# Introduction to Experimental Biophysics

## Biological Methods for Physical Scientists, Second Edition

# FOUNDATIONS OF BIOCHEMISTRY AND BIOPHYSICS SERIES

**Introduction to Experimental Biophysics:**
**Biological Methods for Physical Scientists,**
**Second Edition**
Jay L. Nadeau

**Introduction to Single Molecule Biophysics**
Yuri L. Lyubchenko

**Biomolecular Thermodynamics: From Theory to Application**
Douglas Barrick

**Biomolecular Kinetics: A Step-by-Step Guide**
Clive R. Bagshaw

**An Introduction to Biophysics:**
**Quantitative Understanding of Biosystems,**
**Second Edition**
Thomas M. Nordlund and Peter M. Hoffmann

# Introduction to Experimental Biophysics

## Biological Methods for Physical Scientists, Second Edition

Jay L. Nadeau

CRC Press
Taylor & Francis Group
Boca Raton  London  New York

CRC Press is an imprint of the
Taylor & Francis Group, an **informa** business

**Library of Congress Cataloging-in-Publication Data**

Names: Nadeau, Jay L., author.
Title: Introduction to experimental biophysics : biological methods for physical scientists / Jay L. Nadeau.
Other titles: Experimental biophysics | Foundations of biochemistry and biophysics.
Description: Second edition. | Boca Raton, FL : CRC Press, Taylor & Francis Group, [2017] |
Series: Foundations of biochemistry and biophysics
Identifiers: LCCN 2017010261| ISBN 9781138088153 (hardback) | ISBN 1138088153 (hardback) |
ISBN 9781498799591 (pbk. ; alk. paper) | ISBN 1498799590 (pbk. ; alk. paper)
Subjects: LCSH: Biophysics--Experiments--Technique.
Classification: LCC QH505 .N247 2017 | DDC 572--dc23
LC record available at https://lccn.loc.gov/2017010261

# Contents

# Detailed Contents

## Chapter 7
## Introduction to Biological Light Microscopy    225
### *Coauthored with Michael W. Davidson*



## Chapter 8
## Advanced Light Microscopy Techniques    279
### *Coauthored with Lina Carlini*



## Chapter 9
## Advanced Topics in Microscopy II: Holographic Microscopy    305
### *Coauthored with Manuel Bedrossian*

# Chapter 13
# Advanced Topics in Gold Nanoparticles: Biomedical Applications    429



# Chapter 14
# Surface Functionalization Techniques    453

# Chapter 17
# Introduction to Nanofabrication    623
### *Orad Reshef*

# Series Preface

Biophysics encompasses the application of the principles, tools, and techniques of the physical sciences to problems in biology, including determination and analysis of structures, energetics, dynamics, and interactions of biological molecules. Biochemistry addresses the mechanisms underlying the complex reactions driving life, from enzyme catalysis and regulation to the structure and function of molecules. Research in these two areas is having a huge impact in pharmaceutical sciences and medicine.

These two highly interconnected fields are the focus of this book series. It covers both the use of traditional tools from physical chemistry, such as nuclear magnetic resonance (NMR), x-ray crystallography, and neutron diffraction, as well as novel techniques including scanning probe microscopy, laser tweezers, ultrafast laser spectroscopy, and computational approaches. A major goal of this series is to facilitate interdisciplinary research by training biologists and biochemists in quantitative aspects of modern biomedical research and teaching core biological principles to students in physical sciences and engineering.

Proposals for new volumes in the series may be directed to Lu Han, senior publishing editor at CRC Press, Taylor & Francis Group (lu.han@taylorandfrancis.com).

# Preface

The second edition has been revised and updated to reflect changes in the fields between 2010 and 2016, with references, suppliers, and software all brought up to date. The study questions at the back of each chapter have been thoroughly revised and expanded, and a solutions manual is available.

The book has also been restructured to make a clear distinction between the basic techniques and more advanced approaches that are usually not accessible to an undergraduate laboratory. The book can be used on two levels: as an introductory course with only the basic techniques covered or as a more advanced course that requires access to more sophisticated equipment. The advanced material may also be used for self-study.

The advanced material is included within selected chapters as callouts, as well as forming the basis of five entirely new chapters: advanced molecular biology techniques (Chapter 4), advanced light microscopy (Chapter 8), holographic microscopy (Chapter 9), biomedical applications of gold nanoparticles (Chapter 13), and microfabrication techniques (Chapter 17).

A large fraction of the basic course material provides the basis for a one-semester or summer course on introductory molecular biology techniques.

This textbook is bundled with a laboratory companion guide. It is structured according to the chapters in the book, although it refers to the first edition of this book. The series of 14 experiments presents a wide variety of techniques that may be performed during a semester-long, three-credit course or during a 1-month intensive.

# Acknowledgments

I thank all of the people who made this book possible. The biggest thanks are to the chapter authors, who provided years of firsthand experience on how to do things right (or wrong). Sections of some chapters were also contributed by colleagues. I am grateful to Chris Ratcliffe of the National Research Council, who wrote the section on solid-phase nuclear magnetic resonance (NMR) in Chapter 16, and to Jonathan Saari of McGill, who contributed the section on time-resolved absorption spectroscopy in Chapter 16.

A special mention also goes to Jenna Blumenthal, who was a senior undergraduate in physics/physiology when she helped to proofread the first edition and prepare the first version of the glossary. Ildiko Horvath of McGill drew some of the illustrations in Chapter 1. Thanks also to my former graduate students Samuel Clarke, Xuan Zhang, and Daniel Cooper, whose thesis material is incorporated into several of the chapters.

Another thanks goes to all of the people who provided figures, both published and unpublished, to help illustrate this work. When approached out of the blue, they responded with data, micrographs, and other material that allowed the illustrations to be as beautiful, relevant, and practical as I hoped they might be.

Thanks to those who helped to proofread, and special hugs to Susan Foster, the world's best copy editor.

Finally, this book would not have been possible without my editor, Lu Han, who helped develop the book's idea, encouraged me throughout its evolution, and solicited the second edition long before I had started to think about it.

# Author

**Jay L. Nadeau** is an associate professor of physics at Portland State University (PSU). Prior to PSU, she was a research professor in the Graduate Aerospace Laboratories (GALCIT) at the California Institute of Technology (2015–2017) and an associate professor of biomedical engineering and physics at McGill University (2004–2015). Her research interests include nanoparticles, fluorescence imaging, and development of instrumentation for the detection of life elsewhere in the solar system.

She has published over 70 papers on topics ranging from theoretical condensed matter physics to experimental neurobiology to the development of anticancer drugs and, in the process, has used almost every technique described in this book. Her work has been featured in *New Scientist*, *Highlights in Chemical Biology*, Radio Canada's *Les Années Lumière*, *Le Guide des Tendances*, and educational displays in schools and museums. Her research group features chemists, microbiologists, roboticists, physicists, and physician–scientists, all learning from each other and hoping to speak each other's language. A believer in bringing biology to physicists as well as physics to biologists, she has created two graduate-level courses: methods in molecular biology for physical scientists and mathematical cellular physiology. She has also taught pharmacology in the medical school and was one of the pioneers in the establishment of multiple mini-interviews for medical school admission.

She retains adjunct positions at McGill and Caltech and has collaborators in industry and academia in the United States, Europe, Australia, and Japan. She has given several dozen invited talks at meetings of the American Chemical Society, the American Geophysical Union, the International Society for Optics and Photonics (SPIE), the Committee on Space Research, the American Association of Physics Teachers (AAPT), and many others. Before her time at McGill, she was a member of the Jet Propulsion Laboratory's Center for Life Detection, and previous to that, a Burroughs Wellcome postdoctoral scholar in the laboratory of Henry A. Lester at Caltech. She earned a PhD in physics at the University of Minnesota in 1996.

# Contributors


**Edward S. Allgeyer**
Department of Chemical and Biological Engineering
University of Maine
Orono, Maine

**Manuel Bedrossian**
Graduate Aerospace Laboratories
California Institute of Technology
Pasadena, California

**Oliver M. Baettig**
Department of Biochemistry
McGill University
Montreal, Quebec, Canada

**Albert M. Berghuis**
Department of Biochemistry
McGill University
Montreal, Quebec, Canada

**Lina Carlini**
Laboratory of Experimental Biophysics
EPFL, Switzerland

**Gary Craig**
Department of Chemical and Biological Engineering
University of Maine
Orono, Maine

**Michael W. Davidson**
National High Magnetic Field Laboratory
Florida State University
Tallahassee, Florida

**Jeremy Grant**
Department of Chemical and Biological Engineering
University of Maine
Orono, Maine

**Sanjeev Kumar Kandpal**
Department of Chemical and Biological Engineering
University of Maine
Orono, Maine

**Thomas Knöpfel**
Riken Brain Science Institute
Saitama, Japan

**Christian Lindensmith**

Jet Propulsion Laboratory

California Institute of Technology

Pasadena, California

**Michael D. Mason**

Department of Chemical and Biological Engineering

University of Maine

Orono, Maine

**Joshua A. Maurer**

Department of Chemistry

Washington University

St. Louis, Missouri

**Orad Reshef**

Department of Physics

Harvard University

Cambridge, Massachusetts

# CHAPTER 1

# Introduction and Background

## 1.1 BASIC BIOCHEMISTRY

### Molecules important to molecular biophysics

The chemicals of life are *organic* compounds, or compounds that contain carbon. Carbon (C, atomic number 6) is one of the few *tetravalent* atoms, meaning that it has four *valence* electrons available to form bonds with other atoms. Each of the four atoms to which it bonds can be different and can include other carbons. Carbon is thus central to the formation of complex, three-dimensional molecules, and it makes up about 10.7% of the atomic ratio of living matter. Other molecules necessary for the building blocks of life are hydrogen (H, atomic number 1, monovalent, 60.5%); oxygen (O, atomic number 8, divalent, 25.7%); nitrogen (atomic number 7, trivalent, 2.4%); phosphorus (P, atomic number 15, trivalent up to hexavalent, 0.17%); and sulfur (S, atomic number 16, divalent, tetravalent, or hexavalent, 0.13%).

The valence of the key elements forms the basis of the *structural model* of organic chemistry that permits us to predict which combinations of atoms will combine to form stable molecules. **Figure 1.1** shows the classes of organic molecules that are most important in biochemistry and their functional groups. If the letter $R$ is used to designate any chemical moiety besides hydrogen, then an *amine* has the general formula $RNH_2$ (for a *primary amine*), $R_2NH$ (for a *secondary amine*), or $R_3N$ (for a *tertiary amine*). A *carboxylic acid* is RCOOH; at physiological pH, it will usually dissociate into a free proton ($H^+$) and a negatively charged ion $RCOO^-$ (called a *carboxylate*). A *ketone* is RCOR where the second R is not an OH group. *Phosphates* in biology have the form $RPO_3^{2-}$; when R is OH, this is referred to as inorganic *phosphate* or $P_i$. *Alcohols* are ROH where R can be nearly anything; any biomolecule with a name ending in *-ol* terminates in an OH group. A *sulfhydryl*, also known as a thiol group, is RSH. Thiols are also known as mercaptans. Finally, an *aromatic* group is a planar ring that may be made of carbon only or of carbon plus oxygen, nitrogen, or sulfur (called *heterocyclic* compounds). The simplest example is the six-carbon benzene ring.

The structural and functional makeup of a cell results from combinations of four basic molecular types, each of which falls into one or more of the categories in

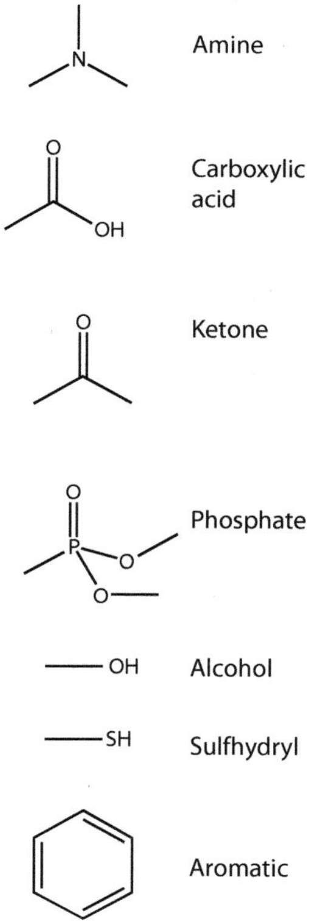

**Figure 1.1 Functional groups seen in biochemistry.**

Amine

Carboxylic acid

Ketone

Phosphate

OH   Alcohol

SH   Sulfhydryl

Aromatic

Figure 1.1; these molecules join end to end (*polymerize*) to achieve their final active form:

- *Amino acids* (polymerize to form *peptides* and *proteins*). There are twenty naturally occurring amino acids, whose structure consists of a central carbon atom with a carboxylic acid on one end and a primary amine on the other, and a *side chain* that branches off the first carbon after the amine. The side chain determines the amino acid's identity and ranges from a hydrogen (glycine) to complex charged or aromatic groups (**Figure 1.2**). Short chains of amino acids are called peptides and may be synthesized by organisms like fungi in order to kill bacteria. The example shown is bacitracin, which is a cyclic peptide active against many bacteria; it is often found in first-aid creams. Some peptides are available from biological suppliers, and custom peptides are also available, though costly. Full-length proteins are encoded genetically and synthesized as a long polypeptide chain. They then fold to form their final *tertiary structure*; the example shown is *green fluorescent protein*, or GFP, which has 238 amino acids. The physics of protein folding still remains largely a mystery. Proteins usually cannot be purchased but must be expressed and purified by the experimenter (**Figure 1.3a**).

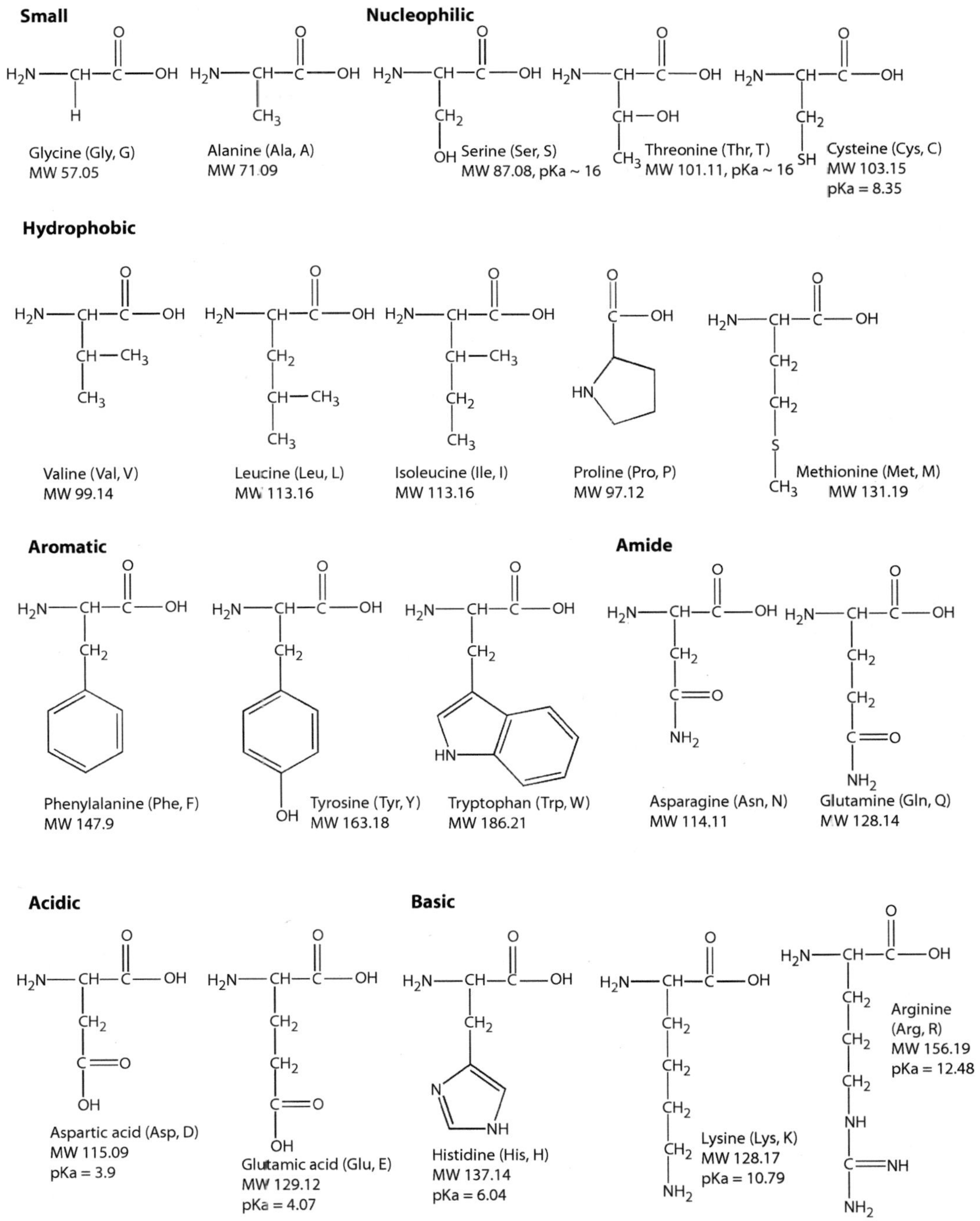

Figure 1.2 The 20 naturally occurring amino acids, showing their one- and three-letter abbreviations, their molecular weights, and their acid dissociation constants (pKa values).

**Figure 1.3 Monomers and polymers of living systems.** Images are not to scale with each other. (a) An amino acid (alanine; side chain CH₃), a peptide (bacitracin), and a protein (GFP). (b) A monosaccharide or simple sugar (glucose), and the polymer of glucose (cellulose). (c) A DNA base (adenine), a nucleotide (deoxyadenosine monophosphate), and a double-stranded oligonucleotide. (d) A fatty acid (oleic acid) and a triglyceride (SOP: steric, oleic, and palmitic acid).

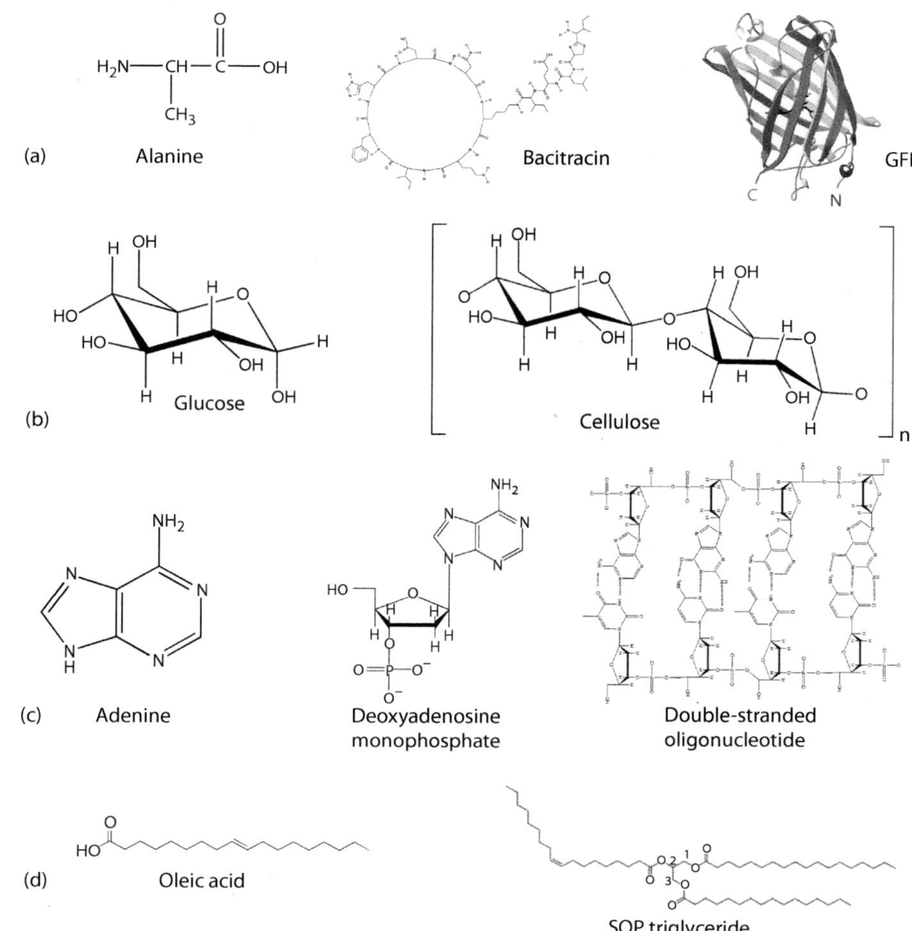


- *Monosaccharides* (polymerize to form *polysaccharides*). Shown in **Figure 1.3b** is glucose (also known as grape sugar or corn sugar), which is the major source of fuel for every living cell on Earth. The active form in biology is right-handed and polarizes light to the right; thus, it is often called simply dextrose, especially in the food industry (see **Advanced Topic 1.1**). Monosaccharides can polymerize to form important storage and structural molecules. Storage molecules include starch and glycogen; the latter is what provides energy after carbo loading. Structural molecules include some of the most abundant natural materials in the world: chitin (a polymer of a glucose derivative found in fungi, arthropods, crustaceans, and insects) and cellulose (a polymer of glucose, the primary component of wood; **Figure 1.3b**).

- *Nucleotides* (polymerize to form *nucleic acids* [DNA, RNA]). DNA is made of four *nitrogenous bases*: adenine, guanine, cytosine, and thymine (abbreviated A, G, C, and T). Adenine is shown in **Figure 1.3c**. A and G are *purines*, while T and C are *pyrimidines* (**Figure 1.4**). When the base is linked to a sugar (in the case of DNA, this sugar is *deoxyribose*; in the case

## ADVANCED TOPIC 1.1:   CHIRALITY

Many organic compounds are not identical to their mirror images. These molecules are called *chiral*, from the Greek *cheir* ("hand"), since human hands are also mirror images of each other but not superposable. In general, any tetrahedral atom with four different groups attached to it will be chiral. This includes all of the amino acids except glycine, all the monosaccharides, and many other compounds (Figure A1.1.1a).

Chirality is of great importance in chemistry and biology for several reasons. First, the chemistry of the right- and left-handed *enantiomers* of the same compound is not identical. Although they have the same molecular weight, solubility properties, index of refraction, and melting and boiling points, they behave differently when they interact with other chiral compounds or with light. The easiest way to observe chirality is to use a polarimeter to observe the rotation of plane-polarized light as it passes through the substance in question. A clockwise rotation is characteristic of a dextrorotatory or right-handed substance; a counterclockwise rotation indicates a levorotatory or left-handed enantiomer.

In biology, one enantiomer or the other is preferred almost exclusively. With a few exceptions in bacteria, sugars in biological systems are D- and amino acids are L- (where D and L refer to structure and not necessarily to the way in which they polarize light). Enzymes are all correspondingly chiral. The opposite-handed compounds have no nutritional value, and large amounts of D-amino acids may be harmful. Some drugs are hazardous only in one enantiomeric form; the best example may be thalidomide, which acts as a sedative and appetite enhancer in its right-handed form but whose left-handed form is highly teratogenic (Figure A1.1.1b). Different enantiomers also often taste and smell different, reflecting their differing interactions with our receptors and enzymes.

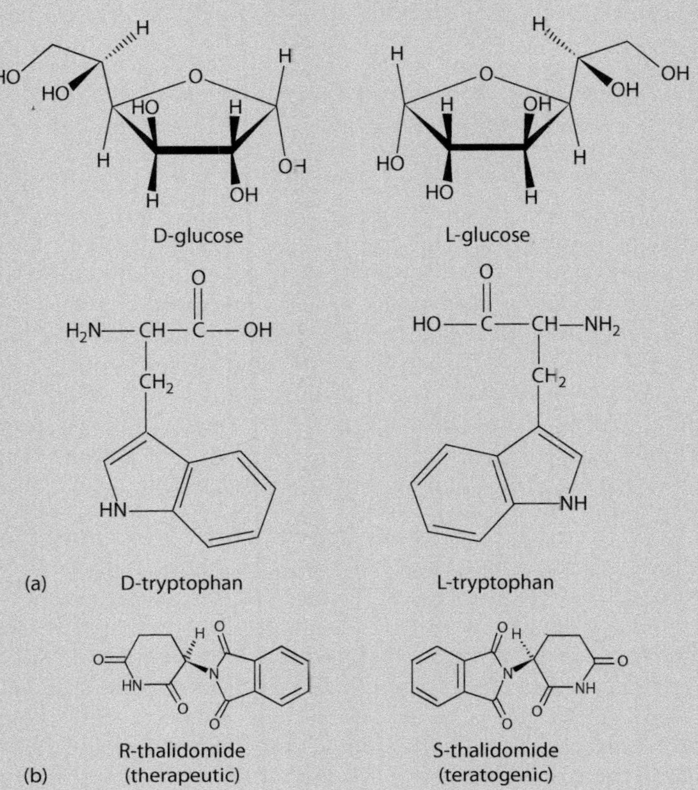


**Figure A1.1.1  Chirality.** (a) Sugars and amino acids are chiral because their mirror images cannot be superposed. (b) Thalidomide is a good example of how different enantiomers react differently with biological systems.

(*Continued*)

### ADVANCED TOPIC 1.1 (CONTINUED):   CHIRALITY

The origin of this exclusiveness, called *homochirality*, is unknown and widely studied because of its implications for the evolution of life on Earth and for the search for life on other planets. It is possible that the "choice" of one enantiomer or another was an evolutionary accident—i.e., an enzyme happened to evolve for an L-amino acid, thereby driving selection for all L-amino acids in the future. If the former is true, then life on other planets would be expected to be homochiral, but not necessarily in the same way as Earth life. Organic molecules that form from abiotic processes, however, should not show this homochirality but instead exist as *racemic mixtures* of both entantiomers. (Indeed, chiral mixtures left to their own devices are found to eventually *racemize*; this fact can be used as a dating technique.)

However, some physicists believe that the observed forms of these molecules are thermodynamically favored, possibly by an asymmetry in the weak force. If this is true, all molecules throughout the universe would be expected to be homochiral, or at least have an *enantiomeric excess*. Thus, finding homochirality on another planet would not be a sign of life. Finding the solution to this problem will have important implications for the design of orbital and landed extraterrestrial missions.

#### SUGGESTED READING

Bakasov, A., Ha, T.K., and Quack, M. (1998). Ab initio calculation of molecular energies including parity violating interactions. *Journal of Chemical Physics* 109, 7263–7285.

Barron, L.D. (2008). Chirality and life. *Space Science Reviews* 135, 187–201.

Borchers, A.T., Davis, P.A., and Gershwin, M.E. (2004). The asymmetry of existence: Do we owe our existence to cold dark matter and the weak force? *Experimental Biology and Medicine* 229, 21–32.

MacDermott, A.J. (2000). The ascent of parity-violation: Exochirality in the solar system and beyond. *Enantiomer* 5, 153–168.

of RNA, *ribose*), it is called a *nucleoside*: e.g., adenine becomes adenosine (in RNA) or deoxyadenosine (in DNA). Addition of one or more phosphate groups makes it a *nucleotide*. Short chains of A, C, G, and T nucleotides form *oligonucleotides* (if there are few, usually 20 or fewer bases) or *polynucleotides* (for longer chains). Oligonucleotides may be purchased from many suppliers and are inexpensive. As provided, they are single-stranded. However, DNA in nature is usually double-stranded, with A being complementary to T and C to G due to complementary hydrogen bonding (**Figure 1.3c**). *Complementary* oligonucleotides can be made to hybridize into double strands by simply heating them to 95°C and then allowing them to cool. However, if they are not fully complementary, the final double-stranded form is much less stable, and the strands can separate at relatively low temperatures. This fact forms the basis of much of molecular cloning and many types of *biosensors*.

- *Fatty acids* (form diglycerides and triglycerides by *dehydration synthesis*). Free fatty acids are molecules with a long carbon chain terminated in a carboxylic acid (**Figure 1.3d**). Fatty acids are crucial components of every cell, as they are the principal constituents of *cell membranes*. Most dietary fats, as well as fats stored in our own bodies, are in the form of triglycerides, which is a *glycerol* head attached to three fatty acid tails. The composition of these tails varies widely and plays an important role in the taste and texture of fatty foods. The number of double bonds in a fatty acid is called its degree of unsaturation and determines its melting temperature. Fully *saturated* fats (no double bonds) are solid at room

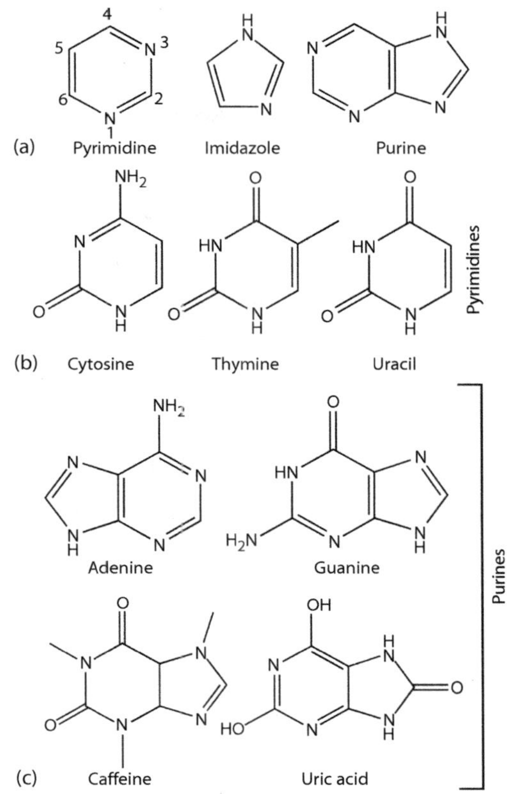

**Figure 1.4 Purines and pyrimidines.** (a) The structure of the pyrimidine ring is shown with its carbon-numbering convention. A pyrimidine ring bound to imidazole makes purine. (b) Cytosine, thymine, and uracil are derivatives of pyrimidine. (c) There is a very large number of biologically relevant purines, including the bases adenine and guanine as well as caffeine, uric acid, and many more.

temperature (butter is >50% saturated) whereas *unsaturated* fats are liquid (canola oil is nearly 95% unsaturated). A mix of different numbers of double bonds in the three chains allows triglycerides to have very complex melting properties. The triglyceride shown in **Figure 1.3d** is one found in cocoa butter, and its melting properties are responsible for the "melt in your mouth, not in your hand" nature of chocolate.

## Making use of functional groups

The different functional groups of the molecules in **Figure 1.1** can be manipulated to create new bonds. Some of these groups are highly reactive, and simple reagents known as *cross-linkers* can catalyze their reactions with a complementary group. For example, a carboxylic acid and an amine can be joined in an *amide bond*; a carboxylic acid and an alcohol can be linked to form an *ester*, or a carboxylic acid and a thiol to a *thioester*; a phosphate can be linked to two other molecules via a *phosphodiester* bond; or two thiols can form a *disulfide* bond (**Figure 1.5a**). Sulfur also forms strong bonds to gold by mechanisms that are still being investigated. These principles can be used to adhere biomolecules to a surface or a particle, a process called *biofunctionalization*; to label a biomolecule with a dye (many dyes are sold that are made prereactive to a specific functional group; see **Chapters 7 and 8**); or simply to join two or more biomolecules (**Figure 1.5b**). Biofunctionalization of nanoparticles will be covered more fully in **Chapters 11 through 13**, and surface functionalization is treated in **Chapter 14**. This is a broad and complex field and is the subject of several excellent review articles and textbooks referenced at the end of each of these chapters.

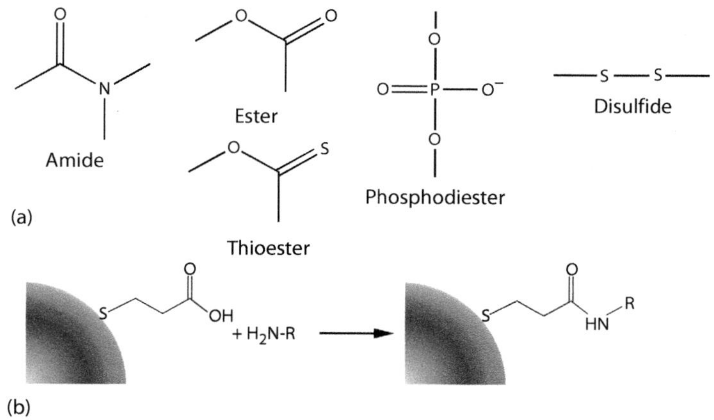

**Figure 1.5 Linking biomolecules.** (a) Types of bonds that can be made by linking amines, carboxylic acids, alcohols, phosphates, and/or thiols. (b) Biofunctionalization example. A gold-covered surface—which may be a nanoparticle, slide, tip, cantilever, etc.—is coated with a molecule containing a thiol group, one or more carbon atoms, and a carboxylic acid. It is then reacted with any molecule containing a primary amine to give an amide bond. Note that all proteins contain both primary amines and carboxylic acids, as they are made up of amino acids.

## 1.2 ENERGIES AND POTENTIALS

### Biologically relevant energy scales

The structural model is empirical; it was developed by August Kékulé, Archibald Scott Couper, and Alexander M. Butlerov independently between 1858 and 1861. It does not provide any mechanistic description of bond formation, which had to wait until the invention of quantum mechanics for the development of a theory of *orbital* formation based upon electron wave functions (see **Advanced Topic 1.2**). Quantum mechanics also describes several other types of interatomic and intermolecular forces besides *covalent bonds,* all of which are equally crucial to biology, and without which molecules such as DNA could not exist. **Table 1.1** lists some examples of these forces and their relative energies. For comparison, $kT$ at room temperature is 2.5 kJ/mol.

### Ionic bonds

An *ionic bond* can be thought of as a covalent bond in which one of the partners is more *electronegative* than the other—that is, it has a stronger affinity for the shared

**Table 1.1**

Types of Interatomic/Intermolecular Interactions and Their Relative Strengths

| Type of Interaction | Example | Bond Energy (kJ/mol) |
| --- | --- | --- |
| Covalent bond | C–C | 200–400 |
| Covalent double bond | C=C | 600–800 |
| Ionic bond | Na–Cl | 700–1000 |
| Hydrogen bond | O–H | 10–40 |
| Ion–dipole interaction | $Na^+$–$H_2O$ | 40–600 |
| Dipole–dipole interaction | HCl–HCl | 5–25 |
| Ion–induced dipole interaction | $Cl^-$–hexane | 3–15 |
| Dipole–induced dipole interaction | $H_2O$–Ar | 2–10 |
| London dispersion interaction | Hexane–octane | .05–2 |
| Cation–pi interaction | Benzene…$Na^+$ | 2–50 |

## ADVANCED TOPIC 1.2:  A QUANTUM MECHANICAL DESCRIPTION
## OF BONDING: MOLECULAR ORBITAL THEORY

The first major conceptual breakthrough in the quantum mechanics of chemical bonding was the idea that bond energy results from exchange (*resonance*) of electrons between two nuclei. For example, for the hydrogen molecule, Heitler and London expressed the wave function of the two electrons as a spin singlet part $\Psi_s$ and spin triplet part $\Psi_t$:

$$\Psi_s(1,2) = N_s\left[\phi_n(r_1)\phi_m(r_2)+\phi_m(r_1)\phi_n(r_2)\right]\chi_s(s_1,s_2)$$
$$\Psi_t(1,2) = N_t\left[\phi_n(r_1)\phi_m(r_2)-\phi_m(r_1)\phi_n(r_2)\right]\chi_t(s_1,s_2),$$

(A1.2.1)

where the $N$'s are normalization constants, the $\phi$'s are spatial wave functions, and the $\chi$'s are spin wave functions. They then calculated energy shift resulting from bonding as a perturbation:

$$\Delta E = \int \Psi^* H \Psi$$

where $H$ is a sum of the interactions between the two electrons (1, 2), the electrons and the nuclei (a, b), and the two nuclei (separated by distance $R$):

$$H = \frac{e^2}{R} + \frac{e^2}{r_{12}} - \frac{e^2}{r_{a2}} - \frac{e^2}{r_{b1}}.$$

(A1.2.2)

The results show that the singlet state has an energy level lower than that of the ground state of a single atom. (It is called the *bonding orbital*.) The triplet state has a higher energy and so is called the *antibonding orbital* (Figure A1.2.1a). The sum of the energies of these orbitals gives an attractive potential that approaches $r^{-6}$ behavior at large distances, but which becomes steeply repulsive at short distances (Figure A1.2.1b). This type of potential is seen in all diatomic molecules, and its general form is also seen in other types of interactions other than covalent.

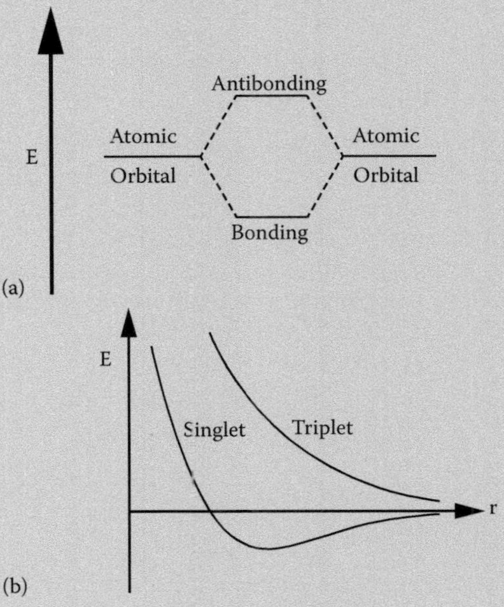


**Figure A1.2.1 Molecular orbital energies for diatomic hydrogen.** (a) Relative energies of atomic hydrogen and of the singlet state of molecular hydrogen (bonding) and triplet state (antibonding). (b) Energy versus distance for the singlet and triplet state of hydrogen using the Heitler–London model.

(*Continued*)

> **ADVANCED TOPIC 1.2 (CONTINUED):   A QUANTUM MECHANICAL DESCRIPTION OF BONDING: MOLECULAR ORBITAL THEORY**
>
> *Molecular orbital calculations* become highly complex for molecules more sophisticated than hydrogen. The Hückel approximation, developed in 1931, lends itself to analytic solutions but is a poor approximation for most problems. More sophisticated quantum chemistry approaches rely upon modern computational power as well as upon development of appropriate approximation techniques; the 1998 Nobel Prize in Chemistry was awarded to John Pople "for his development of the density-functional theory" and to Walter Kohn "for his development of computational methods in quantum chemistry." Many software packages, both open-source and commercial, make use of a variety of approaches to solve these quantum chemistry problems. Density-functional theory is the least computationally intensive; the "functional" refers to the energy of the molecule as a function of electron density as a function of position. Another very common approach to molecular structure calculations is the *Hartree–Fock* approximation. Rather than electron density, Hartree–Fock looks at the wave function of the molecule as a product of the wave functions of each electron. Simplification of this tremendous task can be achieved using semiempirical methods (e.g., spectroscopy data) or ab initio approaches, which use mathematical simplifications to facilitate processing. Improvements to the Hartree–Fock approximation are called post-Hartree–Fock methods. In an advanced version of this course, one or more computational chemistry packages will be provided or suggested to you for the solution of test problems.
>
> **SUGGESTED READING**
>
> Albright, T.A., Burdett, J.K., and Whangbo, M.-H. *Orbital Interactions in Chemistry*. Edn. 2. Wiley-Interscience, 2013.
> Fleming, I. *Molecular Orbitals and Organic Chemical Reactions.* Wiley, 2009 (student edition also available).
> Kotz, J.C., Treichel, P.M., and Weaver, G.C. Bonding and molecular structure: Orbital hybridization and molecular orbitals. *Chemistry and Chemical Reactivity*. Thomson Brooks/Cole, Belmont, CA, 457–466, 2006.

electron. In the extreme case, the electron is almost entirely localized around this partner. Nearly all covalent bonds have some ionic character. An ionic bond can be described using the same quantum mechanical formulations as covalent bonds, with an alteration in the electrostatic term. The potential energy between two ions falls off as $1/r$.

## Ion–dipole interactions

Most atoms do not have permanent dipoles, but many molecules do, meaning that there is an uneven distribution of charge along the molecule. In addition, both atoms and molecules can show induced dipole moments caused by exposure to an electric field (which may come from ions or dipolar molecules). Permanent dipole interactions are stronger than induced dipole interactions and we will consider these first. A *polar* molecule with a dipole moment $m$ interacts with an ion of charge $q$ with the potential

$$V(r) = -\frac{mq}{4\pi\varepsilon_0 r^2}\cos\theta,\qquad (1.1)$$

where

$r$ is the distance between them,

$\varepsilon_0$ is the permittivity of free space, and

$\theta$ is the angle of the dipole (**Figure 1.6a**).

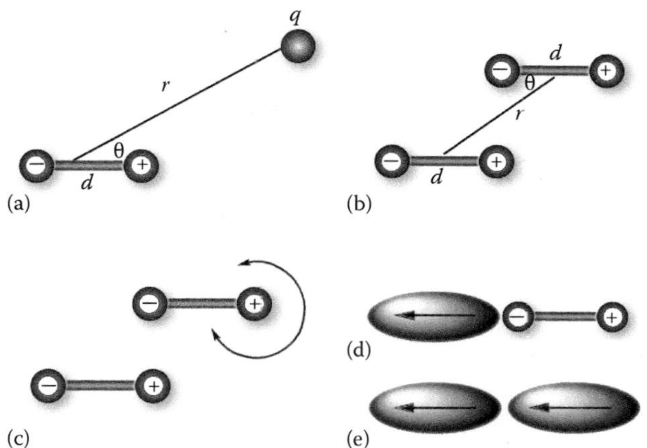

**Figure 1.6 Intermolecular interactions.** (a) A dipole–point charge interaction. When the two are far enough apart, $d$ can be considered negligible relative to $r$. (b) Two parallel dipoles at a fixed angle. This sort of arrangement occurs in a solid. (c) If one or more of the dipoles is completely free to rotate, integration over all angles gives an interaction energy of 0. However, real molecules are limited in their rotations, so dipole–dipole interactions in a liquid or gas are nonzero. (d) A dipole can polarize a nonpolar molecule. (e) Two nonpolar molecules can have induced dipole moments and interact with each other.

If the dipole is free to rotate, thermal averaging results in a potential that falls off more rapidly with distance:

$$V(r) = -\frac{1}{3k_{\mathrm{B}}T}\frac{(mq)^2}{(4\pi\varepsilon_0)^2 r^4},$$

(1.2)

where $k_{\mathrm{B}}$ is the Boltzmann constant and $T$ is the absolute temperature.

A key example of an ion–dipole interaction is the interaction between water and dissolved ions in solution. These relatively strong forces give rise to the *energy of hydration* of these ions, which needs to be overcome if an ion is to be separated from its surrounding water molecules. This energy is why most ions permeate through biological pores and channels in a hydrated state. (For polar solvents other than water, this can be generalized to an *energy of solvation*.)

## Dipole–dipole interactions

The potential between two permanent dipoles $m_1$ and $m_2$ (assuming their own radii are negligible) falls off with distance more quickly than that between a dipole and a charge (**Figure 1.6b**):

$$V(r) = \frac{-2m_1 m_2}{4\pi\varepsilon\varepsilon_0 r^3}$$

(1.3)

This formula is valid for fixed dipoles, as in a solid. However, if the dipoles are completely free to rotate, their attractive and repulsive components cancel, and the net interaction is 0. An important concept is that in liquids and gases, rotation is not completely free but is weighted by the Boltzmann distribution (**Figure 1.6c**). Keesom showed that the average dipole–dipole interaction for rotating molecules at temperature $kT$ is given by

$$V(r) = \frac{-2m_1^2 m_2^2}{3(4\pi\varepsilon_0)^2 k_{\mathrm{B}}T r^6}$$

(1.4)

where

$m_1$ and $m_2$ are the dipole moments,

$r$ is the distance between the dipoles,

$\varepsilon_0$ is the permittivity of free space,

$k_\mathrm{B}$ is the Boltzmann constant, and

$T$ is absolute temperature.

This is the familiar form of the dipole–dipole interaction encountered in energy transfer experiments, e.g., fluorescence resonance energy transfer (FRET; see **Chapter 16**). It is called the *Keesom interaction* and is the strongest of the $1/r^6$ interactions. A derivation is given in **Advanced Topic 1.3**.

A single dipolar molecule can also induce an instantaneous dipole in a nonpolar molecule, with an induced dipole moment $m_\mathrm{i}$ related to the molecule's polarizability $\alpha$ and the applied electric field E as $m_\mathrm{i} = \alpha E$. The average interaction for a dipole and a nonpolar molecule in a liquid or gas (**Figure 1.6d**) is

$$V = \frac{-m^2\alpha}{4\pi\varepsilon_0 r^6} \tag{1.5}$$

This formula does not need to be thermally averaged, because the direction of the induced dipole follows the permanent dipole; thermal fluctuations do not affect it. It is called the *Debye interaction* and is about $1/10$ the strength of the Keesom interaction.

If both molecules are nonpolar, induced dipole–induced dipole interactions can still occur. These are called *London dispersion interactions* and, although weak, are responsible for the only possible interactions between nonpolar species such as noble gases. Their distance dependence is also $1/r^6$, and they are about $1/10$ as strong as the Debye interaction (**Figure 1.6e**).

Collectively, these noncovalent interactions with $1/r^6$ dependence (Keesom, Debye, London) are called *van der Waals interactions* and may be parametrized by a single equation. The attractive $1/r^6$ potential is only valid at relatively long distances relative to the size of the molecule. To better describe what happens at short distances, a repulsive term must be added. The exact form can vary, but the Lennard-Jones potential is often used because it simply describes a very steep repulsion that occurs within a certain radius (called steric hindrance):

$$V = -\frac{A}{R^6} + \frac{B}{R^{12}} \tag{1.6}$$

Values of $A$ and $B$ are determined empirically or computationally for different molecules and can be found in journal articles and books. The arguments relating to dipoles also relate to higher multipoles (**Advanced Topic 1.4**).

## Hydrogen bonds

*Hydrogen bonds* are a special case of fixed dipole–dipole interaction. A hydrogen attached to an electronegative atom (usually oxygen, fluorine, or nitrogen)

### ADVANCED TOPIC 1.3:    DERIVATION OF KEESOM INTERACTION

Derivation of the Keesom interaction is not done in most textbooks. The averaging is nontrivial. First, it must be noted that the averaging is not simply over all angles, which would give a result of 0; the averaging is Boltzmann-weighted, which gives greater weight to the lower-energy configurations.

To derive Equation 1.4, start from looking at the generalization of Equation 1.3 for dipoles each free to rotate in the plane $(\theta_1, \theta_2)$ as well as to twist relative to each other $(\varphi)$. This expression becomes

$$V(r,\theta_1,\theta_2,\varphi)=\frac{-m_1 m_2}{4\pi\varepsilon_0 r^3}\left[2\cos\theta_1\cos\theta_2-\sin\theta_1\sin\theta_2\cos\varphi\right].$$

This formula is a product of a radial part and an angular part

$$V(r,\theta_1,\theta_2,\varphi)\equiv V_0(r)f(\Omega),$$

where

$$V_0(r)=\frac{m_1 m_2}{4\pi\varepsilon_0 r^3}.$$

Also define

$$\beta=\frac{-m_1 m_2}{4\pi\varepsilon_0 r^3(k_\mathrm{B}T)}.$$

This gives the Boltzmann-weighted average as

$$<V>=\frac{V_0(r)\displaystyle\int d\theta_1\,d\theta_2\,d\varphi\sin\theta_1\sin\theta_2\,f(\Omega)e^{\beta f(\Omega)}}{\displaystyle\int d\theta_1\,d\theta_2\,d\varphi\sin\theta_1\sin\theta_2\,e^{\beta f(\Omega)}},$$

where the integral over $\varphi$ goes 0 to $2\pi$ and the integrals over $\theta$ go 0 to $\pi$. Now, using the properties of the natural logarithm, allows us to write

$$<V>=V_0(r)\frac{d}{d\beta}\ln\left[\int d\theta_1\,d\theta_2\,d\varphi\sin\theta_1\sin\theta_2\,e^{\beta f(\Omega)}\right].$$

Since we make the assumption that $\beta f(\Omega)\ll 1$, we can expand the exponential as a Taylor series and keep only the first two terms:

$$e^{\beta f(\Omega)}\sim 1+\beta f(\Omega)+\frac{\left(\beta f(\Omega)\right)^2}{2}.$$

So the integrals we need to evaluate are

$$A=\int d\theta_1\,d\theta_2\,d\varphi\sin\theta_1\sin\theta_2=8\pi$$

(Continued)

**ADVANCED TOPIC 1.3 (CONTINUED):    DERIVATION OF KEESOM INTERACTION**

$$B = \int d\theta_1\, d\theta_2\, d\varphi \sin\theta_1 \sin\theta_2 \beta f(\Omega)$$

$$= \int d\theta_1\, d\theta_2\, d\varphi \sin\theta_1 \sin\theta_2 [2\cos\theta_1 \cos\theta_2 - \sin\theta_1 \sin\theta_2 \cos\varphi] = 0$$

$$C = \frac{1}{2}\int d\theta_1\, d\theta_2\, d\varphi \sin\theta_1 \sin\theta_2 \left[\beta f(\Omega)\right]^2$$

$$= \frac{\beta^2}{2}\int d\theta_1\, d\theta_2\, d\varphi \sin\theta_1 \sin\theta_2 [2\cos\theta_1 \cos\theta_2 - \sin\theta_1 \sin\theta_2 \cos\varphi]^2 = \frac{8}{3}\pi\beta^2$$

This gives

$$<V> = V_0(r)\frac{d}{d\beta}\ln\left[8\pi\left(1+\frac{\beta^2}{3}\right)\right] = V_0(r)\left(1+\frac{\beta^2}{3}\right)^{-1}\frac{2}{3}\beta \sim V_0(r)\left(\frac{2}{3}\beta\right)$$

$$= \frac{-2(m_1 m_2)^2}{3(4\pi\varepsilon_0)^2 (k_B T) r^6},$$

which is the final result.

**SUGGESTED READING**

The original papers reporting this derivation are as follows:
Keesom, W. H., *Physik. Z.* 22, 129, 1921.
Keesom, W. H., *Physik. Z.* 22, 643, 1921.

develops an effective positive charge. This hydrogen is then called the donor or *proton donor*. It interacts with an electronegative atom that doesn't necessarily have any hydrogens and might be of a different species (**Figure 1.7a**). The strongest hydrogen bonds are between HF and F. Hydrogen bonds are crucial in biology: they hold double-stranded DNA into its helical structure and polypeptides into helical protein conformations, and assist in all varieties of specific binding interactions (antibody/antigen, receptor/ligand, DNA/transcription factor, etc.) (**Figure 1.7b**).

### The (strept)avidin/biotin interaction

One of the strongest noncovalent interactions known is the interaction between *avidin* and *biotin*. Avidin is a protein produced in the oviducts of egg-laying animals and deposited into the egg white to protect the developing embryo from bacterial invasion by sequestering biotin, an essential vitamin in the B family that bacteria need to grow (mammals and birds also require biotin; it is possible to acquire a deficiency by eating raw egg whites). In its typical state, avidin is a tetramer in which each subunit binds biotin with equal affinity (**Figure 1.8a**). This gives an overall dissociation constant of $\sim 10^{-15}$ M! A similar protein isolated from a bacterium is called *streptavidin* from its origin (*Streptomyces*), and shows

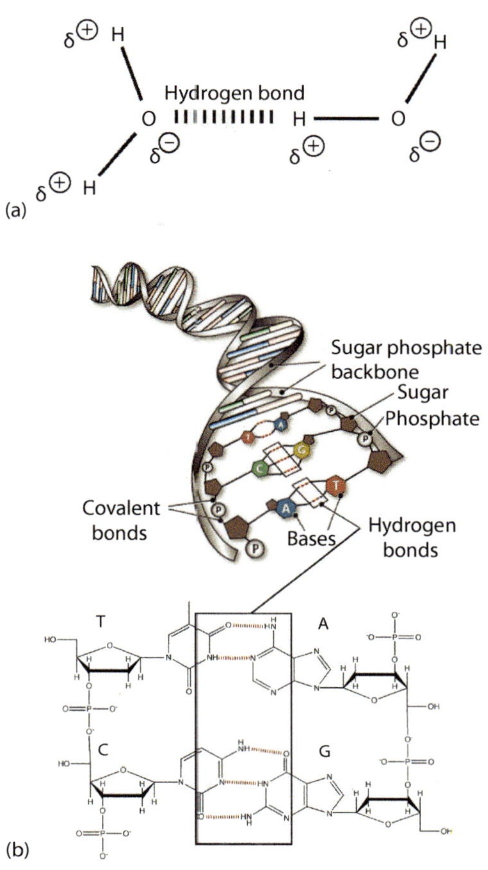



**Figure 1.7 Hydrogen bonding.** (a) Schematic of a hydrogen bond between two water molecules, where hydrogen acts as the proton donor. (b) Hydrogen bonds hold together the helices of the DNA double helix. Two hydrogen bonds link each A–T pair, and three link each G–C pair; no other combinations are possible. Important features of a DNA fragment, such as its melting temperature, are determined by the strength of these bonds. The upper image shows the hydrogen bonds between A–T and G–C in the context of the double helix; the lower inset shows the details of the structure.

similar chemistry but is easier to handle, with less aggregation and unwanted interactions with cells.

This interaction is exploited in biology for just about any type of experiment where specific, irreversible binding is desired. Biotin is a small molecule (molecular weight, 244.31) and may be chemically linked to any other molecule of choice, a process called *biotinylation*. Because it is so small, it very rarely affects the function of the molecule to which it is attached. The biotinylated molecule will then bind avidin or streptavidin specifically with great affinity.

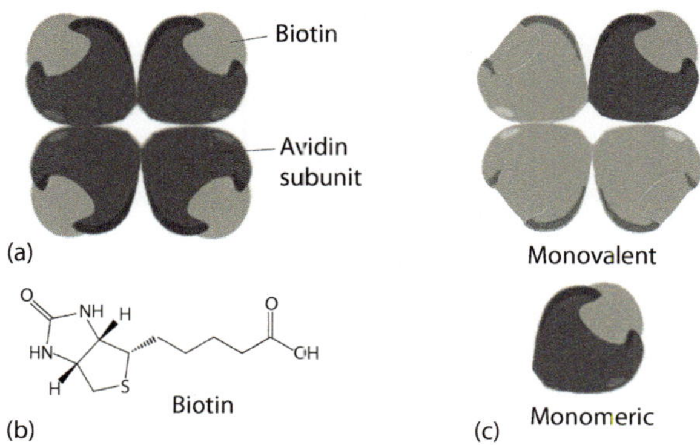



**Figure 1.8 Avidin/biotin.** (a) Schematic of tetrameric structure of avidin/streptavidin, with four binding sites for biotin. (b) Chemical structure of biotin. (c) Monovalent streptavidin has one active site for biotin and three blocked or inactive sites (the protein remains the same size). (d) Monomeric streptavidin is one-quarter the size of the full protein and has only one binding site in consequence.

### ADVANCED TOPIC 1.4:    CATION–PI, ANION–PI, AND PI–PI INTERACTIONS

Some molecules with no permanent dipole moment have a permanent quadrupole moment, and interactions of this moment with charges can be comparable in strength to that of a hydrogen bond. The most notable example is benzene, which has no dipole moment due to symmetry. However, the quantum mechanical delocalization of electrons above and below the plane of the molecule creates a quadrupole moment and a significant attractive potential to positive ions (Figure A1.4.1a). Because two amino acids (tryptophan and tyrosine) contain benzene rings, the *cation-pi* interaction has important implications for protein structure. This interaction is also believed to play an important role in the binding of the *neurotransmitter acetylcholine* to its receptor, and in the ability of nicotine to act as a full agonist for this receptor. (Acetylcholine and nicotine are pictured in Figure A1.4.1b.) More recently, a quadrupole–anion interaction was proposed, and potentially stabilizing anion–pi interactions were found in over 3000 proteins, the majority of the sample tested. Quadrupole–quadrupole interactions between aromatic rings result in *pi stacking*, or arrangement of the rings in energetically favorable conformations. Pi stacking is important in both nucleic acid and protein structure; it also is important for the concentration-dependent optical properties of many dyes in solution, since the majority of fluorescent compounds are aromatic.

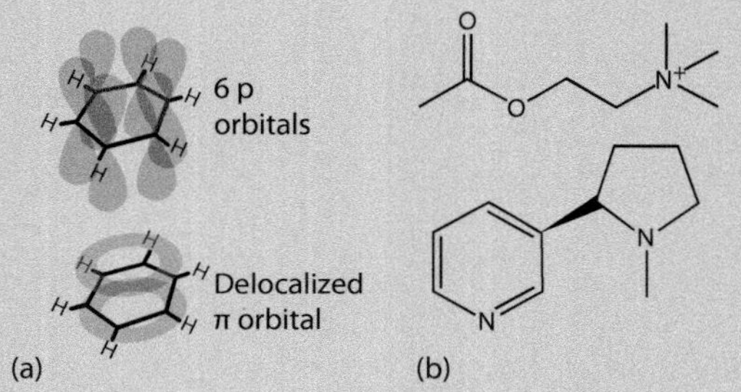

**Figure A1.4.1 The cation–pi interaction.**
(a) Delocalized electron clouds around the benzene plane create a quadrupole moment.
(b) Acetylcholine and nicotine.



## SUGGESTED READING


Jones, G.J., Robertazzi, A., and Platts, J.A. (2013). Efficient and accurate theoretical methods to investigate anion–pi interactions in protein model structures. *The Journal of Physical Chemistry B* 117, 3315–3322.

Lovas, S., He, D.Z., Liu, H., Tang, J., Pecka, J.L., Hatfield, M.P., and Beisel, K.W. (2015). Glutamate transporter homolog-based model predicts that anion–pi interaction is the mechanism for the voltage-dependent response of prestin. *The Journal of Biological Chemistry* 290, 24326–24339.

Wilson, K.A., Wells, R.A., Abendong, M.N., Anderson, C.B., Kung, R.W., and Wetmore, S.D. (2016). Landscape of pi-pi and sugar-pi contacts in DNA–protein interactions. *Journal of Biomolecular Structure and Dynamics* 34, 184–200.

Zhao, Y., Li, J., Gu, H., Wei, D., Xu, Y.C., Fu, W., and Yu, Z. (2015). Conformational preferences of pi-pi stacking between ligand and protein, analysis derived from crystal structure data geometric preference of pi-pi interaction. *Interdisciplinary Sciences* 7, 211–220.

Zhong, W., Gallivan, J.P., Zhang, Y., Li, L., Lester, H.A., and Dougherty, D.A. (1998). From ab initio quantum mechanics to molecular neurobiology: A cation–pi binding site in the nicotinic receptor. *Proceedings of the National Academy of Sciences U S A* 95, 12088–12093.

Variations of avidin and streptavidin have been developed specifically for molecular biology. *Monovalent* streptavidin is still a tetramer but has only one active binding site for biotin. *Monomeric* streptavidin has been dissociated into its components (**Figure 1.8b**). The advantages of both are that they cannot cross-link in reactions. Monomeric streptavidin has the additional advantage of being only one-fourth as large as the tetramer, an advantage for some applications. The disadvantages are much weaker binding (dissociation constants, $\sim 10^{-7}$ M). Other variants have been developed for pH-sensitive or reversible binding.

We will revisit biotin/streptavidin in many chapters of this book. The system is useful for labeling dyes (**Chapters 7 and 8**) and nanoparticles (**Chapters 11 through 13**), and for biofunctionalization (**Chapter 14**).

## 1.3  PRINCIPLES OF SPECTROSCOPY

### What can be measured

The various types of *spectroscopy* used in molecular biophysics probe the interactions described previously, confirming the accuracy of the physical description and allowing for atomic-scale insight into bond lengths and angles, functional groups, molecular symmetry, and electronic structure. All of the methods listed here will be covered in more detail in **Chapter 16**; the following is an introduction to what they are capable of measuring.

Individual atoms and molecules have different quantized electronic energy states in which electrons are permitted to exist. *Transitions* to a higher or lower state require the absorbance or emission of a precise amount of energy. These energies (as well as lifetimes of excited states before they decay to lower-energy states) are characteristic of the atomic system in question. Complex molecules have other types of transitions as well, resulting from the spatial degrees of freedom of the atoms with respect to one another. These are rotations and vibrations about the bonds, giving rise to rotational and vibrational transitions (**Figure 1.9**).

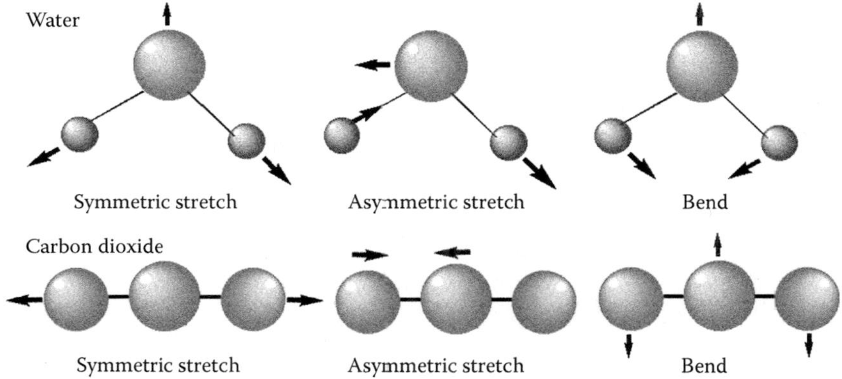


**Figure 1.9 Vibrational modes of a polar triatomic molecule (water) and a nonpolar, linear triatomic molecule (carbon dioxide).** Note that the symmetric stretching mode of $CO_2$ does not change the dipole moment, so it does not appear on an infrared spectrum.

The allowed rotational and vibrational states are quantized and can be estimated with quantum mechanical rigid rotor and harmonic oscillator approximations, respectively; we will return to this in Chapter 16.

## How transitions are measured

When a molecule is exposed to electromagnetic radiation, its dipole interacts with the electric field, and energy can be transferred if the dipole moment's mode and the electric field have the same frequency and phase. If enough energy is transferred, a transition can occur. Calculation of the transition dipole moment between two states can predict which transitions will be experimentally observed; these selection rules will be revisited in more detail in Chapter 16.

The energy needed to excite electronic transitions is comparable to that of photons in the x-ray and ultraviolet-visible (UV-Vis) range. Vibrational transitions have energies comparable to that of infrared (IR) light. Finally, rotational transitions are the least energetic and can be measured using microwaves (Figure 1.10). Irradiating a sample to excite it and observing the resulting transitions is called spectroscopy, and the transitions are spectroscopic transitions. Some common forms of spectroscopy, which we will cover in Chapter 16, are as follows:

- *UV-Vis absorbance spectroscopy*, which provides characteristic signals for molecules that have electronic transitions within this energy range. These molecules are usually aromatic, and if the signal is in the visible, the solution will appear colored. Metals also provide useful signals in this range.

- *Fluorescence spectroscopy* is the complement of absorption spectroscopy as it measures transitions from the excited state to the ground state. Absorbance of a visible or UV photon can excite a molecule to a higher electronic level; relaxation to the ground state causes the emission of a less energetic photon. The photons emitted can vary slightly in energy as relaxation can occur into different vibrational levels of the ground

**Figure 1.10 Energies of electronic, vibrational, and rotational transitions (a) relative to each other and (b) relative to the electromagnetic spectrum.**

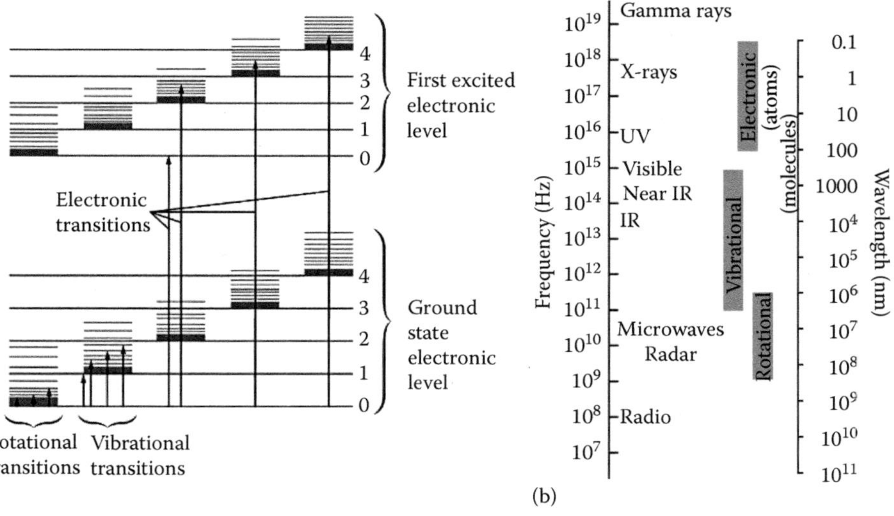

electronic state; this can be used to probe vibrational energies. Many organic molecules are fluorescent in the UV or visible, especially those with aromatic rings.

- Any vibrational modes of the molecule that result in a change in the permanent dipole moment are IR active. (It can be inferred from **Figure 1.9** that water has a very complex IR spectrum.) The energies of these transitions are characteristic for each type of covalent bond. Thus, *IR spectroscopy* can be used to give a "fingerprint" of the types of bonds present in the sample. It can be used to measure very complex materials. *IR fluorescence* can also be used to probe vibrational levels.

- When an intramolecular bond breaks, the result may be an unpaired electron, known as a *free radical*. Free radicals are highly reactive, and specific species, such as oxygen radicals, have been implicated in cell death, cancer, and aging. They are also of interest because when exposed to microwave radiation, an unpaired electron can move between parallel and antiparallel states in a magnetic field. This is the basis for *electron paramagnetic resonance* (EPR) spectroscopy.

## 1.4 CELLS

Life can be thought of as a controlled series of reactions in which the building block molecules are taken up, assembled, recycled, and broken down in precisely controlled order and proportion. The overall process is called metabolism, where building-up processes are anabolism and breaking-down ones are catabolism. A lack of key elements or disruption in the balance of assembly and breakdown leads to death. The first requirement for life is thus that it be enclosed within some kind of container to allow for nonequilibrium concentrations of key elements. The second is that this container must be able to assimilate all of the needed building blocks. Hence the *cell*, the unit upon which all Earth life is based. A cell is a water-impermeable lipid container bearing a series of passive and active (i.e., energy-dependent) transport systems for import and export of key molecules.

A single cell placed in a surrounding "infinite" sea with a certain nutrient concentration will be limited in the extent of its reactions by the rate of diffusion of nutrients to its surface. The poorer the sea, the smaller the cell must be to meet its requirements. In ocean water, calculations have shown this limit to be approximately a cubic micron.

Once inside the cell, the nutrients also must diffuse to specific sites to undergo reactions. This limits the size of the cell unless it develops subcellular containers dedicated to specific metabolic processes. We can distinguish two distinct domains of life: those that do not have these subcellular compartments (called *organelles*, including a *nucleus*) and those that do. The former are called Prokarya, and the latter, Eukarya. The Prokarya include the kingdoms Bacteria and Archaea, and the Eukarya are Animalia, Plantae, and Fungi.

An example of a prokaryotic cell is molecular biology's key organism, the bacterium *Escherichia coli*, usually just called *E. coli*. The structure of an *E. coli* cell is shown in **Figure 1.11a,b**; it consists principally of a *cell wall*, *cell membrane*, circular DNA *chromosome* packaged into a *nucleoid* region, and *ribosomes*. Within its fluid contents or *cytoplasm* are all of the enzymes needed to replicate

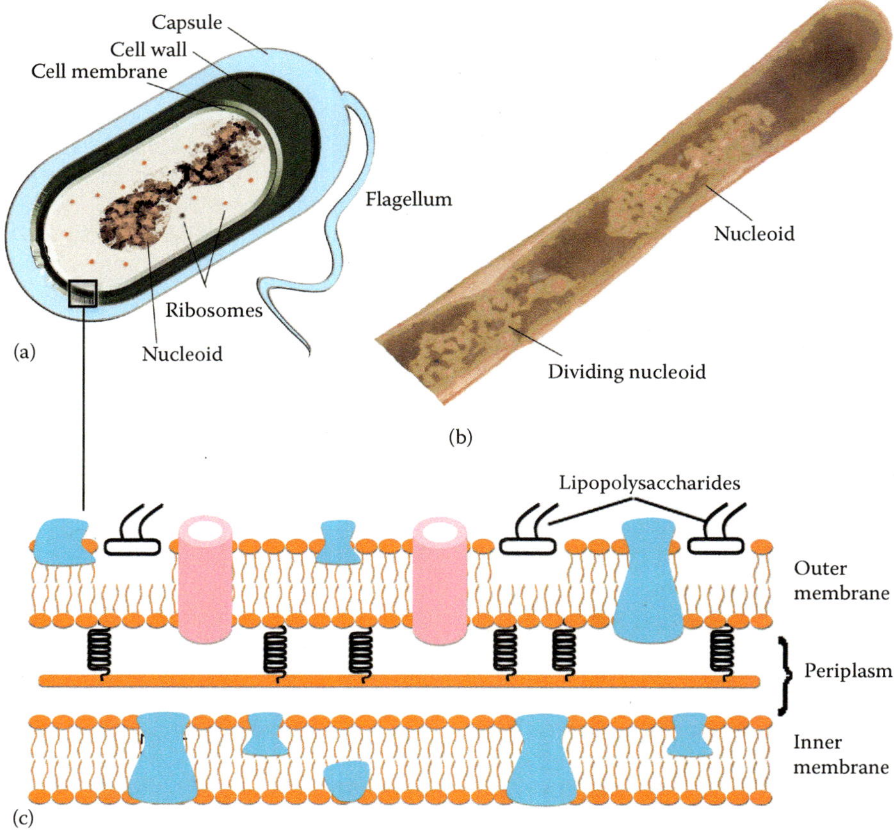



**Figure 1.11 Structure of a typical prokaryote, the bacterium *E. coli.*** (a) Schematic of cross-section through the cell. (b) Transmission electron microscopic image (TEM) of dividing cell. (c) Schematic of inner membrane, periplasm (or periplasmic space), and outer membrane.

DNA and *transcribe* DNA to RNA. RNA is *translated* to protein in the ribosomes. The cytoplasm also contains disequilibrium concentrations of ions, particularly potassium, leading to a nonzero *membrane potential* (see **Chapter 15** for more on the origin and measurement of membrane potentials). The space between the membrane and cell wall is called the *periplasm* and may comprise 40% of the cell's volume; many important reactions, such as neutralization of *antibiotics*, occur in this space. The structure of the *E. coli* membrane and periplasm is characteristic of the class of bacteria called *Gram negative* (**Figure 1.11c**).

A eukaryotic cell is much larger and more complex. Plant and fungal cells are surrounded by rigid cell walls, whereas animal cells have only a lipid membrane. This significantly changes the way in which the different types of cells are handled in the laboratory. Introducing a foreign agent through a rigid cell wall is challenging, and so we will not deal with culture or transfection of fungal or plant cells in this book. Animal cells used for molecular biophysics include those from normal human tissue or from human cancers and rat and mouse cells. These cells are usually immortalized into *cell lines*, usually by infecting with viral agents. This causes the cells to divide essentially forever, so that the experimenter can simply maintain a flask of cells and extract some as needed for each experiment. The cells are seeded onto the substrate of choice and allowed to grow to the desired density before the experiment takes place. Many different cell lines are available from commercial suppliers, along with specialized media that permit their growth. A cell line is chosen based upon the desired experiment: some cell lines simply

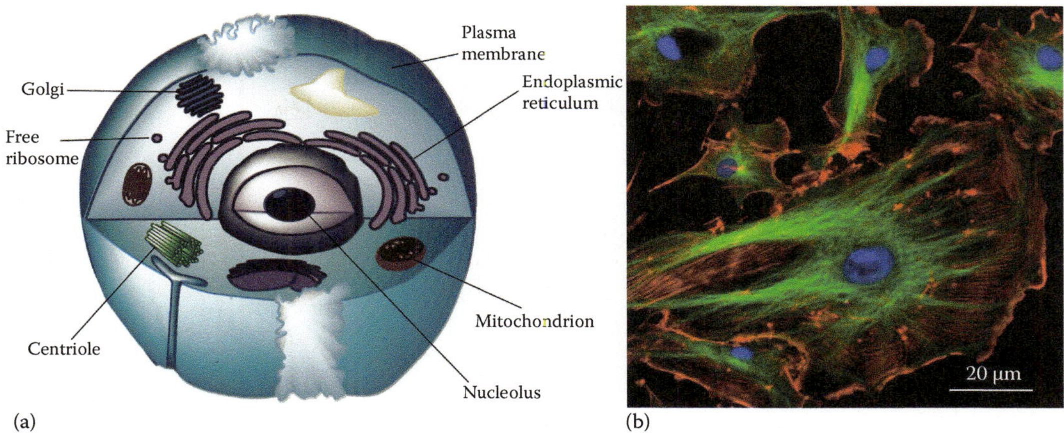



**Figure 1.12 Animal cells.** (a) Schematic of a typical animal cell showing organelles. (b) Light micrograph of cultured cell line showing labeled cytoskeleton (microtubules in green, actin in red) and nuclei (blue).

express foreign DNA well; others express specific receptors or ion channels; some are models for disease processes, especially cancer. The alternative to cell lines is *primary cells*, which are taken directly from an animal and cultured in a dish. This requires obtaining ethical approval for sacrifice of the animals as well as the expertise needed to extract, isolate, and culture the cells. Depending upon the cell type, this can range from easy (e.g., blood or immune cells) to very difficult (e.g., specific neuronal populations). Primary cells are more difficult to manipulate than cell lines, as will be discussed in **Chapter 3**. Several textbooks on cell lines and culture of specific types of primary cells are available.

A typical animal cell is shown in **Figure 1.12**. Organelles include the nucleus, *mitochondria, lysosomes, Golgi apparatus*, and smooth and rough *endoplasmic reticulum* (ER). Ribosomes may exist free or bound to the ER (which is what makes the rough ER "rough" under electron microscopy). The function of these organelles will be discussed in the following sections. Eukaryotic cells are also characterized by a complex *cytoskeleton* of *microfilaments, intermediate filaments*, and *microtubules*. These molecules are often studied in molecular biophysics.

# 1.5  DNA, RNA, REPLICATION, AND TRANSCRIPTION

### The structure and function of DNA and RNA

In cells, nucleic acids are used to encode information on when, where, and in what quantities proteins and other nucleic acids will be made. Their chemistry is what makes this possible. DNA usually exists in cells in a stable double-helical form with a hydrophilic sugar–phosphate backbone surrounding the hydrogen-bonded nitrogenous bases. The phosphate terminus is called the *5' end* of each strand, and the hydroxyl terminus is the *3' end*. The double helix is *antiparallel*: that is, what is considered the top strand runs 5' to 3', and the bottom strand runs 3' to 5'. A only bonds to T and G only to C; these are complements (**Figure 1.13a**). If this helix is separated (*denatured* or melted), free nucleotides that are complementary to the single strand can hydrogen-bond to the single-stranded DNA to regenerate the double helix. Thus, the molecule encodes its own copy

**Figure 1.13 Structure and replication of DNA.**
(a) DNA bases and deoxyribose. (b) Principles of DNA replication. Complementary strands form from each strand of denatured DNA. Each of the resulting copies is then made up of one old strand and one new strand. This is called *semiconservative* replication.

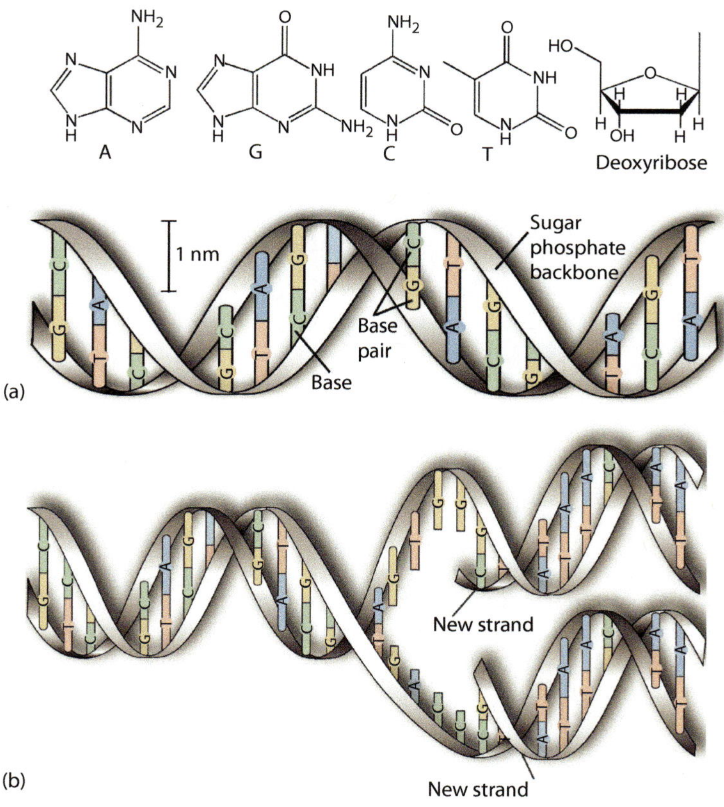


(**Figure 1.13b**). The process is thermodynamically favorable and is accelerated in cells by enzymes that *catalyze* the reactions. Of course, it implies that there must be a certain concentration of free nucleotides available whenever DNA needs to be synthesized. In prokaryotic cells, these are found throughout the cytoplasm, whereas they are compartmentalized in eukaryotic cells according to the need for them. Both prokaryotes and eukaryotes take free nucleotides up from the environment (ingest or eat them); organ meats, seafood, and legumes are good dietary sources. Duplication of DNA is called replication.

RNA is similar to DNA except that each of its sugars contains an extra oxygen (ribose instead of deoxyribose) and it contains the base uracil in the place of thymine, which lacks a carbon (**Figure 1.14a**). The extra oxygen makes RNA less stable, as it is more susceptible to hydrolysis. In every creature on Earth except some viruses, RNA does not encode the genome but serves as a messenger between the stable DNA chromosome and downstream applications. RNA nucleotides can hydrogen-bond to single-stranded DNA in the same way as DNA nucleotides can bind to each other, and thereby, the information contained in DNA is transmitted (**Figure 1.14b**). The process of making RNA from DNA is called transcription, and it relies upon free RNA nucleotides being present. Because RNA is single-stranded, it can have sequences that are self-complementary, causing it to fold back in on itself and form very complex *secondary structures*. These structures are important for many of RNA's roles, including transcription; RNA molecules can also act as enzymes and even catalyze their own replication (**Figure 1.14c**).

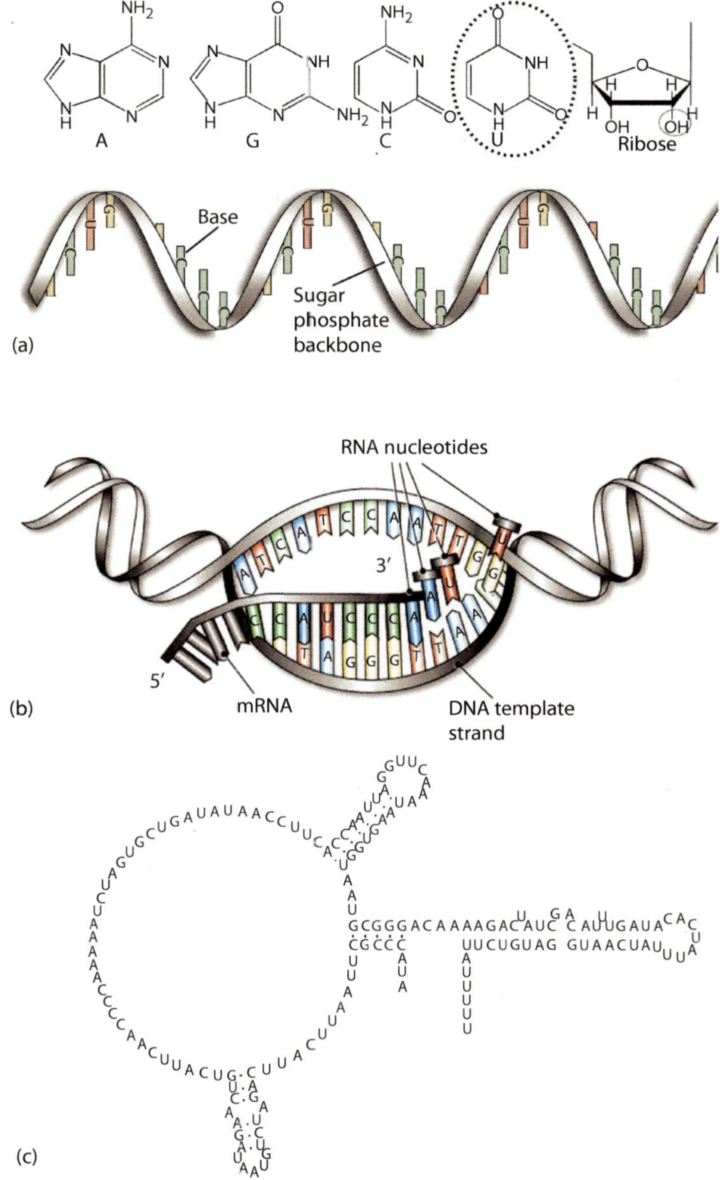


**Figure 1.14  Structure and functions of RNA.** (a) RNA bases and ribose. The circles show the differences with respect to DNA. (b) Principle of transcription. RNA bases can hybridize with single-stranded DNA, creating a single-stranded RNA messenger (messenger RNA [mRNA]) that carries the information. (c) Secondary structure. Since RNA is single-stranded, its self-complementary sequences can bind to each other, creating loops and hairpins. This feature allows RNA to perform roles such as catalysis that DNA cannot. (This image is of a telomerase RNA.)

# Replication

Enzymes serve as *catalysts* to increase the rate of reactions and the local concentration of needed ingredients. Without enzymes, DNA replication and transcription would occur too slowly and haphazardly to be practical for life as we know it. The enzymes involved in DNA processing have been identified, purified, and studied at the atomic level in many organisms, particularly in bacteria, and the physical chemistry of many of the reactions has been elucidated in detail. The protein structure of the enzymes, combined with the physics and chemistry specific to given DNA sequences, allows for careful regulation of when and how

DNA will be replicated, transcribed into RNA, or otherwise modified. We will give an overview here that is focused on prokaryotes (bacteria). Many of the same principles apply in eukaryotes, though the processes are often significantly more complex.

In *E. coli*, there are at least 30 proteins involved in replication. We will see many of these enzymes in later chapters as their purified forms play important roles in molecular biology. For example, *topoisomerase* uncoils DNA from its stably packaged *supercoiled* form. *DNA helicase* breaks open hydrogen bonds to unzip the DNA. *Single-strand binding proteins* (SSBs) stabilize the DNA in a single-stranded configuration so that it does not rewind while transcription is taking place or self-bind as RNA can do. After complementary free nucleotides bind, DNA polymerase joins them to each other with phosphodiester bonds.

Additional complications require even more enzymes. *DNA polymerase* only works in the 5′ to 3′ direction. This means that the parent strand with the free 3′ end elongates continuously; it is called the *leading strand*. The strand with the free 5′ end, called the *lagging strand*, is made in small fragments, which are joined later by *DNA ligase*. In addition, DNA polymerase can only add a base to an existing strand; it cannot start from zero. Thus, a primer (usually made of RNA) and a *DNA primase* are required to start replication. The primers are later removed by *endonucleases*. Replication is shown schematically in **Figure 1.15a**.

Replication occurs in both directions along the DNA strand from a location called a *replication fork*. Multiple replication forks usually occur in an *E. coli* chromosome at the same time. This allows the entire 4.7-megabase genome to be replicated in 20 min even though the rate of replication is only 1000 base pairs/s. **Figure 1.15b** shows a schematic of replication forks along with a micrograph of replication in a bacterium.

DNA replication is extremely accurate thanks to the system of complementary hydrogen bonding, but it is not perfect. Polymerases have proofreading mechanisms

**Figure 1.15 DNA replication in *E. coli*.** (a) Schematic of replication, showing some of the enzymes involved and the synthesis of short fragments on the lagging strand versus continuous synthesis on the leading strand. (b) Schematic and electron micrograph of bidirectional replication of circular bacterial chromosome. Each strand has its own origin, and synthesis proceeds bidirectionally at each origin.

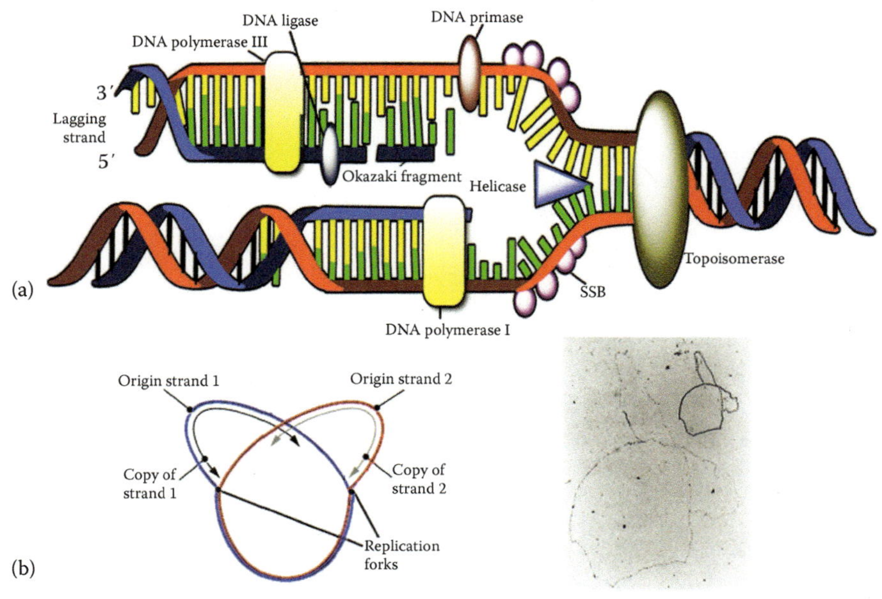

that allow them to trim off misplaced bases; this is called $3' \rightarrow 5'$ *exonuclease* activity and is an important feature of these enzymes that is used in molecular biology. The proofreading ability of DNA polymerase I, operating on the leading strand, is about 20-fold better than that of DNA polymerase III on the lagging strand. Nonetheless, *E. coli* can replicate its genome with only about one error every 10 million base pairs. These errors are called *mutations*, and an organism's mutation rate is crucial for its ability to survive. It shouldn't be too high, or too many individuals will be nonviable mutants, but it must not be 0, or the organism would not be able to evolve.

## Transcription

The generation of a single-stranded RNA molecule from its DNA complement is called transcription. One generally speaks of transcription of *genes*, since as a general rule, each sequence of DNA that encodes for a specific protein is called a gene. (However, the original idea of "one gene, one protein" is an oversimplification, as a single gene can encode for multiple proteins.) The parts of DNA that are directly transcribed into complementary RNA are called coding regions. Noncoding regions include *promoters, enhancers, repressors,* and many others, some still being discovered. (Identifying all of the *regulatory regions* for specific genes is a major challenge in the biology of eukaryotes.) Regulatory regions are usually upstream of the coding region, or toward the 5' end of the DNA molecule, and are involved in controlling the activity of coding regions.

A promoter is a key regulatory region found in every gene; it is what permits the enzyme complex involved in transcription to recognize the gene and bind to it. In *E. coli*, there is only one enzyme that transcribes DNA, called *RNA polymerase*. It binds nonspecifically and weakly to DNA, effectively confining its diffusion to one dimension along the organism's chromosome. When it encounters a promoter, it binds tightly (**Figure 1.16a**). The equilibrium binding constant determines the strength of the promoter and can range from $10^6/$M for weak promoters to $10^9/$M for strong promoters. The first step of transcription is when the promoter DNA and the polymerase form a *closed complex*, in which the DNA is still double-stranded. However, transcription cannot take place until the DNA is made single-stranded so that RNA nucleotides can bind. This occurs in an isomerization reaction, in which the closed complex becomes the *open complex*. During this reaction, a short sequence (12–17 base pairs) of the DNA just upstream to the gene is opened or melted. This open fragment is called the *transcription bubble* (**Figure 1.16b**). Along with DNA melting, the polymerase undergoes a large conformational change during this reaction, developing a hole or pocket allowing the DNA template strand to move inside. The exact details of this reaction remain largely a mystery and may be different for different promoters. The formation of the open complex is the rate-limiting step in the initiation of transcription; rate constants for formation range from $10^{-3}/$s for weak promoters to $0.1/$s for strong promoters.

The binding of free RNA nucleotides can now begin. However, they will be limited to fragments 7–9 base pairs in length (called abortive transcripts) unless the RNA polymerase manages to dissociate from the promoter and bind to the nonspecific DNA downstream. This step is called *promoter clearance* and requires a protein to displace the strong polymerase–promoter binding subunit σ. Transcription has now been initiated, and elongation has begun (**Figure 1.16c**). The RNA transcribed from the template strand is called *messenger RNA*, or mRNA, because it serves to relay the information encoded in the genome to other areas of the cell.

**Figure 1.16 DNA transcription in *E. coli*.** The images of the polymerase are schematics inspired by electron crystallography data of the protein. (a) The first step involves the RNA polymerase binding strongly to a fragment of DNA containing a promoter. A bacterial promoter sequence is shown. (b) The rate-limiting step is the melting of a 12- to 17-base-pair *bubble* to form the open complex. (c) As long as the polymerase is bound to the promoter, only short abortive transcripts can be made. For promoter clearance, the sigma subunit of the polymerase must dissociate, and the enzyme must clear the promoter. (d) Termination occurs at a specific DNA sequence (shown) that leads to a hairpin RNA with a poly-U tail. (Red indicates T or U, blue represents A, yellow represents G, green represents C, and orange indicates any base.)

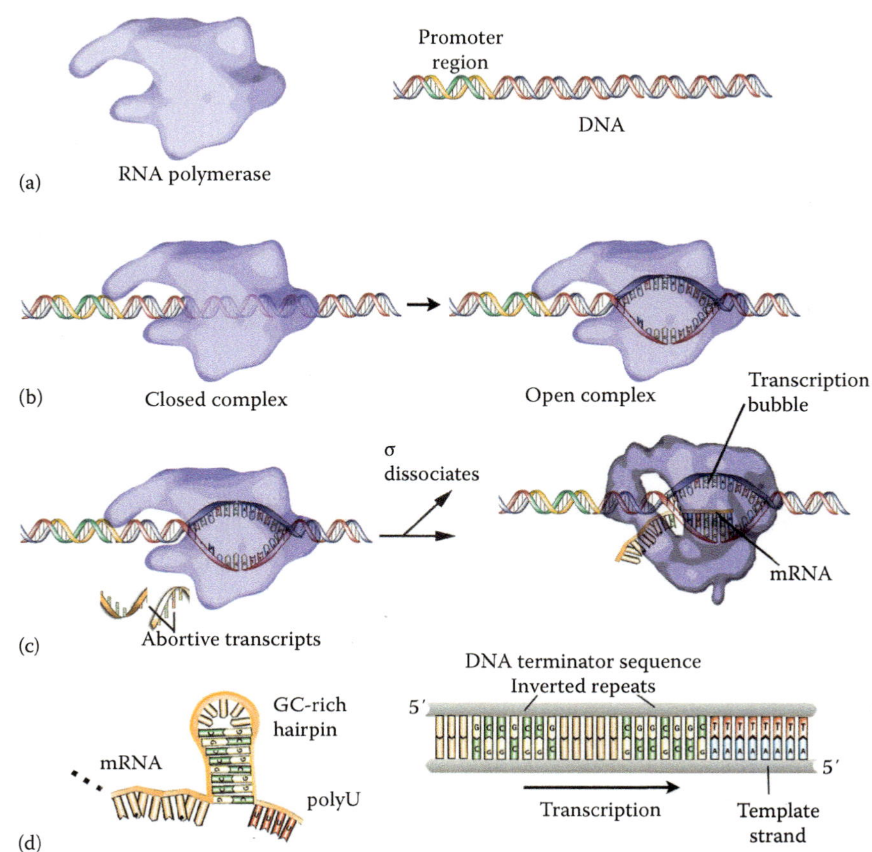



Recent studies have found that mRNA elongation is not a boring, steady process but instead very complex, with different types of paused complexes.

Termination of transcription is assisted by the chemistry of RNA, which allows it to form three-dimensional secondary structures. A *terminator* is a sequence on the DNA that contains two sequences that are inverted repeats, followed by a *poly-A* (multiple adenines). When the RNA is transcribed from the repeats, it bends into a hairpin, probably causing the polymerase to pause. The poly-A creates the weakest possible DNA–RNA hybrid because there are only two hydrogen bonds per base. The combination of these factors causes the mRNA to dissociate from the DNA, and transcription is finished (**Figure 1.16d**).

## 1.6  TRANSLATION AND THE GENETIC CODE

After the synthesis of mRNA is complete, the single strand travels to the ribosomes to be translated into protein. The mRNA is read in units of 3 base pairs, each of which is a *codon* that encodes for a single specific amino acid. At each codon, a *transfer RNA* (tRNA) binds to the mRNA at one end, while bound to the specific amino acid at the other end. The successive tRNAs form a polypeptide chain that emerges from the ribosome (**Figure 1.17**).

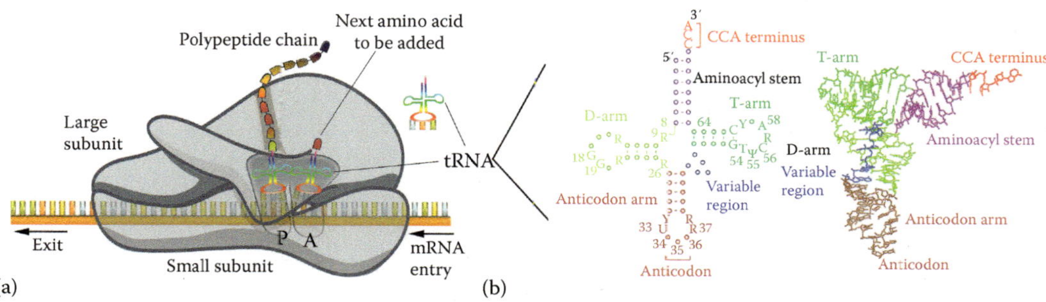


**Figure 1.17  Schematic of translation in *E. coli*.** (a) The ribosome is made up of a large subunit and a small subunit. The *A site* binds aminoacyl-tRNA (tRNA with an amino acid attached); the *P site* binds peptidyl-tRNA (tRNA that's bound to the nascent peptide chain), with an exit tunnel for the new protein to emerge. There is also an *E site* for tRNA preparing to exit, which is not shown here. The mRNA moves through the ribosome being translated one codon at a time. It is possible for more than one ribosome to bind a single mRNA. The tRNA features an anticodon at one end, which is the complementary sequence to the codon in the mRNA, and an attachment site for an amino acid on the other end. (b) Detailed structure of tRNA. (From Hori, H., Methylated Nucleosides in tRNA and tRNA Methyltransferases, *Front. Genet.* 23 May 2014, http://dx.doi.org/10.3389/fgene.2014.00144.)

There are several features of this process that are key in molecular biology and play major roles in the design of most experiments. The first is the genetic code. Because there are 4 unique RNA bases, the number of combinations of $n$ bases is $2^n$; thus, 3 is the smallest number that can encode for the 20 amino acids. However, this creates a lot of degeneracy, as there are now 64 available codons for the 20 amino acids. Three codons are *stop* or termination codons, leaving 61 for the 20 amino acids (**Table 1.2**).

**Table 1.2**

The Genetic Code

| First Nucleotide | Second Nucleotide | | | | Third Nucleotide |
|---|---|---|---|---|---|
| | **U** | **C** | **A** | **G** | |
| U | Phe | Ser | Tyr | Cys | U |
| | Phe | Ser | Tyr | Cys | C |
| | Leu | Ser | STOP | STOP | A |
| | Leu | Ser | STOP | Trp | G |
| C | Leu | Pro | His | Arg | U |
| | Leu | Pro | His | Arg | C |
| | Leu | Pro | Gln | Arg | A |
| | Leu | Pro | Gln | Arg | G |
| A | Ile | Thr | Asn | Ser | U |
| | Ile | Thr | Asn | Ser | C |
| | Ile | Thr | Lys | Arg | A |
| | Met | Thr | Lys | Arg | G |
| G | Val | Ala | Asp | Gly | U |
| | Val | Ala | Asp | Gly | C |
| | Val | Ala | Glu | Gly | A |
| | Val | Ala | Glu | Gly | G |

*Note:* The amino acid abbreviations are given in **Figure 1.2**.

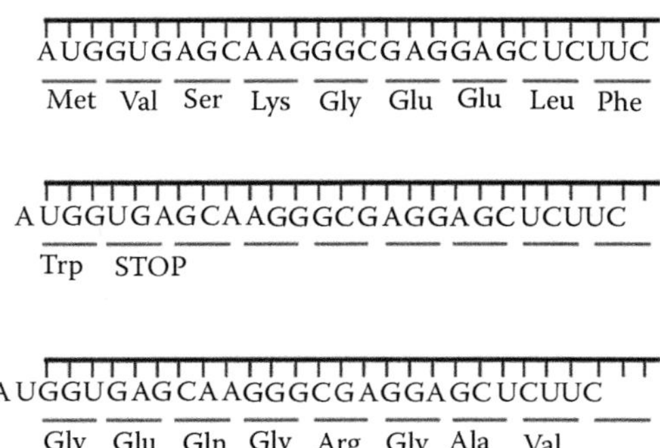


This degeneracy helps to protect organisms from harmful mutations. Of the 64 codons, 32 require only the first 2 base pairs to be specified, and thus are fourfold degenerate; these correspond to the most common amino acids, such as alanine and glycine. Thus, no mutation in the third base would have any effect on these codons. Mutations are discussed further in **Chapter 2**, as they can be used as a valuable tool in the manipulation of DNA. They are also important in determining molecular mechanisms of disease, as mutations leading to large changes in hydrophobicity or charge of a single amino acid can have disastrous consequences to a protein and often to the entire living organism.

The degeneracy of the code is also important when designing genes for expression in different systems. Nearly all organisms use the same genetic code; however, eukaryotes often prefer a different codon than prokaryotes for the same amino acid. Thus, a sequence that uses the bacterial codon would be expressed correctly in mammalian cells, but possibly at a low level. To increase expression levels, silent mutations must be inserted to optimize the codon for the expression system desired. This process is called *codon optimization* and needs to be considered any time genes are being expressed in cells very different from their species of origin.

Another important concept in translation is that of the *reading frame*. Since every three bases code for an amino acid, the exact position of the start point determines the frame in which the sequence is read. A single insertion or deletion will change every subsequent amino acid (**Figure 1.18**). The translation of mRNA does not begin at the beginning of the molecule; most mRNAs contain a 5′ untranslated region (5′ UTR). Instead, an AUG codon (which encodes the amino acid methionine) signals the start of the protein and determines the reading frame. Any additions to the protein must then be in frame with the start codon in order to be read correctly. We will return to this in more detail in **Chapter 2**.

It is important to note that tRNAs are small molecules and thus under the control of the resourceful chemist (**Advanced Topic 1.5**).

## 1.7 PROTEIN FOLDING AND TRAFFICKING

The ribosome produces a one-dimensional polypeptide chain that is biologically inactive. This is referred to as the protein's *primary structure* (**Figure 1.19a**). The *secondary structure* results from hydrogen bonding and creates typical forms such as *alpha helices* and *beta sheets* (**Figure 1.19b**). The *tertiary structure* is the final

## ADVANCED TOPIC 1.5:    UNNATURAL AMINO ACIDS

A tRNA bearing a normal codon can be coupled to something that is not an amino acid at all, or that is some sort of variation of an amino acid (caged, fluorescently tagged, etc.). This is called an *unnatural amino acid* and can be used to study the role of single amino acids in proteins. Many unnatural amino acids are available commercially (Figure A1.5.1).

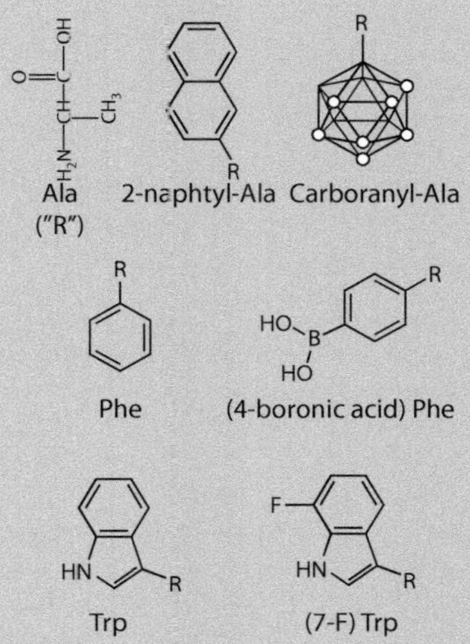

Figure A1.5.1 Some commercially available unnatural amino acids that are derivatives of alanine (Ala), phenylalanine (Phe), or tryptophan (Trp).

## SUGGESTED READING


Adumeau, P., Sharma, S.K., Brent, C., and Zeglis, B.M. (2016). Site-specifically labeled immunoconjugates for molecular imaging—Part 2: Peptide tags and unnatural amino acids. *Molecular Imaging and Biology* 18(2), 153–165.

Brown, K.A., and Deiters, A. (2015). Genetic code expansion of mammalian cells with unnatural amino acids. *Current Protocols in Chemical Biology* 7, 187–199.

Dippel, A.B., Olenginski, G.M., Maurici, N., Liskov, M.T., Brewer, S.H., and Phillips-Piro, C.M. (2016). Probing the effectiveness of spectroscopic reporter unnatural amino acids: A structural study. *Acta Crystallographica Section D, Structural Biology* 72, 121–130.

Hino, N., Hayashi, A., Sakamoto, K., and Yokoyama, S. (2006). Site-specific incorporation of non-natural amino acids into proteins in mammalian cells with an expanded genetic code. *Nature Protocols* 1, 2957–2962.

Lang, K., and Chin, J.W. (2014). Cellular incorporation of unnatural amino acids and bioorthogonal labeling of proteins. *Chemical Reviews* 114, 4764–4806.

Lang, K., Davis, L., and Chin, J.W. (2015). Genetic encoding of unnatural amino acids for labeling proteins. *Methods in Molecular Biology* 1266, 217–228.

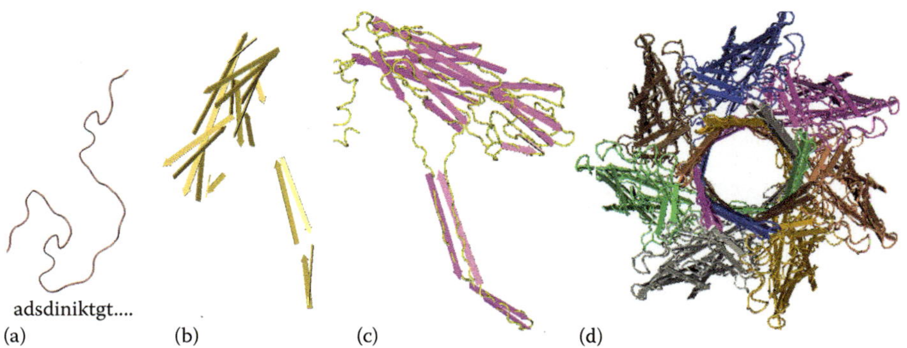



**Figure 1.19 Protein structure.** The example protein shown is alpha-hemolysin, a pore-forming protein from the bacterium *Staphylococcus aureus*. (a) The primary structure is an amino acid sequence, or a polypeptide chain. (b) The secondary structure consists of helices, sheets, and other forms made by hydrogen bonding. (c) The tertiary structure includes all interactions and leads to a specific *native* conformation under the right conditions. (d) Some proteins have quaternary structure. Alpha-hemolysin's final form consists of seven subunits identical to the structure in (c). (It is a homoheptamer.) Each of the colors represents an independent subunit.

three-dimensional form and results from all of the intermolecular interactions among the amino acid residues in the context of the correct temperature, pH, and ionic concentration (**Figure 1.19c**). A correctly folded protein is called the *native* conformation. Some proteins have a *quaternary structure*, which refers to the assembly of multiple subunits that fold independently (**Figure 1.19d**).

How proteins fold is an extremely complex subject, and many sophisticated resources exist on the topic, of which we reference a few at the end of this chapter. It was originally believed that all of the information for the tertiary structure was contained within the primary structure, but this turned out only to be the case for relatively small, soluble proteins. More complex proteins require the assistance of enzymes, called *chaperones*, to fold properly and reach their final destination without aggregating. Chaperones assist proteins in finding their native conformation, assist refolding of misfolded proteins, and bind to the hydrophobic surfaces of proteins in order to prevent aggregation. Other types of *posttranslational processing* of the polypeptide chain may also be necessary to produce an active protein. These include proteolytic cleavage (enzymes called *proteases* trim the ends of the polypeptide chain, or cut it into smaller pieces) and chemical modifications in which new chemical groups are added to specific amino acid residues. Some chemical modifications, such as *phosphorylation* (addition of a phosphate group), are simple and are performed by all organisms. More complex modifications such as *glycosylation* (addition of a large carbohydrate side chain) are only performed in eukaryotic cells.

In *E. coli*, chaperones are located in the cytoplasm, whereas in eukaryotic cells, folding takes place in a specialized membranous organelle called the endoplasmic reticulum. Despite this major difference, many types of eukaryotic proteins can be successfully produced in large amounts in *E. coli* and show normal folding and biological activity; these include human insulin (the first recombinant DNA pharmaceutical), bovine growth hormone (BGH), and many others.

However, many proteins fail to fold properly in *E. coli*, and attempts to optimize expression can be frustrating and fruitless. Reasons for failure include the following:

- The expressed protein contains many cysteine residues. The cytoplasm of *E. coli* is reducing, so disulfide bonds cannot form there. Thus, a protein that requires the formation of one or more disulfide bonds to attain its final tertiary structure may not fold properly.

- The protein has a complex structure with many turns. The formation of turns is a rate-limiting step, and rapid overproduction of complex proteins may lead to aggregation.

- The protein has a need for glycosylation or other complex posttranslational processing.

Failure to fold properly usually results in insoluble aggregates containing misfolded proteins and chaperones. The aggregates are called *inclusion bodies* and can be half as large as a bacterial cell. There are several possible approaches to avoiding inclusion body formation, including simple incubation at lower temperatures or producing lower amounts of protein; these will be discussed more fully in **Chapter 4**. It is also possible to purify protein from inclusion bodies and refold it, although this approach is not often successful. In many cases, *E. coli* cannot be used to express the protein of interest, and a eukaryotic expression system must be used. The most common of these systems are yeast cells and insect cells, and their cultivation and use will be covered in **Chapter 3**. Even in this case, posttranslational modifications may not be identical to those in mammalian systems, so proteins for therapeutic use may still be incorrectly expressed. Mammalian cell systems exist for expression and purification of these proteins, but these are costly and inefficient compared to the other culture types mentioned and should be used as a last resort.

The delivery of proteins to their sites of activity is known as *trafficking*. This is an immensely complex subject, especially in specialized eukaryotic cells such as neurons. More than 100 human diseases are known or believed to be the result of errors in protein trafficking, including some forms of cystic fibrosis, osteogenesis imperfecta, and Alzheimer disease. New proteins and mechanisms involved in trafficking are being continually discovered, many using the techniques of single-molecule biophysics, which we will discuss in later chapters. Here we will mention only the key concepts that have important implications for the design of molecular biology experiments. In *E. coli*, which has no intracellular compartments, trafficking is relatively simple: proteins either are cytoplasmic or are secreted into the periplasmic space or outside the cell. In order to be excreted, the proteins usually do not fold until they have passed through the cell membrane (although there are exceptions to this). In order to prevent premature folding, the proteins are tagged with a signal sequence that marks them as to be excreted. A chaperone protein binds the signal sequence and prevents premature folding, and the polypeptide is secreted through specialized protein pores. The chaperone is then cleaved, and the protein folds (**Figure 1.20**). Numerous signal sequences have been identified in *E. coli* and can be used in molecular biology experiments to tag proteins for excretion.

In eukaryotes, the presence of complex intracellular compartments necessarily makes targeting much more complex. An intracellular organelle, the Golgi apparatus or Golgi complex, is responsible for targeting of proteins to the

**Figure 1.20 Protein trafficking in *E. coli*.** A protein tagged for secretion will have a specific peptide sequence on one end as it emerges from the ribosome (1). It binds the protein SecB (2), which prevents premature folding, and is excreted through a membrane pore made of other proteins SecA, SecY, and SecE (3). Once in the periplasm, the signal sequence is cleaved (4). The protein can then fold correctly. If disulfide bonds must be formed, they are assisted by the dsb proteins. Incorrectly folded proteins aggregate and/or are degraded.

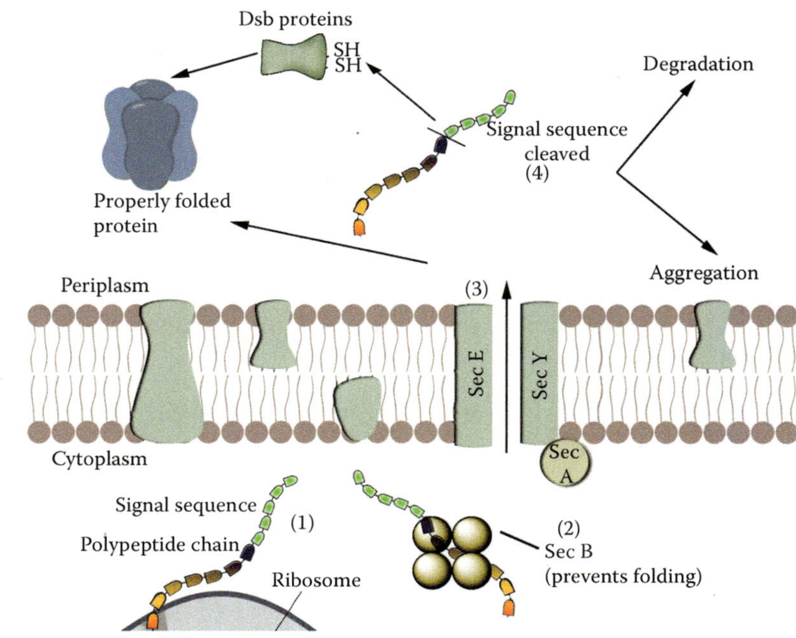


membrane or outside the cell via the *secretory pathway*. The proteins arrive at one face of the Golgi (called the *cis* face) in vesicles after translation in the ER, are processed, and then bud in vesicles from the other side (the *trans* face). Trafficking to other regions, such as the nucleus and mitochondria, is controlled via organelle-specific targeting sequences (**Figure 1.21**).

**Figure 1.21 Protein trafficking in an animal cell.** Proteins destined for the endoplasmic reticulum (ER) or the secretory pathway are tagged with an ER signal sequence. The *resident ER proteins* (needed for ER function) remain in the ER while the others pass to the Golgi. In the Golgi, they are processed further for the lysosomal, membrane, or excretion pathways. Multiple other types of tags allow proteins to target organelles without passing through the secretory pathway: there are nuclear, mitochondrial, and peroxisomal targeting sequences. Of course, many proteins are untagged and remain in the cytoplasm.

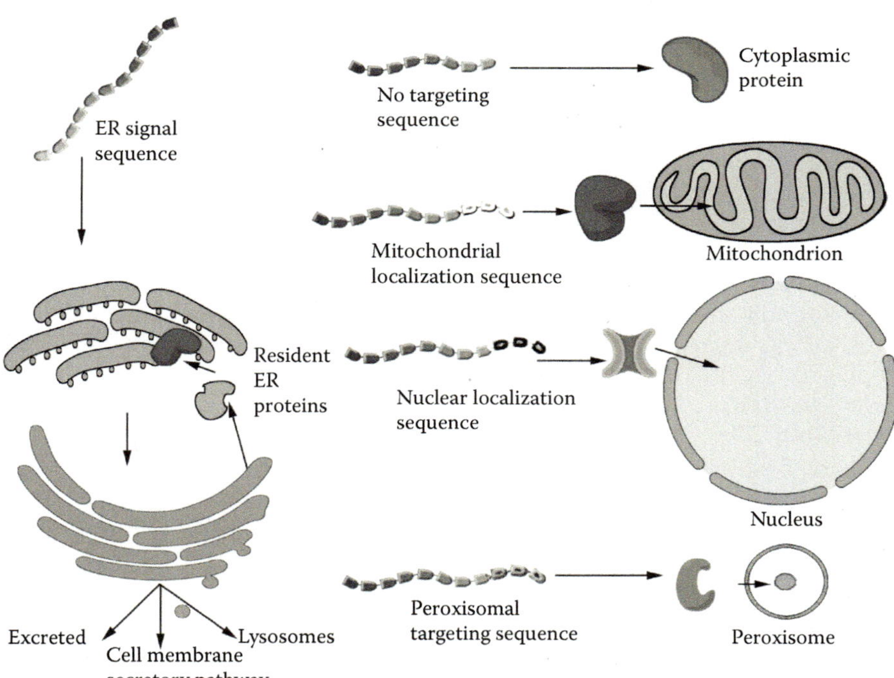

**ADVANCED TOPIC 1.6:    MOLECULAR MECHANICS**

Molecular mechanics is a classical mechanical approach to modeling macromolecules. It creates a force field by summing all of the possible covalent and noncovalent potential energies in a system, where noncovalent includes electrostatic and van der Waals interactions. The simulation must include not only the molecule itself but also the solvent—usually water, though lipid membranes are often commonly needed as well. Empirical data are usually used to input parameters such as bond lengths. Molecular mechanics is usually used in the field of molecular dynamics, or simulations of the motion of macromolecules, including protein folding and conformational changes involved in protein function. The disadvantage of this technique is the immense computational power required; usually, only very short timescales can reasonably be modeled, short even on the timescale of folding or ligand binding. Many of the quantum chemistry software packages also contain molecular mechanics tools, though many do not; dedicated packages are also available.

**SUGGESTED READING**

Haile, J.M. *Molecular Dynamics Simulation: Elementary Methods*. Wiley-Interscience, 1997.
Leach, A. *Molecular Modelling: Principles and Applications*. Edn. 2. Prentice Hall, 2001.

The major implications of trafficking for molecular biology are as follows:

- Mammalian proteins that are targeted to the Golgi or other organelles are likely not to be properly processed or folded in *E. coli*.

- One solution is to target these proteins to the periplasm via a targeting sequence. Many times, this allows membrane proteins to fold properly.

- If this does not work, eukaryotic and possibly mammalian expression systems will be needed to produce these proteins.

- The expression of recombinant proteins may be inhibited even in mammalian systems if the designed protein inhibits processing in the ER or Golgi apparatus. For example, adding GFP to membrane proteins often results in a product that does not escape from the Golgi, hence never reaching its target and causing stress to the cell.

- Targeting sequences can be added to genes that do not otherwise have them, in order to obtain a desired result such as an extracellular protein or a specific organelle label, or in order to study the function of a protein in a particular organelle.

# 1.8  ALTERNATIVE GENETICS

Nearly all forms of known life have their genomic information stored in DNA. RNA acts as a messenger from DNA to protein. The few exceptions are found among the viruses, which have very small genomes—as small as 5 kilobase pairs (kb), compared to the very smallest bacterial genome of nearly 500 kb. The so-called *RNA viruses* use RNA exclusively as their genetic material; this RNA may be single-stranded or double-stranded. Single-stranded RNA viruses may be *positive sense*, meaning that they are transcribed directly into protein by the host cell, or *negative sense*, meaning they are first turned into complementary RNA by a viral enzyme and then transcribed by the host cell. This viral enzyme is an *RNA-dependent RNA polymerase*, meaning a polymerase that makes RNA from RNA. Some of the most dangerous human and animal pathogens are single-stranded RNA viruses: Ebola virus, influenza virus, Lassa fever virus, rabies virus, measles, and mumps are all negative sense; poliovirus, hepatitis A and E, yellow fever, West Nile, and some varieties of common cold viruses are all positive sense.

RNA viruses that use DNA intermediates—that is, turn their RNA genomes into DNA—are called *retroviruses*, and the RNA-dependent DNA polymerase is called *reverse transcriptase*. Once the genome has been turned into double-stranded DNA inside a host cell, it integrates into the host genome, where it is propagated and transcribed with the host cells. It is estimated that 5–8% of the human genome is made up of endogenous retroviral sequences—that is, sequences that have been propagated for a long time after some ancestral infection. The most commonly known pathogenic retrovirus is the human immunodeficiency virus (HIV), the cause of acquired immune deficiency syndrome (AIDS). The use of HIV-based viral vectors as delivery vehicles for genes is described in detail in **Chapter 3**.

Unlike the DNA polymerases found in more complex organisms, RNA-dependent RNA polymerase and reverse transcriptase lack proofreading mechanisms. This accounts for the rapid mutation rates of RNA viruses and retroviruses and hence the difficulty of developing effective vaccines and treatments. The mutation rate is also increased in drug-resistant forms of reverse transcriptase, making future drug failures more likely. These features also influence the way these enzymes must be used in the laboratory; reverse transcriptase can be expected to make approximately 1 error every 2000 base pairs.

## 1.9  WHAT IS CLONING?

For the purpose of this book, we use the term *cloning* to refer not to the cloning of organisms but to molecular cloning. The terms *gene cloning*, *DNA cloning*, *molecular cloning*, and *recombinant DNA* all refer to the same process: the transfer of a DNA fragment of interest from one organism to a self-replicating genetic element, where it is then propagated in a foreign host cell. Like cloning of whole organisms, the output is a series of exact copies of the original. The *host* is nearly always *E. coli*, which is readily grown in the laboratory in large quantities. Over time, a multitude of *cloning strains* of *E. coli* have been developed, designed to take up foreign DNA readily and to replicate it at controlled numbers of copies. These strains are available commercially, and their attributes will be discussed in detail in **Chapter 2**.

Purified forms of the enzymes involved in DNA and RNA replication, transcription, ligation, and other reactions are also available from commercial suppliers. Molecular biology consists of using these enzymes along with purified nucleotides, nucleosides, and other building blocks in order to amplify and manipulate DNA. Manipulations may be done inside *E. coli* cells or *in vitro* (i.e., in a test tube outside a living organism).

## 1.10  DESIGN OF A MOLECULAR BIOLOGY EXPERIMENT AND HOW TO USE THIS BOOK

The design of a molecular biology experiment depends upon the final downstream application or applications (**Scheme 1.1**). The questions to ask are as follows:

- What is the DNA sequence I want to clone?

- What modifications do I want to make to it?

- What is the desired final output (a DNA sequence, cell line expression, primary cell expression, or purified protein)?

- How much of the final output do I need? For example, a crystallization experiment requires at least several milligrams of highly purified protein. On the other hand, a cellular transfection experiment requires only micrograms of DNA, and the production and trafficking of the protein are all done by the cells. The total amount of protein made in this case is very small, however, so transfection of animal cell lines should not be used when purified protein is desired.

- How pure does it need to be? If it needs to be pure, how can I separate it from undesired contaminants?

- How will I confirm the desired output?

In the following chapters, we will proceed through the techniques needed at each step, beginning with obtaining and modifying DNA sequences and proceeding to different methods of expression, purification, and characterization. The chapters are intended to stand alone for each method, so that you can choose which ones to consult based upon your application. **Table 1.3** shows some examples of start-to-finish experiments and their corresponding chapters.

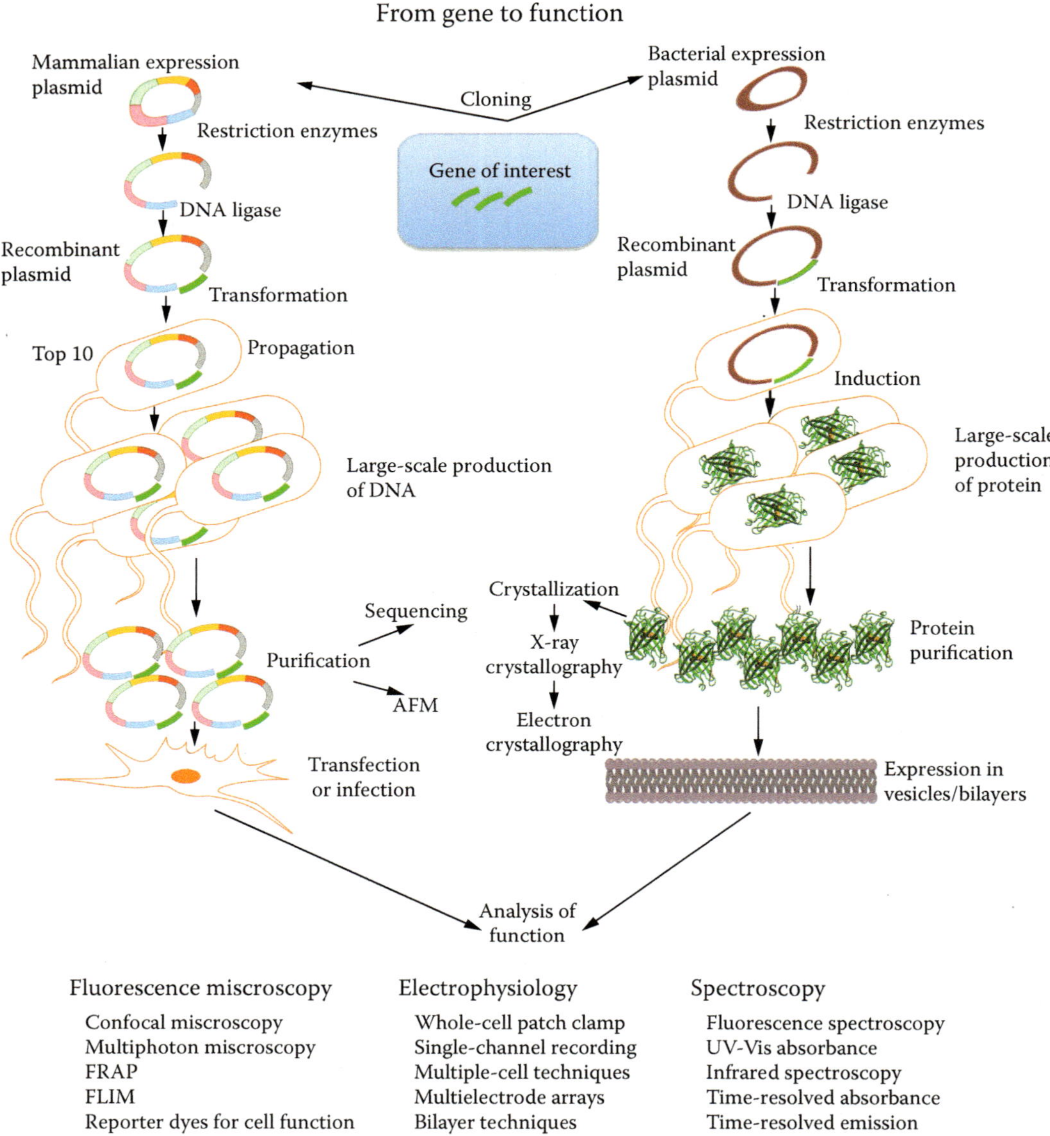


**Scheme 1.1 Design of a molecular cloning experiment based upon desired outcome and downstream analysis techniques.** A fragment of DNA is first inserted into either a mammalian (left) or a bacterial (right) expression vector. In the case of a mammalian vector, the DNA is amplified in *E. coli*. This DNA is purified, and may be sequenced or used in other experiments such as atomic force microscopy. It is then used to transfect mammalian cells, which may be subjected to many of the forms of microscopy, physiology, or spectroscopy listed at the bottom of the scheme. A bacterial expression vector is used when protein is to be purified (unless the protein of interest does not express properly in bacteria; see **Chapter 3**). Bacterial expression vectors can generate many milligrams of protein, which can be purified for spectroscopy or crystallography. Electrophysiological techniques can also be applied to pure protein in lipid bilayers, which ensures that only the protein of interest is being studied.

**Table 1.3**

Guide to Use of This Book

| Desired Application(s) | Chapters to Read |
|---|---|
| Clone a gene, purify the protein, perform UV-Vis and/or infrared spectroscopy | 2, 5, 16 |
| Clone a gene, purify the protein, crystallize, perform x-ray or electron crystallography | 2, 5, 6 |
| Clone a very difficult gene, quantify its expression | 2, 4 |
| Clone a gene, express in cell line, examine by fluorescence microscopy (with additional labels) | 2, 3, 7–8 [11–13] |
| Clone a gene, express in cell line, examine by electron microscopy (with additional labels) | 2, 3, [11–13], 16 |
| Clone a gene, express in cell line, examine by holographic microscopy | 2, 3, 9 |
| Clone a gene, purify the protein, express it in a lipid bilayer, perform electrophysiology | 2, 5, 15 |
| Clone a gene, express in cell line, perform electrophysiology (with additional labels) | 2, 3, 15 [7, 11–13] |
| Clone a gene, purify the protein, biofunctionalize a surface, characterize the surface | 2, 5, 14 |
| Clone a gene, express in cell line, examine cellular toxicity | 2, 3, 10 |
| Clone a gene, express in cell line, pattern cells on microfabricated surface | 2, 3, 17 |

## End-of-Chapter Questions and Problems

*Biochemistry*

1. Where does *valence* come from? Can you tell the valence of an element by looking at its position in the periodic table? What are the valences of the following elements: H, He, O, K, Ar, Cl, Fe?

2. Look up the structural formulas of the following compounds, draw them, and identify their functional groups: cholesterol, vitamin E, dopamine, amphetamine, and aspartame. What are their alternative names, if any?

3. The concept of *acidity constant* $K_a$ or $pK_a$ comes from the equilibrium between the associated form of an acid (HA) and its dissociated form (A$^-$) in water:

$$HA + H_2O \Leftrightarrow H_3O^+ + A^-$$

$$K_a = \frac{[H_3O^+][A^-]}{[HA]}$$

$$pK_a = -\log K_a.$$

pH is simply a special case of the hydronium ion:

$$pH = -\log [H_3O^+].$$

Using the values of $pK_a$ given in **Figure 1.2**, answer the following questions for glutamic acid, histidine, arginine, and lysine. (a) Which of these amino acids is the most acidic? (b) What percentage of each of these amino acids would be in its dissociated form at pH 7.4 (physiological pH)?

4. Draw the following functional groups and discuss their importance: amino, hydroxyl, carboxyl, and phosphate.

5. Match the following (more than one may apply!):

____Triacylglycerols

____Phospholipids

____Sphingomyelins

____Cholesterol

____Glycogen

____Olestra

    a.  Nonpolar lipids

    b.  Found in fat cells

    c.  Look like steroids

    d.  Have a glycerol backbone

    e.  Polar lipids

    f.  Glucose storage in animal cells

    g.  Insulate neuronal axons

    h.  Alter membrane fluidity

    i.  Found in biological membranes

    j.  Artificial fat used as a diet food additive

**6.** Draw and label a segment of lipid bilayer membrane showing the fatty acids, cholesterol, and membrane proteins.

## Energies and Potentials

**7.** What is the speed of a molecule of nitrogen ($N_2$) at room temperature?

**8.** Derive **Equation 1.1** starting from Coulomb's law.

**9.** Derive **Equation 1.2**.

**10.** Polarizability of a nonpolar molecule may be modeled as a spherically symmetric electron cloud surrounding a positively charged nucleus. An externally applied electric field causes a shift in the electron cloud, resulting in an internal field developing at the nucleus to oppose it. Derive a formula for polarizability as a function of electron cloud radius, $a$, based upon this model. If you apply this formula to water, with an assumed radius of ~0.1 nm, what value do you obtain? Compare with the literature value of $1.66 \times 10^{-40}$ mC²/N.

**11.** Which of the following molecules have a permanent dipole moment? Quadrupole moment?

    a.  $H_2O$

    b.  $CO_2$

    c.  $CH_4$

    d.  $N_2$

    e.  CO

    f.  $NH_3$

**12.** Contrary to most other substances, the density of water decreases as temperature decreases between 4°C and 0°C, and then even further upon freezing. Why? How much does the density change between 0°C and 4°C, and upon freezing?

**13.** Calculate the dipole moment of water given that the HOH angle is 104.5°, OH length is 0.98 Å, and partial charges are +0.4 on H and −0.8 on O. Express your answer in debyes (D).

**14.** Take a close look at one of the DNA bases, adenine, and calculate its dipole moment in debyes. Assume the molecule is planar. Then, given the partial charges and $x$–$y$ coordinates in the picture below (given as $q$ in esu, $x$, $y$ in angstroms), calculate the dipoles in the $x$ and $y$ direction and the total moment. Draw an arrow showing the direction of the moment vector. Compare with the measured value of 3.0 D. (ADVANCED: Use molecular modeling software to calculate these partial charges.)

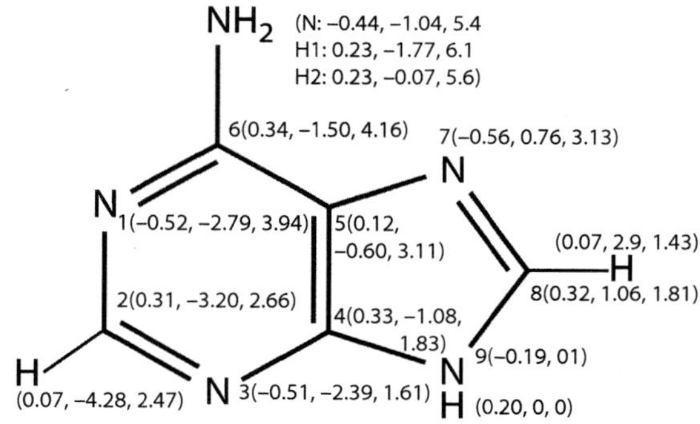


**15.** Calculate the electrostatic potential, $V(r)$, for two water molecules oriented (a) in a configuration permitting hydrogen bonding and (b) in a configuration unfavorable to hydrogen bonding, given that the HOH angle is 104.5°, OH length is 0.96 Å, partial charges are +0.4 on H and −0.8 on O, and hydrogen bond length is 2.5 Å.

16. (ADVANCED) Use molecular modeling software to predict the hydrogen bonds that can occur (a) between adenine and thymine and (b) between guanine and cytosine. Which of these actually occur in DNA?

17. What is the most stable conformation for pi–pi stacking? In which conformation is the interaction repulsive? (ADVANCED: Use modeling software to quantify.)

## Spectroscopy

18. To study crystal diffraction, neutrons of energy approximately 0.33 eV are required. What is the corresponding energy of a photon of the same wavelength? An electron? Is it safe to treat the neutrons nonrelativistically?

19. (a) Show that a hydrogen atom has energy levels $E_n = -E_0/n^2$ where $E_0 = 13.6$ eV. (b) Calculate the wavelengths of the spectral lines corresponding to the $2 \to 1$ transition, the $3 \to 1$ transition, and the $3 \to 2$ transition. Which is the one corresponding to the common telescope filter?

## Cells

20. Estimate the volume of a prokaryotic cell (radius, 1 µm) and of a eukaryotic cell (radius, 10 µm).

21. Estimate the minimum size of a complete living cell containing a genome of 1 megabase (each nucleotide pair is 0.34 nm long) and 15,000 ribosomes each 20 nm in diameter. How is the DNA packaged to fit inside the cell?

22. Estimate the maximum size of a bacterial cell. Hints: for a sphere of radius $R$, the flux across the surface is $Q = 4\pi DRC$, where $D$ is the diffusion coefficient and $C$ is the external concentration of the diffusing molecule; a reasonable metabolic rate is about 2 W/kg; the energy yield of glucose is about 260 kcal/mol; the diffusion coefficient of glucose is $6.7 \times 10^{-10}$ m²/s; and the external concentration is 50 mM. Does your answer seem reasonable? What factors could have made it too high or too low? (ADVANCED: Derive the formula for Q.)

23. If a cell of radius $r$ divides into two cells with the interior volume being conserved, how much new membrane has to be made?

## DNA, RNA, Proteins

24. Estimate how long it would take to replicate the *E. coli* genome from a single replication point. Explain how *E. coli* can divide every 20 min.

25. (a) Estimate how many mRNAs can be made in 20 min if the rate-limiting step is the formation of the open complex ($RP_o$) from the closed complex ($RP_c$). The rate equation is

$$[\text{RNAp}] + [\text{P}] \underset{k_{-1}}{\overset{k_1}{\rightleftharpoons}} [\text{RP}_c] \xrightarrow{k_2} [\text{RP}_o],$$

where [RNAp] and [P] are the concentrations of RNA polymerase and promoters, respectively. First, show that

$$[\text{RP}_o] = [\text{P}_{\text{tot}}](1 - \exp^{-k_{\text{obs}}t})$$

where

$$k_{\text{obs}} = \frac{k_1 k_2 [\text{RNAp}]}{k_1[\text{RNAp}] + k_{-1} + k_2} \equiv \frac{k_2[\text{RNAp}]}{[\text{RNAp}] + k_B^{-1}}$$

and

$$P_{\text{tot}} = [\text{P}] + [\text{RP}_c] + [\text{RP}_o].$$

Now use values of $k_2 = 0.04$/s, $k_B = 10^7$/M, and [RNAp] = 30 nM to get a numerical result. Is this consistent with the replication rate?

(b) Discuss how you would distinguish experimentally whether $k_{-1}$ or $k_2$ serves as the rate-limiting parameter.

(c) If the half-life of an mRNA molecule is 3 min before it is degraded by enzymes, what is the approximate equilibrium number of mRNAs in the cell?

26. Estimate the average translation rate (codons/s) in a cell if a cell is about 30% by weight protein and divides every 20 min. Then assuming 15,000 ribosomes, what is the translation rate per ribosome?

27. How many different polypeptides of 50 amino acids could, in principle, be produced? How

many different secondary structures could be produced from a given sequence, if each secondary domain is 10 amino acids long and can be either an alpha helix or beta sheet?

28. Match the following:

    a. Replication fork

    b. Origin of replication

    c. Lagging strand

    d. Leading strand

    e. DNA helicase

    f. Single-strand binding protein

    g. DNA gyrase

    h. Primase

    i. DNA polymerase III holoenzyme

    j. DNA polymerase I

    k. DNA ligase

    __Proofreads newly synthesized DNA

    __Joins lagging strand edges

    __Is an RNA polymerase

    __Prevents unwound DNA from re-forming base-paired helix

    __Is synthesized discontinuously

    __Relieves stress induced by positive supercoiling

    __Removes RNA primers

    __Requires ATP to add negative supercoils to DNA

    __First place where primosome functions

    __Polarity of synthesis is opposite to replication fork movement

    __Unwinds DNA at replication fork

    __Synthesizes most of DNA during replication

29. Discuss some consequences of errors in DNA replication. Some types of mutations are deletions, substitutions, insertions, and frameshifts. Give short descriptions of each and discuss how they might affect the downstream protein.

30. Why is thymine used in DNA instead of uracil?

31. Which is faster, DNA polymerase I or III? What are their respective rates of replication? Why is this the case?

32. Describe one of the experimental methods that gave evidence that the genetic code is a triplet code.

33. Choose a protein–cofactor, protein–drug, or similar interaction of your choice (e.g., acetylcholine in the acetylcholine receptor). Go to the protein database (PDB) to look it up and import it into Protopedia. Then illustrate the polar, charged, and hydrophobic residues. Indicate which areas of the protein interact with the drug or cofactor, and what types of noncovalent interactions dominate. Discuss what else the protein interacts with and the biological implications.

## Background Reading

## Books

Alberts, B., Johnson, A., Lewis, J., Raff, M., Roberts, K., and Walter, P. *Molecular Biology of the Cell.* Garland Science, 2008.—This is an outstanding textbook on cell biology that should be read by anyone with an interest in the biological sciences. Garland Science, New York, NY.

Atkins, P., and De Paula, J. *Atkins' Physical Chemistry.* Oxford University Press, 2006.—An excellent, mathematically intensive introduction to physical chemistry. Oxford University Press, Oxford, UK.

Carey, F., and Sundberg, R. *Advanced Organic Chemistry: Part A: Structure and Mechanisms.* Springer Science, 2007.—One of the few texts to cover advanced concepts in organic chemistry. Springer Science, Berlin, Germany.

Hettema, H. *Quantum Chemistry: Classic Scientific Papers.* World Scientific Books, 2000.—English translations of (originally German) papers by Born, London, and others, covering atoms, bonds, spectroscopy, and intermolecular interactions. World Scientific Books, Hackensack, NJ.

Kaplan, I.G. *Intermolecular Interactions*. Wiley, 2006.—Excellent coverage of the mathematics needed for intermolecular interaction calculations. Includes dozens of model potentials. John Wiley & Sons, Hoboken, NJ.

Lehninger, A., Nelson, D., and Cox, M. *Lehninger Principles of Biochemistry*. W. H. Freeman, 2008.—Classic biochemistry text with superb illustrations and problems. WH Freeman, New York, NY.

Pavia, D.L., Lampman, G.M., Kriz, G.S., and Vyvyan, J.A. *Introduction to Spectroscopy*. Brooks Cole, 2008.—Classic introduction to spectroscopy by the author of several chemical methods textbooks. Brooks Cole, Boston, MA.

Solomons, T.W.G., and Fryhle, C.B. *Organic Chemistry*. John Wiley & Sons, 2007.—Classic organic chemistry text with emphasis on applications to biology and very clear presentation. John Wiley & Sons, Hoboken, NJ.

Stone, A.J. *The Theory of Intermolecular Forces*. Clarendon Press, 1996.—A comprehensive and modern coverage of intermolecular forces in biophysical chemistry. Clarendon Press, Oxford, UK.

Tinoco, I., Sauer, K., Wang, J.C., and Puglisi, J.D. *Physical Chemistry: Principles and Applications in Biological Sciences*. Prentice Hall, 2002.—An excellent, highly physical introduction to biological concepts. Prentice Hall, Upper Saddle River, New Jersey.

Tuszynski, J.A. *Molecular and Cellular Biophysics*. Chapman & Hall, 2008.—A one-of-a-kind introduction to biological concepts from a physicist's point of view. Chapman & Hall, New York, NY.

## Journal articles

Alfasi, S., Sevastsyanovich, Y., Zaffaroni, L., Griffiths, L., Hall, R., and Cole, J. (2011). Use of GFP fusions for the isolation of *Escherichia coli* strains for improved production of different target recombinant proteins. *Journal of Biotechnology* 156, 11–21.

Bayer, E.A., Skutelsky, E., and Wilchek, M. (1979). The avidin–biotin complex in affinity cytochemistry. *Methods in Enzymology* 62, 308–315.

Bratthauer, G.L. (2010). The avidin–biotin complex (ABC) method and other avidin–biotin binding methods. *Methods in Molecular Biology* 588, 257–270.

Izrailev, S., Stepaniants, S., Balsera, M., Oono, Y., and Schulten, K. (1997). Molecular dynamics study of unbinding of the avidin–biotin complex. *Biophysical Journal* 72, 1568–1581.

Koch, A.L. (1996). What size should a bacterium be? A question of scale. *Annual Review of Microbiology* 50, 317–348.

Kyratsous, C.A., Silverstein, S.J., DeLong, C.R., and Panagiotidis, C.A. (2009). Chaperone-fusion expression plasmid vectors for improved solubility of recombinant proteins in *Escherichia coli*. *Gene* 440, 9–15.

LaVallie, E.R. (2001). Production of recombinant proteins in *Escherichia coli*. *Current Protocols in Protein Science*, 00, 5.1, 5.1.1–5.1.8.

Mamat, U., Wilke, K., Bramhill, D., Schromm, A.B., Lindner, B., Kohl, T.A., Corchero, J.L., Villaverde, A., Schaffer, L., Head, S.R., Souvignier, C., Meredith, T.C., and Woodard, R.W. (2015). Detoxifying *Escherichia coli* for endotoxin-free production of recombinant proteins. *Microbial Cell Factories*, 14, 57.

McClure, W.R. (1980). Rate-limiting steps in RNA chain initiation. *Proceedings of the National Academy of Sciences U S A* 77, 5634–568.

Prescott, D.M., and Kuempel, P.L. (1972). Bidirectional replication of the chromosome in *Escherichia coli*. *Proceedings of the National Academy of Sciences U S A* 69, 2842–2845.

Rothman-Denes, L.B. (2013). Structure of *Escherichia coli* RNA polymerase holoenzyme at last. *Proceedings of the National Academy of Sciences U S A* 110, 19662–19663.

Shimada, K., and Koga, H. (2009). High-throughput production of the recombinant proteins expressed in *Escherichia coli* utilizing cDNA resources. *Methods in Molecular Biology* 577, 83–96.

Sivashanmugam, A., Murray, V., Cui, C., Zhang, Y., Wang, J., and Li, Q. (2009). Practical protocols for production of very high yields of recombinant proteins using *Escherichia coli*. *Protein Science* 18, 936–948.

Tougu, K., and Marians, K.J. (1996). The interaction between helicase and primase sets the replication fork clock. *Journal of Biological Chemistry* 271, 21398–213405.

Walls, D., and Loughran, S.T. (2011). Tagging recombinant proteins to enhance solubility and aid purification. *Methods in Molecular Biology* 681, 151–175.

Wilkinson, D.L., and Harrison, R.G. (1991). Predicting the solubility of recombinant proteins in *Escherichia coli*. *Biotechnology (N Y)* 9, 443–448.

Young, R., and Bremer, H. (1976). Polypeptide-chain-elongation rate in *Escherichia coli* B/r as a function of growth rate. *Biochemical Journal* 160, 185–194.

## Online resources and software

### *Databases*

Entrez. Search engine and database for biomedical journals (PubMed), DNA/RNA sequences (Nucleotide), protein sequences (Protein), structures (Structure), and more. The first place to go to find the sequence of a gene or protein.

SRS. Sequence retrieval system.

NIH Center for Molecular Modeling. National Institutes of Health site with links to software and databases of relevance to biochemistry and modeling. http://cmm.cit.nih.gov/modeling/

## Software

Wikipedia maintains a list of computational chemistry software packages and their capabilities here: https://en.wikipedia.org/wiki/List_of_quantum_chemistry_and_solid-state_physics_software

The following are some commonly used commercial quantum chemistry packages:

Amsterdam Density Functional (ADF). Semiempirical, Hartree–Fock, and density functional theory.

Gaussian. All methods. Frequently updated. Very commonly used.

Jaguar. Ab initio and Density Functional Theory (DFT); focus on metal-containing systems.

SCIGRESS. Many methods; no Hartree–Fock.

Spartan. All methods. Easy-to-use Graphical User Interface (GUI).

TURBOMOLE. Ab initio methods, Hartree–Fock and post-Hartree–Fock.

The following are some commonly used quantum chemistry packages that are free for academic use but not necessarily open source:

DALTON. Hartree–Fock, post-Hartree–Fock, and density functional theory.

DIRAC. Hartree–Fock, post-Hartree–Fock, and density functional theory.

GAMESS. All methods.

The following are some commonly used open-source quantum chemistry packages:

MONSTERGAUSS. Started as open-source answer to Gaussian, with other features. Hartree–Fock.

PSI4. Hartree–Fock and density functional theory.

OpenAtom. Molecular mechanics and density functional theory.

# CHAPTER 2

# Basic Molecular Cloning of DNA and RNA

## 2.1 INTRODUCTION

Modern molecular biology is about manipulating small amounts of invisible, highly sensitive molecules where there is often no direct evidence that what is in the tube is what you think it is. It begins usually with a catalog, a list of naturally occurring enzymes that have been identified, purified, and in some cases mutated in order to facilitate the manipulation of DNA and RNA molecules. The currency of most cloning experiments is the *plasmid*, a circular piece of DNA found in bacteria that usually ranges from 2,000 to 14,000 base pairs (or 2–14 kb) in length. (For comparison, the genome of *Escherichia coli* is 4.6 Mb.) (See **Practical Tips 2.1**.) A plasmid usually contains all of the following features:

- An *origin of replication* (ORI). This is a 50- to 100-base-pair sequence to which host (*E. coli*) enzymes bind and signal the replication of the entire plasmid. This allows the plasmid to be replicated in *E. coli* cells, so that the experimenter can produce as much of it as desired.

- A promoter. As discussed in **Chapter 1**, a promoter is a sequence that permits RNA polymerase to bind, and thus the gene sequence that is downstream of this promoter to be expressed. Depending upon the experiment desired, this promoter can be weak or strong; inducible (i.e., requiring a nutrient or chemical to turn on) or constitutive (always on); and bacterial or eukaryotic. If a plasmid has a bacterial ORI but a mammalian promoter, the DNA of the plasmid itself is replicated in *E. coli*, but the genes it encodes are only expressed if the plasmid is put into mammalian cells.

- One or more genes of interest. These occur downstream of the promoter.

- A *selectable marker*, which is almost always a gene encoding for resistance to a specific antibiotic under the control of a bacterial promoter. All bacteria containing the plasmid will be resistant to the antibiotic, and those without it almost always will not be, allowing the experimenter to selectively amplify bacteria containing the plasmid in culture medium that contains that antibiotic. The most commonly used selection antibiotics in cloning are ampicillin (Amp), kanamycin (Kan), and tetracycline (Tet).

## PRACTICAL TIPS 2.1:    PLASMIDS

Bacteria reproduce asexually, with cell division giving rise to two genetically identical daughter cells each containing a circular genome (chromosome) on the order of a few hundred thousand to a few million base pairs in size. If they had no means of exchanging genetic information, their evolution would be restricted to the rate of random mutations, much too slow to adapt to environmental changes.

However, bacterial evolution is driven not by changes to the primary genome, but to what is called the *mobilome*: pieces of DNA that are separate from the primary chromosome but can interact with it as well as being carried from one bacterium to another via horizontal gene transfer (HGT). (The passage of DNA to a daughter cell is referred to as vertical gene transfer.) The size and composition of the pieces and the means of HGT vary widely; some are due to viruses and are carried by infection, and others are spread by direct *conjugation* of two bacteria or even by uptake from the environment (*transformation* or *transduction*, which we will take advantage of in the laboratory). A *plasmid* is a small (1–20 kb) circular piece of DNA that is usually passed between organisms by conjugation (certain plasmids, called conjugative plasmids, are able to initiate this process) (**Figure P2.1.1**).

Unless it bears a gene conferring resistance to an antibiotic or the ability to metabolize a common element in a restricted medium, the presence of a plasmid lowers the bacterium's fitness, as it takes energy to replicate and maintain. The term *plasmid* was coined in 1952 by Joshua Lederberg, who won the 1958 Nobel Prize in Physiology or Medicine for the discovery that bacteria can exchange genes. Since then, it has been found that HGT is responsible for crucial differences in bacterial properties, such as the variation in virulence among *E. coli* O157 strains and multidrug resistance in species of *Salmonella*, *Staphylococcus*, and others. However, only since the availability of full genome sequencing has HGT's importance in overall bacterial evolution become appreciated.

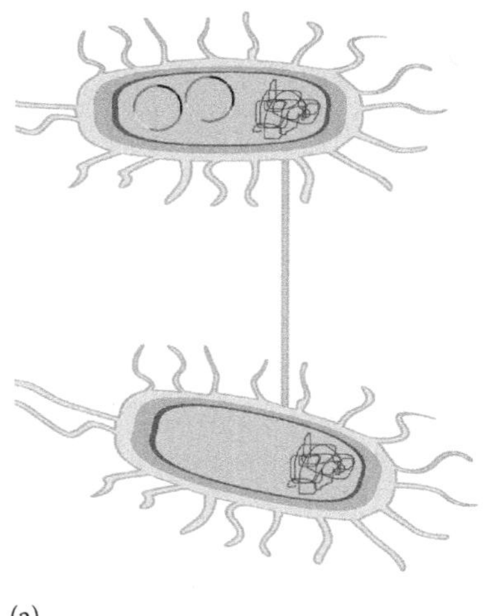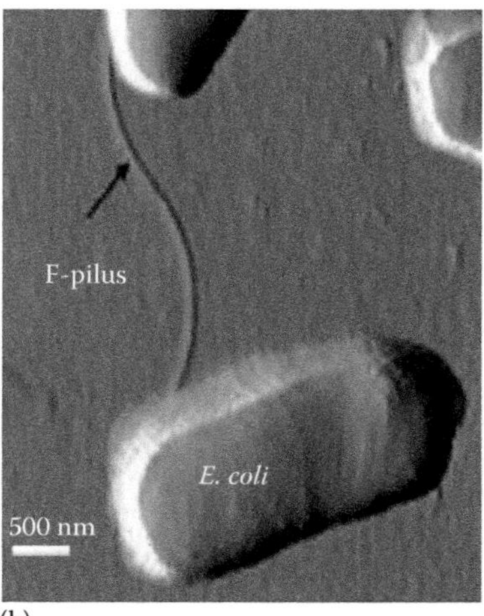


**Figure P2.1.1 Plasmids and horizontal gene transfer.** (a) Schematic of a bacterial cell containing plasmids (top) connected by a conjugation pilus to a cell with only genomic DNA and no plasmids (bottom). The plasmids themselves give rise to the pilus, which is a protein bridge permitting DNA to be exchanged. (b) Electron micrograph of conjugating *E. coli*.

(a)                    (b)

### RECOMMENDED REVIEW ARTICLES

### Classic

Meynell, E., Meynell, G.G., and Datta, N. (1968). Phylogenetic relationships of drug-resistance factors and other transmissible bacterial plasmids. *Bacteriological Reviews* 32, 55–83.
Richmond, M.H. (1965). Penicillinase plasmids in *Staphylococcus aureus*. *British Medical Bulletin* 21, 260–263.

(*Continued*)

**PRACTICAL TIPS 2.1 (CONTINUED): PLASMIDS**

**Modern**

Ahmed, N., Dobrindt, U., Hacker, J., and Hasnain, S.E. (2008). Genomic fluidity and pathogenic bacteria: applications in diagnostics, epidemiology and intervention. *Nature Reviews Microbiology* 6, 387–394.

Bower, D.M., and Prather, K.L. (2009). Engineering of bacterial strains and vectors for the production of plasmid DNA. *Applied Microbiology and Biotechnology* 82, 805–813.

Nikaido, H. (2009). Multidrug resistance in bacteria. *Annual Review of Biochemistry* 78, 19–146.

Hundreds to thousands of copies of a plasmid can be made inside a single bacterium, allowing for rapid and efficient amplification of this specific sequence. These plasmids can then be isolated from the bacterial cells, resulting in ultrapure solutions of plasmid in water at concentrations up to several milligrams per milliliter.

The trick for the molecular biologist is to insert a sequence of interest into the correct plasmid for the desired experiment. This chapter will illustrate each of the steps in amplifying, purifying, and screening plasmid DNA, and will illustrate design of a ligation experiment with an example that includes key troubleshooting steps. At the end of this section, you should be comfortable with basic plasmid manipulations and be able to design your own cloning experiment. We then survey several other cloning techniques of special interest to molecular biophysics, including some approaches to cloning large fragments of DNA and methods of mutating and rearranging genes.

## 2.2 OBTAINING AND STORING PLASMIDS

Many hundreds of plasmids are available commercially; the primary suppliers are listed at the end of this chapter. Some plasmids are empty except for a resistance gene and a promoter; these are known as *cloning vectors* (**Figure 2.1a**). More complex plasmids are intended for expression of the protein in *E. coli*, in which case they are called *bacterial expression vectors* (**Figure 2.1b**). Expression vectors for eukaryotic cells (yeast, mammalian cells, plants, etc.) contain an entire expression sequence that permits the gene to express in these cells (**Figure 2.1c**). Plasmids for expression of genes in bacteria other than *E. coli* and its relatives also usually use such an expression sequence, and *E. coli* is used as an intermediate because of its ease of use for cloning; this will be discussed further in **Chapter 3**.

It is often possible to purchase a plasmid containing your gene of interest and use it directly in experiments. For example, if you want to express green fluorescent protein–labeled *actin* (AcGFP) in mammalian cells, the plasmid in **Figure 2.1c** will do the trick. However, sometimes the commercial plasmid contains the wrong promoter, the wrong resistance gene, or other undesirable features. In this case, the gene of interest can be removed from its original plasmid and put into a plasmid with the necessary features. For example, if you wanted to express AcGFP in mammalian cells only in the presence of tetracycline, then the constitutive promoter in **Figure 2.1c** would not work. You would cut out the AcGFP gene from this plasmid and place it into a plasmid containing a tetracycline-inducible promoter. This is an example of a *molecular cloning* experiment; we will go through this example in detail later in the chapter.

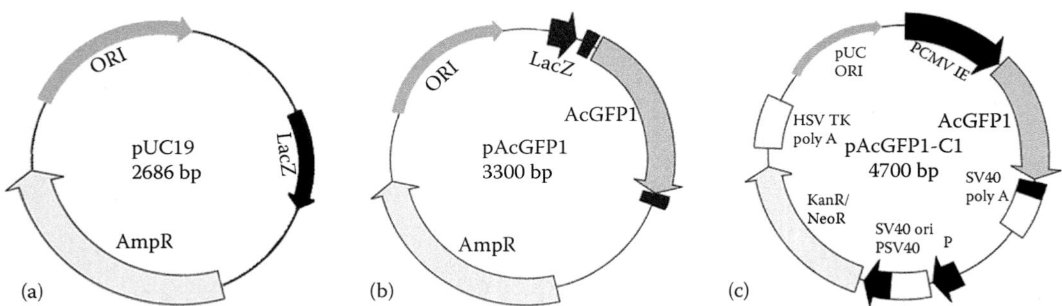

**Figure 2.1 Plasmids for cloning and expression.** (a) A commonly used basic cloning plasmid vector, pUC19. It has an origin of replication (ORI) and encodes a resistance gene for ampicillin (AmpR) as well as a fragment of the gene for β-galactosidase (LacZ) that can be used as a colorimetric screen for the plasmid in some strains of bacteria. (b) A bacterial expression vector, pAcGFP1, made from a pUC19 base. The AcGFP1 gene is inserted immediately downstream of the LacZ. This means that the bacteria will make a protein that is the fusion of β-galactosidase and AcGFP1, which encodes a green fluorescent protein. (c) Plasmids for expression in mammalian cells are more complex. This example vector encodes the same AcGFP1 fluorescent protein as in (b). However, two things are necessary for expression in mammalian cells. The first is a mammalian *promoter* sequence, the cytomegalovirus immediate early promoter (PCMV IE). This will permit expression in nearly all mammalian cell types at a high level. After the fluorescent protein gene is a polyadenylation sequence (SV40 polyA), which permits correct processing of the mRNA. Instead of AmpR, this plasmid contains a resistance gene that is useful in both bacteria and mammalian cells: KanR/NeoR. (Kanamycin is used in bacteria, neomycin or G418 in mammalian cells.) This gene is preceded by two promoters, a bacterial promoter (P) and PSV40, which permits expression in mammalian cells. Another polyadenylation sequence is needed for correct processing of this second gene.

Another very common type of situation is that you wish to express a gene that has been described in the literature but that is not available commercially. In this case, the procedure is to contact the author and request a sample of the plasmid described in his/her published work. Authors in most molecular biology journals are required to make these plasmids available upon request. This sequence may then be used as is, or the gene or pieces of the gene may be removed from the host plasmid and placed into a vector of your choice.

Whether commercial or from an individual laboratory, the plasmid will come in one of three forms:

- As a solution at a given concentration in water or simple buffer

- *Lyophilized* (dry) in a given amount (micrograms)

- As a bacterial *stab* (a culture of *E. coli* containing the plasmid "stabbed" into nutrient *agar*)

Dissolved plasmids should be frozen at −20°C. Lyophilized plasmids may be stored at 4°C until they are resuspended in water; then they should be frozen. Bacterial stabs should be stored at 4°C. In all cases, the plasmid should be amplified before any further experiments are done, to make sure enough is available and to store stocks in a stable form for future use.

## 2.3 SELECTION OF AN APPROPRIATE *E. COLI* AMPLIFICATION STRAIN; TRANSFORMATION OF *E. COLI* WITH PLASMID

### Transformation

If the plasmid comes by itself and not inside cells (i.e., in solution or lyophilized), it will first need to be *transformed* into an appropriate *amplification strain* of *E. coli*. These are procedures that modern molecular biology suppliers have made routine.

*Transformation* refers to the process of inserting the plasmid into *E. coli* cells, where it can be replicated. *E. coli* will not simply take up plasmid DNA from its environment; it must be made *competent* to do so. Amplification strains of *E. coli* are sold in small vials (aliquots) of *chemically competent* or *electrically competent* cells. Each vial is designed for a single cloning experiment. A small amount of plasmid DNA is placed into the vial, and the cells are exposed to heat (for chemically competent cells) or an electric field (for electrically competent cells). Each of these procedures is thought to result in the opening of minute pores in the *E. coli* membrane that permit the plasmid DNA to enter. No special equipment apart from a heat plate or heated water bath is needed to use chemically competent cells. (The "chemically" refers to how they are made, not how they are transformed.) However, to use electrically competent cells, an *electroporator* is needed. This is a specialized instrument that applies a specific voltage across *electroporation cuvettes*, which are plastic cell holders containing two parallel metal plates (**Figure 2.2**).

Which amplification strain should be chosen? There are a few common ones that can be used for nearly all plasmids. However, if the plasmid is especially large or if it shows signs of instability (see the end of **Section 2.4** for further discussion), a particularly stable strain may be used (**Table 2.1**).

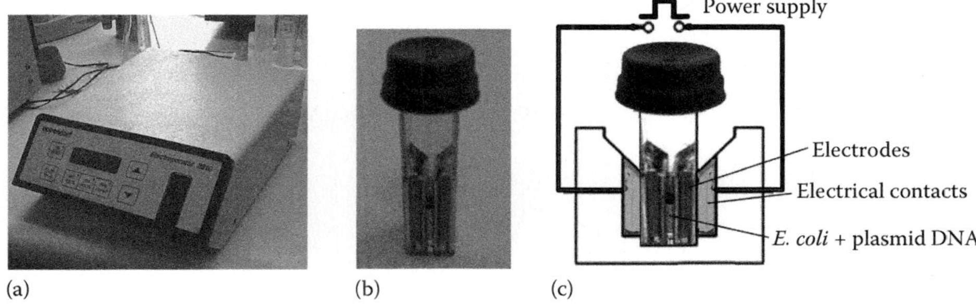


**Figure 2.2 Electroporation.** (a) Electroporator. (b) Electroporation cuvette (holds about 1 mL, path length usually 1 cm). (c) Schematic of how it works. A critical value of the electric field is required to open pores in the cell (usually 10 kV/cm for bacteria). The solution containing the *E. coli* must be low in electrolytes (salt), or the electrodes will arc, making a loud noise and killing all of the cells.

**Table 2.1**

Commercially Available Expression Strains of *E. coli* and Their Recommended Uses

| Strain | Uses | Transformation Efficiency |
|---|---|---|
| TOP10 | Routine cloning. Very commonly used. | $10^9$ (chemical) $10^{10}$ (electro) |
| DH10B | Routine cloning. Very commonly used. | . $10^9$ |
| DH5α | Routine cloning. Very commonly used. | $10^9$ |
| Stbl2, Stbl3, Stbl4 | For unstable plasmids. Stbl4 can be used for very large plasmids (>200 kb). | $10^8$–$10^9$ |
| Mach1 | Very fast-growing for rapid turnover times. | $10^9$ |
| MegaX DH10B Electrocomp | Highest transformation efficiency (tricky ligations, etc.). | $>3 \times 10^{10}$ |
| INV110 | Prepares unmethylated DNA. | $10^6$ |

## Selection

After transformation, the cells are plated onto a *selective plate*. This is a petri dish containing *nutrient agar* plus the antibiotic whose resistance gene is expressed in the plasmid. (Recipes for different nutrient media are given in **Chapter 3**.) Good plating technique (**Figure 2.3a**) ensures the growth of single bacterial *colonies* (**Figure 2.3b**). Each colony is made up of genetically identical cells, so it is good microbiological practice to always work with cells from a single, distinct colony. When too many bacteria are plated at once, this results in a *lawn* in which individual colonies cannot be distinguished (**Figure 2.3c**). It is bad practice to take bacteria from a lawn. Similarly, cells should not be allowed to grow for too long, or the selective antibiotic will begin to degrade, and colonies of nonresistant (thus non-plasmid-containing) bacteria will begin to grow. These are called "satellite" colonies and should be ignored. Note that if the original plasmid was supplied as a stab, the material from the stab is plated directly onto a plate in the same fashion as the transformed bacteria. A plate is stable at 4°C for a week or so unless

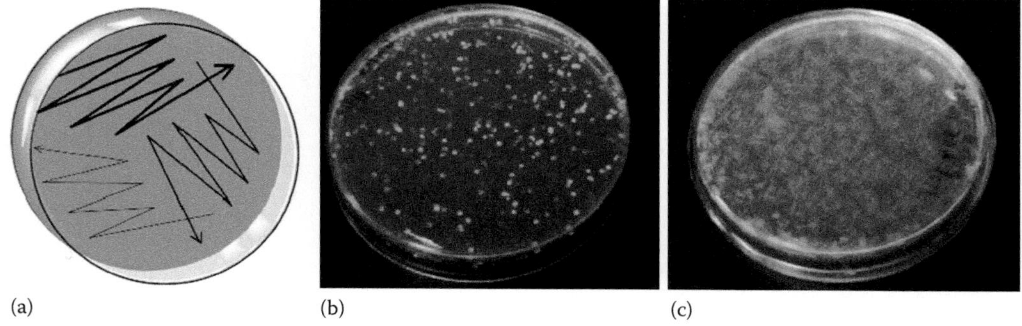

(a)   (b)   (c)

**Figure 2.3 Plating bacteria.** (a) Method of streaking a droplet of transfected cells or a smear from a stab so as to obtain single colonies. Each successive streak contains a lower concentration of bacteria. (b) Plate showing single, discrete colonies. (c) Bacterial lawn with no distinguishable colonies.

it becomes contaminated with environmental fungi or other bacteria; this can be identified as colonies of an unexpected shape, size, or color on the plate.

## Transformation efficiency

The fraction of the transformed competent cells that should be plated to get a good number of colonies, not a lawn, can be estimated by adjusting the amount of DNA used and using the published *transformation efficiency* ($T$) of the strain. This is defined as the number of colonies obtained per microgram of DNA. Typical DNA concentrations used per reaction are 1–10 ng; more than 10 ng usually does not result in more colonies. Thus, the number of colonies obtained ($N$) is given by

$$N = T\,(transformants\,/\,\mu g\,) \times DNA\,transformed\,(\mu g\,) \times fraction\,plated \qquad (2.1)$$

If your transformation efficiencies are much lower than the published values, your technique is probably faulty. Treat the cells with care!

Competent cells are rather costly, and most labs prepare their own by setting aside one aliquot from a commercial batch and amplifying it. This is a recommended procedure but can be tricky. See **Practical Tips 2.2** for recipes and tips for the preparation of chemically and electrically competent cells. Transformation efficiencies should be determined for each batch by using a standard plasmid, such as pUC19, at a known concentration. For routine cloning experiments, efficiencies of $10^6$ transformants/μg are acceptable. However, for *ligation reactions, mutagenesis,* or other low-efficiency operations, values of $10^9$–$10^{11}$ are desirable. Some labs use homemade cells for routine operations and commercial cells for difficult cloning procedures.

# 2.4  PLASMID AMPLIFICATION AND PURIFICATION

## Amplification

Once single colonies have been obtained, they may be "picked" with a sterile pipette tip or toothpick and used to inoculate 3–5 mL of nutrient medium containing the selective antibiotic. The cells are grown for 12–16 h, at which point the culture should be visibly turbid. (If not, something has gone wrong, and the transformation should be repeated.) This volume of culture may be used to purify a small amount of DNA for screening purposes. This scale of purification is called a *miniprep*.

For plasmids of known sequence that will be used in experiments, larger amounts are desirable. In this case, the seed culture is diluted 1:100 into 50–100 mL (a *midiprep*) or 150–250 mL (a *maxiprep*) in fresh medium containing antibiotic. The cultures are grown for a further 12–16 h, and the cells are *pelleted* (centrifuged to the bottom of a tube) for DNA extraction.

## Purification

Traditional DNA purification methods involve ultracentrifugation on a CsCl density gradient. First the *E. coli* are broken open or *lysed*, and centrifuged to remove the major contaminants such as the cell wall. The lysate is then mixed with CsCl and a fluorescent dye (ethidium bromide [EtBr]) that intercalates into the DNA double helix. After 20 h of centrifugation, the plasmid DNA forms a distinct band that is identified by the dye fluorescence and removed with a needle.

**PRACTICAL TIPS 2.2:    PREPARATION OF ELECTRICALLY AND CHEMICALLY COMPETENT CELLS**

**TIPS**

- Start with fresh, discrete single colony.
- Make sure there is no detergent on the glassware. (Rinse well with distilled water.)
- Watch your cell densities carefully. Bacterial concentrations are estimated by measuring the optical density at 600 nm ($OD_{600}$) in a UV–Vis spectrophotometer. Do not let $OD_{600}$ go above log phase (different for different spectrometers; do a growth curve to determine exact value).
- Treat the cells very gently. Centrifuge at the lowest possible speed; do not pipette them up and down.

**FOR ELECTRICALLY AND CHEMICALLY COMPETENT CELLS**

**The Night Before**

Inoculate 100 mL of a very rich nutrient medium (e.g., 2x YT, see **Appendix 2**) with 10 µL of commercial competent cells. No antibiotics are added since there is no plasmid! Incubate 37°C for 14–16 h, shaking at 200 rpm.

**The Next Morning**

Keep *everything* on ice from now on. All pipettes, glassware, etc. should be chilled. You can even do the preparation in a cold room if you have access to one.

(1) Dilute the overnight culture 1:10 in fresh medium (for 1 L total volume). Place in shaker at 37°C and read $OD_{600}$ every 20 min until it reaches mid-log phase (usually ~$OD_{600}$ = 0.6).
(2) Pellet the cells by centrifugation for 20 min at 4°C and approximately 4000 g.

**FOR ELECTRICALLY COMPETENT CELLS**

(3) Resuspend each pellet in 35–40 mL of sterile cold water and transfer to a 50 mL centrifuge tube. Remove the supernatant, resuspend the pellet in fresh cold water, and repeat.
(4) Repeat these wash steps with 10% glycerol in sterile cold water. On the final wash, pool all of the cells together.
(5) Estimate the pellet volume and resuspend in an equal amount of 10% glycerol (or slightly less if very high levels of competence are desired).
(6) Prepare a slush of dry ice/ethanol.
(7) Distribute the cells into Eppendorf tubes in 50 µL aliquots and quick-freeze them in the slush.
(8) Transfer to a precooled cardboard box and store at –80°C for a year or more.

**FOR CHEMICALLY COMPETENT CELLS**

(3) Resuspend the pellet in 1/2 the original volume of sterile, cold 100 mM $CaCl_2$. Incubate on ice 20 min.
(4) Pellet the cells, resuspend in 1/10 the original volume of sterile, cold 100 mM $CaCl_2$. Incubate on ice 60 min.
(5) Add cold, sterile glycerol to a final concentration of 15%.
(6) Prepare a slush of dry ice/ethanol.
(7) Distribute the cells into Eppendorf tubes in 50 µL aliquots and quick-freeze them in the slush.
(8) Transfer to a precooled cardboard box and store at –80°C for a year or more.

The genomic DNA bands at a lower density than plasmid DNA. Although this method is time consuming and the reagents are toxic, it can be repeated to yield extremely pure plasmid DNA, is less costly than commercial kits for large-scale preparations, and is still widely used by those who need large amounts (hundreds of micrograms to milligrams) of pure plasmid DNA. Some references given at the end of the chapter provide protocols for these methods.

For routine applications, most laboratories use commercial plasmid purification kits that are based upon *exchange chromatography*. A resin or membrane is provided in a column along with a selection of buffers. The composition of the initial buffers favors binding of the DNA to the resin, usually based upon the molecule's negative charge. Once it is bound, it can be washed to remove impurities. The last buffer favors DNA dissociation, and the pure product is collected. Anion or cation exchange chromatography techniques are ubiquitous in biochemistry for purification of molecules, and we will not discuss them in detail here except to refer to basic textbooks and papers cited at the end of the chapter. The size of the column varies according to the amount of DNA to be purified; its binding capacity can be as high as 10 mg of DNA for a *gigaprep*. Most commercial kits yield plasmid DNA at least as pure as a single round of CsCl centrifugation, which is sufficient for nearly all applications. Some kits also contain an extra wash step for removal of the *E. coli* lipopolysaccharide (LPS), or *endotoxin*. These endotoxin-free or "endo-free" kits are recommended when the DNA is going to be used in eukaryotic cells, since LPS is toxic.

## Measuring concentration and purity of extracted DNA

Plasmid DNA is eluted from the purification columns in a supercoiled form. Its concentration and purity can be determined using ultraviolet–visible (UV–Vis) absorbance spectroscopy; nucleic acids absorb most strongly at 260 nm ($A_{260}$), while proteins absorb at 280 nm and nonspecific turbidity can be measured further toward the visible (usually $A_{320}$) (Figure 2.4). Concentration is calculated using Beer's law:

$$c = \frac{A}{\varepsilon \ell},$$

(2.2)

where $A$ is $A_{260}$(sample) − $A_{260}$(blank) or $A_{260}$ − $A_{320}$, $\ell$ is the cuvette path length (usually 1 cm), and $\varepsilon$ is the extinction coefficient. The concentration of the sample

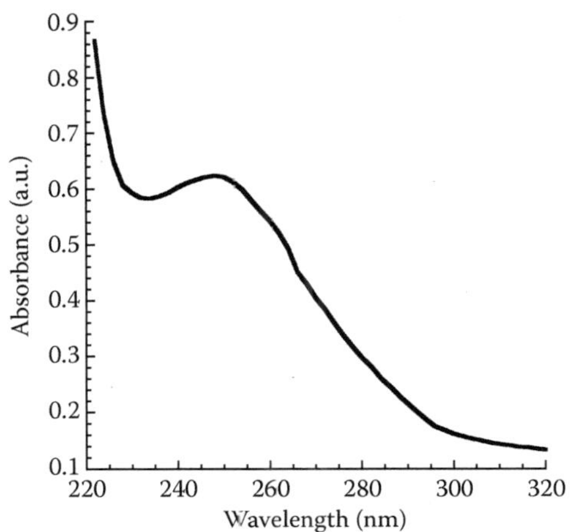


**Figure 2.4 Absorbance spectrum of purified plasmid DNA.**

should be adjusted so that $A_{260}$ is between 0.05 and 1.0, as more concentrated samples fall outside the linear concentration-versus-absorbance range, and more dilute samples are prone to error. For most maxipreps and minipreps, this means that the sample should be diluted 50–100 times. Values of $\varepsilon$ are determined by the length and sequence of the DNA but can be approximated as 6600 M$^{-1}$ cm$^{-1}$ for double-stranded DNA (or RNA) and 8500 M$^{-1}$ cm$^{-1}$ for single-stranded DNA (or RNA). (Note that absorbance techniques do not work well with quick and dirty miniprep kits, whose products are too full of contaminants to yield reliable values.)

The purity of the DNA can be estimated by the ratio $A_{260}/A_{280}$, since the major contaminants are usually proteins, which absorb strongly at 280 nm. A value of 2.0 indicates pure nucleic acid; a good value for a very clean midiprep or maxiprep is 1.8. Anything less than 1.6 is cause for concern. Again, this technique should not be used for most kit-based minipreps. This also cannot identify nucleic acid contaminants such as genomic DNA or cellular RNA. Finally, it is important to note that a preparation can contain pure plasmid DNA and still not be what was wanted. Some plasmids, particularly large ones, have a tendency to *recombine*, and entire large pieces of the sequence may be missing. It is also possible for bacteria to "expel" toxic genes, retaining only the antibiotic resistance. For these reasons, it is always necessary to screen any amplified plasmid by the technique of *restriction mapping*.

## 2.5  PLASMID RESTRICTION MAPPING AND AGAROSE GEL ELECTROPHORESIS

### Restriction enzymes

Restriction mapping is one of the most useful tools in molecular biology. It resulted from the identification and exploitation of the way in which DNA is exchanged and

**Table 2.2**

Some Restriction Enzymes, Illustrating Types of Ends Produced, Recognition Sequences, and Different Types of Degeneracy

| Enzyme Name (Organism) | Recognition Sequence and Cut Pattern | Notes |
| --- | --- | --- |
| Sma I (*Serratia marcescens*) | 5′—CCC↓GGG—3′<br>3′—GGG↑CCC—5′ | Isoschizomer of Xma I, produces blunt end |
| Xma I (*Xanthomonas malvacearum*) | 5′—C↓CCGGG—3′<br>3′—GGGCC↑C—5′ | Isoschizomer of Sma I, produces sticky end |
| Bam HI (*Bacillus amyloliquefaciens* H) | 5′—G↓GATCC—3′<br>3′—CCTAG↑G—5′ | Compatible sticky ends with Bg III |
| Bg III (*Bacillus globigii*) | 5′—A↓GATCT—3′<br>3′—TCTAG↑A—5′ | Compatible sticky ends with Bam HI |
| Bse YI (*Bacillus* sp. 2521) | 5′—C↓CCAGC—3′<br>3′—GGGTC↑G—5′ | Nonpalindromic |
| Bpm I (*Bacillus pumilus*) | 5′—CTCGAG(N)$_{16}$↓—3′<br>3′—GACCTC(N)$_{14}$↑—5′ | Cuts downstream of recognition site |
| Not I (*Nocardia otitidis-caviarum*) | 5′—GC↓GGCCGC—3′<br>3′—CGCCGG↑CG—5′ | 8-base-pair recognition site |

*Note:*  The name of the enzyme is an abbreviation of the scientific name of the organism from which it was isolated; if more than one from the same species is used, it is numbered II, III, etc. Note that Sma I and Xma I recognize the exact same site; they are thus called isoschizomers of each other. However, Sma I produces a blunt end, while Xma I yields sticky ends. Bam HI and Bg III are not isoschizomers of each other but produce compatible sticky ends. Also note that most restriction sites are palindromic (they read the same way 5′ to 3′ as 3′ to 5′), but Bse YI is an exception to this rule. Not all enzymes cut exactly where they recognize; "N" refers to any nucleotide.

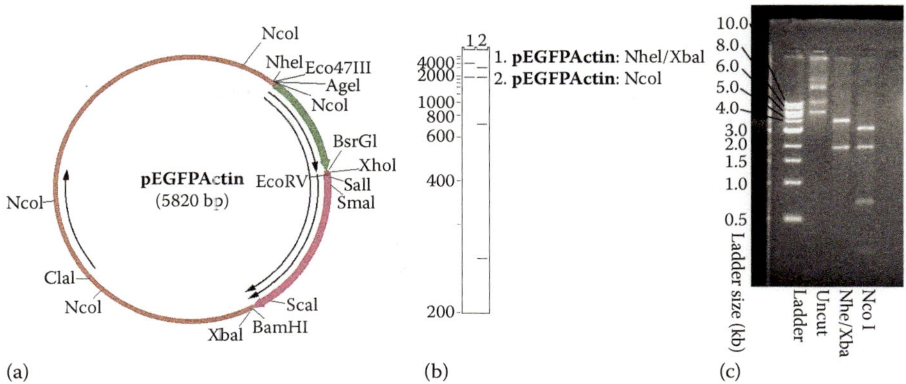


**Figure 2.5  Restriction analysis by mapping and agarose gel electrophoresis.** (a) Plasmid map of the plasmid pEGFP–actin, showing the gene coding regions and many of the restriction sites. Different computer programs are available to find these restriction sites based upon the plasmid sequence published by the manufacturer or on public databases. (b) Predicted appearance of the gel fragments if this plasmid were to be cut with (1) the enzymes Nhe I and Xba I or (2) Nco I. Even though Nco I is not a good enzyme for cloning, because it cuts the plasmid in too many places, it is useful for screening as it provides a good characteristic pattern with four different-sized fragments. (c) Actual appearance of the gel with plasmid cut as in (b). The leftmost lane contains the ladder; the brightest band (3.0 kb) contains 125 ng of DNA, while the other bands each contain ~40 ng. The next lane contains uncut plasmid, showing an unpredictable pattern. The third and fourth lanes are the restriction digests corresponding to (b).

eliminated in nature. Plasmids can only replicate in bacteria if they are circular; if cut open, they are simply degraded. A series of enzymes have evolved in bacteria to protect them against invading foreign DNA by recognizing and cleaving specific DNA sequences. These are known as restriction endonucleases or *restriction enzymes*, and are isolated and purified by molecular biology suppliers and sold in catalogs according to their target sequence. Most restriction enzymes identify and cut a 4- or 6-base-pair sequence; a few recognize 8- or 10-base-pair sequences and are especially useful when cloning using large plasmids. Sometimes, the recognized sequence can be degenerate (**Table 2.2**). Over a hundred of these enzymes are commonly used.

These specific sequences recognized by restriction enzymes, known as *restriction sites*, are a key element in plasmids used for cloning. The occurrence of restriction sites for the most common enzymes is used to create a unique *restriction map* of each plasmid (**Figure 2.5a**).

## Screening purified DNA

The plasmid map can be used in several ways. For screening, a small amount of plasmid (20–100 ng) is cut with one or more enzymes, resulting in two or more fragments of specific sizes. These fragments are then separated by a technique called *agarose gel electrophoresis*, which is based upon two principles: the negative charge of DNA and the ability of 0.5–2% solutions of agarose to gel into a hydrogel with pore sizes that limit the diffusion of linear DNA molecules according to size. Thus, if DNA is placed into the gel in an electrolyte solution and an electric field is applied, the fragments will migrate toward the cathode as a function of size. The precise physics behind the migration has not been determined (see **Advanced Topic 2.1**), but linear DNA of molecular weight MW migrates at a speed proportional to $\log^{-1}(\text{MW})$ (**Figure 2.5b**). The DNA is visualized using a fluorescent *intercalating* dye, usually EtBr (**Figure 2.5c**). Commercial *ladders* are available

**ADVANCED TOPIC 2.1:   HOW GEL ELECTROPHORESIS WORKS**

Many molecules have a net positive or negative charge at physiological pH; DNA has a net negative charge. Therefore, application of an electric field to a solution of DNA molecules will cause the molecules to migrate toward the positive electrode. The *electrophoretic mobility* of a molecule is defined as

$$\mu = \frac{v_{\text{migration}}}{E} = \frac{Ze}{f}, \tag{A2.1.1}$$

where
   $Z$ is the magnitude of charge on the molecule,
   $e$ is the elementary charge,
   $E$ is the magnitude of the applied field, and
   $f$ is the frictional coefficient (kg/s).

The trick in gel electrophoresis is to control the coefficient $f$ (and associated parameters) to such an extent that the desired mass resolution is obtained—that is, if a field is applied to the gel for a specific amount of time, molecules will be separated according to their molecular weights to a visible extent without moving off the end of the gel. The desired resolution varies greatly—from a single base pair for sequencing applications to hundreds of base pairs for screening plasmids and ligation products. In order to obtain this resolution, gels of different charges and pore sizes must be created. A gel is a three-dimensional network of pores that creates an effectively long path for the diffusing molecules, as well as physical and charge barriers to molecule movement. Mixing different concentrations of agarose will alter this network of pores and make the gel suitable for larger or smaller DNA molecules.

The precise hydrodynamics of gel electrophoresis have not been worked out.

that provide standardized weight markers at 100- or 1000-base-pair increments or some other calibrated values (**Figure 2.5c**). The positions of the fragments on the ladders can be compared with the restriction map to make sure that they are consistent. The brightness of each piece corresponds to the mass of DNA present, allowing for a rough estimate of concentration by comparing the brightness of the screened plasmid's bands with the known amount of DNA supplied in the ladder.

The $\log^{-1}$ (MW) rule does not apply to DNA in other configurations. Linear DNA moves more slowly than supercoiled DNA and more rapidly than circular DNA. Plasmids purified using kits consist primarily of DNA in a supercoiled form, with some nicked or open circular molecules present. An unmodified maxiprep sample usually shows two or three bands whose sizes cannot be determined accurately (**Figure 2.5c**). It is thus standard practice to cut the plasmid with an enzyme that cuts only once—that is, to linearize it—before performing electrophoresis. Running a gel is a standard procedure that should be done with every DNA preparation. Recipes for gels and buffers are given in the Appendix.

## Separation of restriction fragments for ligation

Restriction enzyme digestion is of much greater value than simply mapping. Passing the cut fragments of DNA through agarose separates but does not damage them, and selected fragments can easily be removed from the gel with a razor blade or plastic knife and purified from the agarose with commercial cleanup kits.

These fragments can then be pasted together in a specific, oriented fashion with another enzyme, DNA ligase. This is a *ligation* reaction.

The key to the specificity of ligations is that each restriction enzyme has its own distinct pattern of DNA cleavage. Rather than cut the double-stranded DNA molecule flush, the enzymes usually leave an overhang on the top or bottom strand; this overhang is known as a *sticky end* because it is readily ligated to its complement when DNA ligase is added. Overhangs that are not complementary to one another are incompatible sticky ends and will not ligate. Thus, the position and direction of a ligation can be controlled by matching compatible sticky ends (**Figure 2.6**).

Some restriction enzymes do cut flush, creating what is called *blunt ends*. The efficiency of ligation of blunt ends is much lower than that of sticky ends, and they do not permit control of directionality. For these reasons, sticky-ended restriction enzyme cloning is preferred whenever possible (**Figure 2.6**).

It is important to note that a ligation is something performed with very small amounts of DNA (approximately 100–200 ng) and then transformed into competent cells and amplified in the same way as any other plasmid. Large amounts of DNA, enough to use in downstream experiments, cannot be ligated with DNA ligase. This means that at least one of the fragments in the ligation reaction must contain a resistance gene. This piece is the vector, and the fragment ligated to it is called the *insert*.

The whole amount of DNA in a ligation reaction is usually transformed into highly competent cells; note that this is about tenfold as much as is used for ordinary plasmid amplification. This is necessary because the efficiency of DNA ligase is extremely low; less than one in a million plasmids cut with two enzymes and religated with a separate insert will ligate successfully. However, we can again make use of the remarkable amplification ability of *E. coli* to isolate the desired product. So long as all of the undesired products are cut open, only the fully ligated product will be replicated. Specific controls ensure that empty vectors or other undesired products will not grow (**Figure 2.7**).

```
     5'--TCTGAC          GATCATGCAT  --3'
     3'--AGACAGCTAG          TACGTA  --5'
     (a) Compatible sticky


     5'--TCTGACGATC          ATGCAT  --3'
     3'--AGACAGCTAG          TACGTA  --5'
     (b) Blunt


     5'--TCTGAC          GACCATGCAT  --3'
     3'--AGACAGCTAG          TACGTA  --5'
     (c) Incompatible sticky
```

**Figure 2.6 Types of ends produced by restriction enzyme cuts.** (a) A sticky end has an overhang on the 5′ or 3′ end of the double-stranded DNA. Compatible sticky ends are produced when each end of the plasmid has been cut by the same restriction enzyme. It can also occur when two enzymes with compatible sticky ends are used; such ends are tabulated by manufacturers for ease of reference. Compatible sticky ends adhere readily when DNA ligase is added. (b) Blunt ends are produced by some enzymes, and every blunt end is compatible with every other blunt end. However, their efficiency of ligation is much lower than that of sticky ends. (c) Incompatible sticky ends, even with a single-base-pair mismatch as shown here (highlight), are very unlikely to ligate and thus will leave the plasmid nicked and unable to grow.

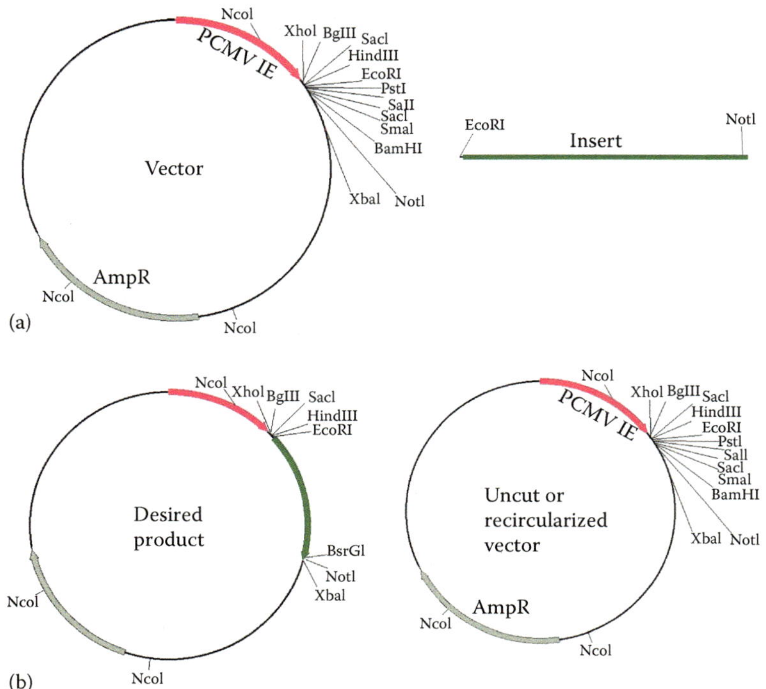



**Figure 2.7  Controlling for unwanted products in ligations.** (a) An example ligation showing a cloning vector with Eco RI and Not I sites. If cut with these two enzymes, the vector should link to an insert cut with these same enzymes. The vector alone should not grow if cut with both enzymes, since it has incompatible sticky ends. The insert itself cannot grow on selection plates, because it has no AmpR gene. (b) The desired product will contain the vector, minus some restriction sites, plus the insert. However, if one or more of the chosen enzymes do not cut the vector completely, undesired products may result. Uncut vector has escaped the action of both enzymes and thus will grow even in the absence of ligase. A control for the presence of uncut vector is thus a reaction containing no ligase. If growth is seen on these plates, the ligation has almost certainly failed, as uncut vector will grow much more efficiently than ligated vector, even if present in very small amounts. Recircularized vector can occur if one enzyme fails to cut—this leaves the vector with compatible sticky ends. A control for this condition is a ligation reaction with no insert but with ligase. It can also be seen that the undesired products contain restriction sites that should be absent from the ligated product: all of those between Eco RI and Not I. This can be used to eliminate the uncut/recircularized vector by cutting with one of these enzymes after performing the ligation reaction (after ensuring that the chosen enzyme does not cut the insert!).

# 2.6  AN EXAMPLE CLONING EXPERIMENT

## Determining a cloning strategy

The following example illustrates how to remove a gene (actin–GFP or AcGFP) from a mammalian expression vector with a constitutive promoter (pEGFP–actin) and place it into an empty vector with a tetracycline-inducible promoter (pTre2) (**Figure 2.8a**). The empty vector contains an area rich in restriction sites; this is called a *multiple cloning site* (MCS) or *polylinker* and is engineered to facilitate cloning. The first thing to appreciate is that the only two restriction sites on the 3′ end of EGFP–actin are Xba I and Bam HI, and that both of these produce sticky ends. Then inspect the target, pTre2, for these sites in the polylinker: both are present, but Bam HI is near the 5′ end of the polylinker, whereas Xba I is the last site, meaning that using this enzyme would leave all of the other elements of the polylinker free to use for the 5′ end. You therefore want to cut the 3′ end of both the vector and insert with Xba I.

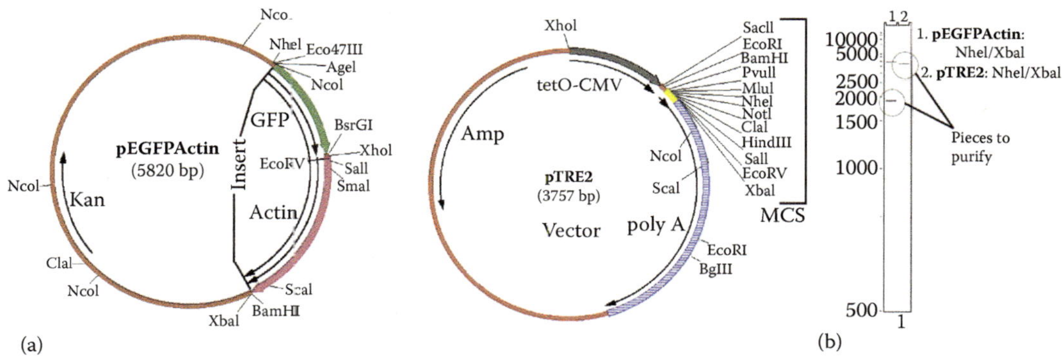

**Figure 2.8 Example ligation.** (a) The insert is a piece of 1.9 kb taken from a kanamycin-resistant plasmid of 5.8 kb. The target vector is 3.8 kb, amp resistant, and has a multiple cloning site (MCS) containing many restriction enzyme sites. (b) Expected appearance of the gel after cutting with the enzymes suggested in the example, and desired pieces to choose.

On the 5′ end of the insert, the enzymes available are Nhe I, Eco 47III, Age I, and Nco I. Now look again at pTre2. Nco I cannot be used, since it cuts into the poly-A sequence of the plasmid. Age I is absent from the pTre2 polylinker. Eco 47III is a blunt-ended cutter, so it could potentially be matched with either of the blunt-cutting enzymes in pTre2: Pvu II or Eco RV. However, the enzyme Nhe I is also present in the polylinker, and it is a sticky end, so it is the ideal choice. (Also note that Eco RV is very close to Xba I; you would have to verify that they would be able to cut together.)

The strategy is thus to cut out the insert with Nhe I and Xba I and put it into the vector, also cut with the same two enzymes.

## Digestion and purification of fragments

Digest a significant amount (1–3 µg) of each plasmid and run them on a 0.8% agarose gel. The digest of pEGFP–actin will show two bands: a larger one corresponding to the vector, and a smaller one corresponding to the insert. Physically cut out the smaller piece and purify it. The digest of pTre2 will show only one band; the small piece between the two enzymes is too small to be resolved on 0.8% agarose and will run off the bottom of the gel. This is good, because it is no longer available to ligate. Cut out the band and purify it (**Figure 2.8b**).

## Determination of parameters for optimal ligation

A ligation reaction is usually performed at a 3:1 insert: vector *molar* ratio, not mass ratio. In this case, the molecular weight of the insert is approximately 2 kb, and that of the target vector is approximately twice that, or 4 kb. So a 3:1 molar ratio is a 3:2 mass ratio; i.e., in a ligation of 100 ng of DNA, you would want 60 ng of the insert (EGFP–actin) and 40 ng of the vector (pTre2). This ratio allows for maximization of ligation changes with minimal probability of *concatemers* (two or more of the inserts end to end).

To estimate the concentration of the DNA fragments after gel extraction and purification, you need to know (1) how much by mass $m_{DNA}$ was in the original digest; (2) the final volume $V$ it is dissolved in; and (3) a rough estimate of the efficiency $E$ of your extraction (usually considered to be around 0.8 if all of the band was used and nothing was lost, lower if you remove only part of the band, some falls on the floor, etc.). The concentration of each fragment is then given by

$$[DNA]\left(\frac{\mu g}{\mu L}\right) = \frac{MW(purified\,fragment)}{MW(total\,plasmid)} \times m_{DNA}(\mu g) \times \frac{E}{V(\mu L)} \qquad (2.3)$$

In this case, the vector is essentially the entire plasmid, so the first ratio in **Equation 2.3** is 1. So if you initially digested 2 µg of pTre2, and purified it into 50 µL, it would have a concentration of about 32 µg/mL. For the insert, the ratio is 2:5.8 (see **Figure 2.9**), so the same digest and purification would give an insert concentration of only 11 ng/mL. The ligation should then contain approximately 1.3 µL of vector and 5.5 µL of insert. Control ligations should be prepared containing the vector only with no ligase and the vector only with ligase.

After the ligation is transformed and plated, the controls should be clean (no colonies), whereas the reaction should show 100–200 colonies. If the controls contain an equal number of colonies as the reaction, screening is not worthwhile; throw out the plates and start again. A ligation is screened by picking 6–12 (or more!) colonies into 3–5 mL volumes of medium plus antibiotic, letting the cultures grow, and performing minipreps. The minipreps are screened by restriction digest and should show the presence of the new product (**Figure 2.9**).

For an example as simple as this one, with two sticky ends, 6–12 minipreps should be more than sufficient to obtain a positive clone. If by some chance, the vector-only control plate is empty and yet the ligation shows only recircularized vector, you have done something wrong, probably in the transformation of the control reaction. Cut the vector again and repeat the experiment. If it does not work a second time, there may be something wrong with one or more of the enzymes. Screening gels can be performed to test the function of each enzyme in turn; for the aforementioned example, the enzyme Xho I is in a useful location for this. Cut the vector with Xho I and Xba I in one small-scale digest, and with Xho I and Nhe I in a second digest. When run on a gel, each should show two fragments of the same size, or something is wrong. Check the expiration dates on the enzymes or order more of the enzyme that did not work.

It is a fortunate ligation where the restriction sites match as perfectly as in this example. In many cases, only one end finds a match, or in other cases, neither end matches. There are multiple ways around this problem, and a few of the especially clever ones are shown in **Practical Tips 2.3**. As you become adept at molecular cloning, you will discover which ones work best for you and develop your own.

**Figure 2.9 Expected outcome of example ligation.** (a) Plasmid map. (b) Result of digestion with different enzymes. Any of these could be used for screening.

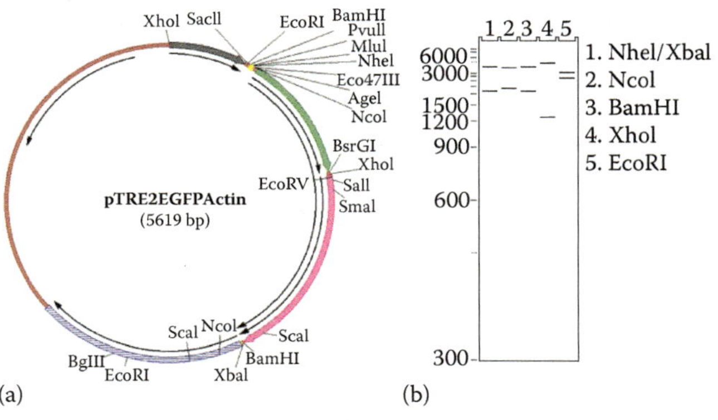

## PRACTICAL TIPS 2.3:    SOME CLONING TRICKS

There are several ways to get around incompatible restriction sites. Here are a few tested favorites.

1. Subcloning

    The principle is very easy: your insert's restriction sites are not compatible with those of your target vector, but they do match another cloning vector, which also matches your target. Cut the insert out, put it into the "middleman," and then cut it out again and put it into your target.

2. Blunt-ending

    One or both of the ends of your insert are incompatible with your target vector. DNA polymerase can "fill in" the sticky end to create a blunt end, permitting the ligation. This is especially recommended if only one end does not match, as double-blunt ligations are difficult (**Figure P2.3.1a**).

3. Linkers

    This approach requires three- or four-way ligation, but it works well. If the restriction sites in the insert are incompatible with those in the vector, you can order single-stranded oligonucleotides that anneal to leave sticky ends that match both vector and insert. The oligos are added to the ligation in excess to facilitate ligation (**Figure P2.3.1b**). You can develop a library of such linkers that can be used in multiple cloning experiments.

4. Partial digestion

    If a particular enzyme cuts where you want it to, but also somewhere else (e.g., inside the gene), you can let that particular enzyme incubate for only a short period of time (10–15 min) and then run the product on an agarose gel. By chance, some of the fragments you want will show up. Remove them from the undesired fragments and purify. You have to be careful to identify the fragments correctly (**Figure P2.3.2**)!

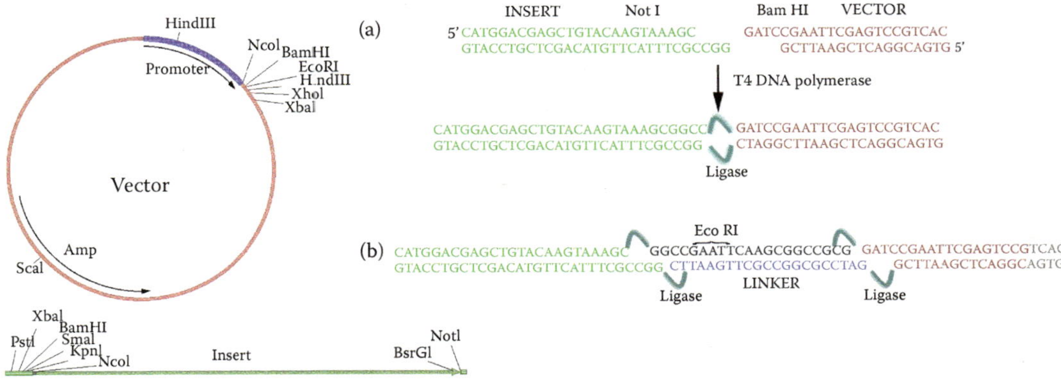


**Figure P2.3.1 Cloning tricks.** The vector and insert shown match at the 5′ end (e.g., with Nco I) but not at the 3′ end. In order to clone them into each other, the insert can be cut with Nco I and Not I, and the vector with Nco I and Bam HI. (These are just examples; any other choice at the 3′ end would also work.) (a) The incompatible ends of the vector and insert can be blunt-ended to make them compatible using DNA polymerase. The blunt-ending must be done before digesting with Nco I, so that the Nco I end remains sticky. This is now a one-blunt, one-sticky ligation; this type of ligation works quite well. (b) Linker oligonucleotides can also be ordered and used in the ligation. They should be long enough to anneal well (~20 base pairs) and can be ordered such that the sticky ends preexist as shown; there is no need to digest. Other sites can be added into the "filler" region for ease of screening or future cloning (shown is Eco RI).

*(Continued)*

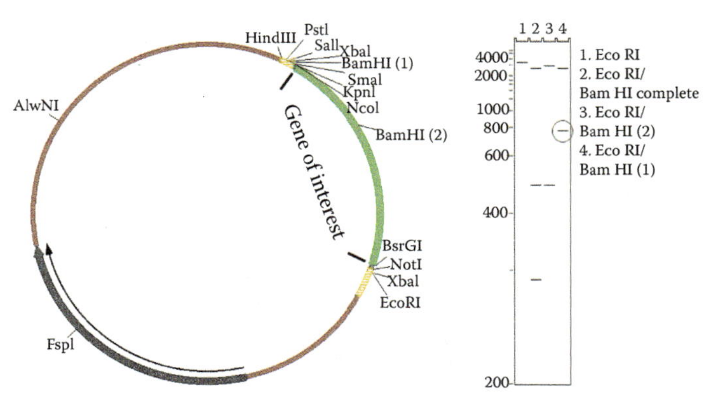

**PRACTICAL TIPS 2.3 (CONTINUED):   SOME CLONING TRICKS**

**Figure P2.3.2 Partial digestion.** A gene of interest is between Eco RI and Bam HI sites, but there is an additional Bam HI site within the gene. The plasmid may be fully digested with Eco RI and then digested very briefly with Bam HI (10 min or less). The resulting gel will contain a mix of fragments of the gene cut at both Bam HI sites (lane 2), cut only with the unwanted site (lane 3), and cut only at the wanted site (lane 4). Identify the desired piece at close to 800 bare pairs in size and purify it; this will be the whole gene cut with Eco RI and Bam HI.



## 2.7  CLONING BY THE POLYMERASE CHAIN REACTION

The polymerase chain reaction, or PCR, is based upon the ability of double-stranded DNA to serve as its own template, and its invention revolutionized molecular biology. PCR can amplify a precise target sequence of double-stranded DNA by repeated cycles of heat denaturation of the double strand, binding of specific *primers*, and polymerization using a heat-stable DNA polymerase (**Figure 2.10**).

The applications of PCR are vast, and here we will only focus upon when it should be used in routine cloning of average sized (<5 kb) genes or gene fragments. There are circumstances in which restriction enzyme cloning is inconvenient and others

**Figure 2.10 PCR, showing the temperatures involved for thermal cycling that permits denaturation of template DNA, primer binding, and polymerase-based extension of strands.**

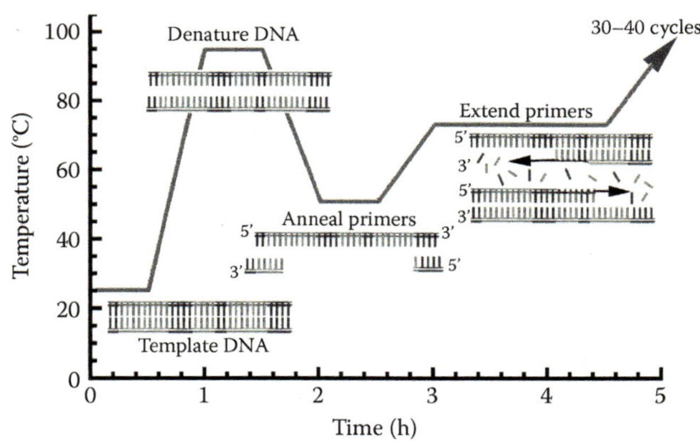

in which it does not work at all; in these cases, PCR is often a good solution. These situations include

- When the sequence of the desired gene is known, but the identity and sequence of the surrounding plasmid are unknown

- When the gene has an unwanted stop codon that is not preceded by any restriction sites

- When only part of the gene is desired, but there are no restriction sites

- When the restriction sites would place the gene out of frame with respect to a desired fusion tag or protein

- When the restriction sites of the gene and the target vector do not match at all

PCR allows for free choice of restriction sites and reading frames by selective primer design. All of the ingredients are available commercially; the trick in a PCR cloning experiment is to choose primers appropriate for the desired application. In general, the primers should be 18- to 24-base-pair oligonucleotides aimed at the beginning (*forward*) and end (*back*) of the sequence to be amplified. Any restriction site of choice can be added by adding the restriction enzyme sequence onto this oligonucleotide. Since this sequence will not bind the template, it should be in addition to the 18–24 base pairs complementary to the sequence (**Figure 2.11**). After the PCR reaction is performed, the resulting amplified fragment is run on an agarose gel, purified, and then digested with the selected enzymes.

Special plasmids are designed to accept PCR fragments without the need for enzyme digestion. The PCR reaction is run, and a portion of it is added directly to the plasmid, without the need for gel purification. The principle behind this is that the most commonly used polymerase, Taq, leaves a single *A* overhang on each end of the product. The supplied linearized vector then has a single *T* to match. This is called *TA* cloning. The drawbacks are that you cannot choose the plasmid and that additional cloning steps may be needed to remove your product from this plasmid to another, with no guarantee that the TA plasmid restriction sites will work well for this.

There are many subtleties to primer design that we will not go into detail about here, except to mention that the use of software to aid in primer choice is highly recommended. Several excellent, free resources exist for designing primers. Some important factors include matching primer melting temperatures, selecting the

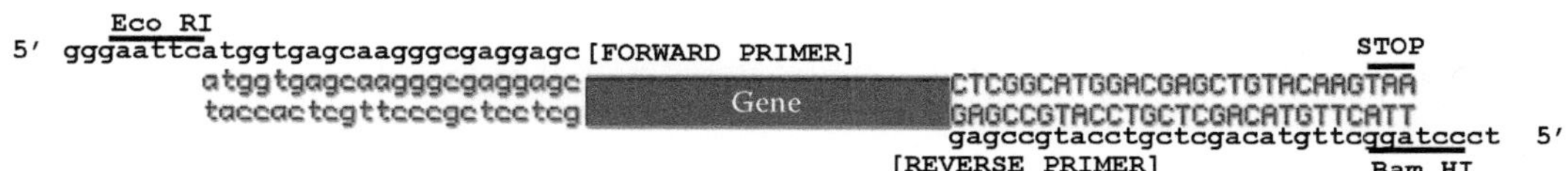


**Figure 2.11 Example forward and reverse primers for a gene.** An Eco RI restriction site is added to the 5′ end, and a Bam HI site to the 3′ end. Note that the stop codon has been removed from the PCR product. Two "extra" base pairs have been inserted on the ends of the restriction sites to permit the restriction enzymes to cut efficiently. Some enzymes require more than this.

PCR annealing temperature, and ensuring that the primers do not bind to one another (*primer dimer*).

A PCR reaction then involves placing a small amount of the template DNA (10–100 ng), the primers, free nucleotides (A, C, G, and T), and polymerase into a thin-walled tube that permits rapid heating and cooling. The tube is placed into a *thermal cycler* or *PCR machine* and subjected to 20–30 cycles at temperatures that allow for denaturation but not boiling, annealing of the specific primers, and elongation.

Many sophisticated variations on PCR exist. An example is *real-time PCR*, which allows for the (usually fluorescent) quantification of the amount of DNA produced at each amplification cycle. The results can be plotted on a standard curve and the amount of DNA carefully quantified. This requires a special thermal cycler; these are often available in university shared facilities.

## 2.8 SEQUENCING

Cloning by restriction enzyme digestion usually does not lead to any changes, or mutations, in the target gene. However, it is possible for restriction enzymes to remove one or more excess base pairs and for ligation to still take place. It is always good practice to *sequence* the area right around the restriction digest after a cloning experiment. After PCR cloning, sequencing is mandatory, as all DNA polymerases make errors and the presence of one or more mutations is likely.

DNA sequencing involves the determination of the identity of each of the base pairs in a target region. Sequencing technology has become orders of magnitude faster and cheaper over the past three decades. As of 2016, a large number of full eukaryotic genomes have been sequenced, including those of mammals (human, cow, rat, mouse, cat, dog, tiger, elephant); birds (chicken, duck, eagle, pigeon—a very large number of bird genomes were reported in 2014); other animals (puffer fish, mosquito, alligator); plants (corn, rice, poplar, watermelon, papaya); and many protists. Dozens of bacterial and archaeal genomes have also been sequenced. Sequencing is performed on automatic *sequencers* that work on essentially the same principle as PCR. The DNA is denatured in the presence of primers, and elongation begins. However, into the mix of abundant free A, C, G, and T are added *dideoxynucleotides*, each fluorescently labeled with a different color. (This is called dye-terminator sequencing.) The dideoxynucleotides lack a 3′ OH group, so when they are added to a strand, elongation terminates—no new nucleotides can bind. Since most of the nucleotides in the reaction are ordinary, the point at which a deoxynucleotide will be added is random, leading to an eventual series of fragments of all different sizes that terminate in one of four colors. By resolving the size of these fragments with single-base-pair resolution, the position of each A, C, G, or T in the sequence can be known (**Figure 2.12**). Size resolution used to be done on meter-sized *sequencing gels*; modern sequencers use capillary electrophoresis.

Very few labs have their own sequencers. However, the service is available from key biological suppliers and on the majority of university campuses, and is inexpensive. What needs to be provided is high-quality template DNA of the

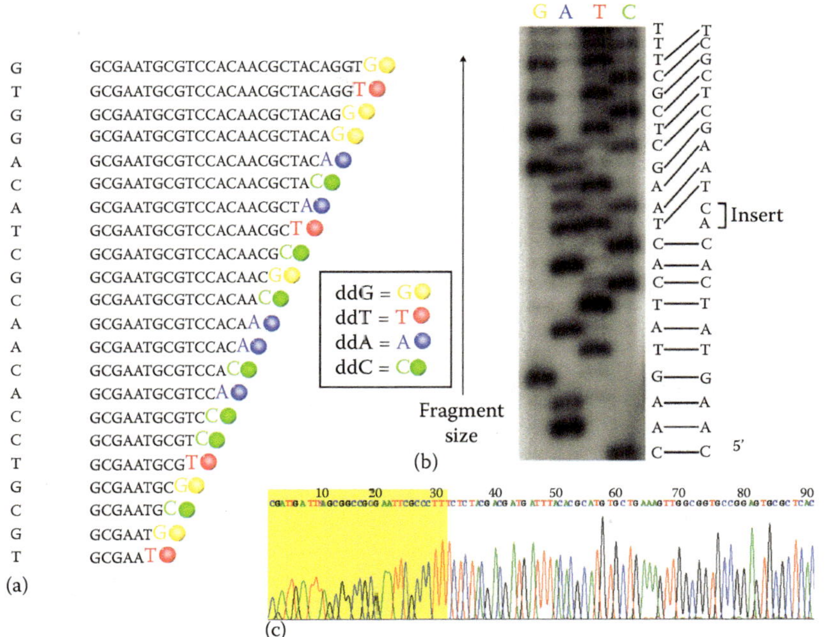

**Figure 2.12 Sequencing.** (a) PCR in the presence of dideoxynucleotides of four colors produces randomly sized fragments that terminate in one of the colors. By separating the fragments by size, the identity of the ends can be determined, and so by proceeding base by base, the entire sequence is known. (b) Appearance of a sequencing gel using only one color in four lanes. This sequence shows the identification of an insertion mutation in the gene for hemoglobin, leading to the disease beta-thalassemia. In a four-color gel, all of these lanes would be shown in a single column. (c) Appearance of sequencing data as they are usually presented.

finished cloning product (not miniprep DNA!) and one or more sets of *sequencing primers*. A sequencing primer is not subject to many of the concerns of a PCR primer, since it is not required to melt and bind repeatedly. However, it should begin 25–40 base pairs upstream of the region of interest, since the first few dozen bases cannot be read precisely. A good sequence is usually obtained for no more than 900 base pairs downstream of the primer. If the region to be sequenced is longer than this, multiple primers are needed.

## 2.9  RNA METHODS

There are some circumstances in which the basic coding material—that is, DNA—is not the template from which you wish to work, and you must work instead with RNA. One situation is when you wish to know gene expression levels in a cell, tissue, or organism. Measuring the amount of RNA that has been transcribed (messenger RNA [mRNA]) is a good way of determining the activity of a specific gene. Looking at the mRNA is also the way to determine the sequence of an unknown eukaryotic gene. Since eukaryotic DNA contains many noncoding *introns*, its sequence does not reveal the sequence of amino acids in the resulting protein. Looking at the mRNA allows the experimenter to reconstruct and express a complementary DNA, or *cDNA*, that contains only the coding region (**Figure 2.13**). Isolation of *total RNA*, not just mRNA, is also a useful technique for a wide variety of applications. It allows for ribosomal RNA *fingerprinting*, which can be used to study environmental communities, and for study of *noncoding RNA* sequences and *microRNAs*. However, total RNA should not be used in cases where mRNA is the desired template, since mRNA makes up only 1–5% of the total RNA in a cell.

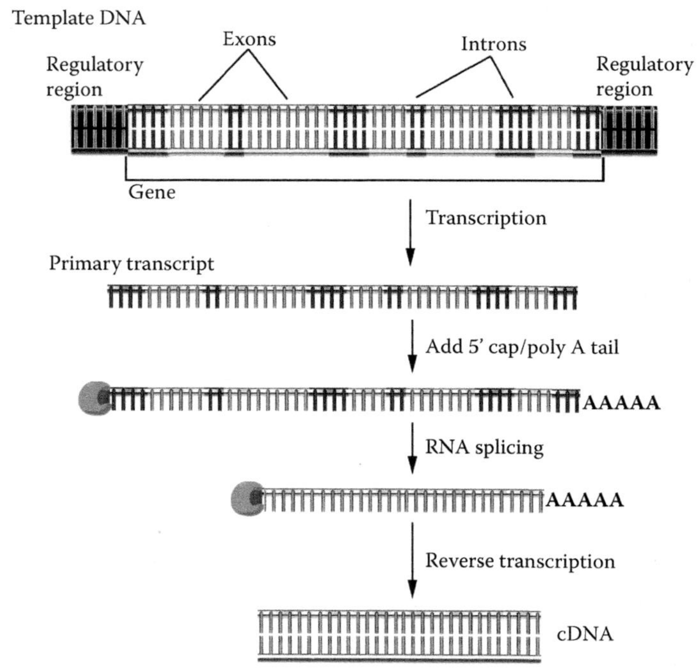


**Figure 2.13 Noncoding DNA sequences are called introns.** After transcription of DNA to RNA, the RNA is spliced, and the noncoding segments are removed. Reverse transcription of the mRNA then yields the DNA sequence of only the coding region of the gene, called the cDNA.

Single-stranded RNA can be manipulated in a similar fashion to double-stranded DNA, with certain exceptions. The most important exception is that, while enzymes that degrade DNA (*DNases*) are heat labile and require divalent cations, *RNases* cannot be destroyed by autoclaving, require no cofactors, and are ubiquitous in the environment. (All organisms produce them; common sources in the laboratory are bacteria and culture media, and human sweat, saliva, and tears.) RNA work thus requires dedicated pipettes and tips, specially prepared solutions, and meticulous technique. Resources are available from molecular biology suppliers to aid in RNase decontamination. One of the most common is to add diethyl pyrocarbonate (DEPC) to any solutions used for RNA work. DEPC covalently modifies histidines and deactivates nucleases. (Note that it should not be used with amine-containing buffers, such as Tris.) However, it is toxic and can inhibit some downstream applications, so final purified RNA should not be dissolved in DEPC water. Additional products with names such as *RNaseZap* and *RNase AWAY* are also sold for cleaning benchtops and glassware or plasticware.

RNA is usually extracted from a target cell culture or clinical or environmental sample. Efficient extraction used to be a challenge, but now kits are available for total RNA and mRNA extraction from all types of samples: plants, animal cells, blood, soil, fresh or fixed tissue samples, and so on. Yield and purity of isolated RNA can be measured by $A_{260}$ and agarose gel electrophoresis like with DNA. The difference is that the gel should be *denaturing*, which causes the RNA to lose its complex secondary structure so it can migrate according to its linear size. This is accomplished by adding formaldehyde to the agarose and by preheating the RNA sample at 65–70°C for 5–15 min. Intact total RNA and mRNAs will show discrete bands, whereas degraded RNA will be a smear (**Figure 2.14**).

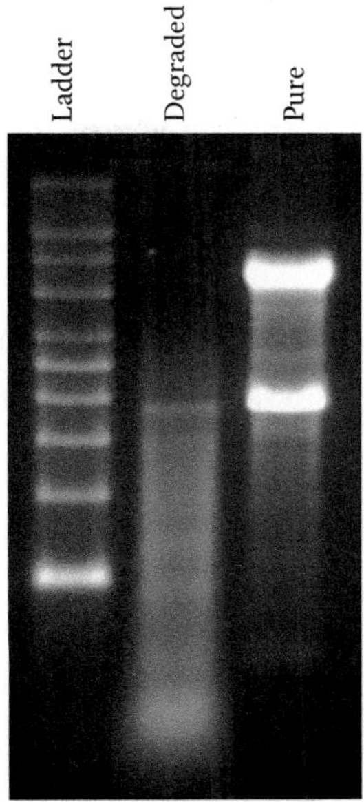



**Figure 2.14 Total RNA run on a denaturing agarose gel, showing the appearance of degraded versus high-quality pure RNA.** The two bands in the good-quality sample reflect 28S and 18S ribosomal RNAs (rRNAs).

## 2.10 SOUTHERN AND NORTHERN BLOTS

In diagnostics, genomics, and other applications, it is often useful to detect a specific sequence from a sample of DNA or RNA, and to know how many times that sequence appears. A straightforward technique to do this makes use of the techniques described previously, and is called a *Southern blot* (DNA) or *Northern blot* (RNA).

The Southern blot is an eponym, named for its inventor, Edwin Southern. The Northern blot is so named as it is the "opposite" of the Southern blot. Both techniques use the complementarity of nucleic acids in order to detect specific sequences in a sample using *probe hybridization*. For a Southern blot, the DNA is cleaved with restriction enzymes and run on an agarose gel. The DNA is then transferred to a membrane that is more stable than the gel and that permits ready access by a probe; this membrane should be positively charged, usually nylon or nitrocellulose. The membrane is placed over the gel, and pressure is applied by suction or a stack of paper towels. Capillary action pulls the DNA onto the membrane, where it binds tightly by electrostatic interactions. The DNA is permanently affixed to the membrane by baking in an oven at 80°C or by exposure to UV light. Once the transfer is complete, the membrane is exposed to a labeled oligonucleotide, which is a target sequence of DNA ~25 base pairs in length that has been tagged by a radioactive, luminescent, or colorimetric tag. These probes can be purchased in the same fashion as primers, with a wide variety of modifications. Any DNA fragment that possesses the complementary sequence will bind to the probe, giving a signal (**Figure 2.15**). Both luminescent and radioactive signals can be quantified using x-ray film, which is the standard technique. As sequencing is becoming increasingly rapid and inexpensive, Southern blotting is disappearing. It is still used in the confirmation of genomic insertion of target sequences into *transgenic* animals, where sequencing the entire genome would be impractical; it is virtually never used anymore for studying plasmids.

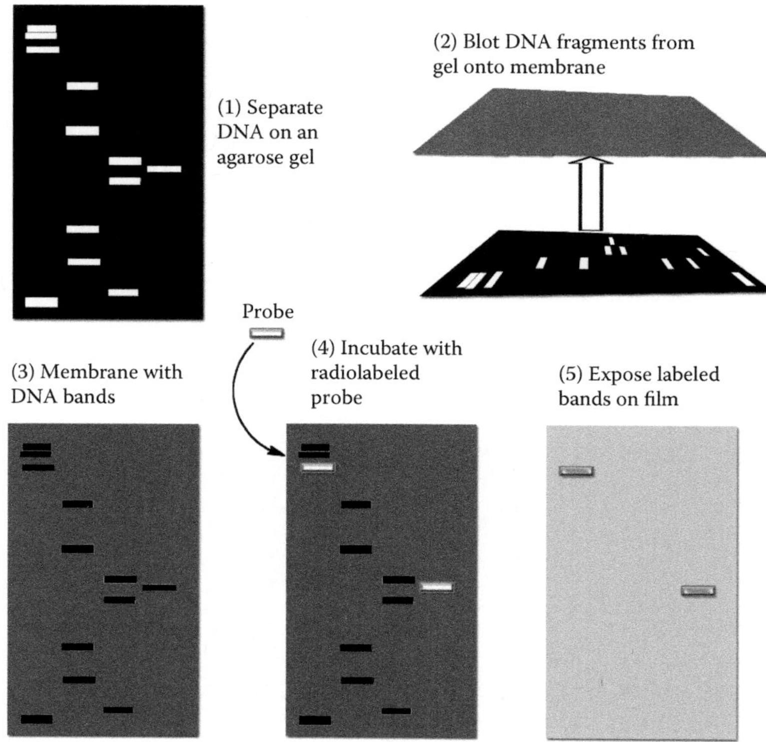



**Figure 2.15 Southern blotting.** Northern blotting is similar but uses RNA instead of DNA.

Northern blotting is similar, except the gel must be denaturing as discussed previously, because the target is RNA. As in Southern blotting, the probes may be DNA or RNA. (The use of an RNA probe to hybridize to DNA is sometimes also done and is called a *reverse Northern blot*.) Although rt-PCR has largely replaced Northern blotting in medicine, most molecular biologists consider the Northern blot the gold standard for the direct measurement of mRNA expression levels. It is also used to study RNA degradation, half-lives, and *alternative splicing*.

## 2.11  SITE-DIRECTED MUTAGENESIS

Polymerases make mistakes, inserting the wrong base pair (a *transition* if it substitutes a purine for a purine or a pyrimidine for a pyrimidine, a *transversion* if it substitutes a purine for a pyrimidine or vice versa); failing to insert one or more base pairs (a *deletion*); or inserting one or more extra base pairs (an *insertion*). All of these are called mutations, and the mutation rate varies among polymerases: from about $4 \times 10^{-7}$ errors per base pair per cycle for *high-fidelity* polymerases, to $2 \times 10^{-5}$ per base pair per cycle for typical commercial Taq polymerase, to $5 \times 10^{-3}$ per base pair per cycle for *error-prone* polymerase. Because of the degeneracy of the amino acid code, a single nucleotide error, called a *point mutation*, may not change the resulting protein; this is a *silent mutation*. If a different amino acid is substituted, it is called a *missense mutation*. These may change the protein's function very little or tremendously, depending upon the location and the degree of change. The insertion of a stop codon leads to a truncated protein and is called a *nonsense mutation* (**Figure 2.16**). The insertion or deletion of 1 or 2 base pairs

Missense mutation

```
      Thr   Pro   Glu   Glu
   ...ACT  CCT  GAG  GAG ...   Wild-type
                                hemoglobin

   ...ACT  CCT  GTG  GAG ...   Sickle-cell
      Thr   Pro   Val   Glu     anemia
```
(a)

Nonsense mutation

```
      Pro   Gln   Gln   Val
   ...CCT  CAA  CAA  GTC...   Wild-type
                              dystrophin

   ...CCT  CAA  TAA        DMD
      Pro   Gln  STOP
```
(b)

**Figure 2.16 Example of point mutations causing human disease.** (a) A missense mutation in the gene for the hemoglobin β chain results in a change in the quaternary structure of hemoglobin, and sickle cell trait or sickle cell disease. This particular mutation is the most common cause of sickle cell. (b) A nonsense mutation in the dystrophin gene causes Duchenne muscular dystrophy (DMD). About 35–40% of cases of DMD result from nonsense mutations.

leads to a *frameshift mutation*, changing the encoding of all of the amino acids downstream of the error.

Deliberate insertion of a specific point mutation into a gene is called *site-directed mutagenesis* and is a common and easy way to study protein function. There are several different ways to perform the mutagenesis; the simplest is to PCR a plasmid encoding a gene with *mutagenic primers* containing the desired error in the middle of the primer. This single error does not prevent the primer from binding; the DNA produced is nicked and is repaired by the bacteria into which it is transformed. Kits for site-directed mutagenesis are available from several suppliers, providing a low-error polymerase to permit PCR of the entire plasmid. It is crucial to use very clean DNA for these experiments.

Mutagenesis techniques are ubiquitous. Silent mutations can be used for codon optimization to allow a gene to express better in a system of interest. For example, GFP was initially cloned from a jellyfish. In order to get optimal expression in mammalian cells, many point mutations were made to optimize codon usage. Silent mutations can also be used to create a restriction enzyme site at a location of choice; by changing one or more bases, the sequence of a restriction site can often be generated (**Figure 2.17**).

Missense mutagenesis allows for the substitution of one amino acid for another and so is one of the key techniques in studying protein structure and function at the molecular and atomic level. Since different amino acid side chains have different charges, a single point mutation allows a positive or negative charge to be exchanged for its opposite or for a neutral molecule at a chosen site anywhere in a protein. For example, it was used to show that a single negative charge in the *transmembrane domain* of potassium-selective channels is crucial for channel opening and closing.

Another ubiquitous application of site-directed mutagenesis is the targeted insertion of cysteines, which have a reactive –SH group. In this case, it is not key amino acids of the protein that are targeted but, rather, key structural regions, as the idea is not to disrupt function but to insert a tag. After the protein is expressed, a tag bearing an *alkylating agent*, such as a maleimide, iodoacetamide, or similar agent, will react with the thiol group and form a covalent bond. (These are discussed more

```
       Thr   Pro   Gly   Ser
   ...ACT  CCT  GGG  TCC  ...


   ...ACT  CCT  GGA  TCC  ...
       Thr   Pro   Gly   Ser
                            Bam HI
```

**Figure 2.17 Creating a restriction site with a silent point mutation.**

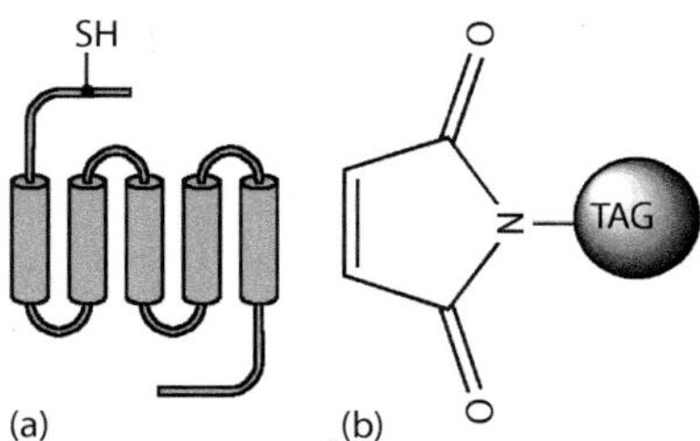



**Figure 2.18 Attaching cysteines to proteins for targeted labeling.** (a) An amino acid mutated to a cysteine at a targeted location in a protein provides a reactive –SH group for binding. (b) A maleimide (or other alkylating agent) attached to a tag will form a thioester bond with the cysteine and thus label the protein at the chosen site. (Note that sometimes, native cysteines must be mutated out of proteins in order to specifically target a site.)

fully in **Chapters 7 and 14**.) Thiol-reactive compounds bearing fluorescent dyes, luminescent probes, spin probes, or other tags can be added to proteins expressing inserted cysteine(s) and will attach to the specified location(s) (**Figure 2.18**). The biggest caveat with this technique is that it is necessary to ensure that the cysteine-substituted protein has normal function before it can be used as a model of the wild type. For some proteins, such as ion channels, measurement of function can be relatively straightforward (e.g., current–voltage relationships); for other proteins, this is more difficult.

Another method that makes use of cysteine's reactivity is the scanning cysteine accessibility mutagenesis (SCAM) method. Amino acids that are expected to move significantly during protein function are mutated to cysteines; a thiol-active probe is added in each state of the protein to confirm or deny the hypothesis by showing whether or not it is accessible to the cysteine in question. Recent developments allow this method to be applied in lipids as well as in aqueous solution.

## 2.12 SUMMARY

A plasmid and a few kits are all that are needed to get started with cloning. Even if the plasmid is provided with all of the genes and promoters in the right place, it is still necessary to learn how to amplify and purify it to ensure a lifetime supply. Many labs have their own preferences of kits, cloning strains, media recipes, and other basic supplies and procedures. For example, if there are a bunch of electroporators on the benchtops, chances are the lab uses electrocompetent cells and probably knows how to grow them. There are a vast number of variations possible for all of the steps and recipes, and for the most part, they all work. Part of getting comfortable with molecular biology is accepting its black-magic nature and realizing that most of the steps in an experiment cannot and should not be quantified, unless you want to spend years growing a single gene. Beware of following the advice of too many people at once—every biologist does things a little bit differently.

The more advanced techniques often take years to master and often form the basis of someone's entire thesis or research program. It is important to select these techniques with care and sometimes to set up collaborations with those who are already experts in the field of choice. Core facilities at universities can also be of tremendous help in performing large or complex cloning experiments.

## End-of-Chapter Questions

1. Assuming a 50% recovery, how many bacteria would be needed to isolate 1 mg of a 2 kb plasmid? Estimate the volume in which this number of bacteria would grow.

2. In **Equation 2.2**, the calculated DNA or RNA concentrations will be in molar. In order to convert these values to mg/mL, what molecular weight should be used? Why?

3. Assuming that sequences of any of the 4 base pairs occur by chance in a length of DNA, what is the expected frequency of a specific 6-base-pair sequence? An 8-base-pair sequence?

4. How would you estimate the relative efficiency of uncut versus ligated plasmid?

5. From **Figure 2.5c**, estimate the minimum amount of DNA that can be visualized on an agarose gel under the conditions of the figure. Estimate the relative fluorescence intensity of the different bands of the pictured restriction digest.

6. What can go wrong in a ligation? What is the significance of the various controls?

7. Why do we use a 3:1 insert-to-vector ratio? What possible products could we get other than the desired one? Estimate their probability.

8. What is blue/white selection, and how does it work?

9. You read a journal article with the title "New Cool Fluorescent Protein is Twice as Bright and Twice as Pretty as GFP." You want to try it, so you send an e-mail to the authors. You do not get a reply to your message, but 2 days later, you get an envelope containing a piece of filter paper with a 1 cm circle on it, and the map in the next column.

   The filter paper does not come with any instructions, but you remember that you only need nanograms of a plasmid to get colonies. You soak the circled part in an Eppendorf tube with some water for half an hour, transform it into chemically competent cells,

and plate on Amp plates. Sure enough, the next day, you have colonies!

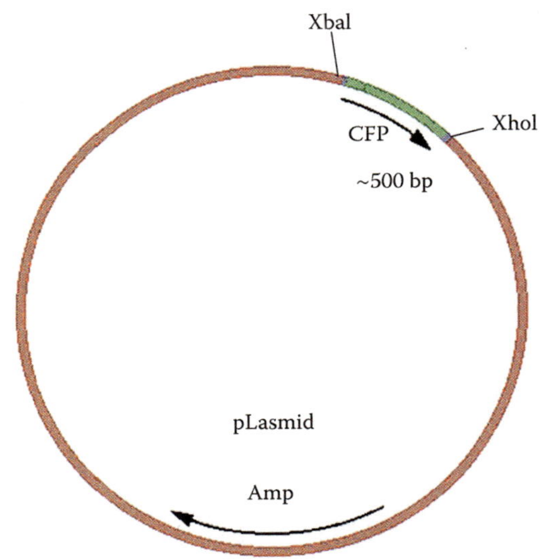


(a) What's the very first thing you should do?

(b) What's the first screening test you should perform?

(c) You transfect some of the DNA into some mammalian cells you have around, but you do not see any fluorescence. Give at least two explanations for why this could be true.

(d) You cannot find pLasmid in any catalogs, and when you e-mail the authors, they do not answer, so you decide to cut the gene out and put it into your favorite vector. Interpret the possible meanings of the gels shown at the end of the problem.

(e) How will each of these results affect your cloning strategy?

(f) At what point would you just give up and PCR it out? How would you design primers?

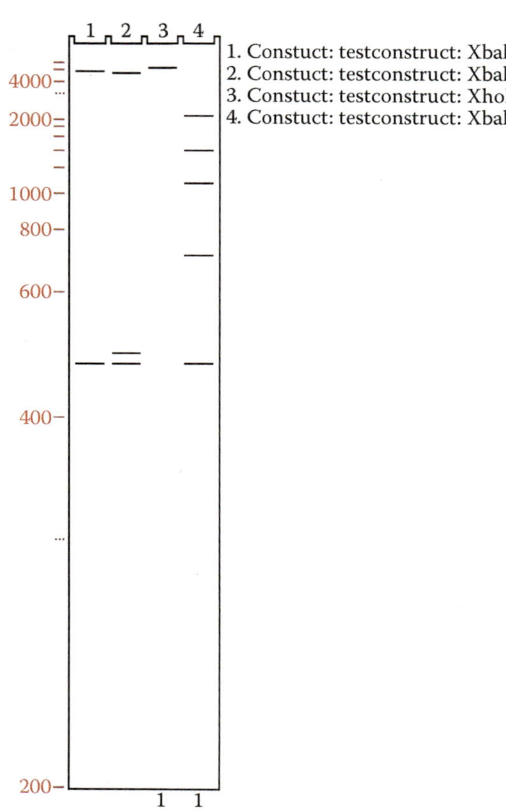



**10.** Someone in Siberia finally manages to unearth a frozen *Smilodon fatalis* (sabertooth cat) from at crevasse. Preliminary sequencing indicates that the reason for the sabertooth's sabers is overexpression of a growth gene in tooth enamel. You want to make a cute little housecat-sized *Smilodon*. How do you approach the problem?

**11.** The discoverers of the *Smilodon* are impressed by its big, pouncy rear legs and suspect that the cat overexpressed a specific ryanodine receptor related to muscle growth. Similar receptors are known in existing cats. You are given a fragment of the *Smilodon*'s leg muscle and asked to clone the gene. How do you go about it?

**12.** Design PCR primers for the feline *enamelin* gene to be inserted into the Bam HI site of a cloning vector. Give the sequence and melting temperature of your chosen primers. What is the size of the protein? Is there any problem amplifying the entire sequence with a single set of primers?

**13.** Do not try this, but imagine you wanted a GFP tattoo. How would you do it?

**14.** What is the pET expression system, and how does it work?

**15.** The restriction enzyme Sal I cuts the sequence GTCGAC to leave a 4-base, 5′ overhang. The restriction enzyme Xho I cuts the sequence CTCGAG to leave a 4-base, 5′ overhang. If you mix a Sal I–digested insert with a Xho I–digested plasmid vector and perform a ligation, which enzymes (if any) will be able to cut the ligation product? What about a Psp XI site ligated to Xho I?

**16.** Discuss the evolutionary significance of restriction enzymes.

**17.** Detail the steps involved in purification of DNA.

**18.** Describe the steps involved in PCR. One of the key discoveries facilitating PCR was the isolation of polymerase from *Thermus aquaticus*, which can survive temperatures up to 80°C. Why is this important for molecular biology?

**19.** *Salmonella* bacteria are used to determine whether chemicals are mutagenic. How does this work?

**20.** You have a membrane protein with a single extracellular domain (P2X purinoreceptor). You want to make a mutant with a single cysteine right in the middle of the extracellular loop, so you can bind a dye to it. How do you go about this?

**21.** Pick a DNA sequence from GenBank, and answer these questions: (a) What gene is this, and from which species? (b) What is its corresponding mRNA sequence? (c) Its protein sequence? (d) Using at least five homologues, what species is closest to this one? With what percent homology? (e) What is the protein structure (if known)? (f) What is the protein's pattern of distribution in the selected species?

**22.** You have the following sequence for a bacterial ion channel. What is the gene, and what does it do? What restriction sites are found in it? Codon-optimize it for expression in a human cell line. Point out which codons have changed, and comment.

```
1 atgctcaaag gattcaagga gtttctcgcg
cggggtaata tcgtcgacct ggctgtcgcg

61 gtggtaatcg gcacagcgtt cacggcgttg
gtcaccaagt tcaccgacag catcattacg

121 ccgctgatca accggatcgg cgtcaacgca
cagtccgacg tcggcatctt gcggatcggt
```

```
181 atcggcggtg gtcagaccat tgacttgaac
gtcttgttgt cggcagcgat caactttttc

241 ctgatcgcgt tcgcggtgta cttcctagtc
gtgctgccct acaacacact acgcaagaag

301 ggggaggtcg agcagccggg cgacacccaa
gtcgtgctgc tcaccgaaat ccgcgatctg

361 ctcgcgcaaa cgaacgggga ctcgccgggg
aggcacggcg gccgtgggac accatcgcca

421 accgacgggc ctcgcgcgag cacagaatcg caatag
```

**23.** What is TOPO cloning, and how does it work?

**24.** You want to make a fusion between GFP and a bacterial ion channel, and express it in *E. coli*. Discuss the cloning strategy; where the GFP could/ should go, and if this affects the strategy; and different ways in which this could go wrong.

**25.** When PCR primers fail to produce the expected product, what are two (contradictory) strategies that might help?

**26.** Discuss the advantages and disadvantages of cloning by PCR.

**27.** PCR has been suggested as a technique for detecting traces of microbial life on nearby planets, such as Mars. Do you think this is a good technique? Discuss the pros and cons.

**28.** Why is it important to heat-inactivate restriction enzymes before ligation? Some enzymes do not heat-inactivate. (Name one.) What do you do in this case?

**29.** (ADVANCED) Derive **Equation A1.1**.

**30.** (ADVANCED) Describe how gel electrophoresis is used to sequence DNA.

## Background Reading

### Books

Nicholl, D.S.T. *An Introduction to Genetic Engineering*. Edn. 3. Cambridge University Press, 2008. *An excellent introductory text written for scientific nonexperts, with extensive glossary.* Cambridge University Press, Cambridge, UK.

Primrose, S.B., Twyman, R.M., and Old, R.W. *Principles of Gene Manipulation*. Edn. 6. Blackwell Publishers, 2002. Narrative history of molecular biology along with well-illustrated description of methods and concepts. *Principles of Gene Manipulation and Genomics*, by the same author, includes genomic techniques. Blackwell Publishers, Hoboken, NJ.

Sambrook, J., and Russell, D. *Molecular Cloning: A Laboratory Manual*. Edn. 4. Cold Spring Harbor Laboratory Press, 2012. Indispensable benchtop companion. New revised version contains all modern methods. An absolute must in any lab. Cold Spring Harbor Laboratory Press, Cold Spring Harbor, NY.

Watson, J.D., Myers, R.M., Caudy, A.A., and Witkowski, J. A. *Recombinant DNA: Genes and Genomes—A Short Course*. Edn. 3. W. H. Freeman, 2007. Clearly written narrative explaining the theory and practice of molecular biology. WH Freeman, New York, NY.

### Book series and methods journals

These series and periodicals provide up-to-date, detailed protocols for laboratory use.

*Methods in Molecular Biology*, Humana Press (many indexed online at Springer Protocols, https://www.springer.com/gp /eproducts/springerprotocols)

*Journal of Biological Methods* (http://www.protocol-online.org /prot/Molecular_Biology/)

*Cold Spring Harbor Protocols* (http://cshprotocols.cshlp.org/)

*Journal of Visualized Experiments* (http://www.jove.com/)

*Nature Protocols* (http://www.nature.com/nprot/index.html)

*Current Protocols* (http://onlinelibrary.wiley.com/book/10.1002 /0471142727)

### Journal articles

Boyer, H.W. (1971). DNA restriction and modification mechanisms in bacteria. *Annual Review of Microbiology* 25, 153–176.

Dieffenbach, C.W., Lowe, T.M., and Dveksler, G.S. (1993). General concepts for PCR primer design. *PCR Methods and Applications* 3, S30–S37.

Froger, A., and Hall, J.E. (2007). Transformation of plasmid DNA into *E. coli* using the heat shock method. *Journal of Visualized Experiments* (6), e253, doi:10.3791/253 (2007).

Glauner, K.S., Mannuzzu, L.M., Gandhi, C.S., and Isacoff, E.Y. (1999). Spectroscopic mapping of voltage sensor movement in the Shaker potassium channel. *Nature* 402, 813–817.

Lehman, I.R. (1974). DNA ligase: Structure, mechanism, and function. *Science* 186, 790–797.

Nathans, D., and Smith, H.O. (1975). Restriction endonucleases in the analysis and restructuring of DNA molecules. *Annual Review of Biochemistry* 44, 273–293.

Zhou, M.Y., Clark, S.E., and Gomez-Sanchez, C.E. (1995). Universal cloning method by TA strategy. *Biotechniques* 19, 34–35.

Zhou, M.Y., and Gomez-Sanchez, C.E. (2000) Universal TA cloning. *Current Issues in Molecular Biology* 2, 1–7.

Zhu, Q., and Casey, J.R. (2007). Topology of transmembrane proteins by scanning cysteine accessibility mutagenesis methodology. *Methods* 41, 439–450.

## Websites and software

### *Some major suppliers of kits and enzymes*

(Note that this is a very small sample of what is available. A custom search will allow you to identify the best products for your experiments and geographical area. Many universities and institutes have on-site supply cabinets from these and other companies, which can help in your choice of reagents. Also see the list of suppliers in Cell biology: Table of suppliers. *Nature* 446, 941–942, 19 April 2007, doi:10.1038/446941a.

Thermo Fisher Scientific (purchasers of Invitrogen). Major supplier of competent cells, enzymes, gel and blotting supplies, PCR and rt-PCR supplies, custom primers, custom oligonucleotides with or without tags, kits (especially cloning kits), sequencing, and more.

Takara Clontech. Major supplier of mammalian expression vectors of many different types, including inducible vectors. A large selection of fluorescent proteins in different kinds of vectors with different targeting sequences (nucleus, membrane, mitochondria, etc.).

New England Biolabs. The major resource for restriction enzymes. Also DNA-modifying enzymes, PCR supplies, competent cells, and a few cloning vectors.

Qiagen. Major supplier of kits for DNA and RNA processing.

Roche. Supplies and instruments for PCR, real-time PCR, sequencing, etc.

Genscript. Custom synthesis of large DNA sequences (several kilobase pairs), as well as other services such as mutagenesis, protein expression, custom plasmid preparation, and cloning services for BACs and other large vectors.

EMD Millipore. Instrumentation and supplies from multiple manufacturers.

Promega. Major supplier of PCR reagents, including master mixes for real-time PCR, rt-PCR, and high-fidelity PCR. Also competent cells, cloning vectors, restriction

enzymes, microarrays, and instrumentation. Site contains user guides and protocols for many molecular biological techniques.

### *Molecular biology news/support sites*

This is a sampling of what was hot in early 2016. Search for any molecular biology keywords to update the list.

> http://www.addgene.org/plasmid-reference/index/
>
> http://www.bio-protocol.org/
>
> http://openwetware.org/
>
> http://biowww.net/protocols/
>
> http://www.123genomics.com/protocols.html
>
> http://bitesizebio.com/
>
> https://www.thermofisher.com/us/en/home /support/scientific-tools-and-tips.html
>
> http://www.cellbiol.com/
>
> http://biocurious.com/

### Software

> Lists of software for Macintosh, Windows, and Linux: http://cellbiol.com/molecular_biology_software _download.php
>
> Also try this list: http://omictools.com/gene-design -category

### Free software

> Primer3. A standard program for use in designing PCR primers. Very widely used. The code is available from SourceForge, or the program may be used in a web browser as Primer3Plus.
>
> Gene designer 2.0. Design tools including codon optimization, from a company that provides custom synthesis.
>
> NEB cutter (http://tools.neb.com/NEBcutter2). Plasmid mapping, from New England Biolabs.
>
> Benchling. Browser-based tools for collaborative molecular biology.
>
> Molecular Cloning Designer Simulator. All-in-one package.

DNAWorks. Optimization of gene synthesis.

EMBOSS. Software package for Linux.

## Commercial software

DNA star. A family of software tools for sequence analysis and experimental design.

Gene Construction Kit (Textco BioSoftware). Plasmid mapping software, showing restriction sites, reading frames, and cloning project history.

Geneious. Integrated platform covering DNA and protein tools.

# CHAPTER 3

# Expression of Genes in Bacteria, Yeast, and Cultured Mammalian Cells

## 3.1 INTRODUCTION

A vast number of experiments begin by inserting a plasmid into a type of cell in which it will express—this may be *Escherichia coli* or other bacteria, yeast cells (the simplest eukaryotes), or mammalian cells of many different types. Once it is expressed in the appropriate host, the protein encoded by the plasmid may be evaluated for function biochemically or by imaging; its effect on the health of the cells may be determined by toxicity assays or by the molecular investigation of cell death pathways; tracking of one or more protein molecules may be studied; promoter function can be measured as a function of time, cell phenotype, or other factors; and effects of genetic changes to the protein on any of the factors above may be evaluated by performing the same experiments with a series of different plasmids encoding proteins with targeted mutations.

The introduction of foreign DNA or RNA into mammalian cells is called *transfection* when it is done by nonviral methods, and *infection* or *transduction* when it is performed with a virus. The corresponding process in bacteria and yeast is usually called *transformation*. Transformation was discussed in Chapter 2 for cloning strains of *E. coli*; in this chapter, we extend the discussion to other types of microorganisms, such as Gram-positive bacteria and yeast. We then discuss mammalian cell culture and some of the myriad of ways in which foreign genes can be introduced into mammalian cells *in vitro*. Some of these methods are routine and others much more difficult, requiring specialized cloning or instrumentation. The choice of method depends on the downstream application and is often key to the success of an experiment. We then address some particular transfection techniques for difficult cells: electroporation, microinjection, the "gene gun," optical and magnetic transfection, and viral vectors.

After reading the appropriate sections of this chapter, you should be able to prepare and perform a straightforward transfection experiment of bacteria, yeast, or an immortalized mammalian cell line. You should have a reasonable appreciation of when each type of cell can be used, and the advantages and disadvantages of each. You should also understand under what circumstances a more challenging approach is needed, and possess sufficient resources to go into the literature to design an experiment involving primary cultures, explants, stable transfections, and/or viruses.

# 3.2 EXPRESSING GENES IN MICROORGANISMS

### *E. coli*

Some experiments are readily performed by expressing the genes of interest in *E. coli*. These include studies of gene regulatory networks such as the "repressilator," where the interactions among mutually repressing or enhancing promoters can be studied. The advantage of working with *E. coli* is that the cells are easy to grow and transform, and may be produced in large amounts and readily transferred onto any substrate of interest (microscope slides, chips, microfluidics, and so on). Selection for successful transformation is easy and efficient using antibiotics. The genotypes and phenotypes of *E. coli* are also well characterized for many different strains, so that unexpected interactions between the introduced plasmid and the host can be minimized.

Transformation of *E. coli* strains is readily performed by heat shock or electroporation as described in **Chapter 2**. The only significant challenge in these systems is choosing and obtaining an appropriate strain for the experiment of interest. For many experiments, such as those involving fluorescent labeling, motility/flagellar studies, or other types of imaging, the choice of strain is not critical, and any cloning strain may be used. For experiments on gene regulation or expression levels, it is important that naturally occurring proteins in the cell do not interfere with the promoter on the introduced plasmid. For example, if a gene of interest is expressed from a *lac* promoter, it may be desirable to use a strain that produces no natural *lac* genes (a so-called *lac*⁻ strain). Databases of available strains allow searches by gene product, mutation, and other features and can be consulted when specialized strains are needed.

## Other bacterial strains

For some types of biophysical experiments, expression of genes in bacteria other than *E. coli* is desired. For instance, fluorescent markers can aid in visualization of processes such as bacterial adhesion and biofilm formation, which are studied in specific environmental organisms or pathogens that are known to form biofilms. Motility experiments are also often performed with strains that are more motile than *E. coli*. The expression of proteins from *Gram-positive* bacteria is also frequently very limited in *E. coli*, so studies of these proteins should be performed in the native host or in another Gram-positive strain.

The first step in a bacterial expression experiment is to choose a strain. Many *type strains* can be purchased commercially; some of these have been completely sequenced. Microorganisms may also be environmental or clinical isolates. Strains can be propagated by *passaging* cells from a mature culture into fresh culture medium, being careful to avoid cross-contamination with other bacteria from the environment or a neighboring experiment. Passaging should only be done five or fewer times; at this point, a fresh culture from a petri dish or a frozen bacterial stock should be used (see **Practical Tips 3.1**).

Plasmids that replicate in *E. coli* do not necessarily do so in other bacteria, as the origins of replication are different. A plasmid with ORIs for both *E. coli* and another type of microorganism is known as a *shuttle vector,* and is the most common type of plasmid used in non–*E. coli* cloning experiments. The advantage of this type of vector is that cloning manipulations—such as insertions, deletions, mutagenesis,

## PRACTICAL TIPS 3.1:   GROWING AND MAINTAINING BACTERIA

*General notes.* Always use good sterile technique. Plates showing growth of mold or unknown bacterial colonies should be discarded into an appropriate biohazard disposal. As with cloning strains, treat the bacteria with appropriate microbiological technique to prevent contamination—use single colonies and seed them into sterilized media and glassware. For aerobic strains, make sure that the flask is at least two to three times the size of the volume of medium it contains, and that the culture is shaken rapidly (150–200 rpm) to permit oxygenation.

As discussed in **Chapter 2**, bacteria may be grown on solid media (Petri dishes or containing agar) or in liquid culture. Different bacteria have different nutrient requirements; sometimes prepared media are available for a specific strain, and other times specific media must be made. These are usually rich media that contain many more nutrients than the bacteria actually need. The so-called *minimal media* can also be prepared that contain the minimal nutrients with which the strain can grow. These can be used to determine phenotypes of the bacteria, to select for strains, or in cases where the rich media interfere with the experiment in some way (for example, some vitamins in rich media may quench the emission of fluorescent probes). Bacteria also have different ideal incubation temperatures, and an appropriate incubator must be chosen. **Table P3.1.1** shows some media appropriate for common strains and cloning protocols. It is important to note that these are not selective media for microbiology, but rich media intended to allow for

**Table P3.1.1**

Examples of Growth Media for Commonly Used Bacterial Strains

| Medium | Example Strain(s) and Temperature | Recipe |
|---|---|---|
| Lysogeny broth (LB) (also called Luria–Bertani) | *Nonfastidious* organisms<br>*E. coli* (37°C)<br>*B. subtilis* (30°C)<br>*Staphylococcus epidermidis* (37°C) | 10 g Bacto tryptone<br>5 g Bacto yeast extract<br>10 g NaCl<br>$H_2O$ to 1 L |
| Brain–heart infusion (BHI) | *Fastidious* organisms<br>*Staphylococcus aureus* (37°C)<br>*Streptococcus* sp. (37°C)<br>*Meningococcus* sp. (37°C) | Available commercially as a powder |
| Trypticase soy broth (TSB) | General purpose for many strains; 5% sheep's blood may be added for fastidious strains | 15 g Tryptone<br>5 g Soytone<br>5 g NaCl<br>$H_2O$ to 1 L |
| Terrific broth (TB) | Highly enriched medium used to support organisms containing large or toxic plasmids; helps increase plasmid yield | 12 g Bacto tryptone<br>24 g Bacto yeast extract<br>4 mL glycerol<br>$H_2O$ to 900 mL<br>100 mL 0.17 M $KH_2PO_4$ and 0.72 M $K_2HPO_4$ (prepared separately and added after cooling) |
| Super optimal broth (SOB) | High-nutrient forms of LB used to support *E. coli* after plasmid transformation | 20 g Bacto tryptone<br>5 g Bacto yeast extract<br>0.5 g NaCl<br>0.186 g KCl<br>$H_2O$ to 1 L |
| SOB with catabolite repression (SOC) | Used to support *E. coli* after plasmid transformation | From an SOB base, add concentrated solutions of $MgCl_2$ (final concentration 10 mM) and glucose (20 mM) |

*(Continued)*

---

### PRACTICAL TIPS 3.1 (CONTINUED):   GROWING AND MAINTAINING BACTERIA

easy cultivation of these strains. Avoidance of contamination by unwanted environmental or laboratory strains is critical.

A culture provided as a frozen stock should be stored in a cryogenic refrigerator (–70 to –80°C). It may be scratched with a sterile toothpick or pipette tip while still frozen; the tip is used to streak a plate, and the rest of the culture is returned to the freezer. Cultures provided in liquid may be streaked directly.

Petri dishes represent the best method for short-term growth and maintenance of strains. Streak a plate as described in **Chapter 2** in order to obtain discrete colonies. Each bacterial strain will produce colonies of a specific color and morphology, allowing you to determine if any contaminants are in the culture. Grow the plate at growth temperature until the colonies form; do not let it go longer or the cells will die. An agar plate may then be stored in the refrigerator for a month or more. The night before an experiment, pick a single colony and seed it into 3–5 mL of growth medium. The following day, this culture may be further seeded into a larger volume and grown to the desired optical density for the experiment. It may also be passaged into a volume of fresh medium for the following day, although the risk of contamination is greater with this method than if single colonies are picked each day.

For long-term storage of bacterial cultures, grow a reasonable amount (10–20 mL) of broth culture overnight (do not overgrow). Distribute 1 mL aliquots into sterile cryotubes. Add 0.2 mL of clean dimethylsulfoxide (DMSO) or 1 mL of sterile glycerol, and mix by inverting several times. The cultures should be viable for at least 1 year at –70°C (if the toothpick method fails to yield colonies, a whole aliquot of frozen cells may be thawed and added to 5–10 mL of liquid medium).

---

and so on—can be performed in the well-characterized, easy-to-use *E. coli*. A few shuttle vectors are available commercially with broad host ranges, typically Gram-negative or Gram-positive bacteria, and with antibiotic selection markers for both *E. coli* and the target strain (**Figure 3.1**). Individual researchers have also reported shuttle vector constructions for many strains, including some *extremophiles* that grow in exceptionally hot (thermophiles) or cold (psychrophiles) conditions. For most bacterial expression experiments with noncloning strains, it is useful to contact these laboratories to request the plasmids for the species or strain of interest.

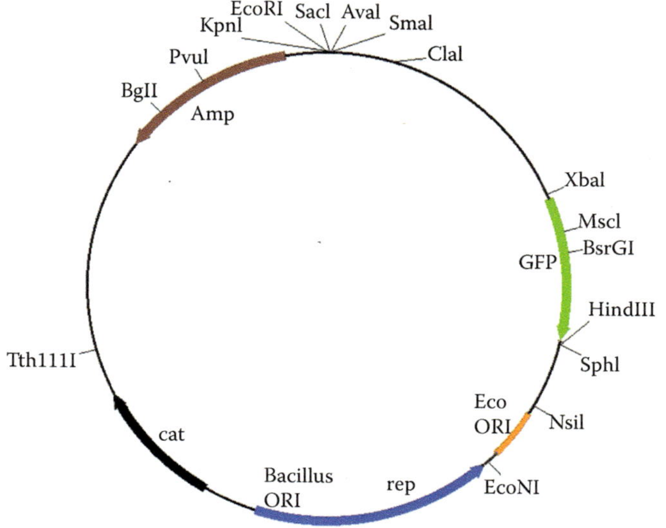


**Figure 3.1** ***E. coli*–Gram-positive shuttle vector.** The plasmid encodes GFP from a constitutive *Bacillus* promoter that drives expression in Gram-positive bacteria. Both an *E. coli* and *Bacillus* origin of replication (ORI) are present; the latter is followed by the *rep* gene encoding a replication initiation protein. The selectable marker *cat* (chloramphenicol) can be used in both strains; Amp resistance is only expressed in *E. coli*.

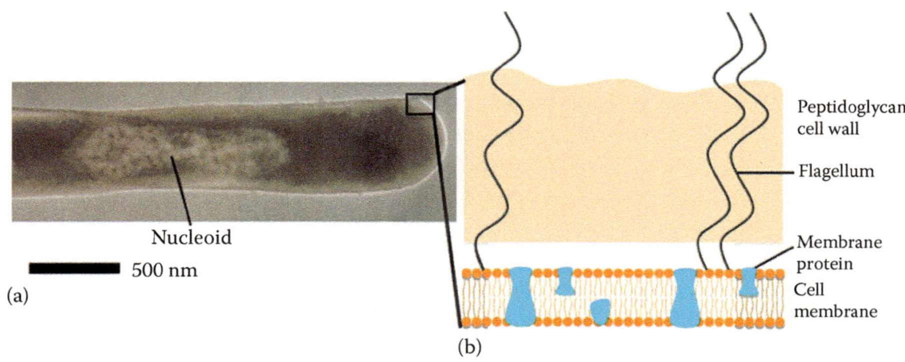


**Figure 3.2 *B. subtilis*, a commonly used Gram-positive strain.** (a) Pseudocolored TEM image of a single bacterial cell, showing the DNA in a nucleoid region. (b) Schematic of Gram-positive cell wall. Unlike Gram negatives, only one membrane is present.

Once the right plasmid has been obtained and the gene of interest cloned into it, it must be delivered to the bacterial cells. Some strains show "natural competence," meaning that they can take up foreign DNA without any special chemical treatment. An example of a naturally competent strain is *Bacillus subtilis*, a large (1 × 5 μm), highly motile, nonpathogenic Gram-positive strain (**Figure 3.2**). It can be readily used in transformation experiments, with its natural competence being enhanced by growth under starvation conditions. Certain mutations that make the strain require a particular nutrient, e.g. tryptophan, make induction of starvation easier. **Practical Tips 3.2** gives a recipe for production and transformation of competent *B. subtilis*.

Most other Gram-positive strains are trickier to transform because of the thick peptidoglycan cell wall that acts as a barrier to DNA entry. Digestion of the cell walls can produce living cells called *protoplasts* (if Gram positive) or *spheroplasts* (if Gram negative), which may be transformed using electroporation or chemical methods such as polyethylene-glycol mediated DNA delivery; references to these methods are given at the end of the chapter. It can be tricky and time-consuming to generate the protoplasts and to reconstitute them into intact cells afterward, but the transformation efficiency of these methods is 100-fold or more greater than that of natural competence. Protoplasts and spheroplasts have other uses in biophysics as well, such as in electrophysiology (**Chapter 15**). **Practical Tips 3.3** gives methods for *E. coli* and *B. subtilis* that are specifically designed for DNA transfer.

## Yeast cells

Yeast cells combine the ease of culture that characterizes microorganisms with the ability to express and correctly post-translationally modify complex proteins. They are therefore useful for certain types of studies on transfected mammalian proteins, such as high-throughput screening of mutants. Yeast cells also possess complex features associated with eukaryotes, such as a cytoskeleton, that have been the object of many biophysical studies. They are also commonly used in gene network studies, as many details of their regulatory networks have been elucidated. In these studies, transfection may be needed to express reporters, such as a target protein linked to a fluorescent protein.

There are two primary types of yeast: *fission yeast*, which reproduce by symmetric cell division (**Figure 3.3a**); and *budding yeast*, which reproduce by the production of a small daughter cell or bud (**Figure 3.3b**). A typical example of a fission yeast used in many experiments is *Schizosaccharomyces pombe*. *S. pombe* was originally isolated from millet beer and is also found in grape juice and kombucha tea; it can even be used to make bread, although the dough does not rise as predictably as

**PRACTICAL TIPS 3.2:    EXPRESSING GENES IN *BACILLUS SUBTILIS***

*General notes. B. subtilis* demonstrates "natural competence" when it is limited in key nutrients. This can be exploited for cloning by using particular strains deficient in a key nutrient, usually the amino acid tryptophan (strains deficient in the *trp* gene cannot synthesize tryptophan and thus require it in the medium). By carefully controlling nutrients as described next, the bacteria can be made to take up DNA.

**PROTOCOL USING NATURAL COMPETENCE IN *B. SUBTILIS***

**Media**

10× Spizizen salts

- Ammonium sulfate, $(NH_4)_2SO_4$ (20 g, 150 mM)

- Anhydrous potassium hydrogen phosphate, $K_2HPO_4$ (139.7 g, 800 mM)

- Potassium dihydrogen phosphate, $KH_2PO_4$ (60 g, 440 mM)

- Tri-sodium citrate (10 g, 35 mM)

- Distilled water to 1 L, autoclave, then add 1 mL of 1 M $MgSO_4$

**Medium A: Growth Medium**

For 100 mL, with the balance brought up by $H_2O$:

- 10× Spizizen salts, 10 mL

- 0.5% glucose

- 50 µg/mL DL-tryptophan

- 50 µg/mL uracil

- 0.02% casein hydrolysate

- 0.1% yeast extract

- 8 µg/mL arginine

- 0.4 µg/mL histidine

- 1 mM $MgSO_4$

**Medium B: Starvation Medium**

For 100 mL, with the balance brought up by $H_2O$:

- 10× Spizizen salts, 10 mL

- 0.5% glucose

- 5 µg/mL DL-tryptophan

- 5 µg/mL uracil

- 0.01% casein hydrolysate

*(Continued)*

---

**PRACTICAL TIPS 3.2 (CONTINUED):   EXPRESSING GENES IN *BACILLUS SUBTILIS***

- 0.1% yeast extract

- 1 mM $MgSO_4$

- 2.5 mM $MgCl_2$

- 0.5 mM $CaCl_2$

### A. PREPARING COMPETENT *B. SUBTILIS*

1. Begin with a fresh plate of cells on nonselective medium.

2. Grow in medium A to stationary phase at 37°C (approximately 1.5 h after the end of exponential phase; you may have to measure this).

3. For each transformation desired, dilute 0.05 mL of culture into 0.45 mL of prewarmed medium B and continue incubating for 1.5 h at 37°C.

4. Pellet the cells; remove and discard 0.4 mL of the supernatant.

5. Resuspend the pellet in the remaining 0.1 mL and add ~0.6 µg of DNA in 1–10 µL of $H_2O$.

6. Incubate for 30 min at 37°C with shaking; lay tube on side for proper aeration.

7. Plate on nutrient agar (LB) containing the appropriate antibiotic for plasmid selection.

### B. GLYCEROL STOCKS OF COMPETENT *B. SUBTILIS*

1. Prepare competent cells to step 4 above.

2. Remove supernatant.

3. Resuspend in 0.5 mL sterile 60% glycerol.

4. Store at −80°C until use.

5. Before use, spin down to pellet and remove all glycerol; add 0.1 mL of prewarmed medium B and proceed with step 5.

### SUGGESTED READING

Anagnostopoulos, C., and Spizizen, J. (1961). Requirements for transformation in *Bacillus subtilis*. *Journal of Bacteriology* 81, 741–746.

Spizizen, J. (1958). Transformation of biochemically deficient strains of *Bacillus subtilis* by deoxyribonucleate. *Proceedings of the National Academy of Sciences U S A* 44, 1072–1078.

---

with ordinary baker's yeast. Its entire genome and proteome are known, and because of its medial fission, it is often used as a model of cell division.

*Saccharomyces cerevisiae*, baker's or brewer's yeast, is a budding yeast and is the most common model organism for probing structure and regulation in eukaryotic cells. It was the first eukaryotic genome sequenced (1996) and is very commonly used in

## PRACTICAL TIPS 3.3:    PREPARATION OF PROTOPLASTS/SPHEROPLASTS

*General notes.* The procedure involves digestion of the cell walls with the enzyme lysozyme (muramidase). Lysozymes are a family of proteins produced by eukaryotes to protect against bacterial invasion by attacking peptidoglycans, a component of cell walls. They act by hydrolyzing the glycosidic bond between N-acetylmuramic acid and N-acetylglucosamine. Lysozyme is found in human milk, saliva, and tears and is important in preventing infections such as diarrhea and conjunctivitis. The most common lysozyme used in research comes from egg white, where it serves to prevent infection of the developing bird embryo.

While the procedures are similar for Gram-negative and Gram-positive cells, the amounts of lysozyme needed vary because Gram positives contain much more peptidoglycan than Gram negatives. The purpose of the sucrose in the solutions is to control the solution's osmolarity; if the solution is too hypo-osmotic, the cells will burst.

### A.  FOR *E. COLI*

### PA Medium

- 10 g casamino acids

- 10 g nutrient broth (Difco)

- 10 g glucose

- 100 g sucrose

- To 1 L with distilled water and sterilize

### Procedure

1. Add 1 mL overnight culture to 200 mL fresh medium (LB).

2. Grow until the cells reach $5 \times 10^8$/mL ($OD_{600} \sim 1$).

3. Centrifuge and resuspend in 3.5 mL of 1.5 M sucrose. Do not chill or wash the cells.

4. Add 1 mL of 30% bovine serum albumin (Povite) and 0.2 mL of a 2 mg/mL lysozyme solution in Tris buffer (pH 8.1).

5. Add 0.4 mL of unbuffered 4% EDTA.

6. After 40–120 s, add 95 mL of PA medium.

7. The cells will become spheroplasts over the course of about 10 min at room temperature. This can be monitored visually (**Figure P3.3.1**).

**Figure P3.3.1 Images of spheroplasts as they form.** (a) Fluorescence. (b) Brightfield.

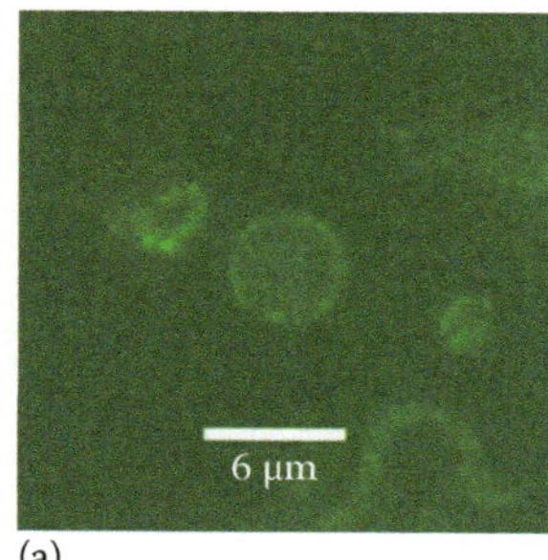

(a)

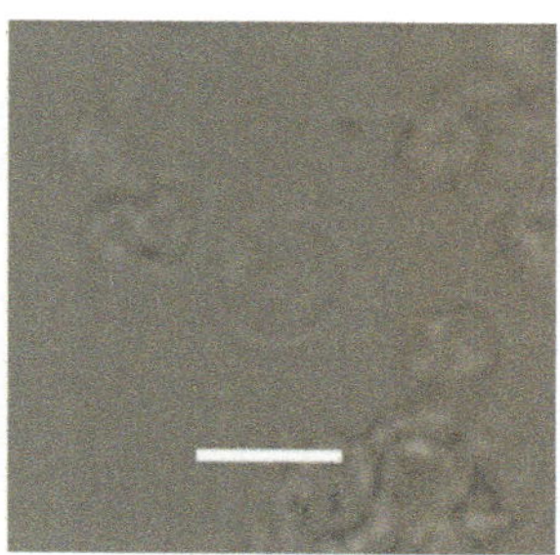
(b)

*(Continued)*

### PRACTICAL TIPS 3.3 (CONTINUED):   PREPARATION OF PROTOPLASTS/SPHEROPLASTS

8. After the process is complete, add 2 mL of 10% $MgSO_4$ to chelate the EDTA and stop the reaction. The addition of 1% protamine sulfate will aid competence; if you are using the spheroplasts for other purposes, it can be omitted.

9. Place the spheroplasts on ice; they will be stable for several hours.

### B. FOR *B. SUBTILIS*

**Penassay Broth (Rich Medium for *B. subtilis* Growth)**

- 10 g peptone

- 1.5 g yeast extract

- 1.5 beef extract

- 3.5 g sodium chloride

- 1.0 g glucose

- 3.68 g potassium phosphate, dibasic

- 1.32 g potassium phosphate, monobasic

- To 1 L with distilled water and sterilize

**Protoplast Medium (PM)**

- A v/v mixture of 2× penassay broth and 1 M sucrose, 0.04 M maleate, 0.04 M MgC12, pH 6.5

**Procedure**

1. Add 1 mL overnight culture to 200 mL fresh penassay broth.

2. Grow until the cells reach $5 \times 10^8$/mL ($OD_{600} \sim 1$).

3. Centrifuge and resuspend in 5 mL of SM and place in a 100 mL flask.

4. Add lysozyme (crystalline) to a final concentration of 2 mg/mL.

5. Incubate with gentle shaking in a water bath or heated room.

6. The cells will become spheroplasts over the course of about 1 h at 37°C. This can be monitored visually (**Figure P3.3.1**).

7. Collect by gentle centrifugation (2000 × *g* for 10 min), wash once in PM, and resuspend in 5 mL of PM.

8. Proceed with PEG-based transformation protocols or other experiments.

### SUGGESTED READING

Benzinger, R., Kleber, I., and Huskey, R. (1971). Transfection of *Escherichia coli* spheroplasts. I. General facilitation of double-stranded deoxyribonucleic acid infectivity by protamine sulfate. *Journal of Virology* 7, 646–650.

Bourne, N., and Dancer, B.N. (1986). Regeneration of protoplasts of *Bacillus subtilis* 168 and closely related strains. *Journal of General Microbiology* 132, 251–255.

Hopwood, D.A. (1981). Genetic studies with bacterial protoplasts. *Annual Review of Microbiology* 35, 237–272.

**Figure 3.3 Yeast cells.**
(a) Fission yeast, *S. pombe*.
(b) Budding yeast, *S. cerevisiae*.
(Images courtesy of Kathryn
Cross and Steve James, National
Collection of Yeast Cultures, IFR,
Norwich, UK.)

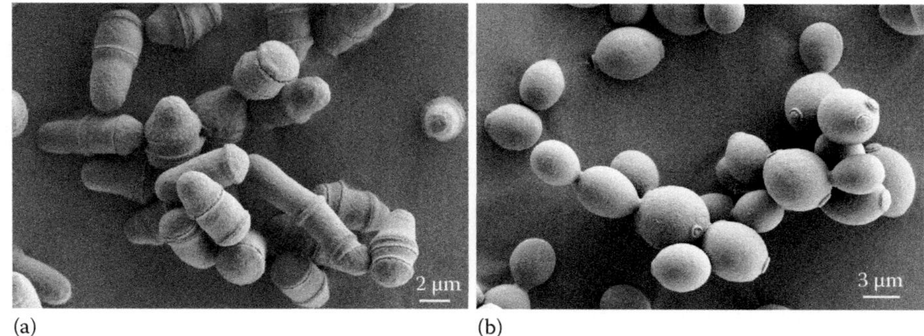


studies of gene regulation. It also displays all of the *motor proteins* of the eukaryotic cytoskeleton, which use ATP to generate mechanical force and which are responsible for many different types of cell motility: contraction (as in muscle cells), vesicle motility, cytoplasmic streaming, chromosome segregation during cell division, and more. Because the cytoskeleton of prokaryotes is much more rudimentary (until recently, it was thought to be completely lacking), yeast cells have provided biophysicists the simplest possible model cell in which to study these processes.

Budding yeast and fission yeast are evolutionarily very different, and media and plasmids must be tailored to the type selected. Plasmids are shuttle vectors as with bacteria—that is, grown and replicated in *E. coli*, but expressed in yeast. Yeast strains and plasmids are available through national repositories in many countries, such as the American Type Culture Collection and several others as listed at the end of the chapter. Individual researchers often also generate large numbers of expression plasmids or libraries, which they then share with the scientific community. References to some key papers are also given at the end of this chapter.

Growth and maintenance procedures for yeast are similar to those for bacteria. There are two standard, easy methods for introducing plasmid DNA into yeast cells: electroporation and the lithium acetate (LiAc) method. A protocol for the latter is given in **Practical Tips 3.4**. Yeast cells may also be made into protoplasts and transformed. The efficiency of all of these methods ranges from ~$10^3$ to $10^5$ transformants/µg DNA.

## 3.3 MAMMALIAN CELL CULTURE

### Introduction to immortalized cell lines

Study of mammalian proteins must usually be done in mammalian cells, as bacteria and even yeast cannot perform the post-translational modifications such as glycosylation or disulfide bond formation that are necessary for correct protein function. This is especially true of membrane-associated proteins such as receptors and channels. Even proteins that express well in bacteria may be easier to study in mammalian cells for a variety of reasons. Mammalian cells are much bigger than bacteria or yeast (10–20 µm or more in diameter as opposed to 1–2 µm), making it possible to image subcellular structures by light microscopy (**Figure 3.4**). They have no cell walls, so the outer membrane is readily accessible to electrodes (such as for patch clamp), dyes, and other interventions. Different cell types rarely cross-contaminate one another, and usually when something is wrong, it is readily observed by eye.

**PRACTICAL TIPS 3.4:   LiAc TRANSFECTION OF YEAST**

*General notes.* This protocol has survived essentially unchanged from when it was first published in 1992. It is easy, with a few pitfalls. Maintaining sterility is critical. PEG and LiAc solutions should be made fresh from 10× stocks just before use. The glucose is added after autoclaving to prevent it from caramelizing; it can be sterilized by ultrafiltration.

*YPAD Medium*

- 20 g peptone

- 10 g yeast extract

- To 950 mL with distilled water, adjust pH to 6.5 and sterilize

- After cooling to 55°C, add glucose to 2% and 13 mL of 2 mg/mL adenine hydrochloride and adjust final volume to 1 L

*YPAD Plates*

- Same as medium, but add 20 g Bacto agar before autoclaving and 1 pellet of NaOH to prevent agar breakdown during heating

- After cooling and adding glucose, add selective antibiotic

- Pour into plates before the medium hardens

*TE-LiAc*

- 0.1 M lithium acetate, pH 7.2

- 0.1 M Tris, pH 7.5

- 50 mM EDTA

*Li-PEG*

- 40% PEG 4000 in TE-LiAc

**PROCEDURE**

1. Add 50 mL overnight culture to 300 mL fresh medium (YPD).

2. Grow to an $OD_{600}$ of 0.8.

3. Centrifuge to pellet (about 2500×$g$, 5 min) and rinse once with 20 mL sterile $H_2O$.

4. Rinse once with TE-LiAc and incubate at 30°C for 10 min; invert the tubes every minute.

5. Centrifuge, remove supernatant, and then resuspend in TE-LiAc.

6. Incubate at 30°C for 10 min.

7. Mix 100 mL of the yeast suspension with 1–5 µg DNA plus 100 µg carrier DNA (salmon sperm DNA, available commercially); pipette up and down once.

8. Add 600 µL Li-PEG and slowly pipette until the solution is homogeneous.

*(Continued)*

---

**PRACTICAL TIPS 3.4 (CONTINUED):    LiAc TRANSFECTION OF YEAST**

9.  Incubate at 30°C for 30 min; invert the tubes every 5 min.

10. Incubate at 42°C for 15 min.

11. Centrifuge for 2 min at 1600×g and remove 600 µL of the supernatant.

12. Add TE (40–400 µL) and plate on selective plate.

13. Yeast cells may be spread onto plates using sterilizable glass or plastic "hockey sticks"; we also recommend glass plating beads.

**SUGGESTED READING**

Gietz, D., St Jean, A., Woods, R.A., and Schiestl, R.H. (1992). Improved method for high efficiency transformation of intact yeast cells. *Nucleic Acids Research* 20, 1425.

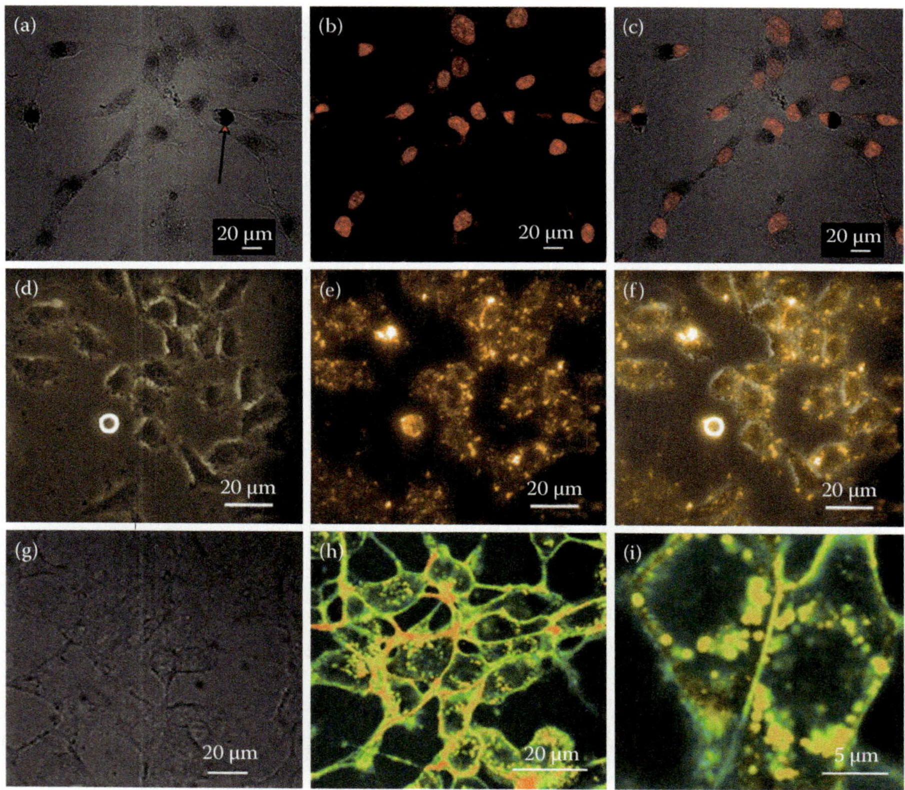


**Figure 3.4 Appearance of some mammalian cell lines under brightfield and fluorescence microscopy.** (a) B16 mouse melanoma cells. The cells produce melanin (arrow), which makes them appear dark under brightfield. (b) Same cells fluorescently labeled with a nuclear-targeting drug, the anticancer drug doxorubicin. (c) Overlay of the fluorescence and brightfield on the right shows the nuclear targeting of the drug. (d) HeLa cervical carcinoma cells, which are epithelial cells with a typical appearance. (e) HeLa cells labeled with orange-emitting quantum dots (QDs; see **Chapter 11**). (f) Overlay of brightfield and fluorescence showing that the QDs surround but do not enter the nucleus. (g) NIH 3T3 fibroblasts. The large, flat appearance is typical of this type of cell. (h) Fluorescence lifetime (FLIM) image (see **Chapter 8**) of NIH 3T3 cells labeled with QDs (not the same cells as in g). The speckled appearance suggests endosomal uptake of the QDs. (i) Close-up of cells under FLIM clearly showing the pattern of endosomes.

On the other hand, growth and culture of mammalian cells require some specific infrastructure, instrumentation, and supplies. There is a requirement for *biosafety containment* for cell lines, often biosafety level 2 (BSL-2). This usually means that a designated laboratory or at least a room must be set aside for mammalian cell work. A specialized tissue culture hood (biosafety cabinet) is needed to protect the cells from contamination and the experimenter from exposure to the cells (**Figure 3.5a**). Most universities provide training in the use of these cabinets, and some require them for all personnel working in a BSL-2 laboratory. Water-jacketed *$CO_2$ incubators* are required to provide a sterile, temperature-controlled environment for cell growth under a $CO_2$-enriched atmosphere (usually 5% $CO_2$; controlling the $CO_2$ in the atmosphere balances the culture's pH when the medium is buffered by sodium bicarbonate). A tank of liquid nitrogen must be kept in the laboratory or nearby to preserve cultures; cells may be stored long-term in liquid nitrogen because the growth of ice crystals is inhibited at temperatures below $-130°C$. Finally, specialized plastic- and glassware is necessary for maintaining healthy cultures. Most mammalian cells must adhere to a substrate in order to grow, and cannot do so on untreated plastic or glass. Flasks, dishes, and plates can be bought *tissue culture treated*, with coatings that permit cell growth, or dishes may be coated with polylysine, laminin, or other substrates by the experimenter (**Figure 3.5b**). Specialized nutrient growth media must be purchased and usually supplemented with fetal bovine serum (FBS) and/or horse serum (HS). Antibiotics are usually added to the media to prevent contamination with bacteria and fungi. Contamination of a laboratory or incubator can be catastrophic for the mammalian cell experiments, and *sterile technique* should be rigorously followed. Those without experience in such techniques should be trained before beginning any cell culture experiment. Cells should be inspected daily for changes in medium color or the presence of cloudiness, which can indicate contamination. Medium may be tested for contamination by placing 1–2 mL into a sterile dish and placing it in the incubator; the growth of bacteria or fungi after 24–48 h indicates contamination.

The easiest mammalian cell types to culture and transfect are immortalized cell lines (introduced in **Chapter 1**). These are cells that continue to divide essentially indefinitely in culture, either because they are of malignant origin or because

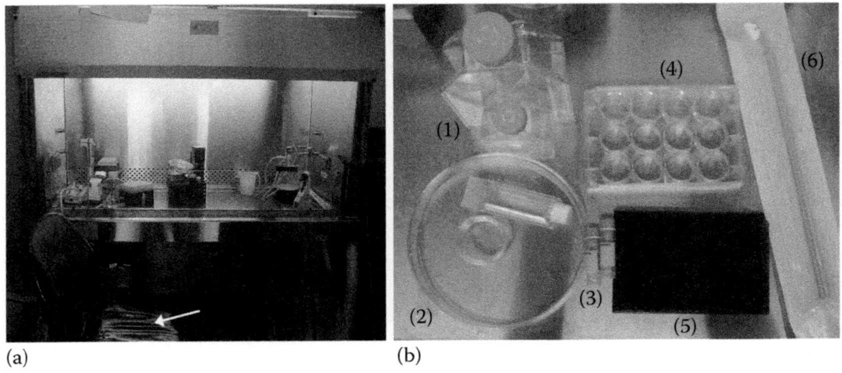


**Figure 3.5 Some of the requirements for tissue culture.** (a) Biosafety hood with glass sash opening just wide enough to admit the experimenter's arms. These hoods provide laminar air flow and overhead UV disinfection. Note that the chair (arrow) is coated with plastic. Some kind of nonporous covering is needed on fabrics to prevent contamination. (b) Some tissue culture supplies. (1) Flasks, (2) culture dishes, (3) cryo vials, (4) 12-well plate, (5) 96-well black plate (black plates are used for measuring fluorescence as they reduce scattering), and (6) cell scraper.

**PRACTICAL TIPS 3.5:    GROWING, PASSAGING, AND STORING ADHERENT MAMMALIAN CELLS**

*General notes.* Each time you begin working with a new cell line, you will become familiar with its appearance, cell morphology, and growth rate. If these parameters change, the cells are either becoming senescent or have become infected. Yeast and bacterial infections are usually obvious because the medium becomes turbid; if the pH indicator phenol red is present, it will change color. However, some infections, such as those due to *mycoplasma*, can be invisible. When a cell line is no longer behaving as it should, or after it has gone through 20 or more passages, a new aliquot of frozen cells should be thawed.

**GROWTH REQUIREMENTS OF MAMMALIAN CELL LINES**

- Physiological parameters: 37°C, pH 7.2–7.5 (phenol red should be red, not purple or yellow).

- Humidity: A tray of sterile water needs to be placed in the incubator.

- Enhanced $CO_2$ concentrations (usually 5%) when bicarbonate is used as a buffer.

- Should be kept in the dark.

- Rich media containing sugars, amino acids, vitamins, and *serum*. The serum is usually from fetal calves (fetal bovine serum), horses (horse serum), or a mix of the two. It is not sold added to cell growth medium, but must be added as a supplement. Its exact composition is unknown; it contains a large number of proteins, hormones, and growth factors that assist in cell attachment to the substrate, inhibit toxins and damaging enzymes (including trypsin), and perform other undefined roles.

- Antibiotics to prevent infection. Unlike in *E. coli* cultures, this is not to select for plasmids, but rather to prevent bacterial and/or fungal contamination. Usually penicillin/streptomycin (pen/strep) is used, although some cell types (e.g., neurons) cannot tolerate it.

**PROCEDURE FOR PASSAGE OF ADHERENT CELLS WITH TRYPSIN**

1. Aspirate cell medium and wash once with phosphate-buffered saline (PBS).

2. Aspirate the PBS and add just enough trypsin to cover the bottom (for a T-25 flask, this is about 5 mL). The trypsin should be prewarmed to 37°C as this will make it work more quickly. Most strongly adherent cells require a 0.25% trypsin solution, but for weakly adherent cells where surface protein integrity is important, 0.05% trypsin should be used with 0.53 mM EDTA.

3. Rock the flask back and forth about 10 times (30 s).

4. Aspirate the trypsin. Observe the cell monolayer (or individual cells under the microscope) to see when they begin to detach. Grossly, this will appear as if the monolayer is sliding down the surface of the flask. As soon as this starts happening, immediately add full culture medium with serum. The serum will inhibit the trypsin and prevent cell damage due to overtrypsinization.

5. Pipette the cells up and down gently at least 10 times to fully resuspend them.

6. Subculture at the desired ratio into a new flask, at least 1:3 of old cells/fresh medium. The ratio chosen depends upon the cell growth rate and the number of days anticipated before a confluent flask is needed again. The remainder of the resuspended cells may be seeded into culture dishes or onto coverslips or multiwell plates for different types of experiments. Cells that are not used are discarded or frozen as below.

*(Continued)*

---

**PRACTICAL TIPS 3.5 (CONTINUED):   GROWING, PASSAGING, AND STORING ADHERENT MAMMALIAN CELLS**

**FREEZING PROTOCOL FOR MAMMALIAN CELLS**

1. The general rule is "freeze slow, thaw fast."

2. After trypsinizing and resuspending, count the cells in a hemocytometer.

3. Centrifuge the cells gently and resuspend in 90% fetal bovine serum (FBS)/10% dimethylsulfoxide (DMSO, tissue culture grade) to concentration of $10^6$–$10^7$ cells/mL.

4. Aliquot into 1 or 1.5 mL cryo vials and place immediately into the –20°C freezer.

5. The next day, place the vials into the –80°C freezer.

6. The following day, transfer them into the liquid nitrogen tank in a cryo storage box.

**THAWING PROTOCOL FOR FROZEN CELLS**

1. Prewarm medium and place the appropriate amount into a tissue culture flask.

2. Remove a vial of cells from liquid nitrogen storage and place them immediately into the 37°C water bath.

3. As soon as any crystals have melted, remove the vial, spray it with 70% ethanol and wipe it down to disinfect, and transfer it into the tissue culture hood.

4. Quickly pipette cells into the warm medium and place into the incubator.

**SUGGESTED READING**

Freshney, R.I. *Culture of Animal Cells: A Manual of Basic Technique*, Edn. 5. WileyLiss, New York, 2005.—This textbook is highly recommended to anyone doing tissue culture.

---

sequences conferring immortality have been introduced by viral infection or fusion with other immortalized cells (the latter are called *hybridomas*). When seeded into a culture flask, the cells will divide until overcrowding leads to nutrient depletion and *senescence*. Before this happens, the cells should be passaged. There are two ways in which cell lines can grow: in suspension or adherent. Suspension cells are usually hematopoietic in origin, such as lymphoma or leukemia cells. Adherent cells will die unless provided with an appropriate substrate for attachment.

Suspension cells are passaged simply by transferring some of the culture into a new flask with fresh medium. Adherent cells may be passaged by dissolving the adhesions with the enzyme *trypsin*, then resuspending the cells and subculturing into a new flask (see **Practical Tips 3.5**). The degree to which adherent cells fill the available space is called their degree of *confluency*; they should be passaged at ~80% confluency to avoid die-off that occurs with crowding (some cell types will never recover from being too crowded). While passaging, some of the cells may be used to seed different types of substrates for experiments. In this way, the cells can be kept going for many months; a fresh culture should be started after a certain number of passages that varies for cell type, usually around 15–20 (although some cell types, such as HEK 293, are quite stable and routinely kept for

**Table 3.1**

Some Commonly Used Cell Lines with Their Name, Organism, and Tissue of Origin, and Useful Phenotypic Features

| Cell Line | Organism | Tissue | Features |
|---|---|---|---|
| HEK 293 | Human | Embryonic kidney | Voltage-gated potassium channels<br>Easy to transfect |
| 293-T | Human | Embryonic kidney | 293 with the addition of the SV 40 T antigen, making them easier to transfect and able to produce very large amounts of transcript; tend to form aggregates (syncytia) |
| CHO | Chinese hamster | Ovary | Typical epithelial cells; easy to transfect |
| COS-7 | African green monkey | Kidney fibroblast | Have the SV 40 T antigen |
| HeLa | Human | Cervical cancer | Human epithelial cells; must be handled with BSL-2 precautions as they contain papovavirus |
| KB | Human | Cervical cancer | HeLa contaminant with higher levels of folate receptor; can be used for folate uptake studies; BSL-2 |
| Jurkat | Human | T-cell leukemia | Suspension cells; express CD3 antigen |
| NIH-3T3 | Mouse | Embryonic fibroblast | Easy to transfect; large, flat cells; excellent for imaging |
| PC12 | Rat | Adrenal gland | Produce catecholamines; express nerve growth factor (NGF) receptors; can be differentiated using NGF |
| Vero | African green monkey | Kidney epithelium | Used for propagating and studying many types of viruses, including measles, polio, rubella, influenza |
| SK-N-SH | Human | Neuroblastoma | Express dopamine β-hydroxylase |
| SNU-5 | Human | Gastric cancer | Express muscarinic acetylcholine receptor, vasoactive intestinal peptide (VIP) receptor |

50 or more passages). It is routine to count the cells with a *hemocytometer* after resuspending them, so that a good estimate of the number of cells seeded into the new culture can be obtained. This way, the cells' growth rate is known, and the ratio of cells to fresh medium can be controlled to time when the cells will be ready for an experiment.

Several hundred cell lines are available commercially, along with media and instructions for growth and passaging. The species and tissue of origin determine their usefulness for a particular application. Many cell lines are useful only because they are easy to transfect. Others possess pathways, receptors, or enzymes from their tissue of origin. Some cell lines are undifferentiated for growth and passaging, but may be made to differentiate in culture into neurons or other specialized cell types. A list of some commonly used cell lines and their features is given in Table 3.1.

## Primary cultures

A primary culture is a culture of cells taken directly from an organism. These are usually animals, commonly rats and mice, but may also come from human tissue (e.g., surgical specimens delivered directly to the laboratory). In either case, the institution's ethics review board must approve any protocols for use of vertebrate animals or human tissue. Except for blood cells, tissues to be made into monolayer cultures must be *dissociated*: that is, the tissue must be disaggregated mechanically and/or enzymatically to allow the cells to form a two-dimensional layer. An enzyme commonly used for this purpose is papain, isolated from papaya, which digests extracellular matrix proteins without damaging the cell bodies. The cells are then plated on an appropriate substrate, which may be more complex than a standard "tissue culture treated" culture dish. Culture media also must be specially prepared according to cell type.

Plated cells may be used immediately or after some degree of maturation. Cells that normally divide in the body will undergo a certain number of generations in culture, but their lifetime will be limited unless they are immortalized. Detailed procedures for immortalization are outside the scope of this chapter, but several companies offer assistance or kits for immortalization of extracted cells. Cells that are *terminally differentiated* and that do not divide in the body, such as neurons and macrophages, do not divide in culture. Thus, the density of the cells in the dish needs to be carefully controlled for each given experiment, and the cultures protected carefully from contamination, dehydration, or loss of nutrients. When well-tended, cultured neurons can survive for months.

In the case of neurons, dissociation causes a loss of axons and dendrites, and so any studies investigating synaptic connections must wait until the processes re-form and the neurons form synapses with their neighbors. This can take from 1 to 2 weeks (**Figure 3.6**). Primary neurons are usually taken from embryonic rats and mice, or from pups immediately after birth, as more mature cells usually do not survive dissociation and plating.

Primary cultures are used when no cell line exists with the genotypic or phenotypic features needed for the experiment. For example, there are no immortalized cell lines that fire sodium/potassium action potentials (although some show calcium spikes). The growth of mature axons and dendrites and the formation of synapses also require a mature neuronal phenotype. If these properties are needed for the experiment, a primary neuronal culture should be used.

Entire textbooks have been devoted to the extraction and cultivation of primary cells. Some cell types, such as macrophages and adrenal chromaffin cells, are relatively easy to extract and grow. Others, such as dopaminergic neurons (which can only be identified in postnatal animals), are very difficult to identify, isolate,

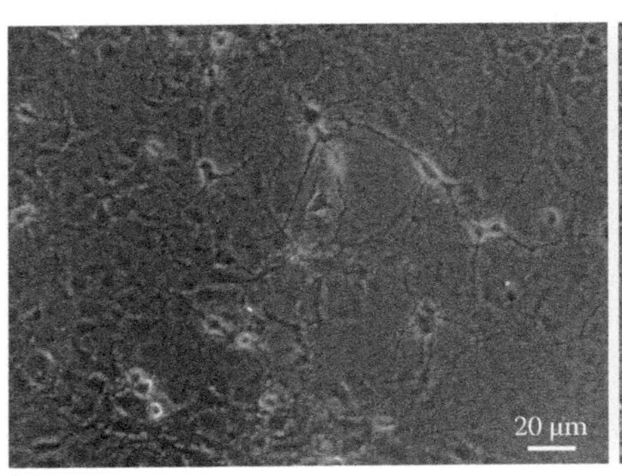

(a)

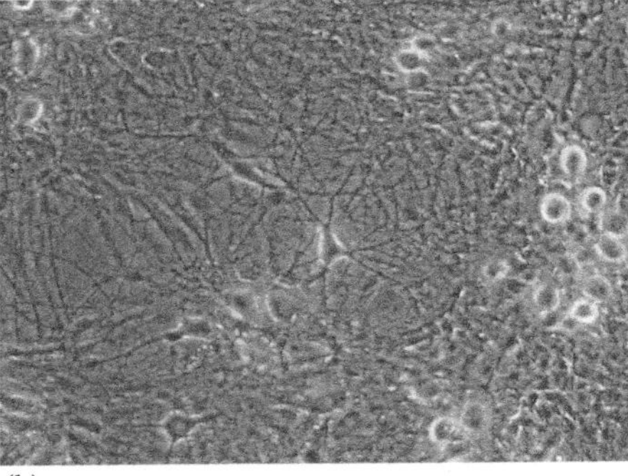
(b)

**Figure 3.6 Development of hippocampal neurons in culture.** (a) One week after plating. (b) Four weeks after plating. Note massive development of processes.

**Figure 3.7 Differentiation of PC-12 cells.** The pictures show the progress of differentiation of the treated cells over time: (a) 0 h, (b) 24 h, (c) 48 h, and (d) 96 h after the initiation of treatment with 100 ng/mL of NGF. Scale bar = 25 μm. (From Vancha, A.R. et al., *BMC Biotechnology* 4, 23, 2004. Used by permission of author and under the terms of the BMC Biotechnology Open Access License.)

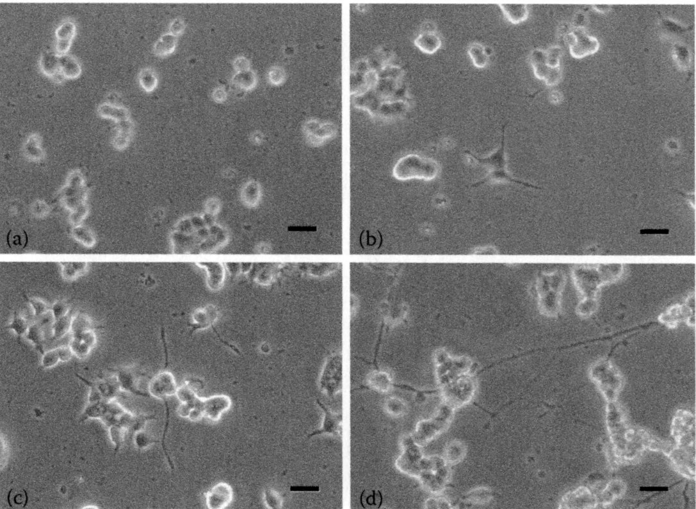

and culture. For any experiment involving primary cultures of anything other than hematopoietic cells or chromaffin cells, we recommend consulting the most recent literature and obtaining expert advice before proceeding.

There are two ways to avoid the need to make primary cultures from animals. The first is to purchase dissociated primary cells from a biological supplier; several companies offer different types of cells from rat, mouse, and human, including bone marrow, blood, neurons, and *glial cells* (the cells that support neurons, particularly *astrocytes*). Many of these cells grow very well; however, difficulties may be encountered with very fragile cell types such as neurons. The commercial cells are costly, and a good deal of waste can occur while determining optimal culture parameters.

The second option is to force a cell line to *differentiate*. A large body of literature exists on this subject, and new cell lines are continually emerging that are designed to be easily differentiated into mature cell types, particularly for human cells where primary material is in short supply. One of the most common examples of this process uses the pheochromocytoma cell line, PC-12. When nerve growth factor (NGF) is added to PC-12 cells every 2 days for a week, the cells cease to divide and extend neurites (**Figure 3.7**). They also become electrically excitable and respond to acetylcholine, thus taking on the properties of cholinergic neurons. The primary disadvantage of this approach is that the cells do not all differentiate identically, so a range of phenotypes is seen that might confuse experimental results. The process of differentiation is also time-consuming, is costly, and can be tricky. These factors need to be weighed against the availability of primary tissue; if the latter is readily available, direct culture of dissociated cells may be the easiest approach.

Cultured cells grow in a one-dimensional monolayer. While this is sufficient for many experiments, in some cases, cell–cell connections in 3D or tissue architecture are needed, such as explants or slices (see **Advanced Topic 3.1**).

Another approach to obtaining desired cell types is to use stem cells (**Advanced Topic 3.2**).

## ADVANCED TOPIC 3.1:    EXPLANTS AND SLICES

An explant is a piece of tissue placed into a culture dish and allowed to grow as is, retaining its native structure, development, and biochemistry. Usually a dissected block of tissue is placed on a porous substrate with an appropriate culture medium underneath. The tissue is then at the air/water interface without being immersed; in this way, it may survive for 2–3 weeks. Blocks of tissue to be cultured may be dissected by hand, being careful not to crush the tissue, and usually cannot be larger than 1–2 mm in any dimension without becoming necrotic in the center.

Explant culture is commonly used for cells that are killed by enzyme digestion, such as smooth muscle cells. It is also used in situations where the tissue architecture is important to the process being studied—for example, in studies of hormone responses in pancreas, placenta, or uterus; for investigations of mechanical stresses or implant integration on cartilage and/or bone; and in some models of cancer growth and invasion. Another use of explants is to elucidate the role of different cell types in a complex tissue, as specific types may be added or subtracted.

Such architecture is becoming increasingly appreciated in other systems as well, where its influence is less obvious, such as in models of infectious disease. For example, cultured human macrophages can be infected with HIV, they must be activated to support viral replication, and their expression of cell-surface molecules is very different from what is seen *in vivo*. Explants of human lymphoid tissue can be used as a model of infection without exogenous activation and with preservation of cell-surface receptors. Such tissue can be obtained from postmortem specimens or from surgical specimens (e.g., tonsillectomy) with the appropriate ethics protocols (Figure A3.1.1).

For neural tissue, slices are usually used in the place of explants. A "slice" is precisely what it sounds like—a section of tissue (brain or spinal cord) several hundreds of microns thick that retains all of the supporting structure and many thousands to millions of synaptic connections. These are made by dissecting out the area of interest, often the hippocampus, and using a vibrating tissue slicer (vibratome) to produce sections of the desired thickness. Brain slices can be used to study complex functions such as networking, learning,

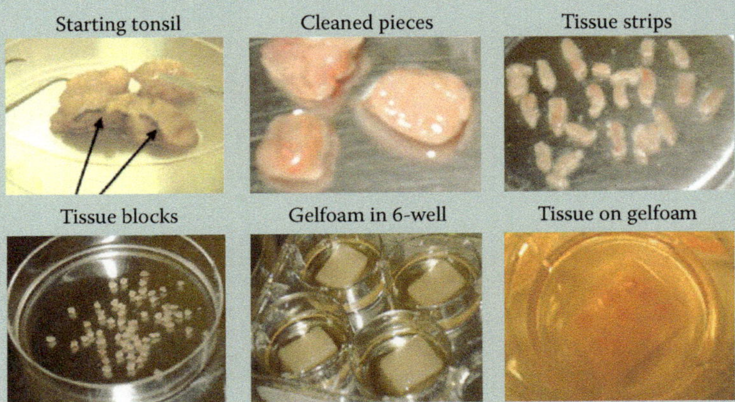


**Figure A3.1.1 Explant.** Example of preparation steps for explants of tonsillar tissue. Preparation begins by removing areas of necrosis or cauterization (arrows). The large cleaned pieces are cut into strips and then appropriately sized blocks for tissue culture, and placed onto supporting media in a 6-well culture plate. (Reprinted by permission from Macmillan Publishers Ltd: *Nature Protocols*, Grivel, J. C., and L. Margolis, Use of human tissue explants to study human infectious agents, 4(2): 256–269, copyright 2009.)

*(Continued)*

## ADVANCED TOPIC 3.1 (CONTINUED):    EXPLANTS AND SLICES

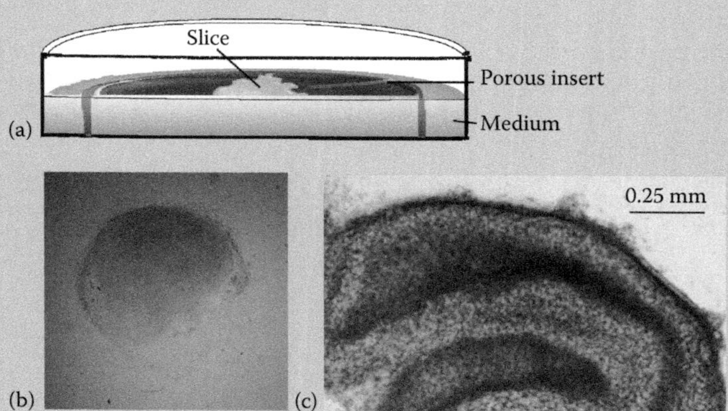


**Figure A3.1.2  Organotypic brain slice.** (a) Schematic of usual method of culturing. An insert bearing a porous membrane is placed into a culture dish above the culture medium. The slice is placed on the insert and thins over time. (b) 2× micrograph of organotypic slice after several weeks in culture; it can be stained (c) to show its architecture. (Courtesy of D. Juncker, McGill.)

and memory. An *acute slice* is a slice used immediately after preparation, bathed in a solution of artificial cerebrospinal fluid (ACSF) for up to several hours. An *organotypic slice* is a slice cultured for weeks to months on a porous substrate, in much the same way as other tissue explants (Figure A3.1.2). Organotypic slices thin over time and grow more complex connections. Although a slice obviously cannot have the same experiences as a mouse or rat, the development of neurons inside an organotypic slice is believed to be similar to that of the same neurons inside an animal at comparable stages. While mouse and rat are the most common animals used, usually in adolescent stages, it is also possible to grow human postmortem tissue for up to several weeks. Organotypic slices are used in studies of mechanisms and treatment strategies for neurotoxin exposure, ischemic stroke (by depriving the slice of oxygen and glucose), epilepsy, Alzheimer's disease, Parkinson's disease, HIV dementia, and more.

## SUGGESTED READING

Grivel, J.C., and Margolis, L. (2009). Use of human tissue explants to study human infectious agents. *Nature Protocols* 4, 256–269.

Humpel, C. (2015). Organotypic brain slice cultures: A review. *Neuroscience* 305, 86–98.

Lein, P.J., Barnhart, C.D., and Pessah, I.N. (2011). Acute hippocampal slice preparation and hippocampal slice cultures. *Methods in Molecular Biology* 758, 115–134.

Marciniak, A., Selck, C., Friedrich, B., and Speier, S. (2013). Mouse pancreas tissue slice culture facilitates long-term studies of exocrine and endocrine cell physiology in situ. *PLoS One* 8, e78706.

Pampaloni, F., Reynaud, E.G., and Stelzer, E.H. (2007). The third dimension bridges the gap between cell culture and live tissue. *Nature Reviews. Molecular Cell Biology* 8, 839–845.

Randall, K.J., Turton, J., and Foster, J.R. (2011). Explant culture of gastrointestinal tissue: a review of methods and applications. *Cell Biology and Toxicology* 27, 267–284.

## ADVANCED TOPIC 3.2:    STEM CELLS

*Stem cells* get their name because they are the cells from which all others "stem." They have two particular properties: they can divide many times without differentiation, but when triggered, they can differentiate into more than one mature cell type. The class of stem cells that can become any cell type except an embryo are called *pluripotent* stem cells. Pluripotent stem cells can be collected most readily by harvesting from early embryos before implantation (*biastocysts*), where they form the embryo's *inner cell mass* (Figure A3.2.1). These embryonic stem cells (ES cells) then must generally be grown on a *feeder layer* of mouse fibroblasts and in medium containing serum; both of these ingredients are necessary to keep them from differentiating. However, these practices were shown to contaminate human-derived cells with a nonhuman cell-surface protein that could cause severe allergic reactions if the cells were used for therapy. Serum-free, feeder-free methods of ES cell cultivation are therefore an important area of research.

Most ES cell work is done in humans and model organisms (mice). Mouse ES cells are available commercially from different strains of mice or may be isolated from blastocysts; a reference to a straightforward protocol is given in the Suggested Readings. The feeder layers, specialized media for growth and differentiation, and other reagents must also all be purchased specifically for ES cell work. Mouse ES cells are used for studies of growth and differentiation as well as for generation of transgenic mice. Because culture conditions are tricky, most universities have core facilities to assist with ES cell work. However, major refinements to techniques in the mid-2010s have made ES cell culture easier and possible for all mouse strains, not only the so-called permissive strains.

Human ES cells come from embryos generated for *in vitro* fertilization and then donated to research. This has generated significant controversy, and specific ethical guidelines are in place for research involving these cells. Between 2001 and 2009, the United States prohibited the generation of new ES cell lines with federal funding. However, new cell lines were produced in other countries, such as Israel and Britain, leading to some improved clones. Various human ES cell–derived products are available commercially, including neural precursors, neural cells (farther along in development than precursors), and *mesenchymal* cells (able to differentiate into osteoblasts, chondrocytes, and adipocytes) (Figure A3.2.2). Very recently,

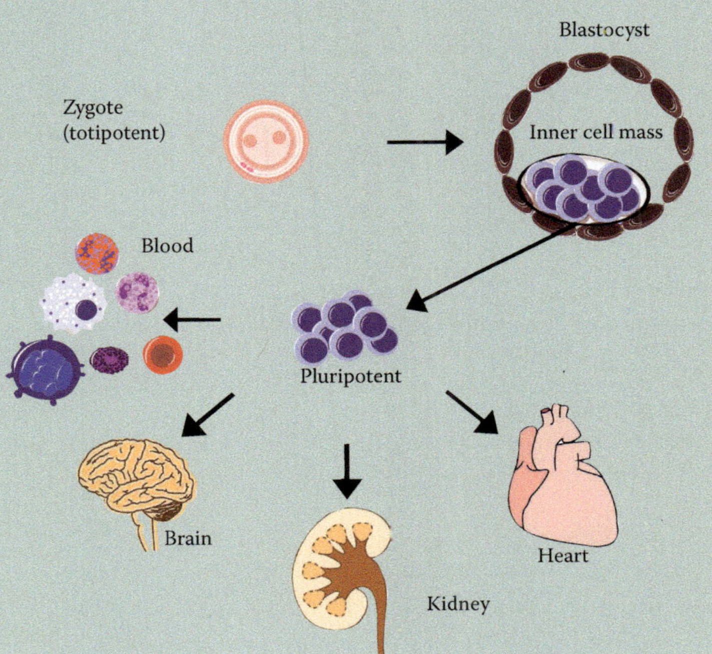


**Figure A3.2.1 Stem cells.** A fertilized egg, or zygote, is totipotent. After passing through several cell divisions (but before implantation), it becomes a blastocyst. The inner cell mass of the blastocyst contains pluripotent stem cells that can be maintained undifferentiated in culture, or made to differentiate into any cell type except an embryo.

*(Continued)*

## ADVANCED TOPIC 3.2 (CONTINUED):  STEM CELLS

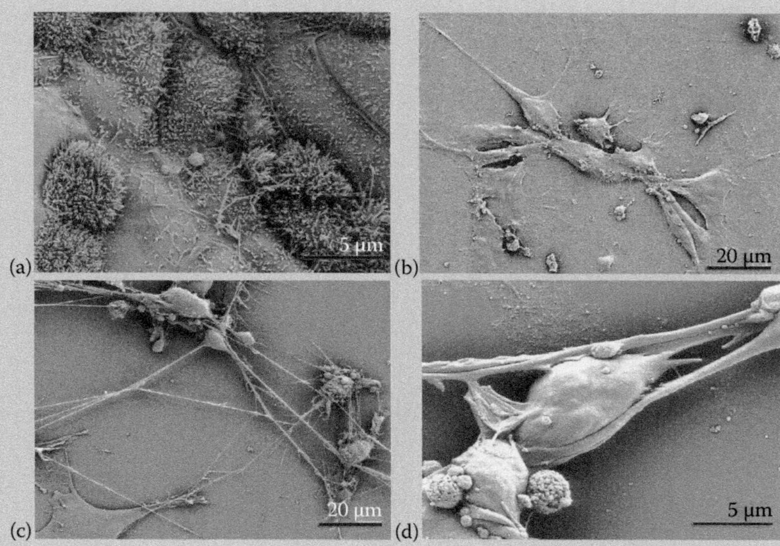

**Figure A3.2.2 Scanning electron micrographs of hESC before, during, and after differentiation into neurons.** (a) hESC showing microvilli. (b) Neural precursor cells differentiated from hESC. (c) Precursors may be further differentiated with complex medium and BDNF into cells with characteristics of the neuronal phenotype, such as complex organization and ion channels. (d) Higher-resolution image of the neuronal cells, showing processes. (Images courtesy of Jennifer Mumaw and Steve L. Stice, University of Georgia.)

protocols have been developed for inducing pluripotency in differentiated cells by genetic manipulation. These cells are not identical to ES cells and are a significant area of study.

*Adult stem cells* are small reserves of cells found in differentiated tissues or organs that are able to differentiate in only a few select ways (they are *multipotent*). The first reserve of stem cells was found in bone marrow: hematopoietic stem cells (HSCs), which give rise to all blood cell types. Mesenchymal stem cells, found also in bone marrow, can form bone, cartilage, fat, and connective tissue. The existence of neural stem cells was only confirmed in the 1990s; these can form neurons, oligodendrocytes, and astrocytes. Other types of adult stem cells include mammary, olfactory, neural crest, and germline.

A significant area of research is in identification and isolation of these cells, which by their nature are small and inconspicuous. They are also difficult to culture. Many labs specialize in the identification and culture of adult stem cells from mice or humans. Procedures for isolating different types of stem cells are under rapid development; references are given at the end of the chapter to illustrate examples of isolation protocols for mesenchymal and neural stem cells from mice.

Adult stem cells have generated a great deal of interest for their potential in regenerative therapies, especially for injuries or diseases of the central nervous system where regeneration is limited: Alzheimer's disease, Parkinson's disease, and spinal cord injury are a few examples. While these therapies have yet to be realized, some clinical applications of adult stem cells are already highly successful. Skin stem cells are used to grow skin for grafting of serious burns, a great improvement over the previous technique of harvesting skin from the patient's nonburned areas. Even more well-known is *bone marrow transplantation*, more accurately referred to as hematopoietic stem cell transplantation. HSCs can be isolated from a donor's bone marrow or, more recently, peripheral blood and injected into a host to treat conditions such as leukemia and aplastic anemia.

### SUGGESTED READING

Bibel, M., Richter, J., Lacroix, E., and Barde, Y.A. (2007). Generation of a defined and uniform population of CNS progenitors and neurons from mouse embryonic stem cells. *Nature Protocols* 2, 1034–1043.

(Continued)

**ADVANCED TOPIC 3.2 (CONTINUED):    STEM CELLS**

Bond, A.M., Ming, G.L., and Song, H. (2015). Adult mammalian neural stem cells and neurogenesis: Five decades later. *Cell Stem Cell* 17, 385–395.

Corti, S., Faravelli, I., Cardano, M., and Conti, L. (2015). Human pluripotent stem cells as tools for neurodegenerative and neurodevelopmental disease modeling and drug discovery. *Expert Opinion on Drug Discovery* 10, 615–629.

Czechanski, A., Byers, C., Greenstein, I., Schrode, N., Donahue, L.R., Hadjantonakis, A.K., and Reinholdt, L.G. (2014). Derivation and characterization of mouse embryonic stem cells from permissive and nonpermissive strains. *Nature Protocols* 9, 559–574.

Ema, H., Morita, Y., Yamazaki, S., Matsubara, A., Seita, J., Tadokoro, Y., Kondo, H., Takano, H., and Nakauchi, H. (2006). Adult mouse hematopoietic stem cells: Purification and single-cell assays. *Nature Protocols* 1, 2979–2987.

Futami, I., Ishijima, M., Kaneko, H., Tsuji, K., Ichikawa-Tomikawa, N., Sadatsuki, R., Muneta, T., Arikawa-Hirasawa, E., Sekiya, I., and Kaneko, K. (2012). Isolation and characterization of multipotential mesenchymal cells from the mouse synovium. *PLoS One* 7, e45517.

Lewandowski, J., and Kurpisz, M. (2016). Techniques of human embryonic stem cell and induced pluripotent stem cell derivation. *Archivum Immunologiae Et Therapiae Experimentalis (Warszawa)* 64, 349. doi:10.1007/s00005-016-0385-y.

Neofytou, E., O'brien, C.G., Couture, L.A., and Wu, J.C. (2015). Hurdles to clinical translation of human induced pluripotent stem cells. *The Journal of Clinical Investigation* 125, 2551–2557.

Pieters, T., Haenebalcke, L., Hochepied, T., D'hont, J., Haigh, J.J., Van Roy, F., and Van Hengel, J. (2012). Efficient and user-friendly pluripotin-based derivation of mouse embryonic stem cells. *Stem Cell Reviews* 8, 768–778.

Rebuzzini, P., Zuccotti, M., Redi, C.A., and Garagna, S. (2016). Achilles' heel of pluripotent stem cells: Genetic, genomic and epigenetic variations during prolonged culture. *Cellular and Molecular Life Sciences* 73(13), 2453–2466. doi: 10.1007/s00018-016-2171-8.

Sreejit, P., Dilip, K.B., and Verma, R.S. (2012). Generation of mesenchymal stem cell lines from murine bone marrow. *Cell and Tissue Research* 350, 55–68.

Stock, P., Bruckner, S., Ebensing, S., Hempel, M., Dollinger, M.M., and Christ, B. (2010). The generation of hepatocytes from mesenchymal stem cells and engraftment into murine liver. *Nature Protocols* 5, 617–627.

Xie, N., and Tang, B. (2016). The application of human iPSCs in neurological diseases: From bench to bedside. *Stem Cells International* 2016, 6484713.

# 3.4 TRANSFECTION OF MAMMALIAN CELLS I: STANDARD TECHNIQUES

## Introduction

Many of the cell lines in Table 3.1 are used simply as transfection hosts. That means that they are not used for any particular receptor or hormone that they express, but simply are able to take up foreign DNA readily, express it well, and possess useful properties for the downstream application (e.g., for fluorescence imaging, they should be large and not exhibit excessive *autofluorescence* [see Chapter 7]). Transfecting these types of cells with moderate-sized plasmids (<15–20 kb) is generally a very simple task, performed using commercial liposomes (described in further detail below).

However, there are occasions in which routine techniques are not sufficient. Many methods of DNA delivery have been developed, including physical means, viral means, and others, each with its own set of advantages and disadvantages. The following questions should be asked to aid method selection:

- What number or fraction of cells need to be transfected? For example, is 1% transfection of $10^6$ cells in a dish sufficient, as it usually is for imaging experiments? Or do nearly 100% of the cells need to express the gene, as is desirable for microarray analysis or for effects on networks? Is it necessary to be able to control the ratio of transfected to untransfected cells?

- How long does the transfection need to last? Will the cells be used for a few hours and then discarded, such as for an imaging or electrophysiology experiment, or do cells need to be kept for days, weeks, or longer?

- What cell type or types need to be used? Will nearly any cell type do, or is the presence of a specific receptor or ion channel or other phenotype required? Does the experiment need to be performed with nondividing primary cells such as neurons or macrophages?

- Does gene expression need to be temporally controlled, reversible, or inducible? If so, how tight must the control be for success of the experiment?

## Cationic liposomes: Easy, transient expression in 40–90% of dividing cells

A very large number of gene-expression experiments are done using *transient transfection* techniques, in which a foreign gene is expressed at high levels from multiple copies of an introduced plasmid. The plasmid does not integrate into the host genome, so it is eventually degraded or diluted through cell division; the gene can be expected to express at high levels for only a few days.

Introducing plasmid DNA into the cytoplasm of cells may be accomplished very efficiently by mixing the DNA with cationic liposomes, which are micron-sized spheres of lipids bearing positively charged head groups (**Figure 3.8a**). They complex spontaneously with the anionic DNA, and the complex is able to cross cell membranes by mechanisms that are not entirely understood, but which involve the endocytic machinery (**Figure 3.8b,c**). The plasmids then must enter the nucleus in order to be expressed, as the necessary enzymes for gene transcription are all nuclear. This can occur via cell machinery or when the nuclear envelope breaks down before mitosis; the latter is one reason why liposome transfection methods work much more effectively with dividing cells (**Figure 3.8d**).

**Figure 3.8 Liposomal transfection.** (a) Negatively charged plasmid DNA complexes with positively charged liposomes. The resulting complex is sometimes called a *lipoplex*. (b) Complex is taken into cells by endocytosis. (c) Entire lipoplex is inside an endosome. It may be delivered to lysosomes and degraded, or instead the lipids can mix (d) and release the DNA, which must travel to the nucleus before it can be expressed.

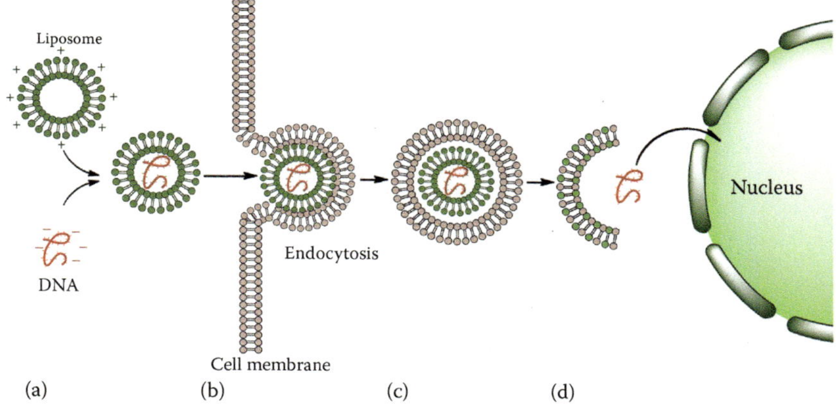

Liposomes may be prepared in the laboratory, but this can be tricky and difficult to keep sterile. Most researchers purchase transfection reagents from one or more commercial suppliers. Different reagents have different levels of toxicity and transfection efficiency in different cell lines, so a reagent should be chosen based upon the target cell type.

A day or two before transfection, the cells should be plated on the appropriate substrate, which may need to be glass for high-resolution imaging, or on specialized coverslips for electrophysiology. They must not be too confluent, or their division will be inhibited. They should be transfected during their exponential growth phase, at roughly 40–80% confluency (use the higher values if the transfection reagent is toxic, because significant cell death will occur). It is important to use high-quality plasmid DNA for transfection, not miniprep DNA. Follow the manufacturer's instructions, which may include some steps in medium without supplemented serum ("serum free"). It is very important to make sure that serum-free medium is indeed used here, as the serum contains proteins that can complex with the liposomes and impede transfection.

For a first round of experiments with an unknown reagent and/or cell line, transfection conditions should be optimized for cell density and amount of DNA and liposomes (see **Practical Tips 3.6** for a sample optimization experiment).

---

**PRACTICAL TIPS 3.6:    OPTIMIZING TRANSFECTION**

*General notes.* All transfection methods lead to some degree of cell death. The degree of cell confluency should be selected to account for this. General health factors will also affect transfection efficiency, and many times transfection fails due to poor cell condition caused by contamination, media lacking supplements or expired media, incubator problems, senescent cells, or other such factors. Transfection efficiencies vary greatly among cell types, so each cell line should be optimized separately. The key variables to adjust are (1) DNA concentration, (2) DNA/liposome ratio, (3) amount of time the cells are exposed to the complexes before washing, and (4) cell density. The following protocol examines all but item 3 in 96-well plates to minimize use of reagents. Item 3 may be examined in separate plates, as desired.

**PROCEDURE FOR OPTIMIZATION**

1. *The day before*: plate cells in a 96-well plate. Use two different dilutions, one to fill the left half of the plate (six columns of eight wells each) and the second to fill the right half (six columns).

2. The following day, prepare the complexes in eight different tubes. Tube 1 should contain plasmid DNA only in the appropriate concentration in a total volume of 150 μL. Tubes 2–7 contain the same amount of DNA and increasing ratios of liposomes (starting ratios are determined based upon the product; the manufacturer will give a suggested range). Tube 8 contains liposomes only, no DNA. All should be in total volumes of 150 μL.

3. Allow the complexes to form. Meanwhile, remove the cells from the incubator and wash/replace medium as necessary.

4. Add nothing to rows 1 and 7. Add 10 μL from each of tubes 1–8 to rows 2 and 8. Add 20 μL to rows 3 and 9, 30 μL to rows 4 and 10, 40 μL to rows 5 and 11, and 50 μL to rows 6 and 12.

5. Incubate for a chosen amount of time before washing and replacing the medium with serum-containing medium. Some reagents do not require this step.

To effectively measure gene expression, a plasmid encoding a *reporter gene* that may be easily quantified should be used. Fluorescent proteins make excellent reporters, as the number of successfully transfected cells can be measured using fluorescence microscopy, and the total amount of protein produced can be quantified using a fluorescence plate reader or spectrometer. However, fluorescent proteins have a low dynamic range; it is hard to tell whether a cell contains 100 or 1000 GFP molecules. A very common reporter gene with a high dynamic range is β-galactosidase (β-gal, or the LacZ gene), which develops a blue color when exposed to the substrate X-gal. Other substrates with luminescent or fluorescent readout are also available (see **Practical Tips 3.7**). These detection methods must be done on fixed and permeabilized cells or cell lysates, which is a disadvantage compared with fluorescent proteins.

*Cotransfection* is the practice of mixing two plasmids together during the same experiment. One can encode a reporter gene, and the other another protein, such as a hormone or ion channel. For reasons not fully understood, essentially all successfully transfected cells will encode both genes, although not necessarily at the same ratio. This is extremely useful for saving cloning steps and minimizing plasmid size, as not everything has to be on the same plasmid.

Commercial suppliers usually provide data on the efficiency of their reagents in different cell types. Specialized reagents are available for difficult-to-transfect cells, including primary and nondividing cells. However, the success rate for liposomal transfection of primary hippocampal neurons is no better than about 1%. For efficient transfection of these cells, a method of delivering DNA directly into the nucleus should be sought (see sections on physical methods and viral vectors).

## Stable transfection: For long-term and/or inducible expression of entire cultures of dividing cells

Stable transfection is the insertion and maintenance of the plasmid sequence into the host genome. It is useful for when all cells in the dish need to express a gene of interest, or when the gene must express for long periods. It is also needed for the use of inducible promoters, such as the tetO promoter that turns off in the presence of tetracycline. In ordinary transient transfections, inducible promoters show very high backgrounds.

A stable transfection begins with ordinary transient transfection, followed by a selection protocol that allows only those few cells that have integrated the plasmid DNA into their genome to survive. Most mammalian expression vectors contain a selectable marker, almost always *neo*, or neomycin resistance (neomycin is an *aminoglycoside* antibiotic found in topical preparations, e.g., Neosporin). Neo can neutralize the toxin geneticin, also called G418. Other common markers are puromycin (*puro*) and hygromycin (*hyg*).

The principle of selecting resistant colonies of mammalian cells is the same as with bacteria, but there are a few complications that make stable transfection somewhat of a challenge. First, each cell line responds differently to the toxin used for selection. A "kill curve" must thus be determined to obtain the optimal concentration for selection; this optimal concentration should be the lowest concentration of drug that begins to give visible cell death in 3 days and kills all the cells in 2 weeks. Second, mammalian cells do not grow in genetically identical isolated colonies the way bacterial cells do. In order to isolate a specific clone,

### PRACTICAL TIPS 3.7:    β-GALACTOSIDASE (LacZ) AS A REPORTER GENE

*General information.* The enzyme β-*galactosidase* is encoded by the bacterial gene *lacZ*. Its physiological role in bacteria is to cleave lactose (milk sugar) into glucose and galactose. It can be used as a reporter gene by expressing it in cells where it is not usually found, and then using its enzymatic activity to indicate that the cells are successfully expressing the delivered gene. Because enzymatic activity can be amplified, this is potentially a much more sensitive reporter than genes such as GFP that must be observed directly and whose signal is thus simply proportional to the number of copies expressed. Because of this, *LacZ* remains a favorite reporter gene for bacteria, yeast, and mammalian cells.

There are a number of ways to indicate β-galactosidase activity in cells. The most commonly used is to provide the enzyme with the substrate *Xgal* (5-bromo-4-chloro-3-indolyl-β-galactopyranoside), which is cleaved to yield galactose and an indoxyl derivative that can dimerize to yield an insoluble blue product (**Figure P3.7.1**).

The procedure for mammalian cells involves fixation, as follows.

### PROCEDURE FOR X-GAL STAINING IN MAMMALIAN CELLS

1. X-gal is kept as a 25 mg/mL stock solution in dimethylformamide (DMF).

2. Protect from light. Dilute just before use.

3. Aspirate off culture media and wash cells with cold PBS.

4. Fix the cells on ice with 3% glutaraldehyde in PBS for 5 min.

5. Rinse the cells three times with PBS, on a rocker for 5 min per wash.

6. Dilute X-Gal stock solution into stain solution to a total concentration of 1 mg/mL X-gal. Typical stain solution: PBS containing 2 µg/mL $MgCl_2$, 1.64 mg/mL potassium ferricyanide, 2.12 mg/mL potassium ferrocyanide.

7. Cover the cells with the X-gal solution and incubate at 37°C for at least 1 h or until they turn blue. The color may continue to develop for up to 20–24 h. Note: this should *not* be done in a tissue culture incubator, as the $CO_2$ will alter the solution pH.

**Figure P3.7.1  (a) Structure of X-gal. The galactose (left) is cleaved at the site of the arrow; the part of the molecule on the right then dimerizes and oxidizes to produce an insoluble blue product (b).** The reaction can be favored by iron ions which are often supplied in LacZ developer kits.

(*Continued*)

---

### PRACTICAL TIPS 3.7 (CONTINUED):  β-GALACTOSIDASE (LacZ) AS A REPORTER GENE

Other substrates for β-galactosidase are available, including fluorescent substrates, chemiluminescent substrates, substrates with increased sensitivity, and those designed for use with live cells. If fixation is not desired or if a more quantitative assay than the blue colorimetric detection is desired, a search for these substrates will provide a list of dozens of choices.

There are several other enzymatic reporter genes that work in a similar fashion. Two of the most popular are *luciferase* (which produces readily quantifiable luminescence) and *chloramphenicol acetyltransferase* (CAT). These reporter genes should be considered when fluorescent proteins are not sensitive enough or quantitative enough for detection, or when fluorescent detection is not possible or not desired. Reporter genes also exist for generating contrast for other types of imaging, such as MRI and PET (see references).

#### SUGGESTED READING

Acton, P.D., and Zhou, R. (2005). Imaging reporter genes for cell tracking with PET and SPECT. *Quarterly Journal of Nuclear Medicine and Molecular Imaging* 49, 349–360.

Alam, J., and Cook, J.L. (1990). Reporter genes: Application to the study of mammalian gene transcription. *Analytical Biochemistry* 188, 245–254.

Gilad, A.A., Winnard, Jr. P.T., van Zijl, P.C., and Bulte, J.W. (2007). Developing MR reporter genes: Promises and pitfalls. *NMR in Biomedicine* 20, 275–290.

Gilad, A.A., Ziv, K., McMahon, M.T., van Zijl, P.C., Neeman, M., and Bulte, J.W. (2008). MRI reporter genes. *Journal of Nuclear Medicine* 49, 1905–1908.

Zhang, Y.Z., Naleway, J.J., Larison, K.D., Huang, Z.J., and Haugland, R.P. (1991). Detecting lacZ gene expression in living cells with new lipophilic, fluorogenic beta-galactosidase substrates. *The FASEB Journal* 5, 3108–3113.

#### WEBSITE

http://openwetware.org/wiki/LacZ_staining_of_cells

---

you have to ensure that you separate the cells enough to ensure a monoclonal population. There are several ways to do this:

- Dilute the cells to a density of 1 cell/200 mL, then plate 100 mL in each well of a 96-well plate. Many wells will contain only one cell and thus be monoclonal (if your genetic marker is visible, e.g., GFP, you will be able to tell this).

- Plate the cells very sparsely on a large plate (10–15 cm). After 2–3 weeks of growth under selective conditions, the cells will grow in "islands" about 1 mm across. If the cells are not strongly adherent, you can detach a few from a colony by pipetting and transfer them to a new dish (with selection, of course). If the cells are more strongly adherent, a cloning cylinder is needed (**Figure 3.9**). This is a ring placed over the colony and held to the dish with sterile grease (silicone high-vacuum grease may be sterilized by autoclaving). The specific colony can then be trypsinized and resuspended. Cloning rings come off easily, so care is needed. An alternative is to apply a trypsin-soaked ring of filter paper to one of the colonies until it detaches.

Because of the difficulties of selection and the slow growth of the diluted cells, establishing a stably transfected cell line takes at least several weeks and is often not successful on the first round. At least 6–10 colonies should be picked and

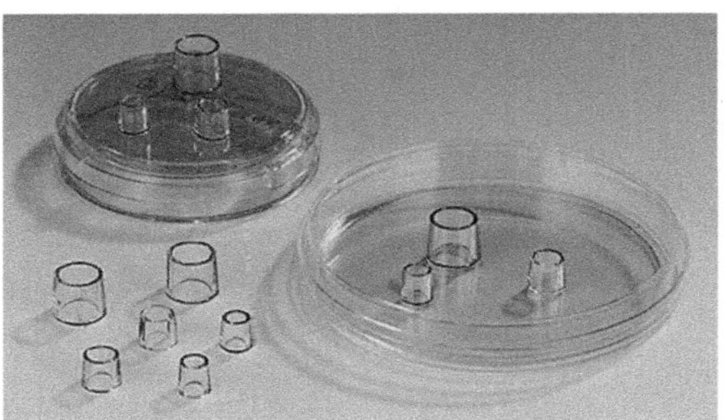

**Figure 3.9 Cloning cylinders.** Different sizes are available for different sized colonies.

amplified, then tested for levels of gene expression. If the gene is not visible, Western blotting should be performed to test expression levels.

As soon as clones with desirable expression levels have been isolated and amplified, stocks should be frozen in liquid nitrogen to ensure a future supply of stably transfected cells. The cells must always be grown in the presence of the selection drug, or else they will sooner or later lose the transfected sequence.

## Example experiment: Transfecting CHO cells with LacZ and GFP

This example illustrates the steps in a routine transfection experiment with two reporter genes with constitutive promoters, CMV-GFP (plasmid size 3 kb total) and CMV-LacZ (5 kb size total). The first step is to obtain the cells, either by purchase or obtaining a flask from a neighboring lab. If the cells were frozen, passage them at least twice before use to ensure that all DMSO is out of the medium and that they are growing well. Set aside ~20 mL of medium as "serum-free" in a sterile tube before adding serum and antibiotics to the main bottle.

It is also necessary to ensure that the DNA is available in sufficient quantity and quality for transfection. Miniprep DNA should not be used. At least 1 µg of DNA per 35 mm dish will be needed at a concentration of 100 ng/mL or more. A midi- or maxi-prep, preferably one made "endotoxin free," is desirable for this.

The next step is to choose a transfection reagent by searching catalogs or BioBars. A good deal of companies report transfection efficiencies for CHO cells, with or without cytotoxicity data. You can see that most reagents work with these cells, so it is not necessary to buy specialized or expensive reagents (a far from exhaustive list of available reagents is given in Table 3.2). Check for pricing and special lab coupons or deals.

Say that for practical reasons (there is some in the lab fridge) you decide to use Lipofectamine. The manufacturer's protocol is quite detailed, but provides quite a wide range of possible DNA and reagent quantities. If you need precise control over efficiencies, perform an optimization procedure as in Practical Tips 3.6. If not, you may choose values that are in the middle of the suggested ranges, e.g., 2.5 µg of DNA and 5–10 µL of Lipofectamine per 35 mm dish.

One to two days before transfection, seed the cells to the desired confluency. They should be 50–80% confluent and 90% viable before transfection. Make at least four dishes: one as a no-DNA control, one as a GFP-only control, one as a LacZ-only control, and one to contain both plasmids. When the cells are ready, proceed according to the manufacturer's instructions:

**Table 3.2**

Example of Some Commercial Transfection Reagents as They Relate to CHO Cells

| Reagent | Company | CHO Data | Notes |
|---|---|---|---|
| Xfect | Clontech | 80% transfection/80% viability | No need for serum-free medium |
| Polyfect | Qiagen | COS-7, NIH/3T3, HeLa, 293, and CHO cells; 50–60% reported efficiency for CHO | Company provides searchable database of results for cell types and transfection reagents |
| Superfect | Qiagen | 75–80% efficiency | |
| Lipofectamine LTX Plus | Invitrogen | Number not given; recommendations provided for "+++" efficiency | Company provides database of results and protocols; OptiMEM reduced serum medium recommended for diluting DNA |
| TransPass D2 | New England Biolabs | Luciferase expression reported; D2 reagent tested specifically for CHO | Serum-free medium not needed, but no antibiotics in transfection medium |

- Start by taking the transfection reagent out of the refrigerator and into the tissue culture hood. Do not even *dream* of opening this vial outside of the hood. Do not touch the cells yet; leave them in the incubator. Put the tubes of DNA in the hood as well.

- For each dish, dilute 2.5 µg of DNA into 0.5 mL of Opti-MEM without serum or antibiotics. Use four different tubes for the four different conditions. For the tube containing both plasmids, use a 1:1 *molar* ratio of each (i.e., 5:3 LacZ/GFP weight ratio for the example given) with the total amount of DNA kept at 2.5 µg.

- Pipette the chosen amount of Lipofectamine into each tube. Be careful not to cross-contaminate any of the reagents. Mix gently and let complexes form in the tissue culture hood for 25 min.

- Take the cells out of the incubator near the end of the incubation period. Remove the old growth medium and add fresh, complete medium (with serum).

- Add the DNA–liposome complexes, rock gently, and return to the incubator.

Other reagents will require slight variations in medium conditions, and many require the complexes to be washed off the cells after some amount of time (1–24 h) to prevent cytotoxicity; this is not the case with Lipofectamine LTX, although if a good deal of toxicity is observed, the medium may be replaced after 2–4 h and replaced with fresh medium.

GFP should begin to become visible after 24 h or less (**Figure 3.10a**). If you wish to track the level of expression, be sure to keep the dish sterile as it is transported to the microscope. When the desired level of GFP expression is seen in the GFP-only control and/or the cotransfected dish, fix the cells and proceed with LacZ analysis as in **Practical Tips 3.7**.

The biggest pitfalls in this type of simple transfection have to do with the health of the cells. If the cells are sick, the transfection will finish them off and very few viable

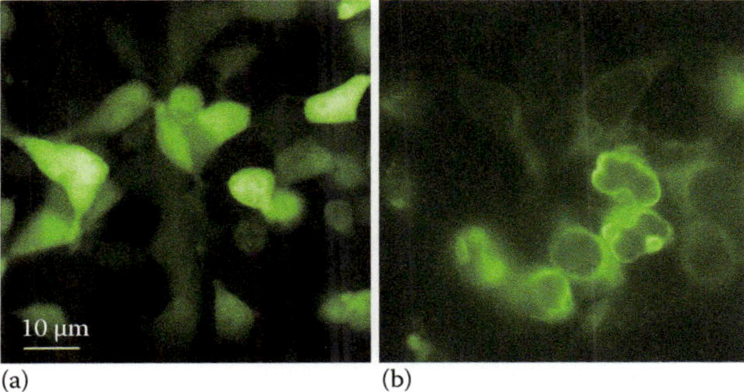


**Figure 3.10 Good and bad transfections.** (a) Good GFP transfection after 24 h, showing cells with varying levels of fluorescence. (b) Bad transfection. Note sick, blebby cells, and greatest expression of GFP in dead/dying cells.

cells will be seen (**Figure 3.10b**). Bacterial and fungal contaminations from the DNA solution and/or liposomes are also common. Please do not do this to your poor cells.

If chronic, unexplained toxicity is seen, you can use a less transfection reagent for the same amount of DNA, or try a different transfection reagent. Some are designed to be particularly gentle to cells. It might also be worthwhile to redo the DNA prep, perhaps with a different kit. Protein contamination and endotoxin can cause serious cell problems. Cells might also have been plated too sparsely, and an increase in confluency can lead to better survival as well as to a greater absolute number of transfected cells. The manufacturers of the different reagents also often provide troubleshooting tips that may be specific to your cell type.

It is also important to remember that your gene product might be toxic. Some fluorescent proteins are more damaging than others, and some cells are particularly sensitive even to wild-type GFP, especially when it is expressed at very high levels. Another thing that may sound trivial—but which causes a lot of grief in fluorescent protein experiments—is that a weak illumination source in the fluorescence microscope might only allow you to see the brightest cells, and these are usually dead. Pulling out or switching filters, changing the lamp, or using a different microscope can reveal unsuspected healthy green cells.

## Electroporation of cell cultures

Electroporation of mammalian cells works in a similar fashion as with bacteria: the cells are placed in a cuvette with parallel metal plates, and a pulse is applied, which transiently disrupts the cell membrane to allow passage of plasmid DNA into the cell. This technique was first reported over 30 years ago, but is more difficult with mammalian cells than with bacteria, and requires a more expensive, tunable electroporator. The pulse duration and strength that will allow nucleic acid uptake without killing the cells has to be determined empirically for each cell type, and substantial cell death nearly always occurs at even the optimum parameters. We do not recommend this technique for animal cells, although many researchers use it with plant cells. It may be of use with stem cells; some references are given at the end of the chapter. Caution is advised.

## Microinjection of DNA and RNA: For a few select cells or constructs that are difficult to transfect

Microinjection is direct delivery of DNA or RNA (or other materials) into a cell cytoplasm or nucleus with a sharp needle. DNA for transfection must be delivered to the cell nucleus; if injected into the cytoplasm, it will simply degrade. There are some disadvantages to using this method that make it often not a technique of choice. It is labor-intensive, as it requires treating one cell at a time. It also demands significant skill, especially when the injection is done into the cell nucleus, and micromanipulators for precise control of needles are costly. The survival rate of sensitive cell types such as neurons is low, usually 5–15% with the best technique. However, it is in widespread use for certain applications using large, well-isolated cells, particularly those with pronounced nuclei—an example is motor neurons. Some immortalized cell lines that meet these criteria are Chinese hamster ovary (CHO) and MDBK cells; these can be used for practice, although they are usually not microinjected with plasmid DNA since they are easy to transfect.

There are certain advantages to microinjection compared with other transfection methods that should be considered for certain types of experiments. Particular cell types can be targeted, which is important in mixed cultures, especially those including neurons and glia. The DNA can be nicked or linear, not necessarily supercoiled, and very large sequences can be delivered. Precise control of DNA concentration is possible, and since the concentration injected determines the time course of expression, it is possible to mix two plasmids and different concentrations to permit one to express before the other. Perhaps most importantly, a large number of plasmids (five or six) can be cotransfected at once with guarantee that they all make it into the cell at a precise ratio. This is key in some neuroscience experiments, where expression of receptors and channels requires several different subunits.

A typical microinjection setup consists of a microscope, a micromanipulator for holding and moving the needles, and a microinjector (**Figure 3.11**). Commercial microinjectors can adjust injecting pressure, inject in discrete pulses or a continuous stream, and use a high-pressure pulse to clean needle blockages. For cells that are not adherent, an additional pipette is needed to hold them in place. Depending upon the size of the cells, magnification of 20× to 63× is needed (we

**Figure 3.11 Microinjection setup, showing microscope, needle holder (arrow) attached to micromanipulator, and Eppendorf microinjector.** (Image courtesy of Sam Johnson, Duke Light Microscopy Core Facility.)

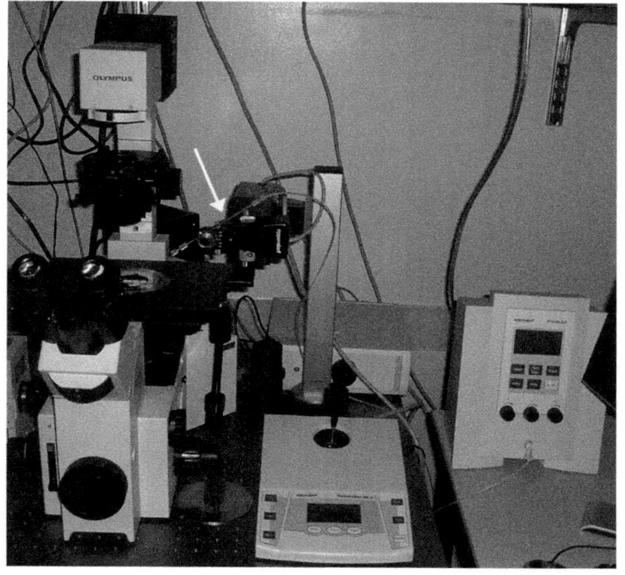

recommend 63× for neurons). Those doing electrophysiology will already have the infrastructure for micromanipulation preparing needles. The needles are made from 1 mm outer diameter thin-walled borosilicate capillaries pulled to a sharp tip using a pipette puller (see **Chapter 15** for more on micropipette pulling). The needles are backfilled with the DNA solution, usually at 50–150 ng/µL.

Optimizing microinjection requires becoming proficient at locating the needle under the microscope, approaching and touching the cell (usually over the nucleus), and rapidly delivering the solution. The pipette should be in the cell for no more than 1–2 s, and the cell should not swell excessively or burst with injection. Depending upon the cell type, some swelling may be observed. Injection volume should be no more than 10–50 fL for a nuclear injection and up to 10 times that much for a cytoplasmic injection. Once these techniques have been mastered, the injection time, pressure, and other variables on the microinjector can be refined.

Usually the greatest hurdle to successful microinjection is needle blockage. The best way to avoid this is to have very clean solutions and to use a microinjector that allows "cleaning." A single needle can be used a limited number of times before it needs to be changed; this needs to be determined empirically for the experimental conditions (type of needle, type of cell, etc.). Needles should also not be stored for more than a few days, as dust and debris from the air accumulate on them. If necessary, a small amount of bovine serum albumin can be added to the pipette to improve flow. There are other methods for preventing blockage that are usually not needed with DNA solutions, but are discussed in the chapter on quantum dots (**Chapter 11**).

Neuronal cultures are particularly tricky to microinject because the cells are round and sticky, and the cells are extremely sensitive. Cells that are grown in serum-free cultures or with a minimal number of glia are usually not healthy enough to survive microinjection. If the cells tear when the needle is removed, this is another sign that the culture is not healthy enough to withstand the procedure. Good cells for microinjection can be selected by their appearance under phase contrast microscopy; the best are cells that are well stuck down and not rounded (**Figure 3.12a,b**). Thinner needles are needed to inject these cells than for flatter cells; however, the needles must not be too thin or the cell will burst. A successful microinjection will lead to a cell with normal morphology that continues to adhere to the dish for up to several weeks. If the cell is irremediably damaged by the microinjection, it will show changes under phase contrast microscopy that indicate cell death, such as crenellation of the membrane. In cells expressing a fluorescent gene, the fluorescence of damage cells will appear granular (**Figure 3.12c,d**).

Microinjection has significant value apart from transfection, as many other molecules besides plasmid DNA may be microinjected: RNA, antibodies, proteins, quantum dots, and so on. For molecules that do not give a visible sign of their presence, the addition of a fluorescent marker such as 0.01% fluorescein-dextran indicates which cells have been treated. This is very useful for experiments using invisible RNAs, such as in RNA interference (RNAi) experiments. The key when microinjecting anything is to ensure that the solution is clean, nonsticky, and free of harmful substances. Preservatives such as sodium azide are highly toxic and must be avoided. All solutions for microinjection should be centrifuged before use to eliminate aggregates that might block the needle. Some molecules, such as quantum dots, are particularly sticky and require the addition of something

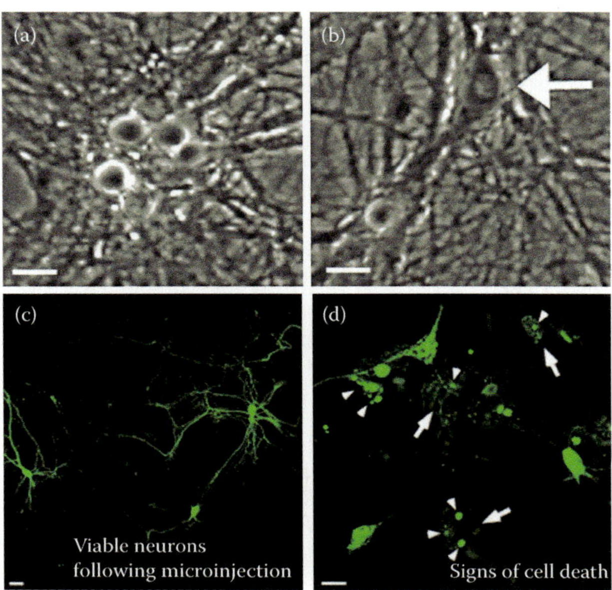


**Figure 3.12 Microinjection of neurons.** Scale bar = 20 μm. (a) Rounded cells that appear "haloed" under phase contrast do not microinject well. (b) Flattened cells with a visible nucleus (arrow) are better candidates. They must be tightly adhered to the glial underlayer. (c) Successful microinjection of GFP plasmid showing neurons expressing fluorescence throughout the cell bodies and processes. The cells appear even without bubbles or blebbing. (d) Signs of cell death include broken or patchy labeling, loss of processes (axons and dendrites), blebs, and cell shrinkage. ([c, d] Reprinted from *Journal of Neuroscience Methods*, 175, Lappe-Siefke, C., C. Maas, and M. Kneussel, Microinjection into cultured hippocampal neurons: a straightforward approach for controlled cellular delivery of nucleic acids, peptides and antibodies, 88–95, Copyright 2008, with permission from Elsevier.)

to improve flow, such as polyethylene glycol (see **Chapter 11**). More specialized transfection methods have been developed for specific applications or reagents. Some of the more common ones are discussed in **Advanced Topic 3.3**.

## 3.5  GENE DELIVERY USING VIRUSES

A large amount of work has been done on the development of vehicles to carry plasmid DNA into nondividing cells, particularly neurons. We have introduced many of them above, at all different levels of complexity from liposomes to ultrafast laser systems. Although many of these techniques can be made to work with refinement and patience, neurotropic *viral vectors* are by far the most efficient technique if large numbers of transfected cells are needed. Transfection via viruses is called, not surprisingly, *infection*. The drawbacks of using viruses for gene delivery are that cloning of the constructs can be difficult, their capacity is often limited, cotransfection does not work (one virus inhibits infection by a second), and viruses often require biosafety containment. However, commercial suppliers and researchers have worked to improve nearly all of these factors over the past several years, and it is now possible for the average biophysics lab to generate certain types of viruses for gene transfer without any particular new expertise or equipment. Once the system is set up and the conditions for infection are optimized, the infection procedure is very easy and results are highly reproducible.

## ADVANCED TOPIC 3.3:    SPECIALIZED PHYSICAL TRANSFECTION METHODS MAGNETOFECTION AND NANOTUBE SPEARING

The principle behind magnetofection is too complex plasmid DNA with a magnetic particle, either with or without liposomes, add this complex to cells, and then apply a magnetic field to pull the complexes toward (or into) the cells (Figure A3.3.1). The particles are superparamagnetic iron oxide coated with cationic polymers, usually polyethyleneimine (PEI). These may be synthesized without much difficulty, and permanent magnets such as those of neodymium–iron–boron (Nd–Fe–B) create a sufficient field for the purpose when placed under a culture dish. Alternatively, reagents for commercial magnetofection are now available commercially; they include the paramagnetic "ferrofluid" particles as well as magnets for different sizes of culture plates.

Advantages of magnetofection are that it is rapid, efficient (requiring very little vector), and nearly uniform over the dish (the rate of transfection will follow the magnetic field lines). Levels of gene expression can be up to a thousand-fold greater than what is seen with typical liposomal transfection. This is therefore a great method for when it is important to express the gene of interest at high levels in nearly every cell. It is also good for difficult cells, such as primary cells of different kinds, and for delivering difficult reagents, such as RNAi.

The term "magnetofection" is trademarked; the trademark is owned by Christian Bergemann of Germany. Several companies sell proprietary reagents for magnetofection, which may work better than homemade PEI particles for some applications.

Another emerging technique in gene delivery is "carbon nanotube spearing." Single-wall carbon nanotubes with ferromagnetic nickel in the tips respond to magnetic fields, and can be accelerated with different amounts of force into cells and even cell nuclei. The magnetic field required can be generated by a simple stir plate as found in almost every laboratory. The plasmid DNA is covalently bound to the carboxylates of the nanotubes via carbodiimide coupling (see Chapter 10). As with all transfection techniques, some cell types are much easier to transfect than others. Efficient transfection of cortical neurons

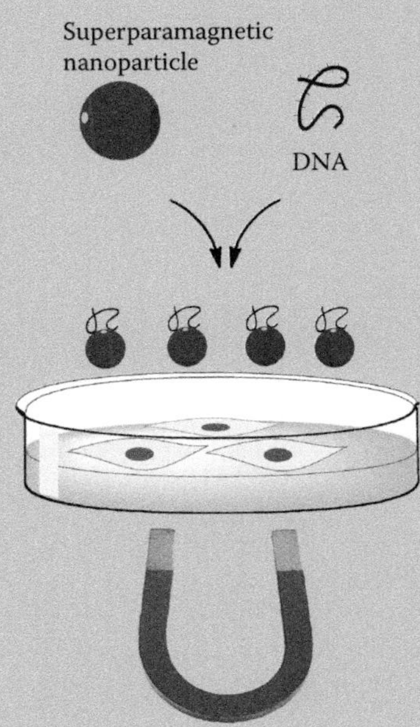


**Figure A3.3.1 Magnetofection.** Superparamagnetic iron oxide nanoparticles ("ferrofluid") are mixed with plasmid DNA, sometimes also with cationic liposomes. They are placed into a dish of cells exposed to a magnet for a specific amount of time, usually minutes. The magnet pulls the DNA onto the cells and facilitates transfection.

*(Continued)*

**ADVANCED TOPIC 3.3 (CONTINUED):   SPECIALIZED PHYSICAL TRANSFECTION METHODS MAGNETOFECTION AND NANOTUBE SPEARING**

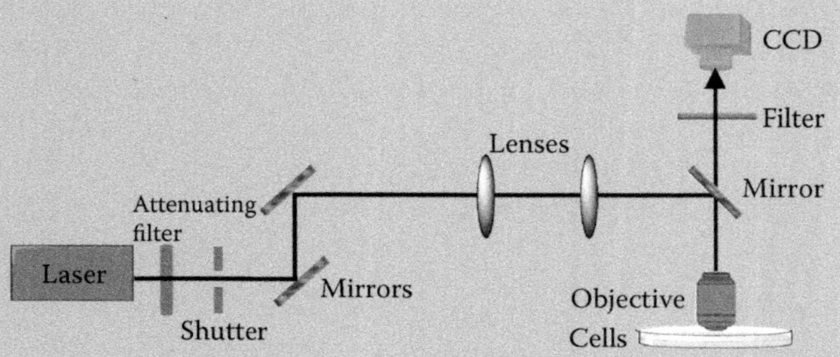


**Figure A3.3.2 Schematic of a simple optofection system.** A laser (Ti:sapphire) is attenuated as needed and focused to a diffraction-limited spot within the dish of cells, allowing for targeted transfection of a selected cell.

requires two steps: agitation of the nanotubes for 20 min in a rotating magnetic field to cause initial cell spearing, then application of a static field for an additional 20 min to pull the tubes deeper into the cells. Gene expression is seen in 1 to 2 days. Unlike other magnetic nanomaterial-based methods, carbon nanotube spearing does not rely upon endocytosis but delivers the cargo directly into the cell cytoplasm or nucleus. This accounts for the increased efficiency and the low amount of plasmid DNA needed, as most endocytosed DNA is destroyed in the endosomes. However, this is a new technique and the extent of cell damage has not been fully explored. The author has not tried it, but recommends it for the curious.

## OPTICAL TRANSFECTION

Optical transfection or "optofection" is performed by using a focused pulse of light to open transient pores in cell membranes. When a sufficient quantity of plasmid DNA is present in the medium, transfection may result. Optofection has been demonstrated using a wide variety of laser light sources, from 405 to 800 nm, both continuous-wave and pulsed. The most convenient method for many labs is to make use of the 800 nm focused light of a *laser tweezers* system. This same light, when focused through a high numerical aperture objective, can function to optofect cells. With the increased availability of femtosecond lasers, the application of femtosecond pulses to cells has become an increasingly popular optofection method. The single pulses deliver nanojoules of energy and thus cause minimal cell damage.

The setup of an optofection system is very simple, involving a laser, mirror, beamsplitter, and microscope with an appropriate objective lens (Figure A3.3.2). An electronic shutter should be used to control the exposure time, and an attenuator may be needed to control laser power; a power density of several TW/cm$^2$ is sufficient to puncture any cell membrane.

Advantages of optofection are that it is "hands-off" and potentially less damaging to cells than other methods. Disadvantages are that it is inefficient and requires huge amounts of plasmid; both of these reasons are because the naked plasmid DNA is readily degraded. If optical transfection is to become a useful method, the optimal way of packaging or complexing the DNA will need to be determined to prevent the excessive waste and possible toxicity associated with application of large amounts of DNA to cultures.

## GENE GUN

Also called "biolistics," a gene gun is a device similar to an air pistol that accelerates micron-sized heavy metal particles (usually gold or tungsten) toward a Petri dish or tissue explant (Figure A3.3.3). When the particles are coated with plasmid DNA, the DNA is carried into the cell and can be expressed. The coating is readily accomplished by electrostatic assembly; the gold particles are mixed with a positively charged molecule such as *spermidine*, then with the plasmid DNA, which adheres spontaneously to the "bullets."

*(Continued)*

**ADVANCED TOPIC 3.3 (CONTINUED):    SPECIALIZED PHYSICAL TRANSFECTION METHODS MAGNETOFECTION AND NANOTUBE SPEARING**

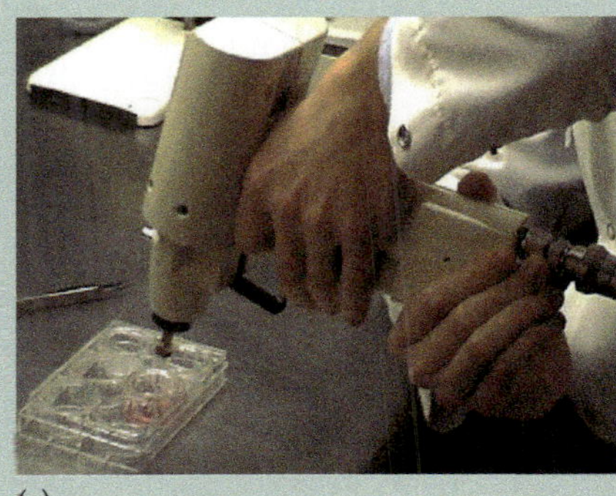 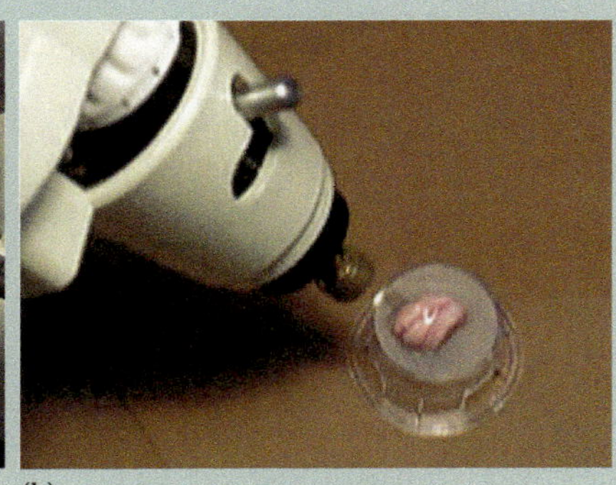

**Figure A3.3.3  Use of the gene gun.** (a) Shooting cells in a 6-well plate. (b) Gene gunning a fish brain. This gene gun has a custom barrel designed for brains and slices. (Images courtesy of J. O'Brien, MRC Laboratory of Molecular Biology, Cambridge, UK, http://www.genegunbarrels.com.)

Gene guns are available commercially, usually helium-propelled. A functional gene gun can also be made using a solenoid valve triggered by a relay switch, with the pressure delivered by a helium tank. Care must be taken to use clean helium, or fungal infection of the cells can result. Some references to do-it-yourself gene guns are included at the end of the chapter.

Gene-gunning has been used in some studies using mammalian cells in culture, but is much more widely used in plants, since the plant cells' cellulose cell walls prevent excessive damage from the "shooting." Even organelles such as chloroplasts can be targeted, which is important for delivering organelle-mediated resistance genes into plants. The gene gun can also be used with yeast and other fungi and worms of different kinds: the nematode *Caenorhabditis elegans* is commonly transfected this way.

Tissue explants and brain slices may also be readily transfected with the gene gun, as the tissue architecture protects the individual cells, and deeper transfection can be obtained than is possible with other methods. For these applications, it may be desirable to modify the commercial gene gun's barrel to target a smaller area; such modified barrels are now available commercially through independent companies. Successful slices show well-defined cells with excellent morphology that can be used for studies of dendritic spines or three-dimensional architecture (Figure A3.3.4).

More recently, the gene gun has been used to deliver proteins rather than nucleic acids. This may be desirable if a protein of interest expresses poorly in a cell type or if it is not available as a plasmid.

**Figure A3.3.4  Neurons transfected with the gene gun.** (a) GFP-transfected pyramidal neurons in a 60 μm thick mouse hippocampal brain slice. (b) Purkinje cell from a cerebellar organotypic slice expressing GFP. (Images courtesy of J. O'Brien, MRC Laboratory of Molecular Biology, Cambridge, UK, http://www.genegunbarrels.com.)

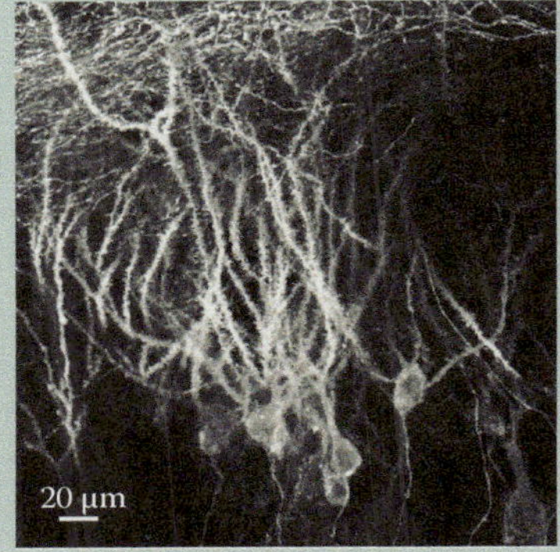 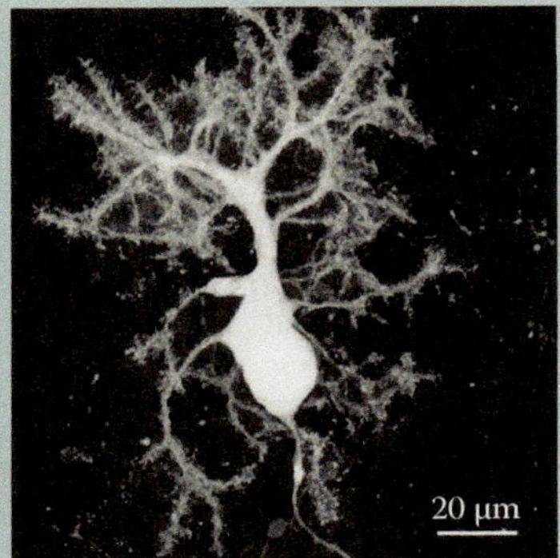


*(Continued)*

**ADVANCED TOPIC 3.3 (CONTINUED):    SPECIALIZED PHYSICAL TRANSFECTION METHODS MAGNETOFECTION AND NANOTUBE SPEARING**

### Sonotransfection

Sonotransfection is an emerging technique that uses ultrasound to deliver nucleic acids. The application of ultrasound disturbs the area around the cell membrane, leading to the formation of small pores. These are insufficient for transfection unless "microbubbles" are added to the medium. Microbubbles are gas pockets, usually of perfluorocarbon, surrounded by a lipid shell, which resonate in the presence of ultrasound. The size of the microbubbles, and hence their resonant frequency, must be adjusted to match the frequency of the ultrasound being used.

### SUGGESTED READING

Arsenault, J., Nagy, A., Henderson, J.T., and O'brien, J.A. (2014). Regioselective biolistic targeting in organotypic brain slices using a modified gene gun. *Journal of Visualized Experiments*, e52148.

Cai, D., Mataraza, J.M., Qin, Z.H., Huang, Z., Huang, J., Chiles, T.C., Carnahan, D., Kempa, K., and Ren, Z. (2005). Highly efficient molecular delivery into mammalian cells using carbon nanotube spearing. *Nature Methods* 2, 449–454.

Clark, I.B., Hanania, E.G., Stevens, J., Gallina, M., Fieck, A., Brandes, R., Palsson, B.O., and Koller, M.R. (2006). Optoinjection for efficient targeted delivery of a broad range of compounds and macromolecules into diverse cell types. *Journal of Biomedical Optics* 11, 014034.

Dib-Hajj, S.D., Choi, J.S., Macala, L.J., Tyrrell, L., Black, J.A., Cummins, T.R., and Waxman, S.G. (2009). Transfection of rat or mouse neurons by biolistics or electroporation. *Nature Protocols* 4, 1118–1126.

Feril, L.B., Jr. (2009). Ultrasound-mediated gene transfection. *Methods in Molecular Biology* 542, 179–194.

Hernot, S., and Klibanov, A.L. (2008). Microbubbles in ultrasound-triggered drug and gene delivery. *Advanced Drug Delivery Reviews* 60, 1153–1166.

Hochbaum, D., Ferguson, A.A., and Fisher, A.L. (2010). Generation of transgenic C. elegans by biolistic transformation. *Journal of Visualized Experiments* 23;(42) (2010). pii: 2090. doi: 10.3791/2090.

Huang, R.Y., Chiang, P.H., Hsiao, W.C., Chuang, C.C., and Chang, C.W. (2015). Redox-sensitive polymer/SPIO nanocomplexes for efficient magnetofection and MR imaging of human cancer cells. *Langmuir* 31, 6523–6531.

Karimi, M., Solati, N., Ghasemi, A., Estiar, M.A., Hashemkhani, M., Kiani, P., Mohamed, E., Saeidi, A., Taheri, M., Avci, P., Aref, A.R., Amiri, M., Baniasadi, F., and Hamblin, M.R. (2015). Carbon nanotubes part II: A remarkable carrier for drug and gene delivery. *Expert Opinion on Drug Delivery* 12, 1089–1105.

Kesharwani, P., Gajbhiye, V., and Jain, N.K. (2012). A review of nanocarriers for the delivery of small interfering RNA. *Biomaterials* 33, 7138–7150.

Lei, M., Xu, H., Yang, H., and Yao, B. (2008). Femtosecond laser-assisted microinjection into living neurons. *Journal of Neuroscience Methods* 174, 215–218.

Liu, W.M., Xue, Y.N., He, W.T., Zhuo, R.X., and Huang, S.W. (2011). Dendrimer modified magnetic iron oxide nanoparticle/DNA/PEI ternary complexes: a novel strategy for magnetofection. *Journal of Controlled Release* 152 Suppl 1, e159–160.

Ma, Y., Zhang, Z., Wang, X., Xia, W., and Gu, H. (2011). Insights into the mechanism of magnetofection using MNPs-PEI/pDNA/free PEI magnetofectins. *International Journal of Pharmaceutics* 419, 247–254.

Martin-Ortigosa, S., and Wang, K. (2014). Proteolistics: A biolistic method for intracellular delivery of proteins. *Transgenic Research* 23, 743–756.

Munsell, E.V., Ross, N.L., and Sullivan, M.O. (2016). Journey to the center of the cell: Current nanocarrier design strategies targeting biopharmaceuticals to the cytoplasm and nucleus. *Current Pharmaceutical Design* 22, 1227–1244.

Mykhaylyk, O., Antequera, Y.S., Vlaskou, D., and Plank, C. (2007). Generation of magnetic nonviral gene transfer agents and magnetofection in vitro. *Nature Protocols* 2, 2391–2411.

Nguyen, T.H., Shamis, Y., Croft, R.J., Wood, A., Mcintosh, R.L., Crawford, R.J., and Ivanova, E.P. (2015). 18 GHz electromagnetic field induces permeability of Gram-positive cocci. *Scientific Reports* 5, 10980.

O'brien, J.A., and Lummis, S.C. (2006). Biolistic transfection of neuronal cultures using a hand-held gene gun. *Nature Protocols* 1, 977–981.

(*Continued*)

## ADVANCED TOPIC 3.3 (CONTINUED):    SPECIALIZED PHYSICAL TRANSFECTION METHODS MAGNETOFECTION AND NANOTUBE SPEARING

O'brien, J.A., and Lummis, S.C. (2013). Biolistic transfection of neurons in organotypic brain slices. *Methods in Molecular Biology* 940, 157–166.

Plank, C., Zelphati, O., and Mykhaylyk, O. (2011). Magnetically enhanced nucleic acid delivery. Ten years of magnetofection-progress and prospects. *Advanced Drug Delivery Reviews* 63, 1300–1331.

Sikorskaite, S., Vuorinen, A.L., Rajamäki, M.L., Nieminen, A., Gaba, V., and Valkonen, J.P. (2010). HandyGun: An improved custom-designed, non-vacuum gene gun suitable for virus inoculation. *Journal of Virological Methods* 165, 320–324.

Stevenson, D.J., Gunn-Moore, F.J., Campbell, P., and Dholakia, K. (2010). Single cell optical transfection. *Journal of the Royal Society, Interface* 7, 863–871.

Taylor, T., Bose, I., Luckie, T., and Smith, K. (2015). Biolistic transformation of a fluorescent tagged gene into the opportunistic fungal pathogen *Cryptococcus neoformans*. *Journal of Visualized Experiments*.

Waleed, M., Hwang, S.U., Kim, J.D., Shabbir, I., Shin, S.M., and Lee, Y.G. (2013). Single-cell optoporation and transfection using femtosecond laser and optical tweezers. *Biomedical Optics Express* 4, 1533–1547.

Viral vectors used in gene transfer experiments bear little resemblance to their infectious parent viruses. Most of their genome is removed, including all of the sequences that permit viral replication. The space in the genome usually reserved for these genes can then be used for cloning; depending on the original genome size, this may be a few kb up to several hundred kb. Viral vectors are also usually *pseudotyped*, meaning that their outer protein *capsid* comes from a different virus than the rest of the genes expressed. It is this capsid that determines *tropism*, or which cells will be infected. Usually a capsid is selected that will allow for broad tropism, essentially any cell type; occasionally one that targets a specific cell type will be used, although this usually decreases infection efficiency overall (**Figure 3.13**). These vectors have often been developed for *gene therapy* of genetic diseases, and the gene therapy literature is often the best source for description of these vectors.

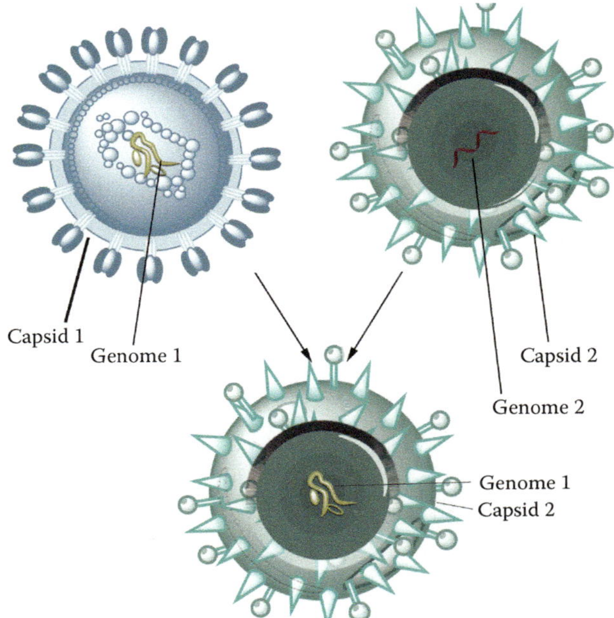


**Figure 3.13 Pseudotype virion is formed when the genome of one virus is inside the capsid of another.** This can occur in nature when two types of virus infect a single cell. In that case, the virus's tropism will be changed by the new capsid, but the daughter virions will have the properties of the parent genome, not the capsid. For viral vectors, which do not reproduce, pseudotyping serves to control the tropism.

There are many classes of viruses, some with highly complex, double-stranded DNA genomes and others with small RNA genomes < 10 kb. Some references are given at the end of the chapter to books on general virology and viral vectors. For this chapter, we will discuss one example of a viral vector whose components are available commercially, which works well in primary nondividing cells, and with which the author has a good deal of experience: *lentivirus*. Lentivirus allows for stable, low-level expression of genes in nondividing cells for weeks or longer, and so is ideal for long-term experiments, developmental studies, and organotypic slices. We will conclude with a brief discussion of other types of viruses whose constructs are available commercially.

## Lentivirus

Lentiviruses are members of the viral family of retroviruses (introduced in **Chapter 1**). Retroviruses transcribe their RNA genomes to DNA, the reverse of what is seen in all other organisms, using the enzyme reverse transcriptase. Reverse transcriptase is error-prone, accounting for the high mutation rate of retroviruses. All retroviruses integrate into the host genome after transcription into DNA (the DNA is called a *provirus*). Different viruses show different preferences for integration sites—some home to promoters, and others to sites of active transcription. Because the sequence now forms an integral part of the host genetic material, it is passed down indefinitely during cell division (**Figure 3.14**).

Reverse transcription occurs in the cytoplasm, and most retroviruses require the nuclear envelope to dissolve (i.e., the initiation of cell division) before they can access the host DNA. The lentiviruses are an exception to this, as they possess enzymes that facilitate the transport of the *integration complex* across an intact nuclear membrane: the complex consists of the DNA provirus, the enzyme *integrase* that catalyzes insertion into the host genome, an accessory protein, and viral protein matrix. Their target cells are usually those of the immune system, such as lymphocytes and macrophages, and the diseases they cause are related to a slow depletion of these cells in the host and resulting immunodeficiency. The human immunodeficiency virus (HIV) and the feline immunodeficiency virus are two examples and are the basis for most of the lentiviral vectors developed for gene transfer.

A detailed understanding of the biology of HIV was necessary before this dangerous pathogen could be made into a safe vector. The ability of the virus to infect

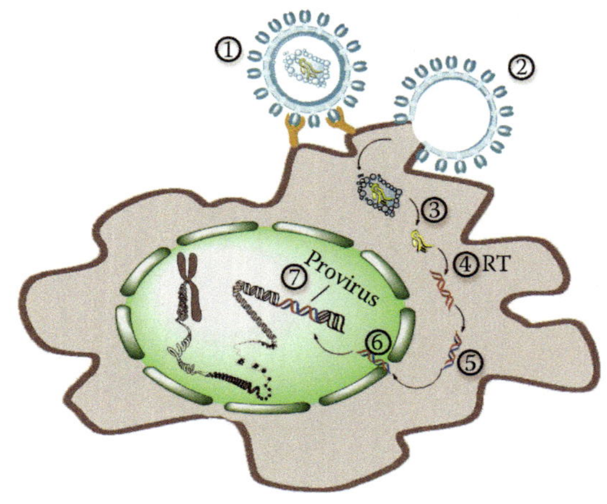


**Figure 3.14 Retrovirus lifecycle.** The virus gains entry to the cell via specific receptors (1). Its entire contents, including packaging proteins, are injected into the cell (2). The viral RNA is then liberated from the packing material (called uncoating) (3) and reverse-transcribed to yield a viral DNA–RNA hybrid (4), then double-stranded viral DNA (5). The DNA traffics to the nucleus (6) and integrates into a certain spot in the host genome; this integrated sequence is the provirus (7).

nondividing cells needed to be preserved while removing all genes necessary for replication and several responsible for virulence. Today's generation of commercial HIV-based vectors do not require high biosafety containment (they are designated BSL-2), although their ability to infect certain specific cell types such as macrophages is greatly reduced compared to vectors containing more HIV genes.

Lentiviral vectors are nearly always generated in a transient transfection system in which three (or four) separate plasmids are transfected into a *packaging cell line*. The requirement of the cell line is that it be easily transfected at a high level; HEK 293T cells are nearly always the ones used. The plasmids contain no overlapping sequences, so it is essentially impossible for them to recombine into an infectious particle.

Key structural or enzymatic components of the virus are located on each plasmid; these components are described in **Table 3.3**. The vector plasmid contains the cloned gene of interest along with sequences necessary for the vector to infect the target cell. The limits on insert size are stringent—usually 7 kb or less. Since cotransfection is not possible, all of the genes of interest need to be encoded

**Table 3.3**

HIV Genes and Their Functions and Roles in Vectors for Gene Transfer

| Gene or Sequence | Function | Role in Gene-Transfer Vector |
| --- | --- | --- |
| LTR | Promoters for gene expression | Necessary for function |
| Ψ | Packaging sequence | Necessary for packaging RNA into viral particle |
| RRE | Interacts with rev (stands for "rev—responsive element"); needed for transport and processing of RNA | Useful |
| PPT | Needed for priming of plus-strand synthesis | Essential |
| Gag/Pol | Encodes structural proteins and enzymes | Essential |
| Vif | Affects viral particle assembly and infectivity in certain cell types | Can be deleted |
| Vpu | Downregulates CD4 and MHC class I molecules; stimulates release of virions from infected cell | Deleted for safety; not needed |
| Vpr | Needed for infection of nondividing cells | Needed to target certain cell types; not necessary for vector infection of nondividing cells, usually deleted for safety |
| Tat | Necessary for high-level expression from the viral LTR | Not needed since other promoters are used |
| Rev | Necessary for expression of unspliced and singly spliced mRNAs | Usually needed |
| Nef | Virulence factor | Deleted for safety; not needed, but absence may decrease expression levels |

*Note:* Compare with **Figure 3.23**. Some of the genes are not present in lentiviral vectors for safety reasons or simply to increase cloning capacity.

within this limit. They may either be fused together, creating a single RNA that is translated into a *fusion protein*, or be transcribed separately from a single promoter by placing an internal ribosomal entry site (IRES) in between the genes (**Figure 3.15a**). The packaging plasmid(s) encode(s) the enzymes required to generate virus particles and all of the structural proteins *except* for the outer envelope (**Figure 3.15b**). A third plasmid, the envelope plasmid, encodes an envelope gene of a different virus for pseudotyping (**Figure 3.15c**). The most commonly used envelope is the G glycoprotein of the vesicular stomatitis virus (VSV-G), which is able to target nearly all cell types and is very stable, permitting easy ultracentrifugation and concentration to high titers.

When the three plasmids are transfected together, they transiently produce viral proteins and a genome containing the cloned insert with a "packaging" ($\Psi$) sequence. The result is infectious but "single-use" particles that are capable only of transducing the gene of interest (**Figure 3.16**).

A typical lentivirus production is done in sufficient quantity to be able to generate several hundred microliters of a solution containing $10^8$–$10^9$ viral particles/mL. This quantity of particles requires a relatively large volume of packaging cells: we routinely use four to six 25 cm dishes of 293 T cells per experiment. Of course, this requires large amounts of the DNA of the three (or four) component plasmids and a large amount of transfection reagents. Because commercial liposomes are expensive, some researchers choose to use calcium phosphate transfection for virus production, although this method requires a very large amount of plasmid DNA (up to several mg per preparation). The transient transfection is done in the same way as any other, and the viral particles are released into the growth medium of the cells. Viral particles may be "harvested" 36–48 h after transfection by simply removing the medium with a pipette. If fresh medium is added to the cells, at least one more harvest may be performed the following day. The viral particles in the medium are concentrated either by ultracentrifugation (no pellet is visible) or by the use of a low-speed centrifugal filter device. The latter provides only 10–100 fold concentration, yielding probably $10^6$ particles/mL, but this concentration is adequate for any *in vitro* experiment. Extremely high concentrations are only needed for animal

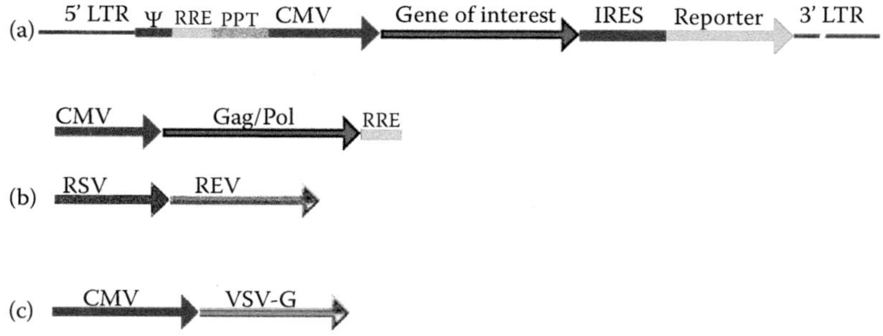



**Figure 3.15 Example of a typical replication incompetent, HIV-based lentiviral vector encoded on four separate plasmids.** The genes and sequences are described in **Table 3.3**. Several other variations of this gene arrangement exist. (a) Gene of interest, with or without a separate reporter, is cloned into a plasmid containing the viral long terminal repeats (LTRs), PPT, RRE, the Ψ sequence, and a strong mammalian-cell promoter such as CMV. This plasmid can get large, 15 kb or more, which means there are few unique restriction sites and amplification in *E. coli* can be unstable. (b) Packaging plasmids. Using two separate plasmids and two different promoters (RSV and CMV) increases safety. Commercial lentiviral vectors contain only the *gag-pol* and *rev* genes of HIV. (c) Envelope plasmid. This is usually VSV-G.

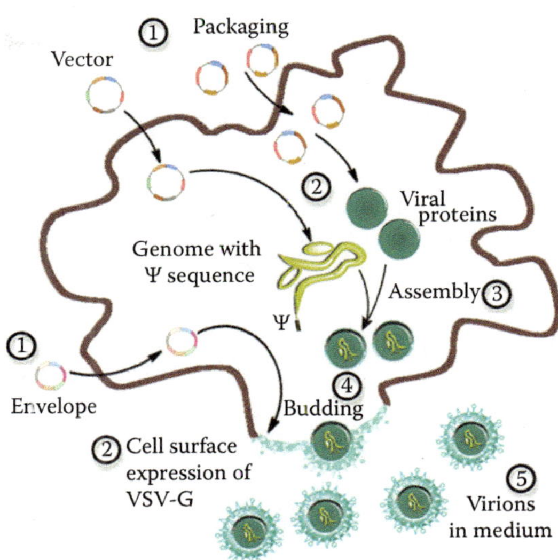



**Figure 3.16 Schematic of lentivirus generation in packaging cell line.** The vector, packaging, and envelope plasmids are all transfected at once (1). The vector plasmid gives rise to multiple copies of genomic RNA encoding for the cloned insert, and tagged with the Ψ sequence that marks it for packaging; the packaging plasmids give rise to viral proteins; and the envelope protein gives rise to VSV-G that is expressed on the cell surface (2). The viral core proteins assemble with the genome (3) and then bud through the VSV-G expressing membrane (4), leading to mature, infectious but replication-incompetent virions in the culture medium where they can be harvested by pipetting (5).

injections, where the volume injected is limited to a few microliters to avoid tissue damage caused by the fluid.

The concentrated viral particles are stored frozen or applied immediately to target cells at concentrations that need to be determined empirically for each cell type. Expression of reporter genes is usually seen 48 h later in nondividing cells. Given the many steps involved—infection, reverse transcription, transport to the nucleus, integration, and gene expression—it is fruitless to examine the cultures before then, especially using fluorescent proteins that have to be expressed at relatively high levels (100–1000 molecules) before they can be seen. Once expression is obtained, it is very stable. We have seen infected neurons survive and express reporter genes for up to 6 weeks.

A transfection rate of 70–90% can be achieved with primary neurons if the culture type and infection conditions are chosen carefully. However, problems with infection can be seen in typical dissociated neuronal cultures, which contain a good deal of glial cells: astrocytes of varying morphologies, oligodendrocytes, microglia, and others. Glia spread out to form a continuous underlayer with individual cells often more than 100 μm across. They are necessary for the health and survival of neurons, and part of their role is to act as a "sponge" for excess potassium, reactive oxygen species, and other toxins. They also have a great affinity for VSV-G pseudotyped lentivirus: glial cells are always infected more readily than neurons (**Figure 3.17**).

Methods for limiting glial infection and maximizing neuronal infection include

- Infecting the culture early after plating (1–2 days in culture vs. weeks). Because glial cells divide, there are more glia as the culture ages, so infecting early may allow a greater fraction of virus to adsorb onto

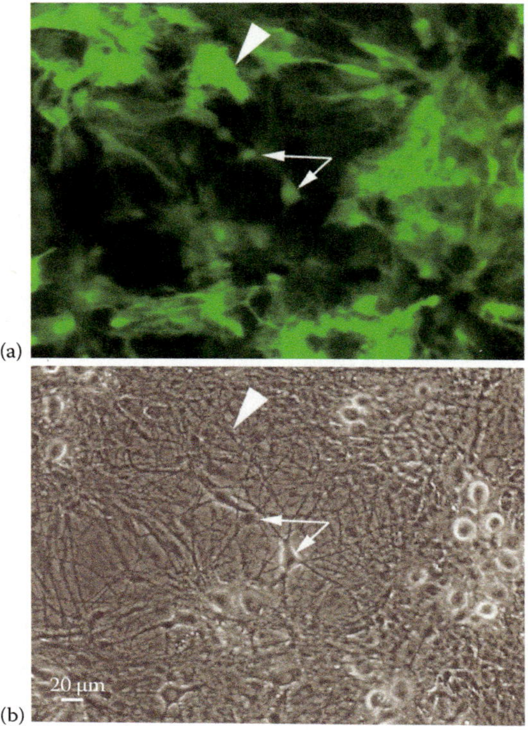


**Figure 3.17 Mature hippocampal culture (3 weeks after plating) infected with lentivirus encoding GFP 7 days previously.** (a) Fluorescence image, showing some cells that are clearly neurons (arrows) and others that are clearly glia (arrowhead). However, much of the fluorescent signal is from mixed cell types and difficult to interpret. (b) Phase contrast image of the same culture. The neurons appear three-dimensional and are often easy to see; the astrocytes are essentially invisible (arrows and arrowhead pointing to the same regions as in a).

neurons. It is also possible that some of the neurons can undergo a single division after culturing, improving their infectability. However, any infected glial cells will continue to divide and express reporter, so this is not always a very effective solution.

- Using cultures such as "Banker" cultures where the glia are grown on a separate coverslip that is inverted over the neurons. This coverslip can be removed before virus is added, and replaced after 24–48 h when all the viruses have had time to enter the neurons.

- Limiting the growth of glia using serum reduction or other methods. This is effective, but neurons grown in such conditions are less healthy overall than neurons with an astrocyte layer (**Figure 3.18**).

Lentivirus infection also works well in organotypic slices (**Figure 3.19**), but is not recommended for acute slices because such slices do not live long enough for gene expression. When high-level, rapid expression is needed, another type of vector should be used.

## Some other types of viruses used as vectors

*Adenoviruses* are a family of double-stranded DNA viruses with over 50 subtypes. Most of them cause relatively minor infections that are more common in children under 2 years, such as respiratory infections, gastroenteritis, and croup. A significant fraction of common colds, especially spring and summer colds, are caused by adenoviruses. Adenoviral vectors are replication-deficient and can

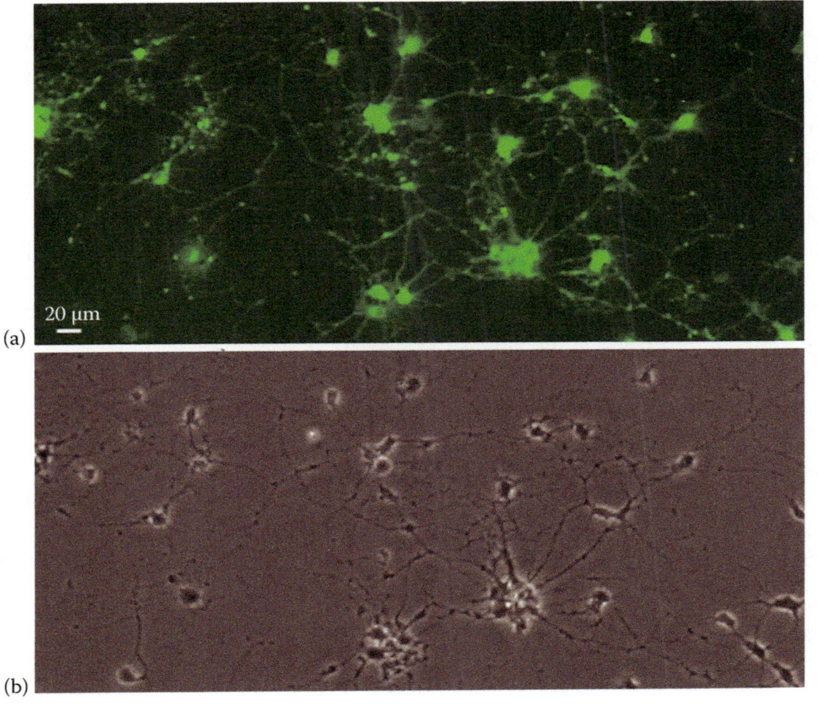

**Figure 3.18  Serum-free hippocampal culture with inhibited glial growth.** These cells are 10 days after plating and 9 days after infection with GFP-encoding lentivirus. (a) Fluorescence and (b) phase contrast. Note the visibility of the neuronal axons and dendrites in this type of culture.

hold inserts up to 8 kb. As gene therapy vectors, adenoviruses have gone all the way to clinical trials; however, the immune response they cause can make patients sick or serve as a significant barrier to gene delivery (essentially just causing a stuffy nose). In the laboratory, adenovirus is great for high-level, fast, transient expression. We recommend it for infecting acute slices and for neuronal cultures where rapid expression is desired (e.g., Purkinje cell cultures or other types of cultures that are usually used within a day or two of preparation). They may also be used for traditional neuronal cultures, but they are significantly more toxic than lentivirus. There is a lot of batch-to-batch variation in adenovirus toxicity as well, which often makes repeat experiments unreliable.

The biggest barrier to use of adenovirus is that cloning the constructs is very tricky. The genome is too large to have unique restriction sites, so constructs are

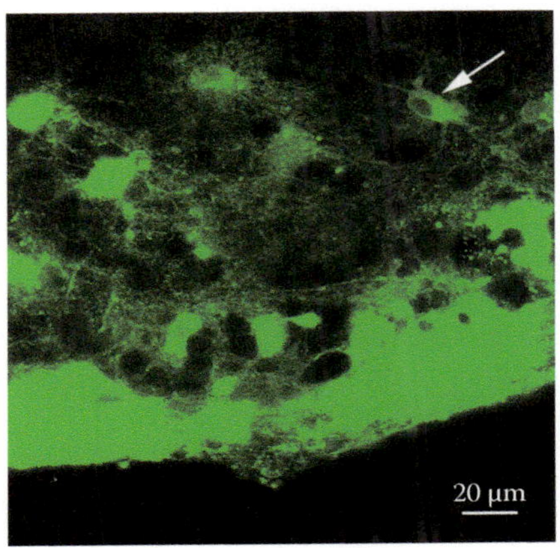

**Figure 3.19  GFP-expressing neurons in a 3-week-old organotypic hippocampal slice, 2 days post infection.** The arrow indicates a neuron.

usually made by *homologous recombination*. A gene is cloned into a plasmid bearing some sequence homology with the adenoviral genome; when these are cotransfected into packaging cells, recombination will occur with some (low) frequency and produce particles if you are lucky. This is a hit-or-miss procedure and can be frustrating, involving viral plaque purification, which is a cumbersome process. Recent developments have produced the "AdEasy" system, in which the plasmids are cotransfected into *E. coli*, where the rate of homologous recombination is much higher than in mammalian cells. No plaque purification is necessary; recombinants are screened by restriction enzyme digestion and then transfected into 293 cells. Kits are available for producing adenoviruses this way from a number of suppliers. It is also possible to purchase stock adenoviruses for hundreds of common genes or to request custom synthesis.

Another barrier to use of adenovirus is that it only infects certain cell types efficiently—those that bear the coxsackievirus-adenovirus receptor (CAR), to which the viral fiber or "knob" binds. Efforts have been made to create adenoviruses with either broader tropism or cell-specific tropism by removing the outer targeting fiber and replacing it with a selected targeting sequence. Other methods have also been used to improve adenoviral transfection of nontarget cells, such as the use of magnetofection with adenoviruses; however, these approaches can lead to significant cell death.

*Adeno-associated virus* (AAV) is a small parvovirus with a single-stranded DNA genome of only 4.8 kb. It is not known to cause disease or severe immune response, but is able to integrate into the host genome at a specific site (called AAVS 1). However, in AAV vectors, the integrative capacity is removed, along with essentially all the other viral genes to make space for cloning genes of interest. AAV cannot replicate on its own, so traditionally adenoviruses were used as a *helper virus* to generate infectious particles. More recently, the adenoviral genes permitting replication were identified and cloned onto a plasmid. Kits for producing AAV are "helper free," requiring only the plasmids and the packaging cell line.

The advantages of AAV are that it is easy to manipulate, is low in toxicity, and readily infects cultured neurons as well as neurons *in vivo*. In fact, many of the gene therapy applications for which it has been designed target neurodegenerative conditions such as Parkinson's disease and Alzheimer's disease, and the virus is administered directly into the brain. The greatest disadvantage is the small cloning capacity. It also will be lost from rapidly dividing cells because it does not integrate into the genome, although its expression in nondividing cells may be stable for weeks.

Other viruses commonly used for gene transfer are nonlentiviral retroviruses such as the moloney murine leukemia virus (MMLV). These vectors are very easy to prepare and are the most commonly used of all the viral vectors, but they only infect dividing cells. Herpes simplex virus (HSV) is also a good vector for some applications, as it has a very large cloning capacity—up to 150 kb. It also has a broad host range and is neurotropic, capable of transducing neurons (and other nondividing cells). Drawbacks are the difficulties of working with its large genome and with large inserts. Several HSV vector systems have been developed that are both safe (containing essentially no viral genome) and relatively easy to work with. Very large titers (up to $10^{10}$/mL) can also be obtained, making this a good choice for animal or explant injections.

It is important to keep in mind that these vectors nearly always require BSL-2 containment. If *oncogenes* or genes for toxins are cloned into viral vectors, the viruses should be handled under BSL-2 with additional precaution rules ("BSL-2+"). The biosafety office of your institution can provide information as to particular requirements for this level of containment and recommend practices for decontamination and disposal of media and plates used to make viral particles. In some places, AAV and MMLV without toxic inserts may be considered BSL-1.

## 3.6 SUMMARY

If all you need is a cell to express your gene, think of a bacterium or yeast cell first. These cells are easy and inexpensive to grow, do not require specialized infrastructure, and will express nearly all genes except some membrane proteins. If their rigid cell walls are a problem, spheroplasts or protoplasts can be made with only a membrane.

When only mammalian cells will do, make sure you can handle them properly. If you do not have the equipment to incubate them and keep them sterile, they will only die horribly and provide no useful data. Some buildings do not have the ventilation infrastructure to permit the installation of hoods; if this is the case, tissue culture must be done elsewhere. In this case, be very careful carrying cells around. Shaking, exposure to the sun, extremes of temperature, and other discomforts can kill them quickly, and of course there is always the risk of contamination. My group keeps sterile Tupperware and injects $CO_2$ through a filter into the box before adding dishes of cells to be transported.

When the infrastructure is available, maintaining and transfecting immortalized cell lines is easy. A huge number of these lines are available, which you can choose based upon their immediate availability, their gene expression profiles, or simply their size and shape (we will revisit cell shape in **Chapter 7** when we discuss microscopic techniques). Not all cell lines are equally good for every application, and researchers usually maintain several of them at a time for different uses.

Sometimes not even a cell line is enough, and primary cells, stem cells, or explants are needed. In these cases, the difficulty of transfection goes up tremendously, especially for notoriously sensitive and difficult cell types such as neurons. In these cases, you should rely upon the expertise of lab members and/or collaborators for a choice of tissue culture and transfection system. Learning any of these primary culture techniques or transfection methods from scratch is extremely time-consuming and difficult. It is also often possible to share animals to minimize their use, for example, to obtain mouse embryo cortex from collaborators who are only using the hippocampus.

Although this chapter is long, it has only scratched the cell-culture surface. There are a few extremely important expression systems that are simply outside the scope of this book, most notably for the biophysicist *Xenopus* oocytes. These are large (1 mm) eggs from the African clawed frog, *Xenopus laevis*. Because of their large size, they are excellent for microinjection and electrophysiology. However, their use requires the presence of an animal facility to house the frogs and surgically remove the eggs. Care of the frogs is nontrivial, as they are subject to bacterial and fungal infections, injury, and poisoning by chlorine in tap water. Some references to the subject are given at the end of the chapter.

For gene delivery into cells, liposomal transfection works for essentially everything except primary nondividing cells and stem cells. There are many choices for transfecting these tricky and sensitive cells, but all of them are difficult and many are extremely costly. For those who already have the apparatus for electrophysiology, microinjection is a good approach as only the microinjector must be purchased. Similarly, if optical tweezing or femtosecond lasers are being used in the laboratory, it may be worth trying to set up a system for optical transfection, although these systems have been sparsely described in the literature and some instrument development will certainly be required. If slices or explants are used, the gene gun is a good bet, but we certainly do not recommend it for dissociated cultured neurons.

Viral vectors can be efficient means of carrying genes into essentially any cell type and are capable of infecting nearly every cell in a dish. They are a good option for anyone who is prepared to learn the techniques of production, purification, and manipulation and who has the biosafety training and containment facilities necessary.

Here as well there are many methods we have not covered. The first method of transfection developed was the calcium phosphate precipitation method, still in use in some labs especially for viral vector production. This is a straightforward method and protocols are available in essentially every lab for those who might wish to try it. Its disadvantages are that it is time-consuming and uses a very large amount of plasmid DNA per transfection. There are also a wide variety of emerging nanoparticle techniques, which may be searched using "nonviral gene delivery" or "nanoparticle gene transfer" or similar keywords. Mastery of these techniques usually requires some polymer chemistry expertise, and their host range and efficiencies are at best what is seen with liposomal transfection. A few references to emerging methods are given at the end of the chapter for the curious.

## End-of-Chapter Questions

1. Bacterial growth is usually measured as the absorbance or optical density (OD) at 600 nm. Why? What do you suppose is the relationship between OD and number of cells/mL? How could you measure this?

2. If a cell's doubling time is 12 h and its radius is 10 μm, how many cells should be seeded into a 35 mm dish to achieve 40% confluency in 24 h? In 48 h? What if the cell's doubling time is 6 h?

3. What assumption does the calculation in Problem 2 make? Can you think of when it might be wrong?

4. For the cells above, what dilution factor corresponds to the number of cells desired? For example, you begin with a T25 flask that is fully confluent and resuspend all of the trypsinized cells in 5 mL of medium. How much volume of this suspension should be added to 2 mL in the 35 mm dishes above?

5. For a typical transfection experiment using 1 μg DNA per 35 mm dish, how many plasmids is this per cell?

6. Estimate the volume of a cell and a cell nucleus. For an injection of 10 fL, what fraction of the nuclear volume is this? The cell volume? Do you expect to see a bulge or swelling of the cells with a 10 fL injection? A 100 fL injection?

7. (ADVANCED) What is the pattern of transfection you expect to see with magnetofection when the field source is a 35 mm permanent magnet placed under a dish of the same size?

8. (ADVANCED) Femtosecond lasers are usually used for optical transfection of single cells, whereas nanosecond lasers are used for a group of cells. Why? Discuss how laser repetition rate and power might affect the mechanisms of transfection.

9. (ADVANCED) When continuous wave lasers are used for optofection, a red dye is usually added to the medium. Why?

10. Estimate the number of virions that could be made in a 10 cm dish of fully confluent packaging cells. How many 10 cm dishes would be needed to yield 100 μL of virus at $10^{10}$ particles/mL?

11. Why does retroviral infection take so long? Using the principles introduced in **Chapter 1**, estimate the time

that each step takes. What is the rate-limiting step or steps? Could you think of how to speed this up? (Hint: early lentiviral vectors were mixed with dNTPs before adding them to cells)

12. Describe the differences between primary, secondary, and transformed cells.

13. Why are each of the following added to cell culture: glutamine, serum, penicillin/streptomycin? Can cells be grown in serum-free medium, and why would this be desirable?

14. Discuss when and why you would have to use a primary neuronal culture rather than a neuronal cell line.

15. Since retroviruses are RNA viruses, why do we not have to work with RNA to produce them?

16. A common concept in virology is "multiplicity of infection" (MOI). Adding $10^6$ viruses to a dish of $10^6$ cells gives an MOI of 1; adding $10^7$ viruses to a dish of $10^6$ cells gives an MOI of 10. For each of these cases, what fraction of cells will remain uninfected?

17. How would you pick a cell line for each of the following experiments: expression of a potassium channel; testing a new breast cancer drug for efficacy; testing a food additive for toxicity?

18. Oh no… your cell culture is contaminated. This is the most common type of problem in cell culture experiments. Discuss how you would (a) know it is contaminated, (b) identify the problem, and (c) fix it.

19. Why is phenol red added to culture media? What does it mean if the medium turns purple? Yellow? What should you do in each case?

20. Why are cell lines grown under a $CO_2$-enriched atmosphere?

21. Explain how Dolly the sheep was cloned. Why is it harder to clone a cat than a sheep?

22. (ADVANCED) Explain how embryonic stem cells are used to make transgenic or "knockout" mice. Why is it harder to make transgenic rats than transgenic mice? What is one way of generating transgenic rats?

23. (ADVANCED) How are hematopoietic stem cells identified? Mesenchymal stem cells? What types of cells can each differentiate into?

24. (ADVANCED) Discuss the effects of applying ultrasound (US) to a dish of cells. What parameters of the US need to be optimized in order to obtain maximum transfection and minimum cell killing? What processes are responsible for cell death?

25. The book *The Immortal Life of Henrietta Lacks* describes how the first immortal cell line—the HeLa cell line—was isolated from an aggressive cervical cancer in a patient named Henrietta Lacks. Why was this cell line successful when so many other attempts had failed? What do we know now that allows other cell lines to be immortalized? Discuss the ethical implications of sequencing the HeLa cell line.

26. Why can mammalian cell lines not be used to clone DNA the way *E. coli* are used?

27. You go to the Antarctic, find a new bacterium, and you would like to label it with GFP. What are the steps to take?

28. Discuss the concept of "cancer stem cells" and their implications for therapy.

29. Describe the general approach to expressing foreign genes in *Saccharomyces cerevisiae*. What are the advantages of using this organism?

30. What are the advantages and disadvantages of stable transfection vs. lentiviral infection for creating a cell line expressing a certain gene?

## Background Reading

### Books

Banker, G. and Goslin, K. (eds.) *Culturing Nerve Cells,* Edn. 2. Massachusetts University Press, 1998.—The bible of neural cell culture.

Freshney, R.I. *Culture of Animal Cells: A Manual of Basic Technique*, Edn. 5. Wiley-Liss, 2005.—A must-read for anyone doing cell culture.

Kaplitt, M.G. and Loewy, A.D. (eds.) *Viral Vectors: Gene Therapy and Neuroscience Applications.* Academic Press, 1995.—A discussion of different viral vectors with a neuroscience slant.

Maurizio, F. (ed.) *Lentivirus Gene Engineering Protocols.* Humana Press, 2003.—Contributed chapters with detailed protocols for production and use of lentiviruses for different systems.

Friedmann, T. and Rossi, J. (eds.) *Gene Transfer: Delivery and Expression of DNA and RNA, a Laboratory Manual*. Cold Spring Harbor University Press, 2007.—A detailed, practical guide to many forms of gene delivery. Highly recommended.

Heiser, W.C. (ed.) *Gene Delivery to Mammalian Cells: Nonviral Gene Transfer Techniques*. Humana Press, 2004.—A practical, protocol-based guide to gene transfer using nonviral means.

Lasic, D.D. *Liposomes in Gene Delivery*. CRC Press, 1997.—Detailed description of DNA and liposomes and their interactions.

## Journal articles

### *Bacteria and yeast*

Averhoff, B., and Muller, V. (2010). Exploring research frontiers in microbiology: Recent advances in halophilic and thermophilic extremophiles. *Research in Microbiology* 161, 506–514.

Bakota, L., Brandt, R., and Heinisch, J.J. (2012). Triple mammalian/yeast/bacterial shuttle vectors for single and combined Lentivirus- and Sindbis virus-mediated infections of neurons. *Molecular Genetics and Genomics* 287, 313–324.

Barros, M.E.C., Rawlings, D.E., and Woods, D.R. (1985). Production and regeneration of thiobacillus-ferrooxidans spheroplasts. *Applied and Environmental Microbiology* 50, 721–723.

Benzinger, R. (1978). Transfection of enterobacteriaceae and its applications. *Microbiological Reviews* 42, 194–236.

Cho, S.N., Hwang, J.H., Park, S., Chong, Y.S., Kim, S.K., Song, C.Y., and Kim, J.D. (1998). Factors affecting transformation efficiency of BCG with a Mycobacterium Escherichia coli shuttle vector pYUB18 by electroporation. *Yonsei Medical Journal* 39, 141–147.

Cui, B.T., Smooker, P.M., Rouch, D.A., and Deighton, M.A. (2015). Enhancing DNA electro-transformation efficiency on a clinical *Staphylococcus capitis* isolate. *Journal of Microbiological Methods* 109, 25–30.

Cuperus, J.T., Lo, R.S., Shumaker, L., Proctor, J., and Fields, S. (2015). A tetO toolkit to alter expression of genes in *Saccharomyces cerevisiae*. *ACS Synthetic Biology* 4, 842–852.

Gietz, R.D., and Schiestl, R.H. (2007a). Frozen competent yeast cells that can be transformed with high efficiency using the LiAc/SS carrier DNA/PEG method. *Nature Protocols* 2, 1–4.

Gietz, R.D., and Schiestl, R.H. (2007b). High-efficiency yeast transformation using the LiAc/SS carrier DNA/PEG method. *Nature Protocols* 2, 31–34.

Gietz, R.D., and Schiestl, R.H. (2007c). Large-scale high-efficiency yeast transformation using the LiAc/SS carrier DNA/PEG method. *Nature Protocols* 2, 38–41.

Hopwood, D.A. (1981). Genetic-studies with bacterial protoplasts. *Annual Review of Microbiology* 35, 237–272.

Iizasa, E., and Nagano, Y. (2006). Highly efficient yeast-based in vivo DNA cloning of multiple DNA fragments and the simultaneous construction of yeast/*Escherichia coli* shuttle vectors. *Biotechniques* 40, 79–83.

Jarrell, K.F., and Sprott, G.D. (1984). Formation and regeneration of halobacterium spheroplasts. *Current Microbiology* 10, 147–152.

Johnston, C., Martin, B., Fichant, G., Polard, P., and Claverys, J.P. (2014). Bacterial transformation: Distribution, shared mechanisms and divergent control. *Nature Reviews Microbiology* 12, 181–196.

Jones, M.J., Donegan, N.P., Mikheyeva, I.V., and Cheung, A.L. (2015). Improving transformation of *Staphylococcus aureus* belonging to the CC1, CC5 and CC8 clonal complexes. *PLoS One* 10 (3): e0119487. https://doi.org/10.1371/journal.pone.0119487.

Lin, H.Y., Lin, S.E., Chien, S.F., and Chern, M.K. (2011). Electroporation for three commonly used yeast strains for two-hybrid screening experiments. *Analytical Biochemistry* 416, 117–119.

Lofblom, J., Kronqvist, N., Uhlen, M., Stahl, S., and Wernerus, H. (2007). Optimization of electroporation-mediated transformation: *Staphylococcus carnosus* as model organism. *Journal of Applied Microbiology* 102, 736–747.

Low, K.B., and Porter, D.D. (1978). Modes of gene transfer and recombination in bacteria. *Annual Review of Genetics* 12, 249–287.

Mann, W., and Jeffery, J. (1986). Yeasts in molecular-biology—Spheroplast preparation with *Candida utilis*, *Schizosaccharomyces pombe* and *Saccharomyces cerevisiae*. *Bioscience Reports* 6, 597–602.

Matsushima, P., and Baltz, R.H. (1985). Efficient plasmid transformation of *Streptomyces ambofaciens* and *Streptomyces fradiae* protoplasts. *Journal of Bacteriology* 163, 180–185.

Mayrhofer-Iro, M., Ladurner, A., Meissner, C., Derntl, C., Reiter, M., Haider, F., Dimmel, K., Rossler, N., Klein, R., Baranyi, U., Scholz, H., and Witte, A. (2013). Utilization of virus phi Ch1 elements to establish a shuttle vector system for halo(alkali)philic archaea via transformation of *Natrialba magadii*. *Applied and Environmental Microbiology* 79, 2741–2748.

Mercer, A.A., and Loutit, J.S. (1979). Transformation and transfection of *Pseudomonas aeruginosa*—Effects of metal-ions. *Journal of Bacteriology* 140, 37–42.

Miteva, V., Lantz, S., and Brenchley, J. (2008). Characterization of a cryptic plasmid from a Greenland ice core Arthrobacter isolate and construction of a shuttle vector that replicates in psychrophilic high G+C Gram-positive recipients. *Extremophiles* 12, 441–449.

Monk, I.R., Shah, I.M., Xu, M., Tan, M.W., and Foster, T.J. (2012). Transforming the untransformable: Application of direct transformation to manipulate genetically *Staphylococcus aureus* and *Staphylococcus epidermidis*. *Mbio* 3 (2). pii: e00277–11 (2012). doi: 10.1128/mBio.00277-11.

Riveline, D., and Nurse, P. (2009). 'Injecting' yeast. *Nature Methods* 6, 513–U564.

Rivera, A.L., Magana-Ortiz, D., Gomez-Lim, M., Fernandez, F., and Loske, A.M. (2014). Physical methods for genetic transformation of fungi and yeast. *Physics of Life Reviews* 11, 184–203.

Ruan, Y.L., Zhu, L.J., and Li, Q. (2015). Improving the electro-transformation efficiency of *Corynebacterium glutamicum* by weakening its cell wall and increasing the cytoplasmic membrane fluidity. *Biotechnology Letters* 37, 2445–2452.

Saunders, J.R., and Saunders, V.A. (1999). "Introduction of DNA into bacteria," in *Methods Microbiology, Vol 29: Genetic Methods for Diverse Prokaryotes,* eds. M.C.M. Smith and R.E. Sockett., 3–49. Elsevier, Amsterdam, The Netherlands.

Serratosa, C., Lopez, R., Garcia, E., and Ronda, C. (1984). Conditions to enhance transfection in *Streptococcus pneumoniae*. *Virus Research* 1, 443–453.

Stanislauskiene, R., Gasparaviciute, R., Vaitekunas, J., Meskiene, R., Rutkiene, R., Casaite, V., and Meskys, R. (2012). Construction of *Escherichia coli–Arthrobacter–Rhodococcus* shuttle vectors based on a cryptic plasmid from *Arthrobacter rhombi* and investigation of their application for functional screening. *FEMS Microbiology Letters* 327, 78–86.

Strathern, J.N., and Higgins, D.R. (1991). Recovery of plasmids from yeast into *Escherichia coli*—Shuttle vectors. *Methods in Enzymology* 194, 319–329.

Strike, P., Humphreys, G.O., and Roberts, R.J. (1979). Nature of transforming deoxyribonucleic-acid in calcium-treated *Escherichia coli*. *Journal of Bacteriology* 138, 1033–1035.

Suarez, J.E., and Chater, K.F. (1980). Polyethylene glycol-assisted transfection of *Streptomyces* protoplasts. *Journal of Bacteriology* 142, 8–14.

Tripp, J.D., Lilley, J.L., Wood, W.N., and Lewis, L.K. (2013). Enhancement of plasmid DNA transformation efficiencies in early stationary-phase yeast cell cultures. *Yeast* 30, 191–200.

## Mammalian cells: general and nonviral methods

http://www.jove.com/science-education/5068/an-introduction-to-transfection

Ahmed, M., and Narain, R. (2011). The effect of polymer architecture, composition, and molecular weight on the properties of glycopolymer-based non-viral gene delivery systems. *Biomaterials* 32, 5279–5290.

Bishop, C.J., Abubaker-Sharif, B., Guiriba, T., Tzeng, S.Y., and Green, J.J. (2015). Gene delivery polymer structure–function relationships elucidated via principal component analysis. *Chemical Communications (Cambridge)* 51, 12134–12137.

Capecchi, M.R. (1980). High-efficiency transformation by direct micro-injection of DNA into cultured mammalian-cells. *Cell* 22, 479–488.

Colosimo, A., Goncz, K.K., Holmes, A.R., Kunzelmann, K., Novelli, G., Malone, R.W., Bennett, M.J., and Gruenert, D.C. (2000). Transfer and expression of foreign genes in mammalian cells. *Biotechniques* 29, 314.

Doerr, A. (2005). A synthetic solution to gene delivery. *Nature Methods* 2, 808.

Doulatov, S.R. (2003). Plasmid-based reporter genes: Assays for green fluorescent protein. *Methods in Molecular Biology* 235, 297–304.

Felgner, P.L., Gadek, T.R., Holm, M., Roman, R., Chan, H.W., Wenz, M., Northrop, J.P., Ringold, G.M., and Danielsen, M. (1987). Lipofection—A highly efficient, lipid-mediated DNA-transfection procedure. *Proceedings of the National Academy of Sciences of the United States of America* 84, 7413–7417.

Fornaguera, C., Grijalvo, S., Galan, M., Fuentes-Paniagua, E., De La Mata, F.J., Gomez, R., Eritja, R., Caldero, G., and Solans, C. (2015). Novel non-viral gene delivery systems composed of carbosilane dendron functionalized nanoparticles prepared from nano-emulsions as non-viral carriers for antisense oligonucleotides. *International Journal of Pharmaceutics* 478, 113–123.

Guignet, E.G., and Meyer, T. (2008). Suspended-drop electroporation for high-throughput delivery of biomolecules into cells. *Nature Methods* 5, 393–395.

Haladjova, E., Toncheva-Moncheva, N., Apostolova, M.D., Trzebicka, B., Dworak, A., Petrov, P., Dimitrov, I., Rangelov, S., and Tsvetanov, C.B. (2014). Polymeric nanoparticle engineering: From temperature-responsive polymer mesoglobules to gene delivery systems. *Biomacromolecules* 15, 4377–4395.

Halterman, M.W., Giuliano, R., Dejesus, C., and Schor, N.F. (2009). In-tube transfection improves the efficiency of gene transfer in primary neuronal cultures. *Journal of Neuroscience Methods* 177, 348–354.

Huang, L., and Li, S. (1997). Liposomal gene delivery: A complex package. *Nature Biotechnology* 15, 620–621.

Jiang, M., and Chen, G. (2006). High Ca2+-phosphate transfection efficiency in low-density neuronal cultures. *Nature Protocols* 1, 695–700.

Kim, T.K., and Eberwine, J.H. (2010). Mammalian cell transfection: The present and the future. *Analytical and Bioanalytical Chemistry* 397, 3173–3178.

Kirtane, A.R., and Panyam, J. (2013). Polymer nanoparticles: Weighing up gene delivery. *Nature Nanotechnology* 8, 805–806.

Lappe-Siefke, C., Maas, C., and Kneussel, M. (2008). Microinjection into cultured hippocampal neurons: A straightforward approach for controlled cellular delivery of nucleic acids, peptides and antibodies. *Journal of Neuroscience Methods* 175, 88–95.

Liu, M. (2003). Plasmid-based reporter genes: Assays for beta-galactosidase and alkaline phosphatase activities. *Methods in Molecular Biology* 235, 289–296.

Luo, D., and Saltzman, W.M. (2000). Synthetic DNA delivery systems. *Nature Biotechnology* 18, 33–37.

Mathupala, S., and Sloan, A.A. (2009). An agarose-based cloning-ring anchoring method for isolation of viable cell clones. *Biotechniques* 46, 305–307.

Muramatsu, T., Mizutani, Y., Ohmori, Y., and Okumura, J. (1997). Comparison of three nonviral transfection methods for foreign gene expression in early chicken embryos in ovo. *Biochemical and Biophysical Research Communications* 230, 376–380.

Templeton, N.S. (2001). Developments in liposomal gene delivery systems. *Expert Opinion on Biological Therapy* 1, 567–570.

Van Gaal, E.V., Van Eijk, R., Oosting, R.S., Kok, R.J., Hennink, W.E., Crommelin, D.J., and Mastrobattista, E. (2011). How to screen non-viral gene delivery systems in vitro? *Journal of Controlled Release* 154, 218–232.

Won, Y.W., Lim, K.S., and Kim, Y.H. (2011). Intracellular organelle-targeted non-viral gene delivery systems. *Journal of Controlled Release* 152, 99–109.

Yang, Y., Zhao, H., Jia, Y., Guo, Q., Qu, Y., Su, J., Lu, X., Zhao, Y., and Qian, Z. (2016). A novel gene delivery composite system based on biodegradable folate-poly (ester amine) polymer and thermosensitive hydrogel for sustained gene release. *Scientific Reports* 6, 21402.

Zhang, Y., and Yu, L.C. (2008). Single-cell microinjection technology in cell biology. *Bioessays* 30, 606–610.

## Viruses

Farson, D., Harding, T.C., Tao, L., Liu, J., Powell, S., Vimal, V., Yendluri, S., Koprivnikar, K., Ho, K., Twitty, C., Husak, P., Lin, A., Snyder, R.O., and Donahue, B.A. (2004). Development and characterization of a cell line for large-scale, serum-free production of recombinant adeno-associated viral vectors. *Journal of Gene Medicine* 6, 1369–1381.

Graham, F.L., and Prevec, L. (1995). Methods for construction of adenovirus vectors. *Molecular Biotechnology* 3, 207–220.

Grieger, J.C., Choi, V.W., and Samulski, R.J. (2006). Production and characterization of adeno-associated viral vectors. *Nature Protocols* 1, 1412–1428.

Jiang, W., Hua, R., Wei, M.P., Li, C.H., Qiu, Z.L., Yang, X.F., and Zhang, C. (2015). An optimized method for high-titer lentivirus preparations without ultracentrifugation. *Scientific Reports* 5, Article number: 13875 (2015). doi:10.1038/srep13875.

Khan, I.F., Hirata, R.K., and Russell, D.W. (2011). AAV-mediated gene targeting methods for human cells. *Nature Protocols* 6, 482–501.

Luo, J., Deng, Z.L., Luo, X., Tang, N., Song, W.X., Chen, J., Sharff, K.A., Luu, H.H., Haydon, R.C., Kinzler, K.W., Vogelstein, B., and He, T.C. (2007). A protocol for rapid generation of recombinant adenoviruses using the AdEasy system. *Nature Protocols* 2, 1236–1247.

Mcclure, C., Cole, K.L.H., Wulff, P., Klugmann, M., and Murray, A.J. (2011). Production and titering of recombinant adeno-associated viral vectors. *Jove-Journal of Visualized Experiments* 57:e3348 (2011). doi: 10.3791/3348.

Meghrous, J., Aucoin, M.G., Jacob, D., Chahal, P.S., Arcand, N., and Kamen, A.A. (2005). Production of recombinant adeno-associated viral vectors using a baculovirus/insect cell suspension culture system: From shake flasks to a 20-L bioreactor. *Biotechnology Progress* 21, 154–160.

Naldini, L., Blomer, U., Gallay, P., Ory, D., Mulligan, R., Gage, F.H., Verma, I.M., and Trono, D. (1996). In vivo gene delivery and stable transduction of nondividing cells by a lentiviral vector. *Science* 272, 263–267.

Sanber, K.S., Knight, S.B., Stephen, S.L., Bailey, R., Escors, D., Minshull, J., Santilli, G., Thrasher, A.J., Collins, M.K., and Takeuchi, Y. (2015). Construction of stable packaging cell lines for clinical lentiviral vector production. *Scientific Reports* 5, Article number: 9021 (2015) doi:10.1038/srep09021.

Thomas, C.E., Ehrhardt, A., and Kay, M.A. (2003). Progress and problems with the use of viral vectors for gene therapy. *Nature Reviews Genetics* 4, 346–358.

Verma, I.M. (1998). Hot papers—gene therapy—In vivo gene delivery and stable transduction of nondividing cells by a lentiviral vector by L. Naldini, U. Blomer, P. Gallay, D. Ory, R. Mulligan, F.H. Gage, I.M. Verma, D. Trono—Comments. *Scientist* 12, 10–10.

Wang, Q.Z., Dong, B., Firrman, J., Roberts, S., Moore, A.R., Cao, W.J., Diao, Y., Kapranov, P., Xu, R.A., and Xiao, W.D. (2014). Efficient production of dual recombinant adeno-associated viral vectors for factor VIII delivery. *Human Gene Therapy Methods* 25, 261–268.

Zheng, Y.X., Yu, F., Wu, Y.M., Si, L.L., Xu, H., Zhang, C.L., Xia, Q., Xiao, S.L., Wang, Q., He, Q.C., Chen, P., Wang, J.Y., Taira, K., Zhang, L.H., and Zhou, D.M. (2015). Broadening the versatility of lentiviral vectors as a tool in nucleic acid research via genetic code expansion. *Nucleic Acids Research* 43, E73–U42.

Zolotukhin, S., Potter, M., Zolotukhin, I., Sakai, Y., Loiler, S., Fraites, T.J., Chiodo, V.A., Phillipsberg, T., Muzyczka, N., Hauswirth, W.W., Flotte, T.R., Byrne, B.J., and Snyder, R.O. (2002). Production and purification of serotype 1, 2, and 5 recombinant adeno-associated viral vectors. *Methods* 28, 158–167.

Zufferey, R., Nagy, D., Mandel, R.J., Naldini, L., and Trono, D. (1997). Multiply attenuated lentiviral vector achieves efficient gene delivery in vivo. *Nature Biotechnology* 15, 871–875.

## Xenopus

Goldin, A.L. (1992). Maintenance of Xenopus laevis and oocyte injection. *Methods in Enzymology* 207, 266–279.

Gurdon, J.B., and Wickens, M.P. (1983). The use of Xenopus oocytes for the expression of cloned genes. *Methods in Enzymology* 101, 370–386.

Heikkila, J.J. (1989). Use of Xenopus oocytes to study the expression of cloned genes and translation of mRNA. *Biotechnology Advances* 7, 47–59.

Heikkila, J.J. (1990). Expression of cloned genes and translation of messenger RNA in microinjected Xenopus oocytes. *The International Journal of Biochemistry* 22, 1223–1228.

Hitchcock, M.J., Ginns, E.I., and Marcus-Sekura, C.J. (1987). Microinjection into Xenopus oocytes: Equipment. *Methods in Enzymology* 152, 276–284.

Krafte, D.S., and Lester, H.A. (1992). Use of stage II-III Xenopus oocytes to study voltage-dependent ion channels. *Methods in Enzymology* 207, 339–345.

Nowak, M.W., Gallivan, J.P., Silverman, S.K., Labarca, C.G., Dougherty, D.A., and Lester, H.A. (1998). In vivo incorporation of unnatural amino acids into ion channels in Xenopus oocyte expression system. *Methods in Enzymology* 293, 504–529.

Romero, M.F., Kanai, Y., Gunshin, H., and Hediger, M.A. (1998). Expression cloning using Xenopus laevis oocytes. *Methods in Enzymology* 296, 17–52.

Soreq, H., and Seidman, S. (1992). Xenopus oocyte microinjection: From gene to protein. *Methods in Enzymology* 207, 225–265.

Yang, X.C., Karschin, A., Labarca, C., Elroy-Stein, O., Moss, B., Davidson, N., and Lester, H.A. (1991). Expression of ion channels and receptors in Xenopus oocytes using vaccinia virus. *FASEB Journal* 5, 2209–2216.


## Equipment and supplies

*Some major suppliers*

The suppliers listed in **Chapter 2** are the primary suppliers of media, supplements (serum, antibiotics), transfection reagents, viral vector kits, and general cell-culture supplies. We list here only the suppliers unique to **Chapter 3**.

*American Type Culture Collection (ATCC)*: The source of cell lines, hybridomas, yeasts and fungi, bacteria, and viruses. They have almost everything.

*UK Stem Cell Bank*: Stem cells. May not be available for export.

*Clonexpress*: Human neuronal and glial precursor cells and cell lines.

*Marker Gene Technologies*: Substrates for marker genes such as β-galactosidase.

*Midsci*: Reagents and kits as well as tissue culture flasks and dishes, biosafety hoods, and cabinets.

*Labconco*: Many types of hoods and enclosures for cell culture, chemistry, nanoparticles, and more.

*Thermo Scientific*: Key supplier of hoods and incubators.

*Biorad*: The supplier of the gene gun.

*Chemicell*: Kits and magnets for magnetofection.

*Eppendorf*: Microinjectors, both manual and automated.

*Applied Viromics*: Viral vectors—adenovirus, AAV.

*Addgene*: Source of lentiviral plasmids. Guide to lentiviral vectors here: https://www.addgene.org/viral-vectors/lentivirus/lenti-guide/

*Clontech*: Viral vectors—adenovirus, AAV, lentivirus.

# CHAPTER 4

# Advanced Topics in Molecular Biology

## 4.1 INTRODUCTION

Using restriction enzymes, ligase, and simple polymerase chain reaction (PCR) is sufficient for many problems but becomes unwieldy or even impossible under certain situations. If the insert to be cloned is too large to purify using standard gel techniques, or if it is unstable or recombines, some other approaches can make it tractable.

Recent techniques have simplified the approach to genetic screening, cloning, and mutagenesis in single cells, cell populations, and animals. Most notably, CRISPR is a rapidly emerging technique for whole-genome editing with a myriad of applications.

The goal of this chapter is to give a practical introduction to some of these techniques so that interested readers will know where to search further. While detailed discussions of all of the techniques are outside the scope of this book, I try whenever possible to give object lessons from real laboratories in order to give an idea of the relative difficulty of different approaches.

## 4.2 CLONING TECHNIQUES FOR LARGE CLONING PROBLEMS AND MULTIPLE INSERTS

### Phage vectors

A *bacteriophage* or phage is a virus that infects bacteria (**Figure 4.1**). The λ phage is a specific phage of *Escherichia coli* that has a significant number of applications in molecular cloning. In this section, we will only discuss its use as a cloning vector; some other applications will be mentioned in later chapters. While λ is not the only phage used as a cloning vector, it is the most common.

Phage are grown by simply incubating them with the appropriate strain of *E. coli*. The mixture can then be mixed with agar and plated onto a petri dish. Empty areas in the bacterial lawn will appear, which are called *plaques*; each contains a genetically identical population of phage, in the same way that bacterial colonies contain genetically identical cells (**Figure 4.2**).

The way phage may be used as cloning vectors is to take advantage of the way they replicate their genomes, and of the fact that most of the genome of a phage can be removed without affecting its ability to make viral particles. Phages have

**Figure 4.1  Life cycle of lytic bacteriophage λ in _E. coli_**. (a) The phage consists of a protein coat and a DNA genome. (b) The viral particle binds specifically to its bacterial target and injects its DNA. (c) Once inside the cell, the DNA circularizes and encodes for proteins to make new viruses and for copies of itself. (d) The DNA and proteins are packaged into new viral particles. (e) The bacterial cell lyses, releasing many phage particles.

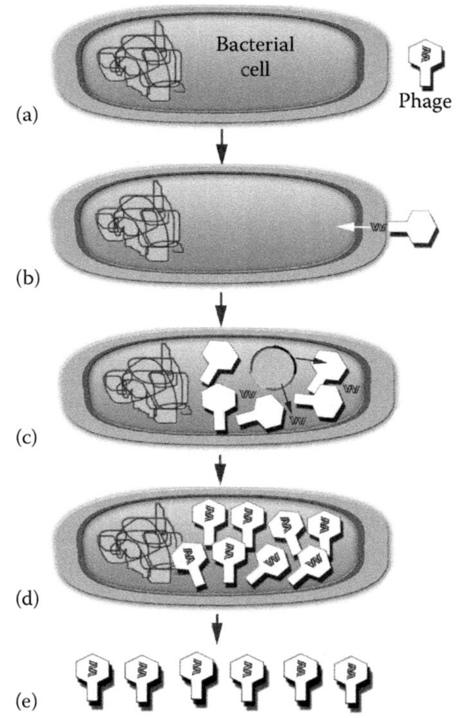


**Figure 4.2  Preparing phage.**

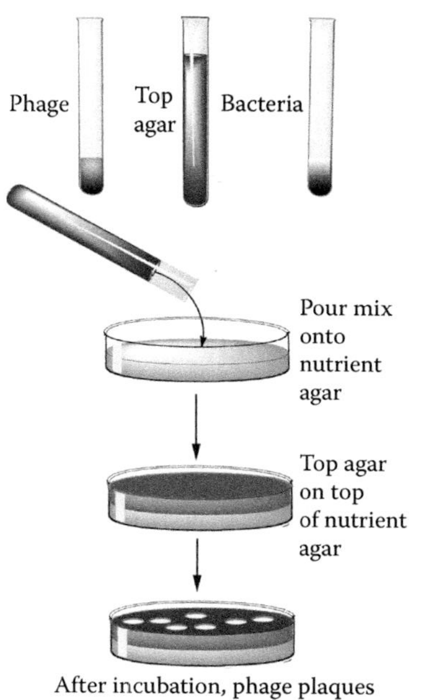

genomes of 38–53 kb in size, or up to five times the size of a typical plasmid vector. This DNA is packaged inside the head of the virus, and if it is significantly larger or smaller than the wild-type genome, the phage probably will not grow. Commercial suppliers thus provide phage DNA as the essential *arms* plus the nonessential *stuffer* containing restriction enzyme sites. Digestion of the stuffer with the appropriate enzymes allows insertion of DNA to be cloned (**Figure 4.3a**). Ligation results in a long concatemer of DNA attached by cleavable cos sites. This long strand is cleaved by specific packaging protocols before being packaged into phage (**Figure 4.3b**). The phage are then grown on a plate, and the plaques are screened with antibodies or labeled DNA probes (**Figure 4.3c**).

This type of replication and screening is easier, stabler, and more efficient than handling the corresponding number of plasmid vectors. Phage infect *E. coli* much more efficiently than plasmids transform them. The advantages of λ vectors come, therefore, in situations when many DNA fragments need to be cloned, screened, and stored for long periods. An example of such an application is in the creation of *DNA libraries*, which are many genes or genome fragments from an organism representing a significant fraction of the genome or *transcriptome.*

For large-scale production of DNA or for expression of proteins, plasmids are superior to phage in both efficiency and ease. Modern kits have allowed researchers to have the best of both worlds by creating λ vectors that can undergo *in vivo excision* to become plasmids. A special strain of *E. coli* in which the phage genome cannot replicate is used to grow the plasmid from one selected plaque or from a mass of different plaques. The resulting plasmids can be bacterial or mammalian expression vectors.

*Phage display* is a technique where a gene or part of a gene is fused to the gene for the phage coat protein in the expression vector. When the gene is transcribed, the protein appears on the phage surface. This allows for efficient high-throughput

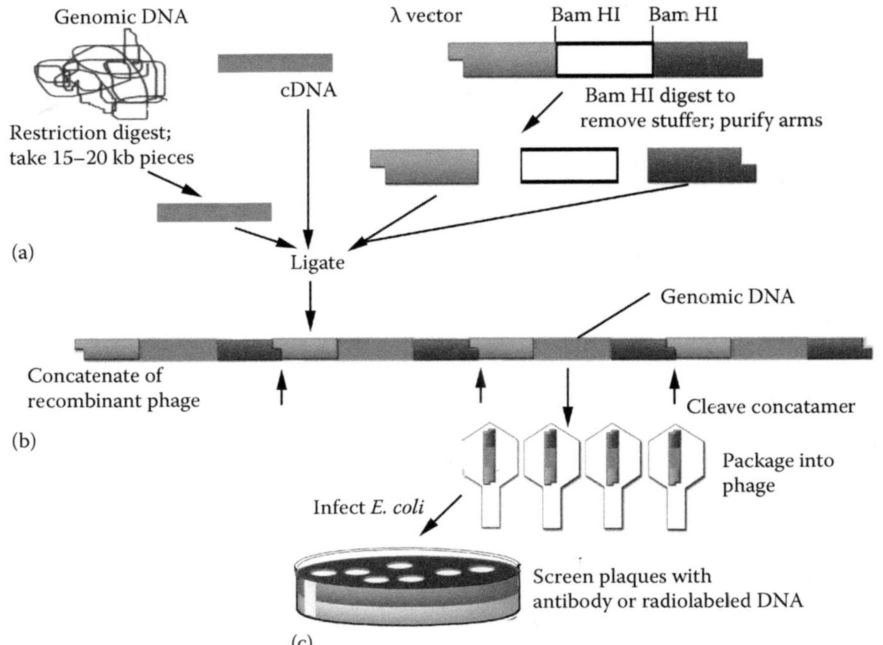


**Figure 4.3  Using phage λ as a cloning vector.** (a) Most commercial vectors have a unique restriction site that releases the stuffer, commonly Bam HI; sometimes there are two different sites so that the arms are incompatible with the stuffer once digested. The genomic DNA is then partially digested with any enzyme that creates Bam HI–compatible ends; Sau3A I is often used as it recognizes a 4-base-pair sequence and so occurs frequently. (b) Ligation leads to a concatemer of phage genomes with the structure left arm–genomic fragment–right arm. When the concatemers are cleaved and packaged, they can be used to infect *E. coli*. (c) Each plaque then contains a genetically identical population of phages containing a specific fragment of the inserted DNA.

screening of the phenotypes of expressed proteins—for example, binding of drugs to expressed antibodies. Many companies offer phase display library construction and screening for in-house or custom libraries.

## Cosmids

A hybrid of phage and plasmid technology is the *cosmid*, which is a plasmid containing phage cos sites so that it can concatemerize to form libraries (**Figure 4.4**). Its cloning capacity is three to five times greater than that of a plasmid, since it does not include any of the arm DNA that leads to the production of viral particles; the entire phage genome space can be taken up by the insert. Replication incompetent *helper* phage, which are essentially viral empty shells, are used to package the concatemer DNA and infect *E. coli*. This DNA then replicates in the bacteria like a plasmid. No viral particles are produced, and no plaques form. This type of vector is excellent for large inserts (30–40 kb).

## Bacterial artificial chromosomes and yeast artificial chromosomes

A bacterial artificial chromosome (BAC) is a solution for very large cloning projects. Based upon the F' (for "fertility") plasmid, an unusual vector that *E. coli* use to exchange DNA, a BAC is essentially a plasmid vector that can hold up to 700 kb of insert DNA, or about 70-fold that of a normal plasmid. BACs are handled in much the same way as ordinary plasmids, although it is important that the *E. coli* in which they are replicated be deficient in homologous recombination machinery, so that the plasmid does not recombine (a so-called recA⁻ strain). Transfection efficiencies of BACs into competent cells are low, as are yields of total DNA in midipreps and maxipreps. To improve yield, *E. coli* should be grown in rich medium (2× YT), and about twofold larger culture volumes should be used

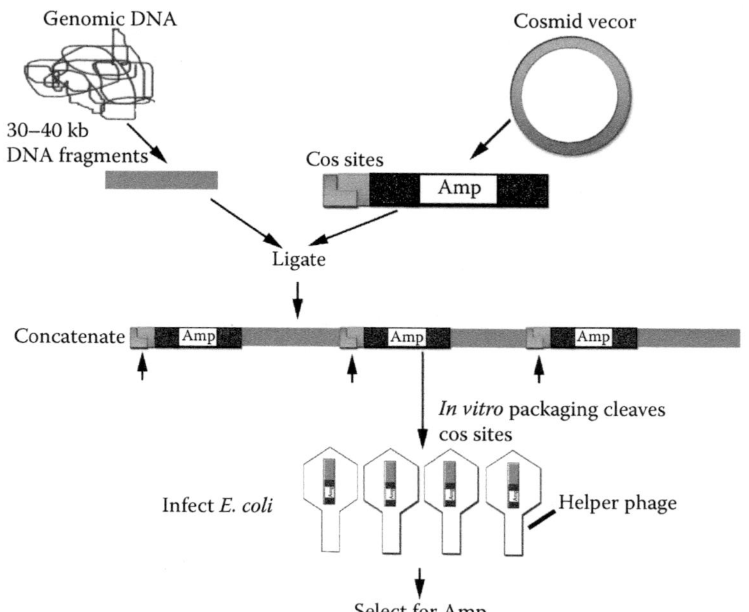


**Figure 4.4 Cosmids.** The cloning steps are the same as with phage vectors, but no plaques of phages are produced; instead, the cosmid is replicated in the *E. coli* like a plasmid.

than for ordinary plasmid preps. Because of the large size of the insert, ordinary agarose gels will not resolve fragments well. A technique called *pulsed-field gel electrophoresis*, where the voltage is periodically switched in direction, is needed to separate pieces larger than ~20 kb. This requires specific instrumentation and special agarose.

There are some particular cases in quantitative biology in which very large inserts are desirable. Perhaps the most important is *genome mapping* of complex genomes (from bacteria to humans). The genome is broken into pieces that are cloned into BACs and then sequenced, and the sequences rearranged using *genome assembly* algorithms. BAC libraries are increasingly becoming available; the National Institutes of Health (NIH) recently funded the BAC resource network to generate at least 10 genome-sized libraries.

BACs may also be used to study genetic regulatory elements, such as promoters, in their native configuration. These elements may often be many kilobase pairs upstream or downstream from the gene, so that very large fragments are needed to capture them all. Some BACs have been modified to permit easy rearrangement of pieces without the need for restriction enzymes.

Yeast artificial chromosomes (YACs) are based upon the large eukaryotic genome of *Saccharomyces cerevisiae* and may be megabases in size. YACs contain all three essential structural components of a yeast chromosome: an autonomously replicating sequence required for replication, a *centromere*, and two *telomeres*.

For such large inserts, restriction enzyme cloning would be nearly impossible—but it is not necessary. What is special about *S. cerevisiae* is that its recombination pathways for repair of double-strand breaks are very efficient, so recombinant molecules may be effectively generated by homologous recombination.

## 4.3 MULTIPLE MUTAGENESIS: WHEN POINT MUTATIONS ARE NOT ENOUGH

Making point mutations is tedious, and when many selected ones are desired—for example, for codon optimization—the easiest procedure is to determine the desired sequence and have the DNA synthesized by a commercial supplier. Prices for synthesis are dropping rapidly, and the selected DNA can be delivered in a plasmid of choice.

Another approach to making multiple mutations is to use the techniques of *mutagenic PCR* and *directed evolution*. If there is not enough information available about the target protein to identify potentially beneficial mutations with confidence, such as in the case of complex or novel proteins or if the target protein must be extensively modified for its intended purpose, then many hundreds of random mutations need to be generated and screened. Directed evolution involves methods for introducing random mutations to a target gene at a controlled rate, resulting in a sizable library of candidate mutant genes, along with methods for retrieving the desired mutants. In this way, directed evolution emulates natural evolutionary processes—natural selection accelerated to a fraction of the time. An advantage of directed evolution is that no prior knowledge of the protein's structure or function is required. The creation of an overwhelming library of unexpressed mutant genes in itself does little to benefit the researcher, but by ligating the diverse library into the appropriate expression vectors, they can be expressed, and monoclonal populations can be screened and isolated. The process can then be iterated, each time selecting the best clones from the previous round of best clones. High-throughput screening techniques are essential for the success of this process, due to the requisite sorting and selection from libraries of thousands of candidate mutant proteins.

The process of directed evolution begins by amplifying the gene of interest while concurrently introducing random mutations. This can be done through mutagenic PCR (an *in vitro* process) or *in vivo* through the use of bacteria or mammalian cells. Mutagenic PCR is perhaps the simplest method and allows the introduction of mutations at a rate that varies depending on reaction conditions such as the concentration of Mn and pH. It can be performed using the same polymerase for standard PCR reactions but, by tweaking the reaction conditions necessary for the proper function of the polymerase, cause it to make occasional "mistakes" in replication. To encourage nucleotide misincorporation, a higher ratio of two nucleotides is used: for example, five times as much CTP and TTP as ATP and GTP. Mutagenic PCR is straightforward to perform, as kits are available from major suppliers that include simple recipes for fixed rates of mutation, i.e., $x$ number of mutations per 1000 base pairs per cycle of amplification. Propagated through many cycles, a library of many thousands of mutant genes results. Of course, each target will be different, and the procedure will most likely need to be optimized for each new experiment.

Mutations in the nucleic acid sequence of the gene result in deviations from the wild-type amino acid sequence, which in turn determine the protein's structure and, subsequently, its function. While some mutations of this variety will cause undesirable loss of function in the protein, others will ideally result in enhanced function or novel activity. Most experts suggest that it is desirable to aim for 30% inactive clones in a mutant library, though this figure varies depending on target gene and targeted property.

Another complementary technique is known as *DNA shuffling* and involves the combination of different pieces of successful mutant genes in order to further enhance them. References to this technique are given at the end of the chapter. With the ability to evolve powerful molecular machines for enhanced function or novel uses, the way is paved for many exciting opportunities in protein design.

## 4.4 REVERSE TRANSCRIPTASE PCR AND QUANTITATIVE REAL-TIME PCR

### Reverse transcriptase PCR

DNA is generated from RNA using *reverse transcriptase PCR*, or rt-PCR. The enzyme *reverse transcriptase*, which evolved in retroviruses, is used to transcribe the RNA template into DNA. DNA primers then allow the target sequence to be amplified many times as in ordinary PCR. Modern methods allow both reactions to take place in a single tube using both enzymes (reverse transcriptase and DNA polymerase) in the same buffer and with the same site-specific primers. It is also possible to separate the rt step from the PCR step for increased flexibility for use in specialized applications such as rapid amplification of cDNA ends (RACE) or differential display reverse transcription (DDRT). These techniques are outside the scope of this discussion, but references are provided at the end of the chapter for many sophisticated PCR and rt-PCR applications.

## Quantitative real-time PCR

Quantitative real-time PCR (also called Q-PCR, RTQ-PCR, and qPCR) is a type of PCR that quantifies the amount of nucleic acids in a sample (either DNA or RNA) while amplifying a defined sequence of interest, called the amplicon. This quantification is based on a fluorescent reaction that occurs every time a nucleotide is added during the annealing portion of the PCR reaction. This differs from traditional PCR, which amplifies a product that can be roughly quantified only as an end product. Real-time detection allows for more accuracy, with precision up to a twofold DNA concentration change.

Quantitative PCR can be used for fast detection of a given gene, for example, to determine if a certain bacterial strain or species is present in an environmental sample. It is also useful in gene expression studies, to observe if expression levels change for a given environmental condition. It is especially useful in environmental microbiology, where it is much faster than culture-based techniques and also allows for the monitoring of bacteria that cannot be cultured.

A qPCR assay must first be validated for a given set of primers and template. This entails finding the optimal parameters for the best possible performance of the primer amplification. Specifically, you must determine the optimal annealing temperature, confirm binding specificity of the primer, and construct a standard curve to which future assays can be compared.

The first step in designing a validated qPCR protocol is to find the optimal annealing temperature for the given set of primers and template. This is done by running a temperature gradient protocol on the qPCR machine. This is crucial for primer specificity, as a temperature that is too low may result in nonspecific binding, whereas a temperature that is too high will reduce the final yield. During a temperature gradient, each reaction tube will be subjected to a different annealing temperature. This range should be designed to encompass temperatures both below and above the melting temperature $(T_\mathrm{m})$ of the primer. The optimal annealing temperature will result in the lowest number of cycles of PCR needed before the output signal crosses the threshold value, called $C_\mathrm{t}$. A low $C_\mathrm{t}$ means that the most efficient detection of amplicon occurs at this temperature, since the primer annealed at the lowest level of template.

In addition to a temperature gradient, the product specificity of the PCR reaction must be confirmed both by melt curve analysis and by running the end product on an agarose gel. These measures will confirm that the amplicon is indeed the only sequence being amplified in the reaction. If one product is being amplified, the melt curve will have a single peak, which should correspond to a single product on the gel.

The next step in protocol design is to create a standard curve against which future samples will be compared. **Figure 4.5** is a serial dilution curve for a specific gene sequence showing relatively poor sensitivity—the 10-fold dilutions show unequal spacing and even some overlap. (The threshold is chosen at the linear part of the curve to avoid artifacts.) Optimization of qPCR to improve these results can be tricky. If primer sequences can be changed, this is an option. If they cannot, then reaction conditions and temperatures need to be optimized for maximum resolution and reproducibility. Efforts spent to debug qPCR in the beginning of the experiment have significant payoff later on!

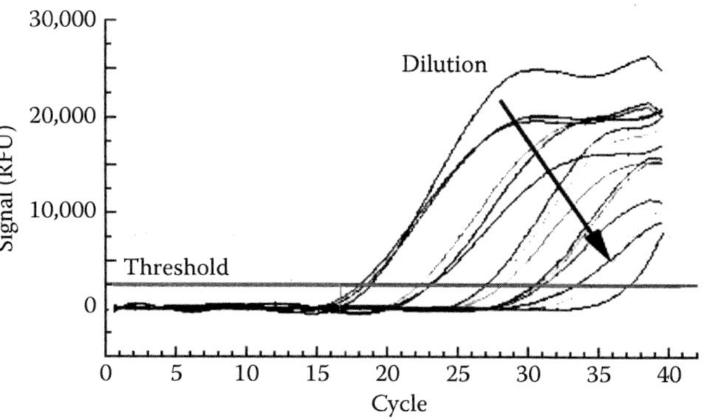


**Figure 4.5 qPCR.** qPCR standard curve showing serial 10-fold dilutions. The *y*-axis is the signal, and the *x*-axis is the cycle of PCR. The cycle at which the signal appears over threshold is plotted versus the number of copies of the target sequence present to give a standard curve against which an unknown sample may be compared.

The techniques of qPCR are also applicable to rt-PCR, called rt-qPCR. (It is important not to confuse "real-time" and "reverse transcriptase," a common error!) rt-qPCR is used to quantify RNA rather than DNA and so is useful as a measure of gene expression as well as for measuring *microRNAs*, which may be biomarkers of diseases such as cancer. Some references to the applications of this technique are given at the end of the chapter.

A new approach to DNA quantification, enabled by modern microfluidics, is called *digital PCR* (dPCR). In dPCR, a sample is diluted into many individual PCR reactions, called *partitions*. Some of the partitions are positive, containing the target molecule, and others are negative. During amplification, dye-labeled probes indicate the target sequence; negative partitions contain no signal. After the PCR is run, the fraction of negative reactions is used to generate a count of the number of target DNAs in the sample. This eliminates the need for standards, can indicate small changes, and works in complex mixtures. To perform dPCR, a special dPCR machine is required. These are available commercially. Specialized chips are used for each reaction.

## 4.5 MICROARRAYS

A microarray can be thought of as a more sophisticated, higher-throughput form of Southern blot. A typical microarray chip or "gene chip" consists of dozens to hundreds of probes immobilized on a solid surface or on beads. Commercially available arrays represent all of the genes expressed in a tissue (e.g., mouse leukocytes, human brain, bovine kidney) or an entire organism (bacterium, parasite). This allows for *gene expression profiling* in samples from the corresponding tissue or organism. Fluorescently labeled messenger RNA (mRNA) or cDNA is allowed to hybridize with the array; the degree of hybridization is measured by the brightness of the fluorescence, giving a semiquantitative measure of gene expression for all of the genes in the array (**Figure 4.6**).

There are many possible experiments in which this is desirable; a few examples are

- To monitor changes in gene expression during stages of development of an organism

- To monitor gene expression in cancer cells, in order to distinguish different types of cancer for prognostic or therapeutic aims, and understand cancer progression

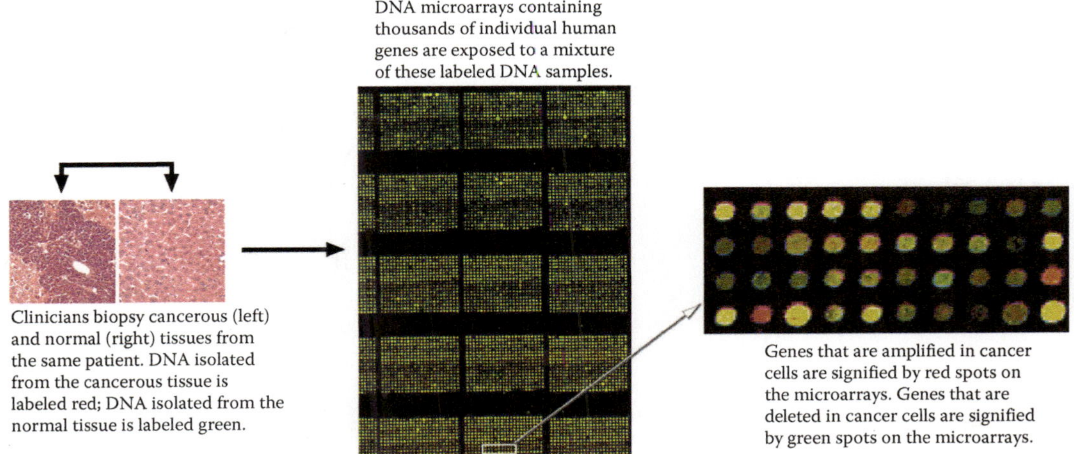



**Figure 4.6 Microarray.** An example of a 40,000 probe spotted oligo microarray with an enlarged inset to show detail. (Image courtesy of MolecularStation.)

- To identify genes in pathogenic organisms that vary in expression with life cycle or upon exposure to drugs, in order to find possible new targets for drugs

- To identify pathogens in environmental or tissue samples without the need for culturing

- To quantify the effect of *knockout* or suppression of a single gene on many others

Some references at the end of the chapter are provided illustrating each of these examples.

Along with commercially available arrays, custom arrays are also available containing probes of the experimenter's choice. *Array synthesizers* have also become reasonably priced in recent years, allowing laboratories to spot their own arrays. The choice of genes to probe can be made using genomics databases; some of these databases represent ongoing projects to classify genes according to functional categories, particularly in the human and mouse.

More recently, protein microarrays have been developed for studying the *proteome*. The human genome's 20,000 coding genes result in at least 70,000 different proteins due to alternative splicing and posttranslational modifications. Standard protein arrays make use of full-size proteins, but these are difficult to generate and have short shelf lives; alternatives include protein fragments and peptides. Arrays may be generated from proteins produced in *E. coli* (called *purified protein arrays*) or from cell lysates using cells from the target organism (called *natural protein arrays*). *Cell-free protein arrays* can also be made, using *in vitro* transcription; this approach does not require purification and leads to greater stability. Protein arrays can be made using a DNA template array in this fashion; such arrays are available commercially. Protein microarrays are used to study protein–protein interactions, most often antibody reactivity to proteins that occurs in cancer, infectious disease, and *autoimmune* disorders.

**Figure 4.7 Mechanism of RNA interference.** Precursor RNA (pre-RNA) is exported to the cytosol and processed by Dicer to small RNA duplex. The guide strand of this duplex forms a complex called RISC to digest the target mRNA for degradation.

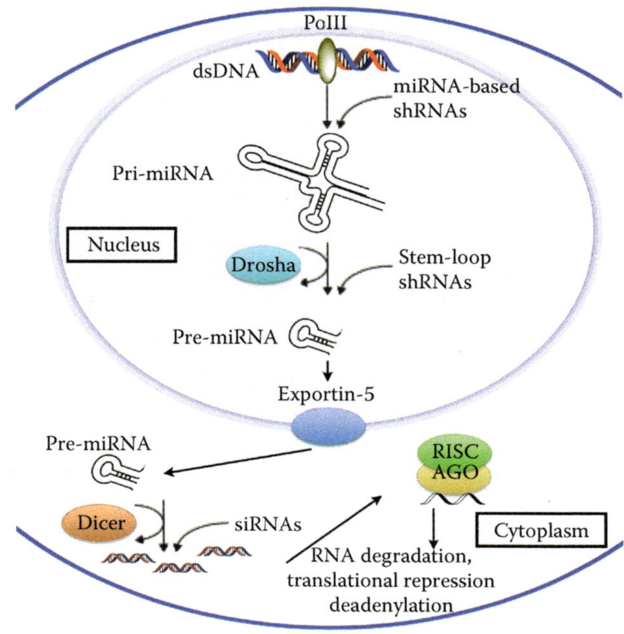


# 4.6 SMALL INTERFERING RNA*

## General principles

*RNA interference* was discovered in 1998 as an endogenous process in *Caenorhabditis elegans*. Specific short sequences of double-stranded RNA, called *small interfering RNA* (siRNA), were found to silence gene expression by degrading corresponding mature mRNA. The mechanism is shown in **Figure 4.7**. The siRNA is transcribed by polymerase II as stem-loop primary miRNA (pri-RNA). pri-miRNA is processed in the nucleus to precursor RNA (pre-RNA), which is later transported to the cytosol (mediated by the exportin-5 complex) and cut by Dicer to siRNA. The guide strand of the siRNA then complexes with proteins to form an *RNA-induced silencing complex* (RISC). When RISC finds and binds the complementary mRNA strand, it is activated to cleave the target mRNA, which downregulates gene expression.

This process was later found to be ubiquitous among eukaryotes. Only a few siRNA molecules are sufficient for full silencing. This has become a powerful tool in molecular biology, since short RNA sequences may be synthesized rapidly and inexpensively. Synthetic strands were first shown to work as siRNA in mammalian cells in 2001; since then, thousands of *in vitro* and *in vivo* siRNA sequences have been made.

There are three ways to introduce exogenous siRNA to cells: (1) direct transfection, (2) transfection of a plasmid containing the pre-miRNA sequence, and (3) infection by a virus coding a pre-miRNA sequence. The first two methods are transient, while the use of viruses permits long-term silencing. However, using viruses is more labor-intensive and takes longer than transient transfection, since the viral vectors must be constructed, grown, and purified before they are ready to use.

---

* This section was adapted from the PhD thesis of Xuan Zhang, McGill University, 2015.

One of the highly practical aspects of RNAi is that many RNAi libraries have been developed. Each library is composed of either plasmids containing pre-miRNA sequences or siRNA sequences ready for use. The libraries may cover the whole human or mouse genome, with three to five constructs from each gene, or may be targeted to a specific subset of genes. (Examples include ion channels, apoptosis, transcription factors, and many more.) Design of custom libraries is also increasingly available and inexpensive. If in the form of plasmids, the library is packaged into viruses and integrated into target genome. Tens of thousands of genes may be reliably targeted using this method.

## Example experiment: Mechanisms of drug resistance

Gold nanoparticles bound to doxorubicin (Au–Dox; see **Chapter 13**) have different effects on cells than doxorubicin alone. An siRNA screen may be used to identify genes that confer Au–Dox resistance on cancer cells. Cells are infected with viruses encoding siRNAs, and then a lethal dose of Au–Dox is applied. Survivors are collected and sequenced, and the knocked-down genes are identified.

Step 1 is to choose a library. Because Au–Dox resistance may be conferred by genes of many possible classes (examples might be particle trafficking, necrosis, cytoskeleton, and metabolic processes), a genome-wide screen was used. Because the cell type used in the experiments was B16, a murine cell line, the genome-wide siRNA library should come from mouse. Taking these aspects into consideration, we chose a second-generation *short hairpin RNA* (shRNA) library developed by the Elledge and Hannon laboratories. shRNAs are artificial RNA molecules with tight hairpin turns that can be used in RNAi experiments. Rather than transient transfection of the RNAi directly into cells—which does work, but is somewhat tricky and only lasts a few days—stable expression of inhibiting RNAs can be achieved using viral expression. In this case, the RNAs are delivered as DNA inside ordinary plasmid vectors and then transcribed. This facilitates the construction of plasmid libraries and also permits delivery via viral vectors, which can lead to stable expression of the RNAs for the lifetime of the cells.

The particular library we chose is composed of more than 85,000 plasmid constructs targeting 30,000 mouse genes. Each construct is stably integrated into cultured cells using a *self-inactivating retroviral vector*, pSM2. The transcription of priRNA is driven by U6 promoter, and a mammalian selectable marker (puromycin) is included to allow for selection of infected cells. The subsequent knockdown of gene expression is achieved through the endogenous RNAi machinery. The plasmids contain a *bar code* so that the integrated sequences may be readily identified in the genome by sequencing (**Figure 4.8**).

A general flowchart for the siRNA experiment is shown in **Figure 4.9**. B16 cells were seeded onto 10 cm dishes the day before infection at a density that permitted efficient lentiviral infection without cell overgrowth. Four dishes were used per experiment: three were treated with virus cocktail (containing a random mix of genes), and one dish was mock-infected (medium only) as control. Puromycin selection was carried out for 5 days to allow the genes time to express; cells were passaged into 15 cm dishes and treated with Au–Dox at the EC50 for 24 h. The surviving cells were left growing on the same plate under puromycin selection for a week to form colonies. The cells from each colony were then collected separately, and the DNA was extracted using a blood and tissue DNA extraction kit.

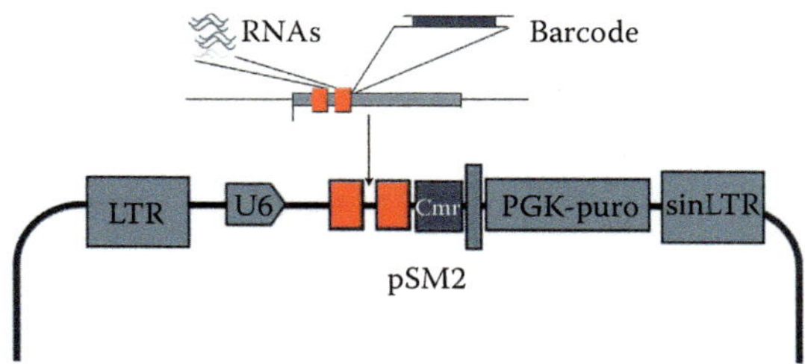



**Figure 4.8 Vector for siRNA experiments.** A selectable marker for mammalian cells (puromycin, puro) is under the control of the PGK promoter. The U6 promoter drives expression of the bar-coded RNAs. The LTRs are the long terminal repeats of the lentiviral backbone. The plasmids also contain bacterial selectivity markers for amplification (not shown).

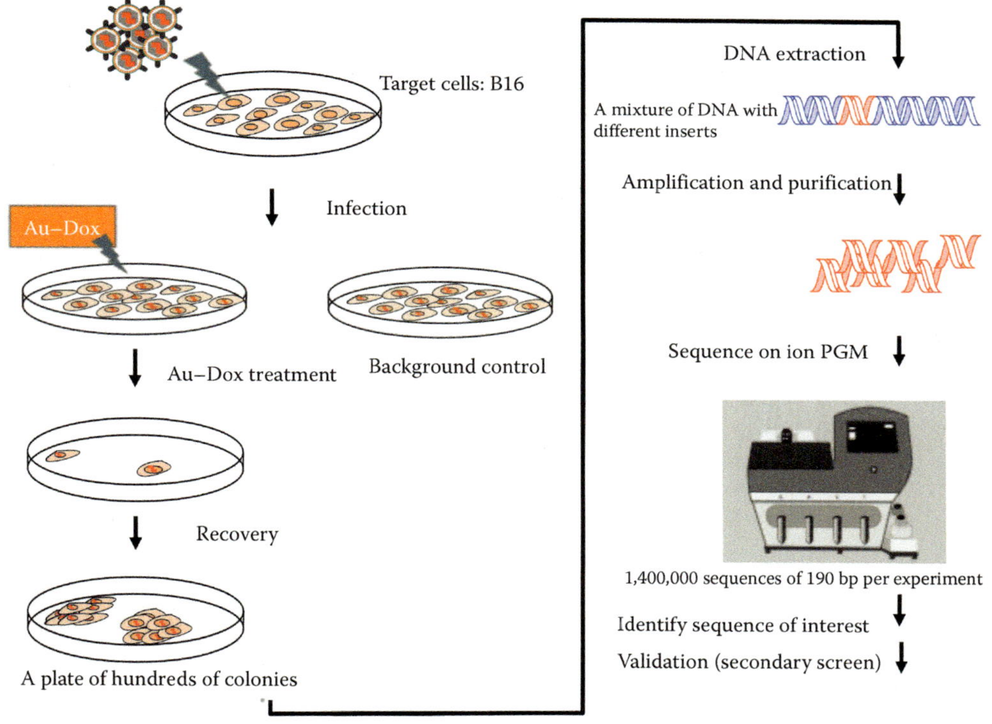



**Figure 4.9 Genome-wide screening using an siRNA library.** The genetic information of shRNA was integrated into the genome of cultured B16 cells via retroviral infection. The cells were then subjected to Au–Dox treatment at a predetermined concentration, and the surviving cells were collected after colony formation. DNA was extracted for amplification of integrated genetic information. Sequences were then read by the Ion Personal Genome Machine.

In order to determine which genes from the library were represented in surviving cells versus controls, shRNA inserts were recovered by PCR with genomic DNA from both nontreated cells and those treated with Au–Dox. Different amounts of genomic DNA were tested, and the number of amplification cycles was optimized to determine the best PCR conditions. The PCR products were purified and quantified using a double stranded DNA (dsDNA) dye specific for the purpose (Quantifluor), and then 100 pg of DNA (~$1.2 \times 10^8$ molecules) was subjected to enrichment and used for sequencing.

The sequencing results were aligned with the original library and extracted for construct information. All three replicates were pooled together. Due to the high occurrences of false-positive effects of siRNA experiments, only the genes represented by multiple shRNA constructs (≥3) were considered. These are shown in Table 4.1.

## Data analysis

A variety of genomics tools are available to assist in interpreting the results of the screening. Some of the ways in which they may be used are illustrated here using our example experiment.

The Database for Annotation, Visualization and Integrated Discovery (*DAVID*) was used first to rank the overall importance of gene groups represented by the genes in Table 4.1. Two clusters that grouped similar annotation terms were found to be significantly enriched, including one protein cluster associated with the sterile alpha domain (SAM) and another one associated with nuclear chromosomes and non-membrane-bounded organelles. However, no individual gene was significantly enriched.

The results were then examined using *PANTHER* (Protein Analysis through Evolutionary Relationships), a classification system for proteins and their genes. Proteins are classified according to family (evolutionarily related proteins) and subfamily (proteins in the same family with similar functions); molecular functions (for example, phosphatase); biological processes at the cellular or organismal level (for example, apoptosis); and pathway (which specifies relationships between interacting molecules). A free account allows data upload. The results of PANTHER analysis of the genes in Table 4.1 are shown in Figure 4.10. Of the 72 genes in Table 4.1, 59 were recognized by PANTHER. By the protein family (Figure 4.10a), transporters (15.2%) and nucleic acid binding proteins (13.6%) were the major groups. Enzyme modulators, hydrolases, and receptors were each 9.1% in this category. By molecular function (Figure 4.10b), proteins that had binding (29.7%), catalytic (28.1%), or transport (17.2%) activity dominated. When categorized by biological process (Figure 4.10c), most of the proteins were involved in metabolism (29.9%) and cellular processes (16.5%). It is noteworthy that localization (15.5%), which covered almost all the transporters categorized by the protein family, was the third group of this category. It confirmed the importance of transportation. Interestingly, 17 pathways (Figure 4.10d) were identified by the list, but none of them was significantly enriched.

To gain further insights into the protein–protein interaction network, Ingenuity software's Pathway Analysis (IPA) was used. This is a commercial, web-based software that allows for complex omics analysis of data sets; use of the tools may be purchased by the data set. We used IPA to identify pathways involved in Au–Dox resistance as well as possible *in vivo* toxicity profiles of Au–Dox. The first trial run was performed using the 72 genes in Table 4.1, of which 65 were recognized. Top canonical pathways identified were mitotic roles of polo-like kinase and semaphoring signaling in neurons. Meanwhile, tumor necrosis factor was identified as the top upstream regulator, leading the top networks (Figure 4.11a), that both involved inflammatory response and cell death and survival. The top toxicity that could be caused by Au–Dox was suggested to be increased liver steatosis.

**Table 4.1**

Genes Identified in Preliminary Screen as Possible Contributors to Au–Dox Resistance

| Gene (CSHL) | Occurrence | Current Accession |
|---|---|---|
| Mus musculus a disintegrin-like and metalloprotease | 11 | NM_175496 |
| Mus musculus solute carrier family 25 | 8 | NM_175194 |
| Mus musculus protein phosphatase 1, regulatory (inhibitor) subunit | 7 | NM_011625 |
| Mus musculus transient receptor potential cation channel, subfamily | 7 | NM_138301 |
| Mus musculus potassium inwardly-rectifying channel, subfamily J | 6 | NM_008426 |
| Mus musculus budding uninhibited by benzimidazoles 1 homolog | 5 | NM_009772 |
| Mus musculus UDP-Gal:betaGlcNAc beta 1,4-galactosyltransferase, | 5 | NM_020579 |
| Mus musculus nudix (nucleoside diphosphate linked moiety X)-type | 5 | NM_027722 |
| Mus musculus gamma-aminobutyric acid (GABA-A) receptors subunit | 4 | NM_008066 |
| Mus musculus gamma-aminobutyric acid (GABA-A) receptor, subunit | 4 | NM_008067 |
| Mus musculus interferon-induced protein with tetratricopeptide | 4 | NM_008332 |
| Mus musculus dual-specificity tyrosine-(Y)-phosphorylation | 4 | NM_010092 |
| Mus musculus kinesin family member 11 (Kif11), mRNA | 4 | NM_010615 |
| Mus musculus tyrosine 3-monooxygenase/tryptophan 5-monooxygenase | 4 | NM_011740 |
| Mus musculus adducin 3 (gamma) (Add3)s mRNA | 4 | NM_013758 |
| Mus musculus solute carrier family 24 (sodium/potassium/calcium) | 4 | NM_133221 |
| Mus musculus pleckstrin homology domain containing, family K member | 4 | NM_133244 |
| Mus musculus acetyl-Coenzyme A dehydrogenases long-chain (Acadl) | 3 | NM_007381 |
| Mus musculus Eph receptor A8 (Epha8), mRNA | 3 | NM_007939 |
| Mus musculus flavin containing monooxygenase 3 (Fmo3)s mRNA | 3 | NM_008030 |
| Mus musculus a disintegrin and metalloproteinase domain 17 | 3 | NM_009615 |
| Mus musculus activation-induced cytidine deaminase (Aicda), mRNA | 3 | NM_009645 |
| Mus musculus caspase 1 (Casp1), mRNA | 3 | NM_009807 |
| Mus musculus interferon (alpha and beta) receptor 2 (Ifnar2), mRNA | 3 | NM_010509 |
| Mus musculus v-maf musculoaponeurotic fibrosarcoma oncogene family, | 3 | NM_010756 |
| Mus musculus similar to ribosomal protein L27 [Rattus norvegicus] | 3 | NM_011289 |
| Mus musculus ubiquitin specific protease 12 (Usp12), mRNA | 3 | NM_011669 |

*(Continued)*

**Table 4.1 (Continued)**

Genes Identified in Preliminary Screen as Possible Contributors to Au–Dox Resistance

| Gene (CSHL) | Occurrence | Current Accession |
|---|---|---|
| Mus musculus sema domains immunoglobulin domain (Ig), transmembrane | 3 | NM_013660 |
| Mus musculus Unc-51 like kinase 2 (C. elegans) (Ulk2), mRNA | 3 | NM_013881 |
| Mus musculus aldo-keto reductase family 1, member E1 (Akr1e1), | 3 | NM_018859 |
| Mus musculus elongation of very long chain fatty acids (FEN1/Elo2, | 3 | NM_019423 |
| Mus musculus tissue-type vomeronasal neurons putative pheromone | 3 | NM_019918 |
| Mus musculus C-type (calcium dependent, carbohydrate recognition | 3 | NM_019948 |
| Mus musculus SAR1a gene homolog 2 (S. cerevisiae) (Sara2), mRNA | 3 | NM_025535 |
| Mus musculus RIKEN cDNA 5730410I19 gene (5730410I19Rik), mRNA | 3 | NM_025666 |
| Mus musculus testis expressed gene 12 (Tex12)s mRNA | 3 | NM_025687 |
| Mus musculus proteasome (prosome, macropain) 26S subunit, | 3 | NM_025894 |
| Mus musculus down-regulator of transcription 1 (Dr1)s mRNA | 3 | NM_026106 |
| Mus musculus RIKEN cDNA 9130427A09 gene (9130427A09Rik), mRNA | 3 | NM_026240 |
| Mus musculus lipases gastric (Lipf), mRNA | 3 | NM_026334 |
| Mus musculus solute carrier organic anion transporter family, | 3 | NM_026413 |
| Mus musculus RIKEN cDNA 1810055E12 gene (1810055E12Rik), mRNA | 3 | NM_026437 |
| Mus musculus exosome component 6 (Exosc6), mRNA | 3 | NM_028274 |
| Mus musculus prestin (motor protein) (Pres), mRNA | 3 | NM_030727 |
| Mus musculus brain and acute leukemia, cytoplasmic (Baalc), mRNA | 3 | NM_080640 |
| Mus musculus serine/threonine kinase 17b (apoptosis-inducing) | 3 | NM_133810 |
| Mus musculus coiled-coil-helix-coiled-coil-helix domain containing | 3 | NM_133928 |
| Mus musculus DNA segment, Chr 10, University of California at Los | 3 | NM_133996 |
| Mus musculus protein phosphatase 1s regulatory (inhibitor) subunit | 3 | NM_138605 |
| Mus musculus Rho GTPase activating protein 4 (Arhgap4), mRNA | 3 | NM_138630 |
| Mus musculus myeloid/lymphoid or mixed lineage-leukemia | 3 | NM_139311 |
| Mus musculus olfactory receptor 843 (Olfr843), mRNA | 3 | NM_146567 |
| Mus musculus olfactory receptor 1248 (Olfr1248), mRNA | 3 | NM_146791 |
| Mus musculus solute carrier family 6, neurotransmitter transporter | 3 | NM_148931 |

(Continued)

**Table 4.1 (Continued)**

Genes Identified in Preliminary Screen as Possible Contributors to Au–Dox Resistance

| Gene (CSHL) | Occurrence | Current Accession |
|---|---|---|
| Mus musculus RIKEN cDNA 4930438M06 gene (4930438M06Rik), mRNA | 3 | NM_172721 |
| Mus musculus monoamine oxidase B (Maob), mRNA | 3 | NM_172778 |
| Mus musculus RIKEN cDNA 4732496O08 gene (4732496O08Rik), mRNA | 3 | NM_172877 |
| Mus musculus sex comb on midleg-like 4 (Drosophila) (Scml4)s mRNA | 3 | NM_172938 |
| Mus musculus RIKEN cDNA A630047E20 gene (A630047E20Rik), mRNA | 3 | NM_173032 |
| Mus musculus polo-like kinase 4 (Drosophila) (Plk4), mRNA | 3 | NM_173169 |
| Mus musculus TGFB-induced factor 2 (Tgif2), mRNA | 3 | NM_173396 |
| Mus musculus RIKEN cDNA 2700007P21 gene (2700007P21Rik)s mRNA | 3 | NM_173750 |
| Mus musculus RIKEN cDNA 1810063B05 gene (1810063B05Rik)s mRNA | 3 | NM_174987 |
| Mus musculus RIKEN cDNA 3830422N12 gene (3830422N12Rik), mRNA | 3 | NM_174993 |
| Mus musculus RIKEN cDNA A130007F10 gene (A130007F10Rik), mRNA | 3 | NM_175539 |
| Mus musculus RIKEN cDNA 9330155M09 gene (9330155M09Rik), mRNA | 3 | NM_177254 |
| Mus musculus RIKEN cDNA B130016O10 gene (B130016O10Rik), mRNA | 3 | NM_177355 |
| Mus musculus Scm-like with four mbt domains 2 (Sfmbt2), mRNA | 3 | NM_177386 |
| Mus musculus formin homology 2 domain containing 1 (Fhod1), mRNA | 3 | NM_177699 |
| Mus musculus potassium channel tetramerisation domain containing 12 | 3 | NM_177715 |
| Mus musculus ATP-binding cassette, sub-family A (ABC1), member 13 | 3 | NM_178259 |
| Mus musculus synovial sarcoma translocation gene on chromosome | 3 | NM_178750 |

Because 65 genes represented a relatively short list for analysis on IPA, the analysis was rerun with all of the genes that bore two or more constructs on the siRNA screen. The top canonical pathways changed in this case to transcriptional regulatory network in embryonic stem cells, macropinocytosis signaling, and natural killer cell signaling, while previously identified canonical pathways ranked 10th and 11th on the list. Interestingly, the top upstream regulator leading the top network (**Figure 4.11b**) in this list was MYD88, which was also a key factor in the top network in the previous analysis. Nervous system development and function and cellular growth and proliferation were also associated with the network. In addition to increased liver steatosis, increased heart failure was on the top toxicity list as well.

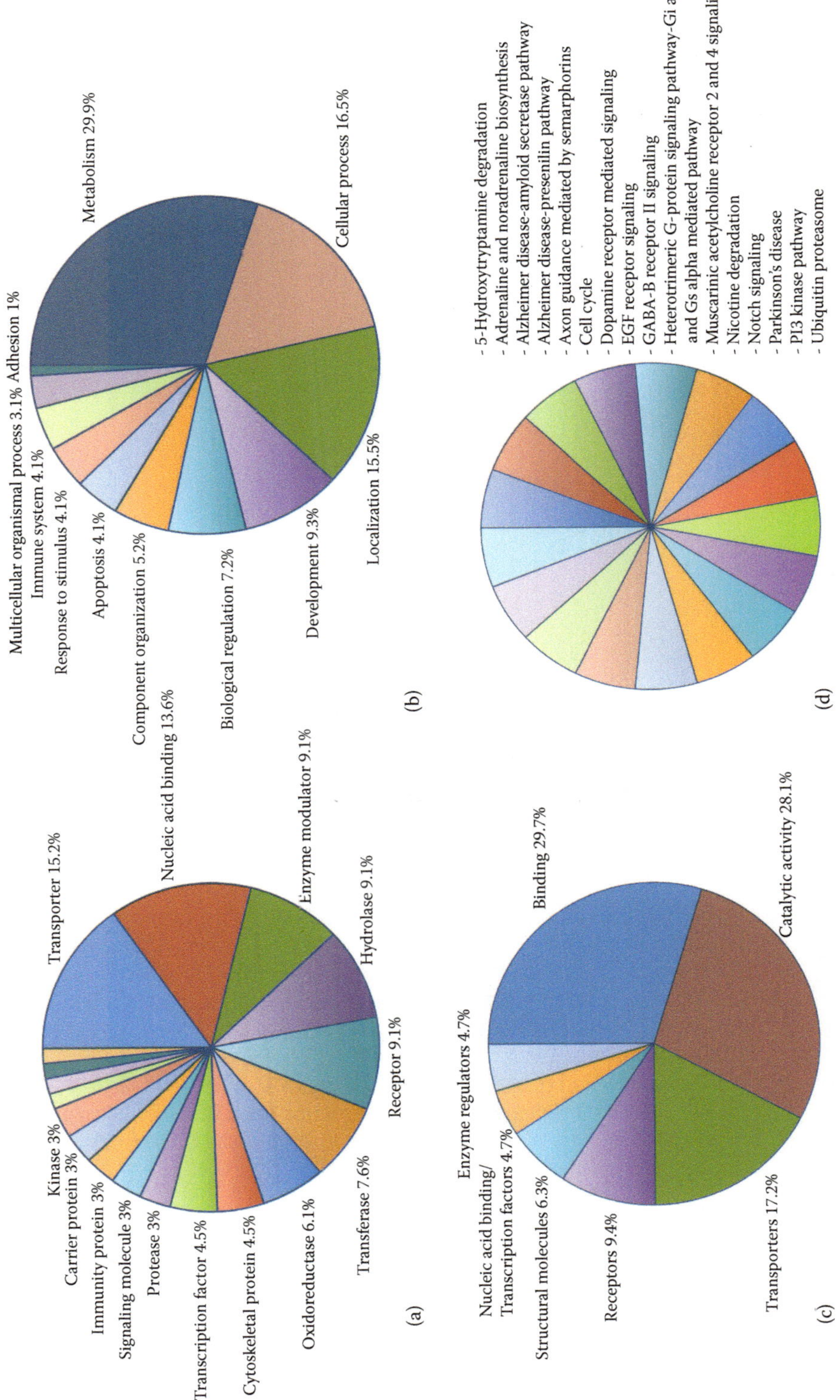


**Figure 4.10  Genes involved in the resistance of melanoma cells to Au–Dox classified using PANTHER by (a) protein family, (b) biological process, (c) molecular function, and (d) pathway.**

**Figure 4.11 Network map of validated screen hits produced by IPA.** (a) Using only the 72 genes of **Table 4.1**. (b) Using all of the genes that showed at least 2 hits in the siRNA screen.

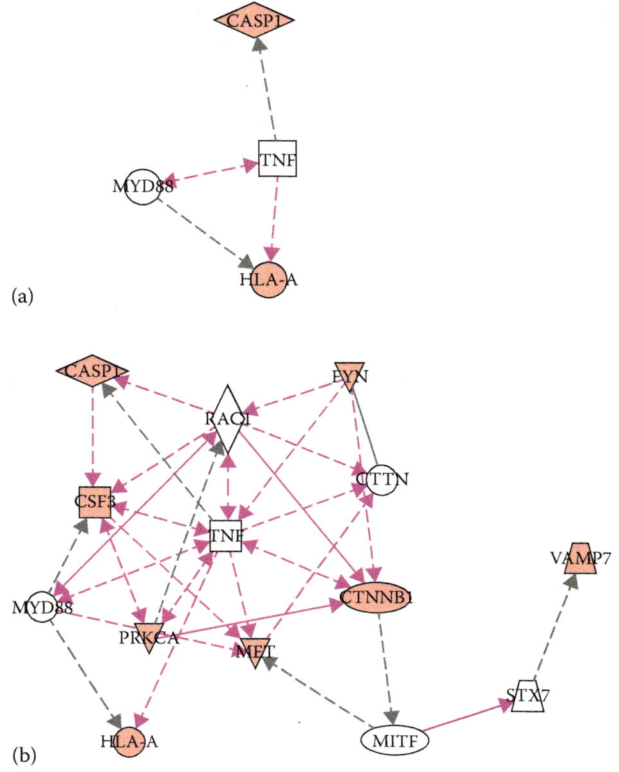


## Secondary screening

The greatest disadvantage of siRNA is off-target effects. Genes of similar sequences can be targeted by the same siRNA, giving the wrong gene identification. Therefore, a secondary screen must always be performed to validate the results. The results of the PANTHER, DAVID, and IPA analyses can be used to identify several (10–20) genes of particular interest. Custom siRNAs should then be ordered for those genes and transiently transfected into cells to verify that they do indeed confer resistance. This is different from library screening because in this case, transfection of the RNA is performed directly. There are certain tricks to doing this that will facilitate the task. Because RNAs are tricky to handle, and because so little is needed per sequence, the easiest way to do this screen is often to order a 96-well plate where each well contains a specific amount of a selected custom siRNA. Predefined plates are also available for those cases in which a gene family target has been identified. The RNA is transfected into cells using a transfection reagent specifically shown to work well for RNAi.

## 4.7  CRISPR

### General principles

*CRISPR* stands for *clustered regularly interspaced short palindromic sequences* and refers to arrays of short repeated DNA sequences separated by unique spacers. CRISPR is found on both chromosomal and plasmid DNA of prokaryotes and

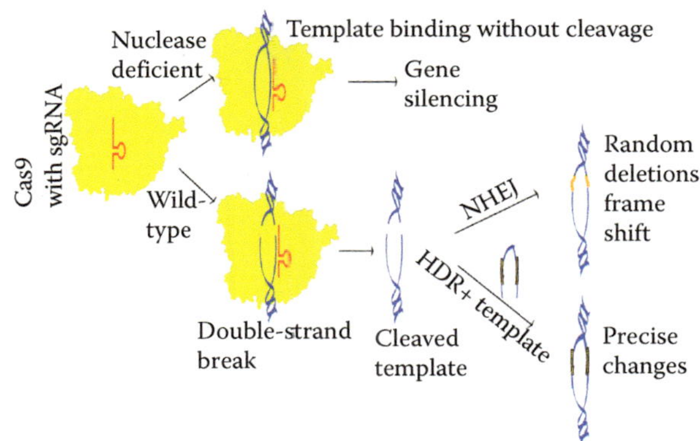


**Figure 4.12 Schematic of CRISPR genome editing.** All approaches start with a custom sgRNA and Cas9. Wild-type Cas9 creates double-strand breaks in the target DNA and then moves away. NHEJ repair mechanisms lead to random errors that may inactivate or knock out the gene. Providing a template for repair allows the HDR system to make precise insertions or knock-ins. Nuclease-deficient Cas9 does not cut the DNA but remains bound, inhibiting transcription and silencing the gene.

constitutes an *adaptive immune system* in these single-celled organisms, defending against viral infection by recognizing and destroying specific nucleic acid sequences with the aid of *CRISPR-associated proteins* (Cas), which act as *endonucleases*.

CRISPR sequences were discovered in 1987, but applications of CRISPR-Cas to genome editing only began to appear in 2013. Since then, there has been an explosion of literature on the topic, because CRISPR sequences may be used as a powerful tool to edit any genome in any organism in a wide variety of fashions: gene silencing, knock-ins, activation, and more. This approach can thus replace transfection or infection of cells with a mutant gene, generation of mutant mice using modification of embryonic stem cells, or the use of *zinc finger nucleases*.

How does it work? Many applications are based upon a specific Cas endonuclease, *Cas9*, which creates a single double-strand break in DNA. The precise site where the break is made can be selected by making a custom *single guide RNA* (sgRNA). The sgRNA is generated specifically for each experiment and has an approximately 20-base sequence specific to the target DNA. Once the sgRNA and the Cas9 are introduced into the cell, they locate the target and create a double-strand break. The repair mechanisms native to the eukaryotic cell are then exploited to achieve the desired result. If only the sgRNA and Cas9 are provided, the cell will repair the break with *nonhomologous end joining* (NHEJ), which often disrupts the gene by deleting a few base pairs. Alternatively, another piece of DNA can be delivered at the same time. If it has homology to the target, *homology directed repair* (HDR) will insert the precise sequence into the desired location, leading to the ability to add, modify, or delete specific base pairs as wanted. A *nuclease-deficient Cas9* (dCas9) is also available; it will bind the target and block transcription, thus silencing the gene, but will not cleave the DNA. dCas9 coupled to a fluorescent protein may be used for fluorescent tagging of specific genome sequences. **Figure 4.12** gives a schematic of all of these approaches.

The applications—real and envisioned—of CRISPR are immense. The following are simply a few examples:

- Correction of *monogenetic* diseases in animal models

- Genome editing of organisms that have been traditionally difficult to modify, from bacterial strains to large mammals

- Use of dCas9 in conjunction with chemical or light induction to control gene expression

- *Epigenome* editing to target complex conditions such as Down syndrome and cancer

- Targeting of cancer-related genes in animal models to inhibit tumor growth

- Studying basic cellular biological processes by editing specific genes in cultured cells, such as those affecting cellular organelles

- High-throughput screening of genomic functions by performing genome-wide knockouts in cultured cells

- Modifying crops and/or insect pests in order to reduce agricultural losses without the use of toxic pesticides

Some references to the latest review articles are given at the end of this chapter. Given the rapid rate of the field's development, many more will no doubt become available in the near future.

## Practical considerations

Cas9 is not specific to each experiment, although varying Cas9s from different organisms are available. The Cas9 may be delivered to cells as a DNA plasmid, as RNA, or as a protein. The sgRNA must be specific to each experiment. The sgRNA can be transfected into cells as RNA, or may be encoded on a DNA plasmid (**Figure 4.13**). When it is transfected as RNA, it is usually ordered as DNA oligonucleotides and

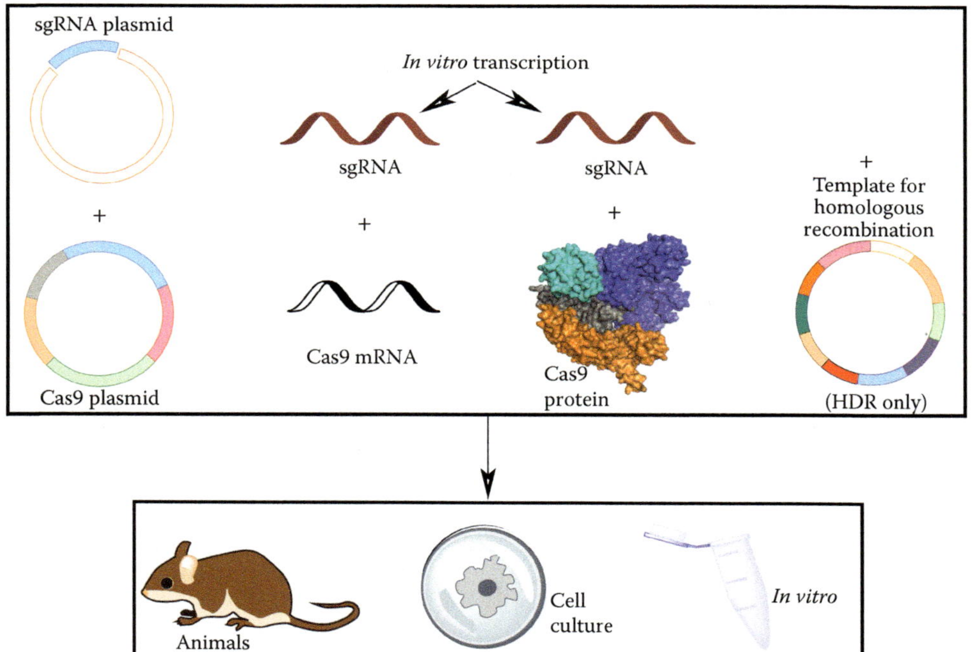


**Figure 4.13  How CRISPR is implemented.** An sgRNA may be delivered as a plasmid, or *in vitro* transcribed to RNA before delivering it to its target. Cas9 may be delivered as a plasmid, as RNA, or as protein. For HDR-based approaches, a template for homologous recombination must also be provided. These elements are delivered into cells or animals or used to edit DNA *in vitro*.

then generated by means of an sgRNA kit (*in vitro transcription*). The choice of which approach to use depends upon the application. For cultured cells where plasmid vectors with appropriate promoters are available, the simplest approach is to use a plasmid-based system. For cells that are difficult to transfect, Cas9/sgRNA may be encoded in a lentiviral or adeno-associated virus (AAV) vector; plasmids for production of these systems are available from plasmid repositories. However, many organisms do not have plasmid vectors with appropriate promoters. In this case, delivery of RNA/RNA or protein and RNA is necessary. These will lead to only transient genomic effects but have been used to edit the genomes of organisms that, until very recently, had never been genetically engineered at all.

The genomic target cannot be just any 20-base-pair sequence. It should ideally be unique in the genome and must be located immediately 5′ of a *protospacer adjacent motif* (PAM). This motif is required for target binding and depends upon the exact species of Cas9. Table 4.2 gives the motifs for the different Cas9s in common use. Cas9 will bind any sequence with a PAM but will only cut if the sgRNA sequence matches the target. Cutting at an undesired location is called *off-target activity* and is one of the key disadvantages of CRISPR. Software tools are available to help design sgRNAs by locating PAM and target sequences within a sequenced genome and predicting on- and off-target activity. Alternatively, a *validated* sgRNA that has already been used successfully may be selected; these are often available as plasmids immediately upon publication of the associated paper. Using validated sgRNAs can save a lot of time but of course reduces the originality of the experiment.

In order to use HDR, a repair template containing the desired edit flanked by a region of homology is required. The template can be RNA or a DNA plasmid (Figure 4.14). The template *cannot* contain a PAM. It is important to keep in mind that the efficiency of HDR is always low, even in CRISPR applications. The resulting cells will be a mix of unedited, NHEJ-repaired, and HDR-repaired genotypes and so must be screened and selected for the desired edit. (See Section 4.7, "Validating CRISPR" section.) Some laboratories are working on ways to increase HDR efficiency and/or suppress NHEJ to make this type of editing easier. Validation of HDR-introduced changes is facilitated by the introduction of a restriction site, as shown in Figure 4.14.

CRISPR has become widely used, so a large number of easy-to-use reagents, guides, and software tools are readily available through all of the major molecular biology suppliers.

**Table 4.2**

PAM Sequences for Different Cas9 Variants

| Species/Variant of Cas9 | PAM Sequence |
| --- | --- |
| Streptococcus pyogenes (SP); SpCas9 | NGG |
| SpCas9 D1135E variant | NGG (reduced NAG binding) |
| SpCas9 VRER variant | NGCG |
| SpCas9 EQR variant | NGAG |
| SpCas9 VQR variant | NGAN or NGNG |
| Staphylococcus aureus (SA); SaCas9 | NNGRRT or NNGRR(N) |
| Neisseria meningitidis (NM) | NNNNGATT |
| Streptococcus thermophilus (ST) | NNAGAAW |
| Treponema denticola (TD) | NAAAAC |

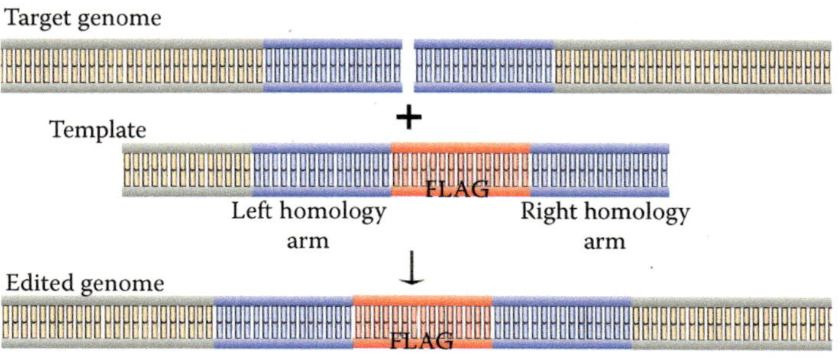


**Figure 4.14 HDR templates.** A template contains left and right arms of sequences homologous to those on the target genome, and an insert containing the edit of interest (*FLAG*). After the target genome is cut by Cas9, recombination inserts the edit into the genome. This is a low-efficiency process, and the length of the homology arms must be carefully chosen for each experiment.

## Caveats

Off-target effects could potentially lead to serious complications, especially in clinical applications. Most seriously, if a disrupted gene is an oncogene or tumor suppressor gene, carcinogenesis could result.

One way to try to prevent off-target cleavage is to use a modified Cas9. Modified Cas9s called *nickases* have been developed, which, instead of generating a double-strand break, cut only one strand to leave a *nick* in the DNA. A second nickase is required to obtain a double-strand break on the opposite strand, so the likelihood of off-target cutting is greatly reduced because two off-target events would be needed before the modification occurs. This is called a *dual-nickase system* and should be used whenever off-target events must be avoided.

It is also possible that for a given target, there is not a PAM sequence positioned correctly in the genome. In that case, another Cas9 will need to be used. The identification of Cas9s from species other than *Streptococcus pyogenes*, as well as the generation of variants, helps the genome coverage, but some targets still remain inaccessible.

It is also important to note that overexpression or underexpression of Cas9 can cause experiments not to work. A Cas9 gene under the control of an inducible promoter can be helpful in these cases.

## Validating CRISPR

After sgRNA and Cas9 have been delivered to cells, there will be a range of genotypes in the cell population as in any transfection experiment. Some cells will escape modification completely and remain wild type. Some will be *heterozygous* for the changes, and others *homozygous*. In homozygotes, changes made by NHEJ might be different at each allele.

Methods of screening depend upon the method used. For editing based upon NHEJ repair, a *mismatch cleavage assay* gives a measure of the percentage of mutated alleles in a cell population.

For HDR edits that introduce a novel restriction site, PCR of the target region followed by digest of the products by the restriction enzyme and agarose gel electrophoresis is the simplest approach to validation. If there is no unique site, or for NHEJ edits, PCR amplification plus sequencing is necessary to obtain a quantitative picture of the edits.

## End-of-Chapter Questions

1. Describe why phage cloning is more efficient than using plasmids when making a cDNA library.

2. How many clones would be needed to cover the human genome? How do you ensure that each region is covered?

3. Each phage virion should contain only one genomic DNA molecule. How can you be sure this is the case?

4. Phage have been used to assemble templates for inorganic materials. Describe how this works.

5. In viral cloning, it is common to read of *cis* and *trans* elements. What does this mean, and what is its relevance to phage cloning? What is a *helper* phage?

6. What is the difference between a genomic DNA library and a cDNA library? When would each be chosen?

7. Why do YACs have a ColE1 plasmid origin of replication?

8. Pick a specific 20-base-pair sequence and find out.

How many times it occurs in the human genome. What if one mismatch is allowed? two?

9. Human embryos have been edited with CRISPR. Why would such editing be useful, but why has it not yet been implemented (apart from ethical considerations)?

10. Discuss how you would use CRISPR to (a) make a hypoallergenic peanut; (b) treat cystic fibrosis; and (c) create a pig whose heart could be transplanted directly into humans.

11. Reticuline epimerase (RE) is an enzyme recently discovered from poppy that is a key step in the biosynthesis of morphine. How would you (a) obtain a genomic clone of the RE gene; (b) make an opium poppy microarray; and (c) determine whether growth in cold versus warm temperatures affects morphine yield?

12. Describe how you could determine what parts of the poppy produce morphine.

13. Could you make *E. coli* that produce morphine? How about yeast?

## Background Reading


### Books

Cowell, I.G., and Austin, C.A. (eds.). *cDNA Library Protocols (Methods in Molecular Biology)*. Humana Press, Cold Spring Harbor Laboratory Press, Cold Spring Harbor, New York, 1997.

Doudna, J., and Mali, P. *CRISPR-Cas: A Laboratory Manual*. Cold Spring Harbor Laboratory Press, Cold Spring Harbor Laboratory Press, Cold Spring Harbor, New York, 2016.

Dymond, J.S. Explanatory chapter: Quantitative PCR. In Lorschs, J. (ed.). Laboratory Methods in Enzymology: DNA. pp. 279–289, 2013. Academic Press, Cambridge, MA.

Frei, J.C., and Lai, J.R. Protein and antibody engineering by phage display. In Pecoraros, V.L. (ed.). Peptide, Protein and Enzyme Design., pp. 45–87, 2016. Academic Press, Cambridge, MA.

Jia, Y.B. Real-time PCR. In Conns, P.M. (ed.). Laboratory Methods in Cell Biology: Biochemistry and Cell Culture. Methods in Cell Biology Vol. 112, pp. 55–68, 2012. Academic Press, Cambridge, MA.

Lundgren, M., and Charpentier, E. *CRISPR: Methods and Protocols (Methods in Molecular Biology)*. Humana Press, Cold Spring Harbor Laboratory Press, Cold Spring Harbor, New York, 2015.

McCubbin, A.G., and Roalson, E.H. Construction of bacterial artificial chromosome libraries for use in phylogenetic studies. In Zimmer, E.A., and Roalsons, E.H. (eds.). Molecular Evolution: Producing the Biochemical Data, Part B. pp. 384–400, 2005. Academic Press, Cambridge, MA.

Sioud, M. *siRNA and miRNA Gene Silencing: From Bench to Bedside (Methods in Molecular Biology)*. Humana Press, Cold Spring Harbor Laboratory Press, Cold Spring Harbor, New York, 2008.

### Journal articles

Phage, BACs, YACs, and Cosmids

Bates, P. (1987). Double cos-site vectors—Simplified cosmid cloning. *Methods in Enzymology* 153, 82–94.

Christensen, A.C. (2001). Bacteriophage lambda–based expression vectors. *Molecular Biotechnology* 17, 219–224.

Dilella, A.G., and Woo, S.L.C. (1987). Cloning large segments of genomic DNA using cosmid vectors. *Methods in Enzymology* 152, 199–212.

Giraldo, P., and Montoliu, L. (2001). Size matters: Use of YACs, BACs and PACs in transgenic animals. *Transgenic Research* 10, 83–103.

Hall, R.N., Meers, J., Fowler, E., and Mahony, T. (2012). Back to BAC: The use of infectious clone technologies for viral mutagenesis. *Viruses—Basel* 4, 211–235.

Kasai, K., and Saeki, Y. (2006). DNA-based methods to prepare helper virus-free herpes amplicon vectors and versatile design of amplicon vector plasmids. *Current Gene Therapy* 6, 303–314.

Lech, K., Brent, R., and Irwin, N. (1987). Lambda as a cloning vector. *Current Protocols in Molecular Biology* 1.10, 1.10.1–1.10.11.

Molek, P., and Bratkovic, T. (2015). Bacteriophages as scaffolds for bipartite display: Designing Swiss army knives on a nanoscale. *Bioconjugate Chemistry* 26, 367–378.

Monaco, A.P., and Larin, Z. (1994). YACS, BACs, PACs and MACs—Artificial chromosomes as research tools. *Trends in Biotechnology* 12, 280–286.

Muyrers, J.P.P., Zhang, Y.M., and Stewart, A.F. (2001) Techniques: Recombinogenic engineering—New options for cloning and manipulating DNA. *Trends in Biochemical Sciences* 26, 325–331.

Narayanan, K., and Chen, Q.W. (2011). Bacterial artificial chromosome mutagenesis using recombineering. *Journal of Biomedicine and Biotechnology*.

Paschke, M. (2006). Phage display systems and their applications. *Applied Microbiology and Biotechnology* 70, 2–11.

Qi, H., Lu, H.Q., Qiu, H.J., Petrenko, V., and Liu, A.H. (2012). Phagemid vectors for phage display: Properties, characteristics and construction. *Journal of Molecular Biology* 417, 129–143.

Van Dorst, B., Mehta, J., Rouah-Martin, E., Blust, R., and Robbens, J. (2012). Phage display as a method for discovering cellular targets of small molecules. *Methods* 58, 56–61.

Zhou, F.C., and Gao, S.J. (2011). Recent advances in cloning herpes viral genomes as infectious bacterial artificial chromosomes. *Cell Cycle* 10, 434–440.

## Multiple mutagenesis, directed evolution

Arnold, F.H. (2015). The nature of chemical innovation: New enzymes by evolution. *Quarterly Reviews of Biophysics* 48, 404–410.

Bassalo, M.C., Liu, R.M., and Gill, R.T. (2016). Directed evolution and synthetic biology applications to microbial systems. *Current Opinion in Biotechnology* 39, 126–133.

Lapa, S.A., Chudinov, A.V., and Timofeev, E.N. (2016). The toolbox for modified aptamers. *Molecular Biotechnology* 58, 79–92.

Merola, F., Erard, M., Fredj, A., and Pasquier, H. (2016). Engineering fluorescent proteins towards ultimate performances: Lessons from the newly developed cyan variants. *Methods and Applications in Fluorescence* 4, 012001.

Pakulska, M.M., Miersch, S., and Shoichet, M.S. (2016). Designer protein delivery: From natural to engineered affinity-controlled release systems. *Science* 351, 1279.

Peisajovich, S.G. (2012). Evolutionary synthetic biology. *ACS Synthetic Biology* 1, 199–210.

Rosenfeld, L., Heyne, M., Shifman, J.M., and Papo, N. (2016). Protein engineering by combined computational and in vitro evolution approaches. *Trends in Biochemical Sciences* 41, 421–433.

Swint-Kruse, L. (2016). Using evolution to guide protein engineering: The devil IS in the details. *Biophysical Journal* 111, (1), 10–8.

Tiwari, V. (2016). In vitro engineering of novel bioactivity in the natural enzymes. *Frontiers in Chemistry* 4.

Xiong, A.S., Peng, R.H., Zhuang, J., Davies, J., Zhang, J., and Yao, Q.H. (2012). Advances in directed molecular evolution of reporter genes. *Critical Reviews in Biotechnology* 32, 133–142.

## Advanced PCR techniques

Botes, M., de Kwaadsteniet, M., and Cloete, T.E. (2013). Application of quantitative PCR for the detection of microorganisms in water. *Analytical and Bioanalytical Chemistry* 405, 91–108.

Bower, N.I., and Johnston, I.A. (2010). Targeted rapid amplification of cDNA ends (T-RACE)—An improved RACE reaction through degradation of non-target sequences. *Nucleic Acids Research* 38 (21), e194.

D'Haene, B., Vandesompele, J., and Hellemans, J. (2010). Accurate and objective copy number profiling using real-time quantitative PCR. *Methods* 50, 262–270.

Dellett, M., and Simpson, D.A. (2016). Considerations for optimization of microRNA PCR assays for molecular diagnosis. *Expert Review of Molecular Diagnostics* 16, 407–414.

Devonshire, A.S., Sanders, R., Wilkes, T.M., Taylor, M.S., Foy, C.A., and Huggett, J.F. (2013). Application of next generation qPCR and sequencing platforms to mRNA biomarker analysis. *Methods* 59, 89–100.

Friedrich, S.M., Zec, H.C., and Wang, T.H. (2016). Analysis of single nucleic acid molecules in micro- and nano-fluidics. *Lab on a Chip* 16, 790–811.

Gadkar, V.J., and Filion, M. (2013). Quantitative real-time polymerase chain reaction for tracking microbial gene expression in complex environmental matrices. *Current Issues in Molecular Biology* 15, 45–57.

Gadkar, V.J., and Filion, M. (2014). New developments in quantitative real-time polymerase chain reaction technology. *Current Issues in Molecular Biology* 16, 1–5.

Griffiths, K.R., Burke, D.G., and Emslie, K.R. (2011). Quantitative polymerase chain reaction: A framework for improving the quality of results and estimating uncertainty of measurement. *Analytical Methods* 3, 2201–2211.

Huggett, J.F., Cowen, S., and Foy, C.A. (2015). Considerations for digital PCR as an accurate molecular diagnostic tool. *Clinical Chemistry* 61, 79–88.

Ivanova, A.V., and Ivanov, S.V. (2002). Differential display analysis of gene expression in yeast. *Cellular and Molecular Life Sciences* 59, 1241–1245.

Kaminski, T.S., Scheler, O., and Garstecki, P. (2016). Droplet microfluidics for microbiology: Techniques, applications and challenges. *Lab on a Chip* 16, 2168–2187.

Leenen, F.A.D., Vernocchi, S., Hunewald, O.E., Schmitz, S., Molitor, A.M., Muller, C.P., and Turner J.D. (2016). Where does transcription start? 5′-RACE adapted to next-generation sequencing. *Nucleic Acids Research* 44, 2628–2645.

Ling, D., and Salvaterra, P.M. (2011). Robust RT-qPCR data normalization: Validation and selection of internal reference genes during post-experimental data analysis. *PLoS One* 6, e17762.

Marr, C., Zhou, J.X., and Huang, S. (2016). Single-cell gene expression profiling and cell state dynamics: Collecting data, correlating data points and connecting the dots. *Current Opinion in Biotechnology* 39, 207–214.

Sager, M., Yeat, N.C., Pajaro-Van der Stadt, S., Lin, C., Ren, Q.Y., and Lin, J. (2015). Transcriptomics in cancer diagnostics: Developments in technology, clinical research and commercialization. *Expert Review of Molecular Diagnostics* 15, 1589–1603.

Schaefer, B.C. (1995). Revolutions in rapid amplification of cDNA ends—New strategies for polymerase chain-reaction cloning of full-length cDNA ends. *Analytical Biochemistry* 227, 255–273.

Sedlak, R.H., and Jerome, K.R. (2014). The potential advantages of digital PCR for clinical virology diagnostics. *Expert Review of Molecular Diagnostics* 14, 501–507.

Stahlberg, A., and Bengtsson, M. (2010). Single-cell gene expression profiling using reverse transcription quantitative real-time PCR. *Methods* 50, 282–288.

Stahlberg, A., and Kubista, M. (2014). The workflow of single-cell expression profiling using quantitative real-time PCR. *Expert Review of Molecular Diagnostics* 14, 323–331.

Sturtevant, J. (2000). Applications of differential-display reverse transcription-PCR to molecular pathogenesis and medical mycology. *Clinical Microbiology Reviews* 13, 408.

Sunohara, M., Kawakami, M., Kage, H., Watanabe, K., Emoto, N., Nagase, T., Ohishi, N., and Takai, D. (2011). Polymerase reaction without primers throughout for the reconstruction of full-length cDNA from products of rapid amplification of cDNA ends (RACE). *Biotechnology Letters* 33, 1301–1307.

VanGuilder, H.D., Vrana, K.E., and Freeman, W.M. (2008). Twenty-five years of quantitative PCR for gene expression analysis. *Biotechniques* 44, 619–626.

Whiley, H., and Taylor, M. (2016). Legionella detection by culture and qPCR: Comparing apples and oranges. *Critical Reviews in Microbiology* 42, 65–74.

*Microarrays*

Brown, P.O., and Botstein, D. (1999). Exploring the new world of the genome with DNA microarrays. *Nature Genetics* 21, 33–37.

Bryant, P.A., Venter, D., Robins-Browne, R., and Curtis, N. (2004). Chips with everything: DNA microarrays in infectious diseases. *The Lancet Infectious Diseases* 4, 100–111.

Gresham, D., Dunham, M.J., and Botstein, D. (2008). Comparing whole genomes using DNA microarrays. *Nature Reviews Genetics* 9, 291–302.

Hartmann, M., Roeraade, J., Stoll, D., Templin, M.F., and Joos, T.O. (2009). Protein microarrays for diagnostic assays. *Analytical and Bioanalytical Chemistry* 393, 1407–1416.

Hoheisel, J.D. (2006). Microarray technology: Beyond transcript profiling and genotype analysis. *Nature Reviews Genetics* 7, 200–210.

Horton, J.D., Shah, N.A., Warrington, J.A., Anderson, N.N., Park, S.W., Brown, M.S., and Goldstein, J.L. (2003). Combined analysis of oligonucleotide microarray data from transgenic and knockout mice identifies direct SREBP target genes. *Proceedings of the National Academy of Sciences of the United States of America* 100, 12027–12032.

Kawasaki, E.S. (2006). The end of the microarray Tower of Babel: Will universal standards lead the way? *Journal of Biomolecular Techniques* 17, 200–206.

Kononen, J., Bubendorf, L., Kallioniemi, A., Barlund, M., Schraml, P., Leighton, S., Torhorst, J., Mihatsch, M.J., Sauter, G., and Kallioniemi, O.P. (1998). Tissue microarrays for high-throughput molecular profiling of tumor specimens. *Nature Medicine* 4, 844–847.

Rhodes, D.R., Yu, J., Shanker, K., Deshpande, N., Varambally, R., Ghosh, D., Barrette, T., Pandey, A., and Chinnaiyan, A.M. (2004). Large-scale meta-analysis of cancer microarray data identifies common transcriptional profiles of neoplastic transformation and progression. *Proceedings of the National Academy of Sciences of the United States of America* 101, 9309–9314.

Steelman, C.A., Recknor, J.C., Nettleton, D., and Reecy, J.M. (2006). Transcriptional profiling of myostatin-knockout mice implicates Wnt signaling in postnatal skeletal muscle growth and hypertrophy. *FASEB Journal* 20, 580–582.

Tanaka, T.S., Jaradat, S.A., Lim, M.K., Kargul, G.J., Wang, X., Grahovac, M.J., Pantano, S., Sano, Y., Piao, Y., Nagaraja, R. et al. (2000). Genome-wide expression profiling of mid-gestation placenta and embryo using a 15,000 mouse developmental cDNA microarray. *Proceedings of the National Academy of Sciences of the United States of America* 97, 9127–9132.

Wang, B., Howel, P., Bruheim, S., Ju, J., Owen, L.B., Fodstad, O., and Xi, Y. (2011). Systematic evaluation of three microRNA profiling platforms: Microarray, beads array, and quantitative real-time PCR array. *PLoS One* 6, e17167.

Wang, D., Coscoy, L., Zylberberg, M., Avila, P.C., Boushey, H.A., Ganem, D., and DeRisi, J.L. (2002). Microarray-based detection and genotyping of viral pathogens. *Proceedings of the National Academy of Sciences of the United States of America* 99, 15687–15692.

Yu, X., Petritis, B., and LaBaer, J. (2016). Advancing translational research with next-generation protein microarrays. *Proteomics* 16, 1238–1250.

Zarate, X., and Galbraith, D.W. (2014). A cell-free expression platform for production of protein microarrays. *Methods in Molecular Biology* 1118, 297–307.

### RNAi

Blow, N. (2008). RNAi technologies: A screen whose time has arrived. *Nature Methods* 5, 361–368.

Chang, K., Marran, K., Valentine, A., and Hannon, G.J. (2012). RNAi in cultured mammalian cells using synthetic siRNAs. *Cold Spring Harbor Protocols* 2012, 957–961.

Fellmann, C., Zuber, J., McJunkin, K., Chang, K., Malone, C.D., Dickins, R.A., Xu, Q., Hengartner, M.O., Elledge, S.J., Hannon, G.J. et al. (2011). Functional identification of optimized RNAi triggers using a massively parallel sensor assay. *Molecular Cell* 41, 733–746.

Hammell, C.M., and Hannon, G.J. (2016). Inducing RNAi in *Caenorhabditis elegans* by Injection of dsRNA. *Cold Spring Harbor Protocols* 2016 (1):, pdb. prot086306.

Housden, B.E., and Perrimon, N. (2016). Comparing CRISPR and RNAi-based screening technologies. *Nature Biotechnology* 34, 621–623.

Kaur, P., Nagaraja, G.M., and Asea, A. (2011). Combined lentiviral and RNAi technologies for the delivery and permanent silencing of the hsp25 gene. *Methods in Molecular Biology* 787, 121–136.

Luo, J., Emanuele, M.J., Li, D., Creighton, C.J., Schlabach, M.R., Westbrook, T.F., Wong, K.K., and Elledge, S.J. (2009). A genome-wide RNAi screen identifies multiple synthetic lethal interactions with the Ras oncogene. *Cell* 137, 835–848.

Poulin, G.B. (2011). A guide to using RNAi and other nucleotide-based technologies. *Briefings in Functional Genomics* 10, 173–174.

Seo, M., Lee, S., Kim, J.H., Lee, W.H., Hu, G., Elledge, S.J., and Suk, K. (2014). RNAi-based functional selection identifies novel cell migration determinants dependent on PI3K and AKT pathways. *Nature Communications* 5, 5217.

Siolas, D., Lerner, C., Burchard, J., Ge, W., Linsley, P.S., Paddison, P.J., Hannon, G.J., and Cleary, M.A. (2005). Synthetic shRNAs as potent RNAi triggers. *Nature Biotechnology* 23, 227–231.

Stephan, J.P. (2014). Using RNAi screening technologies to interrogate the extrinsic apoptosis pathway. *Methods in Enzymology* 544, 129–160.

Zuber, J., McJunkin, K., Fellmann, C., Dow, L.E., Taylor, M.J., Hannon, G.J., and Lowe, S.W. (2011). Toolkit for evaluating genes required for proliferation and survival using tetracycline-regulated RNAi. *Nature Biotechnology* 29, 79–83.

### CRISPR

Dominguez, A.A., Lim, W.A., and Qi, L.S. (2016). Beyond editing: Repurposing CRISPR-Cas9 for precision genome regulation and interrogation. *Nature Reviews Molecular Cell Biology* 17.

Gasiunas, G., and Siksnys, V. (2013). RNA-dependent DNA endonuclease Cas9 of the CRISPR system: Holy Grail of genome editing? *Trends in Microbiology* 21, 562–567.

Heidenreich, M., and Zhang, F. (2016). Applications of CRISPR-Cas systems in neuroscience. *Nature Reviews Neuroscience* 17, 36–44.

Jusiak, B., Cleto, S., Perez-Pinera, P., and Lu, T.K. (2016). Engineering synthetic gene circuits in living cells with CRISPR technology. *Trends in Biotechnology* 34, 535–547.

Lander, E.S. (2016). The heroes of CRISPR. *Cell* 164, 18–28.

Lopes, R., Korkmaz, G., and Agami, R. (2016). Applying CRISPR-Cas9 tools to identify and characterize transcriptional enhancers. *Nature Reviews Molecular Cell Biology* 17, 597–604.

Luo, M.L., Leenay, R.T., and Beisel, C.L. (2016). Current and future prospects for CRISPR-based tools in bacteria. *Biotechnology and Bioengineering* 113, 930–943.

Mei, Y., Wang, Y., Chen, H.Q., Sun, Z.S., and Ju, X.D. (2016). Recent progress in CRISPR/Cas9 technology. *Journal of Genetics and Genomics* 43, 63–75.

Mohanraju, P., Makarova, K.S., Zetsche, B., Zhang, F., Koonin, E.V., and van der Oost, J. (2016). Diverse evolutionary roots and mechanistic variations of the CRISPR-Cas systems. *Science* 353, 556.

Nunez, J.K., Harrington, L.B., and Doudna, J.A. (2016). Chemical and biophysical modulation of Cas9 for tunable genome engineering. *ACS Chemical Biology* 11, 681–688.

Peng, R.X., Lin, G.G., and Li, J.M. (2016) Potential pitfalls of CRISPR/Cas9-mediated genome editing. *FEBS Journal* 283, 1218–1231.

Raitskin, O., and Patron, N.J. (2016). Multi-gene engineering in plants with RNA-guided Cas9 nuclease. *Current Opinion in Biotechnology* 37, 69–75.

Sorek, R., Lawrence, C.M., and Wiedenheft, B. CRISPR-mediaed adaptive immune systems in bacteria and archaea. In Kornbergs, R.D. (ed.). *Annual Review of Biochemistry*, Vol 82. pp. 237–266, 2013.

Tsai, S.Q., and Joung, J.K. (2016). Defining and improving the genome-wide specificities of CRISPR-Cas9 nucleases. *Nature Reviews Genetics* 17, 300–312.

Tycko, J., Myer, V.E., and Hsu, P.D. (2016). Methods for optimizing CRISPR-Cas9 genome editing specificity. *Molecular Cell* 63, 355–370.

Unniyampurath, U., Pilankatta, R., and Krishnan, M.N. (2016). RNA interference in the age of CRISPR: Will CRISPR interfere with RNAi? *International Journal of Molecular Sciences* 17 (3), 291.

White, M.K., and Khalili, K. (2016). CRISPR/Cas9 and cancer targets: Future possibilities and present challenges. *Oncotarget* 7, 12305–12317.

Wright, A.V., Nunez, J.K., and Doudna, J.A. (2016). Biology and applications of CRISPR systems: Harnessing nature's toolbox for genome engineering. *Cell* 164, 29–44.


## Suppliers

Antibody Design Laboratories. Phage vectors, helper phage.

Agilent. Products for phage cloning, mutagenesis, microarrays.

Creative Biolabs. Phage display, antibody libraries.

Lucigen. BACs and vectors for cloning of large/difficult inserts.

Cambridge Protein Arrays. Tools for proteomics.

Bio-Rad. Instruments and support for qPCR, dPCR.

Qiagen. RNAi tools, CRISPR tools, Pathway Analysis, and much more.

Thermo Fisher Scientific. Specialized transfection reagents, antibiotics for selection, plates, instruments, and more.

New England Biolabs. Molecular biology reagents, including restriction enzymes, all reagents for CRISPR, many more.

GenScript. Many gene and protein services, including cloning and mutagenesis, sgRNAs, CRISPR-edited cell lines. Publishes *CRISPR Handbook*.

Dharmacon (GEBiosciences). RNAi, CRISPR, custom RNA and DNA, transfection reagents.

Clontech. Kits for CRISPR experimental design and validation, custom RNA and DNA, inducible expression systems, reagents for qPCR, rt-PCR, rt-qPCR.

AddGene. Plasmid repository from thousands of research labs. Publishes guide to CRISPR (http://blog.addgene.org/how-to-design-your-grna-for-crispr-genome-editing).

GeneON. Molecular biology supplies including a wide variety of master mixes for qPCR, rt-qPCR.

## Software and databases

IPA: http://www.ingenuity.com/products/ipa

PANTHER: http://www.pantherdb.org/help/PANTHERhelp.jsp

DAVID: https://david.ncifcrf.gov/

Broad Institute sgRNA tool: http://portals.broadinstitute.org/gpp/public/analysis-tools/sgrna-design

MIT CRISPR tool: http://crispr.mit.edu/

CRISPR outcome analysis tool: http://crispresso.rocks

Sequence matcher: http://www.vmatch.de/

Another sequence aligner: http://bio-bwa.sourceforge.net/

# CHAPTER 5

# Protein Expression Methods

Joshua A. Maurer

## 5.1 INTRODUCTION

Obtaining pure protein samples is a critical first step for many experiments. While a small number of highly abundant proteins can be purified directly from natural sources, most interesting proteins must be obtained by heterologous expressions or total chemical synthesis. The purification of a protein from a natural source requires large amounts of tissue containing the desired protein and extensive knowledge of the protein's physicochemical properties or the use of immunoaffinity methodologies. As such, natural source isolation of the quantity of protein necessary for biochemical or biophysical studies is limited to a small number of proteins and will not be discussed here. Total chemical synthesis and expressed chemical ligation (ECL) technologies provide intriguing methods for the preparation of modified proteins, but these methods require techniques that are beyond the scope of many laboratories. A brief discussion of these techniques and their applications is provided at the end of the chapter. Generally, proteins are prepared by overexpression and purification using heterologous expression systems; as such, this will be the focus of this chapter. Heterologous expression has the advantage that it can be used for the preparation of small or large amounts of purified proteins, including modified and engineered proteins. Moreover, it can be used to express proteins from pathogenic organisms without the extensive biosafety precautions that would be required for handling the source organism and allows for the functional analysis of open reading frames.

## 5.2 EXPRESSION SYSTEMS

The first step in the heterologous expression and purification of proteins is choosing an appropriate expression system. There are five major classes of expression systems that can be employed for protein preparation: bacteria, yeast, insect cells, mammalian cells, and *in vitro* systems. Each of these systems has an array of advantages and disadvantages that are briefly discussed here. The choice of a protein expression system needs to be undertaken early during project development, since it can greatly effect cloning (see **Chapter 2**) and project design.

*Bacteria.* Bacterial expression systems are some of the most common heterologous expression systems employed. These are the easiest expression systems to use and have the advantage of producing extremely large quantities of protein (>100 mg).

However, expression of functional mammalian proteins in these systems is not always possible. The failure of these systems to give functional proteins is a result of bacteria lacking the machinery to carry out standard eukaryote posttranslational modifications, a lack of the extensive folding machinery and chaperon proteins found in eukaryotic cells, and for transmembrane proteins, a difference in lipid bilayer thickness.

*Yeast.* Yeast expression systems are the simplest eukaryotic systems to employ. However, while yeasts carry out some posttranslational modifications and have a subset of the folding machinery found in mammalian cells, they do not always yield properly folded and packaged proteins. Additionally, they are not capable of expressing transmembrane proteins.

*Insect cells.* Insect cell expression systems are more finicky than yeast or bacterial expression systems; however, these systems provide a powerful tool for the expression of large quantities of functional eukaryotic proteins. Insect cells have been used to express fully functional modified mammalian proteins and mammalian membrane proteins, such as the human glycine receptor.

*Mammalian cells.* Mammalian expression systems are capable of expressing small quantities of eukaryotic proteins. While these systems provide the most complete set of enzymes and chaperons for the production of fully functional and fully posttranslationally modified mammalian proteins, scaling up protein expression in these systems is extremely challenging. As such, mammalian expression systems are typically only employed for studies that require extremely small quantities of protein.

*In vitro. In vitro* expression or cell-free expression systems come in two varieties: translation-only systems and transcription–translation system. The translation only system takes RNA, which can be produced from a *in vitro* transcription kit, to protein, while the transcription–translation kits are capable of taking vector DNA directly to protein. Cell-free lysates for protein expression are available from a wide variety of organisms including *Escherichia coli*, wheat germ, and rabbit reticulocytes, which allow for the production of fully functional and modified proteins. Although *in vitro* expression typically gives extremely low protein yields, it is useful for the production of small amounts of protein and can be used to easily produce radiolabeled proteins through the incorporation of radiolabeled amino acids (typically $^{35}$S-labeled methionine or cysteine).

While it is important to determine the appropriate expression system for a project early during project development, there are a number of cloning vectors that are compatible with multiple expression systems (Section 5.4, "Selecting an Expression Vector" section). For most genes, it makes sense to initially attempt heterologous expression in bacterial systems, since these systems are easy to manipulate and give large quantities of proteins. If expression of functional protein fails in *E. coli*, other expression systems can subsequently be explored. The exception to this strategy is proteins that are known to contain significant posttranslational modifications or eukaryotic membrane proteins, in which case insect cells or *in vitro* systems provide good starting points depending on the amount of protein required for subsequent assays.

## 5.3 IDENTIFICATION OF A DNA SOURCE

There are a number of different ways to obtain a gene of interest. If the protein has been previously studied, it may be possible to obtain the genetic material from another research group. In this case, the gene may or may not be in an appropriate expression vector depending on the types of experiments previously performed with the protein. Many genes can be purchased in plasmids used for sequencing or cloning from repositories discussed in Chapter 2; some others are also given at the end of this chapter. These genes typically need to be subcloned into expression vectors in order to produce protein.

If genes are not available, either from an academic group or commercially, they can be cloned out of genomic DNA or cDNA libraries. For bacterial proteins, cloning genes from genomic DNA is a straightforward process, and typically, genes can be directly inserted into expression vectors without the use of intermediate cloning vectors. Isolated genomic DNA for a large number of bacterial species can be obtained from the American Type Culture Collection (ATCC) or directly from bacteria genome sequencing projects. For eukaryotic proteins, genes must be fished out of cDNA libraries, since the genomic DNA contains introns that are typically spliced out of the messenger RNA (mRNA) prior to expression in the host and may not be processed properly during heterologous expression. A variety of cDNA libraries are available in the public domain and commercially; however, it is critical to make sure that the cDNA library contains full-length mRNA transcripts. Many libraries contain only short stretches of cDNA encoding for a small portion of the mRNA, since it can be challenging to obtain full-length transcripts during reverse transcription of mRNA.

# 5.4  SELECTING AN EXPRESSION VECTOR

Vectors used for protein expression share many of the same basic components as the vectors discussed in Chapter 2. All vectors must contain an origin of replication (ORI), selection marker (typically Amp, Kan, or Tet), and a multiple cloning region (MCS) into which the gene is inserted (Figure 5.1). In addition, expression vectors must contain an appropriate promoter for protein expression and typically contain tags for purification and antibody recognition.

## Promoters

There are two main classes of promoters used for protein expression: constitutively active promoters and inducible promoters. Constitutively active promoters

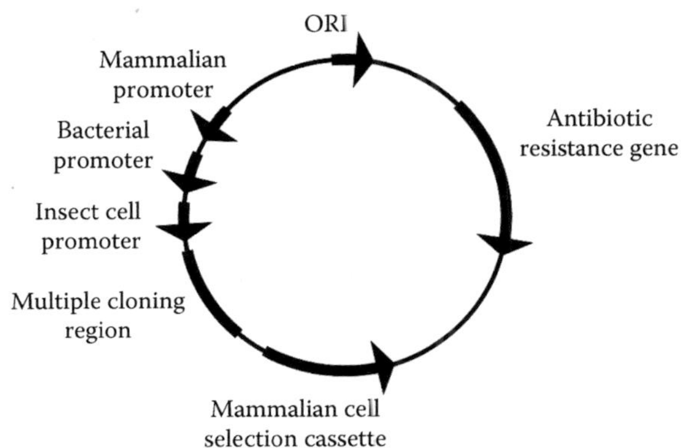


**Figure 5.1 An expression vector with the required components labeled.** All expression vectors must contain an origin of replication (ORI), an antibiotic resistance gene for bacterial selection, a multiple cloning region, and at least one promoter. The example shown contains promoters for expression in multiple hosts (similar to Merk's pTriEx vectors and Qiagen's pQE-TriSystem vector) and a mammalian selection cassette.

produce protein at all times after a gene is introduced into the expression system, while inducible promoters produce protein only upon application of a stimulus. Typically, inducible promoters are used in protein overexpression, since these allow for control over the start of protein expression. By controlling the start of protein expression, extremely strong promoters, typically viral promoters, can be employed. Strong promoters divert most of the host organism's energy into heterologous protein expression. As a result, these promoters are not viable for constitutive expression, since they would compromise the ability of the host to replicate. Another advantage of inducible promoters is that they allow for the expression of proteins that would be toxic to the host strain at high levels.

While there are a wide variety of promoters available for expression in each of the expression systems, in practice, only a limited number of promoters are employed for typical protein expression. The most common promoters for each of the expression systems discussed above are summarized in **Table 5.1**.

Expression of proteins in *E. coli* under control of the T7 promoter is the most common system employed. This system places the gene of interest under the control of the promoter region from bacteriophage T7, which is not recognized by natural *E. coli* mRNA polymerase. For overexpression using this system, the T7 RNA polymerase is introduced into the host bacterial strain under the control of a lac operon, typically the lacUV5 promoter. The T7 RNA polymerase gene (also called DE3) is typically inserted directly into the host cell genomic DNA. Many common bacterial expression strains already contain the DE3 gene, and the gene can be easily inserted into any *E. coli* strain using a commercially available phage-based kit. The T7 gene can also be introduced on a plasmid or through phage infection.

In a T7 expression system where the DE3 gene is under the control of a *lac operon*, expression is induced through the addition of isopropyl β-D-1-thiogalactopyranoside (*IPTG*). Induction occurs as a result of IPTG binding to the Lac repressor protein and inducing a conformational change that prevents it from binding to the Lac promoter (**Figure 5.2**). Once the Lac repressor protein is removed, the *E. coli* strain makes mRNA for the T7 RNA polymerase using *E. coli* RNA polymerase, which in turn produces mRNA for the protein of interest.

In a system where tighter control over protein expression is desired (such as the expression of toxic proteins), the T7 promoter in the plasmid, containing the gene of interest, can also be placed under the control of a lac operon. This double-lock system displays lower levels of background expression compared to systems containing the T7 promoter alone. Background expression in the standard system results from an *expression leak* of the lacUV5 promoter, which produces an

**Table 5.1**

Promoters Used for Protein Expression

| Host | Common Promoters |
| --- | --- |
| Bacterial | T7, lacUV5, tac, araBAD |
| Yeast | LAC4, AOX1, GAP |
| Insect cells | Hr5/ie1, p10, gp64, polh, Ac5, MT |
| Mammalian cells | CMV, SV40, RSV, U6, E1b, EF-1α |
| *In vitro* systems | T7, SP6 |

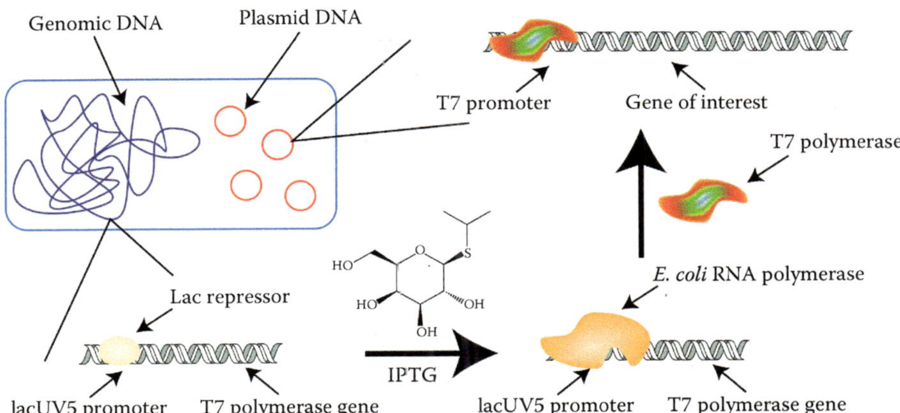



**Figure 5.2 Schematic representation of how the T7 expression system can be used to produce a protein of interest in *E. coli* carrying the DE3, T7 RNA polymerase gene.**

extremely low level of T7 RNA polymerase. In the double-lock system, the activity of this polymerase is further blocked by a second lac operon.

## Protein tags

Typically, protein expression vectors include tags for purification and immunohistological identification of the heterologously expressed protein. In most cases, tags are appended to the N-terminus or C-terminus of the protein during subcloning into the protein expression vector. If C-terminal tags are to be appended to the target protein, then the protein must be inserted into the vector without a stop codon. When the vector contains N-terminal tags, care must be taken to insert the gene in frame with the start codon before the N-terminal tags. If a vector contains C-terminal tags that one does not want to append to the protein of interest, it is typically sufficient to leave the native stop codon in place during the subcloning. However, if the vector contains N-terminal tags that one does not want to append to the protein of interest, these tags must be removed from the vector during subcloning. While tags are typically inserted at only either the N-terminus or the C-terminus of the protein, it is possible to insert tags for purification or immunohistological identification into the core of the protein. This strategy has been successfully applied to purification tags inserted into the extracellular loops of multipass transmembrane proteins.

Protein tags are commonly divided into two classes: *purification tags* and *immunohistochemistry tags*. However, many tags can be used for both purification and immunohistochemical identification. Some of the most common tags are listed as follows, and more details on their use are found in other subsections of this chapter.

*His-tag.* A short polypeptide repeat of six to nine histidine residues that can be used for either purification or immunohistochemical analysis. The polyhistidine repeat motif can be used to bind a protein containing the tag to either a solid support–bound transition metal complex for protein purification or a fluorescent or enzymatically linked transition metal complex for protein identification. (More details may be found in the sample experiment in **Section 5.12**, "Example Experiment: Expression and Purification of Fluorescent Protein Dronpa" section.) Additionally, a wide range of antibodies with varying specificities are available commercially for traditional immunohistochemical analysis.

*GST-tag.* A GST-tag appends glutathione S-transferase to the protein of interest. Typical GST-tags are on the order 220 amino acids (much larger than standard peptide tags) and can be used for protein purification by binding of the tagged protein to a solid support containing immobilized glutathione. Antibodies are available for immunohistochemical analysis of proteins containing GST-tags.

*CBD-tag.* CBD-tag can be used to refer to two different protein tags: the cellulose-binding tag and the chitin-binding tag. There are several different cellulose-binding tags available, which are based on cellulose-binding domains and can be used for protein purification through protein interactions with cellulose. The chitin-binding domain tag is a 53-amino-acid affinity tag capable of binding to beads coated with chitin (poly-N-acetylglucosamine). Antibodies are available for immunohistochemical analysis of proteins containing both types of CBD-tags.

*Epitope tags.* In addition to the standard purification tags described previously, a number of different epitope tags are available. While these tags are primarily used as antibody recognition sites, protein purification methods have also been developed around many of these tags. Some of the most common tags include the HA-tag (YPYDVPDYA), Myc-tag (EQKLISEEDL), FLAG-tag (DYKDDDDK), HSV-tag (QPELAPEDPED), S-tag (KETAAAKFERQHMDS), and Strep-tag (WSHPNFRK).

In many vectors, *protease cleavage sites* are located immediately downstream from the protein tags. These cleavage sites allow for removal of the protein tags following protein expression and purification and can yield tag-free proteins for biological analyses in which the protein tags might interfere with the analysis. Typical protease cleavage sites include thrombin, enterokinase, and factor Xa. These will be discussed more thoroughly in Chapter 6.

## Vector selection

In selecting an expression vector, the most important things to consider are expression host and protein purification. A number of vectors exist that contain promoters for expression in multiple expression hosts, such as Merk's pTriEx vectors and Qiagen's pQE-TriSystem vector that can be used for expression in insect cells, mammalian cells, and bacteria. These vectors represent a good choice for protein expression when it is unclear which expression system will be needed to obtain functional protein.

Most vectors come with a variety of the protein tags described in Section 5.4, "Protein Tags" section; however, it is important to select a vector that contains all of the tags needed for purification, biochemical, and biophysical analysis. Additional tags can be added later by DNA manipulation, but it is easier to start with a vector containing all of the desired tags. Generally, it is possible to start with a vector containing all the desired tags, due to the large number of commercially available expression vectors.

Another factor to consider in the selection of an expression vector is antibiotic resistance. Most commercial vectors encode resistance to ampicillin or kanamycin for bacterial selection, and often, vectors are available in both variants. In choosing an antibiotic resistance for expression in *E. coli*, it is important to consider the *E. coli* host strain that will be used for expression, since many *E. coli* strains used for expression already have antibiotic resistance due to genetic manipulation of the host strains. (See below for a discussion.) For expression in mammalian cells, it is often useful to create stable cell lines, which requires the inclusion of a selection cassette within the vector. Common section cassettes for mammalian cell lines are used to select for cells resistant to blasticidin, G418 (Geneticin), Zeocin, and Hygromycin B.

## 5.5  SUBCLONING INTO AN EXPRESSION VECTOR

Once the gene source has been identified and an expression vector has been selected, the gene must be subcloned into the expression vector. Chapter 2 provides details on how to carry out subcloning, but the exact cloning method to be employed will depend on the gene source selected and the expression vector. Expression vectors that are compatible with traditional cloning techniques, ligation-independent cloning (LIC) techniques, and TOPO cloning are available. For direct cloning, without the use of an intermediate cloning vector, LIC cloning and TOPO cloning provide rapid means of directly inserting genes to be overexpressed into an expression vector.

## 5.6  SELECTION OF AN EXPRESSION STRAIN OR CELL LINE

A wide variety of different expression strains and cell lines exist for protein production, with different strains or lines providing application-specific advantages. Here, the common expression strains and cell lines used for each of the various expression systems outlined in Section 5.2, "Expression Systems" section are briefly reviewed, focusing on bacterial expression systems.

### Bacterial strains

Numerous bacterial strains have been optimized for protein overexpression. Many of these strains have been genetically engineered and mutated to enhance protein yields and facilitate the expression of eukaryotic proteins. Some of the most commonly used strains are described here.

*BL21(DE3).* The most common bacterial strain used for protein expression. This strain typically gives high protein yields in routine protein expression and is deficient in Lon and OmpT proteases.

*Origami(DE3).* The Origami *E. coli* strains encode for mutations in thioredoxin reductase and glutathione reductase, which alter the reducing environment of the bacterial cytoplasm to enhance disulfide bond formation when overexpressing eukaryotic proteins. Origami is available as either a k12-derived strain or a BL21-derived strain (Origami B).

*Rosetta(DE3).* Rosetta and Rosetta 2 are *E. coli* strains derived from BL21 that are designed to enhance eukaryotic expression through the incorporation of tRNAs for rarely used *E. coli* codons. The Rosetta strain contains tRNAs for the condons AUA, AGG, AGA, CUA, CCC, and GGA, and the Rosetta 2 contains all of the addition tRNAs found in Rosetta plus tRNA for the CGG codon. The Rosetta strains generally grow rapidly and give high protein yields.

*Turner(DE3).* The Turner bacterial strain contains a *lac Y* deletion mutation that allows for uniform entry of IPTG into all cells within a bacteria population.

As a result, IPTG concentration in these strains can be used to control protein expression level. This makes possible low-level protein expression, which can increase the amount of properly folded overexpressed eukaryotic protein by avoiding protein aggregation and misfolding.

*Rosetta-gami(DE3)*. The Rosetta-gami strains of *E. coli* incorporate the mutations associated with the Rosetta and Origami strains and, in the case of Rosetta-gami B(DE3), the mutations associated with Turner. While these strains offer an impressive array of tools for eukaryotic protein expression, they are also very slow growing and tend to give much lower proteins yields than classical strains, such as BL21(DE3).

### Yeast

While a variety of yeast expression systems are available, the most common expression systems make use of either *Kluyveromyces lactis* or *Pichia pastoris*. Both of these expression systems are scalable for large-scale protein expression.

### Insect cells

The most common expression strains for insect cells are Sf21 cells, Sf9 cells, High Five, and S2 cells. The Sf21 and Sf9 cell lines are derived from *Spodoptera frugiperda* (fall armyworm) ovarian tissue, with the Sf9 cell line having been selected from the Sf21 cell line for faster growth rate and higher cell densities. The High Five cells are a clonal isolate derived from the *Trichoplusia ni* (cabbage looper moth) cell line, and S2 cells are derived from *Drosophila*. Typically, Sf9 or High Five cells are used in protein expression, since these lines grow rapidly and produce high-density cultures.

### Mammalian cells

Many common mammalian cell lines can be used for protein expression. Typical cell lines employed for protein expression include CHO (Chinese hamster ovary) cells, NIH 3T3 cells, HeLa cells, Jurkat cells, and HEK 293 cells. Overexpression is possible in both adherent and nonadherent cell lines. However, maximum protein expression is typically obtained from cells that can be cultured in suspension, since higher cell densities can be achieved in these systems.

## 5.7 PROTEIN EXPRESSION

The exact details of plasmid transformation and protein expression vary from expression system to expression system. Furthermore, there are significant variations even within expression systems for yeast, insect cells, mammalian cells, and *in vitro* expression systems. However, protein overexpression in bacteria is relatively well defined. As such, *E. coli* is the most common heterologous expression system employed for protein overexpression and allows even novice researchers to obtain high levels of expression.

To express proteins in *E. coli*, the vector containing the protein of interest under the control of a T7 promoter is transformed into the desired expression strain. Following transformation, a bacterial culture is grown overnight to *stationary* phase (**Figure 5.3**). The stationary-phase culture is then diluted at least 1:10 into

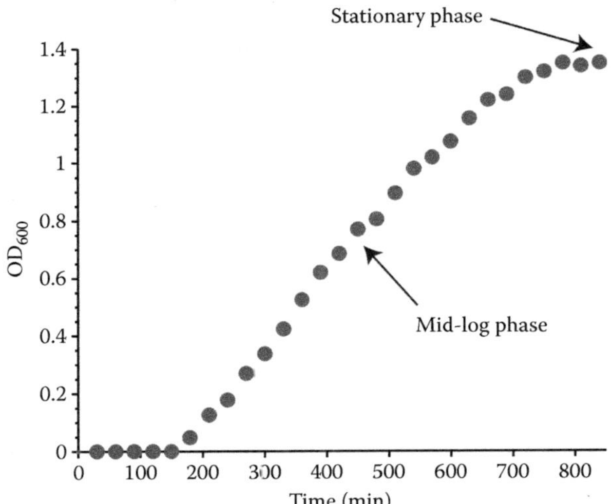


**Figure 5.3  A bacterial growth curve in LB media for the DBC8404 strain of *E. coli*, which is a DE3 variant of the CA8404 strain that contains deletions of the *cya* and *crp* genes.**

fresh media and allowed to grow to "mid-log" phase. At this point, IPTG is added to the culture to induce protein expression, and expression is allowed to proceed for several hours.

*E. coli* growth under standard conditions is logarithmic, with a growth curve of the shape shown in **Figure 5.3**. In the initial portion of the growth curve, the bacteria double with a fixed time constant (approximately 20 min for the BL21 strain). This time constant is associated with the bacterial strain, the type of growth medium employed, and the aeration of the growth medium. In the later portion of the growth curve, bacterial growth slows as the number of bacteria increases and the amount of available nutrients in the medium decreases. This decrease in doubling rate continues until the nutrients are essentially exhausted and the culture reaches stationary phase. Bacteria are healthiest at the midpoint of this curve, known as mid-log phase. As a result, protein expression by the addition of IPTG is typically induced during mid-log phase.

Bacterial growth is easily monitored during expression by determining the optical density of the culture at 600 nm (OD$_{600}$). This is typically accomplished by reading the density of the culture using a conventional ultraviolet–visible (UV–Vis) spectrometer and a disposable plastic cuvette. As the density of the culture increases, the apparent absorbance measured by the UV–Vis spectrometer also increases due to an increase in the scattering of light at 600 nm. The increase in apparent absorbance of the culture is directly proportional to the number of bacteria in the culture. For typical UV–Vis spectrometers, mid-log phase occurs at optical densities between 0.5 and 0.8 absorbance units, and stationary phase occurs at optical densities between 1 and 1.6 absorbance units. However, the exact absorbances associated with these phases is dependent on both growth conditions and instrument configuration.

For optimal bacterial growth, it is essential to employ rich growth media that is heavily aerated. Aeration is typically achieved by growing cultures in sparsely filled Erlenmeyer flasks (500 mL of media in a 2 L flask) using a shaking incubator that contains a platform rotating between 250 and 400 rpm. For protein expression, aerations can be improved by using baffled growth flasks (Erlenmeyer flasks with three to four indentions in the flask bottoms). While low-salt growth media, such as Luria-Bertani (LB) medium, are typically employed for plasmid purification, rich growth media are typically employed for protein expression. Common growth medium formulations for protein expression include 2× LB (LB media at twice its normal concentration), 2× YT, and terrific broth (TB). Some of these media formulations, like 2× LB and 2× YT, simply contain extra nutrients and salts compared to classic growth media formulations, while other media formulations such as TB contain alternate carbons sources such as glycerol that can be metabolized by the bacteria when the glucose in the media is exhausted. In addition to these common growth media formulations, a number of media formulations designed for overnight growth and expression have recently been developed, including Merk's Overnight Express and MP Biomedicals' CircleGrow.

Further improvements in aeration and growth control can be obtained using commercially available bacterial fermenters, which are computer controlled and allow for precise control over culture conditions including agitation, pH, temperature, and oxygen concentration. These fermenters are available for growing cultures as small as 1.5 L to cultures over 3000 L and are used for the industrial preparation of heterologously expressed proteins. Using a fermenter, growth conditions can be monitored in real time, and mass flow controls can be used to provide continuous or batch feeding of the bacterial cultures. Moreover, fermenters are commonly used for expression in yeast, insect, and mammalian cell lines, since expression conditions for these systems must be more precisely controlled than expression conditions for *E. coli*.

At mid-log phase, protein expression is induced by the addition of IPTG to a final concentration between 0.1 and 2 mM. Lower IPTG concentrations are typically employed for proteins that tend to misfold and are designed to slow protein expression. Rapid protein expression can lead to protein misfolding and the formation of insoluble protein aggregates called inclusion bodies. The Turner bacterial strain provides for the best control of protein expression using IPTG. Induction temperature provides another method to control the rate of protein expression, with the optimal temperature for protein expression typically being between 26°C and 37°C. Finally, induction time provides another method of controlling protein expression. Typical induction times for protein expression range from 2 to 12 h.

Unfortunately, there is not one magic set of protein expression conditions that work for every protein. Each of the expression parameters described previously must be optimized individually for every protein. When not using the Turner bacterial strain, the most critical parameters to optimize are expression temperature and induction time. Typically, this is done by carrying out a series of small-scale protein expressions (25 mL) in TB media with induction at 30°C and 37°C for 2, 4, and 6 h. These expressions can then be analyzed by *sodium dodecyl sulfate–polyacrylamide gel electrophoresis* (SDS-PAGE) to determine the extent of protein expression.

## 5.8 CHECKING PROTEIN EXPRESSION (AND PURITY) USING SDS-PAGE

SDS-PAGE provides an easy and convenient method for checking protein expression levels during test expressions and determining protein purity during

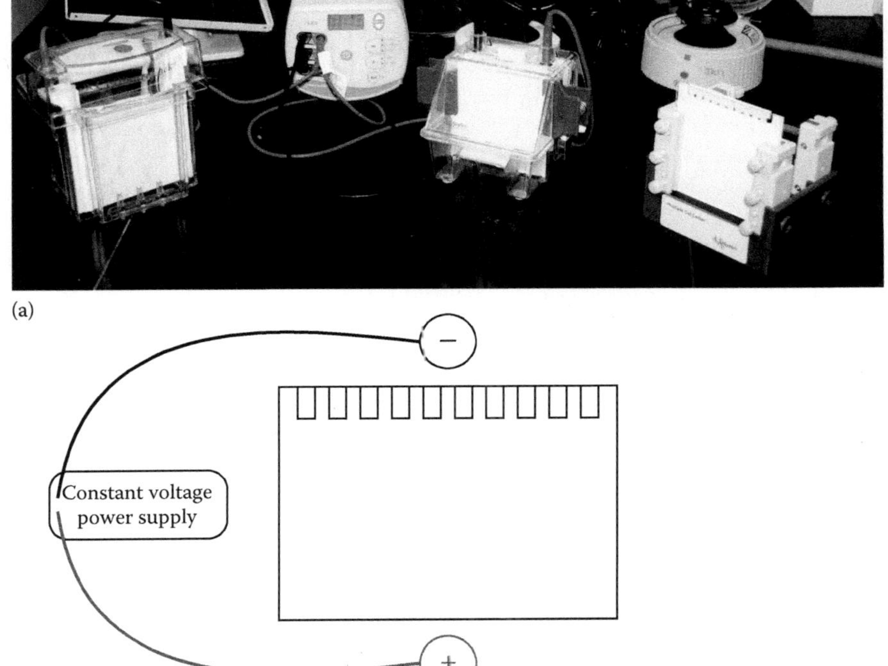


**Figure 5.4 (a) Picture of two different styles of SDS-PAGE minigel running equipment and a gel-pouring apparatus (right). (b) Schematic of an SDS-PAGE setup showing how the voltage is applied across the gel.**

and after protein purification. SDS-PAGE separates proteins by molecular weight using a setup that is similar to agarose gel electrophoresis for DNA as discussed in Chapter 2. To carry out SDS-PAGE analysis, a protein sample is first mixed with a loading buffer containing SDS and glycerol. This mixture is then loaded onto a *polyacrylamide gel*, which is run vertically at a constant voltage with a running buffer containing SDS (Figure 5.4).

## Protein separation

Separation of DNA or RNA in agarose gel electrophoresis is achieved due to the negative charge of the DNA backbone, which results in each base pair being associated with a −2 charge. While proteins are typically charged in solution, their charge can be either positive or negative depending on their amino acid composition. However, upon addition of SDS to a protein mixture, SDS *denatures* the protein and binds to the protein with a defined ratio of 1.4 g of SDS per gram of protein. As a result, the protein becomes negatively charged, with the total number of charges proportional to the molecular weight of the protein. This allows SDS-PAGE to be used for molecular weight determination within 5–10% based on the migration of protein standards.

Denaturation of proteins prior to electrophoresis is achieved by mixing protein samples with an SDS-loading buffer. In addition to SDS, most loading buffers also contain a reducing agent, either dithiothreitol (DTT) or β-mercaptoethanol, to reduce disulfide bonds to free thiols. To ensure that the proteins in a sample are fully denatured, samples are typically heated to 95°C or boiled for 5 min prior to loading. In the analysis of protein expression conditions, pelleted cells can be directly mixed with loading buffer and boiled. This process breaks up the cells and solubilizes all of the cellular proteins.

Similar to the case with agarose, polyacrylamide gels serve as a size sieve, with small molecules migrating through the gel faster than larger molecules. As a result, the density of polyacrylamide within a gel directly affects its molecular weight resolution. Gels containing a single polyacrylamide percentage (typically between 8% and 20%) are easily prepared by polymerization of commercially available acrylamide/bis-acrylamide solutions. However, acrylamide solutions must be handled with extreme care as unpolymerized arylamide is a neurotoxin. These gels provide good resolution over a relatively narrow molecular weight range. To obtain ideal separation for a single protein, the percent acrylamide and ratio of acrylamide to bis-acrylamide can be tailored within the gel. If gels capable of separating a broader molecular weight range are required, *gradient gels* are typically employed. While these gels can be prepared using a gradient mixer, they are also commercially available from a number of vendors.

In addition to acrylamide percentage, the running buffer system employed also affects the molecular weight resolution in SDS-PAGE. For most proteins, a Tris–glycine or bis-Tris–glycine running buffer provides good resolution. However, for very small proteins and peptides, a tricine buffer system is used, and for very large proteins, a Tris acetate buffer system can be employed.

## Protein visualization

There are two principal methods for protein visualization: total protein stains and immunohistochemical methods. Total protein stains allow for the visualization of an expressed protein relative to all of the other proteins in a sample, while immunohistochemical methods indicate if a particular protein is present in a protein sample. These methods are often used in combination with parallel gels being run during protein expression and purification experiments. Gels stained for total protein content are required to confirm protein purity, following protein purification, and can be used to check protein expression levels relative to other proteins in the cell during expression optimization (**Figure 5.5**). Gels visualized with an antibody specific to the protein of interest can be used to rapidly confirm protein identity.

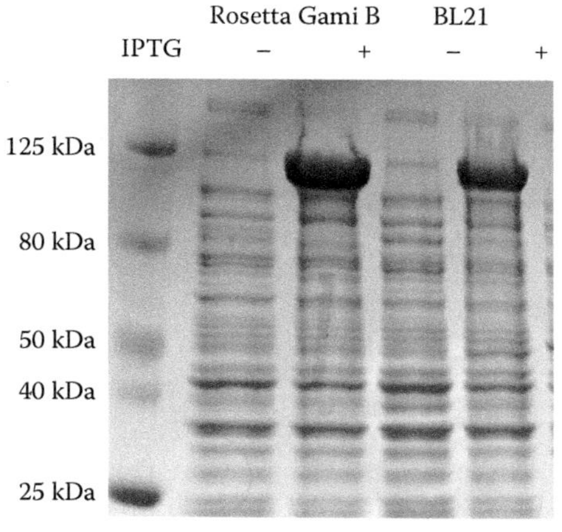


**Figure 5.5 Example of a gel stained with Coomassie blue.** The gel shows a test expression of β-galactosidase (110 kDa) in the Rosetta-gami B (DE3) and BL21 strains of *E. coli*. In this test expression, induction with IPTG gives a significant amount of the β-galactosidase protein in both stains.

Visualization of total protein content can be achieved using a variety of different staining techniques. The classical protein stain, which is still heavily used, is Coomassie blue. Coomassie has a sensitivity of ~8 ng, and staining of polyacrylamide gels is easy and inexpensive (**Figure 5.5**). A number of fluorescent dyes for rapidly staining gels with increased sensitivity have been developed, including SYPRO Orange (4–8 ng), SYPRO Red (4–8 ng), Oriole (<1 ng), Krypton (<1 ng), and Flamingo (<1 ng). Many of these dyes can be visualized and documented using a standard UV transilluminator and gel documentation system for the visualization of DNA gels, as discussed in **Chapter 2**. While these dyes provide increased sensitivity relative to Coomassie, they are also much more expensive. The classic method for visualization of extremely low protein levels in polyacrylamide gels is silver staining, which has a sensitivity rivaling the fluorescent protein dyes. In silver staining, $Ag^+$ ions are allowed to bind to proteins immobilized in the gel. After washing the gel to remove unbound silver ions, the ions are chemically reduced to metallic silver.

*Western blotting* is used for immunohistochemical visualization of proteins separated by a polyacrylamide gel. The Western blot is a close relative to Southern and Northern blots, described in **Section 2.10**, "Southern and Northern Blots" section. The first step in Western blotting is transferring the protein from the gel to a polymer membrane, typically either polyvinylidene fluoride (PVDF) or nitrocellulose. This is accomplished by "sandwich" electrophoresis, in which a sandwich is created containing filter paper, the polyacrylamide gel, the polymer membrane, and more filter paper. Electrophoresis is then used to transfer the proteins from the gel to the polymer membrane, using either a semidry blotting apparatus or a wet blotting apparatus (**Figure 5.6**). After the transfer is complete, the membrane is blocked with a solution of bovine serum albumin (BSA) or powered nonfat milk reconstituted in buffer. The purpose of the blocking step is

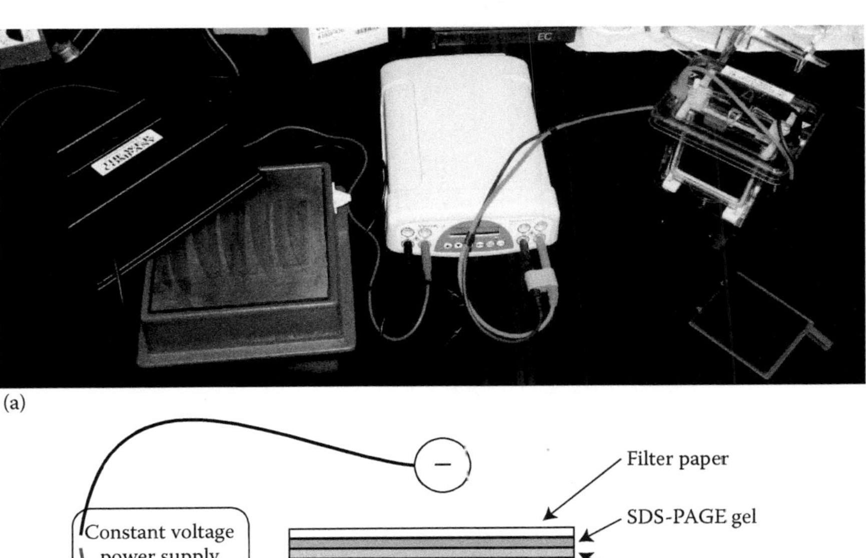



Figure 5.6 (a) Picture of a semidry Western blotting apparatus (left) and a wet Western blotting apparatus (right). (b) Schematic of a Western blot showing how voltage is applied across the sandwich electrophoresis.

to coat the entire membrane with protein, thus avoiding nonspecific absorption of the antibodies used for protein visualization. The blocked membrane is then probed with a primary antibody specific for the protein being expressed. In many cases, the antibody is specific for an epitope tag, which was appended to the protein during subcloning. The primary antibody is then probed with a labeled secondary antibody, specific for the primary antibody. For maximum sensitivity, the secondary antibody is typically labeled with either alkaline phosphatase (AP) or horseradish peroxidase (HRP), both of which catalyze a chemiluminescent reaction that can be visualized using a high-sensitivity camera or black-and-white film.

In addition to total protein stains and Western blotting, a number of kits are now available for the direct visualization of His-tagged proteins on polyacrylamide gels, such as Invitrogen's InVision and Thermo's Pierce 6xHis Protein Tag Stain. These reagents allow for the visualization of His-tagged proteins in polyacrylamide gels without Western blotting and are compatible with total protein staining. This technology is based on a His-tag's ability to coordinate and bind metal ions and is similar to His-tag affinity purification as discussed in **Section 5.12**, "Example Experiment: Expression and Purification of Fluorescent Protein Dronpa" section.

## 5.9  PROTEIN ISOLATION AND PURIFICATION

After establishing protein expression conditions and confirming protein expression, protein must be isolated and purified. Protein isolation and purification involve cell lysis followed by chromatographic separation and buffer exchange. The degree of purification required depends on the downstream application. Proteins being prepared for biochemical assays may only need to be 80–90% pure, while proteins being prepared for crystallographic (see **Chapter 6**) or spectroscopic applications (see **Chapter 16**) often need to be greater than 99% pure.

### Native versus nonnative purification

The first step in protein purification is choosing a general strategy for the purification. Proteins can be purified in either their native state or a denatured state. Native protein purification typically gives properly folded and functional proteins following purification, while nonnative purification results in unfolded proteins that must be refolded prior to carrying out biochemical or biophysical experiments. In most cases, native protein purification is preferred, due to the obvious advantage that it directly yields functional proteins. However, native protein purification requires careful section of buffers, detergents, and chromatography conditions to prevent protein denaturation and may even require testing protein function at various stages during the purification to ensure that the protein remains in an active conformation. Additionally, native purification must typically be carried out directly following protein expression, since it can be difficult to determine appropriate storage conditions of cell pellets or lysed protein mixtures.

Nonnative protein purification typically gives much higher yields, since the harsh denaturants and detergents used in the purification are capable of solubilizing inclusion bodies. Furthermore, the protein purification may be stopped and the intermediate material stored at −80°C at any stage of the purification. Therefore, in situations where protein refolding is possible, nonnative protein purification may be advantageous. Protein refolding is typically accomplished through removal of denaturants by dialysis, during affinity chromatography (on *column refolding*), or by rapid dilution. There is no way of predicting the exact conditions for protein refolding a priori. In consequence, typically, an experimenter must try hundreds of different conditions to determine how to properly refold a protein, which requires a high-throughput screen for protein activity.

## Preparation of protein lysate

For nonnative protein purification, cell lysis is accomplished with ionic detergents, such as SDS, in the presence of a concentrated denaturant, typically either 6 M guanidinium hydrochloride or 8 M urea, and a reductant, typically either DTT or β-mercaptoethanol. These harsh solubilization conditions tend to fully solubilize all of the proteins in the cells. However, if any cellular debris remains, it can be removed by centrifugation. The concentrated denaturant or detergent is typically maintained in all of the buffers during the initial chromatographic steps of the purification.

For native protein purification, cell lysis must be carried out under milder conditions than for nonnative protein purification. Cells are typically lysed using mechanical force, nonionic detergents, or a combination of the two. Small-scale mechanical disruption (cell pellets under 10 g) is typically carried out in low-ionic-strength buffers and in the presence of protease inhibitors using a *probe sonicator*. Care must be taken during the sonication to keep the cell suspension cold, since during sonication, the cell suspension can become hot, resulting in protein denaturation. For large-scale mechanical disruption, a *French press* is employed to disrupt the plasma membrane by passing it through a narrow valve under high pressure. In addition to mechanical disruption, high concentrations of nonionic detergents (typically 1%), such as n-octyl-β-D-glucoside (OG) and n-dodecyl-β-D-maltoside (DDM), can be used to disrupt the plasma membrane. A number of kits containing nonionic detergent mixes are also commercially available for cell disruption. In all cases after disruption, the insoluble material must be removed by centrifugation prior to column chromatography.

The purification of membrane proteins represents a unique challenge that is first encountered during the lysis step. For membrane protein purification, the protein itself must be solubilized in a detergent that allows the membrane protein to retain its folded structure. This is particularly challenging for multi-subunit membrane proteins, such as ion channels, since protein multimerization must also be maintained during solubilization. It is difficult to predict which detergents will be successful for membrane protein solubilization, and often, many detergents must be screened. Nonionic detergents, such as OG and DDM, along with lipid mimetic detergents, such a FOS-Choline-16, have been successful for the purification of many bacterial ion channels. After initial solubilization, the detergent must be maintained in all of the purification buffers at a low concentration (typically 0.1–0.01%) in order to keep the membrane protein soluble. In some purification strategies, one detergent is used for solubilization, and a second detergent is used during purification. This strategy is typically employed when an expensive detergent such as FOS-Choline-16 is necessarily for solubilization and has been successful for bacterial ion channels, for example, *MscS* and *bCNG*.

There are two approaches for lysis that can be employed for membrane protein purification. The first approach is to solubilize the protein and plasma membrane in high concentration of nonionic detergent, typically by sonication. This is analogous to the method described previously for native protein purification. The second approach uses a membrane preparation to provide partial purification prior to chromatography. In this approach, the cells are first disrupted mechanically, using either a sonicator or a French press. The process of mechanical disruption shears the plasma membrane into *small unilamellar vesicles* (SUVs; 100–200 nm) containing the membrane proteins. Following disruption, the insoluble cellular debris is pelleted via standard centrifugation that does not pellet the unilamellar vesicles. The supernatant is then pelleted by ultracentrifugation to obtain a membrane pellet of the SUVs produced during mechanical disruption. This membrane pellet can then be resuspended in nonionic detergent and the proteins subjected to chromatographic purification—or if only crude protein purification is required, the membrane pellet can be used directly in biophysical or biochemical assays. This second approach provides a significant degree of purification without chromatographic separation, since most of the proteins in the cell are not membrane proteins, and upon successful overexpression, most of the membrane proteins in the cell are the protein of interest. For example, a crude membrane preparation containing an ion channel may be used in lipid bilayer electrophysiology (see **Chapter 15**) if the channel of interest can be identified with certainty.

# 5.10 CHROMATOGRAPHY

While there are a wide variety of techniques that can be used to enrich samples in the protein of interest, including batch purification and spin column purifications, chromatographic separation is the best method for large-scale protein purification. *Column chromatography* has been used for the purification of microgram quantities of protein to the purification of gram quantities of protein and is a scalable method. The exact columns used and the number of columns required to obtain a pure protein sample are dependent on the protein, the tags used to label the protein, the strain or cell line employed for protein expression, and the required protein purity for the final application.

## Chromatography systems

Protein chromatography can be carried out using simple gravity-flow methods or complex instrumentation and at low, medium, or high pressure. The simplest systems employ a simple chromatography column connected to a fraction collector. A slightly more complicated low-pressure system uses a peristaltic pump to provide a constant flow rate and an inline UV detector to detect protein elution from the column (**Figure 5.7a**). Protein elution is typically detected by absorption at 220 nm of the peptide backbone, absorption at 280 nm of aromatic amino acids, or both. Medium-pressure systems, such as GE Healthcare's fast protein liquid chromatography (FPLC) and Bio-Rad's medium-pressure liquid chromatography (MPLC), allow for higher flow rates and computer control, and incorporate an injection valve or autoinjector.

Standard high-pressure liquid chromatography (HPLC) systems can also be used for protein purification and typically incorporate all of the advantages of the MPLC systems (**Figure 5.7b**). In addition, HPLC systems can be configured with a diode array UV detector in place of a single- or dual-wavelength UV detector. If an HPLC is used for protein purification, it should be plumbed with PEEK tubing (instead of stainless tubing), and a 1000 psi back-pressure regulator should be inserted between the pump (or pumps) and the column when using media that can only

 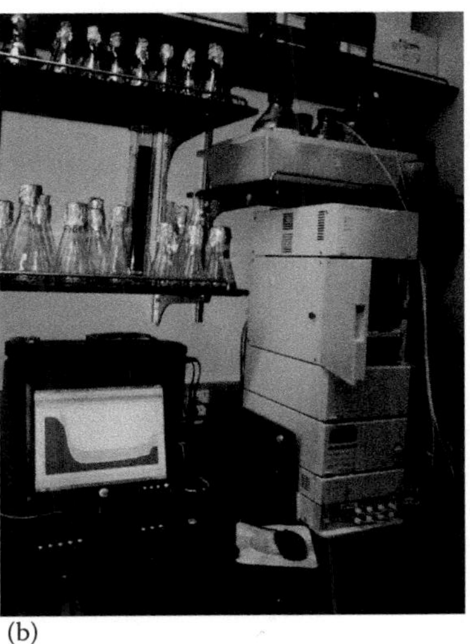

(a)    (b)

Figure 5.7  (a) Picture of a low-pressure peristaltic pump–based protein purification system. (b) Picture of an HPLC system configured for protein purification.

tolerate low pressure. The back-pressure regulator will provide a much smoother pump curve, since the flow rate of HPLC pumps operating at low pressures tends to oscillate greatly with each pump stroke.

MPLC and HPLC systems typically contain either two pumps that can be used for the creation of binary gradients or a low-pressure mixing valve that can be used to mix four to six solutions to create complex buffers (sometimes referred to as inline buffer preparation). Additionally, MPLC and HPLC systems can be configured with a variety of other detectors depending on the application, including conductivity meters and fluorescence detectors. Furthermore, some MPLC and HPLC systems incorporate valves that allow for two-dimensional chromatography. In two-dimensional chromatography, the effluent of one column, at a specific time or absorbance, can be directed onto a second column. This allows for a protein to be purified using two columns in a single chromatographic run.

## Affinity chromatography

Most protein purification schemes begin with an *affinity chromatography* step. In affinity chromatography, the protein is specifically bonded to the column due to a characteristic of the protein and then eluted from the column in a competition experiment. A classic example of affinity chromatography is the purification of the nicotinic acetylcholine receptor from *Torpedo* electroplax. In this purification, the receptor's affinity for α-bungarotoxin, a competitive antagonist, is used for protein purification by immobilizing α-bungarotoxin on a column and then passing the crude protein mixture over the column. In this process, only the nicotinic acetylcholine receptors stick to the column, and the nonspecific bond proteins can be eluted. The nicotinic acetylcholine receptor can subsequently be eluted from the column by addition of free α-bungarotoxin or another competitive agonist or antagonist of the receptor. While this is an extremely effective method for the purification of nicotinic acetylcholine receptors, it is not general and cannot be applied to all proteins.

The development of affinity tags has led to general methods for the rapid enrichment and purification of heterologously expressed proteins. The most

**Figure 5.8 Schematic representation of protein attachment to metal chelate resin via a His-tag.**

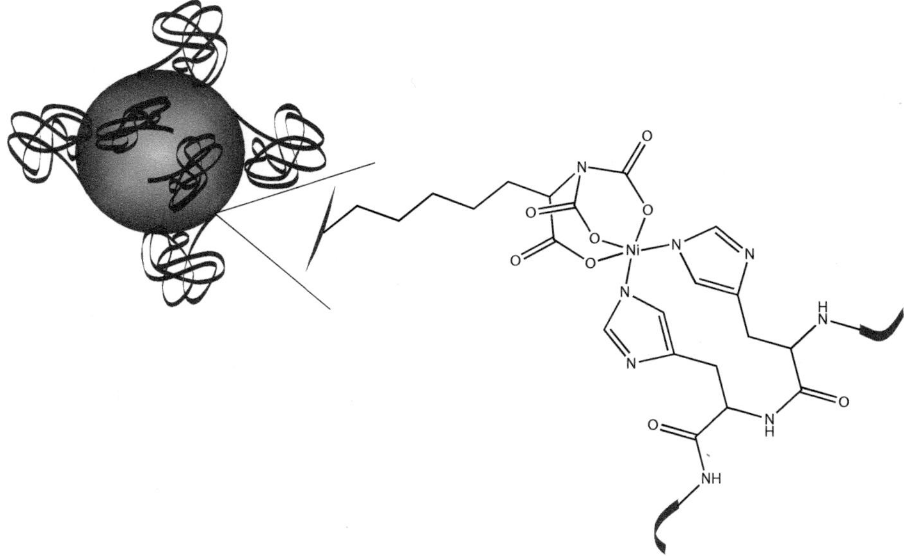

common of these tags is the His-tag. In His-tag protein purification, the protein is bound to a column displaying either immobilized nickel (II) ions or cobalt (II) ions and subsequently competed off with high concentrations of imidazole (**Figure 5.8**). A wide variety of metal chelate resins are available for protein purification, ranging from low-pressure/low-flow-rate resins to high-pressure/high-flow-rate resins. For many applications, a single affinity purification step employing an affinity tag produces protein that is sufficiently pure for subsequent analysis.

Ion exchange chromatography represents a second type of affinity purification and is typically used as a polishing step to further purify proteins subjected to affinity tag purification. In ion exchange chromatography, the protein is bonded to a negatively or positively charged column through electrostatic interaction. How tightly a protein binds to a specific column depends on its isoelectric point (pI) and the charge of the column. Proteins that do not denature at a pH above their pI can be purified using anion exchange chromatography, and proteins that do not denature at a pH below their pI can be purified using cation exchange chromatography. In ion exchange chromatography, proteins are eluted either by increasing the ionic strength of the eluent or by changing the pH toward the pI of the protein, thus making the protein neutral. As with metal chelate resins, a wide variety of resins are available for protein purification, ranging from low-pressure/low-flow-rate resins to high-pressure/high-flow-rate resins.

## Size exclusion chromatography

Size exclusion chromatography is typically employed as the final step in a protein purification scheme and is sometimes referred to as gel permeation chromatography (GPC). Just as it sounds, in size exclusion chromatography, beads containing holes of defined sizes are used to separate proteins based on size and shape. Small proteins enter the holes in the beads, while larger proteins are too big to fit in the holes and are excluded from the beads (**Figure 5.9**). As a result, in size exclusion chromatography, large proteins are eluted first followed by smaller proteins and buffer molecules. GE Healthcare's Sephadex, a polydextran gel, is

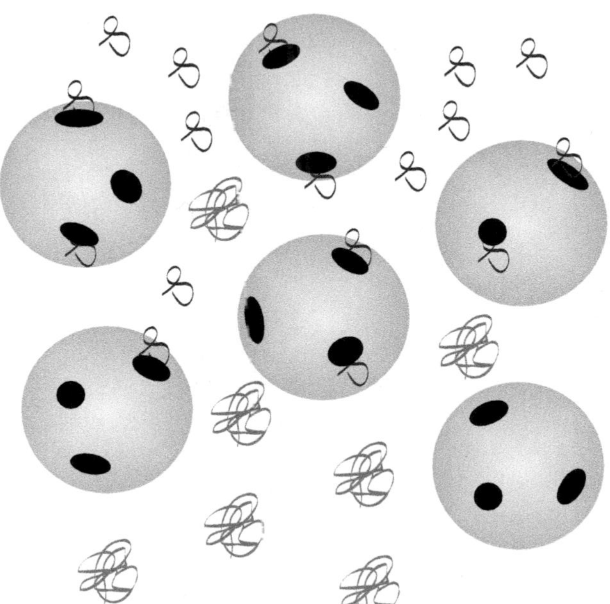

**Figure 5.9 Schematic representation of how proteins are separated during size exclusion chromatography.**

the most common size exclusion resin used for protein purification and comes in a variety of sizes, pressures, and flow rates. The Sephadex used for size exclusion chromatography is Sephadex G-xx, where xx represents the size of holes within the beads, with a larger number corresponding to bigger holes. For example, Sephadex G-50 is useful for separating proteins between 1.5 and 30 kDa, and Sephadex G-200 is useful for separating proteins between 5 and 800 kDa. As with SDS-PAGE, size exclusion chromatography can be used to estimate the molecular weight of a protein (assuming the protein has a globular shape) by comparison with known molecular weight standards.

# 5.11  BUFFER EXCHANGE AND CONCENTRATION

The final step in preparing a protein for a biochemical or a biophysical assay is typically buffer exchange and protein concentration. This step is especially important if an affinity chromatography step was the last step in protein purification, since elution typically involves high concentrations of salt or ligands (imidazole in the case of His-tag purification). Since high concentrations of salt or ligand can interfere in many assays, it is often necessary to exchange the purified protein into an appropriate buffer for the assay. Additionally, it may be necessary to increase the protein concentration, if the purified protein is too dilute for the desired application.

## Buffer exchange

There are two main methods to achieve buffer exchange: dialysis and gel filtration. In dialysis, the protein is loaded into a dialysis bag or cassette, which is then placed into a chamber containing the desired buffer (**Figure 5.10**). The dialysis bag or cassette consists of a membrane containing holes large enough to allow buffer molecules to exchange with the external solution but too small for the protein

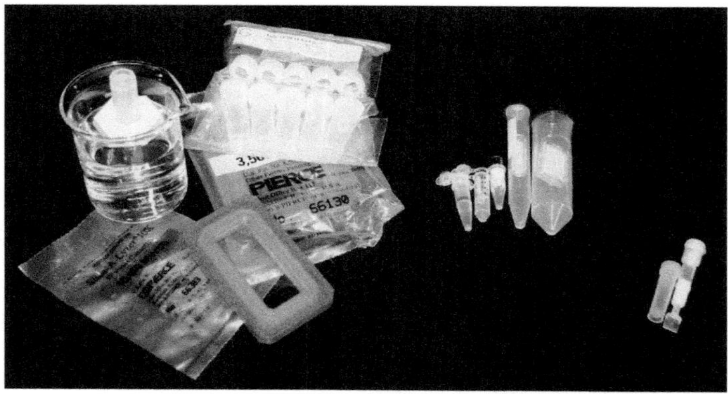

**Figure 5.10 Picture of dialysis devices (left), centrifugal concentrators (center), and a spin gel filtration column (right).**

to escape. Typically, the buffer in the dialysis bag and the outer chamber are allowed to come to equilibrium, and then the outer chamber is replaced with new buffer. This process can be repeated multiple times to achieve nearly complete buffer exchange. A wide variety of dialysis products are available, ranging from traditional dialysis bags to dialysis cassettes and dialysis caps for small-scale buffer exchange. Dialysis products also come in assortment of molecular weight cutoffs (MWCO), where the MWCO corresponds to the molecular weight of the largest globular protein that could pass through the membrane. It is typically best to use the largest MWCO possible for the protein of interest, since the larger the holes, the more rapid the buffer exchange.

Gel filtration provides a rapid method for buffer exchange using size exclusion chromatography. Typically, gel filtration is carried out using either Sephadex G-10 or G-50 size exclusion media depending on the size of the protein. For gel filtration, the protein should be *excluded* from entering the beads, which results in the buffer molecules becoming trapped in the beads and the protein being eluted in the buffer used to equilibrate the size exclusion media. Gel filtration can be carried out on short columns or using spin columns similar to those employed for DNA minipreps. (See Chapter 2 and Figure 5.10.)

## Protein concentration

The classical methods for protein concentration are protein precipitation using trichloroacetic acid (TCA) and lyophilization. However, these methods tend to cause protein denaturation, resulting in protein that is not useful for most biochemical or biophysical studies. To prevent protein denaturation during concentration, a number of methods have been developed for protein concentration based on dialysis. The most common of these methods is centrifugal concentration. In this method, the protein solution is centrifuged through a dialysis membrane with a cutoff below the molecular weight of the protein (Figure 5.10). During centrifugation, buffer passes through the membrane, leaving a concentrated protein solution. The downside to centrifugal concentration is that typically, up to 50% of the protein is irreversibly adsorbed to the dialysis membrane during concentration despite manufacturer claims to the contrary. Another option for protein concentration using dialysis is to dialyze the protein against a solution containing a high concentration of a high-molecular-weight compound that will osmotically absorb water across the dialysis membrane.

# 5.12 EXAMPLE EXPERIMENT: EXPRESSION AND PURIFICATION OF FLUORESCENT PROTEIN DRONPA

You wish to make several milligrams of the photoswitchable fluorescent protein Dronpa. This is available commercially in a bacterial expression vector without any tags. The first step is to subclone the gene into the appropriate expression vector with a tag or tags of choice. A simple affinity tag will be fine; there is no need to use antibodies to detect this protein, because it is so fluorescent. Because this is a soluble protein related to the green fluorescent protein family, there is no reason to think that it will not express well in bacteria. So you decide to use the expression vector pEColi-Nterm-6xHN (Clontech), which has a His-tag followed by an enterokinase cleavage site for future removal of the protein (**Figure 5.11**). Ensure that the Dronpa gene is cut out of its parent vector with a restriction enzyme that puts it in frame with the tag. If such a site does not exist, use polymerase chain reaction (PCR) to amplify the Dronpa fragment in such a way that it can be inserted in frame.

This vector expresses from the T7 *lac* promoter and has ampicillin resistance. So selection will be done using ampicillin, and induction using IPTG. It is important to be aware of the difference between plasmid selection and protein induction. Before induction, the bacteria are exposed to ampicillin to select for cells containing the plasmid, but the plasmids are not producing significant amounts of Dronpa yet, since the lac promoter has not been turned on. Only with induction using IPTG will the fluorescent protein be made and become visible.

Because Dronpa is expected to be nontoxic, an ordinary BL21 strain is chosen for expression. Transform the bacteria with the new plasmid, pEColi-Nterm-6xHN-Dronpa, and plate on a selective (ampicillin) plate. The next day, it will have plasmid-containing colonies; use it right away or store it at 4°C for up to a month. Two days before the desired purification, seed one or more colonies into 3–5 mL of the chosen medium (we like TB) with ampicillin. After 8–12 h, seed ~1 mL of this culture into a larger culture (100–1000 mL), depending upon the total amount of protein desired.

It is a good idea to do a test induction at this stage. We like to do this with a single 50 mL culture. Add the IPTG and return to the shaking incubator at 37°C. (This is such an easy protein that we will not vary the temperature.) Every 60 min, remove 1 mL of the culture for analysis. Because Dronpa is highly fluorescent, there is no need to use SDS-PAGE with it; the level of protein expression can simply be measured by the fluorescence of the removed sample. Continue for at least 16 h and determine the optimal induction time as measured by the greatest fluorescence. We usually induce Dronpa for 12–16 h; the bacteria look green!

Now you are ready for the real thing. First determine which affinity columns you will use and the size needed. Clontech provides different column types specifically for use with its pEcoli vectors, some of which maximize binding capacity and others that maximize purity. Aiming for the latter, we choose the

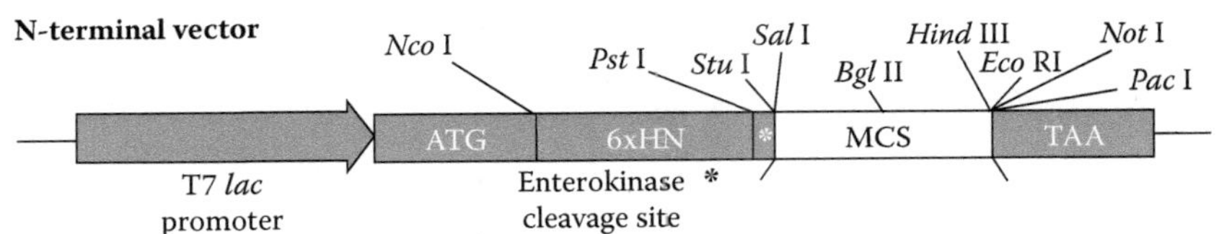


Figure 5.11 Multiple cloning site (MCS) region of pEcoli-Nterm-6xHN (Clontech), showing a His-tag, enterokinase cleavage site, and restriction sites for cloning.

HisTALON columns. These are sold either as an active resin or prepacked into columns (**Figure 5.12**). Checking the manufacturer's specifications shows that 1 mL of resin will bind 5–10 mg of protein. Choose the column size accordingly depending upon the amount of protein desired. These are gravity-flow columns, meaning that no pump system is needed to drive the protein through the column.

Two days before the experiment, seed a colony or colonies into 5 mL of ampicillin-containing medium and then into a larger flask after 8–12 h. How much total culture to prepare? Expression levels vary a lot by protein type. Expect tens to hundreds of milligrams of protein per liter of culture; for a small-scale purification (e.g., one 3 mL column), a liter of culture is a good place to start, either as one single culture or divided into two flasks of 500 mL each. Allow these to grow to mid-log, and then induce for the optimal time as determined previously. Induction frequently ends at an inconvenient time of day, and the bacterial cells may be pelleted and drained, and the pellet stored at −80°C until you are ready to use them. Once the purification begins, however, you are committed; do not stop in the middle, or proteases and other contaminants will destroy your protein! Also, be careful which tubes or bottles you store at −80°C. Some can tolerate that temperature, and say so on their specifications; others will crack.

While the protein is inducing, prepare the buffers according to the manufacturer's specifications. There will be an equilibration buffer for preparing the column; a wash buffer that will remove impurities but not detach the His-tagged protein; and an elution buffer to wash the tagged protein off the column. For the HisTALON columns, the recipes are as follows:

*Equilibration:* 50 mM Na phosphate, 0.3 M NaCl, pH 7.4.

*Elution:* 50 mM Na phosphate, 0.3 M NaCl, 150 mM imidazole, pH 7.4.

*Wash:* 50 mM Na phosphate, 0.3 M NaCl, 10 mM imidazole, pH 7.4 (can be made by mixing 34 mL:0.66 mL equilibration buffer: elution buffer).

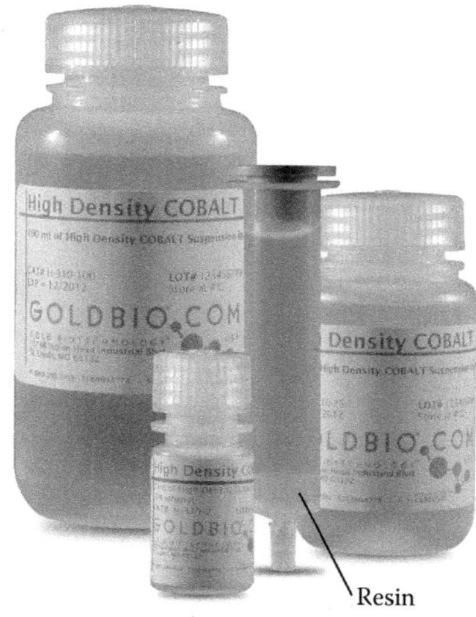


**Figure 5.12 Appearance of resin and gravity-flow column.** The resin near the end is a certain fraction of the column volume; the resin volume determines the binding capacity. (Photographed by Amanda Sneed. Courtesy of Gold Biotechnology, Inc.)

Because Dronpa is a soluble protein, we wish to get the soluble fraction from the induced *E. coli* pellet and remove all of the membranes and cell walls. This may be done using a probe sonicator or French press as described previously, but the HisTALON kit recommends the use of their special xTractor lysis buffer. Use 2 mL of this buffer per 100 mg of cell pellet, and pipette up and down to suspend the pellet well in the buffer. Then add DNase I to digest DNA and incubate on ice for 15 min with intermittent mixing.

Regardless of the method of cell lysis used, centrifugation of the lysate at 10,000 g for 20 min at 4°C is used to remove the undesirable material and produce a *cleared lysate*. This lysate contains your protein, so in the case of Dronpa, it should be powerfully, almost overwhelmingly, green. If it is not, something is wrong. It is important to note that the cell lysis step can be tricky, and is one of the major reasons for low protein yields. Practicing with an easy protein such as Dronpa can help you troubleshoot these methods.

While the lysate is centrifuging, the column may be equilibrated. All of the buffers should be degassed, and everything should be brought to room temperature. Pass 5–10 column volumes of the equilibration buffer over the column, allowing the buffer to drip into a waste receptacle.

Now you are ready to apply the cleared lysate to the column, if and only if said lysate does not contain any reagents that would damage the column or interfere with its function. For HisTALON, these include chelators such as EDTA or some chelators that are present in protease inhibitors, and strong reducing agents such as DTT. If any of these are present, run the lysate over a desalting column before applying it to the TALON column.

After applying the lysate to the column, make sure that what drips out the end of the column is not green. If it is, this means that your protein is failing to bind. There are various reasons this may be true—the His-tag may not have been expressed properly, you made have accidentally cut it out while cloning, or it may be inaccessible to the resin; in the latter case, purifying under denaturing conditions may be needed.

If all goes well, the green will stay inside the column at this step. Now wash with 8 column volumes of equilibration buffer followed by 7 volumes of wash buffer, allowing these buffers to run to waste. Finally, prepare an array of eight to sixteen 1.5 mL Eppendorf tubes. Add a column volume of elution buffer and watch for the migration of the green. Add another volume. Keep doing this until the green starts to come out the bottom of the column; collect this green fraction in 1 mL increments in the Eppendorf tubes.

All of the protein should be eluted within 5–8 column volumes. Again, having a brightly colored protein helps a lot. If the protein is slow to elute, just keep adding more elution buffer. The fractions will show a bell-shaped curve of fluorescence (or absorbance). They should be analyzed by absorption and emission spectroscopy (**Figure 5.13**) and by SDS-PAGE.

When the protein is not so brightly colored, the elution step can be tricky. In this case, it will be necessary to collect more elution fractions, perhaps all of them, and analyze them all by absorption spectroscopy (peak at 280 nm), by the *Bradford assay*, by functional assays, and/or by SDS-PAGE.

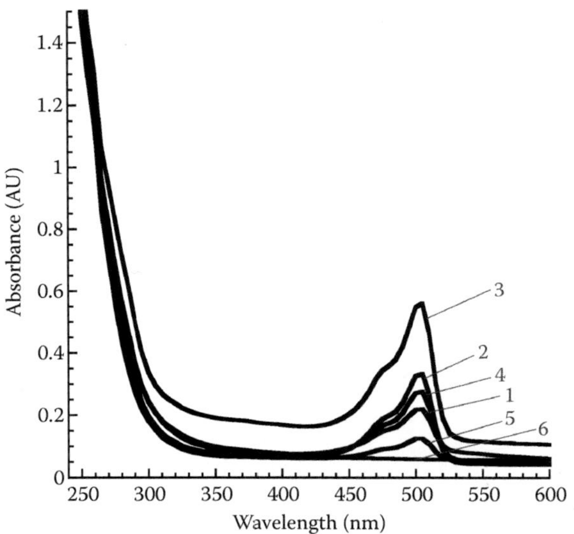

**Figure 5.13 Absorbance spectra of different elution fractions of Dronpa (1–6).** The fractions were collected based upon their appearance—that is, no fractions were saved until the eluate appeared bright green. The total absorbance at 280 nm can indicate if there are other proteins present by comparing to the concentration of Dronpa calculated using the published extinction coefficient $\varepsilon_{503} = 95,000 \ M^{-1} \ cm^{-1}$.

## 5.13 CONCLUSIONS AND FINAL REMARKS

A protein that is soluble, brightly colored or fluorescent, and nontoxic to *E. coli* may be purified by a rank beginner, usually without too many issues. Most problems encountered are those of laziness—failure to measure $OD_{600}$ of the cultures or to determine optimum induction times. Incomplete bacterial lysis can also be a serious problem, and the techniques should be mastered with an easy protein if possible. Sometimes, a particular type of column simply does not work for the experiment or experimenter in question. There are many different types of affinity columns made for His-tags, and changing the style or brand can make all the difference.

On the other end of the spectrum, many proteins are extremely challenging to express and purify, and biochemists can spend their careers determining and refining methods for these molecules. For any experiment involving proteins that cannot be expressed in *E. coli*, the advice of an expert is advised. Many commercial suppliers now offer yeast and insect cell expression, so that only the plasmid need be supplied, and a cell pellet containing the expressed protein is provided. This does not eliminate all of the challenges, of course, but it saves your lab from having to establish eukaryotic expression systems.

Once the protein is purified, it may be used in a wide variety of biophysical and biochemical applications. It might be crystallized (Chapter 6), functionalized onto a surface (Chapter 14), or used in electrophysiology (Chapter 15) and various forms of spectroscopy (Chapter 16). Even if you never attempt a highly challenging purification, it is good to know the basic techniques, which have been made very straightforward with the commercial availability of plasmids, bacterial strains, and purification columns.

## End-of-Chapter Questions

1. Which expression host would you choose for expression of (a) small quantities (milligrams) of a bacterial protein; (b) small quantities of a human transmembrane protein; (c) large quantities (kilograms) of a soluble human protein; (d) large quantities of a human protein that requires substantial posttranslational glycosylation?

2. Describe the advantages and disadvantages of expression using *E. coli*, yeast, and mammalian cells.

3. The first step in seeing whether a protein can be produced in *E. coli* is to see whether the protein is soluble in the bacteria, or whether it forms inclusion bodies. If it is not soluble, is there anything to try before you give up and use a different expression system?

4. It is very useful to use *E. coli* for protein expression, but prokaryotes lack certain posttranslational modification mechanisms. List three of these mechanisms and discuss how the lack can be compensated in order to use *E. coli* to produce therapeutic proteins for human use. Discuss these proteins, specifically insulin, erythropoietin, and interleukin-2.

5. When would you choose Coomassie blue staining for a gel, and when would you choose an immunohistochemical method?

6. What is the isoelectric point (pI) of a protein, and what determines it? How is it measured?

7. How are strong and weak anion and cation exchangers defined? Under which circumstances is each used? Give an example of each type of exchanger.

8. Imagine that you do not know the isoelectric point of a particular protein, so you are not sure whether to use a cation- or anion exchange column. What can you do, and what results would you expect?

9. Explain what type of ion exchange resin should be used to purify the following proteins, assuming a buffer at physiological pH (7.4): human acidic fibroblast growth factor (pI = 4.6); human basic fibroblast growth factor (pI = 8.2); cytochrome c (pI = 10.6); transferrin (pI = 5.9).

10. Imagine that you are too broke to buy restriction enzymes off the shelf. Could you imagine purifying one or more of these from bacteria? If so, how? Detail the purification steps. Pick a specific enzyme as an example if you wish.

11. A published procedure for purification of Bam HI (Jack et al., *Nucleic Acids Research*, vol. 19, no. 8, 1825) uses the following columns for purification, in this order: phosphocellulose with an NaCl gradient; hydroxyapatite with a potassium phosphate gradient; heparin sepharose with an NaCl gradient; and Q sepharose with an NaCl gradient. What is the rationale for these choices and, what do you expect to see after each step?

12. To find out whether the cystic fibrosis transmembrane conductance regulator (CFTR) protein really is an ion channel, Bear et al. purified the protein and observed its behavior when placed in phospholipid vesicles. Give a brief description of the protocol. Do you expect this to work for other transmembrane proteins?

13. Strychnine is a poison because it binds strongly to the neuroinhibitory glycine receptor. Discuss how you might use this face to purify glycine receptor proteins.

14. Which of the following proteins can be effectively separated using a size exclusion column with a molecular weight range of 25–100 kDa? What would be the order of elution? Myosin (molecular weight = 200,000); β-galactosidase (molecular weight = 116,250); myoglobin (molecular weight = 17,000); ovalbumin (molecular weight = 45,000).

15. Describe the role of the following in protein purification: lysozyme; DTT; β-mercaptoethanol; nitrocellulose membrane.

16. You have purified a new protein but do not know anything about it yet. Size exclusion chromatography of the native protein gives a molecular weight of 300 kDa. When 6 M guanidine hydrochloride is added, there is a single peak at 60 kDa. When both 6 M guanidine hydrochloride and 10 mM β-mercaptoethanol are added, there are now two peaks: one at 15 kDa and one at 45 kDa. What can you infer about the protein?

**17.** What problems can lead to protein flow-through in a column?

**18.** How does ammonium sulfate precipitate proteins? How about urea?

**19.** Give a brief description of how size exclusion chromatography works and its major advantages. This technique can suffer from broad peaks, aggregation, and interference by salts. Discuss how the following can cause problems and how they can be optimized: flow rate; pH; salt concentration; column length.

**20.** The image below shows a size exclusion chromatography calibration curve. For what molecular weights is this column appropriate? Indicate the permeation limit and exclusion limit.

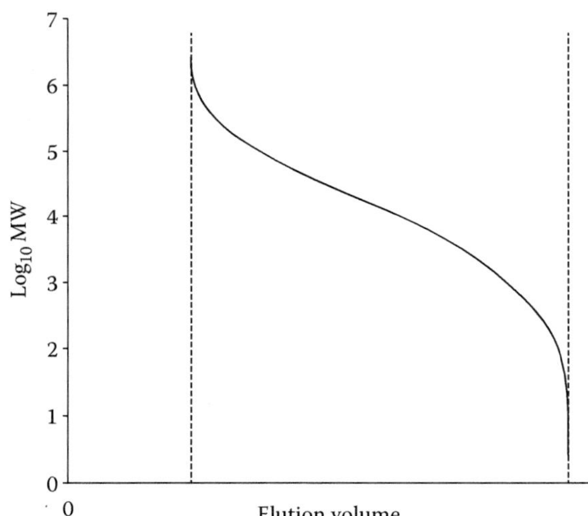


**21.** The chromatography resolution equation expresses resolution as a function of three variables: column efficiency, column selectivity, and solute retention. Identify each of these factors in the equation below and explain how each can be changed experimentally. Which of them do you expect to have the greatest influence on the ability to separate peaks?

$$R \sim \frac{\sqrt{N}}{4} \frac{(\alpha - 1)}{\alpha} \frac{k}{k+1}$$

**22.** You are purifying a new protein and run it through a few logical steps: precipitation with ammonium sulfate; desalting on a PD-10 column; an ion exchange step; and an affinity chromatography step. Given the chart below, calculate the specific activities and determine whether all of these steps were necessary.

| Step | Total Protein (mg) | Total Activity (U) | Specific Activity (unit?) |
|---|---|---|---|
| Before purification | 100 | 100,000 | |
| Precipitation | 50 | 60,000 | |
| Desalting | 40 | 48,000 | |
| Ion exchange | 30 | 36,000 | |
| Affinity | 3 | 30,000 | |

**23.** You are trying to characterize a new protein, and find that it can only be extracted from the cells using SDS. You have it sequenced and find that it is 413 amino acids long, but your measured value of the molecular weight is 50 kDa. What kind of protein is this?

**24.** Give examples of ionic, nonionic, and zwitterionic detergents and discuss the pros and cons of each for membrane protein solubilization. How would you pick the best detergent for your application?

**25.** An important blood test for diabetics is the measurement of glycosylated hemoglobin levels, which give a measure of the degree to which the hyperglycemia has been controlled. Discuss how affinity chromatography could be used for this test.

**26.** Suppose you have a mixture of the following proteins: myoglobin (pI = 7.0, MW = 16.7 kDa); bovine serum albumin (BSA; pI = 4.7, MW = 66.2 kDa); lactate dehydrogenase (LDH; pI = 8.5, MW = 35.0 kDa); lysozyme (pI = 9.3, MW = 14.4 kDa); aprotinin (pI = 10.5, MW = 6.5 kDa). Indicate the order in which they will elute from (a) a size exclusion column; (b) a cation exchange column in acetate buffer, pH 4.76, eluted with NaCl; (c) an anion exchange column in Tris, pH 8.0.

**27.** How can size exclusion chromatography be used for protein folding/refolding studies?

## Solutions

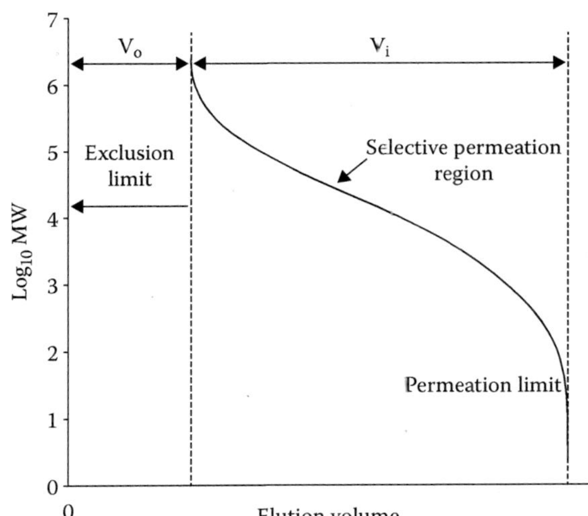


## Background Reading

### Books

Ausubel, F.M. et al. *Short Protocols in Molecular Biology*. Edn. 5. John Wiley & Sons, Inc., Hoboken, NJ, 2002.—An invaluable resource for molecular biology protocols including protein analysis.

Freifelder, D. *Physical Biochemistry: Applications to Biochemistry and Molecular Biology*. Edn. 2. W. H. Freeman and Company, New York, 1982.—An excellent textbook on physical separation methods and applications.

NovaGen. *pET System Manual*. 11th edition. Merk, 2006.—A valuable resource for bacterial protein expression.

*Recombinant Protein Purification Handbook: Principles and Methods*. GE Healthcare.—A good over view of different chromatographic strategies.

Simpson, R.J. *Basic Methods in Protein Purification and Analysis: A Laboratory Manual*. Cold Springs Harbor Laboratory Press, Cold Spring Harbor, NY, 2009.—A good resource for protein purification and analysis protocols.

Taylor, G. *Current Protocols in Protein Science*. John Wiley & Sons, Inc. Hoboken, NJ—A books series that is an invaluable reference for modern protein expression and purification protocols.

### Journal articles

Arnold, F.H., and Haymore, B.L. (1991). Engineered metal-binding proteins: Purification to protein folding. *Science* 252, 1796–1797.

Bear, C.E., Li, C.H., Kartner, N., Bridges, R.J., Jensen, T.J., Ramjeesingh, M., and Riordan, J.R. (1992). Purification and functional reconstitution of the cystic fibrosis transmembrane conductance regulator (CFTR). *Cell* 68, 809–818.

Bock, H.G., Skene, P., Fleischer, S., Cassidy, P., and Harshman, S. (1976). Protein purification: Adsorption chromatography on controlled pore glass with the use of chaotropic buffers. *Science* 191, 380–383.

Brondyk, W.H. (2009). Selecting an appropriate method for expressing a recombinant protein. *Methods in Enzymology* 463, 131–147.

Cheeseman, I.M., and Desai, A. (2005). A combined approach for the localization and tandem affinity purification of protein complexes from metazoans. *Science's STKE* 2005, pl1.

Cregg, J.M., Tolstorukov, I., Kusari, A., Sunga, J., Madden, K., and Chappell, T. (2009). Expression in the yeast *Pichia pastoris*. *Methods in Enzymology* 463, 169–189.

Dietrich, A., Meister, M., Spicher, K., Schultz, G., Camps, M., and Gierschik, P. (1992). Expression, characterization and purification of soluble G-protein beta gamma dimers composed of defined subunits in baculovirus-infected insect cells. *FEBS Letters* 313, 220–224.

Drew, D., Lerch, M., Kunji, E., Slotboom, D.J., and De Gier, J.W. (2006). Optimization of membrane protein overexpression and purification using GFP fusions. *Nature Methods* 3, 303–313.

Drew, D., Slotboom, D.J., Friso, G., Reda, T., Genevaux, P., Rapp, M., Meindl-Beinker, N.M., Lambert, W., Lerch, M., Daley, D.O. et al. (2005). A scalable, GFP-based pipeline for membrane protein overexpression screening and purification. *Protein Science* 14, 2011–2017.

Feng, Y., Zhang, M., Zhang, L., Zhang, T., Ding, J., Zhuang, Y., Wang, X., and Yang, Z. (2012). An automatic refolding apparatus for preparative-scale protein production. *PLoS One* 7, e45891.

Greene, P.J., Heyneker, H.L., Bolivar, F., Rodriguez, R.L., Betlach, M.C., Covarrubias, A.A., Backman, K., Russel, D.J., Tait, R., and Boyer, H.W. (1978). A general method for the purification of restriction enzymes. *Nucleic Acids Research* 5, 2373–2380.

Hannig, G., and Makrides, S.C. (1998). Strategies for optimizing heterologous protein expression in *Escherichia coli*. *Trends in Biotechnology* 16, 54–60.

Hays, F.A., Roe-Zurz, Z., and Stroud, R.M. (2010). Overexpression and purification of integral membrane proteins in yeast. *Methods in Enzymology* 470, 695–707.

Kamionka, M. (2011). Engineering of therapeutic proteins production in *Escherichia coli*. *Current Pharmaceutical Biotechnology* 12, 268–274.

Lathrop, J.T., and Hammond, D. (2007). The Bead blot: A method for selecting small molecule ligands for protein capture and purification. *Nature Protocols* 2, 3102–3110.

Ma, H., Mclean, J.R., Chao, L.F., Mana-Capelli, S., Paramasivam, M., Hagstrom, K.A., Gould, K.L., and Mccollum, D. (2012). A highly efficient multifunctional tandem affinity purification approach applicable to diverse organisms. *Molecular and Cellular Proteomics* 11, 501–511.

Maine, G.N., Li, H., Zaidi, I.W., Basrur, V., Elenitoba-Johnson, K.S., and Burstein, E. (2010). A bimolecular affinity purification method under denaturing conditions for rapid isolation of a ubiquitinated protein for mass spectrometry analysis. *Nature Protocols* 5, 1447–1459.

Meij, P., Vervoort, M.B., Meijer, C.J., Bloemena, E., and Middeldorp, J.M. (2000). Production monitoring and purification of EBV encoded latent membrane protein 1 expressed and secreted by recombinant baculovirus infected insect cells. *Journal of Virological Methods* 90, 193–204.

Middelberg, A.P. (2002). Preparative protein refolding. *Trends in Biotechnology* 20, 437–443.

Morr, J., Rundstrom, N., Betz, H., Langosch, D., and Schmitt, B. (1995). Baculovirus-driven expression and purification of glycine receptor alpha 1 homo-oligomers. *FEBS Letters* 368, 495–499.

Nallamsetty, S., and Waugh, D.S. (2007). A generic protocol for the expression and purification of recombinant proteins in *Escherichia coli* using a combinatorial His6-maltose binding protein fusion tag. *Nature Protocols* 2, 383–391.

Newstead, S., Kim, H., Von Heijne, G., Iwata, S., and Drew, D. (2007). High-throughput fluorescent-based optimization of eukaryotic membrane protein overexpression and purification in Saccharomyces cerevisiae. *Proceedings of the National Academy of Sciences of the United States of America* 104, 13936–13941.

Pihlasalo, S., Auranen, L., Hanninen, P., and Harma, H. (2012). Method for estimation of protein isoelectric point. *Analytical Chemistry* 84, 8253–8258.

Rege, K., and Heng, M. (2010). Miniaturized parallel screens to identify chromatographic steps required for recombinant protein purification. *Nature Protocols* 5, 408–417.

Sorensen, H.P., Sperling-Petersen, H.U., and Mortensen, K.K. (2003). Dialysis strategies for protein refolding: Preparative streptavidin production. *Protein Expression and Purification* 31, 149–154.

Spirin, A.S. (2004). High-throughput cell-free systems for synthesis of functionally active proteins. *Trends in Biotechnology* 22, 538–545.

Structural Genomics Consortium, Architecture et Fonction des Macromolécules Biologiques, Berkeley Structural Genomics Center, China Structural Genomics Consortium, Integrated Center for Structure and Function Innovation, Israel Structural Proteomics Center, Joint Center for Structural Genomics, Midwest Center for Structural Genomics, New York Structural GenomiX Research Center for Structural Genomics, Northeast Structural Genomics Consortium et al. (2008). Protein production and purification. *Nature Methods* 5, 135–146.

Tratner, I., De Togni, P., Sassone-Corsi, P., and Verma, I.M. (1990). Characterization and purification of human fos protein generated in insect cells with a baculoviral expression vector. *Journal of Virology* 64, 499–508.

Tsai, A., and Carstens, R.P. (2006). An optimized protocol for protein purification in cultured mammalian cells using a tandem affinity purification approach. *Nature Protocols* 1, 2820–2827.

Tschantz, W.R., Pfeifer, N.D., Meade, C.L., Wang, L., Lanzetti, A., Kamath, A.V., Berlioz-Seux, F., and Hashim, M.F. (2008). Expression, purification and characterization of the human membrane transporter protein OATP2B1 from Sf9 insect cells. *Protein Expression and Purification* 57, 163–171.

Waugh, D.S. (2011). An overview of enzymatic reagents for the removal of affinity tags. *Protein Expression and Purification* 80, 283–293.

Waugh, D.S. (2005). Making the most of affinity tags. *Trends Biotechnol* 23(6), 316–320.

Wu, W.Y., Mee, C., Califano, F., Banki, R., and Wood, D.W. (2006). Recombinant protein purification by self-cleaving aggregation tag. *Nature Protocols* 1, 2257–2262.

Zerbs, S., Frank, A.M., and Collart, F.R. (2009). Bacterial systems for production of heterologous proteins. *Methods in Enzymology* 463, 149–168.


## Online resources

*Some major suppliers of products for expression and purification products*

(Note that this is a very small sample of what is available. A custom search will allow you to identify the best products for your experiments and geographical area.)

Merck: http://www.merck-chemicals.com/life-science-research. Major supplier of expression vectors (pET series), expression strains and cell lines, protein standards, antibiotics, in vitro expression systems, media, and protein purification kits.

Invitrogen: http://www.invitrogen.com. Expression strains and cell lines, expression vectors, gel and blotting supplies, media, protein stains, and purification kits.

GE Healthcare: http://www.gelifesciences.com. Protein purification systems, gel and blotting supplies, protein standards, and protein purification resins.

Bio-Rad: http:// www.bio-rad.com. Protein purification systems, gel and blotting supplies, protein standards, dialysis supplies, and protein purification resins.

Applied Biosystems: http://www.appliedbiosystems.com. High-pressure affinity chromatography resins (POROS), PEEK columns.

Qiagen: http://www.qiagen.com. Expression vectors and affinity purification resins.

New England Biolabs: http://www.neb.com. Expression vectors and cell lines (yeast expression systems).

Thermo Scientific: http://www.thermofisher.com. Dialysis supplies, gel and blotting supplies, protein standards, and protein purification resins.

*Sources of genes*

The American Type Culture Collection (ATCC): http://www.atcc.org.

Open Biosystems—now part of GE Healthcare: http://dharmacon.gelifesciences.com/about-us/about-open-biosystems/

AddGene: http://www.addgene.org. A plasmid-sharing service with over 45,000 different plasmids as of early 2016.

# CHAPTER 6

# Protein Crystallization

Oliver M. Baettig and Albert M. Berghuis

## 6.1 INTRODUCTION

When we think of crystals, we usually think of diamonds (covalent crystals) or maybe salt (ionic crystals). However, proteins also crystallize (macromolecular crystals). These crystals have highly divergent physical characteristics but share one significant and important property: they are always solids with a repeating orderly pattern. In this chapter, we will focus on the crystallization of proteins. In other words, we will describe ways of getting proteins to align themselves in an ordered pattern. It is this pattern formation that ultimately allows crystallographers to solve protein structures and analyze their functions. A large and well-ordered protein crystal diffracts better than one that is small and/or less ordered. Thus, obtaining a large (~0.2 mm in all dimensions) and high-quality crystal is critical when studying protein structures using diffraction methods such as x-ray crystallography. However, despite significant efforts in optimizing and automating protein crystallization, obtaining a well-diffracting crystal for structural analysis continues to often be the bottleneck in structure determination of proteins. The intent of this chapter is to provide an insight into protein crystallization so that you will know where to begin after purification of a target protein.

The term crystallization describes the process of crystal formation (nucleation) and its subsequent growth. Hemoglobin was the first protein documented to be crystallized, in 1840 by the chemist Hünefeld. After adding a drop of blood to a microscope slide, he added the coverslip but serendipitously neglected to seal it. Under the microscope, he then observed the formation of "red bodies," which he aptly called "blood crystals." Today, we know he observed the first macromolecular crystals and reported the first published method of crystallization.

Deliberate, methodical attempts at crystallizing macromolecules for the purpose of structure determination began in Cambridge in the 1950s and continue to the present day. The first protein structure solved by x-ray analysis was myoglobin from sperm whale in 1958. Being able to study the three-dimensional structure of proteins revolutionized the field of molecular cell biology. For the first time, a protein's function could be linked to its three-dimensional structure. Today, protein structures are central in characterizing binding interactions, in designing and improving new drugs, in engineering novel protein functions, and more. New techniques to study crystals are also rapidly being developed, some of which allow for time-resolved analysis of changes in protein structures required for biological activity.

Theoretically, macromolecular protein crystallization is straightforward: purify soluble protein, allow the protein solution to supersaturate, and under the right conditions, protein crystals will nucleate and grow. The difficult part is to purify the protein properly and to find the right crystallization conditions (buffer, salt, temperature, etc.). The principles behind purification of proteins for most molecular biology applications are the same as those for purification for the purpose of crystallization. However, for crystallization, the sample needs to be as pure and homogeneous as possible.

Much effort has been invested in improving and rationalizing approaches to protein crystallization in order to increase the success of producing a well-diffracting protein crystal. Some references are given at the end of the chapter to key resources on the subject. However, despite a half century of studying and improving methods of protein crystallization and an exponential growth in the number of proteins crystallized during that time, protein crystallization still remains a process of trial and error. Many crystallographers feel that protein crystallization is more an art than a science.

Obtaining a protein crystal is only the first step in the pursuit of a protein structure determination. Once suitable protein crystals are obtained, diffraction patterns must be collected, and the *phase problem* must be solved to transform the data and allow for creation of a molecular model. These procedures are outside the scope of this chapter, and we recommend that a protein crystallographer be consulted to help with the design and implementation of any new crystallization experiment. As you will find out in the following sections, crystallizing a protein is seldom straightforward, and the possibilities of what to do next can be very daunting.

When protein structures are solved, they are deposited online in the Protein Data Bank. Crystallographers usually obtain protein structures in order to answer biological questions that are related to the structures. If the structure of your protein of interest has been deposited, you might not need to solve it again. It is also useful to find structural information of related, homologous proteins, which will be of help when solving a new structure. Thus, it is a good first step to see if the protein of interest or a related protein structure has already been deposited before starting the journey of macromolecular crystallization. **Practical Tips 6.1** gives some suggestions for how to tell if the existing data are of high enough quality for your application.

## 6.2  CRYSTALLIZATION OF MACROMOLECULES

### General concerns and motivations

Three-dimensional structure determination is essential to understanding protein function. If no structure for your protein of interest is available, you may have interest in determining it yourself. The most common technique for high-resolution determination of protein structure is x-ray crystallography, which requires an ordered three-dimensional lattice of macromolecules. It is this ordered repeat of identical structures that yields distinct diffraction patterns, which then can be transformed into electron density maps. The maps finally allow building and refinement of a protein structure model. However, without an ordered lattice, or in other words, a high-quality crystal, none of the above are possible. Thus, crystallizing the protein of interest is the first step toward structure determination.

---

**PRACTICAL TIPS 6.1:    JUDGING THE QUALITY OF A PROTEIN STRUCTURE**

The protein database (PDB) archive holds three-dimensional structures of proteins and other biological macromolecules. The deposited structures have been determined by methods such as x-ray crystallography, nuclear magnetic resonance spectroscopy, or cryo-electron microscopy. For x-ray structures, there are two major numbers to consider when judging their quality: resolution and $R$-factor. High-resolution structures are 2 Å or less, while structures with 3 Å or more are considered low-resolution structures. The $R$-factor is a percent number that describes how well a calculated structure matches the collected data, using the formula

$$R = \frac{\sum \left| F_{\text{obs}} - F_{\text{calc}} \right|}{\left| F_{\text{obs}} \right|} \times 100\%,$$

where $F_{\text{obs}}$ are empirical values extracted from experimental data (diffraction patterns) and $F_{\text{calc}}$ are calculated values obtained from the structure model. The $R$-factor is dependent on the resolution, so higher-resolution structures generally have a better (lower) $R$-factor. A 2 Å resolution structure, for example, should have an $R$-factor of about $20 \pm 2.5\%$, whereas a 3 Å structure typically has a higher $R$-factor, around $23 \pm 3\%$. Finally, there is a second $R$-factor reported, referred to as $R_{\text{free}}$. The $R_{\text{free}}$ is there to monitor if the $R$-factor was artificially lowered by overfitting the data. This number is higher than the regular $R$-factor, but ideally by no more than ~5%.

---

One fundamental assumption made about protein crystals is that the structure of the protein in the crystal is comparable to the one in solution. This is generally believed to hold true for theoretical and empirical reasons. The solvent content of protein crystals is similar to that in living cells (40–60%), so it is not expected that the crystals exist in a highly unnatural environment. More importantly, there is very good experimental evidence that crystallized and freely soluble proteins share similar structures: (1) other structural methods, such as nuclear magnetic resonance spectroscopy, have confirmed the structures determined by x-ray crystallography; (2) different crystal forms of the same protein have yielded largely similar crystal structures; and (3) enzymes have been shown to retain their activity in their crystallized form. Thus, crystal structures are probably a valid depiction of *in vivo* structures.

However, crystal structures do sometimes depict artifacts, such as more flexible loops adopting strange conformations due to biologically irrelevant crystal contacts. It is important to remember that x-ray crystallography provides a snapshot of a dynamic structure, so one crystal typically shows only one of several conformations. Finding more than one crystallization condition is desirable because it might capture the protein in different conformations.

Protein crystallization makes use of the metastable nature of supersaturation. A solution is supersaturated if it contains a higher concentration of solubilized molecules, in our case proteins, than would usually be possible in a given solvent at a given temperature. One common way of creating a supersaturated sodium chloride solution, for example, is to dissolve generous amounts of salt in water while heating it. Supersaturation will be reached once the solution is allowed to cool down. Supersaturation is a metastable state, and small disturbances such as vibration or addition of some more salt will force the dissolved salt out of solution, a phenomenon called precipitation.

During crystal formation, supersaturation needs to be reached, followed subsequently by slow precipitation of protein molecules. For crystals to form, this

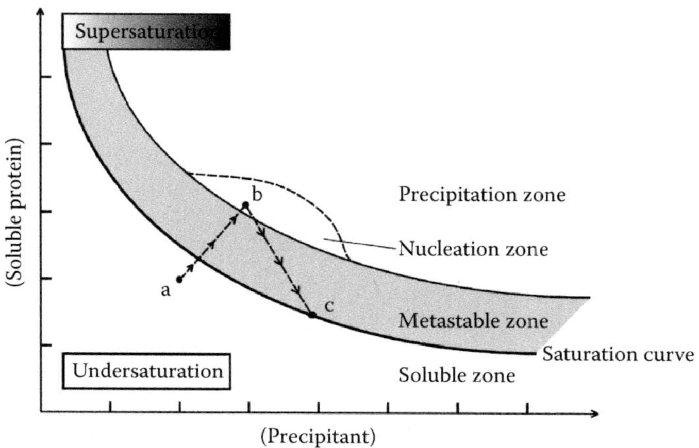


**Figure 6.1 Schematic representation of a two-dimensional phase diagram where the solubility of a protein is related to its concentration as a function of the concentration of the precipitant in solution.** The saturation curve divides the undersaturated state (soluble zone) from the supersaturated state (metastable, nucleation, and precipitation zones) of the protein in solution. (a) At the start of protein crystallization, proteins are usually soluble. Protein and precipitant concentration will increase slowly due to water evaporation until the solution reaches supersaturation. (b) Under the right conditions, a crystal may then nucleate, allowing for growth to occur. (c) The crystal will grow until the protein concentration reaches the saturation curve. (Adapted from Chayen, N. E., and E. Saridakis, Protein crystallization: From purified protein to diffraction-quality crystal, *Nat Methods* 5, 147–153, 2008.)

precipitation must be conducive to the configuration of an ordered arrangement of the molecules. When working with organic molecules such as protein, one needs to find ways other than heating that allow solutions to reach supersaturation. Also, if the process occurs too fast, it will result in disordered precipitate rather than crystals. However, if the right conditions are found, protein molecules will spontaneously order themselves and form what is called a nucleus. Thus, during crystallization, the protein molecules transition from solute to supersaturation, to nucleation and subsequent growth. **Figure 6.1** illustrates this phase transition.

There are several methods to create a supersaturated protein solution. Most methods have in common that the solvent (generally water) is slowly removed from the crystallization mixture. Common methods of crystallization used today include *vapor diffusion, interface diffusion, microbatch,* and *dialysis.*

## Vapor diffusion

Vapor diffusion is the most commonly used method for protein crystallization. Two primary setups for vapor diffusion exist: *sitting drop* and *hanging drop* (**Figure 6.2**). Each method has its advantages and drawbacks; a comparison of the two methods is given in **Table 6.1**. Both make use of a sealed chamber to allow solvent exchange between two solutions: drop and reservoir. The drop contains protein, while the reservoir does not. The reservoir solution (100–1000 µL or less), also referred to as the crystallization solution, usually contains three constituents: buffer, *precipitant,* and salt. (The following section describes in detail what these individual parameters are thought to do.) The drop (1–10 µL or less) is a mixture of the reservoir solution and protein sample, also called *mother liquor.* The ratio of this mixture can vary but is often 1:1. Upon creating a sealed environment, volatile constituents in the drop, usually water, will slowly leave the drop to

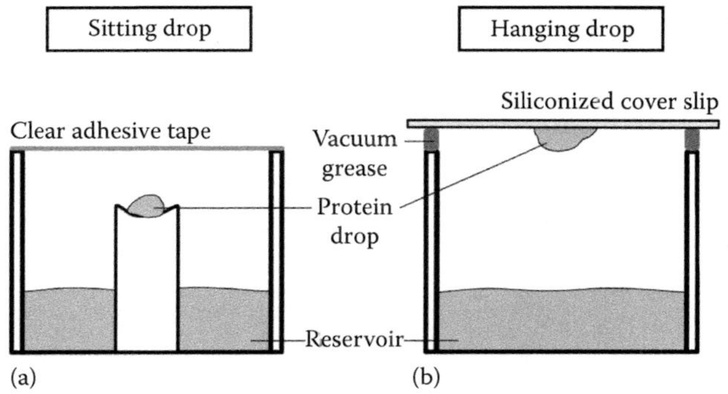



**Figure 6.2 Schematic of sitting drop and hanging drop vapor diffusion experiments.** (a) In the sitting drop method, the protein drop is elevated/separated from the reservoir solution by a pedestal. The chamber is sealed with clear adhesive tape or film. (b) In the hanging drop method, the protein drop is suspended from a coverslip above the reservoir solution. The chamber is sealed with vacuum grease between the well and the coverslip. Both methods allow water to diffuse from the protein drop to the reservoir solution, causing the solutions to equilibrate.

**Table 6.1**

A Comparison of Sitting Drop and Hanging Drop Methods

| | **Sitting Drop** | **Hanging Drop** |
|---|---|---|
| Materials required | Purified soluble protein<br>Crystallization solutions<br>Sitting drop 24- or 96-well plate<br>Clear adhesive tape | Purified soluble protein<br>Crystallization solutions<br>Hanging drop 24-well plate<br>Coverslips (ideally siliconized)<br>Vacuum grease |
| Ease of setup | Fast to set up because the plates allow for automation. Very easy to seal. (~15 min for 96 wells when automated, otherwise up to 60 min.) | Plates have to be pregreased, and every coverslip should be individually cleaned of dust before use. (~45–90 min for 24 wells) |
| Drop size | Small (normally ~2 µL but can vary between 1 and 5 µL) for 96-well plate. (24-well plates can have drops up to 10 µL.) | Potential to be much larger (usually ~4 µL but can be anywhere from 1 to 10 µL). |
| Quantity of reagents | Uses ~100 µL of crystallization solution and ~1 µL of protein solution per well (96-well plates). | Uses 500–1000 µL of crystallization solution and ~2 µL of protein solution per well. |
| Visibility under the microscope | Somewhat tedious to observe because drops are small. Some plates have imperfections that might be seen as potential crystals. Also, condensation on the tape can become a major problem. | Much easier to observe since the protein drop is right below the coverslip. |
| Crystal manipulation | Crystals tend to grow on walls or the bottom of the drop well. Also, plate walls in general can make it difficult to access the crystals at the bottom of the drop. Also, the tape needs to be cut to gain access to the crystal, and resealing the well might not work very well. | Crystals usually grow unattached in solution. Most importantly, crystals of individual drops can be easily manipulated. Once the coverslip is removed there are no walls to hinder access to the crystal. Resealing crystal drops after manipulation is easier. Only the coverslip is manipulated, so the other crystal conditions of the plate are not exposed to unfavorable conditions (temperature and vibrations). |
| Several drops in one well | Sitting drop plates with one to four drop wells per reservoir well exist. | Possible to have several drops on one coverslip. However, each additional drop makes it more difficult and time consuming to set up. |
| Best suited for | *Sparse matrix screening* because it is fast and easy to set up and uses less protein. | Crystal optimization because it makes crystal manipulation such as *microseeding, soaking,* etc. easier. |

travel to the reservoir solution. This process continues until the concentrations of the precipitant, salt, and so on in the drop solution approach their respective concentrations in the reservoir solution. Slowly, the concentrations of all nonvolatile components within the drop effectively increase.

Most vapor diffusion experiments make use of a wide assortment of plastic plates available from different suppliers (**Figure 6.3**). In the sitting drop variant, plates range from 24 to 384 wells. Most sitting drop plates are 96-well plates. Each individual well is subdivided into a reservoir well and one or more smaller wells for the protein drops. To seal the well, clear adhesive tape or sealing film is used.

Most hanging drop plates are 24-well plates, but 48- and 96-well plates also exist. In the hanging drop experiment, the protein drop is placed on a coverslip that is then placed inverted over the reservoir. Consequently, the protein drop is hanging upside down over the reservoir solution. The seal between well and coverslip is provided by vacuum grease. Ideally, siliconized coverslips are used because the hydrophobicity of the silicon prevents drops from spreading. Nicely formed drops help later with the visualization of crystals under the microscope and also contribute to the success and reproducibility of the experiment. These coverslips may be purchased, or they may be easily prepared by dipping ordinary glass coverslips into a silane solution such as Sigmacote. (See **Chapter 14** for more on glass functionalization and silanes.)

Regardless of the method used, proper sealing is important for crystallization to work and to be reproducible. If using the sitting drop method, it is advisable to buy the adhesive tape or sealing film from a distributor that also sells crystal plates to ensure highly water-impermeable material. When using the hanging drop method, sealing can be tricky at first. **Practical Tips 6.2** gives some tips on sealing these plates. It is also important to remember that most plates are made of polystyrene, which is slightly porous, so even properly sealed crystallization experiments will dry out in the long run (several months).

In theory, it should be possible to reproduce crystal growth in a particular condition using any vapor diffusion plate as well as to switch from sitting drop to hanging drop. However, this theory might not necessarily transfer to practice. The different kinetics in two given plates may prevent crystallization conditions from being transferred from one plate to another, even between two sitting drop plates!

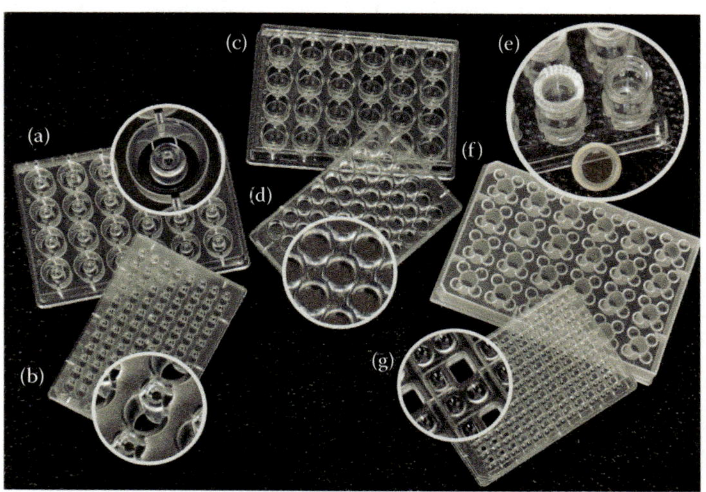

**Figure 6.3 Examples of different crystallization plates for vapor diffusion.**
(a) A 24-well sitting drop plate. (b) A 96-well sitting drop plate. (c) A 24-well hanging drop plate. (d) A 48-well hanging drop plate. (e) A 24-well hanging drop plate with screw caps. (f) A 24-well sitting drop plate with 4 drop wells per reservoir well. (g) A 96-well sitting drop plate with 3 drop wells per reservoir well.

---

**PRACTICAL TIPS 6.2:    HOW TO SEAL HANGING DROP PLATES**

The plates for hanging drop vapor diffusion have to be greased along the rim of each well with vacuum grease. Attaching a 100 µL pipette tip to a large (8–30 mL) surgical syringe works well as a grease dispenser. Parafilm can be used to attach and seal the tip to the syringe. Additionally, trimming the point of the tip slightly will help in dispensing the grease on the rim of the crystallization plates. When greasing the plates, it is advantageous to leave one or more little gaps on the rim of the well, i.e., do not grease the whole rim. It is suggested to first add the crystallization solution to the reservoir. Then, mix the drop in the middle of a coverslip by adding the reservoir solution to the protein solution. Carefully place the coverslip with the drop upside down onto the greased rim. Gently push down on the edges of the coverslip. (Avoid pushing in the middle, as the coverslip might break.) While doing this, the gaps in the grease will allow air to escape and the pressure inside the well to equilibrate. Finally, finish the sealing by turning the coverslip until the seal is complete.

---

## Interface diffusion

As the name suggests, this method utilizes the free diffusion of a concentrated protein sample with a crystallization solution. The two solutions are layered directly on top of each other in small sealed capillaries, allowing the solution to diffuse to equilibrium and creating a transient supersaturation condition where crystal nucleation and subsequent growth can occur.

## Microbatch

In this method, the protein sample is mixed with the crystallization solution containing the usual constituents (precipitant, buffer, salt). However, instead of creating a sealed environment where water slowly diffuses from the protein drop to the reservoir solution, here, water is simply allowed to evaporate. To minimize the evaporation rate, the protein drop is covered with an oil such as paraffin, silicon oil, or a mixture of the two. Silicon oil is more permeable than paraffin, and thus, by mixing the two, the evaporation rate can be controlled. Microbatch setups can be very effective in combination with *sparse matrix screens* (see **Section 6.6**) to test hundreds of different conditions. Microbatch methods have been shown to be very successful at producing crystallization *hits*. However, since a microbatch is not a closed system and the time between a crystal forming and the drop drying out can be very short, observation and timely diffraction measurements of those conditions can be demanding. Additionally, some prefer not to manipulate crystals through a layer of oil.

## Dialysis

In dialysis, one makes use of the ionic strength of solutions. The protein sample is added to a microdialysis chamber, called a *button*, which is sealed with a semipermeable membrane. The button is then placed in crystallizing solution, allowing small molecules to diffuse in and out while preventing diffusion of the larger protein molecules. Dialysis is usually used to either decrease or increase the ionic strength. Since proteins can be forced out of solution by a change in ionic strength, dialysis often allows for crystallization of a protein without need of a precipitant. If needed, change in ionic strength can always be combined with the effects of a precipitant by including a precipitant that can pass through the membrane. High-molecular-weight precipitants such as polyethylene glycol (PEG) do not work with this method.

# 6.3 PREPARATION OF PROTEINS FOR CRYSTALLIZATION

The purification methods described in Chapter 5 can be used as a starting point to prepare a protein sample for crystallization. The protein should be as homogenous as possible and might therefore need to be purified more stringently than for most other types of experiments. Threshold values for crystallization experiments are usually reported as protein purity ≥ 95% and *monodispersity* ≥ 85%. If the values of your sample are lower, it is strongly suggested to pursue further optimization of the protein production and purification process before a crystallization attempt.

## Protein purity

Protein purity is probably the most important factor in protein crystallization. During the crystallization process, a lattice of ordered molecules is created, where the same molecule is incorporated in an identical manner along all three axes. The purer the sample, the easier it is to achieve this, mainly because fewer impurities can become incorporated into the ordered lattice.

Proteins may be purified using the techniques discussed in Chapter 5. It is important to note that simple affinity-column purification is rarely enough for crystallization-quality protein. An affinity column followed by an ion exchange column is very often used to generate the required purity. The following tips can also help:

- Collect protein in small fractions. Keep only those fractions at ≥50% of the maximum peak; discard the others.

- Discard any fractions that contain significant impurities, even if they also contain a good deal of target protein.

- After affinity purification, the protein may be passed over a size exclusion column (see Chapter 5). This will also remove aggregates.

- The use of two different ion exchange columns may yield dramatic improvement.

- If all else fails, consider double-tagging your protein (e.g., with His- and *Strep*-tags) and using two subsequent rounds of affinity chromatography. This should yield purities of ≥98%. Kits for double-purification are available from some manufacturers; the difficulty is in cloning both tags onto the protein of interest. This requires going all the way back to the cloning step and should be considered a last resort.

Polyacrylamide gel electrophoresis (PAGE) and high-resolution gel chromatography (both discussed in Chapter 5) are probably the most commonly used methods to gauge protein purity; mass spectroscopy can also be of use. When using sodium dodecyl sulfate–polyacrylamide gel electrophoresis (SDS-PAGE), gels should be stained with a more sensitive method than Coomassie blue staining. Possible staining methods include silver nitrate staining and zinc–imidazole staining, both of which are classic techniques for which references are provided at the end of the chapter.

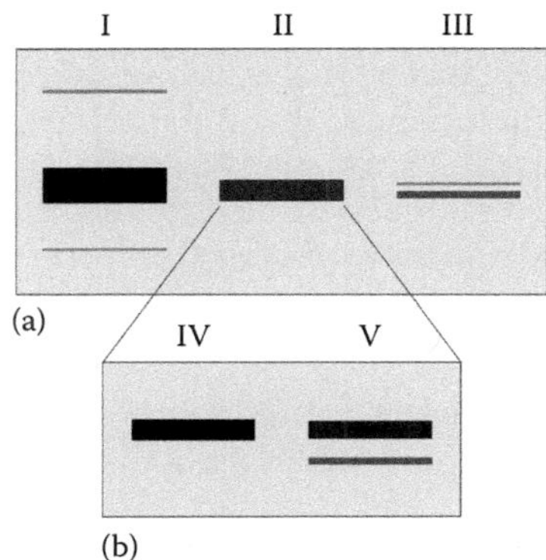

**Figure 6.4 Schematic representations of PAGE gels.** (a) A hypothetical SDS-PAGE gel that is loaded with different amounts of protein to analyze the purity of the protein sample. Loading the gel with the same protein sample of differing quantities can provide information about impurities. Comparing lane *I* to lane *II* shows how overloading the gel can highlight small impurities that otherwise might not be detected. Comparing lane *III* to lane *II* shows how underloading the gel can identify split bands indicative of impurities of similar molecular weight as the protein of interest. (b) A hypothetical nondenaturing PAGE gel. A protein sample that runs as a single band on an SDS-PAGE gel can run as either a single band (lane *IV*), indicating homogeneity, or multiple bands (lane *V*), indicating heterogeneity.

Loading higher and lower concentrations of protein sample is important to adequately determine the purity of your sample (**Figure 6.4a**). Loading large amounts of protein sample ensures that there are no other protein bands present indicative of contaminants. Conversely, loading very little of the protein sample enables you to discriminate distinct split bands that are very close together. In addition to a denaturing SDS-PAGE gel, you should also run a nondenaturing PAGE. Instead of separating denatured proteins according to their mass-to-charge ratio, the nondenaturing PAGE does not use detergent and therefore can separate proteins of similar mass but different shape, providing more detailed information about the homogeneity of the sample (**Figure 6.4b**).

## Monodispersity

For crystallization trials, the protein needs to be stable and monodisperse at high concentrations. Although crystals themselves are aggregates, they are large organized aggregates. Formation of randomly oriented oligomeric aggregates in solution is not conducive to protein crystallization. Thus, identifying a suitable buffer condition (mainly pH and salt concentration) will help ensure that the sample does not aggregate and precipitate prematurely.

It is common practice in crystallography to determine the monodispersity of the protein sample in different buffer systems. *Dynamic light scattering* (DLS) provides a method to perform broad-scale buffer screens. DLS is also called photon correlation spectroscopy or quasi-elastic light scattering and is performed on a specialized instrument called a *particle sizer* or *protein size analyzer*. These instruments use a laser or laser diode, usually at 633 or 830 nm, to perform Rayleigh scattering off a suspension of particles. The intensity variations of the scattered light caused by Brownian motion of the particles can be used to infer the particle size distribution by relating the decay of the intensity autocorrelation function to the diffusion coefficient, and then the diffusion coefficient to particle size using the Stokes-Einstein equation. Depending upon the model, particle sizers have nominal sensitivities of subnanometers up to several micrometers. In practice, the smaller particles are often difficult to observe as their scattering intensity is much less than that of the larger particles. Most core facilities and also many individual labs have particle sizers; this is a useful instrument for many applications in which particle size and aggregation need to be measured. (See **Chapters 11** and **12** for discussions of nanoparticles and particle sizing.)

DLS is very sensitive, and any form of impurity or dust will drastically affect the results, so producing conclusive results can be challenging. Nonetheless, this method is fast and uses low quantities of protein (<10 µg), allowing for the effective screening of many different conditions. Investing some time at this stage can save time down the road when setting up crystal trays. As a starting point, buffer concentrations are usually around 25 mM. The results of a DLS experiment are represented as a histogram of particle sizes. **Figure 6.5** illustrates protein distribution histograms that are desirable or undesirable for crystallization.

Proteins are dynamic in nature; this will affect protein stability, and a once-monodisperse sample will not remain so forever. Some proteins are not amenable to freezing or have a short shelf/fridge life. Similarly, the aggregation state might change over time, and it is therefore good practice to use the purified protein for crystallization as soon as possible. Adding ligands, cofactors, or metals can sometimes help stabilize proteins.

Aggregation can also occur due to highly charged affinity tags. If in doubt, use a tag with a cleavable sequence (called, for example, a "cleavable His-tag" and available in many cloning vectors). A protease is required to cleave the tag; kits are available for cleavage and removal of tags that are very easy to use. (See **Practical Tips 6.3.**)

**Figure 6.5 Schematic illustration of hypothetical DLS histograms from protein samples.** (a) A unimodal peak, indicative of a monodisperse sample. This is the most desirable result. (b) A broad unimodal peak, indicative of a generally monodisperse sample. This may work. (c) A broad shoulder distribution, indicative of protein unfolding, proteolysis, or small oligomerization. Bad. (d) A bimodal distribution indicative of aggregation. Also not workable.

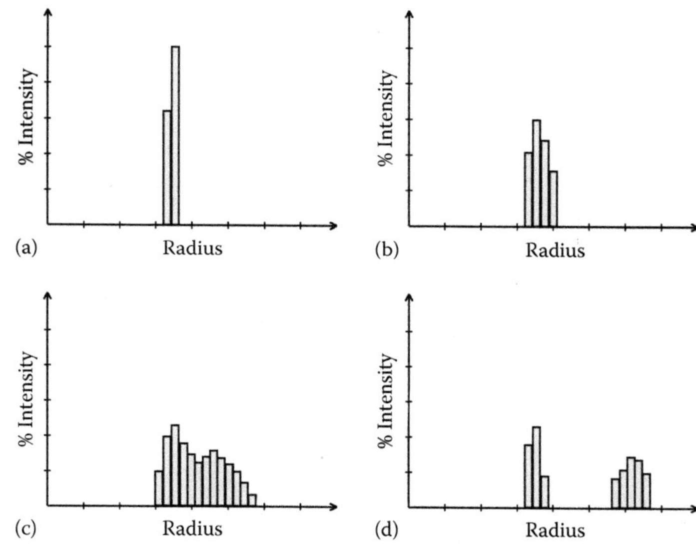

## PRACTICAL TIPS 6.3:   CLEAVABLE AFFINITY TAGS

Affinity tags can introduce highly charged, flexible extensions to a protein that interfere with crystallization and maybe change function. For these reasons, many plasmid vectors come equipped with a short sequence recognized by a *protease* that comes just after the affinity tag (for N-terminal tags) or just before the tag (for C-terminal tags). For example, the cleavage site for *enterokinase* (ETK) is four aspartates followed by a lysine (($Asp)_4Lys$). An N-terminal *cleavable His-tag* would then be $(His)6$-AlaAlaGly-(($Asp)_4Lys$, where the AlaAlaGly is inserted as a spacer and/or cloning site (Pst I). Even if your His-tag vector does not come with the ETK site, this can be easily inserted as an oligomer (see **Chapter 1**).

Cleavage of the tags is easy, but the protease must be removed afterward. This may involve the use of antibodies or additional affinity columns and may be costly. This is the only real drawback to this method, except that some cloning may be needed at the start of the experiment. Using cleavable tags is a good idea in crystallization experiments in the author's opinion, and should be done from the outset. We have seen a protein refuse to crystallize with one type of His-tag (($His$-$Arg)_6$), although literature results reported excellent crystallization with another type (($His)_6$).

The following is a protocol for enterokinase cleavage of an affinity tag. It should be done immediately following protein purification on the affinity column.

### Reagents Needed

- Enterokinase

- Enterokinase buffer: 500 mM Tris-HCl, pH 8.0, 2.0 mM $CaCl_2$, 1% Tween

- Enterokinase removal kit (containing affinity agarose and spin columns) or enterokinase removal affinity agarose

- 20× wash buffer: 0.4 M Tris-HCl, pH 7.4, containing 1 M NaCl and 0.04 M $CaCl_2$ (other buffers may also be used; keep pH below 8.5)

### Procedure

- Purify the protein and dilute it to 1.5 mg/mL in enterokinase buffer.

- Add enterokinase at a concentration of 0.02 units/mg protein. Incubate on a rocker for 16 h. Do not vortex!

- Just before the incubation is ready, prepare the enterokinase removal reagents. Each 50 μL of affinity agarose can remove up to 1 μg of enzyme. The actual amount of enzyme you will have added depends upon its activity. For example, if your enzyme had an activity of 20 units/mg, then for each milligram of protein to be purified, you had to add 1 μg of enzyme.

- Wash the needed amount of affinity agarose by adding 10 bed volumes of 1× wash buffer, and then sedimenting it by centrifugation for 2 min at 1000g and discarding the supernatant. Wash twice; the third time, resuspend in 1 bed volume of wash buffer.

- Transfer the agarose to a clean Eppendorf tube or spin filter.

- Add the protein solution and rock for 45 min.

*(Continued)*

---

**PRACTICAL TIPS 6.3 (CONTINUED):    CLEAVABLE AFFINITY TAGS**

- Spin through the spin filter and collect the filtrate (or if you have no spin filters, centrifuge and collect the supernatant).

- Analyze the protein by SDS-PAGE.

Although the antibody method is perhaps the easiest, for purifying greater amounts of protein, it is more cost-effective to use methods that remove the protease using an affinity column. A good enzyme to use for this is *thrombin*, for which affinity columns are available commercially. The cleavage site is Leu-Val-Pro-Arg-Gly-Ser. Another option is to use a tagged protease, which may be removed using an ordinary affinity column. Tagged proteases are also available commercially; an example is the tobacco etch virus (TEV) protease, for which the cleavage site is Glu-Asn-Leu-Tyr-Phe-Gln-Gly.

*Suppliers*
Sigma-Aldrich: enterokinase and removal kit
Invitrogen: His-tagged TEV protease
GE Healthcare: HiTrap Benzamidine FF columns to remove thrombin

**SUGGESTED READING**

LaVallie, E.R., McCoy, J.M., Smith, D.B., and Riggs, P. (2001). Enzymatic and chemical cleavage of fusion proteins. *Current Protocols in Molecular Biology*, Chapter 16:Unit16 4B.

---

## Protein quantity

To set up one 24-well plate with a protein sample drop volume of 2 µL means that you need about 75 µL of protein per plate. One screen normally contains 96 conditions (equal to four 24-well plates). Protein concentration for crystallization is usually ~10 mg/mL. Thus, to set up one full screen, you need at least 3 mg of pure protein. This would need to be doubled if another condition, such as temperature, is going to be varied. This is a large amount of protein, equivalent to at least 1 L of bacterial culture under optimal conditions, and frequently more. It is a good idea to make sure that you will have enough protein to proceed before beginning a crystallization screen.

## Protein variability

Even for one specific protein, every batch is slightly different. Mixing different protein batches is highly discouraged since homogeneity, dispersity, and stability are the most critical properties of protein samples. Being able to quantitatively reproduce those purification standards is one of many challenges in crystallization and is very important for successful reproduction of a given crystal condition.

Interprotein variability is tremendous. Not all proteins crystallize equally well. Even proteins that behave well in solution can prove next to impossible to crystallize. Flexible areas that interfere with the formation of an ordered lattice might be the reason that no crystals can be obtained. In addition, many biomolecular studies focus on heterocomplexes made of two or more proteins. These complexes tend to have weak interactions and so are extremely hard if not impossible to crystallize. The focus of this chapter from here on will be predominantly on globular, monomeric or homomeric, soluble proteins, which tend to be easier to crystallize.

Proteins that are difficult to solubilize appropriately pose the greatest difficulties for crystallization experiments. Membrane proteins fall into this category and are thus exponentially more difficult to crystallize than cytosolic, water-soluble proteins. These represent a special challenge for crystallographers and will be discussed in **Advanced Topic 6.1**.

## 6.4 COMPONENTS OF CRYSTALLIZATION SOLUTIONS

Having the perfect protein solution is not enough. It must be put into the right conditions and exposed to a reservoir solution and its components that permit crystallization. The components of the reservoir solution commonly include a precipitating agent or precipitant, a buffer, and a salt. The buffer need not be identical to that used for the protein solution. This makes the possible combinations vast and complex, and often, hundreds of different solutions must be tried before good crystals are obtained. Before we discuss how to do these large screens, we will examine each ingredient and its role in the crystallization process.

### Precipitant

After preparing a protein that is as pure and monodisperse as possible, finding the right precipitant is probably the most important factor in protein crystallization. Precipitants force proteins out of solution by lowering their solubility. Thus, one way to identify potentially effective precipitants is to look at all conditions that produce precipitate.

Commonly used precipitants include PEG, *2-methyl-2,4-pentanediol (MPD)*, and ammonium sulfate (**Figure 6.6**). Salts such as ammonium sulfate lower the solubility by dehydrating the protein solution. Organic solvents such as MPD are thought to hydrogen-bond with water molecules. Polymers such as PEGs also withdraw water molecules from the solution just as salts and organic solvents do, but have added benefits: they greatly increase the viscosity of the solution and thus decrease diffusion rates, and they might confer different levels of *cryo-protection* (see **Section 6.8**), which will subsequently help with data collection. Precipitants can also be used effectively in combinations: two different PEGs with diverse molecular weights, PEG combined with MPD, etc. **Practical Tips 6.4** gives a protocol for preparing a PEG solution.

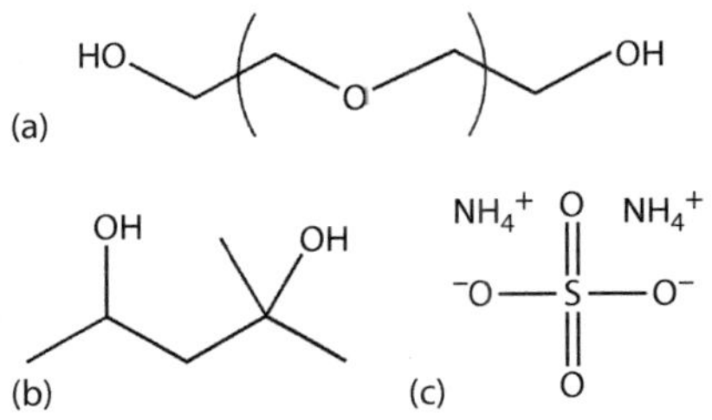

Figure 6.6 **Chemical structures of common precipitants.** (a) PEG. (b) MPD. (c) Ammonium sulfate.

---

**PRACTICAL TIPS 6.4:   PREPARING A PEG SOLUTION**

PEG solutions are usually given by percentage (%w/v), which is the weight of solute (PEG) in a given volume ($H_2O$). For example, 50% PEG 4000 is 50 g of PEG 4000 in 100 mL of $H_2O$. To solubilize PEG, start with a volume of $H_2O$ approximately half of what is needed (i.e., ~50 mL). Then add the PEG slowly to the water while stirring and heating the solution constantly to about 50°C (do not boil the sample). When all the PEG has been solubilized, measure the volume of the solution in a graduated cylinder and adjust it to the desired volume (100 mL). As with all solutions used for crystallization, this solution should then be filtered.

---

### Buffer

The choice of buffer to use in the crystallization solution is as important as that of the precipitant. A typical concentration for a crystallization buffer is ~100 mM (recall that in the protein solution, the value is more like 25 mM), and these buffers may be the same or different as those used with the protein. There are several dozen common buffers used in crystallization experiments. Their names, structures, and associated pH ranges are given in Table 6.2. Manufacturers' catalogs provide even more extensive lists of buffers for crystallization.

Vapor diffusion will result in the mixing the buffer of the protein sample with the buffer of the crystallization solution. This alters the overall pH and thus provides a way to control and alter the ionization state of individual protein molecules, subsequently affecting protein solubility. However, small alterations in the pH of the buffer in the crystallization solution can have significant results on solubility and thus will change the phase transition the protein goes through. Thus, during fine-screening, it is usually the buffer pH that is varied rather than buffer concentration.

### Salt

Salt can act as a precipitant by dehydrating proteins. It also has another function: ionic shielding of surface charges of protein can enable intermolecular interactions among proteins. Changing the concentration of the salt often has a less dramatic effect than changing precipitant concentration or pH. A certain amount of salt is usually needed to permit crystallization (0.1–0.25 M), while adding no salt often prevents the formation of crystals. Adding too much salt, however, will simply force the protein out of solution, a process called *salting out*. Instead of just changing the concentration of the salt, one should also consider changing the type of salt used during fine-screening. There are more than 20 salts commonly used in crystallization experiments, including ammonium salts, potassium salts, sodium salts, and salts of heavy metals including cadmium, mercury, and cobalt.

## 6.5 OTHER FACTORS AFFECTING CRYSTALLIZATION

Everything from your building's airflow to the clothes you wear can affect a crystallization experiment. We discuss a few of the important variables in this section.

**Table 6.2**

Some Buffers Commonly Used in Crystallization Experiments, with Their Working pH Ranges

| Buffer Name | pH Range | Structure |
| --- | --- | --- |
| Citric acid | 2.2–6.5 | 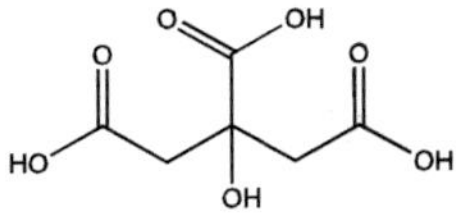 |
| Sodium acetate | 3.6–5.6 | 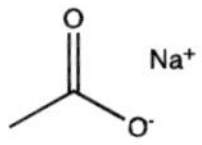 |
| MES<br>2-[N-Morpholino]ethanesulfonic acid | 5.5–6.7 | 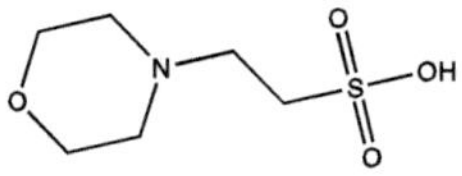 |
| Bis-Tris<br>Bis[2-Hydroxyethyl]<br>  iminotris[hydroxymethyl] methane | 5.5–7.5 | 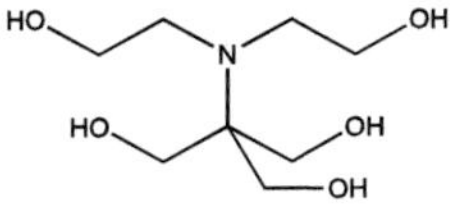 |
| PIPES<br>Piperazine-N,N′-bis[2-ethanesulfonic<br>  acid] | 6.1–7.5 | 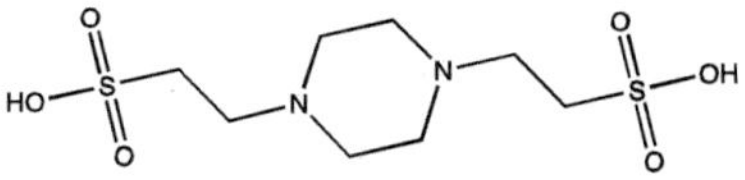 |
| MOPS<br>3-[N-Morpholino]propanesulfonic<br>  acid | 6.5–7.9 | 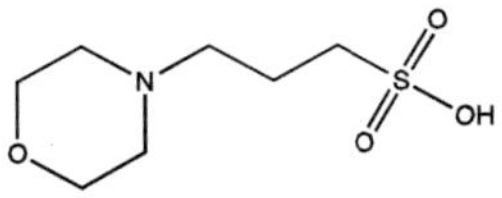 |
| HEPES<br>N-[2-Hydroxyethyl]piperazine-<br>  N′-[2-ethanesulfonic acid] | 6.8–8.2 | 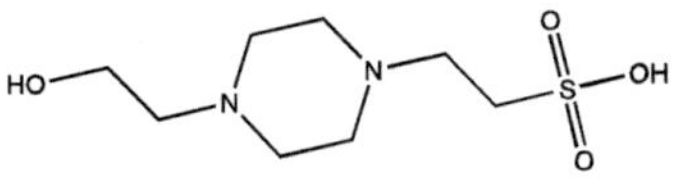 |
| Tris<br>Tris(hydroxymethyl)aminomethane | 7.0–9.0 | 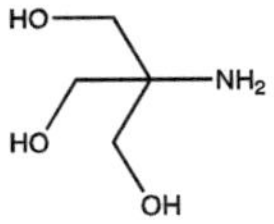 |
| TEA<br>Triethanolamine | 7.3–8.3 | 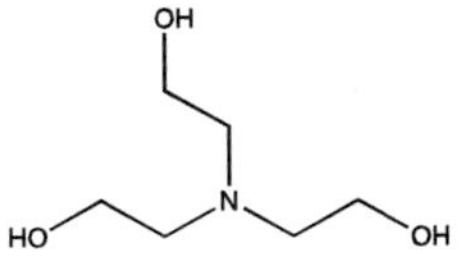 |
| Glycine<br>N,N-bis(2-Hydroxyethyl)glycine | 8.6–10.6 | 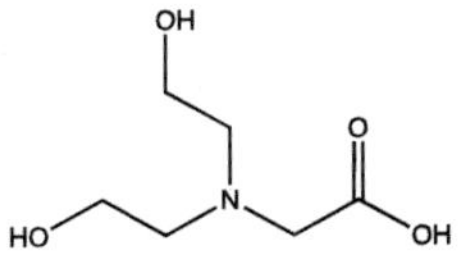 |
| CAPS<br>3-[Cyclohexylamino]-1-propanesulfonic<br>  acid | 9.7–11.1 | 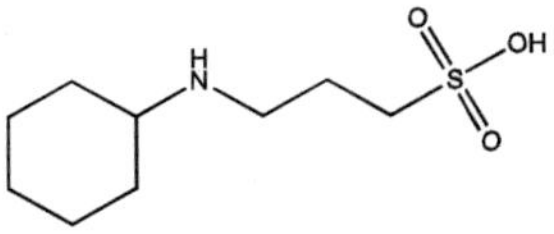 |

**Figure 6.7 Schematic representation of a two-dimensional phase diagram illustrating how changing the drop ratio of protein sample to reservoir solution can affect the phase transition.**
(a) Protein reaches the nucleation zone; crystals can grow.
(b) Protein concentration is too high so that the protein never travels through the nucleation zone. The protein drop will show precipitate. (c) Initial protein concentration is too low so that it never can reach the nucleation zone; protein drop will remain clear.

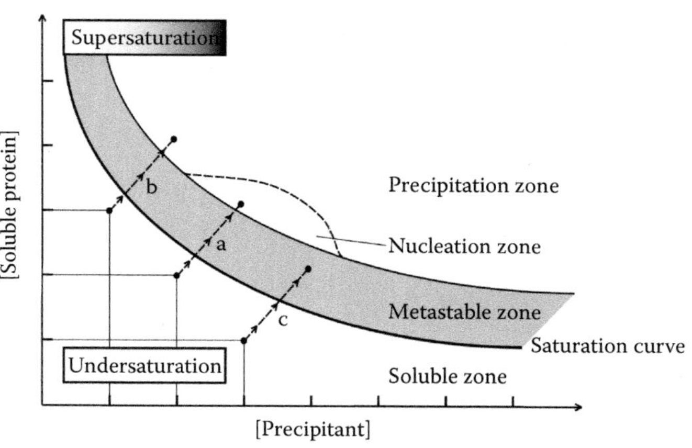



## Protein concentration

Finding the ideal concentration of your protein sample for crystallization is crucial. It is good to know how concentrated the protein sample can be while remaining homogeneous and monodispersed. On average, crystallographers use protein samples between 5 and 15 mg/mL. There are commercially available tests that aid in estimating the optimal protein concentration, marketed under the names "precrystallization test" and "prescreening test." However, the best indication can be derived from observation of the sparse matrix screens, as discussed in the following section.

Altering protein concentrations and drop sizes will affect phase transitions by changing the starting point. **Figure 6.7** illustrates how such a change can affect nucleation and crystal growth.

## Diffusion rate

There are two distinct diffusion phenomena in most crystallization methods: diffusion of components inside the drop and diffusion of water (evaporation) from the drop. Diffusion of molecules inside a drop affects nucleation rates and growth. Higher viscosity slows down diffusion inside the drop and prevents unwanted convection and crystal sedimentation. Using high-molecular-weight PEGs, glycerol, or ethylene glycol is one way to increase viscosity. However, changing the viscosity of your crystallization condition by changing or adding chemicals might impede crystallization altogether. As an alternative, it has been shown that one can promote the growth of single high-quality crystals by using gels such as agarose to increase the viscosity without affecting the chemical composition.

Water diffusion from the drop affects the kinetics of the phase transition. Lowering the evaporation rate allows the crystal to grow more slowly, which in turn should increase the crystal quality. (Evaporation kinetics can also affect the nucleation rate.) Overlaying the reservoir solution with an oil such as paraffin is a common way of trying to control evaporation in a vapor diffusion experiment. Changing the drop size, drop ratio, and to some degree, reservoir volume will also affect evaporation rates. The reason that for some proteins, crystal growth seems to depend upon the brand or model of plate is most like the fact that plates have an effect on evaporation rates.

## Temperature

Crystallization experiments are temperature sensitive since temperature affects solubility, pH, and diffusion rates. It is important to keep crystallization plates in temperature-controlled environments, such as incubators. In initial screening, it is less important which temperature is used, but it is important to know what that temperature is, so that you can reproduce the growth condition. Conventionally, crystallographers tend to set up crystallization trials at two temperatures (4°C and 22°C). Crystals often grow at one or the other temperature but not both.

Handling of plates inevitably alters the temperature. Removing a plate from 4°C or exposing it to the heat of the microscope lamp can drastically change the microenvironment, sometimes leading to the disappearance of protein crystals.

## Vibrations

Vibrations cannot be easily controlled, but they significantly affect crystallization. The more vibration there is, the more nucleation events occur, as well as dissolution of preformed nuclei. Both of these events are bad, since the ideal case is that of single nuclei. The rule of thumb is, the less nucleation, the better, since it usually allows for the growth of bigger and better crystals. It is advisable to protect crystallization plates by putting them on foam matting. Once set up, the less disturbance the plates get, the better. Vibrations should also be minimized when setting up plates in the first place (i.e., do not jostle, jolt, or shake them).

## Mechanical contaminants

Mechanical contaminants such as dust should be avoided for several reasons. It is not unusual to erroneously identify contaminants such as lint as crystals during the first crystallization trials. More importantly, contaminants often act as nucleation points. This is a problem because a crystal that forms around a piece of dust is unlikely to give a good diffraction pattern, and more than one crystal is likely to form, often leading to clusters of crystals that cannot be separated. This is one of the reasons why all solutions should be filtered before use. In addition, the work area must be dust free, crystallization wells/coverslips should be cleaned with pressurized air before adding solutions, and experimenters should wear clothing that does not shed. Certain clothes, particularly sweaters, tend to shed a lot of lint and should be avoided or covered by a clean, lint-free lab coat.

## Solution quality

The quality/purity of the chemicals and, most importantly, the water used in solutions has a vast effect on crystallization. It is good practice to note where all the chemicals came from because it might be necessary to use the same chemical again in order to reproduce a crystal. The source of water is especially important. It is not unusual that in order to reproduce a successful crystallization experiment, a researcher is compelled to go back to the original source of water.

Before setting up crystallization trials, all solutions prepared in the lab, including the protein sample, should be filtered through a 0.22 μm filter. This removes the mechanical contaminants as discussed previously and also semisterilizes the solution, inhibiting bacterial growth. This is important because many solutions will be used over several months or even longer. Many components of crystallization conditions are organic, so they cannot be autoclaved without destroying them. Filter sterilization is usually adequate, but if needed, one can add small amounts of sodium azide to prevent bacterial contamination, with the caveat that the azide may (like so many other things) affect crystallization. After filtration, solutions should always be stored properly. This includes keeping the protein sample on ice or refrigerated, as well as protecting some organic solutions such as PEGs from light.

## 6.6 CRYSTALLIZATION STRATEGIES

The previous sections discussed the major factors that influence protein crystallization. Given the number of different possible variables, it may seem miraculous that anyone ever finds the right parameters. Indeed, it is sometimes the labor of many months or years to get good crystals, especially of a protein that has never been crystallized before. Fortunately, existing kits and reagents are available to enable rapid and efficient screening of hundreds of conditions at a time. This section will look at how to screen for and identify preliminary conditions suitable for crystal growth and how to optimize subsequent crystallization experiments to improve the quality of the crystals.

### Initial screening

If you are trying to crystallize a protein for which protocols are already found in the literature, this gives you a starting point. However, there are no guarantees that what worked for another lab will work for you. Differences in chemical reagents including water, protein sequence, purification procedures, and other factors can all combine to make reproducing published results extremely difficult. If you have no luck with reproducing a published method, and more importantly, if you are working with a new protein, the best approach is to try a large number of conditions to see what may work, and then zero in on the promising ones and refine them.

*Sparse matrix screens* are considered by most crystallographers to be a good starting point to identify potential crystallization conditions. Each screen often contains 96 or 48 different conditions to fit into one 96-well plate. The concept was developed in the 1990s and is often referred to as "brute-force screening," since different solution environments are randomly explored to find crystal hits. Several different screens are commercially available that attempt to efficiently span the parameter space by varying salt, precipitant, and buffer. Table 6.3 gives an example of the components of one commercial screen. Alternative forms of initial screens also exist and can be found in the textbooks and manufacturer sites referenced at the end of the chapter.

Once a plate is finished, it should be stored at a secure, vibration-safe, and temperature-stable location. You should also consider duplicating crystallization trials at two different temperatures to check for temperature dependence. With vapor diffusion methods, it is advisable to observe the crystal drops within 24 h of creating them and then every 2 days thereafter. After 2 or 3 weeks, it is usually sufficient to look at the drops once a week. Also, some crystallographers look at their plates immediately after they set them up to identify precipitation, physical contaminants, or imperfections in the plate.

**Table 6.3**

Example of a Sparse Matrix Screen

|  | Precipitant | Buffer | Salt |
|---|---|---|---|
| 1 | 30% MPD | 0.1 M Na acetate, pH 4.6 | 0.02 M Ca chloride |
| 2 | 0.4 M K Na tartrate | None | None |
| 3 | 0.4 M ammonium phosphate | None | None |
| 4 | 2.0 M ammonium sulfate | 0.1 M TRIS HCl, pH 8.5 | None |
| 5 | 30% MPD | 0.1 M Na HEPES, pH 7.5 | 0.2 M Na citrate |
| 6 | 30% PEG 4000 | 0.1 M TRIS HCl, pH 8.5 | 0.2 M Mg chloride |
| 7 | 1.4 M Na acetate | 0.1 M Na cacodylate, pH 6.5 | None |
| 8 | 30% 2-propanol | 0.1 M Na cacodylate, pH 6.5 | 0.2 M Na citrate |
| 9 | 30% PEG 4000 | 0.1 M Na citrate, pH 5.6 | 0.2 M ammonium acetate |
| 10 | 30% PEG 4000 | 0.1 M Na acetate, pH 4.6 | 0.2 M ammonium acetate |
| 11 | 1.0 M ammonium phosphate | 0.1 M Na citrate, pH 5.6 | None |
| 12 | 30% 2-propanol | 0.1 M Na HEPES, pH 7.5 | 0.2 M Mg chloride |
| 13 | 30% PEG 400 | 0.1 M TRIS HCl, pH 8.5 | 0.2 M Na citrate |
| 14 | 28% PEG 400 | 0.1 M Na HEPES, pH 7.5 | 0.2 M Ca chloride |
| 15 | 30% PEG 8000 | 0.1 M Na cacodylate, pH 6.5 | 0.2 M ammonium sulfate |
| 16 | 1.5 M Li sulfate | 0.1 M Na HEPES, pH 7.5 | None |
| 17 | 30% PEG 4000 | 0.1 M TRIS HCl, pH 8.5 | 0.2 M Li sulfate |
| 18 | 20% PEG 8000 | 0.1 M Na cacodylate, pH 6.5 | 0.2 M Mg acetate |
| 19 | 30% 2-Propanol | 0.1 M TRIS HCl, pH 8.5 | 0.2 M ammonium acetate |
| 20 | 25% PEG 4000 | 0.1 M Na acetate, pH 4.6 | 0.2 M ammonium sulfate |
| 21 | 30% MPD | 0.1 M Na cacodylate, pH 6.5 | 0.2 M Mg acetate |
| 22 | 30% PEG 4000 | 0.1 M TRIS HCl, pH 8.5 | 0.2 M Na acetate |
| 23 | 30% PEG 400 | 0.1 M Na HEPES, pH 7.5 | 0.2 M Mg chloride |
| 24 | 20% 2-Propanol | 0.1 M Na acetate, pH 4.6 | 0.2 M Ca chloride |
| 25 | 1.0 M Na acetate | 0.1 M Imidazole, pH 6.5 | None |
| 26 | 30% MPD | 0.1 M Na citrate, pH 5.6 | 0.2 M ammonium acetate |
| 27 | 20% 2-propanol | 0.1 M Na HEPES, pH 7.5 | 0.2 M Na citrate |
| 28 | 30% PEG 8000 | 0.1 M Na cacodylate, pH 6.5 | 0.2 M Na acetate |
| 29 | 0.8 M K Na tartrate | 0.1 M Na HEPES, pH 7.5 | None |
| 30 | 30% PEG 8000 | None | 0.2 M ammonium sulfate |
| 31 | 30% PEG 4000 | None | 0.2 M ammonium sulfate |
| 32 | 2.0 M ammonium sulfate | None | None |
| 33 | 4.0 M Na formate | None | None |
| 34 | 2.0 M Na formate | 0.1 M Na acetate, pH 4.6 | None |
| 35 | 1.6 M Na/K phosphate | 0.1 M Na HEPES, pH 7.5 | None |
| 36 | 8% PEG 8000 | 0.1 M TRIS HCl, pH 8.5 | None |
| 37 | 8% PEG 4000 | 0.1 M Na acetate, pH 4.6 | None |
| 38 | 1.4 M Na citrate | 0.1 M Na HEPES, pH 7.5 | None |

*(Continued)*

**Table 6.3 (Continued)**

Example of a Sparse Matrix Screen

| | Precipitant | Buffer | Salt |
|---|---|---|---|
| 39 | 2% PEG 400/2.0 M ammonium sulfate | 0.1 M Na HEPES, pH 7.5 | None |
| 40 | 20% PEG 4000/20% 2-propanol | 0.1 M Na citrate, pH 5.6 | None |
| 41 | 20% PEG 4000/10% 2-propanol | 0.1 M Na HEPES, pH 7.5 | None |
| 42 | 20% PEG 8000 | None | 0.05 M K phosphate |
| 43 | 30% PEG 1500 | None | None |
| 44 | 0.2 M Mg formate | None | None |
| 45 | 18% PEG 8000 | 0.1 M Na cacodylate, pH 6.5 | 0.2 M Zn acetate |
| 46 | 18% PEG 8000 | 0.1 M Na cacodylate, pH 6.5 | 0.2 M Ca acetate |
| 47 | 2.0 M ammonium sulfate | 0.1 M Na acetate, pH 4.6 | None |
| 48 | 2.0 M ammonium phosphate | 0.1 M TRIS HCl, pH 8.5 | None |
| 49 | 2% PEG 8000 | None | 1.0 M Li sulfate |
| 50 | 15% PEG 8000 | None | 0.5 M Li sulfate |

*Source:* Jancarik, J., and Kim, S.-H. (1991). Sparse matrix sampling: A screening method for crystallization of proteins. *Journal of Applied Crystallography* 24, 409–411.
*Note:* Crystal screen from Hampton Research.

You should not expect to have perfect protein crystals from spare matrix screens. While it can happen, it is very unlikely to occur. Careful observing and scoring of crystallization drops are necessary. There are different systems for scoring. The following is a good start:

- Score 0. Clear drop, no precipitation. In sparse matrix screens, an appropriate protein concentration will yield about 50% clear drops. If there is substantially more than this, increase the concentration; if less, decrease it.

- Score 1. *Skin* and "bad" precipitate (**Figure 6.8a through c**). Bad precipitate is characterized as heavy and *amorphous*, and has a brownish appearance, indicating denatured protein.

- Score 2. "Good" precipitate (**Figure 6.8d**). Good precipitate is generally considered to be granular, nonamorphous, and capable of being resolubilized.

- Score 3. Phase separation and quasi crystals (**Figure 6.8e and f**). In phase separation, the protein forms small transparent droplets. Quasi crystals are formations that have a more solid/darker appearance than phase separation (essentially everything that does not appear to be skin/precipitate but is not yet clearly crystalline).

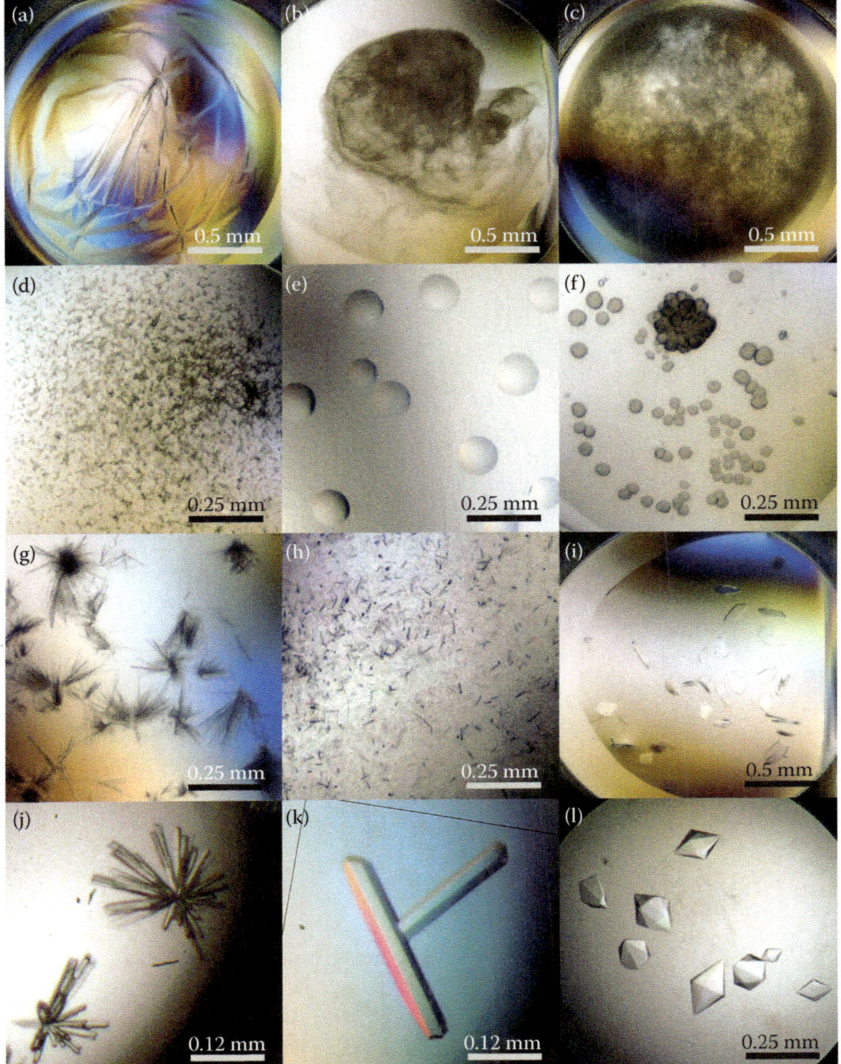

**Figure 6.8 Different crystal drops of increasing quality:** (a) Skin. (b, c) Amorphous ("bad") precipitation. (d) Nonamorphous ("good") precipitation. (e) Phase separation. (f) Quasi crystals. (g) Needles. (h) Small crystals. (i) Plates. (j) Rod clusters. (k) Single rods. (l) High-quality crystals.

- Score 4. Needles, tiny crystals, and thin plates (**Figure 6.8h and i**). Clearly crystalline in nature, but because of their lack in size in one or more dimensions, will not diffract well.

- Score 6. Lager crystals (**Figure 6.8j through l**). Essentially all crystals that might potentially diffract well.

- Star. Marking a star beside all those conditions that show mechanical contaminants or defects in the crystallization well helps with later drop evaluation.

## Pitfalls

The biggest frustration for a starting crystallographer must be the first discovery of a salt crystal, especially because salt crystals tend to be beautifully formed. Learning how to distinguish between a salt and a protein crystal is important. Salt crystals are always more stable than protein crystals due to their ionic intermolecular interactions. Consequently, one of the easiest tests is the "crush test," where the crystals are crushed using a fine, pointed tool. Protein crystals have weak intermolecular forces due to the 40–60% solvent content inside the crystal, and thus will disintegrate when prodded. But then you have ruined what could have been a beautiful crystal!

Alternatively, microscopes with polarizers can show *birefringence* characteristics of protein crystals. Of course, if the protein shows fluorescence, it may be examined under the correct excitation wavelength. Another option is to use a *crystal dye* to try to determine if a crystal is salt or protein. These are simply small-molecule dyes that can penetrate into the solvent tracts of protein crystals, thus dyeing them, but cannot enter salt crystals.

If your ultimate goal is x-ray crystallography, the ultimate test of what kind of crystal you have is to measure an x-ray diffraction pattern (**Figure 6.9**). A diffraction pattern can also tell you how well your crystal diffracts. Thus, if you have access to an x-ray source, it is always worthwhile to collect a diffraction pattern. One never knows how well a certain crystal diffracts from simply looking at it—looks can be deceiving!

Due to their weak intermolecular forces, protein crystals are labile and transient. So once you obtain what you consider a crystal with potential, it is best to obtain a diffraction pattern as soon as possible. It is also a good idea to continue optimizing crystallization conditions to continually obtain more and better crystals.

## Fine-screening

During optimization, the crystal quality is improved by fine-screening the promising conditions. One common procedure is to vary two of the crystallization parameters at a time. Concentration of precipitant and pH tend to be the most influential parameters, so they are a sensible choice to start with. Alternatively, you can alter any of the other described parameters, such as salt, drop size, protein concentration, etc. To vary the temperature, separate plates will need to be set up.

Varying two parameters at a time is used mainly due to convenience since the plates are two-dimensional. The first parameter is varied along the $x$-axis and the second parameter along the $y$-axis. **Figure 6.10** shows an example of a setup for a fine-screening plate. It is very easy to expand the two-dimensional into a three-dimensional as well as a four-dimensional fine screen. However, beware that this will quickly increase the number of plates that need to be set up! Considering that one plate is used for two-dimensional fine-screening, you will need approximately 4 plates to expand to a three-dimensional screen and up to 16 for a four-dimensional screen.

At the beginning of fine-screening it is advisable to set up a broad grid taking "large" steps in all directions (e.g., concentration differences of ~5% or pH steps of 0.5 to 1 unit). This allows the determination of the direction in which optimization can best be achieved and also provides an estimate of how much improvement

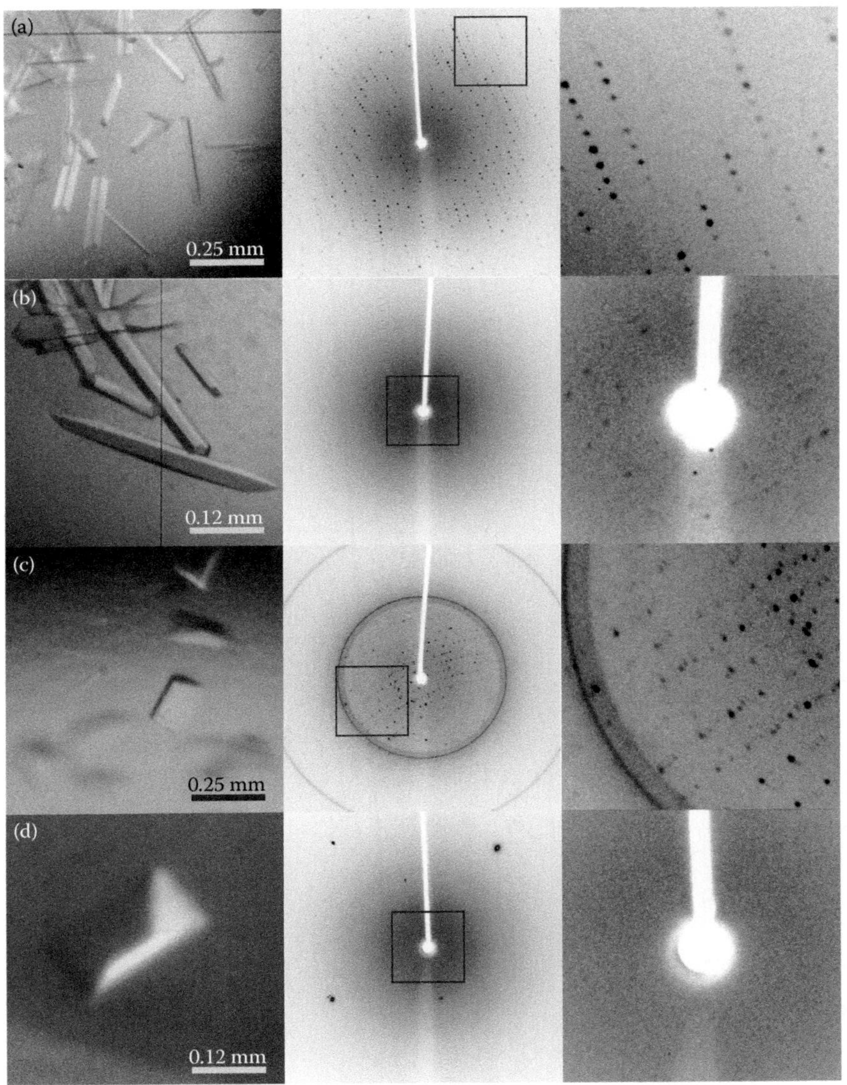

**Figure 6.9 Crystals and their corresponding diffraction patterns.** On the left are pictures of crystals; the middle panels show pictures of whole diffraction patterns; and the right panels are zoom-ins of these diffraction patterns. (a) A high-diffracting crystal (2.5 Å). Well-diffracting crystals have discrete round spots that are regularly spaced from each other. (b) A low-diffracting crystal (10 Å). Low-diffracting crystals only have spots near the center of the beam (low resolution) and no spots on the outside (high resolution). (c) A cracked crystal with ice-ring formation (4 Å). The ring signifies the formation of ice, while the irregularly spaced spots indicate that the crystal cracked. (d) A salt crystal. A few very strong spots are characteristic for salt crystals; there are no spots near the center of the beam.

you can hope to achieve from this optimization. If the results of the first fine screen yield crystals that provide good resolution data sets (3 Å or below), performing a narrow fine screen (concentration difference of ~1% and pH steps of 0.25 unit) should be adequate to further improve the resolution. However, if the resolution is still low (i.e., more than 4–5 Å), more significant changes should be pursued: change of chemicals such as buffer systems, *additive* screens, or *seeding*).

## Additive screens

Additive screens serve as an auxiliary strategy to alterations of common crystallization parameters. It is believed that additives enhance macromolecular crystals by improving crystal contacts. Some additives might achieve this by stabilizing a specific protein confirmation. Others might have a beneficial

Screen conditions in 24-well plate

| | 1 | 2 | 3 | 4 | 5 | 6 |
|---|---|---|---|---|---|---|
| A | NaCl 1 M<br>NaOAc 0.1 M  pH 4 | NaCl 1.2 M<br>NaOAc 0.1 M  pH 4 | NaCl 1.4 M<br>NaOAc 0.1 M  pH 4 | NaCl 1.6 M<br>NaOAc 0.1 M  pH 4 | NaCl 1.8 M<br>NaOAc 0.1 M  pH 4 | NaCl 2 M<br>NaOAc 0.1 M  pH 4 |
| B | NaCl 1 M<br>NaOAc 0.1 M  pH 4.2 | NaCl 1.2 M<br>NaOAc 0.1 M  pH 4.2 | NaCl 1.4 M<br>NaOAc 0.1 M  pH 4.2 | NaCl 1.6 M<br>NaOAc 0.1 M  pH 4.2 | NaCl 1.8 M<br>NaOAc 0.1 M  pH 4.2 | NaCl 2 M<br>NaOAc 0.1 M  pH 4.2 |
| C | NaCl 1 M<br>NaOAc 0.1 M  pH 4.4 | NaCl 1.2 M<br>NaOAc 0.1 M  pH 4.4 | NaCl 1.4 M<br>NaOAc 0.1 M  pH 4.4 | NaCl 1.6 M<br>NaOAc 0.1 M  pH 4.4 | NaCl 1.8 M<br>NaOAc 0.1 M  pH 4.4 | NaCl 2 M<br>NaOAc 0.1 M  pH 4.4 |
| D | NaCl 1 M<br>NaOAc 0.1 M  pH 4.6 | NaCl 1.2 M<br>NaOAc 0.1 M  pH 4.6 | NaCl 1.4 M<br>NaOAc 0.1 M  pH 4.6 | NaCl 1.6 M<br>NaOAc 0.1 M  pH 4.6 | NaCl 1.8 M<br>NaOAc 0.1 M  pH 4.6 | NaCl 2 M<br>NaOAc 0.1 M  pH 4.6 |

Volumes of stock solutions to be added (stock solutions are NaCl, 4 M; NaOAc, 0.5 mM, pH 4; NaOAc, 0.5 M, pH 4.6)

| | 1 | 2 | 3 | 4 | 5 | 6 |
|---|---|---|---|---|---|---|
| A | d.H$_2$O 825 μl<br>NaCl 375 μl<br>NaOAc 300 μl  pH 4<br>NaOAc 0 μl  pH 4.6 | d.H$_2$O 750 μl<br>NaCl 450 μl<br>NaOAc 300 μl  pH 4<br>NaOAc 0 μl  pH 4.6 | d.H$_2$O 675 μl<br>NaCl 525 μl<br>NaOAc 300 μl  pH 4<br>NaOAc 0 μl  pH 4.6 | d.H$_2$O 600 μl<br>NaCl 600 μl<br>NaOAc 300 μl  pH 4<br>NaOAc 0 μl  pH 4.6 | d.H$_2$O 525 μl<br>NaCl 675 μl<br>NaOAc 300 μl  pH 4<br>NaOAc 0 μl  pH 4.6 | d.H$_2$O 450 μl<br>NaCl 750 μl<br>NaOAc 300 μl  pH 4<br>NaOAc 0 μl  pH 4.6 |
| B | d.H$_2$O 825 μl<br>NaCl 375 μl<br>NaOAc 220 μl  pH 4<br>NaOAc 80 μl  pH 4.6 | d.H$_2$O 750 μl<br>NaCl 450 μl<br>NaOAc 220 μl  pH 4<br>NaOAc 80 μl  pH 4.6 | d.H$_2$O 675 μl<br>NaCl 525 μl<br>NaOAc 220 μl  pH 4<br>NaOAc 80 μl  pH 4.6 | d.H$_2$O 600 μl<br>NaCl 600 μl<br>NaOAc 220 μl  pH 4<br>NaOAc 80 μl  pH 4.6 | d.H$_2$O 525 μl<br>NaCl 675 μl<br>NaOAc 220 μl  pH 4<br>NaOAc 80 μl  pH 4.6 | d.H$_2$O 450 μl<br>NaCl 750 μl<br>NaOAc 220 μl  pH 4<br>NaOAc 80 μl  pH 4.6 |
| C | d.H$_2$O 825 μl<br>NaCl 375 μl<br>NaOAc 120 μl  pH 4<br>NaOAc 180 μl  pH 4.6 | d.H$_2$O 750 μl<br>NaCl 450 μl<br>NaOAc 120 μl  pH 4<br>NaOAc 180 μl  pH 4.6 | d.H$_2$O 675 μl<br>NaCl 525 μl<br>NaOAc 120 μl  pH 4<br>NaOAc 180 μl  pH 4.6 | d.H$_2$O 600 μl<br>NaCl 600 μl<br>NaOAc 120 μl  pH 4<br>NaOAc 180 μl  pH 4.6 | d.H$_2$O 525 μl<br>NaCl 675 μl<br>NaOAc 120 μl  pH 4<br>NaOAc 180 μl  pH 4.6 | d.H$_2$O 450 μl<br>NaCl 750 μl<br>NaOAc 120 μl  pH 4<br>NaOAc 180 μl  pH 4.6 |
| D | d.H$_2$O 825 μl<br>NaCl 375 μl<br>NaOAc 0 μl  pH 4<br>NaOAc 300 μl  pH 4.6 | d.H$_2$O 750 μl<br>NaCl 450 μl<br>NaOAc 0 μl  pH 4<br>NaOAc 300 μl  pH 4.6 | d.H$_2$O 675 μl<br>NaCl 525 μl<br>NaOAc 0 μl  pH 4<br>NaOAc 300 μl  pH 4.6 | d.H$_2$O 600 μl<br>NaCl 600 μl<br>NaOAc 0 μl  pH 4<br>NaOAc 300 μl  pH 4.6 | d.H$_2$O 525 μl<br>NaCl 675 μl<br>NaOAc 0 μl  pH 4<br>NaOAc 300 μl  pH 4.6 | d.H$_2$O 450 μl<br>NaCl 750 μl<br>NaOAc 0 μl  pH 4<br>NaOAc 300 μl  pH 4.6 |

**Figure 6.10  Example plate for a two-dimensional fine screen.** NaCl is the precipitant, and Na acetate the buffer.

effect by directly altering protein–protein and/or protein–solvent interactions. To set up a commercially available additive screen, select one of the most promising crystallization conditions and add the additive directly to the drop. (The additive is added only to the drop, not the reservoir solution, unless it is a volatile organic solvent.) Once a potential improvement of crystals is identified, fine-screen the condition again in the presence of the additive. During this fine screen, the concentration of the additive itself can be varied as well as all the other crystallization parameters, even if they already have been optimized before trying the additive screen.

## Seeding

If despite all efforts, you never manage to produce single crystals that diffract well, seeding the crystals might help. There are two general approaches for seeding: microseeding and macroseeding. The principle of the method lies in the separation of the nucleation event from crystal growth.

In microseeding, commercially available Seed Bead kits can be used to grind protein crystals into very small crystals, i.e., seeds, by vortexing. Seeds can be added to a new crystallization trial with the same conditions as before, with the seeds serving as nucleation sites. Seeding can be repeated with the newly formed crystals from the first seeding experiment. Each subsequent cycle of seeding will hopefully produce better-quality crystals. Due to the fact that seeding circumvents nucleation, microseeds can be used to nucleate not only the original crystallization conditions but also other conditions from the sparse matrix screens, with the hope to find new crystal growth conditions.

The major difference between microseeding and macroseeding is that in the latter, the crystal is large enough to see. It essentially involves taking a crystal from one drop and adding it to one with fresh mother liquor. This method might sound straightforward but is rather tedious to accomplish. In macroseeding, the idea is to slightly dissolve the crystal before letting it grow again. Whenever a crystal increases in size, imperfections start to accumulate. Before adding the crystal to the new mother liquor, however, it has to be washed. For the method to succeed, it is critical to find the right wash protocol that removes crystal surface imperfections such as crystalline debris while not completely dissolving the crystal. One way of finding such a protocol is to make a series of reservoir dilutions and then wash different crystals in these solutions before adding them to the mother liquor. Some more detailed references to this method are given at the end of the chapter.

## Improving the protein

If the sparse matrix screen yields only scores of 0–1 (clear drops, skin, and amorphous precipitate), optimizing conditions is not likely to help much, and more desperate measures should be sought. First of all, ensure that your protein purity is adequate and that the protein is well dispersed and has not degraded. If you have not cleaved your affinity tag as in **Practical Tips 6.3**, that is a good place to start.

Another desperate measure is to crystallize a protein in the presence of proteases such as chymotrypsin or trypsin. This is a technique that has found lots of attention in the crystallographic community. The idea is that the protease will cleave the flexible regions, allowing the protein to become more stable and thus increasing the chance of crystal formation. However, the technique should be used with care since the protease might cleave too much of the protein, resulting in a crystal structure that no longer represents the native structure.

## Obtaining different crystal forms of the same protein

Congratulations! You have a well-diffracting crystal. But while one crystal is great, having several is better. Crystals grown under a variety of conditions can sometimes tell more about a protein than those obtained from a single set of parameters. One practical benefit of having more than one crystallization condition is that it will drastically increase the likelihood of collecting a better data set. But the technical benefit is even more compelling. The technique of x-ray crystallography can be thought of as a photo camera in which each crystal structure represents a still shot of one conformation of the protein. Since proteins are inherently flexible, the more still shots exist, the better the ultimate description of the protein structure and its function will be. This holds especially true if the protein binds a ligand. Having two structures, one with the ligand and one without, is greatly desirable.

There are two significantly different ways to obtain ligand-bound crystal structure. Adding ligand to the protein solution before crystallizing is the most common way and is referred to as *cocrystallization*. However, it is not uncommon to obtain crystals without any ligand bound to them when using this method. Alternatively, you can try to *soak* ligands by adding the solubilized ligand to crystallization drops with preformed crystals. However, this technique does not guarantee success either. If successful, some caution is required when analyzing the result. Since the crystal was preformed, it is not certain whether the actual ligand-binding site was accessible when the ligand bound. It is possible for ligands to bind in a biologically irrelevant fashion to a preformed crystal lattice, yielding functionally meaningless data.

The techniques we have discussed in this section represent only a few of the more commonly used methods for optimizing crystallization. The textbooks referenced at the end of the chapter discuss them in greater detail as well as suggest other avenues that can be pursued. Possible avenues of optimization are almost limitless, and so it is important to make educated decisions on what to try next.

# 6.7 EXAMPLE EXPERIMENT: LYSOZYME

Lysozyme was discussed in **Chapter 3** in the context of digestion of bacterial cell walls. This ubiquitous protein has the interesting property of crystallizing more readily than most others and providing excellent diffraction patterns yielding structures with <1 Å resolution. This protocol is an easy and satisfying introduction to crystallography that is guaranteed to give nice crystals.

The purpose of this experiment is to try the sitting and/or hanging drop method. For this, you will need one 24-well plate of each type, or a ComboPlate that can function as both. Note that some sitting drop plates require the separate purchase and assembly of CrystalBridges that sit inside the wells.

Materials needed

- Plates for sitting and hanging drop crystallization (24-well plates)

- Chicken egg-white lysozyme, highly purified

- Sodium acetate (Na acetate)

- Sodium chloride (NaCl)

- Deionized water (d.$H_2O$)

- Clear adhesive tape or film (for sitting drop)

- 18–22 mm coverslips (for hanging drop; size is plate dependent)

- Vacuum grease (for hanging drop)

Stock solutions (for one 24-well plate)

- 15 mg/mL of lysozyme in 25 mM Na acetate, pH 4.6 (~200 µL)

- NaCl 4 M (~15 mL)

- Na acetate 0.5 M, pH 4 (~5 mL)

- Na acetate 0.5 M, pH 4.6 (~5 mL)

- Deionized water (~20 mL)

All stock solutions including the protein sample and water should be filtered with a 0.22 µm filter before use. Small volumes (0.5 mL or less) may be filtered using special Eppendorf tubes with 0.22 µm filters. Create the fine screen according to **Figure 6.10** in twenty-four 1.5 mL Eppendorf tubes. The solution prepared should be enough to set up two plates. This can be either a hanging drop and sitting drop plate, or alternatively, two plates at two different temperatures. If you want to set up more plates, adjust the solutions accordingly.

For the *sitting drop experiment*, install the microbridges if needed. Using pressurized air, remove dust from sitting drop wells. Then add 500 µL of the crystallization solution (fine-screen in the Eppendorf tubes) to the reservoir well. Place 2 µL of the protein solution to the sitting drop well. Then, carefully add 2 µL of crystallization solution from the reservoir to the drop, trying not to create any air bubbles. Repeat this for the other wells. Seal the plate with tape as soon as you are done. It is important not to leave the experiment before all the wells are sealed to minimize evaporation.

For the *hanging drop experiment*, make sure to pregrease the whole plate according to **Practical Tips 6.2**. Using pressurized air, remove dust from a coverslip and place it on a clean area. Now add 500 µL of the crystallization solution to the first reservoir well. Place 2 µL of the protein solution onto the coverslip and then add an equal amount of crystallization solution to the drop. Seal the well and prepare the second well.

Sitting drop plates are generally faster and easier to set up than hanging drop plates. The approach to setting them up is generally the same, with one major difference. In the sitting drop method, all the chambers are filled with crystallization solution before adding the protein solution to the crystallization wells.

**Figure 6.11  A lysozyme crystal.**

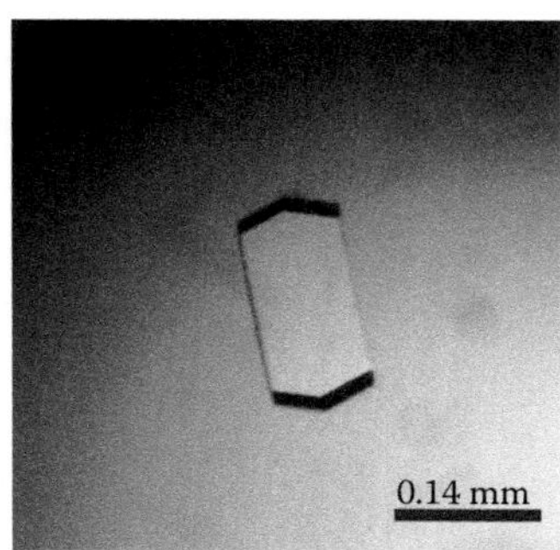


Because the chambers remain open until the last one is set up, it is important to work reasonably fast to minimize evaporation of the solutions. With the hanging drop method, usually, one chamber is filled and sealed at a time.

Once a whole plate is set up, it should be stored in a temperature-controlled, vibration-isolated, and preferably dark place. Crystals should form overnight and can been observed the next day under a microscope (**Figure 6.11**).

Crystallization depends on many factors, and it is critical to control as many as possible to ensure reproducibility. During the crystallization experiment, try to work swiftly but make sure to observe the following key points:

- Label your plate before starting to add solutions. (It is important to label the actual plate with the wells, instead of just labeling the plate lid. This keeps you from mixing up which lid belongs to which plate.)

- Add the crystallization solution to the protein solution. The reverse sequence tends to produce unwanted precipitate.

- Change your pipette tip after every condition.

- Pipette viscous solutions slowly to ensure correct volumes.

- Make sure you have enough protein sample, noting that you will need about 25–50% more than the total quantity in all your drops because there is always some solution on the outside of the pipette tips. (To avoid waste of protein on the outside of the tip, it is good practice to only insert the tip of the pipette into the sample.)

- Avoid transmitting body heat from your fingers to the protein solution. When holding the test tube containing the protein, hold it at the top rather than the bottom. Store the protein on ice when not in use.

- Close the solution tubes tightly when they are not in use to avoid unnecessary evaporation (especially important for commercially available sparse matrix screens).

- Avoid adding dust or other mechanical impurities to the drop. Wear a lab coat and avoid hairy sweaters.

- Pay special attention to sealing the crystallization chamber(s) properly.

- Avoid unnecessary vibration or heat sources.

# 6.8  DATA COLLECTION AND STRUCTURE DETERMINATION USING X-RAY CRYSTALLOGRAPHY

Some references are provided at the end of the chapter for detailed structure determination using x-ray crystallography. This section provides a brief introduction to the key facts needed at the outset.

## Where to do x-ray crystallography

Once protein crystals have been obtained, there are two basic ways of data collection: a synchrotron and a home source. The two major benefits of a synchrotron are that it has a higher beam intensity, which might allow for faster data collection and possibly higher resolution, and that it is frequently possible to tune the wavelength of the x-ray beam, which is needed for more advanced data collections. The benefit of the home source is that it is very convenient. It allows for the easier screening of primary hits, which is the most definite way to determine if a crystal is protein or salt. And since it allows the collection of complete diffraction data sets, it is, in most cases, an ideal place to collect data.

A home source consists of an x-ray source (sealed tube or rotating anode), optional optics for focusing the beam, a detector (CCD or imaging plate), computer and software, and frequently, a crystal cryo-cooling system (**Figure 6.12**). As the home source can be composed from different components, the quality of the data collected will depend ultimately on what is used.

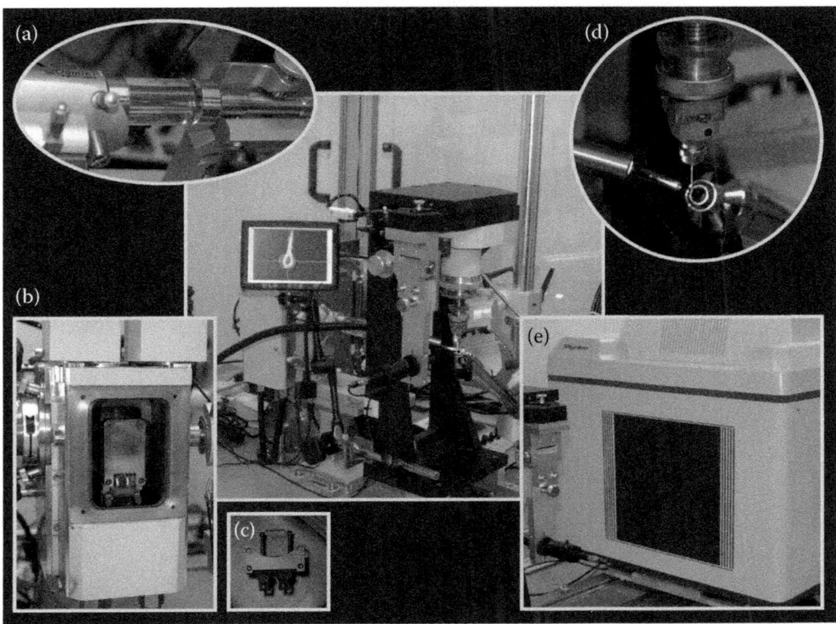


**Figure 6.12 An example of an x-ray diffraction home source.** (a) Optics focus the beam of x-rays (photons) and filter out kβ using a set of mirrors. (b) Chamber containing the rotating copper anode (target) that functions as the source of the x-rays by absorbing electrons and subsequently emitting photons. (c) A cathode filament produces electrons that hit the target. (d) A goniostat is used to center the sample (crystal). Also seen in the picture is the nozzle of the nitrogen cryo-stream and the beam stop to block the direct beam. (e) An image plate functions to record the diffraction patterns.

## Protecting crystals from radiation damage

In crystallography, a complete data set consists of many frames (~200 frames is quite common), where each frame is slightly offset by a small angle (~1°). In the earlier days of crystallography, data sets were collected at room temperature on individual photographic films. Changing films was time consuming, and the collection of a whole data set would often take several days. This usually meant more than one crystal had to be used, since exposing crystals to x-ray radiation creates free radicals that damage the protein, and consequently, the crystal lattice, leading to a decrease in the quality of the data. There are two advances that helped counteract radiation damage. Automated data collection with image plates or the even faster CCD detectors has drastically shortened the length of diffraction experiments. Secondly, collecting x-ray diffractions at −180°C by using a cryo-stream (liquid nitrogen) limits the diffusion rate of radicals.

However, exposing a crystal to cryogenic conditions means that it has to be protected against the formation of ice, which will lead to cracking. Common *cryo-protectants* used for protein crystals are glycerol; low-molecular-weight PEGs; ethylene glycol; sugars (glucose, sepharose); oils (paraffin, Paratone-N); etc. To protect crystals, you can either grow them in conditions that already contain a cryo-protectant, low-molecular-weight PEGs, for example, or soak them in the cryo-protectant after formation. This means taking the crystal from the mother liquor and adding it directly to a solution of cryo-protectant. The risk here is that the crystal might simply dissolve, since the solution will be so different from that under which it was formed. Alternatively, the cryo-protectant can also be directly added to the drop, but this will require time for the solution to equilibrate (minimum of 2 h, but preferably overnight). A third method combines the two previous approaches and entails making a new drop containing cryo-protectant mixed with reservoir solution. The crystal can then be protected by placing into this drop. Several cryo-protectants might need to be tried before one is found that does not damage the crystals. The percentage of a tested cryo-protectant may also be increased gradually. What works best is ultimately situation dependent and can be best evaluated by measuring diffraction patterns of the treated crystals.

Crystals are moved around and transferred from one solution to another using cryo-loops (**Figure 6.13**). Different sizes of loops are available and should be chosen each time according to the size of the crystal being moved. The transfer should be done under the microscope and requires a certain degree of finesse, as it must be dragged through a viscous solution that often contains a lot of junk

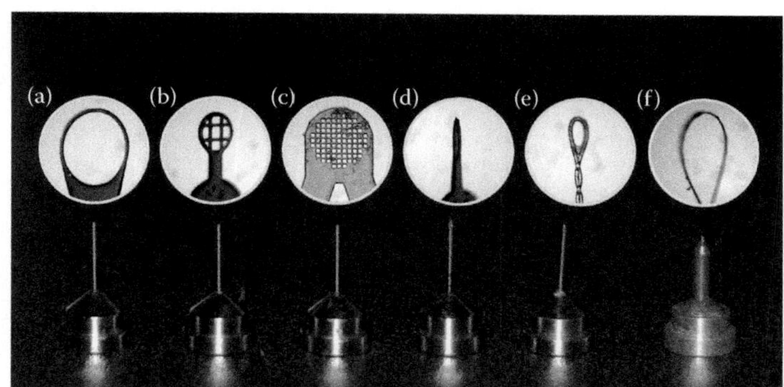


**Figure 6.13 Cryo-loops.** White circles (diameter = 0.6 mm) show magnification of the loops. Different sizes, materials, and forms of loops are available: (a, e, f) loops for mounting single crystals; (b, c) meshes for mounting very small crystals; and (d) elliptical narrow loops for mounting needles (d). All loops are attached by a stem to a magnetic goniometer base.

and pulled through the drop meniscus. Watch out that the drop does not dry out from the heat of the microscope while you are doing this! If using the hanging drop method, one way to limit evaporation is to place the coverslip in a petri dish surrounded with a wet paper towel. If using a sitting drop plate, you may have to periodically shut the scope off and allow the plate to cool if the crystal handling takes a while. Adding some reservoir solution to the drop will help counteract evaporation.

Crystallization still remains a challenge for many proteins, and scientists are continually working to get around the problem. One solution, for proteins that can be crystallized in two dimensions but not three, is to use electron diffraction techniques (see **Advanced Topic 6.1**). A high-cost but potentially revolutionary alternative is to make use of extremely high-intensity laser pulses to determine structure from nanometer- to micrometer-sized crystals (see **Advanced Topic 6.2**).

## 6.9  TROUBLESHOOTING Q AND A

I cannot get a homogenous protein solution—what to do?

Improvement/adaption of the purification steps is probably needed. Consider adding size exclusion columns. Changing the buffer system and/or pH can have a drastic effect on protein stability. More drastic methods involve cleaving of purification tags after purification or the cloning of a new protein construct.

My crystal disappeared.

Protein crystals are metastable and labile. Having a crystal appear and disappear is a good sign in that it is indicative of protein crystals. Making note of the time it took for that crystal to appear and then scheduling the next crystallization trial such that it coincides with data collection should work. Alternatively, the crystal can be flash-frozen and stored under liquid nitrogen. Also, crystals that disappear after a time should be harvested (used for data collection or flash-frozen) as soon as possible since they probably start to degenerate before they disappear completely.

I have too many nucleation sites or none at all.

Vibration and too much or too little mixing of crystallization and protein solution might affect nucleation rate. Consider changing drop ratios or overlay the reservoir solution with oil to alter the phase transition. Lowering the concentration of protein/precipitant might help. Also, seeding separates the nucleation from the growth phase and thus is a very useful way of improving crystal quality. Lastly, temperature has a significant effect on nucleation and crystal growth.

Why do my drops dehydrate?

Check the seal. Not all tapes, films, and plates contain evaporation equally well. When using the sitting drop method, it is very important to equilibrate the pressure in the well before sealing it; otherwise, the coverslip will be pushed up, which will probably destroy the seal.

### ADVANCED TOPIC 6.1:   MEMBRANE PROTEINS

Only a handful of three-dimensional crystals have been made from integral membrane proteins. The approach is to purify large amounts using a bacterial expression system and reconstitute the protein into lipid-containing solutions at high concentration. Even when such purification and reconstitution is possible, it is rare for the proteins to form well-ordered crystals because of flexible structures; some type of anchor is needed to hold them in place. A few references are given here to key papers on the subject, but be forewarned—these techniques are extremely difficult!

More success has been had crystallizing membrane proteins in two dimensions, since in this case, only a single layer of lipid is needed. Electron techniques such as *cryo-transmission electron microscopy* and *electron diffraction* are performed on two-dimensional rather than three-dimensional crystals, so these are the techniques of choice for proteins that cannot be crystallized in 3-D. Some high-resolution structures of membrane proteins have been obtained this way. Several groups are currently working to develop ultrafast electron techniques for direct visualization of protein conformational change.

The membrane protein that usually acts as a test system for 2-D crystallization is *bacteriorhodopsin*, because it is made in such large quantities in the bacterial membrane that it is already crystalline. All that needs to be done to perform 2-D experiments is to isolate the membranes and fuse them into large enough fragments for study (see Practical Tips 6.5).

### SUGGESTED READING

Agah, S., and Faham, S. (2012). Crystallization of membrane proteins in bicelles. *Methods in Molecular Biology* 914, 3–16.

Delmar, J.A., Bolla, J.R., Su, C.C., and Yu, E.W. (2015). Crystallization of membrane proteins by vapor diffusion. *Methods in Enzymology* 557, 363–392.

Johnson, M.C., Dreaden, T.M., Kim, L.Y., Rudolph, F., Barry, B.A., and Schmidt-Krey, I. (2013). Two-dimensional crystallization of membrane proteins by reconstitution through dialysis. *Methods in Molecular Biology* 955, 31–58.

Johnson, M.C., and Schmidt-Krey, I. (2013). Two-dimensional crystallization by dialysis for structural studies of membrane proteins by the cryo-EM method electron crystallography. *Methods in Cell Biology* 113, 325–337.

Kim, J., Kagawa, A., Kurasaki, K., Ataie, N., Cho, I.K., Li, Q.X., and Ng, H.L. (2015). Large-scale identification of membrane proteins with properties favorable for crystallization. *Protein Science* 24, 1756–1763.

Klara, S.S., Saboe, P.O., Sines, I.T., Babaei, M., Chiu, P.L., DeZorzi, R., Dayal, K., Walz, T., Kumar, M., and Mauter, M.S. (2016). Magnetically directed two-dimensional crystallization of OmpF membrane proteins in block copolymers. *Journal of the American Chemical Society* 138, 28–31.

Li, D., Boland, C., Walsh, K., and Caffrey, M. (2012). Use of a robot for high-throughput crystallization of membrane proteins in lipidic mesophases. *Journal of Visualized Experiments*, e4000.

Moraes, I., and Archer, M. (2015). Methods for the successful crystallization of membrane proteins. *Methods in Molecular Biology* 1261, 211–230.

Morrison, V.R., Chatelain, R.P., Tiwari, K.L., Hendaoui, A., Bruhacs, A., Chaker, M., and Siwick, B.J. (2014). A photoinduced metal-like phase of monoclinic VO(2) revealed by ultrafast electron diffraction. *Science* 346, 445–448.

Muller, F.G., and Lancaster, C.R. (2013). Crystallization of membrane proteins. *Methods in Molecular Biology* 1033, 67–83.

Nannenga, B.L., Iadanza, M.G., Vollmar, B.S., and Gonen, T. (2013). Overview of electron crystallography of membrane proteins: Crystallization and screening strategies using negative stain electron microscopy. *Current Protocols in Protein Science*, Chapter 17, Unit17 15.

Siwick, B., and Collet, E. (2013). Physical chemistry: Molecular motion watched. *Nature* 496, 306–307.

Uddin, Y.M., and Schmidt-Krey, I. (2015). Inducing two-dimensional crystallization of membrane proteins by dialysis for electron crystallography. *Methods in Enzymology* 557, 351–362.

Ujwal, R., and Abramson, J. (2012). High-throughput crystallization of membrane proteins using the lipidic bicelle method. *Journal of Visualized Experiments*, e3383.

## ADVANCED TOPIC 6.2:   X-RAY FREE-ELECTRON LASER

As of this writing (mid-2016), structure determination by x-ray free-electron laser (XFEL) is a rapidly emerging technique. XFELs produce extremely intense femtosecond pulses, which allow structure determination from nanocrystals and microcrystals; the crystal is vaporized in the process. By using a stream of crystals, multiple diffraction patterns may be rapidly obtained, a technique called *serial femtosecond crystallography*.

The principle behind XFEL is that the extremely fast pulses permit data collection before the x-rays have rearranged the molecular structure of the crystal. The concept was described in 2000, but implementation has been slow because facilities are costly (~$1 B), with only two currently operational: one at Stanford (California) and one at Riken (Japan). A European XFEL is scheduled to become operational in 2017.

The first studies of protein structure using XFEL used previous knowledge of related structures to analyze the data. However, in 2014, de novo structure determinations began to be published. Software for analysis of XFEL data also began to become available in the early 2010s.

Some protein scientists believe that this is the final, definitive technique for determining protein structures, since lengthy and difficult crystallization procedures are not needed. However, obtaining access to facilities and beam time remain limiting steps for most researchers at the present time.

### SUGGESTED READING


Barends, T.R., Foucar, L., Botha, S., Doak, R.B., Shoeman, R.L., Nass, K., Koglin, J.E., Williams, G.J., Boutet, S., Messerschmidt, M. et al. (2014). De novo protein crystal structure determination from X-ray free-electron laser data. *Nature* 505, 244–247.

Boutet, S., Lomb, L., Williams, G.J., Barends, T.R., Aquila, A., Doak, R.B., Weierstall, U., DePonte, D.P., Steinbrener, J., Shoeman, R.L. et al. (2012). High-resolution protein structure determination by serial femtosecond crystallography. *Science* 337, 362–364.

Chapman, H.N., Fromme, P., Barty, A., White, T.A., Kirian, R.A., Aquila, A., Hunter, M.S., Schulz, J., DePonte, D.P., Weierstall, U. et al. (2011). Femtosecond X-ray protein nanocrystallography. *Nature* 470, 73–77.

Fromme, P., and Spence, J.C. (2011). Femtosecond nanocrystallography using X-ray lasers for membrane protein structure determination. *Current Opinion in Structural Biology* 21, 509–516.

Johansson, L.C., Arnlund, D., White, T.A., Katona, G., Deponte, D.P., Weierstall, U., Doak, R.B., Shoeman, R.L., Lomb, L., Malmerberg, E. et al. (2012). Lipidic phase membrane protein serial femtosecond crystallography. *Nature Methods* 9, 263–265.

Kern, J., Alonso-Mori, R., Hellmich, J., Tran, R., Hattne, J., Laksmono, H., Glockner, C., Echols, N., Sierra, R.G., Sellberg, J. et al. (2012). Room temperature femtosecond X-ray diffraction of photosystem II microcrystals. *Proceedings of the National Academy of Sciences of the United States of America* 109, 9721–9726.

Kern, J., Alonso-Mori, R., Tran, R., Hattne, J., Gildea, R.J., Echols, N., Glockner, C., Hellmich, J., Laksmono, H., Sierra, R.G. et al. (2013). Simultaneous femtosecond X-ray spectroscopy and diffraction of photosystem II at room temperature. *Science* 340, 491–495.

Kirian, R.A., Wang, X., Weierstall, U., Schmidt, K.E., Spence, J.C., Hunter, M., Fromme, P., White, T., Chapman, H.N., and Holton, J. (2010). Femtosecond protein nanocrystallography—Data analysis methods. *Optics Express* 18, 5713–5723.

Neutze, R., Wouts, R., van der Spoel, D., Weckert, E., and Hajdu, J. (2000). Potential for biomolecular imaging with femtosecond X-ray pulses. *Nature* 406, 752–757.

Redecke, L., Nass, K., DePonte, D.P., White, T.A., Rehders, D., Barty, A., Stellato, F., Liang, M., Barends, T.R., Boutet, S. et al. (2013). Natively inhibited *Trypanosoma brucei* cathepsin B structure determined by using an X-ray laser. *Science* 339, 227–230.

Weierstall, U., James, D., Wang, C., White, T.A., Wang, D., Liu, W., Spence, J.C., Bruce Doak, R., Nelson, G., Fromme, P. et al. (2014). Lipidic cubic phase injector facilitates membrane protein serial femtosecond crystallography. *Nature Communications* 5, 3309.

White, T.A., Kirian, R.A., Martin, A.V., Aquila, A., Nass, K., Barty, A., and Chapman, H.N. (2012). CrystFEL: A software suite for snapshot serial crystallography. *Journal of Applied Crystallography* 45, 335–341.

### PRACTICAL TIPS 6.5:    BACTERIORHODOPSIN

Most photosynthetic organisms use *chlorophylls* to harvest light energy from the sun and generate a proton gradient that is used to make ATP. An exception is found in certain archaea of the genus *Halobacterium*, which instead use bacteriorhodopsin. Bacteriorhodopsin is very different in mechanism to chlorophylls, as it acts as a direct proton pump. It is similar in structure to the light-sensing proteins in the vertebrate retina, the *rhodopsins*, as well as having homology to several other proteins in its class, which is the *7TM family*. (It has 7 transmembrane domains.) It is produced in very large amounts in halobacteria, taking up nearly 50% of the surface of the cell in patches called *purple membrane* that are already in a two-dimensional hexagonal crystal lattice. Because of its importance and its ease of production, it often serves as a model system for transmembrane protein crystal experiments. It was the first crystal structure solved by *electron crystallography*. It has the added advantage of being activated by light, so it can be used in time-resolved experiments where a pulse of light determines time 0.

The biggest challenge in obtaining bacteriorhodopsin is finding and maintaining the bacterial strain. *Halobacterium salinarum* strain S9 is a great overproducer of bacteriorhodopsin, as are some strains of *Halobacterium halobium,* but these are not available commercially. The best way to get hold of it is to find a researcher actively using it and obtain a culture. Culturing these strains and getting them to make the purple membrane can be a little tricky as well, as they are extreme halophiles (salt-lovers); their growth medium contains >4 M NaCl. Plates containing such medium must be kept from drying by providing a source of humidity, such as a wet paper towel, or the NaCl will crystallize. Some strains prefer a "coarse" salt; for others, ordinary lab salt will do. Purple membrane is overproduced under anaerobic conditions under white-light illumination. Thus, once the strains have grown (they grow slowly!), shaking is reduced to minimize oxygenation, and the flasks are exposed to incandescent or fluorescent lighting. Many researchers have dedicated incubators with lights installed for this purpose.

Once the strains have grown, the membranes are easily separated, and membrane patches may be fused to yield large 2-D crystals. The references here provide some methods for working with purple membrane. The absorbance peak of bacteriorhodopsin is 568 nm, and it has a striking deep purple color.

### SUGGESTED READING

Kimura, Y., Vassylyev, D.G., Miyazawa, A., Kidera, A., Matsushima, M., Mitsuoka, K., Murata, K., Hirai, T., and Fujiyoshi, Y. (1997). Surface of bacteriorhodopsin revealed by high-resolution electron crystallography. *Nature* 389, 206–211.

Oesterhelt, D. (1976). Bacteriorhodopsin as an example of a light-driven proton pump. *Angewandte Chemie—International Edition in English* 15, 17–24.

Selinsky, B.S. (ed.). *Membrane Protein Protocols: Expression, Purification, and Characterization*. Humana Press, Totowa, NJ, 2003.

Subramaniam, S., and Henderson, R. (1999). Electron crystallography of bacteriorhodopsin with millisecond time resolution. *Journal of Structural Biology* 128, 19–25.

Why does it not work anymore?

Welcome to the art of crystallization. Make sure you use the same solutions/materials as the previous time. The source of the water is especially important. Chemicals to make the solutions should all be of the highest purity, and it is advisable to use the same supplier for a given chemical. If trying to recreate a condition from a sparse matrix screen, prepare the stock solution and final solution the same way as the supplier of the screens. pH is usually given for the buffer alone and not the entire solution. One way to check if your final solution is accurate is to measure the pH of your solution and compare it with that of the original screen solution. If trying to recreate a published crystal experiment, start with a broad-grid fine screen. Also, try to use the same pipettes and take note if the environment is especially humid etc. when making solutions or setting up the experiments. When switching from one plate to another, it is almost always necessary to redo some fine-screening because the kinetics of the phase transition will have changed. The same holds true when using a new protein batch. Lastly, the protein batch might have aged and ceased to be homogenous.

## 6.10  CONCLUSIONS AND FINAL REMARKS

Protein crystallization is an essential but often difficult and time-consuming step in x-ray structure determination. By now, you should have a good idea of all the complexities involved in protein crystallization. Because each protein is unique and behaves differently, there is no clear-cut path to follow when attempting to crystallize a chosen protein. However, with this introduction, you have a good idea where to start and what to look for. Below you will find online sources and other references that will help you to successfully crystallize your protein. Best of luck.

### Problems

1. (Particle sizing) Show that the field autocorrelation function for a monodisperse sample is given by

$$g_1(\tau) = \exp(-\Gamma\tau),$$

where $\Gamma$ is a decay constant. What is the relationship to the diffusion coefficient?

2. (Particle sizing) If the sample is polydisperse, this equation becomes a sum over the distribution of decay rates, $C(\Gamma)$:

$$g_1(\tau) = \int_0^\infty C(\Gamma)\exp(-\Gamma\tau)\,d\Gamma.$$

   Give an expression for $C(\Gamma)$ and describe what is meant by *polydispersity*.

   Given the form of this autocorrelation function, why is it dangerous to particle-size a polydisperse sample?

3. (Birefringence) A birefringent crystal has a refractive index $n_1$ in the $x$ direction and $n_2$ in the $y$ direction. If linearly polarized light is incident on the crystal from below, with its e-vector at an angle of 45° with the $x$-axis, show that the light coming from the upper crystal face is circularly polarized if the crystal thickness $d$ is related to the wavelength of incident light $\lambda$ by

$$t = \frac{\lambda}{4(n_2 - n_1)}.$$

4. Why are crystals, particularly large crystals, so valuable for x-ray diffraction? How does the intensity of the peaks scale with the number of scatterers? Explain the link between crystal appearance and diffraction quality.

5. What would be the benefit of collecting diffraction data at a cryogenic temperature as opposed to room temperature?

6. Structural genomics approaches often screen for the fluorescence changes of their protein when exposed to a collection of small molecules. How can one explain the higher success rate of crystallization when cocrystallizing the protein with small molecules that resulted in wavelength shift?

7. Why are there only a limited amount of DNA-bound protein structures when compared to all the other protein structures deposited in the Protein Data Bank?

8. Why do crystals exhibit different shapes and sizes?

9. What would the phase transition diagram look like for an interface diffusion, microbatch, or dialysis crystallization experiment?

10. Do you expect to see hydrogen atoms in your finalized crystal structure?

11. What structure would you expect the protein to possess in skin or amorphous versus nonamorphous precipitate?

12. Why might it be necessary to modify a perfectly purified protein to make it amenable for crystallization?

13. How can modification of exterior amino acids such as lysines help protein crystallization?

14. Why is it often necessary to simultaneously screen one variable while altering another (i.e., two-dimensional or multidimensional fine screen versus one-dimensional fine screen)?

15. How can polyvalent salts affect the crystallization process differently as compared to monovalent salts?

## Background Reading

### Books

Bergfors, T. *Protein Crystallization*. Edn. 2. International University Line, La Jolla, CA, 2009.

Doublie, S. *Macromolecular Crystallography Protocols, Volume 1: Preparation and Crystallization of Macromolecules*. Humana Press Inc., Totowa, NJ, 2007.

Doublie, S. *Macromolecular Crystallography Protocols, Volume 2: Structure Determination*. Humana Press Inc., Totowa, NJ, 2007.

Drenth, J. *Principles of Protein X-Ray Crystallography*. Edn. 3. Springer Science, New York, 2007.

Iwata, S. *Methods and Results in Crystallization of Membrane Proteins*. International University Line, La Jolla, CA, 2003.

Hunte, C. *Membrane Protein Purification and Crystallization: A Practical Guide*. Edn. 2. Academic Press, San Diego, CA, 2003.

McPherson, A. *Crystallization of Biological Macromolecules*. Cold Spring Harbor Laboratory Press, Cold Spring Harbor, NY, 1999.

Rhodes, G. *Crystallography Made Crystal Clear: A Guide for Users of Macromolecular Models*. Edn. 3. Academic Press, San Diego, CA, 2006.

Rupp, B. *Biomolecular Crystallography: Principles, Practice, and Application to Structural Biology*. Garland Science, New York, 2009.

### Journal articles

Benvenuti, M., and Mangani, S. (2007). Crystallization of soluble proteins in vapor diffusion for x-ray crystallography. *Nature Protocols* 2, 1633–1651.

Bergfors, T. (2007). Screening and optimization methods for nonautomated crystallization laboratories. *Methods in Molecular Biology* 363, 131–151.

Bergfors, T. *Protein Crystallization: Techniques, Strategies, and Tips. A Laboratory Manual*. International University Line, La Jolla, CA, 1999.

Bergfors, T. (2003). Seeds to crystals. *Journal of Structural Biology* 142, 66–76.

Biertumpfel, C., Basquin, J., Suck, D., and Sauter, C. (2002). Crystallization of biological macromolecules using agarose gel. *Acta Crystallographica D Biological Crystallography* 58, 1657–1659.

Bolanos-Garcia, V.M., and Chayen, N.E. (2009). New directions in conventional methods of protein crystallization. *Progress in Biophysics and Molecular Biology* 101, 3–12.

Chayen, N.E. (2006). Methods for separating nucleation and growth in protein crystallisation. *Progress in Biophysics and Molecular Biology* 88, 329–337.

Chayen, N.E. (1997). The role of oil in macromolecular crystallization. *Structure* 5, 1269–1274.

Chayen, N.E. 2004. Turning protein crystallisation from an art into a science. *Current Opinion in Structural Biology* 14, 577–583.

Chayen, N.E., and Saridakis, E. (2008). Protein crystallization: From purified protein to diffraction-quality crystal. *Nature Methods* 5, 147–153.

Collins-Racie, L.A., McColgan, J.M., Grant, K.L., DiBlasio-Smith, E.A., McCoy, J.M., and LaVallie, E.R. (1996). Production of recombinant bovine enterokinase catalytic subunit in *Escherichia coli* using the novel secretory fusion partner DsbA. *Biotechnology (N Y)* 13, 982–987.

D'Arcy, A. (1994). Crystallizing proteins—A rational approach? *Acta Crystallographica D Biological Crystallography* 50, 469–471.

Dong, A., Xu, X., Edwards, A.M., Chang, C., Chruszcz, M., Cuff, M., Cymborowski, M., Di Leo, R., Egorova, O., Evdokimova, E. et al. (2007). In situ proteolysis for protein crystallization and structure determination. *Nature Methods* 4, 1019–1021.

Dunlop, K.V., Irvin, R.T., and Hazes, B. (2006). Pros and cons of cryocrystallography: Should we also collect a room-temperature data set? *Acta Crystallographica D Biological Crystallography* 61, 80–87.

Fernandez-Patron, C., Castellanos-Serra, L., and Rodriguez, P. (1992). Reverse staining of sodium dodecyl sulfate polyacrylamide gels by imidazole–zinc salts: Sensitive detection of unmodified proteins. *Biotechniques* 12, 564–573.

Ferre-D'Amare, A.R., and Burley, S.K. (1994). Use of dynamic light scattering to assess crystallizability of macromolecules and macromolecular assemblies. *Structure* 2, 357–359.

Foresta, K., and Schuttb, C. (1992). Protein engineering for structure determination. *Current Opinion in Structural Biology* 2, 576–581.

Forsythe, E.L., Maxwell, D.L., and Pusey, M. (2002). Vapor diffusion, nucleation rates and the reservoir to crystallization volume ratio. *Acta Crystallographica D Biological Crystallography* 58, 1601–1606.

Garman, E., and Owen, R.L. (2007). Cryocrystallography of macromolecules: Practice and optimization. *Methods in Molecular Biology* 364, 1–18.

Garman, E.F. (2010). Radiation damage in macromolecular crystallography: What is it and why should we care? *Acta Crystallographica D Biological Crystallography* 66, 339–351.

Garman, E.F., and Doublie, S. (2003). Cryocooling of macromolecular crystals: Optimization methods. *Methods in Enzymology* 368, 188–216.

Ireton, G.C., and Stoddard, B.L. (2004). Microseed matrix screening to improve crystals of yeast cytosine deaminase. *Acta Crystallographica D Biological Crystallography* 60, 601–606.

Jancarik, J., Pufan, R., Hong, C., Kim, S.H., and Kim, R. (2004). Optimum solubility (OS) screening: An efficient method to optimize buffer conditions for homogeneity and crystallization of proteins. *Acta Crystallographica D Biological Crystallography* 60, 1670–1673.

Kim, S.-H., and Jancarik, J. (1991). Sparse matrix sampling: A screening method for crystallization of proteins. *Journal of Applied Crystallography* 24, 409–411.

LaVallie, E.R., McCoy, J.M., Smith, D.B., and Riggs, P. (2001). Enzymatic and chemical cleavage of fusion proteins. *Current Protocols in Molecular Biology*, Chapter 16:Unit16 4B.

McPherson, A. (1996). Increasing the size of microcrystals by fine sampling of pH limits. *Journal of Applied Crystallography* 28, 362–366.

McPherson, A. (2001). A comparison of salts for the crystallization of macromolecules. *Protein Science* 10, 418–422.

McPherson, A. (2004). Introduction to protein crystallization. *Methods* 34, 254–266.

McPherson, A., Malkin, A. J., and Kuznetsov, Y.G. (1996). The science of macromolecular crystallization. *Structure* 3, 759–768.

Newman, J. (2004). Novel buffer systems for macromolecular crystallization. *Acta Crystallographica D Biological Crystallography* 60, 610–612.

Saridakis, E., and Chayen, N.E. (2000). Improving protein crystal quality by decoupling nucleation and growth in vapor diffusion. *Protein Science* 9, 755–757.

Stura, E.A., Satterthwait, A.C., Calvo, J.C., Kaslow, D.C., and Wilson, I.A. (1994). Reverse screening. *Acta Crystallographica D Biological Crystallography* 50, 448–456.

Switzer, R.C., 3rd, Merril, C.R., and Shifrin, S. (1979). A highly sensitive silver stain for detecting proteins and peptides in polyacrylamide gels. *Analytical Biochemistry* 98, 231–237.

Vozza, L.A., Wittwer, L., Higgins, D.R., Purcell, T.J., Bergseid, M., Collins-Racie, L.A., LaVallie, E.R., and Hoeffler, J.P. (1996). Production of a recombinant bovine enterokinase catalytic subunit in the methylotrophic yeast *Pichia pastoris*. *Biotechnology (N Y)* 14, 77–81.

Wooh, J.W., Kidd, R.D., Martin, J.L., and Kobe, B. (2003). Comparison of three commercial sparse-matrix crystallization screens. *Acta Crystallographica D Biological Crystallography* 59, 769–772.

## Online resources and software

*Major suppliers of crystallization kits and solutions*

Hampton Research. Major supplier for crystallization screens, chemicals, plates, and other accessories. Hampton Research also has an inclusive list of books about protein crystallization and extensive technical support resources concerning crystal growth and optimization.

Molecular Dimensions. Major supplier specializing in macromolecular crystallization (screens, chemicals, plates, books etc.).

Jena Bioscience. Contains a large section on macromolecular crystallography (freshman kits, screens, plates).

Qiagen. Large selection of screens and other accessories for systematically analyzing and optimizing protein crystallization.

Emerald Biosystems (Rigaku). Specialized company for screens and instruments to create screens.

*Crystallization news and support sites*

Crystallization Guides: http://hamptonresearch.com /growth_101_lit.aspx

Crystallization Tutorial (by Terese Bergfors): http:// xray.bmc.uu.se/terese/tutorials.html

Crystallization Protocols (by Johan Zeelan): http://www.xtal-protocols.de/xtal.html

Crystallization Protocols II (by Enrico Stura): http://www.chem.gla.ac.uk/research/groups/protein/mirror/stura/index2.html

Crystallization Tips and Gallery: http://hamptonresearch.com

*Free software and databases related to protein crystallography*

RCSB Protein Data Bank: http://www.pdb.org/pdb/home/home.do. Contains experimentally determined macromolecular structures.

PDBsum: http://www.ebi.ac.uk/pdbsum. Provides schematic overview of molecular interactions among structural elements of structures deposited on the Protein Data Bank.

Jena Library of Biological Macromolecules: http://jenalib.fli-leibniz.de/. Helps in analyzing and visualizing structures deposited on the Protein Data Bank.

Pfam: http://pfam.xfam.org/. Finds annotated information about protein deposited on the Protein Data Bank.

Marseille Protein Crystallization Database: http://www.cinam.univ-mrs.fr/mpcd/. Lists the crystallization conditions and information for macromolecules with diffraction data and provides statistics. The site can be searched by criteria such as molecular weight, isoelectric point, crystallization methods, conditions, etc.

PyMOL molecular viewer: http://www.pymol.org/. A great visualization system to view structure from the Protein Data Bank.

Xtals: http://xtals.org. Contains useful tools and protocols related to protein crystallization.

ProtParam Tool: http://ca.expasy.org/tools/protparam.html. Predicts physical and chemical parameters for a given protein.

Gridzilla: http://www.code-itch.com/gridzilla/Gridzilla.html. A Graphical User Interface (GUI)-based application to create custom sparse matrix screens for optimization.

XtalPredServer: http://ffas.burnham.org/XtalPred-cgi/xtal.pl. Attempts to predict the ability to crystallize a protein according to sequence. Also, lets one search simultaneously for potential better-suited homologues.

Crystallization Construct Designer: http://xtal.nki.nl/ccd/Welcome.html. Designs sequence truncation to facilitate protein expression and crystallization.

# CHAPTER 7

# Introduction to Biological Light Microscopy

Coauthored with Michael W. Davidson

## 7.1 INTRODUCTION

Light microscopy is an essential technique for the life sciences. It allows for the identification and counting of cells and the observation of cellular growth, development, communication, and death. These processes can all be imaged down to the single-molecule scale thanks to a revolution in labeling techniques, particularly fluorescent labeling. Several excellent, thorough textbooks have been written on light microscopy, and this chapter is not a substitute for any of them. Neither is it a substitute for a complete knowledge of your own instrument, which should be obtained from the literature of the company that made the microscope. All of the large manufacturers provide detailed documentation in their user manuals and websites, often including introductions to the principles of microscopy. What you will find in this chapter is a review of the physics that applies to all light microscopes and their key illumination modes: brightfield, darkfield, phase contrast, differential interference contrast (DIC), epifluorescence, and confocal. The structure of a typical inverted biological microscope is discussed with its practical implications for imaging dishes, plates, and living cells. We then discuss the physics of fluorescence and introduce some of the most common biological fluorescent dyes. At the end of this chapter, you should be comfortable preparing a specimen for a particular type of light microscopy and know why it was prepared the way it was—its substrate of growth, whether it is fixed or live, the presence of one or more labels. You should know how it should be illuminated and which parameters will need to be adjusted to obtain a good image.

Obtaining those good images is then a matter of training, patience, and art. Biological imaging is a lot like wildlife photography, especially when working with living cells. The observer needs to be in the right place at the right time when the cells decide to cooperate.

## 7.2 PHYSICS OF MICROSCOPY: MAGNIFICATION AND RESOLUTION

The compound microscope can produce high magnification ranges by using two lens systems, the objective and eyepiece (or *ocular*), to produce two stages of magnification. The lens nearest the object or specimen, termed the objective, yields a magnified image (the intermediate image) of the object at a plane located at distance $L$ from the objective. Objectives range from 4× to 100×, with a typical research microscope having a turret containing a selection of four to six different magnification objectives. The eyepiece further magnifies the intermediate image by 5× to 30×. Total magnification may then range from 20× to 3000×.

The length $L$ from the specimen to the intermediate image plane, which is positioned in the plane of the eyepiece aperture diaphragm, is known as the *tube length*. Most manufacturers standardized the tube length for many years at 160 mm (**Figure 7.1a**), but many modern objectives have an infinite tube length (known as *infinity-corrected*) in which the image of the specimen is focused with parallel light rays at infinity (**Figure 7.1b**). The reason that this is a significant breakthrough is that infinity-corrected microscopes allow introduction of auxiliary components, such as prisms, polarizers, and epifluorescence illuminators, into the parallel optical path between the objective and the tube lens with only a minimal effect on focus and aberration corrections.

The *resolution* of an optical system is limited by both *aberrations* and *diffraction*. This applies not only to engineered optics but to biology as well. If the human eye was limited only by pupil diffraction, its owner would have a visual acuity of 20/8, or more than twice that considered "normal." In reality, acuity is limited by

**Figure 7.1 Ray traces of the optical train for (a) finite tube length and (b) infinity-corrected microscope systems.**

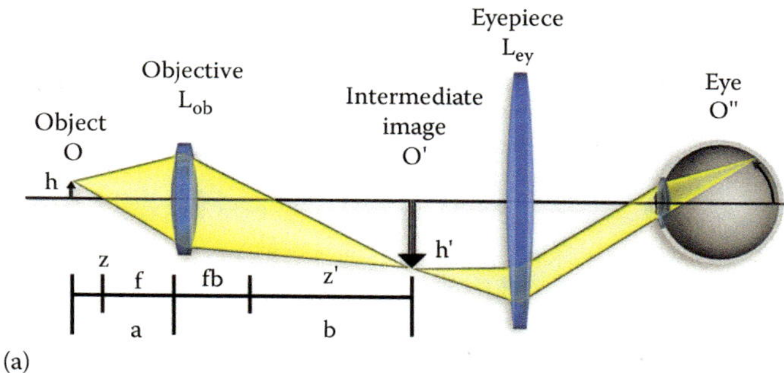


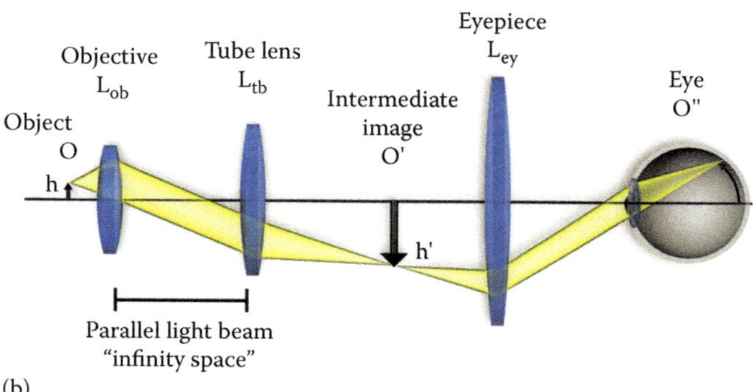

aberrations in the lens, vitreous, and other structures; most of these aberrations increase with age.

The best resolution possible with an ordinary light microscope is diffraction limited. Light passing through an objective results in a ring-shaped diffraction pattern, called the *Airy pattern* or *Airy disk*, consisting of a central bright intensity peak surrounded by a concentric pattern of dark and bright rings having increasingly diminishing intensity. As the separation distance between two points on a specimen is reduced, their Airy disk patterns begin to merge in the intermediate image plane. *Rayleigh's criterion* states that two neighboring points are just distinguishable if the main intensity peak from one point is located on the first dark fringe of the second point's Airy pattern (this is true for the eye and most cameras; some cameras can do better than this). Using this criterion, the minimum detectable separation $d$ between two neighboring points is

$$d = \frac{0.61\lambda}{n\sin\theta} \equiv \frac{0.61\lambda}{NA} \tag{7.1}$$

where $\lambda$ is the optical wavelength (usually averaged at 550 nm for white light), $\theta$ is the half-angle subtended at the specimen by the entrance pupil of the lens, and $n$ is the refractive index of the imaging medium between the objective and the specimen. The product $n\sin\theta$ is defined to be the *numerical aperture* for a lens.

When the imaging medium between the objective and the specimen is air, the practical limit for NA is about 0.95; thus, at the center of the visible spectrum where the wavelength $\lambda = 0.55$ μm, the lateral resolution is limited to: $d \cong 0.35$ μm. With immersion oil of refractive index 1.52 as the operating medium, the maximum numerical aperture is raised to 1.45, which reduces the lateral resolution to: $d \cong 0.2$ μm. The value of oil immersion can be appreciated from these numbers. Other common immersion media in biology are water ($n = 1.33$) and glycerol ($n = 1.473$).

True diffraction-limited imaging is usually not achieved due to lens aberrations. Different objectives have different properties, and often are chosen due to their elimination of one or more types of aberration (see **Practical Tips 7.1**). One of the primary culprits is *spherical aberration*, a monochromatic effect that causes the outer rays exiting from the curved lens surface to converge to a point on a different plane from the rays passing through the lens center, thus limiting resolution. A second common aberration is *chromatic aberration*, which causes rays of different colors (wavelengths) to come to a focus at different points along the optical axis (thus exhibiting color fringing). The third troublesome aberration in microscopy is *field curvature*, manifested by the center of the field of view being in a different focal plane than the periphery.

"Breaking the diffraction limit" refers to achieving greater than 200 nm spatial resolution using optical—usually fluorescence—microscopy. It is possible using a variety of techniques that will be introduced in the next chapter.

## PRACTICAL TIPS 7.1:   OBJECTIVES

A Dutch spectacle maker, Zacharias Janssen, is the reported inventor of the first compound optical microscope in the late sixteenth century. By the seventeenth century, tripod compound microscopes were being sold by several companies and would remain similar in design for the next 200 years. It was not until the late nineteenth century that the light microscopic revolution began with Ernst Abbe, who developed advanced apochromatic objectives that allowed for high-resolution imaging with minimal chromatic and spherical aberration. Microscope objectives are perhaps the most important components of an optical microscope because they are responsible for primary image formation and play a central role in determining the quality of images that the microscope is capable of producing. The objective is the most difficult component of an optical microscope to design and assemble. Modern objectives have three major classes corresponding to their degree of aberration correction: achromats (corrected for chromatic aberrations in two wavelengths and spherical aberrations in green only), fluorites or semi-apochromats (additional spherical correction), and apochromats (almost without chromatic aberration, and corrected for spherical aberration in two to three wavelengths; **Figure P7.1.1**). All of these classes are available in *Plan* versions, which show flat-field correction (i.e., there is a common focus across the entire field of view).

The specifications of each lens can be read off its surface. Apart from the manufacturer and class, other key parameters include magnification, tube length for which it was designed (160 or ∞), optimal coverglass thickness, NA, and whether the objective is dry (no label) or to be used with water (W or WI), glycerol (Gly), or oil (OIL OR OEL or HI) immersion (**Figure P7.1.2**). All dry objectives have an NA < 1.0 and show the greatest aberration caused by the coverslip. Many 40× dry objectives have "correction collars" that can be turned to adjust for coverslip thickness.

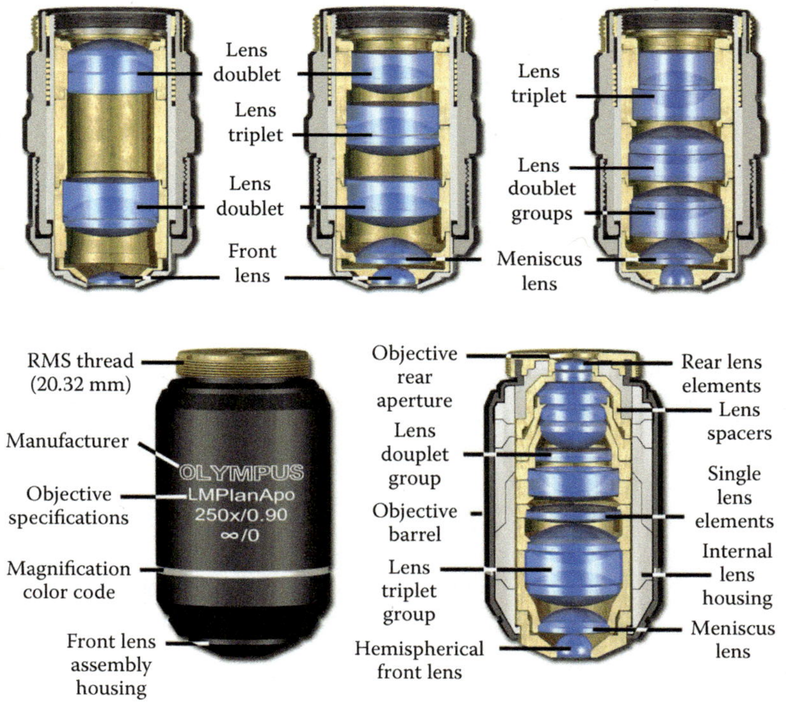


**Figure P7.1.1 The three major classes of objectives.** The achromats have the least amount of correction; the fluorites (or semi-apochromats) have additional spherical corrections; and the apochromats are the most highly corrected objectives available.

**Figure P7.1.2 Reading an objective.** The pictured objective is a long-working-distance dry, infinity-corrected 250× plan apochromat.

*(Continued)*

---

**PRACTICAL TIPS 7.1 (CONTINUED):   OBJECTIVES**

Immersion objectives have higher achievable NA values and less coverslip aberration. Water immersion objectives do not work well with inverted microscopes; they are usually used in upright microscopes, where they dip directly into the culture dish. Oil immersion objectives are very common on both upright and inverted scopes. On the latter, they must be treated with great care, so that immersion oil does not drip down into the objective. Small plastic protective shields can be placed around them to catch excess oil.

Immersion media should *never* be used on dry objectives. This means that if immersion oil is on a sample, you can probably not use a dry objective of more than 10× magnification on that sample without cleaning it, or the objective lens will touch the oil. Dry objectives have no protection against oil penetration and are easily ruined.

There are several other important parameters of objectives that determine their suitability for different applications. The *depth of field d* is the axial range through which an objective can be focused without any appreciable change in image sharpness. It is given by

$$d = \frac{\lambda n}{NS^2} + \frac{ne}{M(NA)},$$

where $\lambda$ is the wavelength of illumination, $n$ is the refractive index of the imaging medium, NA is the objective numerical aperture, $M$ is the objective lateral magnification, and $e$ is the smallest distance that can be resolved by a detector that is placed in the image plane of the objective. Notice that the depth of field goes down rapidly with NA. Practically, this means that the axial position of the objective is very important in achieving clear images in widefield microscopy, and that very thin axial sections are possible with confocal.

The *working distance* is the clearance distance between the closest surface of the cover glass and the objective front lens. Generally, working distance decreases in a series of matched objectives as the magnification and NA increase. Objectives intended to view specimens with air as the imaging medium should have working distances as long as possible, provided that NA requirements are satisfied. Special long-distance objectives are available for certain applications. Immersion objectives, on the other hand, should have shallower working distances to contain the immersion liquid between the front lens and the specimen. Many objectives designed with close working distances have a spring-loaded retraction stopper that allows the front lens assembly to be retracted by pushing it into the objective body and twisting to lock it into place. This allows the objective to be rotated in the nosepiece without dragging immersion oil across the surface of the slide.

# 7.3  ANATOMY OF A BIOLOGICAL MICROSCOPE

## Hardware

New microscopic imaging techniques and technologies are emerging continually, reflecting advances in computers, optics, and biology itself. However, the basic design of the compound microscope remains similar to that of its ancestors of 200 years ago. Microscopes operate on the principle of *conjugate focal planes* that occur along the optical pathway. One set consists of four field planes and is referred to as the field or image-forming conjugate set, while the other consists of four aperture planes and is referred to as the illumination conjugate set. Each plane within a set is said to be conjugate with the others in that set because they are simultaneously in focus and can be viewed superimposed upon one another when observing specimens through the microscope. On the other hand, the reciprocal nature of the two sets of microscope conjugate planes means that an object appearing in focus in one set of planes will not be focused in the other set. The elements of the optical path making up the conjugate planes are the illuminator, condenser, specimen, objective, eyepiece, and detector (a camera or the observer's eye; **Figure 7.2**).

A research grade microscope is much more complex than this simple light path, as it also contains several light-conditioning devices that are often positioned between the illuminator and the condenser, and a complementary detector or filtering device that is inserted between the objective and the eyepiece or camera. The conditioning device(s) and detector work together to modify image contrast as a function of spatial frequency, phase, polarization, absorption, fluorescence, off-axis illumination, and/or other properties of the specimen and illumination technique. Most microscopes used in biology are "inverted," with the objectives placed below the specimen. This allows for micromanipulators, injection needles, perfusion systems, and other hardware to be mounted around the specimen without interfering with the objectives. It causes some issues for imaging that we will discuss below. A diagram of an inverted microscope configured for brightfield, phase contrast, and epifluorescence is shown in **Figure 7.3**.

**Figure 7.2 Relationship between the conjugate plane sets in the optical microscope.** Image-forming conjugate planes reside at the field diaphragm, specimen slide, eyepiece fixed diaphragm, and retina of the eye. Aperture planes occur at the lamp filament, condenser aperture, the objective rear focal plane, and the eyepoint of the eyepiece.

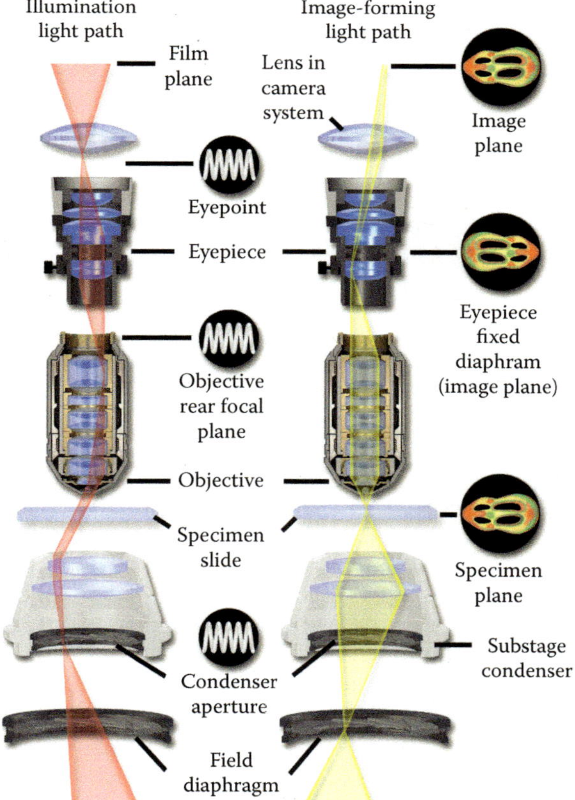

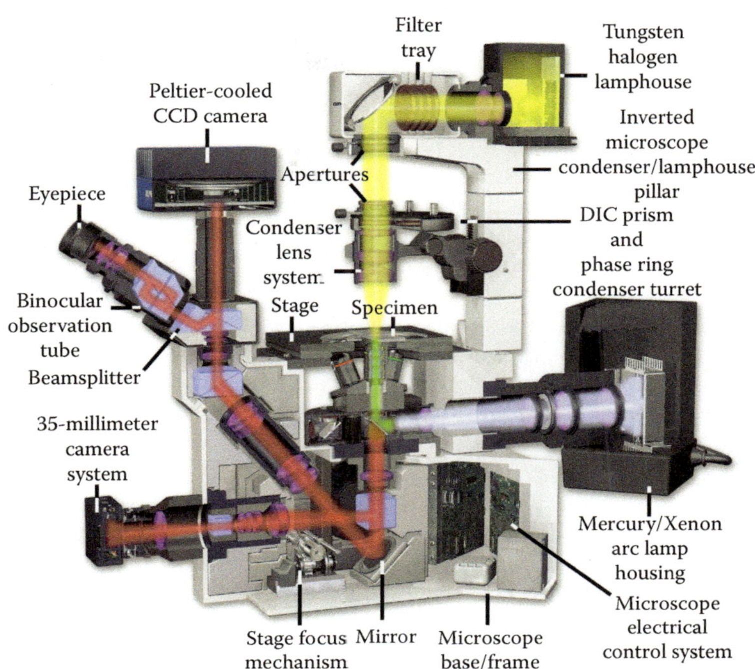


**Figure 7.3 Cutaway view of inverted microscope set up for brightfield and epifluorescence microscopy.**

## Imaging cells on an inverted microscope

On an inverted microscope, the sample must either be able to be flipped over or must have an appropriate transparent substrate for imaging through. This is of course no problem for specimens such as bacteria on a coverslipped slide, but is an issue for either fixed or live cells in typical cultureware. Plastic culture dishes are all right for nearly all 20× objectives, but run into trouble with some 40× objectives and are useless for any higher magnification objectives. They are not only too thick for the working distance of the lens, but they may also create scatter and autofluorescence; multiwell plates are especially bad in this regard. There are several approaches to solving this problem, but they must be considered before the cells are plated:

- Plating cells on coverslips. Tissue-culture treated, standard thickness glass and plastic coverslips are available in a range of diameters. Ordinary glass coverslips may be tissue-culture treated by the experimenter as well. This is the best method if the cells will be fixed before imaging, as the fixed coverslip can then be mounted on a slide using a *mounting medium* to create a highly stable and transportable structure. Slides containing mounted cells may be kept at −20°C and will be stable for many months.

- Plating cells on glass-bottom dishes that are essentially a coverslip fixed into an ordinary tissue culture dish and tissue-culture treated. These are sold ready for use with various sizes of coverslip bottoms. Using these dishes, cells can be taken immediately to the microscope while alive and growing, or they may be fixed and stored for at least several weeks. The authors highly recommend this method and use it for nearly all the experiments in our lab. Note that objective lenses are usually corrected for a specific thickness of coverslip, so ensure that your thickness is appropriate for the lens(es) used.

# 7.4  BRIGHTFIELD IMAGING TECHNIQUES

## Köhler illumination

Illumination is a critical factor in determining the overall performance of the optical microscope. For nearly all imaging techniques, including brightfield, darkfield, polarization, and DIC, the best results are usually achieved by adjusting the illumination system following the principles introduced by August Köhler in 1893. The existence of two interrelated optical paths and two sets of image planes characterizes Köhler illumination and allows the various adjustable diaphragms and aperture stops in the microscope to be used to control both the cone angle of illumination, and the size, brightness, and uniformity of illumination of the field of view. Köhler illumination was developed in order to obtain a bright, evenly illuminated sample without creating artifacts caused by the bulb or filament used for illumination. Its principles are simple: a collector lens on the lamp housing is required to focus light emitted from the various points on the lamp filament at the front aperture of the condenser while completely filling the aperture. Simultaneously, the condenser must be focused to bring the two sets of conjugate focal planes (when the specimen is also focused) into specific locations along the optical axis of the microscope. With the specimen and condenser in focus, the focal conjugates will be in the correct position so that resolution and contrast can be optimized by adjusting the field and condenser aperture diaphragms.

The following procedure to establish Köhler illumination should be done before each experiment, or at least every time the microscope's adjustments have changed:

- Focus the eyes. First adjust the binocular head for your eye separation, with the eyepiece diopters set to 0. Place a specimen on the stage and focus using the 10× objective. Adjust with fine focus on the smallest visible detail. Position the 40× objective and fine focus. Switch to a 4× objective and adjust the eyepiece diopters for best focus. Recheck the focus at 40× and select a working objective.

- Set the condenser focus. Close the field diaphragm to its smallest size. Bring its image into focus with the condenser focus knob. Position the image into the center of the field of view by using the condenser centering mechanism.

- Center the illumination (this step may not be required for pre-centered illumination systems). Remove the diffuser filter from the microscope. Close the aperture diaphragm on the condenser. Use a neutral density filter as a mirror to observe the filament image on the underside of the condenser and focus the filament image by moving the lamp housing in or out until it becomes sharp. After the filament is in focus and centered, replace the diffuser.

- Adjust the condenser diaphragm. Remove one eyepiece and look through the eyepiece tube down toward the back of the objective. Adjust the aperture diaphragm so that it is just inside the exit pupil of the objective.

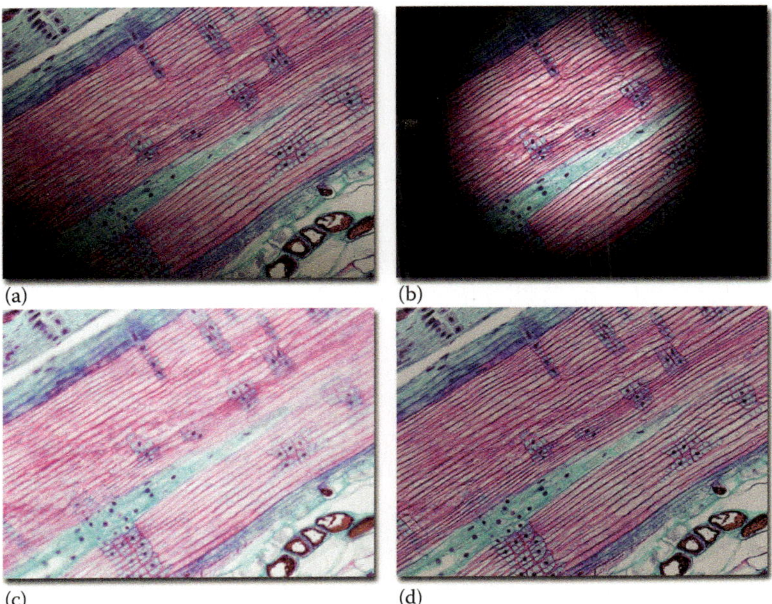

**Figure 7.4  Artifacts in Köhler illumination.** (a) Incorrectly aligned condenser. (b) Condenser position too low. (c) Condenser aperture diaphragm too wide. (d) Condenser aperture diaphragm closed too far.

High numerical aperture, high-magnification objectives should not be used to adjust Köhler illumination. With these lenses, the field diaphragm sometimes cannot be closed to a diameter that can be imaged in the field of view, or the condenser does not produce a diaphragm image that is small enough to be observed.

When the microscope has not been appropriately adjusted for Köhler illumination, certain artifacts can be seen. When the condenser is off-center, a portion of the image will appear dark (**Figure 7.4a**). When the condenser does not focus an image of the lamp filament directly onto the condenser aperture plane, the field diaphragm is out of focus and the illumination is uneven (**Figure 7.4b**). When the condenser aperture diaphragm is open too wide, the sample appears overexposed (**Figure 7.4c**); the opposite is true when it is closed too far (**Figure 7.4d**). It is important to appreciate that similar-appearing artifacts can occur for other reasons, such as filters in the optical pathway causing partial blockage, or incorrect lamp voltage causing images that are too bright or too dim. If an obvious solution cannot be found, all known accessory filters should be removed and Köhler illumination adjusted, then the filters replaced one by one.

## Brightfield and darkfield

Most biological specimens are prepared for examination as thin sections, adherent cells, or smears, and observed with *transmitted light* passed through the specimen and its mounting medium (slide or dish). Opaque samples can be observed with *reflected light,* in which the incident illumination is reflected from the specimen surface into the objective. Most high-end microscopes have shutters for both of these modes. In both cases, specimens often demonstrate very poor contrast between closely adjacent regions, even though the areas differ chemically or morphologically. Consequently, specimen detail is limited by lack of image contrast rather than by diffraction, low numerical aperture, or insufficient magnification. The use of *stains* that fix themselves to different biological structures to improve microscopic contrast is a classic technique. A period of active dye discovery occurred in the late nineteenth century, inspired by Paul Ehrlich and others, leading to a variety of cell and tissue stains, which are still widely used today. There are a few common brightfield stains with which everyone should be familiar:

- Crystal violet/safranin (the *Gram stain*). Crystal violet remains only in the cell walls of Gram-positive bacteria after decolorizing with alcohol. Safranin, a red counterstain, stains the decolorized Gram negatives (**Figure 7.5a**).

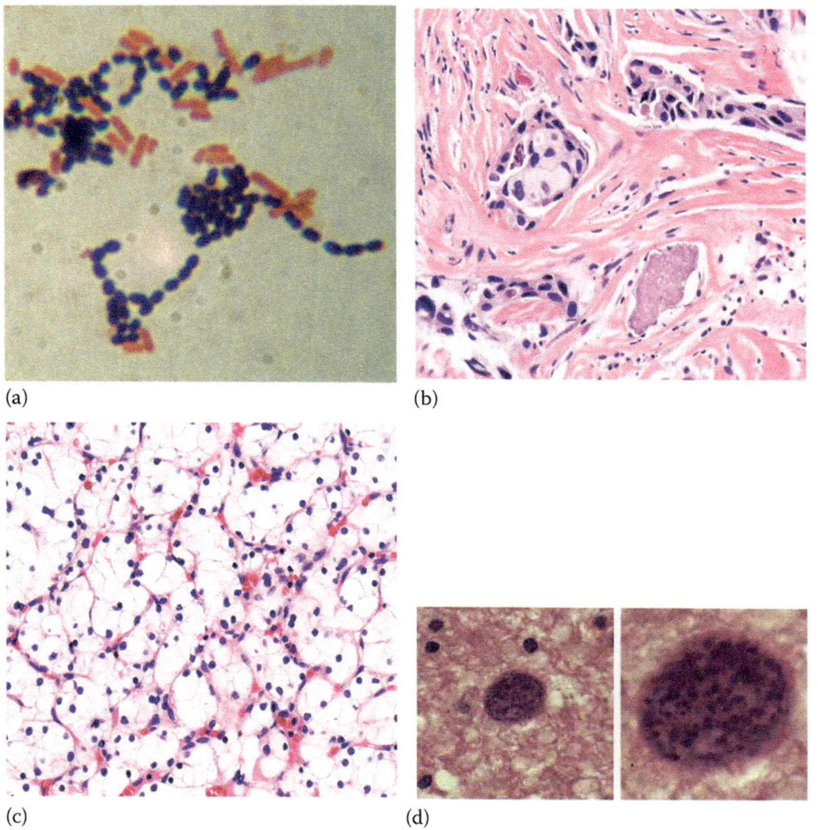

**Figure 7.5 Colorimetric dyes for bacteria and cells.** (a) Gram staining. The rod-shaped bacteria are Gram negative and appear light red. The cocci are Gram positive and appear dark purple. (Courtesy of Rob Hubert, Iowa State University.) (b–d) Hematoxylin/eosin (H&E) staining. Hematoxylin stains nucleic acids, calcium deposits, and bacteria blue. Eosin stains arginine and lysine residues pink (representing most proteins). Clear areas can be water, carbohydrate, lipid, or gas. Stained nuclei are always blue. Protein-rich cytoplasm appears pink. Cytoplasm where protein synthesis is occurring stains purple (pink plus RNAs that stain blue). Protein-poor cytoplasm, filled with carbohydrate, lipid, or water, stains light pink or clear. (b) Breast carcinoma. Note the blue nuclei, pink cytoplasm, and bright-red red blood cells. (c) Clear cell carcinoma of the kidney. Note the very pale staining of the cytoplasm. (d) Brain tissue showing a *Toxoplasma gondii* cyst at two different magnifications. (All images from Wikimedia Commons under the Creative Commons license.)

- Hematoxylin/eosin (H&E). This is one of the most common stain combinations used in histology. Hematoxylin stains nuclei blue, and eosin stains cytoplasm and connective tissue pink. Red blood cells absorb a great deal of eosin and appear bright red (**Figure 7.5b through d**).

- *Acid-fast* (Ziehl–Neelsen) stain. This is used to stain the Mycobacteria, the family of bacteria that includes the causative agent of tuberculosis, as well as a few other structures that have a mycolic acid content, such as bacterial endospores. With this stain, the acid-fast organisms appear red against a blue background.

- Methylene blue. This dye has many uses. It is decolorized by live cells, so it can be used in cell-counting experiments to determine viability. It is also a component of the *Wright's stain* for distinguishing blood cell types and can be used to stain DNA.

One of the easiest ways to identify bacteria is to *heat-fix* them in a smear on a slide by passing the slide through a flame, then stain with one or more stains. The heat fixation permits more stain to be absorbed and fixes the bacteria firmly to the slide, so they may be observed even without a coverslip (see **Practical Tips 7.2**).

*Darkfield* microscopy is a specialized illumination technique that enhances contrast in specimens by illuminating them with oblique light. The zeroth order (direct) light is blocked by an opaque stop in the substage condenser, and light passing through the specimen from oblique angles at all azimuths is diffracted, refracted, and reflected into the microscope objective to form a bright image of the specimen superimposed onto a dark background. Ideal candidates for darkfield illumination are samples that scatter light well: this includes some living organisms such as diatoms and small insects, unstained bacteria and yeast, some cultured cells, and mineral and chemical crystals. Dust also scatters well, so the slide and specimen need to be very clean to avoid artifacts. A sample or substrate that is too thick can also create artifacts; it is important to use a coverslip of standard thickness (1 mm).

Many dissecting/crystallography microscopes come standard with darkfield mode, but most research microscopes do not unless it is requested. For 40× and lower magnification and dry (non-immersion) objectives, it can be achieved simply by placing something opaque over the condenser that blocks the zeroth-order light. This can be a commercial "spider stop" (**Figure 7.6**) or a homemade blocker of the right size made of cardboard, plastic, or even a coin. The diameter needed is different for each objective and can be measured as follows:

- Place a transparent ruler against the bottom of the condenser.

- Remove one eyepiece and observe the ruler at the back focal plane of the objective using a phase telescope or by inserting a Bertrand lens (the condenser aperture and field diaphragm must be fully open).

- The number of ruler divisions visible in the back focal plane will be equal to the size of the stop necessary to block zeroth-order light.

- Rotate through the objectives to obtain the values for each.

### PRACTICAL TIPS 7.2:    PREPARATION OF HEAT-FIXED BACTERIAL SMEARS FOR MICROSCOPIC OBSERVATION

The purpose of heat fixation is to adhere bacteria to the slide and make them take up dyes more readily. The procedure also kills them, so they no longer display motility (this may be an advantage or a disadvantage). The best solution in which to prepare bacterial slides is an isotonic saline solution (0.9% NaCl in distilled water). Do not use rich growth medium, as it will burn during the heat fixation, and it is highly autofluorescent.

If the bacteria are growing in medium, pellet them by centrifugation and rinse twice with 0.9% NaCl, then place a drop (10–50 μL) of the resuspended solution on a slide. If they are growing on a plate, choose a single colony and dissolve it in 0.9% NaCl in an Eppendorf tube or directly into a drop on the slide. A wax pencil may be used to outline different areas of the slide and permit several samples to be placed along its length.

Allow the sample to air dry. It may then be heat-fixed by passing it three to four times through the flame of a Bunsen burner with the bacteria facing upward. Warning: the slide may break if it gets too hot.

Some biosafety departments discourage the use of open flames, and in these cases, a *microincinerator* is used. This is a heating device that has a port for sterilizing inoculating loops as well as an attachment to hold slides for heat fixation (**Figure P7.2.1**).

**Figure P7.2.1 Microincinerator showing slide attachment.** (Photo taken by Tami Port, MS, creator of Science Prof Online educational website.)

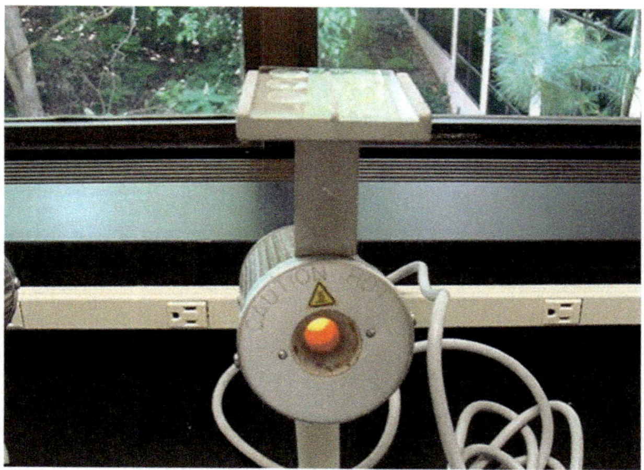

**Figure 7.6** (a) Brightfield (left) and darkfield (right) condensers, showing position of the spider stop. (b) Interchangeable stops are available for the spider housing to accommodate different objectives.

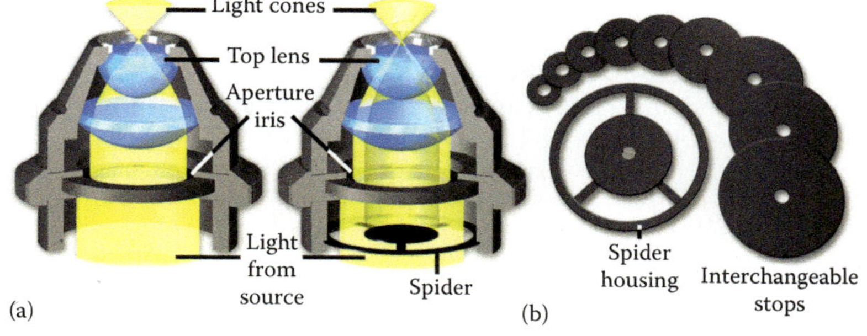

Once the objective is selected and the stop and specimen are in place, the condenser must be carefully focused. It should come very close to the slide, almost touching it. If it is too far away, you will see a dark central region with a bright ring around it.

Special condensers are also made specifically for darkfield. These should be used for highly precise work and/or high NA objectives. There are several varieties, including oil or water immersion darkfield condensers, which require a drop between the top of the condenser and the underside of the specimen. Immersion darkfield condensers have internal mirrored surfaces and can pass highly oblique rays without chromatic aberration, yielding the clearest images and the blackest background.

Darkfield imaging with a high NA condenser is perhaps the easiest way to break the diffraction limit. Noble metal nanoparticles, such as those made of Au, are highly reflective at a specific resonance frequency because of their surface plasmons (see **Chapters 12** and **13**). They can be seen as small diffraction disks under darkfield, and as the particles become smaller, the size of the disks does not change, but its intensity decreases; particles as small as 30 nm can be detected and localized to within nanometer precision. Interparticle interactions can be monitored at the nanoscale by observing interactions between the surface plasmons of two particles, which leads to a redshift of the resonance frequency.

Darkfield microscopy enjoyed a period of great popularity in the late twentieth century, which has been largely eclipsed by other types of brightfield such as phase contrast. It still remains a technique of choice for studying small aquatic organisms such as diatoms and protozoa, where high-resolution images can be spectacular (**Figure 7.7**).

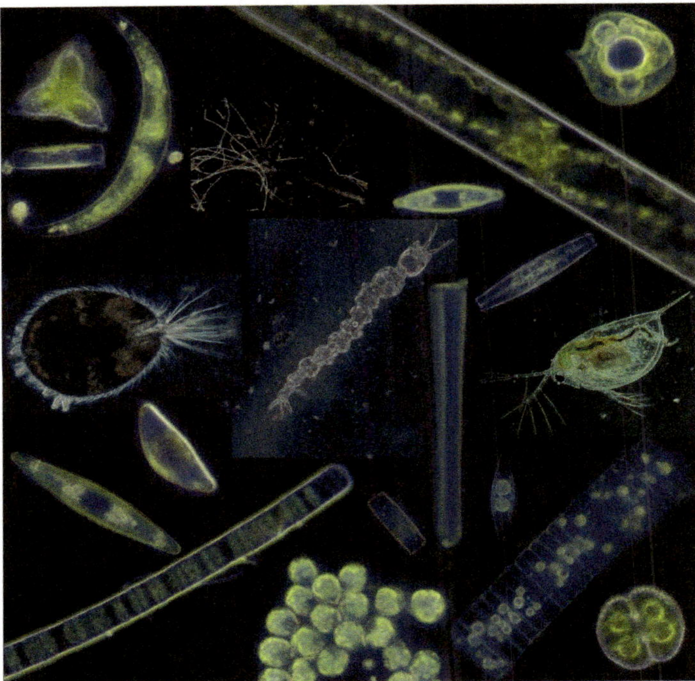

**Figure 7.7 Darkfield images.** Images of freshwater life from Fish Lake, Lagrange County, IN. Magnifications range from 15× to 500×. (Courtesy of Ted Clarke.)

## Phase contrast

Light incident on a specimen emerges as two components: an undiffracted or *surround* or *S-wave*, which has not interacted with the specimen, and a diffracted or *D-wave* that passes through the specimen. The D-wave is phase shifted by an amount δ that depends upon the index of refraction of the medium, *n*, and that of the specimen *n′* along with specimen thickness *t*:

$$\delta = \frac{2n(n-n')t}{\lambda}. \tag{7.2}$$

The product $(n-n')t$ is known as the *optical path difference (OPD)*. The refractive index of a cell is usually ~1.36. The phase shift in water or nutrient medium is usually no more than a quarter wavelength. This type of phase shift is invisible to the eye and to cameras, as it causes no change in intensity or color. Frits Zernike in the early 1940s developed a method that he called *phase contrast* for translating this difference in phase into a difference in amplitude. If the S-wave is sped up (or slowed down) at the same time that the D-wave is slowed down (or sped up) so that their difference in phase becomes λ/2, then the two will destructively (or constructively) interfere and contrast will be created.

To speed up the S-wave, a phase plate is installed with a ring-shaped phase shifter attached to it at the rear focal plane of the objective (**Figure 7.8**). The narrow area of the phase plate is optically thinner than the rest of the plate. As a result, the light passing through the phase ring travels a shorter distance in traversing the glass of the objective than does the D-wave. This causes the details of the specimen to appear dark against a lighter background, and so is called *positive* or *dark* phase contrast.

The opposite is also possible. If the ring phase shifter is thicker than the rest of the plate, the S-wave is slowed, the D-wave arrives at the image plane in phase, and constructive interference takes place. The image appears bright on a darker background for *negative* or *bright contrast*. This is much less frequently used.

**Figure 7.8 Optical train for phase contrast showing the phase ring and annular ring.**

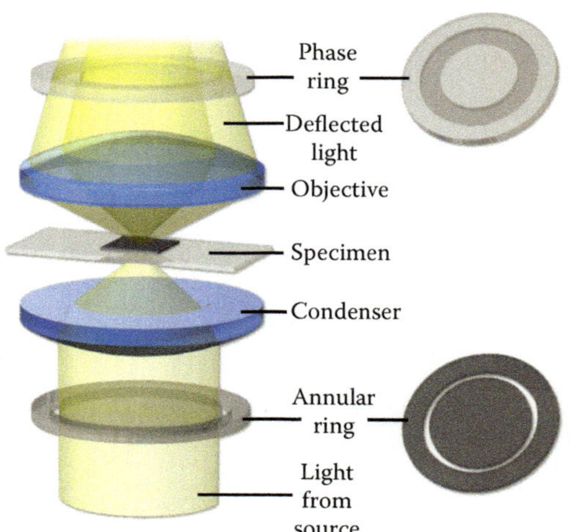

Because the undeviated light of the zeroth order is much brighter than the faint diffracted light, a thin absorptive transparent metallic layer is deposited on the ring to bring the direct and diffracted light into better balance of intensity in order to increase contrast. Also, because the speeding up or slowing down of the direct light is calculated on a 1/4 wavelength of green light, the phase image will appear best when a green filter is placed in the light path (a green interference filter is preferable; note that "average" white light is often taken to be ~550 nm, which is green). Such a green filter also helps achromatic objectives produce their best images, since achromats are spherically corrected for green light.

The ability to do phase contrast should be requested when the microscope is purchased, as the objectives supplied will contain phase plates. They may also be used for brightfield with a small (nearly always negligible) loss of resolution. Of course, new objectives can be purchased later, but these are very costly. A condenser containing a set of phase annuli, one for each phase contrast objective, is also needed. This usually comes as a rotating turret with one position for brightfield (BF), and phase contrast Ph1, Ph2, and so on. The phase outfit also includes a green filter and a phase telescope. The latter is used to enable the microscopist to alight the condenser annulus to superimpose it onto the ring of the phase plate. A set of centering screws in the condenser allows manipulation of the annulus to align it while observing the back focal plane of the objective with the phase telescope.

To set up a phase microscope, first ensure that it is configured for Köhler illumination. Then perform the following steps:

- Obtain a good phase specimen, such as cells, and focus it with the 10× objective.

- Open the condenser aperture diaphragm and swing the turret to the correct position for that objective.

- Place the green filter in the light path.

- Remove one eyepiece and insert the phase telescope.

- Use the centering screws to center the annulus (a Bertrand lens or a pinhole eyepiece, if available, will allow a view of the back focal plane of the objective; **Figure 7.9**).

- Reinsert the eyepiece.

- Rotate the turret to the next phase ring and change objectives. Repeat for each position.

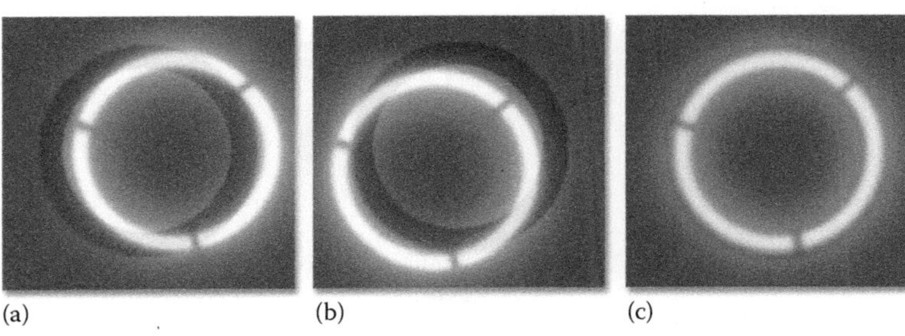

(a)    (b)    (c)

**Figure 7.9 Phase plate and light annulus adjustment.**
(a) and (b) misaligned;
(c) correctly aligned.

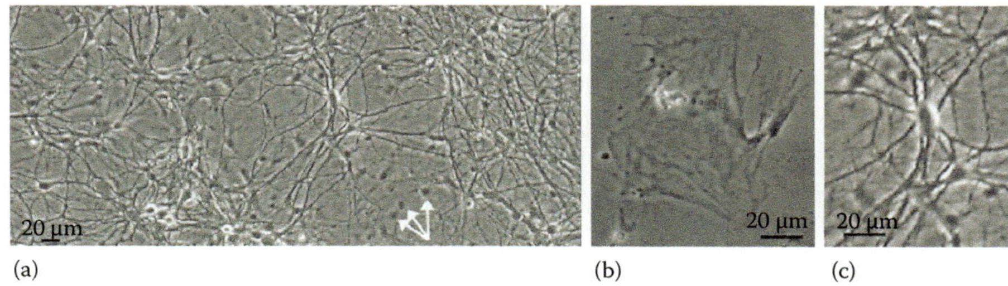

(a)                                                                          (b)                    (c)

**Figure 7.10 Phase contrast images.** Phase contrast is particularly useful in neuronal cultures as it aids in identifying cell types and evaluating cell health. (a) Low-power phase contrast image of a mature neuronal culture. Note the haloes around many of the cells; these cells are rounder than those that appear dark. The arrows show the nuclei of the underlying astrocyte layer. (b) Single astrocyte. Astrocytes are completely invisible under ordinary brightfield. (c) Single neuron, showing the phase contrast halo around the cell body.

Phase contrast is essential for many live-cell techniques, especially those involving cell manipulation such as electrophysiology and microinjection. Under phase contrast, the health of cells can be rapidly ascertained and structures such as the nucleus can be observed (**Figure 7.10**).

## Polarization and DIC

Polarized light microscopy is designed for specimens that are visible due to their optically anisotropic character, such as mineral or protein crystals. The microscope is similar to a brightfield microscope, but is equipped with both a polarizer positioned in the light path somewhere before the specimen, and an analyzer (a second polarizer), placed in the optical pathway between the objective rear aperture and the observation tube. Image contrast arises from the interaction of plane-polarized light with a birefringent specimen to produce two individual wave components that are polarized in mutually perpendicular planes (**Figure 7.11**). If the polarizers are removed, the microscope becomes an ordinary brightfield instrument. Polarizing microscopes also usually come with specially designed objectives that are strain-free and a rotatable stage. They are often low-magnification binocular scopes that can be used for identification and isolation of crystals (see **Chapter 6**).

**Figure 7.11 Principle of polarized light microscopy, showing a birefringent sample between crossed polarizers.**

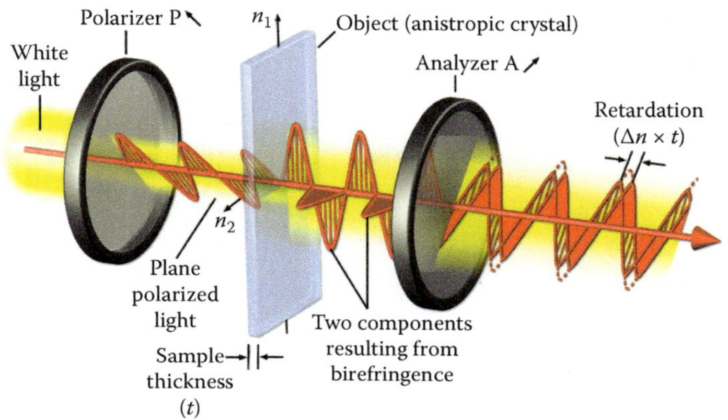

Replacing the prisms of polarizing microscopy with special beamsplitting prisms (Wollaston or Nomarski prisms) is the basic design behind *DIC* microscopy (see **Practical Tips 7.3**). Wavefronts are generated by a beamsplitting prism and then pass through a specimen phase gradient, becoming phase-shifted according to **Equation 7.2**. They are then recombined through differential interference by a

---

### PRACTICAL TIPS 7.3:    BEAMSPLITTING PRISMS

Matched birefringent Wollaston and/or Nomarski prisms are crucial for DIC imaging. They are inserted into the microscope optical pathway with their shear axes oriented at a 45° angle to the polarizer and analyzer. Wollaston prisms are constructed of two precisely ground and polished wedge-shaped slabs of quartz having perpendicular orientations of the optical axis and cemented together at the hypotenuse to generate an optically anisotropic compound plate. Incident linearly polarized wavefronts entering a Wollaston prism are divided into two separate orthogonal waves, termed the *ordinary* and the *extraordinary wave*. Because they are derived from localized areas on a single source, the orthogonal wavefronts generated by a Wollaston prism are coherent, have identical amplitudes, and propagate in the same direction through the lower prism wedge (**Figure P7.3.1a**).

The refractive index difference experienced by the ordinary and extraordinary wavefronts results in varying propagation velocities for the two waves as they travel through the lower and upper sections of a Wollaston prism. In the lower half of the prism, the wavefronts undergo a phase shift that is exactly compensated in the upper half when the geometrical path through the lower and upper halves is identical. However, wavefronts traversing through points away from the center experience a longer journey through either the lower or upper half of the prism, which results in a constant phase shift per unit length in the direction of the shear axis.

The design of a Nomarski prism is similar (**Figure P7.3.1b**). The upper wedge is identical to a conventional Wollaston quartz wedge, but the lower wedge is modified by cutting the quartz crystal in a direction that yields an oblique optical axis with respect to the flat surface of the prism. When the wedges are combined to form a compound birefringent prism, the focal (interference) plane lies at a site several millimeters outside the prism plate. Refraction at the interface between the quartz wedges in a Nomarski prism causes the sheared wavefronts to converge with a crossover point outside the prism. This allows the focal plane to be adjusted over a range of several millimeters, making these prisms easier to use with standard microscope objectives.

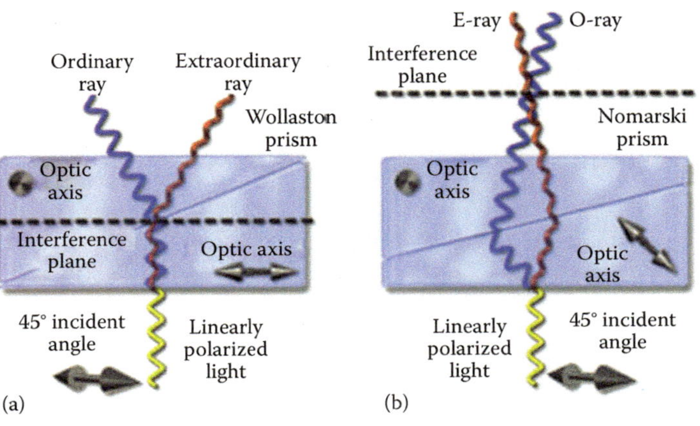


**Figure P7.3.1 Wavefront shear by (a) Wollaston and (b) Nomarski prisms.**

second prism and an analyzer to yield a high-contrast rendition of the gradient. Like phase contrast, the contrast results from OPDs due to differences in refractive index and specimen thickness. However, the mechanism of contrast generation and the appearance of the images is quite different. A DIC image corresponds to the first derivative, rather than the magnitude, of the gradient profile obtained from the specimen OPD. It is important to note that the appearance of "relief" in a DIC image does not necessarily correspond to physical height of the specimen features, but to differences in the *optical thickness*, or the product of the thickness and refractive index difference. It is not possible to separate these two variables using this technique.

Placement of precisely matched optical components in or near conjugate planes is essential to successful DIC (**Figure 7.12**). All of the major manufacturers provide high-quality, precision DIC kits for research-grade microscopes.

Four basic components are required to configure a standard brightfield microscope for DIC:

- *Linear polarizer* to produce the necessary plane-polarized light for interference imaging. It must be inserted into the optical pathway between the microscope light port and the condenser lens assembly.

- *Condenser Wollaston or Nomarski prism* to shear the polarized light emanating from the polarizer into two components. It is placed in or near the conjugate focal plane of the condenser iris diaphragm aperture.

**Figure 7.12 Schematic of DIC optical path.**

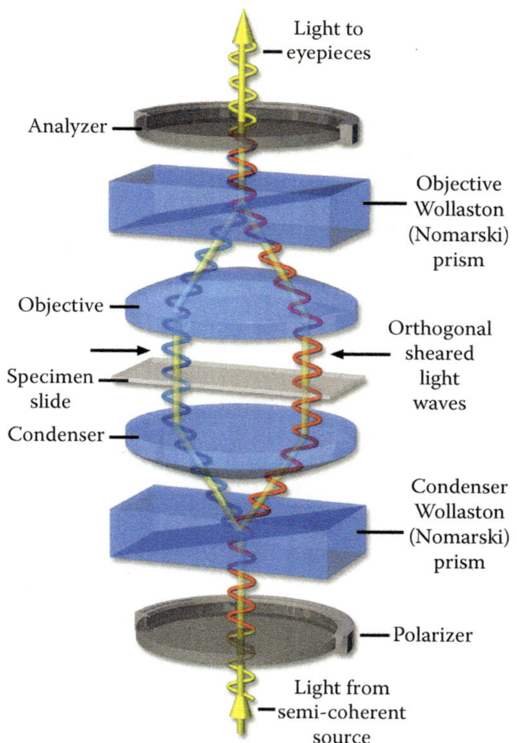

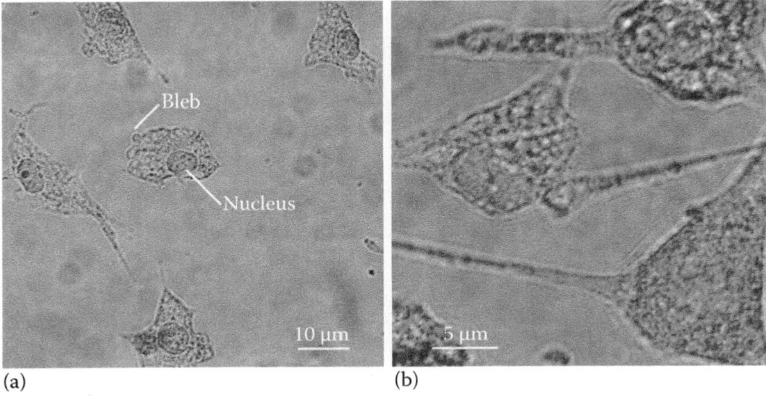

**Figure 7.13  DIC cell images.** These are B16 melanoma cells that have been treated with an anticancer drug to induce cell death. (a) Cell features such as nuclei and membrane blebs (the latter are characteristic of apoptotic cell death) can be easily seen under DIC. (b) High-power image of cells from the same sample.

- *Objective Nomarski prism* to recombine the sheared wavefronts in the conjugate plane of the objective rear aperture. It is positioned behind the objective.

- *Analyzer* to pass the components of circular and elliptically polarized light and allow them to undergo interference to generate the DIC image. It is positioned in the optical pathway before the tube lens.

The major advantage of DIC over phase contrast is that it works better for thick specimens. While phase contrast often produces blurry or confused images from thick specimen because of phase shifts from out-of-focus material, DIC permits optical sectioning of thick cells or tissue and subsequent 3-D image rendering (**Figure 7.13**). For this reason, many researchers employ phase contrast on wide-field microscopes and DIC on confocal laser scanning microscopes. DIC also does not produce the "halos" around cells that are seen with phase contrast, which can be an advantage or a drawback depending upon the information desired.

# 7.5  BASIC FLUORESCENCE MICROSCOPY

## Physics of fluorescent molecules

Biological microscopy relies heavily on fluorescence as an imaging mode, primarily due to its high degree of sensitivity and its ability to specifically target structural components and dynamic processes in fixed or living cells. Many fluorescent probes are constructed around synthetic aromatic organic chemicals designed to bind with a biological macromolecule (for example, a protein or nucleic acid) or to localize within a specific structural region.

What makes a molecule *photoluminescent* (fluorescent or phosphorescent) is the presence of a relatively long-lived excited state from which *radiative recombination* can occur. A *Jablonski diagram* showing the relative energy levels and types of possible transitions is shown in **Figure 7.14**. Most molecules at room temperature are in the ground electronic state ($S_0$). Upon absorption of a photon, an electron passes from $S_0$ to an excited singlet state, usually $S_1$ but sometimes $S_2$. Many different vibrational states of the excited state are allowed, but the electron relaxes to the lowest state of $S_1$ (or $S_2$) via internal conversion within picoseconds. Fluorescence

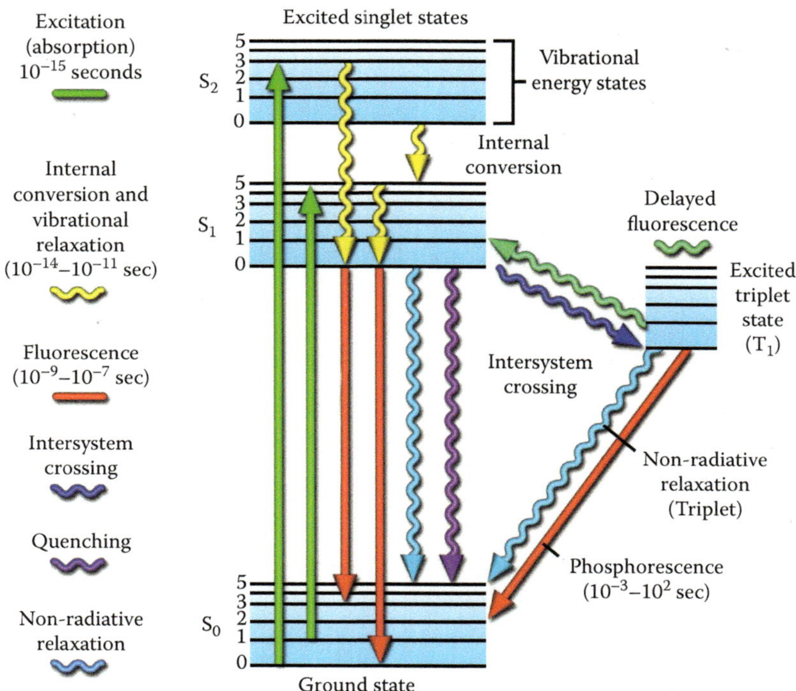



**Figure 7.14 Jablonski diagram of energy levels observed with typical fluorophores and luminescence experiments at biologically relevant temperatures.** Wavy or dashed lines show nonradiative transitions, and solid lines radiative transitions. The typical timescales for each type of transition are also shown.

is a radiative decay between states of the same spin and occurs within $10^{-8}$ to $10^{-5}$ s. The electron may also undergo intersystem crossing to a triplet state; radiative decay from a triplet to a singlet ground state is called *phosphorescence* and has very long lifetimes, up to many seconds.

However, the light emission may be shunted or "stolen" by *nonradiative recombination* of different kinds. Intersystem crossing is often radiationless; the generated triplet state may be highly stable and not produce photons. However, the most common reason for lack of emission is *external conversion* that returns the molecule to its ground state, with the energy of the excited state being lost as heat or transferred to other molecules in the solution by collisions.

In order to achieve the long-lived excited state necessary for luminescence, a molecule must have delocalized electrons. Most aromatics are fluorescent, and an increase in the number of rings and the number of substitutions in the rings increases fluorescence. Rigid structures also demonstrate enhanced emission. The absorbance and fluorescence spectra can be calculated based upon the molecular structure; the fine structure of the absorbance spectrum is given by the vibrational states of $S_1$, and that of the emission spectrum by the vibrational states of $S_0$. These are often mirror images of each other, with the fluorescence spectrum shifted to lower energies by various energy losses (**Figure 7.15**). The amount of shift is known as the *Stokes shift* and is around 20–30 nm for typical fluorophores, although it can range from 10 to 100 nm. Exceptions to the mirror image rule occur when the vibrational states of $S_0$ do not mirror those of $S_1$. This often occurs in pH-sensitive fluorophores, where the proton-dissociated form does not have the same vibrational states as the protonated form. It also occurs with fluorophores whose excited states form complexes with themselves

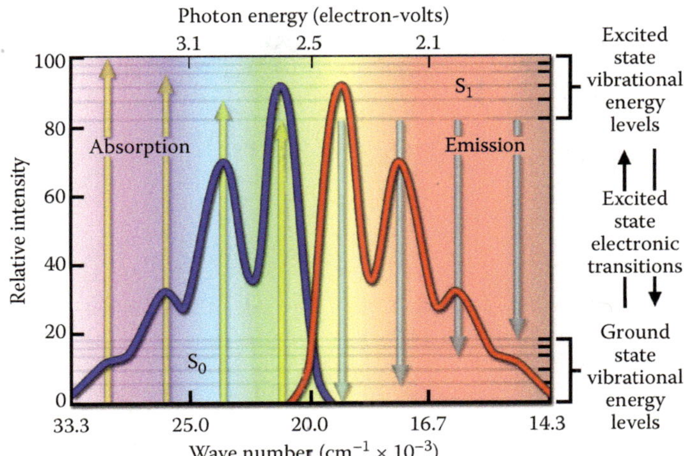


**Figure 7.15 Absorbance and emission spectra for a typical fluorophore, showing correspondence of features with transitions.**

or with other molecules, thereby greatly changing the $S_1$ spectrum relative to the ground state.

Useful fluorophores are catalogued and described according to their absorption and fluorescence emission properties, including the spectral profiles, wavelengths of maximum absorbance and emission, and the fluorescence intensity of the emitted light. There are a few quantitative parameters that need to be known for each fluorophore.

*Molar extinction coefficient* or molar absorptivity $\varepsilon$ (usually measured in units of $M^{-1}\ cm^{-1}$): This is a direct measure of the ability of a molecule to absorb light and provides a relationship between the absorbance (usually at peak) and the concentration $c$ for one or more species in solution:

$$A = \varepsilon c l$$

$$A(\lambda_i) = l \sum_{j=1}^{n} \varepsilon_j(\lambda_i) c_j \quad \text{(for multiple species)} \tag{7.3}$$

The path length $l$ depends upon the optical properties of the spectrometer and is usually 1 cm. The extinction coefficient is determined by measuring the absorbance at the peak characteristic of the absorbing species for a given concentration (**Figure 7.16a**). It is important to realize that most fluorophores have highly nonlinear absorbance spectra as they become more concentrated, so they must be diluted often to nearly colorless solutions before they are within the linear range. Taking spectra across a range of concentrations can determine where this range is for a given molecule.

The value of $\varepsilon$ shows only absorptivity and tells nothing about emission. The *quantum yield Q* of a fluorochrome or fluorophore represents a quantitative measure of fluorescence emission efficiency and is expressed as the ratio of the number of photons emitted to the number of photons absorbed (**Figure 7.16b**). Quantum yields typically range between a value of 0 and 1, and fluorescent molecules commonly employed as probes in microscopy have quantum yields ranging from very low (0.05 or less) to almost 1. The easiest way to determine quantum yield is to compare the integrated fluorescence intensity $I$ of a solution

**Figure 7.16 Absorbance and emission spectra showing measurable parameters.** (a) Absorbance vs. wavelength of three common fluorophores: DAPI, FITC, and Texas Red. The extinction coefficient $\varepsilon$ is taken from the absorbance at peak. (b) Emission spectra of the same example fluorophores. The quantum yield $Q$ comes from the photon intensity integrated over the spectrum.

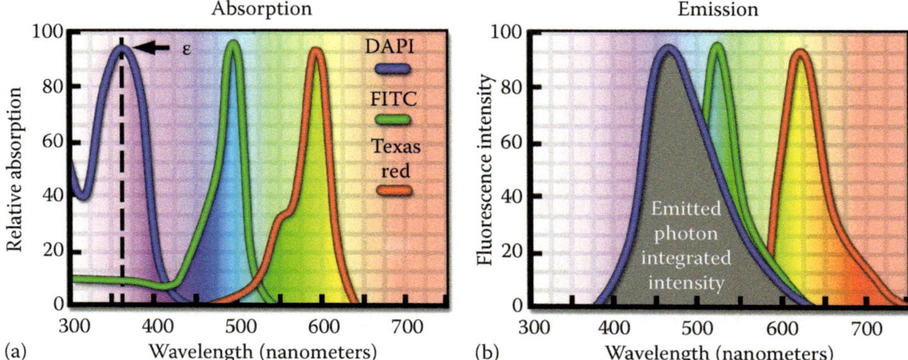


of the fluorophore with that of a reference fluorophore of known quantum yield ("R"). Both fluorophores should be within their linear absorbance range, and both the absorbance and emission spectra should be measured. Then the relative quantum yield is given by

$$\frac{Q}{Q_R} = \frac{I}{I_R}\frac{A_R}{A}\frac{n^2}{n_R^2} \tag{7.4}$$

where $n$ is the refractive index of the solvent. A common reference used is Rhodamine 101, which has a $Q$ value of 1 in ethanol. Other common references are given in **Table 7.1**. The fluorescence intensity, of course, will vary according to the instrument and filters used, but is proportional to the quantum yield, power of excitation $P$, and concentration as

$$I \propto QP(1-10^{-c\ell}) \approx 2.3QPc\ell. \tag{7.5}$$

The greater the power of excitation, the brighter the emission, but the shorter the time before the fluorophore undergoes irreversible *photobleaching*. Also termed fading, photobleaching occurs when a fluorophore permanently loses the ability to fluoresce due to chemical damage and covalent modification. Upon transition from an excited singlet state to the excited triplet state, fluorophores may interact with another molecule to produce irreversible covalent modifications. The triplet state is relatively long-lived with respect to the singlet state, thus allowing

**Table 7.1**

Commonly Used QY Reference Standards

| Dye | Solvent | QY (%) | Excitation (nm) |
|---|---|---|---|
| Rhodamine 101 | Ethanol | 100 | 450 |
| Rhodamine 6G | Water | 95 | 488 |
| Rhodamine B | Water | 31 | 514 |
| Fluorescein | 0.1 M NaOH, 22°C | 95 | 496 |
| Cy 3 | PBS | 4 | 540 |
| Cy 5 | PBS | 27 | 620 |
| Tryptophan | Water | 13 | 280 |

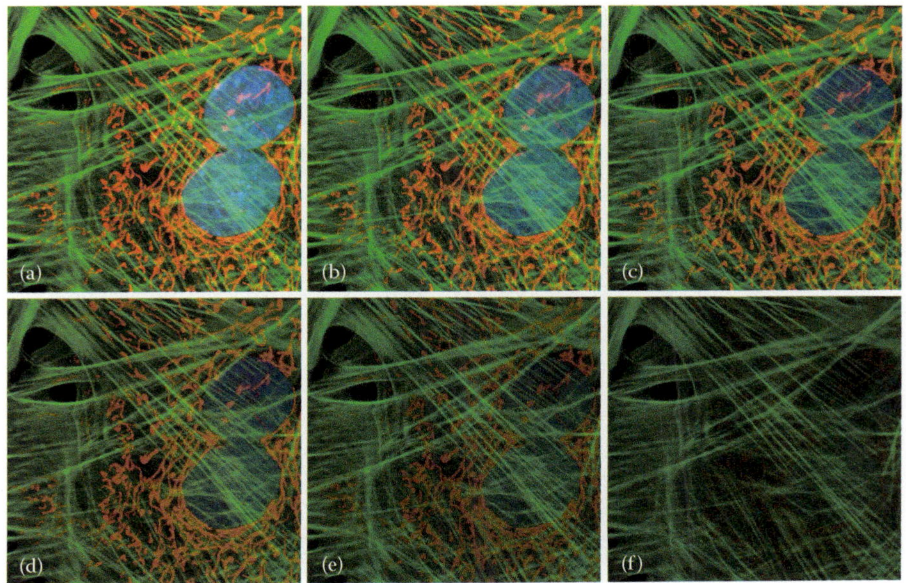

**Figure 7.17 Photobleaching.** Panels (a–f) show increasing amounts of time of light exposure (0–10 min in 2 min intervals) for cells labeled with three fluorophores: one for nuclei (arrow), one for mitochondria (striped arrow), and one for the cytoskeleton (arrowhead). Note that the nuclear stain disappears the most rapidly, and the cytoskeletal stain last.

excited molecules a much longer timeframe to undergo chemical reactions with components in the environment, often water and/or oxygen. The average number of excitation and emission cycles that occur for a particular fluorophore before photobleaching is dependent upon the molecular structure and the local environment. Some fluorophores bleach quickly after emitting only a few photons, while others that are more robust can undergo millions of cycles (**Figure 7.17**). Still others show vastly different stabilities in organic solvents versus water. The best way to reduce photobleaching of sensitive fluorophores is to reduce (or eliminate) oxygen from the solution. However, for live-cell experiments, this is usually not possible. Instead, antioxidants such as vitamin C (water-soluble) or vitamin E (fat-soluble) may be added to the medium in order to scavenge the radicals that damage the fluorophores. Commercial "slo-fade" reagents are also available for both fixed and living specimens.

*Quenching* is a different process usually involving collisional energy transfer with radiationless return of the fluorophore to the ground state. It does not involve chemical modification of the molecule and is reversible. Quenching will be discussed more thoroughly in **Chapter 16**.

## Epifluorescence microscopy

Fluorescence microscopy is usually performed using *reflected* light, even on microscopes where the brightfield examination is done using transmitted light. This mode of fluorescence is also known as incident light fluorescence, episcopic fluorescence, or most commonly *epifluorescence*. The reasons for this are several, primarily that this setup allows the microscope to be readily switched between fluorescence and transmitted light brightfield. Epifluorescence also causes most of the unwanted excitation light to travel away from the objective, and allows for the examination of thick fluorescent specimens, such as tissues, whole animals, and rocks (**Figure 7.18**).

**Figure 7.18 Epifluorescence of thick samples.** Images of Mojave desert gypsum containing red-autofluorescent cyanobacteria, with and without additional fluorescent labels. (a) Autofluorescence of the sandstone (green) showing some bacteria (red), blue excitation 380–460 nm, emission 500–750 nm. (b) Autofluorescence with UV excitation 300–350 nm, emission 420–600 nm. (c) Low-power image of gypsum stained with acridine orange. (d) High-power image of bacteria stained with acridine orange. The green is the dye; the red is the autofluorescence. (e) Gypsum sample stained with wheat germ agglutinin-Alexa 488, showing labeling of fungal cells in green surrounding the red bacteria. (f) Sample stained with calcofluor white, which labels fungi in blue. (g) Sample stained with CFDA, an esterase probe. Green fluorescence indicates regions of bacterial growth. (h) Higher-power image of CFDA-stained cells on rock.

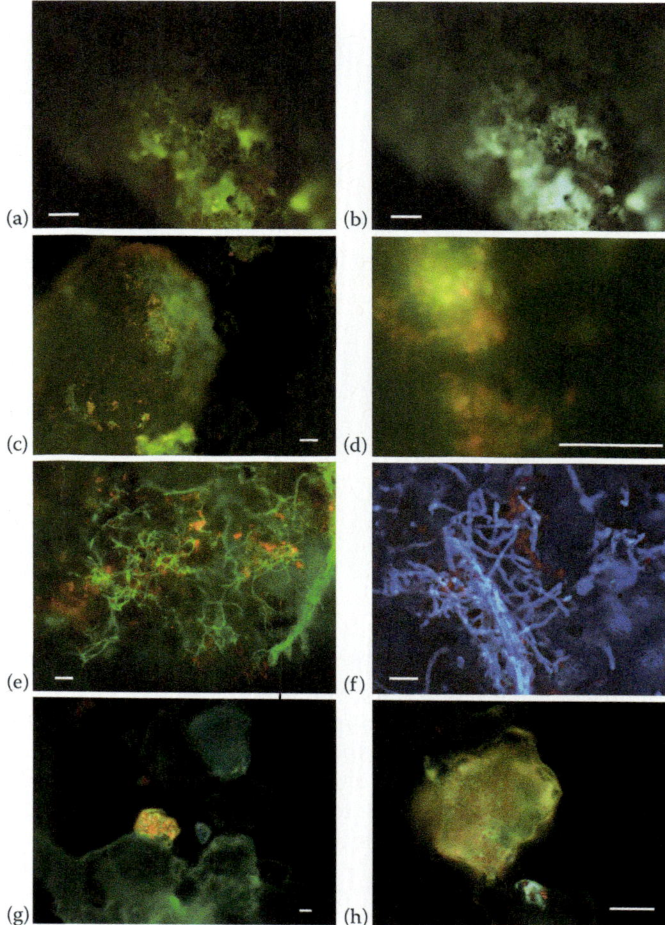

The basic function of any fluorescence microscope is to irradiate the specimen with a desired and specific band of wavelengths and then to separate the much weaker fluorescence emission from the excitation light; the excitation is usually $10^5$–$10^6$ times brighter than the emission. Separation is achieved with filters that block or pass specific wavelength bands in the UV, visible, and near-IR. Epifluorescence microscopes feature an illuminator between the observation viewing tubes and the nosepiece housing the objective (refer again to **Figure 7.3** and to **Figure 7.19**).

The illuminator generates multispectral light through a wavelength-selective *excitation filter*. Wavelengths passed by the excitation filter then reflect from the surface of a dichromatic mirror or *dichroic*, through the microscope objective and to the specimen. The light that emits from the fluorescing specimen radiates spherically in all directions, regardless of the excitation light source direction. Some of the emitted light is gathered by the objective and is sent back to the dichroic. Because the emitted light consists of longer wavelengths than the excitation illumination, it is able to pass through the dichroic and then through the *emission filter* toward the observation tubes. Most of the undesired scattered excitation light reaching the dichroic is reflected back toward the light source,

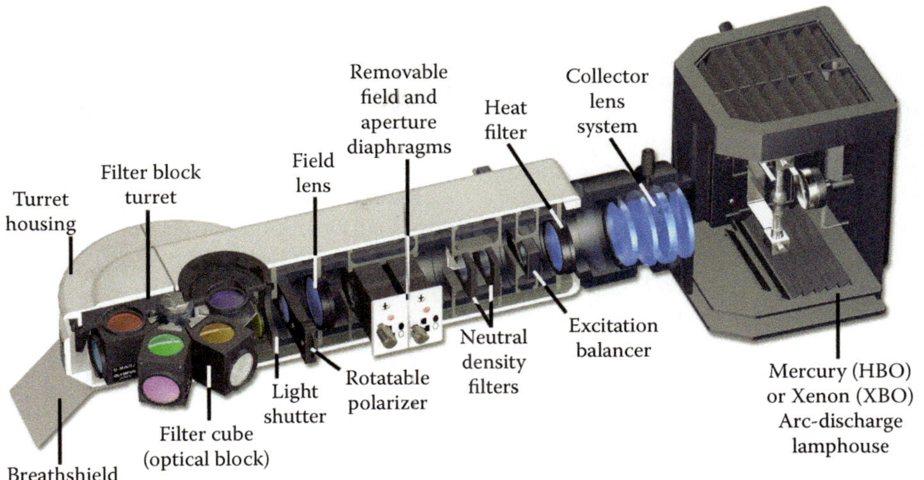



**Figure 7.19 Cutaway image of an illuminator for epifluorescence microscopy, from the lamp source to the filter cubes arranged in a turret.**

and any residual is blocked by the emission filter. The excitation and emission filters can either be placed in front of the light source and eyepiece, respectively, or they can be incorporated into a *filter cube* that also houses the dichromatic mirror (**Figure 7.20a**). Filter cubes are convenient as all elements are held at the correct angles, and excitation and emission filters can be easily replaced, allowing the experimenter to create custom filter sets for different fluorophores. Knowing the absorption and emission spectral profiles for the fluorophores of interest, and how to use these parameters to design or select appropriate filters, is essential to successful fluorescence imaging.

Filters may be long-pass, short-pass (edge filters), and narrow, medium, or wide band-pass filters. For example, the filter in **Figure 7.20b** shows a long-pass emission filter with a cut-on wavelength of 575 nm. The excitation filter in the same set is a narrow band-pass filter with a bandwidth of ~20 nm. The dichroic has transmission regions approximating a medium (455–490 nm) and wide band-pass filter (560–775 nm).

The dichroic or beamsplitter is the most critical component in a filter set, and resembles a long-pass interference filter fabricated to close tolerances with

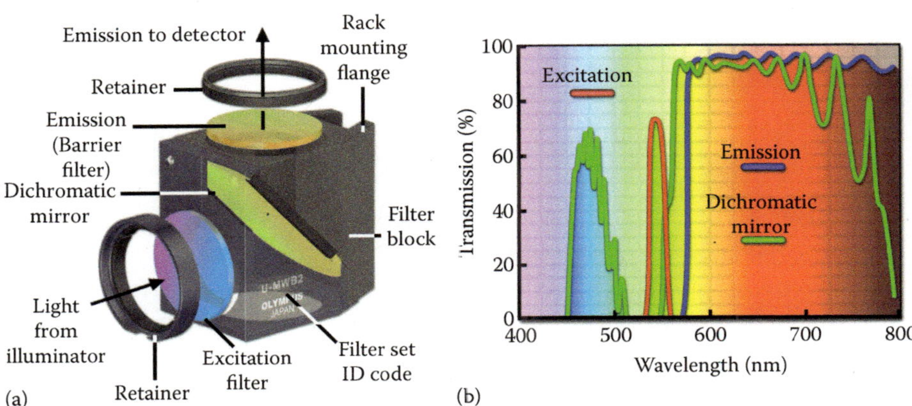



**Figure 7.20 Fluorescence filter cube.** (a) Structure of a cube, showing the position of the excitation and emission filters and dichroic. These elements are held in a block and can usually be replaced easily. (b) Sample spectra of a narrow band-pass excitation filter and long-pass emission.

multiple layers of dielectric materials. The major difference between a dichroic mirror and a standard interference filter is that the mirror is specifically designed for reflection and transmission at defined boundary wavelengths, and must operate at a 45° angle with respect to the microscope and illuminator optical axes. Dichroic mirrors are positioned with the interference coating facing the excitation light source in order to reflect short excitation wavelengths at a 90° angle through the optical train to the specimen. The same mirror must also act as a transmission filter to pass long wavelength fluorescence emission from the objective to the eyepiece or camera. Because the wavelength transition region between almost total reflection and maximum transmission is often limited to 20–30 nm, the dichroic is able to precisely discriminate between excitation and emission wavelengths. If the elements of the filter cube do not match, excitation wavelengths may enter into the image and create a bright background (*bleed-through*), or too much fluorescence might be reflected and lost, making the sample invisible even when it fluoresces brightly. Research-grade fluorescence microscopes usually contain a turret of five or six filter cubes, allowing for a range of choices to optimize excitation and emission parameters. Sliders of emission filters can also be used, allowing the same dichroic to be used with a variety of output filters (**Figure 7.21**).

Light sources for epifluorescence microscopy are continually improving and becoming less costly. In order to generate sufficient excitation light intensity to excite most fluorophores, powerful light sources such as high-energy short arc discharge lamps have traditionally been used. The most common are 50–200 W mercury burners or 75–150 W xenon lamps. These light sources are powered by an external direct current supply that provides enough power to ignite the burner through the ionization of the gaseous vapor and to keep it burning with a minimum of flicker for approximately 200–300 h. After this time period, the light becomes unreliable and the lamp must be replaced.

The spectra of Hg and Xe lamps are shown in **Figure 7.22**. It can be seen that they are not uniform, although the Xe lamp produces a reasonably flat spectrum between 300 and 500 nm. Hg lamps produce certain prominent peaks—e.g., at 313, 334, 365, 406, 435, 546, and 578 nm. At other wavelengths, the intensity is much less bright, although it can still be used to excite most fluorophores. *Excitation balancers* can be used to adjust the intensity at given wavelengths.

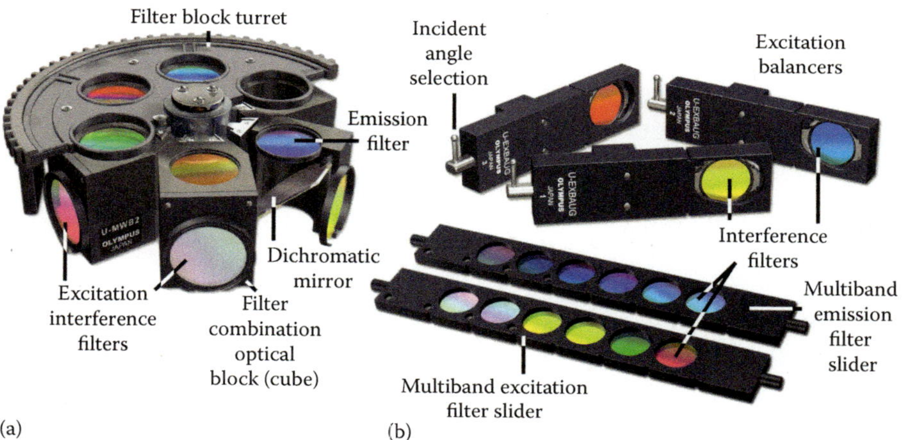


**Figure 7.21 Fluorescence filter choices.** (a) Filter block turret containing multiple filter cubes. The turret can usually be rotated manually as well as by software control. (b) Additional filters can include excitation balancers and interference filters, as well as multiband excitation and/or emission sliders. The latter are useful for ratiometric imaging discussed later in the chapter.

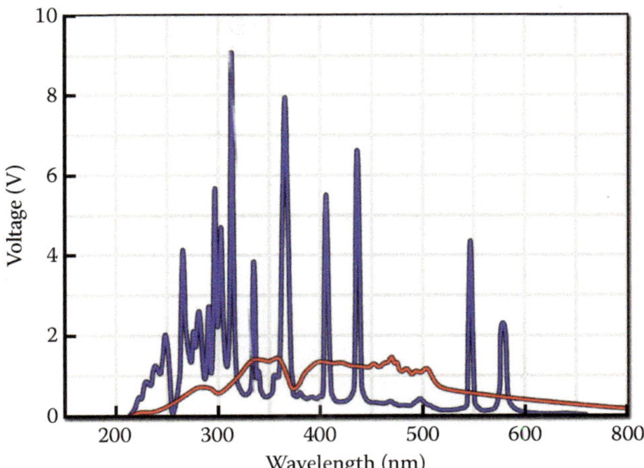


**Figure 7.22 Spectra of Hg and Xe arc lamps.**

The *mean luminance* at the wavelengths of choice is the measure that must be taken into consideration, which is a function not only of source brightness but also arc geometry and the angular spread of the light.

Arc lamps are encased in a lamphouse, usually made of black plastic. The lamphouse protects the experimenter in case of bulb explosion and also has a few other important features. The socket in which the lamp itself is held is equipped with centering screws to permit centering the arc image in the objective rear aperture. Proper centering is essential to avoiding dark and bright spots in the image, and should be one of the first skills learned by the aspiring fluorescence microscopist. The lamphouse also should have an infrared filter to block the far red and infrared wavelengths that generate a tremendous amount of heat. Some designs have a red suppression filter, or a slot for such a filter, to eliminate a reddish background seen through the viewfield in some applications. The lamphouse should also not leak UV as this may damage the experimenter's eyes, and there should be an automatic switch that shuts down the lamp if the housing is opened.

The fact that arc lamps emit a good deal of light in the near-UV contributes to an unfortunate practice of calling them "UV lamps" and/or of referring to all fluorescence excitation as "UV excitation." It is a small minority of useful fluorophores that truly excite in the UV (see **Table 7.2**), and for most uses, Hg and Xe lamps are used because of their bright white light and not for their UV emission, which is filtered out by the excitation filter. Of course, other light sources that produce light more targeted to the fluorophore's excitation wavelength may also be used. Lasers are desirable for their high, adjustable intensity and their monochromaticity. Particularly useful are argon ion lasers, which have strong lines at 488 and 514 nm, krypton/argon lasers, and a variety of helium–neon (HeNe) lasers that can excite in the green to red (543 and 633 nm are common lines used; **Figure 7.23**).

Light-emitting diodes (LEDs) are compact, inexpensive options that work well for many fluorophores. The biggest drawback to the use of LEDs is power output, with tens of milliwatts representing a relatively powerful LED. However, LEDs come in a wide variety of colors with narrow Gaussian outputs. If the LED and fluorophore are selected carefully, nearly all of the LED light can go into exciting the dye.

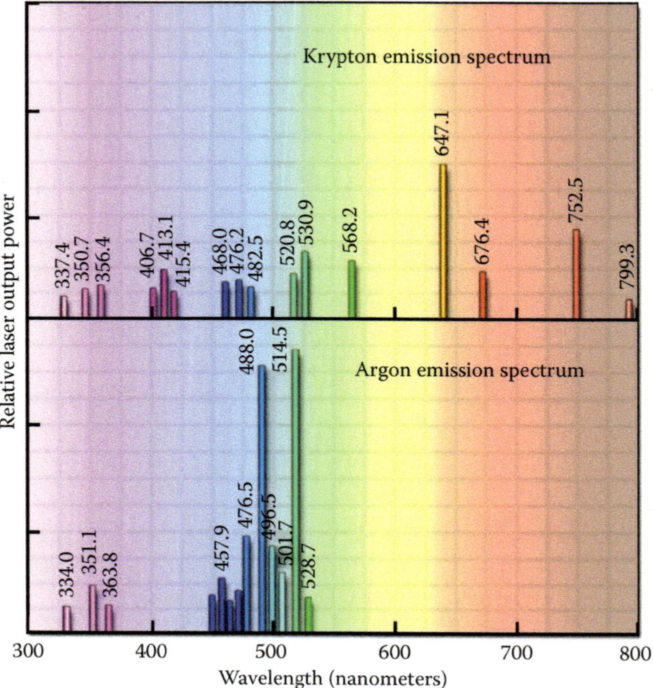

**Figure 7.23 Emission lines of commonly used lasers in confocal microscopy.** For typical Ar lasers, only the 488 and 514 nm lines are powerful enough to excite fluorescence.



We currently use a homemade fluorescence microscope with a single 470 nm LED illuminator for field microbiology applications (see **Practical Tips 7.4**).

## Confocal laser scanning microscopy

In traditional widefield epifluorescence microscopy, a large volume of the specimen is subjected to intense illumination from a high-power source, and a majority of the resulting fluorescence emission directed back toward the microscope is gathered by the objective (depending upon the NA) and projected into the eyepieces or detector (**Figure 7.24 left**). In contrast to this simple concept, the mechanism of image formation in a confocal microscope is fundamentally different. The defining features of confocal microscopy are *point scanning* and the *pinhole aperture*. The laser illumination source is first expanded to fill the objective rear aperture, and then focused by the lens system to a very small spot at the focal plane (**Figure 7.24 right**). The size of the illumination point ranges from approximately 0.25 to 0.8 μm in diameter (depending upon objective NA) and from 0.5 to 1.5 μm deep at the brightest intensity. Spot size is determined by the microscope design, wavelength of incident laser light, objective characteristics, scanning unit settings, and the specimen. The image of an extended specimen is then generated by scanning the focused beam across a defined area in a raster pattern controlled by two high-speed oscillating mirrors driven with galvanometer motors. One of the mirrors moves the beam from left to right along the *x* lateral axis, while the other translates the beam in the *y* direction. After each single scan along the *x*-axis, the beam is rapidly transported back to the starting point and shifted along the *y*-axis to begin a new scan in a process termed *flyback*. During the flyback operation, image information is not collected.

## PRACTICAL TIPS 7.4:   FLUORESCENCE MICROSCOPY IN THE HIGH ARCTIC

The author has been traveling to the Canadian High Arctic (latitude nearly 80°) for the past several years to study the microbial communities in the extreme environments found there, which are considered to be analogs for possible life on colder planets of the solar system, particularly Mars. A limited number of samples can be returned to the laboratory, and for examination of fresh, living specimens, a high-resolution fluorescence microscope is needed that can travel to such a remote location and operate under extreme conditions. It needs to be able to be packed in a suitcase and survive rough handling during flights, endure high winds and dust, operate at temperatures near freezing, and use very little power (there are generators at the station, but for field operation, we rely on a single automotive battery pack for 6–8 h of work). It should also be readily assembled and disassembled for journeys between field sites on foot.

The design we have used was custom-built for our applications by the late Constantine Douketis of CJD Instruments, Inc. It is a "modular" design with interchangeable elements: one element can be an eyepiece or camera; another can be any of several objective lenses; another can be any of a choice of LED illuminators (**Figure P7.4.1**). The pieces are held together by bayonet connectors and can all be easily disconnected and packed into a carrying case. This basic design could be readily adapted by anyone wishing to perform field microscopy.

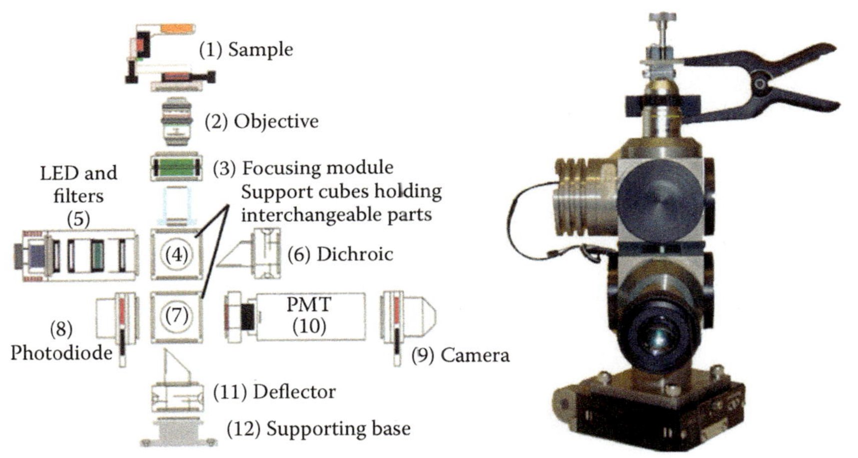


Figure P7.4.1 Schematic (left) and photo (right) of microscope used in High Arctic fluorescence microscopy.

## SUGGESTED READING


Nadeau, J.L., Perreault, N.N., Niederberger, T.D., Whyte, L.G., Sun, H.J., and Leon, R. (2008). Fluorescence microscopy as a tool for in situ life detection. *Astrobiology* 8, 859–874.

Pollard, W., Haltigin, T., Whyte, L., Niederberger, T., Andersen, D., Omelon, C., Nadeau, J., Ecclestone, M., and Lebeuf, M. (2009). Overview of analogue science activities at the McGill Arctic Research Station, Axel Heiberg Island, Canadian High Arctic. *Planetary and Space Science* 57, 646–659.

Rogers, J.D., Perreault, N.N., Niederberger, T.D., Lichten, C., Whyte, L.G., and Nadeau, J.L. (2010). A life detection problem in a High Arctic microbial community. *Planetary and Space Science* 58, 623–630.


As each scan line passes along the specimen in the lateral focal plane, fluorescence emission is collected by the objective and passed back through the confocal optical system. The speed of the scanning mirrors is very slow relative to the speed of light, so the fluorescence emission follows a light path along the optical axis that is identical to the original excitation beam. Return of fluorescence emission through the galvanometer mirror system is referred to as *descanning.*

**Figure 7.24 Schematic of illuminated volume in widefield (left) and confocal (right) microscopy.**

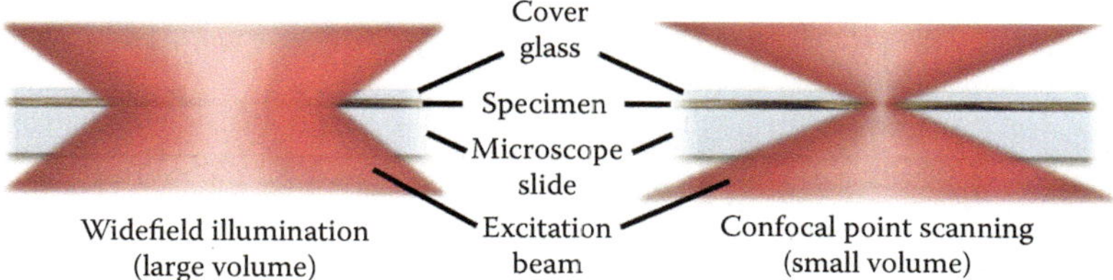


After leaving the scanning mirrors, the fluorescence emission passes directly through the dichromatic mirror and is focused at the detector pinhole aperture. The pinhole acts as a spatial filter at the conjugate image plane and serves to exclude fluorescence signals from out-of-focus features positioned above and below the focal plane, which are instead projected onto the aperture as Airy disks having a diameter much larger than those forming the image. These oversized disks are spread over a comparatively large area so that only a small fraction of light originating in planes away from the focal point passes through the aperture. The pinhole aperture also serves to eliminate much of the stray light passing through the optical system. Several apertures of varying diameter are usually contained on a rotating turret that enables the operator to adjust the pinhole size. The end result of this method of imaging is significantly improved resolution because of the elimination of background light and fluorescence originating from areas above and below the focal plane (**Figure 7.25**).

Most importantly, thin (1–1.5 μm) *optical sections* may be taken through the thickness of the specimen by controlling the fine focus with a stepper motor. This allows for noninvasive imaging throughout a thick cell or tissue; multiple sections may be reconstructed into a 3-D image (**Figure 7.26**).

All modern laser scanning confocal microscope designs are centered on a conventional upright or inverted research-level optical microscope. However, instead of the standard arc lamp, one or more laser systems are used as a light

**Figure 7.25 Images of biological samples under widefield and confocal microscopy.** (a, c, e) Widefield; (b, d, f) confocal.

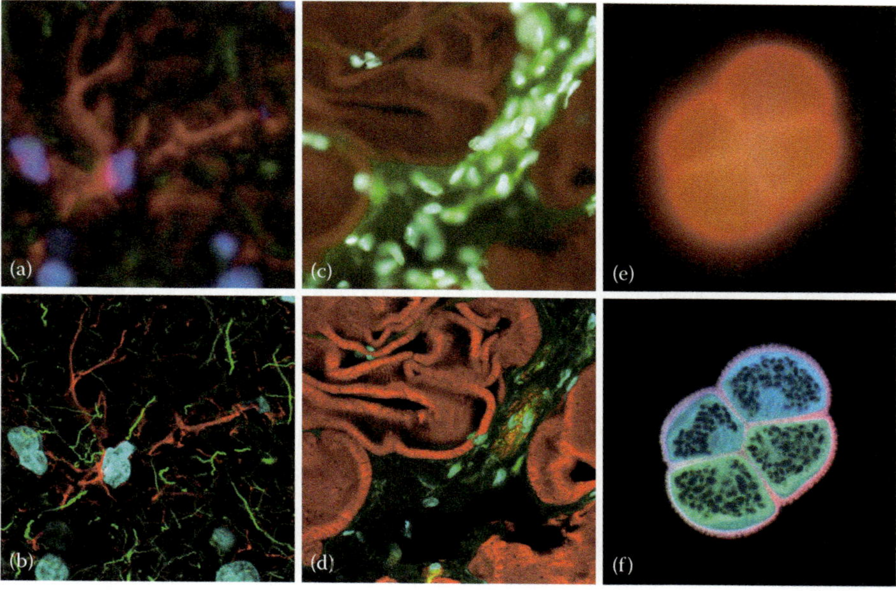

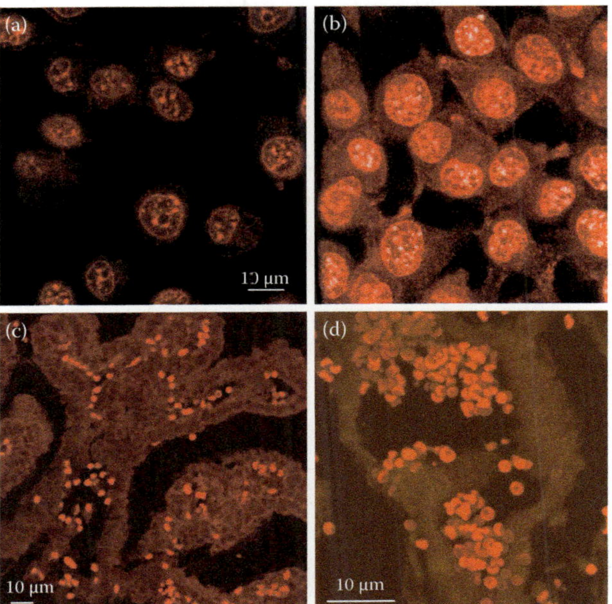

**Figure 7.26  Confocal slices and 3-D reconstructions.**
(a) Single slice through cells labeled with a predominantly nuclear dye. Only the nuclei are visible. (b) 3-D reconstruction showing the shape of the cells and presence of dye throughout the cytoplasm. (c) Single optical slice through a hemorrhagic region of a mouse brain. (d) 3-D reconstruction showing red blood cells inside a leaky capillary.

source to excite the specimen. Image information is gathered point by point with a specialized detector such as a photomultiplier tube (PMT) or avalanche photodiode, and then digitized for processing by the host computer, which also controls the scanning mirrors and/or other devices to facilitate the collection and display of images. The *scan head* is at the heart of the confocal system and is responsible for rasterizing the excitation scans, as well as collecting the photon signals from the specimen that are required to assemble the final image. A typical scan head contains inputs from the external laser sources, fluorescence filter sets and dichromatic mirrors, a galvanometer-based raster scanning mirror system, variable pinhole apertures for generating the confocal image, and PMT detectors tuned for different fluorescence wavelengths (**Figure 7.27**).

PMTs are the ideal photometric detectors for confocal microscopy due to their speed, sensitivity, high signal-to-noise ratio, and adequate dynamic range. High-end confocal microscope systems have several PMTs for simultaneous imaging of different fluorophores in multiply labeled specimens; usually three are assigned to red, green, and blue channels, and a fourth for imaging the specimen with transmitted light using DIC. Signals from each channel can be collected simultaneously and the images merged into a single profile that represents the "real" colors of the stained specimen. If the specimen is also imaged with DIC, the fluorophore distribution in the fluorescence image can be overlaid onto the brightfield image to determine the spatial location of fluorescence emission within the structural domains.

The voltage setting is used to regulate the overall sensitivity of the PMT and can be adjusted independently of the gain and offset values. The latter two controls are utilized to adjust the image intensity values to ensure that the maximum number of gray levels is included in the output signal of the photomultiplier. Offset adds a positive or negative voltage to the output signal and should be adjusted so that the lowest signals are near the photomultiplier detection threshold. The gain circuit multiplies the output voltage by a constant factor so that the maximum signal

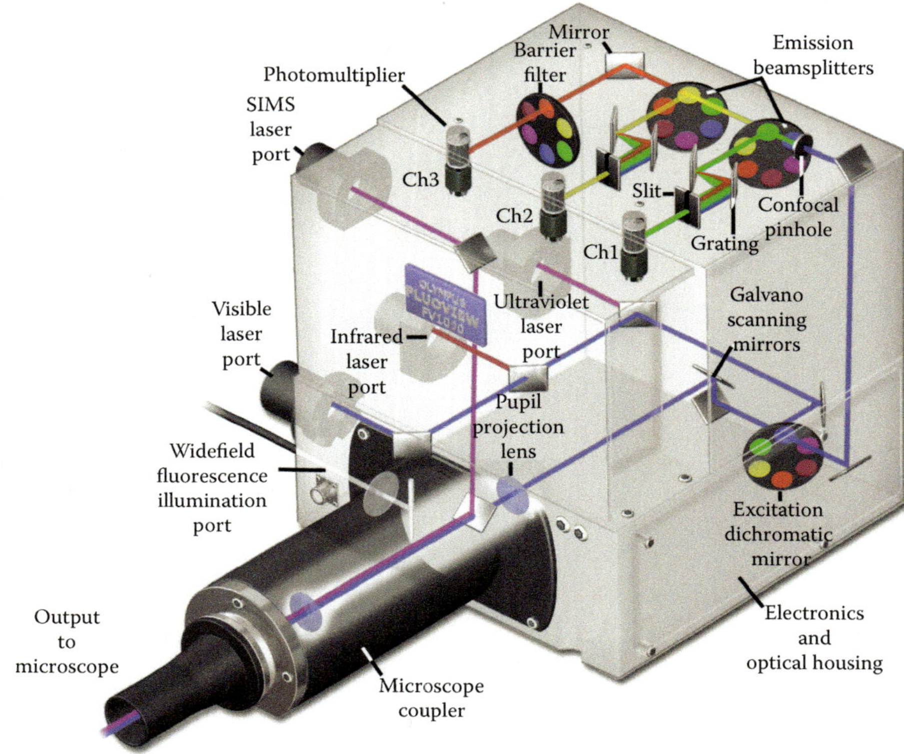



**Figure 7.27 Confocal scan head for a three-channel system.** There is also a port for widefield observation.

values can be stretched to a point just below saturation. In practice, offset should be applied first before adjusting the photomultiplier gain. After the signal has been processed by the analog-to-digital converter, it is stored in a frame buffer and ultimately displayed on the monitor in a series of gray levels ranging from black (no signal) to white (saturation). Photomultipliers with a dynamic range of 10 or 12 bits are capable of displaying 1024 or 4096 gray levels, respectively. Accompanying image files also have the same number of gray levels. However, the photomultipliers used in a majority of the commercial confocal microscopes have a dynamic range limited to 8 bits or 256 gray levels, which in most cases is adequate for handling the typical number of photons scanned per pixel. Effects of offset and gain adjustments on an image are shown in **Figure 7.28**.

Advanced confocal microscopy software packages ease the burden of gain and offset adjustment by using a pseudo-color display function to associate pixel values with gray levels on the monitor. For example, the saturated pixels (255) can be displayed in yellow or red, while black-level pixels (0) are shown in blue or green, with intermediate gray levels displayed in shades of gray representing their true values. When the photomultiplier output is properly adjusted, just a few red (or yellow) and blue (or green) pixels are present in the image, indicating that the full dynamic range of the photomultiplier is being utilized.

Changes to the PMT gain and offset levels should not be confused with post-acquisition image processing to adjust the levels, brightness, or contrast in the final image. Digital image processing techniques can stretch existing pixel values to fill the black-to-white display range, but cannot create new gray levels. As a result, when a digital image captured with only 200 out of a possible 4096

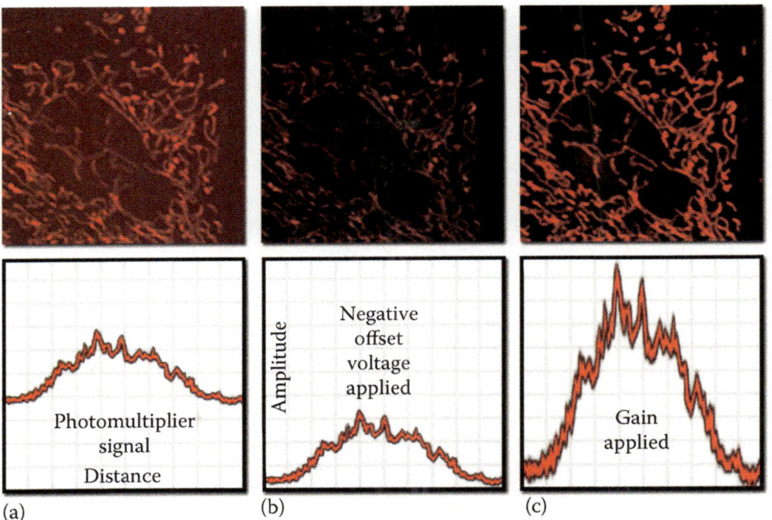



**Figure 7.28 Gain and offset control in confocal PMT detectors.** The sample shows a cell labeled with a mitochondrial dye. (a) Raw image (above) and PMT signal (below). (b) Signal after applying a negative offset voltage to the PMT. (c) Final signal and image after adjusting the gain to fill the entire intensity range.

gray levels is stretched to fill the histogram (from black to white), the resulting processed image appears grainy. In routine operation of the confocal microscope, the primary goal is to fill as many of the gray levels during image acquisition and not during the processing stages.

Confocal is not superior to widefield epifluorescence for all applications. For thin, flat specimens such as bacteria and some cell lines, confocal is unnecessary. For samples with unusual fluorophores or fluorophores that excite away from the common laser lines, arc-lamp excitation may be preferable. Because confocal provides "pseudocolored" images corresponding to assigned channels, it does not necessarily provide a good representation of some multiply labeled samples. Most confocal microscopes do not have UV or violet excitation, although 405 nm diode lasers are becoming increasingly available. This drawback eliminates some very common blue-emitting dyes. Laser radiation can also be harmful to live cells, primarily due to the generation of reactive oxygen species by fluorophores or aromatic groups in the medium (e.g., phenol red, tyrosine). This can be limited by imaging in a minimal medium and adding antioxidants. Finally, confocal microscope systems cost up to an order of magnitude more than comparable widefield microscopes, which often limits their implementation in smaller laboratories. This problem can be easily overcome by cost-shared microscope systems that service one or more departments in a core facility.

## 7.6 FLUOROPHORES FOR CELL LABELING

### Autofluorescence

All cells contain some degree of fluorescence due to the presence of endogenous fluorophores. This is called *autofluorescence* to distinguish it from fluorescence imparted from labels. The degree of autofluorescence varies enormously according to cell type and how the cells are processed; for example, glutaraldehyde should not be used to fix cells for fluorescence microscopy because it generates fluorescent products when it interacts with proteins. A lot of endogenous cellular fluorescence excites in the UV range and emits in the blue, representing molecules such as *nicotinamide adenine dinucleotide* in its reduced form (NADH; absorbance around 340 nm/emission around 460 nm) and tryptophan (excitation peak 230 nm/emission 330 nm in nonpolar solvents; 365 nm in polar solvents such as water). Greenish autofluorescence (500–600 nm emission, excitation at common wavelengths such as 488 and 514) is also very common in cells and is primarily due to endogenous flavoproteins. Red autofluorescence is ubiquitous in plant cells and photosynthetic bacteria due to the presence of chlorophylls; this type of autofluorescence is extremely intense and excites from 350 to 470 nm depending upon the variety. Hemoglobin breakdown products are also highly fluorescent when excited with 488 nm, so that red blood cells usually appear highly fluorescent under Ar laser excitation.

Autofluorescence can be used in certain experiments; for example, the shift of tryptophan with environment is often used to probe protein structural changes, and chlorophyll fluorescence can be used to track swimming bacteria. But for the most part, autofluorescence represents a background signal that is best filtered out. The particular spectrum of autofluorescence of the cells and tissues you are using, as well as your choice of dyes, will determine your ideal filter configurations. In general, narrow-band filters as close as possible to the dye maximum will allow for the greatest signal-to-noise, although if they are too narrow, they cut out a good deal of the dye fluorescence. Sometimes autofluorescence in cells can be reduced by blocking aldehyde groups with sodium borohydride or glycine.

## Traditional organic dyes

Fluorophores chosen for microscopy must exhibit a brightness level and signal persistence sufficient for the instrument to obtain image data that do not suffer from excessive photobleaching artifacts and low signal-to-noise ratios. It is important to keep in mind that some dyes that are popular for widefield applications cannot be used in confocal microscopy because of the more stringent excitation conditions. The structures and physical properties of a few common dyes are given in **Table 7.2**.

Fluorescein is one of the most popular fluorochromes ever designed. This xanthene dye has an absorption maximum at 495 nm, which coincides quite well with the 488 nm spectral line produced by argon-ion and krypton–argon lasers, as well as the 436 and 467 principal lines of the mercury and xenon arc-discharge lamps (respectively). The quantum yield of fluorescein is very high, and a significant amount of information has been gathered on the characteristics of this dye with respect to the physical and chemical properties. On the negative side, its fluorescence emission intensity is heavily influenced by environmental factors such as pH.

Tetramethyl rhodamine (TMR) and the isothiocyanate derivative (TRITC) are frequently employed in multiple labeling investigations in widefield microscopy due to their efficient excitation by the 546 nm spectral line from Hg lamps. They may be used for double-labeling with fluorescein, although there is some overlap. They can also be excited very effectively by the 543 nm line from HeNe lasers, but not by the 514 or 568 nm lines from argon-ion and krypton–argon lasers. When using krypton-based laser systems, Lissamine rhodamine is a far better choice in this fluorochrome class due to the absorption maximum at 575 nm and its spectral separation from fluorescein.

**Table 7.2**

Examples of Some General and Targeted Fluorescent Probes

| Name | Structure | Target | Ex. Peak (nm) | Em. Peak (nm) | $\varepsilon$ (M$^{-1}\chi\mu^{-1}$) |
|---|---|---|---|---|---|
| Fluorescein | 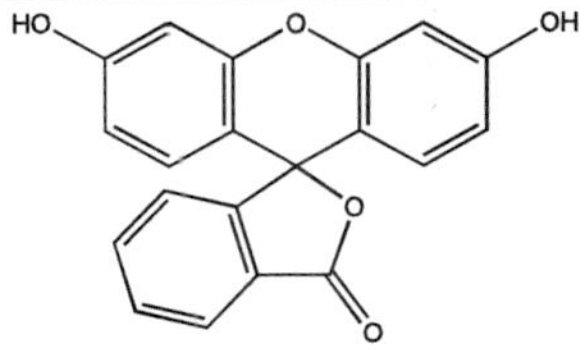 | General | 495 | 520 | 83,000 |
| Rhodamine 6G | 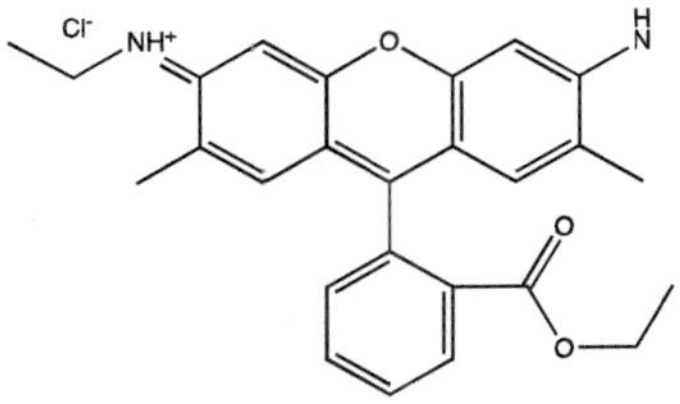 | General | 530 | 556 | 116,000 |
| Alexa 350-succinimide | 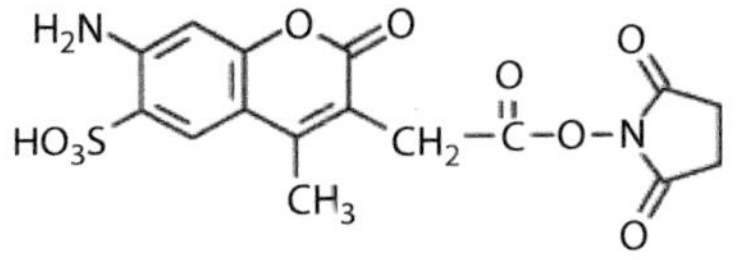 | Amine-reactive | 346 | 442 | 19,000 |
| Texas Red | 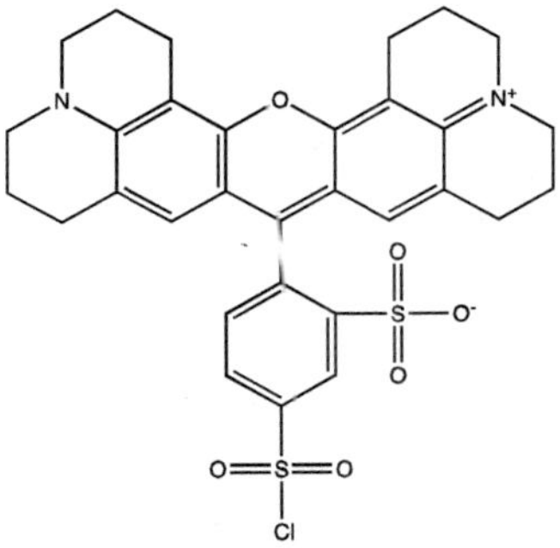 | General | 583 | 603 | 116,000 |
| DAPI | 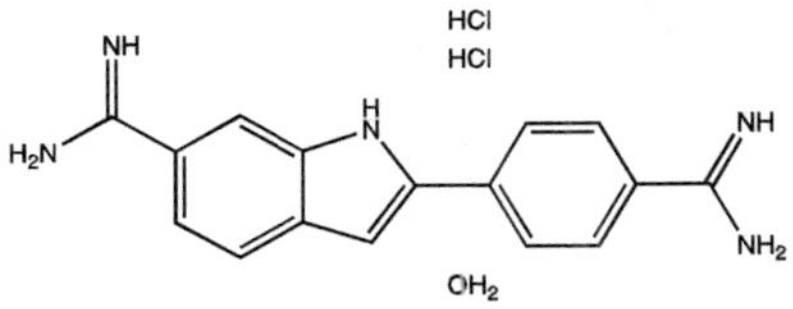 | DNA | 345 | 455 | 27,000 |
| Ethidium bromide | 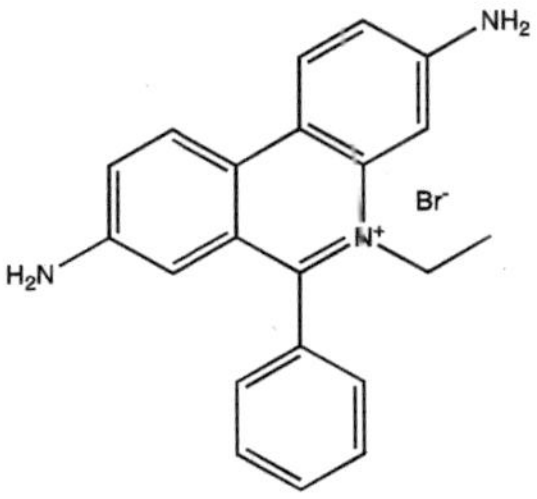 | DNA | 493 | 620 | 5700 |

(Continued)

**Table 7.2 (Continued)**

Examples of Some General and Targeted Fluorescent Probes

| Name | Structure | Target | Ex. Peak (nm) | Em. Peak (nm) | $\varepsilon$ (M$^{-1}$χμ$^{-1}$) |
|---|---|---|---|---|---|
| Fura-2 | 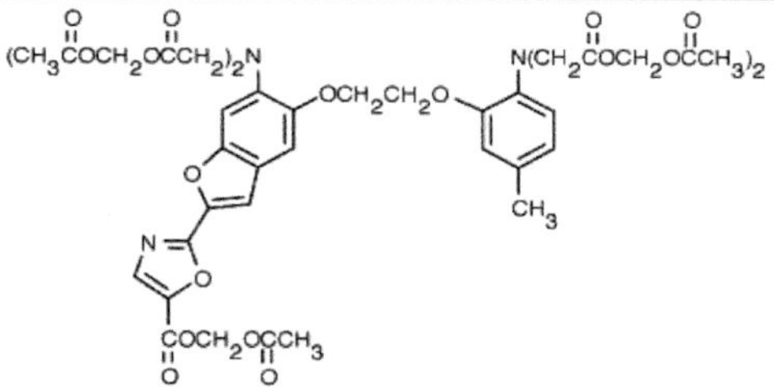 $(CH_3COCH_2OCCH_2)_2N$ … $OCH_2CH_2O$ … $N(CH_2COCH_2OCCH_3)_2$ … $CH_3$ … $COCH_2OCCH_3$ | Ca$^{2+}$ | 363 (0 Ca$^{2+}$) 335 (high Ca$^{2+}$) | 512 (0 Ca$^{2+}$) 505 (high Ca$^{2+}$) | 28,000 (0 Ca$^{2+}$) 34,000 (high Ca$^{2+}$) |

*Note:* More extensive tables are found on the manufacturer's sites.

Several of the acridine dyes, first isolated in the nineteenth century, are useful as fluorescent probes in confocal microscopy. The most widely utilized, acridine orange, consists of the basic acridine nucleus with dimethylamino substituents located at the 3 and 6 positions of the tri-nuclear ring system. In physiological pH ranges, the molecule is protonated at the heterocyclic nitrogen and exists predominantly as a cationic species in solution. Acridine orange binds strongly to DNA by intercalation of the acridine nucleus between successive base pairs, and exhibits green fluorescence with a maximum wavelength of 530 nm. The probe also binds strongly to RNA or single-stranded DNA, but has a longer wavelength fluorescence maximum (approximately 640 nm; red) when bound to these macromolecules. In living cells, acridine orange diffuses across the cell membrane (by virtue of the association constant for protonation) and accumulates in the lysosomes and other acidic vesicles. Similar to most acridines and related polynuclear nitrogen heterocycles, acridine orange has a relatively broad absorption spectrum, which enables the probe to be used with several wavelengths from the argon-ion laser. It is highly phototoxic and will kill cells after several minutes of imaging.

Another popular traditional probe that is useful in confocal microscopy is the phenanthridine derivative, propidium iodide, first synthesized as an antitrypanosomal agent along with the closely related ethidium bromide. Propidium iodide binds to DNA in a manner similar to the acridines (via intercalation) to produce orange-red fluorescence centered at 617 nm. It also has a high affinity for double-stranded RNA. Propidium iodide has an absorption maximum at 536 nm and can be excited by the 488 or 514 spectral lines of an argon-ion (or krypton–argon) laser, or the 543 line from a green HeNe laser. The dye is often employed as a counterstain to highlight cell nuclei during double or triple labeling of multiple intracellular structures. The structurally similar ethidium bromide, which also binds to DNA by intercalation, produces more background staining and is therefore not as effective as propidium.

DNA and chromatin can also be stained with dyes that bind externally to the double helix. The most popular fluorochromes in this category are 4′, 6-diamidino-2-phenylindole (DAPI) and the bisbenzimide Hoechst dyes that are designated by the numbers 33258, 33342, and 34580. These probes are quite water-soluble and bind externally to AT-rich base pair clusters in the minor groove of double-stranded DNA with a dramatic increase in fluorescence intensity. Both dyes are usually used only in widefield as they are too blue for the common laser lines; however, they may be excited by the 351 nm spectral line of high-power argon-ion lasers or the 354 nm line from a HeCd laser, or by a 405 nm diode laser. Similar to the acridines and phenanthridines, these fluorescent probes are popular choices as a nuclear counterstain for use in multicolor fluorescent labeling protocols. The vivid blue fluorescence emission produces dramatic contrast when coupled to green, yellow, and red probes in adjacent cellular structures.

## New-generation fluorescent dyes

Alexa Fluor dyes (registered trademark of Molecular Probes) are sulfonated rhodamine derivatives that exhibit higher Q and enhanced photostability compared with traditional organic fluorophores. They can usually be used in live- and fixed-cell confocal experiments without antifade reagents. They are available in a broad range of fluorescence excitation and emission wavelength maxima, ranging from the ultraviolet and deep blue to the near-infrared regions, many specifically targeted to laser or arc-lamp lines. Alphanumeric names of the individual dyes are associated with the specific excitation wavelength for which the probe is intended. For example, Alexa Fluor 488 is designed for excitation by the 488 nm line of the argon or krypton–argon ion lasers, while Alexa Fluor 568 is matched to the 568 nm spectral line of the krypton–argon laser. There are Alexa Fluor dyes specifically designed for excitation by rarer lines: the blue diode laser (405 nm), the orange/yellow helium–neon laser (594 nm), or the red HeNe laser (633 nm). Other Alexa Fluor dyes are intended for excitation with Hg lamps in the visible (Alexa Fluor 546) or ultraviolet (Alexa Fluor 350), and solid-state red diode lasers (Alexa Fluor 680). Because of the large number of available excitation and emission wavelengths in the Alexa Fluor series, multiple labeling experiments can often be conducted exclusively with these dyes.

The popular family of cyanine dyes, Cy2, Cy3, Cy5, Cy7, and their derivatives, are based on the partially saturated indole nitrogen heterocyclic nucleus with two aromatic units being connected via a polyalkene bridge of varying carbon number. These probes exhibit fluorescence excitation and emission profiles that are similar to many of the traditional dyes, such as fluorescein and tetramethylrhodamine, but with enhanced water solubility, photostability, and higher quantum yields. Most of the cyanine dyes are more environmentally stable than their traditional counterparts, rendering their fluorescence emission intensity less sensitive to pH and organic mounting media. In a manner similar to the Alexa Fluors, the excitation wavelengths of the Cy series of synthetic dyes are tuned specifically for use with common laser and arc-discharge sources, and the fluorescence emission can be detected with traditional filter combinations. The cyanine dyes generally have broader absorption spectral regions than members of the Alexa Fluor family, making them somewhat more versatile in the choice of laser excitation sources for confocal microscopy. For example, using the 547 nm spectral line from an argon-ion laser, Cy2 is about twice as efficient in fluorescence emission as Alexa Fluor 488. In an analogous manner, the 514 nm argon-ion laser line excites Cy3 with a much higher efficiency than Alexa Fluor 546, a spectrally similar probe. Emission profiles of the cyanine dyes are comparable in spectral width to the Alexa Fluor series. Included in the cyanine dye series are the long-wavelength Cy5 derivatives, which are excited in the red region (650 nm) and emit in the far-red (680 nm). The Cy5 fluorophore is very efficiently excited by the 647 nm spectral line of the krypton–argon laser, the 633 nm line of the red HeNe laser, or the 650 nm line of the red diode laser. Because the emission spectral profile is significantly removed from traditional fluorophores excited by ultraviolet and blue illumination, Cy5 is often utilized as a third fluorophore in triple labeling experiments. However, Cy5 is poorly visible to the human eye and thus can only be detected electronically. CCD cameras are very sensitive in the far red, so this is usually not a problem for modern systems. Cy dyes are marketed by a number of distributors.

## Attaching dyes to cell-targeting molecules

Dyes such as propidium iodide or DAPI may be used as is, as they will diffuse into the cell nucleus and label the DNA without significant background in the rest of the cell or the dish. This is not the case for dyes such as fluorescein, rhodamine Alexa Fluors, or Cy dyes, which possess no particular targeting capability on their own. If they are added to a dish of cells, they will produce nothing but an indistinct, nonspecific haze that is no use to anyone. In order to effectively label cells, they must be targeted to a specific cell protein or structure that may be more or less specific depending upon the goals of the experiment. As a result, these dyes are often sold as *conjugates* to useful sequences or proteins, or as *reactive dyes* that the experimenter can use to target a molecule of choice.

One of the most long-standing applications of fluorescent dyes is to the technique of immunofluorescence, where a dye-conjugated antibody binds a cell target, showing the distribution of that antigen. *Indirect immunofluorescence* uses two sets of antibodies: a primary antibody against the antigen of interest, and a secondary dye-coupled antibody that recognizes the primary (**Figure 7.29a**). The secondary antibodies are according to species: that is, if the primary is made in mouse, the secondary may be made in goat and designated "goat antimouse." This minimizes the number of fluorescent antibodies that need to be made, ensuring that a commercial fluorescent secondary is always available for any application. Immunofluorescence is nearly always performed on cells that have been fixed and permeabilized with detergents to permit entry of the antibodies. It is especially useful for determining the distribution of a specific antigen (protein or other) within a cell or tissue, and for confirming the expression of a transfected protein (see **Practical Tips 7.5** for a basic indirect immunofluorescence protocol).

**Figure 7.29 Schematic of immunofluorescence.** (a) Indirect. A primary, unlabeled antibody targets an antigen in or on a cell. A dye-labeled secondary antibody targets the primary; more than one secondary may bind, increasing the signal. (b) Direct. A dye-labeled primary antibody targets a cellular antigen.

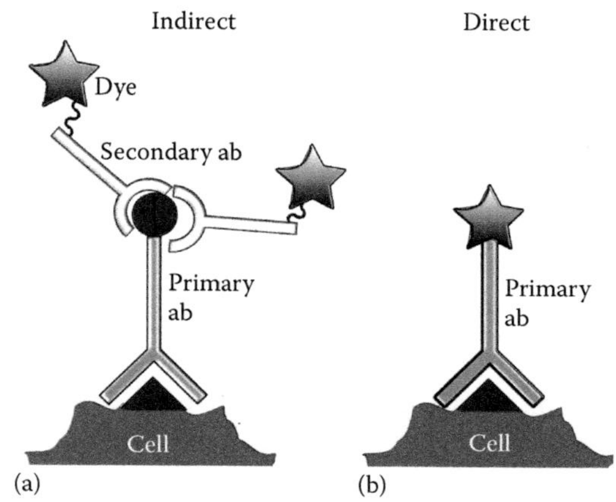

### PRACTICAL TIPS 7.5:   BASIC INDIRECT IMMUNOFLUORESCENCE PROTOCOL FOR INTERNAL ANTIGEN

*General notes.* In all antibody-labeling experiments, negative controls are very important to ensure that any binding seen is truly specific. The antibody concentration usually has to be adjusted carefully in order to obtain good labeling with minimal nonspecific binding; it may need to be varied over a wide range to optimize it. Because antibodies are expensive, labeling is usually done in very small volumes of liquid. The technique of drawing a circle around the specimen with a wax pencil is designed to minimize waste.

### REAGENTS AND SUPPLIES

Cells expressing antigen intracellularly
Primary and secondary antibodies
Optional: DAPI for staining nuclei; mounting medium (we recommend SlowFade Gold from Invitrogen)
Fixation solution
  - 4% paraformaldehyde in PBS, also 0.1 M Tris
Permeabilization solution
  - 0.75 mL (0.5%) Triton-X 100, 0.75 g (0.5%) saporin (optional), 150 mL PBS
Blocking solution
  - 10% FBS, 0.3% Triton-X in PBS
Wash solution
  - 2% FBS, 0.3% Triton-X in PBS

### PROTOCOL

1. Aspirate the culture medium off the cells.
2. Rinse once with PBS.
3. Add fixation solution to cover the cells and incubate for 10 min at room temperature.
4. Remove fixative and incubate in 0.1 M Tris for 5 min.
5. Wash thoroughly in PBS. A good way to do this for cells in dishes is to add PBS and place on a rocker for 3–5 min.
6. Wash three times for 5 min each in permeabilization solution.
7. Wash once for 3 min in PBS.
8. Add enough block solution to cover the cells. Incubate for 1 h at 37°C.
9. Dilute the primary antibody according to the manufacturer's instructions. Keep it on ice. You may wish to make several dilutions.
10. Remove the block solution and add primary antibody to the cells. For a glass-bottom dish, ~30–50 μL should be sufficient to cover the cells. For cells on slides, draw a circle around the area of interest with a wax pencil and place a minimal volume of antibody solution in the circle. In all cases, handle very gently so that the drop does not roll off the area of interest. Place on wet paper towels to minimize evaporation and incubate for 1–1.5 h at 37°C.
11. Prewarm the wash solution to 45°C in a water bath.
12. Wash the cells in warm wash solution three times for 5 min each. The rocking can be done at room temperature.
13. Dilute the secondary antibody in wash solution.
14. Add the minimal volume of secondary to the cells and incubate for 1 h at 37°C.
15. Wash the cells three times for 5 min each in warmed wash solution.

*(Continued)*

## PRACTICAL TIPS 7.5 (CONTINUED):   BASIC INDIRECT IMMUNOFLUORESCENCE PROTOCOL FOR INTERNAL ANTIGEN

16. If DAPI staining is desired, add DAPI at 0.1 µg/mL for 2 min.
17. Wash with PBS.
18. Image immediately, or mount using mounting medium. Slides may be stored at 4°C after mounting.

### SAMPLES TO PREPARE

- The test sample: antigen-expressing cells with both antibodies.
- Negative control: cells that do not express the antigen, labeled with both antibodies. This one is essential.
- Another negative control: add only buffer in the place of primary antibody.

### SUGGESTED READING

Jolly, C., Mongelard, F., Robert-Nicoud, M., and Vourc'h, C. (1997). Optimization of nuclear transcript detection by FISH and combination with fluorescence immunocytochemical detection of transcription factors. *Journal of Histochemistry and Cytochemistry* 45, 1585–1592.

In *direct immunofluorescence*, the primary antibody is directly labeled with a fluorophore (**Figure 7.29b**). Although this requires a specific antibody, it requires fewer labeling steps and may even be used in live cells if the target antigen is on the cell surface; this has been used for single-molecule-tracking experiments using cell membrane receptors. Fluorescent antibodies are available from a variety of suppliers, many of which specialize in this application, or an unlabeled antibody may be conjugated to a reactive dye. Custom antibodies are also available, though they are costly and must be carefully characterized for specificity upon receipt.

Antibodies are very large molecules and not suitable for all applications, especially in living specimens. Other conjugated dyes are available that are designed to be taken up by living cells through receptor-mediated endocytosis or to bind to the surface of living cells. Examples of molecules that are readily available as fluorescent conjugates are as follows: *lectins*, which bind surface carbohydrates (glycoproteins or glycolipids), sometimes with species or cell-type specificity; *annexin V*, which binds to externalized phosphatidylserine, indicating the early stages of cell death; *transferrin*, an iron-binding protein that is readily endocytosed; and various peptides and proteins that are taken up by cells or that pass through the cell membrane. Other conjugates are available that work best with fixed and permeabilized cells; one example is *phalloidin*, which is a mushroom toxin that targets the cytoskeleton. Labeling cells with fluorescent phalloidin provides an image of the actin structure of the cell.

Quite often in molecular biophysics there is no exact fluorescent conjugate that targets the structure or protein of choice. In this case, you can make your own. All of the common dyes are available as conjugates to avidin or streptavidin (see **Chapter 1**) or as reactive intermediates that bind to a specific functional group, for example, a thiol, primary amine, carboxylate, or aldehyde. Streptavidin may be conjugated to anything with a biotin, and kits are available to bind biotin to any protein or small molecule with a reactive group. A very useful class of reactive termini are the amine-reactive molecules such as the succidimydyl esters, which react with primary amines to form amide bonds. The most commonly used is sulfo-*N*-hydroxysulfosuccinimide (*sulfo-NHS*; **Figure 7.30a**). Amine-reactive probes will react with any protein, as multiple amino acid residues within all proteins contain these groups. Thiol-reactive probes will react with cysteine residues, either native or engineered as discussed in **Chapter 2**, as well as other thiols that may be chemically or genetically engineered onto the molecule of interest (**Figure 7.30b**). Most of these labeling reactions are best done with a purified dye and target molecule in a test tube before applying to cells. However, both streptavidin dyes and thiol-reactive dyes may be used to label proteins that are already expressed in cells, provided that these proteins are either biotinylated or contain an easily accessible cysteine residue. The latter can be tricky, as other native proteins may also possess extracellular cysteines. We will return to this subject in **Chapters 11** through **13** when we discuss cell labeling using nanoparticles.

Genetic engineering of proteins can be used to insert a variety of specific tags for dye binding. Some of these include 6-histidine (6His) tags, FLAG tags, and many others. Dyes for specific binding to the most common tags are commercially available, though sometimes difficult to find. A full discussion is outside the scope

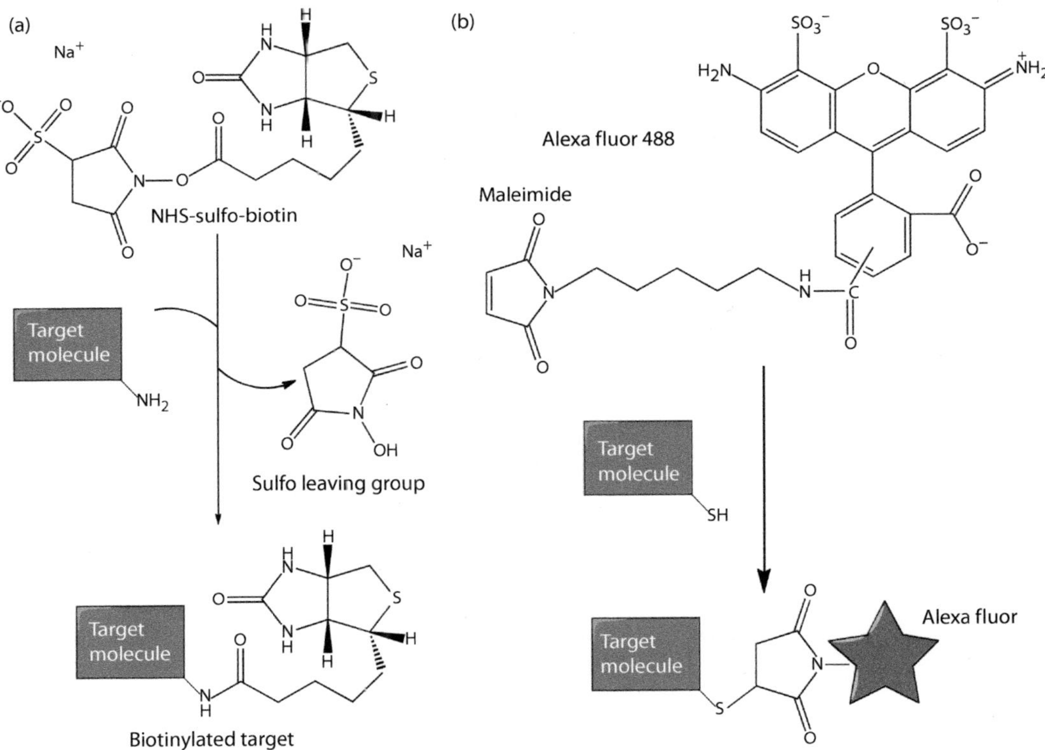


**Figure 7.30 Examples of conjugation strategies with biotin or dyes.** (a) Amine-reactive probes such as sulfo-NHS create an amide bond with primary amino groups; the sulfo group detaches and can be removed by gel filtration. All proteins have primary amines, as do many small molecules. The example shown is biotinylation of a target. (b) Example of a thiol-reactive dye, Alexa fluor 488-maleimide, reacting with a target containing a reactive thiol. Some proteins contain reactive thiols, and many molecules can be engineered to bear a terminal thiol for conjugation purposes. Maleimides create no leaving groups.

of this work, but some references are given at the end of the chapter to some of the techniques.

## Organelle probes

Fluorophores targeted at specific intracellular organelles, such as the mitochondria, lysosomes, Golgi apparatus, and endoplasmic reticulum, are useful for monitoring a variety of biological processes in living cells. In general, organelle probes consist of a fluorochrome nucleus attached to a target-specific moiety that assists in localizing the fluorophore through covalent, electrostatic, hydrophobic, or similar types of bonds. Many of the fluorescent probes designed for selecting organelles are able to permeate the cell membrane, and thus may be delivered to living cells with no special techniques. Organelle probes are useful for investigating transport, respiration, mitosis, apoptosis, protein degradation, acidic compartments, and membrane phenomena in live cells over time courses from milliseconds to days. In many cases, living cells that have been labeled with these probes can subsequently be fixed and counterstained with additional fluorophores in multicolor labeling experiments.

Mitochondrial probes are among the most useful fluorophores for investigating cellular respiration and are often employed along with other dyes in multiple labeling investigations. Specific fluorophores developed by Molecular Probes include the popular MitoTracker and MitoFluor series of structurally diverse xanthene, benzoxazole, indole, and benzimidazole heterocycles that are available in a variety of excitation and emission spectral profiles. The mechanism of action varies for each of the probes in this series, ranging from covalent attachment to oxidation within respiring mitochondrial membranes. MitoTracker dyes are retained quite well after cell fixation in formaldehyde and can often withstand lipophilic permeabilizing agents. In contrast, the MitoFluor probes are designed specifically for actively respiring cells and are not suitable for fixation and counterstaining procedures. Another popular mitochondrial probe, entitled JC-1, is useful as an indicator of membrane potential and in multiple staining experiments with fixed cells. This carbocyanine dye exhibits green fluorescence at low concentrations, but can undergo intramolecular association within active mitochondria to produce a shift in emission to longer (red) wavelengths. The change in emission wavelength is useful in determining the ratio of active to nonactive mitochondria in living cells. Mitochondrial probes are also good for identifying proteins that localize to mitochondria and in identifying oxidative damage due to reactive oxygen species, which causes mitochondrial rounding (**Figure 7.31**).

Labeling lysosomes is generally not difficult, because anything that is endocytosed will eventually end up there, and the low pH environment often causes dyes to shift fluorescence and/or to accumulate once they are protonated and can no longer cross the membrane. Traditional lysosomal probes include the nonspecific phenazine and acridine derivatives neutral red and acridine orange. Fluorescently labeled latex beads and macromolecules, such as dextran, can also be accumulated in lysosomes by endocytosis for a variety of experiments. However, sometimes it is important to find a dye that is essentially nonfluorescent until it is inside lysosomes. The LysoTracker and LysoSensor dyes developed by Molecular Probes contain heterocyclic and aliphatic nitrogen moieties that modulate transport of the dyes into the lysosomes of living cells for both

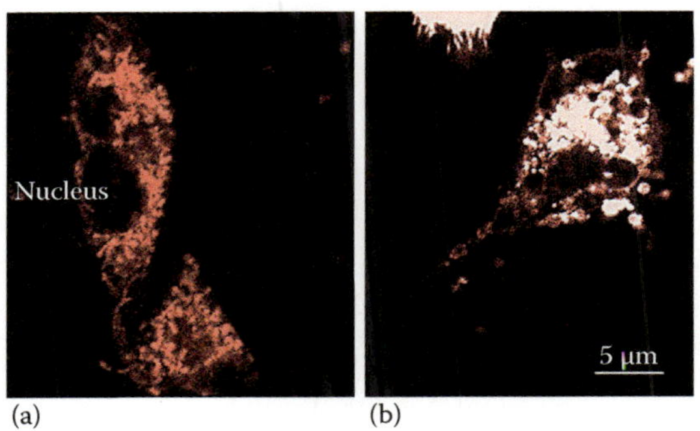


**Figure 7.31 Mitochondrial labeling.** (a) 3-D reconstruction of an A9 cell with normal mitochondria. (b) After application of an oxidative insult, the mitochondria appear round.

short-term and long-term studies. Fluorescence intensity dramatically increases in these dyes upon protonation, making them useful as pH indicators.

Proteins and lipids are sorted and processed in the Golgi apparatus, which is typically stained with fluorescent derivatives of ceramides and sphingolipids. These agents are highly lipophilic and are therefore useful as markers for the study of lipid transport and metabolism in live cells. Several of the most useful fluorophores for Golgi apparatus contain the complex heterocyclic BODIPY nucleus developed by Molecular Probes. When coupled to sphingolipids, the BODIPY fluorophore is highly selective and exhibits a tolerance for photobleaching that is far superior to many other dyes. In addition, the emission spectrum is dependent upon concentration (shifting from green to red at higher concentrations), making the probes useful for locating and identifying intracellular structures that accumulate large quantities of lipids. During live-cell experiments, fluorescent lipid probes can undergo metabolism to derivatives that may bind to other subcellular features, a factor that can often complicate the analysis of experimental data.

The most popular traditional probes for endoplasmic reticulum fluorescence analysis are the carbocyanine and xanthene dyes, DiOC(6) and several rhodamine derivatives, respectively. These dyes must be used with caution, however, because they can also accumulate in the mitochondria, Golgi apparatus, and other intracellular lipophilic regions. Newer, more photostable probes have been developed for selective staining of the endoplasmic reticulum by several manufacturers. In particular, oxazole members of the Dapoxyl family produced by Molecular Probes are excellent agents for selective labeling of the endoplasmic reticulum in living cells, either alone or in combination with other dyes. These probes are retained after fixation with formaldehyde, but can be lost with permeabilizing detergents. Another useful probe is Brefeldin A, a stereochemically complex fungal metabolite that serves as an inhibitor of protein trafficking out of the endoplasmic reticulum.

## Environmental probes

Fluorophores designed to probe the internal environment of living cells have been widely examined by a number of investigators, and many hundreds have been developed to monitor such effects as localized concentrations of alkali and alkaline earth metals, heavy metals (employed biochemically as enzyme cofactors), inorganic ions, thiols and sulfides, nitrite, as well as pH, solvent polarity, and membrane potential. With the possible exception of the pH probes, none of these dyes are easy to use. They require careful calibration, special filters, and often *ratiometric imaging*—that is, the ability to measure the emission (or absorbance) at one specific wavelength relative to that of a second wavelength. A *ratiometric dye* is one that shifts its emission (or absorbance) in response to the environmental situation in question. Ratiometric imaging requires the ability to rapidly change filters, and to adjust for photobleaching in the final data, but is much more sensitive and quantitative than the use of dyes that simply change intensity without spectral shifts.

Divalent calcium ($Ca^{2+}$) is a metabolically important ion that plays a vital role in cellular response to many forms of external stimuli. Because transient fluctuations in $Ca^{2+}$ concentration are typically involved when cells undergo a response, fluorophores not only must be designed to measure localized concentrations within segregated compartments but should also produce quantitative changes when flux density waves progress throughout the entire cytoplasm. Two of the most common $Ca^{2+}$ probes are the ratiometric indicators fura-2 and indo-1, but these fluorophores are not particularly useful in confocal microscopy because they are excited in the UV. Fura-2 requires a dual *excitation* filter, and its emission is usually measured at a constant value around 510 nm. As $Ca^{2+}$ levels rise, emission goes up when the dye is excited at 340 nm, and down when it is excited at 380 nm (**Figure 7.32**).

Fluorophores that respond in the visible range to $Ca^{2+}$ fluxes are, unfortunately, not ratiometric indicators, instead exhibiting an increase or decrease in fluorescence intensity upon ion binding. The best example is fluo-3, a complex xanthene derivative, which undergoes a dramatic increase in fluorescence emission at 525 nm when excited by the 488 nm laser line.

To overcome the problems associated with using dyes that do not shift wavelength, several dyes are often used together for calcium measurements in confocal microscopy. Fura red, a multinuclear imidazole and benzofuran heterocycle, exhibits a decrease in fluorescence at 650 nm when binding $Ca^{2+}$. A ratiometric response to $Ca^{2+}$ fluxes can be obtained when a mixture of fluo-3 and fura red is excited at 488 nm and fluorescence is measured at 525 and 650 nm. Because the emission intensity of fluo-3 increases monotonically while that of fura

**Figure 7.32 Absorbance spectra of fura-2 with emission at 510 nm in the presence of different concentrations of calcium ions.**

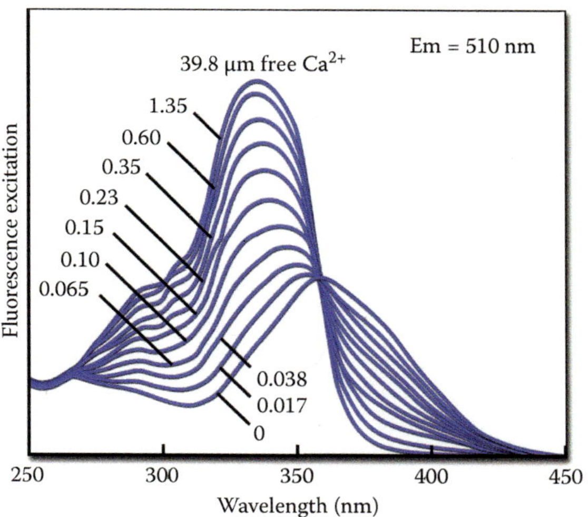

red simultaneously decreases, an isosbestic point is obtained when the dye concentrations are constant within the localized area being investigated. Another benefit of using these probes together is the ability to measure fluorescence intensity fluctuations with a standard FITC/Texas red interference filter combination.

Quantitative measurements of ions other than calcium, such as magnesium, sodium, potassium, and zinc, are conducted in an analogous manner using similar fluorophores. One of the most popular probes for $Mg^{2+}$, mag-fura-2 (structurally similar to fura red), is also excited in the UV and presents the same problems in confocal microscopy as fura-2 and indo-1. Fluorophores excited in the visible light region are becoming available for the analysis of many monovalent and divalent cations that exist at varying concentrations in the cellular matrix. Several synthetic organic probes have also been developed for monitoring the concentration of simple and complex anions.

Important fluorescence monitors for intracellular pH include a pyrene derivative known as 8-hydroxypyrene-1,3,6-trisulfonic acid, trisodium salt (HPTS) or pyranine; the fluorescein derivative (2-carboxy ethyl)-5-(and 6)-carboxy fluorescein (BCECF); and another substituted xanthene termed carboxy semi-naphtorhodafluor (SNARF)-1. Because many common fluorophores are sensitive to pH in the surrounding medium, changes in fluorescence intensity that are often attributed to biological interactions may actually occur as a result of protonation. In the physiological pH range (pH 6.8 to 7.4), the probes mentioned above are useful for dual-wavelength ratiometric measurements and differ only in dye loading parameters. Simultaneous measurements of calcium ion concentration and pH can often be accomplished by combining a pH indicator, such as SNARF-1, with a calcium ion indicator (for example, fura-2).

The *voltage-sensitive dyes* for sensing the membrane potential of excitable cells such as neurons include the ANEPPS class of dyes and the RH dyes. All of these probes show a broad red-orange fluorescence with a large Stokes shift, and most show a shift in absorbance (rather than emission) with membrane potential. These dyes are a story unto themselves and will be covered more fully under the discussion of Optical Recording in **Chapter 15**.

## 7.7 FLUORESCENT PROTEINS

Over the past few years, the discovery and development of naturally occurring fluorescent proteins and mutated derivatives have allowed the investigation of a wide spectrum of intracellular processes. A variety of marine organisms have been the source of more than 100 fluorescent proteins and their analogs, which arm the investigator with a balanced palette of noninvasive biological probes for single, dual, and multispectral fluorescence analysis. Advantages of fluorescent proteins over dyes include the following:

- Fluorescent proteins are genetically encoded, so they may be delivered to selected cells in a dish by microinjection or to regions in animals by injection, viruses, or transgenesis. Stable transfection techniques can lead to uniformly fluorescent populations. Cell- or temporally selective promoters can produce cells or organisms with highly selected fluorescent expression (**Figure 7.33**).

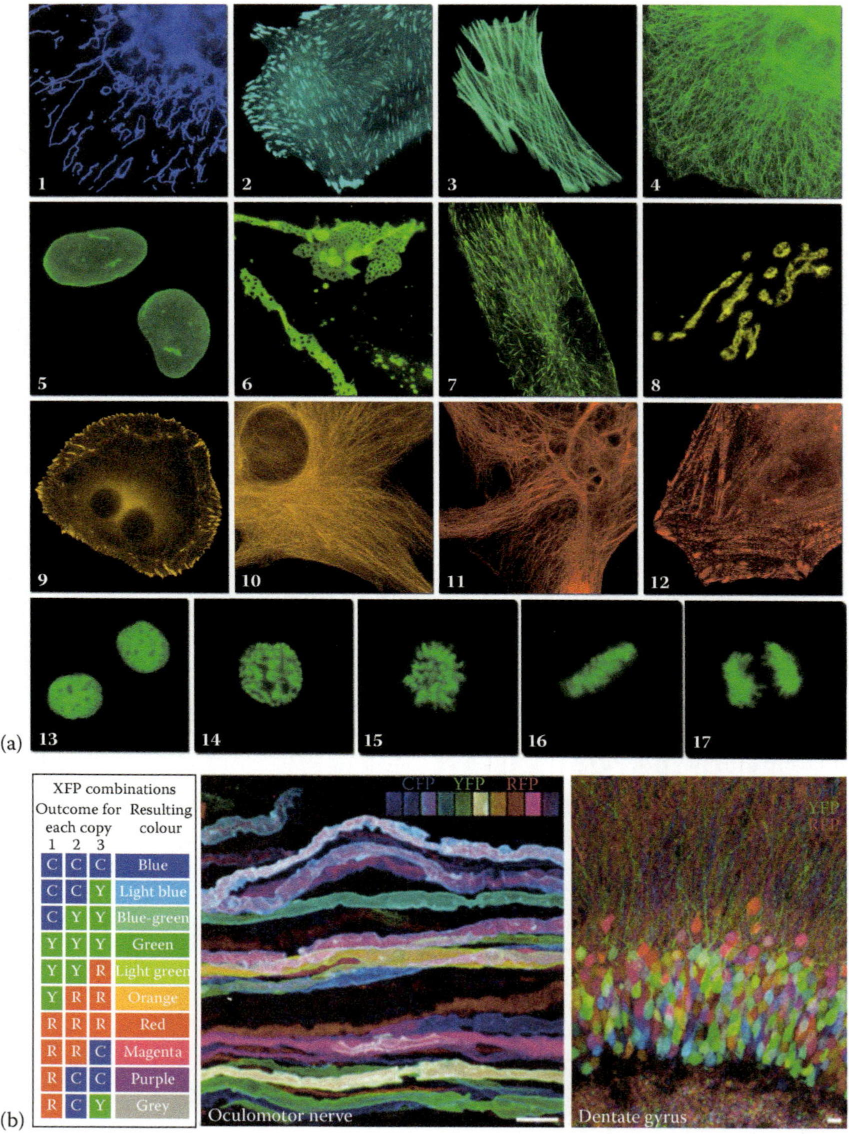



**Figure 7.33 Fluorescent proteins.** (a) Fusions of fluorescent proteins to structure-specific targeting sequences or proteins showing subcellular localization result in targeted labeling of one or more colors. (1) EBFP2-mito-N-7 (human cytochrome C oxidase subunit VIII; mitochondria); (2) mCerulean-paxillin-N-22 (chicken; focal adhesions); (3) mTFP1-actin-C-7 (human β-actin; filamentous actin); (4) mEmerald-keratin-N-17 (human cytokeratin 18; intermediate filaments); (5) sfGFP-lamin B1-C-10 (human lamin B1; nuclear envelope); (6) mVenus-Cx43-N-7 (rat α-1 connexin-43; gap junctions); (7) YPet-EB3-N-7 (human microtubule-associated protein; RP/EB family); (8) mKO-Golgi-N-7 (N-terminal 81 amino acids of human β-1,4-galactosyltransferase; Golgi complex); (9) tdTomato-zyxin-N-7 (human zyxin; focal adhesions); (10) TagRFP-tubulin-C-6 (human α-tubulin; microtubules); (11) mCherry-vimentin-N-7 (human vimentin; intermediate filaments); (12) mPlum-α-actinin-N-19 (human non-muscle; cytoskeleton). 13–17: Fusion of mEGFP with human histone H2B (mEGFP-H2B-N-6). (13) interphase; (14) prophase; (15) prometaphase; (16) metaphase; (17) anaphase. (b) Use of combinatorial expression of three different colors can lead to 10 distinct color combinations. In this "brainbow" mouse, cyan, yellow, and red fluorescent proteins were used along with tissue-specific promoters to map neuronal circuitry. Sections of the oculomotor nerve and dentate gyrus shown here illustrate the varying levels of expression of the fluorescent proteins. (Reprinted by permission from Macmillan Publishers Ltd: *Nature*, J. Livet et al., Transgenic strategies for combinatorial expression of fluorescent proteins in the nervous system, 450, 56–62, copyright 2007.)

- Subcellular localization is easily achieved with organelle targeting sequences that are well understood and do not significantly increase the size of the protein.

- There are no issues with delivering fluorescent proteins to cells, since they are fabricated within the cells themselves.

- Phototoxicity is very low.

- Genetic fusions are readily made by cloning a fluorescent protein sequence onto any other gene. The addition of a fluorescent tag does not affect the function of most membrane proteins.

Disadvantages are mainly related to the limited selection of available proteins:

- Emission spectra are relatively broad.

- Proteins cannot be rationally designed, but must be altered using mutagenesis. This is time-consuming and uncertain.

- Mutants of a given protein are often difficult to distinguish when used together because of spectral overlap.

- Making environmentally sensitive fluorescent proteins is hit-or-miss.

- Most fluorescent proteins form dimers, which can cause artifacts.

- Fusions to soluble proteins often change protein function.

- The proteins are physically large compared to fluorescent dyes, usually in a type of protein structure called a β-*barrel*. Removal of the barrel by genetic engineering eliminates fluorescence.

Fluorescent protein technology is continually improving. The first fluorescent protein discovered was green fluorescent protein (GFP) from the North Atlantic jellyfish *Aequorea victoria*. Early versions of GFP were highly temperature sensitive, excited in the UV, and not mammalian codon optimized. After many rounds of mutagenesis, the still-popular enhanced GFP (EGFP) was developed. Further mutation studies led to blue-shifted variants, cyan fluorescent protein (CFP) and blue fluorescent protein (BFP). CFP is still in wide use and can be excited by several common lines ranging from 405 to 457 nm. BFP has all but disappeared because of its weak fluorescence and light sensitivity.

Another popular fluorescent protein derivative, the yellow fluorescent protein (YFP), was designed on the basis of the GFP crystalline structural analysis to red-shift the absorption and emission spectra. YFP is optimally excited by the 514 nm argon-ion line and provides more intense emission than EGFP, but is more sensitive to low pH and high halogen ion concentrations. With careful use of filters, CFP and YFP may be used together. Further red-shifting of GFP variants has not been achieved, but other proteins have been discovered in the red region, such as the coral protein dsRed. The spectra of the GFP variants and dsRed are shown in **Figure 7.34**.

The "fruit" fluorescent proteins are a new series of mostly monomeric proteins that span the visible spectrum from mCerulean (cyan) through mPlum (far red),

**Figure 7.34 Excitation and emission spectra of four GFP variants (BFP, CFP, GFP, YFP) and dsRed.**

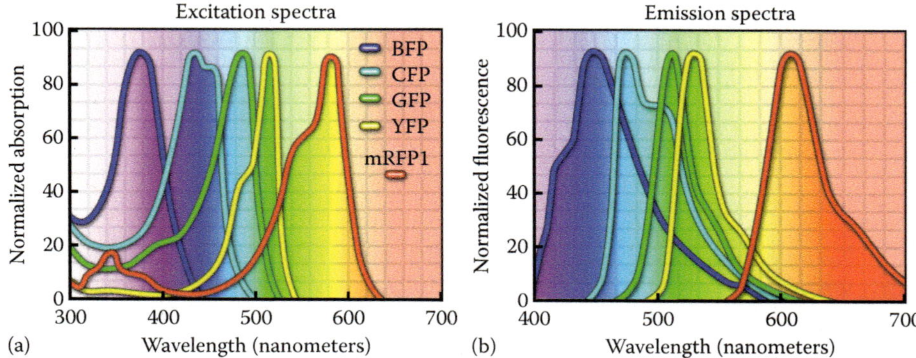



encompassing mHoneydew, mBanana, mCherry, and tdTomato (the "m" stands for monomeric, and "td" for "tandem dimer," a stabilized dimer). These are all now available commercially with or without organelle-localizing tags, and they serve to fill the spectral gaps in the GFP and dsRed families. Some are much brighter than others; for example, tdTomato has a $Q$ nearly sevenfold that of mPlum. The physical properties of the most common commercially available fluorescent proteins are given in Table 7.3.

The physics of fluorescent proteins is still not fully understood, but a few studies in recent years have explored the relationship between the chromophore environment and the emission intensity and wavelength. The extent of p-orbital conjugation contained within the chromophore plays a large role in determining the general

**Table 7.3**

Photophysical Properties of Commonly Used Fluorescent Proteins

| Protein | Color | Ex/Em Maxima | QY | $\varepsilon$ (M$^{-1}$ cm$^{-1}$) |
|---|---|---|---|---|
| EBFP | Blue | 383/445 | 0.31 | 29,000 |
| ECFP | Cyan | 439/476 | 0.15 | 20,000 |
| EGFP | Green | 484/510 | 0.70 | 23,000 |
| EYFP | Yellow | 512/529 | 0.54 | 45,000 |
| Cerulean | Cyan | 433/475 | 0.62 | 43,000 |
| AcGFP1 | Green | 475/505 | 0.82 | 32,500 |
| Superfolder GFP | Green | 485/510 | 0.65 | 83,300 |
| Emerald | Green | 487/509 | 0.68 | 57,500 |
| mWasabi | Green | 493/509 | 0.80 | 70,000 |
| mBanana | Yellow | 540/553 | 0.70 | 6000 |
| Venus | Yellow | 515/528 | 0.57 | 92,200 |
| ZsYellow1 | Yellow | 529/539 | 0.42 | 20,200 |
| mOrange | Orange | 549/565 | 0.60 | 58,000 |
| mTangerine | Orange | 568/585 | 0.30 | 38,000 |
| dsRed-monomer | Red | 557/592 | 0.14 | 27,300 |
| td Tomato | Red | 554/581 | 0.69 | 138,000 |
| mRFP1 | Red | 584/607 | 0.25 | 50,000 |
| mPlum | Far red | 590/649 | 0.10 | 41,000 |

*Note:* Ex/em = excitation/emission; QY = quantum yield. These are only examples—for more, see manufacturers' catalogs and the recent literature.

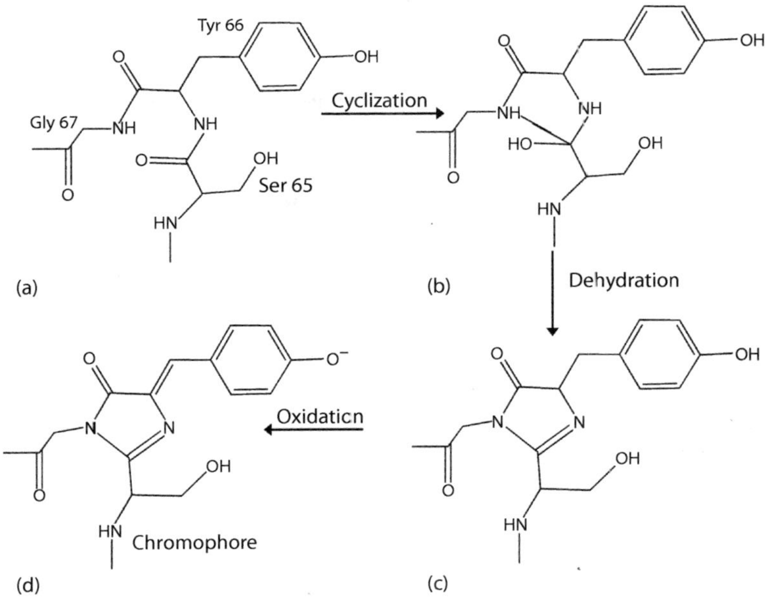

**Figure 7.35 Schematic of chromophore maturation in GFP.** (a) Prematuration EGFP chromophore is a tripeptide amino acid sequence (Ser65–Tyr66–Gly67). (b) First step in maturation is a series of torsional adjustments that relocate the carboxyl carbon of Ser65 so that it is in close proximity to the amino nitrogen of Gly67, allowing for nucleophilic attack on the amide nitrogen of glycine. (c) Dehydration results in formation of heterocyclic ring system. (d) Fluorescence occurs when oxidation of the tyrosine $\alpha$-$\beta$ carbon bond by molecular oxygen extends electron conjugation of the imidazoline ring system to include the tyrosine phenyl ring and its oxygen substituent.

spectral class of a fluorescent protein, such as green or red. Local environmental factors such as charged amino acid residues, H-bonding, and other interactions with the protein can also cause spectral shifts or intensity changes.

GFP is the protein about which the most is known. The fluorophore is made from a tripeptide Ser-Tyr-Gly at positions 65, 66, and 67 in the protein. This is a common motif in proteins but is usually not fluorescent; in GFP, it undergoes a "maturation" process involving bond rearrangement and the formation of a highly conjugated $\pi$-electron resonance system (**Figure 7.35**). One of the steps involves oxygenation of the Tyr, meaning that GFP variants cannot be used under low oxygen conditions (e.g., in anaerobic bacteria). This process also takes several hours, so that the appearance of GFP does not coincide with gene expression, a factor that can affect some experiments.

The ability to alter the photophysics of fluorescent proteins promises to yield a new generation of interactive proteins that signal cellular processes or perform other functions. Fluorescent proteins sensitive to pH and redox potential are already available. Another variant, KillerRed, is able to transfer electrons to molecular oxygen, resulting in reactive oxygen species and cell death. As these proteins become better understood, rational design of desired properties will become possible.

## 7.8  SUMMARY AND REMARKS

Brightfield and fluorescence microscopy are mainstays of modern biology. Mastery of the basic concepts is sufficient for a tremendous array of applications. For thin or transparent samples, widefield microscopy is frequently all that is needed. The optical sectioning capabilities of confocal microscopy make it possible to investigate thicker specimens. Both of these techniques are widely available at nearly all campuses. Shared imaging facilities frequently offer training workshops and courses that instruct users on the techniques and on proper use of available instruments. We encourage our readers to take advantage of available resources in order to generate the highest-quality images possible.

Sometimes the basic techniques are not sufficient, especially for ultraresolution, ultrafast processes, or single-molecule imaging. A large number of new emerging techniques have transcended nearly all of the barriers of light microscopy. Chapter 8 aims to introduce a few of these.

## End-of-Chapter Questions

1. You are buying a new microscope for mammalian cell biology. Which objective lenses should you choose (magnification, numerical aperture [NA])? How does this change if you are doing microbiology? What is the maximum NA of an air objective?

2. Discuss the best way to measure the power (in $W/cm^2$) delivered to a sample for fluorescence microscopy. What variables affect this number?

3. If the power delivered to a fluorescence sample is 0.5 $W/cm^2$, calculate the number of fluorescence photons emitted from the sample as a function of fluorophore properties. What is the numerical value if the fluorophore is rhodamine? Of the photons emitted, estimate what percentage can reasonably be collected.

4. Approximately how many photons of excitation are needed per photon emitted in fluorescence microscopy? Justify your answer.

5. Identify a class of dyes that have been tuned to match the Hg lamp spectra, and estimate how much improvement in imaging results from using one of these dyes for fluorescence imaging with an Hg lamp.

6. Explain the relationship between NA and resolution, and between NA and brightness. Why does increasing magnification not always increase resolution?

7. Given an expression for the lateral ($xy$) resolution of a confocal microscope based upon the Rayleigh criterion. In reality, confocal usually does a bit better than that. Why? How can this be measured?

8. The axial ($z$-axis) resolution of a confocal microscope is often expressed as 1.4 $\lambda n/NA^2$, where $n$ is the index of refraction of the mounting/immersion medium. Derive this equation and explain the assumptions made.

9. What are the advantages of confocal over widefield fluorescence microscopy?

10. Why are some objectives marked "UV"? What hardware would you need to image a fluorophore that excites at 250 nm and emits at 400 nm?

11. Some objectives have a rotating "correction collar." What is this for and how should it be used?

12. How would you choose dyes differently for confocal vs. widefield microscopy?

13. Consider the molecule in Figure 7.15. Using the Boltzmann distribution, estimate whether a significant fraction of these molecules can occupy the first vibrational excited state at room temperature.

14. We mentioned rather blithely in the text that tryptophan fluorescence changes could be used to detect protein interactions. Explain how this could work.

15. What is dye self-quenching? Which types of dyes are particularly susceptible to self-quenching and why? How can this be a problem?

16. Rigid molecules are known to be more fluorescent than those that can twist and bend (compare, for example, fluorene and biphenyl). Why?

17. Figure 7.32 shows the changes in fura excitation spectrum with $Ca^{2+}$ ions. Derive an expression for the ratio of intensities (340 nm/380 nm) as a function of $[Ca^{2+}]$.

18. Discuss the pros and cons of the different $Ca^{2+}$ indicators.

19. The quantum yield of phosphorescence is usually $<10^{-6}$. What does this say about the relative frequency of nonradiative pathways? What can you do to improve phosphorescence quantum yields?

20. Design a filter cube to simultaneously image GFP and the far-red protein mPlum.

**21.** Using **Figure 7.34**, discuss the problems of spectral bleed-through you would expect when using CFP and GFP in a dual-labeling confocal experiment. Suggest parameters and approaches to avoid cross-talk. How about CFP and YFP?

**22.** An optical trap may be constructed using a 1064 nm laser that strikes the sample. What problem(s) would occur if this light is sent to a sample through fluorescence filter cubes? What is the remedy?

**23.** Imaging of living cells over periods greater than about an hour presents certain difficulties: change in pH of the medium, temperature control, phototoxicity, and focal drift. Discuss these and other issues if they occur to you, and suggest some remedies.

**24.** Imaging bacteria over long periods is even more difficult than for mammalian cells, since the small size requires a high magnification, low depth of field objective, and the cells swim and reproduce quickly. Suggest some approaches.

## Background Reading


### Books

Chalfie, M., and Kain S. R. (eds.). *Green Fluorescent Protein: Properties, Applications and Protocols.* John Wiley & Sons, Hoboken, NJ, 2006.—All about GFP in different types of biological systems.

Horobin, R., and Kierman, J. (eds.). *Conn's Biological Stains: A Handbook of Dyes, Stains and Fluorochromes for Use in Biology and Medicine.* Edn. 10. Bios Scientific Publishers, Oxford, UK, 2002.—Classic reference on light microscopic stains and fluorescent probes.

Konnerth, A., and Yuste, R. *Imaging in Neuroscience and Development: A Laboratory Manual.* Cold Spring Harbor University Press, Cold Spring Harbor, NY, 2005.—Laboratory manual based upon imaging course at Cold Spring Harbor. Contains many practical protocols for fluorescent probes. Highly recommended for anyone using neurons or doing calcium imaging.

Mason, W.T. (ed.). *Fluorescent and Luminescent Probes for Biological Activity: A Practical Guide to Technology for Quantitative Real-Time Analysis.* Edn. 2. Academic Press, San Diego, CA, 1999.—Introduction to microscopy, then thorough treatment of the different classes of fluorescent probes. Recommended.

Masters, B.R. *Confocal Microscopy and Multiphoton Excitation Microscopy: The Genesis of Live Cell Imaging.* SPIE—The International Society for Optical Engineering, Bellingham, WA, 2005.—The physics and history of light and confocal microscopy.

Murphy, D.B. *Fundamentals of Light Microscopy and Electronic Imaging.* Wiley-Liss, New York, 2001.—Classic introductory textbook on the subject. Highly recommended.

Paddock, S.W. (ed.). *Confocal Microscopy: Methods and Protocols.* Humana Press, Totowa, NJ, 2010.—Introduction to confocal, plus protocols for specific types of biological samples.

Pawley, J. *Handbook of Biological Confocal Microscopy.* Springer, New York, 2006.—Thorough treatment of confocal microscopy, including components, fluorescent probes, cell and tissue techniques, and many advanced techniques.

### Journal articles

*Darkfield, DIC, and phase contrast*

De Waele, M., Renmans, W., Segers, E., Jochmans, K., and Van Camp, B. (1988). Sensitive detection of immunogold-silver staining with darkfield and epi-polarization microscopy. *Journal of Histochemistry and Cytochemistry* 36, 679–683.

Hayden, J.E. (2002). Adventures on the dark side: An introduction to darkfield microscopy. *Biotechniques* 32, 756–761.

Hill, D.B., Macosko, J.C., and Holzwarth, G.M. (2008). Motion-enhanced, differential interference contrast (MEDIC) microscopy of moving vesicles in live cells: VE-DIC updated. *Journal of Microscopy* 231, 433–439.

Hotani, H., and Horio, T. (1988). Dynamics of microtubules visualized by darkfield microscopy: treadmilling and dynamic instability. *Cell Motility and the Cytoskeleton* 10, 229–236.

Hu, M., Novo, C., Funston, A., Wang, H., Staleva, H., Zou, S., Mulvaney, P., Xia, Y., and Hartland, G.V. (2008). Dark-field microscopy studies of single metal nanoparticles: Understanding the factors that influence the linewidth of the localized surface plasmon resonance. *Journal of Materials Chemistry* 18, 1949–1960.

Lawrence, J.R., Korbera, D.R., and Caldwell, D.E. (1989). Computer-enhanced darkfield microscopy for the quantitative analysis of bacterial growth and behavior on surfaces. *Journal of Microbiological Methods* 10, 123–138.

Reynolds, F.W., and Hesbacher, E.N. (1950). Darkfield microscopy: Some principles and applications. *Journal of Venereal Disease Information* 31, 17–23.

Salmon, E.D., and Tran, P. (2007). High-resolution video-enhanced differential interference contrast light microscopy. *Methods in Cell Biology* 81, 335–364.

Salmon, T., Walker, R.A., and Pryer, N.K. (1989). Video-enhanced differential interference contrast light microscopy. *Biotechniques* 7, 624–633.

Sokabe, M., and Sachs, F. (1990). The structure and dynamics of patch-clamped membranes: A study using differential interference contrast light microscopy. *Journal of Cell Biology* 111, 599–606.

Witlin, B. (1945). Darkfield illuminators in microscopy. *Science* 102, 41–42.

Xiao, L., Qiao, Y., He, Y., and Yeung, E.S. (2010). Three dimensional orientational imaging of nanoparticles with darkfield microscopy. *Analytical Chemistry* 82, 5268–5274.

Yenjerla, M., Lopus, M., and Wilson, L. (2010). Analysis of dynamic instability of steady-state microtubules in vitro by video-enhanced differential interference contrast microscopy with an appendix by Emin Oroudjev. *Methods in Cell Biology* 95, 189–206.

Zernike, F. (1942). Phase contrast, a new method for the microscopic observation of transparent objects. *Physica* 9, 974–986.

## Fluorescence microscopy and dyes

Beatty, K.E., Xie, F., Wang, Q., and Tirrell, D.A. (2005). Selective dye-labeling of newly synthesized proteins in bacterial cells. *Journal of the American Chemical Society* 127, 14150–14151.

Beer, P.D., and Gale, P.A. (2001). Anion recognition and sensing: The state of the art and future perspectives. *Angewandte Chemie—International Edition in English* 40, 486–516.

Brinkley, M. (1992). A brief survey of methods for preparing protein conjugates with dyes, haptens, and cross-linking reagents. *Bioconjugate Chemistry* 3, 2–13.

Chakraborty, A., Wang, D., Ebright, Y.W., and Ebright, R.H. (2010). Azide-specific labeling of biomolecules by Staudinger–Bertozzi ligation phosphine derivatives of fluorescent probes suitable for single-molecule fluorescence spectroscopy. *Methods in Enzymology* 472, 19–30.

de Silva, A.P., Eilers, J., and Zlokarnik, G. (1999). Emerging fluorescence sensing technologies: From photophysical principles to cellular applications. *Proceedings of the National Academy of Sciences of the United States of America* 96, 8336–8337.

de Silva, A.P., Gunaratne, H.Q., Gunnlaugsson, T., Huxley, A.J., McCoy, C.P., Rademacher, J.T., and Rice, T.E. (1997). Signaling recognition events with fluorescent sensors and switches. *Chemical Reviews* 97, 1515–1566.

Ernst, L.A., Gupta, R.K., Mujumdar, R.B., and Waggoner, A.S. (1989). Cyanine dye labeling reagents for sulfhydryl groups. *Cytometry* 10, 3–10.

Fritzsche, M., and Mandenius, C.F. (2010). Fluorescent cell-based sensing approaches for toxicity testing. *Analytical and Bioanalytical Chemistry*.

Griffin, B.A., Adams, S.R., Jones, J., and Tsien, R.Y. (2000). Fluorescent labeling of recombinant proteins in living cells with FlAsH. *Methods in Enzymology* 327, 565–578.

Griffin, B.A., Adams, S.R., and Tsien, R.Y. (1998). Specific covalent labeling of recombinant protein molecules inside live cells. *Science* 281, 269–272.

Hirano, T., Kikuchi, K., Urano, Y., Higuchi, T., and Nagano, T. (2000). Novel zinc fluorescent probes excitable with visible light for biological applications. *Angewandte Chemie—International Edition in English* 39, 1052–1054.

Lin, M.Z., and Wang, L. (2008). Selective labeling of proteins with chemical probes in living cells. *Physiology (Bethesda)* 23, 131–141.

Loudet, A., and Burgess, K. (2007). BODIPY dyes and their derivatives: Syntheses and spectroscopic properties. *Chemical Reviews* 107, 4891–4932.

Miller, E.W., and Chang, C.J. (2007). Fluorescent probes for nitric oxide and hydrogen peroxide in cell signaling. *Current Opinion in Chemical Biology* 11, 620–625.

Minta, A., and Tsien, R.Y. (1989). Fluorescent indicators for cytosolic sodium. *Journal of Biological Chemistry* 264, 19449–19457.

Mujumdar, R.B., Ernst, L.A., Mujumdar, S.R., Lewis, C.J., and Waggoner, A.S. (1993). Cyanine dye labeling reagents: Sulfoindocyanine succinimidyl esters. *Bioconjugate Chemistry* 4, 105–111.

Padmawar, P., Yao, X., Bloch, O., Manley, G.T., and Verkman, A.S. (2005). K+ waves in brain cortex visualized using a long-wavelength K+-sensing fluorescent indicator. *Nature Methods* 2, 825–827.

Panchuk-Voloshina, N., Haugland, R.P., Bishop-Stewart, J., Bhalgat, M.K., Millard, P.J., Mao, F., and Leung, W.Y. (1999). Alexa dyes, a series of new fluorescent dyes that yield exceptionally bright, photostable conjugates. *Journal of Histochemistry and Cytochemistry* 47, 1179–1188.

Stephens, D.J., and Allan, V.J. (2003). Light microscopy techniques for live cell imaging. *Science* 300, 82–86.

Taguchi, Y., Shi, Z.D., Ruddy, B., Dorward, D.W., Greene, L., and Baron, G.S. (2009). Specific biarsenical labeling of cell surface proteins allows fluorescent- and biotin-tagging of amyloid precursor protein and prion proteins. *Molecular Biology of the Cell* 20, 233–244.

Takahashi, A., Camacho, P., Lechleiter, J.D., and Herman, B. (1999). Measurement of intracellular calcium. *Physiological Reviews* 79, 1089–1125.

Tsien, R.Y. (1989). Fluorescent probes of cell signaling. *Annual Review of Neuroscience* 12, 227–253.

Urano, Y., Asanuma, D., Hama, Y., Koyama, Y., Barrett, T., Kamiya, M., Nagano, T., Watanabe, T., Hasegawa, A., Choyke, P.L. et al. (2009). Selective molecular imaging of viable cancer cells with pH-activatable fluorescence probes. *Nature Medicine* 15, 104–109.

Valeur, B., and Leray, I. (2000). Design principles of fluorescent molecular sensors for cation recognition. *Coordination Chemistry Reviews* 205, 3–40.

Weiss, S. (1999). Fluorescence spectroscopy of single biomolecules. *Science* 283, 1676–1683.

White, J.G., Amos, W.B., and Fordham, M. (1987). An evaluation of confocal versus conventional imaging of biological structures by fluorescence light microscopy. *Journal of Cell Biology* 105, 41–48.

Wu, P., Shui, W., Carlson, B.L., Hu, N., Rabuka, D., Lee, J., and Bertozzi, C.R. (2009). Site-specific chemical modification of recombinant proteins produced in mammalian cells by using the genetically encoded aldehyde tag. *Proceedings of the National Academy of Sciences of the United States of America* 106, 3000–3005.

Yang, Y.K., Ko, S.K., Shin, I., and Tae, J. (2007). Synthesis of a highly metal-selective rhodamine-based probe and its use for the in vivo monitoring of mercury. *Nature Protocols* 2, 1740–1745.

## Fluorescent proteins

Barondeau, D.P., Tainer, J.A., and Getzoff, E.D. (2006). Structural evidence for an enolate intermediate in GFP fluorophore biosynthesis. *Journal of the American Chemical Society* 128, 3166–3168.

Betzig, E., Patterson, G.H., Sougrat, R., Lindwasser, O.W., Olenych, S., Bonifacino, J.S., Davidson, M.W., Lippincott-Schwartz, J., and Hess, H.F. (2006). Imaging intracellular fluorescent proteins at nanometer resolution. *Science* 313, 1642–1645.

Campbell, R.E., Tour, O., Palmer, A.E., Steinbach, P.A., Baird, G.S., Zacharias, D.A., and Tsien, R.Y. (2002). A monomeric red fluorescent protein. *Proceedings of the National Academy of Sciences of the United States of America* 99, 7877–7882.

Davidson, M.W., and Campbell, R.E. (2009). Engineered fluorescent proteins: Innovations and applications. *Nature Methods* 6, 713–717.

Day, R.N., and Davidson, M.W. (2009). The fluorescent protein palette: Tools for cellular imaging. *Chemical Society Reviews* 38, 2887–2921.

Frommer, W.B., Davidson, M.W., and Campbell, R.E. (2009). Genetically encoded biosensors based on engineered fluorescent proteins. *Chemical Society Reviews* 38, 2833–2841.

Gurskaya, N.G., Verkhusha, V.V., Shcheglov, A.S., Staroverov, D.B., Chepurnykh, T.V., Fradkov, A.F., Lukyanov, S., and Lukyanov, K.A. (2006). Engineering of a monomeric green-to-red photoactivatable fluorescent protein induced by blue light. *Nature Biotechnology* 24, 461–465.

Heim, R., and Tsien, R.Y. (1996). Engineering green fluorescent protein for improved brightness, longer wavelengths and fluorescence resonance energy transfer. *Current Biology* 6, 178–182.

Lippincott-Schwartz, J., and Manley, S. (2009). Putting super-resolution fluorescence microscopy to work. *Nature Methods* 6, 21–23.

Matz, M.V., Lukyanov, K.A., and Lukyanov, S.A. (2002). Family of the green fluorescent protein: Journey to the end of the rainbow. *Bioessays* 24, 953–959.

McKinney, S.A., Murphy, C.S., Hazelwood, K.L., Davidson, M.W., and Looger, L.L. (2009). A bright and photostable photoconvertible fluorescent protein. *Nature Methods* 6, 131–133.

Ormo, M., Cubitt, A.B., Kallio, K., Gross, L.A., Tsien, R.Y., and Remington, S.J. (1996). Crystal structure of the Aequorea victoria green fluorescent protein. *Science* 273, 1392–1395.

Remington, S.J. (2006). Fluorescent proteins: Maturation, photochemistry and photophysics. *Current Opinion in Structural Biology* 16, 714–721.

Rizzo, M.A., Davidson, M.W., and Piston, D.W. (2009). Fluorescent protein tracking and detection: Applications using fluorescent proteins in living cells. *Cold Spring Harbor Protocols* 2009, pdb top64.

Shaner, N.C., Campbell, R.E., Steinbach, P.A., Giepmans, B.N., Palmer, A.E., and Tsien, R.Y. (2004). Improved monomeric red, orange and yellow fluorescent proteins derived from *Discosoma* sp. red fluorescent protein. *Nature Biotechnology* 22, 1567–1572.

Shaner, N.C., Lin, M.Z., McKeown, M.R., Steinbach, P.A., Hazelwood, K.L., Davidson, M.W., and Tsien, R.Y. (2008). Improving the photostability of bright monomeric orange and red fluorescent proteins. *Nature Methods* 5, 545–551.

Shaner, N.C., Steinbach, P.A., and Tsien, R.Y. (2005). A guide to choosing fluorescent proteins. *Nature Methods* 2, 905–909.

Tsien, R.Y. (1998). The green fluorescent protein. *Annual Review of Biochemistry* 67, 509–544.

Wachter, R.M., King, B.A., Heim, R., Kallio, K., Tsien, R.Y., Boxer, S.G., and Remington, S.J. (1997). Crystal structure and photodynamic behavior of the blue emission variant Y66H/Y145F of green fluorescent protein. *Biochemistry* 36, 9759–9765.

Zacharias, D.A., and Tsien, R.Y. (2006). Molecular biology and mutation of green fluorescent protein. *Methods of Biochemical Analysis* 47, 83–120.

## Multispectral imaging

Dickinson, M.E., Bearman, G., Tille, S., Lansford, R., and Fraser, S.E. (2001). Multi-spectral imaging and linear unmixing add a whole new dimension to laser scanning fluorescence microscopy. *Biotechniques* 31, 1272, 1274–1276, 1278.

Frank, J.H., Elder, A.D., Swartling, J., Venkitaraman, A.R., Jeyasekharan, A.D., and Kaminski, C.F. (2007). A white light confocal microscope for spectrally resolved multidimensional imaging. *Journal of Microscopy* 227, 203–215.

Geladi, P., Burgerb, J., and Lestandera, T. (2004). Hyperspectral imaging: Calibration problems and solutions. *Chemometrics and Intelligent Laboratory Systems* 72, 209–217.

Schultz, R.A., Nielsen, T., Zavaleta, J.R., Ruch, R., Wyatt, R., and Garner, H.R. (2001). Hyperspectral imaging: A novel approach for microscopic analysis. *Cytometry* 43, 239–247.

Zucker, R.M., Rigby, P., Clements, I., Salmon, W., and Chua, M. (2007). Reliability of confocal microscopy spectral imaging systems: Use of multispectral beads. *Cytometry A* 71, 174–189.

## Suppliers and online resources

### *Key microscope suppliers*

All provide a wide selection of advanced research microscopes, including confocal systems.

Nikon. Microscope manufacturer; also host to the site MicroscopyU containing tutorials, galleries, and more. http://www.microscopyu.com/

Olympus. Another key manufacturer with a detailed support site, the Olympus Microscopy Resource Center. http://www.olympusmicro.com/

Leica. Supplier of products for light and electron microscopy, as well as histology systems, software, and cameras.

Zeiss. Microscopes for materials and biological research, including multispectral imaging with deconvolution. Also specialized cameras. Software for free download enables viewing and reconstruction of images taken on Zeiss instruments. Tutorial videos also available.

### *Camera suppliers*

The microscope suppliers listed above also supply cameras. The following suppliers are also recommended.

Andor. Many highly specialized cameras and imaging systems. Also supplies microscopes, including custom designs.

Hamamatsu. Advanced CCD, EMCCD, and CMOS cameras for scientific research.

Qimaging. A variety of cameras for life sciences imaging, including CMOS cameras.

An article on how to choose: http://www.biocompare.com/Editorial-Articles/172456-How-to-Choose-a-Microscopy-Camera/

### *Suppliers of fluorescent probes (a few of many!)*

Invitrogen. Key supplier of dyes, including reactive probes. Maker of Alexa dyes. Also supply quantum dots.

Sigma-Aldrich. Chemical supplier with a limited number of dyes.

Thermo Scientific (Pierce). Dyes, reactive dyes, dye-labeled antibodies. Produce DyLight fluors, an alternative to Alexa or Cy dyes.

AnaSpec. Dyes, quenchers, reactive dyes, antibodies, and more.

Li-Cor Biosciences. Specialists in infrared dyes.

Innova Biosciences. Makers of "lightning link" rapid antibody-labeling kits for many dyes.

Jena Bioscience. Reactive dyes for protein labeling.

Glen Research. Reactive dyes for labeling DNA.

BD Biosciences. Large selection of labeled antibodies.

Santa Cruz Biotechnology. Labeled antibodies.

### *Suppliers of plasmids encoding fluorescent proteins*

Clontech. Key supplier of fluorescent proteins, including mFruits and GFP variants.

BD Biosciences. Organelle-targeted fluorescent protein vectors.

MBL international. Wide array of coral-based fluorescent proteins in all colors, including the photoswitchable Dronpa.

Evrogen. Suppliers of the Tag fluorescent proteins in a wide range of colors, including organelle targeted. Also suppliers of the photoactivatable cytotoxic KillerRed and of photoswitchable proteins.

### *Multispectral imaging*

Gooch and Housego. Makers of the Hsi-300 acousto-optical tunable filter system for hyperspectral imaging.

Perkin Elmer. Makers of the "Nuance" visible multispectral imaging system, optimized for fluorescence or brightfield.

# CHAPTER 8

# Advanced Light Microscopy Techniques

Coauthored with Lina Carlini

## 8.1 INTRODUCTION

Light microscopy is in the middle of a revolution, and one of the greatest challenges for the aspiring microscopist is to appreciate what is available in terms of technology, software, and dyes or labeling techniques. Many of the specialized brightfield and fluorescent techniques described in this chapter require particular adaptations to the microscope and so need to be requested at the time of purchase. Some require new microscopes altogether. Many of these techniques also require specialized cellular labels that show specific optical properties. Despite the impressive number of dyes and proteins available, some cellular processes are still difficult to quantify, especially ionic concentrations and fluxes. The development of new probes is driven by developments in microscopy as well as by the growing understanding of the physics of fluorescence in small molecules and proteins.

Many of the techniques described in this chapter are costly to implement, or require specialized expertise, or both. We make an attempt to note when specialized instrumentation is commercially available, but it is important to bear in mind that commercial instruments are more costly than home-built systems and are less easy to adapt. The nature of the laboratory in which you work, the expertise that is present, and the availability (or not) of shared instrument facilities will determine the choice of whether to buy, borrow, rent, or build a customized microscope.

## 8.2 MULTISPECTRAL TECHNIQUES

Most fluorescence microscopy selects input and output wavelengths using filter cubes, but there are important limitations to this approach. Because each filter has a fixed central wavelength and passband, several filters must be used to image multiple fluorophores of different colors, and the filters are often mechanically interchanged by a rotating turret mechanism. Interference filter turrets and wheels have the disadvantages of limited wavelength selection, vibration, relatively slow switching speed, and potential image shift. Rotation of filter wheels and optical block turrets introduces mechanical vibrations into the imaging and illumination system, which in the best of cases require a damping time and in the worst of cases disturb the experiment (e.g., in electrophysiology or microinjection). Mechanical imprecision in the rotating mechanism can introduce registration errors when sequentially acquired multicolor images are processed. Furthermore, the fixed spectral characteristics of interference filters do not allow optimization for different fluorophore combinations, nor for adaptation to new fluorescent dyes, limiting the versatility of both the excitation and detection functions of the microscope. The *acousto-optic tunable filter* (AOTF) is an electro-optical device that functions as an electronically tunable filter with adjustable bandwidth. It can be used as either a tunable excitation filter or a tunable emission filter for different types of microscopy.

An AOTF designed for microscopy typically consists of a tellurium dioxide or quartz anisotropic crystal to which a piezoelectric transducer is bonded. In response to the application of an oscillating radio frequency electrical signal, the transducer generates a high-frequency vibrational wave that propagates into the crystal. The alternating ultrasonic acoustic wave induces a periodic redistribution of the refractive index through the crystal that acts as a transmission diffraction grating or Bragg diffraction grating to deviate a portion of incident laser light into a first-order beam, which is utilized in the microscope (or two first-order beams when the incident light is nonpolarized). Changing the frequency of the transducer signal applied to the crystal alters the period of the refractive index variation and, therefore, the wavelength of light that is diffracted. The relative intensity of the diffracted beam is determined by the amplitude (power) of the signal applied to the crystal.

In widefield fluorescence microscopy, an AOTF may be used as an emission filter, scanning across the visible range in selected increments from 2 to 50 nm and obtaining an image at each wavelength. This technique is called *hyperspectral imaging* because each image is a four-dimensional "cube," where the "axes" are $x$, $y$, $z$, and $\lambda$. The advantages of this technique over RGB imaging can readily be imagined. It provides a full spectral signature for every pixel in the image, allowing for easy unmixing of multiple fluorophores, even those with highly overlapping spectra (**Figure 8.1**). The disadvantages of hyperspectral imaging are large data set size and very slow acquisition. Depending upon the brightness of a sample and the number of wavelengths desired, collecting and saving a single cube can take up to several minutes. Implementing this technique is as simple

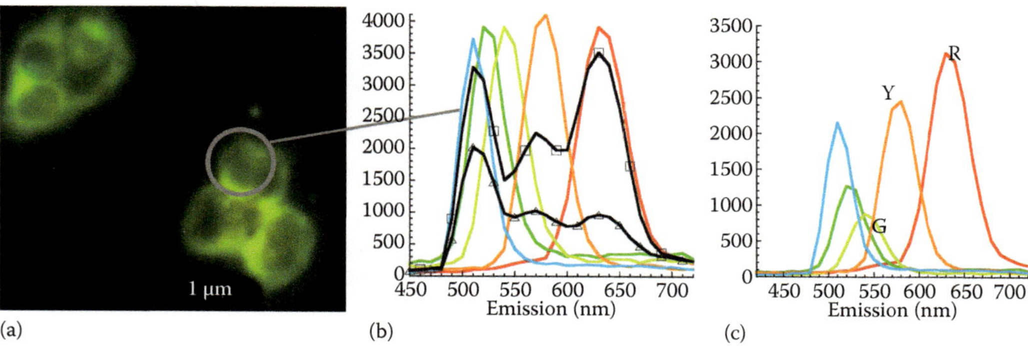


**Figure 8.1 Spectral deconvolution of multiple fluorophores.** (a) Clusters of bacteria labeled simultaneously with five different colored fluorescent probes, of which not all are readily resolvable to the eye. (b) Spectra of probes and labeled cells. The fluorescent labels taken independently are blue (B), cyan (C), green (G), yellow (Y), and red (R). The overlaid black lines are spectra taken with an AOTF from selected cells, one of which is indicated. Regions of interest as small as one pixel are possible. (c) Linear deconvolution of the selected cell spectrum, showing relative amounts of each of the fluorescent probes on this cell.

as purchasing an AOTF and installing it just before the CCD in the optical path. The only advantage to purchasing a complete commercial system is the software, though custom software often performs better if the skills are available to write it.

In confocal microscopy, an AOTF is usually used to modulate the wavelength and intensity of the *excitation*. Different wavelengths of light from multiple laser sources are combined in a single-mode fiber and sent to the AOTF (**Figure 8.2**).

This allows the microscopist control of the intensity and/or illumination wavelength on a pixel-by-pixel basis while maintaining a high scan rate. A variety of advanced microscopy techniques are made simpler using an AOTF; one of the most evident is *fluorescence recovery after photobleaching* (FRAP). Regions of interest can be defined via freehand drawing or selection of geometric shapes, and these regions may be exposed to high intensities to selectively excite or photobleach them (**Figure 8.3**).

In FRAP, fluorescence recovers in the bleached spot as the population of bleached molecules is replaced by diffusion from outside the exposed area. The kinetics of recovery reflect the dynamics of the labeled molecule; when recovery is

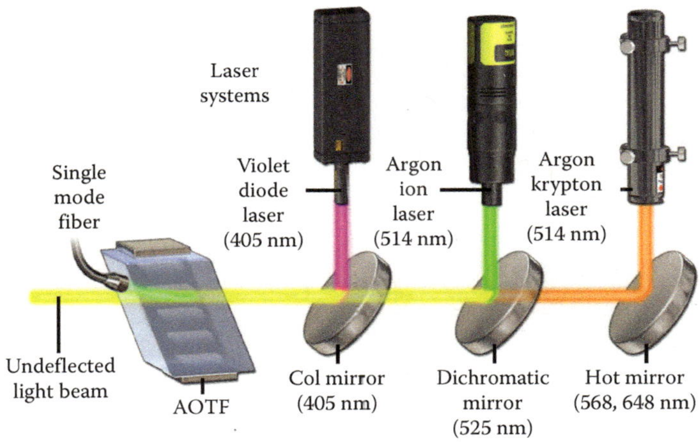


**Figure 8.2 Configuration scheme using an AOTF for excitation intensity and wavelength control in confocal microscopy.**

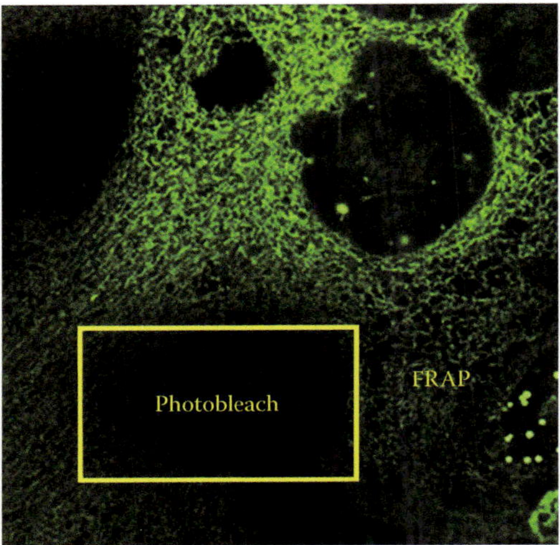


**Figure 8.3 Selectively targeting a region of interest for photobleaching.**

dominated by diffusion, the normalized fluorescence intensity with time $f(t)$ is described by the equation

$$f(t) = e^{-2\tau/t}\left[I_0\left(\frac{2\tau}{t}\right) + I_1\left(\frac{2\tau}{t}\right)\right], \tag{8.1}$$

where $I_0$ and $I_1$ are modified Bessel functions and $\tau$ is the characteristic time scale of the diffusion. For a bleached spot of radius $r$, and diffusion coefficient of the fluorescent molecule $D$, $\tau = \dfrac{r^2}{4D}$ (**Figures 8.4 and 8.5a**).

Since $r$ may be carefully controlled, FRAP is useful for studying

- Diffusional speed of proteins or small molecules
- The presence of connections between/among intracellular organelles or compartments and the exchange speed (for example, whether specific molecules can enter the nucleus from the cytoplasm)

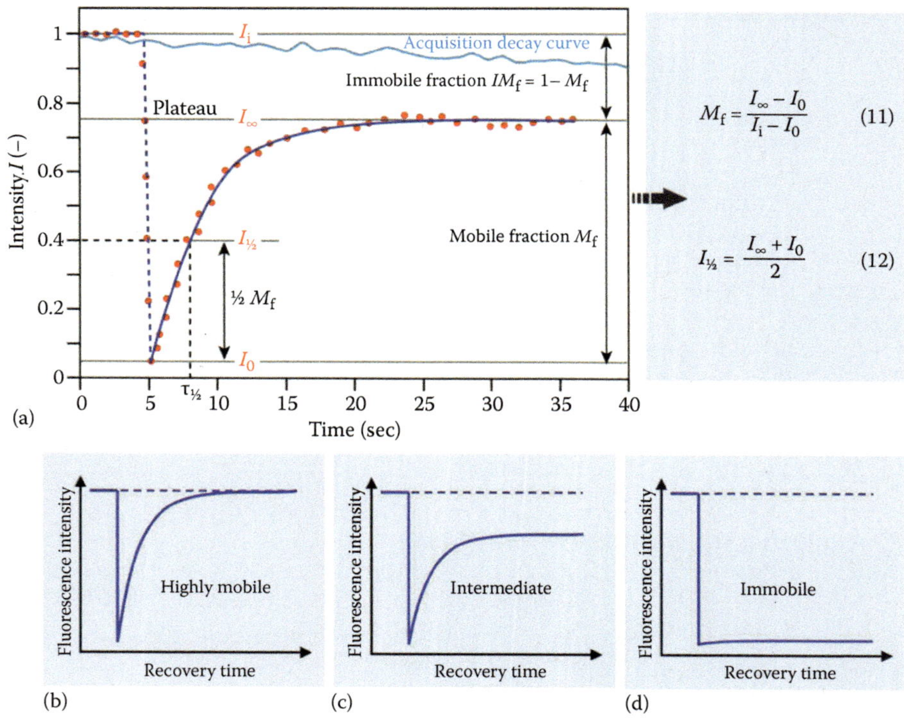


**Figure 8.4 Anatomy of a typical FRAP curve.** (a) From the initial (prebleach) fluorescence intensity ($I_i$), the signal drops to a particular low value ($I_0$) as the high-intensity laser beam bleaches fluorochromes in the Region of interest (ROI). Over time the signal recovers from the postbleach intensity ($I_0$) to a maximal plateau value ($I_\infty$). From this plot and Equations (11) and (12) in panel (a), the mobile fraction ($M_f$), immobile fraction ($IM_f$), $I_{1/2}$, and corresponding time ($\tau_{1/2}$—the time for the exchange of half the mobile fraction between bleached and unbleached areas) can be calculated (light blue line: reference photobleaching curve to correct for fluorescence loss during data acquisition). The information from the recovery curve (from $I_0$ to $I_\infty$) can be used to determine the diffusion constant and the binding dynamics of fluorescently labeled proteins. Based on different recovery profiles, the protein mobility can be classified as (b) highly mobile with virtually no immobile fraction, (c) intermediate mobile with an immobile fraction, or (d) immobile. (Figure and caption from Ishikawa-Ankerhold, H. C. et al., Advanced Fluorescence Microscopy Techniques—FRAP, FLIP, FLAP, FRET and FLIM, *Molecules* 2012, *17* (4), 4047–4132, doi:10.3390/molecules17044047. Reproduced under the Creative Commons license.)

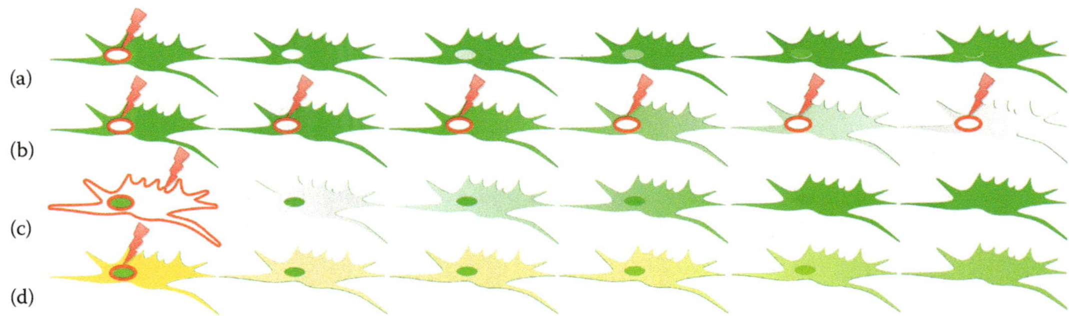

**Figure 8.5 FRAP and its variations.** (a) FRAP, showing bleaching in a region of interest and slow return of fluorescence to that region. (b) FLIP, showing repeated bleaching of a region and diffusion of bleached molecules into the rest of the cell. (c) iFRAP, showing bleaching of the whole cell outside of a region of interest and diffusion of fluorophores from that region into the rest of the cell. (d) FLAP, showing the presence of two fluorophores (red + green = yellow) with selective bleaching of red inside a region of interest. As the molecules diffuse, the region of interest becomes less and less distinguishable from the rest of the cell.

- Binding of proteins to other proteins or to small molecules
- Protein immobilization on cellular structures such as membranes, DNA, or cytoskeleton

There are a number of possible variations on FRAP. In *fluorescence loss in photobleaching* (FLIP), a region of interest is photobleached several times, and the spread of the bleached fluorophores over the entire cell is measured as loss of fluorescence (**Figure 8.5b**). In *inverse FRAP* (iFRAP), a small region is selected to remain intact, and the rest of the cell is photobleached. This is useful for looking at dynamic movement in organelles but requires a good deal of light to photobleach the whole cell (**Figure 8.5c**). *fluorescence localization after photobleaching* (FLAP) is a ratiometric method using two different fluorescent labels. The two labels can be attached to two different target molecules or to the same molecule, but only one of the fluorophores is bleached; the nonbleached population serves as a reference. Then the ratio of the bleached area and of a second nonbleached area is taken to determine the mobility of the targets (**Figure 8.5d**).

FRAP and its variants may be performed on any confocal, so they are popular and used widely. Practical concerns involve data analysis, especially when trying to obtain quantitative results. One major problem is background bleaching of the fluorophores; this can be ameliorated with judicious choices of fluorophores and acquisition protocols. The green fluorescent protein (GFP) family is the most common label used in FRAP. GFPs have intermediate photobleaching resistance; it is important to keep in mind that different colors photobleach at different rates. When small-molecule tags are used, Alexa dyes or quantum dots (QDs) can help minimize background bleaching. Unwanted bleaching can also be limited by acquiring line scans instead of 2-D scans, decreasing laser power, or using faster scans. Some compromise between temporal and spatial resolution will need to be made for each experiment.

It is also important to note that **Equation 8.1** is approximate. Bleaching is not instantaneous, and even small regions of interest are likely to be anisotropic. In addition, factors other than diffusion—such as protein–protein interactions—may affect or even dominate fluorescence recovery. It is important to explore different possible mechanisms when fitting data to a model.

## 8.3 FLUORESCENCE RESONANCE ENERGY TRANSFER MICROSCOPY

This is one of the most common advanced techniques in confocal imaging. It requires no special instrumentation, although an AOTF can be helpful. It uses the principle of dipole–dipole energy transfer between fluorophores to determine distances between structures. It is used as both a spectroscopic and an imaging technique and is discussed fully in **Chapter 16**.

## 8.4 TWO-PHOTON MICROSCOPY

The principle behind this technique is to use a fast-pulsed laser at twice the excitation wavelength of the target fluorophore, so fluorescence is only excited in molecules that absorb two photons simultaneously (**Figure 8.6**). The most common laser used for this is the Ti:sapphire (tunable in the range of ~700–1000 nm), so it can excite fluorophores that excite at ~350–500 nm, such as 4,6-Diamidino-2-phenylindole (DAPI), the $Ca^{2+}$ indicator Indo-1, and several Alexa dyes. It is important to note that the two-photon absorbance curve may be very different from the single-photon curve. The pure electronic transition dominates the one-photon spectrum, whereas specific vibronic transitions are enhanced in a two-photon process by resonant enhancement. Obtaining a curve for two-photon excitation versus emission is important when using a new or uncharacterized probe.

The probability of the two-photon process occurring is proportional to the square the intensity of the excitation, so only those fluorophores right at the focal point

**Figure 8.6 Jablonski diagram comparing one- and two-photon excitation.** Two-photon excitation into the $S_n$ states may be significantly higher than for the corresponding one-photon process, changing the absorption curve.

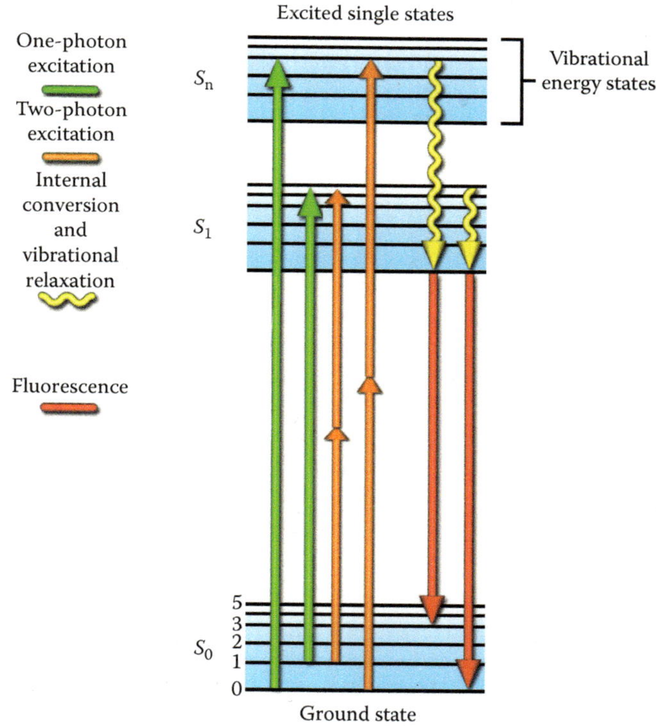

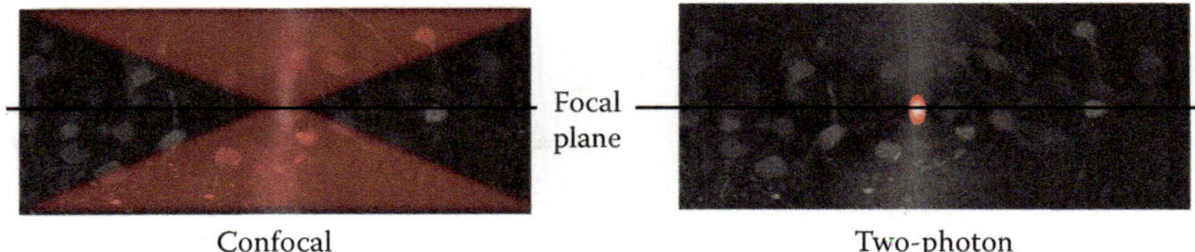



**Figure 8.7 Two-photon versus confocal excitation.** In a thick specimen, confocal excitation light excites a large volume, and a pinhole must be used to exclude light from outside the focal plane. Meanwhile, the out-of-focus areas suffer photobleaching and phototoxicity. With two-photon excitation, only a small (femtoliter-scale) spot at the focal plane is excited, removing the need for a pinhole and minimizing photodamage.

are excited efficiently (**Figure 8.7**). This spares the surrounding specimen from phototoxicity and allows for specifically targeted release of caged compounds. It also allows for imaging in deep specimens, since the penetration depth of near-infrared (near-IR) light is about twice that of visible; this makes it especially valuable in brain-slice or whole-animal experiments. Two-photon excitation is also helpful for FRAP because of the precision of the excitation spot.

The laser is the only needed additional equipment and may be fitted to an ordinary confocal microscope. Pulsed lasers are costly, so only a few individual investigators have two-photon instruments, but most universities have two-photon capabilities in imaging facilities. The recent development of fixed-wavelength femtosecond fiber lasers has made two-photon microscopy less costly; available wavelengths are ~780 nm from frequency-doubled erbium-doped fiber lasers and ~1055 nm from ytterbium (Yb)-doped fiber lasers. **Table 8.1** gives a list of some lasers appropriate for two-photon microscopy.

The biggest drawback of near-IR excitation is that it might "cook" highly pigmented cells, such as *Xenopus* oocytes. (Neurons, in contrast, are highly transparent.) Albino frogs are available for experiments requiring *Xenopus* oocytes; they lay unpigmented eggs that do not heat up with 800 nm light.

Two-photon absorption is a nonlinear optical process and follows different quantum mechanical selection rules than single-photon absorption. Some molecular structures are therefore much more useful as two-photon dyes than others. The probability of absorbing two photons is called the two-photon cross section ($\sigma$) and is usually given in Goeppert-Mayer units (GM), where 1 GM is $10^{-50}$ cm$^4$ s photon$^{-1}$. **Table 8.2** gives the two-photon cross section for some useful probes.

**Table 8.1**

Some Lasers Useful for Two-Photon Microscopy

| Laser Type | Wavelength (nm) | Πυλσε ρατε (MHz) | Πυλσε Length (fs) | Pulse Energy at 10 mW (nJ) | Company |
|---|---|---|---|---|---|
| Ti:sapphire | Tunable (690–1040 or 680–1300) | 80 | 140 or 100 | 0.125 | Coherent |
| Ti:sapphire | 690–1040 or 710–990 | 80 | 100 or 80 | 0.125 | Spectra-Physics |
| Fiber | 1055 | 70 | 70 | 0.14 | Coherent |
| Fiber | 810 | 75 | 140 | 0.13 | IMRA |
| Fiber | 780 | 50 | 100 | 0.13 | IMRA |
| Fiber | 780 | 100 | 90 | 0.10 | Toptica |

**Table 8.2**

Some Dyes and Proteins That Have Been Used for Two-Photon Microscopy, with Their Two-Photon Excitation Wavelengths ($\lambda_{2P}$), Emission Wavelengths ($\lambda_{Em}$), and Two-Photon Cross Sections ($\sigma$)

| Probe | $\lambda_{2P}$ (nm) | $\lambda_{Em}$ (nm) | $\sigma$ (GM) | Ref. |
|---|---|---|---|---|
| *Dyes* | | | | |
| Rhodamine B | 691 | 625 | 194 | Xu, C., and Webb, W. W., Measurement of Two-Photon Excitation Cross Sections of Molecular Fluorophores with Data From 690 to 1050 nm, *J. Opt. Soc. Am. B*, 13, 481–491, 1996. |
| Fluorescein | 1050 | 518 | 0.23 | Ibid. |
| Indo-1 (with $Ca^{2+}$/ without $Ca^{2+}$) | 700 | 400 | 3.5/1.5 | Svoboda, K., and Yasuda, R., Principles of Two-Photon Excitation Microscopy and Its Applications to Neuroscience, *Neuron*, 50, 823–839, 2006. |
| Calcium green (with $Ca^{2+}$/without $Ca^{2+}$) | 820 | 530 | 30/2 | Ibid. |
| SeTau-647 | 875–925 | 695 | 3500 | Podgorski, K. et al., Ultra-Bright and -Stable Red and Near-Infrared Squaraine Fluorophores for in Vivo Two-Photon Imaging, *PLoS One*, 7, e51980, 2012. |
| Quantum dots (CdSe/ZnS) | 700–900 | Varies with size | Up to 10,000 depending upon size | Pu, S. C. et al., The Empirical Correlation between Size and Two-Photon Absorption Cross Section of CdSe and CdTe Quantum Dots, *Small*, 2, 1308–1313, 2006. |
| *Proteins* | | | | |
| GFP | 900–1000 | 510 | 100–200 | Zipfel, W. R. et al., Nonlinear Magic: Multiphoton Microscopy in the Biosciences, *Nat. Biotechnol.*, 21, 1369–1377, 2003. |
| YFP | 930–1000 | 530 | 100–200 | Ibid. |
| mOrange | 640 | 559 | 200 | Drobizhev, M. et al., Two-Photon Absorption Properties of Fluorescent Proteins, *Nat. Methods*, 8, 393–399, 2011. |
| tdTomato | 684 | 581 | 316 | Ibid. |
| dsRed2 | 700 | 582 | 112 | Ibid. |

Fluorescent proteins have a complex structure that is more difficult to model theoretically than a small molecule. The resonant effect is particularly strong with red fluorescent proteins (FPs), making them show two-photon absorption at wavelengths much shorter than twice the one-photon peak. Charges on the chromophore and variations of electric field within the beta-barrel of FPs can all affect two-photon processes and cause surprising differences in relative brightness among the different protein variants, which are not predictable from the one-photon spectra.

It is also possible to have three-photon processes, though these have been much less explored than two-photon process for reasonably obvious reasons. Two-photon microscopy is therefore often called *multiphoton microscopy*. QDs may be excited by three-photon processes relatively easily. The appropriate excitation wavelength is ~1300 nm; size-dependent cross sections may be calculated theoretically.

## 8.5 TOTAL INTERNAL REFLECTANCE MICROSCOPY

This is a technique that allows for high-contrast, high signal-to-noise imaging on surfaces. When light arrives at an interface between two media of different

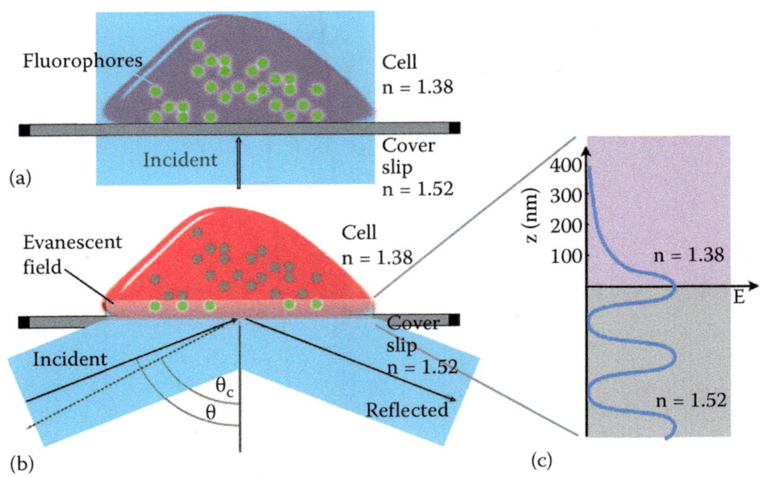



**Figure 8.8 TIRF.** (a) In ordinary epifluorescence microscopy, the incident light (blue) travels through the entire cell and excites all the fluorophores in it (green). (b) In TIRF, the incident light is reflected. Only a thin, decaying evanescent field penetrates the cell, exciting only the fluorophores very near the coverslip surface. (c) Schematic of decaying electric field inside the cell.

refractive indices, the light incident at an angle greater than the critical angle $\theta_c$ undergoes total reflection according to the formula

$$\theta_c = \sin^{-1}\frac{n_2}{n_1}, \tag{8.2}$$

where $n_1$ is the index of the objective and $n_2$ that of the specimen, with $n_1 > n_2$. The field penetrating into the specimen is then only the exponentially decaying evanescent wave, with a depth of only a few hundred nanometers, so only the very surface fluorophores are excited. This allows for an effective optical section that is very thin, and is excellent for seeing vesicle exocytosis and other processes that occur right on the surface (**Figure 8.8**).

Total internal reflectance (TIRF) can be set up on an ordinary inverted fluorescence microscope with a few specialized accessories: a laser light source, which can be an argon, helium–neon, or krypton, or solid-state lasers; an ultrahigh numerical aperture objective (often called a TIRF objective; numerical aperture > 1.4); an appropriately sensitive camera if one is not already in place; ultrahigh refractive index ($n$) coverslips; and ultrahigh-$n$ immersion liquid ($n = 1.78$).

# 8.6 FLUORESCENCE LIFETIME IMAGING (FLIM)

## General principles and use

Fluorescence lifetimes and their meaning and uses are discussed more fully in **Chapter 16**, as lifetime experiments are often performed in pure solutions of fluorophores in order to evaluate photophysical properties. *Fluorescence lifetime imaging microscopy* (FLIM) allows for imaging of fluorescence lifetimes in cells with pixel-by-pixel resolution using highly specialized instruments that have become more widely available over the past few years. Many universities have FLIM systems as part of imaging facilities. They are costly even to build from scratch, as they require detectors with single-photon sensitivities. Such detectors are highly specialized and must be chosen carefully for the intended range of lifetimes needed. A discussion of the hardware is outside the scope of

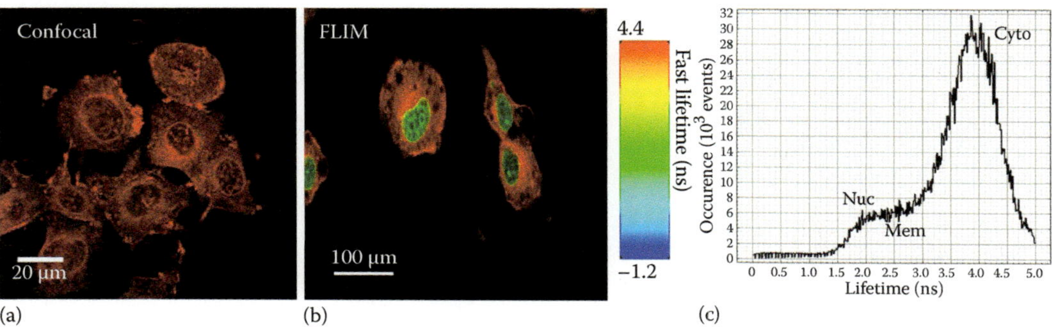

**Figure 8.9 Confocal intensity image versus FLIM image.** (a) Confocal image of cells labeled with gold nanoparticle–doxorubicin conjugate (Au–Dox; see **Chapter 13** for more). The fluorophore is present throughout the cell. (b) FLIM image showing variation of the fluorophore lifetime in different parts of the cell—cytoplasm (red), nuclear membrane (yellow), and nucleus (green). The color scale shows the values of the characteristic lifetime measured by fitting the decay curve to a double exponential. (c) A histogram of the measured lifetimes showing variations in different parts of the cell.

this discussion (see **Chapter 16** for more), but references to suppliers and to descriptions of FLIM instruments are given at the end of the chapter.

FLIM provides photoluminescence decay curves in each pixel of a microscopic image (**Figure 8.9**). The decay curve is a single or multiple exponential with a characteristic lifetime $\tau$. The decay curve is fit (usually automatically within the acquisition software) to

$$I = \sum_{i=1}^{n} A_i exp\left( -\frac{t}{\tau_i} \right), \tag{8.3}$$

and the average lifetime may be given as either an amplitude-weighted lifetime

$$\langle \tau \rangle = \frac{\sum_{i=1}^{n} A_i \tau_i}{\sum_{i=1}^{n} A_i} \tag{8.4}$$

or an intensity-weighted lifetime

$$\langle \tau \rangle = \frac{\sum_{i=1}^{n} A_i \tau_i^2}{\sum_{i=1}^{n} A_i \tau_i} \tag{8.5}$$

With simple fluorophores such as GFP or dyes, the lifetime is fit to a single-exponential decay curve. More complex fits are possible if the sampling statistics are good and the signal is strong, and are needed for more complex fluorophores such as QDs or intrinsic signals from proteins. The value of FLIM is that fluorophore lifetime is sensitive to environmental conditions—pH, redox potential, crowding, and so forth—much more than fluorescence intensity. Thus, FLIM provides a useful tool for probing environment. The type and degree of environmental sensitivity depends upon the fluorophore; intrinsic fluorophores (such as NADH and tryptophan) as well as extrinsic fluorophores show environmental sensitivity. FLIM has been used for

- Quantitative sensing of calcium and other ions by lifetime change of ion-sensitive dyes rather than intensity, so that photobleaching does not need to be accounted for
- Measurement of fluorescent drug release from nanoparticles, capsules, or endosomes due to changes in lifetime between *bound* and *free* drug
- Oxygen concentrations or redox state of cells and organelles, due to changes in fluorophore lifetime with oxygen
- Local changes in pH, viscosity, and other physicochemical parameters
- Interactions between fluorophore-labeled proteins or intrinsically FPs;
- Fluorescence resonance energy transfer (FRET), which may often be quantified more easily using lifetimes rather than intensities
- Diagnostics, where lifetime changes in cells may indicate malignancy or other pathological changes, particularly in pigmented cells such as those of the retina.

FLIM is a relatively new technique, so the extent of changes to lifetimes with environment remains to be explored for a wide range of dyes, FPs, and intrinsic fluorophores. The degree to which lifetime imaging may be useful for ultraresolution techniques also remains largely unknown, as most photoactivated dyes have not been characterized by lifetime analysis. It is only recently that FLIM has become available to a large number of academic researchers, so the number of possible applications is only beginning to be appreciated.

## Example experiment: Measuring lifetimes of QDs inside cells

QDs (see Chapter 11) are fluorescent semiconductor nanoparticles that are exquisitely sensitive to a wide variety of environmental changes. Images of live cells that have taken up QDs show intensity changes (quenching and enhancement) in areas known to have altered pH or redox potential, particularly endosomes and lysosomes. QDs are also sensitive to photooxidation, showing brightening and blue shifting upon exposure to ultraviolet (UV)-to-blue light at typical intensities used for widefield fluorescence excitation. Because of the greater sensitivity of FLIM as compared with intensity, it can be expected that FLIM of cells that contain QDs would show dramatic changes in QD lifetime as the particles are endocytosed and processed in the cells, and upon photooxidation.

However, there are some challenges involved in FLIM using QDs. QDs typically have relatively long lifetimes compared to organic dyes, tens of nanoseconds or even more. However, the lasers used for FLIM have ultrafast repetition rates that usually cannot be adjusted. Typically, in a fluorescence decay experiment, the interval between laser pulses should be four to five times that of the average fluorescence lifetime of the fluorophore, or a limit of 3 ns for an 80 MHz repetition rate. When this is not the case, the signal does not fully decay between laser pulses, leading to erroneous amplitudes (**Figure 8.10**). This needs to be taken into consideration during fitting.

**Figure 8.10 Artifacts caused by too high of a laser repetition rate in lifetime measurements.** The fluorescence has not fully decayed from the first pulse (blue) before the next pulse (green) occurs and excites the sample again. The measured intensity (black) is therefore too high.

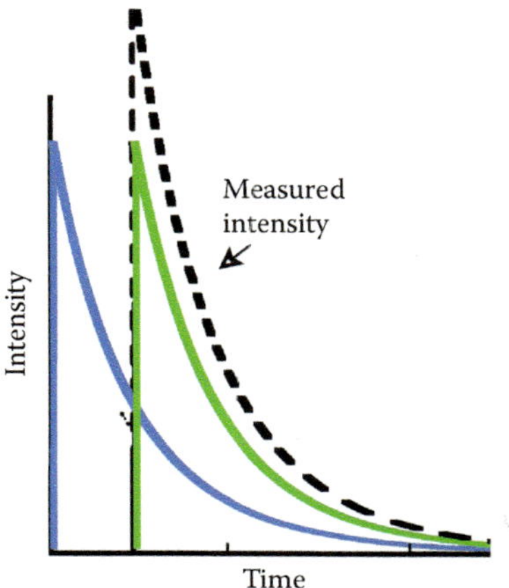


This example illustrates how a FLIM experiment is set up and what can be observed. The procedure is as follows:

- Grow cells and add QDs by a determined protocol to obtain uptake at different concentrations. (See **Chapter 11** for more.)
- Confirm QD uptake using confocal imaging using an Ar ion laser with 488 nm excitation.
- Choose some samples for photooxidation prior to imaging; cells incubated with QDs are exposed to UV illumination from an Hg lamp source through a DAPI filter (excitation, 370–380 nm/emission, 435–485 nm) for 1–2 min.
- Place the sample into the FLIM instrument. Use a 63× water-immersion objective. For the images shown, the parameters were as follows: a Zeiss 710 NLO multiphoton microscope equipped with a Coherent Chameleon Vision II two-photon excitation source; this is a mode-locked Ti:S laser producing pulses of 120 ps full-width half-max at 80 MHz. A reference beam was sent to the SPC-150 *time-correlated single-photon counting* (TCSPC) module (Becker & Hickl). (See **Chapter 16** for more about TCSPC). The fluorescence signal was transmitted to an HPM-100-40 GaAsP hybrid detector (Becker & Hickl) before reaching the TCSPC module. The hybrid PMT and its components are controlled via the DCC-100 detector controller of the FLIM system. (See Becker & Hickl (bh) "Modular FLIM systems for Zeiss LSM 510 and LSM 710 Laser Scanning Microscopes" for details.)
- Because two-photon excitation is being used, the optimal wavelength must be determined. Scan over wavelengths 800–900 nm to optimize excitation. A 565–615 nm bandpass filter is placed before the detector to match the QD emission. Imaging parameters—laser power, transmission, and excitation wavelength—should all be adjusted to minimize cellular autofluorescence and maximize QD signal. Despite the fact that these CdSe/ZnS QDs were optimally excited with 400 nm one-photon excitation, they were poorly excited with 800 nm two-photon excitation. The optical parameters were chosen to be 900 nm excitation, 980 mW, 9% transmission. These parameters gave zero counts from unlabeled cells (not shown).
- All scans lasted 60 s, and images were recorded with 512 × 512 pixel resolution.

The confocal intensity images indicated that the QDs did get into the cells (**Figure 8.11a**). In the lifetime images, there were three distinct regions with different lifetimes: discrete vesicles corresponding to endosomes/lysosomes; a continuous area just inside the cell membrane, corresponding to cytoplasm; and regions outside the cell or adhered to the cell membrane (**Figure 8.11b**). Each region showed a distinct lifetime, with that in the endosomes markedly shorter than in the other regions.

A remarkable increase in lifetimes was observed when cells were exposed to higher QD concentrations. While 20 nM QDs showed the expected decrease in lifetime in the endosomes, cells with 50 nM QDs had enhanced lifetimes in the endosomal/lysosomal regions of the cells (**Figure 8.11c**). Lifetimes in the cytoplasmic region were the same as with the lower concentration.

Photooxidation of endosomes by exposure to UV light led to a similar effect: lengthening of lifetimes in endosomes/lysosomes, but not in the rest of the cell (**Figure 8.11d**).

The accuracy of the fits to the lifetimes is evaluated by measuring the $\chi^2$ value. FLIM software usually provides these values as pixel-by-pixel maps (**Figure 8.12**). Values of $\chi^2 > 2$ indicate poor fits. If poor results are obtained with single exponentials, two or three exponentials may be required. At least 1000, $10^4$, or $10^5$ photons at peak are required for a fit to a single, double, or triple exponential, respectively. Depending upon the brightness of the probe and the acquisition time, obtaining sufficient counts often requires increasing the binning. This risks sampling areas

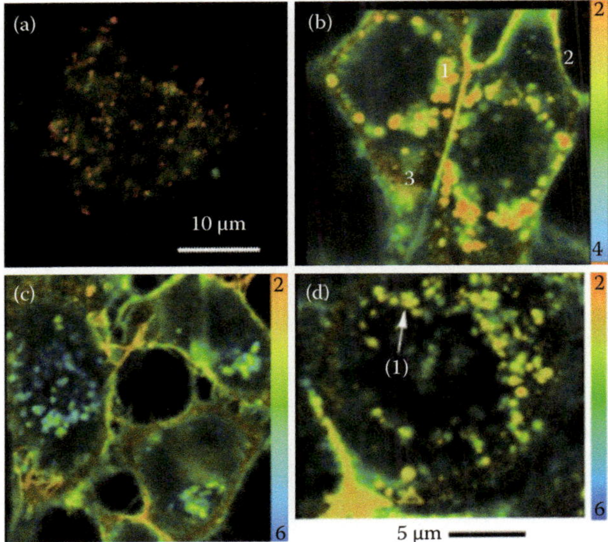



**Figure 8.11 Confocal and FLIM images of fluorescent nanoparticle quantum dots taken up into cultured cells.** (a) Confocal image of QDs (green) and the dye LysoTracker Red (red), showing substantial overlap of the channels (yellow), indicating that most of the QDs are in cell endosomes/lysosomes. (b–d) FLIM images; scale bar applies to all. The color bar on the right-hand side of each image is the lifetime scale in nanoseconds. (b) FLIM with QDs at 20 nM concentration. Three regions of distinct lifetimes were identified: (1) endosomes, (2) membrane, (3) cytoplasm. (c) FLIM image of cells incubated with QDs at 50 nM. (d) FLIM image of cells incubated with 20 nM QDs after photooxidation using mercury lamp illumination through a DAPI filter.

**Figure 8.12 Goodness of fit of lifetime estimates as shown by $\chi^2$.** FLIM images and fits for (a) 20 nM QDs and (b) 50 nM QDs. Pixels painted red represent chi-squared values ranging from 0 to 1.5, green pixels represent values between 1.5 and 2, and those painted blue are chi squared values greater than 2. Each image is fit to a monoexponential function. Note that the poorest fits occur at the higher concentration in the membrane areas of the cell.

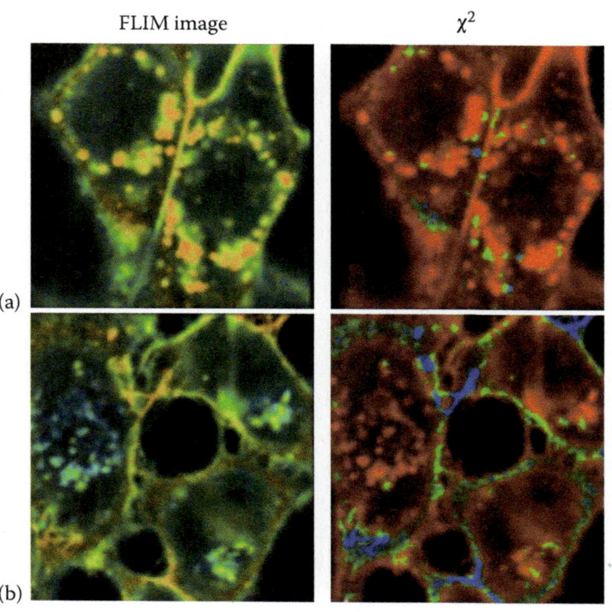


outside the region of interest, but may be the only possible approach to obtaining reasonable fits to highly nonexponential decays.

In order to account for the long lifetimes of QDs, an "incomplete decay model" may be used. Such models have been implemented in Becker & Hickl's image-processing software for over a decade; a reference is given at the end of the chapter. The model calculates the measured amplitude as a sum of the excited fluorescence plus the tails of any decays remaining from previous pulses. Only multiexponential fits are affected: for single exponentials, only the amplitude is changed by incomplete decay, and this is normalized away during fitting. **Table 8.3** shows a comparison of the mean lifetimes obtained from multiexponential fits with or without implementation of incomplete decay correction. It can be seen that the incomplete fit significantly changed the measured mean lifetimes and so was essential to accuracy with these particles.

**Table 8.3**

Mean Lifetimes Measured in Different Regions of Quantum Dot–Labeled Cells Using Multiexponential Fits with and without Incomplete Decay Correction

| Sample | Region or Condition | Exponential Model Fit | $\langle\tau_m\rangle$ (ns) | $\langle\tau_m\rangle$ (ns) Incomplete |
|---|---|---|---|---|
| 20 nM QD | (1) Endosomes | Double | 1.7 | 1.7 |
|  | (2) Membrane | Double | 4.4 | 6.6 |
|  | (3) Cytoplasm | Double | 1.1 | 2.3 |
| 50 nM QD | (1) Endosomes | Double | 3.5 | 4.4 |
|  | (2) Membrane | Double | 2.1 | 3.1 |
|  | (3) Cytoplasm | Double | 1.5 | 3.8 |
| 20 nM QD + light | (1) Endosomes | Single | 2.8 | 3.1 |

*Note:* "+ light" indicates photooxidative pretreatment via exposure to UV light.

# 8.7  FOUR PI MICROSCOPY

Four pi (4Pi) microscopy is an approach to breaking the diffraction limit that uses interference to reduce the axial ($z$-plane) focal volume and thus increase the resolution in $z$. The way this is accomplished is to set up a confocal system having two opposing objective lenses (**Figure 8.13**). The counterpropagating wave fronts from the two lenses form interference fringes at a common focal point. This reduces the axial focal volume (**Figure 8.14**). The resolution that can be obtained in $z$ is ~100 nm.

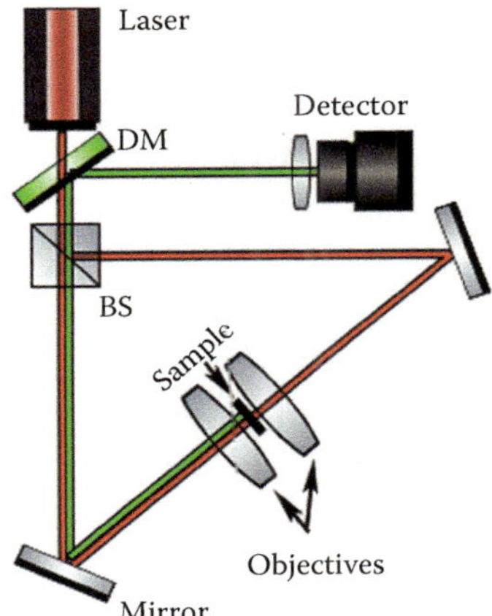


**Figure 8.13  Schematic of a 4Pi microscope.** The laser usually provides two-photon excitation. The excitation light is split into two beams by a beam splitter (BS) and focused by two objectives onto the same point in the sample. Constructive interference between these two beams reduces the size of the focal point. (DM = dichroic mirror.)

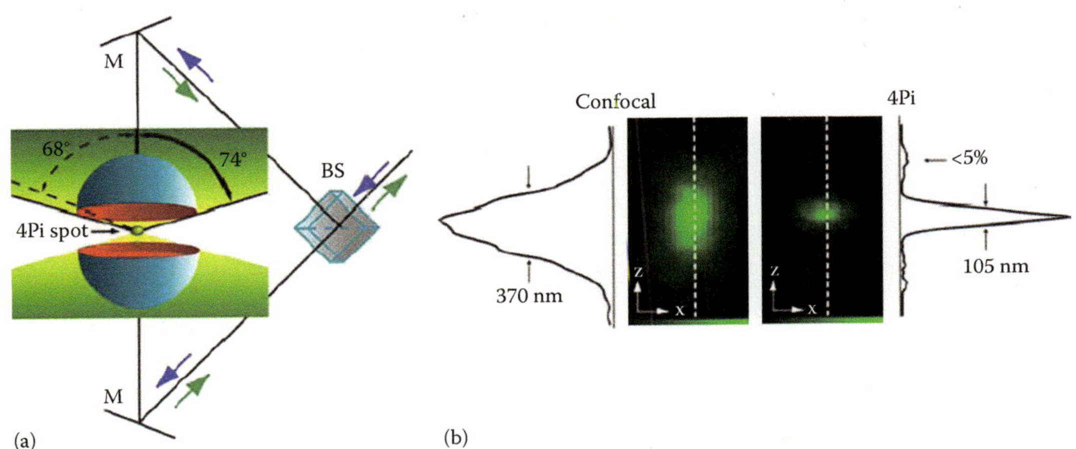


**Figure 8.14  Axial resolution improvement in 4Pi microscopy.** (a) Coherent illumination is divided at a beam splitter (BS) into two counterpropagating beams that are focused onto the sample by opposing lenses. This converts the beams' wave fronts into spherical wave-front caps of semiaperture angle $\alpha$. In 4Pi microscopy of type A, only the excitation light (blue arrows) is coherently added (at the focus), whereas in type C, the spherical wave-front caps of fluorescence emission (green arrows) is also coherently added at a common point of detection. (b) Improvement in measured axial point spread function (PSF) in 4Pi versus confocal microscopy. Both recordings were performed using the same fluorescent objects and objective lenses. The axial full-width half-maximum PSF is 370 nm for the confocal and 105 nm for the 4Pi.

Leica produces a commercial 4Pi module that may be integrated with a confocal microscope. It requires specialized coverslips and very thin specimens and is best suited to cells that have been fixed, though several studies of live cells using 4Pi microscopy have been published.

The biggest drawback of 4Pi microscopy has been the requirement for specimens significantly thinner than the average mammalian cell—tens to hundreds of nanometers as opposed to 10+ μm. This restriction is because of the interferometric nature of the technique. Signals at $z$ coordinates $\pm\lambda/2$ (200–250 nm) give the same interference pattern and so are ambiguous. Point spread function (PSF) deconvolution techniques can help with samples <1 μm thick, but are complex and do not work for thicker samples. A technique first reported in 2016, whole-cell 4Pi single-molecule switching nanoscopy (W-4PiSMSN), makes it possible to apply 4Pi techniques to whole cells. It uses a combination of deformable mirrors in the interferometric arms, software analysis techniques, and axial scanning to produce high-resolution images throughout entire cells.

4Pi microscopy may also be coupled with other interferometric methods, such as spectral self-interference fluorescent microscopy, to further improve spatial resolution.

## 8.8  PHOTOACTIVATED LOCALIZATION MICROSCOPY (PALM) AND STOCHASTIC OPTICAL RECONSTRUCTION MICROSCOPY (STORM)

### Principles of photoactivated localization microscopy/stochastic optical reconstruction microscopy

*Photoactivated localization microscopy* (PALM) and *stochastic optical reconstruction microscopy* (STORM) are fluorescence-based superresolution techniques, which rely on imaging single molecules. The defining difference between these techniques is the probe used, since PALM uses FPs and STORM makes use of synthetic dyes. Both techniques rely on fluorescent probes that can convert between bright and dark states, in order to temporally separate the fluorescence signal of interest and subsequently localize molecular positions. The localizations from many individual molecules (typically acquired over thousands of frames) are summed to yield a superresolution image (**Figure 8.15**).

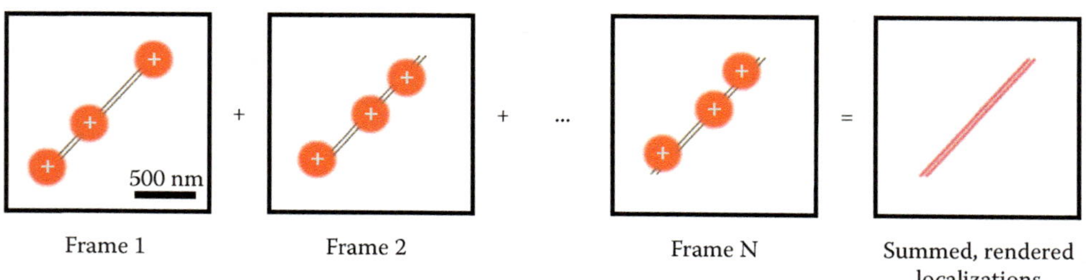



**Figure 8.15  Concept of localization microscopy.** A fluorescent probe is made to photoswitch. Each frame of the acquisition contains single molecules, which are localized and summed to yield a superresolution image.

## Probe requirements

PALM/STORM image quality relies greatly on the performance of the probes used. In addition to possessing the ability to photoswitch or photoconvert, probes should exhibit high, stable photon yields since the localization precision of single molecules is inversely proportional to the number of photons collected to first order. Next, the probes should be as close as possible to the target protein. Otherwise, a linkage error can significantly displace the probe from the protein's true location; this results in errors in the sizes and positions of proteins measured. For example, using a primary and secondary antibody can displace a synthetic dye from the target structure by ~20 nm. One strategy to reduce this linkage error is to use "nanobodies." These are single-domain antibodies, naturally produced in camelids, are only ~2 nm in size, and were only recently put to use for STORM.[6] Finally, probes must yield high molecular densities since this impacts image resolution. In the case of PALM, this means that FPs must efficiently photoactivate or photoconvert. For STORM, this means that synthetic dyes must densely label proteins, and an oxygen-scavenging, thiol-containing buffer must be used. The Nyquist criterion describes the required molecular density to resolve features of a given size (see **Problem 8.11**). Although this equation describes how increasing molecular density enhances resolution and has historically been used as a resolution metric for STORM and PALM, it has recently been shown to overestimate the true spatial resolution of an image. A more quantitative resolution measure is possible based on Fourier ring correlation for PALM and STORM; some references to this technique are given at the end of the chapter to this technique.

To summarize, a probe used for PALM or STORM should

1.  Possess a high photon yield

2.  Be smaller than the desired localization precision of single molecules

3.  Yield high molecular densities

### *Photoswitching mechanisms: FPs*

These proteins are characterized by a chromophore formed in their beta-barrel structure. This chromophore shielding renders the photoswitching performance of FPs less sensitive to their nanoenvironment as compared to synthetic dyes. FPs used for PALM are either *photoactivatable, photoconvertable*, or *photoswitchable*. Photoactivatable and photoconvertable FPs undergo permanent structural changes upon UV activation, such as permanent breakage of their protein backbone. Therefore, in principle, one molecule should be imaged only once after the FP has been photoactivated or photoconverted. In practice, however, these probes have been shown to reversibly switch, which means that a single FP can yield several molecular localizations. This complicates experiments where quantitative molecular counting is required. Several studies have addressed this concern and developed methods to overcome this overcounting problem.

Photoswitchable FPs are designed to undergo reversible structural transformations. Many of these FPs, such as Dronpa, undergo a reversible *cis–trans* isomerization in order to switch. *Cis* and *trans* isomers differ in their protonation states, which is precisely what controls the fluorescence emission of this FP.

**Figure 8.16 Concept of synthetic dye photoswitching.** Triplet state not shown for simplicity. Single-molecule photoswitching is achieved by reduction and oxidation of a dye molecule.

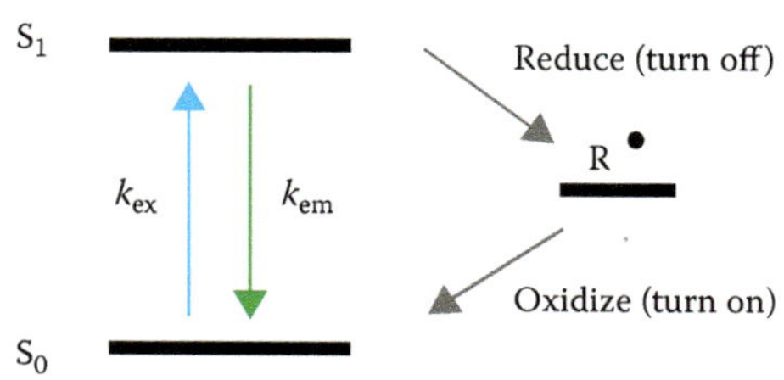


Several exceptions to a *cis–trans* isomerization mechanism have been reported. One example of this is Dreiklang, a photoswitchable FP that switches between hydration states.

### Photoswitching mechanisms: Synthetic dyes

Synthetic dyes can be made to photoswitch using specialized buffers. These buffers introduce a radical ion dark state to the dye, which is typically achieved by reduction. Dark states can last several milliseconds (a few frames), and the dye can return to its ground state (or turn "on") with a complementary oxidizing agent. Accordingly, such buffers contain an oxygen-scavenging system supplemented with reducing and oxidizing (redox) components (**Figure 8.16**). For commonly used Alexa dyes, thiols are efficient reducing agents via the triplet state. The purpose of the oxygen-scavenging system is to do the following:

1. Prevent reactions with molecular oxygen. This extends the lifetime of the radical ion dark state and results in more efficient photoswitching.

2. Minimize the generation of singlet oxygen, a reactive oxygen species (ROS) that is known to cause irreversible photodamage and bleaching via the triplet state.

Several dyes used for STORM have reduced states that strongly absorb in the UV range.[18] As a result, UV illumination allows for these dyes to transition from a dark to a bright state. Conveniently, this is a way to control single-molecule density: increasing the 405 nm laser irradiance during an acquisition increases the molecular density of a given frame. Synthetic dyes can stochastically cycle between bright and dark states multiple times. This reversible photoswitching mechanism is known as the *reducing and oxidizing system* (ROXS). Before performing a STORM experiment, one should consult the PALM/STORM literature to find dye–buffer combinations that work well since many synthetic dyes cannot be made to switch in standard STORM buffers. Failure to photoswitch is usually observed as fast photobleaching. After turning on, single molecules transition to an irreversible "off" state, rather than cycling between bright and dark states.

## 8.9  SUMMARY AND CONCLUSION

The past few years have seen the creation of a large number of techniques for advanced fluorescence microscopy. Many of these, such as 4Pi and PALM/STORM, are superresolution techniques; others, such as FLIM, provide qualitatively new information. Many of these techniques require custom instrumentation that is difficult (though possible) to design in-house. Commercial instrumentation is becoming widely available. Most large universities have FLIM microscopes, and most have the capacity for TIRF and PALM/STORM. Techniques such as FRET microscopy do not require any specialized instrumentation but involve careful selection of probes and labeling techniques. The greatest limitation to PALM/STORM is the discovery/development of photoswitchable dyes and FPs, currently a highly active area of research. These techniques have made new areas of biology possible, such as superresolution imaging of nanometer-scale processes in living cells. Applications to medicine and *in vivo* studies are emerging rapidly. The richness of information available in a fluorescent tag is remarkable, so it is worth learning about all the available techniques before designing a fluorescence microscopy experiment.

## End-of-Chapter Questions

**1.** Give expressions for the Abbe diffraction limit for axial ($z$) and lateral ($xy$) resolution. Why are they different?

**2.** Derive **Equation 8.1**. Hint: start from the general FRAP formula:

$$F(t) = QY \iint_{R^2} I(x,y)C(x,y,t)\,dx\,dy,$$

where QY is the quantum yield, $I$ is the laser excitation, and $C$ is the fluorophore concentration. Assume that $I$ is a uniform disk and that $C(x,y,t)$ is a product of pure diffusion.

**3.** Repeat problem 2 for the case of a Gaussian laser profile.

**4.** Derive **Equations 8.3 through 8.4**. What is the penetration depth of an Ar ion laser (514 nm) if $\theta = 64°$?

**5.** Compare the critical angles and maximum angles of incidence for objectives with (a) 60× magnification, NA = 1.45; (b) 100× magnification, oil immersion, NA = 1.65. Which would be easier to work with and why? Can you estimate the minimum NA for a 60× objective for which TIRF would be possible?

**6.** For a FLIM setup with a laser repetition rate of 80 MHz, what is the longest lifetime that can be expected to completely decay between pulses? What types of molecules fit this limit, and what types do not? How can FLIM be performed on longer-lifetime fluorophores?

**7.** Discuss how the timed release of an encapsulated drug may be monitored by FLIM. Pick a specific example and sketch how the lifetime changes with release.

**8.** Show that a general expression for the resolution of a sample containing photoactivatable fluorophores is

$$\frac{\lambda}{2(NA)\sqrt{1+\dfrac{I}{I_{sat}}}},$$

where $I$ is the intensity of illumination light, $I_{sat}$ is the intensity required for saturating the light-dark transition, and $a$ is a parameter taking into account the shape of the laser beam. How is this ratio maximized?

Hint: write this as a rate equation with two states A ("on") and B ("off"). Assume the following: the rate of transition from A to B depends upon intensity: $k_{AB} = \sigma I$, whereas the transition from B to A is independent of intensity; at time 0, all N fluorophores are in state A; and the total number of fluorophores is constant. Solve the rate equation and take the limit as $t$ goes to infinity to get the result.

**9.** For techniques such as PALM and STORM, what are the key properties of the fluorophores that determine the resolution? Discuss how the following work in photoswitching and what laser lines are required: the cyanine switch; Alexa 647; caged fluorescein.

**10.** Can quantum dots be used for PALM/STORM? If not, could they be engineered to work? Discuss.

**11.** Discuss how density of fluorophores can affect superresolution microscopy. (Hint: use the Nyquist sampling criterion.) Why is this particularly important in PALM/STORM?

**12.** Discuss how the fluorescence intermittency (blinking) of fluorophores can be exploited for superresolution microscopy.

**13.** Which technique would you use to: (a) image red fluorescent protein deep inside a live mouse; (b) observe a fluorescent drug being taken up by endosomes in cancer cells; (c) watch exocytosis of fluorescently labeled dopamine from cultured neurons;

(d) image the structure of nuclear pores in fixed cells; (e) determine how the influenza hemagglutinin protein travels through living cells. Justify your answers.

**14.** Imagine labeling a membrane with a diffuse photoswitchable dye and capturing a live-cell STORM acquisition. What artifacts would be present in the STORM image, and how can adjusting the camera frame rate minimize this effect? Discuss.

**15.** What assumptions are we making when applying the Nyquist criterion to STORM images? How is this problematic in biology?

**16.** In principle, which fluorescent probe would you use to (a) measure protein stoichiometry and (b) determine the diameter of a microtubule? Discuss.

## Background Reading

### Books

Diaspro, A., and Zandvoort, M.A.M.J.v. *Super-resolution Imaging in Biomedicine*. CRC Press, Boca Raton, FL, 2016.

Marcu, L., French, P.M.W., and Elson, D.S. *Fluorescence Lifetime Spectroscopy and Imaging: Principles and Applications in Biomedical Diagnostics*. CRC Press, Boca Raton, FL, 2016.

Novotny, L. and Hecht, B. *Principles of Nano-Optics*. Edn. 2. Cambridge University Press, Cambridge, UK, 2012.

Periasamy, A., and Clegg, R.M. *FLIM Microscopy in Biology and Medicine*. Taylor & Francis, Boca Raton, FL, 2010.

Popescu, G. *Nanobiophotonics*. McGraw-Hill, New York, 2010.

### Journal articles

*General review articles on superresolution*

Galbraith, C.G., and Galbraith, J.A. (2011). Super-resolution microscopy at a glance. *Journal of Cell Science* 124, 1607–1611. http://www.ncbi.nlm.nih.gov/pubmed/21536831

Halpern, A.R., Howard, M.D., and Vaughan, J.C. (2015). Point by point: An introductory guide to sample preparation for single-molecule, super-resolution fluorescence microscopy. *Current Protocols in Chemical Biology* 7, 103–120. http://www.ncbi.nlm.nih.gov/pubmed/26344236

Schermelleh, L., Heintzmann, R., and Leonhardt, H. (2010). A guide to super-resolution fluorescence microscopy. *The Journal of Cell Biology* 190, 165–175. http://www.ncbi.nlm.nih.gov/pubmed/20643879

Wegel, E., Gohler, A., Lagerholm, B.C., Wainman, A., Uphoff, S., Kaufmann, R., and Dobbie, I.M. (2016). Imaging cellular structures in super-resolution with SIM, STED and localisation microscopy: A practical comparison. *Scientific Reports* 6, 27290. http://www.ncbi.nlm.nih.gov/pubmed/27264341

*Hyperspectral imaging*

Gao, L., and Smith, R.T. (2015). Optical hyperspectral imaging in microscopy and spectroscopy—A review of data acquisition. *Journal of Biophotonics* 8, 441–456. http://www.ncbi.nlm.nih.gov/pubmed/25186815

Gowen, A.A., Feng, Y., Gaston, E., and Valdramidis, V. (2015). Recent applications of hyperspectral imaging in microbiology. *Talanta* 137, 43–54. http://www.ncbi.nlm.nih.gov/pubmed/25770605

Lu, G., and Fei, B. (2014). Medical hyperspectral imaging: A review. *Journal of Biomedical Optics* 19, 10901. http://www.ncbi.nlm.nih.gov/pubmed/24441941

*TIRF*

Asbury, C.L. (2016). Data analysis for total internal reflection fluorescence microscopy. *Cold Spring Harbor Protocols* 2016, pdb prot085571. http://www.ncbi.nlm.nih.gov/pubmed/27140913

Axelrod, D. (2001). Selective imaging of surface fluorescence with very high aperture microscope objectives. *Journal of Biomedical Optics* 6, 6–13.

Boehm, E.M., Subramanyam, S., Ghoneim, M., Washington, M.T., and Spies, M. (2016). Quantifying the assembly of multicomponent molecular machines by single-molecule total internal reflection fluorescence microscopy. *Methods in Enzymology* 581, 105–145. http://www.ncbi.nlm.nih.gov/pubmed/27793278

Delgado-Peraza, F., Nogueras-Ortiz, C., Acevedo Canabal, A.M., Roman-Vendrell, C., and Yudowski, G.A. (2016). Imaging GPCRs trafficking and signaling with total internal reflection fluorescence microscopy in cultured neurons. *Methods in Cell Biology* 132, 25–33. http://www.ncbi.nlm.nih.gov/pubmed/26928537

Kudalkar, E.M., Davis, T.N., and Asbury, C.L. (2016). Single-molecule total internal reflection fluorescence microscopy. *Cold Spring Harbor Protocols* 2016, pdb top077800. http://www.ncbi.nlm.nih.gov/pubmed/27140922

Mashanov, G.I., Tacon, D., Knight, A.E., Peckham, M., and Molloy, J.E. (2003). Visualizing single molecules inside living cells using total internal reflection fluorescence microscopy. *Methods* 29, 142–152.

Sako, Y., and Uyemura, T. (2002). Total internal reflection fluorescence microscopy for single-molecule imaging in living cells. *Cell Structure and Function* 27, 357–365. http://www.ncbi.nlm.nih.gov/pubmed/12502890

Steyer, J.A., and Almers, W. (2001). A real-time view of life within 100 nm of the plasma membrane. *Nature Reviews. Molecular Cell Biology* 2, 268–275. http://www.ncbi.nlm.nih.gov/pubmed/11283724

Toomre, D., and Manstein, D.J. (2001). Lighting up the cell surface with evanescent wave microscopy. *Trends in Cell Biology* 11, 298–303. http://www.ncbi.nlm.nih.gov/pubmed/11413041


### FRAP and variants

De Los Santos, C., Chang, C.W., Mycek, M.A., and Cardullo, R.A. (2015). FRAP, FLIM, and FRET: Detection and analysis of cellular dynamics on a molecular scale using fluorescence microscopy. *Molecular Reproduction and Development* 82, 587–604. http://www.ncbi.nlm.nih.gov/pubmed/26010322

Deschout, H., Raemdonck, K., Demeester, J., De Smedt, S.C., and Braeckmans, K. (2014). FRAP in pharmaceutical research: Practical guidelines and applications in drug delivery. *Pharmaceutical Research* 31, 255–270. http://www.ncbi.nlm.nih.gov/pubmed/24019022

Hauser, G.I., Seiffert, S., and Oppermann, W. (2008). Systematic evaluation of FRAP experiments performed in a confocal laser scanning microscope—Part II: Multiple diffusion processes. *Journal of Microscopy* 230, 353–362. http://www.ncbi.nlm.nih.gov/pubmed/18503660

Ishikawa-Ankerhold, H.C., Ankerhold, R., and Drummen, G.P. (2012). Advanced fluorescence microscopy techniques—FRAP, FLIP, FLAP, FRET and FLIM. *Molecules* 17, 4047–4132. http://www.ncbi.nlm.nih.gov/pubmed/22469598

Lippincott-Schwartz, J., Snapp, E., and Kenworthy, A. (2001). Studying protein dynamics in living cells. *Nature Reviews. Molecular Cell Biology* 2, 444–456. http://www.ncbi.nlm.nih.gov/pubmed/11389468

Mueller, F., Mazza, D., Stasevich, T.J., and McNally, J.G. (2010). FRAP and kinetic modeling in the analysis of nuclear protein dynamics: What do we really know? *Current Opinion in Cell Biology* 22, 403–411. http://www.ncbi.nlm.nih.gov/pubmed/20413286

Nouar, R., Devred, F., Breuzard, G., and Peyrot, V. (2013). FRET and FRAP imaging: Approaches to characterise tau and stathmin interactions with microtubules in cells. *Biology of the Cell* 105, 149–161. http://www.ncbi.nlm.nih.gov/pubmed/23312015

Seiffert, S., and Oppermann, W. (2005). Systematic evaluation of FRAP experiments performed in a confocal laser scanning microscope. *Journal of Microscopy* 220, 20–30. http://www.ncbi.nlm.nih.gov/pubmed/16269060

Wachsmuth, M. (2014). Molecular diffusion and binding analyzed with FRAP. *Protoplasma* 251, 373–382. http://www.ncbi.nlm.nih.gov/pubmed/24390250

Watanabe, N., Yamashiro, S., Vavylonis, D., and Kiuchi, T. (2013). Molecular viewing of actin polymerizing actions and beyond: Combination analysis of single-molecule speckle microscopy with modeling, FRAP and s-FDAP (sequential fluorescence decay after photoactivation). *Development, Growth and Differentiation* 55, 508–514. http://www.ncbi.nlm.nih.gov/pubmed/23621590

White, J., and Stelzer, E. (1999). Photobleaching GFP reveals protein dynamics inside live cells. *Trends in Cell Biology* 9, 61–65. http://www.ncbi.nlm.nih.gov/pubmed/10087620


### Multiphoton

Bestvater, F., Spiess, E., Stobrawa, G., Hacker, M., Feurer, T., Porwol, T., Berchner-Pfannschmidt, U., Wotzlaw, C., and Acker, H. (2002). Two-photon fluorescence absorption and emission spectra of dyes relevant for cell imaging. *Journal of Microscopy—Oxford* 208, 108–115.

Drobizhev, M., Makarov, N.S., Tillo, S.E., Hughes, T.E., and Rebane, A. (2011). Two-photon absorption properties of fluorescent proteins. *Nature Methods* 8, 393–399. http://www.ncbi.nlm.nih.gov/pubmed/21527931

Feng, X.B., Ang, Y.L., He, J., Beh, C.W.J., Xu, H.R., Chin, W.S., and Ji, W. (2008). Three-photon absorption in semiconductor quantum dots: Experiment. *Optics Express* 16, 6999–7005.

Mank, M., Santos, A.F., Direnberger, S., Mrsic-Flogel, T.D., Hofer, S.B., Stein, V., Hendel, T., Reiff, D.F., Levelt, C., Borst, A. et al. (2008). A genetically encoded calcium indicator for chronic in vivo two-photon imaging. *Nature Methods* 5, 805–811.

Mullineaux, C.W., and Kirchhoff, H. (2007). Using fluorescence recovery after photobleaching to measure lipid diffusion in membranes. *Methods in Molecular Biology* 400, 267–275. http://www.ncbi.nlm.nih.gov/pubmed/17951740

Mutze, J., Iyer, V., Macklin, J.J., Colonell, J., Karsh, B., Petrasek, Z., Schwille, P., Looger, L.L., Lavis, L.D., and Harris, T.D. (2012). Excitation spectra and brightness optimization of two-photon excited probes. *Biophysical Journal* 102, 934–944.

Niko, Y., Moritomo, H., Sugihara, H., Suzuki, Y., Kawamata, J., and Konishi, G.I. (2015). A novel pyrene-based two-photon active fluorescent dye efficiently excited and emitting in the 'tissue optical window (650–1100 nm)'. *Journal of Materials Chemistry B* 3, 184–190.

Podgorski, K., Terpetschnig, E., Klochko, O.P., Obukhova, O.M., and Haas, K. (2012). Ultra-bright and -stable red and near-infrared squaraine fluorophores for in vivo two-photon imaging. *PloS One* 7, e51980. http://www.ncbi.nlm.nih.gov/pubmed/23251670

Pu, S.C., Yang, M.J., Hsu, C.C., Lai, C.W., Hsieh, C.C., Lin, S.H., Cheng, Y.M., and Chou, P.T. (2006). The empirical correlation between size and two-photon absorption cross section of CdSe and CdTe quantum dots. *Small* 2, 1308–1313.

Svoboda, K., and Yasuda, R. (2006). Principles of two-photon excitation microscopy and its applications to neuroscience. *Neuron* 50, 823–839.

Xing, G.C., Ji, W., Zheng, Y.G., and Ying, J.Y. (2008). Two- and three-photon absorption of semiconductor quantum dots in the vicinity of half of lowest exciton energy. *Applied Physics Letters* 93, 241114.

Xu, C., and Webb, W.W. (1996). Measurement of two-photon excitation cross sections of molecular fluorophores with data from 690 to 1050 nm. *Journal of the Optical Society of America B* 13, 481–491.

Zipfel, W.R., Williams, R.M., and Webb, W.W. (2003). Nonlinear magic: Multiphoton microscopy in the biosciences. *Nature Biotechnology* 21, 1369–1377. http://www.ncbi.nlm.nih.gov/pubmed/14595365

*FLIM*

Bakker, G.J., Andresen, V., Hoffman, R.M., and Friedl, P. In Conn, P.M. (ed.). *Imaging and Spectroscopic Analysis of Living Cells: Optical and Spectroscopic Techniques. Methods in Enzymology.* Academic Press, Cambridge, MA. Vol. 504. pp. 109–125, 2012.

Bastos, A.E.P., Scolari, S., Stockl, M., and de Almeida, R.F.M. In Conn, P.M. (ed.). *Imaging and Spectroscopic Analysis of Living Cells: Optical and Spectroscopic Techniques. Methods in Enzymology.* Academic Press, Cambridge, MA. Vol. 504. pp. 57–81, 2012.

Becker, W. (2012). Fluorescence lifetime imaging—Techniques and applications. *Journal of Microscopy* 247, 119–136.

Blum, C., Cesa, Y., Escalante, M., and Subramaniam, V. (2009). Multimode microscopy: Spectral and lifetime imaging. *Journal of the Royal Society Interface* 6, S35–S43.

Carlini, L., and Nadeau, J.L. (2013). Uptake and processing of semiconductor quantum dots in living cells studied by fluorescence lifetime imaging microscopy (FLIM). *Chemical Communications* 49, 1714–1716.

Chang, C.W., Sud, D., and Mycek, M.A. In Sluder, G., and Wolf, D.E. (eds.). *Digital Microscopy.* Academic Press, Cambridge, MA. Edn. 3. Vol. 81. p. 495, 2007.

Chen, L.C., Lloyd, W.R., Chang, C.W., Sud, D., and Mycek, M.A. In Sluder, G., and Wolf, D.E. (eds.). *Digital Microscopy* Edn. 4. Vol. 114. pp. 457–488, Elsevier, Amsterdam, the Netherlands, 2013.

Chen, Y., Mills, J.D., and Periasamy, A. (2003). Protein localization in living cells and tissues using FRET and FLIM. *Differentiation* 71, 528–541.

Clegg, R.M., Holub, O., and Gohlke, C. (2003). Fluorescence lifetime-resolved imaging: Measuring lifetimes in an image. *Biophotonics, Part A* 360, 509–542.

Festy, F., Ameer-Beg, S.M., Ng, T., and Suhling, K. (2007). Imaging proteins in vivo using fluorescence lifetime microscopy. *Molecular Biosystems* 3, 381–391. <Go to ISI>://WOS:000246804300013

French, T., So, P.T.C., Dong, C.Y., Berland, K.M., and Gratton, E. (1998). Fluorescence lifetime imaging techniques for microscopy. *Methods in Cell Biology* 56, 277–304.

Niehorster, T., Loschberger, A., Gregor, I., Kramer, B., Rahn, H.J., Patting, M., Koberling, F., Enderlein, J., and Sauer, M. (2016). Multi-target spectrally resolved fluorescence lifetime imaging microscopy. *Nature Methods* 13, 257.

Petrasek, Z., Eckert, H.J., and Kemnitz, K. (2009). Wide-field photon counting fluorescence lifetime imaging microscopy: Application to photosynthesizing systems. *Photosynthesis Research* 102, 157–168.

Provenzano, P.P., Eliceiri, K.W., and Keely, P.J. (2009). Multiphoton microscopy and fluorescence lifetime imaging microscopy (FLIM) to monitor metastasis and the tumor microenvironment. *Clinical and Experimental Metastasis* 26, 357–370.

Sarder, P., Maji, D., and Achilefu, S. (2015). Molecular probes for fluorescence lifetime imaging. *Bioconjugate Chemistry* 26, 963–974.

Suhling, K., French, P.M.W., and Phillips, D. (2005). Time-resolved fluorescence microscopy. *Photochemical and Photobiological Sciences* 4, 13–22.

Sun, Y.S., Hays, N.M., Periasamy, A., Davidson, M.W., and Day, R.N. In Conn, P.M. (ed.). *Imaging and Spectroscopic Analysis of Living Cells: Optical and Spectroscopic Techniques*, Vol. 504. pp. 371–391, Academic Press, NY, 2012.

Wallrabe, H., and Periasamy, A. (2005). Imaging protein molecules using FRET and FLIM microscopy. *Current Opinion in Biotechnology* 16, 19–27.

Wang, X.F., Periasamy, A., Herman, B., and Coleman, D.M. (1992). Fluorescence lifetime imaging microscopy (FLIM)—Instrumentation and applications. *Critical Reviews in Analytical Chemistry* 23, 369–395.

Zhang, X., Shastry, S., Bradforth, S.E., and Nadeau, J.L. (2015). Nuclear uptake of ultrasmall gold-doxorubicin conjugates imaged by fluorescence lifetime imaging microscopy (FLIM) and electron microscopy. *Nanoscale* 7, 240–251.

*4Pi*

Baddeley, D., Carl, C., and Cremer, C. (2006). 4Pi microscopy deconvolution with a variable point-spread function. *Applied Optics* 45, 7056–7064.

Bahlmann, K., Jakobs, S., and Hell, S.W. (2001). 4Pi-confocal microscopy of live cells. *Ultramicroscopy* 87, 155–164.

Bewersdorf, J., Bennett, B.T., and Knight, K.L. (2006). H2AX chromatin structures and their response to DNA damage revealed by 4Pi microscopy. *Proceedings of the National Academy of Sciences of the United States of America* 103, 18137–18142.

Bewersdorf, J., Schmidt, R., and Hell, S.W. (2006). Comparison of (IM)-M-5 and 4Pi-microscopy. *Journal of Microscopy* 222, 105–117.

Davis, B.J., Dogan, M., Goldberg, B.B., Karl, W.C., Unlu, M.S., and Swan, A.K. (2007). 4Pi spectral self-interference microscopy. *Journal of the Optical Society of America a-Optics Image Science and Vision* 24, 3762–3771.

Dilipkumar, S., and Mondal, P.P. (2013). Taylor series expansion based multidimensional image reconstruction for confocal and 4Pi microscopy. *Applied Physics Letters* 103.

Egner, A., Jakobs, S., and Hell, S.W. (2002). Fast 100-nm resolution three-dimensional microscope reveals structural plasticity of mitochondria in live yeast. *Proceedings of the National Academy of Sciences of the United States of America* 99, 3370–3375.

Glaschick, S., Rocker, C., Deuschle, K., Wiedenmann, J., Oswald, F., Mailander, V., and Nienhaus, G.U. (2007). Axial resolution enhancement by 4Pi confocal fluorescence microscopy with two-photon excitation. *Journal of Biological Physics* 33, 433–443.

Gugel, H., Bewersdorf, J., Jakobs, S., Engelhardt, J., Storz, R., and Hell, S.W. (2004). Cooperative 4Pi excitation and detection yields sevenfold sharper optical sections in live-cell microscopy. *Biophysical Journal* 87, 4146–4152.

Hell, S., and Stelzer, E.H.K. (1992). Fundamental improvement of resolution with a 4Pi-confocal fluorescence microscope using 2-photon excitation. *Optics Communications* 93, 277–282.

Hell, S.W. (2003). Toward fluorescence nanoscopy. *Nature Biotechnology* 21, 1347–1355.

Hell, S.W., Lindek, S, and Stelzer, E.H.K. (1994). Enhancing the axial resolution in far-field light-microscopy—2-photon 4Pi confocal fluorescence microscopy. *Journal of Modern Optics* 41, 675–681.

Hell, S.W., and Nagorni, M. (1998). 4Pi confocal microscopy with alternate interference. *Optics Letters* 23, 1567–1569.

Hell, S.W., Stelzer, E.H.K., Lindek, S., and Cremer, C. (1994). Confocal microscopy with an increased detection aperture—Type-B 4Pi confocal microscopy. *Optics Letters* 19, 222–224.

Huang, F., Sirinakis, G., Allgeyer, E.S., Schroeder, L.K., Duim, W.C., Kromann, E.B., Phan, T., Rivera-Molina, F.E., Myers, J.R., Irnov, I. et al. (2016). Ultra-high resolution 3D imaging of whole cells. *Cell* 166, 1028–1040.

Rasmussen, A., and Deckert, V. (2005). New dimension in nano-imaging: Breaking through the diffraction limit with scanning near-field optical microscopy. *Analytical and Bioanalytical Chemistry* 381, 165–172.

Schnitzbauer, J., McGorty, R., and Huang, B. (2013). 4Pi fluorescence detection and 3D particle localization with a single objective. *Optics Express* 21, 19701–19708.

Schrader, M., and Hell, S.W. (1996). 4Pi-confocal images with axial superresolution. *Journal of Microscopy—Oxford* 183, 189–193.

Strack, R. Nanoscopy for the whole cell. (2016). *Nature Methods* 13, 709–709.

Vicidomini, G., Hell, S.W., and Schonle, A. (2009). Automatic deconvolution of 4Pi-microscopy data with arbitrary phase. *Optics Letters* 34, 3583–3585.

Vicidomini, G., Schmidt, R., Egner, A., Hell, S.W., and Schonle, A. (2010). Automatic deconvolution in 4Pi-microscopy with variable phase. *Optics Express* 18, 10154–10167.

Zhang, P., Goodwin, P.M., and Werner, J.H. (2014). Interferometric three-dimensional single molecule localization microscopy using a single high-numerical-aperture objective. *Applied Optics* 53, 7415–7421.

## *PALM/STORM*

Andresen, M., Stiel, A.C., Trowitzsch, S., Weber, G., Eggeling, C., Wahl, M.C., Hell, S.W., and Jakobs, S. (2007). Structural basis for reversible photoswitching in Dronpa. *Proceedings of the National Academy of Sciences of the United States of America* 104, 13005–13009.

Annibale, P., Scarselli, M., Kodiyan, A., and Radenovic, A. (2010). Photoactivatable fluorescent protein mEos2 displays repeated photoactivation after a long-lived dark state in the red photoconverted form. *Journal of Physical Chemistry Letters* 1, 1506–1510.

Banterle, N., Bui, K.H., Lemke, E.A., and Beck, M. (2013). Fourier ring correlation as a resolution criterion for super-resolution microscopy. *Journal of Structural Biology* 183, 363–367.

Betzig, E., Patterson, G.H., Sougrat, R., Lindwasser, O.W., Olenych, S., Bonifacino, J.S., Davidson, M.W., Lippincott-Schwartz, J., and Hess, H.F. (2006). Imaging intracellular fluorescent proteins at nanometer resolution. *Science* 313, 1642–1645.

Brakemann, T., Stiel, A.C., Weber, G., Andresen, M., Testa, I., Grotjohann, T., Leutenegger, M., Plessmann, U., Urlaub, H., Eggeling, C. et al. (2011). A reversibly photoswitchable GFP-like protein with fluorescence excitation decoupled from switching. *Nature Biotechnology* 29, 942-U132.

Cordes, T., Vogelsang, J., and Tinnefeld, P. (2009). On the mechanism of Trolox as antiblinking and antibleaching reagent. *Journal of the American Chemical Society* 131(14), 5018–5019.

Dempsey, G.T., Vaughan, J.C., Chen, K.H., Bates, M., and Zhuang, X. (2011). Evaluation of fluorophores for optimal performance in localization-based super-resolution imaging. *Nature Methods* 8, 1027–1036.

Fitzgerald, J.E., Lu, J., and Schnitzer, M.J. (2012). Estimation theoretic measure of resolution for stochastic localization microscopy. *Physical Review Letters* 109, 048102.

Hamers-Casterman, C., Atarhouch, T., Muyldermans, S., Robinson, G., Hamers, C., Songa, E.B., Bendahman, N., and Hamers, R. (1993). Naturally occurring antibodies devoid of light chains. *Nature* 363, 446–448.

Heilemann, M., van de Linde, S., Schuttpelz, M., Kasper, R., Seefeldt, B., Mukherjee, A., Tinnefeld, P., and Sauer, M.

(2008). Subdiffraction-resolution fluorescence imaging with conventional fluorescent probes. *Angewandte Chemie—International Edition* 47, 6172–6176.

Hess, S.T., Girirajan, T.P., and Mason, M.D. (2006). Ultra-high resolution imaging by fluorescence photoactivation localization microscopy. *Biophysical Journal* 91, 4258–4272.

Lee, S.H., Shin, J.Y., Lee, A., and Bustamante, C. (2012). Counting single photoactivatable fluorescent molecules by photoactivated localization microscopy (PALM). *Proceedings of the National Academy of Sciences of the United States of America* 109, 17436–17441.

Li, D., Shao, L., Chen, B.C., Zhang, X., Zhang, M., Moses, B., Milkie, D.E., Beach, J.R., Hammer, J.A. 3rd, Pasham, M. et al. (2015). Advanced imaging. Extended-resolution structured illumination imaging of endocytic and cytoskeletal dynamics. *Science* 349, aab3500.

Nieuwenhuizen, R.P., Lidke, K.A., Bates, M., Puig, D.L., Grunwald, D., Stallinga, S., and Rieger, B. (2013). Measuring image resolution in optical nanoscopy. *Nature Methods* 10, 557–562.

Olivier, N., Keller, D., Gonczy, P., and Manley, S. (2013). Resolution doubling in 3D-STORM imaging through improved buffers. *Plos One* 8(7), e69004.

Ries, J., Kaplan, C., Platonova, E., Eghlidi, H., and Ewers, H. (2012). A simple, versatile method for GFP-based super-resolution microscopy via nanobodies. *Nature Methods* 9, 582–584.

Rust, M.J., Bates, M., and Zhuang, X. (2006). Sub-diffraction-limit imaging by stochastic optical reconstruction microscopy (STORM). *Nature Methods* 3, 793–795.

Stepanenko, O.V., Stepanenko, O.V., Kuznetsova, I.M., Verkhusha, V.V., and Turoverov, K.K. (2013). Beta-barrel scaffold of fluorescent proteins: Folding, stability and role in chromophore formation. *International Review of Cell and Molecular Biology* 302, 221–278.

Thompson, R.E., Larson, D.R., and Webb, W.W. (2002) Precise nanometer localization analysis for individual fluorescent probes. *Biophysical Journal* 82, 2775–2783.

van de Linde, S., and Sauer, M. (2014). How to switch a fluorophore: from undesired blinking to controlled photoswitching. *Chemical Society Reviews* 43, 1076–1087.

van de Linde, S., Loschberger, A., Klein, T., Heidbreder, M., Wolter, S., Heilemann, M., and Sauer, M. (2011a). Direct stochastic optical reconstruction microscopy with standard fluorescent probes. *Nature Protocols* 6, 991–1009.

van de Linde, S., Krstic, I., Prisner, T., Doose, S., Heilemann, M., and Sauer, M. (2011b). Photoinduced formation of reversible dye radicals and their impact on super-resolution imaging. *Photochemical and Photobiological Sciences* 10, 499–506.

Vogelsang, J., Steinhauer, C., Forthmann, C., Stein, I., Person-Skegro, B., Cordes, T., and Tinnefeld, P. (2010). Make them blink: Probes for super-resolution microscopy. *ChemPhysChem* 11, 2475–2490.

Zhou, X.X., Lin, M.Z. (2013). Photoswitchable fluorescent proteins: Ten years of colorful chemistry and exciting applications. *Current Opinion in Chemical Biology* 17, 682–690.

## Online resources and software

### *Resources*

Directory of localization microscopy software: http://bigwww.epfl.ch/smlm/software/

Guide to superresolution from BiteSizeBio: http://bitesizebio.com/13436/overcoming-the-limits-of-light-a-guide-to-super-resolution-microscopy-part-1/

Leica's guide to superresolution microscopy: http://www.leica-microsystems.com/science-lab/a-guide-to-super-resolution-fluorescence-microscopy/

Biocompare 2016 editorial comparing superresolution techniques: http://www.biocompare.com/Editorial-Articles/186077-Super-resolution-Microscopy-2016-Update/

Becker & Hickl *TCPSC Handbook* (free download): http://www.becker-hickl.com/handbookphp.htm

Other Becker & Hickl handbooks: http://www.becker-hickl.com/literature.htm

### *Free software*

FLIMFit. Open-source software for fitting FLIM data. Can import Becker & Hickl and PicoQuant data, as well as other data types with some manipulation sometimes required.

Globals. Free software for phasor plot analysis, a necessary complement to FLIM studies that is not provided in commercial software.

FluorTools. Image analysis tools including DecayFit (FLIM), AniFit (anisotropy), and iSMS (single-molecule FRET).

FRETBursts. Open-source software for single-molecule FRET.

smFRET. Zhuang lab's software for single-molecule FRET analysis: http://zhuang.harvard.edu/smFRET.html.

iSMS. Aarhus University single-molecule FRET package: http://isms.au.dk/about/.

ImageJ Plug-Ins: ThunderSTORM; GraspJ; fairSIM; single-molecule localization microscopy (SMLM) (many plugins for different techniques); NanoJ and NanoJ-SRRF; and more. (See http://imagej .net/List_of_update_sites.)

rapidSTORM. Open-source software for localization microscopy, with 3-D capability.

### *Commercial software*

SymPhoTime. Lifetime acquisition and analysis software from PicoQuant, designed for use with their instruments.

SPC Image. Software for FLIM and FRET from Becker & Hickl.

Zen. Acquisition and analysis software from Zeiss. Other software is also available for many types of microscopy: https://www.zeiss.com/microscopy /us/products/microscope-software.html.

## Instrument manufacturers

Zeiss. Zeiss manufactures the Elyra superresolution system, which permits structured illumination and PALM; both can be used with correlative scanning electron microscopy.

Nikon. Manufactures microscopes for structured illumination (N-SIM); STORM (N-STORM).

Leica. Manufactures commercial stimulated emission depletion microscopes (TCS SP8 STED 3X), including gated and 3-D STED; single-molecule localization microscopes (Leica GSD); and 4Pi.

GEBiosciences. DeltaVision OMX structured illumination microscopy system.

Bruker. Manufacturer of a fast (video-rate) superresolution microscope (Vutara 350).

Becker & Hickl. All things lifetime related. Systems for bulk TCSPC (see **Chapter 16**) as well as FLIM modules specific to Olympus, Zeiss, or Leica microscopes or a general-purpose FLIM module.

PicoQuant. Lifetime systems, including FLIM modules for Olympus, Nikon, Zeiss, and Leica microscopes and complete FLIM microscope systems of varying levels of complexity.

### *Reagent suppliers*

Evrogen. Photoactivatable fluorescent proteins.

Clontech. Some photoactivatable/photoswitchable FPs.

AddGene. Many FP constructs from published literature.

# CHAPTER 9

# Advanced Topics in Microscopy II: Holographic Microscopy

Coauthored with Manuel Bedrossian

## 9.1 INTRODUCTION

Denis Gabor won the Nobel Prize in Physics in 1971 for "invention and development of the holographic method." Holography is an interferometric technique that relies upon coherent interference between a clear reference beam and a so-called object beam transmitted through, or reflected by, a sample of interest. *Digital holographic microscopy* (DHM) adds a lens (usually a microscope objective) in the object path to obtain a magnified image that is interfered with the reference wave; the object wave front of interest is a magnified replica of the sample.

DHM possesses several key advantages over traditional light microscopy (brightfield, phase contrast) for some biological applications. The optical components involved in DHM are relatively simple and low cost compared to those needed for fluorescence microscopy. A DHM instrument can be robust and have no moving parts, making it ideal for field applications, including underwater imaging. A single hologram captures all of the information in a three-dimensional volume, allowing for high throughput and instantaneous imaging of fast three-dimensional processes such as swimming of microorganisms or contractile motion of cardiomyocytes. The data are essentially "precompressed" by being encoded in the *hologram*, which is desirable for any application requiring data relaying. Thus, applications of holography to biological problems have mostly focused upon *in situ* use in remote areas. Depending upon the instrument design, this may be for detection of parasitic diseases; for tracking microbial behavior underwater; or for detection of microorganisms in extreme environments with a view to developing instruments for space flight.

The goal of this chapter is to introduce the physics of hologram acquisition and reconstruction in the context of biological cell imaging. A few of the designs in current use are discussed in detail, showing the advantages and disadvantages of each. Finally, we provide a simple design for a low-cost do-it-yourself DHM and links to sites for reconstruction software.

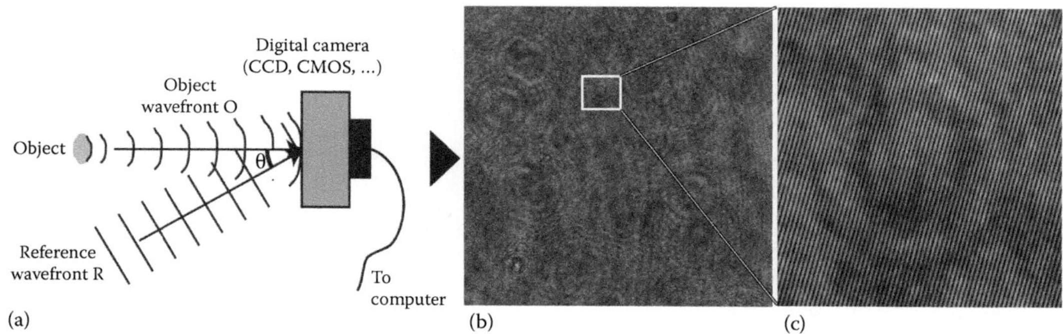


**Figure 9.1 Principle of DHM.** (a) The setup requires two light beams, usually coherent laser beams, to interfere and encode the wave front of the object as a fringe pattern. (b) Appearance of a raw hologram of a bacterial culture. (c) Inset showing a magnified area of the hologram to illustrate the fringes.

## 9.2 PHYSICS OF HOLOGRAPHY

Holography with electron waves has been used for three-quarters of a century; Gabor invented the technique as an attempt to improve the resolution of the electron microscope. It has also been used with x-rays. Holography using visible light was limited until recently by the quality of coherent light sources (lasers) available and the pixel size of CCD cameras. With current technology, it is possible to record high-resolution holograms and reconstruct them in almost real time.

A typical DHM is essentially a Mach–Zehnder interferometer. An object wave front interacts with an object—usually by transmission, though sometimes by reflection. It is then interfered with a slightly off-axis reference wave front. This encodes the complex wave front (amplitude and phase) as intensity modulation (fringes) in the detector plane. The resulting fringe pattern is recorded by a digital camera (**Figure 9.1**).

Because nearly all of the illuminating photons are captured, low-power excitation sources can be used—specifically milliwatt diode lasers, which are available in a large range of wavelengths. For similar reasons, camera sensitivity can be significantly less than for fluorescence imaging; cooled CCDs are not required. Acquisition speed is limited by the speed of the camera. For capture of rapid motion, cameras with full-frame shutter (CCD) are preferable to rolling readout (CMOS). Spatial resolution is determined by the numerical aperture of the system, with a limiting resolution of $\lambda$/numerical aperture (NA) as in conventional light microscopy as long as the camera pixels sample the fringes at the Nyquist frequency or better. Because a single wavelength of illumination light is used, compound objective lenses with chromatic aberration correction are not needed.

## 9.3 RECONSTRUCTING HOLOGRAMS

Following recording, *reconstruction* may be undertaken on a computer as a postprocessing step. A large body of literature exists on reconstruction of amplitude and phase images from digital holograms. As for any interferometric recording, in digital holography, the acquired hologram $h$ is the sum of individual noncoherent (intensity) terms and interference (complex) contributions:

$$h = (R+O)(R+O)^* = R^2 + O^2 + RO^* + R^*O, \qquad (9.1)$$

where the first two terms are intensities of the reference wave front $R$ and the object wave front $O$, and the last two are the interference terms of interest, respectively called the real and the virtual image.

The hologram reconstruction algorithms are very simply described and depend on well-developed fast Fourier transforms (FFT). Reconstruction of a single image $\Gamma(\xi,\eta)$ is described by the equation

$$\Gamma(\xi,\eta) = \Im^{-1}\{\Im(h \cdot R) \cdot G\}, \qquad (9.2)$$

where $h$ is the raw recorded hologram, $R$ is a term describing the reference wave, and $G$ is the Fourier transform of the *propagator*, a refocus term that propagates the solution to the desired plane of reconstruction. A very useful feature of **Equation 9.2** is that although three Fourier transforms are computed for each reconstructed image, two of them can be precomputed and used for multiple reconstructions. The term $G$ is entirely independent of the image and can be calculated and expressed analytically for a number of different wave front propagator choices (typically the Fresnel approximation), simplifying and speeding computation. The term $\Im(h \cdot R)$ is the Fourier transform of the product of the hologram and the reference wave, $R$. It needs to be computed only once per hologram, after which it is multiplied by the $G$ term for a particular focus position, and a single Fourier transform of the product yields the focused image at the desired plane. The intensity image is then simply given by $|\Gamma(\xi,\eta)|^2$, and the phase image is given by $\arctan(\mathrm{Im}[\Gamma(\xi,\eta)]/\mathrm{Re}[\Gamma(\xi,\eta)])$.

Practically speaking, this means that the intensity image can be reconstructed with ordinary FFT algorithms and some additional code for the $R$ and $G$ terms. The first step is to take the hologram and apodize it (**Figure 9.2a**) to reduce aliasing

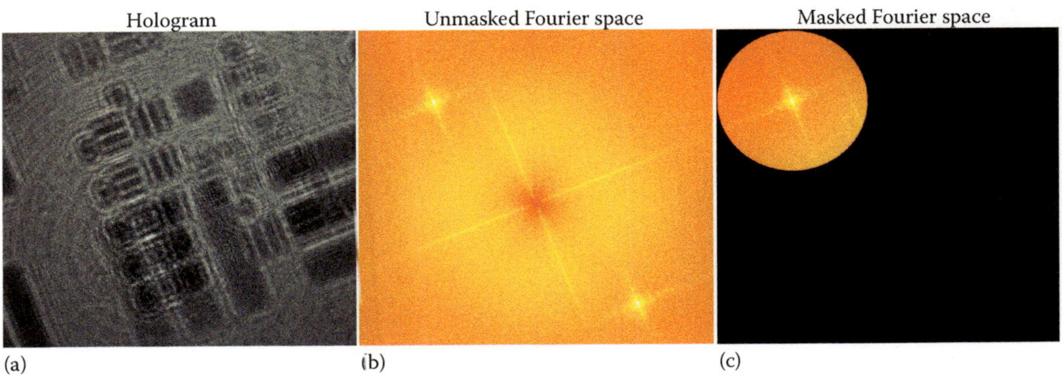


**Figure 9.2 Masking in Fourier space for reconstruction.** (a) The hologram of a standard US Air Force resolution target, apodized to reduce aliasing. (b) Unmasked Fourier transform. (c) The masked Fourier transform should preserve the real or virtual image.

effects due to sharp cutoff at the edges of the hologram. The next step is to take the 2-D FFT of the image (**Figure 9.2b**) and create a spectral mask (**Figure 9.2c**). Here we have chosen the manifestation of the carrier frequency on the top left (corresponding to the image that is physically behind the sensor). Two similar carriers are masked out by this—a direct current (DC) term at the center of the image and one for an identical image in front of the sensor in the lower right. This identical image could have been chosen instead.

The mask here is simply a circle of radius 250; thus, the maximum frequency that can be resolved is nominally 250 cycles/3.45 μm pixel size = 72.5 cycles per micron on the sensor. More complex masks can be used to remove static optical aberrations at this stage, prior to reconstruction. The removal of tilt is accomplished simply by shifting the remaining spectrum to the center.

The reconstruction follows **Equation 9.2** exactly, taking the inverse FFT to get the refocused image. It is important to note that the air force test target shown in **Figure 9.2** is a poor test of anything except spatial resolution and amplitude reconstruction. It is an amplitude object, not a phase object, and its features of interest are all on a single plane. In order to test the performance of the instrument in phase and on multiple $z$-planes, a biological object such as a cell culture should be used.

Reconstruction of quantitative phase and amplitude images may be numerically performed at selected $z$-planes throughout the volume or at all $z$-planes. Amplitude images are equivalent to transmission light microscopy. Phase images have no direct counterpart in light microscopy; Zernicke phase contrast, discussed in **Chapter 6**, is not the same as the quantitative phase images produced by

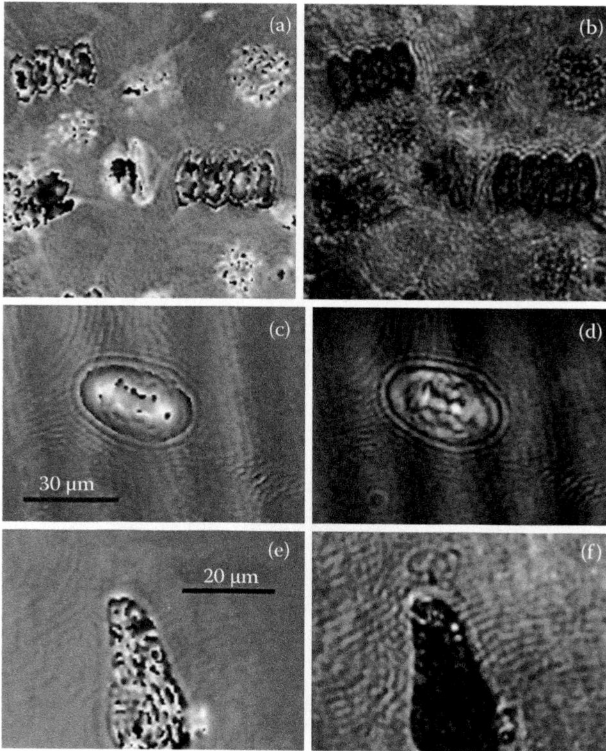

**Figure 9.3 Images of selected biological samples in amplitude and phase.** Reconstruction is in a single $z$-plane. (a) Untreated pond water, phase. (b) Untreated pond water, amplitude. The dark-appearing cells contain chlorophyll. (c) *Paramecium* sp., phase. (d) *Paramecium* sp., amplitude. (e) *Euglena* sp., phase. (f) *Euglena* sp., amplitude. Note the polar flagellum that cannot be seen in phase.

holography, though phase-contrast-like images may be generated from hologram reconstructions. The observed quantitative phase shift seen in a phase image is related to the product of the cell thickness $h$ and the difference in refractive index between the medium ($n_m$) and cell ($n_c$):

$$\Delta\varphi = \frac{2\pi}{\lambda} h(x,y)\left[n_c(x,y) - n_m\right], \tag{9.3}$$

where $\lambda$ is the wavelength of illumination.

Different types of cells show very different contrast in amplitude and phase. Those that are highly pigmented, such as photosynthetic organisms, are often more readily viewed in amplitude. Those that are transparent under brightfield often stand out well in phase. (They are *pure phase objects*.) **Figure 9.3** shows single $z$-plane reconstructions in amplitude and phase for some selected biological samples.

## 9.4  SOURCES OF NOISE

The signal-to-noise ratio (SNR) in raw reconstructions of biological samples is much lower than that seen in fluorescence microscopy. Typical SNR values are 2–4. Small cells, such as bacteria, are difficult to distinguish from background unless they are motile. Sources of noise in DHM include vibrations, laser speckle, temporal phase noise that results from uncorrelated noise between the two fields of the interferometer, as well as noise introduced by the imaging device. Both preprocessing and postprocessing filtering methods are used to reduce noise. The setup may be optimized by reducing the recording distance and using a camera with a smaller pixel size. Other methods to reduce noise include phase error compensation, spatial light modulation (SLMs), and multiple frequency overlapping, which improve the reconstructed phase image of holograms. Noise reduction is also accomplished by filtering certain frequencies, both in the spatial and Fourier planes, with Butterworth filters and masks, respectively. In addition, efficient encoding methods and correlation-based denoising algorithms have been developed to significantly reduce speckle noise. During biological studies, the power of the light source must sometimes be modulated so as to not harm the organisms being observed. This has prompted investigation into the reduction of errors introduced by shot noise that becomes significant at low illumination.

A significant source of noise can also arise from the material used to contain the sample being investigated, which has not yet been well documented. *Polydimethylsiloxane* (PDMS) and other polymer-based microfluidic channels are often used to contain cells for light microscopy, due to PDMS's ease of fabrication into suitable complex structures and relatively low effect on light intensity imaging. However, the complex phase structure of PDMS makes it incompatible with phase imaging. The same is true for other hydrogels such as agarose. Even high-quality plastic sample chambers that yield excellent amplitude images may cause the phase images to appear washed out and low contrast (**Figure 9.4**).

It is also important to note that samples that are dense or turbid will inhibit reconstruction. Hologram reconstruction assumes a form of the propagator that does not account for multiple scattering events that occur when the sample is highly concentrated. This is a fundamental property of holography, which results

**Figure 9.4 Images of *E. coli* in amplitude in phase on a glass slide versus an acrylic slide.** The contrast in phase is markedly reduced by the acrylic.

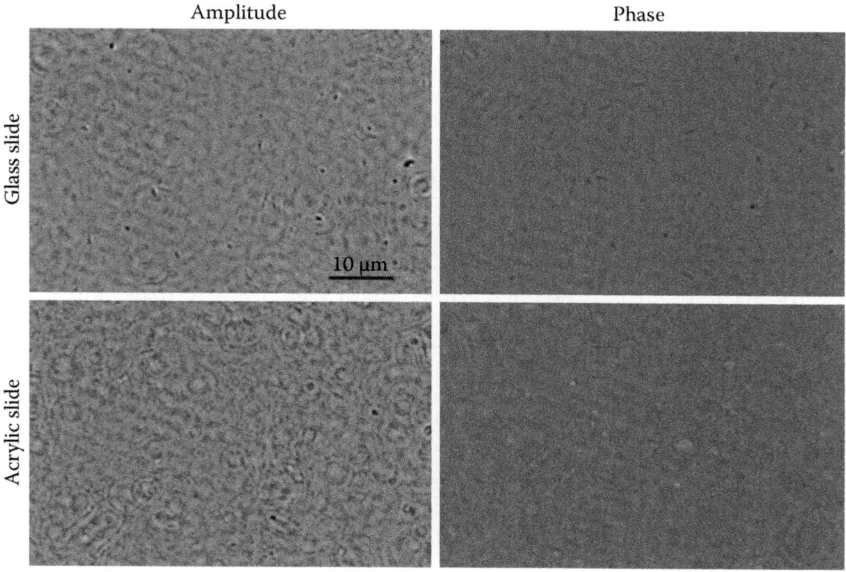


from the particle field itself; when the virtual image and the reference wave are spatially separated from the real image, this noise is reduced. Several papers have investigated the signal-to-noise attainable in DHM as a function of particle size, particle number density, sample chamber depth, and wavelength. An expression for the image intensity $I$ over the noise $\langle I_N \rangle$ is

$$\frac{I}{\langle I_N \rangle} = \frac{\pi \tan^2 \Omega}{\lambda^2 n_s L},$$

(9.4)

where $\lambda$ is the wavelength of light, $n_s$ is the particle number density, $L$ is the chamber depth, and $\Omega$ is the angular aperture. The value $\Omega$ corresponds to the actual angular aperture given by the numerical aperture of the system:

$$NA = n \sin \Omega$$

(9.5)

As a practical guideline, we have found that reconstruction of *Escherichia coli* samples becomes impossible at concentrations greater than about $10^8$ cells/mL in a sample chamber 0.6 mm thick (**Figure 9.5**). It is interesting to note that for a mean cell size of $1 \times 1 \times 2$ µm, $10^8$ cells/mL corresponds to a projection of sqrt(2) $\times 10^{12}$/mm², or essentially close packing. Dilution of samples improved the signal-to-noise in an essential stepwise fashion: that is, below the too-crowded density, further dilution did not help. Samples could be too dilute for the desired applications, as well. The ideal density for tracking experiments in this context was thus ~$5 \times 10^6$–$10^7$ cells/mL for *E. coli*. Significantly larger cells would of course have a lower maximum cell density that permits reconstruction, with a good starting estimate being the density whose projection into a single plane results in a filled field.

Many experiments using DHM go well beyond simply identifying cells. Ordinarily, a refractive index difference of 0.05 is approximately the limit at which a cell of 1 µm

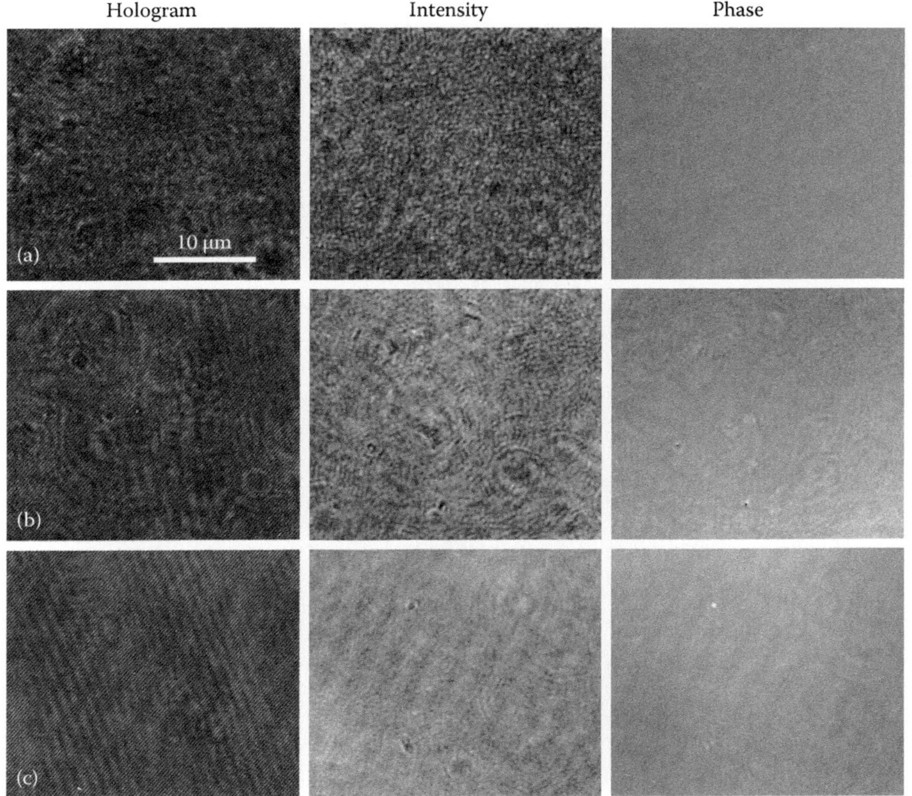



**Figure 9.5 Dilution series showing reduction in background noise with sample dilution.** Raw holograms are shown, along with intensity and phase reconstructions on a single plane. (a) Bacterial culture at $OD_{600}$ = 0.4, corresponding to ~$10^8$ colony-forming units (CFU)/mL. (b) Tenfold dilution of the culture. (c) 100-fold dilution.

thickness becomes invisible in raw images. By careful reduction and subtraction of noise, both during and after acquisition, extremely precise measurements of phase have been used to monitor a variety of biological processes. Changes in the refractive index can occur as cells take up or lose water, thus generating signals under phase imaging. Processes that can be observed this way include cell cycle arrest, initiation of apoptosis, *excitotoxicity*, and neuronal response to glutamate stimulation. However, the phase changes are very small: as little as $4 \times 10^{-4}$ in refractive index. Monitoring of such small changes is only possible by using background phase monitoring and noise subtraction. Phase in multiple areas outside the cell is chosen to be the background, and excessive noise or fluctuations in the background should lead to changes in the experimental technique. At this level of sensitivity, even flowing in a solution such as 10 mM glutamate can affect the phase.

## 9.5  INSTRUMENT DESIGNS

### Mach–Zehnder

Many laboratory DHMs are a Mach–Zehnder design. This technique has the advantages of being straightforward to set up and interface with other instruments, such as confocal microscopes or spectrometers. It has the disadvantage of being misaligned easily—even heavy footsteps can knock the system out of alignment.

**Figure 9.6 Design of a Mach–Zehnder off-axis DHM.** (a) Schematic showing the following: BS, beam splitter; L, lens; M, mirrors; MO, microscope objective; O, object beam; R, reference beam. (b) Photo of an actual breadboard instrument with the solid line indicating the object beam and the dashed line showing the reference beam.

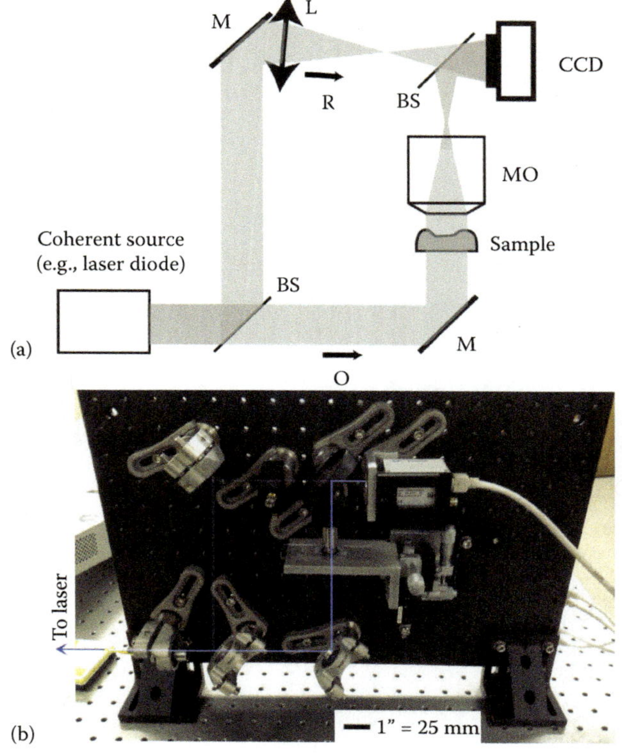


It is thus best used by people who know how to align the system, and is not very good for fieldwork.

**Figure 9.6** shows a Mach–Zehnder DHM specifically designed for imaging motile bacteria. For this purpose, spatial resolution of better than 1 µm was required.

The optical design aimed to achieve this by minimizing on-axis optical interfaces. This reduces coherent parasitic reflections and so allows for diffraction-limited resolution. Antireflection (A/R)-coated plate beam splitters (BSs) were used, and the objective lens was a simple asphere with an antireflective coating. This eliminated the multiple scattering events that can take place inside complex microscope objectives. The lens used had an NA of 0.5 as a magnification $M = 20\times$ single-lens objective. The illumination source was a violet ($\lambda = 405$ nm) laser diode, which sets the theoretical diffraction-limited resolution at $\sim\lambda/\text{NA}$ of $\sim0.8$ µm (0.5 µm under the Rayleigh criterion). The camera is a $2448 \times 2050$ digital CCD camera with 3.45 µm pixel size, delivering up to 22 fps in $1024 \times 1024$ (1 Mpx) mode and 14 fps with $2048 \times 2048$ pixels (4 Mpx). All of these elements led to spatial resolution that had not been previously reported in holographic microscopy, with the ability to image and track low-contrast microorganisms in liquids. Individual bacteria were still difficult to see in raw images but stood out clearly after median subtraction (**Figure 9.7**).

## Common path

Implementing the off-axis geometry as a *common-path* layout makes DHM insensitive to errors in alignment, making it ideal for rough transport and handling

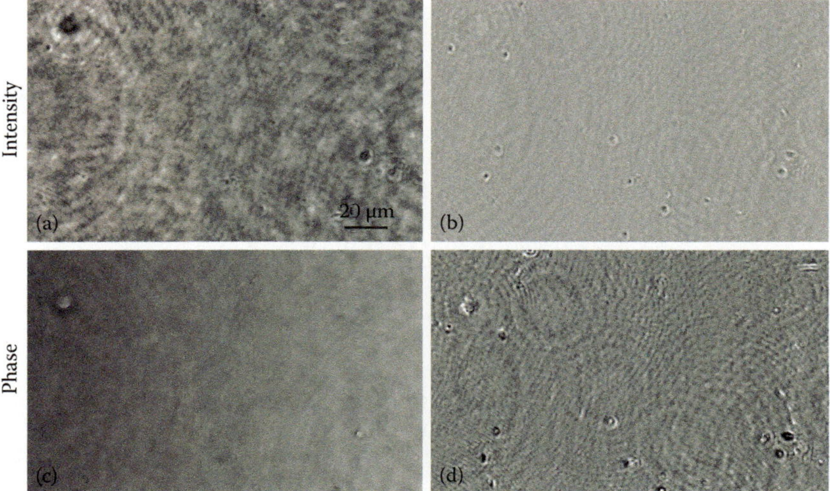

**Figure 9.7 Performance of the Mach–Zehnder off-axis DHM in amplitude and phase.** The images show individual cells of a marine bacterium. (a) Intensity, raw image after reconstruction, no processing. (b) Intensity image after median subtraction over the time series. (c) Raw phase image. (d) Phase image after median subtraction.

conditions. What this means is that instead of passing the illumination light through a BS to create object and reference beams, the sample itself is adjusted to create object and reference channels. Light from the laser is collimated by a lens matched to the numerical aperture of the optical fiber; this homogeneous source then passes through a sample chamber with geometrically separated channels. One channel contains the sample, and the other is empty. Light from both channels passes through a single relay lens and to the CCD (**Figure 9.8**).

The disadvantage of this design is that special sample chambers need to be used in order to ensure that a clean reference channel is present at the precise spacing required by the design (**Figure 9.9**). It is interesting to note that in the case of

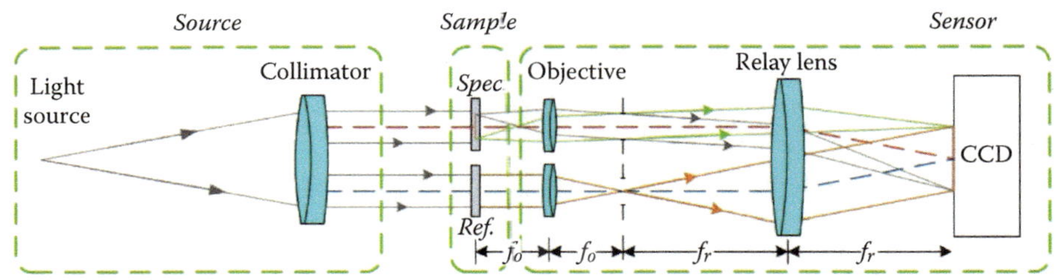


**Figure 9.8 Schematic of a common-path DHM showing four main elements: the source, the sample, the microscope, and the sensor.**

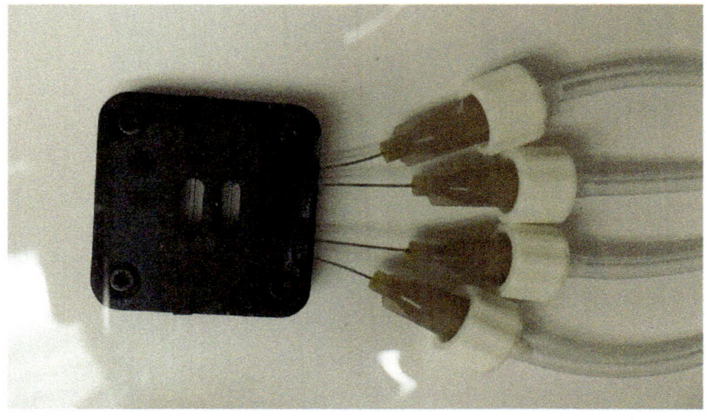

**Figure 9.9 Sample chamber for common-path DHM, showing reference and object chambers at a precise separation distance.**

dilute samples, it is perfectly fine to put sample into both channels—the objects in each channel will appear to be moving in opposite directions and will be in mirror focal planes, but do not interfere with each other (unless the motion becomes confusing). However, in more dense samples such as cell cultures, this does not work. If the reference beam passes through a cell, the image quality will be very poor. This restriction is not a problem for suspension cells, such as bacteria, which may be pipetted into the object channel, but can be a serious problem with adherent cells. The cells must be grown on a substrate with a blank area, or to a low enough density so that the reference beam can pass through an empty spot.

## In-line

An *in-line* DHM has no independent reference beam. Instead, it uses a single beam as both object and reference, and reconstructs using the assumption that the object is a small perturbation (i.e., the first Born approximation holds). This allows the optical train to be very simple. A laser beam passes through a pinhole to create a clean Gaussian beam that acts as a point source (**Figure 9.10**). The tricky part is aligning the pinhole the first time the instrument is set up, though once a procedure is established, this process becomes routine. Pinhole sizes can be 500 nm to several micrometers, and should be on the order of the wavelength of illuminating light. Smaller pinholes provide better resolution but are more expensive and more difficult to align.

However, the in-line makes reconstruction more difficult. When attempting to isolate one of the components of interest—thus enabling numerical propagation of the amplitude and phase information through, for example, a Fresnel or angular spectrum propagator—in-line holographic approaches suffer from the spatial superposition of all four terms in **Equation 9.1**. While the two intensity terms can usually be discarded with prior reference-only and object-only calibration recordings, the complex conjugate term can only be suppressed by using sequential phase-shifting approaches, thus compromising fast real-time capabilities and adding a great deal of complexity. Alternatively, one can choose to live with it, i.e., imaging the superposition of the two complex wave fronts.

However, when dealing with a thick-volume sample and attempting to image multiple planes in the sample through digital focusing, the result

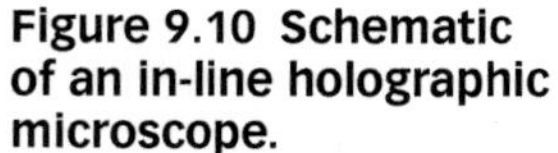

**Figure 9.10 Schematic of an in-line holographic microscope.**

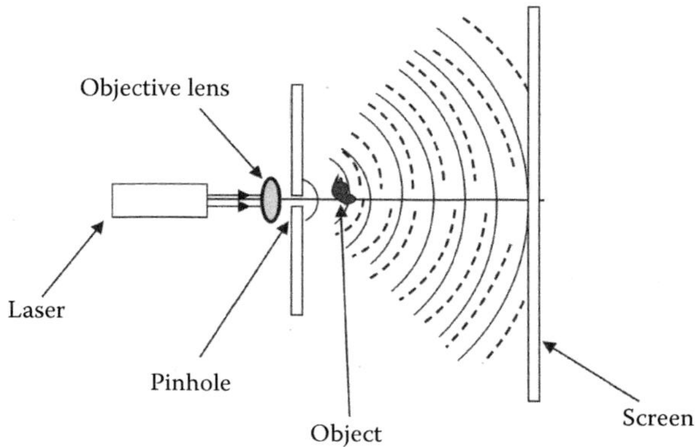

**Figure 9.11 Use of in-line holographic microscope in the field.** (a) Image of swimming trajectories of single-celled eukaryotes from Colour Lake, Nunavut, Canada. (b) The instrument deployed in Colour Lake. The camera and laser are within a waterproof aluminum can, and controlled by a computer on shore connected via cables.

is the superposition of a sharply focused image with a blurred, out-of-focus contribution from the twin image. This dual image arises because the complex conjugate term produces a virtual image that propagates in the opposite direction, i.e., if the real image is in focus at $+d$, the virtual image will be in focus at $-d$: there is therefore always a superposed out-of-focus image at $2d$ when imaging this way. This effect can be actually observed in images taken on such systems, where it clearly translates to a loss of contrast and resolution. Images taken from in-line DHM are nearly always processed before publication, usually by frame-to-frame subtraction, which provides outlines of moving organisms (**Figure 9.11a**). The advantage of such systems is that they are readily made robust enough for field use because of simplicity of their design (**Figure 9.11b**).

## Incoherent DHM

DHM may be performed using LEDs or an array of LEDs. The advantage of this approach is that the instruments are extremely small, low cost, and low power: instruments weighing <50 g with achieved 1–2 µm resolution have been reported. A single amber (591 nm) LED was used for illumination. Arrays of LEDs can improve resolution because the image using each LED is a slightly shifted hologram, allowing for use of a pixel superresolution algorithm to break the diffraction limit. One highly portable design is shown in **Figure 9.12**.

The field of view for these systems is large, 10 times or more the field seen with a typical 20× microscope objective (**Figure 9.13**). The disadvantage is that depth is lost in the use of incoherent light, so the technique is best applied to very thin samples. It is of particular use in field diagnostics, where it may be used to inspect blood smears for parasitic infection or to examine histological specimens. Instruments can be constructed based upon cell phones, which allows nearly everyone to have access to high-resolution microscopy. These instruments are of great value in public health and help to revolutionize and democratize sophisticated imaging.

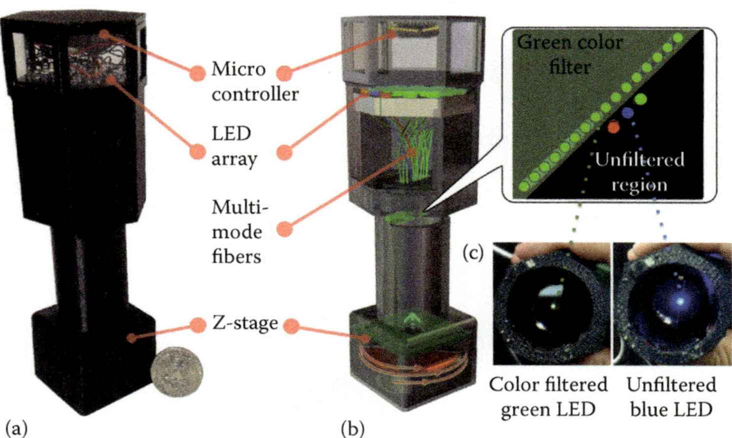



**Figure 9.12  A portable lens-free superresolution color microscope.** (a) A photograph of the microscope that weighs <145 g. (b) A schematic diagram of the microscope. The LED array is separated into two groups; the first group enables pixel superresolution based on source shifting, and it contains 17 green LEDs ($\lambda = 527$ nm) where each LED is butt-coupled to a multimode fiber and its emission passes through a color filter as shown in the inset. The second group of LEDs enables the acquisition of a lower-resolution color image, and it is composed of three LEDs ($\lambda = 470$ nm, 527 nm, and 625 nm). (c) A photograph of 1 of the 17 green LEDs (left) and the blue unfiltered LED (right). (Image courtesy of A. Ozcan, UCLA. Originally published in Greenbaum, A. et al., Field-Portable Pixel Super-Resolution Colour Microscope, *PLoS One* 8, e76475, 2013.)

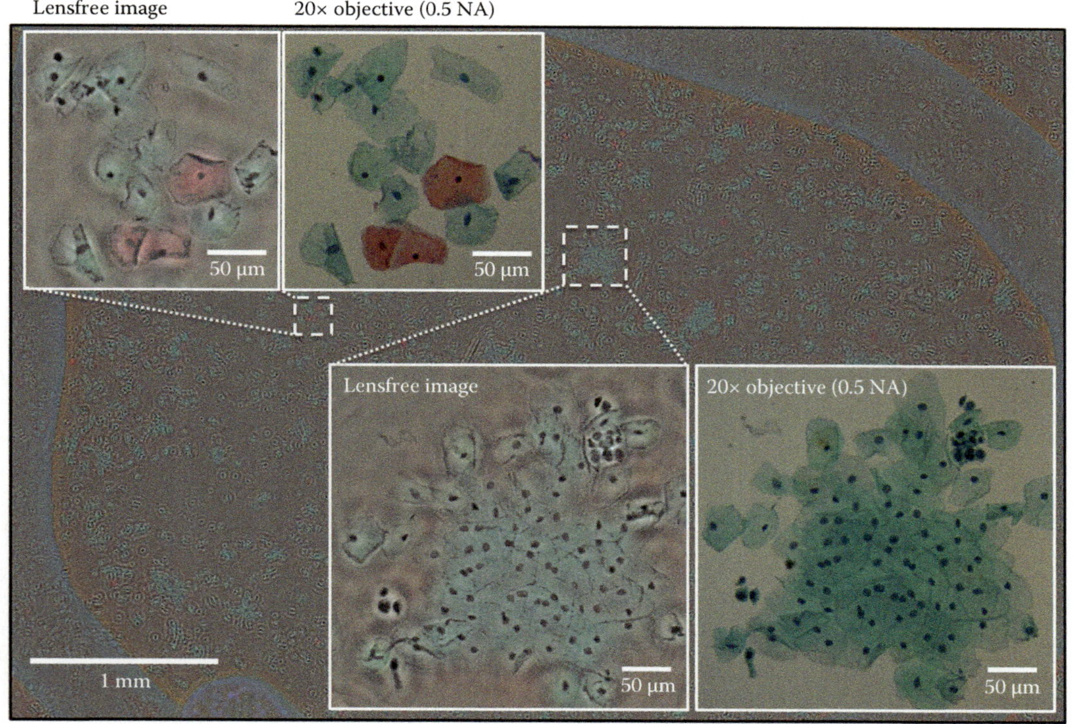



**Figure 9.13  A wide field of view (FOV) (~21 mm²) image of a Pap smear sample taken by the instrument shown in Figure 9.12.** 20× microscopes images (0.5 NA objective lens) of the same specimen are also shown for comparison. (Image courtesy A. Ozcan, UCLA. Originally published in Greenbaum, A. et al., Field-Portable Pixel Super-Resolution Colour Microscope, *PLoS One* 8, e76475, 2013.)

# 9.6  BUILDING A LOW-COST DHM

## Hardware

The instrument shown in **Figure 9.8** costs only about $200 to build, exclusive of the light source, the camera, and the custom 3-D printed piece. However, light sources and high-resolution cameras cost several thousand dollars each. In order to reduce the cost of a DHM, these elements need to be replaced by commercial items that most people or labs would own (e.g., cell phone, digital camera) or by inexpensive replacements.

The most inexpensive design is an in-line setup using an LED with a pinhole attached as an illumination source. This eliminates the need for any optical components such as lenses and BSs. However, spatial resolution will be on the order of 10–20 µm. Some references to low-cost instruments of this kind are given at the end of the chapter.

## Gradient index lens common mode

To achieve micrometer-scale resolution, several low-cost approaches are possible. We have recently reported a lensless design based upon a single-mode optical fiber and gradient index (GRIN) lenses (**Figure 9.14**). The GRIN lenses increase the NA of the fiber output beam. GRIN lenses with diameters on the scale of a millimeter and lengths on the order of 2 mm are readily available at low cost, and can generate a high NA focus at their output for a collimated input. Illuminating both GRIN lenses with a single collimated beam removes the need for a second fiber, making the system *common mode* until the fiber launch point, eliminating potential path length drifts to which a dual-fiber system would be susceptible. From the pair of coplanar output foci, the two beams expand at high NA and pass through the sample, and then both fill the camera array. With no optical elements between the two source points and the detector array (other than the sample), the imaging stage is indeed lensless.

### Parts needed

- Coherent light source (laser)

- Sensor

- GRIN lenses

- Test samples: air force test target, micrometer-sized beads in a slide chamber of some kind (note: do not used curved hanging-drop slides for DHM; the curvature causes lensing)

- Electronics to acquire and process data

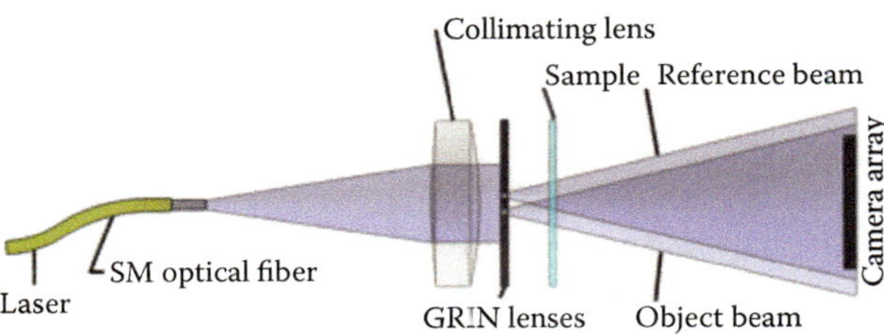


**Figure 9.14 Schematic of a lensless off-axis DHM design.**

For this application of off-axis holography, a coherent light source is needed. Many optical supply distributors carry coherent light sources at many wavelengths. A commonly used coherent light source for a compact implementation of off-axis holography is a laser-coupled diode, often called a "pig-tailed" laser. The light-producing diode is coupled directly to the single-mode fiber-optic cable that routes the light to the desired location and can be terminated with any sort of connector desired. (FC/PC is the most commonly used for laboratory settings.)

The sensor and associated electronics must be chosen carefully with respect to the illumination source wavelength and spacing between GRIN lenses. Important characteristics of the sensor include pixel size, array size, and monochromaticity. The dependence of sensor pixel size on illumination wavelength and distance from GRIN lenses will be discussed in the following section. A monochromatic sensor is needed due to the need for high-resolution imaging at a single wavelength. Polychromatic imaging is done by Bayer interpolation, which uses a mosaic of separate red, green, and blue pixels to interpolate RGB values at each pixel, effectively reducing the spatial resolution of the sensor.

Compatible electronics are needed to communicate with the sensor and record holograms. Most sensors come with software development kits (SDKs) in order to create customized data acquisition (DAQ) sequences as well as prebuilt graphical user interface (GUI) data acquisition software for basic camera operation and image acquisition. Depending on the resolution and frame rate desired of the instrument, an appropriate processor must be used to be able to handle the data transfer rates of real-time image acquisition at high fidelity.

### *Assembly*

Almost every part of a lensless DHM is a readily available commercial optical component, except for the GRIN lens holder. Starting from the illumination source, a collimator lens is placed directly after the illumination light source inlet. (Most common is an FC/PC connector for a fiber-optic cable.) After the collimator are the GRIN lenses. The GRIN lenses can be held in place any number of ways, but a straightforward method is to use a 3-D printer to create a circular mount with two off-axis holes where the GRIN lenses will be held in place. After the GRIN lenses, the sample is to be placed before the sensor array. The GRIN lenses can be selected to provide the desired field of view and magnification. The fringe spacing seen by the sensor array is dictated by the distance between the GRIN lenses and the sensor array.

In order to calculate the fringe spacing at the sensor array, one can see this instrument's resemblance to a Young double-slit experiment setup. In a Young double-slit experiment, the fringe spacing produced is calculated by the following expression:

$$z = \frac{a\lambda_0}{nd},\tag{9.6}$$

where $a$ is the distance between the GRIN lenses and the detector array along the optical axis, $\lambda_0$ is the illumination wavelength, $n$ is the index of refraction through which the light is to propagate (air), and $d$ is the center-to-center distance of the GRIN lenses perpendicular to the optical axis.

A fringe spacing of three to four pixels per fringe is desired to provide the necessary sampling criterion in order to sufficiently sample the fringe at an appropriate spatial frequency as well as to avoid aliasing.

It is worth noting that when constructing one's own DHM, the noise introduced to the instrument is proportional to the distance $a$, as mentioned earlier. Minimizing this distance as much as possible decreases the noise of the instrument, which increases its resolution.

### Testing

In order to build a functional off-axis holographic microscope, a few performance metrics are useful in order to set up the instrument correctly. These performance metrics include fringe visibility and stability.

Fringe visibility is a measure of the amplitude of intensity peak-to-valleys of the interference pattern recorded in the hologram. The intensity of an ideal hologram across the sensor array can be expressed by the following expression:

$$h(x,y) = I_1 + I_2 + 2\sqrt{I_1 I_2}\, \cos\big(\Delta\phi(x,y)\big), \tag{9.7}$$

where $I_1$ and $I_2$ are the intensities of beams 1 and 2, respectively, and $\Delta\phi$ is the local phase difference between the two beams of light.

The ideal fringe visibility is defined as the ratio of the intensity of the interference term with the constant illumination due to the two beams of light.

$$V_{\text{ideal}} = \frac{2\sqrt{I_1 I_2}}{I_1 + I_2} \tag{9.8}$$

In practice, calculating fringe visibility is accomplished by the following expression:

$$V_{\text{real}} = \frac{I_{\text{max}} - I_{\text{min}}}{I_{\text{max}} + I_{\text{min}}}, \tag{9.9}$$

where $I_{\text{max}}$ is the maximum pixel intensity recorded along a single fringe and $I_{\text{min}}$ is the minimum pixel intensity recorded along a single fringe.

Fringe visibility is an important performance metric of any interferometer due to its direct proportionality to the SNR of the reconstructed image.

Fringe stability is used to quantify the temporal stability of fringes. This is done by calculating the fringe visibility of a hologram as a function of time. A completely stable interferometer will have no change in the fringe visibility as a function of time (a flat profile). Calculating the numerical derivative of fringe visibility with respect to time will show the stability of fringes. This is a necessary performance metric in the ability of recording a time series of holograms (e.g., a dynamic process like motility of bacteria).

With any coherent light source, there is a narrow spatial region where the coherence of the light is suitable in creating interference patterns. This spatial region is known as the coherent envelope. The exact definition of the coherent envelope varies but can be regarded as the percentage of the instrument's field of view where the fringe visibility is in the top 30 percentile of the hologram.

Due to the nature of the implementation of off-axis holography, a single hologram has many fringes. In order to calculate the fringe visibility of a single hologram, a recursive algorithm can be used to partition a single hologram into submatrices containing single fringes. The fringe visibility of each fringe can be calculated and stitched together to yield a *visibility map* of the hologram. This visibility map can be used to calculate the mean fringe stability of a single hologram and the fringe stability of a time series of holograms.

## Reconstruction and analysis software

There is little available as far as open-source or commercial code that is tailored to useful holographic reconstruction. Some commercial code is available but is costly and not open source. Some partial algorithms exist for MATLAB and ImageJ but will require tailoring to specific instruments. It is key to check premade software to see whether it was designed for off-axis or in-line holography, which require very different types of deconvolution.

For in-line holography, the HoloJ (holographic analysis with ImageJ) suite of plug-ins is open source and easy to use. The suite handles complex images, including FFTs on complex images, and performs amplitude reconstruction and phase unwrapping.

For off-axis holography, the author's group has begun developing an open-source software package called SHAMPOO, available on GitHub. Please check for updates. References to a few other sources are also given at the end of the chapter.

## End-of-Chapter Questions

1. Using the diagram below, derive the expression for fringe spacing in a basic Young double-slit experiment, where the fringe spacing is as follows:

$$X = \frac{a\lambda_0}{nd}$$

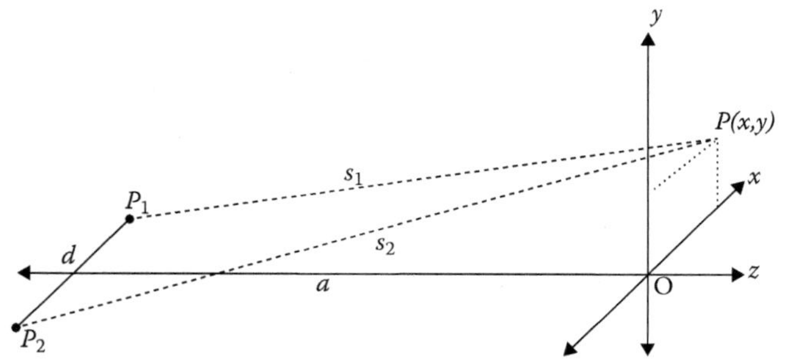


Here, $P_1$ and $P_2$ are the two pinholes of the double-slit setup, $d$ is the distance between them, $a$ is the distance along the optical axis between the two light sources and the observation plane where the interference pattern is created, and $P(x,y)$ is an arbitrary point on the observation plane at coordinate $(x, y, z = 0)$.

2. Calculate the diffraction pattern for light of wavelength $\lambda$ passing through a slit of width $d$ onto a screen a distance $x$ away ($x \gg d$). How can the wavelength be determined by the fringe spacing?

3. Sketch the diffraction pattern for a square pinhole and for a hexagonal pinhole.

4. The fringe intensity for a monochromatic wavelength is given by the following expression:

$$I = I_0\left(1 + \cos\left[\frac{2\pi}{\lambda}\Delta x\right]\right)$$

(a) Calculate the fringe pattern created by two closely spaced wavelengths given by $\lambda + \Delta\lambda$ and $\lambda - \Delta\lambda$. (Hint: change to units of frequency first.)

(b)  Calculate the fringe pattern for a continuous spectral range by integrating from $\nu - \Delta\nu$ to $\nu + \Delta\nu$. Assume a uniform spectral density.

(c)  The visibility of a fringe is given by

$$V = \frac{I_{max} - I_{min}}{I_{max} + I_{min}} \rightarrow \frac{I_{amp}}{I_{avg}}$$

From the expression derived above, determine the allowable path length difference that gives a visibility of >50%.

(d)  An imaging array has a pixel size of 4 μm, and is 1000 × 1000. If we desire the visibility to be >80% at opposite corners of the array, what is the required optical bandwidth for the laser?

5.  A sample moves at a maximum rate of 2 μm/s. If this sample is magnified by a factor of 10 on to the CCD, how quickly must the camera sample in order to freeze the motion of the sample? Assume it will not move by more than 0.1μm from frame to frame.

6.  The depth of focus of an optical system is given by $2\lambda(f/\#)^2$. This is the range over which the brightness of the peak of the point spread function (PSF) drops by 20%. However, in an optical system that images with magnification, the lateral magnification in the reimaged plane is given by $m$, such that $x' = mx$ and $y' = my$. But the longitudinal magnification goes by magnification squared: $z' = m^2z$. At a wavelength of 0.405 um, and an input number of f/1.4, and a magnification of 20 determine the following:

(a)  The depth of focus at the original input plane

(b)  The depth of focus in the reimaged output plane

7.  Sketch the optics of a Michelson interferometer and explain how it works. Give an equation relating the movement of the mirror to the number of fringes crossing the field of view. For an illuminating wavelength of 700 nm, what distance would a mirror have to be moved in order to observe 800 fringes?

8.  You want to put a high-index coating on glass to minimize reflection (antireflective coating). If the material's index is 2.0, what thickness needs to be used for 405 nm light illumination? (The index of glass is 1.5.)

9.  Quantum mechanical beam splitter: Photons entering a beam splitter may be considered quantum mechanically one by one. Give an expression for the transformation representing two photons entering a 50:50 beam splitter.

10.  Give an explanation for the Hong–Ou–Mandel effect and how it is measured.

11.  In a Mach–Zehnder interferometer, a photon passes through a beam splitter and then enters the sample, which we can assume for simplicity is a piece of glass. Write the matrix representation of operator corresponding to the sample.

12.  In a Mach–Zehnder configuration, each beam sees one beam splitter reflection and one transmission. Assuming energy conservation, i.e., $r^2 + t^2 = 1$, where $r$ and $t$ are the field reflection and transmission coefficients, show that the maximum fringe modulation occurs for $R = r^2 = 1/2$, i.e., for an equal split.

13.  Simulate the hologram of a variety of objects of your choice in water, and then reconstruct in amplitude and phase. Examples: a 2-μm-diameter sphere of index $n = 1.4$; a cylinder of circular diameter 2 μm and length 10 μm of index $n = 1.37$; and a cell of radius 10 μm containing a nucleus of 1 μm diameter where the nucleus has an index of 1.8 and the rest of the cell of 1.5.

14.  **Equation 9.3** gives a relationship between observed phase shifts and optical path length (product of index of refraction and sample thickness). A major challenge in biological DHM is deconvolving these values—that is, determining both index and thickness quantitatively. Discuss how this can be done.

## Background Reading

### Books

Françon, M. *Optical Interferometry*. Academic Press, New York, 1966.

Goodman, J. W. *Introduction to Fourier Optics*. Edn. 3. Roberts & Co., Englewood, CO, 2005. Table of contents: http://www.loc.gov/catdir/toc/ecip051/2004023213.html; Contributor biographical information: http://www.loc.gov/catdir /enhancements/fy1202/2004023213-b.html; Publisher description: http://www.loc.gov/catdir/enhancements/fy1202/2004023213-d .html.

Hariharan, P. *Optical Holography: Principles, Techniques and Applications*. Edn. 2, Cambridge University Press, Cambridge, UK, 1996.

Kasper, J. E., and Feller, S. A. *The Complete Book of Holograms: How They Work and How to Make Them*. Dover, Mineola, NY, 2001.

Kim, M. K. *Digital Holographic Microscopy: Principles, Techniques, and Applications*. Springer, New York, 2011.

Saxby, G., and Zacharovas, S. *Practical Holography*. Edn. 4. CRC Press, Boca Raton, FL, 2015.

## Journal articles

Beiderman, Y., Amsel, A. D., Tzadka, Y., Fixler, D., Mico, V., Garcia, J., Teicher, M., and Zalevsky, Z. (2011). A microscope configuration for nanometer 3-D movement monitoring accuracy. *Micron* 42, 366–375.

Chmelik, R., Slaba, M., Kollarova, V., Slaby, T., Lostak, M., Collakova, J., and Dostal, Z. In E. Wolf (ed.), *Progress in Optics*, Vol. 59, pp. 267–335, 2014.

Di, J. L., Li, Y., Xie, M., Zhang, J. W., Ma, C. J., Xi, T. L., Li, E. P., and Zhao, J. L. (2016). Dual-wavelength common-path digital holographic microscopy for quantitative phase imaging based on lateral shearing interferometry. *Applied Optics* 55, 7287–7293.

Dixon, L., Cheong, F. C., and Grier, D. G. (2011). Holographic deconvolution microscopy for high-resolution particle tracking. *Optics Express* 19, 16410–16417.

Garcia-Sucerquia, J. Color digital lensless holographic microscopy: Laser versus LED illumination. *Applied Optics* 55, 6649–6655 (2016). <Go to ISI>://WOS:000381965800017

He, X. F., Nguyen, C. V., Pratap, M., Zheng, Y. J., Wang, Y., Nisbet, D. R., Williams, R. J., Rug, M., Maier, A. G., and Lee, W. M. (2016). Automated Fourier space region-recognition filtering for off-axis digital holographic microscopy. *Biomedical Optics Express* 7, 3111–3123.

Hosseini, P., Zhou, R. J., Kim, Y. H., Peres, C., Diaspro, A., Kuang, C. F., Yaqoob, Z., and So, P. T. C. (2016). Pushing phase and amplitude sensitivity limits in interferometric microscopy. *Optics Letters* 41, 1656–1659.

Indebetouw, G., Tada, Y., and Leacock, J. (2006). Quantitative phase imaging with scanning holographic microscopy: An experimental assesment. *Biomedical Engineering Online* 5, 63.

Jericho, S. K., Klages, P., Nadeau, J., Dumas, E. M., Jericho, M. H., and Kreuzer, H. J. (2010). In-line digital holographic microscopy for terrestrial and exobiological research. *Planetary and Space Science* 58, 701–705.

Kim, S. B., Bae, H., Koo, K. I., Dokmeci, M. R., Ozcan, A., and Khademhosseini, A. (2012). Lens-free imaging for biological applications. *Jala* 17, 43–49.

Kim, T., Zhou, R. J., Goddard, L. L., and Popescu, G. (2016). Solving inverse scattering problems in biological samples by quantitative phase imaging. *Laser and Photonics Reviews* 10, 13–39.

Knox, C. (1966). Holographic microscopy as a technique for recording dynamic microscopic subjects. *Science (New York, N.Y.)* 153, 989–990.

Kuhn, J., Niraula, B., Liewer, K., Wallace, J. K., Serabyn, E., Graff, E., Lindensmith, C., and Nadeau, J. L. (2014). A Mach–Zender digital holographic microscope with sub-micrometer resolution for imaging and tracking of marine micro-organisms. *Review of Scientific Instruments* 85 (12):123113.

Lee, K., Kim, K., Jung, J., Heo, J., Cho, S., Lee, S., Chang, G., Jo, Y., Park, H., and Park, Y. (2013) Quantitative phase imaging techniques for the study of cell pathophysiology: From principles to applications. *Sensors* 13, 4170–4191.

Lee, S. H., Roichman, Y., Yi, G. R., Kim, S. H., Yang, S. M., van Blaaderen, A., van Oostrum, P., and Grier, D. G. (2007). Characterizing and tracking single colloidal particles with video holographic microscopy. *Optics Express* 15, 18275–18282.

Lindensmith, C. A., Rider, S., Bedrossian, M., Wallace, J. K., Serabyn, E., Showalter, G. M., Deming, J. W., and Nadeau, J.L. (2016). A submersible, off-axis holographic microscope for detection of microbial motility and morphology in aqueous and icy environments. *Plos One* 11 (1):e014770.

Liu, P. Y., Chin, L. K., Ser, W., Chen, H. F., Hsieh, C. M., Lee, C. H., Sung, K. B., Ayi, T. C., Yap, P. H., Liedberg, B., Wang, K., Bourouina, T., and Leprince-Wang, Y. (2016). Cell refractive index for cell biology and disease diagnosis: Past, present and future. *Lab on a Chip* 16, 634–644. <Go to ISI>://WOS:000370333200001

Mann, C. J., Yu, L. F., Lo, C. M., and Kim, M. K. (2005). High-resolution quantitative phase-contrast microscopy by digital holography. *Optics Express* 13, 8693–8698.

Marquet, P., Depeursinge, C., and Magistretti, P. J. In M.L. Yarmush (ed.), *Annual Review of Biomedical Engineering*, Vol 15, pp. 407–431, 2013.

Memmolo, P., Miccio, L., Paturzo, M., Di Caprio, G., Coppola, G., Netti, P. A., and Ferraro, P. (2015). Recent advances in holographic 3D particle tracking. *Advances in Optics and Photonics* 7, 713–755.

Mir, M., Bhaduri, B., Wang, R., Zhu, R. Y., and Popescu, G. In E. Wolf (ed.), *Progress in Optics*, Vol 57, ) pp. 133–217, 2012.

Moon, I., Yi, F., and Javidi, B. (2010). Automated three-dimensional microbial sensing and recognition using digital holography and statistical sampling. *Sensors* 10, 8437–8451.

Nadeau, J. L., Bin Cho, Y., Kuhn, J., and Liewer, K. (2016). Improved tracking and resolution of bacteria in holographic microscopy using dye and fluorescent protein labeling. *Frontiers in Chemistry* 4, 17. doi: 10.3389/fchem.2016.00017.

Pan, G., and Meng, H. (2003). Digital holography of particle fields: Reconstruction by use of complex amplitude. *Applied Optics* 42, 827–833.

Popescu, G. In B.P. Jena (ed.), *Methods in Nano Cell Biology*, Vol. 90, p. 87, 2008.

Rappaz, B., Charriere, F., Depeursinge, C., Magistretti, P. J., and Marquet, P. (2008). Simultaneous cell morphometry and refractive index measurement with dual-wavelength digital holographic microscopy and dye-enhanced dispersion of perfusion medium. *Optics Letters* 33, 744–746.

Rosenhahn, A., and Sendra, G. H. (2012). Surface sensing and settlement strategies of marine biofouling organisms. *Biointerphases* 7, 63. doi:10.1007/s13758-012-0063-5.

Wallace, J. K., Rider, S., Serabyn, E., Kühn, J., Liewer, K., Deming, J., Showalter, G., Lindensmith, C., and Naceau, J. (2015). Robust, compact implementation of an off-axis digital holographic microscope. *Optics Express* 23, 17367–17378.

Wang, C., Zhong, X., Ruffner, D. B., Stutt, A., Philips, L. A., Ward, M. D., and Grier, D. G. (2016). Holographic characterization of protein aggregates. *Journal of Pharmaceutical Sciences–USA* 105, 1074–1085.

Xu, W. B., Jericho, M. H., Meinertzhagen, I. A., and Kreuzer, H. J. (2001). Digital in-line holography for biological applications. *Proceedings of the National Academy of Sciences of the United States of America* 98, 11301–11305.

Yu, X., Hong, J., Liu, C., and Kim, M .K. (2014). Review of digital holographic microscopy for three-dimensional profiling and tracking. *OPTICE* 53, 112306–112306.

## Suppliers and online resources

### *Optical and optomechanical parts*

ThorLabs. A huge array of optical parts, breadboards, lasers, etc.

Asphericon. Custom and stock aspheres, optical design, coatings.

Toptica. High-end lasers.

### *Cameras*

Baumer. Many CCD and CMOS camera models.

Allied Vision. High-performance industrial cameras.

### *Manufacturers of holographic microscopes*

4Deep. In-line for use in bodies of water, primarily for research on eukaryotic microorganisms 10–100 μm in size.

LynceeTec. Off-axis for biological applications; several different objective lenses.

Tomocube, Inc. A tomographic holographic microscope.

### *Software*

#### Commercial software

KOALA (LynceeTec). Primarily designed to work with LynceeTec's off-axis instruments, but may be adapted for other instruments provided they use a supported camera driver.

#### Open-source software

HoloVision. http://ostatic.com/holovision.

HoloRec3D. Digital Holography Matlab Toolbox.

HoloJ. Plug-ins for ImageJ.

SHAMPOO: For off-axis holography; written in Python. https://github.com/bmorris3/shampoo.

# CHAPTER 10

# Quantitative Cell Culture Techniques

## 10.1 INTRODUCTION

Sometimes it is important to know exactly how many cells are present in a liquid culture or adherent monolayer. Sometimes only the number of cells relative to a control culture or an earlier time point is what is required. At other times, it is important to distinguish the fractions of live cells versus dead, live versus apoptotic, differentiated versus undifferentiated, or labeled versus unlabeled. In all of these cases, the cells must be counted somehow. There are three general approaches to doing this: directly counting cells by eye; using a measurable parameter (absorbance, impedance) as a surrogate for cell count; or placing the cells into a specialized flow system that permits computerized cell counting. All of these methods are very commonly used, and each has its advantages and drawbacks. In this chapter, we discuss some of the most well-established ways for quantifying populations of bacteria and mammalian cells and provide some of the protocols we have developed that are designed to help sidestep the pitfalls that can occur. At the end of this chapter, you should be comfortable obtaining a bacterial $IC_{50}$ growth curve; performing an end-point mammalian toxicity assay; and setting up a simple fluorescence-activated cell sorting (FACS) experiment. We also briefly introduce some of the emerging techniques, such as real-time impedance measurements for mammalian cells, imaging cytometry, and microfluidic techniques.

## 10.2 QUANTIFYING BACTERIAL GROWTH AND DEATH

### Quantifying bacterial concentrations

The usual method for quantifying bacterial concentration in a solution is by using ultraviolet–visible (UV–Vis) absorbance spectroscopy to measure the absorbance (*optical density*) of the solution at a particular wavelength, usually 600 nm ($OD_{600}$). However, it is important to note that unless calibrated, this method gives only a *relative* measure of bacterial populations. The exact number of cells per milliliter corresponding to a given $OD_{600}$ depends upon the bacterial strain, the growth medium, and the spectrometer; if these data are needed, calibration must be performed by counting the cells.

Bacteria may be counted on a hemocytometer, which is a gridded slide containing a well of calibrated volume specifically designed for counting cells (originally blood cells, hence the name). Special hemocytometers are available for counting

**Figure 10.1 Hemocytometer.**
(a) Appearance of chamber
and coverslips. The coverslips
provided are thicker than normal
and should not be substituted.
(b) Microscopic appearance
of the grids. There are usually
different sizes to permit counting
of different sizes and densities of
cells. (c) Appearance of cells on
grid. (d) Counting procedure. To
avoid double-counting, eliminate
the top and right edges from
all squares counted. (Images
courtesy of Oliver Kim, http://
www.microbehunter.com.)

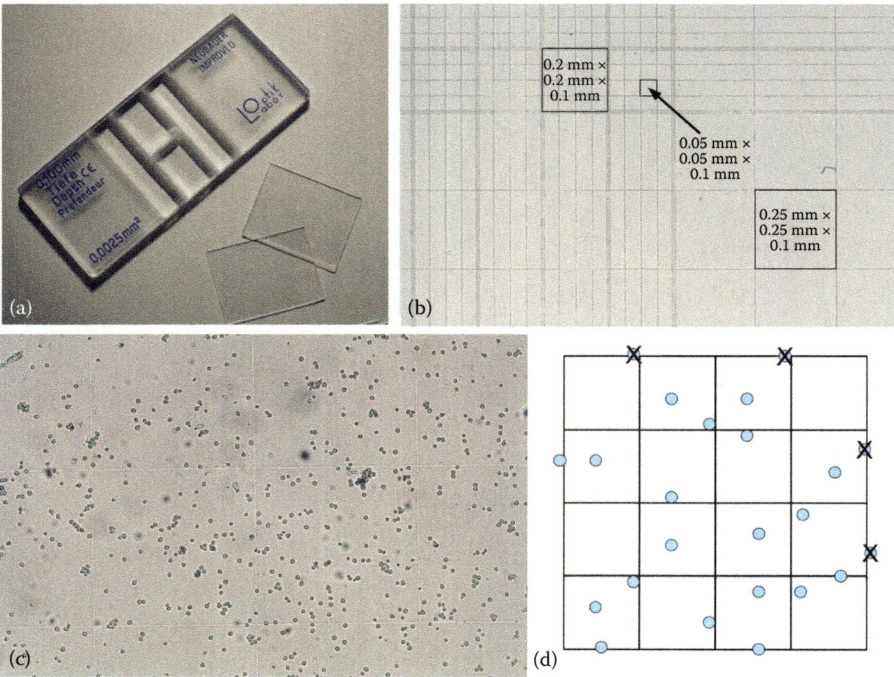


bacteria to account for the cells' small size. The suspension is diluted to yield a concentration where the cells on the grid appear uncrowded yet not too sparse (**Figure 10.1**). Enough grids are then counted by hand to give a reasonable mean value of cells per volume. Naturally, this will not work with motile cells unless the cells are first killed or immobilized!

Bacteria may be visualized using phase contrast microscopy (no stain), brightfield microscopy with a colorimetric dye such as trypan blue, or epifluorescence microscopy using a stain such as DAPI or Live/Dead (SYTO BC/propidium iodide). We highly recommend the latter method if the appropriate microscope is available, since the cells are much easier to see when fluorescently labeled, and fluorescent counts have been shown to correlate somewhat better than brightfield counts with plating methods of enumeration.

A method of obtaining the count of viable bacteria in a culture is by plating and counting colonies, a method called *plate count*. The concept is simple enough: when the solution is sufficiently dilute, discrete colonies will form on plates that each correspond to a single viable cell. In practice, plate counts can be highly variable and tricky, for two main reasons: (1) the bacterial suspension must be spread uniformly over the plate; (2) the resulting plate should have between 25 and 250 colonies, considered the limits of the "countable range" that do not induce too many errors due to cell crowding or low numbers. This requires testing a large number of concentrations, which means a lot of plates. **Practical Tips 10.1** gives a protocol for performing a bacterial plate count.

## Bacterial growth curves

There are many models for bacterial growth, all of which may be derived from the generic growth rate equation for the population size $x$ at time $t$:

---

**PRACTICAL TIPS 10.1:    VIABLE BACTERIA NUMBERS BY PLATE COUNT**

**MATERIALS NEEDED**

Bacterial culture

Sterile $H_2O$ or other diluting medium (e.g., 0.9% NaCl if bacteria are sensitive to osmotic shock)

Nutrient agar plates (5 or 10 for each culture to be tested)

Recommended: 6 mm borosilicate beads (autoclaved in glass tubes of 30 each)

**PROCEDURE**

Make 10-fold serial dilutions ranging from $10^{-3}$ to $10^{-7}$:

- Add 0.9 mL diluting medium to each of eight 1.5 mL tubes.
- Pipette 0.1 mL of sample into the first tube. Vortex.
- Change the pipette tip!
- Pipette 0.1 mL of the dilution into the next tube.
- Continue until all dilutions have been made.
- Begin by pipetting 0.1 mL of the *most dilute* solution onto a plate.
- Add 30 glass beads, cover, and shake vigorously clockwise and counterclockwise in a circular motion for 1–2 min. When the beads no longer move freely, they can be removed and used on the next plate.
- Proceed to the next plate. Ideally, make duplicates of all dilutions.
- Continue until the $10^{-3}$ dilution is reached. (Note that it is actually $10^{-4}$ because only 0.1 mL is plated.)
- Incubate plates until colonies appear. Count those with 25–250 colonies per plate.

---

$$\frac{dx}{dt} = \frac{\beta}{k^n} X^{1-np} (k^n - x^n)^{1+p}, \tag{10.1}$$

where $\beta$, $k$, $n$, and $p$ are parameters that determine the shape of the curve. Integrating gives the generic growth rate model:

$$x(t) = \frac{k}{(1+[1+\beta np(t-t_0]^{-1/p})^{1/n}}, \tag{10.2}$$

where $t_0$ is a constant of integration. Depending upon the strain and conditions, one or more of these parameters can be eliminated. To obtain a physical model, the fewest possible fit parameters should be used. When there are no limits on

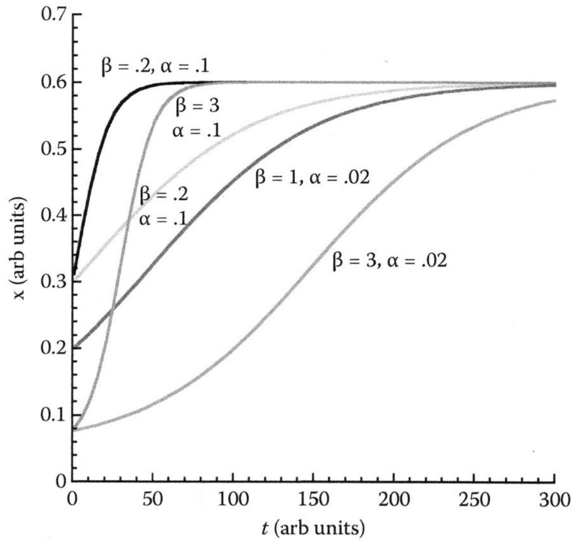

**Figure 10.2 Logistic growth curves.** The population size ($x$, arbitrary units; usually optical density) versus $t$ (usually time in minutes) for different values of $\alpha$ and $\beta$. The values of $x_0$ and $x_{max}$ are fixed.



cell growth, **Equation 10.2** reduces to simple exponential growth. When growth is limited by the availability of a factor, it reduces to the *logistic equation*:

$$x = \frac{kx_0}{x_0 + (k - x_0)\exp(-\beta t)} \equiv \frac{k}{1 + \exp(\alpha - \beta t)}, \tag{10.3}$$

where $\beta$ corresponds to the growth rate, $k$ to the carrying capacity, and $x_0$ to the original population size. This gives the familiar sigmoidal bacterial growth curve showing an exponential growth phase and plateau (**Figure 10.2**). It can be seen that the time at which $x$ reaches half maximum occurs at $t = t_{50}$, given by

$$t_{50} = \frac{\alpha}{\beta},$$

and that at small times, the equation becomes an exponential proportional to $\exp(\alpha t)$, with *doubling time* $\tau = \ln(2)/\alpha$.

Many bacterial growth curves are excellent fits to this equation. Others are more accurately fit by different special cases of the growth model; some common ones are the hyperlogistic, *Gompertz*, hyper-Gompertz, and Bertalanffy–Richards models. The logistic model does a poor job of describing the lag phase at the beginning of growth, and so the fit should be performed after this phase, or another model used to account for it.

## Bacterial inhibition curves and modeling

The addition of antibiotic agents can affect the growth curve in multiple nonlinear ways. Some bacteria in a population may be wholly or partially resistant to the agent; the agent may degrade slowly over time or act once and then disintegrate, and it may kill bacteria outright (*bactericidal* activity) or simply inhibit their growth (*bacteriostatic* activity).

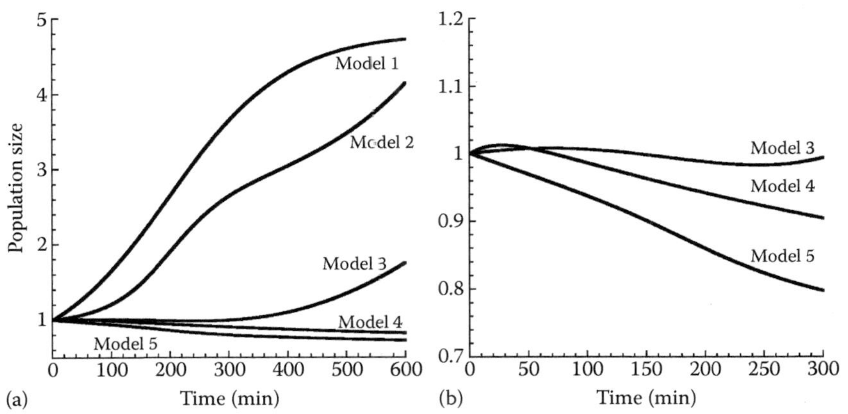



**Figure 10.3 Models of bacterial growth described by Equation 10.4.** (a) All models showing population size normalized to $t = 0$ with parameters chosen to lead to stationary behavior by ~600 min. (b) Zoom-in of early times of decaying models 4, 5, and 6 showing the differences. Note especially the slight increase in model 4 followed by the decay. This is a commonly observed feature of bacterial populations.

general equation for a bacterial growth curve in the presence of a toxic agent can be given in terms of a fraction of cells, $k_{dying}$, killed by the treatment and undergoing exponential decay at a rate $\alpha$; a fraction of cells, $k_{stat}$, prevented from reproducing but not killed outright; and a fraction of cells that continues to grow in a logistic fashion. The sum of these three populations leads to the equation for population size $x$ as a function of time $t$:

$$x(t) = \frac{k - k_{stat}}{1 + \exp(\beta - \alpha t)} + k_{stat} + k_{dying}\,e^{-\alpha t} \tag{10.4}$$

Depending upon the bacterial strain and the type of antibacterial agent(s) used, not all of these parameters are required to fit each curve. In addition, changing signs of certain parameters can be used to define new models using the same equation. For instance, if $\alpha < 0$, $k_{dying}$ becomes $k_{surviving}$, and the model describes a fraction of the surviving cells that begins to grow exponentially; that is, cell death is limited by the amount of the toxin or the damage to the cell.

Five distinct types of curves result from this type of equation:

Model 1: No cell death, classic S-shaped growth curve. $k_{dying} = 0$.

Model 2: Short lag followed by exponential growth. $k_{dying} > 0$, $\beta > 0$.

Model 3: Long lag followed by exponential growth. $k_{dying} > 0$, may have $\beta < 0$ and $\alpha < 0$.

Model 4: Slow rise followed by decay. $\alpha > 0$, $\beta > 0$, $k_{dying} > k$.

Model 5: Exponential decay only. $\alpha > 0$, $k = k_{stat} = 0$

These models are illustrated in **Figure 10.3**.

## IC$_{50}$ and minimum inhibitory concentration

The bacterial density at a chosen time point can be plotted against the log of the concentration of the antibiotic agent applied ($\log(C)$), usually resulting in an S-shaped curve. This curve may be fit to a Hill equation in order to determine the concentration needed to inhibit 50% of growth ($IC_{50}$). The function that describes the curve is

$$x = x_{min} + \frac{x_{max} - x_{min}}{1 + 10^{[Log(IC50) - Log(C)]H}},$$

(10.5)

where $H$ is the Hill coefficient (**Figure 10.4**). Although more rarely used, values of concentrations needed to inhibit arbitrary fractions of growth may also be reported ($IC_x$ where $x$ is anything, e.g., $IC_{20}$, $IC_{80}$). These values are determined from the same curve fit using the equation

$$\log(IC50) = \log(ICx) - \frac{1}{H}\log\frac{x}{100 - x}.$$

(10.6)

Minimum inhibitory concentration (MIC) may also be determined from the growth curves by relating the fractional area under the growth curve (AUC) to two parameters, $P_1$ and $P_2$, and the drug concentration $C$:

$$\frac{AUC}{AUC_0} = \exp\left[-\left(\frac{C}{P_1}\right)^{P_2}\right],$$

(10.7)

where MIC is the intercept of the tangent to the maximum gradient of the curve of AUC versus $\log(C)$:

$$MIC = P_1\exp\left(\frac{1}{P_2}\right)$$

(10.8)

However, the standard definition of the MIC is the concentration at which no bacterial growth is observed on a plate or in medium. This is determined by serial dilutions of the antibiotic agent as given in **Practical Tips 10.2**.

Most of the discussion here can be generalized to other microorganisms such as yeast or protozoa. Adjustments to the exact wavelength used may need to be made, and of course, adjustments to reading times must be used to take into account the different growth rates. Some single-celled organisms grow very slowly, so growth curves are determined by readings every 4–8 h for several days.

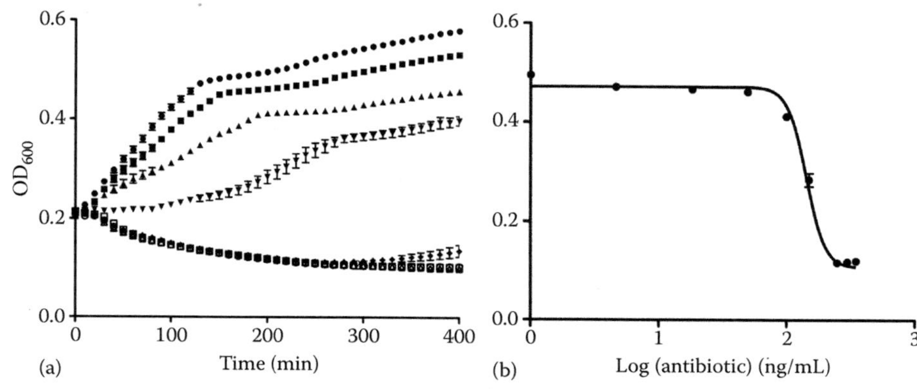


**Figure 10.4 $IC_{50}$ curves.** (a) Samples of growth curves at different antibiotic concentrations. It is important to use a concentration that does not inhibit growth noticeably, as well as one that kills all the bacteria if possible. (b) $IC_{50}$ curve taken at 300 min. The points on the plateaus ensure a good fit.

**PRACTICAL TIPS 10.2:    MIC BY DILUTION ASSAY**

**GENERAL REMARKS**

The minimum inhibitory concentration (MIC) is defined as the concentration of an agent that kills 99.9% of a specific microorganism in a specific growth medium within a certain time point (usually 24 h). This amount of cell killing corresponds to no visible growth in medium and/or no colonies on plates. The following assay uses a 96-well format and so is designed to conserve reagents.

**MATERIALS NEEDED**

Bacterial culture

Antibiotic in appropriate solution

Nutrient medium

Recommended: multichannel pipette

**PROCEDURE**

Make twofold serial dilutions of the drug as follows:

- Prepare a stock solution of the antibiotic in growth medium at twice the highest concentration to be tested. (If drug contains DMSO, final DMSO concentration should not exceed 5%.)

- Add 200 µL of this stock to each well of the first row in a clear 96-well plate.

- Add 100 µL of growth medium to all other wells.

- Pipette 100 µL of the drug suspension into the next set of wells. Mix.

- Continue serially until you reach row 11. Do not add drug to row 12. Change the pipette tips in between steps!

- Add 100 µL of bacterial suspension to all wells (1:2).

- Incubate 24 h or other determined time point. Inspect by eye and/or read optical density.

**SUGGESTED READING**

Motyl, M., Dorso, K., Barrett, J., and Giacobbe, R. (2006). Basic microbiological techniques used in antibacterial drug discovery. *Current Protocols in Pharmacology* Supp 31.

# 10.3  QUANTIFYING MAMMALIAN CELLS

## Counting mammalian cells

Mammalian cells should be counted using a hemocytometer. For adherent cells, trypsinize the cells (as described in **Chapter 3**) and resuspend them in complete culture medium. Centrifuge once gently to pellet, remove the supernatant, and resuspend in complete medium. Dilute as necessary for counting, and place into the hemocytometer. If desired, an equal volume of 0.4% trypan blue solution may be added to the cells to be counted. The dead cells will take up the dye while the live ones exclude it. With some practice, this becomes unnecessary—the dead cells usually have a very recognizable morphology, especially under phase contrast. Do not use any cultures for toxicity assays that have a large fraction of dead cells. The number of cells counted per hemocytometer square should be ~100 for statistical accuracy (choose a square of the appropriate size to ensure this), and they should not be so crowded that they overlap.

Of course, no sooner have you counted the cells that they begin to grow again. There are a few ways to use a known starting point to quantify future growth with more or less precision. A *plating efficiency* is determined by diluting the cells to a point where single cells attach and form single colonies, which can then be counted to give an estimate of the number of cells that are able to grow. It is very sensitive for comparing the differences among different growth conditions: e.g., different media, plates, sera (different lots of serum can make a huge difference!), or presence of a toxic agent. Sometimes, the plating efficiency is referred to as the plating efficiency of the *cells*, and other times of the *condition*, depending upon the experiment. The procedure after cell counting is as follows:

- Mix cells carefully so that they are evenly suspended.

- Calculate the volume required to yield 2, 10, and 20 cells/cm² of plating surface area (for comparison, 100% confluency is ~$10^5$ cells/cm²). A typical 96-well plate has about 0.3 cm² of plating area.

- Plate the appropriate volume in duplicate or triplicate.

- Incubate the cells until individual colonies are visible to the eye but not touching each other (usually a week to 10 days).

- Wash the cells with phosphate-buffered saline (PBS) and fix in 10% formalin for 10 min.

- Aspirate the formalin and add crystal violet to cover the cells for 10 min.

- Rinse with water until the water runs clear.

- Count the colonies and calculate the plating efficiency as number of colonies formed/number of cells plated.

The results of this assay are often dramatic (**Figure 10.5**).

**Figure 10.5 Example of plating efficiencies comparing two cell lines.** (Fan, H. et al., *PNAS* 94, 13181–13186, The R1 component of mammalian ribonucleotide reductase has malignancy-suppressing activity as demonstrated by gene transfer experiments. Copyright 1997 National Academy of Sciences, U.S.A. Image used with permission.)

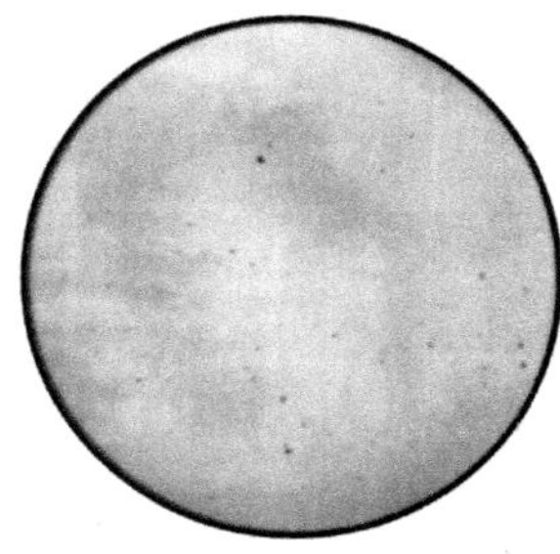
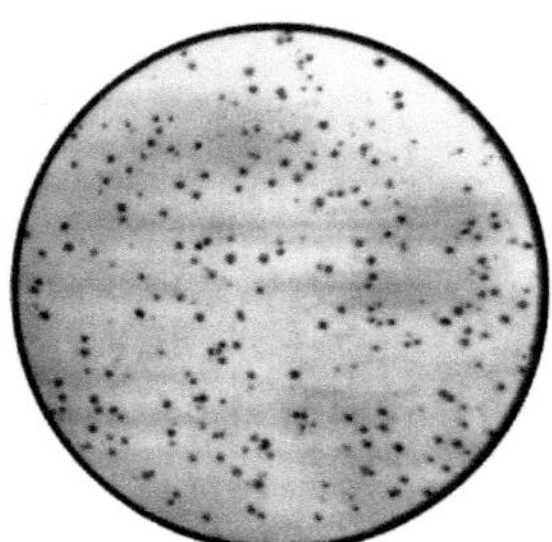

Another frequently measured parameter is a cell line's doubling time. This is usually done by plating the cells at different densities, waiting a specified amount of time to allow the cells to enter the exponential growth phase, and then trypsinizing and counting again. It is important to note that adherent cells have a long lag time, usually nearly 24 h. It is also key to appreciate that cells in monolayer culture show slowed or stopped growth when they are surrounded by other cells, a phenomenon known as *contact inhibition*. This means that the logistic equation is insufficient to describe the growth of these cells once they reach a relatively high density. There are many models available for describing contact inhibition that depend upon the cell type and how they grow. If all cells are equally contact inhibited, a logistic equation with an adjustment parameter may be all that is required. However, if the cells grow in patches, the cells in the middle of the patch will not divide at all, whereas the ones at the edges will grow normally. The situation is further complicated in three-dimensional cases such as tumors. Some references to these various models are given at the end of the chapter for the curious. However, in general, full growth curves are rarely done with mammalian cells. Instead, the density is controlled to maximize signal-to-noise, and cell density is measured at a single point. We call these types of assays end-point assays.

## End-point methods for mammalian cells: The sulforhodamine B assay and other colorimetric methods

The end-point assays described here are designed to quantify the effect of a drug or other treatment on cell proliferation. In order to accurately measure this, the cells must be allowed to go through a certain number of doubling times, usually two to three, before the end-point is reached. A lot can go wrong in this period—contamination, overgrowth, loss of cells due to incautious wash steps, or sometimes just inexplicable cell death. So although these methods sound very easy, they are tricky to get right, often yielding large standard deviations. Lots of repeats and lots of practice should eventually give good curves, along with some of the tips in this chapter.

The author's personal favorite of these assays is the sulforhodamine B (SRB) assay. This is a colorimetric (bright pink) assay that measures the total cell mass in a tissue culture well by binding of the dye to proteins. It is as sensitive as fluorescence-based assays and much easier to use (although the intense color means that care should be taken not to spill any around the lab, or everything will turn pink—benches, walls, even sterile water!). The assay's linear range is approximately 7500 to $1.8 \times 10^5$ cells per well of a 96-well plate, which corresponds to 1% confluence to over 200%. Limits of detection are 1000–2000 cells per well.

The first key to good results is to have the negative control (maximum cell growth) fall near the upper end of the assay's linear range. This maximizes the range of toxicity that can be quantified. If you are not very familiar with your cell line or with growing it in 96-well plates, the assay in **Practical Tips 10.3** helps to determine the starting point.

The protocol for SRB staining then proceeds as follows:

- Materials needed: 0.057% SRB in 1% acetic acid; 1% acetic acid solution in a squirt bottle; 40% tricholoroacetic acid (TCA) solution.

- Plate the cells in the 96-well plate at the determined density.

---

**PRACTICAL TIPS 10.3:  PRETEST FOR SRB ASSAY**

**GENERAL REMARKS**

The idea is to have the untreated control cells show maximum density within the linear range of the test.

**MATERIALS NEEDED**

Cells and culture medium

Appropriate plate to be used in SRB assay (usually a 96-well plate)

SRB reagents (see text)

**PROCEDURE**

Make twofold serial dilutions of the drug as follows:

- Trypsinize the cells and count them.
- Prepare a solution containing $5 \times 10^3$ cells/mL.

Add to wells of a 96-well plate as follows:

| Plating Test | | | | | | | | | | |
|---|---|---|---|---|---|---|---|---|---|---|
| Cells per well | 1000 | 750 | 500 | 350 | 200 | 100 | 50 | 25 | 12 | 0 |
| Cell stock 5000 cells/mL ($\mu$L) | 200 | 150 | 100 | 70 | 40 | 20 | 10 | 5 | 2.5 | 0 |
| Medium ($\mu$L) | 0 | 50 | 100 | 130 | 160 | 180 | 190 | 195 | 197.5 | 200 |

- Allow to grow for the length of the drug assay (e.g., 3 days if you will stop the assay after 72 h).
- Perform the SRB assay as described in the text.
- Plot the resulting curve to determine the linear range.
- Use the high point of the linear range as the starting density in the actual experiments.

- At least 24 h later, incubate the cells with drugs or other agents according to your desired test protocol. Some agents, such as nanoparticles, must be incubated with cells in serum-free medium. If this is the case, remove the medium carefully from the cells by aspiration before adding the serum-free medium.

- Carefully wash away drugs and put back the regular medium.

- Incubate cells at 37°C in 5% $CO_2$ for 72 h or other determined period.

- Add 65 $\mu$L of 40% TCA directly to the medium.

- Aspirate culture medium and rinse wells four times with tap water, shake vigorously to remove residual water, and let air-dry (usually overnight).

- Add 100 μL of SRB stain and incubate for 30 min at room temperature. Make sure to protect SRB from light since it is light sensitive.

- Wash cells four times with 1% acetic acid, using the squirt bottle to fill wells rapidly. Shake vigorously to remove as much acetic acid as possible.

- Leave plates to air-dry. (This takes several hours; if you are in a hurry, place them at an angle in the chemical hood, and the airflow will cause them to dry in 90 min.) Add 100 μL of 10 mM Tris base unbuffered (pH ~7.4). Agitate on rocker till stain dissolves and read optical density at 510 nm. The variations in signal should be obvious to the eye. The plates may be stored for extended periods without loss of signal.

The most common problem with this assay is excessive loss of cells, through incautious pipetting or removal of medium in an excessively rough manner. When adding new medium, try to pipette against the walls of the plate rather than directly over the cells. When removing medium, aspirate rather than inverting the plate to dump the medium out. It is also a good idea to perform multiple repeats of each test case; for example, if eight repeats are performed, the highest and lowest value may be discarded. Sometimes, the wells in the corners of a 96-well plate behave differently from the others, and this might also be taken into account.

There are several other colorimetric end-point assays that are in very common use. In the *MTT* (3-(4,5-dimethylthiazol-2-yl)-2,5-diphenyltetrazolium bromide) assay, MTT is reduced to a purple formazan dye in actively metabolizing cells. Closely related compounds, such as XTT (2,3-bis-(2-methoxy-4-nitro-5-sulfophenyl)-2H-tetrazolium-5-carboxanilide), are also used and work in the same way. The primary difference from SRB is that only actively metabolizing cells, rather than the whole cell mass, are measured. This can sometimes making quantifying the assay a little trickier. However, in general, there is very good agreement between the SRB and MTT assays. One caveat is that some particles and/or compounds can reduce MTT in the absence of cell activity, leading to a false-positive signal. Examples include serum albumin and many antioxidants (i.e., reducing agents).

The end-point assays should be performed with drug concentrations that yield a sigmoidal dose–response curve, the same as **Equation 10.5**. The survival is usually normalized to the control wells and reported as a percentage (**Figure 10.6**).

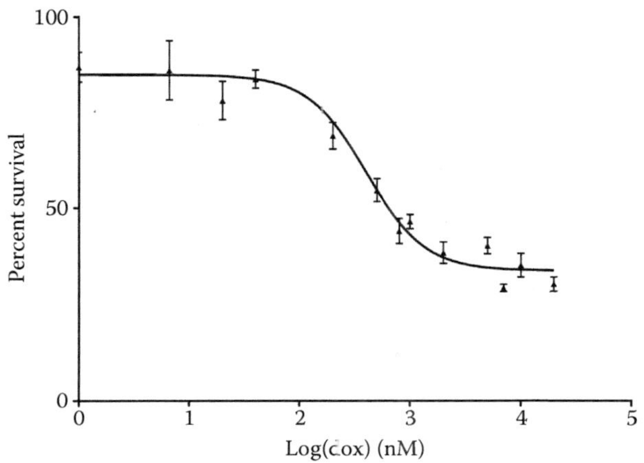


**Figure 10.6 Example of survival curve resulting from SRB assay.** The percent survival of a cell line versus log concentration of the cytotoxic drug doxorubicin (dox) is shown and fit to the Hill equation. Note the error bars, which result from sixfold duplicates of each condition. (The curve does not start from 100%, because the data point with 0 drug is not plotted on this graph.)

It is also possible to quantify mammalian cell growth by growing the cells in colonies and using colony diameter and integrated optical density as measures of growth. The latter has been shown to be proportional to the number of cells. This may be used as either a real-time or an end-point method. As an end-point method, we prefer the colorimetric assays as they are more easily quantified; however, there are some circumstances under which colony-forming assays must be used. Some real-time methods of mammalian cell detection are discussed in **Advanced Topics 10.1 and 10.2.**

### ADVANCED TOPIC 10.1:   ELECTRIC CELL–SUBSTRATE IMPEDANCE SENSING

One of the simplest and most well-established techniques for evaluating cell-surface interactions uses changes in impedance, a method called electric cell–substrate impedance sensing (ECIS). The cells are plated onto a working electrode (usually lithographically patterned and <1 mm in diameter), and an alternating current (AC) is passed between this electrode and a much larger counter electrode (**Figure A10.1.1**). Ordinary culture medium serves as the electrolyte. The cells act as insulators because of their highly resistive membranes, and so cell spreading and the formation of cell–cell contacts are all measurable as changes in impedance per unit area of the cell-covered electrode versus a cell-free electrode.

For disk-shaped cells, only three adjustable parameters are required to fit the measured values to a model: the cell capacitance $C_\mathrm{m}$; the junctional resistance between the cells $R$; and the parameter $\alpha$, given by

$$\alpha = r\sqrt{\frac{\rho}{h}}, \tag{A10.1.1}$$

where $r$ is the radius of the cells, $\rho$ is the resistivity of the culture medium, and $h$ is the cell–substrate separation height (**Figure A10.1.2**). Input values include the electrode surface area, capacitance and resistance of the cell-free electrode, and frequency of the applied AC.

Typical measurements are impedance versus time at a given AC frequency (from 4 to 60 kHz), and impedance as a function of frequency. Changing frequencies allows for observation of scaling behavior of the curves and allows differentiation between changes in cell junctional resistance, capacitance, and/or height from the substrate. All of these parameters are likely to change during toxicity experiments as the cells contract, round, and detach from the surface (**Figure A10.1.3**). It is important to note that if the cell monolayer develops holes, the method no longer works, as large leak currents develop.

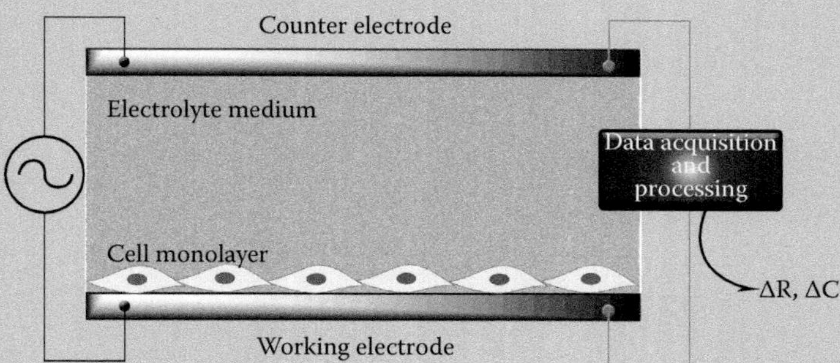


**Figure A10.1.1 Principle of action of ECIS.** The AC can only flow between cells and the substrate or between two cells because the plasma membrane acts as an insulator. A cell monolayer without defects is required.

(*Continued*)

**ADVANCED TOPIC 10.1 (CONTINUED):    ELECTRIC CELL–SUBSTRATE IMPEDANCE SENSING**

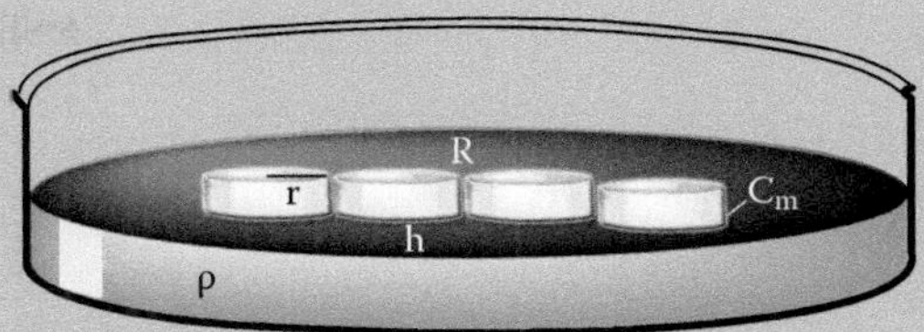


**Figure A10.1.2  Model of a cell monolayer as a collection of disks in contact with one another.**

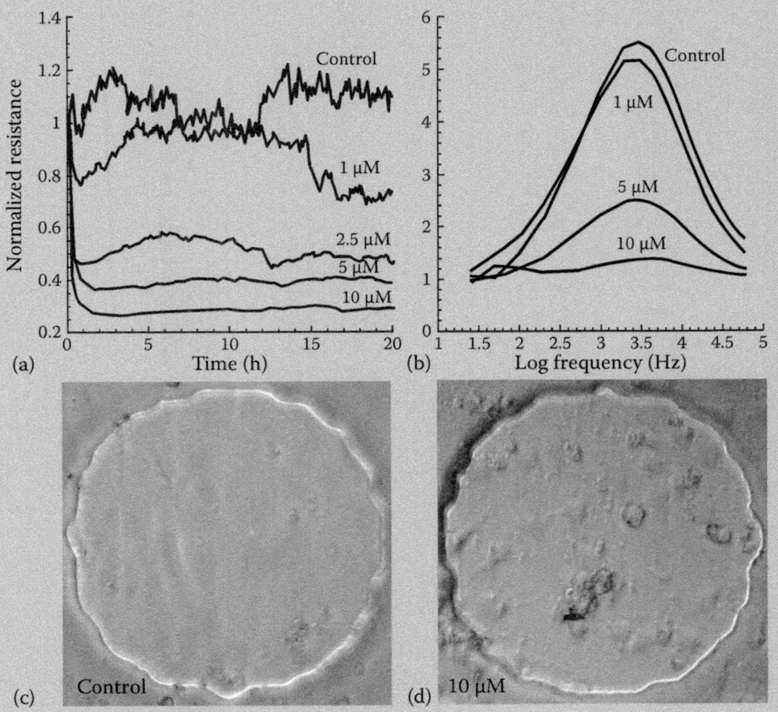


**Figure A10.1.3  Sample ECIS data.** (a) Normalized resistance versus time for cell monolayers exposed to different concentrations of the toxin cytochalasin B, which inhibits cell movement and causes extrusion of nuclei. (b) Normalized resistance versus log(frequency). (c) Image of control cell monolayer on 250 μm electrode. (d) Image of cell monolayer with 10 μM cytochalasin B. (Images and data courtesy Chun-Min Lo, University of South Florida.)

Substrates for ECIS may be made in any micromachining facility, and many researchers use homemade devices. Systems are also now available commercially with multiwell electrodes (Figure A10.1.4). Of course, adherent mammalian cells will not grow well if plated directly on gold, so the electrodes should be coated with an attachment-permitting substrate before the cells are plated. Usually, fibronectin is used for this, although polyelectrolyte multilayers have been suggested as a more stable alternative.

*(Continued)*

## ADVANCED TOPIC 10.1 (CONTINUED):   ELECTRIC CELL–SUBSTRATE IMPEDANCE SENSING

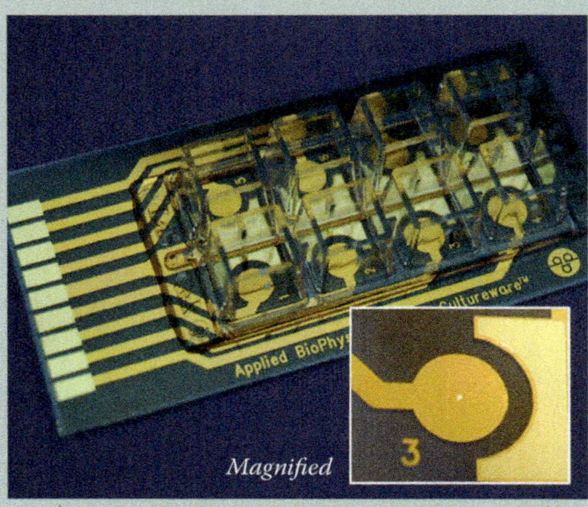


**Figure A10.1.4 Example of multiwell electrode for ECIS.** Each of the eight wells contains a single circular 250-μm-diameter active electrode. (Image courtesy of C. Dehnert, http://www.biophysics.com.)

## SUGGESTED READING

Bogomolova, A., Komarova, E., Reber, K., Gerasimov, T., Yavuz, O., Bhatt, S., and Aldissi, M. (2009). Challenges of electrochemical impedance spectroscopy in protein biosensing. *Analytical Chemistry* 81, 3944–3949.

Giaever, I., and Keese, C. R. (1991). Micromotion of mammalian-cells measured electrically. *Proceedings of the National Academy of Sciences of the United States of America* 88, 7896–7900.

Hug, T. S. (2003). Biophysical methods for monitoring cell–substrate interactions in drug discovery. *Assay and Drug Development Technologies* 1, 479–488.

Keese, C. R., Bhawe, K., Wegener, J., and Giaever, I. (2002). Real-time impedance assay to follow the invasive activities of metastatic cells in culture. *Biotechniques* 33, 842.

Keese, C. R., Karra, N., Dillon, B., Goldberg, A. M., and Giaever, I. (1998). Cell–substratum interactions as a predictor of cytotoxicity. *In Vitro and Molecular Toxicology—A Journal of Basic and Applied Research* 11, 183–192.

Male, K. B., Lachance, B., Hrapovic, S., Sunahara, G., and Luong, J. T. (2008). Assessment of cytotoxicity of quantum dots and gold nanoparticles using cell-based impedance spectroscopy. *Analytical Chemistry* 80, 5487–5493.

Mijares, G. I., Reyes, D. R., Geist, J., Gaitan, M., Polk, B. J., and Devoe, D. L. (2010). Polyelectrolyte multilayer-treated electrodes for real-time electronic sensing of cell proliferation. *Journal of Research of the National Institute of Standards and Technology* 115, 61–73.

Opp, D., Wafula, B., Lim, J., Huang, E., Lo, J. C., and Lo, C. M. (2009). Use of electric cell–substrate impedance sensing to assess in vitro cytotoxicity. *Biosensors and Bioelectronics* 24, 2625–2629.

Rumenapp, C., Remm, M., Wolf, B., and Gleich, B. (2009). Improved method for impedance measurements of mammalian cells. *Biosensors and Bioelectronics* 24, 2915–2919.

Tiruppathi, C., Malik, A. B., Delvecchio, P. J., Keese, C. R., and Giaever, I. (1992). Electrical method for detection of endothelial-cell shape change in real-time—Assessment of endothelial barrier function. *Proceedings of the National Academy of Sciences of the United States of America* 89, 7919–7923.

Xiao, C., and Luong, J. H. T. (2003). On-line monitoring of cell growth and cytotoxicity using electric cell–substrate impedance sensing (ECIS). *Biotechnology Progress* 19, 1000–1005.

### ADVANCED TOPIC 10.2:   QUARTZ CRYSTAL MICROBALANCE AND OPTICAL WAVEGUIDE LIGHTMODE SPECTROSCOPY

There are other biophysical techniques that are less often used than ECIS but that have been reported to be sensitive reporters for cell growth, movement, and/or death. The author has no experience with these, and your mileage may vary! One of these techniques is *quartz crystal microbalance* (QCM), which makes use of the piezoelectric nature of quartz (originally natural quartz, now usually laboratory-grown). A small stress on the crystal leads to a proportional electrical potential on its surface. When driven with alternating current (AC), the crystal produces a standing shear wave with a high quality factor, up to $10^6$ (Figure A10.2.1). The resonance frequency of the wave is shifted by an amount $\Delta f$ when mass $m$ is deposited on the crystal surface; when the shift is small relative to the frequency, the Sauerbrey equation gives

$$\Delta f = -C_f \mathrm{m}$$

where $C_f$ is the crystal calibration constant given by the crystal manufacturer.

QCM is often used for microgravimetry; commercial instruments are sometimes called "thickness monitors." Its use in liquids is much less obvious, as the damping may stop the crystal from oscillating. Even if this fatal flaw is overcome, the resonant frequency is influenced by the solution in ways that remain poorly defined. However, in the 1980s, it was shown that the mass sensitivity in solution approaches that of the Sauerbrey equation, and with proper design, electrical influence of the solution can be avoided. Along with $\Delta f$, measurements of the damping parameter $\Delta R$ are also informative in cell studies. Deposition of a cell monolayer causes the resonant frequency to increase and the damping factor to increase. Cell death results in the opposite, thus leading to a sign change in the $\Delta f/f$ and $\Delta f/\Delta R$ curves (Figure A10.2.2).

Many QCM systems are available commercially, including turnkey systems designed for use in liquids. It is important to note that any mass adsorption will change the signal, so it is crucial to allow media components such as serum to equilibrate before adding cells, and to keep these components constant throughout the experiment.

Other techniques use changes in refractive index to infer the health of the cells. The most common of these techniques is *optical waveguide lightmode spectroscopy* (OWLS), which measures the effective refractive

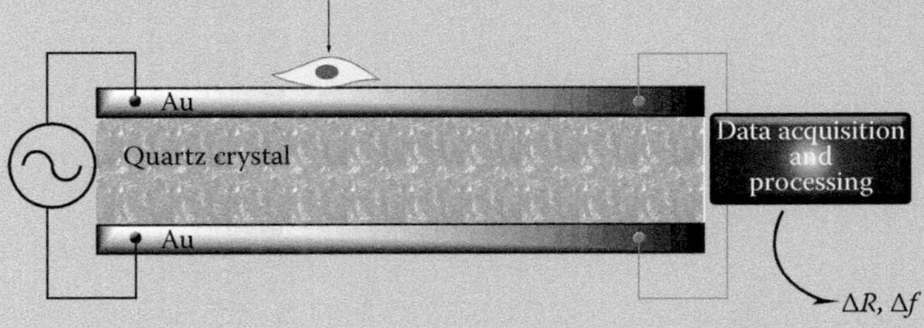


**Figure A10.2.1  Principle of action of QCM.** Mass deposition (e.g., a cell) on an oscillating piezoelectric quartz crystal causes a change in the resonant frequency $\Delta f$. Decay of the oscillations measured by ringdown gives the damping parameter $\Delta R$.

*(Continued)*

**ADVANCED TOPIC 10.2 (CONTINUED):   QUARTZ CRYSTAL MICROBALANCE
AND OPTICAL WAVEGUIDE LIGHTMODE SPECTROSCOPY**

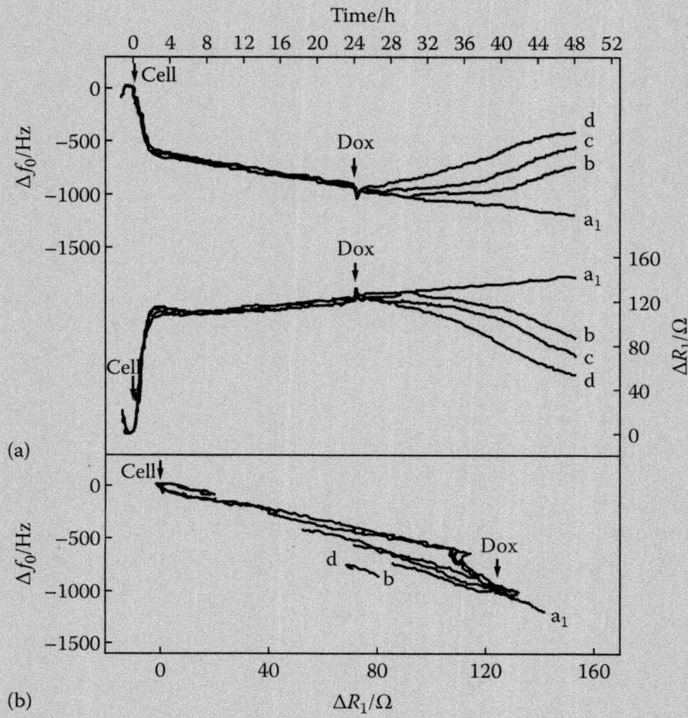

**Figure A10.2.2** Real-time $\Delta f_0$ and $\Delta R_1$ responses to the addition of cells onto QCM Au electrode in the absence (a1) and presence of midway-added 10.0 µg/mL (b), 20.0 µg/mL (c), and 30.0 µg/mL (d) doxorubicin (Dox). (b) The relationships of $\Delta f_0$ versus $\Delta R_1$.

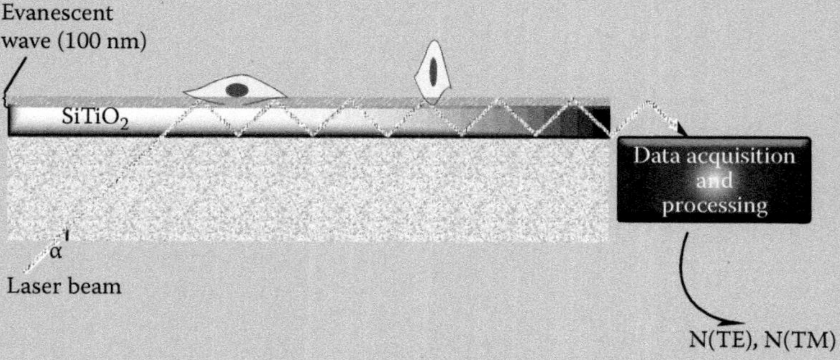


**Figure A10.2.3  Principle of action of OWLS.** A polarized laser beam is coupled into a waveguide. The light probes the sensor surface to the depth of the evanescent field, an area that might contain proteins or cells. The coupling angle $\alpha$ at which light is maximally propagated allows for the determination of the effective refractive indices of the electric and magnetic field components, $n$(TE) and $n$(TM).

*(Continued)*

**ADVANCED TOPIC 10.2 (CONTINUED):   QUARTZ CRYSTAL MICROBALANCE
AND OPTICAL WAVEGUIDE LIGHTMODE SPECTROSCOPY**

index of a thin layer above the surface of a waveguide. The theoretical description of how this is done is given in detail in the end-of-chapter references. In brief, a polarized laser beam is coupled into a waveguide using an optical grating. A phase shift occurs when the polarized light is totally reflected at the interface between the waveguide and the medium. The phase shift is a function of the quantity and polarizability of anything adsorbed to this surface within the penetration depth of the evanescent wave (about 100 nm), and causes light to be maximally propagated at a specific angle, called a *coupling angle* (Figure A10.2.3). From the coupling angles, the refractive mode indices can be calculated. Like QCM, OWLS is not specific, and gives a signal due to adsorption of serum or any other proteins. However, it has been shown that cell growth can be monitored over time using this technique if serum is adequately controlled for. Full OWLS systems are available commercially.

**SUGGESTED READING**

Bruckenstein, S., and Shay, M. (1985). Experimental aspects of use of the quartz crystal microbalance in solution. *Electrochimica Acta* 30, 1295–1300.

Hug, T. S., Prenosil, J. E., Maier, P., and Morbidelli, M. (2002a). On-line monitoring of adhesion and proliferation of cultured hepatoma cells using optical waveguide lightmode spectroscopy (OWLS). *Biotechnology Progress* 18, 1408–1413.

Hug, T. S., Prenosil, J. E., Maier, P., and Morbidelli, M. (2002b). Optical waveguide lightmode spectroscopy (OWLS) to monitor cell proliferation quantitatively. *Biotechnology and Bioengineering* 80, 213–221.

Leclercq, L., Modena, E., and Vert, M. (2013). Adsorption of proteins at physiological concentrations on PEGylated surfaces and the compatibilizing role of adsorbed albumin with respect to other proteins according to optical waveguide lightmode spectroscopy (OWLS). *Journal of Biomaterials Science, Polymer Edition* 24, 1499–1518.

Tan, L., Jia, X. E., Jiang, X. F., Zhang, Y. Y., Tang, H., Yao, S. Z., and Xie, Q. J. (2009). In vitro study on the individual and synergistic cytotoxicity of adriamycin and selenium nanoparticles against Bel7402 cells with a quartz crystal microbalance. *Biosensors and Bioelectronics* 24, 2268–2272.

Yu, H., Eggleston, C. M., Chen, J., Wang, W., Dai, Q., and Tang, J. (2012). Optical waveguide lightmode spectroscopy (OWLS) as a sensor for thin film and quantum dot corrosion. *Sensors (Basel)* 12, 17330–17342.

# 10.4  FLOW CYTOMETRY

*Flow cytometry* is a method of computerized counting and/or sorting of cells based upon their optical properties (scattering and fluorescence) as measured by selected beams of light. It allows for counting of thousands of cells per minute in a controlled cell-by-cell fashion. This single-filing is made possible using *hydrodynamic focusing*, where the cell-containing sample is injected into a central channel that is enclosed by a sheath containing a faster-flowing fluid. The central channel narrows as it approaches the light beam, and a massive drag is exerted upon it by the sheath, resulting in passing of cells single-file through the 70 µm nozzle of the instrument without blockage. Ideally, the flow remains laminar ($R_e < 2300$), and there is no mixing of the sheath fluid with the sample.

Most flow cytometers use at least two lasers to probe the cells, usually green (488 nm) and red (635 nm); advanced systems use even more. One detector is placed line with the light beam(s) to measure forward scatter (FSC) from the cells. Detectors perpendicular to the beam measure side scatter (SSC) and fluorescence; multiple wavelengths of detector are often used for each excitation wavelength (Figure 10.7).

**Figure 10.7 Configuration of a typical multicolor flow cytometer.**

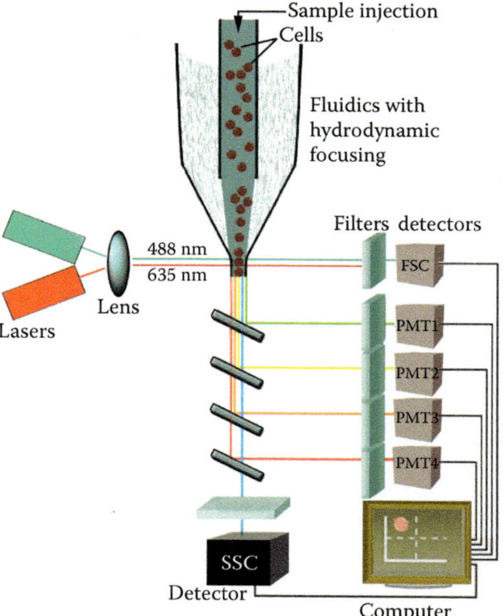


Flow cytometry data may be represented as histograms or as *dot plots*. A histogram quantifies the intensity of a single parameter, be it fluorescence or scattering. Subpopulations are identified as peaks in the histogram (**Figure 10.8a**). A dot plot, in contrast, plots one parameter against another and represents each cell by a dot with a position on the two axes. The density of cells in each region may be color-coded or drawn as a contour plot to indicate relative cell numbers in each region. Dot plots allow for the identification of particular subpopulations—e.g., cells that are red but not green versus cells that are both red and green (**Figure 10.8b**).

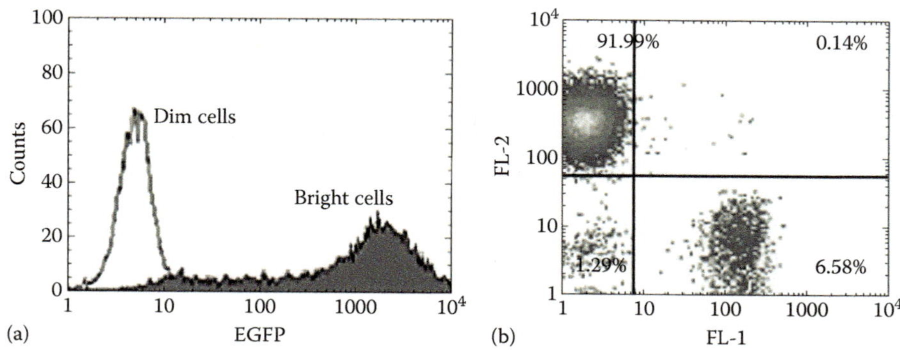


**Figure 10.8 Different ways of presenting flow cytometry data.** (a) Enhanced green fluorescent protein (EGFP) measured by a histogram. The data are plotted on a log scale because of the extremely wide range of protein expression. The median is a better measure than the mean of such a histogram, since a few very bright cells can skew the mean. (b) Dot plot of two fluorophores, FL-1 versus FL-2. Among the cells, 91.99% are positive for FL-2 but not FL-1; 6.58% are positive for FL-1 but not FL-2; 1.29% are negative for both; and 0.14% are positive for both.

Dot plots of FSC versus SSC can be highly informative. FSC correlates with the cell volume, and SSC depends on the inner complexity of the particle (i.e., shape of the nucleus, amount and type of cytoplasmic granules, or the membrane roughness). Thus, scattering alone can be used to distinguish different types of cells in a heterogeneous mixture, such as dead cells versus live cells or different types of white blood cells. Dead cells show lower FSC and higher SSC than living cells (**Figure 10.9a**). Different types of white blood cells show very distinctive scattering properties, since they are of different sizes and granularities (**Figure 10.9b**).

An important principle of flow cytometry data analysis is called *gating*. This is when the cells with desirable values of a given parameter are used in the analysis, while other cells are eliminated. For example, *gating on live cells* would eliminate the cells identified as dead by scattering as in **Figure 10.9a**; *gating on monocytes* would eliminate all but those identified as such in **Figure 10.9b**.

The fluorescent signal depends upon the type of cells and how they have been labeled, transfected, or tagged. The ability to use two or more different fluorescent tags makes this technique ubiquitous in biology. The most common tags include genetically encoded markers (fluorescent proteins and fusions to fluorescent proteins); antibodies to cell-surface markers (these can be used on live cells); and dyes to indicate cell death or apoptosis (propidium iodide, etc.). Most of the environmentally sensitive dyes discussed in **Chapter 7**, such as the calcium-sensitive or voltage-sensitive dyes, may also be used with flow cytometry. FACS is a special case of flow cytometry in which the optical properties are used to sort cells into different containers. The term is often used interchangeably with flow cytometry, although it should not be.

Flow cytometry, and usually FACS, is available at most institutions within core facilities. The amount of assistance with sample preparation and data analysis will vary. An in-depth discussion of the hundreds of possible experiments that are possible using multicolor flow cytometry would take volumes; **Table 10.1** shows some examples, and the next section describes one of them in detail.

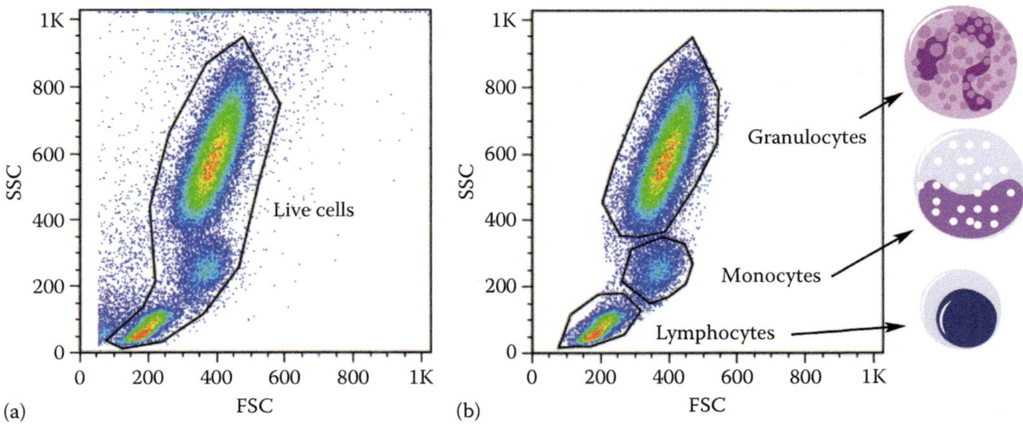


**Figure 10.9 Scattering dot plots.** (a) FSC versus SSC for identifying live versus dead cells in a sample of human blood. (b) FSC versus SSC for identifying different populations of white blood cells in the same sample, with schematic drawings of the cells to show relative size and granularity. (Flow cytometry plots from Melgert, B. N. et al., Pregnancy and Preeclampsia Affect Monocyte Subsets in Humans and Rats, *PLoS ONE* 7 (9), e45229, 2012, doi:10.1371/journal.pone.0045229).

**Table 10.1**

Some of the Many Applications of Flow Cytometry/FACS

| Parameter Measured | Label | Sample Reference |
|---|---|---|
| Differential blood count | None (SSC/FSC) and/or cell-surface marker antibodies (anti-CD4, etc.) | Dauber, J. H. et al., Flow Cytometric Analysis of Lymphocyte Phenotypes in Bronchoalveolar Lavage Fluid: Comparison of a Two-Color Technique with a Standard Immunoperoxidase Assay, *Am. J. Respir. Cell. Mol. Biol.* 7, 531–541, 1992. |
| Environmental community analysis | DNA dyes + autofluorescence | Marie, D. et al., Enumeration and Cell Cycle Analysis of Natural Populations of Marine Picoplankton by Flow Cytometry Using the Nucleic Acid Stain SYBR Green I, *Appl. Environ. Microbiol.* 63, 186–193, 1997. |
| Cell activation | Specific antibodies | Gavasso, S, Flow Cytometry and Cell Activation, *Methods Mol. Biol.* 514, 35–46, 2009. |
| Refractive index of cells/organisms | None | Spinrad, R. W., and J. F. Brown, Relative Real refractive Index of Marine Microorganisms: A Technique for Flow Cytometric Estimation, *Appl. Opt.* 25, 1930, 1986. |
| Cell viability | Uptake or exclusion of dye (SYTOX, acridine orange, propidium iodide, etc.) | Berglund, D. L. et al., A Rapid Analytical Technique for Flow Cytometric Analysis of Cell Viability Using Calcofluor White M2R, *Cytometry* 8, 421–426, 1987. |
| Cell division | BrdU and FSC | Nusse, M. et al., Flow Cytometric Detection of Mitotic Cells Using the Bromodeoxyuridine/DNA Technique in Combination with 90 Degrees and Forward Scatter Measurements, *Cytometry* 10, 312–319, 1989. |
| Drug resistance in cancer cells | Fluorescent drug + P-glycoprotein | Gheuens, E. E. et al., Flow Cytometric Double Labeling Technique for Screening of Multidrug Resistance, *Cytometry* 12, 636–644, 1991. |
| Diagnosis of malaria | Red-cell antigen and propidium iodide | Pattanapanyasat, K. et al., Flow Cytometric Two-Color Staining Technique for Simultaneous Determination of Human Erythrocyte Membrane Antigen and Intracellular Malarial DNA, *Cytometry*, 13, 182–187, 1992. |
| Intracellular ions (Pb, Ca, Mg, etc.) | Fluorescent indicator dyes | Dyatlov, V. A. et al., Lipopolysaccharide and Interleukin-6 Enhance Lead Entry into Cerebellar Neurons: Application of a New And Sensitive Flow Cytometric Technique to Measure Intracellular Lead and Calcium Concentrations, *Neurotoxicology* 19, 293–302, 1998. |
| Membrane potential | Reporter dyes | Novo, D. et al., Accurate Flow Cytometric Membrane Potential Measurement in Bacteria Using Diethyloxacarbocyanine and a Ratiometric Technique, *Cytometry* 35, 55–63, 1999. |
| Gram-positive versus Gram-negative bacteria | Wheat germ agglutinin dye and propidium iodide | Holm, C. et al., A Flow Cytometric Technique for Quantification and Differentiation of Bacteria in Bulk Tank Milk, *J. Appl. Microbiol.* 97, 935–941, 2004. |
| Protein–protein interactions | Beads for immobilization + antibodies | Bridgeman, J. S. et al., Development of a Flow Cytometric Co-Immunoprecipitation Technique for the Study of Multiple Protein–Protein Interactions and its Application to T-Cell Receptor Analysis, *Cytometry A* 77, 338–346, 2010. |
| Continuous measurements of phytoplankton | Chlorophyll and phycoerythrin (native pigments) | Swalwell, J. E. et al., SeaFlow: A Novel Underway Flow-Cytometer for Continuous Observations of Phytoplankton in the Ocean, *Limnol. Oceanogr. Methods* 9, 466–477, 2011. |
| Measurement of bacterial physiological processes | Varies | Ambriz-Avina, V. et al., Applications of Flow Cytometry to Characterize Bacterial Physiological Responses, *Biomed. Res. Int.* 2014: 461941, 2014. |

# 10.5 EXAMPLE EXPERIMENT: DETERMINING LEUKEMIC B CELLS AND T CELLS BY FLOW CYTOMETRY

Along with their differences in size and granularity, white blood cells also show a variety of distinctive cell-surface markers that make antibody-based flow cytometry of tremendous use in identifying and counting them. The clinical uses of this technique are widespread. Before the identification of the human immunodeficiency virus (HIV) as the causative factor of AIDS, the disease could be diagnosed by measuring the total number of lymphocytes that stained positive to the antibody CD4 (CD4+ cells), or the ratio of CD4+ to CD8+ cells. This technique is still used in some developing countries, as flow cytometry is less costly than specific HIV testing. Determination of the counts and ratios of other types of blood cells is useful in the diagnosis of hematologic cancers, such as leukemia. The following example was designed for identification of leukemia from patient samples. It is designed to be used on unfixed blood samples (unless there is a risk of hepatitis virus or HIV contamination, in which case the samples should be fixed in paraformaldehyde). The samples are collected in 8–10 mL tubes with EDTA to prevent clotting, and centrifuged to remove red blood cells. In a normal sample, there will be about $5$–$10 \times 10^6$ white blood cells/mL. In a leukemic sample, this may be much higher.

The concept behind the procedure is very simple: T cells express CD3, whereas B cells express CD20. Of these B cells, normal mature ones should not express CD5, whereas some of the leukemic cells will express CD5. There may be other cells in the sample that express CD5 that are not B cells. Thus, the following populations need to be identified:

- B cells versus T cells (the relative fraction will give a good hint as to the presence of leukemia). For this, a dot plot of CD3 versus CD20 is needed.

- Fraction of B cells that are CD5 positive. For this, a dot plot of CD20 versus CD5 should show the distinction between CD5-positive cells that are B cells, and those that are not.

Since two independent comparisons are being done, only two colors are needed: CD20 should be linked to an antibody of one color (say phycoerythrin [PE]), and CD5 and CD3 to another color fluorescein isothiocyanate (FITC). All of these antibodies are commercially available. Two control antibodies linked to PE and FITC should also be purchased. These are called *isotype* antibodies and should not recognize anything on the target cells (e.g., CD8). They are used to control for nonspecific binding and to set the zero values for flow cytometry.

A cell pellet of $5 \times 10^5$ to $1 \times 10^6$ cells is needed for each sample. Wash the cells with PBS, and then resuspend them to a concentration of $10^6$/mL in flow cytometry buffer (FCB: 3% fetal calf serum in PBS). Prepare 5 samples:

- Isotype controls: isotype-PE and isotype-FITC antibodies.

- FITC-only sample: CD3-FITC antibody only.

- PE-only sample: CD20-PE.

- Double-stained sample for B versus T determination: CD20-PE and CD3-FITC.

- Double-stained sample for leukemia determination: CD20-PE and CD5-FITC.

Add antibodies according to the manufacturer's recommendation and let incubate on ice for 30 min. Centrifuge, wash in FCB, and resuspend in FCB to $10^6$ cells/mL.

The cells are now ready for flow cytometry analysis. Because of the double labeling, some adjustments are necessary to ensure that the overlap of the spectra of FITC and PE does not create channel bleed-through. First, use the double-isotype control to make a plot of FSC versus SSC. Gate on the cells that are consistent with live cells (**Figure 10.10a**). Next, use the double-isotype control to set both fluorescence channels to close to zero (**Figure 10.10b**).

Next, use the single-color samples to minimize channel bleed-through between the FITC and PE signals. This is called adjusting the *compensation* in flow cytometry. Say FITC is assigned to the FL-1 channel, and PE to the FL-2 channel. All of the FITC-positive cells should be negative in FL-2; it is OK if a few of them are also negative in FL-1 (**Figure 10.11a**). Similarly, all of the PE-labeled cells should be negative in FL-1, and most should be positive in FL-2 (**Figure 10.11b**).

**Figure 10.10 Setting the parameters based upon isotype-stained controls.** (a) Gate on the live, unaggregated cells using scattering. (b) Adjust so that the fluorescence in both channels is negligible.

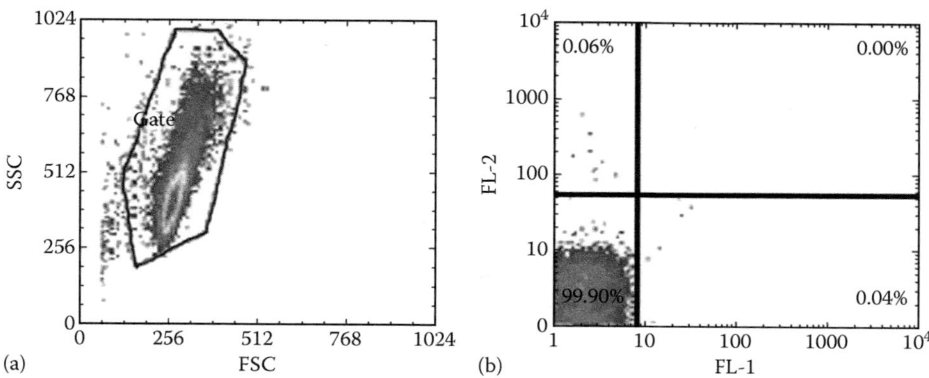

**Figure 10.11 Adjusting compensation for double labeling.** (a) Adjust the FITC channel to be PE-negative. (b) Adjust the PE channel to be FITC-negative.

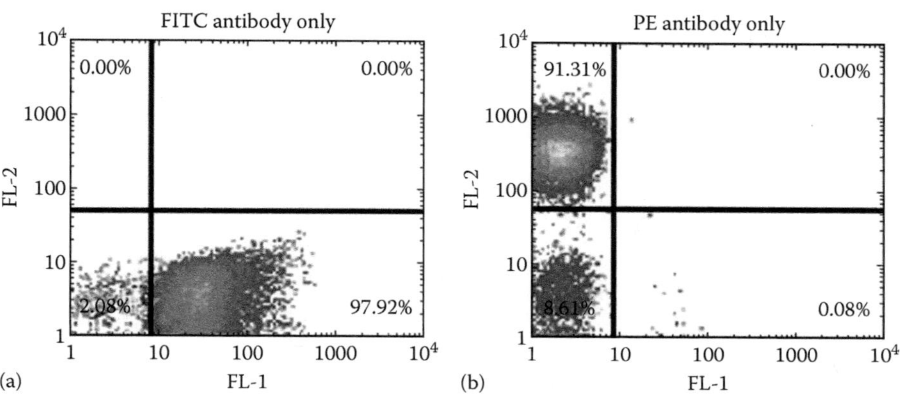

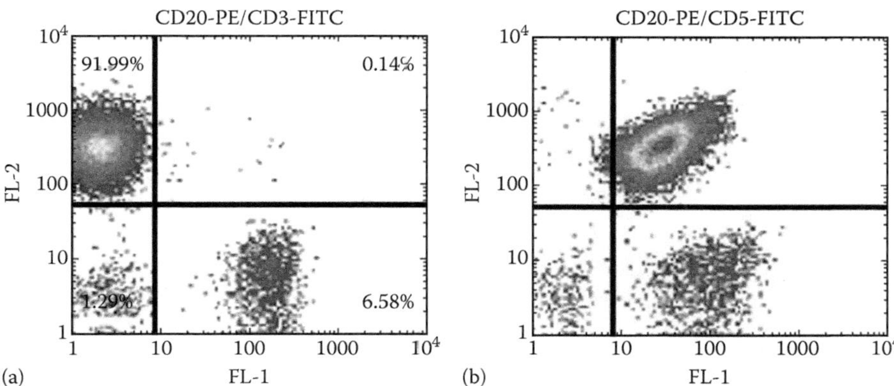


**Figure 10.12  Flow cytometry data from double-labeled cells.** (a) Identification of B cells versus T cells using CD20 versus CD3. Note the abundance of B cells in this leukemic patient. (b) Identification of CD5-positive B cells (those that are positive in both channels). These are leukemic cells. (Images from **Figures 10.10 through 10.12** are courtesy of C. A. Hollmann, McGill University.)

Finally, run the double-labeled samples. The CD20/CD3 sample (**Figure 10.12a**) shows that only 6.58% of the cells are T cells, whereas nearly 92% of the cells are B cells. This is a typical appearance in B-cell leukemia. The double-labeled sample with CD20 and CD5 (**Figure 10.12b**) shows that a vast majority of the cells are positive in both FL-2 and FL-1—that is, they are leukemic B cells.

New instrumentation is available to not only count cells but also image them in fluorescence and brightfield. *Imaging cytometry* is discussed in **Advanced Topic 10.3**.

## ADVANCED TOPIC 10.3:   IMAGING CYTOMETRY

The use of light microscopy to count cells is not new, but recent technological advances have allowed for high-throughput scanning and quantification of cells in multiwell plates, making this approach an easy and flexible method for obtaining real-time growth curves. Imaging cytometers may use phase-contrast microscopy or fluorescence techniques (either wide-field or laser scanning).

Although collection of data with these techniques is straightforward, analysis can be tricky. Software tools to identify cells versus non-cells must be developed; canned routines nearly always do a poor job unless they can be optimized for the specific cell type used and the patterns of growth seen in each type of experiment. Cells in multiwell plates grow differently at the edges of the wells and in corner wells versus central wells.

Two approaches that can be used for quantification are object analysis and field analysis. Object analysis measures the cell counts and is reported as the number of cells per well. Field analysis measures cell confluence and is reported as percentage of the well area covered by cells (% area covered). **Figure A10.3.1** shows the results of both on phase-contrast images of B16 melanoma cells taken using a SoftMax Pro imaging cytometer and analyzed with the included software. The software was trained with shapes and colors of live (countable) and dead (undesirable) cells using a trial-and-error approach until the best cell demarcation was reached. The selection and optimization process was reiterated at least five times or until the analysis led to a majority of live cells being chosen by the software at different cell confluences.

When optimizing protocols for imaging cytometry, results should be compared with those obtained using end-point assays (**Figure A10.3.2**). It is important to keep in mind that cells exposed to a toxic agent may die in different ways, leading to different appearances under light microscopy, and that sometimes, the toxic agent may appear as aggregates that may be mistaken for cells by the software.

*(Continued)*

**ADVANCED TOPIC 10.3 (CONTINUED):   IMAGING CYTOMETRY**

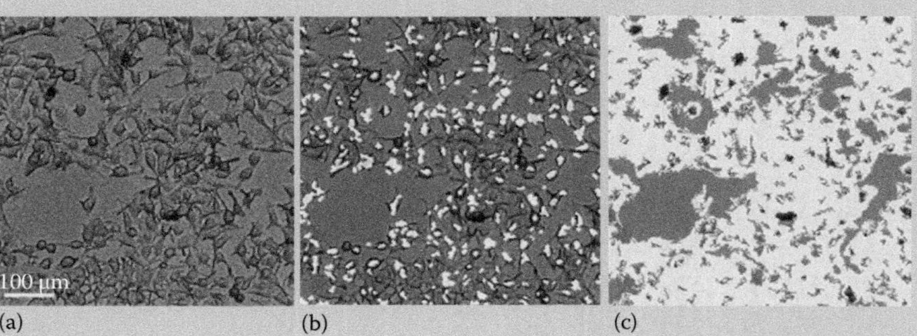


**Figure A10.3.1  Phase-contrast images of cultured B16 cells on a 96 well plate (a) without analysis, (b) with a customized discrete object analysis, and (c) with a customized field analysis from SoftMax Pro software.** The white areas indicate the cell regions that were delineated after analysis. Note the improved detection in panel (c) compared with panel (b). Dead cells (black) were omitted from the count.

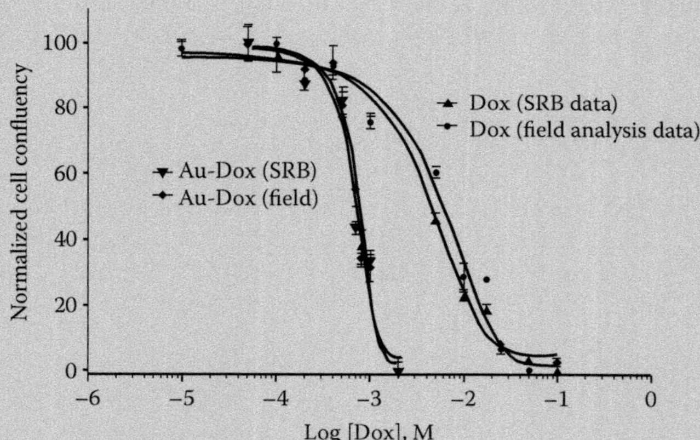


**Figure A10.3.2  Dose–response effect of two toxic agents: the anticancer drug doxorubicin (Dox) and gold nanoparticle conjugates to Dox (Au–Dox) on B16 cell viability determined by SRB assay versus phase-contrast microscopy with field analysis.** The curves are fits to the Hill equation. Each data point is the mean of six replicates ± the standard error of the mean (SEM) and has been normalized to the highest and lowest mean value.

## SUGGESTED READING

Henriksen, M., Miller, B., Newmark, J., Al-Kofahi, Y., and Holden, E. (2011). Laser scanning cytometry and its applications: a pioneering technology in the field of quantitative imaging cytometry. *Methods in Cell Biology* 102, 161–205.

Kim, M. J., Lee, S. C., Pal, S., Han, E., and Song, J. M. (2011). High-content screening of drug-induced cardiotoxicity using quantitative single cell imaging cytometry on microfluidic device. *Lab on a Chip* 11, 104–114.

Krull, D. L., and Peterson, R. A. (2011). Preclinical applications of quantitative imaging cytometry to support drug discovery. *Methods in Cell Biology* 102, 291–308.

*(Continued)*

**ADVANCED TOPIC 10.3 (CONTINUED):   IMAGING CYTOMETRY**

La Muraglia, G. M., 2nd, O'neil, M. J., Madariaga, M. L., Michel, S. G., Mordecai, K. S., Allan, J. S., Madsen, J. C., Hanekamp, I. M., and Preffer, F. I. (2015). A novel approach to measuring cell-mediated lympholysis using quantitative flow and imaging cytometry. *Journal of Immunological Methods* 427, 85–93.

Naoghare, P. K., Kim, M. J., and Song, J. M. (2008). Uniform threshold intensity distribution-based quantitative multivariate imaging cytometry. *Analytical Chemistry* 80, 5407–5417.

Tarnok, A., and Darzynkiewicz, Z. (2013). New insights into cell cycle and DNA damage response machineries through high-resolution AMICO quantitative imaging cytometry. *Cell Proliferation* 46, 497–500.

## 10.6 QUANTIFYING VIRUSES

### Titering viral vectors

With viruses, even more so than bacteria and mammalian cells, it is important to measure the number of *functional* particles per unit volume. Thus, methods like electron microscopy, which allow for direct counting of particles, are of minimal use. In most viral preparations, a large fraction of the particles are noninfectious, either because they are incomplete or because they have been inactivated by purification methods such as ultracentrifugation. It is thus the *infectivity* of the solution to a target cell line that is quantified. This unit is referred to as infectious units (i.u.) or *transducing units* (TUs) per unit volume. Another unit often used in virology is the *multiplicity of infection* (MOI or $m$), which refers to the average number of (functional) viral particles per target cell assuming that the distribution of particles within cells follows a Poisson distribution. That is, if $m$ particles/cell are added, the probability $P(n)$ of a cell containing $n$ particles is given by

$$P(n) = \frac{m^n e^{-m}}{n!}. \tag{10.9}$$

Of course, deviations from this distribution can occur if a single infectious event either blocks or encourages further infections of the same cell. Retroviruses especially are seen to deviate from Poisson behavior, with more copies per infected cell and fewer overall transfected cells than would be expected from a random distribution. However, for small numbers of viruses per cell (<1), the Poisson assumption usually works well.

Procedures for determining titer depend on the type of virus. Viruses such as bacteriophage (introduced in Chapter 2) destroy their target host cells and form *plaques*, which may be counted in a manner similar to colonies. However, many viral vectors are not cytotoxic or have had their cytotoxicity removed so that the particles act only to transduce genes of interest. In this case, expression of the gene of interest is what is used for titering. Establishing a consistent protocol will allow for good comparison among experiments performed with different stocks of viruses and is key to successful experiments. It is important to use the same protocol any time data will be directly compared, since titers obtained using different methods usually vary a significant amount. It is also important to retiter viruses after long periods of storage, as they may lose infectivity over time.

The following sections give some specific examples of protocols for titering of some common types of viruses. Many of the principles given in specific examples are generalizable to other types of viral vectors having similar properties and expressing similar types of genes. However, it is always recommended to consult the literature for your specific type of viral vector in case there are particular concerns specific to it. Many times, there are specific kits that may be used for rapid titering based upon a viral protein (e.g., anti-Hexon antibody for adenovirus).

## Titering phage by plaque assay

The target cells for phage are bacteria, so the plaques to be counted represent holes in a bacterial lawn. The procedure is simple, but it is important to mix things well, dilute carefully, change pipette tips when necessary, and ensure even distribution of phages in the plates. The following protocol works well:

- Begin with an overnight culture of a susceptible strain of *Escherichia coli* (4–5 mL is sufficient).

- Make serial dilutions of the phage culture by preparing seven tubes each containing 9 mL of growth medium.

- Add 1 mL of phage culture to the first tube and mix it carefully by pipetting up and down or inverting. Proceed with a clean pipette to the following tubes, always putting 1 mL into 9 mL. You should have dilutions of $10^{-1}$ to $10^{-6}$, plus a control with no virus.

- Label six agar plates as follows: $10^{-3}$, $10^{-4}$, $10^{-5}$, $10^{-6}$, $10^{-7}$, control.

- Prepare seven tubes of 2 mL of soft top agar and place them in the 42°C water bath.

- As soon as you remove a tube from the water bath, work quickly! Add 100 µL of the appropriate phage dilution to the tube, and 250 µL of *E. coli*. Vortex to mix and pour into the agar plate labeled for that dilution. (Note: it is important not to add more than about 100 µL of phage culture, or the top agar will not harden. This is why the dilutions are not done at this stage.)

- Swirl the plate to distribute the top agar.

- Repeat until all dilutions have been plated.

- Invert the plates and incubate overnight.

- Count all plaques on plates with a countable number. The plaques are discrete clear areas several millimeters in size (**Figure 10.13**).

- The titer is TU/mL = (number of plaques/plate)(1/mL plated)(dilution factor).

## Titering adenovirus by plaque assay

The principle here is the same as with phage, only the target cells are no longer bacteria but mammalian cells. The handling conditions are thus somewhat different, and the assay takes much longer (at least 2 weeks from start to finish). The agar *overlay* is necessary to prevent viral progeny from leaking from one plaque to the next. Many labs no longer titer adenoviruses in this fashion, as

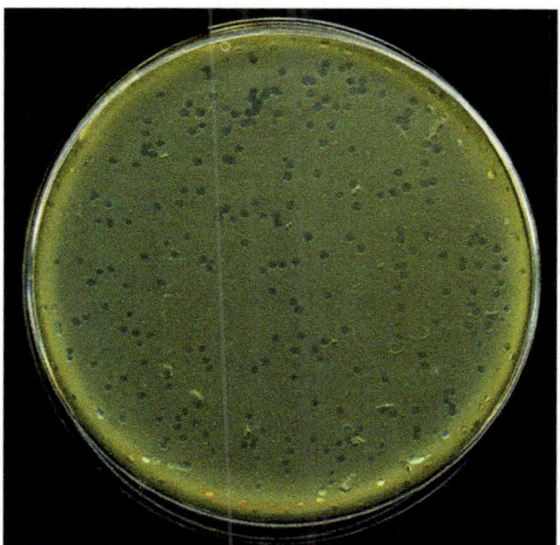

**Figure 10.13  Lambda plaques on a lawn of _E. coli_.**

there are a multitude of new kits available to count both physical particles and infectious particles in easier ways. Real-time polymerase chain reaction (rt-PCR) is also easier and faster than the plaque assay. However, this technique is at least of historical interest and may certainly be used when alternative methods are unavailable. The protocol is as follows:

- In order to make the overlay, some specialized reagents are needed, available from tissue culture suppliers: double-strength (2×) growth medium; 100× stock of neutral red dye; and a 2.8% Bacto agar solution prepared in water, autoclaved, and cooled. Make 36 mL of agar solution for each 100 mL of overlay (25 mL of overlay is needed per six-well plate).

- The day before the experiment, plate 293 cells into six-well plates at 50–70% confluency ($2$–$5 \times 10^5$ cells per well). (If they are too confluent, they will detach too easily from the dish.)

- The day of the experiment, heat the Bacto agar solution in the microwave until fully melted, and then store it in a 56°C water bath to prevent solidification.

- Prepare serial dilutions of adenovirus. The total volume of virus added to the cells should be negligible relative to 2 mL.

- Remove cells from incubator. Do not wash the cells; they are too fragile! Simply remove all but 2 mL of medium from each well, and then add the viral stock. Do each dilution in duplicate. Return the cells to the incubator.

- While the cells are incubating with virus, prepare the overlay. For 100 mL,

  - 50.0 mL 2× growth medium.

  - 2.0 mL 1.0 M HEPES.

  - 1.25 mL 1.0 M $MgCl_2$.

  - 10.0 mL FBS.

  - 1.0 mL 100× penicillin/streptomycin solution.

- 36.0 mL melted 2.8% Bacto agar.

  - Mix well and swirl in a 37°C water bath. It's best to do it in 50 mL aliquots as there is less solidification that way.

- After the desired infection period (usually 6–16 h), remove the cells from the incubator and aspirate the virus-containing medium (very carefully so as not to disturb the cells!).

- Very gently add 4 mL of overlay to each well. Allow to solidify at room temperature (10–20 min), and then return to the incubator. To keep the agar from drying, sterile PBS may be added to the spaces between the wells, and Parafilm wrapped around the edges of the plate. Do not let it get too wet, or the plaques will spread.

- After 3–5 days, add a second layer of the same overlay containing neutral red. Allow it to solidify, and return cells to the incubator. The neutral red is taken up by healthy cells but not dead cells, so the plaques appear clear.

- Start counting plaques the day after the neutral red overlay is added. Plates can be kept and examined for up to a month. A light table can help in visualization of plaques (**Figure 10.14a**).

- TU/mL = (number plaques/plate)(1/mL plated)(dilution factor).

Titering adenovirus can be tricky. Plates can dry out or get too wet and flow, or become too acidic (the overlay turns yellow); the kinetics of plaque formation can be slower or faster than anticipated; or plaques can be very small. It is important to develop a consistent technique for all lab members using adenovirus. If several viral batches are going to be directly compared in experiments, it is best to titer them at the same time to ensure consistent conditions. The kinetics of plaque formation as well as the total final number of plaques help indicate the viral cytotoxicity (**Figure 10.14b**).

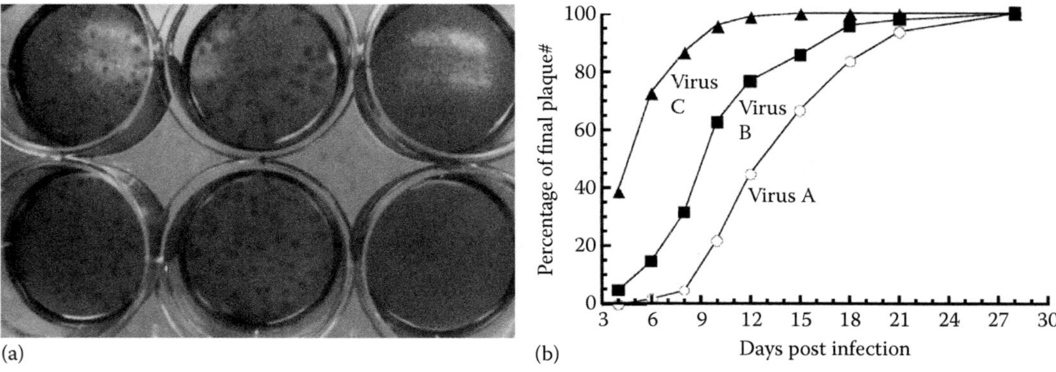


**Figure 10.14 Viral plaques.** (a) Appearance of plaques generated by viral vectors under different conditions, stained with MTT. (Image courtesy of Dr. Marylou Gibson, Virapur, San Diego, CA.) (b) Kinetics of plaque formation for three different adenoviruses, A, B, and C. The more cytolytic the virus, the earlier plaques will form. (With kind permission from Springer Science+Business Media: *Adenovirus Methods and Protocols,* Preparation and Titration of CsCl-Banded Adenovirus Stocks, 2007, A.E. Tollefson et al., Springer eBook, copyright Springer.)

## Titering adenovirus by optical density

The "lazy man's" way to titer adenovirus is simply by measuring the viral DNA content by reading the $OD_{260}$. One OD unit corresponds to ~$10^{12}$ viral particles/mL. However, the great majority of these particles are noninfectious; typical infectious-to-total particle ratios range from 1:20 to 1:100 or more. This method should only be used to confirm or complement methods that measure TU directly.

## Titering lentiviral vectors by flow cytometry

Lentiviruses (discussed in **Chapter 3**) and most other retroviral vectors are noncytotoxic and so do not form plaques. However, when they are used for gene transduction experiments, they usually encode a gene that can be readily quantified by flow cytometry: a fluorescent protein or a cell-surface marker that may be targeted with a fluorescent antibody. When this is the case, infection with serial dilutions followed by flow cytometry is the best method for titering these viruses. The following protocol provides some hints for doing this quantifiably and reproducibly.

- The day before the experiment, seed $10^5$ target cells per well into as many six-well plates as you will need. HeLa or 293 cells are good choices. The number of wells should be as follows: one well for cell count; one well for no-virus control; one well for other negative control, if needed (e.g., AZT, see Optional page); and one well for each dilution ×2, since everything should be done in duplicate.

- The next day, trypsinize one of the wells and do a viable cell count. This will be important for accurately calculating the titer.

- *Optional*: Add 2 mL/mL of 8 mg/mL Polybrene to all wells including control. (Polybrene is reported to aid lentiviral infection. We sometimes use it, usually not.)

- *Optional*: Add a reverse-transcriptase inhibitor such as AZT to one well. This will identify "pseudotransduction," a passive transfer of the transgene that can occur without true viral infection. We do not usually do this, but it can be informative!

- Add the serially diluted virus to duplicate wells. Make sure to change pipette tips in between dilutions! The goal is to have some wells with 1–10% transduced cells. (If there is an AZT control, add the maximum amount of virus to this well.)

- Gently rock the plates to mix, and place in the incubator overnight. The next morning, aspirate the medium and replace it with fresh medium.

- Allow the infection to proceed for 48 h.

- Trypsinize all the wells and place the cells into individual tubes. Pellet the cells and rinse with the appropriate FCB. Resuspend in a total volume of 0.5 mL.

- If the cells express a fluorescent protein, you are ready to go. Keep them on ice and use as soon as possible for flow cytometry. Alternatively, fix them in 1% paraformaldehyde and store for up to several days.

- If the cells express a cell-surface marker, add the appropriate fluorescently labeled antibody and incubate in the dark on ice for 30–60 min. Rinse with PBS containing 1% FBS, and then pellet and fix in 1% paraformaldehyde in PBS.

- Proceed with flow cytometry and collect 5,000–10,000 events. Eliminate the dead cells (very low SSC and FSC). Make sure that the threshold is set so that the control cells are <0.25% "fluorescent." Be aware that with lentiviral transduction; many cells may be only weakly fluorescent (**Figure 10.15**).

- Calculate the titer as

$$TU = \frac{\%FluorescentCells}{100}CV,$$

where $C$ is the cell count in cells/mL obtained in step 2, and $V$ is the volume of virus added in mL.

If flow cytometry is not available, microscopic counts may be made of the cells in the wells or trypsinized and placed into a hemocytometer. These methods work quite well but require some patience, as well as a careful look at the negative control to make sure that autofluorescence is not confused with low levels of expression.

## Titering retroviruses expressing a selectable marker

Retroviral vectors often contain an antibiotic resistance gene for ease of selection of transfected cells. When this is the case, titering may be performed

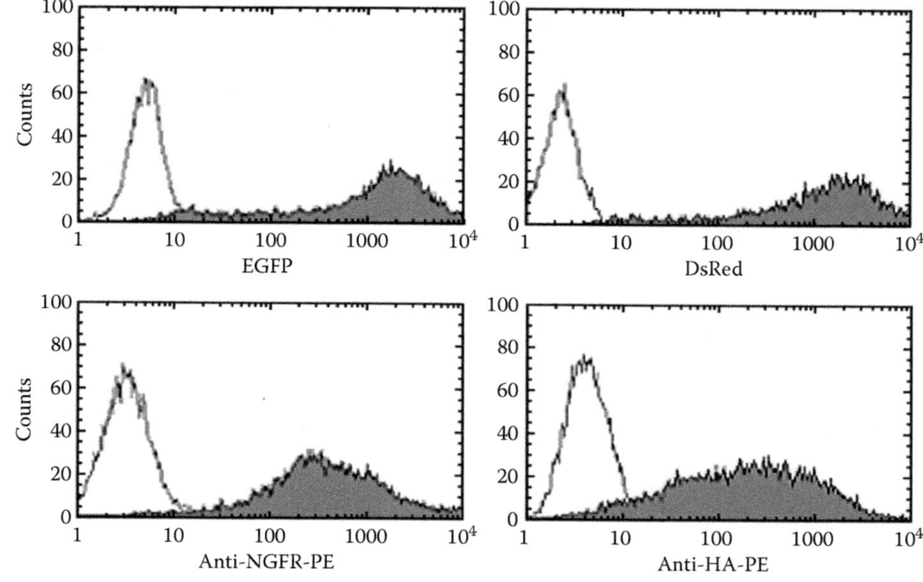


**Figure 10.15 Examples of flow cytometry data for lentivirally transduced cells.** The top two are fluorescent proteins, enhanced green fluorescent protein (EGFP) and DsRed. The bottom two are cell-surface markers, hemagglutinin (HA) and nerve growth factor receptor (NGFR), both tagged with the red dye phycoerythrin (PE). (Image courtesy of Lung-Ji Chang, University of Florida.)

using this gene. This takes significantly longer than fluorescence methods but is quite accurate. The procedure begins the same way as for fluorescent markers, but lower concentrations of virus should be used in order to obtain a small, discrete number of cell colonies (usually 10- to 100-fold less than for the fluorescent reporters). Allow the infection to proceed for only 24 h, and then perform the following steps:

- Trypsinize all the wells; wash with PBS.

- Resuspend in fresh medium.

- Plate each condition at 1:10 and 1:50 dilutions in fresh six-well plates.

- Add the selective antibiotic (e.g., neomycin).

- Incubate for 1–2 weeks. Change medium as needed; fresh antibiotic should be added at this time as well.

- When no colonies are seen on control plates, stain and count as in **Section 10.3**, "Quantifying Mammalian Cells."

- The titer is given by TU = (no. of colonies)(dilution factor)$/V$, where $V$ is the volume of virus added.

### Titering lentivirus using p24

Sometimes a viral vector is made without an easily screenable gene of interest. There are methods of determining infection of target cells in this case using quantitative rt-PCR, but they are tricky and often poorly quantifiable. A good alternative in the case of lentiviral vectors is to measure the number of physical particles by quantifying p24, the HIV capsid protein. There are approximately 2000 molecules (or $8 \times 10^{-17}$ g) of p24 per lentiviral particle. A reasonably well packaged vesicular stomatitis virus G glycoprotein (VSV-G) pseudotyped lentiviral vector will have an infectivity index in the range of 1 TU per 1000 particles (PP) to 1 TU per 100 physical particles, or 10–100 TU/pg of p24. The best way to quantify p24 is by using a commercial enzyme-linked immunosorbent assay (ELISA) test kit that is designed for lentiviral vectors. This approach is similar to that used for screening donated blood for HIV and to assist in diagnosing HIV infection, but with one important difference. The best kits for testing lentiviral vectors eliminate free p24 generated from the transfected packaging cells. This p24 is not associated with viral particles (see **Chapter 3**), so it may be separated before assaying (**Figure 10.16**). This still does not eliminate the problem of p24 associated with defective viral particles, but it is an improvement over general p24 tests. These kits may be used as an adjunct to titering by flow cytometry if you wish to estimate your packaging efficiency.

## 10.7  SUMMARY AND FINAL REMARKS

Quantifying cells usually means just counting them, but this is done very differently based upon their size, growth requirements (adherent vs. suspension), optical properties, and whether their functional status needs to be known. A simple hemocytometer is a tool that every lab should have, and everyone can become

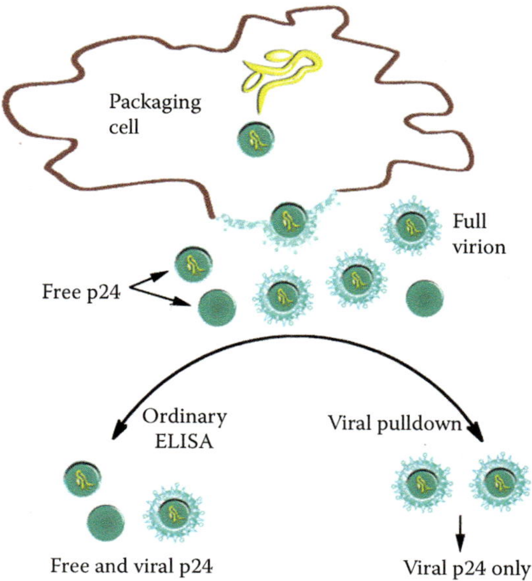



**Figure 10.16 Free p24 versus total p24.** Free p24 escapes from transfected packaging cells along with whole virions. If viral pull-down is used, only the p24 associated with full particles is measured, which gives a better idea of infectious titer.

rapidly familiar with its use. With many experiments involving mammalian cells, this is all that is needed, along with colorimetric assays for determining end-point toxicity. However, sometimes real-time measurements of the effects of a drug or toxin on cells need to be determined. For bacteria, yeast, and many suspension cells, this is as easy as obtaining growth curves with time and fitting these to an appropriate model. For adherent mammalian cells, however, both counting and modeling can be tricky, and it may be worthwhile to investigate some of the emerging optical techniques for this purpose. Although these techniques measure very different parameters from the end-point techniques, the correlation among all of the measures of cell death is quite good—meaning that if cells are going to die, they will change shape, stop moving, and detach from the dish.

Flow cytometry is also a tool that is widely available and usually straightforward to use, and allows for the rapid counting of tens of thousands of cells. Standard protocols for labeling using dyes or antibodies allow for high-throughput discrimination among cell types, identification of mechanisms of cell death, and even monitoring of ion flow. This technique may also be used to titer viruses that express a fluorescent reporter gene or to sort cells depending upon their label.

Some of the techniques in this chapter can be a little difficult to calibrate, but all of them are within the scope of anyone at a research university. For all of the techniques, the biggest pitfalls involve poor cell handling—depositing different numbers of cells per well, losing cells when washing, or even counting the wrong thing. (If your *Staphylococcus aureus* is contaminated with *E. coli*, you are not counting what you think you are counting!) Along with careful mastery of the techniques, we advise you to know your cells and to do frequent visual inspections to identify any issues that may lead to erroneous results.

## End-of-Chapter Questions

1. Derive **Equation 10.3** from **Equation 10.1**. (Hint: let $n = p = 1$.)

2. For a population following **Equation 10.1** with $n = p = 1$, find when the growth rate is maximum.

3. If bacteria were not stopped from reproducing, how long would it be before a single cell (doubling time 0.5 h) generated a mass of bacteria equal to the mass of the earth ($6 \times 10^{24}$ kg)?

4. Experiments with bacterial cells usually are performed with a "mid-log" culture. What does this mean, and why is it done? How long would *E. coli* (doubling time 0.5 h) stay in the logarithmic phase? Assume that carrying capacity is $10^{10}$ cells/mL. If you want to keep the cells in log phase longer, what can you do?

5. Derive the expressions for $t_{50}$ and the doubling time.

6. As mentioned in the text, the logistic model does not capture the lag phase well. Why not? What are some models that fit this phase of bacterial growth more accurately?

7. Derive **Equation 10.5**. What happens when the drug concentration in the solution is not constant?

8. Derive **Equation 10.8**. Explain why **Equations 10.7** and **10.8** make sense relative to the "standard" definition of MIC.

9. What are some pitfalls in fitting the Hill equation? How will missing data points affect the fits?

10. For antibiotic treatment, a target value is set for the ratio of the area under the pharmacokinetic curve to MIC (AUC/MIC). Why is this done, and what does it mean?

11. Propose an equation for inhibition due to two antimicrobial agents. How would you know whether the two agents add additively or synergistically?

12. Under what circumstances should colony-forming assays be done to quantify mammalian cells? Why?

13. Derive a modified logistic equation for cells that are uniformly contact-inhibited.

14. Derive a modified logistic equation for cells that are contact-inhibited and grow in colonies.

15. A dish of cells is exposed to a transforming agent, such as a carcinogen. The fraction of the cells that is transformed is no longer contact-inhibited; the untransformed cells show the same inhibition as in problem 13. Model what happens in the dish.

16. Define and describe the Allee effect. How is it important in tissue culture?

17. (ADVANCED) In ECIS, how do you expect impedance to scale with $C_m$, $R$, and $\alpha$?

18. (ADVANCED) Explain how the Sauerbrey equation results from the modeling of a quartz crystal as a harmonic oscillator in vacuum, with the deposited mass $m$ much less than the mass of the oscillator.

19. Discuss some of the advantages of flow cytometry over other cell analysis techniques.

20. How would you measure the total DNA content of a cell population using flow cytometry?

21. How would you analyze the different proportions of cells in each phase of the cell cycle using flow cytometry?

22. How could you use flow cytometry to monitor the production of a (nonfluorescent) foreign protein in bacterial cells?

23. Flow cytometry leads to an underestimate of the actual size of cells. Can you imagine why this might be? How would you account for this in experiments where absolute cell size is needed?

24. Channel bleed-through is an issue with flow cytometry as it is for fluorescence microscopy, and is controlled for by adjusting the compensation. Imagine that you are using two common fluorophores, fluorescein (FITC) and phycoerythrin (PE). (a) Draw the emission spectra and discuss the degree of overlap. (b) How would you perform compensation using these two dyes?

25. Autofluorescence is another major issue in flow cytometry, especially because many types of cells become more autofluorescent as they die. Discuss ways to compensate for autofluorescence, noting that it is typically spectrally wide, nearly white. Sketch a plot of FL-1 versus FL-2 for a population of autofluorescent cells labeled with FITC and indicate the different populations.

**26.** (ADVANCED) Compare and contrast imaging cytometry and imaging flow cytometry, and discuss when each should be used.

**27.** Infection with the human immunodeficiency virus (HIV) is diagnosed using an ELISA test and a Western blot. Describe how these work and the sensitivity and specificity of each.

**28.** Discuss what should be done to troubleshoot quantitative PCR (qPCR) in the following circumstances: (a) little or no product; (b) nonspecific product; (c) poor yield; (d) primer dimer; (e) multiple peaks on melting curve.

**29.** Explain how qPCR may be used to estimate the relative abundances of major taxonomic groups of bacteria and fungi in environmental communities.

## Background Reading

### Books

(Also refer to the books cited in **Chapter 3**.)

Federico, M. (ed.). *Lentivirus Gene Engineering Protocols*. Humana Press, Totowa, NJ, 2003.

Hawley, T. S., and Hawley, R. G. (Eds.). 2011. *Flow Cytometry Protocols*, 3rd Edition. Methods in Molecular Biology, vol. 699. Humana Press, Totowa, NJ.

Shapiro, H. M. *Practical Flow Cytometry*, John Wiley & Sons, Hoboken, NJ, 2003.

Watson, J. V. *Flow Cytometry Data Analysis: Basic Concepts and Statistics*. Cambridge University Press, Cambridge, MA, 1992.

Wold, W. S. M., and A. E. Tollefson (eds.). *Adenovirus Methods and Protocols*. Humana Press, Totowa, NJ, 20010.

### Journal articles

(For more on flow cytometry applications, see **Table 10.1**.)

Cheng, G., Youssef, B. B., Markenscoff, P., and Zygourakis, K. (2006). Cell population dynamics modulate the rates of tissue growth processes. *Biophysical Journal* 90, 713–724.

Coleman, A. W. (1980). Enhanced detection of bacteria in natural environments by fluorochrome staining of DNA. *Limnology and Oceanography*. 25, 948–951.

Dairkee, S. H., and Glaser, D. A. (1984). A mutagen-testing assay based on heterogeneity in diameter and integrated optical density of mammalian cell colonies. *Proceedings of the National Academy of Sciences of the United States of America* 81, 2112–2116.

Dong, H. P., Kleinberg, L., Davidson, B., and Risberg, B. (2008). Methods for simultaneous measurement of apoptosis and cell surface phenotype of epithelial cells in effusions by flow cytometry. *Nature Protocols* 3, 955–964.

Fehse, B., Kustikova, O. S., Bubenheim, M., and Baum, C. (2004). Pois(s)on—It's a question of dose. *Gene Therapy* 11, 879–881.

Forestell, S. P., Milne, B. J., Kalogerakis, N., and Behie, L. A. (1992). A cellular automaton model for the growth of anchorage-dependent mammalian-cells used in vaccine production. *Chemical Engineering Science* 47, 2381–2386.

Frame, K. K., and Hu, W. S. (1988). A model for density-dependent growth of anchorage-dependent mammalian-cells. *Biotechnology and Bioengineering* 32, 1061–1066.

Fujikawa, H., Kai, A., and Morozumi, S. (2003). A new logistic model for bacterial growth. *Shokuhin Eiseigaku Zasshi* 44, 155–160.

Fujikawa, H., Kai, A., and Morozumi, S. (2004). Improvement of new logistic model for bacterial growth. *Shokuhin Eiseigaku Zasshi* 45, 250–254.

Geraerts, M., Willems, S., Baekelandt, V., Debyser, Z., and Gijsbers, R. (2006). Comparison of lentiviral vector titration methods. *BMC Biotechnol* 6, 34.

Holmes, K., Lantz, L. M., Fowlkes, B. J., Schmid, I., and Giorgi, J. V. (2001). Preparation of cells and reagents for flow cytometry. *Current Protocols in Immunology*. Chapter 5, Unit 5 3.

Houghton, P., Fang, R., Techatanawat, I., Steventon, G., Hylands, P. J., and Lee, C. C. (2007). The sulphorhodamine (SRB) assay and other approaches to testing plant extracts and derived compounds for activities related to reputed anticancer activity. *Methods* 42, 377–387.

Jaroszeski, M. J., and Heller, R. (eds.). *Flow Cytometry Protocols*. Humana Press, Totowa, NJ, 1998.

Labarre, D. D., and Lowy, R. J. (2001). Improvements in methods for calculating virus titer estimates from TCID50 and plaque assays. *Journal of Virological Methods* 96, 107–126.

Lim, J. H. F., and Davies, G. A. (1990). A stochastic-model to simulate the growth of anchorage dependent cells on flat surfaces. *Biotechnology and Bioengineering* 36, 547–562.

Motyl, M., Dorso, K., Barrett, J., and Giacobbe, R. (2006). Basic microbiological techniques used in antibacterial drug discovery. *Current Protocols in Pharmacology* Suppl 31.

Papazisis, K. T., Geromichalos, G. D., Dimitriadis, K. A., and Kortsaris, A. H. (1997). Optimization of the sulforhodamine B colorimetric assay. *Journal of Immunological Methods* 208, 151–158.

Pruitt, K. M., and Kamau, D. N. (1993). Mathematical models of bacterial growth, inhibition and death under combined stress conditions. *Journal of Industrial Microbiology* 12, 221–231.

Riccardi, C., and Nicoletti, I. (2006). Analysis of apoptosis by propidium iodide staining and flow cytometry. *Nature Protocols* 1, 1458–1461.

Ruaan, R. C., Tsai, G. J., and Tsao, G. T. (1993). Monitoring and modeling density-dependent growth of anchorage-dependent cells. *Biotechnology and Bioengineering* 41, 380–389.

Troiano, L., Ferraresi, R., Lugli, E., Nemes, E., Roat, E., Nasi, M., Pinti, M., and Cossarizza, A. (2007). Multiparametric analysis of cells with different mitochondrial membrane potential during apoptosis by polychromatic flow cytometry. *Nature Protocols* 2, 2719–2727.

Vichai, V., and Kirtikara, K. (2006). Sulforhodamine B colorimetric assay for cytotoxicity screening. *Nature Protocols* 1, 1112–1116.

## Websites and software

*Some major suppliers*

ZeptoMetrix. ELISA tests, including p24 ELISA; viral lysates; antibodies.

Applied BioPhysics (ABP). Electric cell–substrate impedance sensing (ECIS) systems.

Biolin Scientific: Q-sense. Quartz crystal microbalance (QCM) systems.

Microvacuum Ltd. Optical waveguide lightmode spectroscopy (OWLS) systems.

Cell Biolabs. Assays for viruses, as well as cell-based assays for migration, adhesion, etc.

Virapur. Viruses and viral assay services.

Becton-Dickinson. Key supplier of flow cytometers and fluorescence activated cell sorting (FACS) systems.

Millipore. Benchtop flow cytometers, affordable for many individual labs.

*Free software*

Cyflogic. Flow cytometry data analysis tool for Windows.

CytoWin. Data analysis tool for flow cytometry data in many formats, including ASCII.

WinMDI. One of the best flow cytometry tools. Windows only.

FCS Assistant/FCSPress. Flow cytometry tools for Macintosh.

*Commercial software*

FCS Express (de Novo software). Complete flow cytometry analysis package.

FlowJo. Complete flow cytometry package for Windows or Macintosh.

MATLAB (TheMathWorks). Highly recommended data analysis tool for nonlinear regression and many forms of statistical analysis.

Maple (MapleSoft). Another highly recommended data analysis package for scientists and engineers.

# CHAPTER 11

# Semiconductor Nanoparticles (Quantum Dots)

## 11.1 INTRODUCTION

Particles of many different materials show unique properties when they are smaller than 10–100 nm in diameter. Recent developments in synthesis procedures have allowed for controlled synthesis of specific sizes and shapes of metals, semiconductors, and insulators within this size range and even smaller. The physics and chemistry of these materials have been investigated for several decades and have provided significant new insights into quantum mechanics. More recently, techniques of water solubilization have permitted these materials to be used in biological environments, yielding a whole new set of surprises. The physics of interactions between nanomaterials and biological structures, and the reaction of living systems to nanosized materials, have only begun to be understood. This chapter introduces some of the most common semiconductor nanomaterials and their synthesis methods, with a focus on methods accessible to most laboratories. We then discuss different ways to make nanoparticles biologically compatible and to conjugate them to antibodies or other biological molecules. The most well-developed techniques for use of nanoparticles, such as single-particle tracking (SPT), fluorescence resonance energy transfer (FRET), and electron microscopic imaging, are then discussed. Finally, we introduce some of the ways nanoparticles can be used to develop sensors for biological targets.

## 11.2 QUANTUM DOT PROPERTIES AND SYNTHESIS

### Physics of quantum dots

A *quantum dot* (QD) is a nanoscale particle of a semiconductor material that (more or less) retains the lattice structure of the bulk semiconductor. For commonly used materials, this lattice structure may be hexagonal (wurtzite) or cubic (zinc blende) (**Figure 11.1**). In order to understand the physics of QDs, it is important to be familiar with the basics of the band theory of semiconductors, including the concepts of Bohr radius, exciton, bandgap, and effective mass. Some values for the relevant materials are given in **Table 11.1**. For our purposes, we will discuss only *colloidal* QDs, or nanocrystals synthesized and dispersed in liquid; there are also many types of lithographically patterned QDs, of less use to biology.

**Figure 11.1 Nanoscale semiconductor lattices.**
(a) Illustration of a wurtzite lattice, with a high-resolution TEM image of a single CdSe nanocrystal of diameter ~5 nm (TEM image courtesy of Prof. Paul O'Brien). (b) Illustration of a zinc blende lattice, with a high-resolution TEM image of a single lead sulfide (PbS) nanocrystal. (TEM image courtesy of John Watt, Victoria University of Wellington.)

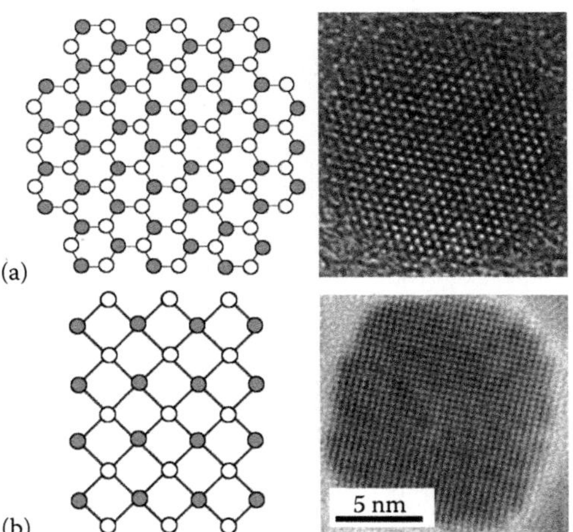

**Table 11.1**

Parameters for Some II–VI Semiconductor Materials

| Material | $E_{gap}$ (eV) | $m_e$ | $m_h$ | Lattice at RT | Nearest-Neighbor Distance (nm) | Density (g/cm³) |
|---|---|---|---|---|---|---|
| ZnS | 3.68 (zb) 3.91 (w) | 0.4 | ? | zb or w | 0.541 (zb) 0.234 (w) | 4.11 (zb) 3.98 (w) |
| ZnSe | 2.82 | 0.21 | 0.6 | zb | 0.246 (zb) | 5.26 (zb) |
| ZnTe | 2.39 | 0.2 | 0.2 | zb | 0.264 (zb) | 5.65 (zb) |
| CdS | 2.50 | 0.21 | 0.8 | w | 0.252 (w) | 4.87 (zb) 4.82 (w) |
| CdSe | 1.714 | 0.13 | 0.45 | w | 0.263 (w) | 5.66 (zb) 5.81 (w) |
| CdTe | 1.474 | 0.11 | ? | zb | 0.281 (zb) | 5.86 (zb) |

*Note:* Effective masses are given relative to the electron mass $m_0$; "lattice at RT" refers to the stable lattice form at room temperature (w, wurtzite; zb, zinc blende).

The most striking property of QDs is a bright and size-dependent *photoluminescence*. For a semiconductor nanoparticle to be a QD, it must have a bandgap energy comparable to energies of photons of visible light. Numerous narrow-bandgap semiconductors, which may belong to group II-VI, group IV-VI, or group III-V, fit this criterion. (The elements in group VI are called *chalcogens* or "ore formers": they include O, S, Se, and Te.) Excitation by a photon slightly more energetic than the bandgap leads to the creation of an electron–hole pair, or exciton. When the exciton recombines, it emits a photon slightly redder than that which was absorbed (**Figure 11.2**).

As these semiconductors are reduced to the nanoscale, their excitons are confined to nearly infinite potential wells at the edges of the particle; this adds a particle-in-a-box energy to the bandgap energy, broadening the bandgap. The density of states of the excited electron is also affected by the confinement. In a 3-D bulk semiconductor, the electron density of states is continuous and proportional to the square root of the energy $E$:

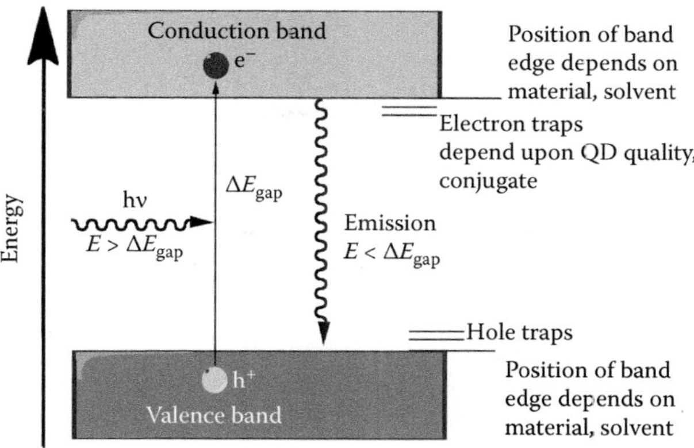



**Figure 11.2 Schematic of QD photoluminescence.** A photon more energetic than the bandgap excites an exciton, which recombines to give a slightly redder photon. The amount of redshift is determined by the QD electronic fine structure, and can also be influenced by ligands, solvents, and surface defects.

$$\rho_{3D}(E) = \left(\frac{m_e}{\hbar^2}\right)^{3/2} \frac{\sqrt{2E}}{\pi^2} \tag{11.1}$$

where $m_e$ is the effective electron mass. In a 0-D system (QD; confined in all three dimensions) this becomes discrete, quantized levels with quantum numbers $n$, $m$, $l$:

$$\rho_{0D}(E) = 2\sum_{n,m,l} \delta(E - \varepsilon_{n,m,l}) \tag{11.2}$$

These states are represented schematically in **Figure 11.3**. Because their discrete energy states recall those of an atom, QDs are sometimes called "artificial atoms."

For the lowest 1s state (the position of the first exciton absorbance peak), the total exciton energy is given by

$$E = E_{gap}(\infty) + \frac{h^2}{8R^2}\left(\frac{1}{m_e} + \frac{1}{m_h}\right) - \frac{1.8e^2}{4\pi\varepsilon R} \tag{11.3}$$

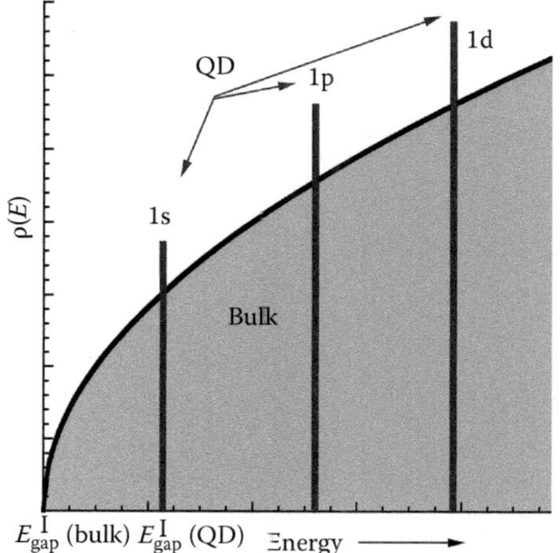



**Figure 11.3 Electron density of states (DOS) ρ(E) in bulk semiconductors versus QDs.** In the bulk, the DOS is a continuum beginning at the bandgap energy and proportional to $E^{1/2}$. In a QD, the lowest (1s) state is found at the bandgap energy (which is shifted from the bulk bandgap energy). Higher energy states exist at discrete levels.

where $\varepsilon$ is the crystal dielectric constant, $m_e$ and $m_h$ are the effective masses of the electron and hole, and $R$ is the radius of the particle. When the particle radius is smaller than the Bohr radius (a few nanometers for most materials), the quadratic term dominates, leading to significant shifts in the first exciton peak with changes in particle size. This is called the *strong confinement regime* and corresponds to different colors for different particle sizes. The peak of the first exciton can be observed as a feature in the nanoparticle absorbance spectrum (**Figure 11.4a**).

For CdSe, the most commonly used QD material, particle sizes of 2–6 nm show photoluminescence emission that spans the visible range. A cadmium-free alternative, InP, spans a similar range. CdTe, with a slightly smaller bandgap, can range from midvisible to the near infrared (IR). PbS and PbSe emit in the near- to mid-IR range, and CdS covers the near ultraviolet (UV) (**Figure 11.4b**). All of the colors may be excited with a single source, which for CdSe, CdTe, and InP can be near UV or violet (**Figure 11.5**).

**Figure 11.4 QD colors.**
(a) Absorbance (left) and emission (right) spectra of five distinct colors of CdSe/ZnS QDs: blue (B), green (G), yellow (Y), orange (O), and red (R). Note the exciton peaks in the absorbance spectra and the increasing absorbance at higher energies than the bandgap. (b) CdS, CdSe, and PbS QDs span the UV to near-IR emission ranges.

(b)

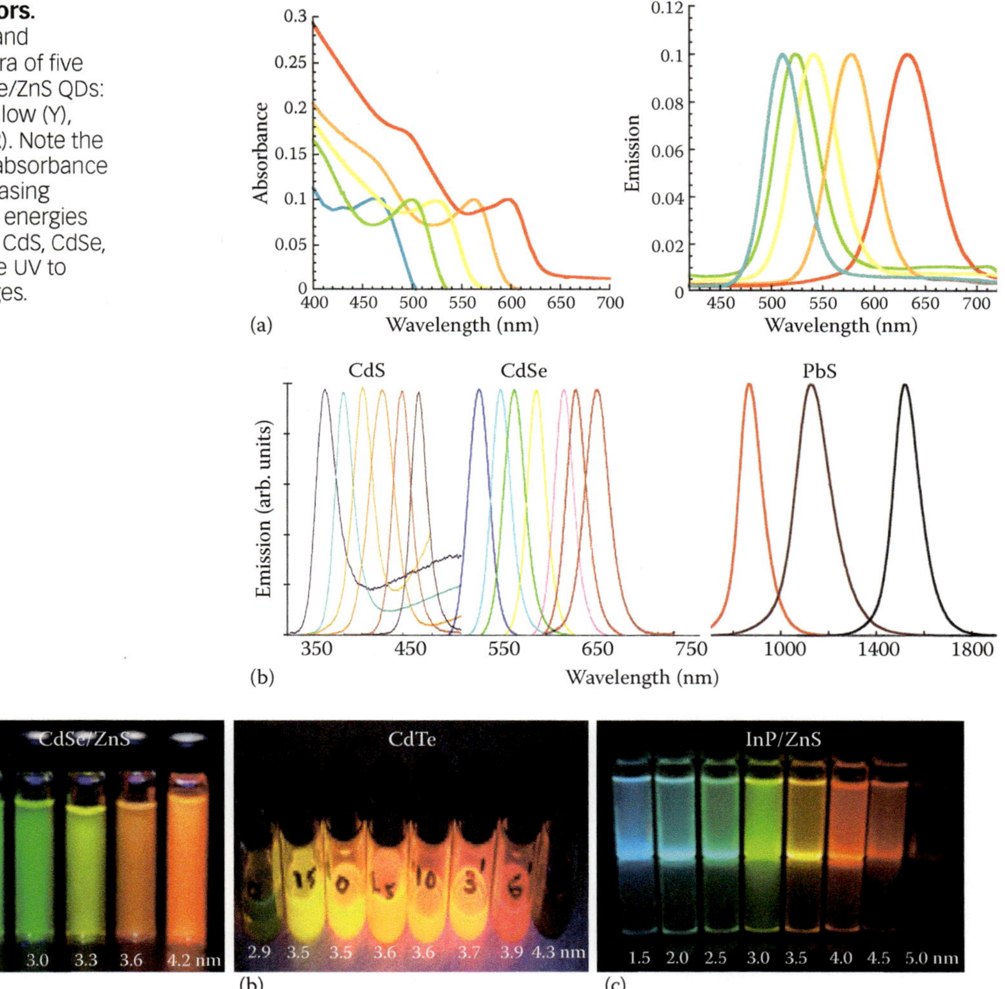


**Figure 11.5 Appearance of solutions of quantum dots under UV wand (365 nm) excitation.** (a) CdSe/ZnS. (b) CdTe. (c) InP/ZnS.

A *bare-core* QD made simply of its core semiconductor material is very sensitive to interactions with surrounding solvents and ions, particularly water and oxygen. The electron and hole from the exciton can transfer to the solvent, oxidizing or reducing the dot. These interactions, particularly oxidation, usually reduce fluorescence and eventually break down the nanoparticle by a process called *anodic decomposition*. The exciton can be protected or *passivated* by the overlay of a material with a higher bandgap energy. Provided that this material does not alloy well with the semiconductor core, its presence will affect the emission of the QD very little. The most common material used as a cap for CdSe and InP is ZnS. This cap can be as thin as a single atomic monolayer, and particles capped in this fashion are called *core–shell QDs*. Increasing the shell thickness does not improve the quality of the dot; it may even cause cracking due to lattice mismatch.

Interestingly, CdTe nanoparticles can be highly fluorescent in water, whereas CdSe particles cannot. Most biological applications of CdTe QDs use bare-core QDs for this reason. However, the particles are still sensitive to oxidation and may break down. This has important implications for toxicity, since release of $Cd^{2+}$ ions is highly damaging to cells, and means that toxicity studies using CdSe/ZnS cannot be directly compared with those using CdTe. Synthesis of core–shell CdTe/ZnS QDs has been reported, but core–shell CdTe has not yet become a standard biological reagent.

## Synthesis of QDs

At least several dozen methods of synthesis for common types of QDs have been published. It is possible to generate CdSe and CdTe directly in aqueous solution, with no need for further solubilization steps before they are used in biological systems; some references to these methods are given at the end of the chapter. However, the quality of these particles is generally lower than that obtained from synthesis in organic solvent, and none of these methods have been widely adopted. The standard procedure is *organometallic synthesis* at high temperatures in organic solvents, a procedure involving toxic materials and some risk of explosion, as solvents are often heated past their flash point. The highest-quality CdSe particles are obtained using dimethylcadmium as a precursor, but this is a highly toxic chemical that most labs choose not to handle. Excellent particles can be obtained using inorganic precursors instead, such as cadmium oxide.

The principle of synthesis is based upon the LaMer model of colloid growth, which states that nucleation and growth are separate phases. If all particles are nucleated at once, the resulting colloid will be monodisperse. (By this we mean they vary in size by no more than ±5%.) In order to achieve this nearly instantaneous nucleation, the (cold) selenium or tellurium precursor is rapidly injected into a hot, stirred solution of the cadmium precursor. The high temperature ensures complete dissolution of the reagents, and the rapid addition with stirring creates a supersaturated solution with concentrations above nucleation threshold. Nuclei form, the concentration in the solution drops below threshold (aided by the cooler temperature of the injected solution), and no new nuclei are able to form unless new precursors are added (**Figure 11.6a**).

The particles then grow homogeneously, with the time of growth determining the size of the particles. If the reaction is allowed to go long enough, *Ostwald ripening* occurs; this is the well-known process by which smaller particles dissolve and their material is added to larger particles (**Figure 11.6b**).

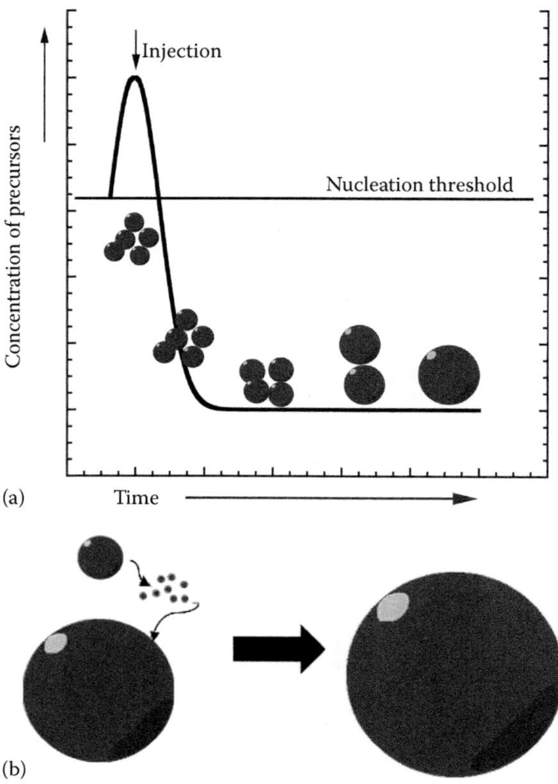



**Figure 11.6** (a) Nucleation and growth of nanoparticles by the LaMer model. (b) In Ostwald ripening, smaller particles dissolve and are added to larger particles. This is a process that occurs in many crystalline systems; the grittiness of old ice cream is due to Ostwald ripening of ice crystals within the cream.

The achievable particle sizes and crystal structures depend upon the temperature of synthesis, the precursors used, the reaction solvent, and the ratio of surfactants to precursors. The trick to high-quality colloidal nanoparticles is to minimize the width of the size distribution and maximize the surface passivation to increase fluorescence. Early papers determined ideal parameters empirically, followed later by theoretical studies of particle size distributions under diffusion, reaction, and reaction–diffusion control. The key parameter in a diffusion-controlled reaction is the activation energy for growth versus dissolution; this is strongly influenced by solvent. The most commonly used solvent is *tricoctylphosphine oxide* (TOPO), although several others are also widely found in the literature (**Table 11.2**). The degree of coordination ability of the solvent can vary from strongly *coordinating* to noncoordinating, referring to the degree to which the solvent complexes to the QD atoms, either the metal or the chalcogen or both. Nanoparticles synthesized in coordinating solvents will be capped with the solvent; those made in noncoordinating solvents require a *surfactant* such as oleic acid to coat the particle surface.

Preparation of QDs requires facilities for anaerobic synthesis and a good deal of patience and quality control, as well as facilities for characterization by electron microscopy, zeta potential measurements, and others. However, making the dots in house is the only way to be certain of their exact chemical structure. Bare-core CdSe and CdTe are easier to synthesize than core–shell structures. The synthesis procedures that the author has used for many years for CdSe, CdTe, and CdSe/ZnS are given in **Practical Tips 11.1**.

**Table 11.2**

Some Solvents Used for Synthesis of CdSe QDs and Their Properties

| Solvent | Properties | Comments |
| --- | --- | --- |
| Trioctylphosphine oxide (TOPO) | Coordinating to metal sites via oxygen. | Reported in original papers. Very commonly used. Results in hexagonal CdSe. Elongated QDs result from extended growth times. |
| Triphenylphosphine (TPP) | Nonsurfactant solvent. Coordinating to chalcogen via the P atom. | Results in a distorted crystal phase intermediate between hexagonal and cubic. |
| Oleylamine (OLA) | Primary amine. Coordinating to metals via N and to chalcogen via weak H bonds. | Hexagonal CdSe. Elongated and branched QDs result from extended growth times. |
| Hexadecylamine (HDA) | Primary amine. Coordinating to metals via N and to chalcogen via weak H bonds. | Hexagonal CdSe. Elongated and branched QDs result from extended growth times. |
| Dioctylamine (DOA) | Secondary amine. Weakly coordinating to metals via N. | Results in cubic CdSe. |
| Octadecylamine (ODE) | Noncoordinating. | Results in cubic CdSe. Commonly used. |

## Determination of QD size and concentration

**Equation 11.3** can be used for an approximate measure of QD emission peak versus size. However, significant variations from the particle-in-a-box model do occur, especially as the QDs become redder and thus more weakly confined. The best way to estimate the QD size, size distribution, and extinction coefficient is to take an absorbance spectrum and then calculate the diameter from parametrized curves from size-versus-wavelength data that have been determined empirically for different types of nanoparticles. Transmission electron microscopy (TEM) is also highly recommended for confirming particle size. The relationship between particle diameter, $D$, and emission at the exciton peak, $\lambda$, can be expressed as

CdSe

$$D=(1.6122\times10^{-9})\lambda^4-(2.6575\times10^{-3})\lambda^3+(1.6242\times10^{-3})\lambda^2-(0.4277)\lambda+41.57$$

CdTe

$$D=(9.8127\times10^{-7})\lambda^3-(1.7147\times10^{-3})\lambda^2+(1.0064)\lambda-194.84$$

CdS

$$D=(-6.6521\times10^{-8})\lambda^3+(1.9557\times10^{-4})\lambda^2-(9.2352\times10^{-2})\lambda+13.29$$

**PRACTICAL TIPS 11.1:    PREPARATION OF CdSe AND CdTe QDs**

## GENERAL SETUP

- A Schlenk line with a flow of inert gas (nitrogen or argon) is essential.

- The synthesis is performed in a three-necked flask in which one neck serves to monitor temperature, one to flow inert gas, and one to inject and remove solutions (**Figure P11.1.1**).

- The three-necked flask is placed in a high-quality heating mantle. The ability to accurately measure and control temperature is necessary for reproducible synthesis of high-quality particles.

- The elemental precursors (Se and Te) are purchased as a mesh: Te, powder 200 mesh; Se, powder 100 mesh.

- The given quantities are appropriate for a 50–100 mL three-neck flask. The synthesis may be scaled up as desired.

- Do not attempt this synthesis without the appropriate setup! Many of these reagents must be handled with great care. Oxygen must not be allowed to contact the hot reaction mixture. Dimethylzinc must also be handled under inert atmosphere, or it will combust.

### CdTe and CdSe

### Several Hours or the Day Before

CdTe: Prepare the tellurium precursor (TOPTe) by mixing 0.02 g (0.16 mmol) of Te with 0.415 g (1.12 mmol) trioctylphosphine (TOP) and 2 mL octadecene (ODE) under $N_2$ in a sealed vial containing a stir bar. Stir vigorously until the solution becomes light yellow (at least 1 h). The solution may need to be sonicated to aid solubility; it also helps to warm it to 45°C.

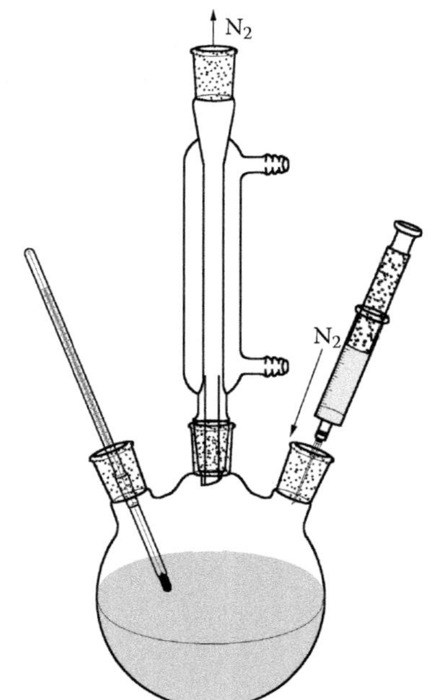


**Figure P11.1.1 Three-necked flask for QD synthesis, showing thermometer, syringe for injection, and connection to glassware for purging with inert gas.**

(*Continued*)

---

**PRACTICAL TIPS 11.1 (CONTINUED):    PREPARATION OF CdSe AND CdTe QDs**

CdSe: Prepare the selenium precursor (TOPSe) by mixing 0.016 g of Se with 0.5 mL TOP and 2 mL ODE under $N_2$ in a sealed vial containing a stir bar. Stir vigorously for 10–30 min; solubility is better than with Te.

**The Next Day**

***CdTe***

Add 0.026 g (0.20 mmol) of CdO and 0.179 g (0.63 mmol) of oleic acid (OA) to a three-neck flask containing 10 mL of ODE. Degas for 5 min and heat under an inert atmosphere to 220°C until the solution becomes colorless. When this occurs, increase the temperature to 310°C. Formation of a grey $Cd^0$ precipitate will be evident after heating for 10–20 min at 310°C. Immediately after this happens, rapidly inject the TOPTe precursor. Turn down the heat mantle and allow the temperature to drop and stabilize at 270°C for the growth of the nanoparticles. Aliquots of the reaction mixture may be withdrawn at various time points and injected into cold toluene to quench further growth and provide a size series of nanoparticles.

***CdSe (size series)***

Add 0.05 g of cadmium oxide (CdO) and 2 mL of OA to a three-neck flask containing 20 mL of ODE. Degas and heat under an inert atmosphere to 250°C. Turn off the heat mantle and rapidly inject the Se precursor with a syringe to initiate the growth of the CdSe core nanoparticles. Withdraw aliquots of the reaction mixture at various time points and inject into cold toluene to quench further growth and provide a size series of nanoparticles.

***CdSe/ZnS (one single size)***

In a separate flask, prepare the ZnS precursor by combining 0.75 mL of TOP, 0.2 mL hexamethyldisilathiane $((TMS)_2S)$, and 0.3 mL dimethylzinc $(Zn(CH_3)_2)$ under nitrogen. Dilute to 5 mL with ODE.

Prepare CdSe as above; after injection of the selenium precursor, monitor the QD fluorescence with a UV wand. When the desired color is observed, inject half of the ZnS precursor right away, and the rest dropwise over 5 min.

---

InP

$$D = \left[ \frac{3.6}{\dfrac{hc}{\lambda} - 1.35} \right]^{0.74} \tag{11.4}$$

The molar extinction coefficients, $\varepsilon$, are then given by (in units of $M^{-1}\ cm^{-1}$)

$$
\begin{aligned}
&CdSe \\
&\varepsilon = 5857 D^{2.65} \\
&CdTe \\
&\varepsilon = 10043 D^{2.12} \\
&CdS \\
&\varepsilon = 21536 D^{2.3} \\
&InP \\
&\varepsilon = 46595 D^{2.88}
\end{aligned}
\tag{11.5}
$$

These empirical fits can be used to calculate QD size and concentration from an absorbance spectrum intensity and position at the first exciton peak.

## Solubilization and biofunctionalization of QDs

The optical properties of QD photoluminescence make these particles highly promising for biological labeling. Compared to typical organic dyes, QDs show narrow emission spectra, resistance to photobleaching, high quantum yields, and large effective Stokes shifts. The absorbance spectra are broad, allowing multiple colors to be excited with a single wavelength. Their spectral properties also make them ideal FRET donors. (They are poor acceptors because of their broad absorbance; it is difficult to find a wavelength at which they do not excite.) (See **Chapter 16** for a discussion of FRET.)

However, use of the particles in aqueous solutions and biological systems has proved to be challenging. In order to make them water soluble and hence useful in biology, the particles must be attached to or encased in highly polar, water-soluble molecules. One or more of the following problems result from all solubilization procedures: loss of QD photoluminescence due to energy or electron transfer to the solubilizing molecules; instability of the bond between the QD and the molecules, leading to loss of solubility and changes in fluorescence over a time course of hours to days; large increases in QD effective radius, making the particles too large to label small biological structures; and access of water and oxygen to the QD surface, leading to QD breakdown and the release of toxic ions, particularly $Cd^{2+}$. As a result of these issues, many dozens of solubilization and functionalization schemes have been designed for nanoparticles: coating with polymers, Si shells, lipid micelles, bidentate thiols, and other compounds (**Table 11.3**).

Each functionalization technique leads to QDs that show unique biomolecular interactions, from fluorescence stability to toxicity to subcellular localization. Thus, there is no "best" approach to nanoparticle functionalization, but the method should be chosen according to the goals of the experiment and its required sensitivity and duration as well as its target cells. One of the easiest methods, which results in the smallest particles, is simple *cap exchange* with an *alkanethiol* of chosen chain length. This results in carboxylate-capped QDs that may then be conjugated to other molecules via carbodiimide coupling (see **Chapter 14**) (**Figure 11.7a**). These QDs are not stable for more than a few days, however. The use of a construct with more than one sulfur group per molecule (a *bidentate thiol*) can improve stability (**Figure 11.7b**). Either of these constructs is appropriate for many types of experiments, including photophysics and cell labeling, but the resulting fluorescence is usually too weak to permit SPT.

Alkanethiol-solubilized QDs also show nonspecific binding to many biological surfaces, including most cell lines. Because of this, they are unsuitable for use in any animal experiments or cell experiments where nonspecific binding is undesirable. In order to reduce QD stickiness and to permit flow through the bloodstream, a polymer such as polyethylene glycol (PEG) is needed. Two relatively simple approaches have been shown to work well for these purposes: one that uses a bidentate thiol–PEG (**Figure 11.8a**) and the other that uses a peptide–PEG, where several cysteines on the peptide bind the QD surface tightly (**Figure 11.8b**). The former molecule must be synthesized, whereas the latter may be purchased. **Practical Tips 11.2** gives recipes for solubilization of CdSe/ZnS and CdTe using the molecules in **Figures 11.7a** and **11.8b**.

It is important to note that *methods designed for CdSe/ZnS do not necessarily work for any other type of nanoparticle!* The core QD material has a profound but poorly

**Table 11.3**

Some of the Methods Used to Solubilize QDs with the Classic References Reporting Their Synthesis and/or Use

| Solubilization Method | Mechanism of Action and Comments | Reference |
| --- | --- | --- |
| Short-chain carboxy alkanethiols (MAA, MPA) | S–S bond (for core–shell) or S–Cd bond (for core); easiest method; negatively charged at basic pH. | Chan, W. C. W., and S. M. Nie, Quantum Dot Bioconjugates for Ultrasensitive Nonisotopic Detection, *Science* 281, 2016–2018, 1998. |
| Longer-chain carboxy alkanethiols | Same as short-chain. Requires higher pH for solubility than shorter chains; more stable. | Hanaki, K. et al., Semiconductor Quantum Dot/Albumin Complex is a Long-Life and Highly Photostable Endosome Marker, *Biochem. Biophys. Res. Commun.* 302, 496–501, 2003. |
| Zwitterionic thiols (cysteine, penicillamine) | Same as above. Zwitterionic at physiological pH. | Choi, H. S. et al. Renal Clearance of Quantum Dots, *Nat. Biotechnol.* 25, 1165–1170, 2007. Breus, V. V. et al., Zwitterionic Biocompatible Quantum Dots for Wide pH Stability and Weak Nonspecific Binding to Cells, *ACS Nano* 3, 2573–2580, 2009. |
| Bidentate thiols | S–S or S–Cd bond with two interactions per molecule. More stable than univalent thiols. | Mattoussi, H. et al., Self-Assembly of CdSe–ZnS Quantum Dot Bioconjugates Using an Engineered Recombinant Protein, *J. Am. Chem. Soc.* 122, 12142–12150, 2000. |
| Lipid micelles | Hydrophobic interactions; TOPO not removed. | Dubertret, B. et al., In Vivo Imaging of Quantum Dots Encapsulated in Phospholipid Micelles, *Science* 298, 1759–1762, 2002. |
| Triblock copolymer | Hydrophobic interactions; TOPO not removed. | Gao, X. et al., In Vivo Cancer Targeting and Imaging with Semiconductor Quantum Dots, *Nat. Biotechnol.* 22, 969–976, 2004. |
| Thin silica shell | S–S or S–Cd bond with cross-linked shell. | Bruchez, M. Jr. et al., Semiconductor Nanocrystals as Fluorescent Biological Labels, *Science* 281, 2013–2016, 1998. |
| Tridentate phosphines | S–P bond. Causes less fluorescence loss than S–S bond. | Kim, S., and M. G. Bawendi, Oligomeric Ligands for Luminescent and Stable Nanocrystal Quantum Dots, *J. Am. Chem. Soc.* 125, 14652–14653, 2003. |
| Thiol-terminated peptides | Single S–QD bond. | Akerman, M. E. et al., Nanocrystal Targeting in Vivo, *Proc. Natl. Acad. Sci, U S A* 99, 12617–12621, 2002. |
| Polyhistidine or polycysteine peptides | Multiple His– or Cys–metal interactions. | Slocik, J. M. et al., Monoclonal Antibody Recognition of Histidine-Rich Peptide Encapsulated Nanoclusters, *Nano Lett.* 2, 169–173, 2002. Pinaud, F. et al., Bioactivation and Cell Targeting of Semiconductor CdSe/ZnS Nanocrystals with Phytochelatin-Related Peptides, *J. Am. Chem. Soc.* 126, 6115–6123, 2004. Dif, A. et al., Interaction Between Water-Soluble Peptidic CdSe/ZnS Nanocrystals and Membranes: Formation of Hybrid Vesicles and Condensed Lamellar Phases, *J. Am. Chem. Soc.* 130, 8289–8296, 2008. |

*Note:* See the end of the chapter for additional references.

understood influence on the solubilization properties. For CdTe, alkanethiol cap exchange does work but is less stable than with CdSe. (See **Practical Tips 11.2** for a recipe.) Interestingly, CdTe particles can be solubilized with short-chain amines such as cysteamine, whereas CdSe cannot. As of this writing, InP remains difficult to solubilize using biologically appropriate reagents. A method for capping with a thin silica shell has been published, but its utility has not yet been demonstrated in cells. Some references to solubilization procedures for these other types of QDs are given at the end of the chapter.

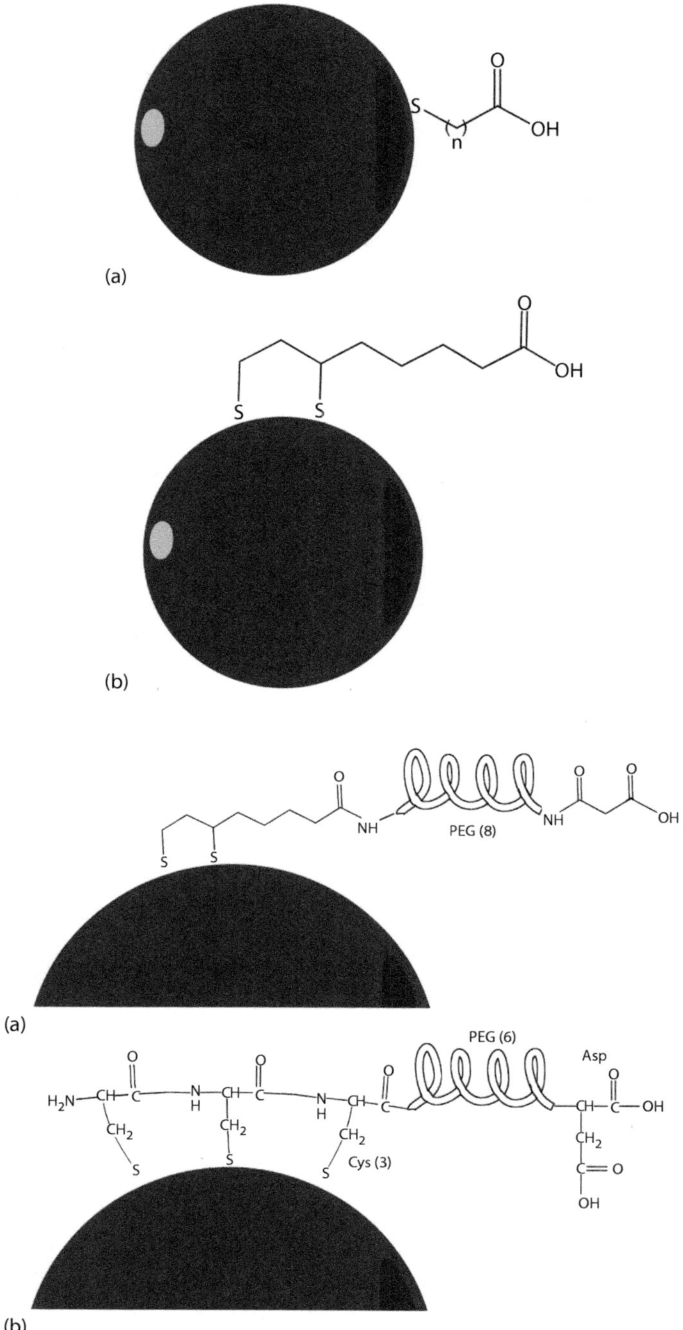

**Figure 11.7 Common ligands for solubilizing QDs.**
(a) Alkanethiols, which are a single thiol followed by a carbon chain (of variable length) and terminating in a carboxylic acid. These are negatively charged at basic pH. (b) Bidentate thiol. The two thiols make a stronger QD–ligand bond and thus a more stable cap than the alkanethiols.

**Figure 11.8 Ligands for biocompatible QDs with minimal nonspecific binding.**
(a) Bidentate thiol–PEG–carboxylic acid. (b) Peptide (with three Cys residues)–PEG–Asp.

Once the QDs are solubilized, they may be used as is or conjugated to other biomolecules such as peptides, proteins, antibodies or antibody fragments, signaling molecules, and many others. Each type of conjugate is different and may be hindered by complications such as aggregation or fluorescence quenching; thus, preparing conjugates to untested molecules requires a reasonable amount of characterization and testing. Instrumentation for particle sizing, absorbance

---

### PRACTICAL TIPS 11.2:    SOLUBILIZING CdSe/ZnS AND CdTe QDs

After solubilization, the QDs should be stored in the dark at room temperature. If precipitation is observed, discard them and try again.

#### CdSe/ZnS: USING MERCAPTOACETIC ACID OR MERCAPTOPROPIONIC ACID

Begin with 1 mL of concentrated QDs (optical density > 5 at exciton peak) in ODE/hexanes. Dilute to 5 mL in methanol and then add a large excess of mercaptoacetic acid (MAA) or mercaptopropionic acid (MPA) (approximately 1 M final concentration). Adjust the pH to ~10 using tetramethylammonium hydroxide pentahydrate (TMAH) and leave in the dark with stirring for 72 hours at room temperature. Precipitate with ethyl acetate, remove the supernatant, and then resuspend in distilled water or buffer.

#### CdSe/ZnS: USING THE PEPTIDE IN FIGURE 11.8B

The peptide N-CysCysCys(PEG6)Asp-C can be custom-synthesized by several suppliers. PEG6 is a hexamer of polyethylene glycol (PEG). Begin with 1 mL of concentrated QDs in hexane (optical density ≥ 5 at exciton peak). Precipitate them with a large volume excess of anhydrous ethanol followed by centrifugation at 10,000 rpm in a tabletop centrifuge for 5 min. Redissolve the resulting pellet in 900 µL of pyridine. Separately, dissolve the peptide in 100 µL of dimethyl sulfoxide (DMSO) to a final concentration of 10 mM. Mix the QDs in pyridine and the peptide in DMSO, and incubate at room temperature for 10 min. Add drop by drop a solution of TMAH (1 M in methanol) until the solution becomes cloudy, and then centrifuge as above. Discard the supernatant and dry the QDs in air for 20 min. The pellet can be suspended in 1 mL of distilled water. For purification from excess peptide and reaction side products, the sample may be cleaned using centrifugation filters with a 50 kDA molecular weight cutoff.

#### CdTe: USING MPA (SMALL-SCALE PREPARATION; USE AS QUICKLY AS POSSIBLE)

Add 400 µL of toluene, 500 µL of 200 mM phosphate-buffered saline (PBS) or borate buffer (pH 9), and 1 µL of MPA to 100 µL of concentrated hydrophobic QDs. Mix by shaking vigorously. The QDs will move from the organic phase to the aqueous phase, which can be extracted using a pipette. QDs may be isolated from excess thiol by several cycles of concentration and dilution using a centrifuge filter with a 10 kDa molecular weight cutoff.

---

and fluorescence spectroscopy, zeta potential, and electron microscopy is the minimum required.

## Commercial QDs

If the only concern is to have the brightest, stablest QDs possible, and knowledge of their exact chemistry is not necessary, then several types of QDs are available commercially (Table 11.4). Commercial QDs in water with carboxylate caps are not recommended for further functionalization by either the author or the manufacturer, as methods such as EDC coupling do not work on these particles; their coatings are proprietary, so determination of the exact chemistry is not possible. The particles should be used as is, or conjugates should be made by linking a biotinylated biomolecule of choice to QD–streptavidin, or an antibody of choice to QD–protein A.

A specific type of "sandwich" construct consisting of a primary antibody to a cellular target, biotinylated *Fab fragments* (the antigen-binding fragment of

**Table 11.4**

Some Commercially Available QDs

| Type of QD | Comments | Manufacturer |
| --- | --- | --- |
| CdSe/ZnS–polymer–streptavidin | Very stable, high quantum yield<br>Relatively large<br>Can make new constructs using biotin<br>525–625 nm emission wavelengths | Invitrogen |
| CdSe/ZnS–polymer–protein A | For making conjugates to antibodies of choice | Invitrogen |
| CdSe/ZnS–polymer–antibody | Selected antibodies for targeting | Invitrogen |
| CdSe, CdSe/ZnS, InP, CdS, others including doped | In organic solvent | NN-Labs |
| CdSe/ZnS, Mn-doped ZnSe | Water-soluble; not recommended for conjugation | NN-Labs |
| CdSe/ZnS, InP, PbC | Organic solvent or water | Sigma-Aldrich |

antibodies; see **Chapter 14**), and finally QD–streptavidin has been used in many biological studies (**Figure 11.9a**). The large size of the sandwich is offset by its great chemical and optical stability relative to less fully functionalized QDs. The first demonstration of the use of this construct was in tracking of single glycine receptors in neurons. Individual receptors were followed for 40 min, and the diffusion coefficients in the membrane could be estimated. Protocols for preparation of these sandwich conjugates, optimization of targeting to cell-surface proteins, and image acquisition and analysis have been published and are referenced at the end of the chapter.

Another approach, which minimizes the size of the construct somewhat, is to biotinylate a ligand of choice in order to attach it to QD–streptavidin (**Figure 11.9b**). Biotinylation of proteins is straightforward using a number of commercially available reactive biotins such as biotin hydrazide (aldehyde reactive), biotin sulfosuccinimidyl ester (amine reactive), biotin maleimide (thiol reactive), and others. The conjugated protein (or other molecule) is also likely to retain its biological activity unless the QD is conjugated directly to the active site.

An interesting and useful twist to the biotinylation approach is that a specific peptide sequence (GLNDIFEAQKIEWHEAR) serves as a *biotin acceptor* for the *Escherichia coli* biotin ligase, BirA (the boldfaced lysine is biotinylated). Thus any protein can be biotinylated at a specific location by inserting this sequence during

**Figure 11.9 Sandwich constructs using QD–streptavidin.** (a) Biotinylated Fab fragments bind to QD–streptavidin, and a cell-targeting antibody recognizes the Fab fragments as well as the cellular target. (b) Any protein may be biotinylated and then bound to QD–streptavidin. The biotinylation reaction may be done in a test tube after protein synthesis, or performed by the *E. coli* or mammalian cells that express the protein.

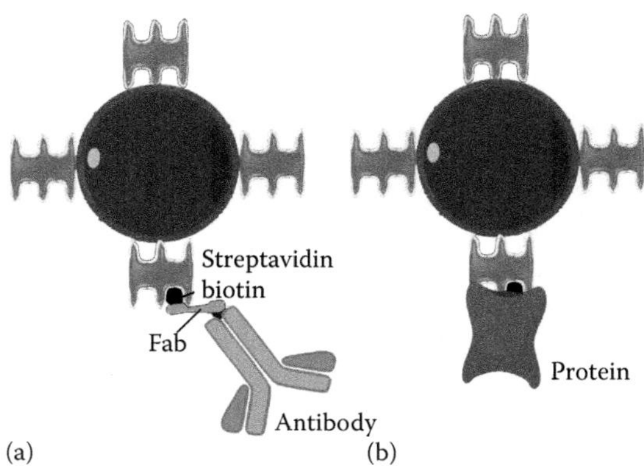

molecular cloning, making it possible to biotinylate membrane proteins *in situ*. Native mammalian biotin protein ligases also exist so that the proteins will be expressed biotinylated in mammalian cells, and commercial cloning vectors are available that link the target sequence to a protein of choice. However, the tag is significantly longer than that needed for biotinylation by bacterial enzymes (72 amino acids) and thus may disrupt the function of complex proteins.

## 11.3 QD APPLICATIONS

### Single-particle tracking

SPT observes the diffusion of single particles in a medium such as the cytoplasm or cell membrane for periods ranging from seconds to hours. The trajectories and diffusion coefficients of the particles, which can be labeled receptors or other proteins, gives insight into the life cycle of the molecules and the diffusion properties of the medium. The major limitation of SPT is photobleaching. The longer the fluorophore lasts, the longer it can be tracked. SPT was therefore one of the first applications of QDs and perhaps the one in which the most progress has been made. Commercial QDs are especially suitable for this purpose, as QDs conjugated to small molecules, or coated with a minimal layer of solubilizing ligands without conjugation, are much more sensitive than the heavily passivated sandwich constructs to chemical and photoinduced degradation. The sandwich constructs are particularly useful for cell-surface proteins such as ion channels or receptors. The antibody (or conjugated protein) is simply added to a dish of cells at a low concentration (nanomolar), allowed to bind for several minutes, and the target is tracked. Several interesting insights into neuroscience have been obtained using this method, as the long-lived QD fluorescence has allowed for the observation of phenomena such as receptor clustering on a long timescale (many minutes to hours).

Regardless of conjugate, all commonly used QDs show fluorescence intermittency, or "blinking"—stochastic periods of emissive ("on") and dark ("off") states (**Figure 11.10**). This poses a problem for SPT, since a particle that is dark cannot be tracked. (Interestingly, organic fluorophores also show intermittency, but this is usually not observed, since the single molecules are not bright enough to track

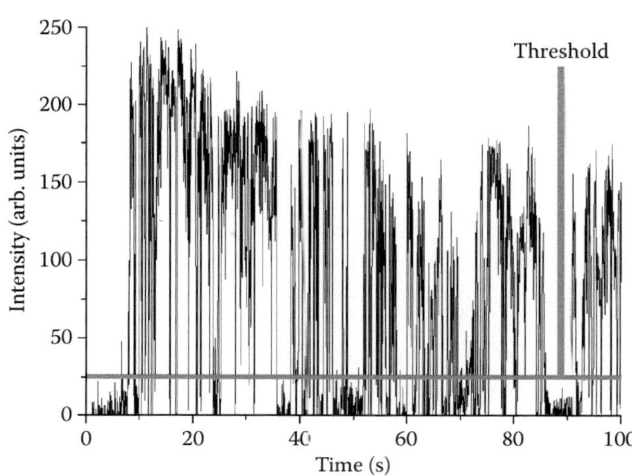


**Figure 11.10 QD blinking.** Typical intensity-versus-time trace taken over 100 s. A threshold is chosen below which the QD is considered "off."

and the blinking occurs on rapid timescales.) Currently, the best approaches to dealing with blinking are analytic—that is, using data analysis techniques to separate the intensity fluctuations due to particle photophysics from those due to motion. A technique called $k$-space image correlation spectroscopy ($k$ICS) can be used to extract accurate diffusion coefficients and QD blinking probability distribution functions.

QDs made of CdSe with thick CdS shells have been reported to show greatly reduced blinking, and completely nonblinking nanocrystals of an alloyed core–shell material (CdZnSe/ZnS) have also been described. The particles are only slightly larger than ordinary core–shell QDs. In both cases, fluorescence imaging was performed in organic solvents in the original studies, and biological applications of these materials remain to be established.

## QD delivery to living cells

Not every experiment is quite so easy as applying an antibody conjugate and waiting for it to bind a cell-surface protein. If the protein or process of interest is in the interior of the cell, the QD conjugate must be taken up or delivered to the cell cytoplasm, nucleus, or wherever else the target might be. If the cells do not need to be alive for the labeling, then they may be fixed and permeabilized as was discussed for antibody labeling in Chapter 7. Different cell types have different requirements for this labeling; some cell types such as primary neurons show particular difficulties. Practical Tips 11.3 gives a protocol for QD–antibody labeling that works particularly well with neurons.

Many cultured cell lines take up QD–COOH or QD–NH$_2$ constructs nonspecifically when the QDs are added at low nanomolar concentrations; however, some cell lines show little or no uptake. (Most cell lines show some degree of nonspecific uptake at high QD concentrations, hundreds of nanomolar, and more, but use of these high concentrations is wasteful, usually toxic, and not recommended.) Nonspecific uptake results in endosomal localization of QDs. Efficient endocytosis of QDs can also be achieved by targeting a cell-surface receptor that shows high internalization rates. When the receptor is internalized, the QD will come along. Endosomal uptake leads to a characteristic "speckled" appearance of the cells. This type of internalization has been demonstrated using many different molecules and receptors: some examples are as follows: QD–dopamine taken up by dopamine receptor–bearing cells (Figure 11.11); glucose-conjugated QDs taken up by yeast cells; transferrin-conjugated QDs taken up by pancreatic cancer cells; nerve growth factor (NGF) QDs taken up by neural cells; QD–folate taken up by cancer cells that overexpress the folate receptor; and QD–antibodies internalized by different types of cells after targeting to surface receptors with high turnover rates.

Endocytosis is sufficient for toxicity studies, useful for tracking of endosomes (e.g., for studies of retrograde transport), and may be sufficient for drug delivery studies depending upon the nature of the drug and its cellular target. If the drug is cleaved from the QD inside the endosome, it may diffuse out, or reactive species created by QD activation may rupture the endosome and release its contents. However, there are cases where endosomal delivery is not sufficient, such as when nuclear delivery or targeted intracellular labeling is desired. In these situations, there are methods of functionalizing QDs to permit escape from endosomes.

### PRACTICAL TIPS 11.3:  LABELING CULTURED NEURONS WITH QD–ANTIBODY CONJUGATES

Reprinted from Pathak, S. et al., Quantum Dot Applications to Neuroscience: New Tools for Probing Neurons and Glia, *J. Neurosci.* 26 (7), 1893–1895, 2006. With permission.

#### STEP 1. PREPROCESSING AND FIXING

1. Remove media from wells by gently aspirating.
2. Wash cells with warmed PBS.
3. Fix cells with electron-microscopy grade 4% paraformaldehyde in PBS for 10 min at room temperature.
4. Wash cells 3× with PBS.
5. Permeabilize cells with 0.2% Triton X-100 in PBS for 5 min.
6. Wash cells 3× for 5 min with PBS.
7. Incubate with 10% horse serum in PBS for 30 min at room temperature.
8. Rinse with PBS.
9. Apply streptavidin/biotin blocking kit to block endogenous biotin.

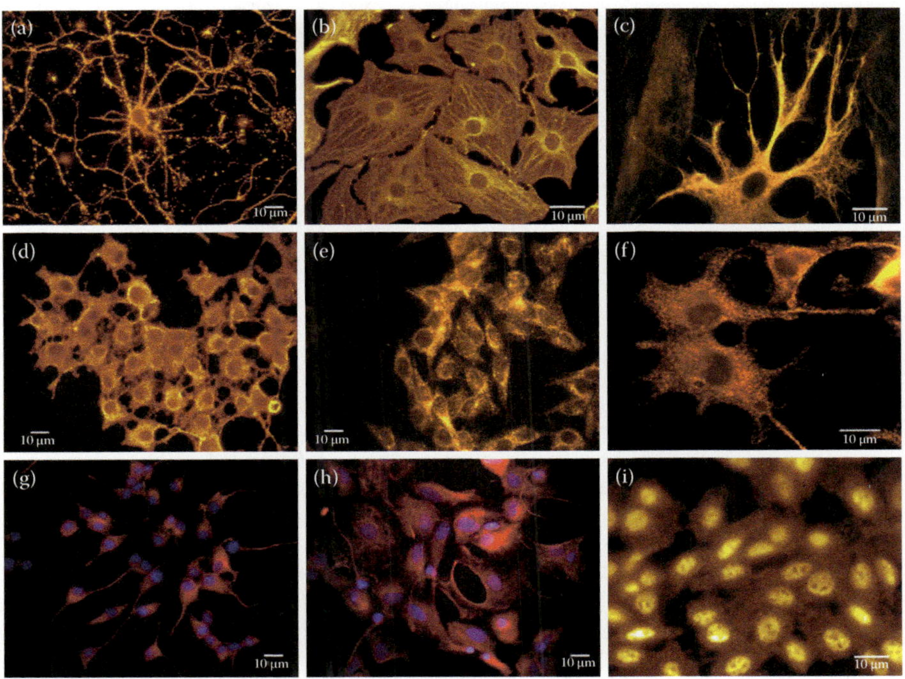

**Figure P11.3.1**  (a) Primary cortical neurons specifically labeled for β-tubulin. (b,c) Primary cortical astrocytes specifically labeled for glial fibrillary acidic protein (GFAP). (d,f) PC12 cells labeled for β-tubulin. (e) r-MC1 neural retinal Muller glial cells specifically labeled for GFAP. (g) PC12 cells labeled for β-tubulin using standard immunocytochemistry. (h) Primary spinal cord astrocytes labeled for GFAP using standard immunocytochemistry. (i) An example of artifactual nonspecific labeling in r-MC1 Muller cells with anti-GFAP-conjugated 605 nm quantum dots. In this case, putative nonspecific electrostatic interactions between quantum dots and cellular proteins led to intense nuclear staining and mild cytoplasmic staining using other quantum dot conjugation protocols described for mammalian cells.

(Continued)

**PRACTICAL TIPS 11.3 (CONTINUED):   LABELING CULTURED
NEURONS WITH QD–ANTIBODY CONJUGATES**

**STEP 2. INCUBATION WITH BIOTINYLATED CELL-TARGETING MOLECULE**

1. Rinse with PBS.
2. Add biotinylated molecule of interest (e.g., antibodies). You may need to use a kit to biotinylate antibodies; these are available from several suppliers.
3. Incubate 2 h at room temperature.
4. Remove antibodies by gently aspiration and rinse 3× with PBS.

**STEP 3. QUANTUM DOT INCUBATION**

1. Add streptavidin-conjugated QDs
2. Incubate 1 h at room temperature.
3. Rinse 3× with PBS.
4. Mount with 90% glycerol in PBS.

Results are shown in **Figure P11.3.1**.

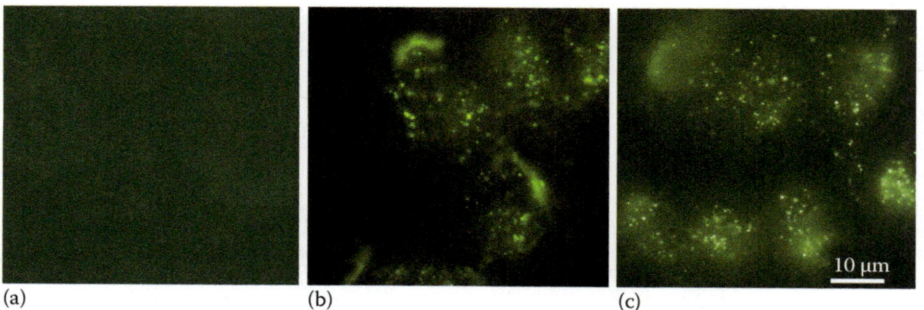


**Figure 11.11 Specific and nonspecific uptake of QDs and conjugates by dopamine receptor–transfected PC12 cells.** (a) PC12 cells alone under broadband blue excitation (400–450 nm) and long-pass emission (460–800 nm). (b) Unmodified QD–MPA added to the same cells at a concentration of 5 nM and exposed for 2 h, showing a typical endosomal uptake pattern: speckled near the center of the cell with no nuclear entry. (c) QD–dopamine conjugates applied to the same cells at 5 nM for 1 h, also showing intracellular vesicles consistent with endosomal uptake.

Some peptide sequences, both naturally occurring and synthetic, associate with the cell surface and facilitate QD endocytosis, or in some cases, cell penetration independent of endocytosis. The highly positively charged TAT peptide (containing a polyarginine sequence) can deliver QDs and QD–protein complexes into cells. This peptide may be attached to a QD by a terminal thiol using simple methods of cap exchange. The peptide is simply mixed with alkanethiol-solubilized QDs, and a certain number of the alkanethiols are replaced by the peptide. Alternatively, in order to mediate tight binding of the peptide to the QDs, several positively charged His residues can be added to the peptide during synthesis. This construct can also be added to solubilized QDs simply by incubating for several hours (**Figure 11.12a**).

The commercially available peptide Chariot (KETWWETWWTEWSQPKKKRKV) can also be used to deliver QDs to cells. It must first complex noncovalently with a

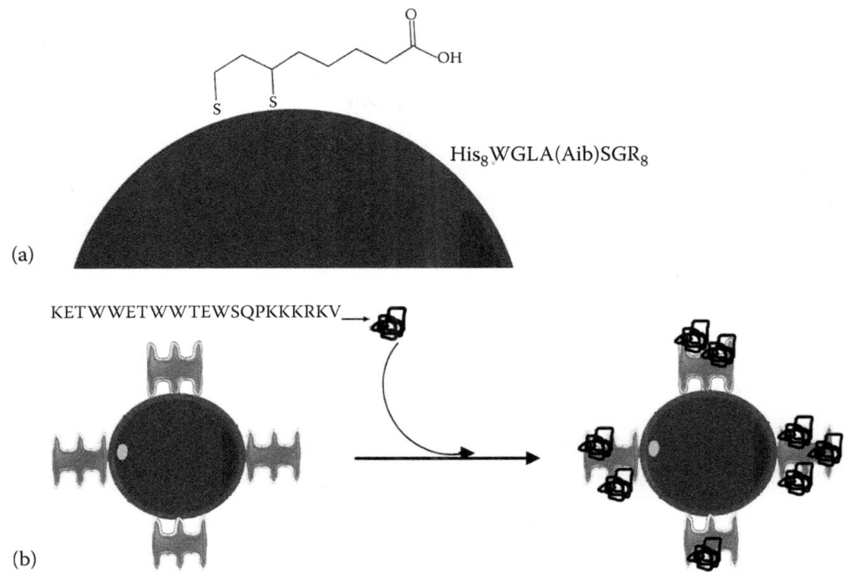



**Figure 11.12 Peptide-mediated delivery of QDs into cells.** (a) TAT peptide consisting of eight His residues to mediate QD binding, a spacer sequence (Aib is α-amino isobutyric acid), and the eight Arg residues of the TAT peptide that mediate cell entry. The histidines coordinate directly to the Zn atoms of the QD; dihydrolipoic acid (DHLA) solubilization groups remain on the particle. (b) The CHARLOT peptide complexes with proteins to mediate cell delivery. In the case of commercial streptavidin-coated QDs, the peptide binds to the streptavidin coat noncovalently and shuttles the entire construct across the cell membrane.

protein, which can be the streptavidin of commercial QDs, and is then able to deliver its cargo across the cell membrane (**Figure 11.12b**). Organelle-specific targets can be conjugated to the QDs before delivery and may achieve specific, nonendosomal localization; however, these QDs do not enter the nucleus efficiently.

Certain hyperbranched copolymers are endosomolytic (endosome disrupting), and result in QD release into the cytoplasm after uptake (**Figure 11.13**). Another possible approach is to use patterned striations of alternating anionic and hydrophobic groups on the surface of the QD. Gold nanoparticles show spontaneous formation of such patterns, approximately 6 Å wide, which have been shown to permit penetration directly into cells without membrane disruption.

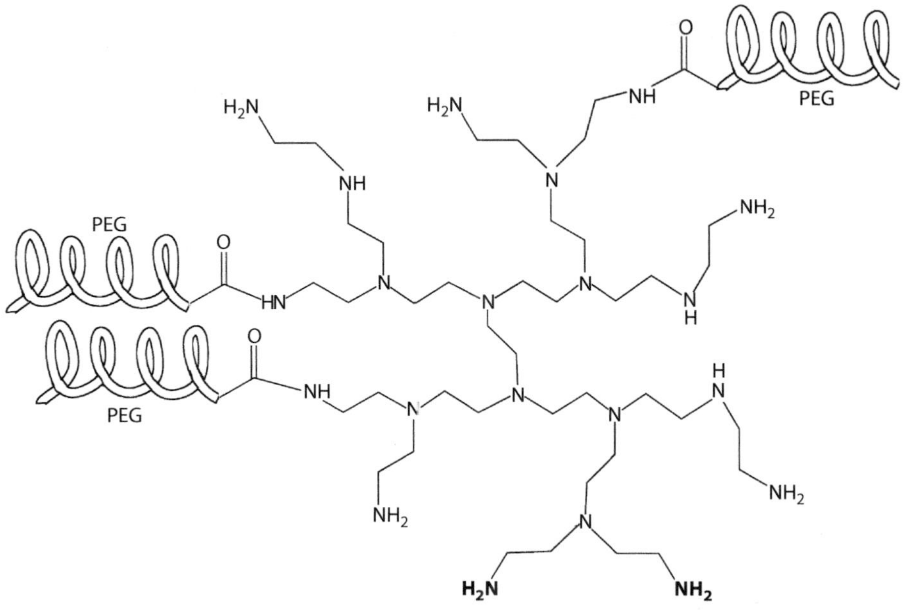

**Figure 11.13 PEG-grafted polyethylene imine (PEI), which serves to disrupt endosomes.** This copolymer can be added to the surface of QDs by ligand exchange.

Other alternatives include electroporation and microinjection of QDs, carried out in a similar fashion as described for nucleic acids in Chapter 3. The biggest risk with both of these methods is QD aggregation, which will block injection pipettes and damage cells. For microinjection, PEGylation of the QDs or encasing them in lipid micelles can help prevent this. Aggregation during electroporation is not reduced by coating the QDs with PEG or cross-linked shells, indicating that it results from mechanisms other than loss of stabilizing ligands on QDs due to the electric field. Successful microinjection leads to a uniform labeling of the cell cytoplasm in contrast to the speckled appearance seen with endosomal uptake.

For microinjection of QDs, we recommend a *closed tip* method. This is a way of eliminating the air interface at the tip of the microinjection needle by pulling an extremely long, fine needle that is closed at the end. The needle is then backfilled and placed into the microscope, and the tip is carefully broken to yield an opening of the appropriate size for injection. Learning how to break the tips requires some practice! Find an empty spot on the cell coverslip (a plastic dish will not work). Lower the needle very slowly until a small piece breaks off, and then raise it and try to inject a cell. Break off more as needed. Usually, the tips can be broken twice before they become too wide to use.

## Multicolor labeling and avoidance of autofluorescence

There are numerous biological applications where a maximum number of distinguishable fluorescence colors is desirable. Multiple structural features of a single cell may need to be distinguished; this is probably the most developed of all QD applications. Several colors of QDs, each conjugated to its own antibody, ligand, or other cell-targeting molecule, can all be added to the cell at once and excited with a single wavelength. Narrow bandpass filters or spectral deconvolution can provide quantitative measures of the amount of each color of QD in a given region. Usually, such experiments are done with fixed and permeabilized cells so as to facilitate delivery of the QDs.

Because all cells show some degree of *autofluorescence*, distinguishing fluorophores is not always easy. The many colors of QDs available allow for avoidance of autofluorescence in many cases, but in more stubborn cases, more sophisticated tricks may be used. Some cyanobacteria show such intense autofluorescence that imaging them with fluorescent labels proved to be nearly impossible until QDs were developed. The autofluorescent pigments can be photobleached more quickly than the QDs, allowing the cells to be observed after deliberate light exposure. QD fluorescent lifetimes are also much longer than those of typical autofluorescent pigments (hundreds vs. tens of nanoseconds). By using a technique called *time-gated imaging*, in which a gated CCD with nanosecond resolution is used to exclude short-lived fluorescent signals, the autofluorescence can be eliminated from the image (**Figure 11.14**).

The many colors of QDs available also make these particles useful for multiplexing studies, such as in microarrays. Arrays of DNA and RNA oligonucleotides and of proteins have all been made using QDs. Uniform spotting of the QDs on the array is important; smaller spot sizes (<100 μm), including nanoarrays, have been reported to be more easily prepared than larger spot sizes. Over a dozen QD colors can be used simultaneously in highly multiplexed arrays.

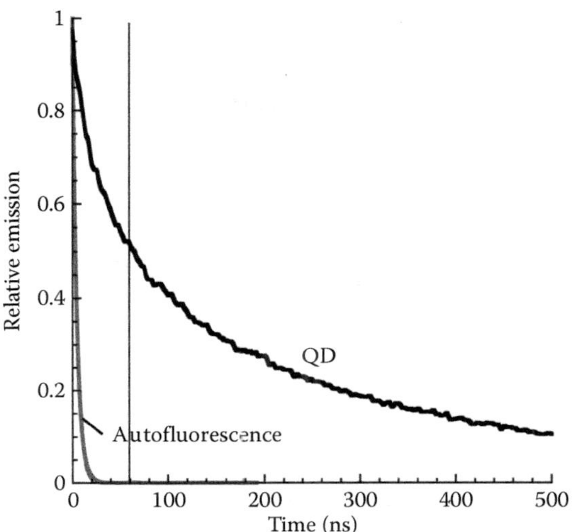

**Figure 11.14 Time-gated imaging.** Shown are typical fluorescence decay curves (normalized intensity vs. time) for a QD and cellular autofluorescence. Collecting the signal after 50 ns will successfully eliminate the autofluorescence while retaining a large fraction of the QD signal.

## Correlated fluorescence and electron microscopy

Although not so electron-dense as metal nanoparticles, CdSe possesses sufficient electron density to permit visualization using TEM (**Figure 11.15a**). Unstained QDs can be difficult to distinguish from structures like ribosomes, but standard methods can be used to improve contrast. Silver enhancing can be used to observe single-molecule QD tags. This is a standard TEM technique usually used to enhance immunogold labeling by exploiting the ability of metals to transfer electrons to silver ions and thus precipitate metallic silver. Kits for silver enhancement can be purchased from any electron microscopy supplier. Because semiconductors are capable of the same reaction, QDs are also enhanced by silver; the stain may even distinguish photooxidized from undamaged QDs.

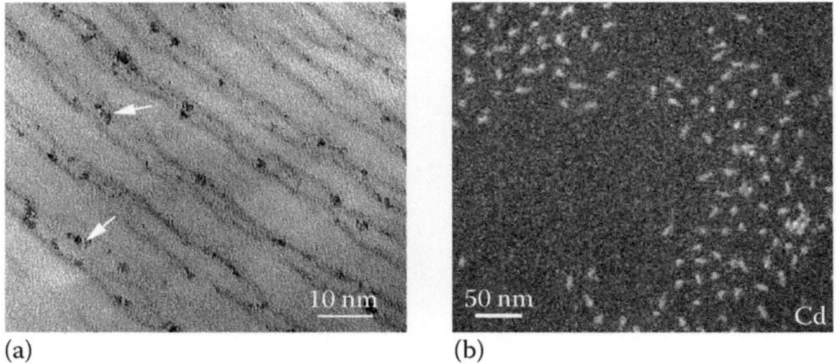


**Figure 11.15 QDs and TEM.** (a) Standard TEM image of a thin section of QDs embedded in a lipid bilayer. The section is stained with uranyl acetate and osmium tetrachloride, so that the lipid layers appear gray and the QDs are nearly black. The identity QDs (indicated by arrows) can be confirmed using energy-dispersive x-ray spectroscopy. (b) Energy-filtered TEM (EFTEM) image of QDs in a cell. (Reproduced from Nisman, R. et al., Application of Quantum Dots as Probes for Correlative Fluorescence, Conventional, and Energy-Filtered Transmission Electron Microscopy, *J. Histochem. Cytochem.* 52 (1), 13–18, 2004. With permission.)

An excellent technique for unambiguous detection of QDs is energy-filtered TEM (EFTEM), also called electron spectroscopic imaging (ESI). EFTEM generates maps of elemental distributions and can distinguish elements that have different ionization edges. The $M_{IVV}$ ionization edge of cadmium permits it to be distinguished from nitrogen. A map of Cd distribution in a sample shows the location of the QDs (**Figure 11.15b**).

Correlative fluorescence and electron microscopy (EM) may also be performed using QDs. Because the different colors of QDs are not only different sizes but also usually different shapes (the reddest emitting particles are rod-shaped when synthesized in coordinating solvent), fluorescence images can be compared with EM ultrastructure. At least three colors/sizes of QDs may be distinguished this way (**Figure 11.16**). An important caveat is that QD fluorescence is irreversibly destroyed by osmium tetroxide, a typical TEM stain. Thus, osmication cannot be used on samples for correlated imaging.

Semiconductor nanoparticles may also be used for biological labeling using scanning electron microscopy (SEM). Under SEM, the particles appear bright relative to surrounding biological tissue (**Figure 11.17a**). However, it is easy to mistake particles for cellular structures such as villi with typical SEM, especially because individual QDs are too small to resolve.

There are methods to scan through thick biological samples without the need for thin sectioning. Scanning transmission electron microscopy (STEM) tomography of uranyl acetate–stained, Epon-embedded tissue slices as thick as 1 µm may be examined with this technique. Under these conditions, nanoparticles appear bright (**Figure 11.17b,c**). Another approach is to use ion abrasion in conjunction

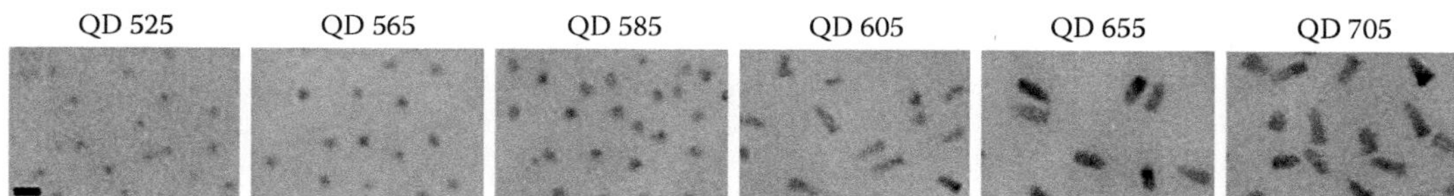


**Figure 11.16 Appearance of commercial QDs of different colors under TEM (scale bar = 10 nm).** The number after the QD indicates its emission peak wavelength. Note the increasing elongation in the reddest QDs. (Reprinted by permission from Macmillan Publishers Ltd., *Nat. Methods*, Giepmans, B. et al., Correlated Light and Electron Microscopic Imaging of Multiple Endogenous Proteins Using Quantum Dots, 2, 743–749, copyright 2005.)

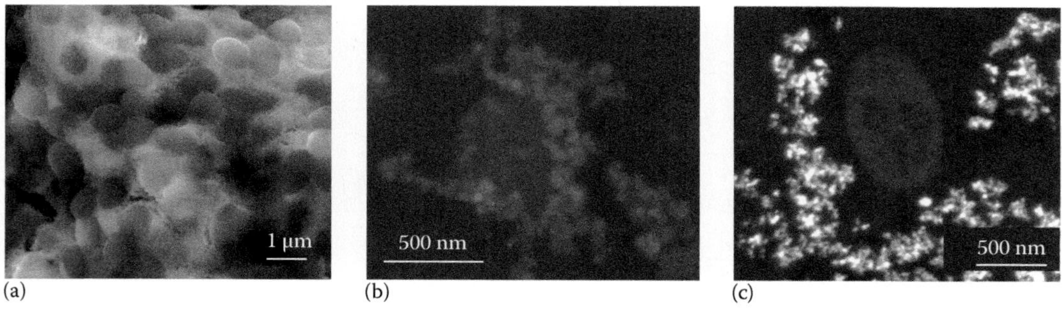


**Figure 11.17 QDs and electron microscopy.** (a) Scanning electron micrograph of unstained *Staphylococcus aureus* bacteria coated with CdSe/ZnS QDs. (b,c) STEM images of *E. coli* cells with CdTe QDs (b) and $TiO_2$ nanoparticles (c). The samples are stained with uranyl acetate and osmium tetroxide. (Images courtesy of R. Mielke.)

with SEM, a technique called ion abrasion SEM (IA-SEM). This allows thin sections of samples to be milled away by a focused ion beam, and successively imaged by the electron beam, providing a set of slices that can be reconstructed into an image of the entire thickness of the sample. It requires an instrument capable of housing both beams.

Finally, with the appropriate instrumentation, electron microscopic techniques can be applied to living, or at least fully hydrated, samples. Environmental scanning electron microscopy (ESEM) is performed in a chamber with up to 8 mbar of water pressure, preserving hydration. Bacteria and plant cells are readily imaged with ESEM, as they do not collapse when medium is removed. Mammalian cells pose a greater challenge; however, this technique is ideal for some types of tissue culture, such as organotypic neuronal slices. Very recently, TEM suppliers have introduced environmental chambers for TEM and STEM of samples in liquid. Nanoscale resolution of live cells is possible with these new techniques, although they remain uncommon and expensive. Some researchers custom-design environmental chambers permitting varying degrees of water vapor during TEM imaging.

## QDs as biosensors

The band edge structure and absorbance/emission spectra of QDs can be exploited to create particles that change their optical properties under specific conditions. The first uses of QDs as environmental sensors made use of their sensitivity to pH. When not highly passivated with large protein layers, QDs show significant changes in emission intensity with pH, becoming weaker as pH falls (**Figure 11.18a**). This effect is reversible, and sensitive enough to be able to be used to detect uptake of QDs into subcellular vesicles such as endosomes and

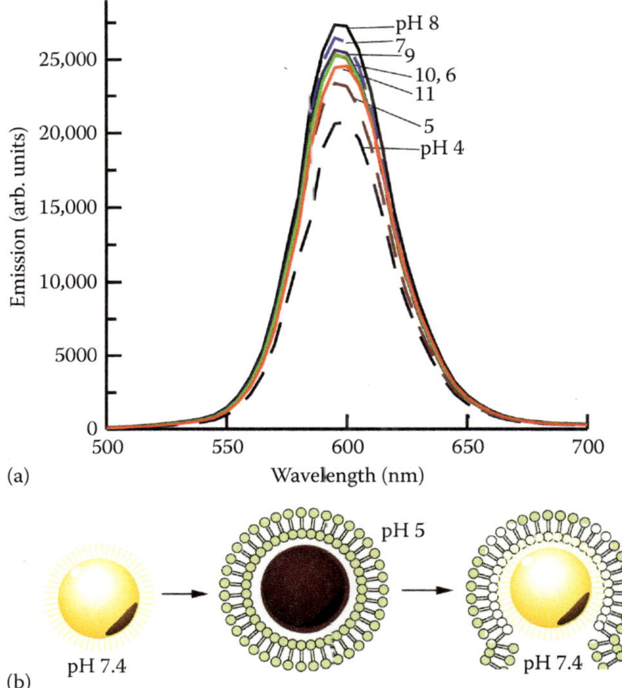


**Figure 11.18 QDs as pH indicators.** (a) pH dependence of fluorescence intensity of mercaptopropionic acid–capped CdSe/ZnS QDs. (b) Schematic of use of QDs as indicators of vesicle uptake and release. When inside the vesicle at pH 5, the QDs are significantly dimmer than when in the extracellular space or cytoplasm at pH 7.4.

synaptic vesicles, which have an acidic internal pH (around 5). The QDs are bright before uptake, dimmer inside the vesicle, and then bright again when the vesicle opens (**Figure 11.18b**).

Conjugation of molecules that transfer energy or electrons to QDs can be used to create sensitivity to essentially any selected environmental condition, provided that the desired chemistry can be found or synthesized. The principle behind the design of such probes is the ability of QDs to interact with biomolecules in energy or electron transfer processes—usually FRET.

The spectral properties of semiconductor QDs make them ideal FRET donors. For FRET to occur, the emission spectrum of a donor fluorophore must overlap the absorbance spectrum of an acceptor; with their broad absorbance and narrow, tunable emission, QDs have the promise to allow FRET with minimal donor–acceptor spectral overlap using almost any fluorophore as the acceptor. If the acceptor changes its absorbance spectrum upon analyte binding, a sharp fluorescence shift can be expected (**Figure 11.19**). This principle has been used to make highly sensitive pH sensors and sensors for maltose, for specific DNA sequences, and for a variety of different ions and a few small molecules.

As will be discussed in **Chapter 16**, FRET is highly distance dependent. A typical dipole–dipole interaction falls off as $r^{-6}$. This principle can be used to create sensors by attaching a QD and a FRET acceptor to complementary molecules or

**Figure 11.19 Use of FRET for specific sensing.** (a) The absorbance of QDs (QD Abs) is broad, but the emission is narrow (QD Em). In the case of a hypothetical nonfluorescent dye whose absorbance redshifts upon addition of analyte, FRET becomes possible with QDs as donors only as the analyte is added. (b) In this case, it can be expected that increasing analyte concentration will lead to increasing amounts of QD quenching. (c) In the case of a hypothetical red fluorescent dye whose absorbance but not emission changes with analyte addition, the same conditions apply, but (d) the emission reflects QD quenching and dye enhancement with analyte addition.

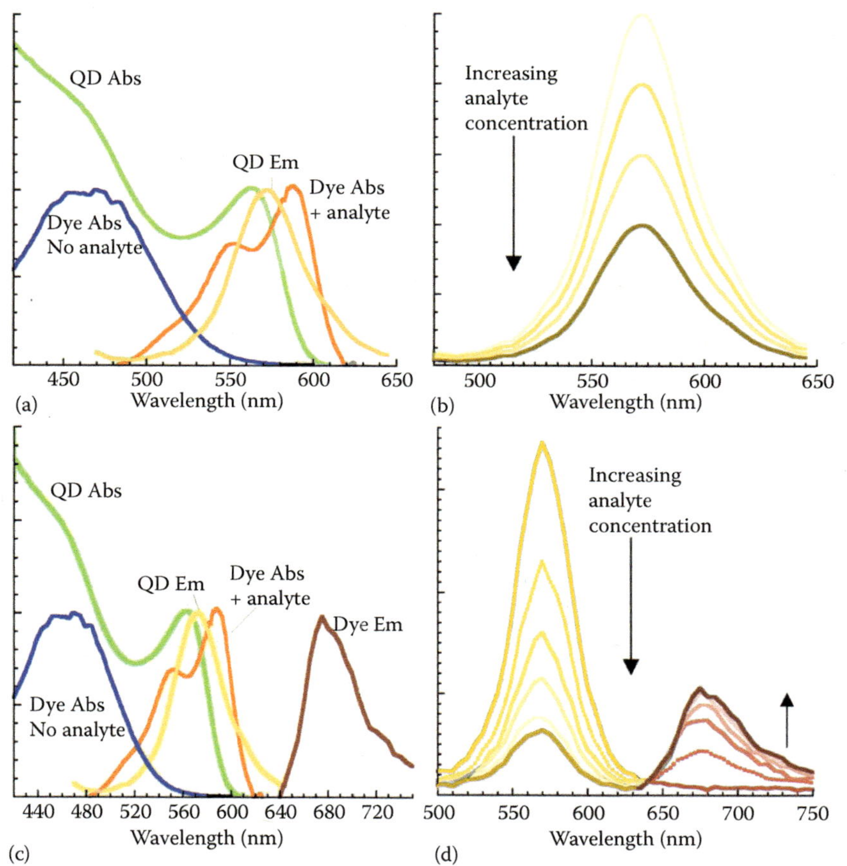

to different parts of a protein. When the molecules bind or the protein changes conformation, the QD and acceptor get close enough to demonstrate FRET (**Figure 11.20**). An important recent discovery is that the QD–dye energy transfer interaction is not a dipole–dipole interaction, but a dipole–surface interaction. This type of interaction falls off as $r^{-4}$. This means that FRET using QDs can be seen over longer distances than is possible with two dye molecules.

Principles similar to FRET, chemiluminescence resonance energy transfer (CRET) or bioluminescence resonance energy transfer (BRET), can be used to create "self-illuminating" QDs that turn on in the presence of a specific chemical signal rather than by light excitation. In this case, the QDs act as the energy acceptors; the donor is a molecule that generates luminescence in response to a biological or chemical signal. Because of their broad absorbance spectra, QDs can act as acceptors for essentially any luminescence that emits 10 nm or more to the blue of their own emission. The most well-known implementation of BRET uses the widely known *Renilla luciferase*, which emits bioluminescence at 480 nm when activated with the substrate coelenterazine. A QD with approximately six luciferase molecules attached will respond to coelenterazine by emitting its typical QD spectrum, which can be over 100 nm redder than that of luciferase (**Figure 11.21**).

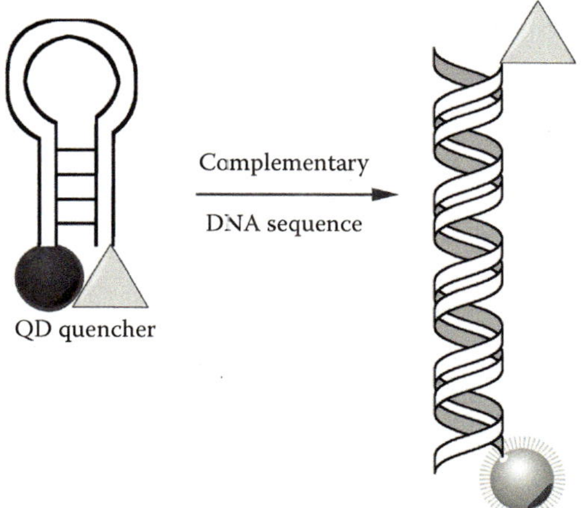


**Figure 11.20 The use of a QD as a FRET-based *molecular beacon*.** A loop sequence is made that is complementary to a sequence to be sensed. To this is added self-complementary ends that will cause the DNA to form a secondary structure with the ends close to each other. A QD and a FRET quencher are put on each end, so in the single-stranded form, there is no fluorescence. When a complementary sequence binds and creates a double helix, the QD is separated from its quencher, and fluorescence is seen.

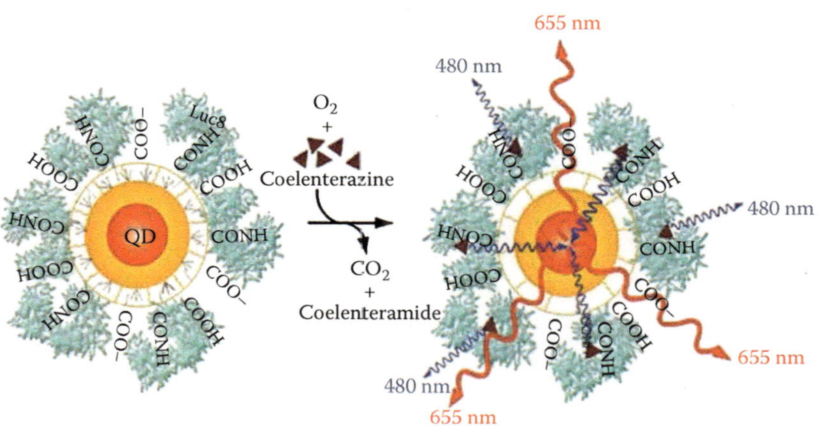


**Figure 11.21 Schematic of a QD coupled to a BRET donor, luciferase (Luc8).** Coelenterazine is oxidized by luciferase, resulting in luminescence at 480 nm. Energy from the luminescence is transferred to the QDs, resulting in QD emission, in this case red QDs emitting at 655 nm. (Reprinted by permission from Macmillan Publishers Ltd., *Nat. Biotechnology*, So, B. M. et al., Self-Illuminating Quantum Dot Conjugates for *in Vivo* Imaging, 24, 339–343, copyright 2006.)

Other processes besides FRET are possible and potentially even more useful. Direct electron transfer from or to the QD will turn the particle off, since it eliminates the electron–hole pair needed for radiative recombination. Not only does each QD have a specific bandgap energy, but also, the energetic positions of the conduction and valence bands vary according to the material. This determines which molecules each QD is able to donate electrons to or accept electrons from. The band edge energies vary with solution composition and pH as well as particle size; approximate values for bulk semiconductors in aqueous solution at neutral pH are shown in **Figure 11.22a**. It is important to remember that as the bandgap widens as the particle shrinks, the band edges shift in proportion to the reduced masses of the electron and hole. The electron, which is much lighter, therefore shifts significantly more. This is shown for CdSe QDs of different sizes in **Figure 11.22b**.

Recall that to calculate the shifts due to decreasing particle size, the electron (conduction band) and hole (valence band) energies will shift in proportion to their effective masses (**Equation 11.1**). The effective mass values are determined by the material and given in standard solid-state physics textbooks; some particularly useful ones are provided in **Table 11.1**. Note, for example, that the hole effective mass in CdSe is nearly 3.5 times that of the electron.

Electron transfer is even more distance dependent than FRET, showing exponential falloff with distance. This has important practical applications: a 1 Å motion of an attached electron donor will lead to a 30% fluorescence change, whereas the same motion of a FRET donor would lead to only a 1% change. However, transfer of electrons from the QD (oxidation) in aqueous environments will degrade the particle and irreversibly quench it, whereas transfer of electrons to the QD (reduction) is a reversible process. Thus, sensors made with electron donor-QD conjugates are possible. The neurotransmitter dopamine is an electron donor, and QD–dopamine conjugates are sensitive to redox potential, becoming brighter in oxidizing conditions when the dopamine is oxidized to a nondonor form. Other electron donors have been attached to QDs and displaced by target molecules to create a signal; this has yielded sensors for maltose, thrombin, and fatty acids. The full potential of FRET and electron transfer for creation of QD sensors has yet to be realized.

**Figure 11.22 Bandgap energies and energetic positions of the band edges (vs. normal hydrogen electrode [NHE]) in aqueous solution for commonly used semiconductor materials**. (a) Values for bulk materials. The positions of the band edges determine the ability of the electron to reduce or the hole to oxidize other molecules. Some common redox couples are shown on the same scale for comparison. (b) Nanoparticles show corresponding energy shifts according to **Equation 11.3**. Shown are energy levels for a red QD (emission peak, 620 nm), a green QD (560 nm), and a blue QD (520 nm).

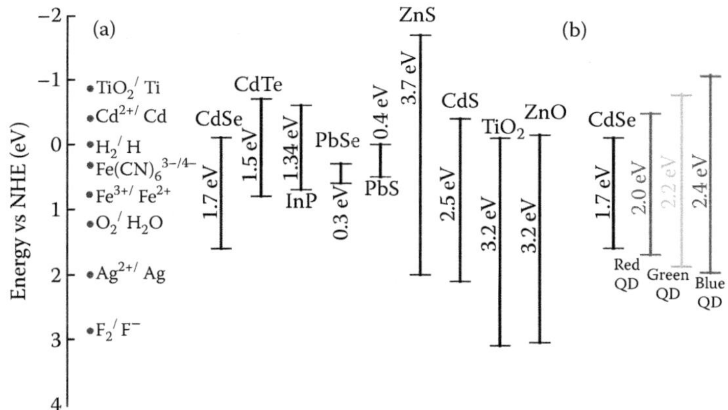

## 11.4 EXAMPLE EXPERIMENT: CONJUGATION OF QDs TO DOPAMINE AND QUANTIFYING THE EFFECTS ON FLUORESCENCE PER MOLECULE BOUND

This example will show how to characterize QD fluorescence before and after conjugation to biomolecules, and introduces one of our favorite protocols for determining the number of biomolecules bound per QD. Dopamine (DA) is a small molecule (molecular weight [MW], 1811.64), so up to a couple of hundred dopamines may fit per QD.

The number of ligands per QD is quantified using an assay based upon the dye o-phthaldialdehyde (OPA). OPA in the presence of beta-mercaptoethanol (βME) reacts rapidly with primary amino groups in order to form highly fluorescent products. The fluorescence emission of OPA is in the blue region (ca. 460 nm) and sufficiently far away from the emission of nearly all QDs in this study to avoid interference. Because DA contains a primary amine, it causes OPA to react and fluoresce with an intensity that is linear in emission intensity over a wide range (**Figure 11.23**).

Thus, in order to estimate the number of DA molecules that bind to the QDs, OPA is reacted with the coupling reaction in the presence and absence of EDC. There should be a reduction in the intensity of the OPA signal in the presence of EDC resulting from the conversion of the DA amino groups to an amide bond with the carboxylate group on the capping molecule and the amount by which the OPA signal differs from the EDC-negative samples can be used to calculate the number of DA ligands that have bound to the surface of the QDs.

The procedure is as follows:

- Begin with carboxylate-functionalized QDs (**Practical Tips 11.2**). Make sure that excess capping molecules (e.g., mercaptoacetic acid [MAA]) are removed by dialysis or filtration before doing the conjugation, or

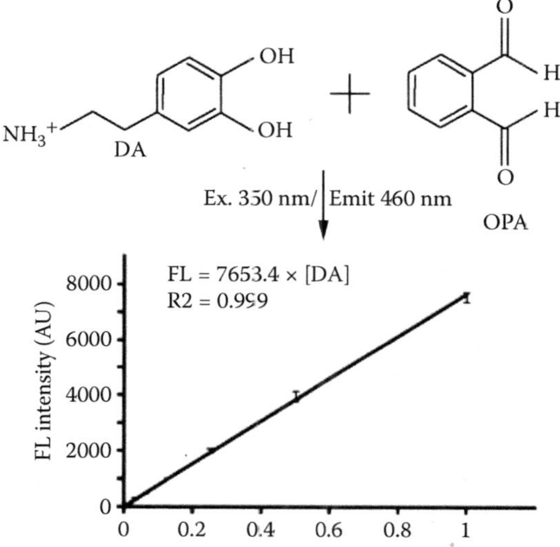


**Figure 11.23 The fluorescent emission of OPA increases linearly with the amount of DA in solution.** Data are an average of four experiments, with error bars indicating the standard error of the mean. The resulting linear fit can be used to determine free dopamine concentrations in solution.

the conjugate will stick to these molecules instead of to the QDs. For small volumes, we like plastic "dialysis cassettes" that contain a small window of dialysis membrane.

- Prepare stock solutions of OPA by dissolving 1 mg/mL of OPA in phosphate buffer and adding 2 mL/mL βME.

- Prepare a standard curve of OPA fluorescence versus dopamine concentration as in **Figure 11.23**.

- In multiple tubes or a multiwell plate, prepare samples containing 1 µM QDs. Add the same amount of dopamine to all tubes. (We find that a 5000:1 ratio works well.)

- Prepare a fresh EDC solution and add different concentrations to the different tubes to span the range of 0 EDC per QD to 16,000 EDC per QD.

- After the desired amount of time (which can range from immediately to 1 h), remove 25 µL of each reaction and add to 250 µL of the OPA stock solution. Let incubate for 1 min and read the fluorescence with excitation/emission at 350/460 nm (**Figure 11.24**).

- Calculate the number of dopamines bound per QD based upon the OPA emission relative to the EDC-free emission based upon the standard curve (**Figure 11.25**).

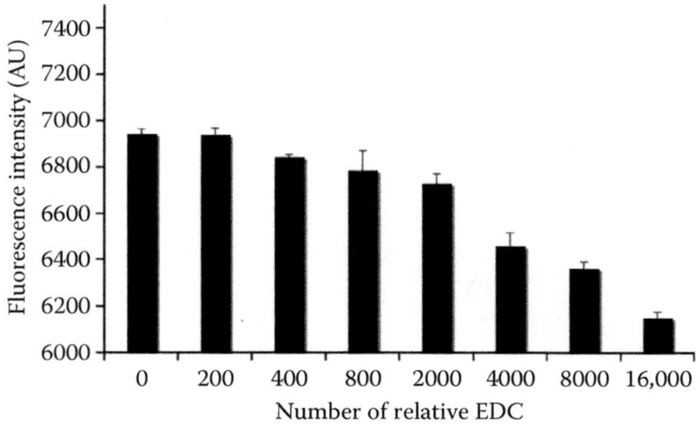

**Figure 11.24 Reaction of red QDs with OPA.** For a fixed ratio of QD to DA (1:2500), increasing levels of EDC lead to a reduction in OPA intensity due to covalent binding of the DA to the surface of the QDs.

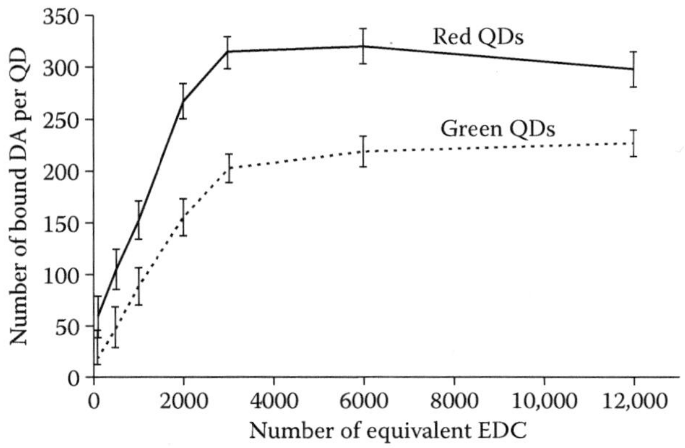

**Figure 11.25 Number of bound DA molecules as a function of the number of equivalent EDC molecules added to the coupling reaction for red and green QDs as measured by OPA assay.** Data are an average of four experiments, with error bars indicating the standard error of the mean.

- Multiple time points can be used to examine the reaction kinetics, if desired. We find that the EDC coupling reaction is essentially complete by 20 min (**Figure 11.26**).

- Clean the coupled QDs by passing through a spin column.

- The QD fluorescence as a function of the number of bound QDs can now be measured by exciting and emitting at the usual QD wavelengths (**Figure 11.27**).

The number of bound dopamines can be compared with the theoretical limiting number of solubilizing ligands. If the QD is modeled as a sphere using the TEM sizing parameters, the surface area of green QDs and red QDs are 32.35 nm$^2$ and 611.77 nm$^2$, respectively. Since the sulfur group on a short-chain thiol has a molecular area of 0.025 nm$^2$, a green QD has a theoretical capacity to bind approximately 647 MPA molecules, and a red QD has a capacity for 1375 MPA molecules on the surface of the particle through Zn–S bonds. The results in **Figure 11.25** suggest that the number of available MPA molecules on the surface of the QDs is considerably less than what the physical size permits, probably due to a combination of steric hindrance and surface defects.

The emission spectra shown in **Figure 11.27** can be fit to models of quenching. It can be seen that they are close to linear and so are a good fit to the Stern–Volmer model. This will be discussed in **Chapter 16**.

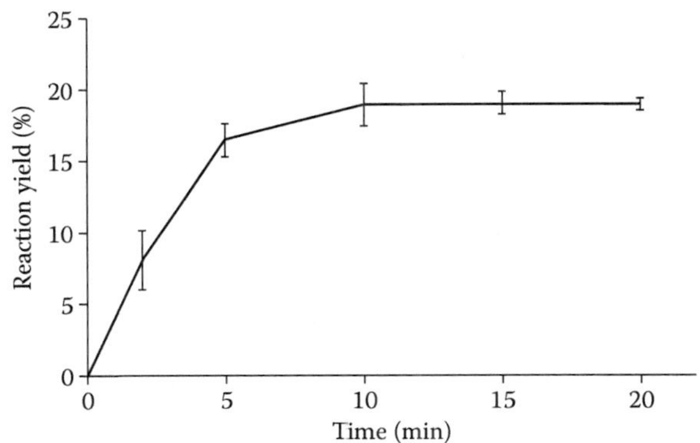

Figure 11.26  **Timescale of the EDC coupling reaction as measured by OPA assay.**

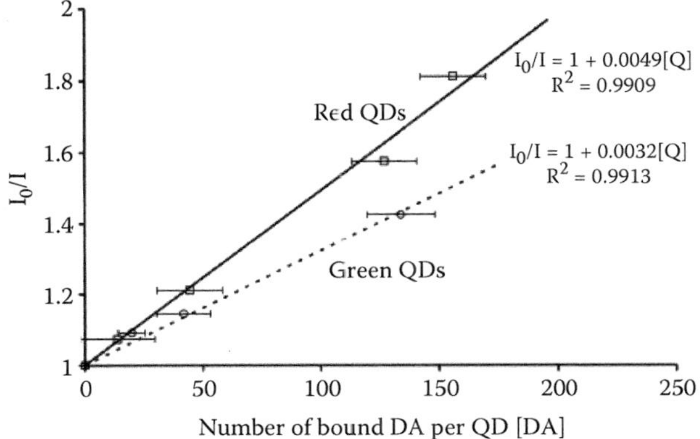


Figure 11.27  **Relationship between the number of DA particles bound to the surface of the QD and the resulting emission intensity $I$ relative to the intensity of unconjugated QDs $I_0$.** For both red and green QDs, the relationship can be fitted to the linear equations depicted on the graph. Data are an average of four experiments, with error bars indicating the standard error of the mean.

# 11.5 SUMMARY AND REMARKS

The use of fluorescent nanoparticles for biological targeting, imaging, and sensing is still in its relative infancy. Many of the techniques described here using QDs were originally developed using dyes, fluorescent beads, or other markers. Many experiments lend themselves to the use of QDs, especially those in which long-lasting signals or single-molecule measurements are needed. For other experiments, dyes might be easier and preferable. Many techniques used with dyes have not yet been adapted to the use of QDs. In some cases, this is due to difficulties with QDs (for example, fluorescence *in situ* hybridization [FISH] has posed significant problems), but in many other cases, the field is open to experimentation. Searching older literature for protocols using microbeads, microspheres, or targeted dyes can provide ideas for labeling techniques that can be readily adapted for QDs.

## End-of-Chapter Questions

**1.** Approximate how many atoms are in a 3-nm-diameter CdSe QD. How would you express a concentration of 1 mg/mL CdSe as a "molarity" of QDs?

**2.** In a 3 nm QD, what fraction of the atoms is on the surface?

**3.** How is it determined that the QD lattice structure is wurtzite? What other structure is possible? Could the two coexist?

**4.** What is the expected position of the first exciton peak for a CdSe QD 3 nm in diameter? What emission wavelength (approximately) will this correspond to? What if the dot is 6 nm in diameter?

**5.** Repeat the calculations above for a CdTe QD 3 nm in diameter.

**6.** If the emission spectrum in no. 4 is ± 10 nm, what is the size variation of the particles?

**7.** A laser shines a photon of energy $E$ onto a QD of diameter $D$ and effective well depth $V$. What minimum diameter must the QD have to ensure that the excited electron remains confined?

**8.** What are trap states? Why are they a problem, and how are they usually dealt with? Do you expect to see emission from trap states?

**9.** Using atomic absorption spectroscopy (AAS), it is possible to determine the elemental composition of the particles; heavier ions are easier to determine (e.g., Cd, Zn). Given a total mass of particles $m$, mass of Cd $m_C$, and mass of Zn $m_Z$, derive an expression for the number of shells on a 3 nm core CdSe/ZnS QD. Each single shell has a thickness of 0.31 nm.

**10.** Calculate the density of states per unit volume at a given energy for (a) bulk 3-D materials; (b) 2-D materials (*quantum well*; one $k$-space component fixed); (c) 1-D materials (*quantum wire*); and (d) 0-D materials (QDs).

**11.** Compare the size dependence of the energies for a quantum well, wire, and dot.

**12.** Derive the dispersion relation for an electron in the conduction band of a semiconductor by Taylor expansion. What is the effective mass? What key parameter determines its value?

**13.** Discuss how the distance dependence of energy transfer between and dye and a nanoparticle can be quantified.

**14.** Solubilization with alkanethiols such as beta-mercaptoethanol will reduce quantum yield of CdSe but increase quantum yield of CdTe. Explain.

**15.** A molecular beacon is made with a QD on one end and fluorescein on the other. Pick the best color of QD for this purpose. Assume that energy transfer obeys an $r^{-6}$ dependence, with $R_0 = 0.57$ nm. In the hairpin configuration (left panel of **Figure 11.20**), the dye and QD are 0.2 nm apart. If each DNA base is 0.3 nm apart, plot the expected fluorescence as a function of the beacon length.

**16.** Blinking statistics in QDs are said to follow *power-law statistics*. What does this mean, and what are some of the difficulties involved in confirming this experimentally?

**17.** Discuss ways to reduce QD blinking, both chemically and by altering the structure of the QD.

**18.** So-called Type II QDs have both the valence and conduction bands in the core lower (or higher) than in the shell. As a result, one carrier is mostly confined to the core, while the other is mostly confined to the shell. Discuss how this alters the photoluminescence properties and what these QDs might be good for.

**19.** Other types of photoluminescent nanoparticles, which emit light by different mechanisms than QDs, are also useful in biology. One example is lanthanide-doped upconversion nanoparticles. Discuss how these work and their advantages and disadvantages.

**20.** A label for cancer cells was created by binding QDs to the fluorescent anticancer drug doxorubicin (Dox). The absorbance and emission spectra of Dox are shown below. What color QD should be chosen to create an effective FRET pairing? What should the wavelength of excitation be? As the particles are taken up by cells, how would you expect the emission to change?

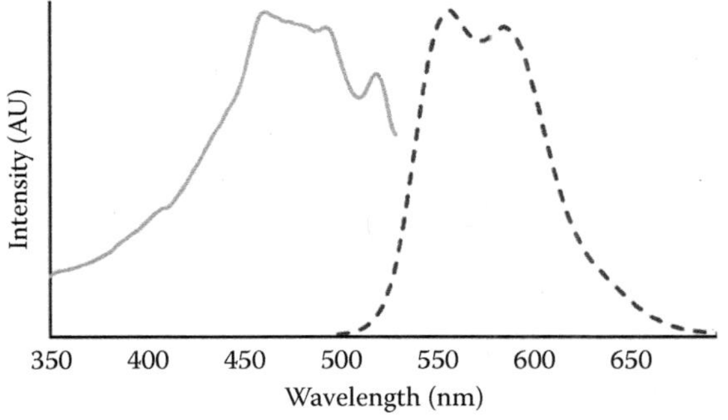

**21.** Describe how a QD sensor can be made to detect a single-point mutation in DNA.

**22.** QDs have been designed to emit white light. How does this work?

**23.** QDs have been proposed as a sensor of electrical potential based upon the quantum-confined Stark effect (QCSE). Discuss this effect and its magnitude.

**24.** QDs have been proposed as a sensor of electrical potential across cell membranes by an energy or electron transfer mechanism. Discuss how this could work and its challenges.

**25.** QDs are often used to transfer electrons to $TiO_2$ nanoparticles to improve visible light collection for solar cells. Draw the positions of the band edges of the two materials and discuss how this works.

**26.** CdTe is a relatively soft material with both a large deformation potential and a large difference in lattice constant relative to capping materials (ZnS, CdS). The strain created by making CdTe/ZnS and CdTe/ZnS QDs can be used to tune the bandgap and obtain a wide range of emission colors. Explain.

## Background Reading

### Books

It is important to distinguish between lithographically patterned QDs and colloidal nanocrystal QDs when browsing for books. More books have been written on the former. I only include the latter here.


Fontes, A., and Santos, B.S. *Quantum Dots: Applications in Biology*. Humana Press, Totowa, NJ, 2014.

Klimov, V. *Nanocrystal Quantum Dots*. CRC Press, 2010.

Pinaud, F. *Peptide-coated Quantum Dots: Applications to Biological Imaging of Single Molecules in Live Cells and Organisms*. VDM Verlag Dr. Müller, 2008.

Rogach, A. *Semiconductor Nanocrystal Quantum Dots: Synthesis, Assembly, Spectroscopy and Applications*. Springer-Verlag, New York, 2008.

## Journal articles

Many thousands of articles have been published on QDs. The following list is by no means exhaustive but is designed to provide a few classic papers on QD synthesis and basic physics, some useful methods and review articles, and a sampling of biological applications. For brevity, we have focused on CdSe, CdTe, and InP; literature on other materials can be found by searching on similar key words.

### *Synthesis*

Al-Salim, N., Young, A.G., Tilley, R.D., McQuillan, A.J., and Xia, J. (2007). Synthesis of CdSeS nanocrystals in coordinating and noncoordinating solvents: Solvent's role in evolution of the optical and structural properties. *Chemistry of Materials* 19, 5185–5193.

Bae, W.K., Nam, M.K., Char, K., and Lee, S. (2008). Gram-scale one-pot synthesis of highly luminescent blue emitting Cd1-xZnxS/ZnS nanocrystals. *Chemistry of Materials* 20, 5307–5313.

Chang, J.Y., Wang, S.R., and Yang, C.H. (2007). Synthesis and characterization of CdTe/CdS and CdTe/CdSe core/shell type-II quantum dots in a noncoordinating solvent. *Nanotechnology* 18, 345602.

Ivanov, S.A., Piryatinski, A., Nanda, J., Tretiak, S., Zavadil, K.R., Wallace, W.O., Werder, D., and Klimov, V.I. (2007). Type-II core/shell CdS/ZnSe nanocrystals: Synthesis, electronic structures, and spectroscopic properties. *Journal of the American Chemical Society* 129, 11708–11719.

Li, L., and Reiss, P. (2008). One-pot synthesis of highly luminescent InP/ZnS nanocrystals without precursor injection. *Journal of the American Chemical Society* 130, 11588.

Liu, Z.P., Kumbhar, A., Xu, D., Zhang, J., Sun, Z.Y., and Fang, J.Y. (2008). Coreduction colloidal synthesis of III–V nanocrystals: The case of InP. *Angewandte Chemie—International Edition* 47, 3540–3542.

Narayanaswamy, A., Feiner, L.F., Meijerink, A., and van der Zaag, P.J. (2009). The effect of temperature and dot size on the spectral properties of colloidal InP/ZnS core–shell quantum dots. *ACS Nano* 3, 2539–2546.

Narayanaswamy, A., Feiner, L.F., and van der Zaag, P.J. (2008). Temperature dependence of the photoluminescence of InP/ZnS quantum dots. *Journal of Physical Chemistry C* 112, 6775–6780.

Park, J., Lee, K.H., Galloway, J.F., and Searson, P.C. (2008). Synthesis of cadmium selenide quantum dots from a non-coordinating solvent: Growth kinetics and particle size distribution. *Journal of Physical Chemistry C* 112, 17849–17854.

Washington, A.L., and Strouse, G.F. (2008). Microwave synthesis of CdSe and CdTe nanocrystals in nonabsorbing alkanes. *Journal of the American Chemical Society* 130, 8916–8922.

Xie, R., Battaglia, D., and Peng, X. (2007). Colloidal InP nanocrystals as efficient emitters covering blue to near-infrared. *Journal of the American Chemical Society* 129, 15432.

Xu, S., Ziegler, J., and Nann, T. (2008). Rapid synthesis of highly luminescent InP and InP/ZnS nanocrystals. *Journal of Materials Chemistry* 18, 2653–2656.

### *Physics*

Aldana, J., Wang, Y.A., and Peng, X.G. (2001). Photochemical instability of CdSe nanocrystals coated by hydrophilic thiols. *Journal of the American Chemical Society* 123, 8844–8850.

Anderson, N.A., and Lian, T.Q. Ultrafast electron transfer at the molecule–semiconductor nanoparticle interface. In *Annual Review of Physical Chemistry*, pp. 491–519, 2005.

Boles, M.A., Ling, D., Hyeon, T., and Talapin, D.V. (2016). The surface science of nanocrystals. *Nature Materials* 15, 141–153.

Boulineau, R.L., and Osborne, M.A. (2013). Direct object resolution by image subtraction: A new molecular ruler for nanometric measurements on complexed fluorophores. *Chemical Communications* 49, 5559–5561.

Bruhn, B., Qejvanaj, F., Sychugov, I., and Linnros, J. (2014). Blinking statistics and excitation-dependent luminescence yield in Si and CdSe nanocrystals. *Journal of Physical Chemistry C* 118, 2202–2208.

Burda, C., Green, T.C., Link, S., and El-Sayed, M.A. (1999). Electron shuttling across the interface of CdSe nanoparticles monitored by femtosecond laser spectroscopy. *Journal of Physical Chemistry B* 103, 1783–1788.

Ding, T.X., Olshansky, J.H., Leone, S.R., and Alivisatos, A.P. (2015). Efficiency of hole transfer from photoexcited quantum dots to covalently linked molecular species. *Journal of the American Chemical Society* 137, 2021–2029.

Empedocles, S.A., and Bawendi, M.G. (1997). Quantum-confined stark effect in single CdSe nanocrystallite quantum dots. *Science* 278, 2114–2117.

Empedocles, S.A., Norris, D.J., and Bawendi, M.G. (1996). Photoluminescence spectroscopy of single CdSe nanocrystallite quantum dots. *Physical Review Letters* 77, 3873–3876.

Fernee, M.J., Tamarat, P., and Lounis, B. (2014). Spectroscopy of single nanocrystals. *Chemical Society Reviews* 43, 1311–1337.

Fomenko, V., and Nesbitt, D.J. (2008). Solution control of radiative and nonradiative lifetimes: A novel contribution to quantum dot blinking suppression. *Nano Letters* 8, 287–293.

Hagfeldt, A., and Gratzel, M. (1995). Light-induced redox reactions in nanocrystalline systems. *Chemical Reviews* 95, 49–68.

Houel, J., Doan, Q.T., Cajgfinger, T., Ledoux, G., Amans, D., Aubret, A., Dominjon, A., Ferriol, S., Barbier, R., Nasilowski, M. et al. (2015). Autocorrelation analysis for the unbiased determination of power-law exponents in single-quantum-dot blinking. *ACS Nano* 9, 886–893.

Javier, A., Magana, D., Jennings, T., and Strouse, G.F. (2003). Nanosecond exciton recombination dynamics in colloidal CdSe quantum dots under ambient conditions. *Applied Physics Letters* 83, 1423–1425.

Ji, B., Giovanelli, E., Habert, B., Spinicelli, P., Nasilowski, M., Xu, X., Lequeux, N., Hugonin, J.-P., Marquier, F., Greffet, J.-J. et al. (2015). Non-blinking quantum dot with a plasmonic nanoshell resonator. *Nature Nanotechnology* 10, 170–175.

Klimov, V.I., Ivanov, S.A., Nanda, J., Achermann, M., Bezel, I., McGuire, J.A., and Piryatinski, A. (2007). Single-exciton optical gain in semiconductor nanocrystals. *Nature*, 447 441–446.

Klimov, V.I., Mikhailovsky, A.A., Xu, S., Malko, A., Hollingsworth, J.A., Leatherdale, C.A., Eisler, H.J., and Bawendi, M.G. (2000). Optical gain and stimulated emission in nanocrystal quantum dots. *Science* 290, 314–317.

Kuno, M., Fromm, D.P., Hamann, H.F., Gallagher, A., and Nesbitt, D.J. (2000). Nonexponential "blinking" kinetics of single CdSe quantum dots: A universal power law behavior. *Journal of Chemical Physics* 112, 3117–3120.

Maity, P., Debnath, T., and Ghosh H.N. (2015). Ultrafast charge carrier delocalization in CdSe/CdS quasi-type II and CdS/CdSe inverted type I core–shell: A structural analysis through carrier-quenching study. *Journal of Physical Chemistry C* 119, 26202–26211.

Mondal, N., and Samanta, A. (2016). Ultrafast charge transfer and trapping dynamics in a colloidal mixture of similarly charged CdTe quantum dots and silver nanoparticles. *Journal of Physical Chemistry C* 120, 650–658.

Mongin, C., Garakyaraghi, S., Razgoniaeva, N., Zamkov, M., and Castellano, F.N. (2016). Direct observation of triplet energy transfer from semiconductor nanocrystals. *Science* 351, 369–372.

Nadeau, J.L., Carlini, L., Suffern, D., Ivanova, O., and Bradforth, S.E. (2012). Effects of beta-mercaptoethanol on quantum dot emission evaluated from photoluminescence decays. *Journal of Physical Chemistry C* 116, 2728–2739.

Osad'ko, I.S., Eremchev, I.Y., and Naumov, A.V. (2015). Two mechanisms of fluorescence intermittency in single core/shell quantum dot. *Journal of Physical Chemistry C* 119, 22646–22652.

Petryayeva, E., Algar, W.R., and Medintz, I.L. (2013). Quantum dots in bioanalysis: A review of applications across various platforms for fluorescence spectroscopy and imaging. *Applied Spectroscopy* 67, 215–252.

Rogez, B., Yan, H., Le Moal, E., Leveque-Fort, S., Boer-Duchemin, E., Yao, F., Lee, Y.-H., Zhang, Y., Wegner, K.D., Hildebrandt, N. et al. (2014). Fluorescence lifetime and blinking of individual semiconductor nanocrystals on graphene. *Journal of Physical Chemistry C* 118, 18445–18452.

Schmidt, R., Krasselt, C., Goehler, C., and von Borczyskowski, C. (2014). The fluorescence intermittency for quantum dots is not power-law distributed: A luminescence intensity resolved approach. *ACS Nano* 8, 3506–3521.

Smith, A.M., and Nie, S. (2010). Semiconductor nanocrystals: Structure, properties, and band gap engineering. *Accounts of Chemical Research* 43, 190–200.

Stewart, M.H., Huston, A.L., Scott, A.M., Oh, E., Algar, W.R., Deschamps, J.R., Susumu, K., Jain, V., Prasuhn, D.E., Blanco-Canosa, J. et al. (2013). Competition between Forster resonance energy transfer and electron transfer in stoichiometrically assembled semiconductor quantum dot–fullerene conjugates. *ACS Nano* 7, 9489–9505.

Weinberg, D.J., He, C., and Weiss, E.A. (2016). Control of the redox activity of quantum dots through introduction of fluoroalkanethiolates into their ligand shells. *Journal of the American Chemical Society* 138, 2319–2326.

Yun, C.S., Javier, A., Jennings, T., Fisher, M., Hira, S., Peterson, S., Hopkins, B., Reich, N.O., and Strouse, G.F. (2005). Nanometal surface energy transfer in optical rulers, breaking the FRET barrier. *Journal of the American Chemical Society* 127, 3115–3119.

Zenkevich, E., Stupak, A., Goehler, C., Krasselt, C., and von Borczyskowski, C. (2015). Tuning electronic states of a CdSe/ZnS quantum dot by only one functional dye molecule. *ACS Nano* 9, 2886–2903.

## Applications

Akerman, M.E., Chan, W.C., Laakkonen, P., Bhatia, S.N., and Ruoslahti, E. (2002). Nanocrystal targeting in vivo. *Proceedings of the National Academy of Sciences of the United States of America* 99, 12617–12621.

Breus, V.V., Heyes, C.D., Tron, K., and Nienhaus, G.U. (2009). Zwitterionic biocompatible quantum dots for wide pH stability and weak nonspecific binding to cells. *ACS Nano* 3, 2573–2580.

Bruchez, M., Jr., Moronne, M., Gin, P., Weiss, S., and Alivisatos, A.P. (1998). Semiconductor nanocrystals as fluorescent biological labels. *Science* 281, 2013–2016.

Chan, W.C.W., and Nie, S.M. (1998). Quantum dot bioconjugates for ultrasensitive nonisotopic detection. *Science* 281, 2016–2018.

Choi, H.S., Liu, W., Misra, P., Tanaka, E., Zimmer, J.P., Itty Ipe, B., Bawendi, M.G., and Frangioni, J.V. (2007). Renal clearance of quantum dots. *Nature Biotechnology* 25, 1165–1170.

Dif, A., Boulmedais, F., Pinot, M., Roullier, V., Baudy-Floc'h, M., Coquelle, F.M., Clarke, S., Neveu, P., Vignaux, F., Le Borgne, R. et al. (2009). Small and stable peptidic PEGylated quantum dots to target polyhistidine-tagged proteins with controlled stoichiometry. *Journal of the American Chemical Society* 131, 14738–14746.

Dif, A., Henry, E., Artzner, F., Baudy-Floc'h, M., Schmutz, M., Dahan, M., and Marchi-Artzner, V. (2008). Interaction between water-soluble peptidic CdSe/ZnS nanocrystals and membranes: Formation of hybrid vesicles and condensed lamellar phases. *Journal of the American Chemical Society* 130, 8289–8296.

Dubertret, B., Skourides, P., Norris, D.J., Noireaux, V., Brivanlou, A.H., and Libchaber, A. (2002). In vivo imaging of quantum dots encapsulated in phospholipid micelles. *Science* 298, 1759–1762.

Gao, X., Cui, Y., Levenson, R.M., Chung, L.W., and Nie, S. (2004). In vivo cancer targeting and imaging with semiconductor quantum dots. *Nature Biotechnology* 22, 969–976.

Hanaki, K., Momo, A., Oku, T., Komoto, A., Maenosono, S., Yamaguchi, Y., and Yamamoto, K. (2003). Semiconductor quantum dot/albumin complex is a long-life and highly photostable endosome marker. *Biochemical and Biophysical Research Communications* 302, 496–501.

Hashemian, Z., Khayamian, T., Saraji, M., and Shirani, M.P. (2016). Aptasensor based on fluorescence resonance energy transfer for the analysis of adenosine in urine samples of lung cancer patients. *Biosensors and Bioelectronics* 79, 334–340.

Keller, A.M., Ghosh, Y., DeVore, M.S., Phipps, M.E., Stewart, M.H., Wilson, B.S., Lidke, D.S., Hollingsworth, J.A., and Werner, J.H. (2014). 3-dimensional tracking of non-blinking 'giant' quantum dots in live cells. *Advanced Functional Materials* 24, 4796–4803.

Kim, S., and Bawendi, M.G. (2003). Oligomeric Ligands for luminescent and stable nanocrystal quantum dots. *Journal of the American Chemical Society* 125, 14652–14653.

Kovtun, O., Sakrikar, D., Tonalinson, I.D., Chang, J.C., Arzeta-Ferrer, X., Blakely, R.D., and Rosenthal, S.J. (2015). Single-quantum-dot tracking reveals altered membrane dynamics of an attention-deficit/hyperactivity-disorder-derived dopamine transporter coding variant. *ACS Chemical Neuroscience* 6, 526–534.

Lee, J.Y., Kim, J.S., Park, J.C., and Nam, Y.S. (2016) Protein–quantum dot nanohybrids for bioanalytical applications. *Wiley Interdisciplinary Reviews—Nanomedicine and Nanobiotechnology* 8, 178–190.

Ma, L., Tu, C., Phuong, L., Chitoor, S., Lim, S.J., Zahid, M.U., Teng, K.W., Ge, P., Selvin, P.R., and Smith, A.M. (2016). Multidentate polymer coatings for compact and homogeneous quantum dots with efficient bioconjugation. *Journal of the American Chemical Society* 138, 3382–3394.

Mattoussi, H., Mauro, J.M., Goldman, E.R., Anderson, G.P., Sundar, V.C., Mikulec, F.V., and Bawendi, M.G. (2000). Self-assembly of CdSe–ZnS quantum dot bioconjugates using an engineered recombinant protein. *Journal of the American Chemical Society* 122, 12142–12150.

Petryayeva, E., Algar, W.R., and Medintz, I.L. (2013). Quantum dots in bioanalysis: A review of applications across various platforms for fluorescence spectroscopy and imaging. *Applied Spectroscopy* 67, 215–252.

Pinaud, F., King, D., Moore, H.P., and Weiss, S. (2004). Bioactivation and cell targeting of semiconductor CdSe/ZnS nanocrystals with phytochelatin-related peptides. *Journal of the American Chemical Society* 126, 6115–6123.

Slocik, J.M., Moore, J.T., and Wright, D.W. (2002). Monoclonal antibody recognition of histidine-rich peptide encapsulated nanoclusters. *Nano Letters* 2, 169–173.

## Equipment and supplies

*Some major suppliers*

See **Table 11.4** for suppliers of quantum dots and other nanoparticles. A key source for most of the chemicals used in synthesis is Sigma-Aldrich. The following suppliers supply reagents for synthesis, solubilization, and biofunctionalization of QDs that are not readily available from the primary chemical companies.

Peprotech. Custom peptides. (Also available through many universities; your mileage may vary.)

Nanocs. PEGylation reagents; conjugation reagents; conjugation and PEGylation services.

Creative PEGWorks. All sorts of PEG thiols, including branched and custom PEGs and PEGylation services.

Nanoscience Instruments. Many different molecules for surface functionalization.

Chemglass. Glassware for QD synthesis, including round-bottom flasks, three-neck flasks, etc.

Safety Emporium. Vacuum and gas manifolds.

Welch Vacuum. Vacuum manifolds and accessories—pumps, traps, etc.

Harvard Apparatus. Spin filters that work well with QDs. Many of the small filters do not work and cause aggregation. We swear by their Micro SpinColumns!

Pierce. Dialysis filters and other supplies.

# CHAPTER 12

# Gold Nanoparticles

Edward S. Allgeyer, Gary Craig, Sanjeev Kumar Kandpal,
Jeremy Grant, and Michael D. Mason

## 12.1 INTRODUCTION

Unlike semiconductor-based nanoparticles (quantum dots; Chapter 11), noble
metal particles are not a recent innovation. They have been used as pigments in
exotic paints and glasses for centuries, although the physics behind their optical
properties have only recently become well understood. The first solutions to
Maxwell's equations describing the interaction between light and small spherical
particles were derived by Gustav Mie roughly a century ago, and later refined
by Gans and Happel. Like many nanomaterials, as a noble metal nanoparticle's
physical dimension is reduced to sizes below 100 nm, the optical and electronic
properties can deviate dramatically from those of the bulk material. However,
it was not until recently that dramatically improved analytical methods made it
possible to directly correlate growth conditions and chemistries with nanoparticle
size and geometry. Similarly, the emergence of surface science as a prominent
field over the past decades has brought new attention to these already existing but
somewhat overlooked materials. This resurgence of research in the area of noble
metal nanoparticles has led to countless new materials and applications.

This chapter introduces one of the most widely implemented noble metal
nanoparticles, gold, including a brief description of the relationship between
particle size, morphology, and photophysical properties, as well as suggestions
as to how to use them in varying light microscope conformations. Although gold
nanoparticles are also widely used in electron microscopy, we do not cover this
here, but instead provide some references to this well-established field. We then
discuss methods for gold nanoparticle synthesis, surface functionalization, and
several of the best developed optical imaging techniques that make use of these
nanoparticles. Finally, we introduce some emerging imaging modalities and
potential applications that take advantage of the more unusual properties of gold
nanoparticles.

Gold particles are probably the simplest nanomaterials to synthesize. At the end
of this chapter, you should be prepared to undertake particle synthesis yourself,
as well as straightforward biofunctionalization. It is the imaging side of gold
that is trickier than with quantum dots, and we will give a background that will
hopefully help you choose your imaging methods according to the available
instrumentation and the needs of your particular experiment.

## 12.2 THE PHYSICS OF SCATTERING AND SPHERICAL METAL NANOPARTICLES

### General theory for all particles

Investigations of the interactions of light with matter have been ongoing for over three centuries, with major developments (quantum electrodynamics) as recently as the late 1940s. The interaction of light with metal nanoparticles is known to have been under investigation as early as the 1850s by Michael Faraday. As mentioned, German physicist Gustav Mie was the first to derive the extinction spectra (extinction is scattering plus absorption) of spherical particles of an arbitrary size from Maxwell's equations at the beginning of the nineteenth century, which have been used extensively since. These classical references, as well as a concise history of the *optical theorem* in scattering that covers roughly a century's worth of work, are given in the references at the end of the chapter. The description presented here provides some theoretical insight to enable selecting a specific nanoparticle system for a particular experiment and instrumental setup.

The physical picture of scattering is not particularly complex and does not require specification of material, shape, or size. As visible radiation is incident upon matter—whether it is dense, as in a solid, or dilute, as in a gas—the electric field interacts with the charged particles composing the atoms or molecules. The incident electric field causes a separation of charge, or an induced dipole, and this dipole responds to the oscillating field by itself oscillating. As accelerating charges isotropically radiate electromagnetic radiation, radiation secondary to the incident light is emitted from the oscillating dipoles. The secondary radiation is what is termed the *scattered radiation*. However, the energy carried in incident radiation can be dissipated, or absorbed, by the atoms or molecules it interacts with in nonradiative ways. This loss of radiation is termed *absorption*. The total attenuation of incident radiation, called *extinction*, is the sum of the scattered and absorbed light and is often expressed in terms of scattering and absorption cross sections, such as

$$\sigma_{\text{total}} = \sigma_{\text{scattered}} + \sigma_{\text{absorption}} \equiv \sigma_{\text{extinction}} \tag{12.1}$$

where the total cross section also equals the commonly used extinction cross section. The extinction cross section gives the rate at which energy is removed from an incoming beam of light due to scattering and absorption in a closed surface surrounding the scatterer via the so- called optical theorem. As the magnetic field (inherent in any light–matter interaction) has little effect on charged particles moving at nonrelativistic speeds, it is generally neglected in the physical picture of scattering. However, note that this is not true for extremely high-power sources, such as high-power lasers, which can accelerate charges to near-relativistic speeds. It is of note that there are a variety of semiredundant ways to present extinction measurements. Twelve of these are shown in Table 12.1. The extinction cross section of a homogeneous sample multiplied by the number density gives the extinction constant. The extinction constant, related to the molar absorptivity or (molar) extinction coefficient via Avogadro's number, can be used in the familiar Beer–Lambert law to compute the attenuation of light traveling through a sample (neglecting multiple scattering events).

Using Mie theory, the absorption, scattering, and extinction cross sections for any spherical particle can be calculated, provided sufficient knowledge of

**Table 12.1**

Different Ways to Present Extinction Data

| | |
|---|---|
| 1. Transmittance $T$ | $T = I/I_0$ |
| 2. Extinction $E$ | $E = 1 - T$ |
| 3. Extinction | Extinction $= \log(I/I_0)$ |
| 4. Optical density OD | $OD = \log(I_0/I)$ or $\ln(I_0/I)$ |
| 5. Absorptance $a$ | $a = I_a/I_0$ |
| 6. Absorbance $A$ | $\log_{10}(I_0/I_a)$ |
| 7. Absorption/extinction/scattering constant $\gamma_a, \gamma_e, \gamma_s$, cm$^{-1}$ | $I = I_0 \exp(-\gamma_i Z)$ |
| 8. Absorption coefficient $k$ ($n = n_r + ik$), absorption index $\kappa$ | $(4\pi_k/\lambda) = \gamma_a$ $n_r + ik = n_r(1 + i\kappa)$ |
| 9. Absorption/extinction/scattering cross sections $\sigma_{abs}/\sigma_{ext}/\sigma_{sca}$ [cm$^2$] | $\gamma_i = N\sigma_i$ where $N$ is the number density and $\gamma_i$ is used in definition 7 |
| 10. Absorption/extinction/scattering efficiencies | Efficiency$_i = \sigma_i/\pi R_{GEO}^2$ where $R_{GEO}$ is the geometric cross section of the particle |
| 11. Imaginary part of the dielectric function | $\varepsilon_2 = 2n_r k$ |
| 12. Optical conductivity | $\Sigma = \omega\varepsilon_2/4\pi$ |

*Source:* From Kreibig, U., and M. Vollmer, *Optical Properties of Metal Clusters*, Springer Series in Materials Science 25, ISBN 3-540-57836-6.

the wavelength dependence of the dielectric constants of the particle and the surrounding medium. A detailed description of Mie theory and many facets of scattering phenomenon can be found in the textbooks referenced at the end of the chapter. Briefly, according to Mie theory, the extinction and scattering cross sections integrated over all angles $(\theta,\phi)$ can be expressed as

$$\sigma_{ext} = \frac{2\pi}{k^2} \text{Re}(a_n + b_n) \tag{12.2}$$

$$\sigma_{sca} = \frac{2\pi}{k^2} \sum_{n=1}^{\infty} (2n + 1)\left(\left|a_n\right|^2 + \left|a_n\right|^2\right) \tag{12.3}$$

where $k = 2\pi/\lambda_0$. The absorption cross section is determined according to **Equation 12.1** as $\sigma_{abs} = \sigma_{ext} - \sigma_{sca}$. The coefficients $a_n$ and $b_n$ in **Equations 12.2** and **12.3** are given as

$$a_n = \frac{\mu m^2 j_n(mx)[x j_n(x)]' - \mu_1 j_n(x)[mx j_n(mx)]'}{\mu m^2 j_n(mx)[x h_n^{(1)}(x)]' - \mu_1 h_n^{(1)}(x)[mx j_n(mx)]'} \tag{12.4}$$

$$b_n = \frac{\mu_1 j_n(mx)[x j_n(x)]' - \mu j_n(x)[mx j_n(mx)]'}{\mu_1 j_n(mx)[x h_n^{(1)}(x)]' - \mu h_n^{(1)}(x)[mx j_n(mx)]'} \tag{12.5}$$

where the $h_n$'s and $j_n$'s are spherical Hankel and Bessel functions of the first kind, respectively; $m$ is the relative index of refraction ($m = n_p/n_m$); $\mu$ and $\mu_1$ refer to the magnetic permeability of the medium and the particle, respectively; and $x = ka$ is the size parameter. While somewhat cumbersome to evaluate, the terms in the infinite series of **Equation 12.3** converge quickly. There are several software packages that capitalize on this fact and yield accurate numerical results.

## Simplifications for nanosized particles

While Mie theory makes it possible to obtain exact extinction spectra for arbitrary spherical particles, the mathematical formalism itself allows for little direct insight into absorption and scattering phenomena. Fortunately, for subwavelength-sized metal nanoparticles, a small extension of the general idea of scattering allows for significant simplification in the mathematical formalism. Within the framework of the Rayleigh approximation, a particle is considered small relative to the wavelength of light when

$$\frac{2\pi a}{\lambda_0} n_p \ll 1, \tag{12.6}$$

where $a$ is the radius of the particle, $n_p$ is the complex index of refraction of the particle, and $\lambda_0$ is the vacuum wavelength. In this instance, the electric field can be taken as constant over the entire particle's volume. When an external electric field is applied to a metallic nanoparticle, the conduction electrons are displaced, and subsequent oscillations of the applied electric field cause the conduction electrons to coherently oscillate on the surface. The dipolar oscillations of the conduction electrons are termed *plasmons*. The frequency of plasmon oscillation is a function of the size, shape, charge distribution, and effective electron mass. Taking the electric field inside the particle to be constant and in phase with the external electric field, a nanoparticle can be treated with reasonable accuracy as a dipole emitter. This approach is referred to as the *dipole approximation*. For relatively small spherical nanoparticles, higher-order multipoles are excited but do not radiate. However, as particle size is increased, higher-order multipoles begin to radiate, and more complex extinction spectra are observed.

Below the Rayleigh limit, it can be shown that for a spherical dielectric nanoparticle, the absorption and scattering cross sections, integrating over all scattering angles, are

$$\sigma_A = \frac{8R^3}{\pi\lambda_0}\left(\frac{m^2-1}{m^2+2}\right), \quad \sigma_S = \frac{8\pi R^6}{3}\left(\frac{2\pi n_m}{\lambda_0}\right)^4\left(\frac{m^2-1}{m^2+2}\right)^2, \tag{12.7}$$

where $m = n_{particle}/n_{medium}$ is the index of refraction of the particle relative to that of the medium, $\lambda_0$ is the vacuum wavelength, and $R$ is the particle radius. Differing forms of these equations are found in the literature and can be derived readily, noting the relationship between the complex dielectric constant and the complex index of refraction ($\varepsilon = n^2$) and the definition $m$. Assuming that the dielectric properties of the particles are equivalent to the bulk values—reasonable except for radii below 5 nm—then the most prominent feature of **Equation 12.7** is the strong dependence on size $R$. This is summarized in **Figure 12.1**, where the total extinction cross section is shown to increase rapidly with $R$ across the entire visible spectrum.

It is important to note that $\sigma_A$ increases as $R^3$, while $\sigma_S$ is proportional to $R^6$ (**Figure 12.2a**). This implies that for most nanoparticles below approximately 100 nm in diameter, absorbance dominates over scattering. Because of this, the absorption cross section is often directly equated to the total extinction cross section, $\sigma_E$, which is the quantity measured by ultraviolet–visible (UV–Vis) "absorbance" techniques, and presented in virtually all reports. While this appears mathematically reasonable, it is not always representative of experimental systems. In fact, many synthesis strategies (see **Section 12.3**, "Synthesis of Gold Nanoparticles" section)

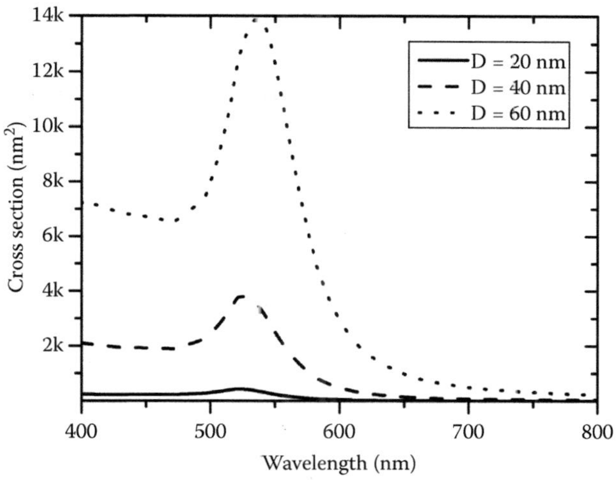

**Figure 12.1 Comparison of calculated extinction spectra of 20-, 40-, and 60-nm-diameter gold nanocrystal suspensions in water.**

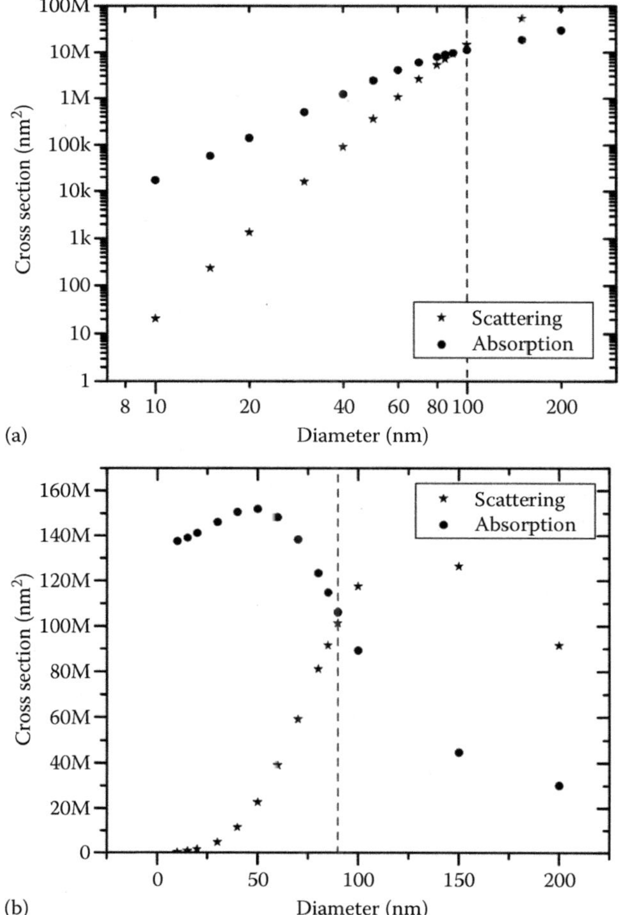

(a)

(b)

**Figure 12.2 (a) Integrated absorption and scattering cross sections for individual spherical gold nanoparticles versus nanoparticle radius, in water, as calculated using Mie theory. (b) Same data as (a) normalized by the relative number density.**

control the final particle size ($R$) by changing the relative concentration of the reducing agent while keeping the concentration of available gold ions fixed. As the number of gold atoms is limiting in batch synthesis, the relative number density ($N_i$) of particles must also scale with particle volume according to

$$N_i = \left( \frac{R_m}{R_i} \right)^3,$$ 
(12.8)

where $R_m$ is the maximum radius when comparing a series of samples, and $R_i$ is the radius of the particle of interest where $R_i \rightarrow R_m$. The measured extinction spectra can then be normalized by multiplying the extinction cross section by the relevant number density: $N_i \sigma_{E_i}$. The results for the normalized individual scattering and absorption terms are shown in **Figure 12.2b**. The approximate nanoparticle radius at which the number density–normalized scattering is equivalent to absorption is obtained by simply setting the individual cross sections of **Equation 12.7** equal and solving for $R$. This is shown as the crossing points of the lines shown in **Figure 12.2**. For sensing or labeling applications based primarily on total absorbance (for example, *in vitro* detection or brightfield imaging) sizes to the left of this line are likely best, whereas for scattering applications (darkfield imaging) sizes to the right of this line may generate better total signal.

Though informative, the simplified form for the scattering cross section shown in **Equation 12.7** assumes that it is reasonable to integrate over all scattering angles, although a typical experimental geometry almost certainly precludes this possibility. A somewhat more general form of the scattering equation, but still within the Rayleigh limit of the dipole approximation and summed over all polarizations, is

$$\frac{d\sigma_S}{d\Omega} = \frac{R^6}{2} \left( \frac{2\pi n_m}{\lambda_0} \right)^4 \left( \frac{m^2 - 1}{m^2 + 2} \right)^2 (1 + \cos^2 \theta),$$
(12.9)

where $\theta$ represents the angle of light scattering relative to the direction of incident light propagation, and $d\Omega = \sin\theta d\theta d\varphi$ is the surface area element subtended by a cone of scattered light at an angle $\theta$ around the light propagation axis. Integrating the last term in **Equation 12.9** from $\theta_{min} \rightarrow \theta_{max}$ gives the angular dependence of scattering relative to the desired experimental geometry. Rigorously speaking, the angle-dependent term is also size dependent, where particles above the Rayleigh limit show increased forward scattering and require application of the exact Mie theory via **Equation 12.3**.

The resulting forward scattering and backscattering fractions are plotted relative to the particle radius in **Figure 12.3a**. Here, the effects of numerical aperture (NA) of the light collection optic for three common objective NAs are also compared. NA = $n \sin\theta_{max}$ is used to determine $\theta_{max}$, which is the maximum angle of the cone of the imaging optic. The potential impact on optimal signal can be compared for transmission (forward scattering) and reflection (backscattering) microscope geometries (**Figure 12.3b**). For smaller nanoparticles (<100 nm), both forward scattering and backscattering geometries are expected to produce approximately equivalent scattering signals. It should be noted, however, that this analysis does not consider the sample environment around the particle, which may exhibit differing scattering fractions and subsequently affect the background signal to varying degrees. The sample environment itself—tissue, cell membrane—should be considered when determining an optimal nanoparticle size and instrumental geometry.

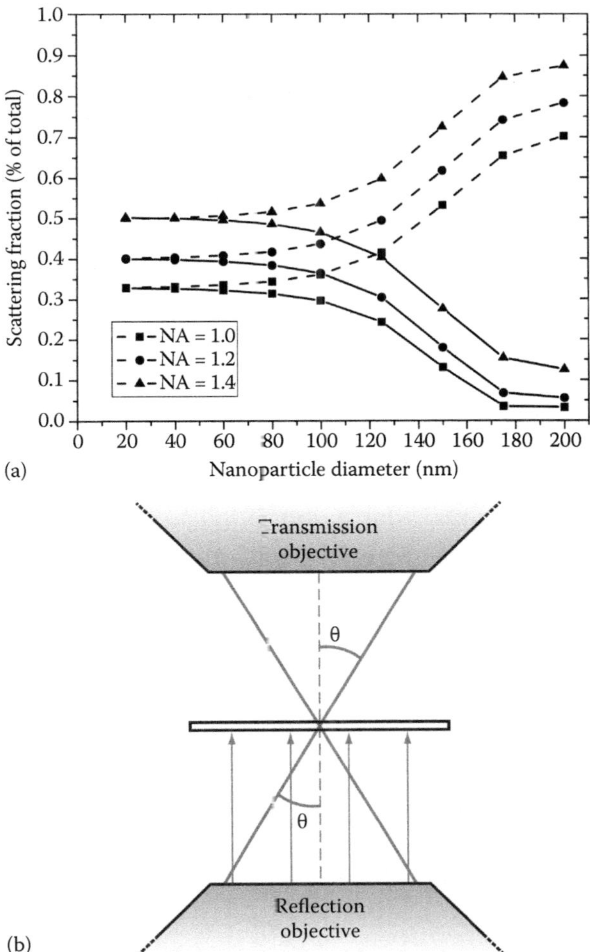



(a)

(b)

**Figure 12.3** (a) Forward scattering (dashed lines) and backscattering (solid lines) fractions for three different NA objectives. (b) Schematic of typical microscopic excitation/collection geometries.

Of particular interest for metal nanoparticles, a relatively strong peak appears in both the absorption and scattering spectra, as seen in **Figures 12.4** and **12.1**. According to **Equations 12.7** and **12.9**, the absorption and scattering signals would be expected to decrease continuously with wavelength at a rate proportional to $1/\lambda$ and $1/\lambda^4$, respectively. Despite this fact, however, a strong peak in the signal is still observed for most nanoparticles.

This apparent inconsistency is due to a wavelength dependence of the complex dielectric constant, $\varepsilon_p \rightarrow \varepsilon_p(\omega)$. This dependence is strong in metals and is shown in **Figure 12.5b** for bulk gold. Recognition that the dielectric function is wavelength dependent is commonly attributed to Drude; however, his initial free-electron model did not produce good predictions in the visible and near-UV regions. It was later corrected to accommodate quantum mechanical effects. (References to these calculations are given at the end of the chapter.) For particles larger than 10 nm in diameter, the effective dielectric constant can be accurately estimated using the bulk values and so is not described in further detail here. As a particle's size approaches the mean free path of the conduction electrons, the conduction electrons' oscillations are somewhat damped, resulting in a size-dependent dielectric constant.

**Figure 12.4 Comparison of calculated Mie theory absorption, scattering, and extinction cross sections for an 85-nm-diameter gold nanoparticle in water.**

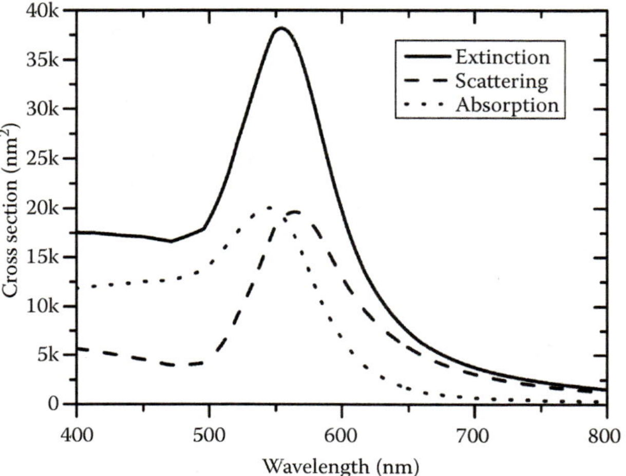

Using bulk values of the index of refraction—more formally, the dielectric constant—for gold, we expect to observe a peak in the predicted spectra according to **Equation 12.7** when the denominator in the last term approaches 0; $m^2 + 2 \rightarrow 0$. In media such as water $\left(\varepsilon_m = n_m^2 \approx (1.33)^2\right)$, this occurs when $n_p^2 = -2n_m^2 \approx -3.5$, or in terms of the dielectric constant, when $\varepsilon_p = -2\varepsilon_m \approx -3.5$. The absorption and scattering equations (**Equation 12.7**) are evaluated using the imaginary and real components of the dielectric constant, and are used in a similar fashion here. As an example, for gold nanospheres with diameters in the range 10–50 nm, comparing the data in **Figure 12.5a and b**, Re($\varepsilon_p = -3.5$) is seen to be satisfied at around $\lambda = 510$ nm. This is the so-called resonance condition, and the wavelength (or frequency) where this occurs is referred to as the *plasmon resonance frequency* ($\omega_p$). For particles larger than ~50 nm in diameter, the dipole approximation loses its validity as multipoles become

**Figure 12.5 (a) Real and imaginary parts of the index of refraction and (b) real and imaginary dielectric functions for bulk gold.**

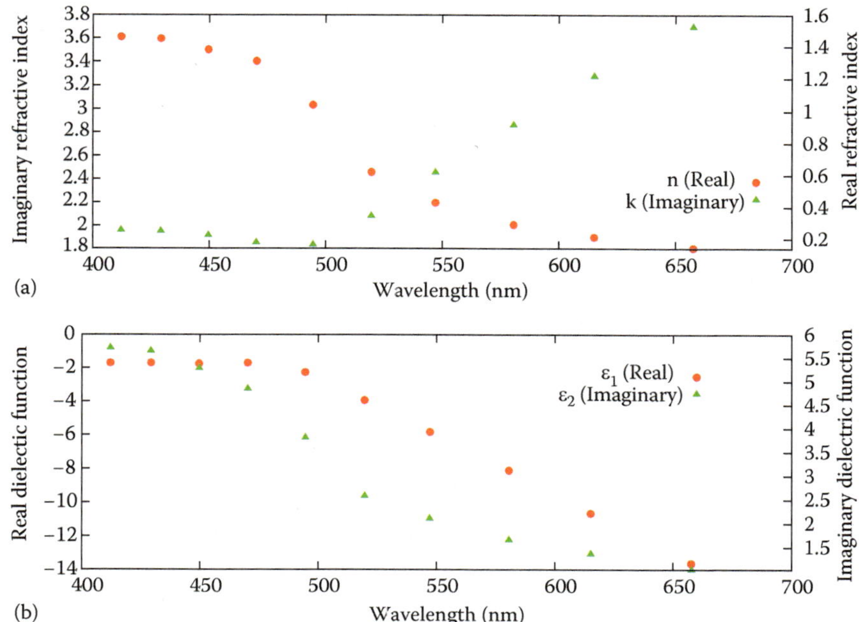

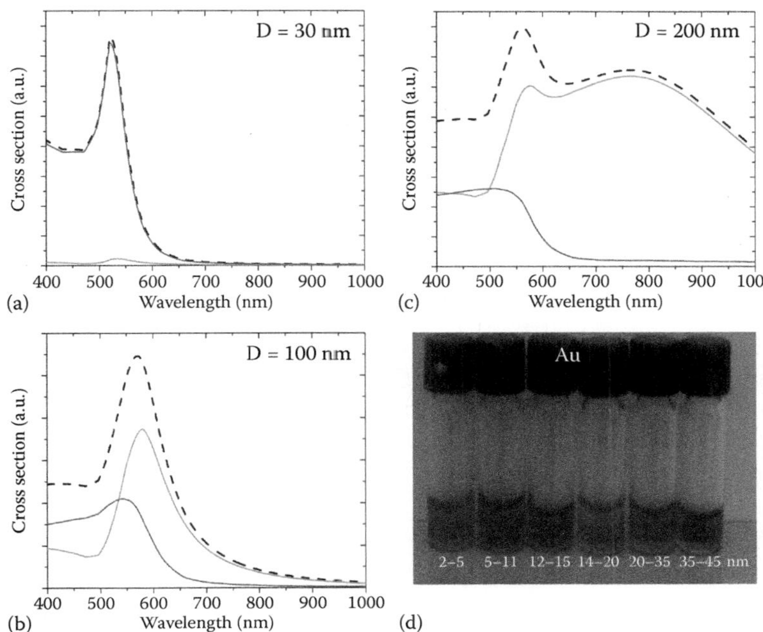

**Figure 12.6 Size changes and absorbance.** (a–c) Calculated cross sections (dashed, $\sigma_E$; dark gray, $\sigma_A$; light gray, $\sigma_S$) for three sizes of gold nanoparticles in water: (a) 30 nm, (b) 100 nm, and (c) 200 nm. (d) Image of different sizes of Au nanoparticles in solution.

supported. This results in the appearance of multiple resonance frequencies, which contribute to the observed spectral structure. In the 50–100 nm range, this gives the appearance of a broadened and slightly redshifted single peak. As the size is increased further, multiple peaks become apparent. This is shown in **Figure 12.6a through c** and has the net effect of redshifting the *average* absorption and scattering wavelength as nanoparticle size is increased. In effect, the apparent *color* of a metallic nanoparticle, under white-light illumination, can be tuned as a function of size (**Figure 12.6d**). Like semiconductor nanocrystals, this size dependence has led to the widespread investigation of metal nanoparticles as probes in biology.

When selecting gold nanoparticles for experiments, some of the key parameters to consider are

- Instrument geometry (reflection/transmission)

- Sample geometry and composition (possible background signal)

- Labeling strategy (large particles may not be well suited for cellular uptake)

- Desired signal "color"

These will all be addressed more fully in the following sections as we discuss synthesis, labeling, and imaging strategies.

## 12.3 SYNTHESIS OF GOLD NANOPARTICLES

While the earliest known applications of gold nanoparticle date back several centuries, direct correlations between synthesis conditions and nanoparticle geometry were not possible until many years later. In fact, much of the base knowledge and instrumentation necessary to directly characterize nanoparticles—transmission electron microscopy (TEM), atomic force microscopy (AFM), and scanning tunneling microscopy (STM)—arose from the semiconductor industry, where the study of optical and electronic properties of semiconductor materials at decreasing length scales is of great interest. Due in large part to the development of these characterization techniques, new solution-phase chemistries have been developed that provide effective size and shape control of the growth of nanocrystals, including noble metals. The study and application of metal nanoparticles have seen a consequent resurgence, and these particles are currently being investigated for a broad range of technological and biological applications. Some references are given at the end of the chapter to emerging uses of gold nanoparticles, including subwavelength optical devices, solar cells, flexible electronics, conducting inks and adhesives, surface-enhanced spectroscopies (such as Raman), chemical catalysis, and biological labeling and detection.

Owing to the sheer number of potential metal nanoparticle–based applications, reasonable-quality metal nanoparticles can now be acquired from commercial sources. While this is of great benefit to many researchers, it should be noted that the overall selection of available nanoparticle sizes and geometries is limited. In addition, the quality (size and shape distribution) and consistency of these materials may not be adequate for more sensitive experiments. Of particular concern may be the addition of stabilizers, either on the particle surface or in solution. Stabilizers extend the shelf life of nanoparticle colloidal suspensions but may complicate downstream applications. Many researchers find it beneficial to perform the synthesis in house, allowing for reduced cost, better quality control, and the investigation of nonstandard geometries.

The most common solution-phase synthesis strategies, which we will focus on here, are relatively straightforward, require little in the way of specialized equipment or materials, and hence are accessible to virtually any interested researcher. Generally, these involve the aqueous-phase reduction of metal salts ($HAuCl_4$, $AgNO_3$, $AgCl_2$) with a moderate-to-strong reducing agent ($NaBH_4$, $Na_3C_6H_5O_7$). There are literally hundreds of synthetic routes leading to metal nanoparticles; synthesis can be performed in either aqueous or organic phases, or both.

Owing to poor solubilization, relative mass, and reduced availability of suitable organic-phase reducing agents, larger metal nanoparticles are usually not well supported in organic solvents. In contrast, a wide range of nanoparticle sizes and shapes can be produced in aqueous systems and later transferred into an organic phase if so desired. Phase-transfer catalysts are usually cetyl-trimethylammonium bromide (CTAB) or tetra-octylammonium bromide (TOAB). In early reports, aqueous-phase reducing agents were limited to small inorganic compounds ($NaBH_4$ or $Na_3C_6H_5O_7$), but they have since been expanded to include polymers, amines, organic acids, intermediate radicals, and even biological molecules. As a result, numerous "green" synthesis strategies are available. A simple method for making 15 nm gold particles is given in **Practical Tips 12.1**.

The generally accepted mechanism for growth of spherical metal nanoparticles was first described by Turkevich and later detailed by Frens. The mechanism involves an initial nucleation phase followed by a gradual layer-by-layer growth phase. During the nucleation phase, the reducing agent, through the transfer of $n$ charges, converts the dissolved metal ion to an uncharged atom: $M^{n+} \rightarrow M^0$. Since uncharged metal atoms are unstable in solution, upon collision with other metal

---

### PRACTICAL TIPS 12.1:    SYNTHESIS OF 15 NM SPHERICAL GOLD NANOPARTICLES

This is one of the more common and straightforward synthetic procedures for gold nanoparticles. In this case, gold nanoparticles stabilized by citrate may be prepared from hydrogen tetrachloroaurate ($HAuCl_4$) in water by reduction with sodium citrate at 85°C. The experimental apparatus consists of a clean 500 mL round-bottom flask heated by means of an oil bath and magnetically stirred at an intermediate speed to ensure uniform temperature throughout. In order to obtain a particle diameter of roughly 15 nm, 20 mg of $HAuCl_4$ is dissolved in 100 mL of 18 M$\Omega$ $H_2O$. Sodium citrate ($Na_3C_6H_5O_7$), 100 mg, is dissolved in 2 mL of water and added to the solution. After several minutes, the solution will turn wine-red, indicating that the reaction is near completion. Solutions are then allowed to heat for 10 more minutes to ensure that all gold has reacted.

---

atoms, they begin to form a second phase, minimizing the overall energy of the system. Over time, the abundance of individual $M^0$ atoms is reduced, while that of the nucleates is increased. Gradually, it becomes more likely that a $M^0$ atom will collide with a nucleate than with another $M^0$ atom, and the growth phase proceeds. Due to the crystalline nature of metals, atoms are added in a layer-by-layer fashion. The number of atoms required for each layer increases as the particle grows. If the rate of addition of atoms is constant across all particles, then the growth rate (per layer added) is greater for smaller particles. Additionally, under some conditions, smaller particles can dissolve, and their constituent atoms can be added to the surface of larger particles. This process is referred to as Ostwald ripening (see Chapter 11), which results in relatively narrow size distributions (Figure 12.7).

Ultimately, particle growth is controlled by the relative concentration of metal ions and reducing agent, temperature, and in some cases, mixing. The resulting nanoparticle surface contains a number of only partially reduced metal ions, which are positively charged. Due to strong Coulombic attraction, any negatively

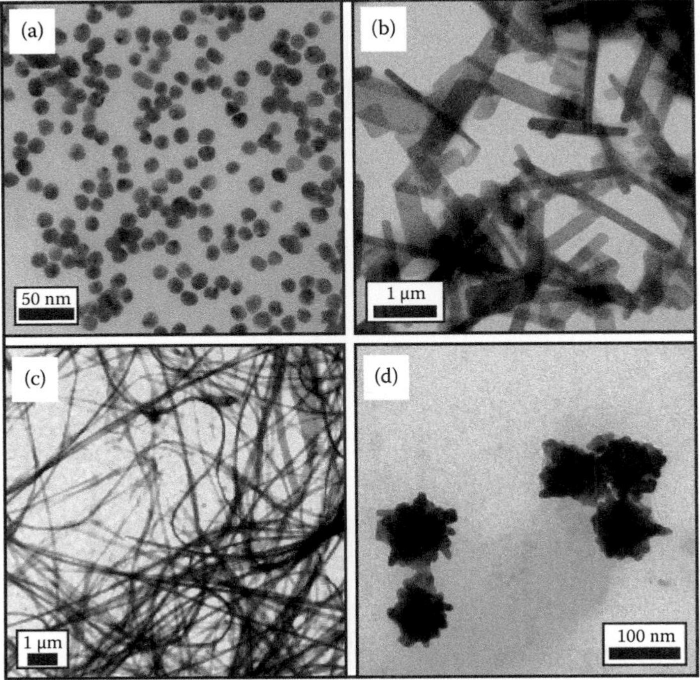

**Figure 12.7 TEM images of (a) nanospheres, (b) nanorods, (c) nanowires, and (d) nanostars.**

charged counterions (such as citrate) are drawn to the surface, where they become tightly held. The net result is a negatively charged, roughly spherical, particle surface (zeta potential, ~30–50 mV) that imparts significant electrostatic stability to this colloidal system.

When preparing nanoparticles for *in vivo* use, it is of the utmost importance that they be well characterized and uniform in size. One important tool for evaluating and purifying nanoparticle dispersions is size exclusion chromatography (SEC). SEC is a technique similar to high-performance liquid chromatography (HPLC), in which particles are separated by their hydrodynamic volume. This is particularly useful when determining the efficiency of ligand coupling to a nanoparticle. **Figure 12.8** shows elution profiles of two samples: PEGylated gold and PEGylated gold that has been coupled to an antibody. The profile of the sample with no antibody is a simple Gaussian, primarily a result of the particle size distribution. The curve for the antibody-linked nanoparticles shows a shift in the primary peak. This is because of the increased size of the conjugate when compared to the unlinked nanoparticle. Two secondary peaks are also evident, one for antibody dimers and the other for unbound antibody. Collection of the eluate in fractions allows for highly characterized antibody conjugates to be obtained.

SEC is a highly effective technique, but it may not be appropriate for large samples. When scaling up for *in vivo* use or other high-volume applications, standard size-exclusion columns do not provide enough throughput. Instead, for large samples, a combination of vacuum filtration and centrifugation provides for the removal of excess reactants and aggregates. Centrifugation is useful for creating injection-sized doses (mouse, ~200 μL).

In addition to spherical metal nanoparticles, other nanoparticle geometries are readily obtained with the use of templating agents. The most common of these are rods—though wires, trigonal prisms, nanoboxes, nanoshells, and even stars have also been produced. Due to the strong correlation between photophysical properties and nanoparticle size, particles having more complicated geometries consequently produce interesting optical signals. The use of these templating agents is generally straightforward, and references to descriptions of their synthesis are given at the end of the chapter. **Practical Tips 12.2** gives a method for making nanostars.

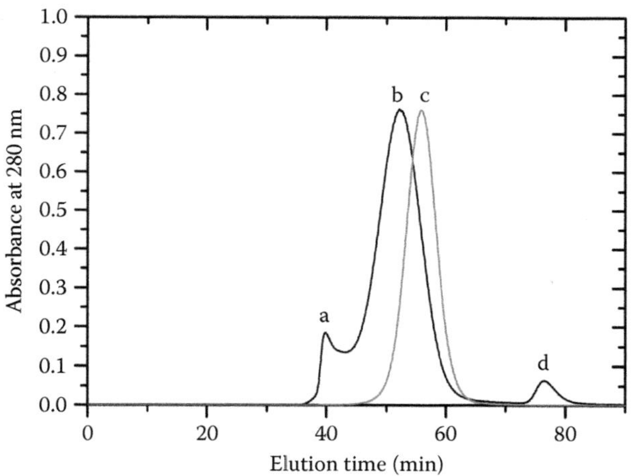


**Figure 12.8 SEC elution profiles of PEGylated gold nanoparticles (gray) and antibody-conjugated nanoparticles (black).** (a) Aggregates, (b) the desired conjugate, (c) unconjugated particles, and (d) free antibody.

## PRACTICAL TIPS 12.2:    SYNTHESIS OF GOLD NANOSTARS

This protocol describes the process for making gold nanostars from the gold spheres described in **Practical Tips 12.1**. Spheres are first synthesized for use as seeds for star growth. A templating agent, polyvinylpyrollidone (PVP), is used, which preferentially, though weakly, physisorbs to the planar surfaces of the gold seed crystals, leaving the edges and vertices exposed. Gold ions (from $HAuCl_4$) are then added as a growth medium, which add to the particle only at the exposed vertices. Growth continues until the gold ions are exhausted. The relative definition of the nanostar shape can be controlled by varying the core size and the concentration of available gold ions.

The protocol is as follows:

- To 200 mL of 20 nm gold spheres, add 4 mL of 10 mM PVP (MW = 10,000) solution in water.
- Stir overnight at room temperature.
- Pellet by centrifugation at 12000g for approximately 40 h.
- Decant the supernatant by vacuum aspiration and resuspend the seeds in 20 mL of anhydrous ethanol.
- While preparing the seeds, prepare the growth solution by dissolving 1.5 g of PVP in 15 mL dimethyl-formamide (DMF) in a 100 mL round-bottom flask. Dissolution will require sonication for a few minutes.
- Add 168 µL of 25 mM gold chloride to the DMF/PVP solution.
- Now add 25 µL of the seed solution and allow it to react at room temperature without stirring. After approximately 30 min, the solution will start to turn blue, an indicator that large particles are forming.
- Let the reaction run overnight.
- Once the stars are fully formed, they should be washed for characterization, tagging, and/or PEGylating. Washing is done by first placing the star solution in 1.5 mL microcentrifuge tubes and spinning the stars into a pellet at 6000 rpm for 20 min. Once centrifugation is complete, 0.75 mL of the DMF/PVP solution is decanted and replaced with 0.75 mL of anhydrous ethanol and vortexed or sonicated until the stars are resuspended. Repeat the centrifugation step. After this second spin, all of the supernatant is decanted and replaced with ethanol and vortexed/sonicated to resuspend.

A series of three such nanostar preparations are shown in **Figure P12.2.1**. A sample extinction spectrum is also shown, which, for this system, demonstrates a broad extinction (largely scattering) profile.

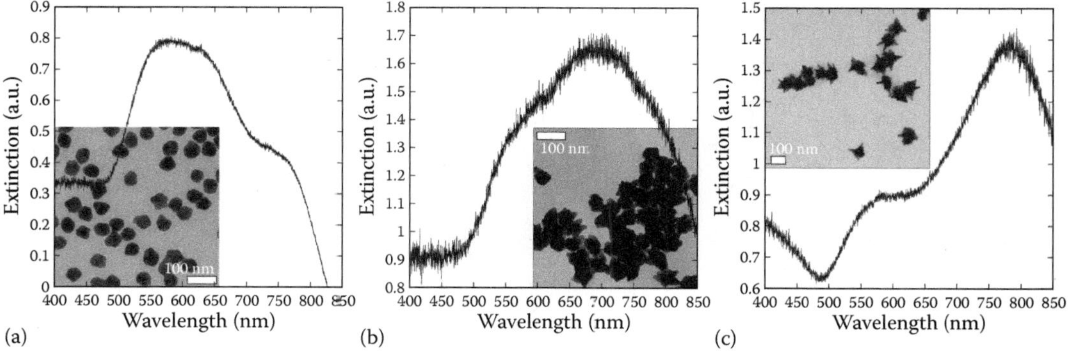

**Figure P12.2.1    TEM series of gold nanostars of increasing size and definition of the star structure (a–c), and corresponding UV-Vis extinction.**

### SUGGESTED READING

Nehl, C.L., Liao, H., and Hafner, J.H. (2006). Optical properties of star-shaped gold nanoparticles. *Nano Letters* 6(4), 683–688.

# 12.4 CHARACTERIZATION AND SURFACE MODIFICATION OF GOLD NANOPARTICLES

## Recommended characterization techniques

Detailed characterization of gold nanoparticles is required before they are used in any quantitative experiment. Unlike typical molecular fluorophores, homogeneous colloids of uncapped metal nanoparticles are inherently unstable and prone to changes over time. Furthermore, even the best-understood synthetic strategies are highly dependent on reaction conditions and often give rise to a distribution of nanoparticle sizes and shapes. In the case of metal nanoparticles, it is possible to produce similar optical spectra from dramatically different geometries. Day-to-day reproducibility can only be ensured under the most controlled conditions, and simple visual inspection of solution color or UV–Vis spectra are not reliable metrics as to the success of a reaction or the quality of a sample. It is recommended that storage time be limited and that detailed characterization be performed just prior to further use—that is, capping or bioconjugation. Preuse characterization should include correlation between multiple independent methods.

While there are a number of common methods for nanoparticle characterization, perhaps the most widely used include UV–Vis absorption (extinction), TEM, and dynamic light scattering (DLS). Example data from these three techniques are shown in **Figure 12.9**. Though individually, these methods are wholly inadequate, collectively, they provide a meaningful and quantitative picture of the nanoparticle under study.

As UV–Vis absorption spectroscopy is both ubiquitous and cost effective, it is typically performed first, providing an ensemble snapshot of the sample. TEM can be used to qualitatively confirm the geometries present, though in a nonstatistical fashion, as only several hundred particles out of many millions are visualized. DLS generates a near-ensemble picture of the nanoparticle size distribution based on single-particle mobility, where individual nanoparticles are assumed to be spheres. (See **Chapter 5** for a discussion of this technique.) It is important to be particularly careful when visualizing DLS data, however, as statistically weighted distributions are common and can vary the apparent size of a single sample dramatically depending on the weighting factor—that is, intensity, volume, or particle number (**Figure 12.9c**).

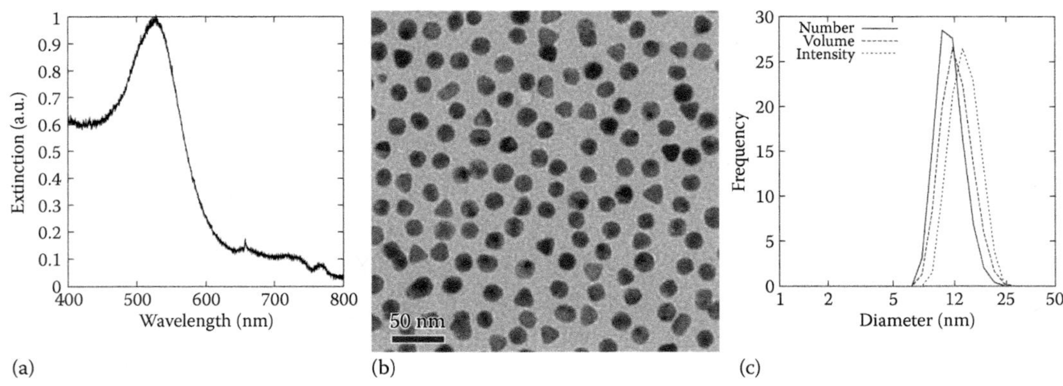


**Figure 12.9 Typical metallic nanoparticle characterization methods.** (a) UV–Vis spectroscopy. (b) TEM imaging. (c) DLS.

## Surface stabilization and biocompatibility

In biological solutions, gold nanoparticles may be subjected to changes in concentration, pH, or temperature, all of which are known to cause agglomeration. Ideally, the particles would remain dispersed (nonagglomerating) and be structurally stable in solution as well as in complex biological systems. There are several approaches to achieve this stability. One general method involves coating the particles with a polymer in order to stabilize them by steric hindrance.

There are a variety of organic surface modifiers available to coat metallic nanoparticles and sterically prevent them from sticking to each other when they collide. Surface modifiers can either be electrostatically adsorbed onto the surface of the particle or covalently bound. Electrostatically adsorbed surface modifiers, such as sodium citrate, are useful during particle synthesis but do not impart enough stability to make the particles viable in biological systems. Polyethylene glycol (PEG) molecules of different chain lengths and functional groups are available for stabilizing nanoparticles and for attachment of targeting moieties. The best way to adsorb PEG onto gold is to use a PEG-thiol that makes use of the strong gold–sulfur bond to create a very stable system that will remain intact even inside cells. A PEG–dithiol linker, with two gold–sulfur bonds per molecule, yields a highly stable particle that exhibits almost no agglomeration (**Figure 12.10a**). Particles stabilized with the PEG-dithiol can be concentrated by a factor of 5000 by spin centrifugation without agglomeration (**Figure 12.10b,c**). In addition to improving stability, PEG-thiols may reduce the toxicity seen with certain sizes and shapes of gold and silver particles (**Figure 12.10d**).

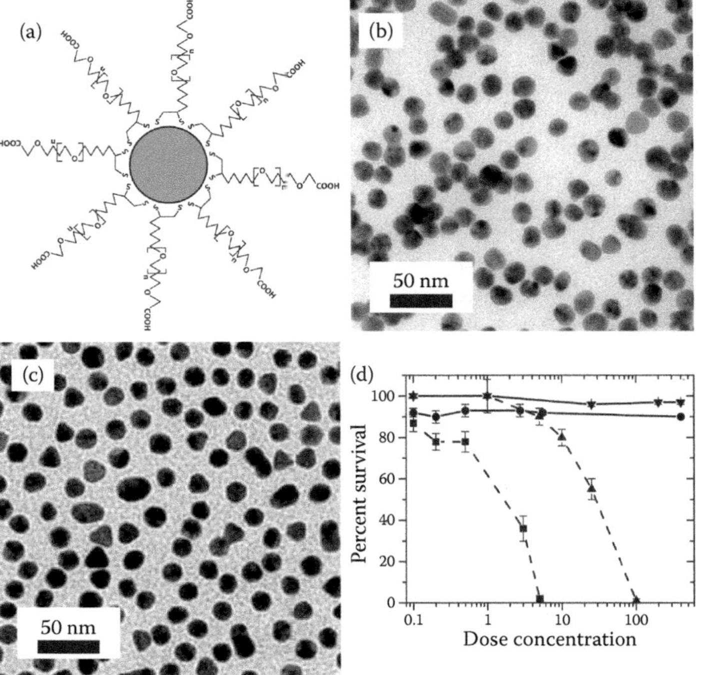


**Figure 12.10 Biofunctionalization of gold particles.** (a) Cartoon of a PEG-coated nanosphere, (b) gold nanoparticles without a PEG coating, (c) gold nanoparticles with a PEG surface coating, and (d) comparison of marine mammalian cell survival with increasing exposure to unprotected silver (■), unprotected gold (▲), PEGylated silver, (●) and PEGylated gold (▼) nanoparticles.

**Figure 12.11 Depiction of potential gold nanoparticle/ antibody targeting schemes.**

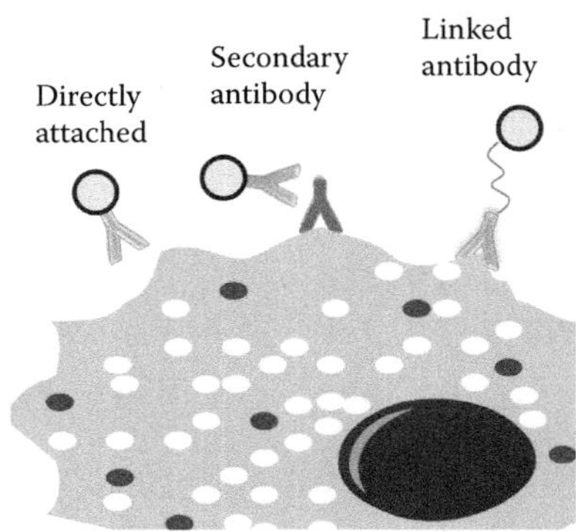


---

## PRACTICAL TIPS 12.3:　BIOFUNCTIONALIZATION OF AU NANOPARTICLES

### PEGYLATION

Gold nanoparticles can be readily functionalized with PEG thiols (monothiol or dithiol) by simple ligand exchange. Citrate-stabilized gold nanoparticles agglomerate immediately upon transfer into salt-containing buffer; thus, the exchange from citrate to PEG must be performed in pure water. In a typical reaction, 500 µL of the gold nanoparticle dispersion is added to 2 mg of the dry PEG ligand and vortexed. The reaction between the PEG dithiol and the nanoparticles happens within several minutes, and an increase in the hydrodynamic radius of the particle can be observed by DLS. After mixing, it is allowed to sit for ~2 h at room temperature. Following this, the nanoparticle dispersion should be transferred into a 1.5 mL Eppendorf tube and centrifuged at 16,000$g$ for 30 min until a pellet is formed. The supernatant is removed, and 1 mL of 18 M$\Omega$ water is added. This process is repeated three times to remove any unbound PEG ligand. After the third washing and decanting, the particles are resuspended into 100 µL of salt-containing medium (0.1 M 2-($N$-morpholino)ethanesulfonic acid [MES] buffer, pH = 5).

### ANTIBODY ATTACHMENT

Functionalization of the PEGylated nanoparticles with antibodies can be performed using standard NHS/ EDC coupling chemistry. In this method, the carboxy groups on the particle surface are activated with EDC hydrochloride and NHS in 0.1 M MES buffer (pH 5). Ten milligrams of NHS and 10 mg of EDC are first dissolved into 100 mL of MES buffer. Twenty-five microliters of this solution is then added to 100 µL of the PEGylated particle dispersion and allowed to sit for 30 min at room temperature. Samples are then subject to repeated washing (four times) with 300 µL of MES (pH = 5) at 0°C in centrifuge filters (Millipore Ultrafree 0.5, 30 kDa cutoff), spinning down at 11,500$g$ to no more than 25 µL each time. Antibodies can be attached by adding the concentrated activated nanoparticles to a dilute solution containing the antibody. First, 15 µg (~100 pmol) of the antibody is diluted to a volume of 300 µL with cold PBS buffer (0°C) at pH 7.4. The solution is then added to the centrifuge filter containing the 25 µL solution of activated gold particles and mixed well. This mixture is then spun down to 25 µL followed by rinsing once with 300 µL of PBS (all at 0°C). Finally, enough PBS to resuspend to a final volume of 25–50 µL should be added. After mixing well, the solution is transferred into an Eppendorf tube and swirled gently overnight at room temperature. After this, all unbound NHS-activated carboxy groups will have reacted or been hydrolyzed.

### SUGGESTED READING

Eck, W., Craig, G., Sigdel, A., Ritter, G., Old, L.J., Tang, L., Brennan, M.F., Allen, P.J., and Mason, M.D. (2009) PEGylated gold nanoparticles conjugated to monoclonal F19 antibodies as targeted labeling agents for human pancreatic carcinoma tissue. *ACS Nano* 2(11), 2263–2272.

## Targeting schemes

Gold nanoparticle systems are being designed for cellular uptake studies, for basic research, and to aid in the diagnosis and treatment of disease. As previously described in Chapter 8, there are many different schemes by which targeting moieties, such as antibodies, can be attached to nanoparticles. Commonly, proteins are linked to the terminus of carboxy-PEG molecules using N-hydroxysuccinimide/1-ethyl-3-(3-dimethylaminopropyl) carbodiimide (NHS/EDC) coupling chemistry. Antibodies may be attached to gold particles as discussed in Chapter 8 for quantum dots, and may be designed to target either a cell antigen directly or another antibody (Figure 12.11). Practical Tips 12.3 gives recipes for PEGylation of and antibody conjugation to gold nanoparticles.

## 12.5 APPLICATIONS FOR COLORIMETRIC DETECTION AND MICROSCOPY

### Metal nanoparticles as local sensors

According to Equation 12.7, the intensity and wavelength of the plasmon resonance peak are expected to show a strong dependence on the index of refraction ($n_m$) of the surrounding matrix. Several interrelated factors can influence local index of refraction: temperature, concentration, composition, viscosity, density, and pH, to name a few. This suggests that by monitoring the intensity and scattering (or extinction) spectrum of either individual gold nanoparticles or suspensions of particles, one can gain insight into the physical properties near the particle.

A simple example of this is shown in Figure 12.12, where an aqueous suspension of 15-nm-diameter gold nanoparticles is used in an absorbance (extinction) measurement to monitor an increase in glucose content from nearly 0 to approximately 20% by weight. Interestingly, not only does the peak intensity shift with glucose content, as predicted according to Equation 12.7, but also, a second peak grows in. The presence of the second peak is not expected to arise from a change in the local optical properties, but rather, indicates the effect glucose has on the stability of the system, as the appearance of the second peak is likely due

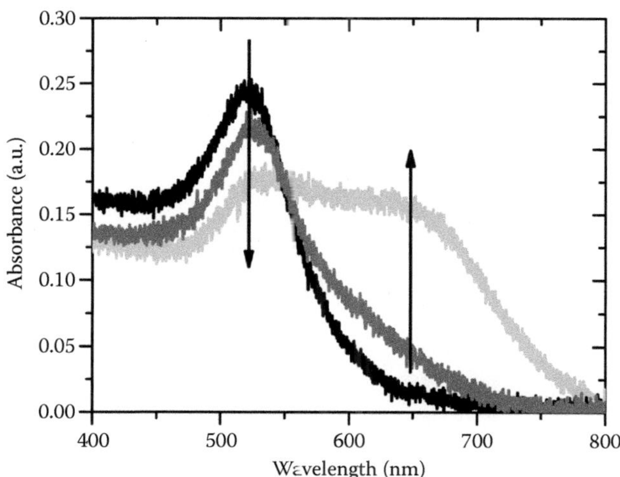

**Figure 12.12 Normalized extinction spectra versus glucose content ranging from nearly 0 (black) to ~20 wt% (light gray).**

to aggregate formation. While not generally desirable, the formation of aggregates can itself be a very sensitive indicator and has, in fact, been used in several biomolecular sensing applications.

## Darkfield microscopy

For larger gold nanoparticles (>20 nm), individual nanoparticles can be visualized in thin tissue sections or even in live cells using conventional darkfield transmission microscopy provided that the samples are sufficiently transparent. In this configuration, the collected signal is based on forward scattering; hence, larger particles should be used whenever possible (**Figure 12.5**). For particles smaller than 10 nm, light scattering from cellular structures tends to dominate, making it difficult to visualize individual particles. Unfortunately, larger nanoparticles are typically not taken into the cell by endosomal routes. While electroporation or microinjection techniques have been used successfully, and are described in **Chapter 8**, another strategy is to begin with smaller particles, which are taken into the cell by endocytosis and subsequently form small aggregates. To achieve this, a surface coating imparting intermediate or short-lived stability should be employed on the nanoparticles. A sample darkfield image of an NIH3T3 cell is shown in **Figure 12.13**, where nontargeted 15 nm gold particles stabilized with noncovalently bound (physisorbed only) PEG were incubated for 24 h in culture. Although the starting nanoparticles showed only a very weak scattering signal, individual diffraction-limited features are clearly visible in the image. Most features show no marked shift in scattering color, suggesting that only some minimum amount of aggregation has occurred. A few particles, however, do exhibit a significant color change and signal increase, visible in **Figure 12.13** as the brightest grayscale features, indicating more prominent aggregation. If single-particle integrity is not required, this simple approach can prove useful.

Darkfield microscopy typically makes use of broadband (white-light) illumination. With the use of filters, narrow wavelength bands can be selected to coincide with the peak plasmon wavelength of a nanoparticle probe. Since cellular features exhibit broad scattering spectra, the use of a filter increases the effective contrast of individual particles, making them visible in (or on) tissue sections (**Figure 12.14**).

Whether the instrumental approach is darkfield (transmission) or confocal (reflection), it is important to take care when selecting the appropriate wavelength

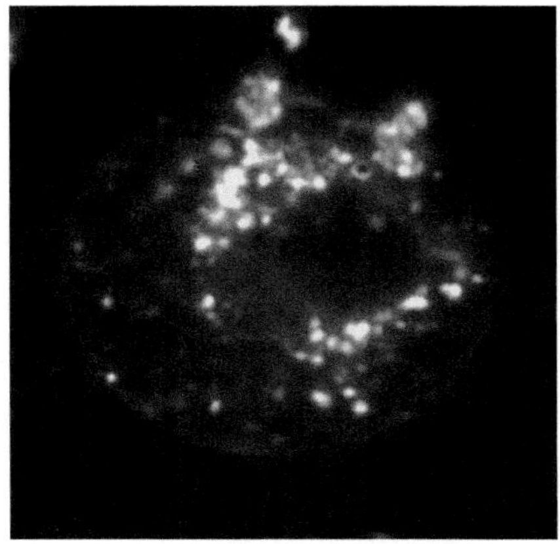

**Figure 12.13 Live cell imaged using 15 nm gold nanoparticles in a transmission darkfield microscope.**

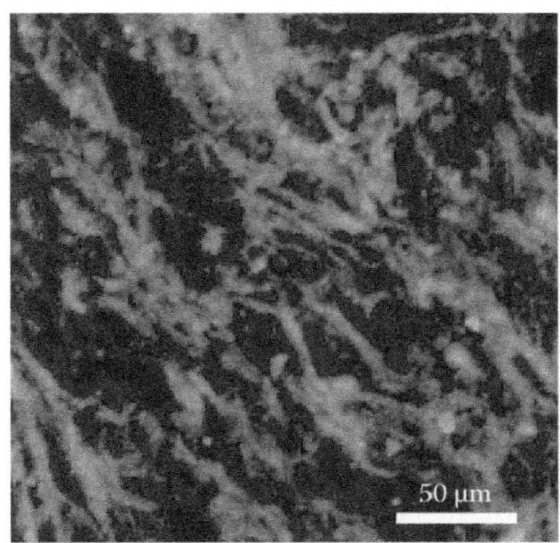

**Figure 12.14  Darkfield transmission image of F19-conjugated 15 nm gold nanoparticles targeting responsive stromal cells in a pancreatic cancer tissue thin section.** This image was obtained using a Moticam 1280 × 1024 camera and a 550/20 nm bandpass filter.

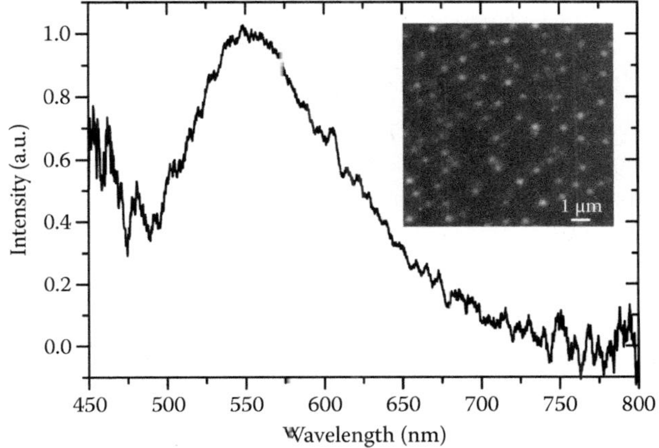

**Figure 12.15  Darkfield transmission image of 15 nm gold nanoparticles in a polymer matrix dispersed on cover glass and corresponding scattering spectrum.**

band (or laser line) for imaging. As the most common way of characterizing colloidal gold is UV-Vis absorbance, the tendency is to use the peak observed in these measurements to determine the optimal excitation wavelength. However, we must recall again that UV-Vis spectrophotometers typically measure *extinction*. Referring again to **Equation 12.1** and **Figure 12.6**, it can be appreciated to what extent the peak in the extinction spectrum is not necessarily the peak scattering wavelength. When possible, it is recommended that the peak scattering wavelength be determined using the same instrument as that used for final applications. In many cases, this can be accomplished with the use of a single focusing lens placed in the camera port and an inexpensive monochromator or charge-coupled device (CCD) array detector system. An example of such a spectrum, and the corresponding image of nanoparticles dispersed on glass, is shown in **Figure 12.15**.

## Prospects for high-speed imaging

Over the past decade, the sensitivity of array detectors (CCD cameras) has made it possible to study the relative motion of individually labeled features, and in some cases single molecules, in biological systems. Particle-tracking experiments have the potential to provide valuable insight into the dynamics of biological

**Figure 12.16 Scattering intensities versus excitation intensity.**

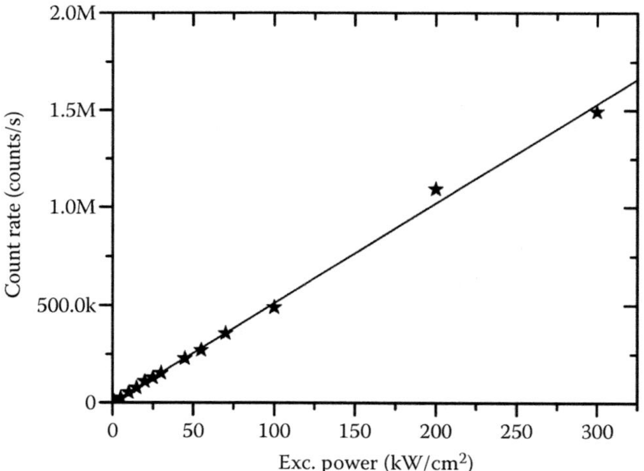

processes. Unlike imaging in fixed systems where spatial resolution is predicated upon image contrast, temporal resolution is determined primarily by available signal (contrast) per unit time. The greater the rate at which photons are collected, the greater the potential for increased image acquisition rates.

To achieve higher photon count rates, one would seemingly only need to increase the intensity of the excitation source. In the case of fluorescence, however, a maximum in the emission rate is quickly reached, as determined by the lifetime of the excited state—typically on the order of nanoseconds. Furthermore, at elevated excitation intensities, the probability of photobleaching is enhanced.

In contrast, scattering does not rely on a photon absorption event and consequently does not suffer the same limitations. Nanoparticles are far more robust than organic molecules and will continue to efficiently scatter radiation up to the melting point of the metal. As can be seen in **Figure 12.16**, as the excitation intensity is increased, the scattered signal from the nanoparticle system increases linearly without saturation. This means that high-speed scattering images can be readily obtained by simply increasing the intensity of the light source, where the limiting factor is not the probe but potential photodamage to the sample.

## Confocal microscopy

In thick or opaque samples, reflectance (backscattering) geometries are typically employed. When imaged in a backscattering geometry, metal nanoparticles produce strong signals (**Figure 12.17**) and may have several potential advantages over commercial organic fluorophores and semiconductor quantum dots. By tuning the size of the particles, they can be engineered for peak scattering at longer wavelengths, allowing for the use of red or near-infrared (IR) light sources that exhibit greater penetration into biological specimens and decreased probability of generating background autofluorescence. Furthermore, when compared to semiconductor materials (Cd, Se, Pb) or organic dyes, noble metals are relatively nontoxic, making them well suited for studies in living systems with or without a surface-capping layer.

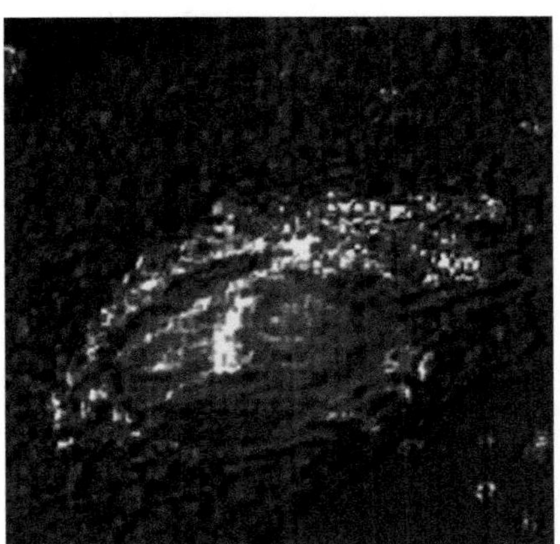

**Figure 12.17  830 nm excitation confocal reflectance image of an NIH3T3 cell labeled with gold nanoparticles.**

## 12.6  SAMPLE EXPERIMENT: LABELING CELLS WITH LECTIN-TAGGED GOLD NANOPARTICLES

*Lectins* are sugar-binding proteins that bind specifically to a glycoprotein or glycolipid. They are ubiquitous in the natural world and show differing degrees of specificity for species or classes of cells or organisms. Some lectins are used in blood typing, others for protein purification (*concanavalin A* is very widely used), and others can be used to tag bacteria. *Wheat germ agglutinin* (WGA) is specific for Gram-positive over Gram-negative bacteria, and is also taken up by a wide variety of mammalian cells, including neurons, where it is synaptically transported from one cell to another and so can be used to trace neural pathways. We will use this one in this particular example because of its wide usefulness, though the procedure described should apply to any lectin.

The biggest problems with lectins are that they tend to aggregate, as the name *agglutinin* might imply. To prepare lectin-conjugated nanoparticles, the pH of the nanoparticle solution must first be adjusted to the isoelectric point of the lectin to be used (8.5 for WGA). The amount necessary to stabilize the particle solution can be calculated using the hydrodynamic radius and concentration of the nanoparticles and varies slightly for different types of lectin. For example, the hydrodynamic radius of WGA, $r_{WGA}$, has been reported in the literature to be ~2.4 nm. The number of WGA molecules that can fit on a gold nanoparticle is then given by

$$n_{WGA} = \frac{4\pi(r_{Au} + r_{WGA})^2}{\pi r_{WGA}^2} \tag{12.10}$$

For a 15 nm gold particle, this value is about 68 WGA molecules/particle. A slight excess over this may be used, but a significant excess will probably lead to aggregation. After adding the lectin to the solution, add approximately 0.05% PEG (MW, 20,000) by weight to stabilize the lectin-nanoparticle solution for washing. Centrifuge the solution at 4°C, 40,000g, for 30 min; aspirate the supernatant; and resuspend the nanoparticles into phosphate-buffered saline (PBS).

To label cells, add aliquots of lectin-conjugated nanoparticles to tissue culture flasks at a rate of 1–5 mL (100 µg/mL, 15 nm radius) gold nanoparticles to a volume of 5 mL of media in standard tissue culture flasks, or several drops to a 35 mm petri dish. Evidence of cellular uptake can be seen by darkfield or confocal microscopy the next day.

# 12.7 APPLICATIONS IN SURFACE–ENHANCED RAMAN SCATTERING

### Introduction to Raman scattering

The scattering cross section includes both the traditional view of scattering (presented previously) and an additional phenomenon termed *Raman scattering*, first predicted by Adolf Smekal and first observed in molecular liquids by Sir Chandrasekhara Raman. In the scattering theory described previously, light is scattered elastically—that is, the energy of the incident photon is conserved upon scattering. In contrast, during Raman scattering, the photon energy is either increased or, more commonly, decreased—hence, it is inelastic scattering. In general, the change in the scattered energy results from an interaction of the incident electric field with that of a molecule, resulting in a change in the rotational, vibrational, or electronic energy of the molecule. As this energy can be either absorbed by the molecule or from the molecule, the resulting scattered photon can lose or gain a small amount of energy equal to that of the quantized transition. In liquids, under excitation with visible radiation, vibrational transitions are most probable and yield a spectrum containing structural and compositional information about the scattering molecule. Raman spectroscopy, like vibrational IR, can be used to generate a detailed "fingerprint" of the analyte molecule.

Raman scattering is a generally weak phenomenon when compared to elastic scattering and can be extremely weak when compared with other optical processes such as fluorescence. Under the best circumstances, a free molecule in solution will scatter fewer than 1 in $10^7$ photons via the Raman mechanism. Interestingly, however, when that same molecule is adjacent to a metal surface, either by physisorption or the formation of chemical bonds, the apparent Raman signal can be enhanced many orders of magnitude. The effect is called *surface-enhanced Raman* (SER or SERS [surface-enhanced Raman scattering] in the case of spectroscopic measurements) and has made it possible to extend the applications of vibrational Raman spectroscopy into a multitude of new research and even commercial areas. Specifically, SER is useful for the detection and identification of low-concentration analytes such as environmental contaminants or pathogens. Although a roughened metal surface is usually used as the Raman-enhancing surface, in recent years, Raman techniques have emerged that employ colloid metal nanoparticles for this purpose.

In general, we can describe the measured intensity from a particle due to light scattering (or any linear process) as follows:

$$I_S = NI_0\sigma_S,\qquad(12.11)$$

where $N$ represents the number of scattering particles present, $I_0$ the excitation intensity, and $\sigma_S$ the wavelength-dependent scattering cross section. In the case of Raman-active molecules deposited on a metal sphere, this equation becomes

$$I_{S,\text{Total}} = I_0 N(\sigma_S + N'\sigma_{\text{Raman}}) + I_{\text{BG}}\qquad(12.12)$$

where $N'$ represents the average number of Raman-active molecules per nanoparticle, and $I_{\text{BG}}$ represents the contribution from all other background sources of light. Since the values of $\sigma_{\text{Raman}}$ are typically very small, and $N'$ is limited by the nanoparticle surface area, it seems reasonable to expect that the inelastic Raman term would be swamped by the elastic term. In the case of broadband illumination, this would be true. However, under monochromatic (laser) illumination, the picture becomes different (**Figure 12.18**). In this case, we expect the elastic scattering from the nanoparticle to exist only at the laser wavelength, illustrated as the shaded circle in **Figure 12.18**, whereas the inelastic Raman scattering is shifted to lower energies (Stokes shifted). Rigorously speaking, a shift to higher energies is also possible (anti-Stokes), though with lesser probability. With the addition of a suitable long-pass filter, the weak Raman signal can be collected independently of the elastic scattering signal and the laser source. This setup is equivalent to that used in laser scanning (confocal) fluorescence imaging.

With the use of this somewhat common setup, the total scattering equation (**Equation 12.12**) can be simplified to include only the Raman-scattering terms. However, Raman scattering from all molecules in the excitation (confocal) volume must be considered. Hence, the scattered intensity becomes

$$I_{\text{Raman,Total}} = I_0 NN' \sigma_{\text{surf,Raman}} + I_0 N_{\text{BG}} \sigma_{\text{BG,Raman}} \tag{12.13}$$

where $N_{\text{BG}}$ and $\sigma_{\text{BG,Raman}}$ are the number of background molecules and their respective Raman cross sections. Though the Raman cross section of the background molecules can be small, in realistic biological systems, their number relative to $NN'$ can be very large. Again, it appears that our desired signal would be swamped by an unwanted background signal.

Finally, we must consider the nature of the SER effect itself, which is believed to result from two factors: an electric field enhancement in the presence of a metallic surface and a chemical enhancement from the interaction of the Raman reporter with its substrate. The dominant enhancement is the electric field enhancement effect, which can be considered qualitatively as an increase in the amplitude of radiation field, $E_0(\lambda)$, experienced by the surface-bound molecule, with the intensity being proportional to the square of the electric field. Based on the dipolar description of scattering, an oscillating incident field ($E_0$) causes the

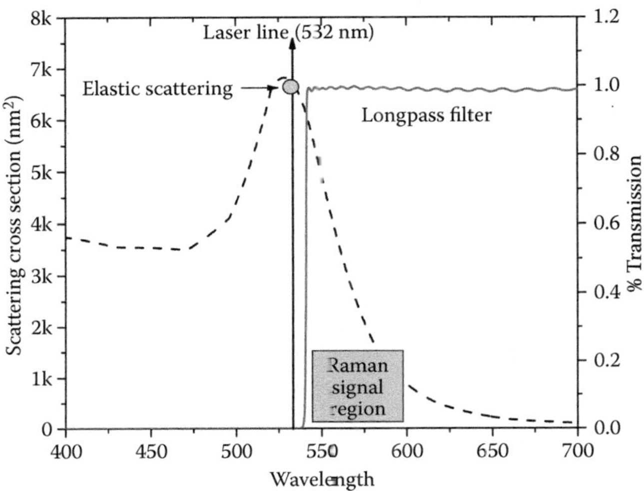


**Figure 12.18 Comparison of scattering and Raman versus illumination and filter.**

electrons in the particle to oscillate, generating an additional radiator ($E_\text{p}$). The exciting field experienced by the surface molecules is the superposition of these two fields: $E_\text{M} = E_0 + E_\text{p}$. We can describe the enhancement relative to the incident field as

$$A(\lambda_\text{L}) = \frac{E_\text{M}(\lambda_\text{L})}{E_0(\lambda_\text{L})} \approx \frac{m^2-1}{m^2+2}\left(\frac{r}{r+d}\right)^3, \qquad (12.14)$$

where $\lambda_\text{L}$ is the laser wavelength, $r$ is the radius of the surface (e.g., a nanoparticle), $d$ is the distance from the surface to the bound molecule, and $m$ is the relative index of refraction defined in **Equation 12.5**. The Raman-scattering field, now at a slightly shifted wavelength, $\lambda_\text{S}$, is also enhanced in a similar fashion, giving its own enhancement factor, $A(\lambda_\text{S})$. For small, tightly bound molecules on larger spheres, $d \ll r$, so the last term in **Equation 12.14** is approximately unity. The total electric field enhancement factor, $G_\text{E}$, can then be written as

$$G_\text{E} = \left|A(\lambda_\text{L})\right|^2 \left|A(\lambda_\text{s})\right|^2 \approx \left|\frac{m^2-1}{m^2+2}\right|^4 \qquad (12.15)$$

An additional enhancement is sometimes observed when an electronic coupling between surface molecules and metal conduction band electrons occurs. This is dependent upon the chemical nature of the surface-bound molecule and is termed the chemical effect, which we represent as the chemical enhancement factor $G_\text{C}$. The total Raman-scattered intensity is then

$$I_\text{Raman,Total} = I_0 G_\text{E} G_\text{C} N N' \sigma_\text{surf,Raman} + I_0 N_\text{BG} \sigma_\text{BG,Raman}, \qquad (12.16)$$

where the enhancement term affects only those molecules directly adjacent to the nanoparticle surface, as the distance-dependent term in **Equation 12.14** drops off at a rate proportional to $(1/d^3)^4 = 1/d^{12}$ for larger distances. If the wavelength of the incident light source ($I_0$) is selected to be at the nanoparticle scattering peak, we observe a huge increase in Raman signal from each of the $N'$ surface-bound molecules on each of the $N$ particles present, relative to that of the background molecules. This suggests that it may be possible to observe otherwise very weak Raman signals, even in complex biological systems, and suggests Raman-based nanoprobes as an alternative to fluorescent molecules or quantum dots.

One of the perceived advantages of Raman-based probes is the highly identifiable fingerprint-like spectra that are observed. In contrast to fluorescence, Raman signatures arise from very small changes in vibrational energy, and as a result, give rise to very narrow peaks (~20/cm). Furthermore, since Raman is a vibrational technique sensitive to composition, most molecules generate multiple distinguishable peaks.

An example is shown in **Figure 12.19**, where the solution-phase Raman spectra of three different probe systems—Raman-active molecules bound to gold nanoparticles—are compared. Note the total spectral range shown as compared to that for typical fluorescent molecules. Additionally, it becomes apparent that probes can be readily identified based on the position of multiple peaks, information not typically available in fluorescence. These features make Raman-active nanoprobes ideal candidates for the simultaneous labeling of large

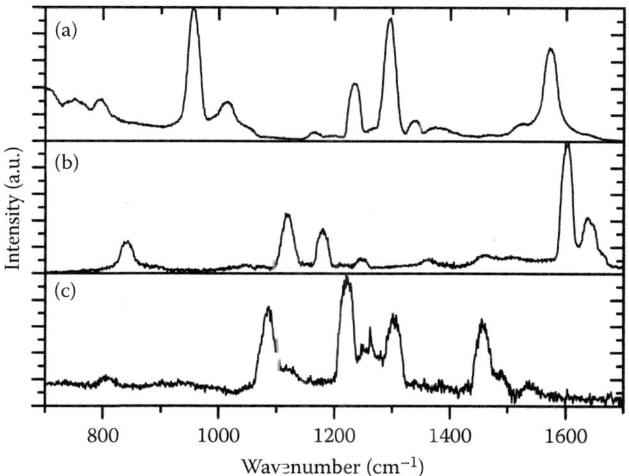

**Figure 12.19 Comparison of three Raman-active probes based on spherical gold nanoparticle cores tagged with (a) naphthalenethiol, (b) bis(4-pyridyl)ethylene, and (c) mercaptopyridine.**

numbers of biological structures or species and thereby possibly overcoming one of the key limitations in fluorescence imaging.

While this area of research—referred to as spectral multiplexing—has garnered significant interest, applications are still forthcoming. This is due to several practical challenges that still exist. In order to obtain a Raman image, a spectrum must be obtained at each image pixel. To obtain detailed spatial information, this can mean upward of a million pixels. For reasonable image acquisition times, where collected signal is dispersed over multiple wavelengths using a monochromator, the Raman signal rates must be very large. Raman signals can be increased orders of magnitude by simply increasing the incident laser power; however, the efficient conversion of absorbed radiation into heat demonstrated by some nanoparticles can become an issue. Furthermore, especially at elevated local temperatures, the adsorption of undesirable molecules from the environment onto the particle, and desorption from the particle, can lead to single-particle Raman signals that fluctuate over time. While this may be useful in determining the presence of molecules in solution, it is a generally undesirable quality for a nanoprobe intended for imaging and is similar to the fluorescence intermittency observed in fluorescent probes. Some of these issues are addressed in the following sections.

## Protected Raman-active nanospheres

In an attempt to address the apparent environmental sensitivity of Raman-active nanoprobes, several surface entrapment strategies have been implemented. In addition to protecting the nanoparticle surface from adsorption of foreign molecules, these methods also eliminate desorption, effectively fixing the number of Raman-active surface species. One relatively popular early approach involves the growth of thin layers of silica (glass) on the nanoparticle surface. A TEM image of such a system is shown in **Figure 12.20**, where the dark spherical gold cores are clearly visible surrounded by a hazy layer indicating the presence of the silica.

Formation of a silica shell on a gold particle is done by chemisorption of an amino- or mercapto-silane. The concentration of the Raman reporter must be reduced in order to accommodate the silane molecules. This method does result in stable Raman-active probes. In fact, in some cases, these probes can be boiled

**Figure 12.20 TEM image of silica-coated Raman nanoprobes.**

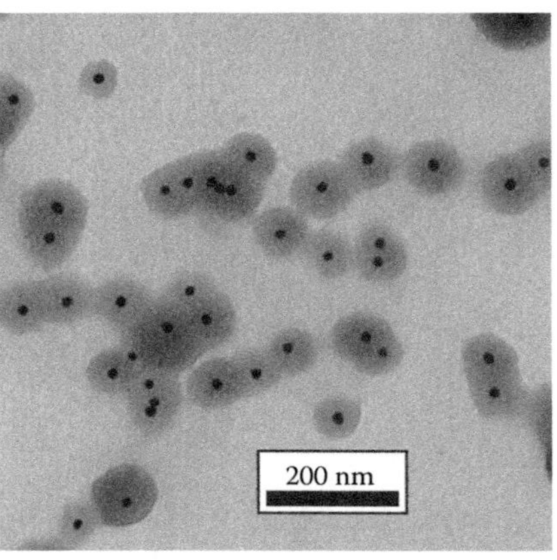


---

### PRACTICAL TIPS 12.4:    RAMAN TAG ADDITION

Preparation of SERS nanoparticles involves surface modification with 5 kDa α-methoxy-omega-mercapto poly(ethylene glycol) (PEG) and 4-mercaptopyridine (4Mpy) and can be done in aqueous solution or in ethanol. Gold nanoparticles are prepared and washed to remove excess reactants and templating agents. After the final washing, the resulting nanoparticles are then surface-modified with a 1:3 molar ratio of the 4Mpy and PEG. Under vigorous stirring, 750 μL of 22 mM 4Mpy is slowly added and stirred for 15 min. Following this, 3.5 mL of 12 μM PEG is slowly added while stirring vigorously for another 15 min. Finally, the mixture is washed and resuspended in water, leaving the SERS nanoparticles ready for characterization.

---

in acid without significant loss in Raman signal. However, the resulting signals are significantly reduced relative to the unprotected species, and the silica growth chemistry itself is exceedingly challenging to perfect. An additional disadvantage to this method is the increase in particle size, which places limitations on practical uses in biological systems.

An alternative approach involves the co-chemisorption of Raman-active molecules with polymer molecules such as PEG–thiol (or dithiol). In these systems, the radius of the particle (typically measured as the hydrodynamic radius) is only slightly increased. Thus, the smaller Raman reporter molecules remain trapped below the dynamic polymer surface, and the overall solubility and long-term stability of the particle system is improved. Possibly the greatest single advantage is the relative simplicity of the chemistry, which proceeds rapidly and is self-driven by the gold–thiol bonds formed. This approach has gained momentum and is recommended to any interested in the Raman imaging method. **Practical Tips 12.4** gives our favorite recipe for Raman tagging of gold nanoparticles.

## SERS nanoparticles: Beyond spheres

Spherical gold particles are by far the most widely studied and are used in commercial and research applications. In all likelihood, this is due to their commercial availability and the ease with which they can be synthesized, purified, and characterized. There are, however, a number of more interesting structures, the synthetic methods for which were found by sheer accident. These structures include rods, platelets, prisms, truncated octahedra, "dog bones," and most recently, nanostars.

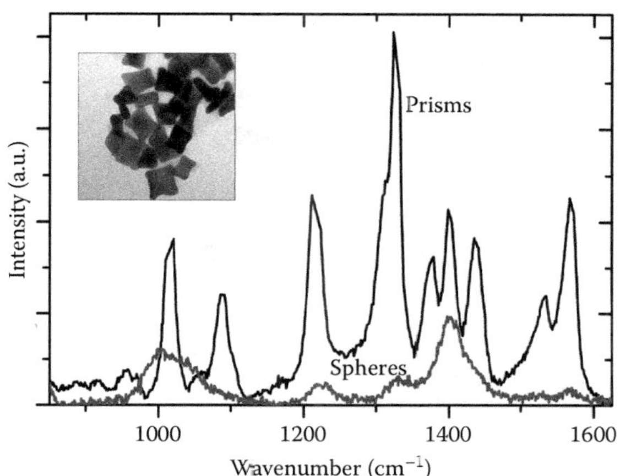



**Figure 12.21 TEM and Raman spectra of mercaptopyridine on approximately 35 nm trigonal gold prisms.**

The local electric field enhancement seen on a metal surface is related to the local radius of curvature of the surface, being amplified near sharp features—this is the so-called lightning rod effect, described in the Raman literature both phenomenologically and theoretically. This effect means that particles with sharp edges or vertices should provide an enhancement of the SERS signal over what is seen with equivalently sized spheres. Prismatic gold nanoparticles have both sharp edges and vertices and are compared with similarly sized spheres in **Figure 12.21**, confirming the effect. While prismatic nanoparticles are a potential improvement over spherical particles, the additional chemical steps involved in their synthesis (not described here) as well as the relative size heterogeneity of the typical samples make these particles less popular as Raman reporters.

Nanostars have shown more promise, especially because their synthesis is easy (**Practical Tips 12.2**). In the context of SERS, these particles exhibit all of the perceived benefits of spherical gold particles, but are composed of numerous sharp features and hence take full advantage of the lightning rod effect. While the synthesis of these structures does require a nonaqueous route, they are readily transferred into aqueous suspension following centrifugation and washing, and even exhibit modest stability without the use of stabilizers such as PEG. Raman tag addition and PEGylation follow the same protocols as that described for gold spheres.

Owing to the relative complexity of the templating and growth mechanism and the dependence on the initial gold core crystalline structure, individual nanostar structures vary greatly. This might seem to suggest that the resulting single-particle Raman signals would also exhibit significant variance. In fact, this is generally not the case.

In **Figure 12.22a**, Raman-active nanostars ~93 nm in size, tagged with mercaptopyridine and protected with PEG–thiol, have been imaged in a sample scanning confocal microscope under 632 nm excitation. After passing through a 647 nm long-pass filter, the resulting Raman signal is integrated to obtain the single-pixel intensities shown. For comparison, in the same instrument, under the same excitation power, Alexa Fluor 647 molecules were imaged as shown in **Figure 12.22b**. Alexa Fluor 647 molecules are among the most efficient and longest-lived fluorophores available. It is evident from the figure that the measured Raman signal is of approximately the same intensity as that measured for the fluorophores. Furthermore, signal intermittency—clearly visible as the

**Figure 12.22 Comparison of gold nanoparticles with commercial fluorophores.** (a) Confocal Raman images of 93 nm Raman-active nanostars tagged with 4-mercaptopyridine with 40 µW at the sample and (b) confocal fluorescence image of Alexa Fluor 647 images under identical conditions as (a) with 45 µW at the sample.

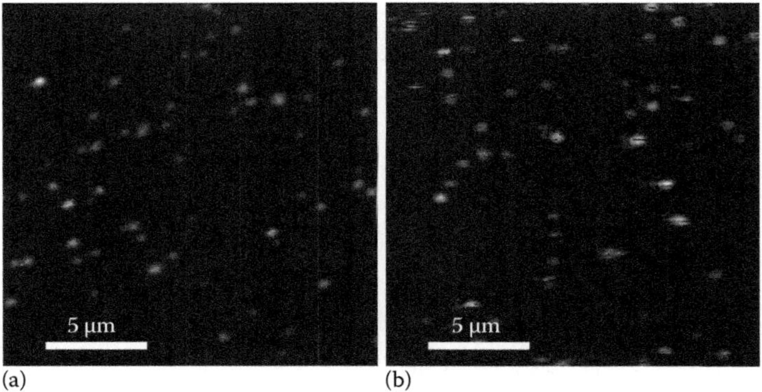



dark horizontal lines across the fluorescence features—is not observed in the Raman image. A more detailed description can be found in references given at the end of the chapter, which also compare the effect of increased excitation intensity, photobleaching, and signal variance. In all cases, the Raman nanostars appear to perform as well as or better than the state-of-the-art fluorophores. It is apparent that this geometry offers great potential for advancing Raman imaging as a powerful tool for the biological sciences.

## 12.8  GOLD NANOPARTICLES AS PHOTOTHERMAL TRANSDUCERS

As the source of light irradiating a gold nanoparticle is brought closer to its absorption maximum, light is absorbed by the particle, and the energy is transferred into electron oscillations and lattice expansion. As this is nonradiative, the temperature of the nanoparticle and its surrounding medium rise. Predicting the change in temperature of the nanoparticle and its surrounding medium is commonly accomplished by solving the heat transfer system of equations, which yields the result

$$\Delta T_{\text{medium}} = \frac{V_{\text{np}}Q}{4\pi r K_{\text{medium}}} \tag{12.17}$$

where $T_{\text{medium}}$ is the temperature of the surrounding medium, $V_{\text{np}}$ is the volume of the nanoparticle, $r$ is the distance from the nanoparticle, $K_{\text{medium}}$ is the thermal conductivity of the medium, and $Q$ is the heat generation rate. Note that the temperature decreases as $1/r$ from the nanoparticle. The heat generation rate is given by

$$Q = \frac{\sigma_{\text{abs}}I}{V_{\text{np}}} \tag{12.18}$$

where $I$ is the laser flux intensity and $\sigma_{\text{abs}}$ is the absorption cross section. Substitution of **Equation 12.18** into **Equation 12.17** gives the change in temperature as a function of laser intensity and distance from the nanoparticle as

$$\Delta T_{\text{medium}} = \frac{\sigma_{\text{abs}}I}{4\pi r k_{\text{medium}}} \tag{12.19}$$

Thus, radiating an area populated with metallic nanoparticles will heat the immediate environment surrounding the nanoparticles but leave areas not well populated at ambient temperatures. In this way, local environments can be selectively heated. Nanoparticle-facilitated local heating has recently drawn interest at the interface of biology and physics—for example, in the targeted killing of cancer cells by local hyperthermia.

## 12.9 CONCLUSION

Recent advances in nanoparticle synthesis have led to strategies for the facile and controlled synthesis of several new gold nanoparticles sizes and geometries. Like other nanomaterials, gold nanoparticles have optical properties different from those of the bulk and can be made to produce unique measurable signals in a range of experimental setups. Gold nanoparticles offer some distinct advantages over other probe systems, and though the physical phenomena behind signal generation may differ, they can be viewed in many standard microscopes, including brightfield, fluorescence, and confocal instruments. Gold nanoparticles can generate long-lived signals free of intermittency or photobleaching and demonstrate cellular toxicity well below that of other probe systems. Owing to the general use of the strong bond between gold and sulfur, standard thiol-based surface modification and NHS/EDC coupling strategies can be employed. Emerging spectroscopic techniques, such as Raman imaging, exploit the engineered optical properties of gold nanoparticles. This and other surface-based methods provide the opportunity for new directions in labeling and imaging. Though not specifically discussed in this chapter, gold nanoparticles have been used as contrasting agents in electron microscopy, which can now be directly correlated with optical measurements. For labeling and sensing measurements, where larger probe sizes are acceptable, gold nanoparticles provide an excellent platform with great flexibility.

### End-of-Chapter Problems

1. (a) Predict the wavelength corresponding to the extinction cross section maximum for a 45-nm-diameter gold nanoparticle in water. (b) For the same particle as (a), predict the maximum scattering wavelength using the scattering cross section.

2. Find the laser power needed to raise the temperature 100 K at a distance of 10 nm from a 100-nm-diameter gold nanoparticle, in water. (b) Repeat this calculation at a distance of 1000 nm for the same nanoparticle as part (a).

3. Consider a thin sample with negligible absorption and background generation for imaging with a single wavelength excitation source. If this sample is populated with 20-nm-diameter gold nanoparticles, would a scattering transmission or reflection mode be preferred, and why? What wavelength would be best for excitation?

4. (a) How many gold atoms are in a gold nanoparticle that is 50 nm in diameter? (b) Roughly what percent of these atoms are on the surface of the particle?

5. Consider Equation 12.9. Multiplying both sides by $d\Omega$ and integrating from 0 to $2\pi$ for $d\varphi$ and from 0 to $\pi$ for $d\theta$ gives the total scattering cross section as presented in Equation 12.7. (a) Verify the total scattering cross section presented in Equation 12.7. (b) Derive the total scattering cross section, using Equation 12.9, for the case of a 1.4-NA objective lens. (c) Using the bulk index of refraction for gold, compute the total cross section for a

25-nm-diameter gold nanoparticle in water excited by 632 nm incident radiation using **Equation 12.7**. (d) Using your result from part (b), compute the scattering cross section for the same gold particle as (c) under the same excitation conditions. (e) What percent of the total light scattered from the gold nanoparticle of part (c) would not be collected by the 1.4-NA objective lens?

## Background Reading

### Books

*Classic and modern texts that cover topics mentioned in this chapter*

Bohren, C.F., and Huffman, D.R. *Absorption and Scattering of Light by Small Particles*. John Wiley & Sons, Inc., New York, New York, 1983.

Born, M., and Wolf, E. *Principles of Optics*. Edn. 7. Cambridge University Press, Cambridge, UK, 1999.

Hecht, E. *Optics*. Edn. 4. Addison Wesley, New York, 2002.

Hulst, H.C.v.d. *Light Scattering by Small Particles. Dover,* New York, 1981.

Jackson, J.D. *Classical Electrodynamics*. Edn. 3. John Wiley & Sons, Inc., New York, 1999.

Jonasz, M., and Fournier, G.R. *Light Scattering by Particles in Water*. Academic Press, London, 2007.

Kokhanovsky, A.A. *Light Scatering in Media Optics Problems and Solutions*. Edn. 3. Springer-Verlag, Berlin, 2004.

Kreibig, U., and Vollmer, M. *Optical Properties of Metal Clusters*, vol. 25. Springer-Verlag, Berlin, 1995.

Maier, S.A. *Plasmonics: Fundamentals and Applications*. Springer Science, New York, New York, 2007.

McHale, J.L. *Molecular Spectroscopy*. Prentice-Hall, Inc., Upper Saddle River, NJ, 1999.

Nehl, C.L., Laio, H., and Hafner, J.H. Plasmon Resonant Molecular Sensing with Single Gold Nanostars, Proceedings of Spie-the International Society for Optical Engineering, Article No. 6323, 2006.

Verwey, E.J.W., and J.T.G. Overbeek. *Theory of the Stability of Lyophobic Colloids*. Elsevier, Amsterdam, 1948.

### Journal articles

*Articles on topics covered by this chapter*

Aherne, D., Gara, M., Kelly, J.M., and Gun'ko, Y.K. (2012). From Ag nanoprisms to triangular AuAg nanoboxes. *Advanced Functional Materials* 20, 1329–1338.

Alivisatos, P.A., Gu, W., and Larabell, C. (2005). Quantum dots as cellular probes. *Annual Review of Biomedical Engineering* 7, 55–76.

Allgeyer, E.S., Pongan, A., Browne, M., and Mason, M.D. (2009). Optical signal comparison of single fluorescent molecules and Raman active gold nanostars. *Nano Letters* 9, 3816–3819.

Basiruddin, S., Saha, A., Pradhan, N., and Jana, N.R. (2012). Advances in coating chemistry in deriving soluble functional nanoparticle. *Journal of Physical Chemistry C* 114, 11009–11017.

Bishop, P.T. (2002). The use of gold mercaptides for decorative precious metal applications. *Gold Bulletin* 35, 89–98.

Boyack, R., and Ru, E.C.L. (2009). Investigation of particle shape and size effects in SERS using T-matrix calculations. *Physical Chemistry Chemical Physics* 11, 7398–7405.

Bruzzone, S., and Malvaldi, M. (2009). Local field effect on laser-induced heating of metal nanoparticles. *Journal of Physical Chemistry C* 113, 15805–15812.

Byrne, J.D., Betancourt, T., and Brannon-Peppas, L. 2008. Active targeting schemes for nanoparticle systems in cancer therapeutics. *Advanced Drug Delivery Reviews* 60, 1615–1626.

Carotenuto, G., Pepe, G. P., and Nicolais, L. (2000). Preparation and characterization of nano-sized Ag/PVP composites for optical applications. *European Physical Journal B* 16, 11–17.

Chylek, P. (1977). Light scattering by small particles in an absorbing medium. *Journal of the Optical Society of America* 67, 561–563.

Costanzo, P.J., Liang, E., Patten, T.E., Collins, S.D., and Smith, R.L. (2005). Biomolecule detection via target mediated nanoparticle aggregation and dielectrophoretic impedance measurement. *Lab on a Chip* 5, 606–612.

De, M., Ghosh, P.S., and Rotello, V.M. (2008). Applications of nanoparticles in biology. *Advanced Materials* 20, 4225–4241.

Derjaguin, B.V., and Landau, L. (1941). Theory of the stability of strongly charged lyophbic sols and the adhesion of strongly charged particles in solution fo electrolytes. *Acta Physicochimica USSR* 14, 633–662.

Eck, W., Craig, G., Sigdel, A., Ritter, G., Old, L.J., Tang, L., Brennan, M.F., Allen, P.J., and Mason, M.D. (2008). PEGylated gold nanoparticles conjugated to monoclonal F19 antibodies as targeted labeling agents for human pancreatic carcinoma tissue. *ACS Nano* 2, 2263–2272.

Faraday, M. (1857). The Bakerian lecture: Experimental relations of gold (and other metals) to light. *Philosophical Transactions of the Royal Society of London* 147, 145–181.

Fuller, S.B., Wilhelm, E.J., and Jacobson, J.M. (2002). Ink-jet printed nanoparticle microelectromechanical systems. *Journal of Microelectromechanical Systems* 11, 54–60.

Hardman, R. (2006). A toxicologic review of quantum dots: Toxicity depends on physicochemical and environmental factors. *Environmental Health Perspectives* 114, 165.

Hughes, M.D., Xu, Y.J., Jenkins, P., McMorn, P., Landon, P., Enache, D.I., Carley, A.F., Attard, G.A., Huthcings, G.J., King, F. et al. (2005). Tunable gold catalysts for selective hydrocarbon oxidation under mild conditions. *Nature Letters* 437, 1132–1135.

Jain, P.K., Qian, W., and El-Sayed, M.A. (2006). Ultrafast cooling of photoexcited electrons in gold nanoparticle–thiolated DNA conjugates involves the dissociation of the gold–thiol bond. *Journal of the American Chemical Society* 128, 2426–2433.

Jana, N.R., Gearheart, L., and Murphy, C.J. (2001). Seed-mediated growth approach for shape-controlled synthesis of spheroidal and rod-like gold nanoparticles using a surfactant template. *Advanced Materials* 13, 1389–1393.

Johnson, P.B., and Christy, R.W. (1972). Optical constants of the noble metals. *Physical Review B* 6, 4370.

Kalele, S., Gosavi, S.W., Urban, J., and Kulkarni, S.K. (2006). Nanoshell particles: Synthesis, properties and applications. *Current Science* 91, 1038.

Kelly, K.L., Coronado, E., Zhao, L.L., and Schatz, G.C. (2003). The optical properties of metal nanoparticles: The influence of size, shape, and dielectric environment. *Journal of Physical Chemisty B* 107, 668–677.

Keren, S., Zavaleta, C., Cheng, Z., de la Zerda A., Gheysens, O., and Gambhir, S.S. (2008). Noninvasive molecular imaging of small living subjects using Raman spectroscopy. *PNAS* 105, 5844–58412.

Khlebtsov, N.G., and Dykman, L.A. (2012). Optical properties and biomedical applications of plasmonic nanoparticles. *Journal of Quantitative Spectriscopy and Radiative Transfer* 111, 1–35.

Khoury, C.G., and Vo-Dinh, T. (2008). Gold nanostars for surface-enhanced raman scattering: Synthesis, cheraterization and optimization. *Journal of Physical Chemistry C* 112, 18849–188512.

Kirchner, C., Liedl, T., Kudera, S., Pellegrino, T., Javier, A.M., Gaub, H.R., Stoelzle, S., Fertig, N., and Parak, W.J. (2005). Cytotoxicity of colloidal CdSe and CdSe/ZnS nanoparticles. *Nano Letters* 5, 331–338.

Law, M., Greene, L.E., Johnson, J.C., Saykally, R., and Yang, P. (2005). Nanowire dye-sensitized solar cells. *Nature Letters* 4, 455–459.

Lin, A.W.H., Lewinski, N.A., West, J.L., Halas, N.J., and Drezek, R.A. (2005). Optically tunable nanoparticle contrast agents for early caner detection: Model based analysis of gold nanoshells. *Journal of Biomedical Optics* 10, 064035.

Lo, C.T., Chou, K.S., and Chin, W.K. (2001). Effects fo mising procedures on the volume fraction of silver particles in conductive adhesives. *Adhesion Science and Technology* 15, 783–792.

Lord, S.J., Lee, H.-l.D., and Moerner, W.E. (2012). Single-molecule spectroscopy and imaging of biomolecules in living cells. *Analytical Chemistry* 82, 2192–2203.

Maxwell, D.J., Taylor, J.R., and Nie, S. (2002). Self-assembled nanoparticle probes for recognition and detection of biomolecules. *Journal of the American Chemical Society* 124, 9606–9612.

Mie, G. (1908). Beitrage zur optik truber medien, speziell kolloidaler metallosungen, *Annaler der Physik* 25, 377–445.

Moskovits, M. (1985). Surfaced-enhanced spectroscopy. *Reviews of Modern Physics* 57, 783.

Mulvaney, S.P., Musick, M.D., Kearting, C.D., and Natan, M.J. (2003). Glass-coated, analyte-tagged nanoparticles: A new tagging system based on detection with surface-enhanced Raman scattering. *Langmuir* 19, 4784–4790.

Murphy, C.J., Gole, A.M., Stone, J.W., Sisco, P.N., Alkilany, A.M., Goldsmith, E.C., and Baxter, S.C. (2008). Gold nanoparticles in biology: Beyond toxicity to cellular imaging. *Accounts of Chemical Research* 41, 1721.

Nair, L.S., and Laurencin, C.T. (2007). Silver nanoparticles: Synthesis and therapeutic applications. *Journal of Biomedical Nanotechnology* 3, 301–316.

Nehl, C.L., Liao, H., and Hafner, J.H. (2006). Optical properties of star-shaped gold nanoparticles. *Nano Letters* 6, 683–688.

Newton, R.G. (1976). Optical theorem and beyond. *American Journal of Physics* 44, 6312.

Otsuka, H., Nagasaki, Y., and Kataoka, K. (2003). PEGylated nanoparticles for biological and pharmaceutical applications. *Advanced Drug Delivery Reviews* 55, 403–419.

Pissuwana, D., Valenzuelaa, S.M., and Cortie, M.B. (2006). Therapeutic possibilities of plasmonically heated gold nanoparticles. *Trends in Biotechnology* 24, 62–67.

Pradhan, N., Pal, A., and Pal, T. (2002). Silver nanoparticle catalyzed reduction of aromatic nitro compounds. *Colloids and Surfaces A* 196, 247–257.

Qian, X., Peng, X.-H., Ansari, D.C., Yin-Goen, Q., Chen, G.Z., Shin, D.M., Yang, L., Young, A.N., Wang, M.D., and Nie, S. (2008). In vivo tumor targeting and spectroscopic detection with surfaced-enhanced Raman nanoparticle tags. *Nature Biotechnology* 26, 83.

Roberts, S. (1960). Optical properties of copper. *Physical Review* 188, 15012.

Ryder, A.G. (2005). Surface enhanced Raman scattering for narcotic detection and applications to chemical biology. *Current Opinion in Chemical Biology* 9, 489–493.

Salata, O. (2004). Applications of nanoparticles in biology and medicine. *Journal of Nanobiotechnology* 2, 3–8.

Schütz, G.J., Pastushenko, V.P., Gruber, H.J., Knaus, H.-G., Pragl, B., and Schindler, H. (2000). 3D imaging of individual ion channels in live cells at 40 nm resolution. *Single Molecule* 1, 25–31.

Shiraishi, Y., Maeda, K., Yoshikawa, H., Xu, J., Toshima, N., and Kobayashi, S. (2002). Frequency modulation response of a liquid-crystal electro-optic device doped with nanoparticles. *Applied Physics Letters* 81, 2845–2847.

Stoller, P., Jacobsen, V., and Sandoghdar, V. (2006). Measurement of the complex dielectric constant of a single gold nanoparticle. *Optics Letters* 31, 2474.

Sun, W., Wang, G., Fang, N., and Yeung, E.S. (2009). Wavelength-dependent differential interference contrast microscopy: Selectively imaging nanoparticle probes in live cells. *Analytical Chemistry* 81, 9203–9208.

Takuro Niidome, M.Y., Okamoto, Y., Akiyama, Y., Takahashi, H., Kawano, T., Katayama, Y., andNiidome, Y. (2006). PEG-modified gold nanorods with a stealth character for in vivo applications. *Journal of Controlled Release* 114, 343–347.

Uppal, M.A., Kafizas, A., Lim, T.H., and Parkin, I.P. (2012). The extened time evolution size decrease of gold nanoparticles formed by the Turkevich method. *New Journal of Chemistry* 34, 1401–1407.

Xia, Y., Yang, P., Sun, Y., Wu, Y., Mayers, B., Gates, B., Win, Y., Kim, F., and Yan, H. (2003). One-dimensinnal nanostructures: Synthesis, characterization, and applications. *Advanced Materials* 15, 353.

Zhang, T., Stilwell, J.L., Gerion, D., Ding, L., Elboudwarej, O., Cooke, P.A., Gray, J.W., Alivisatos, A.P., and Chen, F.F. (2006). Cellular effect of high doses of silica-coated quantum dot profiled with high throughput gene expression analysis and high content cellomics measurements. *Nano Letters* 6, 800.

Zheng, J., Zhang, C., and Dickson, R.M. (2004). Highly fluorescent, water-soluble, size-tunable gold quantum dots. *Physical Review Letters* 93, 077402(1)–077402(4).

Zhou, J., Ralston, J., Sedev, R., and Beattie, D.A. (2009). Functionalized gold nanoparticles: Synthesis, structure and colloid stability. *Journal of Colloid and Interface Science* 331, 251–262.

## *Some articles on topics not covered in this chapter*

Brandenberger, C., Muhlfeld, C., Ali, Z., Lenz, A.G., Schmid, O., Parak, W.J., Gehr, P., and Rothen-Rutishauser, B. (2012). Quantitative evaluation of cellular uptake and trafficking of plain and polyethylene glycol–coated gold nanoparticles. *Small* 6, 1669–1678.

Chhabra, R., Sharma, J., Wang, H., Zou, S., Lin, S., Yan, H., Lindsay, S., and Liu, Y. (2009). Distance-dependent interactions between gold nanoparticles and fluorescent molecules with DNA as tunable spacers. *Nanotechnology* 20, 48, 485201.

Connor, E.E., Mwamuka, J., Gole, A., Murphy, C.J., and Wyatt, M.D. (2005). Gold nanoparticles are taken up by human cells but do not cause acute cytotoxicity. *Small* 1, 325–327.

Curley, S.A., Cherukuri, P., Briggs, K., Patra, C.R., Upton, M., Dolson, E., and Mukherjee, P. (2008). Noninvasive radiofrequency field–induced hyperthermic cytotoxicity in human cancer cells using cetuximab-targeted gold nanoparticles. *Journal of Experimental Therapeutics and Oncology* 7, 313–326.

He, W., Ladinsky, M.S., Huey-Tubman, K.E., Jensen, G.J., McIntosh, J.R., and Bjorkman, P.J. (2008). FcRn-mediated antibody transport across epithelial cells revealed by electron tomography. *Nature* 455, 542–546.

Kim, J.H., Kim, K.W., Kim, M.H., and Yu, Y.S. (2009). Intravenously administered gold nanoparticles pass through the blood–retinal barrier depending on the particle size, and induce no retinal toxicity. *Nanotechnology* 20, 505101.

Kong, T., Zeng, J., Wang, X., Yang, X., Yang, J., McQuarrie, S., McEwan, A., Roa, W., Chen, J., and Xing, J.Z. (2008). Enhancement of radiation cytotoxicity in breast-cancer cells by localized attachment of gold nanoparticles. *Small* 4, 1537–1543.

Lasne, D., Blab, G.A., Berciaud, S., Heine, M., Groc, L., Choquet, D., Cognet, L., and Lounis, B. (2006). Single nanoparticle photothermal tracking (SNaPT) of 5-nm gold beads in live cells. *Biophysical Journal* 91, 4598–4604.

Leenders, M., Gerwin, C., and Sheng, Z.H. (2004). Multidisciplinary approaches for characterizing synaptic vesicle proteins. *The Current Protocols in Neuroscience*, 28:2.7:2.7.1–2.7.18.

Levy, R., Shaheen, U., Cesbron, Y., and See, V. (2010). Gold nanoparticles delivery in mammalian live cells: A critical review. *NanoReviews* 1, 1–18.

Mendoza, K.C., McLane, V.D., Kim, S., and Griffin, J.D. (2012). In vitro application of gold nanoprobes in live neurons for phenotypical classification, connectivity assessment, and electrophysiological recording. *Brain Research* 1325, 19–27.

Pissuwan, D., Cortie, C.H., Valenzuela, S.M., and Cortie, M.B. (2012). Functionalised gold nanoparticles for controlling pathogenic bacteria. *Trends in Biotechnology* 28, 207–213.

Spin, J.M., and Atkinson, D. (1995). Cryoelectron microscopy of low density lipoprotein in vitreous ice. *Biophysical Journal* 68, 2115–2123.

Svarovsky, S., Borovkov, A., and Sykes, K. (2008). Cationic gold microparticles for biolistic delivery of nucleic acids. *Biotechniques* 45, 535–540.

Zhang, X.D., Guo, M.L., Wu, H.Y., Sun, Y.M., Ding, Y.Q., Feng, X., and Zhang, L.A. (2009). Irradiation stability and cytotoxicity of gold nanoparticles for radiotherapy. *International Journal of Nanomedicine* 4, 165–173.

Zheng, J., Zhang, C., and Dickson, R.M. (2004). Highly fluorescent, water-soluble, size-tunable gold quantum dots. *Physical Review Letters* 93, 077402.

## Suppliers and software

### *Some major suppliers of chemicals*

Fisher Scientific. Major supplier of laboratory chemicals and supplies.

Sigma-Aldrich. Major supplier of laboratory chemicals, including metal salts.

Iris Biotech. Supplier of various types of specialty chemicals including PEG molecules with various chain lengths and functional groups.

VWR International. Supplier for tissue culture supplies and general laboratory supplies.

### Software

MiePlot. Free software package for simulating scattering and extinction spectra of simple nanoparticle geometries.

MATLAB. Versatile scripting language with easy to use image handling and processing tools.

### *Recommended spectroscopy vendors*

Princeton Instruments. High-end, high-quality monochrometers and CCD cameras for spectroscopy.

Ocean Optics. Affordable vendor of spectroscopy equipment particularly simple robust UV–Vis spectrometers and light sources.

# CHAPTER 13

# Advanced Topics in Gold Nanoparticles: Biomedical Applications

## 13.1 INTRODUCTION

The goal of this chapter is to introduce the concepts and challenges involved in using metal nanoparticles—particularly gold—as *nanomedicines*. A large number of cell and animal studies have suggested that gold nanoparticles might be highly effective at augmenting cancer therapy by delivering *chemotherapeutic* agents to tumors, providing localized heating (*hyperthermia therapy*), amplifying ionizing radiation dose (*radiotherapy*), or some combination of all three. The first half of this chapter summarizes these studies, with a particular focus on radiation therapy, which can be tricky to perform on cells.

As of mid-2017, very few nanoparticle formulations are in clinical trials. When a nanoparticle is designed to be used in humans, it is subject to much more rigorous control than a laboratory formulation, and it must be able to be fabricated in large, reproducible batches by commercial laboratories. Most basic researchers do not have these issues in mind when they are working with nanoparticles in the laboratory. Most researchers also do not have all of the skills or instruments to perform the panel of tests in cells and animals needed before an agent is submitted for approval as an investigational drug to the Food and Drug Administration (FDA). The second half of the chapter details the steps my laboratory has taken toward approval of a particular gold nanoparticle conjugate, with the aim of providing the reader with an idea of what characterization is necessary before a nanoparticle agent might be taken toward the clinic.

## 13.2 THE USE OF GOLD IN MEDICINE

Approval of nanoparticle-based drugs by the FDA is hindered by the lack of consistent data or guidelines. One of the only solid nanoparticles to have made headway in approvals for nanomedicine is gold (another is iron oxide). Gold is one of the most well-studied nanoparticle delivery vehicles for drugs, particularly anticancer drugs conjugated to the nanoparticle surface. One reason for the popularity of gold is due to its ease of synthesis and conjugation to nearly any molecule with an active group. In addition—perhaps surprisingly, since it is a heavy metal—gold is of low toxicity *in vivo* and has been used for centuries to

**Table 13.1**

Gold Nanomedicines in Clinical Trials as of Mid-2016

| Name | Composition | Company | Disease Target | Clinical Phase |
|---|---|---|---|---|
| Aurimune (CYT-6091) | TNF-$\alpha$ bound to PEGylated Au nanoparticles | CytImmune | Head and neck cancer | Phase II |
| AuroShell | Gold nanoshells for photothermal therapy (AuroLase) | Nanospectra | Lung cancer (primary or metastatic) | Pilot study in humans/ veterinary clinical trials |
| NANOM-FIM | Gold/silica nanoparticles for photothermal therapy | Ural Medical University | Atherosclerosis | Phase I completed |
| CNM-Au8 | Gold nanoparticles | Clene Nanomedicine | Demyelinating disorders | Phase I |

treat inflammatory conditions and cancer. Colloidal gold nanoparticles are FDA-approved for treatment of rheumatoid arthritis, and several gold chemotherapy agents are in clinical trials or seeking approval (Table 13.1).

Not only is gold of low toxicity; conjugation to Au may reduce the side effects of chemotherapeutic agents. For example, in the Phase I trial of CytImmune's gold–*tumor necrosis factor* (TNF) conjugate, the dose-limiting hypotension seen with TNF alone did not occur with the conjugate. The major side effects of gold therapy are itching and dermatitis, though there is the possibility of manifestations of kidney toxicity such as glomerulonephritis.

*Biodistribution* and *clearance* must be considered when designing nano-medicines, both for effective delivery and for preventing toxicity. Nanoparticles are cleared from the body through three primary mechanisms: renal, hepatic, and through the *reticuloendothelial system* (RES). Liver clearance is often very slow, and there remains concern about toxicity of Au nanoparticles to both the liver and kidney, depending upon particle size and surface conjugate. The optimal surface chemistry for *in vivo* use is an object of intense research in all drug discovery labs using Au nanoparticles. Nearly all Au nanoparticles for drug use are coated with a thiol–polyethylene glycol (PEG) shell; adjustment of the size of the PEG affects organ accumulation and clearance speed. Un-PEGylated particles usually cleared too rapidly for accumulation in tumors but have been investigated for some applications, particularly in radiotherapy when radiation is administered immediately after gold injection.

## 13.3 ACTIVE AND PASSIVE TARGETING OF AU NANOPARTICLES

The reason that nanoparticles—of gold or other substances—are so useful for drug delivery is that objects <200 nm in size accumulate passively in tumors due to the so-called *enhanced permeability and retention (EPR) effect* (not to be confused with electron paramagnetic resonance) (Figure 13.1). This occurs because malignant tissues have leaky and disorganized vasculature caused by their rapid growth. (The same is true of infected tissue, so similar arguments apply to antimicrobial treatments.)

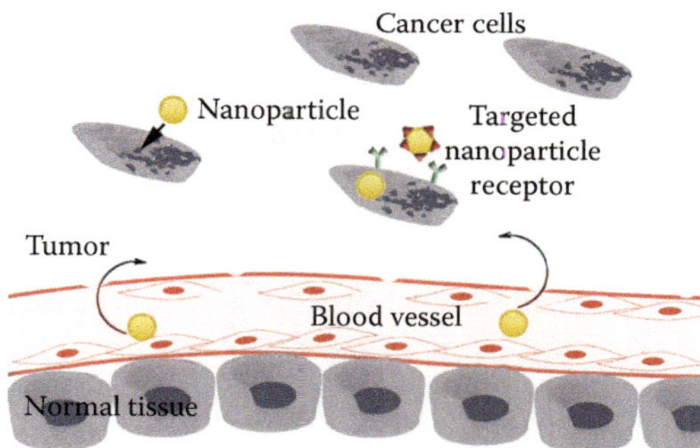


**Figure 13.1 Tumor targeting with nanoparticles.** Leaky blood vessels cause nanometer-sized particles to accumulate at tumor sites via the enhanced permeability and retention effect. Uptake by cancer cells may be further optimized by conjugation of nanoparticles to ligands that bind more strongly to malignant cells than to healthy cells.

There have been many strategies to try to optimize nanoparticle delivery to cancer by taking advantage of the EPR effect. Some anticancer drugs are now used almost exclusively in liposomal formulations that work by EPR; examples are Doxil (doxorubicin), Caelyx (PEGylated doxorubicin), and Abraxane (paclitaxel). In animal studies, Au nanoparticles of different sizes and geometries have been shown to successfully concentrate in tumors, but subtle differences in rate and site of accumulation depend upon particle size and charge and must be carefully considered for each application. Larger particles may enter tumors more effectively than smaller ones, for example, but may not penetrate as deeply into the individual tumor cells.

The EPR effect remains controversial. It may be significant in some human cancers, but not all, and it does not target *micrometastases*, which do not yet have their own vasculature. One of the key questions in nanomedicine is whether active targeting of nanoparticles to improve upon the EPR effect will significantly improve targeting to tumors over nontarget organs. Specific targeting has been attempted using streptavidin–biotin, antibodies, aptamers, and specific ligands for cancer cell receptors. While many of these conjugates are taken up readily by cells, most cells also take up unconjugated nanoparticles, and cell studies do not take into account circulation or filtration by liver, kidney, or RES. This means that targeting studies must be done using animals—a costly, time-consuming, and for some, unethical approach. Biodistribution in animals is measured by injecting nanoparticles and then sacrificing the animals at particular time points and examining organs for the presence of gold. Instrumental neutron activation analysis (INAA) and Inductively coupled plasma mass spectrometry (ICP-MS) give quantitative measures of metal concentration; the limit of detection for gold using ICP-MS is as low as 0.001 µg/kg, or one part per trillion. But the sample preparation is time-consuming and costly, and this method tells nothing about the distribution of the particles within the organs (e.g., inside the cells vs. in the blood vessels).

Au nanoparticles are not simply passive carriers of drugs; they may also act as active agents in one or more ways. They can increase the ability of the drug to generate reactive oxygen species (ROS), making the gold-bound drug more toxic than the free drug; they can absorb near-infrared (near-IR) light, heating the cancer cells to levels that kill them (photothermal therapy), and they can increase the effective dose of ionizing radiation delivered during radiation therapy by the photoelectric effect. A very large body of literature exists on all of these approaches. In the following sections, I will discuss two of them in detail: the use of gold for radiotherapy and the use of gold for *photothermal therapy*.

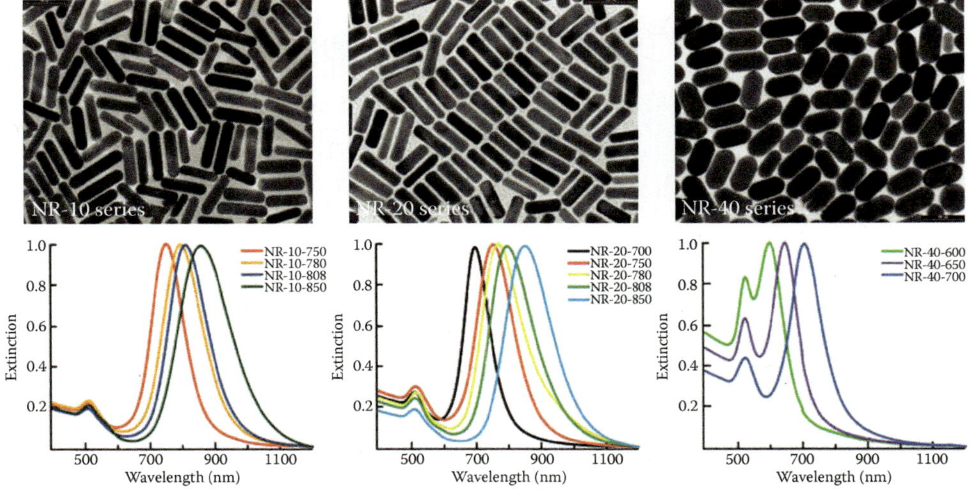



**Figure 13.2 Effects of size and aspect ratio on absorbance spectra of gold nanorods.** These nanorods are available commercially (Image credit: Nanoseeds).

## 13.4 THE USE OF GOLD IN PHOTOTHERMAL THERAPY

Hyperthermia therapy is a minimally invasive treatment in which the temperature is increased locally (up to 44°C) to kill malignant cells. Methods to locally heat the tumor region include high-intensity focused ultrasound (HIFU), microwave heating, magnetic hyperthermia, and photothermal therapy. In photothermal therapy, a light source (usually IR) is used to deliver heat to the tumor. Such approaches are difficult to target, but delivery of nanomedicines to the tumor could improve the local heating profile. Most studies have looked at gold nanoparticles and nanorods for this purpose, because exposure of Au nanoparticles to IR light causes a local temperature increase due to surface plasmon resonance. By modifying the size and shape of these nanoparticles, the resonance peak can be tuned to different wavelengths in the IR (**Figure 13.2**).

## 13.5 THE USE OF GOLD IN RADIATION THERAPY*

### Principles of radiation therapy

Radiation therapy or radiotherapy (XRT) is a critical component of the modern approach to curative and adjuvant treatment of cancers. XRT controls the growth of cancerous cells by bombardment with ionizing radiation, causing DNA damage by direct ionization or through generation of free radicals by ionization of water or oxygen molecules. Sufficient damage to DNA in this fashion can arrest cell growth and prevent metastasis. The primary drawback is collateral damage: there is little distinction in absorption between healthy and malignant tissues, and thus, doses must be limited in order to mitigate

---

* Sections adapted from D. Cooper et al., Gold Nanoparticles and Their Alternatives for Radiation Therapy Enhancement, *Front. Chem.* 2, 86, 2014 Oct 14, doi: 10.3389/fchem.2014.00086.

unwanted damage to the tumor surroundings. *External beam radiotherapy* (EBRT) utilizes x-ray beams produced by orthovoltage units, or linear accelerators that may be spatially oriented and shaped using multileaf collimators in order to maximize the specificity for the target. Distinct energy ranges are available for different EBRT targets: 40–100 kV (kilovoltage or *superficial* x-rays) for skin cancers or other exposed structures, and 100–300 kV (orthovoltage) and 4–25 MV (megavoltage or *deep* x-rays) for subsurface tumors. Techniques such as three-dimensional conformal and intensity-modulated radiation therapies have vastly improved the targeting capabilities of external beam therapy, but naturally, there is still a strong desire to be able to further reduce the doses required for effective treatment. The SI derived unit for *absorbed dose* is the gray (Gy), equivalent to 1 J of energy deposited by ionizing radiation per kilogram of matter ($1\ \mathrm{Gy} = 1\ \mathrm{J/kg} = 1\ \mathrm{m^2/s^2}$). (See **Practical Tips 13.1**.)

*Brachytherapy*, or internal radiotherapy, utilizes a radioactive source to provide a steady or pulsed dose of radiation to a small tissue volume. It is typically used for cervical, prostate, breast, and skin cancers. Radioactive sources include $^{125}$I and $^{103}$Pd, which produce γ-rays of ~20–35 keV; $^{192}$Ir (γ-rays, 300–610 keV); $^{137}$Cs (γ-rays, 662 keV); $^{60}$Co (γ-rays, 1.17 and 1.33 MeV); $^{198}$Au (γ-rays, 410–1009 keV); $^{226}$Ra (γ-rays, 190–2430 keV); and $^{106}$Ru, which decays primarily through $\beta^-$ emission at 3.54 MeV. Seeds of the listed materials can provide doses of up to 12 Gy/h (high-dose-rate [HDR] brachytherapy), though typical low-dose-rate (LDR) treatments amount to around 65 Gy over 5–6 days.

Heavy elements can be potent *radiosensitizers*. It has been demonstrated that platinum-containing DNA-cross-linking drugs such as cisplatin can enhance the effects of ionizing radiation through the *high-Z effect*, or what has come to be known as *Auger therapy*. Heavy elements have significantly higher photoelectric cross sections than soft tissue for sub-million-electron-volt energies, approximated for *x-ray energies* by the equation

$$\sigma_{pe} \propto \frac{Z^n}{E^3} \tag{13.1}$$

where $\sigma_{pe}$ is the cross section, $E = h\nu$ is the photon energy, $Z$ is the atomic number, and $n$ varies between 4 and 5 depending on the value of $E$. The photoelectric effect dominates below the electron rest energy of 511 keV, beyond which inelastic Compton scattering becomes more prevalent. As the photon energy decreases, it is no longer able to eject inner-shell electrons, producing the characteristic sawtooth pattern with K, L, and M edge structures. When ionized by x-ray or γ-ray energy, mid- to high-Z elements (roughly Br and up) can produce a cascade of low-energy Auger electrons that can locally enhance the effective radiation dose. Dense inorganic nanoparticles can also provide radiation dose enhancement that depends upon the composition and size of the particles, uptake of particles into cells, and energy of the applied radiation.

It is important to note that radiation does not kill cells outright—it stops them from dividing. Measures of cell death such as 3-(4,5-dimethylthiazol-2-yl)-2,5-diphenyltetrazolium bromide (MTT) and sulforhodamine B (SRB) assays are therefore inappropriate for testing cell death due to radiation. The correct way to measure radiation effectiveness is via a *clonogenic assay* (**Practical Tips 13.2**).

### PRACTICAL TIPS 13.1:   RADIATION THERAPY IN CELLS

Radiation therapy may be tested *in vitro* using sources that mimic clinical external beam therapy or brachytherapy. For x-irradiation, usually a specialized *in vitro* or small animal setup is used. Traditional isotope irradiators contain radioactive cesium or cobalt, but these pose security risks and have largely been replaced by x-ray tube sources.

Such sources are available as self-contained units so that the operator is not exposed to radiation; users do not need to wear dosimetry badges. The x-ray tube produces a highly homogenous beam with a given energy profile (though the beam intensity may fall off at the edges). Given the distance between the source and the sample and the presence or absence of filters, precise dose rates with known energy profiles can be delivered. **Figure P13.1.1** shows a commercial cell/small animal irradiator and the setup with T25 flasks of cells to be irradiated. Such sources are usually administered by divisions of radiation safety, who can provide training in their use as well as assistance with accurate dosimetry. Instruments such as the one shown are programmable for highly repeatable experiments.

For brachytherapy experiments, access to clinical sources is frequently available after hours in hospitals. Collaborations with clinicians and medical physicists are necessary to obtain access and training. Clinical isotope sources, such as $^{192}$Ir plaques used to treat ocular melanoma, may also be purchased directly from manufacturers. These are costly, however, and the half-life of the source must always be kept in mind when purchasing. A source that will decay before it is used is a waste of money.

Performing and analyzing radiation therapy experiments can be tricky. It is important that the number of cells at each stage in the cell cycle be consistent in each treatment group and from experiment to experiment. Nondividing cells are much less sensitive to radiation than dividing cells, so allowing cells to become too confluent—and thus not dividing—can skew the results enormously. Experiments should be performed at least three times.

It is also important to mention that cells have repair mechanisms that address the strand breaks caused by radiation. Allowing cells to stand for 6–24 h maximizes the repair. This may be used as an additional test case, but it might also confuse results if treated cells are allowed to sit before trypsinization and plating.

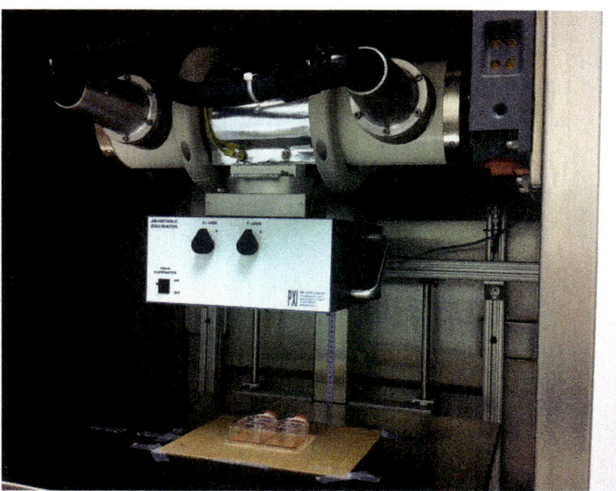

**Figure P13.1.1  Small animal irradiator (Faxitron) with two side-by-side T25 flasks to be irradiated.** The distance between the x-ray source and the cells affects the dose.

*(Continued)*

> ### PRACTICAL TIPS 13.1 (CONTINUED): RADIATION THERAPY IN CELLS
>
> The survival fraction $S$ versus dose $D$ is usually fit to a linear quadratic equation
>
> $$S = \exp[-\alpha D - G\beta D^2],$$
>
> where $\alpha$ and $\beta$ are the fit parameters. $G$ is a time factor; for a single acute dose, $G = 1$. This model takes into account two mechanisms of killing—a *single-hit* kill (linear) and a *multiple-hit* kill (quadratic)—while minimizing the number of fit parameters. The time factor $G$ takes into account repair that can occur between hits.
>
> #### SUGGESTED READING
>
> Brenner, D.J. (2008). Point: The linear-quadratic model is an appropriate methodology for determining iso-effective doses at large doses per fraction. *Seminars in Radiation Oncology* 18, 234–239.
>
> Garcia, L.M., Leblanc, J., Wilkins, D., and Raaphorst, G.P. (2006). Fitting the linear-quadratic model to detailed data sets for different dose ranges. *Physics in Medicine and Biology* 51, 2813–2823.
>
> Hall, E.J. *Radiobiology for the Radiobiologist.* JB Lippincott Company, Philadelphia, PA, 2000.

## Gold nanoparticle–assisted radiation therapy

Au nanoparticles have been under investigation for several years as possible agents for selective amplification of radiation dose in tumors, a concept called *gold nanoparticle–assisted radiation therapy*, or GNRT.

The earliest studies used bulk or microsized gold to enhance radiation dose. Although this could be effective *in vitro* at a range of energies, micron-sized particles are not taken up well *in vivo*, even after intratumoral injection. Later experiments focused on Au nanoparticles or nanoclusters (1.9 nm diameter). When injected intravenously, these ultrasmall particles rapidly accumulated in cancer tissue, with 2.7 g Au/kg body weight resulting in 7 mg Au/g in tumor almost immediately after injection. Irradiation was performed about 60 s after injection, and with typical 250 kVp x-ray therapy, 1-year survival was 86% (compared to 20% with x-rays alone and 0% with gold alone). This result was followed by theoretical and experimental papers examining the mechanism of Au nanoparticle action as well as attempting to optimize Au particle concentration and size and the energy and dose of applied x-rays.

## Improving GNRT by targeting

A significant problem with ultrasmall, nontargeted nanoparticles is rapid excretion by the kidneys. The amount of Au needed in animal studies (>2 g Au/kg body weight) represents a very large amount of Au for human use. This is impractical and costly and may cause toxicity. Achieving therapeutic levels in tumor with less delivered total Au is needed. In addition, irradiation in the mouse studies was performed immediately after particle injection. This is not practical in the clinic and may not work well in humans. Particles with longer circulation times, which can be delivered in multiple doses, are desirable for clinical applications. Optimizing the size, surface chemistry, and targeting of the Au nanoparticles may improve circulation times and accumulation in specific tumors.

**PRACTICAL TIPS 13.2: CLONOGENIC ASSAY**

Performing a clonogenic assay with mammalian cells is conceptually similar to bacterial plating and colony counting. However, it is more difficult to make reproducible and is much more tedious, taking 1–2 weeks for colonies to grow. This means that effort should be made to make each experiment count by plating cell dilutions that yield a *countable* number of colonies—about 50–200 per 10 cm plate. Plates with more or fewer do not yield good statistics. Accurate cell counting and dilution are keys to success with this protocol. Familiarize yourself with the hemocytometer (or other cell counting device) before attempting this assay.

**REAGENTS NEEDED**

*At the Start*
- Fully supplemented culture medium (with serum)
- Trypsin (see **Chapter 3**)
- PBS
- 10 cm culture plates

*After Colony Growth*
- Colony staining/fixation solution (6% glutaraldehyde/0.5% crystal violet in $H_2O$ or 10% formalin/0.5% methylene blue)
- Automatic colony counter (optional but recommended)

**PREEXPERIMENT: PLATING EFFICIENCY**

The plating efficiency (PE) gives you a control survival value—that is, the fraction (or percentage) of cells that grow into colonies per cell plated. For most immortalized cell lines, PE should be 0.7–1.0. If it is much lower, check your reagents and technique. This may be an issue of senescence (cells have been maintained too long), an error with the initial counting, or a fundamental property of the cell line. Some cell types, especially primary cells, do not grow well into colonies when plated sparsely. These cell types require supplementation with conditioned medium or a feeder layer. These techniques are outside the scope of this book, but if your PE is <0.1–0.2, think about the cell type and whether it can grow into colonies well. PE values are often given in the literature and from suppliers.

- Start with a flask of cells at 50–80% confluence (not fully confluent).
- Wash in PBS and trypsinize. When the cells detach, resuspend them in serum-containing medium and pipette (gently) up and down to obtain a uniform suspension.
- Count.
- Dilute the cells to a concentration that will yield 50–200 cells per dish, plus a concentration somewhere on either side of this. A decent rule of thumb is to double the number of cells plated for each 2 Gy radiation dose.
- Plate the dilutions in duplicate.
- Let colonies grow for 7–14 days (until large enough to count).
- Aspirate medium and stain with fixation/stain for 30–60 min.
- Wash with water and air-dry.
- Count colonies. (A colony has >50 cells.)
- PE = colonies counted/cells plated (×100 if expressed as a percentage).

**ADDING TREATMENTS**

- Perform the aforementioned steps, but add flasks that have been treated with the test conditions (e.g., 0–12 Gy radiation in steps of 2 Gy).
- Use multiple dilutions for each treatment to account for the range of possible treatment effects.

*(Continued)*

---

**PRACTICAL TIPS 13.2 (CONTINUED):   CLONOGENIC ASSAY**

- Any cells that are not trypsinized immediately should be kept on ice, or the experiments should be performed quickly enough so that repair does not take place before trypsinizing the cells. (This may mean doing different doses on successive days.)
- Survival fraction = colonies counted/(cells plated × PE).

**SUGGESTED READING**

Franken, N.A., Rodermond, H.M., Stap, J., Haveman, J., and van Bree, C. (2006). Clonogenic assay of cells in vitro. *Nature Protocols* 1, 2315–2319.

Puck, T.T., and Marcus, P.I. (1956). Action of x-rays on mammalian cells. *Journal of Experimental Medicine* 103, 653–666.

---

The increased metabolic rate of tumors relative to normal tissue results in a high demand for glucose. Several studies have used thioglucose-conjugated Au nanoparticles (**Figure 13.3a**) in order to increase uptake by cancer cells. One study using ~14 nm Au demonstrated significantly increased uptake of thioglucose-conjugated particles by an ovarian cancer cell line after 8–96 h of incubation. A significant increase in inhibition was seen in the presence of 5 nM particles using 90 kVp or 6 MV x-rays; dose enhancement was significant relative to control beginning at 5 Gy and extending to 20 Gy, where all cells were inhibited even in the absence of particles. Another study compared cysteamine and thioglucose-coated 15 nM Au nanoparticles in breast cancer and normal breast cell lines. Cysteamine-coated particles were taken up threefold to fourfold more efficiently than glucose-coated particles. However, when applied to cells at concentrations that led to similar intracellular Au concentrations, glucose-coated particles led to increased radiosensitization relative to cysteamine-capped particles. Interestingly, radiosensitization by Au was not seen in a nonmalignant breast cell line, although the cells grew at the same rate as the cancer cells and took up an equal number of particles.

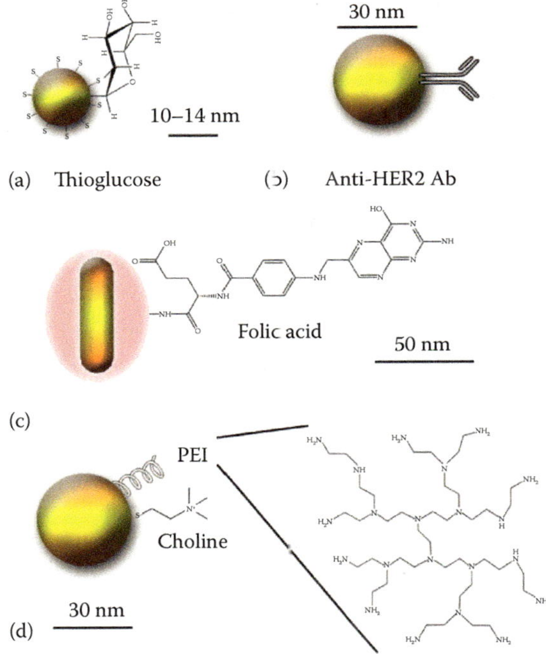


**Figure 13.3 Approaches to creating tumor-targeted Au nanoparticles.** Molecules not to scale. (a) Thioglucose-conjugated Au nanoparticles. (b) Au nanoparticles conjugated to Herceptin (anti-*HER2* antibody). (c) Au nanorods conjugated to folic acid. (d) PEI-coated Au nanoparticles conjugated to choline.

The use of larger Au particles (57 and 84 nm) coated with thioglucose has also been studied. These particles were taken up in equal numbers by HeLa cells. Surprisingly, unconjugated particles showed a greater radiosensitizing effect than thioglucose-conjugated particles, which might result from absorbance of singlet oxygen by the thioglucose shell.

Another study used the humanized anti-*HER2* antibody trastuzumab (Herceptin), PEGylated and conjugated to 30 nm Au particles, for delivery to MDA-MB-361 breast cancer (**Figure 13.3b**). Both *in vitro* and *in vivo* studies were performed. *In vitro*, an effective dose enhancement factor of 1.6 was seen in the presence of 2.4 mg/mL particles using 100 kVp x-rays. Delivery to MDA-MB-361 xenografts was done intratumorally, with ~0.8 mg total Au used (4.8 mg/g tumor). 11 Gy of 100 kVp image-guided x-ray irradiation was performed 24 h after injection. This subtoxic dose led to a 46% reduction in tumor size relative to irradiation alone, with no damage to normal tissue or systemic toxicity.

*Folic acid* (folate) is another nutrient for which the need is increased in cancer cells. Conjugation to folate has been used for a wide variety of targeting applications for cancer and inflammatory diseases. Intraoperative tumor imaging using folate targeting has recently moved to the clinic for ovarian cancer. In terms of GNRT, silica-modified Au nanorods (~50 nm long) conjugated to folate have been tested *in vitro* (**Figure 13.3c**). The rods were shown to be taken up by MGC803 human gastric carcinoma cells. Six grays of x-irradiation led to a 60% decrease in cell viability in the presence of 12.5 μM rods relative to cells without Au.

Cancers are also often distinguished by a lower pH than healthy cells due to hypoxia and resulting anaerobic metabolism within tumors. The pH-sensitive pHLIP peptide was used in one study to target Au nanoparticles to mice bearing *HeLa* tumors. Although radiation was not performed, accumulation of Au in tumors sufficient for radiotherapy enhancement was demonstrated, with the stated goal of using the construct for this purpose.

Prostate cancer is an excellent target for nanoparticle-enhanced radiation, since it is often treated by brachytherapy and is accessible to intratumoral injection. *Choline* is a ubiquitous molecule in all cells for which overactivity of processing enzymes (choline kinase) has been found in prostate tumors. One study reported development of Au nanoparticles conjugated with polyethylene imine (PEI) and choline for the purpose of targeting prostate cancer for GNRT (**Figure 13.3d**). While no radiation experiments were reported, favorable pharmacokinetics were shown in mice.

These studies illustrate that targeted GNRT remains an area requiring substantial further investigation. While one or more targeting agents may be conjugated to Au nanoparticles, and while these may improve delivery *in vitro* and even *in vivo*, it is not fully established whether these formulations improve tumor response to radiation therapy. The possibility that an organic shell can absorb ROS deserves further inquiry. The size of the nanoparticles used, the density of targeting ligands, and the delivery method (intravenous [IV], intratumoral, concentration, timing) all need to be optimized. The good news is that many of these formulations use FDA-approved ingredients, some of which are currently in the clinic for imaging. Optimization in animal studies should thus lead to rapid clinical translation.

## Improving GNRT by addition of photothermal therapy

Even though hyperthermia can kill cells on its own, it is more often used in combination with other treatments such as radiotherapy or chemotherapy; such combinations are in clinical trials. An increase in nuclear damage is one of the mechanisms through which cells are radiosensitized after hyperthermia. In addition, the higher temperature causes dilation of the blood vessels, increasing oxygenation of the tumor. Since oxygen is a potent radiosensitizer, it can increase the damage to the tumor through generation of free radicals.

Delivery of gold followed by heating and ionizing radiation has proved to be a promising approach in preclinical studies. Gold nanoshells with a 120 nm silica core and a 12–15 nm shell were used in one study to treat a murine *xenograft* model of human colorectal cancer. Localized hyperthermic treatment followed 5 min later by a 10 Gy x-ray dose were given 20–24 h after IV delivery of the nanoshells. The tumor volume doubling time was significantly greater for the treated mice. Two mechanisms were identified as contributing to the treatment's efficacy: an increase in perfusion resulting in a decrease in tumor hypoxia, and vascular collapse in the tumor due to accumulation of nanoparticles around the blood vessels. Another study confirmed these results using similar gold nanoshells in two murine breast cancer models.

Another study used gold nanorods modified with silica and conjugated to folic acid to test the effects of photothermal and radiation therapy on MGC803 gastric cancer cells. The two treatments were tested separately and not combined. For the radiation treatment, a 6 MeV source was used to deliver doses of up to 10 Gy. Cell survival with radiation decreased in a concentration-dependent fashion with nanoparticle addition; the particles were nontoxic in the absence of radiation. Photothermal therapy consisted of 3 min of irradiation with a 30 mW, 808 nm laser source. Apoptosis was seen after treatment in the presence of the gold particles.

A recent study calculated the radiation dose required to control 50% of tumors (TCD50) in a mouse *squamous cell carcinoma* model. Gold nanoparticles were delivered intratumorally, and 24 h later, the tumor was heated to 48° for 5 min at 1.5 W/cm$^2$ followed by x-ray irradiation at 100 kVp (7.5 Gy/min). TCD50 was reduced from 55 Gy to less than 15 Gy

These studies illustrate one of the biggest problems of the approach, which is the need for simultaneous delivery of heating and radiation, which poses logistic problems in the clinic. Other drawbacks include a lack of specificity and the difficulty of heating deep tumors.

# 13.6 EXAMPLE: HOW TO MAKE A NANOMEDICINE—THE CASE OF AU–DOX

My laboratory has been working for several years to create a gold nanoparticle-doxorubicin conjugate (Au-Dox) for clinical use. The following sections explain a little about the motivation behind this formulation, and then summarize all of the testing steps we have taken so far to develop it for the clinic. The goal is to illustrate the types of physical, *in vitro*, and *in vivo* characterization that are needed before a nanoparticle conjugate becomes a potential drug candidate.

## Why Au–Dox?

Doxorubicin (Dox) is a chemotherapeutic drug used for nearly all *solid cancers*. Its major drawback is damage to the heart (cardiotoxicity), which may appear years after treatment and which is related to the cumulative dose of the drug that was administered during treatment. Delayed effects may cause death by heart failure years after treatment. Various clinical trials are in progress to test the efficacy of cardioprotective treatments such as beta-blockers, antiangiotensin agents, or iron chelators, but this life-threatening side effect remains a major barrier to treatment. This is especially true because recurrent cancers develop Dox resistance, but higher doses cannot be given without risking serious side effects.

Nanoparticle delivery of Dox can help to prevent cardiotoxicity because of better tumor targeting via the EPR effect, which means that less total drug can be given. Both PEGylated and non-PEGylated liposomal Dox reduce cardiotoxicity. Nanoparticle–Dox may also overcome resistance mechanisms in a variety of ways.

A proposed nanoparticle construct should thus focus on maximizing Dox accumulation in tumor versus nontarget organs, especially the heart, and on optimizing anti–cancer cell activity. Dox has multiple modes of action, including intercalation into DNA. It is well established that at low concentrations (<3 mM), induction of apoptosis is the primary mechanism of Dox action. At higher concentrations, however, other mechanisms take place to which apoptosis-inhibited cells are not resistant. These mechanisms remain somewhat controversial. Thus, comparison of different sizes and stabilities of nanoparticle constructs against cancer cells is of great interest.

## Physical characterization

In creating a nanomedicine, attention must be paid to reproducibility and stability of each ingredient, not simply the immediate physical properties. This often requires preplanning for months or years so that long-term stability can be monitored.

### Au particles alone

The first step in the synthesis is the fabrication of the Au particles capped with a free carboxylate group for conjugating Dox. In this example, we fabricated Au–tiopronin (**Figure 13.4a**). Tiopronin is FDA-approved as a drug to treat *cystinuria*, so its use in the formulation is expected to be nontoxic and to pose few regulatory problems. This is an important consideration, since some capping groups for Au nanoparticles, such as cetyltrimethylammonium bromide (CTAB), are highly toxic. Tiopronin is negatively charged at physiological pH, as confirmed by *zeta potential* measurements ($-16.79 \pm 1.94$ mV). A size distribution of $2.7 \pm 0.5$ nm was obtained from electron microscopy, giving a mean number of atoms per particle of 169 (**Figure 13.4b**). Fourier transform infrared (FTIR) spectra (see **Chapter 16**) show stretches at 3300 nm (O=H of carboxylic acids) and 1700 nm (C=O of carboxylic acids), which further confirmed the successful synthesis of the nanoparticles (**Figure 13.4c**). By dissolving 33 mg of gold–tiopronin particles in 1 L of $H_2O$, the concentration was measured to be 16.6 ppm by flame *atomic absorption spectroscopy*. Therefore, 50.3% w/w of each nanoparticle was the gold core and the 49.7% w/w was the tiopronin ligands, giving a chemical formula of $Au_{169}Tiopronin_{207}$. Absorbance spectroscopy showed no distinct peak at any

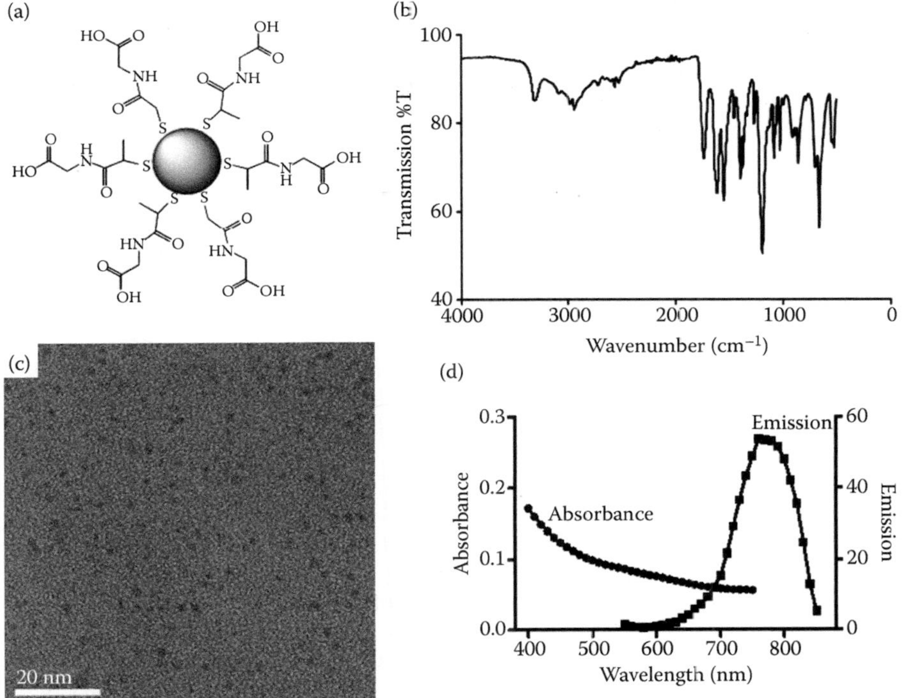

**Figure 13.4 Characterization of tiopronin-capped ultrasmall gold nanoparticles.** (a) Schematic. (b) IR spectrum. (c) Representative TEM image indicating uniform distribution and 2.7 ± 0.7 diameter. (d) Absorbance and fluorescence emission spectra (arbitrary units).

wavelength, consistent with subplasmonic particles. The particles were fluorescent in the near-IR range with a quantum yield of approximately $1.62 \times 10^{-3}$ (**Figure 13.4d**). Particles were also characterized by electron microscopy with selected area electron diffraction (SAED) and energy-dispersive x-ray spectroscopy (EDS).

Multiple repetitions of the synthesis were performed, determining that the size distribution was reproducible to better than 5%. The ligand–Au atom ratio similarly varied by <5%. The maximum concentration in water where no precipitation was apparent was 100 mM. Aliquots of this concentration were saved at –20°C, 4°C, and room temperature and tested periodically for aggregation, which could be observed by eye as precipitation and change in the solution color. Saved aliquots were also used to prepare Au–Dox and compared with fresh particles. In this way, it was determined that the Au alone was stable at 4°C or –20°C for at least 2 years.

It is also important to determine stability under a range of pH values. The particles were suspended in different buffers with pH values 5–10 and stored in parallel with the water samples. The quantum yield of fluorescence was altered by pH, but otherwise, the particles were stable in this range as measured by lack of precipitation.

## *Au–Dox*

Particles were conjugated to Dox using carbodiimide coupling (see **Chapter 14**) (**Figure 13.5a**). Reaction by-products were removed using filtration or dialysis, which is crucial to prevent toxicity. Different pH values in the conjugation reaction and different conjugation times were investigated to optimize coupling. The amount of bound Dox per particle could be estimated from ultraviolet–visible (UV–Vis) absorbance spectra; this had to be corrected for the fact that the presence of Au particles caused an approximately 50% reduction in the absorbance peak of Dox,

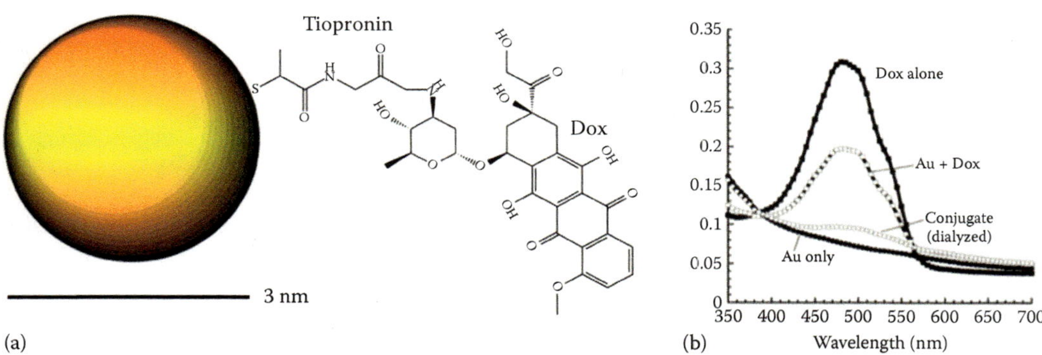



**Figure 13.5 Au–Dox conjugates.** (a) Schematic of Au–Dox with elements to scale. (b) Absorbance spectra of 50 mM doxorubicin alone (filled squares), showing reduction in absorbance when mixed with Au particles (Au + Dox) (open squares). This relative absorbance is used to calculate the number of doxorubicin molecules remaining on the dialyzed conjugate (open circles).

indicating an electronic interaction (**Figure 13.5b**). Taking this into account, we estimated $20 \pm 5$ Dox molecules per Au nanoparticle remaining after dialysis. Because of the quenching effect, stability of the conjugates could be monitored using UV–Vis spectroscopy. This showed no change in absorbance after 24 h at 37°C at pH values 4–8: that is, the particles were stable under physiological conditions. This suggested that the Dox should stay on the nanoparticles when taken up by the cells, unless other mechanisms took place inside cells that were not seen in a test tube.

The particles showed some batch-to-batch variability, the cause of which was never completely clear. Testing for efficacy on cell lines was the only reliable quality check for the conjugates. At 37°C, conjugates were stable for <24 h. When frozen at –20°C, some batches showed nearly normal efficacy after 1 year. Others were completely inactive at this time, though these could usually be identified by visual signs of precipitation. Dox itself also showed variability from assay to assay. It is sensitive to light and heat and can adhere to some plastics, so it is just as important to test the control free drug as the conjugate!

A key element in creating a nanomedicine for efficacy studies is to ensure that the preparation is sterile and free of endotoxin. A good start to ensure sterility is to prepare all particles and conjugates in a tissue-culture hood using sterile reagents. Filter sterilization using a 0.2 μm filter also helps to remove bacteria but is not a substitute for sterile technique during the preparation. Sterility is tested by incubating a sample at 37°C in Dulbecco's Modified Eagle Medium (DMEM) and examining for bacterial growth after 1–2 days; by endotoxin testing; and by streaking a plate and observing for colonies (see **Practical Tips 13.3**).

## Efficacy against cultured cancer cells

All cell lines tested took up Au–Dox readily. The cell line we tested the most thoroughly was B16, a murine melanoma cell line that forms highly aggressive tumors in mice. Uptake of Au–Dox could be easily observed using fluorescence microscopy (see **Chapter 7**) to observe the native Dox fluorescence (**Figure 13.6a through d**). Transmission electron microscopy was needed to show subcellular distribution of the gold. Clusters of Au particles were seen primarily in endosomes, but some Au was also observed in cytoplasm and nucleus (**Figure 13.6e and f**).

### PRACTICAL TIPS 13.3:    STERILE NANOPARTICLES

Preparing nanoparticle conjugates for animal and human use imposes requirements on sterility that are often difficult to achieve, at least at first. Nanoparticle synthesis is often done in a chemical fume hood rather than a biosafety cabinet, protecting the operator but not the particles. Reagents used for synthesis and conjugation—even distilled water—often contain significant amounts of endotoxin. Any time a formulation is intended for preclinical use, it should be tested for bacterial and endotoxin contamination. If contaminated, new synthesis procedures should be put in place so that the formulation is clean as synthesized.

#### TEST FOR BACTERIAL CONTAMINATION

Prepare petri dishes containing nutrient agar (as discussed in **Chapter 2**) with no antibiotics. Have at least two dishes for each condition to be tested. Dissolve the nanoparticles to a concentration of 1 mg/mL in sterile water, and then prepare serial dilutions of this solution also in sterile water, down to at least a 1:1000 dilution.

Also prepare a positive control containing a test strain (usually *Escherichia coli*) at 10–100 colony-forming units (CFU/mL). An additional control may be prepared by deliberately spiking the nanoparticle solution with *E. coli*. This is especially important for nanoparticles such as Ag, which may have an effect on bacterial growth.

Plate duplicates of the negative control (water alone), positive control, spiked control as appropriate, and diluted nanoparticle solutions, and streak as in **Figure 2.3**. Incubate for 72 h at 37°C, and count colonies. Any colonies on the nanoparticle samples indicate contamination, which should be quantified as the number of CFUs per milliliter.

#### TEST FOR ENDOTOXIN

This is a crucial test for any drug to be used in animals or humans, since endotoxin can lead to severe immune responses, obscuring any positive or negative effects of the nanoparticles. Testing is performed using commercial limulus amebocyte lysate (LAL) assay kits. Internal standards are included within the kits for calibration; it is important to perform all controls because some nanoparticles interfere with readout, especially in colorimetric assays. The limit for clinical use is 5 endotoxin units (EU) per kilogram body weight per hour for intravenously delivered drugs.

Using the ToxinSensor chromogenic LAL assay kit from Genscript, the value for Au–tiopronin was 0.014 EU/mL in a 0.1 mM solution, which is acceptable. However, Au–Dox prepared normally gave a value of 1.28 EU/mL in a solution containing 1 mM Dox. This was too high, so synthesis procedures had to be revised. Synthesis was performed in a biosafety cabinet; all media were prepared using endotoxin-free water; and new glassware was purchased specifically for the preclinical formulation. In this way, levels were reduced to 0.029 EU/mL of endotoxin in an Au–Dox solution containing 1 mM of Dox.

#### SUGGESTED READING

This protocol was used in our lab in order to meet the Nanotechnology Characterization Laboratory's standards. It has been adapted from NCL Method STE-2.2.

Crist, R.M., Grossman, J.H., Patri, A.K., Stern, S.T., Dobrovolskaia, M.A., Adiseshaiah, P.P., Clogston, J.D., and McNeil, S.E. (2013). Common pitfalls in nanotechnology: Lessons learned from NCI's Nanotechnology Characterization Laboratory. *Integrative Biology (Cambridge)* 5, 66–73.

Toxicity studies were performed using the MTT and/or SRB assays and real-time growth curves (see **Chapter 10**). These showed that the Concentration needed to inhibit the growth by half ($IC_{50}$) in B16 cells was nearly 20-fold lower with Au–Dox than Dox. Au particles themselves showed no significant toxicity at up to 30-fold the concentrations used in the conjugates. At some point, the Au nanoparticles induced autophagy in the cells, probably due to simple overloading of the endosomes, but this occurred at extremely high concentrations (~100 mM). We also found interesting results in cardiomyocytes using real-time growth curves (see **Chapter 10**). Cardiomyocytes in culture were more sensitive than B16 cells to Dox alone but showed resistance to Au–Dox (**Figure 13.7**).

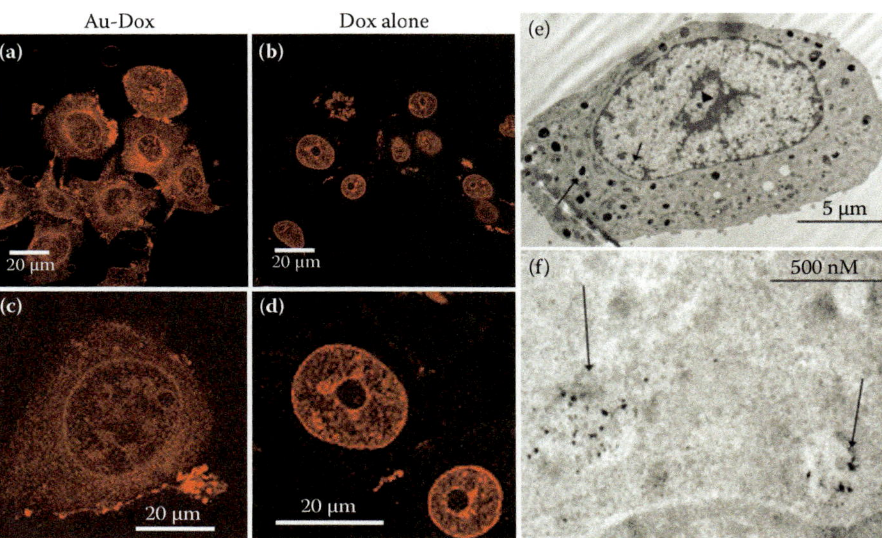


**Figure 13.6 Uptake of Au–Dox and Dox alone by cancer cells.** (a) Au–Dox shows distribution throughout the cell, including in the nucleus and in clusters on the cell membrane. (b) Dox alone, showing nuclear localization. (c) Zoom of a single Au–Dox cell. (d) Zoom of two Dox-alone cells. (e) TEM image of a cell exposed to Au–Dox showing the presence of gold (dark areas; see arrow). (f) Close-up of endosomes under TEM shows individual Au particles.

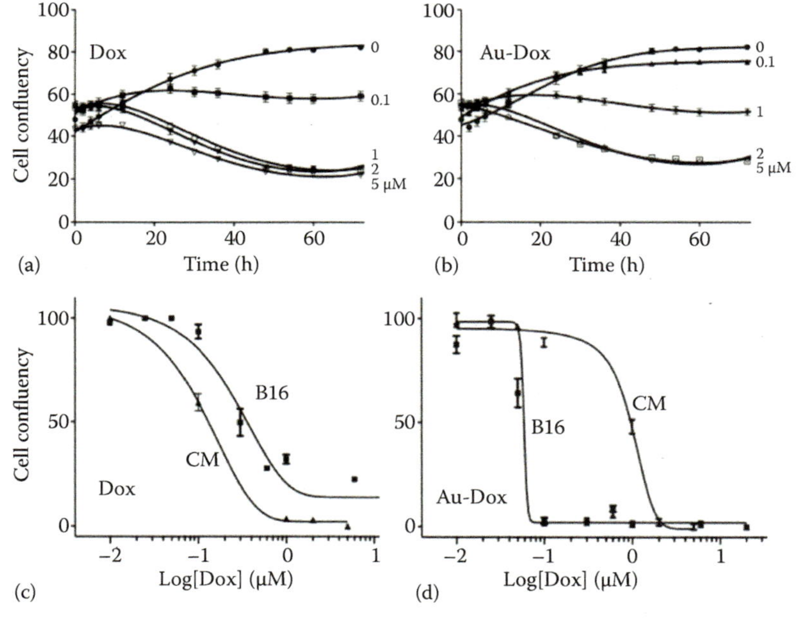


**Figure 13.7 Cytotoxicity measured by real-time growth curves in cells exposed to Au–Dox.** (a) Real-time plot of cardiomyocyte confluence after incubation with doxorubicin at 0, 0.1, 1, 2, and 5 µM. (b) Real-time plot of cardiomyocyte confluence after incubation with Au–Dox at 0, 0.1, 1, 2, and 5 µM. Each data point is the mean of six replicates with error bars indicating standard error of the mean (SEM). (c) Dose–response for Dox alone at 48 h comparing B16 cells and cardiomyocytes (CM). (d) Dose–response for Au–Dox at 48 h comparing B16 cells and CM.

A useful feature of Dox fluorescence is that its lifetime is environmentally sensitive. Several studies have shown different lifetimes of Dox in the cytoplasm and nucleus of cells, and different (longer) lifetimes of encapsulated versus free Dox. Fluorescence lifetime imaging microscopy (FLIM) (see **Chapter 8**) could thus be used to show uptake of Au–Dox and to demonstrate that at least some of the conjugate remains intact inside cell nuclei (**Figure 13.8a through c**). Imaging indicated no nuclear entry of intact Au–Dox in cardiomyocytes (**Figure 13.8d through f**).

A variety of assays were performed in order to further clarify the mechanisms of Au–Dox action. A fluorescent ROS indicator was used in combination with flow cytometry (see **Chapter 10**) to indicate the fraction of ROS-positive cells. Two cell lines were compared: ordinary HeLa cells versus HeLa cells stably transfected with an antiapoptotic protein, Bcl-2. After 1 h, ~28% of Bcl-2 cells with Au–Dox showed substantial ROS, compared with ~41% of untransfected HeLa cells with Au–Dox (**Figure 13.9**).

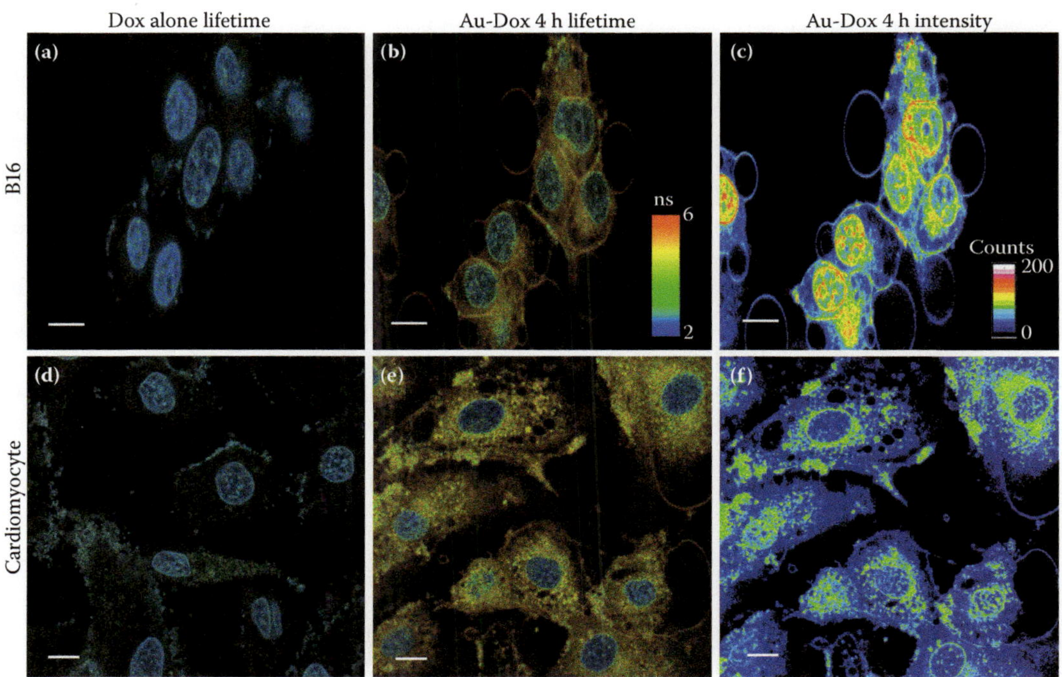


**Figure 13.8  FLIM images of (a–c) B16 melanoma cells and (d–f) cardiomyocytes incubated with Dox and Au–Dox.** The 2–6 ns scale bar applies to panels (a), (b), (d), and (e). (a) Dox alone in B16 cells, showing nuclear localization with short lifetime. (b) Au–Dox in B16 cells after 4 h incubation, showing a long lifetime in cytoplasm, shorter lifetime in nucleus (but longer than for Dox alone), and an intermediate lifetime in the nuclear membrane. (c) Intensity image of panel (b) showing concentration of particles in endosomes and in nuclear and nucleolar membranes. (d) Dox alone in cardiomyocytes, showing nuclear localization with short lifetime. There is a small amount of cytoplasmic Dox with a longer lifetime visible as green. (e) Au–Dox in cardiomyocytes cells after 4 h incubation, showing a long lifetime in cytoplasm, very short lifetime in nucleus (same as for Dox alone), and an intermediate lifetime in the nuclear membrane. (f) Intensity image of panel (e) showing concentration of particles in endosomes and in nuclear membrane with essentially no nuclear entry.

**Figure 13.9 Toxicity of Au–Dox is correlated with ROS production as measured by a fluorescent reporter dye, measured by flow cytometry.** Cells in the right upper quadrant are positive for both doxorubicin and ROS. (a) Wild-type HeLa cells show a significant number of double-positive cells. (b) Overexpression of the antiapoptotic protein Bcl-2 reduces the generation of ROS and the amount of cell death.

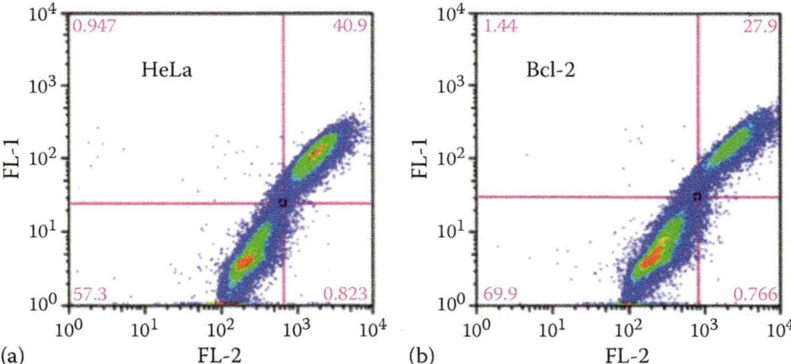

## *In vivo* studies

After establishing effectiveness in cell lines, we compared the efficacy of Au–Dox to Dox alone using two mouse models of melanoma. The first was B16, a murine cell line that creates aggressive tumors in *immunocompetent* mice, and the cell line that was used in most of the *in vitro* studies. The use of immunocompetent mice is important in case the nanomedicine involves some immune-mediated response. Intratumoral and IV delivery were investigated. Intratumoral injection was highly effective. Au–Dox produced >70% reduction in tumor growth in B16 tumors that persisted for at least 19 days; Dox alone was able to suppress tumors for a shorter time before growth accelerated. Due to their aggressiveness and large size, B16 tumors were impossible to eliminate entirely (**Figure 13.10a**). The second model was SK-MEL-28, a human melanoma cell line that forms tumors in nude mice; Au–Dox was able to completely suppress growth of tumors (duration of experiment > 6 months), whereas Dox alone showed early inhibition of tumors followed by rapid tumor growth (**Figure 13.10b**).

The tumors were removed, embedded in paraffin, and thin-sectioned for examination with different types of stains, including hematoxylin and eosin (H&E) (see **Chapter 7**). H&E sections showed a great reduction in tumor size of Au–Dox

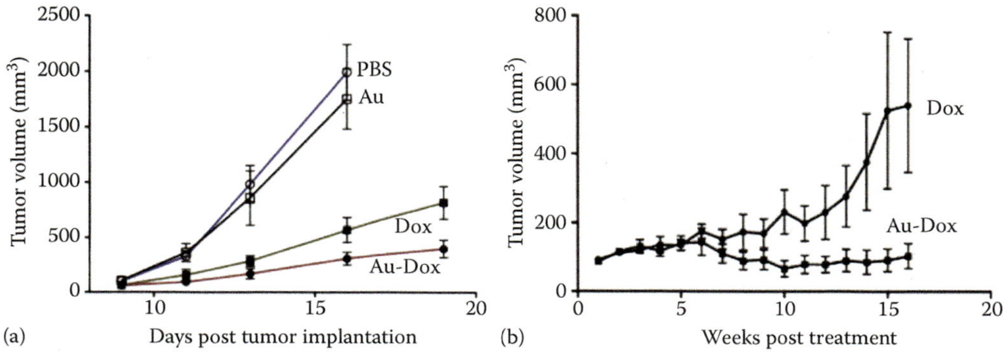

**Figure 13.10 Efficacy of intratumoral injections in B16 melanoma xenografts and SK-MEL-28 xenografts.** (a) B16 tumor volume versus time for intratumoral injections of Dox alone, Au–Dox, phosphate-buffered saline (PBS), or Au nanoparticles alone. The injections were given every 2–3 days; each injection of Au–Dox or Dox alone contained 0.2 mg/kg of doxorubicin. The injections containing Au alone contained the same amount of Au as in the Au–Dox case. The error bars represent standard deviations of 12 mice in each group. (b) SK-MEL-28 tumor volume over 100 days; clinical end points were not reached. The error bars represent standard deviations of 12 mice in each group.

tumors, with large amounts of necrosis (**Figure 13.11a through d**). Microscopic examination showed some apoptotic bodies in both the tumors treated with Dox and Au–Dox, but the bulk of the tumor mass was necrotic (**Figure 13.11e and f**).

Light and electron microscopy showed uptake of particles into cells and nuclei (**Figure 13.12**).

Possible toxicity to nontarget organs was assessed by IV injection of 79.2 mg of Au–tiopronin (corresponding to 29.6 mg of Au) followed by histological examination of the liver and kidneys after 24 h and 7 days. The kidneys were especially of interest following a report of renal toxicity from Au–tiopronin. We examined 12 slides from each animal; there were no visible signs of toxicity in the liver and kidneys in any of the animals (not shown). Biodistribution of particles was also measured in tumor-bearing and wild-type mice by digesting organs and performing atomic absorption spectroscopy. Au was found mostly in the liver and kidney and not at all in the brain; when tumors were present, ~30% of the total Au was distributed in the tumor (**Figure 13.13**).

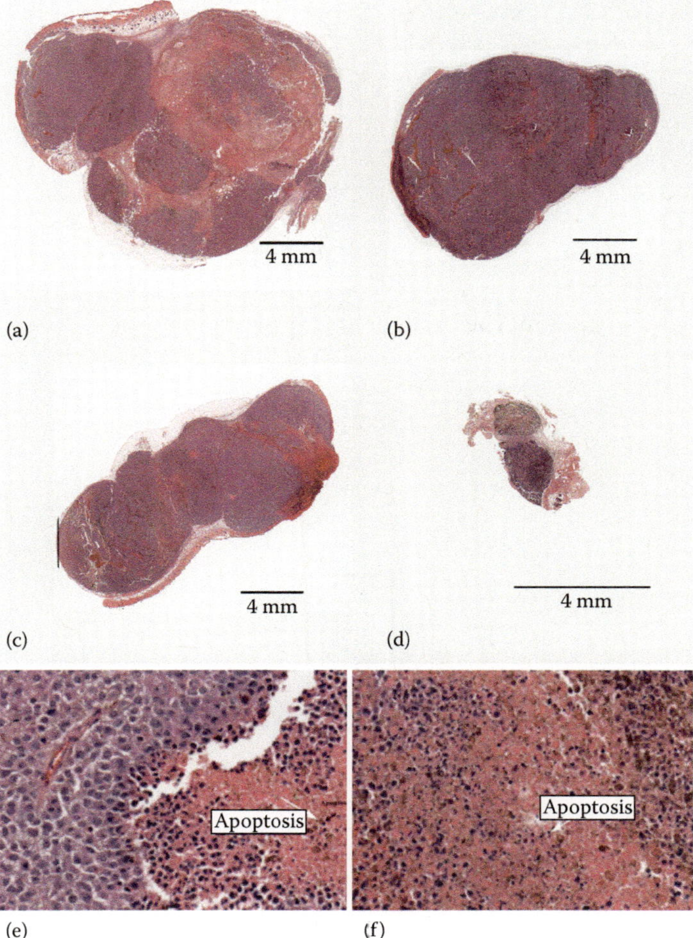


**Figure 13.11** Hematoxylin and eosin (H&E) histology of tumors after intratumoral injection of (a) PBS, (b) unconjugated Au nanoparticles, (c) doxorubicin, and (d) Au–Dox conjugates. The pink areas indicate necrosis, and the purple areas are viable, growing tissue. (e) Microscopic examination of a Dox-injected sample shows viable tumor next to a necrotic region. (f) Microscopic examination of an Au–Dox tumor shows a necrotic mass with occasional apoptotic bodies.

**Figure 13.12 Light and EM microscopy of histology sections from B16 tumors after intravenous injection of Au–Dox.** (a) Silver staining. The black arrows indicate the Au. The white arrowhead shows nonspecific dark areas that are found in controls. The inset shows a high-magnification image of a single cell. (b) Low-magnification of STEM section, showing tumor cells, adipocytes (arrowhead), and deposits of Au (arrow). (c) Higher-magnification image of B16 cell, showing intracellular and intranuclear Au. (d) STEM image showing Au within a capillary (arrows). (e,f) Cell nuclei in tumors with (e) and without (f) Au–Dox to illustrate the difference in contrast.

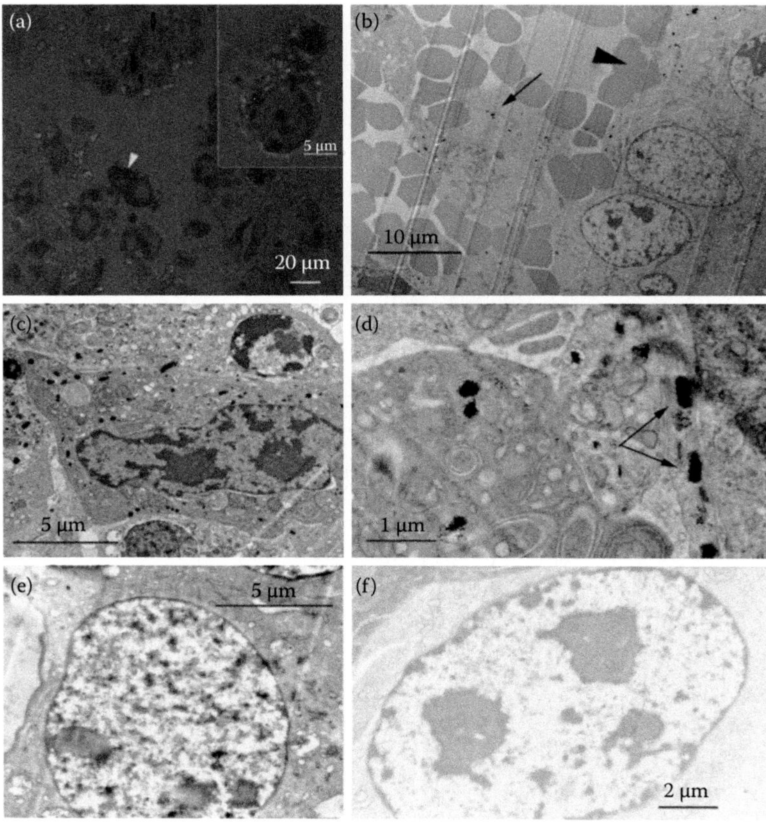

**Figure 13.13 Biodistribution of Au and Au–Dox in mice with B16 melanoma xenograft tumors (*n* = 27) at 1 h and 24 h postinjection.** The Au levels were measured by atomic absorption spectroscopy (AAS) in parts per billion (ppb); error bars give the standard deviation.

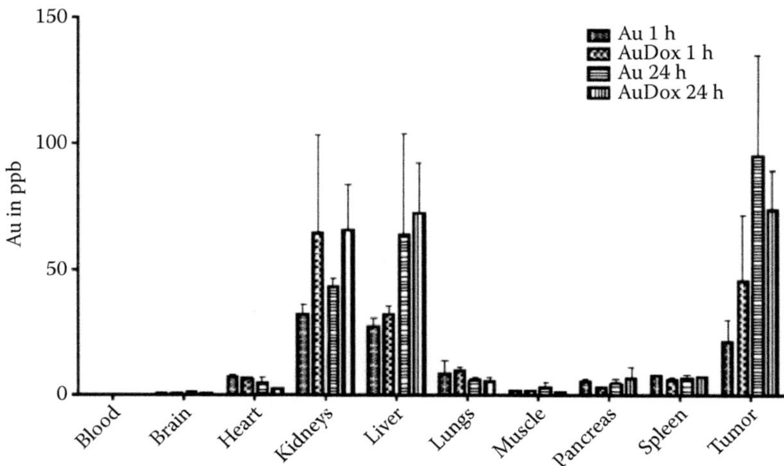

## The nanotechnology characterization laboratory assay cascade

After procedures for reproducible, sterile synthesis are established and efficacy is shown in cells and at least one animal model, a nanomedicine may be submitted to the Nanotechnology Characterization Laboratory (NCL) for further testing. The NCL was established by the National Cancer Institute under guidance from the National Institute of Standards and Technology (NIST) and the FDA, with the goal of standardizing *preclinical* efficacy and toxicity tests of nanomedicines. Potential nanomedicines are selected through a proposal process, which requires preliminary efficacy studies as well as a clear plan to clinical use. Once selected, a nanomaterial will be subjected to a thorough, standardized assay cascade that includes physical characterization, *in vitro* characterization, and *in vivo* characterization in rodents. Many of the components of the cascade are assays that are difficult for individual laboratories to perform because they are expensive or complicated, such as *pharmacokinetic/pharmacodynamic* (PK/PD) studies and immunological testing. This is therefore an extremely valuable resource for nanomedicine development, though it is worth noting that nanomedicines may be stopped in the assay cascade if they show properties incompatible with the testing, or if they are contaminated. It is important not to apply to the program prematurely, but to have a reproducible, fully characterized synthesis procedure that another laboratory will be able to imitate. The NCL also makes protocols available for its assays, which aid in standardizing procedures performed in individual labs.

## Good laboratory practice, good manufacturing practice, and scale-up

A significant question when developing a drug is whether the drug can be readily produced in large amounts in a highly reproducible fashion—not only that the drug itself should behave reproducibly but also that any possible contaminants are also reproducible. When the time comes to ask this question, a laboratory that produces drugs by good laboratory practice (GLP) and/or good manufacturing practice (GMP) should be consulted.

There is a significant difference between GLP and GMP, which should be kept in mind before a lab is consulted. GLPs are designed to provide the FDA with a clear and auditable record of open-ended research studies. GMPs are used for products that are already FDA regulated, and demonstrate that each batch of the product is manufactured according to predefined manufacturing criteria. Thus, in general, for research studies, GLP should be used.

GLP/GMP facilities that can produce nanoparticles are few at the present, but their numbers are growing. Some universities have GMP facilities associated with them, but others do not; in this case, researchers can contact other institutions to inquire about the feasibility and cost of producing their formulations by GMP. The costs are relatively high, and so funding sources specifically for the purpose often need to be solicited. Pharmaceutical companies and venture capitalists are possible sources of funding for GLP/GMP production.

For Au nanoparticles, the only significant barrier is the large volume of $H_2O$ required for purification of the particles by dialysis. Alternative methods of purification, such as centrifugation, need to be substituted.

## Steps toward approval: the investigational new drug

In order to use a drug in humans for limited clinical trials in the United States, it must obtain FDA Investigational New Drug (IND) approval. IND does not permit the drug to be sold but establishes that the product will not expose humans to unreasonable risks in a clinical trial. Data that must be submitted include results of animal studies (including pharmacokinetics/pharmacodynamics); manufacturing data (e.g., that the drug is made by GMP); and details of the clinical trial. A researcher will usually have to have pharmaceutical industry support at this stage and must certainly have clinicians involved who are able to conduct limited clinical trials on appropriate patient groups. Because submission of an IND is complex, the FDA has set up pre-IND consultation programs for different classes of drugs. For example, the Office of Oncology Products deals with anticancer drugs and has three subcategories: biological oncology products, drug oncology products, and hematology products. Medical Imaging Products is a separate division.

*Drug or device?* PEGylated Au nanoparticles for photothermal therapy are classified by the FDA as a device, not as a drug. This permits an accelerated and less costly approval process. Whether other formulations will receive similar approval is unclear.

## End-of-Chapter Questions

**1.** Pick a nanomedicine from the literature and discuss all of the characterization steps that you can find that have been published. Discuss which further steps are needed toward use in the clinic.

**2.** Discuss the issues involved in producing completely reproducible gold nanoparticles. Are there alternatives to chemical synthesis methods?

**3.** If you were going to create a nanomedicine, what would be your first step in choosing the ingredients? Does using a multifunctional nanoparticle pose more issues than a particle with a single functionality? Would having multiple functions be worth it?

**4.** Discuss the EPR effect and its role in human cancers.

## Background Reading

### Gold in medicine

Akhter, S., Ahmad, M.Z., Ahmad, F.J., Storm, G., and Kok, R.J. (2012). Gold nanoparticles in theranostic oncology: Current state-of-the-art. *Expert Opinion on Drug Delivery* 9, 1225–1243.

Beretta, G.L., and Cavalieri, F. (2016). Engineering nanomedicines to overcome multidrug resistance in cancer therapy. *Current Medicinal Chemistry* 23, 3–22.

Dawidczyk, C.M., Russell, L.M., and Searson, P.C. (2014). Nanomedicines for cancer therapy: State-of-the-art and limitations to pre-clinical studies that hinder future developments. *Frontiers in Chemistry* 2, 69.

Dreaden, E.C., Austin, L.A., Mackey, M.A., and El-Sayed, M.A. (2012). Size matters: Gold nanoparticles in targeted cancer drug delivery. *Therapeutic Delivery* 3, 457–478.

Pillai, G., and Ceballos-Coronel, M.L. (2013). Science and technology of the emerging nanomedicines in cancer therapy: A primer for physicians and pharmacists. *SAGE Open Medicine* 1, 2050312113513759.

### Targeting

Kasten, B.B., Liu, T., Nedrow-Byers, J.R., Benny, P.D., and Berkman, C.E. (2013). Targeting prostate cancer cells with PSMA inhibitor-guided gold nanoparticles. *Bioorganic and Medicinal Chemistry Letters* 23, 565–568.

Razzak, R., Zhou, J., Yang, X., Pervez, N., Bedard, E.L., Moore, R.B., Shaw, A., Amanie, J., and Roa, W.H. (2013) The biodistribution and pharmacokinetic evaluation of choline-bound gold nanoparticles in a human prostate tumor xenograft model. *Clinical and Investigative Medicine* 36, E133–E142.

Roa, W., Xiong, Y.P., Chen, J., Yang, X.Y., Song, K., Yang, X.H., Kong, B.H., Wilson, J., and Xing, J.Z. (2012). Pharmacokinetic and toxicological evaluation of multi-functional thiol-6-fluoro-6-deoxy-D-glucose gold nanoparticles in vivo. *Nanotechnology* 23 (37):375101.

Simpson, C.A., Agrawal, A.C., Balinski, A., Harkness, K.M., and Cliffel, D.E. (2011). Short-chain PEG mixed monolayer protected gold clusters increase clearance and red blood cell counts. *ACS Nano* 5, 3577–3584.

## Photothermal therapy and GNRT

Al Zaki, A., Joh, D., Cheng, Z.L., De Barros, A.L.B., Kao, G., Dorsey, J., and Tsourkas, A. (2014). Gold-loaded polymeric micelles for computed tomography–guided radiation therapy treatment and radiosensitization. *ACS Nano* 8, 104–112.

Butterworth, K.T., McMahon, S.J., Currell, F.J., and Prise, K.M. (2012). Physical basis and biological mechanisms of gold nanoparticle radiosensitization. *Nanoscale* 4, 4830–8.

Chang, M.Y., Shiau, A.L., Chen, Y.H., Chang, C.J., Chen, H.H.W., and Wu, C.L. (2008). Increased apoptotic potential and dose-enhancing effect of gold nanoparticles in combination with single-dose clinical electron beams on tumor-bearing mice. *Cancer Science* 99, 1479–1484.

Chattopadhyay, N., Cai, Z.L., Kwon, Y.L., Lechtman, E., Pignol, J.P., and Reilly R.M. (2013). Molecularly targeted gold nanoparticles enhance the radiation response of breast cancer cells and tumor xenografts to X-radiation. *Breast Cancer Research and Treatment* 137, 81–91.

Chithrani, D.B., Jelveh, S., Jalali, F., van Prooijen, M., Allen, C., Bristow, R.G., Hill, R.P., and Jaffray, D.A. (2010). Gold nanoparticles as radiation sensitizers in cancer therapy. *Radiation Research* 173, 719–728.

Cho, S.H., Jones, B.L., and Krishnan S. (2009). The dosimetric feasibility of gold nanoparticle–aided radiation therapy (GNRT) via brachytherapy using low-energy gamma-/x-ray sources. *Physics in Medicine and Biology* 54, 4889–4905.

Geng, F., Song, K., Xing, J.Z., Yuan, C.Z., Yan, S., Yang, Q.F., Chen, J., and Kong, B.H. (2011). Thio-glucose bound gold nanoparticles enhance radio-cytotoxic targeting of ovarian cancer. *Nanotechnology* 22 (28):285101.

Hainfeld, J.F., Dilmanian, F.A., Zhong, Z., Slatkin, D.N., Kalef-Ezra, J.A., and Smilowitz, H.M. (2010). Gold nanoparticles enhance the radiation therapy of a murine squamous cell carcinoma. *Physics in Medicine and Biology* 55, 3045–3059.

Hainfeld, J.F., Slatkin, D.N., and Smilowitz, H.M. (2004). The use of gold nanoparticles to enhance radiotherapy in mice. *Physics in Medicine and Biology* 49, N309–N315.

Herold, D.M., Das, I.J., Stobbe, C.C., Iyer, R.V., and Chapman, J.D. (2000) Gold microspheres: A selective technique for producing biologically effective dose enhancement. *International Journal of Radiation Biology* 76, 1357–1364.

Huang, P., Bao, L., Zhang, C.L., Lin, J., Luo, T., Yang, D.P., He, M., Li, Z.M., Gao, G., Gao, B. et al. (2011). Folic acid–conjugated silica-modified gold nanorods for X-ray/CT imaging-guided dual-mode radiation and photo-thermal therapy. *Biomaterials* 32, 9796–9809.

Kong, T., Zeng, J., Wang, X.P., Yang, X.Y., Yang, J., McQuarrie, S., McEwan, A., Roa, W., Chen, J., and Xing, J.Z. (2008). Enhancement of radiation cytotoxicity in breast-cancer cells by localized attachment of gold nanoparticles. *Small* 4, 1537–1543.

Leung, M.K.K., Chow, J.C.L., Chithrani, D.B., Lee, M.J.G., Oms, B., and Jaffray D.A. (2011). Irradiation of gold nanoparticles by x-rays: Monte Carlo simulation of dose enhancements and the spatial properties of the secondary electrons production. *Medical Physics* 38, 624–631.

Tawagi, E., Massmann, C., Chibli, H., and Nadeau, J.L. (2015). Differential toxicity of gold–doxorubicin in cancer cells vs. cardiomyocytes as measured by real-time growth assays and fluorescence lifetime imaging microscopy (FLIM). *Analyst* 140, 5732–5741.

Van den Heuvel, F., Locquet, J.P., and Nuyts, S. (2010). Beam energy considerations for gold nano-particle enhanced radiation treatment. *Physics in Medicine and Biology* 55, 4509–4520.

Wang, S., Huang, P., Nie, L., Xing, R., Liu, D., Wang, Z., Lin, J., Chen, S., Niu, G., Lu, G. et al. (2013). Single continuous wave laser induced photodynamic/plasmonic photothermal therapy using photosensitizer-functionalized gold nanostars. *Advanced Materials* 25, 3055–3061.

Yao, L., Daniels, J., Moshnikova, A., Kuznetsov, S., Ahmed, A., Engelman, D.M., Reshetnyak, Y.K., and Andreev, O.A. (2013). pHLIP peptide targets nanogold particles to tumors. *Proceedings of the National Academy of Sciences of the United States of America* 110, 465–470.

Zhang, X.D., Guo, M.L., Wu, H.Y., Sun, Y.M., Ding, Y.Q., Feng, X., and Zhang L.A. (2009). Irradiation stability and cytotoxicity of gold nanoparticles for radiotherapy. *International Journal of Nanomedicine* 4, 165–173.

## Au–Dox

Kumar, S.A., Peter, Y.A., and Nadeau, J.L. (2008). Facile biosynthesis, separation and conjugation of gold nanoparticles to doxorubicin. *Nanotechnology* 19, 495101.

Tawagi, E., Massmann, C., Chibli, H., and Nadeau, J.L. (2015). Differential toxicity of gold–doxorubicin in cancer cells vs. cardiomyocytes as measured by real-time growth assays and fluorescence lifetime imaging microscopy (FLIM). *Analyst* 140, 5732–5741.

Zhang, X., Chibli, H., Kong, D., and Nadeau, J. (2012). Comparative cytotoxicity of gold–doxorubicin and InP–doxorubicin conjugates. *Nanotechnology* 23, 275103.

Zhang, X., Chibli, H., Mielke, R., and Nadeau, J. (2011). Ultrasmall gold–doxorubicin conjugates rapidly kill apoptosis-resistant cancer cells. *Bioconjugate Chemistry* 22, 235–243.

Zhang, X., Shastry, S., Bradforth, S.E., and Nadeau, J.L. (2015a). Nuclear uptake of ultrasmall gold–doxorubicin conjugates imaged by fluorescence lifetime imaging microscopy (FLIM) and electron microscopy. *Nanoscale* 7, 240–251.

Zhang, X., Teodoro, J.G., and Nadeau, J.L. (2015b). Intratumoral gold–doxorubicin is effective in treating melanoma in mice. *Nanomedicine* 11, 1365–1375.

## Suppliers

Sigma-Aldrich. Bare and functionalized Au nanoparticles and nanorods.

Nanopartz. Bare and functionalized Au nanoparticles and nanorods, including proprietary *in vivo* coatings for animal experiments and a wide choice of functionalities. May be ordered certified sterile and/or endotoxin-free.

nanoComposix. Many shapes, sizes, and functionalities of Au nanomaterials, including custom synthesis; certified endotoxin-free available; GMP production available.

Innova Biosciences. Au nanoparticles, conjugation kits with different reactive groups, and conjugated Au nanoparticles.

Nanocs. Wide range of Au nanomaterial sizes and surface coatings.

# CHAPTER 14

# Surface Functionalization Techniques

## 14.1 INTRODUCTION

In numerous applications, it is desirable to have specific biomolecules arrayed on or tethered to a physical surface. This surface may serve as a transducer of signals by being conductive or piezoelectric, or by being small enough to change in mass or conformation upon analyte binding. A device that converts a biological binding event into a physicochemical signal is called a *biosensor*, and may be used for detection of ionic, small molecule, or large protein analytes in diagnostics or biochemistry; for studies of protein function; for environmental monitoring; or for numerous other basic and applied applications.

The challenge facing biosensors at every scale, from centimeter to nanometer, is to create an array of active molecules that bind to their targets without binding to nontarget molecules. Difficulties are encountered in various areas:

- Loss of analyte-binding capacity of immobilized protein. Proteins may array in the "wrong" orientation and be unable to bind their targets, or may be sterically hindered by overly dense layers.

- Loss of signal-transducing activity of protein. Proteins that undergo conformational changes upon analyte binding may be hindered by immobilization. Some proteins, such as membrane proteins, cannot function at all on a rigid surface because their conformational changes require a fluid environment. Such proteins must be immobilized onto lipids or lipid-like layers.

- Nonspecific binding. Analyte may bind a surface in areas other than the protein, causing false-positive signals or high background. Passivation of the surface to prevent this binding remains an important challenge.

- Lack of spatial specificity. It is often desirable to pattern different functionalities onto different areas of a slide or chip, often at the submicron scale. To accomplish this, it is necessary to control the surface chemistry and/or functionalization methods in a precisely controlled fashion.

A large body of literature exists in all of these areas. Many studies are contradictory, and others work only for the precise system described. Any new protein immobilization experiment is therefore a challenge and should be undertaken only after a thorough review of the relevant recent literature.

However, there are a few basic techniques that are relatively straightforward to master and that have very wide applicability, and we will present these here. At the end of this chapter, you should be able to prepare a functionalized silicon dioxide/glass or Au surface and immobilize an antibody or streptavidin, characterize the resulting layer, and quantify specific versus nonspecific binding. It is important to note that the exact chemicals mentioned in this chapter are examples, and the final choice of chain lengths, passivation strategies, and so on is up to the experimenter. In the final sections, we will discuss immobilization strategies for membrane proteins such as ion channels and present some ways in which this may be done.

## 14.2 PREPARING MONOLAYERS USING FUNCTIONAL SILANES OR THIOLS

### Silanes

A *silane* is analogous to an alkane, except that in the place of a central carbon atom, it contains a Si. Alkoxysilanes contain either methoxy ($-OCH_3$) or ethoxy ($-OCH_2CH_3$) groups attached directly to the Si atom. This functionality will react with anything possessing free hydroxyl groups: glass, metal oxides, silicon dioxide, or mica. The other side of the silane has a carbon chain of varying length terminated in a reactive group such as an amine, thiol, or epoxide. **Figure 14.1** shows some commonly used silanes for biological applications and their nomenclature.

The value of silanes is in their ability to form uniform, essentially defect-free monolayers when applied in the liquid or vapor phase to an appropriate surface (**Figure 14.2**). These monolayers prevent nonspecific binding to the surface and provide a uniform substrate for attachment of proteins. While surface silanization is relatively straightforward, surface defects and deposition of multilayers of silane can occur. Two things are essential for preparing a uniform silane monolayer:

- The surface must be very clean. If possible, silanization should be performed in a clean room.

- The silane must be fresh. Silanes will polymerize after long storage with oxygen exposure. Store them under inert gas and replace them as necessary. If the silane has been sitting around for a while, or if it appears viscous or ropy when pipetted, it should be discarded.

The first step in silanization is to clean the surface of organic contaminants and to generate reactive hydroxyls. This is usually performed with a strong oxidant; many variations are possible, several of which are called *piranha solution* or *piranha etch* and which contain large fractions of hydrogen peroxide. This should be performed immediately before silanization using fresh solutions, and extreme care is needed when handling piranha solution. A protocol for piranha cleaning is given in **Practical Tips 14.1**.

For liquid phase assembly, the cleaned substrate is then immersed in a silane solution for 1–24 h to permit formation of the silane monolayer. While a disordered

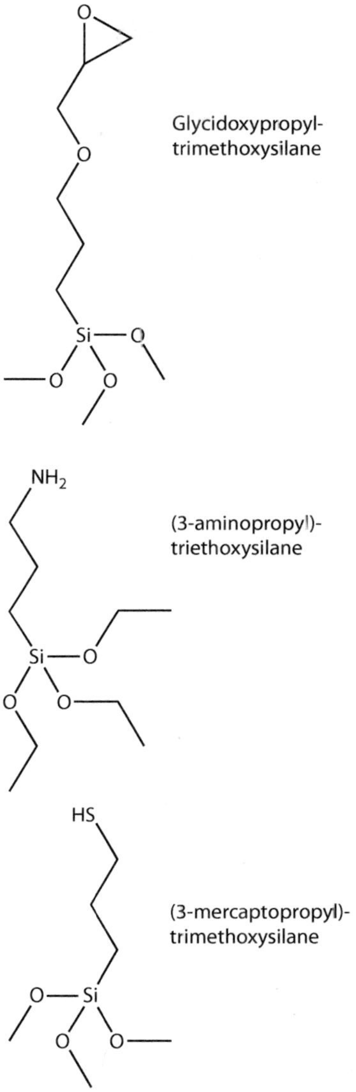
Glycidoxypropyl-
trimethoxysilane
$NH_2$
(3-aminopropyl)-
triethoxysilane
HS
(3-mercaptopropyl)-
trimethoxysilane
Si
O

**Figure 14.1 Some common silanes showing epoxide, amine, and thiol termini.**

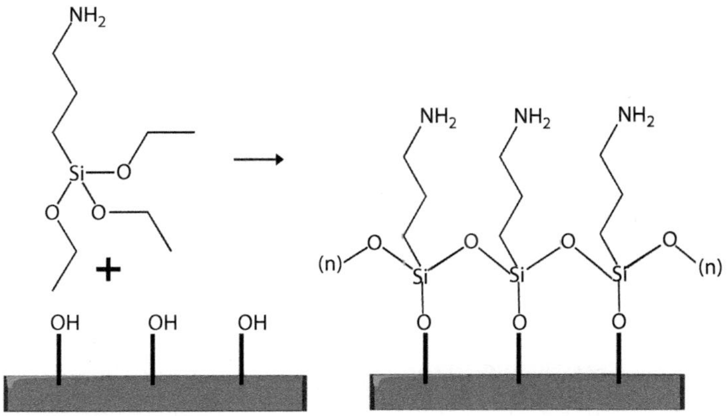
$NH_2$
Si
O
+
OH
(n)

**Figure 14.2 Formation of silane monolayer on -OH-terminated surface.**

---

### PRACTICAL TIPS 14.1:   CLEANING SURFACES WITH PIRANHA SOLUTION

**GENERAL NOTES**

Many different types of strongly oxidizing solutions are called piranha. The following is an example. All piranha solutions must be handled with great caution, using the appropriate personal protective equipment and a chemical fume hood. They should never be allowed to contact organic solvents, or an explosion may result!

**PROTOCOL**

- Add 2 volumes of sulfuric acid ($H_2SO_4$) to a container deep enough to completely submerge the sample.
- Add the sample to the bottom of the container.
- Add 1 volume of hydrogen peroxide ($H_2O_2$, 30%).
- The solution will become hot.
- Agitate gently for 10–15 min.
- Remove the sample and wash with copious amounts of distilled water.
- Dispose of the piranha solution after cooling in an appropriate acid-waste container. Do not reuse.

---

layer is formed almost immediately, the formation of an ordered monolayer is complex. There is significant disagreement in the literature about the optimal incubation time, with many papers claiming that a full 24 h is needed for formation of a stable, uniform layer. Different solvents may also be used; the following are our recommendations for biomolecule immobilization:

- Using mercaptopropyltrimethoxysilane (MPTS). Air will cause oxidation of thiols, so best results are obtained by using MPTS at 2% in dry toluene under an inert atmosphere (e.g., in a glove bag or glove box) at room temperature. The substrate is then rinsed in toluene, and for best results, the protein cross-linking is performed immediately. If the substrate must be stored, it can be dried under a stream of nitrogen and placed in a capped container. Heat treatment of the substrate is not recommended, as this can lead to disulfide bond formation in the terminal thiols with corresponding loss of reactivity.

- Using (3-aminopropyl)triethoxysilane (APTS or APTES). Immerse the substrate in a 5–10% APTS solution in distilled water. Rinse with distilled water followed by absolute ethanol. Then cure in a vacuum oven overnight at 80°C to encourage cross-linking of silanes. The substrate may be stored if kept clean or used immediately for functionalization.

- Using glycidoxypropyltrimethoxysilane (GOPS). Immerse the substrate in a 10:85.5:4.5 GOPS/ethanol/water solution. Rinse in water followed by absolute ethanol and dry in a vacuum oven at 50°C.

Many other silanes apart from these examples are available. When determining the reaction conditions, their solubility and terminal groups must be considered.

It is always best to follow a protocol found in the literature that has been optimized for biomolecule attachment whenever possible.

## Alkanethiol self-assembled monolayers

A large variety of alkanethiols are available in a range of chain lengths from single carbon up to 16 carbons and more. These may be homobifunctional (thiols on both ends) or heterobifunctional (thiol on one end, another reactive group on the other). Chains of any length will produce ordered *self-assembled monolayers* (SAMs) on Au substrates due to the strong Au–S interaction (adsorption energies are typically 35–45 kcal/mol) (**Figure 14.3**). Due to van der Waals interactions between the chains, longer alkanethiols (more than 10 carbons) produce highly crystalline layers. Shorter chains show less crystallinity but may be equally functional.

The preparation of alkanethiol SAMs is easy and rewarding. The only drawback to this technique is that it requires an Au (or sometimes and Ag) substrate, so the metal must be patterned in the desired region of the device or sensor for this to be useful. For sensors requiring an electrode, this method is ideal, but it is of course of lesser use for general functionalization of surfaces.

The Au surface used must be very smooth and very clean. If it is polycrystalline, it may be polished using alumina slurries followed by ultrasonic cleaning. However, we recommend the preparation of smooth Au surfaces by evaporation or sputtering onto optically flat glass or silicon wafers. Since Au adheres poorly to silicon, the ideal substrate is a 10/10/100 nm Ti/Pd/Au or 10/10/100 Cr/Pd/Au multilayer. The Pd blocks diffusion of the alkanethiol through the pores in the Au and prevents it from interacting with the underlayer.

When multilayer metal deposition is not available, the poor man's approach to a smooth Au surface on silicon or glass is to first silanize the surface with MPTS and then deposit the Au by evaporation. This type of substrate is preferred in biosensor applications involving immersion into electrolytes, as it avoids the corrosion seen with metals. The only caveat with this method is that the thiols on the MPTS must be protected from oxidation, as Au will not adhere to disulfides. This means that the sample should be treated with Au immediately after preparation. Also, if the

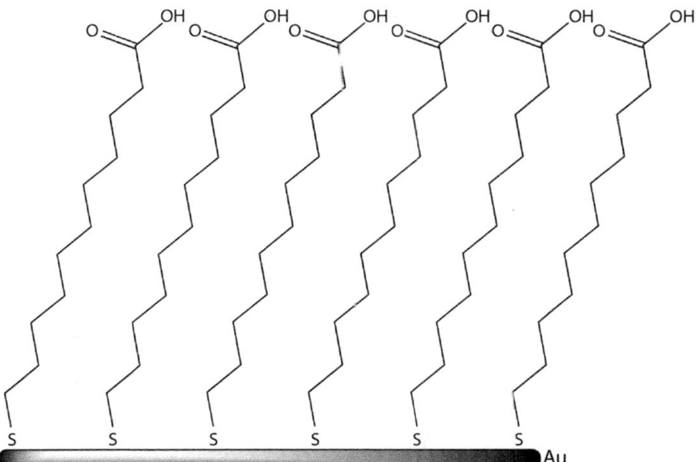


**Figure 14.3 Formation of alkanethiol SAMs on Au.** Shown is an 11-mercaptoundecanoic acid SAM, showing highly ordered crystalline packing. The spacing between the anchor points on the Au is 4.97 Å, larger than the ideal distance needed to maximize van der Waals interactions between the alkanethiol chains. This causes the chains to tilt at an angle of 20–45°, and in general, the longer the chain, the more ordered the structure.

sample is heat-cured, it must be kept at a temperature <115°C. Finally, electron beam evaporation is preferred to thermal deposition, as less heat is generated. If thermal evaporation must be used, the proper choice of evaporation boat is essential to minimize excess heat.

When metal evaporation is not available, it is possible to make a conductive Au surface suitable for functionalization via a solution-based process, using deposition of Au nanoparticles followed by "seeding" to create a conductive surface (see **Chapter 12** for Au nanoparticle synthesis and **Practical Tips 14.2** for a protocol for solution-based deposition onto MPTS).

Au films can be stored for long periods dry but must be cleaned immediately before functionalization. In typical chemistry laboratories, piranha solution, as discussed in **Practical Tips 14.1**, is used for cleaning. However, if you have access to a nanofabrication facility, an Ar or $O_2$ plasma etch is the best way to clean Au.

Assembly of SAMs is then as simple as immersing the substrate in a 1–2 mM solution of the alkanethiol in the appropriate solvent (absolute ethanol for chain lengths ≤18 carbons; a nonpolar solvent such as hexane for longer chains). However, there are a few extra steps than can improve success:

- If possible, recrystallize the alkanethiol to eliminate any disulfides. The presence of disulfides will slow self-assembly.

- Degas the solvent to eliminate oxygen (which can cause disulfide bond formation).

- Use clean glass or plasticware; we recommend disposable polypropylene or glass that has been cleaned thoroughly in a clean room.

Preliminary assembly of the thiols on the Au surface will occur within a few minutes, but it is recommended to allow the substrate to equilibrate in the alkanethiol solution overnight (12–24 h) to improve the quality of the SAM.

---

**PRACTICAL TIPS 14.2:    CREATING A CONDUCTIVE SURFACE USING Au NANOPARTICLES**

**GENERAL NOTES**

Please see **Chapter 12** for a full discussion on the synthesis and properties of Au nanoparticles. The purpose of this protocol is to bind Au particles to a functionalized surface and then grow them by reduction of extra Au ions until they are large enough to form a conducting surface (**Figure P14.2.1**).

**PROTOCOL**

- Prepare a substrate with amino- or thiol-terminated silanes.
- Prepare a colloidal gold solution of citrate-capped particles of ~20–30 nm size.
- Immerse the substrate into the Au nanoparticle solution for 2 h.
- Rinse with distilled water.
- Immerse in a solution containing 1.3 mg hydroxylamine and 10 mg $HAuCl_4$ per 100 mL of water. Agitate for 30 min.

*(Continued)*

**PRACTICAL TIPS 14.2 (CONTINUED): CREATING A CONDUCTIVE SURFACE USING Au NANOPARTICLES**

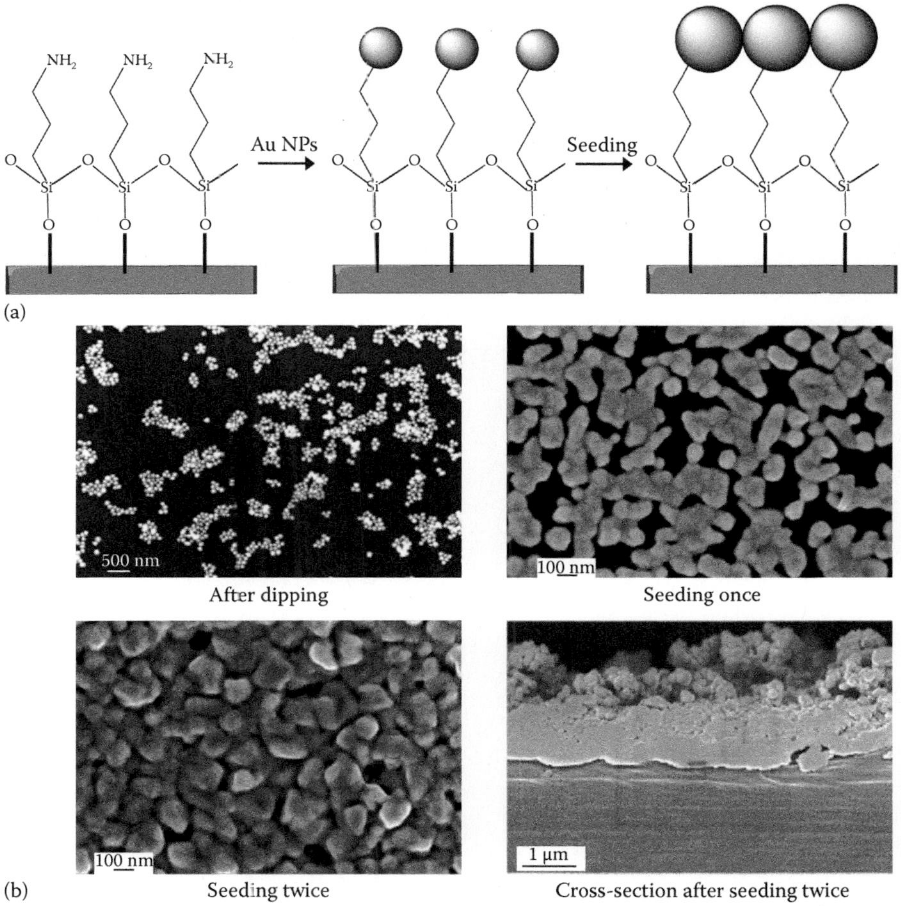

**Figure P14.2.1** (a) Schematic of process for binding Au particles to a surface and then seeding until they become large enough to form a conducting layer. (b) SEM images of surfaces before and after seeding. (Adapted with permission from Supriya, L., and Claus, R. O., Solution-Based Assembly of Conductive Gold Film on Flexible Polymer Substrates, *Langmuir* 20, 8870–8876. Copyright 2004 American Chemical Society.)

- Rinse with water and repeat the immersion twice.
- A multimeter can be used to monitor conductivity of the film, which should approach that of bulk gold.
- Optional: inspect the surfaces by SEM or AFM to observe the growth of the particles (**Figure P14.2.1**).

## SUGGESTED READING

Hassler, B.L., Amundsen, T.J., Zeikus, J.G., Lee, I., and Worden, R.M. (2008). Versatile bioelectronic interfaces on flexible non-conductive substrates. *Biosensors and Bioelectronics* 23, 1481–1487.

Supriya, L., and Claus, R.O. (2004). Solution-based assembly of conductive gold film on flexible polymer substrates. *Langmuir* 20, 8870–8876.

It is often desirable to make "mixed" SAMs containing different chain lengths, often with different functionalities. The shorter chain may be chosen to have a nonreactive group such as a hydroxyl in order to act as a passivating surface. This can improve protein binding and ligand recognition by minimizing steric hindrance (**Figure 14.4a**). It can also allow for automatic insertion of a reactive

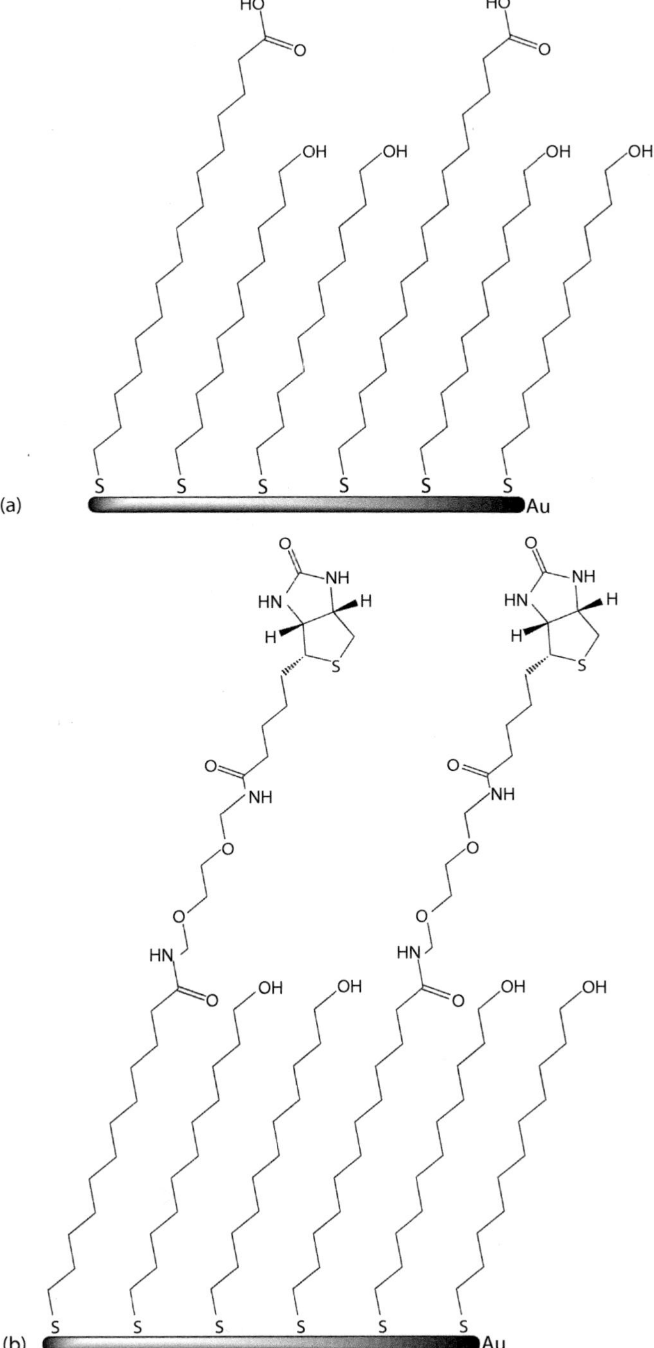

**Figure 14.4 Mixed SAMs.** (a) A mixture of 2:1 11-mercaptoundecanol and 16-mercaptohexadecanoic acid. The –OH group is hydrophilic but will not react with cross-linkers. (b) A mixture of 2:1 11-mercaptoundecanol and a biotinylated dodecanethiol. Usually, fewer biotins are desired per unit area for optimal spacing.

element such as a biotin at desired ratios (**Figure 14.4b**). For assembly of streptavidin, the ideal mixture of biotinylated to nonbiotinylated alkanethiols is 1:9. Formation of mixed SAMs is done in the same fashion as for single-component SAMs; however, the importance of the compounds' purity becomes more acute, so it is highly recommended that they be recrystallized or vacuum-distilled before use. An uncontrolled percentage of unreactive disulfides in either or both of the components will lead to highly irreproducible results!

The most common problem is the formation of "islands" of one or the other component. To avoid this, the two components should not differ greatly in charge or hydrophobicity. They should also be very well mixed before incubation, by sonication if possible.

After SAM assembly, the substrate should be rinsed thoroughly with ethanol and then water. If the assembly has worked properly, a SAM with a charged head group (such as an amine or carboxylate) will result in the surface becoming hydrophilic—that is, water will spread over it easily rather than beading up. If this is not the case, SAM assembly has failed, and the sample should be discarded.

## Some special considerations

The following are a few tricks and/or pitfalls that often occur in the construction of different types of functionalized chips and biosensors.

*Site-specific patterning on Si/SiO$_2$.* An oxide layer is necessary to permit silanes to react. All silicon surfaces exposed to air have a thin oxide layer called *native oxide*, which can be sufficient for silanization, although it is recommended to use a deposited SiO$_2$ layer of micron thickness. If target features on a device have thick oxide layers, the rest of the chip can be stripped of oxide using HF, thus permitting silanization of only those features. This is a very commonly used method for site-specific functionalization, which we will discuss in more detail in the example experiment in **Section 14.5**, "Example Experiment: Preparing a Silane–Biotin–Streptavidin Sandwich on SIO$_2$ Features on a SI Chip" section.

*Silanized surfaces other than silicon or glass (polymers).* In some cases, sensors or devices need to be mechanically flexible. In this case, instead of glass or silicon, a variety of polymers can be used as a substrate for silanization. Two examples of useful polymers are Kapton (**Figure 14.5**) and polyethylene ($-_n[CH_2–CH_2]_n-$). The only difficulty is that these polymers do not have reactive hydroxyl groups; they need to be plasma-etched to generate carbon radicals and then exposed to air to result in reactive species (carbonyl, carboxyl, etc.) that can react with silanes. This requires access to an Ar plasma etcher. Once silanization is accomplished, Au may be evaporated or deposited as Au nanoparticles as in **Practical Tips 14.2**. The latter is recommended as it is more easily controlled.

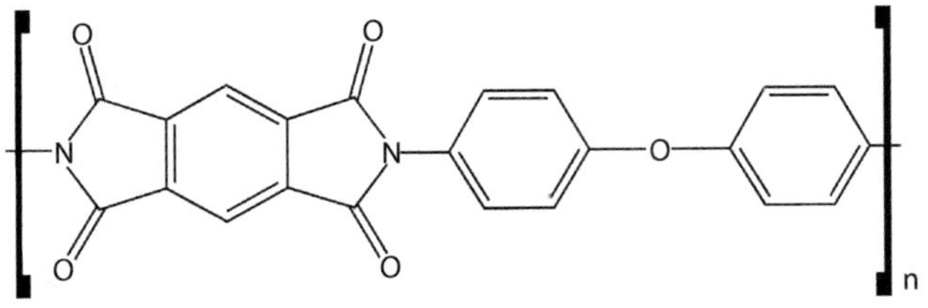

**Figure 14.5 Structure of Kapton.**

**Figure 14.6 The uses
of critical point drying.**
(a) Cantilever sensors are
positioned as "diving boards"
over large etched wells. These
features will break and curl if
passed through an air–liquid
interface. (b) Critical point drying
of bacterial cells can show
internal structure of intact cells.
This image shows two dividing
*Bacillus subtilis* cells after critical
point drying.

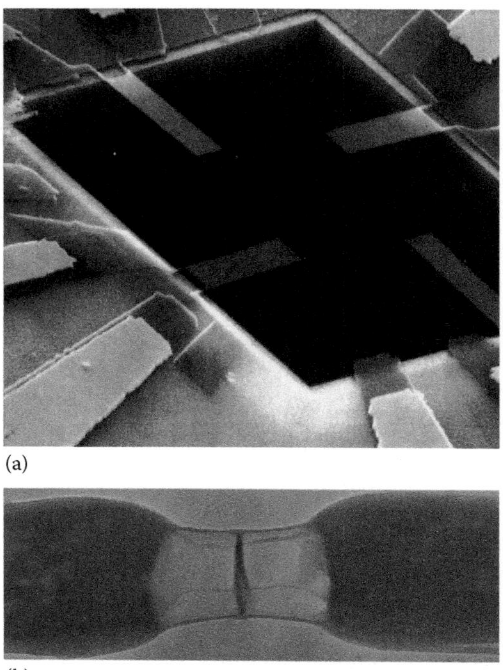

(a)

(b)

*Fragile features.* Many micromachined devices on silicon wafers have features such as cantilevers that break readily under a liquid–air interface (**Figure 14.6a**). For these types of samples, a *critical point dryer* can be used to take the samples from liquid to air without a phase boundary. This is a specialized instrument that is often found in microfabrication core facilities, and may be the only way to treat some types of samples without breaking them. Critical point drying can also be used on biological samples such as bacterial cells, as it dries them without deformation and renders them transparent (**Figure 14.6b**).

## 14.3 TECHNIQUES FOR CHARACTERIZING SURFACE MONOLAYERS

Making a surface monolayer is easy; characterizing one can be the labor of a lifetime. Alkanethiol monolayers are about a nanometer in thickness and essentially transparent in the ultraviolet (UV) to visible. Their structure and chemical composition, uniformity, and roughness all need to be known. Fortunately, the techniques for their characterization are well developed, though quite specialized and often expensive. Many of the techniques discussed in this section require dedicated instruments developed for studying the physics of thin films. They are usually not found in individual labs but can be found all together in core facilities dedicated to microfabrication, advanced materials, or nanotechnology. Often they are located inside clean rooms, which aids in sample handling but requires the user to be trained in appropriate procedures. Despite the initial difficulties, characterizing your surfaces well is highly recommended, as it is the only way to protect against irreproducible results.

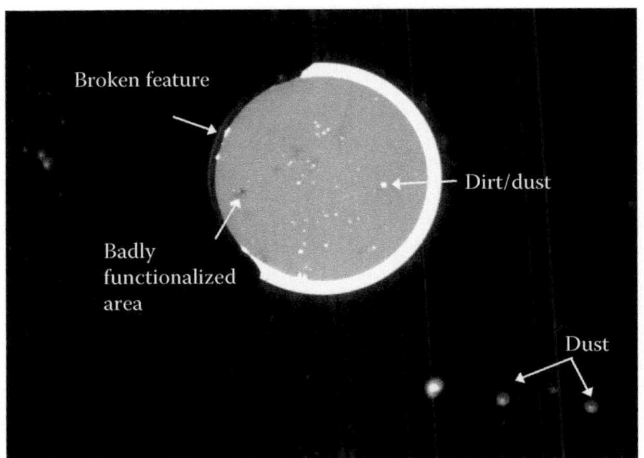



**Figure 14.7  Use of reactive dye to visualize coverage.** Defects are clearly identified.

## Interaction with reactive dyes

This is an easy method to test the success of reactive monolayer formation that requires no specialized equipment. It can provide semiquantitative data about the number of reactive sites on the surface as well as a visual depiction of the surface quality. The only disadvantage of this technique is that it spoils the sample for any future experiments, so duplicate samples must be prepared. The procedure is simple: buy a reactive dye that reacts with the silane terminus in question. (See **Chapter 7** for a discussion of reactive probes; an example would be tetramethylrhodamine maleimide [TMR maleimide] to react with the sulhydryls of MPTS.) Prepare a solution of an excess of reactive dye (10–50 μM), immerse the substrate for 30–60 min, and rinse copiously with water and then with the solvent in which the dye dissolves, as appropriate (for TMR, this would be ethanol). Then record the UV–visible absorbance spectrum of the sample; at any particular wavelength $\lambda$, the number of dye molecules per unit area $n$ will be proportional to the ratio of the absorbance to the extinction coefficient at that wavelength:

$$n = 6 \times 10^{20} \frac{A(\lambda)}{\varepsilon(\lambda)} \tag{14.1}$$

Choosing the correct wavelength can be complicated for some dyes that change their behavior upon close-packing. In these cases, standard absorbance curves should be prepared and a wavelength chosen that does not show strong concentration dependence. This phenomenon is found in TMR; a good wavelength for this dye is 570 nm. In addition, background subtraction can be performed with a slide that has no dye if very precise results are desired.

This technique can also be used to monitor the formation of surface monolayers in real time. It is also good for visual inspection of the samples using the dye fluorescence, as any breakage, defects, or dust will be readily apparent (**Figure 14.7**).

## Ellipsometry

*Ellipsometry* is performed using a specialized instrument called an *ellipsometer* and permits the measurement of film thickness, refractive index, extinction coefficient, anisotropy, and roughness. Nearly all experiments involving monolayer

**Figure 14.8 Sample ellipsometry data.** The data are for the surfactant dioctadecyldimethylammonium bromide (DODAB) at the air–water interface. The angle of incidence was 54°, close to the Brewster angle for water. Measurements on pure water were used as a background. (a) Raw data (points) with fits to an oscillator model (lines). (b) Resulting optical parameters extracted from the model. (Images courtesy Dr. E. Teboul, Horiba Jobin Yvon, Inc.)

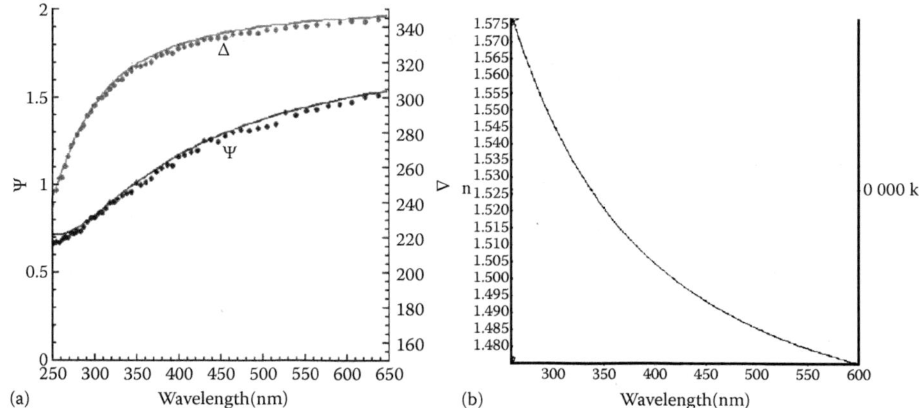

deposition should use this technique for film characterization. The principle of ellipsometry is that the polarization state of light reflected off a sample will change according to the ratio of the Fresnel coefficients for p- and s-polarized light, $R_p$ and $R_s$:

$$\frac{R_p}{R_s} = \tan(\Psi)e^{i\Delta}\tag{14.2}$$

The values measured by ellipsometry are $\Psi$ (amplitude ratio) and $\Delta$ (phase difference). In a typical experiment, data are taken at different wavelengths and angles of incident light. The wavelengths chosen should be roughly comparable to the film thickness; i.e., UV–visible wavelengths are used for samples ranging from 0.1 to 1000 nm.

Optical constants are inferred from ellipsometric data by fitting to a model, as they are directly calculable only in the case of infinitely thick materials (which usually do not interest the makers of SAMs!). Sample models are usually included with ellipsometry software or can be easily entered as approximations of thickness and complex refractive index ($\tilde{n} = n + ik$). The model is used to calculate the predicted $\Psi$ and $\Delta$ from the Fresnel equations and iterate until it fits the experimental data. For alkanethiols, the samples are essentially transparent in the visible. A good approximation is to set $k = 0$ and $n = 1.5$, making estimates of film thickness easy (**Figure 14.8**).

For samples that are nonuniform, *imaging ellipsometry* can give a picture of the optical properties of the film with $x$–$y$ resolution of up to 2 μm. What are needed for this technique are a long-working-distance, high numerical aperture objective lens and scanner in place of the iris of a typical ellipsometer, and a CCD camera in the place of the photodiode.

## Contact angle

A *contact angle goniometer* measures the static (i.e., the drop is still) contact angle $\alpha$ between a drop of liquid (often water) and a surface (**Figure 14.9a**). The solid–liquid (SL), solid–vapor (SV), and liquid–vapor (LV) interfacial tensions $\gamma$ are then related to $\alpha$ by the Young equation:

$$\cos\alpha = \frac{\gamma_{SV} - \gamma_{SL}}{\gamma_{LV}}\tag{14.3}$$

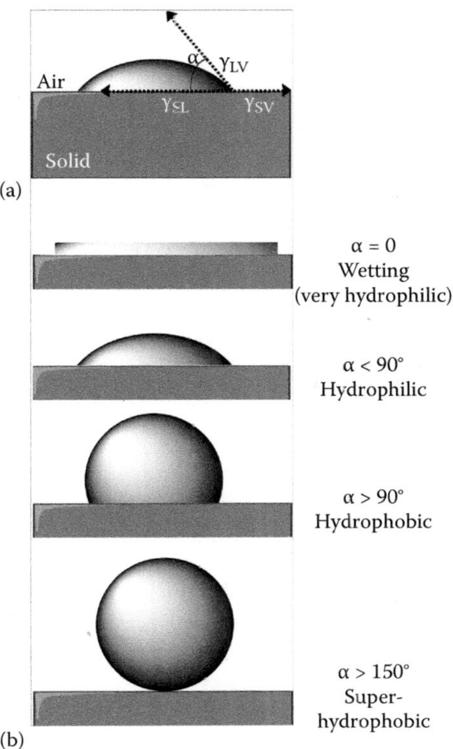



**Figure 14.9 Contact angle measurements.** (a) The contact angle α is at the solid–liquid–air interface. (b) Appearance of drops of water on different types of surfaces. In the absence of a goniometer, the shape of the drops can be used for a qualitative appraisal of hydrophilicity or hydrophobicity of a surface.

More qualitatively, the contact angle gives a measure of the degree of wetting of the surface. If the drop is water, this is a measure of the surface's hydrophobicity or hydrophilicity (**Figure 14.9b**). A good idea of the value can be had by eye, but measurement is greatly aided with a goniometer that digitizes and analyzes images of the drop. The software determines the contact angle by fitting the shape of a drop to a formula (either a sphere or the Young–Laplace equation) and then calculating the slope of the tangent. It is interesting to note that the phenomenon of wetting is extremely complex, and interesting discoveries continue to be made in the study of contact angles between water and surfaces.

## X-ray photoelectron spectroscopy

When an element containing different molecular orbitals of binding energies $B_e$ is bombarded with x-rays of energy $h\nu$, electrons of characteristic kinetic energies $K_e$ are emitted according to the equation

$$K_e = h\nu - B_e - \phi, \tag{14.4}$$

where $\phi$ is the work function. An x-ray photoelectric spectroscopy (XPS) spectrum gives the number of electrons emitted as a function of $K_e$, showing peaks that are characteristic of elements in different chemical states. For example, a survey XPS of a biotinylated alkanethiol SAM on Au (same molecule as **Figure 14.4b**) shows peaks corresponding to C, O, Au, N, and S (**Figure 14.10a**). A detailed spectrum of the S region indicates that there are actually two distinct peaks, one corresponding to the thiolate bound to the Au and the other present in the biotin ring (**Figure 14.10b**).

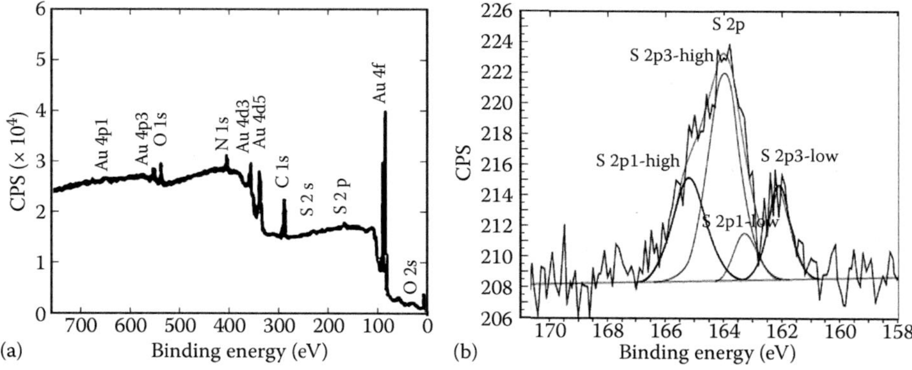

**Figure 14.10 Example of XPS data on a SAM made of the molecule pictured in Figure 14.4b.** (a) Peaks of the key elements in the SAM can be observed: C, N, O, S, and Au. The atomic orbital from which the electrons originate can also be identified as noted. (b) The two different chemical states of sulfur in the SAM can be identified as the splitting of the S 2p peaks into "high" and "low." S 2p-low corresponds to the thiolate bound to Au, while S 2p-high corresponds to the unbound groups within the biotin.

XPS measures only the surface. For samples thicker than 10 nm, successive layers may be removed (for example, by Ar ion sputtering) and XPS repeated at increasing depth. However, the sputtering can affect certain elements preferentially, changing the composition. The lateral resolution (spot size) of an XPS is usually quite large, in the millimeter range. If micron-scale resolution is needed, for example, for composite SAMs, micro-XPS instruments are available with smaller spot sizes.

## Scanning probe microscopy

The techniques of *scanning probe microscopy* (SPM) use a nanometer-scale tip to physically scan across a sample. The most common SPM variant is *atomic force microscopy* (AFM), where the tip is usually of silicon nitride and attached to a cantilever. Interactions of the tip with the sample influence the movement of the cantilever according to Hooke's law, and this deflection is recorded by detectors (usually photodiodes) (**Figure 14.11**). This allows for a measurement of the interaction forces between the tip and the surface, which may be one or more of many types: electrostatic, van der Waals, chemical binding, etc. The sensitivity of the instrument can be adjusted for this wide range of forces by changing the spring constant of the cantilever (which is provided by the manufacturer).

**Figure 14.11 Schematic of operation of atomic force microscope.** Minute deflections of the cantilever are detected by reflecting a laser spot off the cantilever onto a photodiode array.

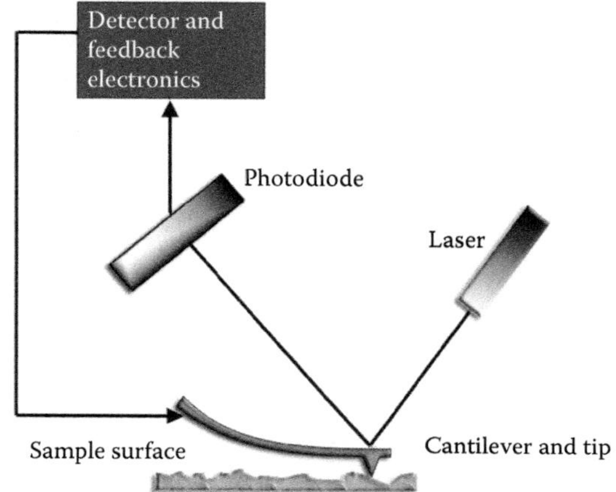

Depending upon the sample, different AFM modes may be chosen. In *contact mode*, the cantilever is moved up and down as it scans while in contact with the surface to maintain the tip at a constant force. This is best used on hard surfaces that will not stick to the tip. In *tapping mode*, the tip is driven up and down by an oscillator. Changes in the phase and amplitude of the oscillation amplitude indicate the surface–probe interaction. Only intermittent contact is made with the sample, so that less damage is done than in contact mode. This is the mode usually used for soft biological specimens, such as unfixed cells. In *noncontact mode*, the tip is driven at a smaller amplitude than with tapping mode or is held farther away, so that it never contacts the surface. Less damage is done to the sample. Hard samples may be imaged in any of these modes with similar results; soft samples show variations depending upon whether the probe penetrates the meniscus of water on the surface and whether it is able to get close enough to measure short-range forces without sticking to the sample.

Other types of AFM include magnetic force microscopy (MFM), in which a magnetically sensitive probe is used to probe the sample's magnetic field; chemical force microscopy (CFM), in which the tip is functionalized to measure the specific force between the molecule on its surface and the sample; and thermal scanning microscopy (TSM), where a resistive tip that acts as a tiny thermometer measures the surface thermal conductivity. In addition, most microscopes can also work in *phase mode* to plot the phase difference between the measured modes.

Researchers who specialize in AFM usually perform many different techniques using specialized tips and in different media (vacuum, air, liquid). Ultrasharp tips may be fabricated using carbon nanotubes, or tips specially functionalized for measuring particular chemical interactions. These techniques are outside the scope of this chapter; some references are provided for those wishing to delve more deeply into the subject.

Standard contact and tapping modes are usually available in thin-film or microfabrication labs, and are among the best ways to characterize monolayers. Vibration isolation is critical, so the microscopes are usually contained in a specialized facility. AFMs for surface characterization are often designed to take full 200 mm wafers, and measure surface roughness with ~5 nm lateral/0.01 nm vertical resolution. Three-dimensional topographical maps of the surface are made by plotting sample height versus tip position. Islands, multilayers, and other surface defects are readily visible in both the images and profiles (**Figure 14.12**).

Another type of SPM is *scanning tunneling microscopy (STM)*, which is based upon the principle of quantum tunneling. When a metal tip is brought very close to a conducting or semiconducting surface, application of a voltage bias $V$ will allow electrons to tunnel through the vacuum or air between them. The resulting *tunneling current $I(V)$* varies as the probe is moved over the surface and is translated into an image. A theoretical expression for this current was calculated by John Bardeen and is given by

$$I = \frac{2\pi e}{\hbar} \sum_{\mu\nu} f(E_\mu)[1 - f(E_\nu + eV)]|M_{\mu\nu}|^2 \delta(E_\mu - E_\nu), \qquad (14.5)$$

where $f(E)$ is the Fermi function, $M$ are the tunneling matrix elements between state $\mu$ (the probe) and $\nu$ (the surface), and $E_\mu$ is the energy of state $\mu$ in the absence

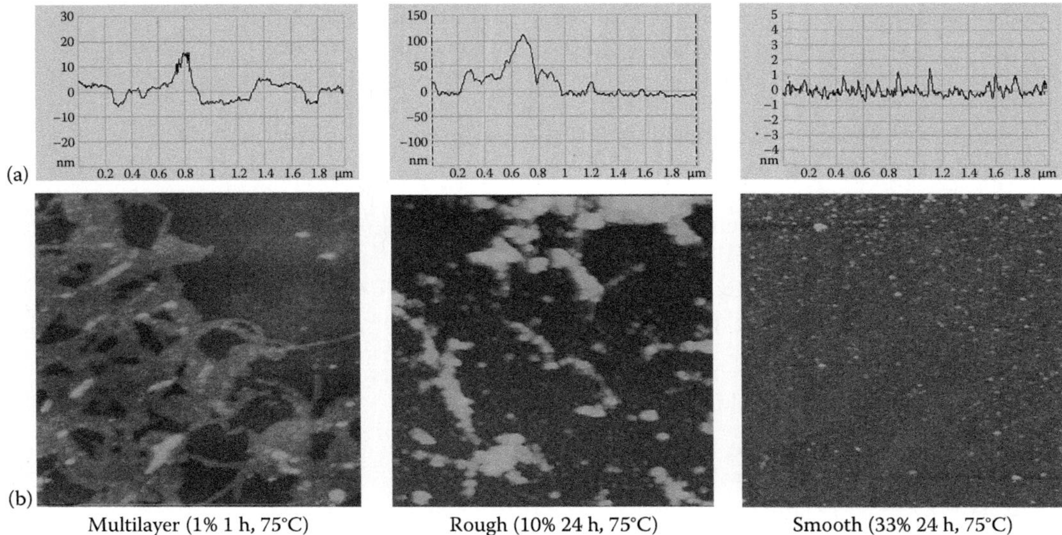



**Figure 14.12 Characterizing APTES deposition using tapping-mode AFM.** (a) The profiles and (b) the topographic images of multilayer, rough, and smooth films made under different conditions. (Reprinted with permission from Howarter, J. A., and J. P. Youngblood, Optimization of Silica Silanization by 3-Aminopropyltriethoxysilane, *Langmuir* 22, 11142–11147. Copyright 2006 American Chemical Society).

of tunneling. In the limit of small $V$ and low temperature (room temperature and below), this reduces to

$$I = \frac{2\pi}{\hbar} e^2 V \sum_{\mu\nu} \left| M_{\mu\nu} \right|^2 \delta(E_\nu - E_F)\delta(E_\mu - E_F), \qquad (14.6)$$

where $E_F$ is the Fermi energy. There are several common ways to approximate the tip wave function; calling it a sphere works well. In the limit of a point source (atomically sharp tip), the current is simply proportional to the value of the sample density of states at the position $r_0$ of the tip:

$$I \propto \sum_\nu \left| \psi_\nu(r_0) \right|^2 \delta(E_\nu - E_F) \qquad (14.7)$$

This simple approximation allows us to appreciate two things:

- The best resolution is obtained with the sharpest possible tip.

- The distance dependence of the current will be exponential, since the wave function through a tunneling barrier decays exponentially.

Scanning tunneling microscopy (STM) is a challenging technique for both of these reasons. The tips must be extremely sharp (and clean), and the surfaces extremely clean as well. Typical tip–sample distances are several angstroms. Data interpretation can be clouded by the smallest errors in distance measurement.

However, there is no technique that can produce such high resolution (up to 10 Å in depth) or atomic-level insight (**Figure 14.13**).

STM has two primary modes. In *constant height* mode, the tip is rastered over the sample, and the changes in current versus position are measured. This gives a measure of charge density and is the fastest mode, but it can only be used on very flat surfaces. In *constant current* mode, the height of the tip required to keep the current constant is what is measured. In this mode, feedback electronics must adjust the height based upon the current to keep $I(V)$ constant. This uses piezoelectric controllers and is slower than constant current mode. The result is a contour of constant transconductance across the sample. Another type of measurement, *scanning tunneling spectroscopy*, holds the tip in place and scans through voltage. This does not give a map of the surface, but rather, the $I(V)$ curve for a single position.

STMs are highly specialized instruments and usually found in the labs of researchers who specialize in the technique, or sometimes in shared nano-fabrication facilities. We highly recommend consulting an expert before undertaking an STM experiment. As with AFM, many variations on STM exist. STM may also be used with a biomolecule instead of air or vacuum in the gap. Tunneling of electrons through the molecule gives a measure of its redox properties (see **Advanced Topic 14.1**).

Another type of SPM is *near-field scanning optical microscopy* (NSOM). It is used more for creation of patterned surfaces than for characterization and will be discussed briefly under Micropatterning.

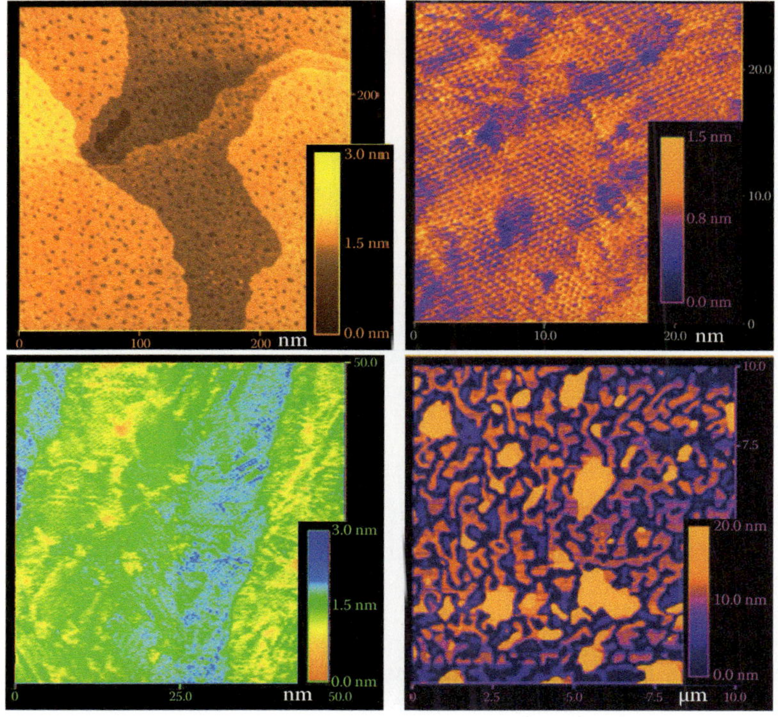

**Figure 14.13 Scanning tunneling microscopy of self-assembled monolayers.** The first three images are depth-coded pictures of a $C_{11}$ alkanethiol SAM. The bottom right image is of a liposome-generated SAM on a Si substrate modified with vinyl silane. (Images courtesy of Amy Blum, McGill University.)

**ADVANCED TOPIC 14.1: STM ON PROTEINS**

Redox-active proteins have attracted a lot of recent attention as light-harvesting devices for conversion of solar energy to electricity. The efficiency of energy-to-light conversion in protein systems is widely studied both theoretically and experimentally. The theoretical approach is based upon solving a set of quantum mechanical transport theory equations for electron current and power conversion efficiency as functions of the molecular energy levels, the coupling to electrodes, and the incident photon energy. Experimentally, STM has been applied to a wide variety of redox proteins, and the resulting $I$–$V$ data are fit to quantum mechanical transport theory describing tunneling across a metal–protein–metal interface. The protein contributes a set of discrete energy states with which the tunneling electrons may interact (Figure A14.1.1). These data yield the tunneling elements $M_{\mu\nu}$.

**SUGGESTED READING**

Marcus, R.A., and Sutin, N. (1985). Electron transfers in chemistry and biology. *Biochimica et Biophysica Acta* 811, 265–322.

Wigginton, N.S., Rosso, K.M., and Hochella, M.F., Jr. (2007). Mechanisms of electron transfer in two decaheme cytochromes from a metal-reducing bacterium. *Journal of Physical Chemistry B* 111, 12857–12864.

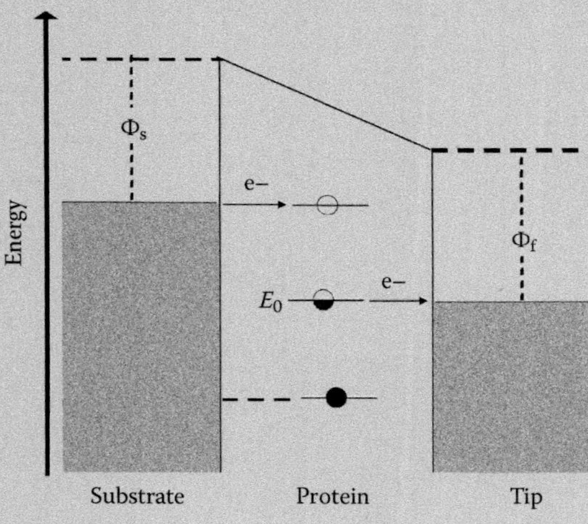


**Figure A14.1.1 Schematics of tunneling as measured by STM.** In the presence of a redox-active protein, the tunneling gap is no longer a vacuum but contains states with which the electrons can interact. $E_0$ is the redox potential, and the $\Phi_s$ are work functions.

## Other methods

Other methods for thin-film characterization include Fourier transform infrared spectroscopy (FTIR, discussed in Chapter 16), scanning and transmission electron microscopy, x-ray and electron diffraction, profilometry, ion beam analysis, secondary ion mass spectrometry, and electrical characterization techniques. Some references are given at the end of the chapter to textbooks and review articles that cover one or more of these techniques in depth.

# 14.4 FUNCTIONALIZATION OF MODIFIED SURFACES USING CROSS-LINKERS

## Types of cross-linkers

Chemical cross-linkers are used to covalently link two reactive groups: in our case, the reactive group of the chosen silane or alkanethiol to a molecule or object of choice: a protein, nanoparticle, whole cell, etc. The choice of cross-linker depends upon the terminal group of the surface molecules and on which reactive groups are present on the molecule to be bound. All proteins have primary amines and carboxylates, and many have cysteine sulfhydryls (these may be engineered). The same applies to cells, which display surface proteins. For smaller molecules (peptides and oligos), it is easier to synthesize or order them with one reactive group already on the end, such as a maleimide. However, if that has not been done and you have the molecule with an amino or carboxyl group, these smaller molecules may be attached using cross-linkers as well.

There are over 100 different cross-linkers, which fall into two major classes. *Homobifunctional* cross-linkers attach two of the same group (e.g., amine to amine). *Heterobifunctional* cross-linkers link two different groups (e.g., amine to carboxylate). Homobifunctional cross-linkers risk causing aggregation by self-polymerization if they are added to a solution of molecules, but can be used to modify the surface first, before addition of the functionalizing molecule. Choice of a cross-linker depends upon many variables, including

- Solvent (aqueous vs. nonaqueous).

- Reaction conditions such as pH (e.g., if you're linking quantum dots, pH < 6 is undesirable).

- Resulting covalent bond desired, e.g., cleavable such as an ester, or noncleavable?

- Length of spacer desired between the two molecules to be conjugated. Some cross-linkers attach the two directly (*zero-length* cross-linkers), while others insert a spacer arm.

- Presence of cells. When cell surfaces are labeled, the cross-linker should be membrane impermeant.

Specialized cross-linkers also exist for photoactivatable groups or for creating photocleavable linkages. There are also many nonspecific cross-linkers, which interact with all reactive groups. We will not discuss either of these here.

Table 14.1 shows the structure, reactivity, and conditions of use for several of the most commonly used cross-linkers. The carbodiimides are especially useful for linking to carboxylates, as there are not many reagents available that interact with these groups. For the amines and sulfhydryls, there are more choices.

The general protocol for all of these cross-linkers is to dissolve them in the appropriate solvent and apply them immediately to the functionalized surface. The following represent two of the most common examples:

- Glutaraldehyde. After silanization with APTS and curing, immerse the sample into a 2.5% solution of glutaraldehyde in carbonate buffer for 2 h. Rinse thoroughly with phosphate-buffered saline (PBS), and then add the protein or other functionalizing molecule and incubate 8 h or overnight. Rinse in PBS and immerse in 100 mM glycine to block any unreacted aldehyde groups, and then rinse and store in PBS.

**Table 14.1**

Some Examples of Cross-Linkers Used to Connect Two Reactive Groups

| Homobifunctional Cross-Linkers | | | | | |
|---|---|---|---|---|---|
| **Name** | **Structure** | **Reactivity** | **Solvent** | **Space (Å)** | **Bond** |
| Glutaraldehyde | | Amine/hydrazide | $H_2O$ | ? | Amide or hydrazone |
| Bismaleimidohexane (BMH) | | Sulfhydryl | $H_2O$ pH 6.5–7.5 | 16.1 | Thioether |
| Bis(sulfosuccinimidyl) suberate (BS³) | | Amine | $H_2O$ | 11.4 | Amide |
| Dimethyl pimelimidate (DMP) | | Amine | $H_2O$ pH 8–9 | 9.2 | Amidine |

| Heterobifunctional Cross-Linkers | | | | | |
|---|---|---|---|---|---|
| **Name** | **Structure** | **Reactivity** | **Solvent** | **Space (Å)** | **Bond** |
| 1-Ethyl-3-[3-dimethylaminopropyl] carbodiimide (EDC) | | Amine/carboxylate | $H_2O$ pH 4.7–7.5 | 0 | Amide |
| Dicyclohexyl carbodiimide (DCC) | | Amine/carboxylate | $H_2O$ or 80% dimethylformamide (DMF) | 0 | Amide |
| Diisopropyl carbodiimide (DIC) | | Amine/carboxylate | Organic | 0 | Amide |
| m-Maleimidobenzoyl-N-hydroxysulfosuccinimide (sulfo-MBS) | | Amine/sulfhydryl | $H_2O$ (without sulfo: dimethyl sulfoxide [DMSO]) | 9.9 | Amide/thioether |
| N-(γ-maleimidobutyryloxy) sulfosuccinimide (sulfo-GMBS) | | Amine/sulfhydryl | $H_2O$ (without sulfo: DMSO) | 6.8 | Amide/thioether |

*Note:*  *Space* gives the length of the spacer arm remaining after conjugation; *bond* is the identity of the bond in the resulting product.

- N-γ-maleimidobutyryl-oxysuccinimide ester (GMBS). After silanization with MPTS, immerse into 2 mM GMBS. Sulfo-GMBS is soluble in buffer; GMBS alone must be dissolved in dimethylformamide (DMF) and the reaction performed in ethanol. Rinse with PBS, and then add the protein or other molecule for 1 h. Rinse thoroughly and proceed with experiments or store.

The concentration of protein or molecule used in these protocols depends, of course, on its size. For typical proteins of 50–150 kD, about 0.05–0.5 mg/mL is a good place to start. Your mileage may vary, and duplicates should usually be prepared the first time to optimize the conditions.

There are many subtleties involved in choice and use of cross-linkers, especially when unwanted polymerization or steric hindrance occurs. If your problem is particularly tricky or if you encounter any difficulties with these standard approaches, please consult some of the texts referenced at the end of the chapter.

## Controlling protein orientation

Chemical cross-linking methods targeting amines and carboxylates, and usually sulfhydryls, lead to a mix of random protein orientations, with more or less denaturation depending upon the nature of the protein. Sometimes this is undesirable, and a layer of "right-side-up" proteins is needed. There are several methods for controlling protein orientation. If the protein is being prepared in house, the easiest way to do this is to place an affinity tag on the end of the protein that you wish to stick to the surface, and reacting the tag with special silanes or alkanethiols designed to bind with it (**Figure 14.14**).

For example, a His tag complexes specifically with a nickel complex with nitrilotriacetic acid ($Ni_2$+-NTA), and NTA-terminated alkanethiols are commercially available. NTA silanes may be made by reacting MPTS with NTA maleimide (**Figure 14.15a**). The complexation with $Ni^{2+}$ and histidines (**Figure 14.15b**) is reversible, which may be a drawback or an asset depending upon the application.

Another variation on this approach is to add a site-specific biotinylation tag (discussed in **Chapter 7**) and biotinylate the protein, and then bind it to a streptavidin-functionalized surface. Still another method is to make a site-specific fusion protein between the protein of interest and a *capture protein* that will react with specific ligands. For example, the protein *cutinase* (a fungal enzyme) forms covalent linkages to phosphonates, so a phosphonate-terminated monolayer can be used to covalently capture a fusion protein (**Figure 14.16**).

Antibodies represent a special case in this discussion. It is crucial that they be oriented properly as they are useless when their ligand-binding domain is not accessible to the solution. Antibody structure allows this to be done in a relatively

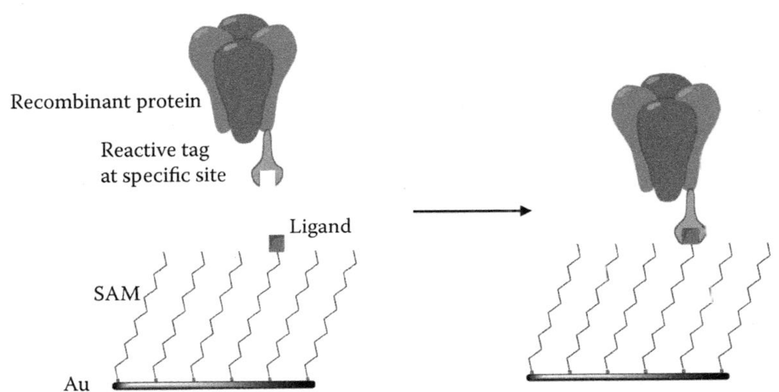


**Figure 14.14 Schematic for controlled orientation immobilization of proteins.** The protein is expressed with a specific reactive tag, such as a His tag, a site that may be biotinylated, or a capture protein. The SAM is prepared bearing a specific ligand for the tag at a desired ratio, and the two are mixed.

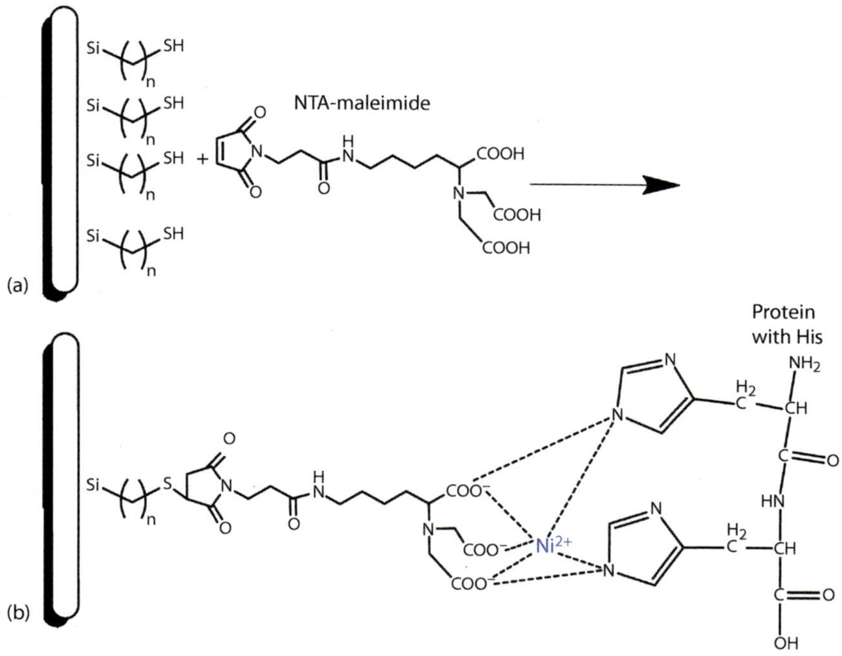

**Figure 14.15 Oriented binding using His tags.**
(a) An NTA-terminated silane can be produced by reacting thiol-terminated silane with NTA maleimide. (b) The His tag on the protein complexes with the NTA in the presence of nickel ions. This is not a covalent bond.

straightforward fashion, using either the entire immunoglobulin molecule (IgG) or the antigen-binding fragments (Fab) only. An IgG consists of two Fab fragments and an Fc fragment; the latter is usually glycosylated (**Figure 14.17a**). The entire molecule may be immobilized specifically by oxidizing these regions to reactive aldehydes, and then coupling to a biotin hydrazide compound and

**Figure 14.16 A genetically encoded capture protein that forms covalent linkages.**
(a) The protein cutinase. (b) Its ligand, phosphonate.

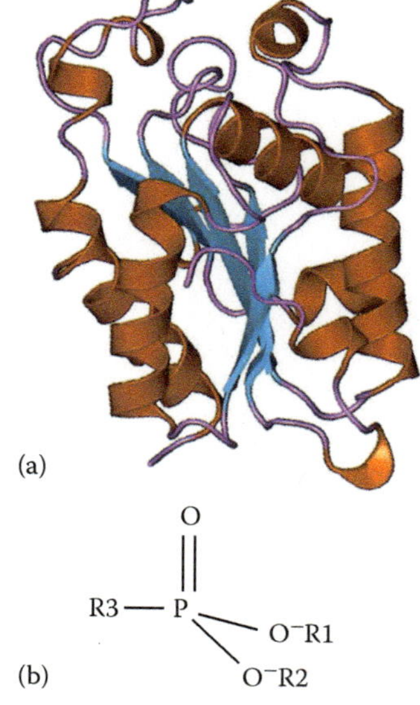

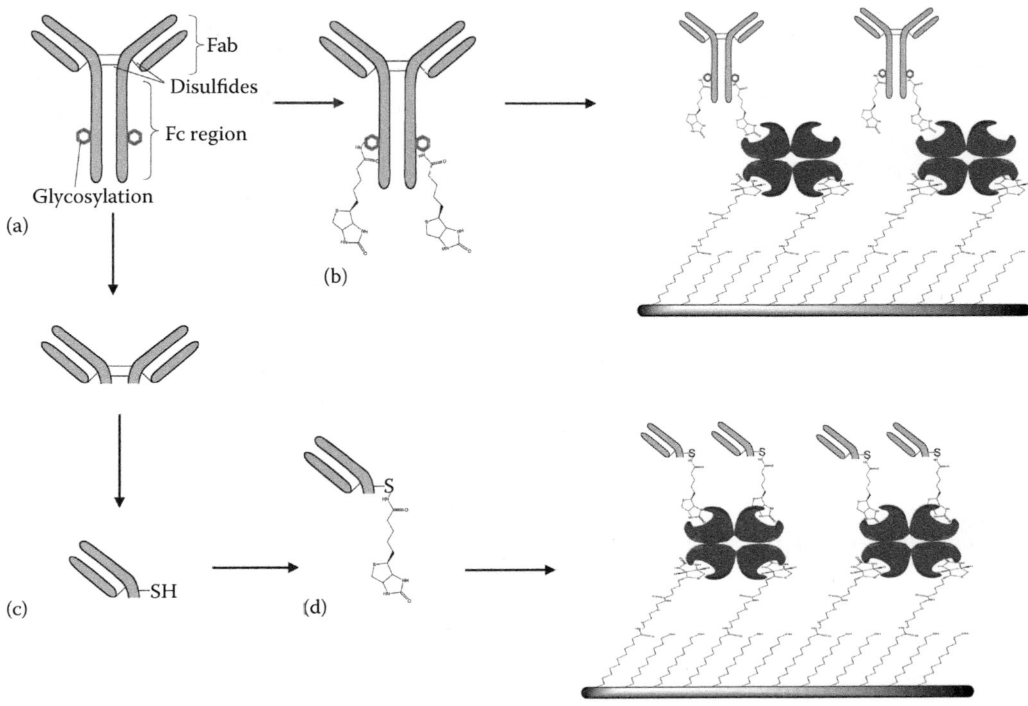

**Figure 14.17 Oriented immobilization of antibodies.** (a) Schematic of IgG structure. (b) The carbohydrate moiety on the Fc fragment may be oxidized and then biotinylated, leading to an ordered array of IgG on the surface. (c) The Fc region may also be specifically removed by protease. (d) Gentle reduction leads to free sulfhydryls that may be biotinylated and the Fab fragments assembled on a streptavidin surface.

assembling onto a streptavidin-coated surface (**Figure 14.17b**). The procedure is as follows:

- Begin with 3–5 mg/mL IgG in 100 mM sodium acetate, pH 5.5.

- Add 20 mM sodium metaperiodate and incubate at just above freezing temperature for 1 h.

- Stop the reaction with 30 mM glycerol.

- Filter and dialyze against sodium acetate buffer.

- Couple to biotin hydrazide or related compound according to the manufacturer's instructions. The use of N-(aminooxyacetyl)-N'-(D-biotinoyl) hydrazine (*ARP*) will lead to greater stability if storage is desired.

- React with a streptavidin surface.

Alternatively, the Fc region may be removed entirely by protease digestion and the Fab fragments separated and purified (**Figure 14.17c**). A protocol for isolation of Fab fragments is given in **Practical Tips 14.3**. Because of the disulfide bonds that connect the antibody fragments, the resulting Fabs contain terminal sulfhydryls, which may be modified with biotin maleimide. An ordered, densely packed layer of Fab fragments is then made on a streptavidin surface (**Figure 14.17d**).

### PRACTICAL TIPS 14.3:    GENERATING FAB FRAGMENTS FROM IgG

**GENERAL NOTES**

The proteolytic enzyme pepsin is used to digest the Fc fragments. The use of immobilized pepsin allows for better control than performing the reaction in solution. Pepsin–agarose beads are available from several manufacturers. Some manufacturers supply complete Fab preparation kits.

**REAGENTS**

- IgG in solution or lyophilized
- Slurry of 4–6% cross-linked beaded agarose functionalized with pepsin
- 20 mM sodium acetate, pH 4–4.5
- 100 mM phosphate buffer, pH 7.2
- 2-mercaptoethylamine (MEA), 100 mM in phosphate buffer
- Ethylenediaminetetraacetic acid (EDTA)
- Desalting column (e.g., PD-10 or Sephadex G-25)

**PROTOCOL: DIGESTION OF Fc**

- Optional: deglycosylate the IgG before use to improve pepsin digestion. This may be done using *PNGase.*
- For lyophilized IgG, dissolve at up to 10 mg/mL in sodium acetate buffer. If the IgG is in solution, or has been treated with PNGase, dialyze it against this buffer overnight.
- For each 1 mL of IgG to be digested, place 0.125 mL of agarose gel (i.e., 0.25 mL of a 50% slurry) into a 15 mL tube. Equilibrate the gel with the sodium acetate buffer by washing twice with 4 mL, centrifuging, and discarding the supernatant. After washing, resuspend in 0.5 mL sodium acetate.
- Add the IgG to the equilibrated beads. Incubate for the appropriate time (to be determined for each antibody; start with 4 h) in a shaking water bath at 37°C.
- Centrifuge and keep the supernatant.
- Dialyze against 100 mM phosphate buffer (pH 7.2) using a 50,000-molecular-weight cutoff filter to remove the digested fragments.
- Optional: undigested IgG may be removed using a *protein A* column.

**PROTOCOL: PREPARING FAB MONOMERS (FAB') FROM LINKED DIMERS (F(AB')$_2$)**

- Add MEA to the dialyzed antibody in phosphate buffer to a final concentration of 50 mM and EDTA to a final concentration of 5 mM. The EDTA helps to keep the sulfhydryls in their reduced form.
- Incubate for 1.5 h at 37°C.
- Allow to cool to remove temperature, and then remove the MEA by passing over a desalting column equilibrated with reducing buffer.
- Elute the protein from the column with reducing buffer, and determine which fractions contain antibody by measuring $A_{280}$.
- Fragments should be used immediately.

**SUGGESTED READING**

Cass, T., and Ligler, F.S. (eds.). *Immobilized Biomolecules in Analysis: A Practical Approach.* Oxford University Press, New York, 1998.

Peluso, P., Wilson, D.S., Do, D., Tran, H., Venkatasubbaiah, M., Quincy, D., Heidecker, B., Poindexter, K., Tolani, N., Phelan, M., Witte, K., Jung, L.S., Wagner, P., and Nock, S. (2003). Optimizing antibody immobilization strategies for the construction of protein microarrays. *Analytical Biochemistry* 312, 113–124.

# 14.5 EXAMPLE EXPERIMENT: PREPARING A SILANE–BIOTIN–STREPTAVIDIN SANDWICH ON SIO₂ FEATURES ON A SI CHIP

This example will illustrate some of the steps and issues involved in a simple biofunctionalization experiment of microtoroids on a chip. The goal of the experiment is to create a streptavidin-functionalized surface that can be used to detect biotinylated substrates of any type.

## Observing and cleaning the substrate

The example substrate is an array of disk-shaped microresonators with a thick $SiO_2$ surface on a Si wafer. Each is about 100 μm in diameter. They can be observed by scanning electron microscopy (SEM) (**Figure 14.18**).

Because the wafer is opaque, they may be observed by epifluorescence when labeled, but not by transmission light microscopy. The goal is to deposit a functional monolayer of streptavidin onto the oxide surface, without any protein binding to the Si. Although streptavidin may be bound directly to the $SiO_2$ or covalently linked to the termini of a silane, it is better to make a silane–biotin–streptavidin "sandwich" as this ensures the correct orientation of each streptavidin molecule. There are thus three functionalization steps in this example: silanization, biotinylation, and binding of streptavidin (**Figure 14.19**).

Before functionalization begins, the first step is cleaning to remove any native oxide from the surrounding Si wafer. This may be done by HF etch or Ar plasma in the clean room. (*Note*: HF is very dangerous and should only be handled in laboratories set up for its use with the appropriate gloves and other protective

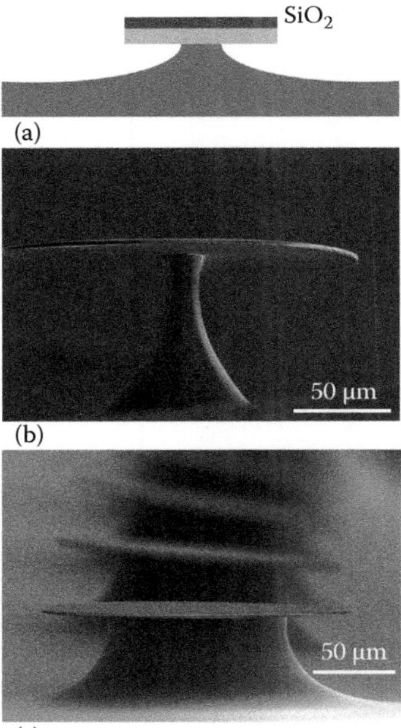


**Figure 14.18 Example of a substrate for biofunctionalization.** (a) Schematic, showing selective deposition of oxide layer on features. (b) SEM image of a single feature. (c) SEM image of array of features on the chip.

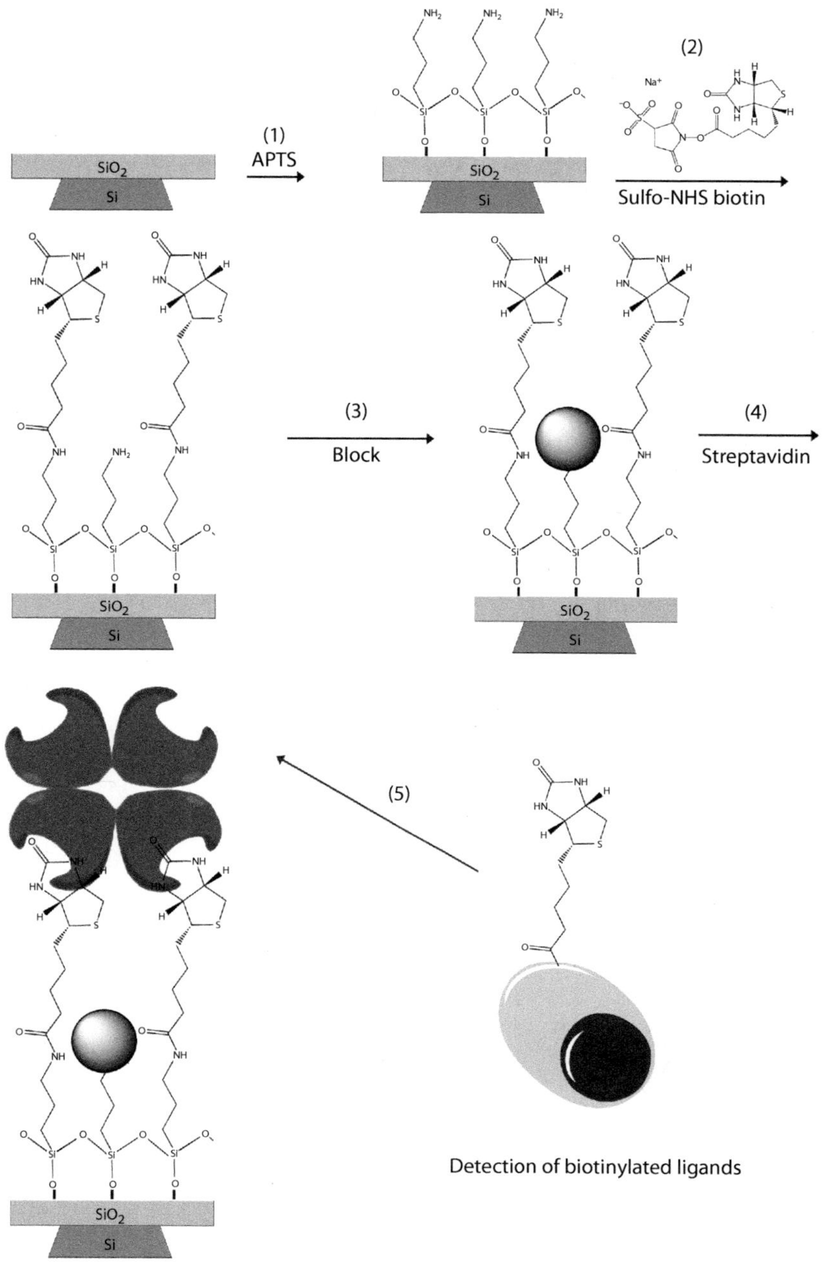

**Figure 14.19 Schematic of steps for functionalizing SiO$_2$ features on a Si chip.** (1) Silane, in this case APTS, binds specifically to areas containing oxide. (2) Amine-reactive biotin (in this case sulfo-NHS-biotin) binds to the amino termini of the silane. (3) Any remaining amines are blocked with a nonspecific protein such as BSA. (4) Streptavidin now binds only to the biotin sites in an ordered fashion. (5) Two sites on each streptavidin are left free, so any biotinylated molecule, cell, etc. will bind specifically to the surface.



equipment.) Because native oxide re-forms very quickly, this should be done immediately before proceeding with the rest of the experiment. Immediately after HF treatment and rinsing, samples are immersed in piranha solution to activate the oxides. They are then rinsed thoroughly in distilled water and dried with a nitrogen gun. It is often helpful to have similarly treated SiO$_2$ chips that are treated in parallel with the devices, since the device surface area is too small for some measurement techniques such as contact angle. These are just wafer fragments with an oxide layer that are handled in parallel with the devices.

## Silanization

Silanization with APTS may be performed in water, but we recommend doing it in anhydrous acetone or toluene in a glove bag or box. Immerse the samples into a 1–2% APTS solution for 30–60 min, and then rinse in acetone (or toluene/acetone) and bake for several hours at 80°C in a vacuum oven. To obtain a thin, smooth layer, it is important not to use too high of an APTS concentration and to limit the reaction time to ≤1 h.

The silane layer should be tested using contact angle; this can be used to test a variety of silanization conditions on sample chips. The contact angle of $SiO_2$ is ~56°; an APTS layer will increase this to 77–81° depending upon the quality of the layer, with higher values indicating better silane coverage. Ellipsometry should also be performed, using refractive index values of $n_{air} = 1.0$, $n_{silica} = 1.465$, and $n_{APTS} = 1.465$. The resulting thickness value should be ~1.5 nm. If it is much higher than this, multilayers have been deposited, and the experiment should be repeated with different silane concentrations, incubation times, or both.

Optional methods of characterization at this stage include AFM, XPS, and binding fluorescent dyes. These are not strictly necessary but may be used to troubleshoot if the contact angle or thickness measurements are not as predicted. AFM can show multilayer island or rough film formation (refer again to Figure 14.12), and XPS can give a measure of the extent of reacted silane. If all silane molecules have formed siloxane linkages, the C/N ratio should be 3:1. Unreacted APTS has a value of 9:1.

## Biotinylation and blocking

In the next steps, the reagents are expensive and sold in very small quantities. It is therefore impractical and wasteful to immerse the substrates in the solutions. Instead, a small drop of the solution is added just to the target area, and a coverslip is added to ensure even distribution of the solution (Figure 14.20). The sample may also be kept in a humidified chamber during incubation to prevent drying out. In the case of test samples, mark the target region with a grease pencil to help hold the solution and to indicate where the functionalized spot is.

Biotinylation of the surface can be performed with amine-reactive sulfo-NHS-biotin. It must be freshly dissolved in $H_2O$ at a concentration of about 1 mg/mL. Allow it to incubate for at least 2 h, and then rinse in $H_2O$. The contact angle should decrease, reflecting the hydrophilicity of biotin.

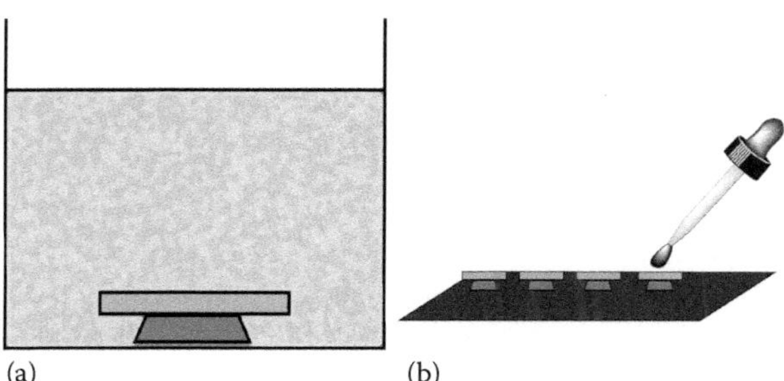


**Figure 14.20  Small versus large volumes on small features.** (a) When using a reagent such as silane, which is inexpensive, the entire chip may be immersed. (b) For reagents such as sulfo-NHS-biotin, they should be applied in small amounts only to the feature with a dropper or pipette. A coverslip may be applied to prevent evaporation if this is possible to do without breaking the features.

Blocking with nonspecific proteins is now necessary to fill any cavities in the biotin surface. This is because any protein, including streptavidin, will adhere nonspecifically to the free amine groups of APTS. This may be easily done with a 1% solution of bovine serum albumin (BSA) in PBS for 1 h; alternatively, you can use a commercial SuperBlock solution. These solutions are designed for sensors and are free of naturally occurring biotin and other contaminants. It may be useful to perform AFM at this stage.

### Assembling streptavidin, final characterization, and using the sensor

After blocking, streptavidin assembles rapidly, and the sample will be ready to use after 15–20 min incubation with streptavidin in PBS; a good starting concentration is 0.01 mg/mL, or about 18 µM. Rinse well to assure that nonspecifically bound streptavidin is removed. If a fluorescent dye–conjugated streptavidin is used, the samples may be characterized by epifluorescence microscopy. This shows rapidly if there are any systematic problems with the samples such as debris, failed functionalization, or nonspecific binding to the Si surface caused by insufficient removal of native oxide (**Figure 14.21**).

The substrate may be further characterized and/or used as a sensor by comparing binding of nonbiotinylated to biotinylated analytes. Our sample microtoroids may be used as high-quality-factor whispering-gallery resonators for biosensing. Alternatively, the signal may be transduced via fluorescence changes. The binding of fluorescently labeled, biotinylated *Escherichia coli* (*E. coli*) to the surface results in alterations of many of the cavity's optical properties (**Figure 14.22**).

**Figure 14.21 Characterizing the functionalized surface using epifluorescence.** In this case, a streptavidin–Alexa dye conjugate was used in the final step to visualize the surface. (a) Examples of good and bad surfaces. (b) Fluorescence intensity of the surface (as measured by fluorescence microscopy, relative units) as a function of streptavidin concentration (µM).

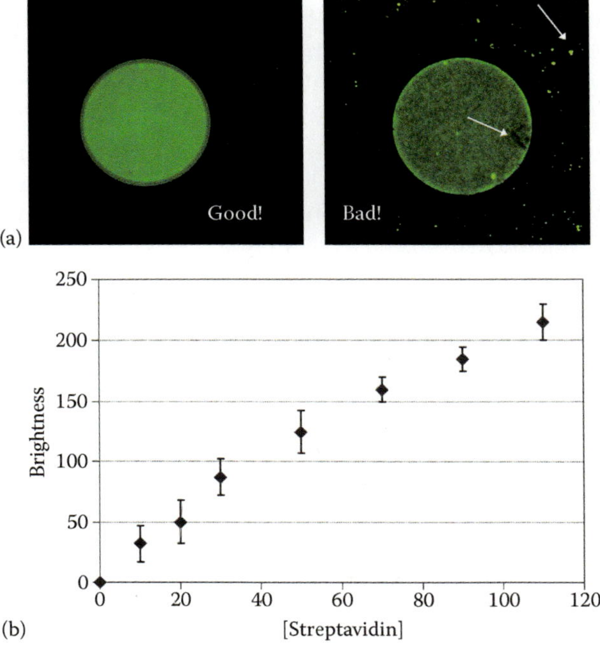

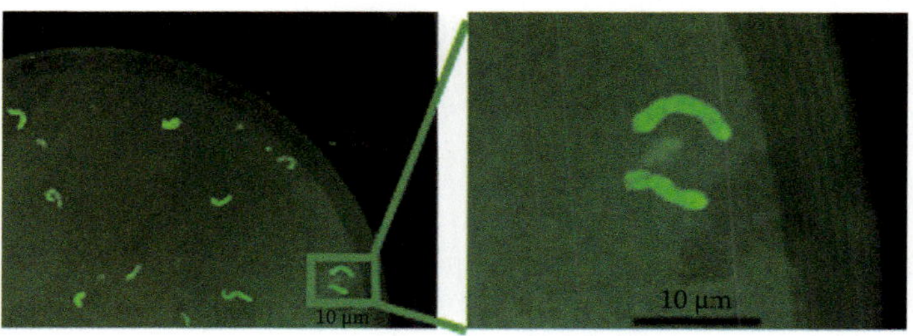

**Figure 14.22 Specific binding of biotinylated *E. coli* to the functionalized surface.** Nonbiotinylated *E. coli* did not bind.

## Variations on a theme

The biotin–streptavidin–biotin sandwich is a particularly useful construct that has many different variations allowing for reversible and patterned functionalization. There are a few variations on streptavidin that are useful to keep in mind, including the monovalent and monomeric forms discussed in **Chapter 1**: *captavidin* has nitrated tyrosine groups within its biotin-binding pockets. This makes it release biotin at pH values above 9, so it can be used as a reversible sensor by binding at a lower pH and releasing all ligands at a high pH. *Neutravidin* has a lower isoelectric point and sometimes reduces background staining.

Hundreds of biotin derivatives are available, many of them expressly designed for use in biosensors. *Desthiobiotin* is a biotin analogue that binds more weakly than wild-type biotin and thus can be used for reversible binding. *Photobiotin* is a biotin with a light-activatable nitrophenyl azide group; it creates an arylnitrene after exposure to UV–blue light that reacts with nucleic acids and proteins via linkages that are not fully understood (**Figure 14.23a**). Alternatively, biotins with photocleavable *protecting groups* may be synthesized, such as

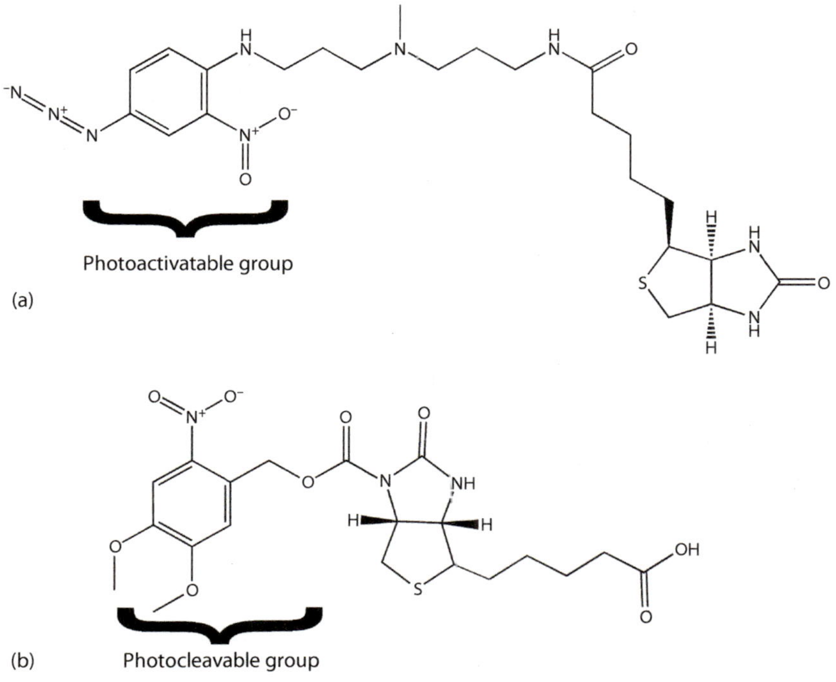

**Figure 14.23** (a) Photobiotin. (b) NVOC–biotin.

nitroveratryloxycarbonyl (NVOC)–biotin (Figure 14.23b). Selective light exposure activates the biotin by releasing the protecting group. These photoactivatable reagents are extremely useful for *micropatterning*.

## Micropatterning

Micropatterning is a method of functionalizing some areas or features of a chip or surface, while other areas remain unfunctionalized or functionalized in a different fashion. There are many ways to do this, quite a few of which involve photoactivatable reagents. A full discussion of these methods is outside the scope of this chapter, though several of them are worth mentioning.

- Use of a *photomask* to selectively activate a photosensitive ligand. Photomasks can be made in *photolithography* facilities, which are available at most universities, or may be ordered commercially from electronics companies. The spatial resolution of these procedures is usually on the micron scale but can be hundreds of nanometers (Figure 14.24).

- *Microcontact printing* involves use of a pattern stamp to deliver biomolecules to a substrate. It can be used with proteins, biotin, and more. The stamps are made of soft materials, such as agarose or *polydimethylsiloxane* (PDMS), poured into a mold (Figure 14.25). Resolution is ~1 mm.

- *Nanografting* involves using an AFM tip to selectively remove molecules from a SAM by probing hard enough to detach the molecule. Resolution is in nanometers.

- *NSOM* (also called SNOM) is a type of SPM in which a probe with dimensions less than a wavelength of light is used to illuminate a sample's near field. It can be used to both create and characterize patterns on a nanometer scale. NSOM capability is often integrated into a standard fluorescence microscope.

If photolithography facilities are not available, grids for transmission electron microscopy (TEM) made of metals such as nickel or copper may be used as

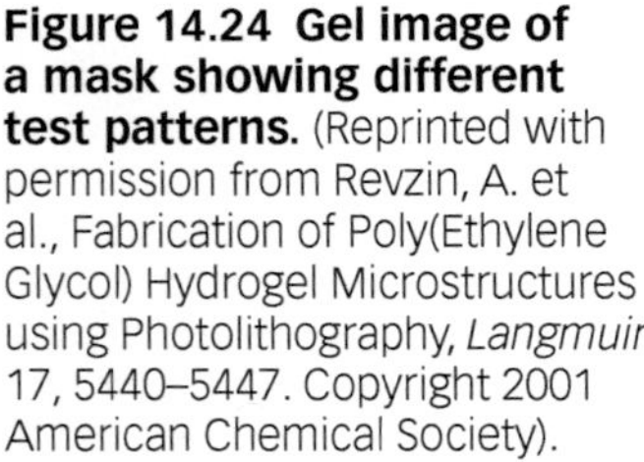

**Figure 14.24 Gel image of a mask showing different test patterns.** (Reprinted with permission from Revzin, A. et al., Fabrication of Poly(Ethylene Glycol) Hydrogel Microstructures using Photolithography, *Langmuir* 17, 5440–5447. Copyright 2001 American Chemical Society).

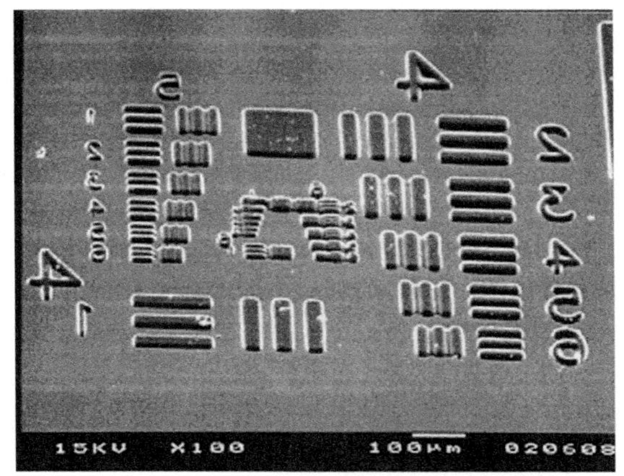

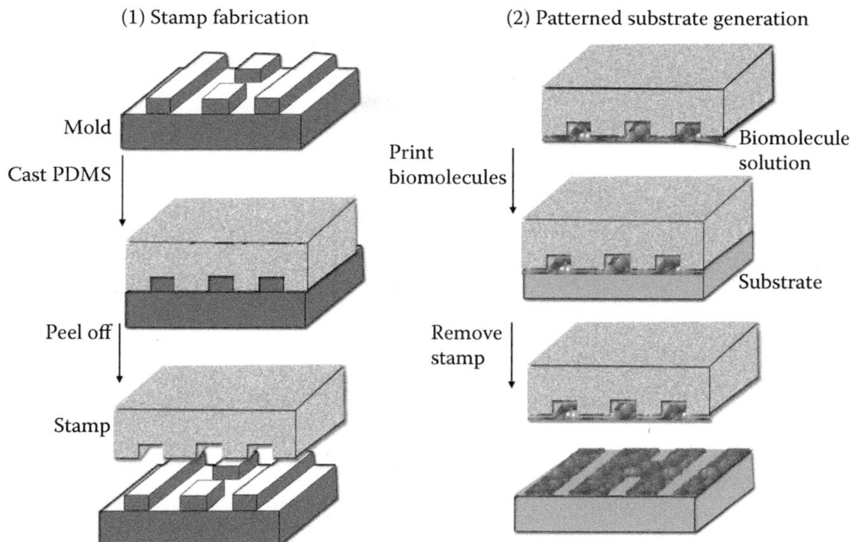



**Figure 14.25 Schematic of the steps in microcontact printing.**

photomasks. While there is little control over the shape and size of features using this method, it is inexpensive and easy.

## 14.6  PREVENTING NONSPECIFIC BINDING OF BIOMOLECULES

As any contact lens wearer can painfully attest, proteins adhere to and build up on any clean surface. In the sample experiment, we used BSA before streptavidin in order to block any stray, nonbiotinylated silane termini. This method is easy and works, but is not ideal for all experiments. BSA is a large molecule, which can cause steric hindrance or interfere with the function of some biosensors. This method also relies on nonspecific, noncovalent interactions and so is not desirable for any surface whose exact chemistry needs to be known, and protein–protein interactions can occur (e.g., between the BSA and streptavidin). Therefore, under many circumstances, the nonspecific adherence of proteins to both the specifically functionalized and/or nonfunctionalized surfaces of a solid surface needs to be prevented.

As we mentioned in **Chapter 11** in the context of nanoparticles, the polymer polyethylene glycol (PEG) greatly reduces surface stickiness. As few as four to six PEG chains are needed for this effect. *Polyethylene oxide* (PEO) and phospholipids also prevent nonspecific protein binding. Over the past few years, PEG silanes and thiol-reactive PEGs have become widely available with many different reactive termini: amines, thiols, biotin, etc. Most nonspecific binding in surface-immobilization experiments can be eliminated by using a reactive PEG–silane or PEG–alkanethiol on the areas intended for further functionalization, and a nonreactive PEG on the areas to be passivated (**Figure 14.26**).

It is also worth mentioning that recent research has shown that the mild nonionic surfactant, n-dodecyl-β-D-maltoside, prevents nonspecific binding on PDMS surfaces. This is important since many microfluidic channels and devices are made

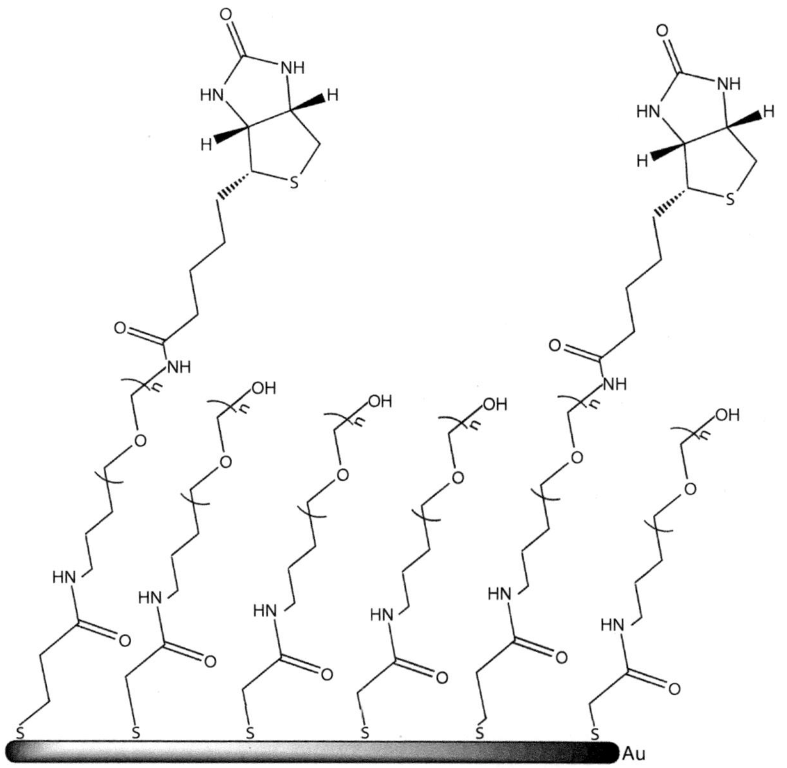

**Figure 14.26 Schematic of SAM on Au made from a reactive thiol–PEG (biotin–PEG mercaptopropionyl) and an unreactive thiol–PEG.**

of this material. References to techniques of functionalization and passivation of PDMS are given at the end of the chapter.

## 14.7 TESTING THE FUNCTION OF IMMOBILIZED PROTEINS

Congratulations! Your protein is on a surface. Now you have to figure out if it functions the way it should. The following are some examples of how this can be done.

### Specific binding: Quantity and kinetics

Many proteins bind a specific ligand, such as biotin in the case of the biotin-streptavidin system. The amount of ligand bound can be measured as a function of concentration, but more informative if possible is to measure binding kinetics. This requires some mechanism of real-time signal transduction from the sensor, which we will not go into here. (If you have made a biosensor, we'll assume you know how it works.) If such capability is available, then for a given concentration of analyte [A], the rate of change of the concentration of detector molecule [B] and of the complex [AB] is given as a function of the association constant $k_a$ and the dissociation constant $k_d$:

$$\frac{d[AB]}{dt} = k_a[A][B] - k_d[AB] \tag{14.8}$$

In the absence of competing interactions or depletion, $\ln(d[AB]/dt)$ will be linear with time. The mechanism(s) of deviations from linearity can be investigated by altering the parameters of the kinetic experiment. Of course, streptavidin–biotin represents a special case since $k_d$ is nearly 0.

## Enzymatic function

In many cases, the immobilized protein is an enzyme. In this case, the reaction of the enzyme with its substrate to give a final product can be quantified by measuring the kinetics of production of the product. Depending upon the product, this is often possible to do using colorimetric or fluorescent reporter kits for the product that require no specialized equipment. For example, the enzyme glucose oxidase produces hydrogen peroxide, which is easily detectable with commercial kits.

## Electrochemistry

Redox-active proteins immobilized on an electrode are often studied with electrochemistry. A current-versus-voltage relationship will give the redox potentials and indicate whether these are changed by immobilization.

## Ion channel function

Ion channels in supported bilayers represent a special case that in principle is very simple. A single ion channel opening event can lead to the flow of $10^8$ ions, making them one of nature's highest-gain proteins. Thus, a measurement of direct current through the immobilized channels on an electrode (such as Au) should give a large-enough signal to be measurable. The problem is that the reservoirs under immobilized channels are often femtoliters in size. A single opening event could deplete the electrolytes in such a small reservoir. In this case, impedance rather than direct current is used to measure channel function.

See **Advanced Topic 14.2** for the special case of membrane proteins.

**ADVANCED TOPIC 14.2:　ASSEMBLING MEMBRANE PROTEINS ON SURFACES**

**THE PROBLEM**

Lipid membranes are remarkably well behaved on solid supports. Bilayers will form spontaneously from vesicles on solid surfaces such as glass, mica, silica, and gold, a phenomenon that has led to an immense body of literature characterizing these bilayers, patterning them onto substrates for potential sensor use, and investigating the function of proteins assembled within them. Bilayers result from a vesicle opening out, so the orientation is predictable, and are assembled directly on the rigid substrate with only the upper surface exposed to the solution. Less than 1 nm of water is found between the bilayer and the surface (**Figure A14.2.1**).

Supported bilayers show remarkable properties that depend upon lipid composition, including formation of rafts or islands of different lipids and complex phase behavior. They are extremely stable, lasting sometimes for months, and can be cooled to cryogenic temperatures without loss of their natural motion, so their properties have been probed with high-resolution electron microscopy and electron diffraction. **Practical Tips 14.4** gives a protocol for preparation of supported bilayers on glass slides.

*(Continued)*

**ADVANCED TOPIC 14.2 (CONTINUED):    ASSEMBLING MEMBRANE PROTEINS ON SURFACES**

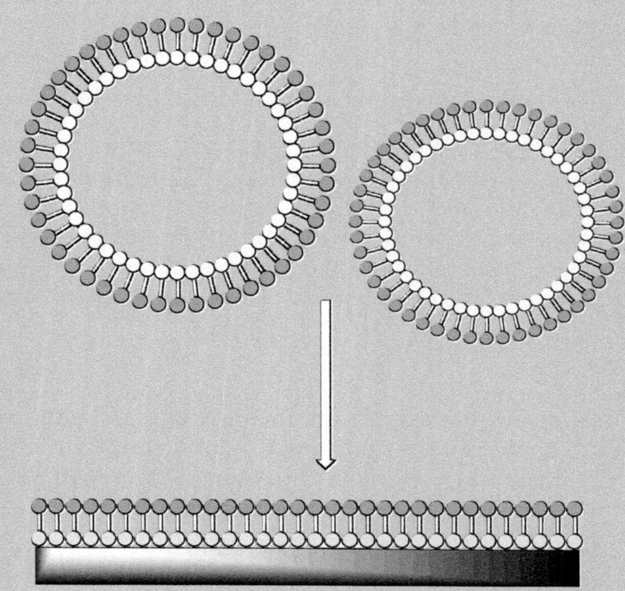

**Figure A14.2.1 Rupture of lipid vesicles to form supported bilayers.** This happens spontaneously on many solid surfaces such as glass and gold. The inside of the vesicle is color-coded to show how oriented rupture occurs, so that if proteins are assembled in the vesicle in an oriented fashion, they will retain this orientation in the supported bilayer.

The trouble starts when proteins are added. Most transmembrane proteins have hydrophilic regions on the intracellular side. When assembled on a solid surface, these regions "hit bottom," and the protein is denatured. Protein mobility is also greatly affected by the solid surface, which may affect function. A few transmembrane proteins remain functional on solid supports, most notably the *photosynthetic reaction centers* and *rhodopsin*. For most proteins, however, some method of cushioning the protein or tethering the lipid above the surface is needed to preserve function. This section is a brief introduction to these methods, which vary widely in complexity and success. Caveat emptor!

## CUSHIONED BILAYERS

The easiest solution to the problem of proteins denaturing on a rigid surface is to assemble the bilayer and protein on a nanoporous or microporous layer that serves as a cushion between the solid surface and the lipid (Figure A14.2.2). The cushion provides all of the physical support of a solid surface, but is gentler to the protein and permits the normal flow of small molecules, including ions and ligands, from the "intracellular" side. There are a variety of materials that work well as cushions, most of which are very easy to assemble onto surfaces. Some examples are as follows:

- Polysaccharides such as agarose or chitosan. Gel-grade agarose may be used as a supporting layer. We recommend low-melting-point agarose as it is easier to handle. A solution of 0.5–1% agarose in electrolyte solution is microwaved and spread on the glass surface while still hot; an additional glass slide placed on top during cooling ensures flatness. Chitosan is a polysaccharide derived from the chitin of shellfish shells that has been shown to form a more stable supporting layer than agarose. Preparation of chitosan supporting layers is simple, involving simple dissolving of the polysaccharide in an aqueous solution and application of several microliters to the surface with a micropipette.
- Bacterial S-layers. The cell surface of many bacteria and archaea consists of a crystalline layer of a single protein or glycoprotein, called an S-layer. When isolated, the proteins will recrystallize in suspension or on solid surfaces or lipids. Recrystallization is begun by injection of protein onto

*(Continued)*

### ADVANCED TOPIC 14.2 (CONTINUED):    ASSEMBLING MEMBRANE PROTEINS ON SURFACES

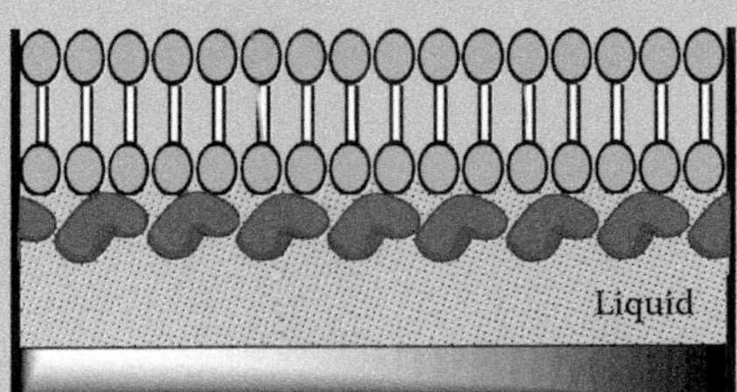


**Figure A14.2.2 Schematic of microporous supporting layer.** A liquid reservoir is present between the solid support and the lipid bilayer; the latter is supported by a polymer, polysaccharide, or crystal. The porosity of the layer ranges from angstroms to several nanometers, permitting ion flow between the bilayer and the electrolyte with normal kinetics, and often the flow of larger molecules such as ligands.

a silanized substrate at the air–buffer interface, where it is then allowed to assemble overnight. For extra stability, the proteins may be cross-linked with glutaraldehyde. Obtaining the proteins may be the most difficult step. Some are now available commercially, although they are sometimes already cross-linked. A reference is given at the end of the chapter to a textbook covering all aspects of S-layer biotechnology.

- Polymers. Several different polymers have been used as supporting layers. These include polyethyleneimine and PEG. Coating of glass or silica substrates is done simply by dipping them into the polymer solution.

From a practical point of view, the drawback of soft supports is that bilayers do not usually self-assemble spontaneously on them, and must be formed by *Langmuir–Blodgett techniques* or by application with a paintbrush. Langmuir–Blodgett application requires specialized equipment and can be tricky.

From a scientific point of view, the problem is that very few of these supporting layers have been well characterized. The relationships between physical characteristics, such as lipid mobility, and the function of membrane proteins assembled into supported layers have not been fully explored. The presence of large, fatal bilayer defects is common and poorly understood. Reports of protein function have been contradictory, sometimes suggesting no function or altered function, especially of complex proteins such as ligand-gated ion channels.

When characterization is reported, the parameter measured is usually the diffusion coefficient of the lipid, more rarely of the protein. These measurements can be made by incorporating a certain percentage of fluorescently tagged lipids into the bilayer and then using fluorescence recovery after photobleaching (FRAP; see Chapter 8) to measure how rapidly the bleached version recovers. Fluorescently labeled proteins can also be used in the same way, although these results are much less frequently reported. The friction between the protein and the supporting layer can be estimated using the extension of the Saffmann–Delbrück equations developed by Evans and Sackmann. When the friction from the support is nonnegligible compared with that from the surrounding lipid, the diffusion coefficient of the protein is given by

$$D_{\text{L}}^{-1} = \frac{4\pi\eta_{\text{m}}}{kT}\left[\frac{\varepsilon^2}{4} - \varepsilon\frac{K_1(\varepsilon)}{K_0(\varepsilon)}\right]$$

$$\varepsilon = R\sqrt{\frac{\eta_{\text{w}}}{\eta_{\text{m}}d_{\text{w}}}}$$

(A14.2.1)

*(Continued)*

## ADVANCED TOPIC 14.2 (CONTINUED):   ASSEMBLING MEMBRANE PROTEINS ON SURFACES

**Figure A14.2.3 Appearance of agarose-supported bilayers visualized with 1% of a fluorescent lipid (BODIPY).** (a) Immediately after formation, showing a normal pattern of bilayer formation where the center is black because it is thinner than the wavelength of light. (b) The same bilayer after 2 days, showing a remaining small black spot in the interior but the appearance of both light and dark areas all around it. The dark areas may represent areas where the lipid adheres to the agarose.

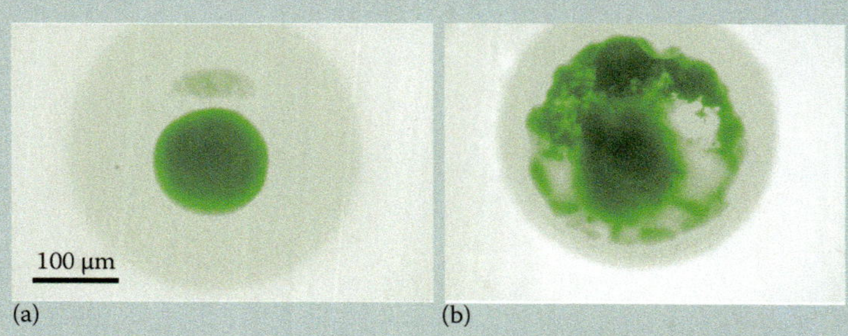


where $\eta_m$ and $\eta_m$ are the viscosities of the membrane and supporting layer, respectively; $R$ is the radius of the (cylindrical) protein; $d_w$ is the supporting layer thickness; and $K_{0,1}$ are Bessel functions.

More sophisticated techniques are sometimes used to look beyond the bulk diffusion coefficient and identify domains or "barriers" in the lipid membrane. The membrane can stick to the supporting layer, creating barriers to diffusion of both lipids and proteins (**Figure A14.2.3**). *Fluorescence correlation spectroscopy* can be used to identify these regions.

### TETHERED BILAYERS

*Tethered bilayers* use a long-chain molecule to hold the bilayer several nanometers above the solid support. This molecule is part of the lipid bilayer—that is, it is a phospholipid with a long tail and anchoring group that attaches to the surface. Thiolipids, for example, can be used to tether to gold (**Figure A14.2.4a**). The

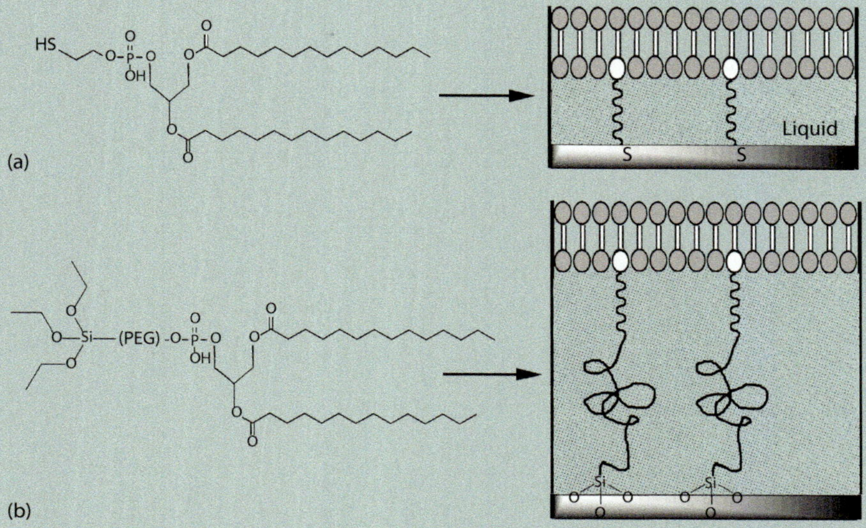

**Figure A14.2.4 Tethered bilayers.** (a) A phospholipid attached to an alkanethiol chain. When mixed with ordinary phospholipids, it forms a bilayer where the alkanethiols hold the membrane just above the surface, forming a femtoliter reservoir. (b) A phospholipid attached to a PEG–silane. The idea is the same as in (a), but in this case, the tether is longer and the resulting reservoir therefore larger.



*(Continued)*

**ADVANCED TOPIC 14.2 (CONTINUED):    ASSEMBLING MEMBRANE PROTEINS ON SURFACES**

resulting is a femtoliter-volume reservoir filled with aqueous electrolyte solution. This prevents transmembrane proteins from denaturing when they assemble.

For some proteins, such a reservoir is still insufficient in volume. If the intracellular hydrophilic domain is large, the protein can still denature on such a support. A way to increase the size of the reservoir is a "hybrid" tether–polymer approach, in which a phospholipid is attached to a large polymer such as a PEG, and then to a thiol or silane for anchoring (Figure A14.2.4b).

The disadvantages of tethered bilayers are that they are fragile, the synthesis of the compounds must usually be done in house, and they do not necessarily permit physiological levels of protein mobility. Proteins in such layers often show a wide range of diffusion coefficients for reasons that have not been fully explained.

## IMPROVING LIPID STABILITY

Even with a supporting layer, the lipid itself is unstable to mechanical stresses, drying, and oxidation. There are several ways to create more stable lipid bilayers, with the warning that all of them may affect your protein function. Some examples are as follows:

- The use of *ether lipids* instead of ester lipids. When the glycerol of the lipid molecule is bound to its alkyl chain via an ether instead of an ester, the lipid is more stable against oxidation. Archaea are distinguished by the ether lipids in their cell walls, and both archaeal and artificial ether lipids are available commercially.
- The use of *photopolymerizable lipids*. Diacetylene lipids, available commercially, are cross-linked with UV (254 nm) light, which is provided from a mercury or xenon lamp attached to a microscope. Passed through a high-power objective lens, the light may be focused onto a targeted area. This may be used to leave some areas fluid for protein assembly.
- The use of *cross-linkable solvents*. Para- or meta-divinylbenzene can be used as a solvent in which to dissolve the lipid. It can be photopolymerized in a similar manner as the diacetylene lipids, eliminating the solvent as a source of leakage. This improves stability to a lesser extent than a cross-linkable lipid but may be kinder to proteins.

## SUGGESTED READING

Fruh, V., Ijzerman, A.P., and Siegal, G. (2011). How to catch a membrane protein in action: A review of functional membrane protein immobilization strategies and their applications. *Chemical Reviews* 111, 640–656.

Kusters, I., Mukherjee, N., de Jong, M.R., Tans, S., Kocer, A., and Driessen, A.J. (2011). Taming membranes: Functional immobilization of biological membranes in hydrogels. *PLoS One* 6, e20435.

McGillivray, D.J., Valincius, G., Heinrich, F., Robertson, J.W.F., Vanderah, D.J., Febo-Ayala, W., Ignatjev, I., Losche, M., and Kasianowicz, J.J. (2009). Structure of functional staphylococcus aureus alpha-hemolysin channels in tethered bilayer lipid membranes. *Biophysical Journal* 96, 1547–1553.

Sarmento, M.J., Prieto, M., and Fernandes, F. (2012). Reorganization of lipid domain distribution in giant unilamellar vesicles upon immobilization with different membrane tethers. *Biochim Biophys Acta* 1818, 2605–2615.

### PRACTICAL TIPS 14.4:   PREPARATION OF SUPPORTED BILAYERS ON GLASS SLIDES

**GENERAL NOTES**

This protocol will work with many different lipids as long as they are not too negatively charged (e.g., ≥20% phosphatidylserine). Different buffers may also be used, as long as the ionic strength is sufficient (50–100 mM salt).

**MATERIALS NEEDED**

- Appropriate lipid or mixture of lipids in chloroform
- Appropriate buffer (e.g., 20 mM phosphate buffer; 50 mM NaCl, pH 7.2)
- 18 mm glass coverslips
- Piranha etch
- Nitrogen tank or house nitrogen
- Vesicle extruder or bath sonicator

**PROCEDURE**

- Clean coverslips with piranha etch. They should be highly hydrophilic, with a contact angle of close to 0. They may also be cleaned by plasma if available. One of the most common causes of failure is insufficiently cleaned glass!
- Prepare small unilamellar vesicles (SUVs). This can be done by extrusion if you have a vesicle extruder. If not, the following works very well:
  - Using a glass pipette or needle (no plastic!); transfer 2 mg of lipid(s) into a glass vial or test tube.
  - Dry down the lipid solution under a stream of nitrogen in a fume hood. If possible, finish with 60 s of centrifugal evaporation (e.g., on a SpeedVac) to remove any residual chloroform.
  - Add 2 mL of buffer solution and vortex to resuspend. The solution will be cloudy; these are *multilamellar* vesicles, which contain many interleaved lipid rings like the layers of an onion (**Figure P14.4.1a**).
  - Place in a bath sonicator and sonicate for 10–30 min, or until the solution is very nearly clear. Now the vesicles are *unilamellar* (**Figure P14.4.1b**).
  - Store at 4°C; use within 1 week.
- Pipette 30 µL of vesicle solution in the bottom of a glass dish and place a clean coverslip on top of the droplet.
- Wait ~30 s to allow the lipid bilayer to form on the surface of the glass coverslip.
- Gently fill the dish with double-distilled $H_2O$ (dd$H_2O$). You may pinch the coverslip with tweezers and agitate it gently to remove excess vesicles.
- Put another coverslip over the first.

**Figure P14.4.1**
**(a) Multilamellar vesicle.**
**(b) Unilamellar vesicle.**

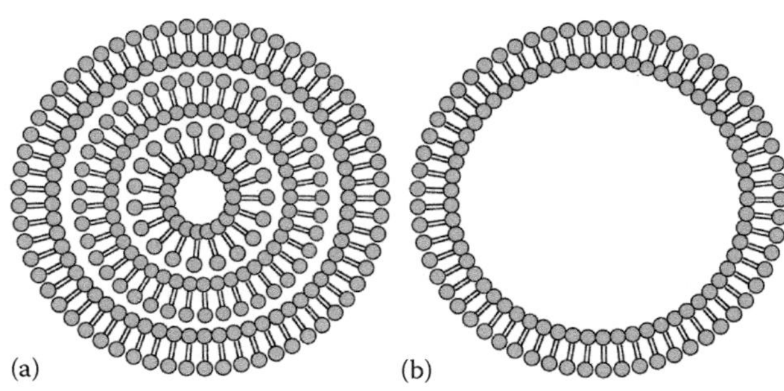


(*Continued*)

> **PRACTICAL TIPS 14.4 (CONTINUED):   PREPARATION OF SUPPORTED BILAYERS ON GLASS SIDES**
>
> **SUGGESTED READING**
>
> Cremer, P.S., and Boxer, S.G. (1999). Formation and spreading of lipid bilayers on planar glass supports. *Journal of Physical Chemistry B* 103, 2554–2559.

## 14.8  CONCLUSION AND FINAL REMARKS

Nearly every biophysicist will need to biofunctionalize a surface at some point. The applications are myriad: investigation of the function of single proteins; creation of electrical or optical biosensors of different sizes, shapes, and materials; controlling the pattern and direction of cell growth; and many more. Tremendous progress has been made over the past few years in these techniques, and usually, it is possible to obtain the desired surface functionality without chemical synthesis. Silanes, PEGs, alkanethiols, cross-linkers, reactive biotins, fluorescent labels, and all the other tools needed are now available commercially and are, for the most part, easy to use. In many cases, the simple examples presented in this chapter will be all you need. However, there are cases where more specialized approaches may need to be pursued, especially when functionalizing features at the nanoscale or when working with a protein that has never been immobilized before. In these cases, a consultation of the recent literature can be invaluable for discovery of new and emerging techniques.

### End-of-Chapter Questions

1.  Derive **Equation 14.1**.

2.  If the spacing between Au atoms is 2.88 Å, why is the spacing between sulfur anchors 4.97 Å?

3.  Estimate the predicted thickness of (a) a monolayer of MPTS and (b) a monolayer of 11-mercaptoundecanol.

4.  Ellipsometry measurements vary significantly according to substrate. Why?

5.  Describe how ellipsometry may be used to evaluate the complex refractive index of a material. How can the real and imaginary parts be determined independently?

6.  Draw a schematic of a water drop in oil with a silica surface and indicate approximate contact angle(s).

7.  Assume you can create a surface with a gradient of free energy along one axis. Draw the contact angle of a water droplet with this surface, and describe the behavior you may expect to see.

8.  XPS must be performed in vacuum. Give at least two reasons why.

9.  Check whether your institution has a site-licensed XPS fitting software, such as CasaXPS. If so, download and fit a test data set and explain the fitting process.

10. Derive **Equation 14.6** from **Equation 14.5**.

11. Derive **Equation 14.7** from **Equation 14.6**, given that a general expression for $M_{\mu\nu}$ is given by the integral over the surface of the gap

$$M_{\mu\nu} = \frac{\hbar^2}{2m} \int dS \left( \psi_\mu^* \nabla \psi_\nu - \psi_\nu \nabla \psi_\mu^* \right)$$

12. Calculate the probability that an electron will tunnel from a metal to an STM probe if the work function is 4.0 eV and the gap is 3 Å.

13. (ADVANCED) Discuss how the energy levels of a redox-active protein determine tunneling current in STM, and give an expression for the tunneling current as a function of applied voltage.

14. What is the isoelectric point (pI) of avidin? Streptavidin? Which is preferable for reducing nonspecific interactions, and why?

**15.** Integrate **Equation 14.8** in the case where the analyte concentration can be considered to be constant.

**16.** How would the $\ln(d[AB]/dt)$ plots be expected to change if (a) the analyte can bind nonspecifically and irreversibly to the surface as well as to B; (b) other proteins can bind nonspecifically to B; and (c) B is heterogeneous because of how it is immobilized?

**17.** Think of some ways that protein function can be measured apart from the ones mentioned.

**18.** The *Vroman effect* refers to a phenomenon where abundant, easily adsorbed proteins adhere to a surface first, and then are displaced by less abundant proteins with stronger binding constants. In blood plasma, the first protein adsorbed is fibrinogen; it is later replaced by other, less abundant proteins. Consider a solution consisting of only two proteins in solution above a sensor surface: fibrinogen and kininogen. The association and dissociation constants of fibrinogen are $k_a$ and $k_d$, respectively. The association constant for kininogen

is $k_s$ when it binds to the bare surface, and $k_c$ when it binds to a site containing fibrinogen, and it binds irreversibly. Write differential equations for this system and solve them with the assumption that fibrinogen binding is fast relative to kininogen binding.

**19.** Discuss how a biosensor for a specific protein might be constructed based upon the Vroman effect.

**20.** Describe a scheme for conjugating the following sets of surfaces/molecules to each other: (a) an amino-terminated silane to a carboxyl-terminated nanoparticle in aqueous buffer; (b) an amino-terminated nanoparticle to an amino-terminated silane in aqueous buffer; and (c) a dendrimer to an amino-terminated surface in dimethylformamide (DMF). Which types of aqueous buffers are appropriate?

**21.** (ADVANCED) Describe a possible design for an ion-channel-based biosensor and the type of readout that could be used. What are the advantages and disadvantages of using ion channels as sensors?

## Background Reading

### Books

There are few books currently available on the subject of biofunctionalization. The following are all highly recommended for the practicalities of surface chemistry.

Auciello, O., and Krauss, A.R. *In Situ Real-Time Characterization of Thin Films*. John Wiley & Sons, New York, 2001.

Bonnell, D. *Scanning Probe Microscopy and Spectroscopy: Theory, Techniques, and Applications*. Wiley-VCH, New York, 2001.

Cass, T., and Ligler, F.S. (eds.). *Immobilized Biomolecules in Analysis: A Practical Approach*. Oxford University Press, New York, NY, 1998.

Hermanson, G.T. *Bioconjugate Techniques*. Edn 3.. Academic Press, San Diego, CA, 2013.

Niemeyer, C.M. (ed.). *Bioconjugation Protocols: Strategies and Methods*. Edn. 2. Humana Press, Totowa, NJ, 2011.

The following books deal with characterization techniques.

Alford, T.L., Feldman, L.C., and Mayer, J.W.. *Fundamentals of Nanoscale Film Analysis*. Springer, New York, 2007 (paperback version 2010).

Auciello, O., and Krauss, A.R. (eds.). *In Situ Real-Time Characterization of Thin Films*. John Wiley & Sons, New York, 2001.

Bonnell, D. (ed.). *Scanning Probe Microscopy and Spectroscopy: Theory, Techniques, and Applications*. Wiley-VCH, New York, 2001.

Jena, B.P., and Horber, J.K.H. (eds.). *Force Microscopy: Applications in Biology and Medicine*. John Wiley & Sons, Hoboken, NJ, 2006.

Hinrichs, K., and Eichorn, K.-J., (eds). *Ellipsometry of Functional Organic Surfaces and Films*, Springer Series in Surface Sciences, Springer, Berlin, Germany, 2013.

### Journal articles

This is by no means an exhaustive list but should give some idea of the methods, applications, and in some cases history of the techniques in this chapter.

Almeida, N.F., Beckman, E.J., and Ataai, M.M. (1993). Immobilization of glucose oxidase in thin polypyrrole films: Influence of polymerization conditions and film thickness on the activity and stability of the immobilized enzyme. *Biotechnology and Bioengineering* 42, 1037–1045.

Alonso, J.M., Reichel, A., Piehler, J., and del Campo, A. (2008). Photopatterned surfaces for site-specific and functional immobilization of proteins. *Langmuir* 24, 448–457.

Arumugam, S., and Popik, V.V. (2012). Attach, remove, or replace: Reversible surface functionalization using thiol-quinone methide photoclick chemistry. *Journal of the American Chemical Society* 134, 8408–8411.

Aspnes, D.E. (1985). The accurate determination of optical properties by ellipsometry. In: Paliks, E.D. (ed.). *Handbook of Optical Constants of Solids*. Academic Press, Orlando, 89–112.

Bailey, J.E., and Cho, Y.K. (1983). Immobilization of glucoamylase and glucose oxidase in activated carbon: Effects of particle size and immobilization conditions or enzyme activity and effectiveness. *Biotechnology and Bioengineering* 25, 1923–1935.

Banuls, M.J., Puchades, R., and Maquieira, A. (2013) Chemical surface modifications for the development of silicon-based label-free integrated optical (IO) biosensors: A review. *Analytica Chimica Acta* 777, 1–16.

Bardeen, J. (1961). Tunnelling from a many-particle point of view. *Physical Review Letters* 6(2), 57–59.

Ben Ali, M., Bessueille, F., Chovelon, J.M., Abdelghani, A., Jaffrezic-Renault, N., Maaref, M.A., and Martelet, C. (2008). Use of ultra-thin organic silane films for the improvement of gold adhesion to the silicon dioxide wafers for (bio)sensor applications. *Materials Science and Engineering C-Biomimetic and Supramolecular Systems* 28, 628–632.

Bernard, A., Renault, J.P., Michel, B., Bosshard, H.R., and Delamarche, E. (2000). Microcontact printing of proteins. *Advanced Materials* 12, 1067–1070.

Bhushan, B., Tokachichu, D.R., Keener, M.T., and Lee, S.C. (2005). Morphology and adhesion of biomolecules on silicon based surfaces. *Acta Biomaterialia* 1, 327–341.

Brock, A., Chang, E., Ho, C.C., LeDuc, P., Jiang, X., Whitesides, G.M., and Ingber, D.E. (2003). Geometric determinants of directional cell motility revealed using microcontact printing. *Langmuir* 19, 1611–1617.

Bumm, L.A., Arnold, J.J., Charles, L.F., Dunbar, T.D., Allara, D.L., and Weiss, P.S. (1999). Directed self-assembly to create molecular terraces with molecularly sharp boundaries in organic monolayers. *Journal of the American Chemical Society* 121, 8017–8021.

Cattani-Scholz, A., Pedone, D., Blobner, F., Abstreiter, G., Schwartz, J., Tornow, M., and Andruzzi, L. (2009). PNA-PEG modified silicon platforms as functional bio-interfaces for applications in DNA microarrays and biosensors. *Biomacromolecules* 10, 489–496.

Chen, X., Sevilla, P., and Aparicio, C. (2013). Surface biofunctionalization by covalent co-immobilization of oligopeptides. *Colloids and Surfaces B: Biointerfaces* 107, 189–197.

Choi, H.J., Kim, N.H., Chung, B.H., and Seong, G.H. (2005). Micropatterning of biomolecules on glass surfaces modified with various functional groups using photoactivatable biotin. *Analytical Biochemistry* 347, 60–66.

Cohen-Karni, T., Timko, B.P., Weiss, L.E., and Lieber, C.M. (2009). Flexible electrical recording from cells using nanowire transistor arrays. *Proceedings of the National Academy of Sciences of the United States of America* 106, 7309–7313.

Coq, N., van Bommel, T., Hikmet, R.A., Stapert, H.R., and Dittmer, W.U. (2007). Self-supporting hydrogel stamps for the microcontact printing of proteins. *Langmuir* 23, 5154–5160.

Csucs, G., Michel, R., Lussi, J.W., Textor, M., and Danuser, G. (2003). Microcontact printing of novel co-polymers in combination with proteins for cell-biological applications. *Biomaterials* 24, 1713–1720.

Delaittre, G., Goldmann, A.S., Mueller, J.O., and Barner-Kowollik, C. (2015a) Efficient Photochemical approaches for spatially resolved surface functionalization. *Angewandte Chemie International Edition* 54, 11388–11403.

Delaittre, G., Greiner, A.M., Pauloehrl, T., Bastmeyer, M., and Barner-Kowollik, C. (2015b). Correction: Chemical approaches to synthetic polymer surface biofunctionalization for targeted cell adhesion using small binding motifs. *Soft Matter* 11, 2314.

Dunn, B., and Zink, J.I. (2007). Molecules in glass: Probes, ordered assemblies, and functional materials. *Accounts of Chemical Research* 40, 747–755.

Eteshola, E., Keener, M.T., Elias, M., Shapiro, J., Brillson, L.J., Bhushan, B., and Lee, S.C. (2008). Engineering functional protein interfaces for immunologically modified field effect transistor (ImmunoFET) by molecular genetic means. *Journal of the Royal Society Interface* 5, 123–127.

Foley, J., Schmid, H., Stutz, R., and Delamarche, E. (2005). Microcontact printing of proteins inside microstructures. *Langmuir* 21, 11296–11303.

Hansen, A.G., Boisen, A., Nielsen, J.U., Wackerbarth, H., Chorkendorff, I., Andersen, J.E.T., Zhang, J., and Ulstrup, J. (2003). Adsorption and interfacial electron transfer of *Saccharomyces cerevisiae* yeast cytochrome c monolayers on Au(111) electrodes. *Langmuir* 19, 3419–3427.

Hassler, B.L., Amundsen, T.J., Zeikus, J.G., Lee, I., and Worden, R.M. (2008). Versatile bioelectronic interfaces on flexible non-conductive substrates. *Biosensors and Bioelectronics* 23, 1481–1487.

Hirsch, J.D., Eslamizar, L., Filanoski, B.J., Malekzadeh, N., Haugland, R.P., and Beechem, J.M. (2002). Easily reversible desthiobiotin binding to streptavidin, avidin, and other biotin-binding proteins: Uses for protein labeling, detection, and isolation. *Analytical Biochemistry* 308, 343–357.

Hodneland, C.D., Lee, Y.S., Min, D.H., and Mrksich, M. (2002). Selective immobilization of proteins to self-assembled monolayers presenting active site–directed capture ligands. *Proceedings of the National Academy of Sciences U S A* 99, 5048–5052.

Holden, M.A., and Cremer, P.S. (2003). Light activated patterning of dye-labeled molecules on surfaces. *Journal of the American Chemical Society* 125, 8074–8075.

Howarter, J.A., and Youngblood, J.P. (2006). Optimization of silica silanization by 3-aminopropyltriethoxysilane. *Langmuir* 22, 11142–11147.

Huang, B., Wu, H., Kim, S., and Zare, R.N. (2005). Coating of poly(dimethylsiloxane) with n-dodecyl-beta-D-maltoside to minimize nonspecific protein adsorption. *Lab on a Chip* 5, 1005–1007.

Jackman, R.J., Wilbur, J.L., and Whitesides, G.M. (1995). Fabrication of submicrometer features on curved substrates by microcontact printing. *Science* 269, 664–666.

Jang, C.H., Tingey, M.L., Korpi, N.L., Wiepz, G.J., Schiller, J.H., Bertics, P.J., and Abbott, N.L. (2005). Using liquid crystals to report membrane proteins captured by affinity microcontact printing from cell lysates and membrane extracts. *Journal of the American Chemical Society* 127, 8912–8913.

Johansson, F., Carlberg, P., Danielsen, N., Montelius, L., and Kanje, M. (2006). Axonal outgrowth on nano-imprinted patterns. *Biomaterials* 27, 1251–1258.

Jonkheijm, P., Weinrich, D., Schroder, H., Niemeyer, C.M., and Waldmann, H. (2008). Chemical strategies for generating protein biochips. *Angewandte Chemie International Edition in English* 47, 9618–9647.

Kumar, S., Raj, S., Kolanthai, E., Sood, A.K., Sampath, S., and Chatterjee, K. (2015). Chemical functionalization of graphene to augment stem cell osteogenesis and inhibit biofilm formation on polymer composites for orthopedic applications. *ACS Applied Materials and Interfaces* 7, 3237–3252.

Le Brun, A.P., Holt, S.A., Shah, D.S., Majkrzak C.F., and Lakey, J.H. (2011). The structural orientation of antibody layers bound to engineered biosensor surfaces. *Biomaterials* 32, 3303–3311.

Lee, S.C., Keener, M.T., Tokachichu, D.R., Bhushan, B., Barnes, P.D., Cipriany, B.R., Gao, M., and Brillson, L.J. (2005). Protein binding on thermally grown silicon dioxide. *Journal of Vacuum Science and Technology B* 23, 1856–1865.

Lei, J., and Ju, H. (2012). Signal amplification using functional nanomaterials for biosensing. *Chemical Society Reviews* 41, 2122–2134.

Li, Y., Maynor, B.W., and Liu, J. (2001) Electrochemical AFM "dip-pen" nanolithography. *Journal of the American Chemical Society* 123, 2105–2106.

Libertino, S., Giannazzo, F., Aiello, V., Scandurra, A., Sinatra, F., Renis, M., and Fichera, M. (2008). XPS and AFM characterization of the enzyme glucose oxidase immobilized on $SiO_2$ surfaces. *Langmuir* 24, 1965–1972.

Lisa, M., Wilde, L.M., Farace, G., Roberts, C.J., Davies, M.C., Sanders, G.H.W., Tendler, S.J.B., and Williams, P.M. (2001) Molecular patterning on carbon based surfaces through photobiotin activation. *The Analyst* 126, 195–198.

Luo, R., Tang, L., Wang, J., Zhao, Y., Tu, Q., Weng, Y., Shen, R., and Huang, N. (2013). Improved immobilization of biomolecules to quinone-rich polydopamine for efficient surface functionalization. *Colloids and Surfaces B: Biointerfaces* 106, 66–73.

Marcus, R.A., and Sutin, N. (1985). Electron transfers in chemistry and biology. *Biochimica et Biophysica Acta* 811, 265–322.

McInnes, J.L., Forster, A.C., Skingle, D.C., and Symons, R.H. (1990). Preparation and uses of photobiotin. *Methods in Enzymology* 184, 588–600.

Mrksich, M., Dike, L.E., Tien, J., Ingber, D.E., and Whitesides, G.M. (1997). Using microcontact printing to pattern the attachment of mammalian cells to self-assembled monolayers of alkanethiolates on transparent films of gold and silver. *Experimental Cell Research* 235, 305–313.

Nielsen, C.H. (2009). Biomimetic membranes for sensor and separation applications. *Analytical and Bioanalytical Chemistry* 395, 697–718.

Orth, R.N., Clark, T.G., and Craighead, H.G. (2003). Avidin–biotin micropatterning methods for biosensor applications. *Biomedical Microdevices* 5, 29–34.

Pallavicini, P., Dacarro, G., Galli, M., and Patrini, M. (2009). Spectroscopic evaluation of surface functionalization efficiency in the preparation of mercaptopropyltrimethoxysilane self-assembled monolayers on glass. *Journal of Colloid and Interface Science* 332, 432–438.

Patolsky, F., Timko, B.P., Yu, G.H., Fang, Y., Greytak, A.B., Zheng, G.F., and Lieber, C.M. (2006). Detection, stimulation, and inhibition of neuronal signals with high-density nanowire transistor arrays. *Science* 313, 1100–1104.

Pauloehrl, T., Delaittre, G., Winkler, V., Welle, A., Bruns, M., Borner, H.G., Greiner, A.M., Bastmeyer, M., and Barner-Kowollik, C. (2012). Adding spatial control to click chemistry: Phototriggered Diels–Alder surface (bio)functionalization at ambient temperature. *Angewandte Chemie International Edition in English* 51, 1071–1074.

Peluso, P., Wilson, D.S., Do, D., Tran, H., Venkatasubbaiah, M., Quincy, D., Heidecker, B., Poindexter, K., Tolani, N., Phelan, M., Witte, K., Jung, L.S., Wagner, P., and Nock, S. (2003). Optimizing antibody immobilization strategies for the construction of protein microarrays. *Analytical Biochemistry* 312, 113–124.

Pieters, B.R., and Bardeletti, G. (1992). Enzyme immobilization on a low-cost magnetic support: Kinetic studies on immobilized and coimmobilized glucose oxidase and glucoamylase. *Enzyme and Microbial Technology* 14, 361–370.

Piner, R.D., Zhu, J., Xu, F., Hong, S., and Mirkin, C.A. (1999). "Dip-Pen" nanolithography. *Science* 283, 661–663.

Quere, D. (2002). Surface chemistry—Fakir droplets. *Nature Materials* 1, 14–15.

Raffa, V., Vittorio, O., Pensabene, V., Menciassi, A., and Dario, P. (2008). FIB-nanostructured surfaces and investigation of bio/nonbio interactions at the nanoscale. *IEEE Transactions on Nanobioscience* 7, 1–10.

Ramachandran, K.B., and Perlmutter, D.D. (1976). Effects of immobilization on the kinetics of enzyme-catalyzed reactions. I. Glucose oxidase in a recirculation reactor system. *Biotechnology and Bioengineering* 18, 669–684.

Reich, C., and Andruzzi, L. (2010). Preparation of fluid tethered lipid bilayers on poly(ethylene glycol) by spin-coating. *Soft Matter* 6, 493–500.

Revzin, A., Russell, R.J., Yadavalli, V.K., Koh, W.G., Deister, C., Hile, D.D., Mellott, M.B., and Pishko, M.V. (2001). Fabrication of poly(ethylene glycol) hydrogel microstructures using photolithography. *Langmuir* 17, 5440–5447.

Rozkiewicz, D.I., Kraan, Y., Werten, M.W., de Wolf, F.A., Subramaniam, V., Ravoo, B.J., and Reinhoudt, D.N. (2006). Covalent microcontact printing of proteins for cell patterning. *Chemistry* 12, 6290–6297.

Santer S., Zong Y., Knoll W., and Ruhe J. (2005). On the formation of molecular terraces. *Langmuir*, 21, 8250–8254.

Sassolas, A., Leca-Bouvier, B.D., and Blum, L.J. (2008). DNA biosensors and microarrays. *Chemical Reviews* 108, 109–139.

Schmalenberg, K.E., Buettner, H.M., and Uhrich, K.E. (2004). Microcontact printing of proteins on oxygen plasma-activated poly(methyl methacrylate). *Biomaterials* 25, 1851–1857.

Shen, K., Qi, J., and Kam, L.C. (2008). Microcontact printing of proteins for cell biology. *Journal of Visualized Experiments* 22, 1065.

Sundberg, S.A., Barrett, R.W., Pirrung, M., Lu, A.L., Kiangsoontra, B., and Holmes, C.P. (1995). Spatially-addressable immobilization of macromolecules on solid supports. *Journal of the American Chemical Society* 117, 12050–12057.

Supriya, L., and Claus, R.O. (2004). Solution-based assembly of conductive gold film on flexible polymer substrates. *Langmuir* 20, 8870–8876.

Tersoff, J., and Hamann, D.R. (1985). Theory of the scanning tunneling microscope. *Physical Review B* 31, 805–813.

Trilling, A.K., Beekwilder, J., and Zuilhof, H. (2013). Antibody orientation on biosensor surfaces: A minireview. *Analyst* 138, 1619–1627.

van Noort, D., Rumberg, J., Jager, E.W.H., and Mandenius, C.F. (2000). Silicon based affinity biochips viewed with imaging ellipsometry. *Measurement Science and Technology* 11, 801–808.

Wang, S., Su, P., Hongjun, E., and Yang, Y. (2010). Polyamidoamine dendrimer as a spacer for the immobilization of glucose oxidase in capillary enzyme microreactor. *Analytical Biochemistry*, 405(2), 230–235.

Weiss, P.S. (2008). Functional molecules and assemblies in controlled environments: Formation and measurements. *Accounts of Chemical Research* 41, 1772–1781.

Wigginton, N.S., Rosso, K.M., and Hochella Jr., M.F. (2007). Mechanisms of electron transfer in two decaheme cytochromes from a metal-reducing bacterium. *Journal of Physical Chemistry B* 111, 12857–12864.

Xu, H., Ling, X.Y., van Bennekom, J., Duan, X., Ludden, M.J., Reinhoudt, D.N., Wessling, M., Lammertink, R.G., and Huskens, J. (2009). Microcontact printing of dendrimers, proteins, and nanoparticles by porous stamps. *Journal of the American Chemical Society* 131, 797–803.

Xu, S., Miller, S., Laibinis, P.E., and Liu, G. (1999). Fabrication of nanometer scale patterns within self-assembled monolayers by nanografting. *Langmuir* 15, 7244–7251.

Yildiz, H.B., Kiralp, S., Toppare, L., and Yagci, Y. (2005). Immobilization of glucose oxidase in conducting graft copolymers and determination of glucose amount in orange juices with enzyme electrodes. *International Journal of Biological Macromolecules* 37, 174–178.

You, C., Bhagawati, M., Brecht, A., and Piehler, J. (2009). Affinity capturing for targeting proteins into micro and nanostructures. *Analytical and Bioanalytical Chemistry* 393, 1563–1570.

Yu, L.M.Y., Leipzig, N.D., and Shoichet, M.S. (2008). Promoting neuron adhesion and growth *Materials Today* 11, 36–43.

Zhang, H., Joubert, J.R., and Saavedra, S.S. (2010). Membranes from polymerizable lipids. *Polymer Membranes/Biomembranes*, 224, 1–42.

## Equipment and supplies

*Some major suppliers*

### Chemicals

Pierce Protein Research Products (part of Thermo Scientific). The ultimate source for cross-linkers. Also purification products, antibodies, and modification kits (e.g., fluorescent labeling, production of Fab).

Nanocs. Surface-reactive, homobifunctional, and heterobifunctional PEGs. Also nanoparticles and conjugation services.

ProChimia. Specialist in thiols for SAMs.

Microsurfaces, Inc. Sells prefunctionalized surfaces on glass, silica, etc.

Xiameter (part of Dow Corning). Many silanes.

### Instruments

Horiba Scientific. All key thin-film characterization instruments. J. A. Woollam Co., Inc. Ellipsometers.

Gaertner Scientific. Ellipsometers and measuring microscopes.

Kruss GmbH. Contact angle and other surface-science instruments.

JEOL. Electron microscopes, XPS spectrometers, scanning probe microscopes.

Park Systems. AFMs.

Thermo Scientific. Many types of spectrometers.

Agilent. Scanning probe microscopes and spectrometers.

Novascan. AFMs and tips.

### Databases

http://www.xpsdata.com. Libraries of XPS spectra and other useful tools.

# CHAPTER 15

# Electrophysiology

Coauthored with Christian A. Lindensmith and Thomas Knöpfel

## 15.1 INTRODUCTION

Electrophysiology is a technique for measuring currents through membranes. Because lipid membranes are impermeable to charged species, currents cannot result from simple diffusion but instead require the presence of *ion channel* proteins in the membrane. Ion channels are pore-forming proteins that allow a flow of ions in the direction established by the cell's electrochemical gradient. They usually display *selectivity*, allowing only one type of ion to pass, although nonselective ion channels also exist that pass either all anions or all cations. Ion channels are present in a wide variety of biological processes that cause rapid changes in cells: transport of nutrients, muscle contraction, the release of insulin in pancreas, and *action potentials* in the neurons of the nervous system and the cardiomyocytes of the heart. Cells that fire action potentials are called *excitable* cells.

Along with selectivity, many ion channels also display *gating*, which means that they are effectively open only in the presence of a stimulus of some kind. *Voltage-gated* ion channels open in response to a particular transmembrane voltage ("membrane potential"); *ligand-gated* ion channels open in response to the binding of a chemical species (**Figure 15.1**). Other types of gating exist as well, such as gating caused by mechanical stress (*mechanosensitive* channels), temperature, or light (channel rhodopsin).

Electrophysiology is a mainstay of biophysics; some may argue that it is what made biophysics exist. There are many reasons to perform electrophysiology, among them are as follows: to examine excitable cells to determine the factors controlling and altering their excitability; to clone and characterize new ion channels; to determine the effect of transfected ion channels on cell properties; to study the mechanisms of selectivity and/or gating; and to examine the effect of chemical species on ligand-gated ion channels for discovery of new drugs. The latter is a field of research unto itself, as drugs that affect ion channels include anesthetic agents, antidepressants, anticonvulsants, and classic drugs of addiction such as nicotine; other ion channel agents are commonly known poisons such as muscarine (found in some poisonous mushrooms), organophosphates, war gases known as "nerve gas," and many more.

The purpose of this chapter is to present the basic physical principles behind the design of an electrophysiology experiment, followed by some practical setup advice for two particular types of recording system: lipid bilayer and mammalian cell patch clamp. The basic principles will enable you to

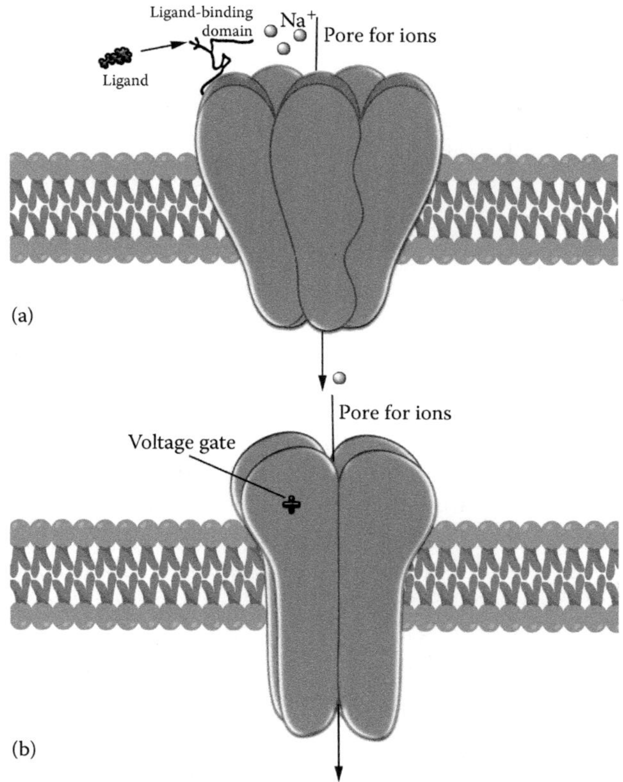



**Figure 15.1 Gated ion channels.** (a) Ligand-gated ion channel has one or more sites where ligands can bind, leading to opening of the channel (for *agonists*) or closing of the channel (for **antagonists**). Many ligand-gated ion channels have multiple binding sites (see the *GABA receptor* for a wonderful example of this). (b) Voltage-gated ion channel has a voltage-sensing region consisting of a string of about a dozen protein-bound positive charges that move from the cytosolic to the exoplasmic surface of the membrane to gate the channel.

appreciate what can be measured in an electrophysiology experiment and which type of experiment should be chosen according to the data desired. How to formulate the appropriate electrolyte solutions for the type of channel and type of experiment will be discussed in some depth. The data acquisition parameters—sensitivity, speed, voltage—will also be introduced. The sections on bilayer recording and mammalian cell recording are designed to be very practical, showing all of the components needed to get started. The sample experiment presents an example of mammalian cells transfected with a foreign potassium channel, and how this channel can be identified and characterized in the cells at the whole-cell level.

Finally, we discuss some advanced and emerging techniques, such as multi-electrode arrays and optical recording. This discussion is intended for those who wish to go beyond single-cell methods in order to examine complex networks of excitable cells, a type of experiment that is still highly challenging. The advantages and drawbacks of existing methods are discussed, with an introduction to some areas of research that are being developed to improve these methods.

At the end of this chapter, you should be aware of what is needed to set up an electrophysiology experiment. If a setup is available, you should be comfortable going to the appropriate literature to design a whole-cell or lipid-bilayer experiment for a particular channel. Actually setting up a rig from scratch will require choosing system components, reading the manufacturer's literature, and understanding the parameters of the instruments. This can be a complex task, although we will try to provide some guidance here as to where the instrumentation can be found and the general principles of its operation.

There are some very common methods that we do not have room to discuss here, especially two-electrode clamp using *Xenopus* oocytes. Some references are given at the end of the chapter, and we encourage you to skip from the first section to these references if you wish to do this type of electrophysiology.

## 15.2  PHYSICAL BASIS AND CIRCUIT MODELS

### Cell circuit models

Much of the electrophysiology that you do will be on cells, so it is useful to understand their electrical properties and what cell characteristics change those properties. Cells also tend to live in and interact with some in an external environment that also plays a role in the electrical properties that we observe.

*Simple vesicle.* We can start by looking at the simplest possible cell—a lipid vesicle filled with internal electrolyte (KCl) and surrounded by external electrolyte, shown in **Figure 15.2**. The two fluids might be the same or different, depending on the desires of the experimenter, but for virtually all biological systems, they will contain various ions. To learn about the electrical properties of this cell, we will want to make electrical contact by sticking an electrode into it. The details will be described later in this chapter, but for now we will assume that the solution outside the cell is grounded, and we can measure the inside of the cell by sticking an electrode into it and connecting our instrumentation between the electrode and the ground. If we have electrically neutral solutions inside and out and put a voltmeter across the connections, we should see zero volts, which is not particularly interesting.

If we could buy a jar of positively charged potassium ions, $K^+$, and change the concentration of these ions inside the cell while keeping the concentration outside the cell the same, then we might see a finite voltage drop across the membrane. In reality, there cannot be a voltage this way, because you cannot buy a jar of a charged species. Real chemicals come as salts of an anion and a cation and give a solution that is electrically neutral—for example, potassium can come as the chloride salt (KCl), gluconate salt (K-gluconate), or any number of other equal mixtures of $K^+$ and an anion $A^-$. In this case, the only way to obtain a difference

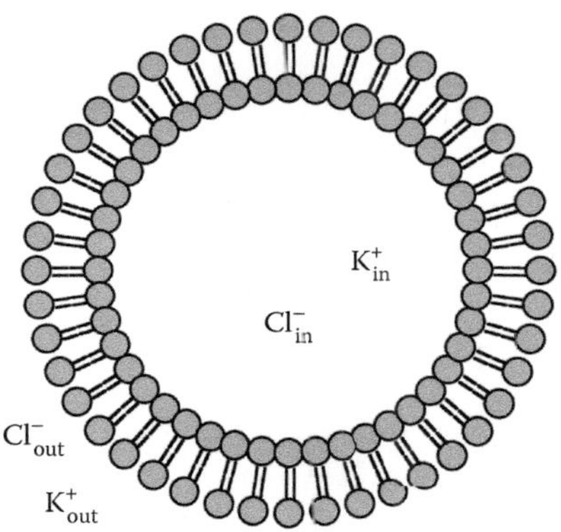


**Figure 15.2  Vesicle with KCl inside and out.** The membrane potential is 0.

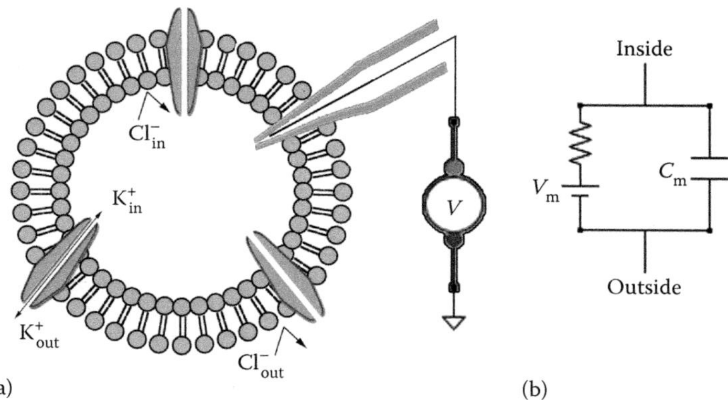


in concentration of one of the charged species across the membrane is to have a pore or channel in the membrane that passes only the anion or only the cation (i.e., it has selective permeability).

*Vesicle with one channel type.* The next simplest cell we can make has only one more element added, but is much more interesting. In **Figure 15.3a**, we have added an ion channel to the membrane of a vesicle containing some concentration of KCl, and immersed in a bath containing a different concentration of KCl. In the vesicle membrane is an ion channel specific for potassium. This "cell" is something we might actually work with—we might make artificial vesicles with just the one type of channel in them to do experiments on that channel in isolation. In most cases, each vesicle will have many (hundreds or thousands) copies of the channel. If we assume that the channel we have inserted has equal ability to pass ions in both directions, then we can treat the channel as a simple resistor.

The cell also has a capacitance. That is because the lipid bilayer is an extremely good insulator. The resistivity of typical recording solutions is tens or hundreds of mΩm, while the lipid has a resistivity of order $10^{11}$ mΩm, making it quite a good insulator. By measuring how the voltage changes with the applied charge, we can determine this capacitance, $C$, with the well-known expression $C = Q/V$, where $Q$ is the charge and $V$ is the voltage difference. If we know the properties of the lipid, we can even estimate the size of the cell from the capacitance—a typical mammalian cell membrane has a capacitance of order 1 μF/cm². The circuit model for this simple cell is then a resistor and capacitor in parallel.

As soon as the channels open, K⁺ ions will flow down their concentration gradient (from the side with more KCl to the side with less). Because only the K⁺ and not the Cl⁻ can move, this results in a charge imbalance. Ions stop moving when the chemical and electrical free energies equal each other, which is when the voltage drop across the membrane reaches the potassium equilibrium potential $E_K$, given by the Nernst equation

$$E_K = \frac{RT}{z_K F} \ln\left( \frac{\left[ K_{in}^+ \right]}{\left[ K_{out}^+ \right]} \right) \tag{15.1}$$

where $R$ is the gas constant (8.31 J K⁻¹ mol⁻¹), $F$ is the Faraday constant (9.648 C mol⁻¹), $z_K$ is the charge of the ion (in this case 1), $[K_{in}]$ is the concentration of K⁺ in the inner solution, and $[K_{out}]$ is the concentration of K⁺ in the solution outside the cell.

The cell is now in dynamic equilibrium. If we use our electrode to try to change the voltage on the cell membrane, currents will flow to try to bring it back to the resting potential. This can be represented in the circuit model of the cell by a battery (**Figure 15.3b**).

*Vesicle with two channel types.* Now that we have something that is starting to exhibit cell-like behavior, we will look at one more idealized case: the same simple vesicle and solutions but with a new type of channel added, this one permeable to Cl⁻ (**Figure 15.4a**). By now you can guess that the circuit for this will look like **Figure 15.4b**. Because the ions have opposite charges, the "batteries" associated with the two ions are drawn with opposite polarity. The Nernst equation is no longer sufficient to determine the resting membrane potential $V_\mathrm{m}$, but the *Goldman–Hodgkin–Katz* (GHK) equation comes to our rescue. This is a generalized version of the Nernst equation (and, indeed, simplifies to it for our earlier case of a single ion) that we can use to predict the resting membrane potential for a cell with multiple ionic species. In the case of monovalent anions A– and cations C⁺, the GHK equation is given in terms of their concentrations and permeabilities $g$ as

$$V_\mathrm{m} = \frac{RT}{F} \ln\left( \frac{\displaystyle\sum_i^N g_{C_i^+}\left[C_i^+\right]_\mathrm{out} + \sum_j^M g_{A_j^-}\left[A_j^-\right]_\mathrm{in}}{\displaystyle\sum_i^N g_{C_i^+}\left[C_i^+\right]_\mathrm{in} + \sum_j^M g_{A_j^-}\left[A_j^-\right]_\mathrm{out}} \right) \tag{15.2}$$

Of course, a separate term with the $z$ factor as in **Equation 15.1** would have to be included for divalent ions. In the case of KCl with K⁺ and Cl⁻ channels, this reduces to

$$V_\mathrm{m} = \frac{RT}{F} \ln\left( \frac{g_K[K^+]_\mathrm{out} + g_{Cl}[Cl^-]_\mathrm{in}}{g_K[K^+]_\mathrm{in} + g_{Cl}[Cl^-]_\mathrm{out}} \right) \tag{15.3}$$

*Real cells.* Real cells start to be much more complicated than the few idealized examples shown above. The ion channels will sometimes have variable resistance that depends on the experimental parameters—it might be voltage dependent, depend on the presence or concentration of some other chemical in the system,

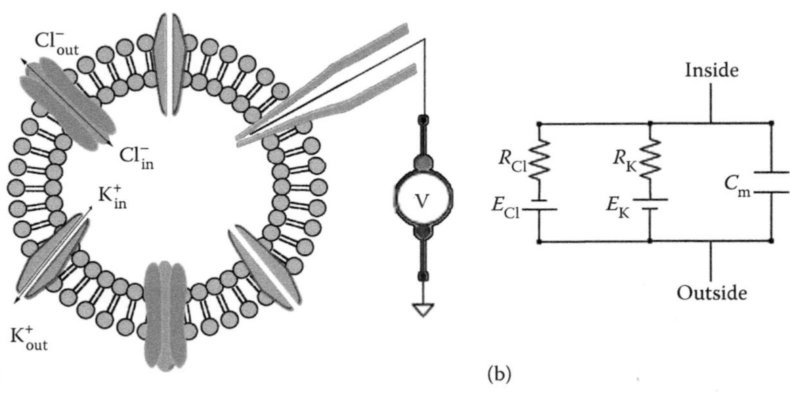


Figure 15.4 (a) Vesicle with KCl inside and out, and two types of channels, one permeable to K⁺ and the other to Cl⁻. (b) Equivalent circuit.

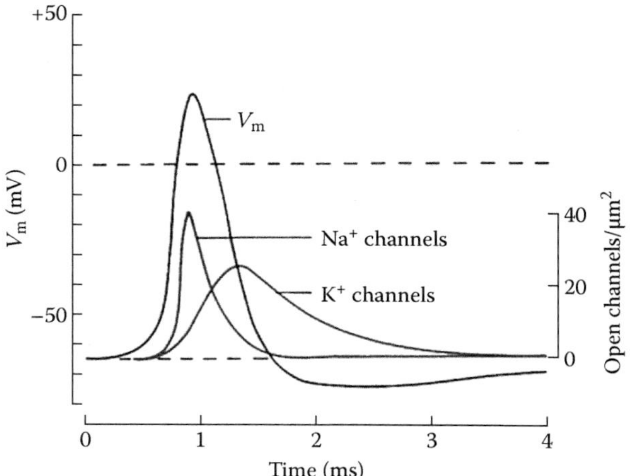



**Figure 15.5 Temporal pattern of opening of Na⁺ and K⁺ channels that result in the neuronal action potential.** The equilibrium potential for Na⁺ is very positive, so opening these channels depolarizes the cell. This depolarization opens voltage-gated K⁺ channels, causing the cell to hyperpolarize once more.

or even on exposure to light. Other ion channels show switching behavior based on the same types of stimuli—the channels will be open (or closed) most of the time, and on exposure to the stimulus will suddenly close (or open), resulting in an abrupt change in the electrical behavior of the cell. Neurons are an excellent example of this—at rest in a dish they have high permeability to K⁺ but much less to Na⁺, and in physiological solutions typically sit at a membrane potential near $E_\mathrm{K}$, or –60 to –70 mV (*polarized*). If we use our electrode to force the membrane potential high (*depolarize* them), then at some threshold voltage suddenly the sodium channels open up and allow a large influx of sodium into the cell, driving the membrane potential rapidly toward $E_\mathrm{Na}$. This is followed by the opening of voltage-gated potassium channels that allow a large flux of potassium out of the cell as the sodium channels inactivate, bringing the membrane potential back down toward $E_\mathrm{K}$. This sequence of events is known as an action potential and is the basis for neural communication (**Figure 15.5**).

The variable nature of ion channels in real cells is a two-edged sword; it can produce unexpected and apparently strange behavior in cells, but it also provides additional variables for an experimenter to work with in deciphering the function of cellular apparatus. You can modify a cell to change how many of a particular ion channel it expresses, alter how a particular type of channel functions, or add completely new channels to the cell. When doing anything of this sort, it can be helpful to keep in mind that there can also be closed-loop feedback systems in the cellular apparatus that the cell uses to try to maintain its normal state against the efforts of even the most determined experimenter.

## Types of recording: Bilayers, single-channel patches, whole cell

There are two main techniques in electrophysiology for exploring the electrical properties of cells and the behavior of ion channels *in vitro*: *patch clamping* and *lipid bilayer recording*. Patch clamping is the direct measurement of electrical properties of cells or pieces of cells by directly contacting them with an electrode. The name comes from two elements of the technique—in many cases, the recording is done using just a "patch" of cell membrane stretched across the orifice of a recording pipette (shown later), and the act of taking data generally

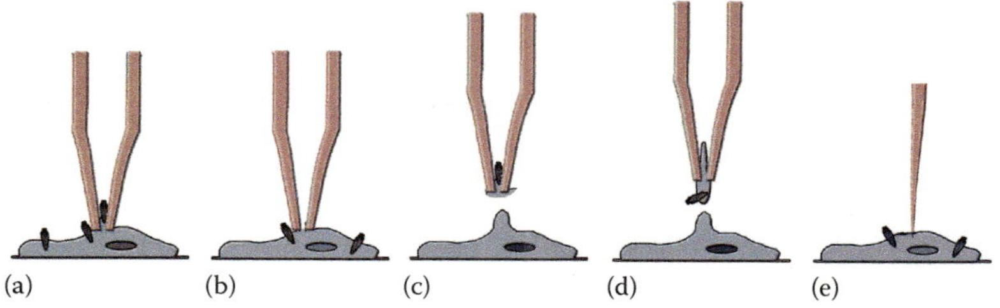

**Figure 15.6 Types of patch clamp.** (a) Cell-attached. (b) Whole-cell. (c) Inside-out. (d) Outside-out. (e) Sharp electrode.

is done by "clamping" the current or voltage at a particular value and measuring the resulting voltage or current at the cell. Details of how to patch clamp will be described below, but a general description of the technique and its variants is given here.

The major variants are shown in **Figure 15.6**. Panel a shows the simplest variant that is generally the starting point for getting to the others: *cell-attached* patch clamping. In cell-attached patch clamping, the pipette is brought into good contact with the cell membrane to form what is known as a gigaohm seal or *gigaseal*. The seal is between the pipette glass and the cell membrane and is (as apparent from the name) on the order of $10^9$ $\Omega$. The high resistance of the seal effectively eliminates noise from currents leaking between the pipette and the bath and allows detection of the current of a single ion channel. In the cell-attached configuration, the communication with the cell is through whatever happens to be located in the patch of the membrane encircled by the end of the pipette. The advantages are that it is easy to do and the cell remains intact. One disadvantage of this technique is that the circuit from the electrode to the ground goes through the cell membrane twice: first at the "patch" and second through the rest of the cell membrane outside the patch, making interpretation of the data more complex for many types of experiments. Also, the pipette contents are not readily changed, so that concentrations of drugs, agonists, etc. cannot be varied during a single experiment, but must be done on successive patches.

**Figure 15.6b** shows different configuration that is only slightly harder to achieve but simplifies the circuit: a whole-cell patch. For a whole-cell patch, the cell membrane is punctured (often by a rapid application of pressure) to create an open hole in the membrane that is encircled by the pipette orifice. This effectively creates a giant "patch" where the resistances of all the channels of the cell appear in parallel in the circuit. Whole-cell recording is not a single-channel technique; the signal is coming from hundreds to thousands of channels. This type of recording is also sometimes done with a *perforated patch*, in which the pipette contains a chemical (such as the antibiotic amphotericin B) that degrades the cell membrane on contact and reduces the resistance from the pipette into the cell. The access resistance into the cell is usually higher this way than with traditional whole-cell recordings, but the advantage is that the cell's contents are not replaced by the contents of the pipette, so that the function of channels remains more physiological. Perforated patch recordings can also be very stable, lasting for hours.

**Figure 15.6c and d** shows true patch configurations, where the pipette has been pulled from the cell in order to break off a patch of membrane that is sealed over the end of the pipette. In the *inside-out patch* of panel c, the pipette is separated from the cell after the gigaseal is made, and a piece of the membrane remains

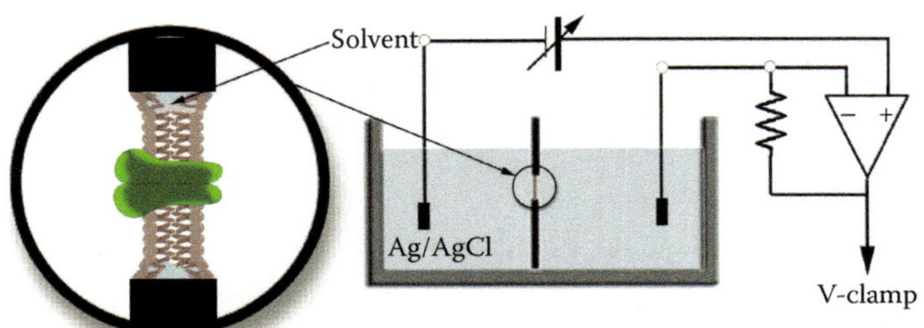



**Figure 15.7 Schematic of a chamber for bilayer recording.** A chamber containing 1–3 mL of electrolyte solution is separated from a similar chamber by a wall containing an orifice on the scale of 0.1 mm in size. The fact that the wall is vertical helps to neutralize any pressure differences that would break the bilayer. Each chamber is connected to an amplifier via an Ag/AgCl electrode, one side to signal and the other to ground. Lipid in solvent is used to block the orifice, which leads to a disordered lipid/solvent annulus on the edges of the orifice, but a solvent-free bilayer near the middle. Ion channels can assemble into this bilayer.

attached with the part that is normally inside the cell exposed to the solution in the bath. The *outside-out patch* of panel d is made by pulling the pipette away from a whole-cell patch and trying to get a piece of the membrane to fold back over the pipette end. The advantage of these two true patch configurations is that a small piece of membrane will contain only a small number of ion channels, sometimes as few as one, which allows observation of the behaviors of a single channel in isolation.

*Sharp electrode* recording is shown in panel e. This is technically not a patch technique, as a much sharper micropipette is simply poked through the cell membrane to access the inside of the cell for recording. Fragile cell types will typically not tolerate sharp electrode recording but are less damaged in patch recording. We will not discuss these in detail in this chapter.

Lipid bilayer recordings are electrically similar to patch clamping, but the physical configuration is drastically different. Two chambers of a bath are separated by a small orifice that is "painted" with lipid to form an artificial cell membrane, as shown in **Figure 15.7**. Ion channels can be implanted in this membrane to create an artificial idealized "cell" that can be used to measure the electrical characteristics of the implanted channels.

## Voltage clamp and current clamp

Once the cell membrane and ion channel configuration have been prepared using one of the techniques above, the recording is done with specialized amplifiers that can drive and detect currents and voltages at levels appropriate for cells. In *voltage-clamp* mode, the amplifier is used to set the cell membrane at a particular voltage, and the resulting currents are measured. In *current-clamp* mode, the

**Figure 15.8 Ideal voltage-clamp circuit.** If $R_s = 0$, the cell membrane is "clamped" to the command voltage $V_{cmd}$.

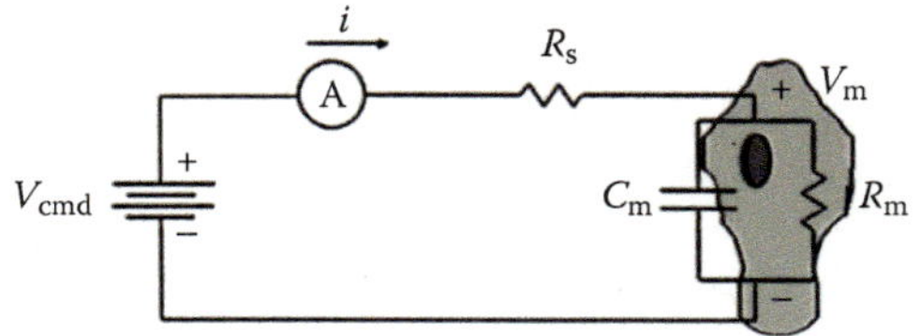

amplifier sets the current through the electrode, and the resulting voltage of the cell membrane is measured.

A conceptualized voltage clamp is shown in **Figure 15.8**. In the ideal case, the circuit consists of only a battery, an ammeter, and the cell. When the battery steps to a command voltage $V_{cmd}$, a pulse of current flows to charge the cell capacitance, then a steady-state current maintains the cell membrane potential $V_m$ at exactly $V_{cmd}$.

In reality, the resistance of the recording electrode combined with that of the access hole into the cell, called $R_s$ or the *series resistance*, can contribute to errors in measured voltage. When it is nonnegligible, the circuit becomes a voltage divider. The voltage of the membrane is no longer equal to the command voltage, but is given by

$$V_m = \frac{V_{cmd} R_m}{R_m + R_s} \tag{15.4}$$

One of the great advantages of patch clamp over sharp electrode recordings is that $R_s$ can be small (~5 M$\Omega$) compared to the resistance of the cell membrane (hundreds of M$\Omega$). One goal of the electrophysiologist is to minimize $R_s$ in every way possible. We will discuss this more in the section on pipettes and the section on amplifiers.

## Issues of space clamp

Since microelectrodes can control membrane voltage only at their tip, the voltage is often less well "clamped" at regions of the cell membrane that are electrically remote from the point of optimal voltage clamp. This is why perfect clamp over the whole plasma membrane, including processes such as axons and dendrites, also called *space clamp,* is difficult to achieve in large cells with processes.

In small, round cells, a quasi-space clamp is routinely achieved. However, in large, branched cells such as neurons, the control of membrane potentials in the branches typically is very limited. This can even result in unclamped action potentials firing in the neuronal processes while the soma is being voltage-clamped (**Figure 15.9**).

There are both experimental and theoretical approaches to this problem. The experimental approaches are to use small round cells whenever possible, to ligate large processes, or to use two-electrode clamp. If none of these are feasible, it may be sufficient simply to understand how the voltage distribution varies throughout the cell and to take this into account in data interpretation. There are software packages available that allow for these calculations using realistic

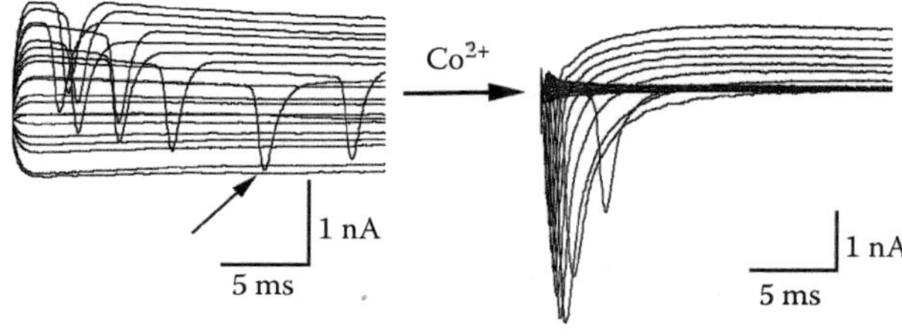


**Figure 15.9 In the left panel, poor space clamp is exhibited in this neuron, manifested by low resistance and unclamped action potentials (arrow).** When membrane resistance is increased by the addition of CoCl$_2$, space clamp improves, and the right panel shows well-behaved current clamp with typical inward Na$^+$ and outward K$^+$ currents.

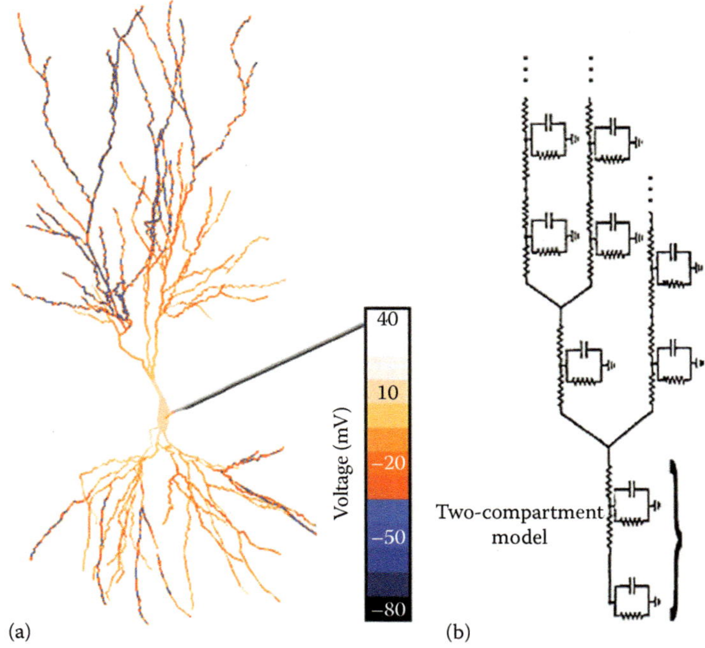



**Figure 15.10 Models of space clamp.** (a) Some software models incorporate realistic neuronal structures and are able to simulate membrane voltage throughout the cell. Here the neuron pictured is firing an action potential, so the membrane potential at the soma is +40 mV. The color scale shows the distribution of voltage through the processes. (b) Compartmental model treats segments of the cell as individual RC circuits separated by resistors. The number of compartments can vary enormously. A simple two-compartment model does well for many cells and helps explain why only partial capacitance and series resistance compensation are possible.

models of neurons that may be imported or drawn (**Figure 15.10a**). Alternatively, a *compartmental model* may be used to model the processes as a series of isopotential compartments. For some minimally branched neurons in culture, a two-compartment model is sufficient, corresponding to the soma and all of the processes (**Figure 15.10b**).

# 15.3  SOLUTIONS AND BLOCKERS

## Internal and external solutions

In any type of electrophysiology, some thought needs to go into formulating the composition of the electrolyte solutions on either side of the membrane. In the case of cells, these will be "internal" and "external" solutions; in all but perforated-patch recordings, the internal solution replaces the internal contents of the cell, so it must be compatible with life. It also must be compatible with gigaseal formation; different solutions will result in various degrees of recording ease. This does not mean that the "easy" solution is the best, however—it may make patching easier because it blocks channels that make the membrane excitable, which is not a good thing for studying excitability. For any experiment, it is usually best to find recipes in the literature for solutions appropriate to the channel being measured. We give some general guidelines here to explain the purpose of the most common ingredients in cellular patch experiments.

*Internal solutions.* The main cationic component of a cell is $K^+$, found in the 100 mM range. The physiological $Cl^-$ range is much lower (5–40 mM), so KCl is often not used as the primary salt. Instead, organic salts such as potassium gluconate, potassium aspartate, or potassium methanesulfonate are used to replace most of the KCl. Both $K^+$ and $Cl^-$ can be replaced with nonphysiological ions designed to disrupt cell function in revealing ways. Using $F^-$ as an anion will inhibit cellular metabolic processes, allowing the effect of these processes on channel function to be elucidated. When $K^+$ is replaced with an impermeant ion, usually $Cs^+$, $K^+$ currents are blocked, allowing for better resolution of other types of channels. It is common to have a control solution and a test solution, where the latter is formulated with $F^-$ or $Cs^+$, and recordings using the two types of solution are compared.

The role of $Ca^{2+}$ is complex and poorly understood in cells and recordings. However, it can be stated with confidence that unless $Ca^{2+}$ is chelated, forming gigaseals is very difficult. A total of 1–10 mM EGTA is usually used to buffer divalent cations, and then $CaCl_2$ added so that a free $Ca^{2+}$ concentration to what is believed to be physiological is reached (10–50 µM).

To avoid rundown of channels that require ATP and/or G proteins, 1–10 mM ATP and/or GTP should be added to the internal solution in recordings where these channels are or may be important. Eventual rundown of $Ca^{2+}$ current is usually seen even with ATP and is difficult to prevent, though creatine phosphate and creatine phosphokinase may help (see references for recipes).

A buffer is used to keep pH constant. One appropriate for the pH range is used, very often HEPES (see **Chapter 6** for a discussion of buffers).

*External solutions.* These will be complementary to the internal solutions, usually containing large amounts of a Na salt and small amounts of $K^+$, so that resting membrane potentials are on the order of –60 to –80 mV. It is not necessary to add ATP or GTP or EGTA to these solutions; some divalent cations ($Ca^{2+}$ and $Mg^{2+}$) at low mM concentrations are usually added. The external solution should be buffered in the same way as the internal solutions.

Usually external solution is made 1 L at a time, and internal solution 100 mL at a time. After preparation, the solutions' pH should be adjusted while checking with a pH meter, usually to 7.1 to 7.3. Solutions are usually too acidic as prepared, so the adjustment is done using the hydroxide of the predominant ion (e.g., KOH or CsOH for internal, NaOH for external). Pellets or concentrated stock solutions can be used for this purpose. After adjusting pH, the osmolarity of the solutions should be matched to prevent cell shrinking or swelling. An *osmometer* measures the freezing point of a solution to determine osmolarity and is much more reliable than osmolarity calculations. Measure the osmolarity of both solutions, and adjust the one that is lower with sucrose. This may take what seems like a lot of sucrose, but be careful not to overshoot the mark or you will have to throw the solution out and start over again. Make sure to calibrate the osmometer with standard solutions as its absolute values are often wrong by a fixed amount.

Because dust can inhibit seal formation, after adjustment of pH and osmolarity, solutions should be filtered through a 0.22 µm filter, either a syringe filter (for internal solutions) or a larger flask-mounted filter for external solutions. Internal solutions that contain ATP and/or GTP should be frozen in single-use, 1 mL aliquots.

In bilayer recordings, there is no need to keep the bilayer alive, but consideration of what currents are going to be measured needs to drive the formulation of the solutions, which in this case are "front" and "back" or "cis" and "trans" rather than internal and external. In many cases, the solutions can be very simple, e.g., 1 M KCl and 10 mM HEPES both front and back for recording from a $K^+$ channel.

## Junction potential

It is important to recall that the GHK equation (Equation 15.2) is an equilibrium equation. Most of the time, ions move rapidly enough to come into equilibrium between the internal and external solution essentially immediately after the recording electrode is dipped into the solution. This is not the case, however, when ions have significantly different *mobilities* and are interacting at a boundary (i.e., the narrow pipette tip). Even $Na^+$ and $Cl^-$ show different mobilities; for large bulky ions often used to replace smaller ions, such as the aspartate or gluconate often used in the place of $Cl^-$, mobilities can be strikingly reduced. The faster ions flow into the bath first, resulting in a net charge relative to the pipette that is maintained at a constant value by the resulting electric field (Figure 15.11). A negative *junction potential* is, by convention, when the bath contains more anions and so is negatively charged with respect to the pipette; a positive junction potential is when the bath is positive relative to the pipette.

An uncorrected junction potential will lead to errors in determining membrane potential. This can lead to erroneous identification of cell type or state of health, and significant issues with calculating IV curves, which will seem to be offset by a certain amount. Most amplifiers allow for junction potential calculation and correction before recording. The correction should be applied just after the pipette is immersed into the bath, before touching the cell.

Junction potentials may be calculated using tools provided by the amplifier manufacturer or found online (references at the end of the chapter). These tools make use of tables of mobilities determined empirically; more and more values are becoming available, and online tools are available for using these data. The solution composition is entered (only the primary species are needed), and a value is calculated using the sign convention discussed above using a generalized version of the Henderson equation that expresses the potential difference between

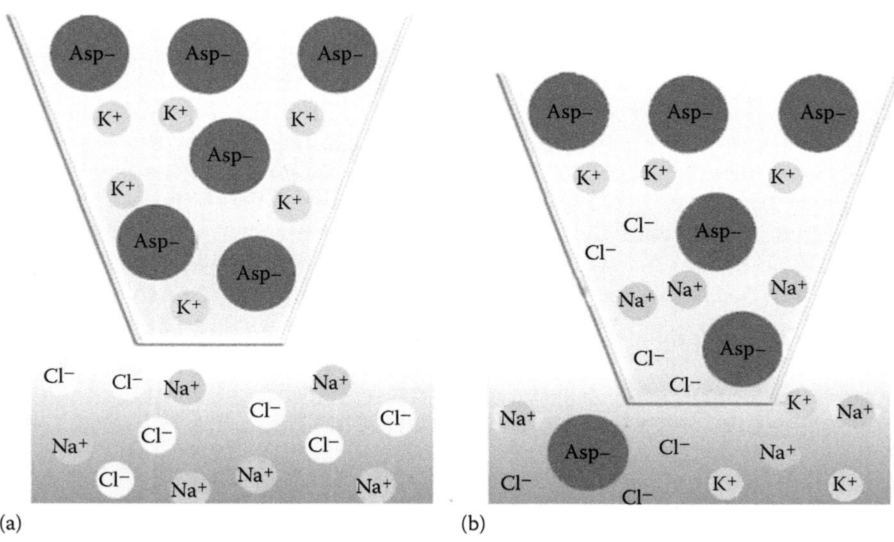



**Figure 15.11 Junction potential.** (a) Pipette and solution before contact, showing different ions. (b) After contact, showing much lower mobility of the large aspartate anion, which leads to a net positive charge in the bath. (Courtesy of N. Blair, http://junctionpotential.blogspot.com)

the solution $V_S$ and the pipette $V_P$ as a function of the mobilities $u$, activities $a$, and valences $z$ of the ionic species present:

$$V_S - V_P = \frac{RT}{F} S_F \ln \left[ \frac{\sum_i z_i^2 u_i a_i^P}{\sum_i z_i^2 u_i a_i^S} \right],$$

where

$$S_F = \frac{\sum_i z_i u_i \left( a_i^S - a_i^P \right)}{\sum_i z_i^2 u_i \left( a_i^S - a_i^P \right)}. \tag{15.5}$$

It is also possible to measure junction potentials experimentally, but we do not recommend this. Recall that if the external solution is changed during the experiment, the junction potential may also change. In this case, do not compensate for the junction potential during the experiment, but note the values for the different solutions and perform the corrections during data analysis. Keep in mind during recordings that as a general rule, the cell membrane potential is more negative than the amplifier indicates if the dominant anion has a lower mobility than the dominant cation in the pipette solution.

## Blockers, agonists, antagonists

One type of ion channel in a cell can be separated and distinguished from the others by the use of agents that block it specifically. Subtracting the current after blockade from that before blockade leads to a clean signal from only the channel in question. The existence and number of highly specific blockers are remarkable and fascinating. One of the earliest discovered was tetrodotoxin (TTX) from pufferfish, which blocks voltage-gated $Na^+$ channels at nanomolar concentrations. Many other blockers come from the venom of exotic organisms such as spiders, snakes, and insects, and some companies are dedicated solely to the discovery and purification of these molecules. Some of the suppliers listed at the end of the chapter provide extensive catalogs of these agents; a few examples are given in Table 15.1.

In the case of ligand-gated ion channels, agonists or antagonists may be used to turn the channels on or off. Some agonists and antagonists bind at the same site as the natural ligand; others bind elsewhere, and electrophysiology is a good way to study this type of interaction. As with the blockers, catalogs of many hundreds of these agents are available. Most bind a wide variety of channels with different affinities, so it is important to recognize all possible binding events that might be occurring in a cell. In a bilayer, of course, there is no such issue. Table 15.2 gives some examples of natural ligands and artificial agonists/antagonists for commonly studied channels.

**Table 15.1**

A Few Specific Blockers (and a Few Activators) of Ion Channels

| Name | Type of Agent | Target |
| --- | --- | --- |
| 1-(Aminomethyl) cyclohexaneacetic acid (Gabapentin) | Blocker (anticonvulsant) | Voltage-gated $Ca^{2+}$ channels |
| $\alpha$-[3-[[2-(3,4-Dimethoxyphenyl)ethyl] methylamino]propyl]-3,4-dimethoxy-$\alpha$-(1-methylethyl) benzeneacetonitrile hydrochloride (Verapamil) | Blocker (vasodilator) | L-type $Ca^{2+}$ channels |
| (2$S$-$cis$)-3-(Acetyloxy)-5-[2-(dimeth ylamino)ethyl]-2,3-dihydro-2-(4-methoxyphenyl)-1,5 -benzothiazepin-4(5$H$)-one hydrochloride (Diltiazem) | Blocker (antihypertensive) | L-type $Ca^{2+}$ channels |
| 5-Nitro-2-(3-phenylpropylamino)benzoic acid (NPPB) | Blocker | $Cl^-$ channels |
| 4-Aminopyridine (4-AP) | Nonspecific blocker | Voltage-gated $K^+$ channels |
| 1,2-$Bis$(2-aminophenoxy)ethane-$N,N,N',N'$-tetraacetic acid tetrakis(acetoxymethyl ester) (BAPTA AM) | Blocker (antithrombotic) | $hK_V1.5$, hERG, and $hK_V1.3$ channels |
| Tertiapin Q | Blocker (bee venom) | ROMK1 ($K_{IR}1.1$) and GIRK1/4 ($K_{IR}3.1/3.4$) channels |
| 6-(1-Piperidinyl)-2,4-pyrimidinediamine 3-oxide (minoxidil) | Activator (hair growth agent) | $K_{ATP}$ channels |
| $N,N,N,N$-Tetraethylammonium chloride (TEA) | Nonspecific blocker | $K^+$ channels |
| Tetrodotoxin (TTX) | Blocker (pufferfish poison) | Voltage-gated $Na^+$ channels |
| Batrachotoxin (BTX) | Activator (poison dart frog toxin) | Voltage-gated $Na^+$ channels |

*Note:* Their effects in humans are given in parentheses if they are used as drugs. There are hundreds more available through suppliers.

## 15.4 INSTRUMENTATION

### Amplifiers

The amplifier is the heart of an electrophysiology experiment. Great demands are placed upon it; currents as small as 1 pA and as large as 1 nA are routinely seen in single-channel and whole-cell experiments, respectively. The amplifier must therefore be low-noise and have various gain levels that permit resolution of currents of different magnitudes. In addition, in order to perform voltage clamp, the amplifier must be able to compensate for and subtract current transients caused by the capacitance of the recording electrode and/or the cell. Special features of voltage-clamp amplifiers include the following:

- "Fast" capacitance compensation. This cancels out the capacitive transient caused by the capacitance of the patch pipette $C_p$, which is usually on the order of 1 pF. Variable gain $A$ in the amplifier scales the applied

**Table 15.2**

A Few Agonists and Antagonists for Some Ligand-Gated Ion Channels

| Name | Type of Agent | Target |
| --- | --- | --- |
| Acetylcholine (ACh) | Natural agonist | nAChR, mAChR |
| α-Bungarotoxin | Irreversible inhibitor | nAChR (a7 subtype) |
| PHA 543613 hydrochloride | Agonist (experimental antischizophrenic) | nAChR (a7 subtype) |
| Carbamoylcholine chloride (carbachol) | Agonist | nAChR, mAChR resistant to acetylcholinesterase |
| Catestatin | Noncompetitive antagonist | nAChR |
| Pilocarpine | Agonist | mAChR |
| Ipratropium bromide | Antagonist (bronchodilator) | mAChR |
| γ-Aminobutyric acid (GABA) | Natural agonist | GABA receptors (all subtypes) |
| Muscimol | Agonist | $GABA_A$ receptor, partial $GABA_C$ |
| (-)-Bicuculline methochloride | Antagonist | $GABA_A$ receptor |
| Diazepam (Valium) | Allosteric modulator at benzodiazepine binding site (sedative, anticonvulsant) | $GABA_A$ receptor |
| Glutamate | Natural agonist | Glutamate receptors (subtypes: AMPA, kainate, NMDA) |
| S-AMPA | Agonist | AMPA receptors |
| (S)-(-)-5-Fluorowillardiine | Agonist | AMPA receptors (more selective than AMPA) |
| CNQX | Antagonist | AMPA and kainate receptors |
| NBQX | Antagonist | AMPA selective |

*Note:* Many of these agents also have activity with other types of channels, which must be considered in whole cell and especially animal and human experiments. Abbreviations: nAChR = nicotinic acetylcholine receptor; mAChR = muscarinic acetylcholine receptor.

voltage before it reaches the injection capacitor so that $(A - 1)C_{inj} = C_p$, cancelling the transient (**Figure 15.12**). The degree of compensation can be adjusted by the experimenter.

- "Slow" capacitance compensation. This works along the same principle, but is used after breaking into a cell to compensate for the cell capacitance, which is on the order of tens to hundreds of picofarads.

- Series resistance compensation. The series resistance $R_s$ described above is measured by the amplifier and compensated by adding a compensatory voltage $I_{inj}R_S$ to the command voltage using a circuit like that shown in **Figure 15.13a**. This cannot be done completely, as this would lead to infinite voltages, and must be done with care to avoid feedback oscillations that can kill the cell. The amount of compensation applied is reported as a percentage; some transient remains (**Figure 15.13b**). Series resistance compensation is most important for sharp, fast currents. A cell

**Figure 15.12 Compensation of pipette capacitance.** (a) Circuit for compensation. The subscripts "p" and "c" refer to pipette and command, respectively. (b) Appearance of signals before and after compensation of a typical patch pipette in a typical bath solution, with a test pulse of 5 mV for 5 ms applied.

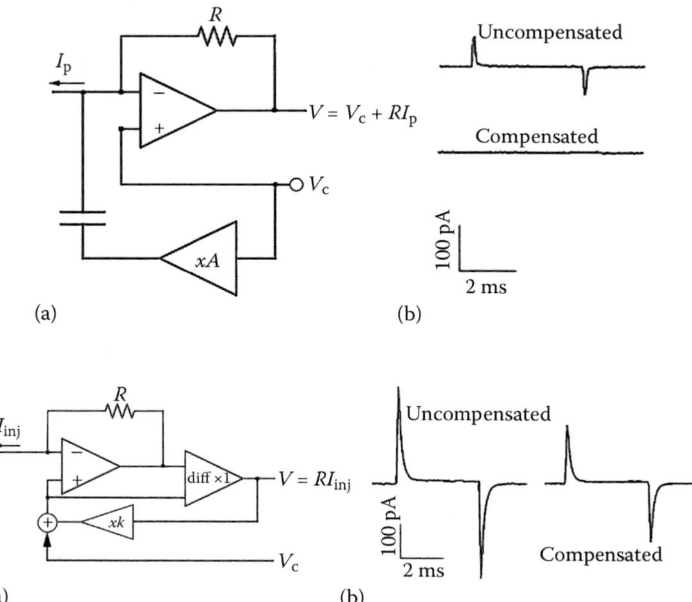

**Figure 15.13 Series resistance compensation.** (a) Circuit for compensation. (b) Appearance of a typical cell under whole-cell recording mode before and after compensation. Test pulse of 5 mV for 5 ms applied. The subscript "inj" is for the current injected through the pipette.

showing sodium and potassium currents with and without capacitance and series resistance compensation is shown in **Figure 15.14**.

- Other on-the-fly methods of noise subtraction that are particularly useful for single-channel recordings, as well as other features such as leak subtraction.

A high-quality amplifier will also do current clamp, and there is a whole new set of demands placed upon a current-clamp amplifier. Recall that in current clamp, current is injected and the cell membrane potential is measured. However, this cannot be done accurately if a variable waveform is injected unless the voltage drop across the recording pipette is compensated—this is especially important for high-resistance intracellular electrodes, which often have resistances on the order of 50 MΩ. This compensation was traditionally done using a Wheatstone bridge, so is still called "bridge balance" on patch-clamp amplifiers. Bridge balance is set by applying a test pulse and adjusting the balance setting until the voltage drop across the pipette disappears (**Figure 15.15a**). This leads to a change in the

**Figure 15.14 Typical hippocampal neuron under voltage clamp, showing steps of −140 to +60 mV in 10 mV increments, before and after series resistance and capacitance compensation.** Note how the Na⁺ currents become sharper and closer together after compensation.

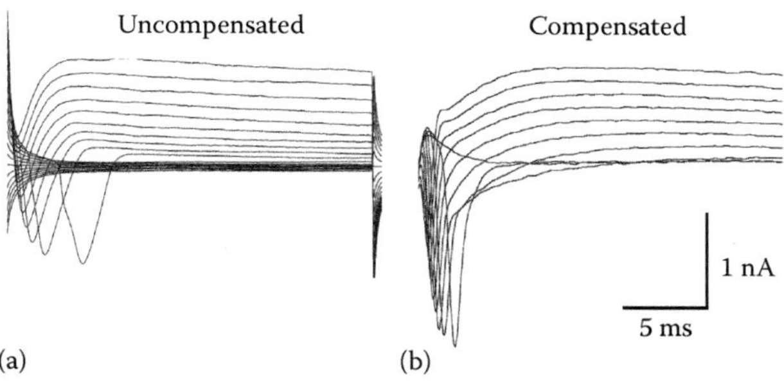

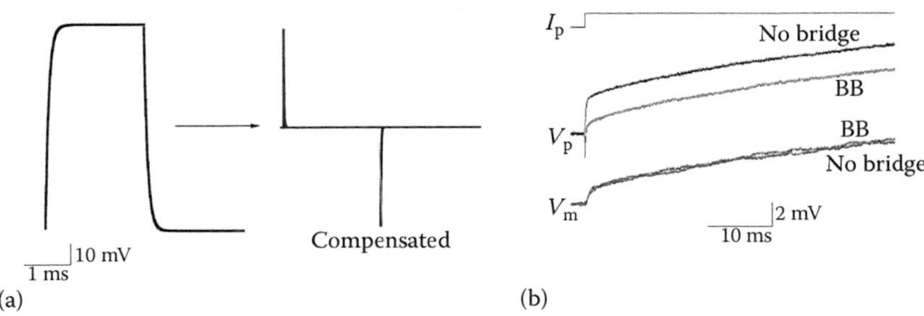



**Figure 15.15 Bridge balance.** (a) Voltage drop across a 50 mΩ pipette under a 2 nA current pulse of 2 ms duration (capacitance 1.4 pF). Compensation removes this voltage drop and should result in sharp transients only. (b) Balancing the bridge in a Purkinje cell. Note the double-exponential charging curve reflecting first rapid charge distribution within the cell, then charging of the whole cell according to $\tau_m$. Balancing the bridge has no effect on the cell ($V_m$), only on the recorded voltage ($V_p$). (Adapted with author's permission from the "Electronics for Electrophysiologists" tutorial by Boris Barbour, which is available online.)

voltage recorded from the cell, but not an actual change in the cell's membrane potential (**Figure 15.15b**).

Different amplifiers do this compensation in different ways. The best is to monitor the voltage drop across a resistor placed in series with the recording pipette, which allows for the real pipette resistance to be monitored and compensated. Another method is to use a separate circuit as a virtual ground, originating from the bath ground electrode. This provides a cancellation current so that the output voltage of the amplifier is directly proportional to the bath current. The biggest problem with the latter method is that the input lead to the virtual ground can pick up noise. Reading the product literature for amplifiers you own or may buy will tell you which method is used by a particular instrument. Many perform the compensation automatically, so there is nothing for the user to do. Others switch between $R_s$ compensation and bridge balance when they are switched between voltage-clamp mode and current-clamp mode, respectively.

There are two primary suppliers of high-end, fully functional, everything-included patch-clamp amplifiers: Molecular Devices (formerly Axon Instruments) and HEKA Instruments. Both of these companies provide high-quality instruments designed particularly for electrophysiology, with various features available (double or triple-cell recording, etc.). Both provide their own software that includes easy setup of protocols for voltage and current clamp, whole-cell vs. patch recording, and "continuous" recording for single-channel currents. They allow for electronically timed application of external solutions with recording of when the application occurred. There is one major functional difference between the two companies: HEKA's amplifiers are completely software controlled, with no knobs or dials to be adjusted on the instrument itself. The amplifier may be placed out of the way and not touched during the experiment. In contrast, the Axon instruments have dials for adjustment of series resistance compensation, capacitance compensation, and other variables. Preference of one over the other is largely a matter of personal taste. The author uses amplifiers from both companies and considers them both roughly equally easy to use.

Other suppliers also make amplifiers, often for specialized applications. Warner Instruments and Harvard apparatus make amplifiers specifically for bilayer recording that are excellent. A few other companies provide lower-cost alternatives, but the noise profiles and cancellation capabilities of these amplifiers may not

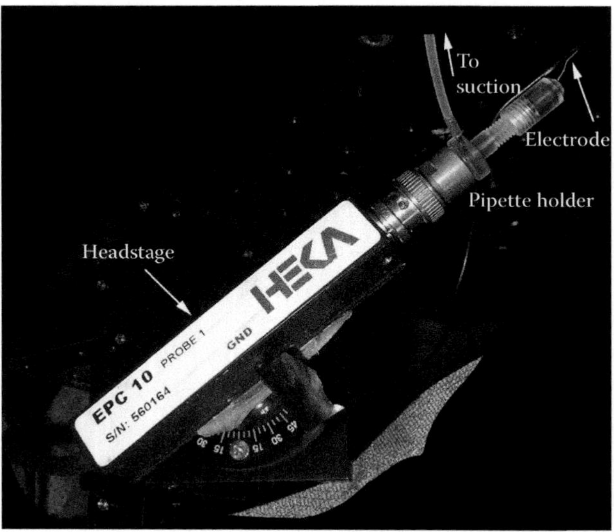



**Figure 15.16 Example of a headstage.** A ground electrode is plugged into the port marked GND. The signal electrode is located within the pipette holder, so the filled pipette can be threaded over it and held stably in place. The suction port allows for formation of gigaseals and breaking through by pulling on an attached syringe.

be sufficient for many experiments. In addition, they usually come without software. Some open-source electrophysiology programs are available; it is also relatively straightforward to write acquisition programs in MATLAB. However, in the authors' experience, the extra setup time and reduced performance of "homemade" systems are rarely worth it. Good electrophysiology starts with a good amplifier. There are also newer, emerging companies that provide alternative designs of the patch-clamp amplifier. The authors have not tried any of these. Before any purchase, it is key to know what voltage and current ranges must be measured and with what tolerance for noise levels, which will help in selection of an instrument.

Patch-clamp amplifiers have a resistive *headstage* that serves to hold the recording electrode in a stable, low-noise fashion. They also have a port for grounding. Most have unity voltage gain, but their current-passing capabilities vary depending upon the magnitude of the resistor in the headstage. This can be chosen according to the desired current measurement. Some headstages have variable resistance, with switching controlled by a computer according to the output gain. An example is shown in **Figure 15.16**. The exact style of pipette holders, grounding wires, and other features vary widely. It is important to note that the input pins for the pipette holder, and sometimes the ground port, are often not of any standard size and may be difficult to find if they need to be replaced or if you wish to attach anything other than the provided pipette holder.

## Grounding and shielding

Electrophysiologists spend a good deal of time creating low-noise setups and troubleshooting noise setups. A good start to all experiments is to find a quiet room, in the basement if possible. Even better, some institutions have electrophysiology rooms that are electrically shielded and on isolated slabs to prevent outside sources of vibration from affecting the experiments. In the absence of this, mount the equipment on a vibration isolation platform or table. Not everyone puts the setup inside a Faraday cage, but doing so makes shielding easier. A Faraday cage can be made from copper wire screwed to an aluminum frame.

Even inside the cage, everything will need to be carefully tied to a common ground and shielded from the electrode. The most common sources of noise are electrical pickup, magnetic pickup, and ground loops. As an example of the scale of electrical pickup, consider a fluorescent light bulb placed 10 cm from the recording electrode and not shielded. It will produce a current

$$I = 2\pi f C V \left( 2\pi f t + \frac{\pi}{2} \right) \tag{15.6}$$

with the capacitance $C$ proportional to the area of the electrode divided by the distance to the lamp. If the area is 0.1 cm$^2$—which is quite small—the noise current will be 6 pA, bigger than the conductance of many single channels.

This type of noise will be visible as sine waves across the signal, and can be shielded by placing a piece of *grounded* metal between the noise source and the electrode. If the shield is not grounded, it will act as an antenna and make things even worse. A good troubleshooting technique is to touch a grounded alligator clip to various areas inside the Faraday cage, examining the signal for improvements in noise characteristics. A good supply of wire braids a few feet long and with strong clips on both ends is very useful for managing noise sources in the system.

Another method of checking for noise sources is simply to turn potentially troublesome things off and see how this affects the signal. If a lamp causes noise, it may be best to just leave it off during the experiment. Computer monitors are a notorious source of noise, and excess computers in the room should be shut down before performing recordings.

Ground loop currents arise when devices or cables are grounded at more than one place, and there is a small potential difference between the two grounds. It can be an informative exercise when setting up a lab to use a voltmeter or oscilloscope to measure the potential differences between the various "grounds" in a room or experiment. Sometimes this is easy to fix, when it results from shields that have been accidentally tied to more than one ground. Simply moving the ground wire to a common source, which is tied to the headstage ground of the amplifier, will cure the problem. Unfortunately, for many electronic devices, the solution is nontrivial and getting rid of ground loops can be difficult. This is especially true with computers, which may need their own ground. The best way to see if an apparatus is causing a ground loop is to shut it off and unplug it.

It is common to spend hours reducing noise in the system, only to come back the next day and discover a huge, hideous, usually periodic (line frequency or its second harmonic) noise signal. Before you give up in despair, check to see what new electronic devices might have been introduced into the room or turned on. Also check all of your ground wires, as these are notoriously disconnected and/or scavenged by other users, who should be trained never to touch an experimental setup without asking its most recent user. If all else fails, blame the air conditioner in the lab next door and try the experiment at midnight—the best data are traditionally taken in the wee hours of Christmas Day, though we hope you are never driven to this.

## Micromanipulators

If you are working with micron-sized cells—or even more difficult, patches pulled from these cells—then a good micromanipulator is the most important of equipment after the amplifier. A good micromanipulator will have sufficient range to allow the patch-clamp electrode to be maneuvered from outside the dish into the dish in a rapid fashion (moving at many mm/s), but then also permit very fine manipulations (fractions of a µm/s) to move the electrode onto the cell. It will not wiggle, drift, shake, or otherwise move in any axis during the duration of the recording, which can be hours. An upward drift of a patch-clamp pipette can pull the electrode right off the cell; downward drift will impale the cell and kill it. Changes in temperature and room vibrations can cause poorer-quality manipulators to become unreliable, at which point they are the bane of the electrophysiologist's existence. Many excellent models are now available, some with spatial resolution down to tens of nanometers. Some are programmable, allowing a "home" position to be defined so that pipettes may be changed and then located again in the microscope very quickly. All of these features are optional, but stability and reliability are crucial to happy cells and long recordings. Good micromanipulators can be motorized, piezoelectric, or hydraulic. Hydraulic manipulators are often best for single-channel current measurements, as they generate no electrical noise.

## 15.5  LIPID BILAYER SETUP

### General principles and use

A typical lipid bilayer setup is designed for easy creation of a solvent-containing lipid bilayer across a relatively large orifice: 0.1–0.25 mm in size, visible to the naked eye. A single channel is incorporated into this bilayer, making this the easiest way to do single-channel recording. There are a couple of caveats:

- In order to be able to use this type of setup, you must have purified protein, since there is no cell machinery available to express the gene. The protein may be in solvent (in the case of simple peptides), but more often must be reconstituted into lipid vesicles (see **Chapter 5**).

- Most bilayer chambers do not permit high-resolution imaging, because they contain large volumes of liquid on both sides. Some highly specialized chambers for imaging exist, but they are much harder to work with than the usual kind, and their limited volumes can cause issues of ion depletion.

- Bilayer experiments on large orifices are not good for studying voltage-activated single channels because of the large capacitance of these bilayers.

- Low noise is crucial for good data. A good Faraday cage and high-quality amplifier are necessary and are only the beginning.

This setup is usually achieved in a *bilayer chamber* that has two equally sized reservoirs for balancing solution height. These are frequently made of Delrin, which is a Dupont trade name for polyoxymethylene plastic (POM). A cup is inserted into the rear reservoir; the cup contains the orifice (**Figure 15.17**).

Each side of the chamber is then filled with an electrolyte solution of the user's choice, and electrical contact to Ag/AgCl electrodes is made via *salt bridges* (**Figure 15.18**; see **Practical Tips 15.1**). Chambers and cups are sold commercially; the reservoirs usually come in sizes from 1 to 3 mL. The cups can be made of different materials, such as polystyrene, polysulfone, or Delrin. We recommend

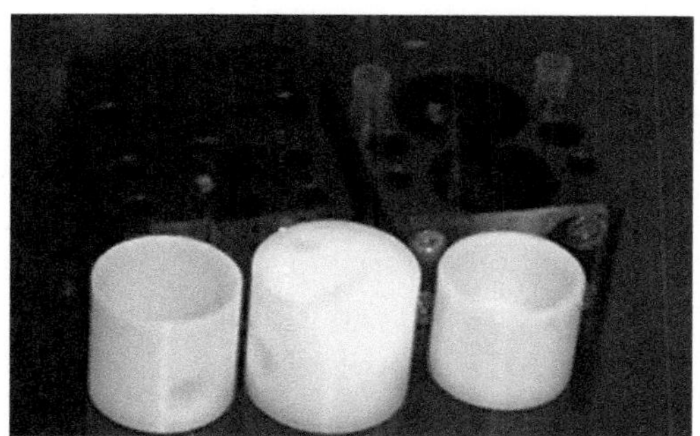

**Figure 15.17 Cups and chambers for bilayer recording.**

**Figure 15.18 Cups and chambers set up for recording.** (Left) Top view, showing placement of cup and salt bridges. The depressions in the cup and chamber are for placement of stir bars. (Right) Side view, showing electrode placement. The small subreservoirs containing the electrodes are filled with 3 M KCl. Make sure that only the AgCl part of the electrode touches the liquid, as bare wire will cause noise.

commercial cups and chambers rather than trying to make homemade bilayer orifices. All of the sizes and materials sold work well, and the author does not have any strong preferences, though some electrophysiologists prefer cups of one material or another.

Complete bilayer systems are also available commercially, providing stirring, illumination, and a Faraday cage. These are great if budget allows, but building a custom cage is often more cost-effective and quite easy. A picture of our custom Faraday cage is shown in **Figure 15.19**. Plastic guides hold the commercial chambers tightly so they do not shift during recording. Holes on the bottom of the chamber permit the use of a magnetic stir plate. The amplifier headstage is located within the Faraday cage and is connected to the electrodes via shielded alligator clips. The chamber lid closes during the recording to isolate the system. If any illumination lamps, stir plates, or other accessories are used, it is very important that they all be grounded to a common ground. The whole setup should be placed on a vibration isolation table or platform.

## Making the lipid bilayer

Once the setup is in place, formation of the bilayer takes a bit of finesse. The first trick is to choose the lipid or lipids to be used in the experiment. The lipids must be "soft" enough to permit channel insertion, yet not so soft that they break too quickly. Different lipids may also be chosen to deliberately study the effects of bilayer composition on channel function. These details are too complex to go into here, and for many, experiments are not needed. The recipe that we use for nearly all recordings is a 3:1 mixture of 1-palmitoyl-2-oleoyl-*sn*-glycero-3-phosphoethanolamine

**PRACTICAL TIPS 15.1:   MAKING ELECTRODES AND SALT BRIDGES FOR BILAYER RECORDINGS**

Salt bridges (or agar bridges) are pieces of bent glass tubing filled with soft agar containing KCl. They form a conductive bridge between the bilayer reservoir and a smaller reservoir containing the Ag/AgCl electrode. The advantage of salt bridges is they minimize junction potential (see text for discussion), and they prevent metal ions from contacting the lipid. Once prepared, agar salt bridges will last for multiple recordings. Discard them when there is no "open hole" signal, or the signal is noisy.

Electrodes are simply pieces of thick silver wire coated with silver chloride except on the very end where the electrical contact is made. The electrodes will last for several days to weeks, until they become oxidized. Silver chloride is purple, whereas silver oxide is whitish. Ag(0), formed by UV decomposition of AgCl, is black. When electrodes develop a layer of oxide or Ag(0), or when data become noisy, discard them and cut a fresh piece of wire. Protect them from sunlight.

To make electrodes:

- Cut ~3 cm of 1.5 mm silver wire.

- Hold it with an alligator clip into a beaker of bleach so that the uppermost part of the wire is not immersed in the bleach.

- Leave it 30 min to overnight.

- Rinse the electrode with water and use.

To make 3M KCl:

- Add 11.18 g KCl to a 50 mL polystyrene tube.

- Dilute to the 50 mL mark with 0.01 M phosphate buffer, pH 7.4.

- Filter through a 0.2 μm filter.

- Keep tightly capped in a tube; beware of "KCl creep" (precipitation of salt around the outside of the vessel).

To make salt bridges:

- Choose a capillary tubing of the appropriate diameter (usually 0.86 mm is used).

- Carefully heat the pieces of tubing in a Bunsen burner to bend them into the appropriate "bridge" shape; this may be V- or U-shaped as long as it bridges between the reservoir and electrode and does not pinch the tubing closed.

- Make a bunch of these pieces of tubing before you get started, at least 20 since quite a few of them will probably fail to fill.

- Prepare a 2% agar solution in 3 M KCl. A total of 3 mL is sufficient. Use a flat vessel to hold the solution; a weigh boat works well.

- Melt the agar on a hot plate or in the microwave until it is fully dissolved and just barely bubbling.

- Quickly dip the bridges into the agar, one end at a time. They should fill swiftly with no bubbles; discard those with large bubbles or rotate the bridge in the agar until the bubble comes out.

- Place into 3M KCl for cooling and storage. Before use, peel off any excess agar that has stuck to the bridge.

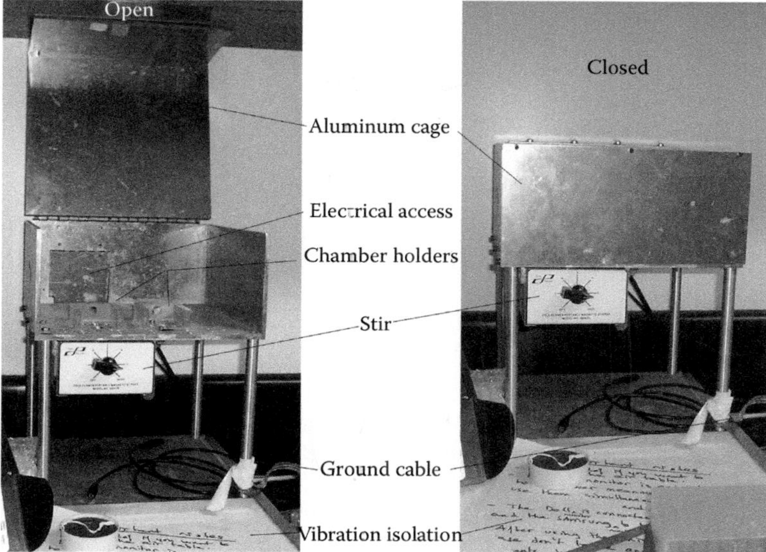



**Figure 15.19 Custom-designed cage for bilayer recording, machined from aluminum.** The cage is about 25 cm across.

(POPE) to 1-palmitoyl-2-oleoyl-*sn*-glycero-3-phospho-L-serine (POPS). The lipids are usually supplied as 10 or 20 mg/mL solutions in chloroform. In order to use them to paint a stable bilayer, they must be put into a solvent that evaporates very slowly—usually decane. To prepare lipids for recording:

- In a chemical fume hood, transfer 25 µL of POPS and 75 µL of POPE to a glass vial using a nonplastic pipette. The solutions in chloroform should be stored at −20°C, under inert gas and with Teflon tape around the edges of the vial. Do not let condensation get in.

- Dry the solutions down completely under a flow of inert gas.

- Resuspend in 80 µL of decane.

- Keep capped; use within a couple of hours as the lipids will oxidize. If the experiment goes on for a significantly longer time than this, make fresh lipid.

The lipid is spread onto the orifice in the cup using a polished glass rod. These are easily made by fire-polishing the tip of a glass Pasteur pipette in a Bunsen burner. Simply hold the rod in the flame until the end seals and becomes smooth. You may wish to make a "hook" or other shape in the rod so that it fits easily into the bilayer chamber; the shape is a matter of personal taste. Make sure that the end of the rod is nicely polished so it does not scratch up the cup, and make a few of them in case some are impractical shapes or for use with different lipids or channels.

When everything is ready, place the cup into the chamber. Some electro-physiologists "prime" the cup by painting around the hole with lipids in chloroform, which are then allowed to evaporate. This is optional. Fill the front and rear reservoirs (called the *cis* and *trans* reservoirs) with equal volumes of the appropriate electrolyte solutions, and place the salt bridges. Usually the *cis* chamber is grounded, and the *trans* chamber is connected to the headstage amplifier, but this is merely a convention. Put the Ag/AgCl electrodes into place. Now attach the amplifier and examine the signal.

## Monitoring bilayer formation electrically

The key to bilayer formation is to add just enough lipid to cover the orifice so that the area in the center thins into a single bilayer. Too little, and there will be huge leak currents; too much, and there will be a thick "glob" of lipid. The only way to tell the difference between a glob of lipid and a bilayer is by monitoring the capacitance. For this, the full electrophysiology setup needs to be put in place and everything turned on.

Arrange all of your solutions and supplies in a convenient way around the bilayer Faraday cage. Place a chamber in the cage and a cup in the chamber; fill each side of the chamber with the "internal" and "external" solutions appropriate to your experiment (many times this is symmetric 1 M KCl with a buffer of some kind). Turn on the amplifier and start the software, attach the silver electrodes and salt bridges, and apply a 5 mV test pulse.

The test pulse with no lipid present should result in a huge leak current, on the order of 20 nA (**Figure 15.20a**). The gain on the amplifier will need to be turned all the way down to see this. It is important to see this leak. Otherwise, something is wrong with the salt bridges (test by dipping the electrode directly into the reservoir); the cup (hole might be blocked with old lipid or just "junk"); the electrodes; or your setup in general (is everything plugged in?).

Once you have your leak or "open hole" current, block the hole with the glass rod and ensure that the current goes away. If it does not, something is still wrong (leakage of electrolyte somewhere, other holes in the cup). Also monitor the noise at this stage. If there is more than about 500 fA of noise, you will have a hard time seeing many channels, and the system needs to be improved. If the noise profile looks good and the hole blocks easily, you are ready to try blocking it with lipid. Some people, at least in the beginning, like to use a dissecting microscope to visualize the orifice and end of the glass rod. Others find this unnecessary. If you do use a microscope, it needs to be on a boom stand for maneuverability.

Dip the glass rod into lipid and sweep it across the hole a few times. If it does not work right away, dip again, but do not use too much lipid. When you succeed in blocking the hole, the current will fall to zero. Turn the gain way up. Notice the small size of the capacitive signal—this is because you have a lipid "glob" (**Figure 15.20b**). Since capacitance goes as $\varepsilon A/d$, where $d$ is the thickness of the layer, thinning of lipid into a bilayer will result in an increase in capacitance, usually observed as a smooth growth of the capacitive transient due to the test pulse to a final value determined by

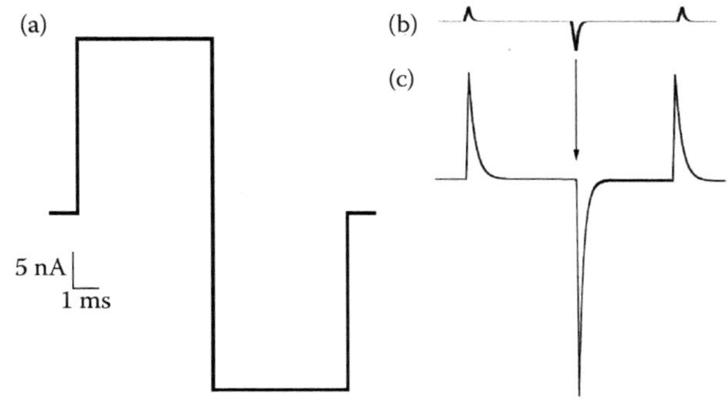


**Figure 15.20 Monitoring bilayer formation with a 5 mV test pulse.** (a) Open orifice current. This may be too big to resolve on the amplifier, and just result in a saturated signal. That is OK. What is bad is if the current is small at this stage. (b) When the hole is blocked, the current falls to nearly 0. Turn the gain up now. (c) Thinning of the central lipid into a bilayer is seen as a smooth growth of the capacitance transient to a stable final value consistent with the size of the orifice (see Problems).

the orifice size (**Figure 15.20c**). Some specialized amplifiers provide audible signals in the form of a tone that changes frequency when the hole is blocked or bilayer is forming, so that the experimenter does not have to look up at the screen.

Usually a bilayer does not form spontaneously from a glob, at least not the first time. It will need to be "teased" by rubbing it with the rod or passing the rod near the edges. Different people have different tricks. You may wish to probe the hole until the leak current appears again, then re-sweep. At some point, the capacitance will grow and should remain stable. If it continues to grow, the bilayer is just going to break. Usually once a bilayer forms, each successive bilayer is easier to make.

## Adding ion channels

When the bilayer has been stable for a few minutes, it might be safe to add channel. The amount of channel added depends upon its identity and purification protocol. Some pore-forming peptides (*ionophores*) such as *alamethicin, valinomycin,* and *gramicidin* are simply dissolved in solvent (water or other) and need to be added at very low levels to get a good recording. Adding too much will just break the bilayer, and you will not be able to re-form it, because the ionophores will still be floating around. They can be difficult to wash out—if you add too much, you may have to take everything apart and wash the cup and chamber before trying again. In general, the desired concentration of peptides is ~1–5 nM. For proteins in lipid vesicles, the amount we usually add is ~1 ng/mL, followed by more if there is no channel seen after 20–30 min.

Add the channel to the appropriate reservoir (usually the front; gramicidin must be added to both sides) and turn on the stir bar. This will cause noise in the current (**Figure 15.21**), so stop the stirring periodically to look at the current for signs of channel insertion and function. There is no need to record the current at this stage; it would be a waste of disk space.

Make sure to monitor the current using the appropriate gain depending upon the size of the single-channel conductance expected. For example, gramicidin will

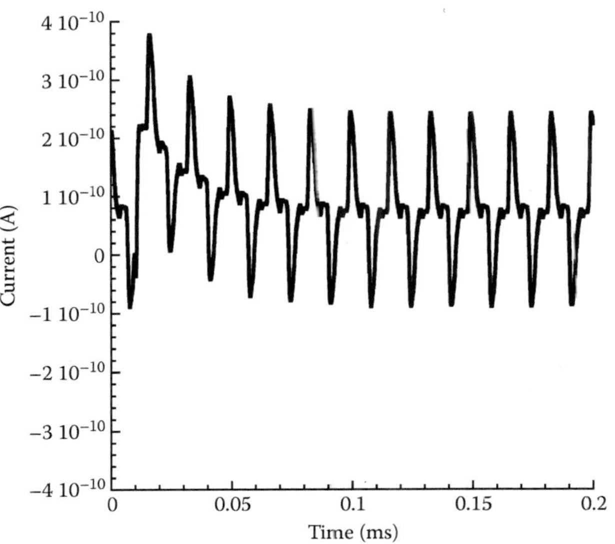

**Figure 15.21 Noise caused by stirring.**

**Figure 15.22 Signal from alamethicin in a bilayer.** Note different subconductance states, which aid in identification of the channel. Also note the large size of these currents. Not all channels are so easy.

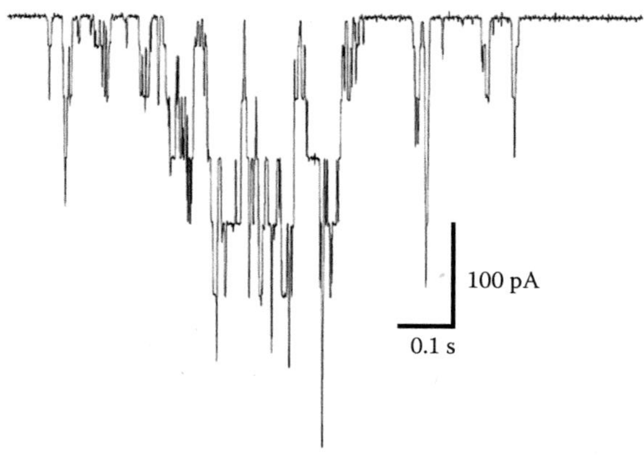

lead to events on the order of 2 pA in size; alamethicin's currents are more on the order of 100s of pAs in 1 M KCl. The appearance of a classic pattern of opening and closing events indicates that your channel has entered the bilayer (**Figure 15.22**). It is up to you to know what this "classic pattern" is for the channel you are working with—otherwise, opening events might be hard to distinguish from noise. It may take 30–60 min for a channel to find the bilayer and insert, so be patient.

Once the channel is in place, you want to record for many minutes to hours to ensure that you have enough statistics for single-channel analysis (discussed in **Section 15.10**). A typical scenario is to record currents for 10 min at a given voltage, then to step through the voltages from negative to positive for 10 min each. **Figure 15.23** shows sample data from glycine receptor purified into vesicles; the current–voltage curve extracted from the data is characteristic of this channel.

**Figure 15.23 Single-channel currents from human glycine receptor (GlyR).** (a) Signals seen at different voltages (indicated). At certain voltages, channel openings are rarer and/or smaller, so analysis must be done accordingly. (b) I–V curve for GlyR. This curve helps to identify the channel.

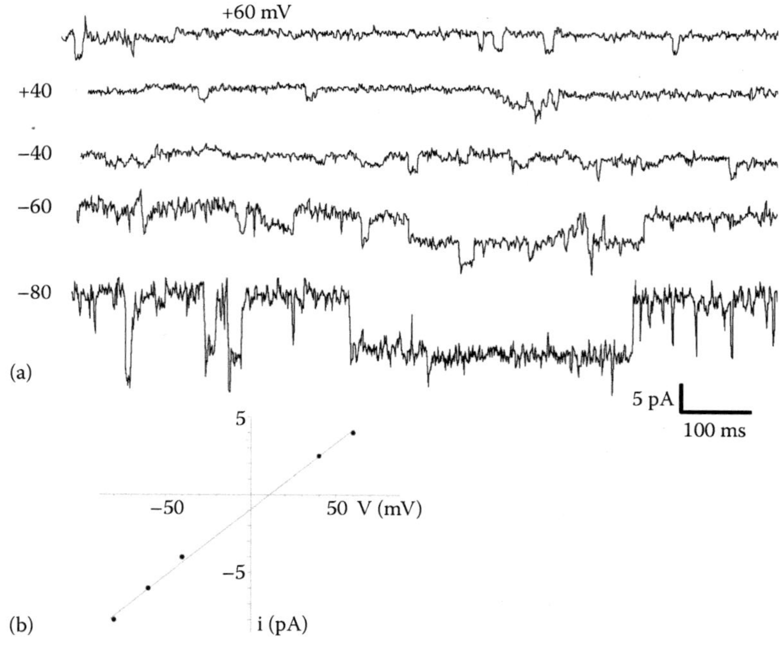

If the bilayer breaks before you have enough data to work with—and it very well might—then that is just sad. Keep trying. Maybe next time it will last longer. If breakage is a persistent problem, consider getting fresh lipid, changing the lipid composition, adding less channel, or giving up for the day and trying again another time. Bilayer recordings can be easy and satisfying, but they can also be frustrating. Another common occurrence is that the channel simply does not insert. This is harder to troubleshoot. If you have assays for the function of your protein, this may be the time to use them. Otherwise, consider adding more protein or dipping the rod directly into the protein solution and then probing or sweeping over the bilayer. The latter method has worked for the author when nothing else would. Your mileage may vary. Also consider changing the lipid if the protein in question has not been proven to work with your choice of lipids.

## 15.6 CELL PATCH-CLAMP SETUP: WHAT IS NEEDED

The setup for recording from mammalian cells is significantly more complex and costly than the lipid bilayer setup. This is primarily because it requires a good-quality microscope for visualizing the cells, preferably with fluorescence; a *pipette puller* for fabricating pipettes for seal formation; and a micromanipulator for manipulating the pipettes into position with submicron precision. The primary advantage of working with mammalian cells is that the cell machinery can be used to express almost any gene with great efficiency. Levels of expression may be controlled by mixing the active gene (containing ion channel) at different ratios with a dummy plasmid (expressing nothing or GFP), so that the number of channels per surface area of the membrane can be controlled for best single-channel results. The cells are small, with capacitance on the order of 10 pF, so fast currents are easily resolved. Many cell lines are nearly round, so problems of space clamp do not occur. The same setup may be used to investigate the complex biology of spontaneously excitable cells, which show highly complex and adaptable patterns of ion channel expression. Whole-cell recordings from neurons can be used to study models of *long-term potentiation* and *long-term depression*, to investigate network properties, and to construct *in vitro* models of genetic or infectious disease or toxin exposure.

*Microscope.* An inverted microscope should be used (see **Chapter 7**) to permit placement of the headstage, micromanipulator, perfusion apparatus, electrodes, etc. Two or even three headstages can be mounted for simultaneous double- or triple-recording; these types of experiments are used to investigate synaptic coupling between neurons. Phase contrast is the recommended brightfield mode for electrophysiology, as its "three-dimensional" appearance allows for easier identification of healthy cells and aiming of the electrode (**Figure 15.24**). It is possible to patch under ordinary brightfield, but it is more difficult. A shutter for switching between fluorescence and phase contrast should be conveniently placed so that fluorescent cells can be quickly identified, then patch-clamped under brightfield so as not to cause phototoxicity.

For most cells, a 20× objective is sufficient for electrophysiology, although some people prefer to use a 40× objective. Bear in mind that increased magnification means a smaller field of view and reduced depth of field. For high-resolution imaging experiments, it may be necessary to patch clamp using a 63× or 100× oil-immersion objective. It can be tricky not to get oil where it is not wanted, so this should be used only if really needed.

**Figure 15.24 Appearance of a neuron with approaching patch pipette under phase contrast.** Use of phase contrast allows for evaluation of cell health (cell is smooth, with a bright halo, and nucleus is invisible; these are all good); it also allows an easier appreciation for when the pipette touches.

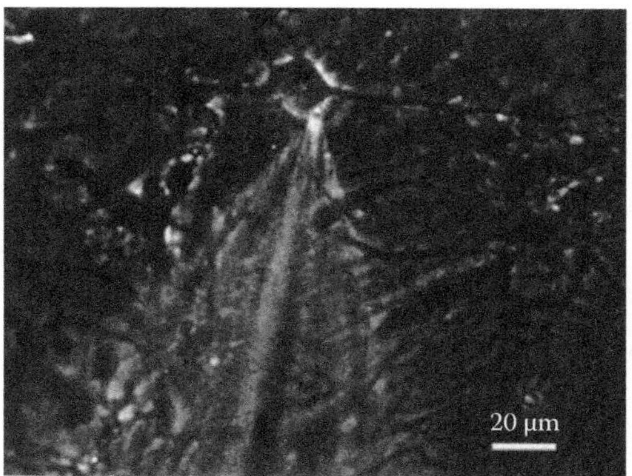


*Micromanipulator.* We have already discussed the importance of a good micromanipulator. The power supply and usually the manipulator controls are placed outside the Faraday cage. The headstage is mounted directly to a micromanipulator mounting plate, which may be made specifically for the given brand of amplifier. This mounting plate allows for precise control of the angle at which the pipette enters the dish. This angle will of course control the x, y, and z components of the travel when the manipulator is moved. There should be three independent axes of movement with total travel of several centimeters.

When first beginning electrophysiology, a significant challenge is learning to place the recording pipettes and locate them under the microscope using the micromanipulator without "crashing" them into the bottom of the dish. If you have not acquired this skill yet, it is good to spend a couple of days practicing before even considering getting a dish of cells. There are various tricks that people use to find the pipette. Some like to turn the magnification all the way down to 4× and locate the pipette, center it in the dish, and then progressively turn up the magnification. Others like to focus up and down to find the pipette, then the bottom of the dish, and so on in succession until the two are close to each other. Sometimes what works for one pipette does not work for the next, for whatever reason. If the pipette is too short or too long, the micromanipulator might reach the end of its travel without the pipette being within the field of view. If this happens, get a new pipette or reset the micromanipulator settings to redefine the zero point.

*Electrodes.* The signal electrode in a patch-clamp recording is located within the pipette. It is a piece of 0.010 inch Ag wire that is chlorided by immersion into bleach for several minutes, then installed directly into the pipette holder (see **Practical Tips 15.2**). Constant replacing of pipettes can wear out the AgCl coating; if this happens, the wire should be re-dipped in bleach, or else the signal will show background drift. Eventually the wire will become bent and/or scratched from use; at this point, replace it with a fresh piece. It is good to have a roll of Ag wire on hand.

The ground electrode is attached with a pin to the headstage, then consists of a shielded wire long enough to reach the bath, and a pellet electrode. The larger the pellet, the lower the resistance. Salt bridges are almost never used in cell recording as they create resistance. However, this does mean that metal ions may enter the bath, which occasionally needs to be considered.

---

**PRACTICAL TIPS 15.2:    WHY SILVER/SILVER CHLORIDE?**

A metal in one of its own salts is a nonpolarized electrode, meaning that it conducts current with very little change in potential. Ag represents a good example of such an electrode for biological applications, since Ag dissolves slowly, is nontoxic, and can be easily formed into electrodes of nearly any shape and readily coated with a chloride later. If both the ground and recording electrodes are made of the same material, the potential between them will be zero. The ion that is actually supporting the current in this case is the anion, $Cl^-$. Because physiological solutions contain a good deal of $Cl^-$, the junction potential of these electrodes is usually negligible. However, the importance of maintaining an AgCl coating can be readily appreciated. Some ions in electrophysiological solutions can destroy the electrode, particularly sulfide, which can convert it to $Ag_2S$. This will cause instability in the potential. If such ions are found in the bath, the electrode should be protected by using a salt bridge.

---

*Pipette holder.* Plastic pipette holders come in different styles and serve three functions: (1) they permit contact between the Ag/AgCl signal electrode and the headstage; (2) they hold the pipette tightly by means of one or two O-rings; and (3) they permit suction to be applied to the pipette for seal formation and breaking. A failure in any one of these functions means a failure of the experiment. Sometimes the signal electrode gets pulled free of the pin in the holder and needs to be soldered into place. O-rings often wear out (more often if pipettes are not fire-polished); this will cause motion of the pipette that can ruin a recording. Change O-rings at least monthly, and choose a model that fits tightly for sensitive experiments. Finally, the application of suction is a key component in electrophysiology. The suction port on the pipette holder is attached to the appropriately sized tubing and then usually to a glass syringe for applying suction. This sounds simple, but do not be fooled. The tubing should be nice and supple but not porous; I like silicone. If the tubing is too stiff, it will make controlling small motions of the pipette difficult and cause motion in the pipette tip when the tubing is moved. It is also important that this stretch of tubing not be too long or too soft, or it will not suck well and may cause the pipette to move. It is good to tape it down to the table every few inches. Finally, the best type of syringe is a heavy glass one, lubricated with a bit of vacuum grease so as to prevent air leakage around the plunger. I use a 30 mL glass syringe; smaller ones work too. Be careful not to break this syringe (putting it down too hard on a metal air table will do this). Many people apply suction with their mouths, using a 1 mL plastic pipette tip inserted into the tubing as a mouthpiece, and claim that the feedback is more sensitive this way. *Chacun à son goût*, but we do not recommend that you put your lips on anything in a lab filled with tetrodotoxin and recombinant viral vectors. If the rig is shared, you could also catch a cold this way.

*Perfusion/drug application.* Once the recording is in progress, the setup is delicate, so adding or removing solutions needs to be done gently to avoid disturbing the seal. The speed with which this needs to be done depends upon the agent applied and the experiment—for example, in an experiment examining drug kinetics, it is important to be able to add and remove drugs quickly. It is also sometimes important to apply drug only to one cell and not its neighbor. Many different systems for these applications are available, ranging from the simple to the highly sophisticated. Some electrophysiologists use *perfusion chambers* where cells on coverslips are exposed to a flow of solution and where the volume can be changed quickly. These can be stand-alone chambers, or inserts that fit into a culture dish. We like the ones designed for use with glass-bottomed culture dishes (**Figure 15.25**).

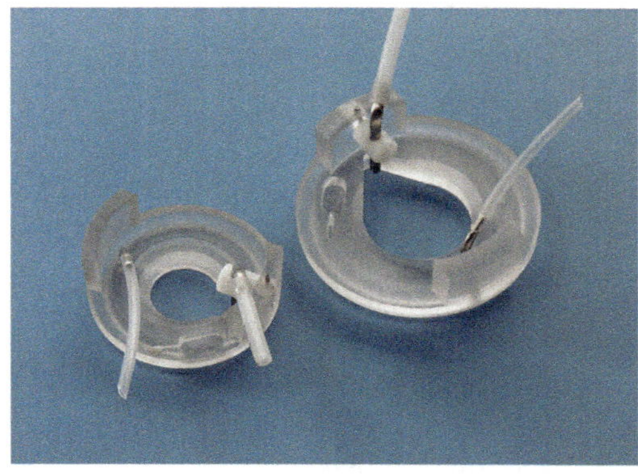

**Figure 15.25 Perfusion inserts for 35 mm culture dishes.** (Courtesy of Warner Instruments, http://www.warneronline.com).

For simple buffer exchange and application of ions, the system pictured in **Figure 15.26** works well. It is simply a series of syringes with stopcocks that can be loaded with different solutions that are then dripped onto the cells. The rate and volume of the application can be controlled by the diameter of the tubing used. To remove solution from the dish, a small nozzle can be attached to a peristaltic

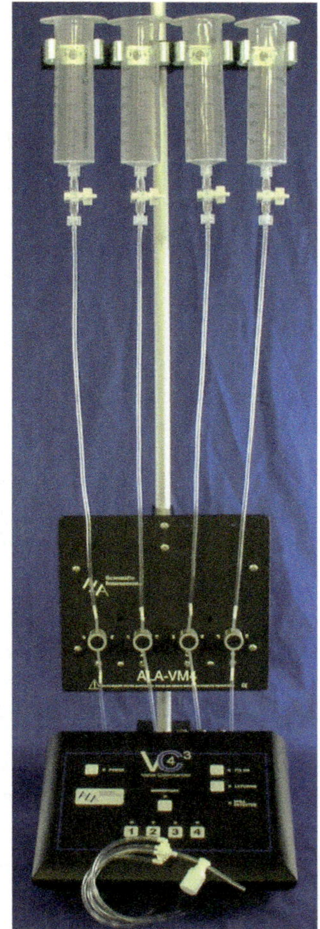

**Figure 15.26 Four-channel gravity perfusion system.** (Courtesy of ALA Scientific Instruments, model VC³-4xG.)

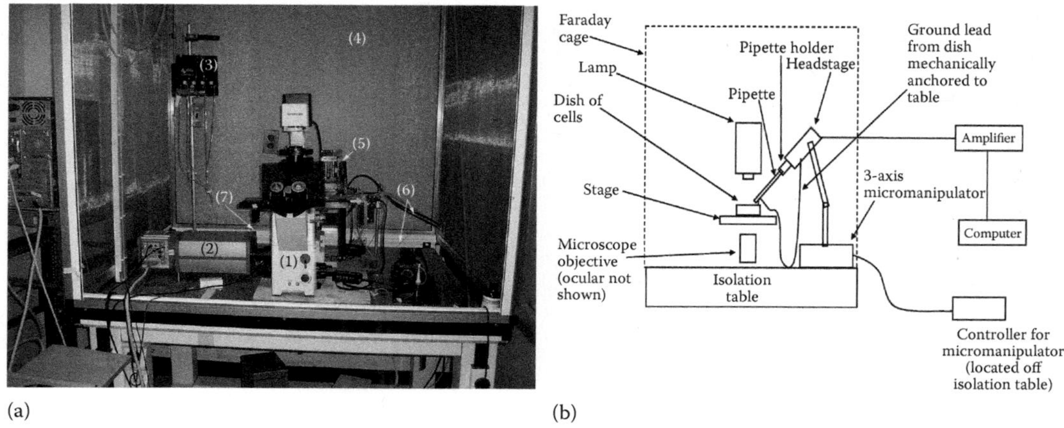



**Figure 15.27 Complete electrophysiology rig for mammalian cell work.** (a) Photo showing (1) microscope; (2) camera; (3) perfusion system; (4) Faraday cage (Cu mesh); (5) headstage; (6) wires to instruments, including tubing for peristaltic pump; and (7) tubing from headstage to syringe. The amplifier is at the lower left off-camera, and the micromanipulator is at the upper right, placed on a shelf at a convenient operator height. All power supplies are on the floor underneath the table. (b) Block diagram.

pump (the kind used in aquariums). Systems of valve control are available for automated regulation.

Perfusion chambers and systems also allow for temperature and gas (usually $CO_2$) control, which are needed in some types of experiments. There are some cellular processes and proteins that function significantly differently at 37°C than at room temperature, so heating is needed; this is also needed for long-term recordings to keep cells healthy. However, most electrophysiology is done at room temperature under ambient air.

A complete system, showing everything inside the Faraday cage, is shown in **Figure 15.27**. Note that the computer, amplifier, micromanipulator, and all power supplies are located outside of the cage.

# 15.7 THE ART AND MAGIC OF PIPETTE PULLING

## Pullers and glass

Patch pipettes are made by pulling on heated thin-walled glass until it stretches and breaks into a desirable shape. Traditionally, gravity was used to do the pulling, and the pipettes were pulled in a painstaking two-stage process involving much skill. Thankfully, horizontal pipette pullers have all but eliminated the old methods, and provide programmable and reproducible heating and pulling. The "desirable shape" part, though, is still up to you, and defining and obtaining it can still be very tricky. The idea is to minimize pipette resistance and capacitance while keeping the open end of a reasonable size relative to the size of the cell. In general, resistance is inversely proportional to the size of the pipette opening, though all bets are off if the pipette is a wonky shape. Patch pipettes should have a blunt shank and an opening of about 1 μm in size (**Figure 15.28**). Sharp electrode recording, in which the pipette tip is pushed through the cell membrane and into the cell, requires a longer and thinner pipette tip (which is easier to make).

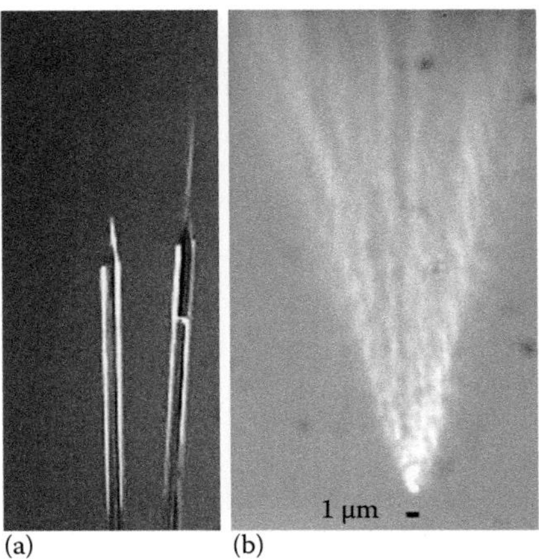



The first step is to buy (or borrow) a puller. Many electrophysiologists do not like to share their pullers, as the heating filament is delicate, the pipette shape is very dependent on consistent behavior of the filament, and all their fine-tuning can be for naught if someone bumps it. Determine what type of glass you will use to make pipettes before buying a puller. Some pullers only work with thin-walled borosilicate glass, which is the standard material for patch pipettes. However, thick-walled glass is desirable for some applications—it seals better and has lower capacitance (it is much harder to make good patch pipettes with it, however). If you want to use this type of glass, make sure that the puller has filaments large enough to accommodate it.

Pullers for all types of glass have platinum–iridium filaments that get up to orange hot, or about 1000°C. This is not sufficient to melt quartz, so if you want to use quartz electrodes, it is necessary to get a laser puller. The cost of the quartz and the puller should be weighed against the advantages of quartz. For high-resolution fluorescence imaging, quartz is highly superior, so it is often used for combined electrophysiology/calcium imaging. It is also less likely to break, so it may be used to patch onto cells that are inside a relatively tough tissue.

Nearly always, it is good to get the pipette glass labeled "with filament." The filament is a fine glass filament running down the channel of the glass that aids in backfilling the pipette with internal solution. Some people do not like filaments for perforated-patch recordings, but this is the only situation I can think of in which you would not want the filament. It prevents bubbles and makes life much simpler. Outer wall thickness is 1.5–2.0 mm; make sure this value matches the size of the pipette holder.

## Making patch pipettes

Once the puller and glass have arrived, read the user's manual and find a suggested protocol for making patch pipettes, which will vary from instrument to instrument. This will provide a good place to start. Pull a test pipette and see how it looks. If it looks roughly correct, and not horrifically long and thin or deformed, fill it with internal solution and place it into the holder and into an empty Petri dish of external solution. Turn on the amplifier and computer/oscilloscope and observe the trace in response to a 5 mV test pulse. A good pipette will show a nice square waveform with a resistance of 3–6 M$\Omega$ (pipettes for cell-attached patch

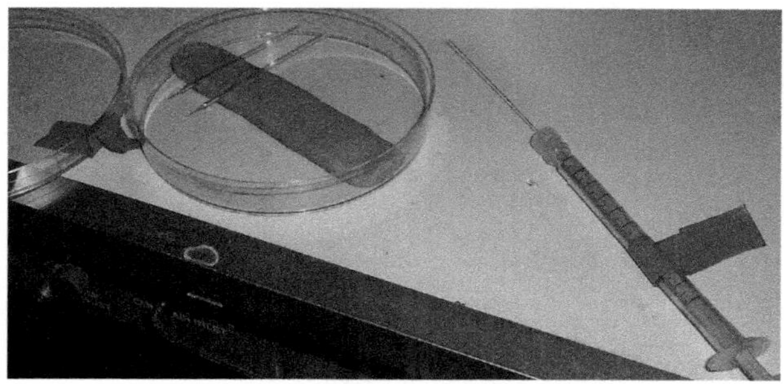

**Figure 15.29 Two pipettes are held stable in modeling clay in a culture dish with a lid (removed for visibility).** A syringe (right) with a long thin needle is used for backfilling the pipettes with internal solution.

recordings may be slightly smaller, with resistances of 6–10 M$\Omega$; pipettes for sharp-electrode recording 40–60 M$\Omega$). Artifacts in the waveform, such as a lack of squareness, can indicate that there is something wrong with the pipette. The lower the resistance, the more effectively $R_s$ can be compensated (note that $R_s$ is not just the pipette resistance but also the access resistance to the cell, which is a combination of the pipette properties and the cell properties, and always larger than the pipette resistance alone). However desirable low-resistance pipettes may be, at some point, the pipette will be too large for the cell to handle and will kill it. Pipettes < 1 M$\Omega$ in resistance will basically never work.

Usually the resistance is too low or too high for your needs at first. The trick is to adjust the resistance without deforming the shank or making it too long. This will require careful iteration of the parameters on the "pull" program—heat temperature and duration, pull strength, and number of cycles before the glass breaks. There are no good general rules for how the shape and resistance will change with these parameters; for example, increasing heat can either increase or decrease pipette resistance. Beware of pipettes that look beautiful on the oscilloscope but that are long, thin, and ugly. Always examine them by eye before bothering to fill them.

It can take several hours or more to calibrate the puller the first time. Each time the heat filament is changed, this will have to be repeated. This is why the filament should be treated with great care. Always place the glass into one side of the puller, clamp it in, and then ease it through the filament so that there is no chance of bumping. Although this may seem obvious, do not let untrained people touch your puller. Placing it in a secured room or office is recommended.

Pipettes pick up dirt and volatiles from the environment quickly, which will inhibit formation of gigaseals. I recommend that they be placed in a covered dish as in **Figure 15.29**, and used within 1 h of pulling—at the very least the same day. Any unused pipettes should be discarded into a sharps waste container (an old plastic kitty-litter container works well).

## Sylgard

Most stray pipette capacitance comes from the thinner region of the pipette that goes into the bath solution. This is because the wall is thin and also because bath solution can creep up it. Both of these can be ameliorated by coating the pipette with a thick layer of an inert hydrophobic substance. There is more than one option for this, but Sylgard (Dow Corning 184) is commonly used. The coating should run from as close to the tip is as feasible (0.5 mm or less) and continue up past the

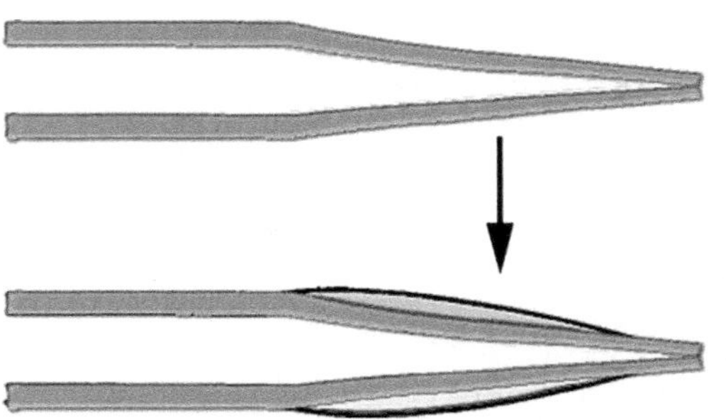

**Figure 15.30 Application of Sylgard should coat the narrow shank of the pipette far enough up so that electrolyte solution does not climb past it, and as far down as possible without blocking the tip.**

taper. It is performed by mixing the Sylgard 10:1 with a curing agent and coating the pipettes using a needle or plastic tool; most people prefer to do this under a dissecting microscope. They are held under a heat gun or into a heated wire coil for several seconds to cure. Make sure that the Sylgard does not block the tip (**Figure 15.30**). For many routine applications such as whole-cell recording, Sylgarding is unnecessary, and a significant fraction of electrophysiologists never do it. It becomes important for very small single-channel events in the low pA range.

Many people also fire-polish their pipette rear opening to prevent scratching of the AgCl electrode coating by a ragged pipette open end. Some fire-polish the tips to improve gigaseal formation, which is significantly trickier. We have never done either of these. Many of the references at the end of the chapter describe the process if you find it desirable.

## Recording artifacts caused by pipette materials

Any ions released into the bath solution or cell may affect the behavior of ion channels. Some types of glass release $Ba^{2+}$ ions, which we will see in the next section can block certain $K^+$ channels with great affinity. Others release heavy metals, which can have a variety of effects on channels. Needles used to fill pipettes may also leach ions. The way to check if this is happening is to add a heavy-metal chelator such as EGTA or EDTA to the internal solution and see whether the current–voltage curve is affected. If so, change glass and/or needles used for filling.

# 15.8  STEP-BY-STEP GUIDE TO PERFORMING A WHOLE-CELL RECORDING

The following steps should be performed in order:

- Pull 8–10 pipettes or as many as your holder will accommodate.

- Prepare external and internal solutions; warm to room temperature or 37°C as appropriate. Solutions may be prepared well in advance and refrigerated. In this case, aliquot out enough solution to use for the upcoming session and allow it to come to temperature while you make other preparations for recording.

- Turn on the amplifier, micromanipulator, computer, oscilloscope, fluorescence lamp, and camera if needed.

- Remove cells from the incubator; rinse two to three times with external solution and fill dish shallowly with external solution (about 1 mL for a 35 mm dish). Filling shallowly helps with resistance and with observation.

- Fill a pipette with internal solution and mount it into the holder. To get the tip to fill properly, it helps to dip the tip briefly into a vial of internal solution first and let it fill by capillary flow before backfilling the pipette with solution using a syringe. Manipulate it into the dish and use a test pulse to check its resistance; it helps to apply a bit of positive pressure to the syringe to keep the electrode clean while dipping.

- Null the junction potential if you wish. Otherwise, this can be done off-line.

- Find the pipette under the appropriate magnification.

- Choose a cell by moving the dish around under the visualized pipette. If cells contain a fluorescent protein, switch back and forth between fluorescence and brightfield. It takes some practice to identify "good" cells of any given cell line or type—there are no general rules, except that clearly broken or floating cells are dead and should not be patched.

- Position the pipette over the chosen cell and set the micromanipulator to the most sensitive setting. Carefully move down toward the cell while keeping an eye on the test pulse (make sure you do not miss the cell). When the pipette just touches the cell surface, the resistance will increase by a few megaohms, usually to ~10–15 M$\Omega$ (**Figure 15.31a and b**). Sometimes the touching can be observed as a dimpling under phase contrast, but quite often it cannot.

- Pull gently on the syringe. Sometimes a gigaseal forms almost immediately, observed as a disappearance of the test pulse current (current should be < 1 pA; **Figure 15.31c**). Sometimes a few episodes of pulling and waiting are needed. The resistance should increase rapidly once it reaches the 100 M$\Omega$ range—if it does not, something is probably

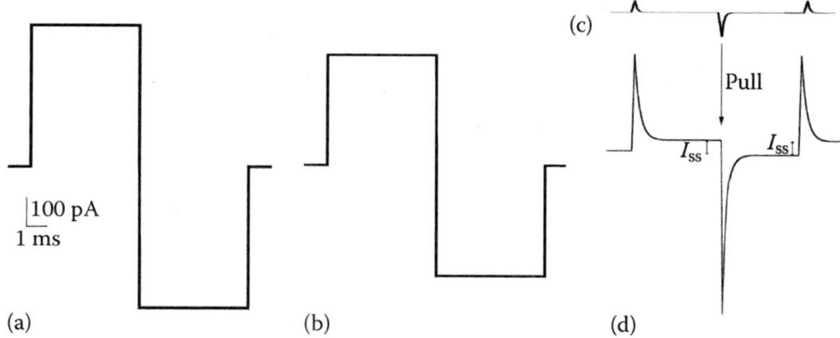



**Figure 15.31 Electrical signals seen during establishment of whole-cell recording mode.** (a) Open-pipette current for a 5 M$\Omega$ pipette. If this rises before the cell is touched, this is probably due to debris and is bad. (b) Touching the cell gently leads to a variable degree of increase in resistance. (c) After this occurs, suction is applied to form the gigaseal. $C_{fast}$ should be compensated at this point. (d) After compensation and stabilization of the patch—perhaps with application of negative potential so as not to shock the cell upon break-in—a few quick tugs leads to the appearance of a transient whose shape depends on the size and leakiness of the cell. Notice that the steady-state current ($I_{ss}$) here is *not* zero (compare with the case of the lipid bilayer); this corresponds to open channels at the holding potential used.

wrong. Hyperpolarizing the cell membrane can help in sealing. Getting the suction right to form a gigaseal can take a good deal of finesse and practice. Be extremely patient while learning to record.

- If the seal fails to form, remove the pipette, replace it with another, and try again with a new cell. Using fresh, clean pipettes is critical to forming gigaseals—do not ever be tempted to use the same pipette on more than one attempt to contact a cell.

- If you fail 10 times in a row, the cells are either too sick, there is something wrong with your technique, there is something wrong with the solutions, or there is something wrong with the hardware (e.g., vibrations, pipette drift, etc.). Figuring out which of these problems is the most acute can be frustrating. If there are other dishes of cells available, try a different dish, or a dish of untransfected cells. Alternatively, if something looks wrong with the rig, try to fix it. Look into the microscope while pulling on the syringe to make sure the pipette is not moving and that there are no obvious vibrations. Setting the tip and a cell in focus in the microscope and then observing quietly for 10–15 min will let you see if the pipette or stage is drifting.

- When the miracle happens and the seal forms, let it rest and stabilize for a few minutes and null the pipette capacitance using the $C_{\text{fast}}$ adjustment.

- To break in, give a couple of brief, sharp tugs on the syringe. When the membrane breaks and you get whole-cell access, a capacitive transient will appear along with a steady-state current that indicates the degree of "leakiness" of the cell (**Figure 15.31d**). All cells show some leak current, but if this is on the order of the size of the capacitive transient, the cell is sick and/or dead.

- Note the cell membrane potential. If it is too depolarized, the cell is probably too sick to work with, even if it did seal well. Acceptable depolarization will vary with cell type and line.

- Adjust the $R_s$ and $C_{\text{slow}}$ compensation to null the transient as much as possible without causing ringing; note their values. A measure of the cell capacitance will probably be used in data analysis to report currents as pA/pF. If $R_s$ is too high, the pipette is too small or the cell is "sticky" and ill. Usually $R_s$ is about twice the value of the resistance of the pipette in the bath.

- Proceed to voltage-clamp protocols. Note capacitance and $R_s$ values at the end; if series resistance climbs during a recording, break-in can be attempted again or the cell discarded. If capacitance decreases, the cell is shrinking and something is wrong.

## 15.9  EXAMPLE EXPERIMENT: WHOLE-CELL RECORDING ON CELLS TRANSFECTED WITH K⁺ CHANNELS AND GFP

Someone has given you a plasmid that he claims encodes the potassium channel *ROMK1*. You want to find out whether this gene expresses in cultured mammalian cells, and whether it makes a channel consistent with ROMK1 using electrophysiology.

The first thing to do is to find out how you will distinguish ROMK1 from background $K^+$ conductances found in most cells. With a bit of research, you find out that ROMK1 is a kidney *inward rectifier* potassium channel—that is, it should conduct more current at negative vs. positive potentials. This should give it a distinctive current-vs.-voltage profile. In addition, it has certain blockers. $Cs^+$ and $Ba^{2+}$ ions will both block ROMK1, as will the snake toxin δ-dendrotoxin. When you see the price of δ-dendrotoxin, you decide to go with $Ba^{2+}$.

The next step is to choose a cell line. Any cell line without great expression of inward rectifiers will do. HEK293 cells are easy to transfect and contain $K^+$ channels but none with the profile of ROMK1. Plate the cells on the appropriate dish or coverslip to use with your microscope and electrophysiology rig. The next day, transfect them with a 1:1 ratio of the ROMK1 plasmid and a visible marker, such as GFP. (You might be tempted to clone the GFP right on to the ROMK1 gene—there are a couple of reasons not to do that at this stage. First, it is more work. Second, the GFP might drastically alter the ion channel behavior.)

While the cells are transfecting, figure out the internal and external recording solutions needed. A general $K^+$ recipe is as follows:

- Internal solution (in mM): 130 KCl, 0.8MgCl₂, 5 EGTA, 5 MgATP, 10 HEPES (pH to 7.2 with KOH)

- External solution: (in mM) 137 NaCl, 10 HEPES, 4.0 KCl, 1.8 CaCl₂, 1.0 MgCl₂, 10 glucose, adjusted to pH 7.4 with NaOH

Adjust the osmolarity of the solutions so that they match. What is the junction potential?

When the cells are ready, examine them under epifluorescence and brightfield. Make sure that there are a significant number of healthy-looking green cells. If all the green cells are dead, the channel might be killing them—especially if you have done a GFP-only control and those cells look fine. If this is the case, consider doing the transfection in the presence of a channel blocker such as $Ba^{2+}$. Alternatively, if you think the cells might be dying from $K^+$ efflux, raise the $K^+$ concentration of the medium.

If and when there are healthy green cells, prepare them for electrophysiology by washing them and replacing the medium with external solution. Pull pipettes and have internal solution ready. The identification of the channel must be done statistically—the presence of a $Ba^{2+}$-blockable current with an I–V curve consistent with ROMK1 must be seen in green cells, but not in non-green cells from the same dish or from GFP-only transfected cells in a control dish. (There is always a risk of having very faint green fluorescence that is not seen in a transfected cell, so one "freak" result from the transfected dish does not ruin the experiment.) Note that the GFP is not linked to the channel, so the distribution of GFP in the cell does not reflect where the channels are going. That is too bad, because if the channels simply do not make it to the membrane for whatever reason, we have no way of knowing this.

Once the whole-cell mode has been established with a cell, step through a voltage protocol appropriate for ROMK1 in both voltage and speed. Check the literature for the parameters needed to see the features of the I–V curve. A protocol

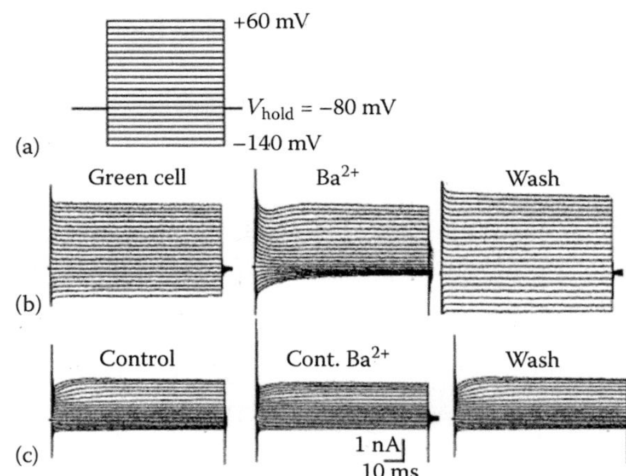



**Figure 15.32 Observing ROMK1 in transfected cells.** (a) Voltage step protocol: holding at –80 mV, stepping from –140 to + 60 mV in 10 mV increments, pulse durations of 100 ms, acquisition at 10 kHz. (b) Sample trace from a green cell, showing K$^+$ currents at all voltages that show preferential block at negative voltages by Ba$^{2+}$. Washout of the ion removes the block. (c) Control cell shows much less overall current at negative voltages, with minimal block by Ba$^{2+}$.

spanning –140 to +60 mV with a few hundred milliseconds duration is a good choice (**Figure 15.32a**). Choose the acquisition bandwidth. Although ROMK1 is not a fast channel, monitoring cell capacitance changes is difficult if bandwidth is too low; 10–20 kHz is appropriate.

Once a full recording of all of the voltage steps has been made, apply the Ba$^{2+}$ either with a pipette or a perfusion system. A good way to do this is to fill one syringe of a perfusion system with a normal recording medium and another with a medium containing 100 mM BaCl$_2$. Wash in the latter to apply the blockage, then wash the Ba$^{2+}$ away with the ordinary medium and repeat the voltage-step protocol. The block should be reversible (**Figure 15.32b**). Control cells should show less K$^+$ conductance at negative voltages and little or no effect of Ba$^{2+}$ (**Figure 15.32c**).

Repeat with a statistically significant number of transfected and control cells. This number depends on the variability observed. If the cells are all essentially the same, four to five cells may be sufficient. If the amount of current seen is widely variable, 20 or more cells may be needed. It may take weeks, and many dishes of cells, to get sufficient data. Patience is a virtue for the electrophysiologist.

Plot the current vs. voltage relationship for all the cells and examine the variability. If a mean of all of the cells makes sense, the curve can be presented as

**Figure 15.33 Mean I–V curves made from multiple ROMK1-transfected and control cells.** Curves are corrected for junction potential. Note that the control cells have some current, so the difference between the ROMK1 cells and control cells gives the correct ROMK1 curve.

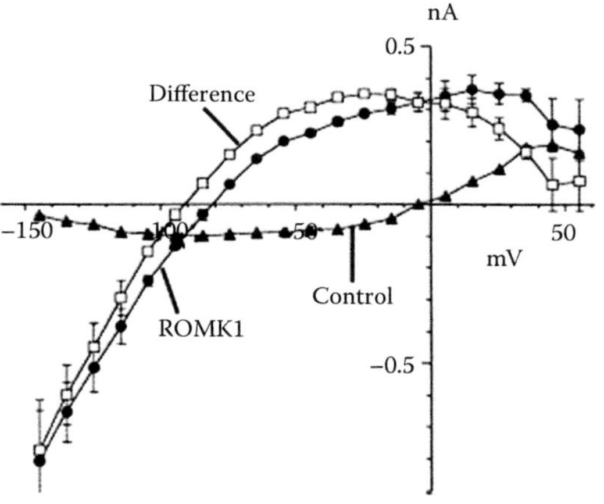

the mean. It is common to report a mean current as pA/pF, which normalizes for cell size. However, in cases where all the cells are roughly the same size, this is not necessary. Compare your I–V curve with that in the literature for ROMK1; it may make sense to subtract the current in the control cells (**Figure 15.33**).

You can now be pretty confident you have seen ROMK1. If further confirmation is desired, more specific toxins could be purchased, or immunofluorescence performed if there is an antibody available (there is). Other electrophysiological measurements may also be done, such as comparisons of the permeability ratios of selected ions, such as $Rb^+/K^+$.

## 15.10  BRIEF INTRODUCTION TO SINGLE-CHANNEL MODELING AND DATA ANALYSIS

### Why do single-channel measurements?

Single-channel measurements can give physical insight into channel properties that cannot be obtained from whole-cell recordings giving average current through hundreds to thousands of channels. The single-channel current is obtained, and the different conductance states can be observed, both of which are important for modeling ion permeation through different types of channels. The *dwell times* of the channel in its different conductance states can be measured and used to infer opening and closing mechanisms (electrophysiology is much too slow to resolve actual opening and closing events: its best resolution is microseconds, whereas protein motion occurs on a picosecond timescale). When drugs are applied, the degree of channel block and the association and dissociation constant of the drug can be observed. All of these methods are statistical, requiring the analysis of hundreds to thousands of events. Whether done in a bilayer system or on cells, one of the key challenges of single-channel recording is keeping the system stable long enough to get sufficient data.

### Analyzing data

Single-channel recordings look like a series of discrete, square current jumps (refer again to **Figures 15.22 and 15.23**). The simplest way to characterize single-channel data is to set a threshold for open vs. closed, usually chosen as $(I_{open} - I_{closed})/2$. A histogram of open vs. closed times is then prepared with all currents above the threshold counted as "open" and those below the threshold as "closed." This method requires a good signal-to-noise ratio, or else false events will occur too frequently. The average frequency of false events $f$ is a Gaussian function of the standard distribution $\sigma_n$ of background noise and the threshold $\phi$:

$$f \sim f_c \exp\left( -\frac{\phi^2}{2\sigma_n^2} \right) \tag{15.7}$$

where $f_c$ is the filter corner frequency. The number of false events that can be tolerated per second of course depends upon the number of real events in the same time period. A channel with infrequent openings imposes greater stringency on the experimenter.

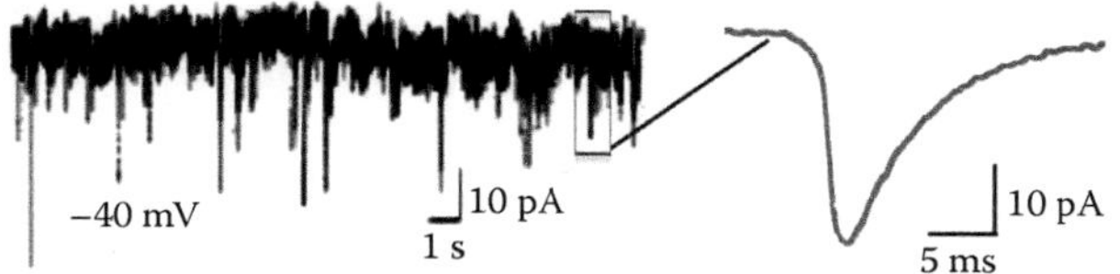

**Figure 15.34 Excitatory postsynaptic currents (EPSCs) are small events that are usually fit with a template, as they have a typical shape as indicated in the expanded view of one event.**

There are other more sophisticated methods of fitting single-channel currents when the threshold method breaks down. One of the most common types of correction needed is to adjust for a drifting baseline. This is easily done by baseline subtraction followed by threshold analysis. Other types of analysis are also possible, for instance, in cases of events that do not reach full amplitude, perhaps reflecting a different channel configuration. In these cases, an event "template" can be defined manually and scrolled over the data to determine whether each blip represents a real event or noise. This technique is usually the one used to characterize postsynaptic currents or potentials, as these small events have a characteristic shape that is not seen in noise (**Figure 15.34**).

Amplitudes and dwell times are usually presented as histograms. **Figure 15.35** shows a histogram of current distributions in alamethicin, which clearly resolves the characteristic open states of the channel (many channels are not nearly so clear-cut.)

A Gaussian or multiple Gaussians are usually fit to these data, describing average currents $I$ with variances $\sigma$ and relative areas of open and closed populations $a$:

$$n = \sum_{i=1}^{n} \frac{a_i}{\sqrt{2\pi\sigma_i^2}} \exp\left[ -\frac{(I_i - I_{0i})^2}{2\sigma_i^2} \right] \tag{15.8}$$

There are certain types of binning errors that need to be considered and avoided when presenting data as histograms that we will not go into here, but which are discussed in the single-channel analysis programs available as well as in textbooks cited at the end of the chapter.

**Figure 15.35 Amplitude histograms of alamethicin, showing the highly characteristic subconductance states of the channel.**

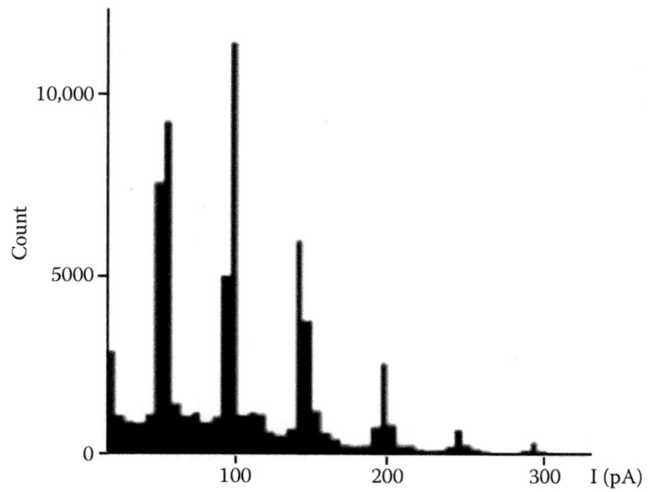

There are many subtleties in the dwell times, some or all of which may be interesting in a given experiment. Dwell times may be analyzed all together, or separated according to amplitude. They may also be examined as a function of time to see if prolonged recording affects the channel. The appearance of *bursts* (**Figure 15.36**) should also be analyzed as to length of burst and length of time between bursts (see section below for importance of bursts).

Time constants can be extracted from dwell-time histograms by fitting to a probability density function *f(t)* either to the histogram or the events directly:

$$f(t) = \sum_{i=1}^{n} \frac{a_i}{\tau_i} \exp\left(-\frac{t}{\tau_i}\right),$$

(15.9)

where $a_i$ are the relative areas whose sum is unity. The mean open time is then given by

$$\sum_i a_i \tau_i.$$

When there is only one time constant, plotting the histogram on a linear time axis is fine (**Figure 15.37**). However, it is easier to appreciate the presence of multiple exponentials when the histograms are plotted on a log-log scale.

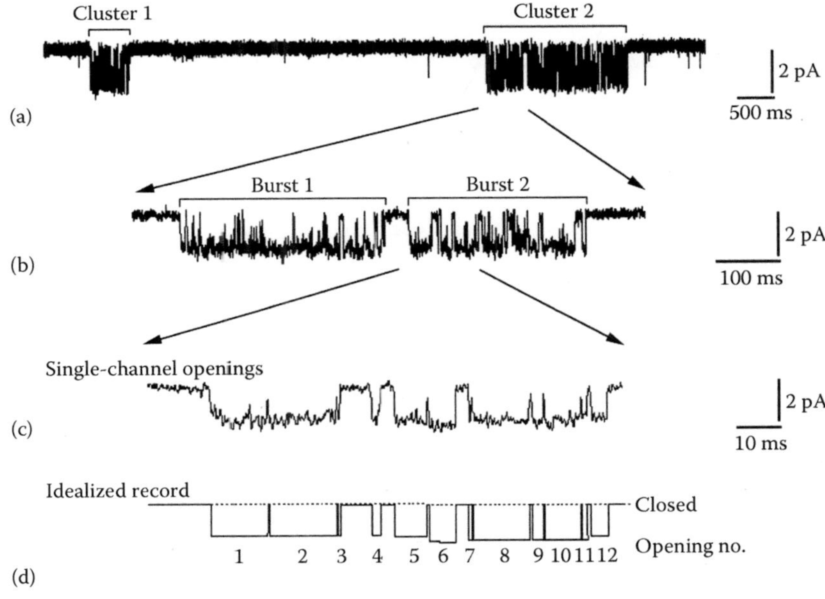


**Figure 15.36 Example of an outside-out patch recording of channel activity for GABA$_A$ receptors expressed in HEK cells in the presence of 1 mM GABA and held at -70 mV, filtered at 3 kHz and sampled at 20 kHz.** Channel openings with a main amplitude of 1.9 pA can be seen as downward deflections. (a) Sample 9 s recording, where two distinct clusters are observed and separated by a long desensitized shut period, interrupted only by a single independent opening. A section of the second cluster is expanded in (b) to show how the cluster is composed of bursts of channel activity. (c) Further time expansion of a section of the second burst illustrates individual channel openings of different duration. (d) Idealized representation after having fit the original channel current with the idealization program, SCAN. (Reprinted by permission from Macmillan Publishers Ltd. *Nature Protocols*, Mortensen, M., and Smart, T.G., Single-channel recording of ligand-gated ion channels, 2: 2826–2841, copyright 2007.)

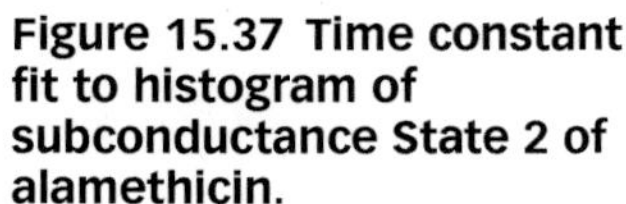

**Figure 15.37 Time constant fit to histogram of subconductance State 2 of alamethicin.**

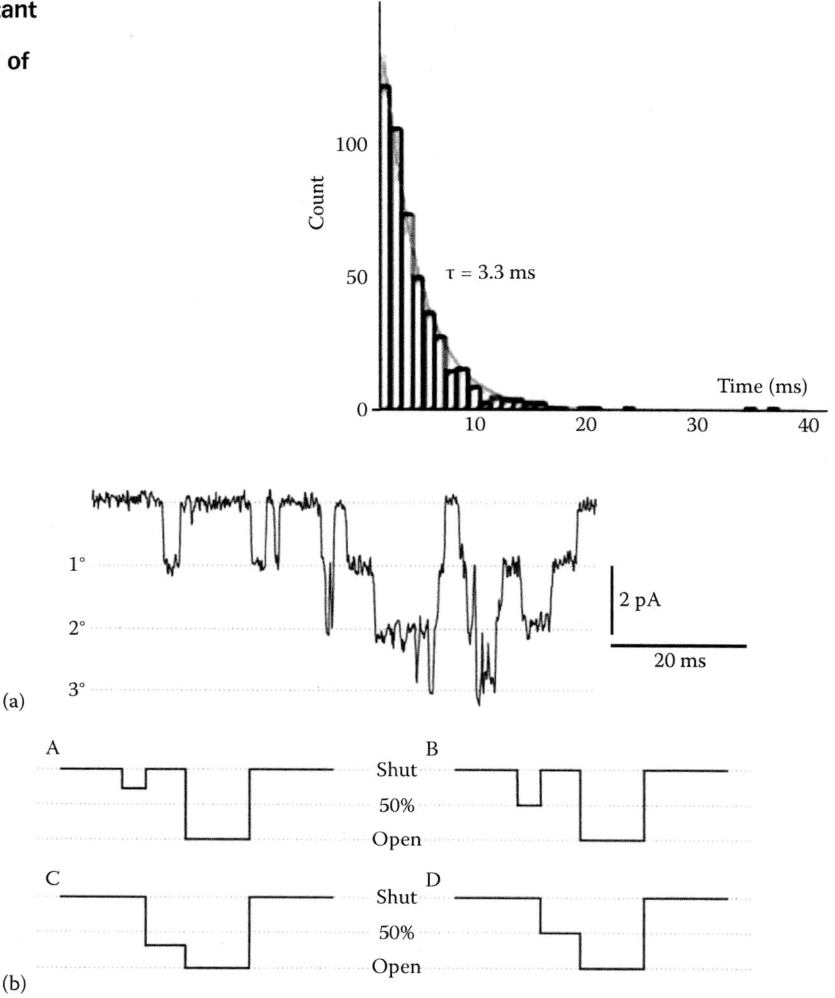



**Figure 15.38** (a) Outside-out patch recording taken from HEK293 cells expressing GABA$_A$ receptors. The currents were activated by 300 µM GABA and display channel stacking ($V_h$: −70 mV, filtered at 3 kHz and sampled at 30 kHz). The main current amplitude level for this receptor is about 1.8–1.9 pS (marked 1°), and the current stacks are multiplications of the main amplitude. Two or three channels being open in the same time window are marked as 2° or 3°, respectively. (b) Schematic single-channel openings displaying low conductance levels that are less than 50% (A) or equal to 50% of the main conductance level (B); or subconductance levels that differ from 50% (C) or are equal to 50% of the main conductance level (D). Low or subconductance levels that are exactly 50% of the main conductance level are probably rarely encountered, but these are shown here as possible scenarios that may need to be addressed when having to distinguish channel stacking from low or subconductance levels. (Reprinted by permission from Macmillan Publishers Ltd. *Nature Protocols,* Mortensen, M., and Smart, T.G., Single-channel recording of ligand-gated ion channels, 2: 2826–2841, copyright 2007.)

Sometimes two or even three channels are found in a bilayer or patch, and it can be nontrivial to tell the difference between one channel with multiple subconductance states and two or more independently opening and closing channels. If a current is exactly twice that of the main current, it is likely to be two channels, whereas if it is not exactly twice, it is likely to be a different state of one channel. Inflections upon opening and closing also give hints as to the presence of multiple channels (**Figure 15.38**).

### Interpreting single-channel data

We do not yet have good models for how channels "open" and "close." Most likely, these processes represent subtle rearrangements of the channel structure rather than large, discrete conformational changes. Until the day when we can write a first-principles equation for conduction, however, we can use discrete-state Markov models to provide a mathematical framework for describing single-channel function. A Markov model postulates specific rate constants $k$ for stochastic, memoryless transition between open (O) and closed (C) states

$$C \underset{k2}{\overset{k1}{\rightleftharpoons}} O. \tag{15.10}$$

This simplest possible model does not account for observed features such as bursting. These begin to appear with a three-state model involving a closed state, open state, and inactivated (I) state:

$$C \underset{k2}{\overset{k1}{\rightleftharpoons}} O \underset{k4}{\overset{k3}{\rightleftharpoons}} I. \tag{15.11}$$

Many attempts have been made to describe complex channels such as the voltage-gated $Na^+$ channel using Markov models. One of the simplest postulates three different closed states, an open state, and an inactivated state. Unfortunately, it is impossible to exclude the existence of other states by fitting the data. Some references are given at the end of the chapter to the use of Markov models in ion channel modeling, as well as some alternatives such as chaos theory.

## 15.11  NETWORKS

Understanding the properties of even the simplest nervous systems requires the ability to understand how networks form, grow stronger and weaker, adapt, and learn. In order to do this, recordings from hundreds or more of excitable cells are required. The ability to do this easily remains one of the holy grails of neuroscience. The Advanced Topics of this chapter introduce just a few of the available techniques and how they are being developed. **Advanced Topic 15.1** discusses multielectrode recording, **Advanced Topic 15.2** discusses optical recording, and **Advanced Topic 15.3** discusses a few more emerging techniques for neuronal recording and simulation.

## 15.12  CONCLUSIONS AND FINAL REMARKS

Patch-clamp recordings are possible to do in any lab that can afford an amplifier, though fine-tuning the system can take a good deal of patience and mastery of some specific skills. If you are starting from scratch, it is always recommended to visit an electrophysiology laboratory to get an idea of how things are set up and why. It is also highly important to know and understand the instrumentation so that the right instruments can be selected and used properly for each application. An amplifier with the wrong noise profile, a micromanipulator that drifts, and a puller that does not handle your pipette glass are all problems that can be avoided.

### ADVANCED TOPIC 15.1:   MULTIELECTRODE ARRAYS

Arrays of four up to thousands of electrodes interfaced with neuronal networks *in vitro* are used to model cellular interactions and to screen neuroactive drugs. The arrays are called *multielectrode arrays* (MEAs) and may be used with sparse cultures, dense cultures, or brain slices for periods of weeks to months or even years (Figure A15.1.1a and b). Ideally, MEAs are able to both stimulate and record, using electrodes that range from millimeters to tens of microns in size. Two major problems hinder MEA recordings: low signal to noise caused by large impedance electrodes; and (even worse) the complex and changing biological–electrode interface seen in neuronal cultures. Glial cells (mostly astrocytes) often "shield" neurons from the electrodes and prevent recording. In addition, the large voltages used for stimulation usually saturate the amplifiers, causing a recording "black out" period for at least several hundred milliseconds. Despite these drawbacks, MEAs are used in many different types of studies, from quantitative network models to *in vitro* studies of phenomena such as epilepsy that can be quantified by observing firing rates across the culture (Figure A15.1.1c and d).

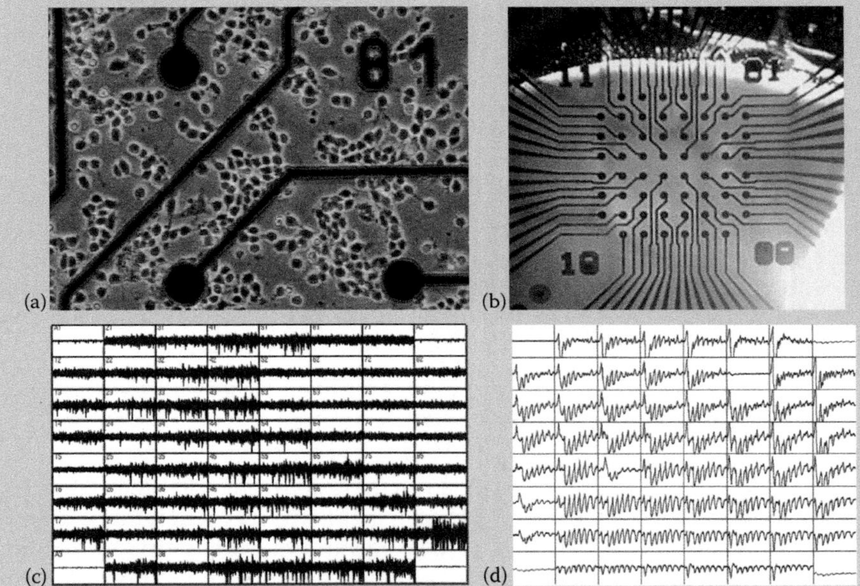

**Figure A15.1.1 EA cultures and recordings.** (a) Dissociated neurons on an MEA. (b) Slice on an MEA. (c) Recording from each of the electrodes in the dissociated culture, showing spikes over the background noise (about 10 mV background). (d) Data from the slice. The slice MEA has protruding electrodes to pierce the dead layer on the slice. (All images courtesy of T. de Marse, University of Florida.)

### SUGGESTED READING

Boehler, M.D., Leondopulos, S.S., Wheeler, B.C., and Brewer, G.J. (2012). Hippocampal networks on reliable patterned substrates. *Journal of Neuroscience Methods* 203, 344–353. http://www.ncbi.nlm.nih.gov/pubmed/21985763

Brewer, G.J., Boehler, M.D., Jones, T.T., and Wheeler, B.C. (2008). NbActiv4 medium improvement to Neurobasal/B27 increases neuron synapse densities and network spike rates on multielectrode arrays. *Journal of Neuroscience Methods* 170, 181–187. http://www.ncbi.nlm.nih.gov/pubmed/18308400

Hales, C.M., Rolston, J.D., and Potter, S.M. (2010). How to culture, record and stimulate neuronal networks on microelectrode arrays (MEAs). *Journal of Visualized Experiments: JoVE* 10.3791/2056. http://www.ncbi.nlm.nih.gov/pubmed/20517199

Killian, N.J., Vernekar, V.N., Potter, S.M., and Vukasinovic, J. (2016). A device for long-term perfusion, imaging, and electrical interfacing of brain tissue in vitro. *Frontiers in Neuroscience* 10, 135. http://www.ncbi.nlm.nih.gov/pubmed/27065793

Newman, J.P., Zeller-Townson, R., Fong, M.F., Arcot Desai, S., Gross, R.E., and Potter, S.M. (2012). Closed-loop, multichannel experimentation using the open-source NeuroRighter electrophysiology platform. *Frontiers in Neural Circuits* 6, 98. http://www.ncbi.nlm.nih.gov/pubmed/23346047

## ADVANCED TOPIC 15.2:   OPTICAL RECORDING

Optical recording uses reporter dyes rather than electrodes to record cellular signals. Reporter dyes are usually but not always fluorescent, and the signals (when speaking about excitable cells) are usually action potentials. The main advantage of optical recording is that it allows for recording from large numbers of cells simultaneously, most notably from networks of neurons. The problems of glial–cell insulation seen with MEAs do not occur with optical recording, since electrodes are replaced by dye molecules placed directly into the cell membrane (voltage indicators) or the cytosol (calcium indicators) of the cells of interest. However, there are all sorts of other issues that make this technique difficult to master. Until recently, only a few laboratories worldwide have been able to generate high-quality optical recordings of voltage signals from neuronal networks. With the development of genetically encoded voltage indicators (GEVIs, see below), voltage imaging in living animals became feasible and practicable.

In the past, most optical voltage recording has been done using voltage-sensitive dyes (VSDs that are low molecular weight organic molecules). The most useful VSDs are the amino-naphthyl-ethenyl-pyridinium (ANEP) dyes, such as the di-4-ANEPPS and di-8-ANEPPS, as well as with dialkylaminophenylpolyenyl-pyridinium dyes also known as RH dyes, after their inventor, Rina Hildesheim. These dyes are hydrophobic, being essentially nonfluorescent in aqueous solution, beginning to fluoresce after embedding within a cell's plasma membrane. Their absorbance and emission spectra are broad, with a large Stokes shift (**Figure A15.2.1**). Once in the cell membrane, they demonstrate an ~10% change in fluorescence intensity per action potential. While they respond almost instantaneously (within microseconds) to a change in membrane potential, the voltage-dependent modulation of their fluorescence intensity is too small (and often too fast) to be directly visualized on an absolute scale. Instead, signals are recorded with devices such as fast CCD or CMOS cameras and digitally processed to display changes of fluorescence relative to baseline fluorescence. In addition, the ANEP dyes are significantly phototoxic and usually lead to cell death within 1–2 h. These and other limitations of classical VSDs motivated the development of GEVIs discussed below.

## GENETICALLY ENCODED VOLTAGE INDICATORS: OPTOGENETIC MONITORING OF MEMBRANE POTENTIALS

Staining of all lipid membranes accessible to the dye solution is the cause of the most serious limitations of classical voltage-sensitive imaging. Nonselective staining causes a large amount of optical noise (see below), and in the case of complex systems, such as mammalian brain tissue, optical signals cannot be easily allocated to individual cells and their intermingled processes. However, as we have seen in the chapters on molecular cloning discussing specific promoters, cellular identity can be established by the genetic

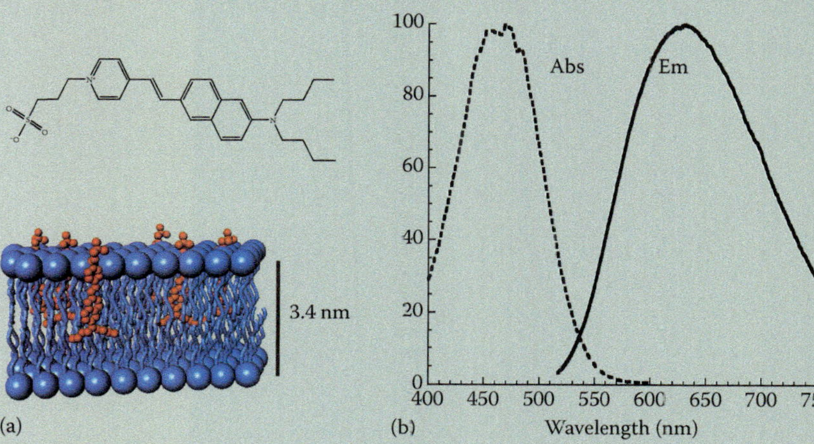



**Figure A15.2.1  i-4-ANEPPS.** (a) Structure of the molecule and model of its insertion into the membrane. (b) Absorbance (Abs) and emission (Em) spectra in lipid.

(*Continued*)

## ADVANCED TOPIC 15.2 (CONTINUED):    OPTICAL RECORDING

fingerprint of expressed genes. For this reason, genes and proteins are important keywords in optical electrophysiology, in particular since the discovery of fluorescent proteins (FPs). Over the last decade, the fusion of optical methods with genetic methods has led to the emerging field of *optogenetics*. In its broadest definition, optogenetics is the use of genetically encoded actuator and reporter proteins that allow the use of light to either control or report on the activity of molecular processes in specific (genetically defined) cell populations.

In this section, we will consider only the optogenetic control and measurement of membrane potentials, where electrodes are replaced by proteins that act as molecular probes. As in a classical electrophysiological current clamp experiment, experimenters like to record membrane potential while having the possibility to control it by injection of current through the membrane. This is achieved with two types of proteins. The first type comprises biosensors engineered from the fusion of FP reporters to voltage sensor proteins, which convert voltage signals into changes in fluorescence output. The second type of protein is derived from light-activated channels and pumps.

Practically all fluorescent protein-based voltage probes follow a common design principle, involving the molecular fusion of a protein that senses voltage with an FP-based reporter. The best studied voltage-sensing membrane proteins are voltage-gated ion channels, so they were the obvious choice as sensor proteins for the first generation of FP-based voltage probes. Two early prototypic voltage sensors (Flash and SPARC) were based on the molecular insertion of an FP into a potassium channel subunit or a sodium channel with the idea that voltage-dependent changes in protein configuration would affect the optical properties of the FP (**Figure A15.2.2a**). A mechanistically more rational concept for voltage-sensitive fluorescent proteins (VSFPs) was based on the fusion of an isolated voltage sensor domain to a pair of FPs that report conformational changes as a change in Förster resonance energy transfer (FRET) efficacy (**Figure A15.2.2b**).

Although the very first-generation GEVIs provided a proof of principle, their application in mammalian systems was limited by poor targeting to the plasma membrane in transfected cells. This issue was solved by using the voltage-sensing domain of the self-contained non-ion channel protein *Ciona intestinalis* voltage sensor-containing phosphatase (ci-VSP; **Figure A15.2.3**).

The resulting (FRET)-based VSFP2 sensors displayed significantly improved targeting to the cell surface and reliable responsiveness to membrane potential signaling from targeted neurons in culture, acute brain slices, and living mice (**Figure A15.2.4**). Like most FRET-based GEVIs, the first VSFP2 voltage reporters were derived from the GFP variants Cerulean and Citrine. Subsequent VSFP2 variants based on different FPs allow the use of longer wavelength light. Further development of the VSFP design and introduction of opsin-based GEVIs resulted in a large toolbox, containing indicators that report with large changes

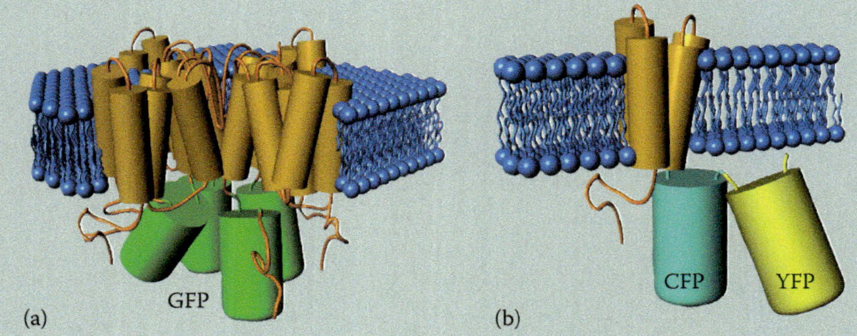


**Figure A15.2.2 Conceptual design of voltage-sensing proteins.** (a) Configuration based. An ion channel fused to one or more fluorescent proteins will show small changes in fluorescence emission as the channel opens and closes. (b) FRET-based. A configuration change in the ion channel caused by changes in voltage will alter the amount of FRET between two different FPs (CFP and YFP) fused to different parts of the channel, resulting in different levels of emission of the two colors.

*(Continued)*

**ADVANCED TOPIC 15.2 (CONTINUED):   OPTICAL RECORDING**

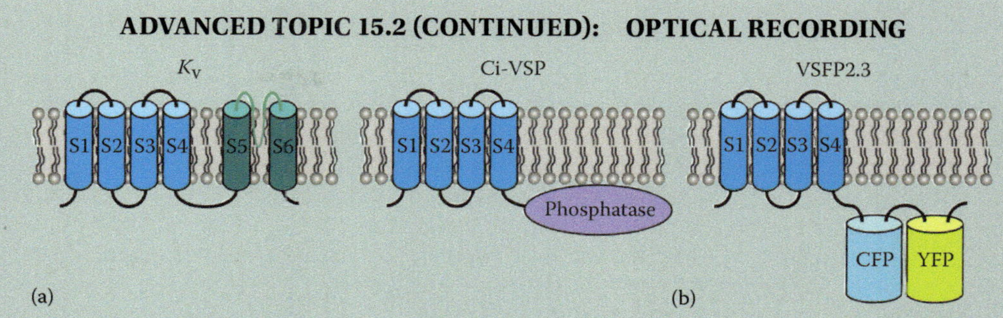


**Figure A15.2.3  Second-generation voltage-sensing proteins contain no ion channels.** (a) Voltage-sensor-containing phosphatase of *C. intestinalis* (Ci-VSP) is homologous to voltage-gated K+ channels, $K_V$. (b) Phosphatase is removed and the voltage sensor is fused to CFP and YFP to create the voltage-sensitive fluorescent protein VSFP2.3.

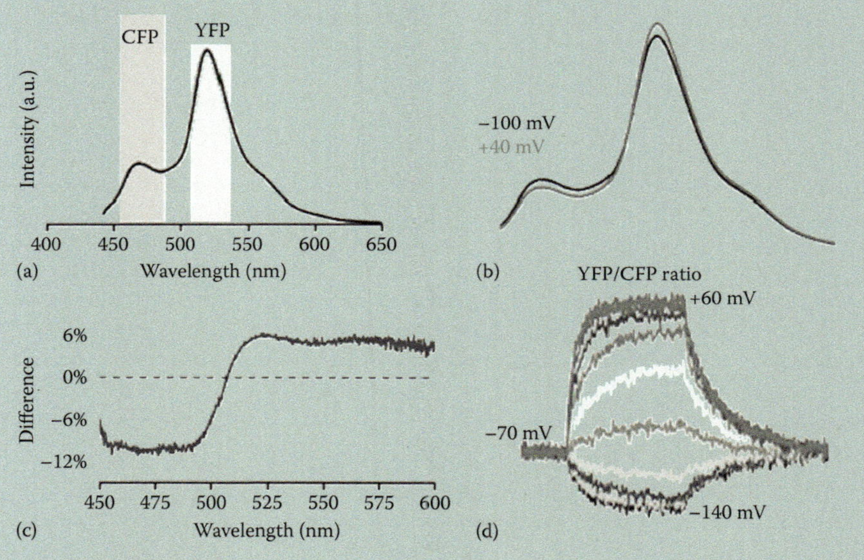


**Figure A15.2.4  Spectral properties of the voltage-sensitive fluorescent protein VSFP2.3.** (a) Overall emission spectrum, showing CFP and YFP bands. (b) Overall spectrum at −100 and +40 mV. (c) Difference vs. wavelength of the spectrum at −100 mV vs. +40 mV. (d) Ratio of emission in the YFP/CFP bands under voltage steps from −140 to +60 mV (holding at −70 mV).

in fluorescence intensity at the speed mammalian of action potentials (1 ms range). Recent reviews that cover this very active field of technical developments and physiological applications are found in the list of references.

## INJECTING CURRENTS USING GENETICALLY ENCODED PROTEINS

GEVIs allow only the *monitoring* of membrane potentials, but it is often desirable to also *manipulate* the membrane potential by activating ion currents through the membrane. While this feature is inherent to the use of intracellular microelectrodes, GEVIs need to be supplemented by specially designed ion channels and pumps to achieve this functionality. As we have seen, currents that flow through an ion channel either depolarize or hyperpolarize the membrane depending on the direction of the driving force of the ionic species that are selectively conducted by the channel. If we have one genetically engineered channel that can inject depolarizing (positive) currents in response to an external signal, and one that can inject hyperpolarizing (negative) currents in response to a different control signal, then we have a system for genetically

*(Continued)*

## ADVANCED TOPIC 15.2 (CONTINUED): OPTICAL RECORDING

controlled stimulation. Such systems have been discovered and developed to a remarkable degree using different wavelengths of light as the control signal.

*Channel rhodopsins* (ChRs) are naturally occurring light-gated channel proteins. They were originally discovered in algae and, as with the GFP-type of proteins, characterized and mutated to result in variants with improved or specialized properties. ChRs allow neuroscientists to optically depolarize (cation channel rhodopsins) or hyperpolarize (anion channel rhodopsins [ACRs]) specific genetically targeted populations of neurons and have been successfully used to study neural circuits at the cellular and system levels.

The first optogenetic tools that were used to optically hyperpolarize membranes involved rat rhodopsin 4 (RO4), which, when expressed in neurons, can couple with endogenous GIRK potassium channels via endogenous Gi/o G-proteins to enable light-induced neural hyperpolarization. Although conceptually elegant, the RO4-based system is difficult to implement, may interact with other processes, and provides only relatively slow current onset and offset kinetics. A subsequent generation of optogenetic hyperpolarizers are based on the light-driven inward chloride pump *halorhodopsin* from *N. pharaonis* (Halo/NpHR) that exhibits onset and offset in tens of milliseconds. Other additions to this toolbox are derived from the green light-driven outward proton pump *archaerhodopsin-3* from *H. sodomense* (Arch) and light-activated Cl conducting channels engineered from ChR2.

### OPTICAL NOISE IN NETWORK RECORDINGS

Optical recordings are typically limited by shot noise due to the quantal nature of light. A sufficient signal-to-noise ratio (S/N) obviously requires a signal that is large relative to the noise. The size of an optical signal is usually given by the change of light intensity (e.g., fluorescence, $F$) over absolute intensity (e.g., $\Delta F/F$). A good probe should have a large $\Delta F/F$ value, but the practically relevant signal scale relates to the level of noise associated with the signal. The number of photons detected during a sampling interval obeys a Poisson distribution, and the shot noise on a signal of $n$ photons is $\sqrt{n}$. The signal-to-noise ratio S/N increases, therefore, with $\sqrt{n}$, and this is why functional fluorescence imaging requires "bright" signals. But what level of brightness is required? In order to achieve a signal-to-noise ratio greater than 1 for a signal of 0.1% $\Delta F/F$, the noise amplitude needs to be minimized to 0.1% of baseline fluorescence, which in turn requires the detection of $n=10^6$ photons per sampling time interval. If this photon flux is sampled from an object area of 1 $\mu m^2$, this value is comparable to the brightness of daylight. In other words, low optical noise at high spatial resolution is not obtained under low light conditions.

### SUGGESTED READING

Abdelfattah, A.S., Farhi, S.L., Zhao, Y., Brinks, D., Zou, P., Ruangkittisakul, A., Platisa, J., Pieribone, V.A., Ballanyi, K., Cohen, A.E. et al. (2016). A bright and fast red fluorescent protein voltage indicator that reports neuronal activity in organotypic brain slices. *The Journal of Neuroscience* 36, 2458–2472. http://www.ncbi.nlm.nih.gov/pubmed/26911693

Acker, C.D., and Loew, L.M. (2013). Characterization of voltage-sensitive dyes in living cells using two-photon excitation. *Methods in Molecular Biology* 995, 147–160. http://www.ncbi.nlm.nih.gov/pubmed/23494378

Akemann, W., Mutoh, H., Perron, A., Rossier, J., and Knopfel, T. (2010). Imaging brain electric signals with genetically targeted voltage-sensitive fluorescent proteins. *Nature Methods* 7, 643–649. http://www.ncbi.nlm.nih.gov/pubmed/20622860

Akemann, W., Song, C., Mutoh, H., and Knopfel, T. (2015). Route to genetically targeted optical electrophysiology: Development and applications of voltage-sensitive fluorescent proteins. *Neurophotonics* 2. http://www.ncbi.nlm.nih.gov/pubmed/26082930

Antic, S.D., Empson, R.M., and Knopfel, T. (2016). Voltage imaging to understand connections and functions of neuronal circuits. *Journal of Neurophysiology* 116, 135–152. http://www.ncbi.nlm.nih.gov/pubmed/27075539

(*Continued*)

**ADVANCED TOPIC 15.2 (CONTINUED):    OPTICAL RECORDING**

Becker-Baldus, J., Bamann, C., Saxena, K., Gustmann, H., Brown, L.J., Brown, R.C., Reiter, C., Bamberg, E., Wachtveitl, J., Schwalbe, H. et al. (2015). Enlightening the photoactive site of channelrhodopsin-2 by DNP-enhanced solid-state NMR spectroscopy. *Proceedings of the National Academy of Sciences of the United States of America* 112, 9896–9901. http://www.ncbi.nlm.nih.gov/pubmed/26216996

Bourgeois, E.B., Johnson, B.N., McCoy, A.J., Trippa, L., Cohen, A.S., and Marsh, E.D. (2014). A toolbox for spatiotemporal analysis of voltage-sensitive dye imaging data in brain slices. *PloS One* 9, e108686. http://www.ncbi.nlm.nih.gov/pubmed/25259520

Briggman, K.L., Kristan, W.B., Gonzalez, J.E., Kleinfeld, D., and Tsien, R.Y. (2015). Monitoring integrated activity of individual neurons using FRET-based voltage-sensitive dyes. *Advances in Experimental Medicine and Biology* 859, 149–169. http://www.ncbi.nlm.nih.gov/pubmed/26238052

Brinks, D., Klein, A.J., and Cohen, A.E. (2015). Two-photon lifetime imaging of voltage indicating proteins as a probe of absolute membrane voltage. *Biophysical Journal* 109, 914–921. http://www.ncbi.nlm.nih.gov/pubmed/26331249

Cohen, A.E., and Hochbaum, D.R. (2014). Measuring membrane voltage with microbial rhodopsins. *Methods in Molecular Biology* 1071, 97–108. http://www.ncbi.nlm.nih.gov/pubmed/24052383

Hou, J.H., Venkatachalam, V., and Cohen, A.E. (2014). Temporal dynamics of microbial rhodopsin fluorescence reports absolute membrane voltage. *Biophysical Journal* 106, 639–648. http://www.ncbi.nlm.nih.gov/pubmed/24507604

Knopfel, T., Gallero-Salas, Y., and Song, C. (2015). Genetically encoded voltage indicators for large scale cortical imaging come of age. *Current Opinion in Chemical Biology* 27, 75–83. http://www.ncbi.nlm.nih.gov/pubmed/26115448

Lin, J.Y. (2011). A user's guide to channelrhodopsin variants: Features, limitations and future developments. *Experimental Physiology* 96, 19–25. http://www.ncbi.nlm.nih.gov/pubmed/20621963

Lin, J.Y. (2012). Optogenetic excitation of neurons with channelrhodopsins: Light instrumentation, expression systems, and channelrhodopsin variants. *Progress in Brain Research* 196, 29–47. http://www.ncbi.nlm.nih.gov/pubmed/22341319

Loew, L.M. (2015). Design and use of organic voltage sensitive dyes. *Advances in Experimental Medicine and Biology* 859, 27–53. http://www.ncbi.nlm.nih.gov/pubmed/26238048

Nagel, G., Szellas, T., Huhn, W., Kateriya, S., Adeishvili, N., Berthold, P., Ollig, D., Hegemann, P., and Bamberg, E. (2003). Channelrhodopsin-2, a directly light-gated cation-selective membrane channel. *Proceedings of the National Academy of Sciences of the United States of America* 100, 13940–13945. http://www.ncbi.nlm.nih.gov/pubmed/14615590

Yamawaki, N., Suter, B.A., Wickersham, I.R., and Shepherd, G.M. (2016). Combining optogenetics and electrophysiology to analyze projection neuron circuits. *Cold Spring Harbor Protocols* 2016, pdb prot090084. http://www.ncbi.nlm.nih.gov/pubmed/27698240

Yan, P., Acker, C.D., Zhou, W.L., Lee, P., Bollensdorff, C., Negrean, A., Lotti, J., Sacconi, L., Antic, S.D., Kohl, P. et al. (2012). Palette of fluorinated voltage-sensitive hemicyanine dyes. *Proceedings of the National Academy of Sciences of the United States of America* 109, 20443–20448. http://www.ncbi.nlm.nih.gov/pubmed/23169660

Zhang, F., Wang, L.P., Boyden, E.S., and Deisseroth, K. (2006). Channelrhodopsin-2 and optical control of excitable cells. *Nature Methods* 3, 785–792. http://www.ncbi.nlm.nih.gov/pubmed/16990810

Zhang, Z., Feng, J., Wu, C., Lu, Q., and Pan, Z.H. (2015). Targeted expression of channelrhodopsin-2 to the axon initial segment alters the temporal firing properties of retinal ganglion cells. *PloS One* 10, e0142052. http://www.ncbi.nlm.nih.gov/pubmed/26536117

Zhang, H., Reichert, E., and Cohen, A.E. (2016). Optical electrophysiology for probing function and pharmacology of voltage-gated ion channels. *eLife* 5. http://www.ncbi.nlm.nih.gov/pubmed/27215841

Zou, P., Zhao, Y., Douglass, A.D., Hochbaum, D.R., Brinks, D., Werley, C.A., Harrison, D.J., Campbell, R.E., and Cohen, A.E. (2014). Bright and fast multicoloured voltage reporters via electrochromic FRET. *Nature Communications* 5, 4625. http://www.ncbi.nlm.nih.gov/pubmed/25118186

We did not discuss here any particular issues involved with patch-clamping neurons, but it is worth mentioning that this can be challenging. Primary neurons are much more fragile than cell lines, so that electrophysiology is significantly more challenging, especially with certain particular cell types such as motor neurons. The neurons must be in a pristine state of health, or they will not form

## ADVANCED TOPIC 15.3:   MORE EMERGING TECHNIQUES
## FOR NEURONAL RECORDING AND STIMULATION

Single-walled carbon nanotubes (SWNTs) have attracted a lot of interest as possible electrodes for stimulation and recording in neuronal cultures. A mesh of SWNTs can be made on a chosen substrate simply by pipetting hepatanal/sarcosine-functionalized NTs onto the surface. Neuronal cell lines and even neurons grow happily on such substrates, and there are many reports of the cells being electrically stimulated by the SWNTs—although what exactly is happening is not entirely clear. Most groups suggest that there is direct coupling between the cell and the NTs, based upon simultaneous patch-clamp and/or high-resolution imaging studies. Several groups have made MEAs consisting of SWNT mesh deposited onto microelectrodes (Figure A15.3.1a).

A few studies suggest that vertically aligned nanotubes might be better as electrical contacts to cells than mesh, as they do not create short-circuit pathways. Models have predicted that electrical recordings using this approach might be as good as those obtained via whole-cell patch clamp, although this has yet to be achieved. A significant problem is that vertically aligned SWNTs are fragile and break under biological conditions; various stabilizers have been tried, such as polypyrrole, which has permitted growth of cells on vertical tubes (Figure A15.3.1b). So far, these techniques are too noisy to permit resolution of action potentials.

For stimulation (but not recording), the photolytic properties of quantum dots can be used. Both PbSe and HgTe particles have been shown to stimulate action potentials in neurons when the particles are excited by light. This is believed to be an electric field rather than an electron-transfer process, as the cells are grown without contacting the nanoparticles.

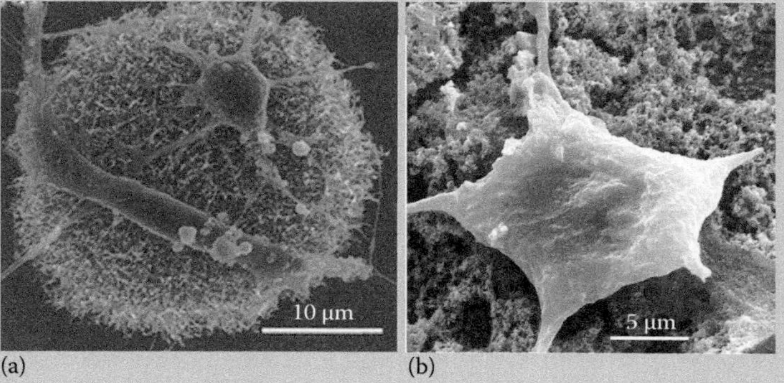


**Figure A15.3.1**   **(a) Neurons and glia on a carbon nanotube mesh.** (From Ben-Jacob, E., and Hanein, Y. (2008). Carbon nanotube micro-electrodes for neuronal interfacing. *Journal of Materials Chemistry* 18, 5181–5186. Reproduced by permission of The Royal Society of Chemistry.) **(b) PC12 cells grown on a carbon nanotube/polypyrrole mesh**. (Reprinted from *Biomaterials*, 31, Lu, Y., Li, T., Zhao, X., Li, M., Cao, Y., Yang, H., and Duan, Y.Y. Electrodeposited polypyrrole/carbon nanotubes composite films electrodes for neural interfaces, 5169–5181, Copyright 2010, with permission from Elsevier.)

## SUGGESTED READING


Burblies, N., Schulze, J., Schwarz, H.C., Kranz, K., Motz, D., Vogt, C., Lenarz, T., Warnecke, A., and Behrens, P. (2016). Coatings of different carbon nanotubes on platinum electrodes for neuronal devices: Preparation, cytocompatibility and interaction with spiral ganglion cells. *PloS One* 11, e0158571. http://www.ncbi.nlm.nih.gov/pubmed/27385031

Fabbro, A., Bosi, S., Ballerini, L., and Prato, M. (2012). Carbon nanotubes: Artificial nanomaterials to engineer single neurons and neuronal networks. *ACS Chemical Neuroscience* 3, 611–618. http://www.ncbi.nlm.nih.gov/pubmed/22896805

Hernandez-Ferrer, J., Perez-Bruzon, R.N., Azanza, M.J., Gonzalez, M., Del Moral, R., Anson-Casaos, A., de la Fuente, J.M., Marijuan, P.C., and Martinez, M.T. (2014). Study of neuron survival on polypyrrole-embedded single-walled carbon nanotube substrates for long-term growth conditions. *Journal of Biomedical Materials Research. Part A* 102, 4443–4454. http://www.ncbi.nlm.nih.gov/pubmed/24677410


(Continued)

**ADVANCED TOPIC 15.3 (CONTINUED):  MORE EMERGING TECHNIQUES FOR NEURONAL RECORDING AND STIMULATION**

Massobrio, P., Massobrio, G., and Martinoia, S. (2016). Interfacing cultured neurons to microtransducers arrays: A review of the neuro-electronic junction models. *Frontiers in Neuroscience* 10, 282. http://www.ncbi.nlm.nih.gov/pubmed/27445657

Singh, N., Chen, J., Koziol, K.K., Hallam, K.R., Janas, D., Patil, A.J., Strachan, A., G. Hanley, J., and Rahatekar, S.S. (2016). Chitin and carbon nanotube composites as biocompatible scaffolds for neuron growth. *Nanoscale* 8, 8288–8299. http://www.ncbi.nlm.nih.gov/pubmed/27031428

gigaseals. To obtain good synaptic signals, the neurons must be at least 1 week (and preferably 2 weeks) in culture, to give time for synapses to develop. Recording synaptic currents can be done on single cells, or it can be done on two or more cells where one is stimulated and the response of the others is recorded. Two-cell recordings are routine; three are done by only a few labs. This requires the ability to manipulate a second (or third!) pipette into the dish and find another cell without disturbing the first, so the difficulties are readily appreciated.

The patch-clamp technique is well established and highly useful for studying ion channel properties and responses of cell currents to drugs, toxins, and drug candidates. It is the best method we have currently to study single ion channels. Where it becomes inadequate is for high-throughput applications or for studies of complex networks. Physical, chemical, and optical methods of recording from networks of cells *in vitro* and *in vivo* are being developed, but have not yet attained the sensitivity attainable with patch clamp. When the experimental challenges are overcome, we will be met with the new difficulties of modeling these incredibly complex connections.

For all of the emerging techniques described above, engineering of new tools is a rapid and ongoing process. It is therefore advisable to consult the most recent literature (or better an expert) before choosing a particular voltage-sensitive dye, optogenetic method, or advanced imaging technique. Research into network recording methods will most likely expand over the next decade with better molecular tools, more advanced optical instrumentation, and fascinating applications aimed at understanding neuronal circuits and hence the core of brain functions.

## End-of-Chapter Questions

1. Write the GHK equation for internal/external solutions made of NaCl and KCl.

2. Relate the GHK equation to the "chord conductance equation" for sodium, potassium, and chloride, which gives the resting membrane potential as

$$V_m = \frac{g_K E_K + g_{Na} E_{Na} + g_{Cl} E_{Cl}}{g_K + g_{Na} + g_{Cl}}.$$

3. What is the resting membrane potential of a cell that has 140 mM KCl in, 10 mM NaCl in, 5 mM KCl out, 145 mM NaCl out, and a permeability to Na$^+$ that is 5% that of its permeability to K$^+$? What if the permeability to Na$^+$ is 50% that of K$^+$? What if the permeability of Cl$^-$ is 45% that of K$^+$, and the internal solution is entirely potassium gluconate rather than KCl? (Assume room temperature.)

4. Suppose an otherwise normal neuron is placed in a bath of high external KCl. What happens to the resting potential? What effect can you expect this to have on the cell?

5. What is the "ideal" capacitance for a bilayer formed across a 0.2 mm orifice? What is a more realistic value and why?

6. Why is ATP put into pipette internal solutions? When is it needed/not needed?

7. Reduce the equation for the junction potential (**Equation 15.5**) in the case of a bi-ionic solution with only monovalent ions.

8. What is the $E_K$ for the solution in the sample experiment at room temperature? At 37°C? What is the junction potential?

9. Why is series resistance compensation more important for Na+ currents than K+ currents?

10. If the standard deviation of the background noise in a single-channel recording, $\sigma_n$, is ¼ of the threshold value, how many false events will occur per second? What if it is 1/6 of the threshold? Assume a recording frequency of 1 kHz.

11. How many photons can be generated by a single GFP molecule before it is bleached and how long can, accordingly, its all-or-none fluorescence be observed at a S/N of 10 at a temporal resolution of 1 ms given a sampling efficacy of 0.1?

12. How is the S/N of an optical signal affected if the signal is carried only by 10% of the chromophores?

13. What are the advantages of red (far red) dyes in optical imaging as compared to green emitting dyes?

## Background Reading

### Books

*Electrophysiology is a well-established field with many comprehensive textbooks on theory and practice available. The following are all highly recommended.*

Bretschneider, F., and de Weille, J.R. *Introduction to Electrophysiological Methods and Instrumentation*. Elsevier, San Diego, CA, 2006.

Hille, B. *Ion Channels of Excitable Membranes*. Edn. 3. Sinauer Associates, Sunderland, MA, 2001.

Liu, X.J. (ed.). *Xenopus Protocols: Cell Biology and Signal Transduction*. Humana Press, Totowa, NJ, 2010.

Molleman, A. *Patch Clamping: An Introductory Guide to Patch Clamp Electrophysiology*. John Wiley & Sons, Hoboken, NJ, 2003.

Molnar, P., and Hickman, J.J. (eds.). *Patch-Clamp Methods and Protocols*, vol. 403. Humana Press, Totowa, NJ, 2010.

Nicholls, J.G., Martin, A.R., Wallace, B.G., and Fuchs, P.A. *From Neuron to Brain: A Cellular and Molecular Approach to the Function of the Nervous System*. Edn 4. Sinauer Associates, Sunderland, MA, 2001.

Sakmann, B., and Neher, E. (eds.). *Single-Channel Recording*. Edn 2. Springer Science + Business Media, New York, 2009.

Walz, W., Boulton, A.A., and Baker, G.B. (eds.). *Patch-Clamp Analysis: Advanced Techniques (Neuromethods)*, vol. 35. Humana Press, Totowa, NJ, 2002.

### Journal articles

*This is a sampling of classic articles as well as new articles that focus on methods. There are thousands of articles reporting results using electrophysiology, and a literature search with the appropriate keywords (ion channel name, etc.) will lead you to them.*

*General articles*

Barry, P.H., and Lynch, J.W. (1991). Liquid junction potentials and small cell effects in patch-clamp analysis. *Journal of Membrane Biology* 121, 101–117.

Becker, J.D., Honerkamp, J., Hirsch, J., Frobe, U., Schlatter, E., and Greger, R. (1994). Analysing ion channels with hidden Markov models. *Pflugers Archiv* 426, 328–332.

Cafiso, D.S. (1994). Alamethicin: A peptide model for voltage gating and protein-membrane interactions. *Annual Review of Biophysics and Biomolecular Structure* 23, 141–165.

Caviedes, P., Ault, B., and Rapoport, S. I. (1990). Replating improves whole cell voltage clamp recording of human fetal dorsal root ganglion neurons. *Journal of Neuroscience Methods* 35, 57–61.

Davie, J.T., Kole, M.H., Letzkus, J.J., Rancz, E.A., Spruston, N., Stuart, G.J., and Hausser, M. (2006). Dendritic patch-clamp recording. *Nature Protocols* 1, 1235–1247.

Edwards, F.A., Konnerth, A., and Sakmann, B. (1990). Quantal analysis of inhibitory synaptic transmission in the dentate gyrus of rat hippocampal slices: A patch-clamp study. *Journal of Physiology* 430, 213–249.

Edwards, F.A., Konnerth, A., Sakmann, B., and Takahashi, T. (1989). A thin slice preparation for patch clamp recordings from neurones of the mammalian central nervous system. *Pflugers Archiv* 414, 600–612.

Hainsworth, A.H., Randall, A.D., and Stefani, A. (2006). Whole-cell patch clamp recording of voltage-sensitive Ca2+ channel currents heterologous expression systems and dissociated brain neurons. *Methods in Molecular Biology* 312, 161–179.

Hamill, O.P., Marty, A., Neher, E., Sakmann, B., and Sigworth, F.J. (1981). Improved patch-clamp techniques for high-resolution current recording from cells and cell-free membrane patches. *Pflugers Archiv* 391, 85–100.

Hodgkin, A.L., and Horowicz, P. (1959). The influence of potassium and chloride ions on the membrane potential of single muscle fibres. *Journal of Physiology* 148, 127–160.

Horn, R., and Lange, K. (1983). Estimating kinetic constants from single channel data. *Biophysical Journal* 43, 207–223.

Horn, R., and Vandenberg, C.A. (1984). Statistical properties of single sodium channels. *Journal of General Physiology* 84, 505–534.

Jackson, M.B. (2001). Whole-cell voltage clamp recording. *Current Protocols in Neuroscience*, Chapter 6: Unit 6 6.

Kagan, B.L., Selsted, M.E., Ganz, T., and Lehrer, R.I. (1990). Antimicrobial defensin peptides form voltage-dependent ion-permeable channels in planar lipid bilayer membranes. *Proceedings of the National Academy of Sciences of the United States of America* 87, 210–214.

Khan, R.N., Martinac, B., Madsen, B.W., Milne, R.K., Yeo, G.F., and Edeson, R.O. (2005). Hidden Markov analysis of mechanosensitive ion channel gating. *Mathematical Biosciences* 193, 139–158.

Kirber, M.T., Singer, J.J., Walsh, J.V. Jr., Fuller, M.S., and Peura, R.A. (1985). Possible forms for dwell-time histograms from single-channel current records. *Journal of Theoretical Biology* 116, 111–126.

Latorre, R., and Alvarez, O. (1981). Voltage-dependent channels in planar lipid bilayer membranes. *Physiological Reviews* 61, 77–150.

Liebovitch, L.S., and Toth, T.I. (1991). A model of ion channel kinetics using deterministic chaotic rather than stochastic processes. *Journal of Theoretical Biology* 148, 243–267.

Major, G. (1993). Solutions for transients in arbitrarily branching cables: III. Voltage clamp problems. *Biophysical Journal* 65, 469–491.

Major, G., and Evans, J.D. (1994). Solutions for transients in arbitrarily branching cables: IV. Nonuniform electrical parameters. *Biophysical Journal* 66, 615–633.

Major, G., Evans, J.D., and Jack, J.J. (1993a). Solutions for transients in arbitrarily branching cables: I. Voltage recording with a somatic shunt. *Biophysical Journal* 65, 423–449.

Major, G., Evans, J.D., and Jack, J.J. (1993b). Solutions for transients in arbitrarily branching cables: II. Voltage clamp theory. *Biophysical Journal* 65, 450–468.

Methfessel, C., Witzemann, V., Takahashi, T., Mishina, M., Numa, S., and Sakmann, B. (1986). Patch clamp measurements on Xenopus laevis oocytes: Currents through endogenous channels and implanted acetylcholine receptor and sodium channels. *Pflugers Archiv* 407, 577–588.

Milne, R.K., Edeson, R.O., and Madsen, B.W. (1986). Stochastic modelling of a single ion channel: Interdependence of burst length and number of openings per burst. *Proceedings of the Royal Society of London B: Biological Sciences* 227, 83–102.

Milne, R.K., Yeo, G.F., Madsen, B.W., and Edeson, R.O. (1989). Estimation of single channel kinetic parameters from data subject to limited time resolution. *Biophysical Journal* 55, 673–676.

Mody, I., Salter, M.W., and MacDonald, J.F. (1988). Requirement of NMDA receptor/channels for intracellular high-energy phosphates and the extent of intraneuronal calcium buffering in cultured mouse hippocampal neurons. *Neuroscience Letters* 93, 73–78.

Mortensen, M., and Smart, T.G. (2007). Single-channel recording of ligand-gated ion channels. *Nature Protocols* 2, 2826–2841.

Neher, E., and Sakmann, B. (1992). The patch clamp technique. *Scientific American* 266, 44–51.

Neher, E., Sakmann, B., and Steinbach, J.H. (1978). The extracellular patch clamp: A method for resolving currents through individual open channels in biological membranes. *Pflugers Archiv* 375, 219–228.

Park, M.R., Kita, H., Klee, M.R., and Oomura, Y. (1983). Bridge balance in intracellular recording; introduction of the phase-sensitive method. *Journal of Neuroscience Methods* 8, 105–125.

Perkins, K.L. (2006). Cell-attached voltage-clamp and current-clamp recording and stimulation techniques in brain slices. *Journal of Neuroscience Methods* 154, 1–18.

Rodighiero, S., De Simoni, A., and Formenti, A. (2004). The voltage-dependent nonselective cation current in human red blood cells studied by means of whole-cell and nystatin-perforated patch-clamp techniques. *Biochimica et Biophysica Acta* 1660, 164–170.

Sakmann, B., Edwards, F., Konnerth, A., and Takahashi, T. (1989). Patch clamp techniques used for studying synaptic transmission in slices of mammalian brain. *Quarterly Journal of Experimental Physiology* 74, 1107–1118.

Sakmann, B., Hamill, O.P., and Bormann, J. (1983). Patch-clamp measurements of elementary chloride currents activated by the putative inhibitory transmitter GABA and glycine in mammalian spiral neurons. *Journal of Neural Transmission, Supplementum* 18, 83–95.

Sakmann, B., and Neher, E. (1984). Patch clamp techniques for studying ionic channels in excitable membranes. *Annual Review of Physiology* 46, 455–472.

Sala, F., and Sala, S. (1994). Sources of errors in different single-electrode voltage-clamp techniques: a computer simulation study. *Journal of Neuroscience Methods* 53, 189–197.

Sengupta, B., Laughlin, S.B., and Niven, J.E. (2010). Comparison of Langevin and Markov channel noise models for neuronal signal generation. *Physical Review E: Statistical Nonlinear Soft Matter Physics* 81, 011918.

Sperelakis, N. (1988). Patch clamp and single-cell voltage clamp techniques and selected data. *Molecular and Cellular Biochemistry* 80, 3–7.

Stuart, G.J., Dodt, H.U., and Sakmann, B. (1993). Patch-clamp recordings from the soma and dendrites of neurons in brain slices using infrared video microscopy. *Pflugers Archiv* 423, 511–518.

Stuhmer, W., Methfessel, C., Sakmann, B., Noda, M., and Numa, S. (1987). Patch clamp characterization of sodium channels expressed from rat brain cDNA. *European Biophysics Journal* 14, 131–138.

Watsky, M.A., and Rae, J.L. (1991). Resting voltage measurements of the rabbit corneal endothelium using patch-current clamp techniques. *Investigative Ophthalmology and Visual Science* 32, 106–115.

Wilson, C.J., and Park, M.R. (1989). Capacitance compensation and bridge balance adjustment in intracellular recording from dendritic neurons. *Journal of Neuroscience Methods* 27, 51–75.

Wonderlin, W.F., Finkel, A., and French, R.J. (1990). Optimizing planar lipid bilayer single-channel recordings for high resolution with rapid voltage steps. *Biophysical Journal* 58, 289–297.

Woody, C.D., and Gruen, E. (1987). Acetylcholine reduces net outward currents measured in vivo with single electrode voltage clamp techniques in neurons of the motor cortex of cats. *Brain Research* 424, 193–198.

Woody, C.D., Baranyi, A., Szente, M.B., Gruen, E., Holmes, W., Nenov, V., and Strecker, G.J. (1989). An aminopyridine-sensitive, early outward current recorded in vivo in neurons of the precruciate cortex of cats using single-electrode voltage-clamp techniques. *Brain Research* 480, 72–81.

Woolley, G.A., and Wallace, B.A. (1992). Model ion channels: Gramicidin and alamethicin. *Journal of Membrane Biology* 129, 109–136.

## Online resources and software

### Free software

*There are a few open-source or free-for-academics acquisition and analysis packages. The following are of particular interest.*

Jclamp (SciSoft). Voltage-clamp acquisition program. Easy to use and has many features. For Windows.

RELACS. Open-source, hardware-independent acquisition program for GNU/Linux.

Spikoscope. For analysis of extracellular and intracellular recordings.

Stimfit. For analysis. Reads Axon, HEKA, and Axograph files. For Windows, Mac, and GNU/Linux.

NeuroMatic. Analysis and acquisition functions for *Igor Pro.*

Serf software suite. For analysis and presentation of many types of biological data, including electrophysiology. Windows.

Strathclyde electrophysiology software. Suite of programs for acquisition and analysis from the University of Strathclyde. Windows.

Cell Electrophysiology Simulation Environment (CESE). Electrophysiology simulation. Windows, Mac, and Unix.

Libraries for electrophysiological data (Sourceforge). Sourceforge project: collection of software to analyze data acquired from electrophysiology experiments.

### Commercial software

*Also see instrument manufacturers for software.*

Axograph. Originally developed for Mac, now also available for PC. Data acquisition and analysis, including its own built-in programming language to create custom analysis modules. Highly recommended for analysis (I have not tried it for acquisition).

CellWorks (NPI Electronic). Software for data acquisition, electrophysiology, and pharmacology. Parent company also makes amplifiers.

SigmaPlot. An electrophysiology module is available for data analysis; supports Axon, HEKA, and Bruxton formats. Recommended.

TAC and DataAccess (Bruxton Corporation). TAC is a single-channel analysis program, recommended. DataAccess allows for importation of all common electrophysiology data and conversion to multiple formats. Also recommended.

Spike2 (Cambridge Electronic Design). Capture and analysis package for spike trains; good for MEAs. Amplifiers also available for specific applications.

## Instrument manufacturers

*These are a few of the key ones, by no means an exhaustive list but certainly enough to get started.*

Molecular Devices. Suppliers of the amplifiers and software formerly sold by Axon instruments. A wide variety of high-end patch-clamp amplifiers specialized for all applications, digitizer boards, and pClamp software.

HEKA Elektronik. Electrophysiology amplifiers; also electrochemistry amplifiers.

ALA Scientific Instruments. A huge variety of electrophysiology instruments and supplies for every conceivable application, including custom designs. Maker of some specialized amplifiers as well as distributor of HEKA amplifiers; filters and conditioners; MEAs; perfusion systems and applicators for small-volume applications; micromanipulators; microforges and bevellers for pipettes; recording chambers; pipette glass and pullers; and much, much more.

Harvard Apparatus. Some amplifiers; vibration isolation platforms; oocyte accessories; lipid bilayer accessories; peristaltic pumps; and many other non-electrophysiology tools and supplies for culturing cells and tissues.

Warner Instruments (part of Harvard apparatus). Lipid bilayer equipment and supplies; amplifiers; oocyte instruments; pipette pullers; micromanipulators; glass; vibration platforms and air tables.

Sutter Instruments. Pipette pullers, including laser puller; filaments; pipette bevellers; glass; microinjectors; micromanipulators; imaging filter wheels and shutters; motorized stages for microscopes; custom services.

Dagan Corporation. Amplifiers (including low-cost patch-clamp amplifiers with choice of headstage(s)); pre-amps; pullers; glass; perfusion systems; recording chambers and temperature controllers.

### *Reagent suppliers*

Tocris Bioscience. Exhaustive catalog of ion channel agents and other drugs.

Alomone Labs. A wide variety of tools including ion channel antibodies, toxins (some fluorescently labeled), agonists and antagonists, and more.

Ascent Scientific. Ion channel agents.

Enzo Life Sciences. Wide variety of ion channel and other neuroscience agents; libraries of blockers/agents for drug discovery or channel characterization.

# CHAPTER 16

# Spectroscopy Tools and Techniques

## 16.1 INTRODUCTION

The previous chapters discussed synthesis, conjugation, and biological labeling using a variety of materials with complex structures and spectra. The purpose of this chapter is to give a practical introduction to the best spectroscopic methods for characterizing these types of samples. The basics of the physics behind each type of measurement will be given if it has not been discussed previously, followed by practical instructions for finding or putting together a source of the instrumentation. Some of the types of spectroscopy covered here—UV–Vis, fluorescence emission—are ubiquitous and found in essentially every laboratory. Some of them are more specialized, but are used every day in laboratories that use them and are found on nearly all university campuses (time-resolved techniques, x-ray spectroscopy, nuclear magnetic resonance [NMR]). Still others require traveling to synchrotron sources and are used rarely and only under special circumstances (specialized x-ray techniques, high-field NMR). After reading this chapter, you should have a good idea of which types of spectroscopy can be used on any type of sample similar to the ones that occurred in **Chapters 1 through 15**, what information this method will provide, and how difficult obtaining the instrumentation is likely to be. It is also a good idea to check your laboratory's list of instruments as well as any shared facilities available to you before deciding which spectroscopic techniques to pursue.

## 16.2 GUIDING PRINCIPLES

As discussed in **Chapter 1**, atoms and molecules have quantized electronic, rotational, and vibrational energy states. Transitions to a higher or lower state require the absorbance or emission of a precise amount of energy, and the energies and lifetimes are characteristic of the system in question. They can identify the specific atoms as well as specific types of bond or ionization state. Depending upon the characteristic energy of what needs to be identified, different types of spectroscopy are used. The guiding principle behind spectroscopy is to provide enough energy to induce specific transitions that allow recognition of the system or quantification of a desired property (such as concentration). The characteristic energies of different types of transitions relative to the electromagnetic spectrum are given in **Figure 16.1**.

**Figure 16.1 Energies
and frequencies of the
electromagnetic spectrum
and molecular transitions.**
This forms the basis of the
different types of spectroscopy.

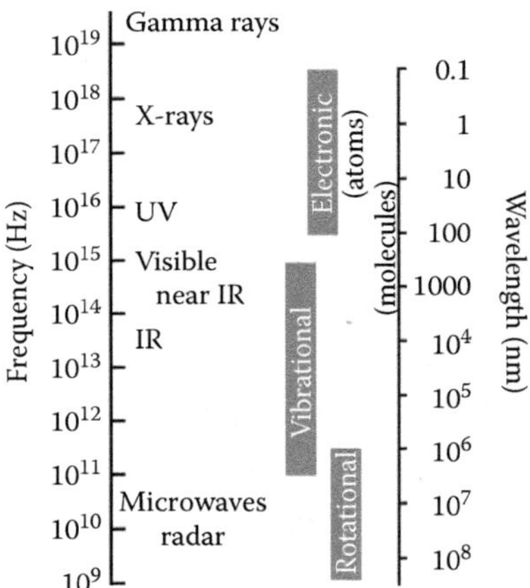



# 16.3  UV–VISIBLE ABSORBANCE SPECTROSCOPY

UV–Vis spectroscopy has energies on the order of

$$E = \frac{hc}{\lambda}\,1.6\times10^{-22}\,\text{kcal}/\gamma \qquad (16.1)$$

or multiplied by Avogadro's number, just under 100 kcal/mole of photons. This is the amount of energy in a covalent bond, so UV–Vis light is sufficient to excite electronic transitions. The energies of allowed transitions for molecules of known structure may be calculated using molecular orbital theory. Even when the structure is not exactly known, however, absorbance peaks can be of use. We have mentioned UV–Vis absorbance spectroscopy many times throughout this book in many different contexts. To recap:

- Proteins have at least four types of transitions that contribute to absorbance in the UV. The aromatic amino acids are the strongest of these. A protein with average amounts of Tyr and Trp will show an absorbance peak at 280 nm, with approximately 1 AU equivalent to 1 mg/mL.

- Nucleic acids show a peak at 260 nm with $\varepsilon \sim 10^4$/M (lower for double-stranded DNA than for single-stranded or free nucleotides). This can be used to quantify the results of DNA preps (**Chapter 2**), and $A_{260}$ may be compared with $A_{280}$ to detect protein contamination.

- The absorbance spectra of fluorescent molecules can be used to determine their energy states, to calculate their concentration, and often to predict emission spectra (**Chapter 7**).

- Absorbance of solutions of cells (optical density) can be used to determine cell number (**Chapter 10**).

- Quantum dots (QDs; **Chapter 11**) and Au nanoparticles (**Chapters 12** and **13**) show distinctive, size-dependent absorbance (or extinction) spectra.

The ubiquity of this technique is apparent, and it is often the first measure used to characterize any nanoassembly, be it crystalline or molecular. Every lab should have a UV–Vis spectrophotometer, sometimes more than one depending upon the features needed by the different types of experiments.

Many different UV–Vis spectrophotometers are available. In general, they can be classified into monochromator-based instruments, or simultaneous instruments that do not pass the light through a monochromator but instead collect all wavelengths on a *diode array* (**Figure 16.2**). The choice of an instrument should be based upon sensitivity and wavelength resolution required (high-end instruments allow for 0.01 nm resolution); speed and kinetic capability; photosensitivity of samples (will they be exposed to light only during data collection, and if so for how long?); desired throughput (some plate readers have 384-well options); portability needed (there are field models available); type of samples used (some instruments can take spectra from solid samples, for example, whereas many cannot, and the size of the sample compartment varies); and other special features such as nitrogen purging or fiber-optic attachment. It is not uncommon to have two spectrophotometers in the lab, one for physicochemical characterization and one for biology. Instruments for biology often run on multiwell format and can accommodate tissue culture plates of all types. There are also instruments aimed at molecular biologists that use extremely small sample volumes, 1 μL or less; some of these even measure a spectrum within a pipette tip and then allow the sample to be moved on to the next stage of an experiment.

Although UV–Vis spectroscopy is one of the easiest quantitative techniques to use and interpret, there are still a few common sources of error that are worth mentioning:

- Most compounds have a range of concentration in which the absorbance is linear. Outside of this linear range, a twofold chance in concentration may lead to essentially no change in absorbance. An example is shown in **Figure 16.3** for the green fluorescent protein variant, Dronpa. Nonlinearities occur at both extremes of concentration. It is especially easy to be deceived with fluorescent compounds, since the concentration needed to get a good fluorescence emission spectrum is often many

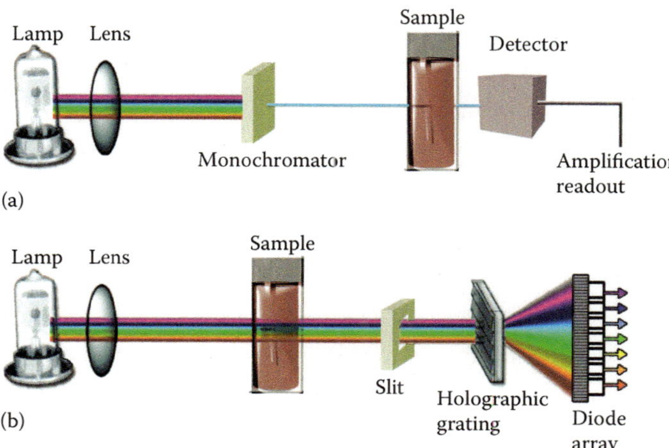


**Figure 16.2 Simplified light paths of typical UV–Vis spectrophotometers.** The lamp usually represents two lamps: a deuterium lamp to provide UV light, and a tungsten–halogen for the visible. (a) Scanning monochromator spectrophotometer. The sample is exposed to one wavelength of light at a time, with the increment controlled by the user and its resolution determined by the instrument. (b) Diode array spectrophotometer. The sample is exposed to all wavelengths of light at once.

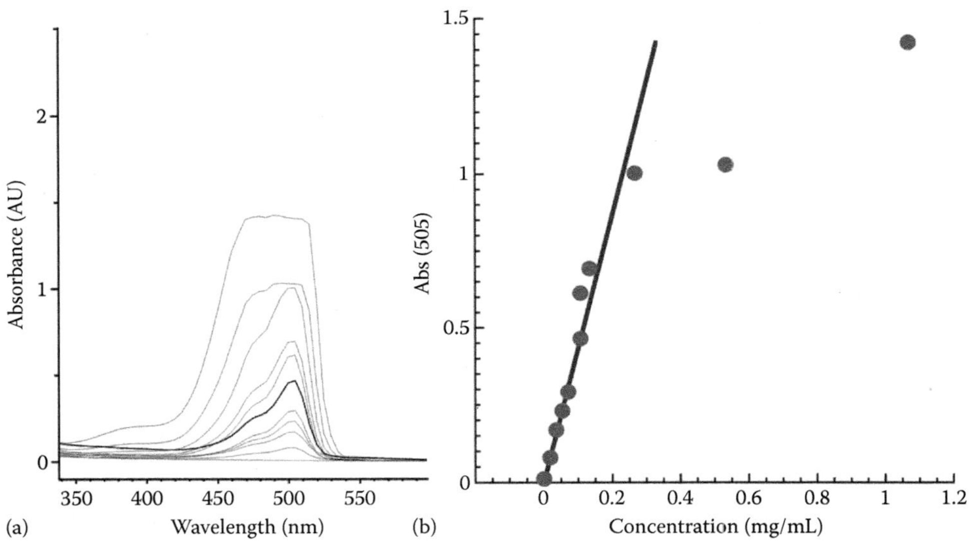

**Figure 16.3 Nonlinear absorbance behavior in a fluorescent protein.** (a) Spectra taken at different concentrations. (b) Spectral peak vs. concentration showing departure from linearity beginning at ~0.3 mg/mL.

orders of magnitude less than what is used for absorbance. For example, using QDs, absorbance spectra are usually collected with samples in the micromolar range. Such a concentration would saturate a fluorescence spectrometer for dots of reasonable quality; only a few nanomolars is needed for emission measurements.

- This sounds obvious but causes a lot of problems. Evaporation of the solvent naturally causes the sample to become more concentrated, changing its absorbance measurements. This can be a serious problem in experiments that go for extended periods, such as bacterial growth curves. Evaporation can make it look as if dead bacteria are growing again, when the increased optical density is just due to loss of water (**Figure 16.4**). Evaporation can occur very rapidly with samples in organic solvent.

- Many glass and plastic cuvettes do not transmit UV light and are designed only for the Vis part of UV–Vis (>380 nm). When measuring UV, use a quartz cuvette or make sure the plastic is intended for UV.

- Solvents, as well, have cutoff points below which they absorb all of the light. For water, this value is 180 nm, so UV spectra are often taken in water. Ethanol is also a good choice, with a cutoff of 205 nm. The value for toluene does not permit resolution of proteins (285 nm), and for acetone the value is 329 nm.

- Absorbance spectrophotometers need regular calibration. If not calibrated, they can show errors in wavelength (e.g., DNA peaks being at $A_{263}$ instead of $A_{260}$), resolution, and sensitivity. Calibration should usually be done by a professional, though kits are available for home testing.

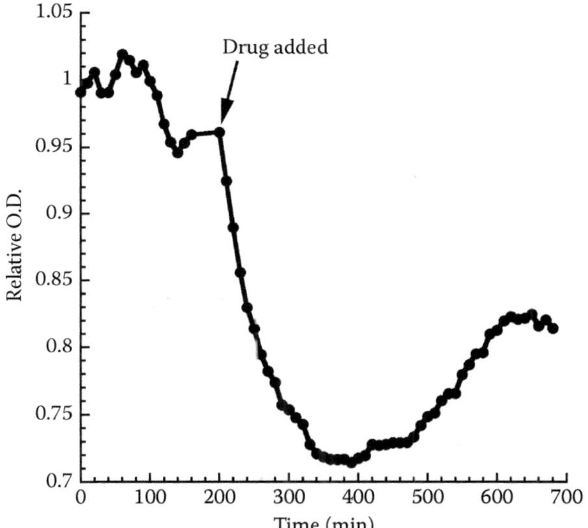



**Figure 16.4 Artifacts of drying.** *E. coli* optical density showing a decrease in cell density following drug addition, then an apparent "recovery" after ~400 min. This was in fact an artifact caused by medium evaporation.

# 16.4 FLUORESCENCE SPECTROSCOPY

In previous chapters, we have made many references to fluorophores, or objects/molecules that fluoresce (or in some cases luminesce). The physics of fluorescence is discussed in Chapter 7. In this section, we will discuss the instrumentation needed to collect basic fluorescence spectra, as well as some tools for characterizing fluorophores using this technique.

## Instrumentation

A fluorescence spectrometer, or spectrofluorometer, consists of the following features:

- A light source for excitation, usually a Xe lamp. Xe lamps have the flattest possible spectrum across the UV–Vis, unlike other lamps such as Hg that show sharp peaks (refer to Chapter 7, specifically Figure 7.22 for the spectrum of Xe vs. Hg lamps).

- A monochromator for the excitation light.

- A slit (usually 0.1 nm increment) to select the incident light. The wider the slit, the more the energy and the bigger the effective volume sampled in the cuvette, at the expense of spectral resolution.

- A beam splitter connecting a reference cell to a reference PMT. A small fraction of the incident beam is deflected here to correct the emission signal for input inhomogeneity.

- The sample compartment. The emission is usually read perpendicular to the excitation beam, but not always.

- Selectable optical filters on the emission path. These are used to eliminate interfering wavelengths (e.g., Rayleigh scattering, harmonics) and for other applications such as polarization.

- A slit for selecting the bandwidth of emitted light.

- An emission monochromator.

- The sample PMT.

A schematic is shown in **Figure 16.5**.

Usually both *excitation* and *emission* spectra are collected for a sample. The excitation spectrum varies the excitation wavelength in designated increments and holds the emission monochromator at a specified point (usually the emission maximum). The emission spectrum holds the excitation wavelength fixed (with a specified bandwidth) and varies the emission filter (**Figure 16.6**).

The properties of the light source, monochromators, and PMTs are more important for fluorescence than for absorbance spectroscopy. That is because for the latter, a blank sample can be used to correct for inhomogeneities in the transmitted light. For fluorescence, however, there is no blank; the measurements are absolute rather than relative. For most applications, it is usually sufficient to have high-quality

**Figure 16.5 Simplified light path for a typical spectrofluorometer.** The lamp is usually a Xe lamp. The light from the sample is usually collected at a right angle, though this is not shown here.

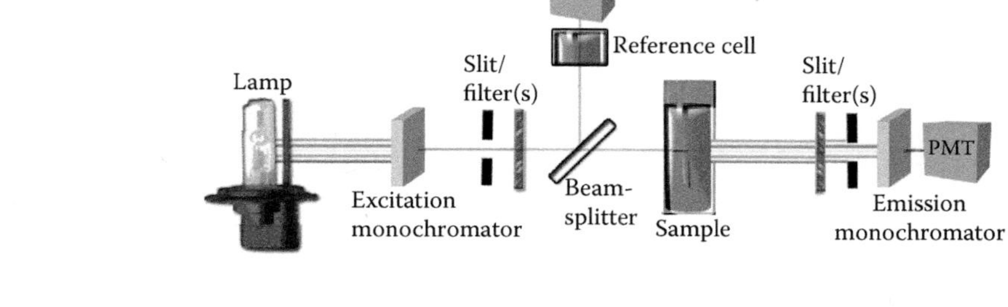


**Figure 16.6 Excitation and emission spectra for the dye acridine orange.** The excitation spectrum was taken with emission held at 540 ± 10 nm, and the emission was excited at 480 ± 10 nm.

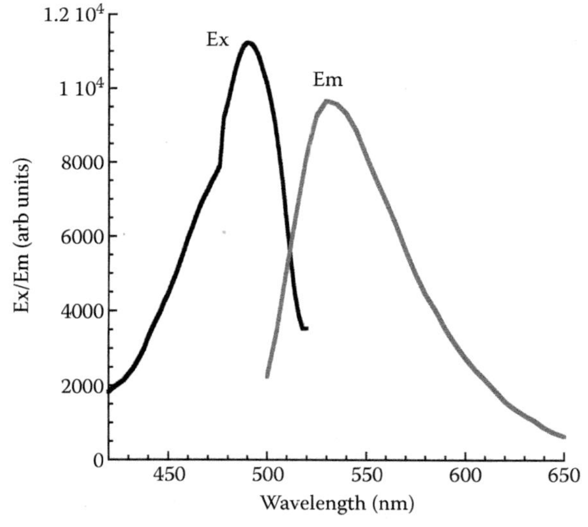

optical components and a reference detector, and to realize that emission spectra taken on one instrument may not completely overlap those taken on another. However, if spectra need to be corrected, there are reasonably straightforward ways to do this. Correcting emission spectra may be done by comparing the measured spectrum with the published standard spectrum of a known substance such as β-carboline, and adjusting for any artifacts. Another way to do this is to measure then intensity of the lamp vs. wavelength, $I(\lambda)$, using the ordinary detection system with no sample. The lamp output vs. wavelength, $L(\lambda)$, should be provided by the manufacturer. The sensitivity of the detector vs. wavelength is then $S(\lambda) = I(\lambda)/L(\lambda)$, and the measured spectrum may be divided by this factor to correct it.

Most spectrofluorometers use PMTs for detection, but they may be used in either *photon counting mode* or *analog mode*. The former makes use of the fact that each photoelectron results in a burst of $10^5$–$10^6$ electrons, which can be detected as a peak on the PMT anode. This is the most sensitive mode and is used by high-end spectrofluorometers. Analog mode measures the average photocurrent and incorporates the noise into the signal, often losing the signal for low-emission samples.

The choice of a spectrofluorometer should be based in a large part on the required sensitivity and time resolution, as well as the ability to measure luminescence and phosphorescence if needed. There are also a number of accessories that may be necessary for certain experiments. Although they may be available as add-ons later, it is good to know how adaptable an instrument is if you foresee using it for a wide range of applications. Some of these accessories include

- Polarizers (for excitation and/or emission); needed for fluorescence polarization experiments, discussed below.

- Sample holders for different geometries. Usually light is collected perpendicular to the sample cuvette, but this does not work for dense or solid samples. Other geometries and holders are often available (**Figure 16.7**).

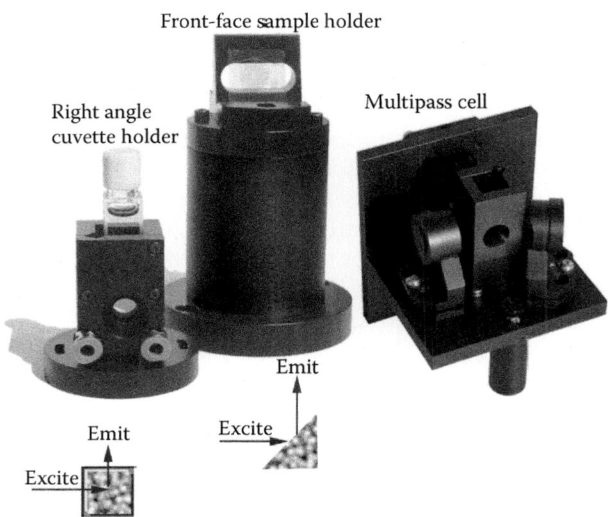


**Figure 16.7 Different sample holders and geometries for fluorescence.** A cuvette holder usually collects at right angles from excitation through a liquid sample. This is not appropriate for samples with very high optical density, including solids, for which a front-face geometry may be used. For very weakly fluorescent samples, a multipass cell uses mirrors to reflect the excitation light multiple times through the sample. (Images of sample holders courtesy of PicoQuant GmbH, http://www.picoquant.com)

- Temperature control. Adapters are available for a wide range of temperatures, including cryogenic temperatures.

- Fiber optics. The ability to use an extended fiber optic to read from an external source—such as an electrochemistry cell—can greatly expand the possible applications.

- The ability to read multiwell plates. Usually, biologists looking to read fluorescence from culture plates (from 6 to 384 wells) use a dedicated instrument called a *microplate reader* or plate reader. However, microplate accessories are also available for some spectrofluorometers.

As spectrofluorometers are costly, most laboratories have only one, and making the right choice for the desired experiments is important. It is a good idea to get demonstrations from various manufacturers when making the instrument selection. We conclude this section with two ways of testing sensitivity for aid in choosing or calibrating instruments:

- *Fluorescein dilution test.* Make serial dilutions of fluorescein until no signal can be detected over the noise. A good instrument should be able to detect 50–100 fM fluorescein. When comparing two instruments, make sure to use the same acquisition parameters and bandwidths.

- *Water Raman test.* This is a test of signal to noise, so is quite valuable for comparing instruments. This peak is seen with 350 nm excitation/397 nm emission (usually with 5 nm band-pass filters on the excitation). The noise can be measured anywhere where there is no Raman signal from water, e.g., at 450 nm. The S/N is given by (peak-background)/sqrt(background).

## Caveats and sources of error

As with absorption, using samples that are too concentrated is an important source of error. Too high of a concentration leads to scattering and a reduction in signal. There are also a few other common types of errors that are not seen with absorption spectroscopy, in particular contaminations in the spectrum due to Rayleigh scattering. Rayleigh scattering is elastic scattering of light, so it refers to photons of the same energy as the excitation light being accidentally collected as part of the emission spectrum. This can occur due to injudicious choice of filters and slits. Because fluorescence is quite weak relative to elastic scattering, it only takes a little bit of Rayleigh scattering to ruin the spectrum. Rayleigh scattering is also seen at the harmonics, usually at twice the wavelength of excitation (e.g., 600 nm for 300 nm excitation; **Figure 16.8**). It is important to note that if an instrument is out of calibration, a Rayleigh peak may not look as if it is at twice the excitation wavelength, so it might be mistaken for something real. Do not make this mistake. Any time a "fluorescence" peak appears to be so bright that it saturates the instrument, suspect Rayleigh scattering. Other sources of error include dust or other scattering particles, which add noise, and contamination of cuvettes with minute amounts of highly fluorescent materials.

Fluorescence spectroscopy can be used to investigate interactions between or among fluorophores, and many inferences about structure and photophysics

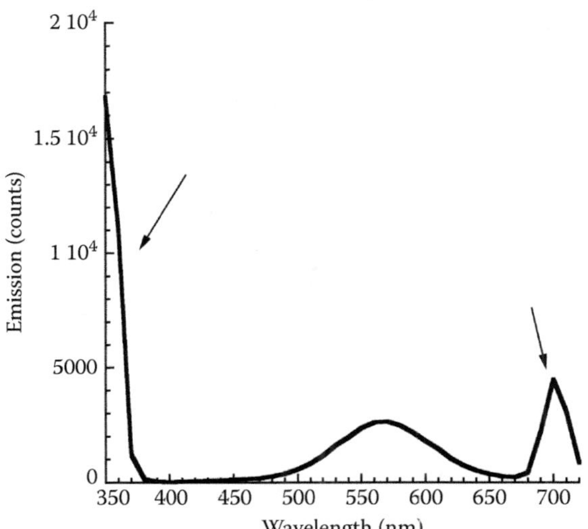

**Figure 16.8 Rayleigh scattering.** Sample of quantum dots excited at 350 nm. Note the remains of the peak around the excitation wavelength, and the secondary peak at 700 nm (arrows). Also note the size of these peaks relative to the desired emission peak at 570 nm, as well as in absolute number of counts.

can be made from carefully done measurements. The following sections give an introduction to a few of the common methods that are relevant to the earlier chapters. For a more thorough treatment, we refer you to the references at the end of the chapter.

## Applications of fluorescence spectroscopy: Quenching

Measurements of *fluorescence quenching* can give insight into a variety of processes that lead to reduced emission from a fluorophore: energy or electron transfer, complex formation, excited-state reactions, and more. *Collisional* quenching occurs when a quencher diffuses to an excited-state fluorophore, causing the latter to return to its ground state without fluorescence (**Figure 16.9a**). *Static* quenching occurs due to the formation of a nonfluorescent complex or conjugate between a fluorophore and a quencher (**Figure 16.9b**). Many different quenchers have been identified, some of which are quite specific to certain fluorophores. Collisional quenching is described by the Stern–Volmer equation, which relates the initial emission intensity $I_0$ to the measured intensity in the presence of quencher $I$ as a function of quencher concentration $[Q]$ and two parameters that may not be

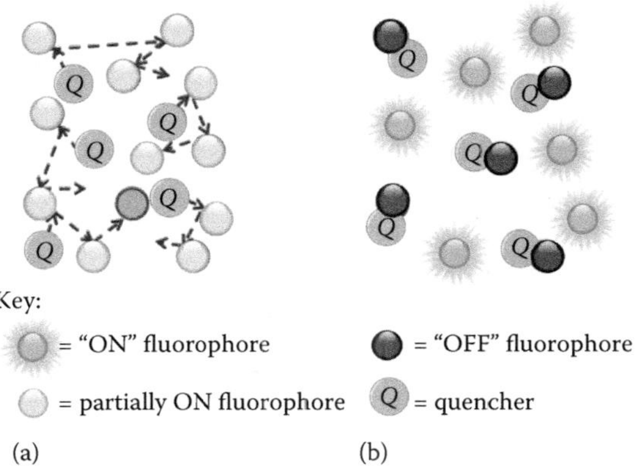



**Figure 16.9 (a) Dynamic quenching vs. (b) static quenching.**

known: the *biomolecular quenching constant* $k_q$ and the fluorophore lifetime $\tau$. The product of these two is the *Stern–Volmer constant* $K_D$:

$$\frac{I_0}{I} = 1 + k_q\tau[Q] \equiv 1 + K_D[Q] \tag{16.2}$$

A Stern–Volmer plot is made by plotting $I_0/I$ vs. $[Q]$. Pure collisional quenching, where the fluorophore has only one population of molecules each of which is equally accessible to the quencher, results in a linear relationship. Deviations from linearity are important and often used in molecular biophysics to identify "hidden" populations of fluorophores—for example, tryptophan residues that are inside a protein. Protein conformational changes that change the accessibility of these residues can give insight into how the protein works.

The quenching constant can also give insight into the quenching process. The value of $k_q$ is the product of the frequency of collisions $k_0$ and the probability that a collision will cause quenching, $f_Q$. If the diffusion coefficients $D$ of the two species, fluorophore and quencher, are known, then $k_0$ can be calculated using the Smoluchowski equation:

$$k_0 = \frac{4\pi N_A}{1000} R(D_f + D_q) \tag{16.3}$$

where $R$ is the sum of the two species' molecular radii, and the factor of 1000 is a correction for units of concentration in M. It is important to note that the diffusion coefficients can be tricky and should be measured whenever possible. The Stokes–Einstein equation is often off by several-fold depending upon the system. It is also important to mention that any quenching measurements should ensure that the reduction in fluorescence intensity does not have another cause, such as increased scattering caused by the quencher molecules.

Quenching has many uses in molecular biophysics apart from simple photophysical studies. Some examples are as follows: gaining insight into protein structure using the accessibility of tryptophan or tyrosine residues; studies of processes such as photosynthesis that involve fluorescent molecules; and the construction of so-called *molecular beacons*. Because of this, there are almost as many commercially available quenchers as there are fluorophores. Some examples of quenchers and their target dyes are given in **Table 16.1**.

Static quenching also results in a linear reduction in emission intensity with concentration. There are several ways to distinguish static vs. dynamic quenching. Raising the temperature usually leads to increased dynamic quenching (due to faster diffusion) and to unchanged or even reduced static quenching (depending upon the stability of the complexes). However, the best way to distinguish the two processes is with lifetime measurements, discussed below.

## Applications of fluorescence spectroscopy: Anisotropy

*Fluorescence anisotropy r* refers to the degree to which the fluorescence emission of a group of fluorophores is polarized when excited with polarized light. It is defined as a function of the intensity observed when the emission polarizer

**Table 16.1**

Examples of Quenchers and Their Target Fluorophores, with References or Supplier Name

| Quencher | Fluorophore | Reference/Manufacturer |
|---|---|---|
| Acrylamide, peroxide, imidazole, succinimide, iodide | Tryptophan (protein) | Lehrer, S. S. (1967). The selective quenching of tryptophan fluorescence in proteins by iodide ion: Lysozyme in the presence and absence of substrate. *Biochemical and Biophysical Research Communications* 29, 767–772.<br>Vanderkooi, J. M., Englander, S. W., Papp, S., Wright, W. W., and Owen, C. S. (1990). Long-range electron exchange measured in proteins by quenching of tryptophan phosphorescence. *Proceedings of the National Academy of Sciences of the U S A* 87, 5099–5103.<br>Walters, J., Milam, S. L., and Clark, A. C. (2009). Practical approaches to protein folding and assembly: Spectroscopic strategies in thermodynamics and kinetics. *Methods in Enzymology: Biothermodynamics A* 455, 1–39. |
| Metal nanoparticles | Organic dyes | Huang, C. C., Chiang, C. K., Lin, Z. H., Lee, K. H., and Chang, H. T. (2008). Bioconjugated gold nanodots and nanoparticles for protein assays based on photoluminescence quenching. *Analytical Chemistry* 80, 1497–1504. |
| Quinones | Chlorophyll | Karukstis, K. K., and Monell, C. R. (1989). Reversal of quinone-induced chlorophyll fluorescence quenching. *Biochimica et Biophysica Acta* 973, 124–130. |
| Disulfides, oxygen, phosphate | Tyrosine (protein) | Lakowicz, J. R., and Maliwal, B. P. (1983). Oxygen quenching and fluorescence depolarization of tyrosine residues in proteins. *Journal of Biological Chemistry* 258, 4794–4801. |
| Deep Dark/Eclipse | Organic dyes | Eurogentec<br>Marras, S. A. E. (2008). Interactive fluorophore and quencher pairs for labeling fluorescent nucleic acid hybridization probes. *Molecular Biotechnology* 38, 247–255. |
| Black Hole | Organic dyes of selected wavelengths | Biosearch Technologies |
| Iowa Black | Organic dyes of selected wavelengths | Integrated DNA Technologies |
| BlackBerry | Long-wavelength dyes | Berry & Associates |

is parallel to the polarized excitation, and the intensity observed when it is perpendicular:

$$r = \frac{I_\parallel - I_\perp}{I_\parallel + 2I_\perp} \qquad (16.4)$$

Why is fluorescence emission polarized? Recall from **Chapter 1** that the probability of occurrence of a transition is related to the transition dipole moment between two states.

If linearly polarized light is used to excite a solution of fluorophores, molecules with transition moments parallel to the polarization plane of the light are excited preferentially, and those with vectors perpendicular are not excited at all. This leads to a biased population of excited molecules, but for small molecules that diffuse more rapidly than the rate of emission, the emission quickly depolarizes and the anisotropy is near 0. For larger molecules or viscous solutions, the anisotropy will be related to the rate of rotational diffusion of the molecules. For large molecules such as proteins, this can provide important structural information.

The *fundamental anisotropy* $r_0$ of a fluorophore refers to the anisotropy that is measured in the absence of rotational diffusion. It is related to the angle $\alpha$ between the absorption and emission dipole moments of the fluorophore:

$$r_0 = \frac{2}{5}\left(\frac{3\cos^2\alpha - 1}{2}\right) \tag{16.5}$$

The derivation of this equation is too complex for this chapter, but is found in the references by Perrin. It can be seen that if $\alpha = 0$ (that is, the moments are colinear), then $r_0 = 0.4$; it can also be seen that fundamental anisotropies can be negative. These values are measured at low temperatures in solvents that form clear glasses, such as propylene glycol, that prevent molecular rotation, and can be found in the literature for a wide range of commonly used dyes. The absorption dipole moment varies with wavelength, so the angle will have different values at different excitation energies.

There are some cases in which $r_0$ has surprisingly low values, especially with benzene compounds. This can be shown to be due to the symmetry of the molecule. Usually a fluorophore has a symmetry axis along which the transition is maximally excited—call this $\Gamma_1$. A perpendicular axis $\Gamma_2$ results in minimum excitation, and the third axis $\Gamma_3$ is perpendicular to the other two. The fundamental anisotropy can then be defined as (for $\alpha = 0$)

$$r_0 = \frac{3}{5}\sum_{i=1}^{3}\Gamma_i^2 - \frac{1}{5} \tag{16.6}$$

So in the case of a completely symmetric molecule (all $\Gamma_i = 1/3$), the anisotropy value is 0. When two values are identical, the value is 0.1. This latter value is seen in symmetrically planar molecules such as benzene derivatives (**Figure 16.10**).

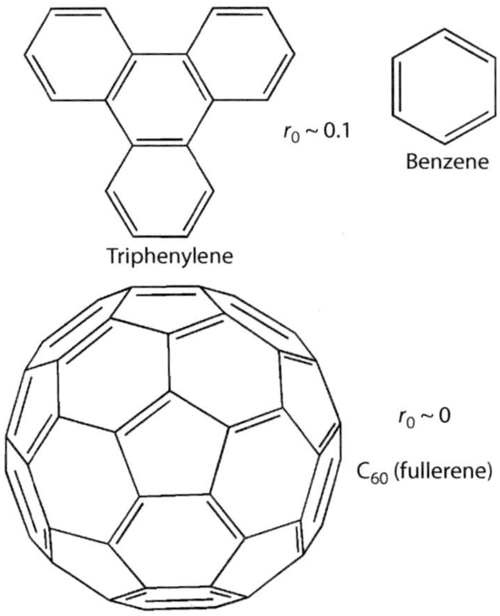

**Figure 16.10 Reduction in fundamental anisotropy due to molecular symmetry.** For molecules with threefold symmetry, such as benzene and triphenylene, $r_0$ is reduced to ~0.1. For molecules that are symmetric about all axes, it is expected that the anisotropy will be 0. This has been experimentally confirmed using $C_{60}$.

Usually anisotropy measurements are made in dilute nonviscous solutions at room temperature. In this case, the rotational motion that occurs between excitation and emission will give a value of $r < r_0$. If the molecule rotating is a sphere, then the time-resolved anisotropy $r(t)$ decays as a single exponential related to the diffusion coefficient $D$:

$$r(t) = r_0 e^{-6Dt} \equiv r_0 \exp\left(-\frac{t}{\tau_{\text{rot}}}\right) \tag{16.7}$$

The value of $\tau_{\text{rot}}$, called the rotational correlation time, can be estimated by the Stokes–Einstein equation to be a function of the molar volume $V$ of the protein and the viscosity of the solution $\eta$:

$$\tau_{\text{rot}} = \frac{V\eta}{RT} \tag{16.8}$$

Now assume that the fluorescence intensity $I(t)$ also shows single-exponential decay behavior with a characteristic lifetime $\tau_f$ (this is true of most simple fluorophores, but not of particles such as QDs; see next section). The steady-state anisotropy is then given by the average of the anisotropy decay over the intensity decay:

$$r = \frac{\displaystyle\int_0^\infty I(t)r(t)\,dt}{\displaystyle\int_0^\infty I(t)\,dt} = \frac{\displaystyle\int_0^\infty I_0 e^{-t/\tau_f} r_0 e^{-t/\tau_{\text{rot}}}\,dt}{\displaystyle\int_0^\infty I_0 e^{-t/\tau_f}\,dt} = \frac{r_0}{1+\left(\dfrac{\tau_f}{\tau_{\text{rot}}}\right)} \tag{16.9}$$

Rearrangement of the right-hand side gives the *Perrin equation*, originally derived in a much different fashion (see references):

$$\frac{r_0}{f} = 1 + \frac{\tau_f}{\tau_{\text{rot}}} \tag{16.10}$$

Perrin plots are used to determine the protein volume by plotting $1/r$ vs. $T/\eta$ or $T\tau_f/\eta$. The plots should be linear up to the denaturation temperature of the protein. **Practical Tips 16.1** gives a protocol for measuring $\tau_{\text{rot}}$ of bovine serum albumin (BSA) by binding it noncovalently to a dye.

Many spectrophotometers come with polarization capability or with polarization accessories that may be added on after the fact. These may be motorized or manual, and involve both an excitation and emission polarizer. For all instruments, an instrument correction factor should be calculated, which is measured using horizontally polarized excitation, making both polarizers perpendicular to the excitation polarization. Any difference in the monitored vertical intensity, $I_{\text{HV}}$, and horizontal intensity, $I_{\text{HH}}$, is thus due to the properties of the detection system

---

**PRACTICAL TIPS 16.1:   HYDRODYNAMIC VOLUME OF BSA USING ANISOTROPY**

*Introduction.* This example experiment makes use of two useful features of the dye 8-anilino-1-naphthalene sulfonic acid (ANS): it binds tightly to bovine serum albumin (BSA), and it is nonfluorescent in water (so there is essentially no background signal). Neither of these features is generalizable to all proteins. Experimental procedure:

- Determine the G factor for your instrument.
- Dissolve 0.4 mg/mL BSA in water; add 8 µM ANS.
- Measure $I_{VV}$ and $I_{VH}$ for temperatures from 25°C to 60°C (~8–10 data points total).
- Calculate $r$ using **Equation 12.11**.
- Plot $1/r$ vs. $T/\eta$, assuming that the viscosity of the solution is the same as that of water at a given temperature.
- Determine the slope of the graph. Compare the calculated hydrodynamic volume with that obtained using x-ray scattering (0.104 m³/mol).

---

(monochromators, filters, etc.; **Figure 16.11**). So the correction factor $G$ is given by $I_{HV}/I_{HH}$, and ideally should be 1. In cases where it is not, the anisotropy can be rewritten as

$$r = \frac{I_{VV} - GI_{VH}}{I_{VV} + 2GI_{VH}} \tag{16.11}$$

where the first subscript represents the orientation of the excitation polarizer, and the second the orientation of the emission polarizer.

There are many other important applications of fluorescence anisotropy that we have not covered here. One key one is in the study of membranes, which are highly spatially anisotropic. Rod-like fluorophores with transition dipoles along their long axis are usually used as probes of these systems; their wobbling can give an idea of the "microviscosity" of the membrane (**Figure 16.12**). Also, for many proteins, the assumption of sphericity is very wrong, and more complex models must be developed. Some references are given at the end of the chapter to more advanced applications, and we will also

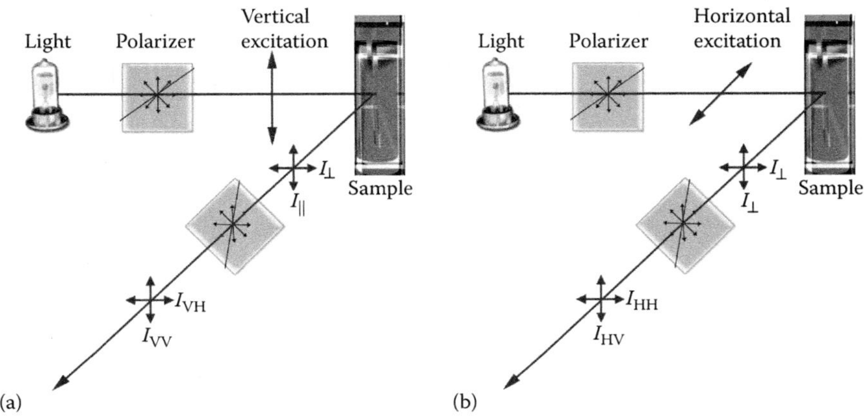


**Figure 16.11 Measuring responses of the instrument to vertically vs. horizontally polarized light.** (a) With vertical excitation, it is difficult to quantify the relationship between $I_\perp$, $I_\parallel$ and the measured $I_{VH}$, $I_{VV}$. (b) With horizontal excitation, both polarizer orientations are perpendicular to the excitation, so both components are equal to $I_\perp$. This means that any differences in $I_{HH}$, $I_{HV}$ are due to the instrument.

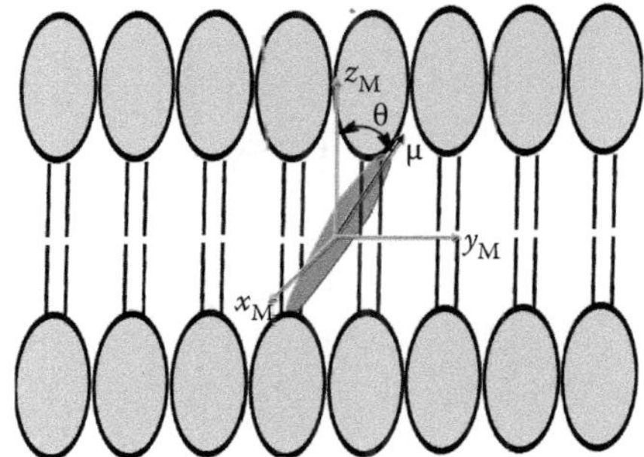

**Figure 16.12 Elongated fluorophore within a membrane.** The membrane has a coordinate system $x_M$, $y_M$, $z_M$, and the fluorophore vector makes an angle $\theta$ with it. The order parameter is a function of this angle and gives information about the potential in which the molecule moves.

return to the subject of anisotropy in the discussion of time-resolved emission spectra (Section 16.5).

## Applications of fluorescence spectroscopy: Energy transfer

*Fluorescence resonance energy transfer* (FRET)—more precisely called simply RET because there is no transfer of a photon—is one of the most commonly used techniques in molecular biophysics. Its principle is simple: when the emission spectrum of a fluorescent *donor* overlaps the excitation of an *acceptor*, then excitation of the donor causes emission at the acceptor wavelength (Figure 16.13). Ideally the donor and acceptor excitation bands should have an area that is free of overlap, because excitation of both at once make the data difficult to interpret. For example, QDs make excellent donors but poor acceptors, since their absorbance spectra are so broad. It is also possible to have a nonfluorescent acceptor; in this case, FRET is measured by donor quenching. This can be more difficult to interpret than FRET using a fluorescent acceptor, since quenching can result from many different mechanisms, including simply photobleaching.

One of the reasons FRET is so useful is its distance dependence. It can be used to measure distances from one residue or protein to another; to observe protein conformational changes in response to a ligand or other signal; or in biosensing, to observe the attachment of a target (Figure 16.14).

The theory of the distance dependence is somewhat complicated, and some references are given at the end of the chapter to the derivations for different systems. For a single donor and acceptor separated by a distance $r$, the rate of energy transfer $k_T$ is given by

$$k_T(r) = \frac{Q_D \kappa^2}{\tau_D r^6} \left( \frac{9000(\ln 10)}{128\pi^5 N_A n^4} \right) J(\lambda) \tag{16.12}$$

where $Q_D$ is the quantum yield of the donor; $n$ is the refractive index (taken as 1.4 in aqueous medium); $\tau_D$ is the donor lifetime; $\kappa$ describes the relative donor and acceptor spatial orientation (taken to be two-thirds for dynamic random averaging); and $J(\lambda)$ is the *overlap integral* describing the degree of overlap between the donor emission spectrum and the acceptor absorption spectrum.

**Figure 16.13 FRET between orange-emitting quantum dots and the membrane probe DiD.** (a) Note the overlap of the QD emission with the "shoulder" of the DiD absorption. The spectra are well separated but the overlap remains significant, so this is a good FRET pair. Excitation at 400 nm will excite the QDs but not the DiD. (b) As DiD concentration increases, excitation at 400 nm leads to quenching of the QD emission and growth of the DiD emission peak at ~675 nm.

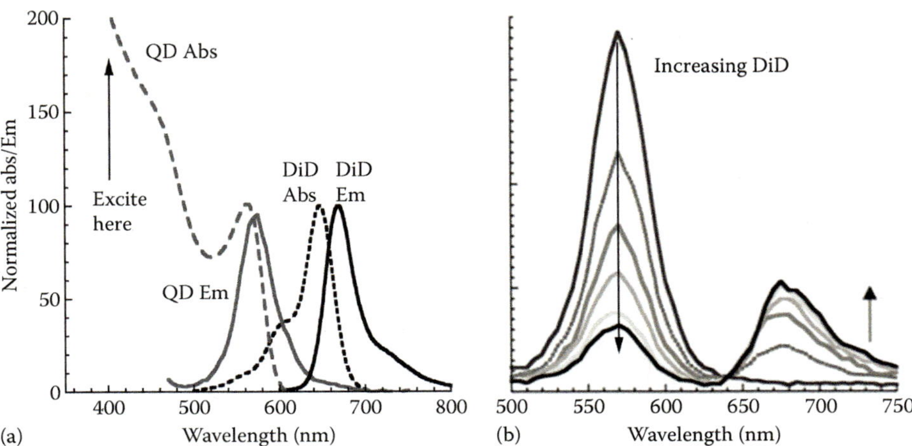


It can be calculated from the fluorescence intensity of the donor $F_D(\lambda)$ (measured from a normalized curve) and the extinction coefficient of the acceptor $\varepsilon_A(\lambda)$:

$$J(\lambda) = \int_0^\infty F_D(\lambda)\varepsilon_A(\lambda)\lambda^4 \, d\lambda, \qquad (16.13)$$

where $J(\lambda)$ is in units of $M^{-1}\,cm^3$.

The $r$-dependence of **Equation 16.12** can be removed by defining a *Förster distance* $R_0$ where the energy transfer rate is equal to the donor decay rate ($k_T(r) = 1/\tau_D$):

$$R_0^6 = Q_D\kappa^2\left(\frac{9000(\ln 10)}{128\pi^5 N_A n^4}\right)J(\lambda) \qquad (16.14)$$

**Figure 16.14 Ways to use FRET.** (a) Receptor and ligand, each labeled with a donor or acceptor, will show FRET when they bind. (b) Receptor that shows significant conformational change when it binds a ligand can be labeled with a donor and acceptor such that FRET only becomes possible when ligand is bound. (c) Tethered ligand can be used that only shows FRET upon binding to its binding site. (d) Activity of proteases and other enzymes can be monitored by placing a donor and acceptor on either side of a sequence cleaved by the enzyme. Enzymatic activity results in a loss of FRET and the appearance of the donor fluorescence peak.

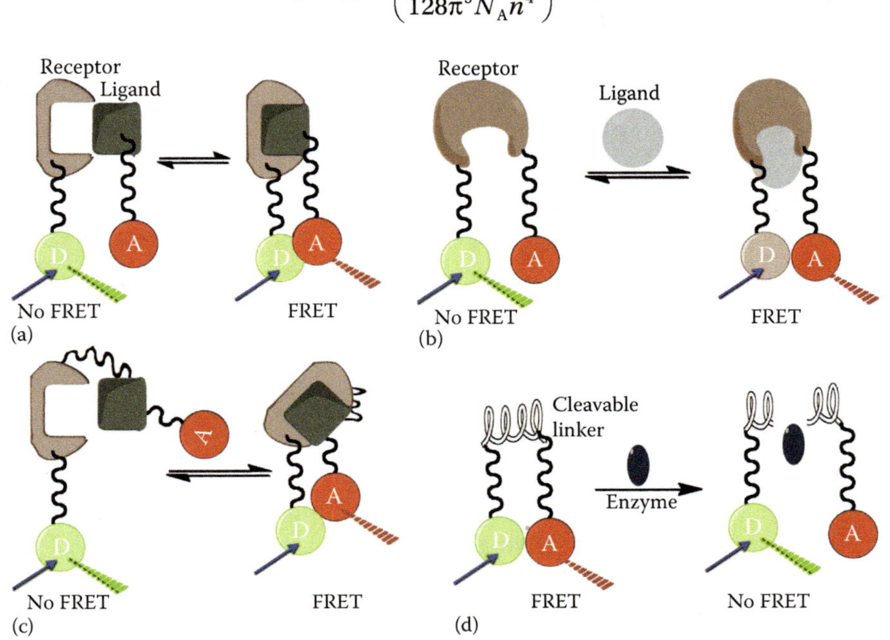

Then the energy transfer rate is given simply as

$$k_{\mathrm{T}}(r) = \frac{1}{\tau_{\mathrm{D}}}\left(\frac{R_0}{R}\right)^6 \tag{16.15}$$

The *FRET efficiency E* then shows dependence upon the rate vs. donor lifetime and on distance:

$$E = \frac{k_{\mathrm{T}}}{k_{\mathrm{T}} + \dfrac{1}{\tau_{\mathrm{D}}}} = \frac{R_0^6}{R_0^6 + r^6} \tag{16.16a}$$

Each pair of fluorophores has a specific Förster distance. These can be found in articles on FRET using specific pairs, and sometimes in reviews. **Table 16.2** gives the values for some common pairs. The distance limit for useful FRET experiments is about 10 nm.

Under realistic experimental conditions, you do not know $r$. What is measured are the extinction coefficients $\varepsilon$ of the donor $D$ and acceptor $A$ at the wavelength used for excitation, and the fluorescence intensity $I$ of the acceptor alone and the acceptor–donor mix. For practical purposes, therefore, the expression for $E$ becomes

$$E = \frac{\varepsilon_{\mathrm{A}}(\lambda_{\mathrm{ex}})}{\varepsilon_{\mathrm{D}}(\lambda_{\mathrm{ex}})}\left[\frac{I_{\mathrm{AD}}(\lambda_{\mathrm{em}})}{I_{\mathrm{A}}(\lambda_{\mathrm{em}})} - 1\right]\left(\frac{1}{f}\right) \tag{16.16b}$$

where $f$ is the fraction of acceptors labeled with donor (often 1).

It is important to remember that the formulas given above apply to two point fluorophores freely rotating. Under other circumstances, the distance dependence

**Table 16.2**

Förster Distances for Some Example Fluorophore Pairs

| Donor | Acceptor | $R_0$ (nm) |
|---|:---:|:---:|
| B-phycoerythrin | Cy5 | 7.2 |
| Fluorescein-5-isothiocyanate (FITC) | Erythrosin-5′-isothiocyanate | 6.2 |
| FITC | Eosin thiosemicarbazide | 6.1–6.4 |
| FITC | Eosin maleimide | 6.0 |
| FITC | Eosin-5-isothiocyanate | 5.4 |
| Lucifer yellow | Eosin maleimide | 5.3 |
| FITC | Tetramethylrhodamine (TMR) | 4.9–5.4 |
| FITC | N-acetylimidazole | 4.6 |
| Dansyl | Octadecylrhodamine | 4.3 |
| Pyrene | Coumarin | 3.9 |
| Monobromobimane | FITC | 3.8 |
| 5-iodo-acetamidofluorescein (IAF) | TMR | 3.7 |
| IAF | DiI | 3.5 |
| Naphthalene | Dansyl | 2.2 |

*Source:* Wu, P., and L. Brand, *Analytical Biochemistry* 218, 1–13, 1994.

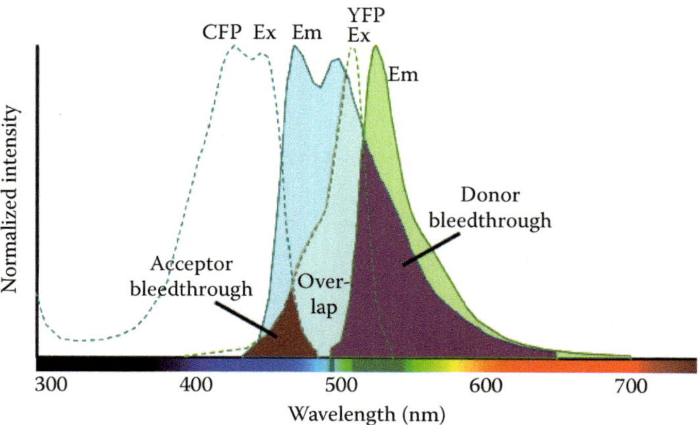


**Figure 16.15 Emission, excitation, and bleed-through with one of the most commonly used FRET protein pairs: CFP as the donor and YFP as the acceptor.**

of energy transfer may be altered. The most striking example of this is in the case of noble metal nanoparticles. A fluorophore dipole freely rotating at some distance from a noble metal surface will show FRET that drops off as $r^{-4}$, with a useful distance of >20 nm. The same applies when the "surface" is a nanoparticle, with the separation between the two larger than the particle. When the particle is of comparable size to the separation, the distance dependence is weaker than $r^{-4}$.

The use of FRET in confocal imaging experiments in cells represents a few particular challenges. The number of wavelengths for excitation and emission is limited, and if transfected cells are being used, the properties of fluorescent proteins must be considered. Proteins that have very different maturation rates and/or brightness values should not be used together. Spectral overlap should be maximized, but this results in some degree of both acceptor and donor spectral bleed-through (**Figure 16.15**).

There are a few techniques that can maximize FRET detection in imaging experiments. If available, spectral imaging using an AOTF or lifetime imaging (**Chapter 8**) will provide unambiguous FRET data. When these techniques are not available, ratiometric imaging with a careful choice of filters is usually used. However, it is necessary to determine the amount of spectral bleed-through under these conditions. Another alternative to reducing this bleed-through is by photobleaching the acceptor to <10% of its value, but this presents difficulties as well: it is slow; the donor may also bleach; and the whole cell must be bleached at once, or else fresh molecules can diffuse into the field of view. FRET experiments using cells therefore need to be designed with a degree of care. Recommended fluorescent protein pairs are given in **Table 16.3**; also keep in mind that new proteins are being discovered and designed on a regular basis, many of them specifically for FRET experiments.

## 16.5 TIME-RESOLVED EMISSION

A time-resolved fluorescence emission experiment in the time domain involves collection of photons emitted from a fluorescent sample at a selected wavelength or wavelengths at specific time intervals, usually nanoseconds or faster for

**Table 16.3**

Fluorescent Proteins That Can Be Used as FRET Pairs

| Donor | Acceptor | $Q_D$ | $\varepsilon_A$ (M$^{-1}$cm$^{-1}$) | $R_0$ (nm) | Brightness Ratio |
|---|---|---|---|---|---|
| EBFP2 | mEGFP | 0.56 | 57,500 | 4.8 | 1:2 |
| ECFP | EYFP | 0.40 | 83,400 | 4.9 | 1:4 |
| Cerulean | Venus | 0.62 | 92,200 | 5.4 | 1:2 |
| MiCy | mKO | 0.90 | 51,600 | 5.3 | 1:2 |
| TFP1 | mVenus | 0.85 | 92,200 | 5.1 | 1:1 |
| CyPet | YPet | 0.51 | 104,000 | 5.1 | 1:4.5 |
| EGFP | mCherry | 0.60 | 72,000 | 5.1 | 2.5:1 |
| Venus | mCherry | 0.57 | 72,000 | 5.7 | 3:1 |
| Venus | td-Tomato | 0.57 | 138,000 | 5.9 | 1:2 |
| Venus | mPlum | 0.57 | 41,000 | 5.2 | 13:1 |

*Source:*   Nikon MicroscopyU, http://www.microscopyu.com.

fluorescence phenomena. (Time-resolved measurements are also done in the frequency domain, but we will not discuss this here.) The result is an exponential decay curve with a characteristic lifetime $\tau$ (**Figure 16.16a**). In the case of complex molecules or nanostructures, the decay may be multiexponential. For example, in a protein, exposed tryptophan residues may show different lifetimes than buried residues, resulting in at least two different $\tau$ values. In this case, the intensity must be fit to a function

$$I = \sum_{i=1}^{n} A_i \, exp\left(-\frac{t}{\tau_i}\right) \tag{16.17}$$

and the average lifetime may be given as either an amplitude-weighted lifetime

$$\tau = \frac{\sum_{i=1}^{n} A_i \tau_i}{\sum_{i=1}^{n} A_i} \tag{16.18}$$

or an intensity-weighted lifetime

$$\tau = \frac{\sum_{i=1}^{n} A_i \tau_i^2}{\sum_{i=1}^{n} A_i \tau_i} \tag{16.19}$$

Most fluorophores have at most two lifetimes, but some cases are even more complex than can be handled by a sum of exponentials. For example, the fluorescence emission from QDs originates from a variety of surface trap states with different lifetimes. Very good quality QDs may show double-exponential lifetimes, but most show triple-exponential decays or more, where there is no good model for the physical meaning of the different time constants. One way to get a good estimate of the average lifetime under these circumstances is to fit the fluorescence intensity $I$ to a "stretched exponential" with a stretching factor $\beta$:

$$I = A \exp\left[-\left(\frac{t}{\tau}\right)^{\beta}\right] \tag{16.20}$$

with the average lifetime then given by (**Figure 16.16b and c**)

$$\frac{\tau}{\beta}\Gamma\left(\frac{1}{\beta}\right) \tag{16.21}$$

The applications of time-resolved measurements are myriad, and we refer you to texts on the subject for details. As a summary,

- Time-resolved measurements allow for discrimination between static and dynamic quenching. Static quenching does not affect the lifetimes, since all of the "off" fluorophores are "off," and all of the "on" fluorophores have normal lifetimes. Dynamic quenching does have an effect on lifetime proportional to the concentration of the quencher.

- Time-resolved anisotropy measurements can give insight into protein structure and conformational change, and the dynamics of nucleic acids and other polymers.

- Time-resolved RET measurements can show a distribution of distances between donors and acceptors, thus giving insight into the conformation(s) of the molecule(s) to which the donor and acceptor are linked.

Time-resolved emission experiments in the time domain are nearly always performed using the technique of *time-correlated single-photon counting* (TCSPC). We refer you in the references to the multiple textbooks devoted entirely to this technique. Its principle is based upon pulsed excitation of the sample with a fast source, usually a femtosecond laser. Each pulse is monitored by a fast photodiode to produce a start or reference signal. Each 50–100 pulses should result in one photon (more than this causes pulse pileup, and the sample should be diluted or a neutral density filter placed to reduce the signal). This single photon is detected by a photon counting board after a specific arrival time measured with respect to the reference pulse. Summing over tens of thousands of pulses yields a probability histogram of counts vs. time (with the time bin determined by the experiment). A schematic of a TCSPC system is given in **Figure 16.17**.

The time resolution of the instrumentation can be estimated by measuring the *instrument response function* (IRF), which is the curve obtained from a zero-lifetime sample (e.g., a scatterer with no emission filter; we use a dilute solution of coffee creamer). Most IRFs are tens of picoseconds to several nanoseconds wide (**Figure 16.18**). It is important to appreciate that the observed signal $I_{obs}(t)$ is a convolution of the measured signal $I(t)$ and the IRF $L(t)$:

$$I_{obs}(t) = \int_{0}^{t} L(t')I(t-t')dt' \tag{16.22}$$

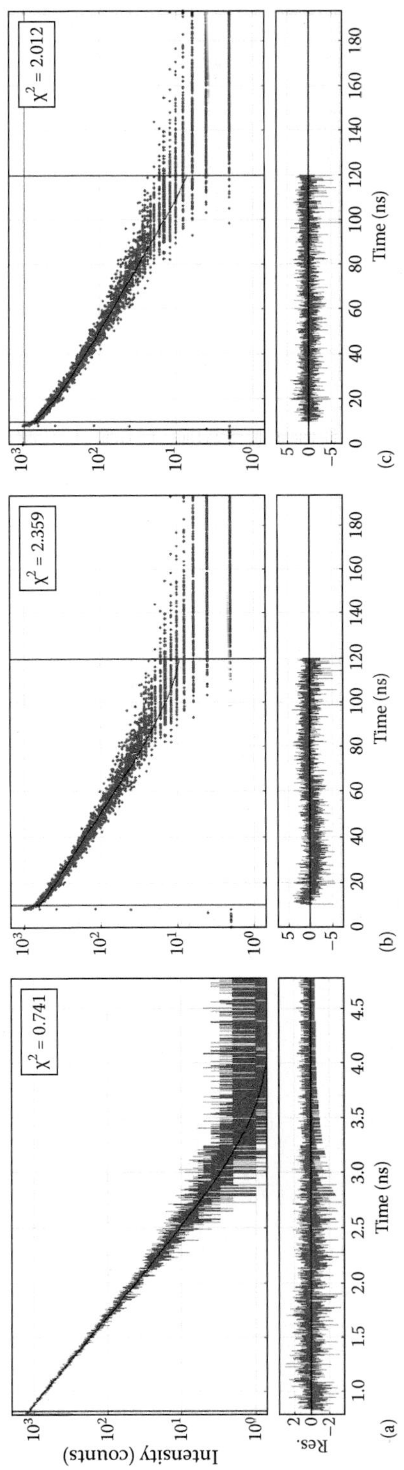

**Figure 16.16  Fluorescence lifetime decays.** (a) Single-exponential decay showing fit and residuals (Res.) for the dye Ru(bpy)3. (b) Lifetime decay of water-solubilized CdTe quantum dots, with triple-exponential fit. (c) Same decay with stretched-exponential fit. Note the small improvement in the residuals and $\chi^2$. Fits to one- and two-exponential models were very poor.

**Figure 16.17 Schematic of a TCSPC system.** fl = focal length. The band-pass filters remove any stray laser light (less than 450 nm and greater than 650 nm is appropriate for the Ti-sapphire laser pictured here).

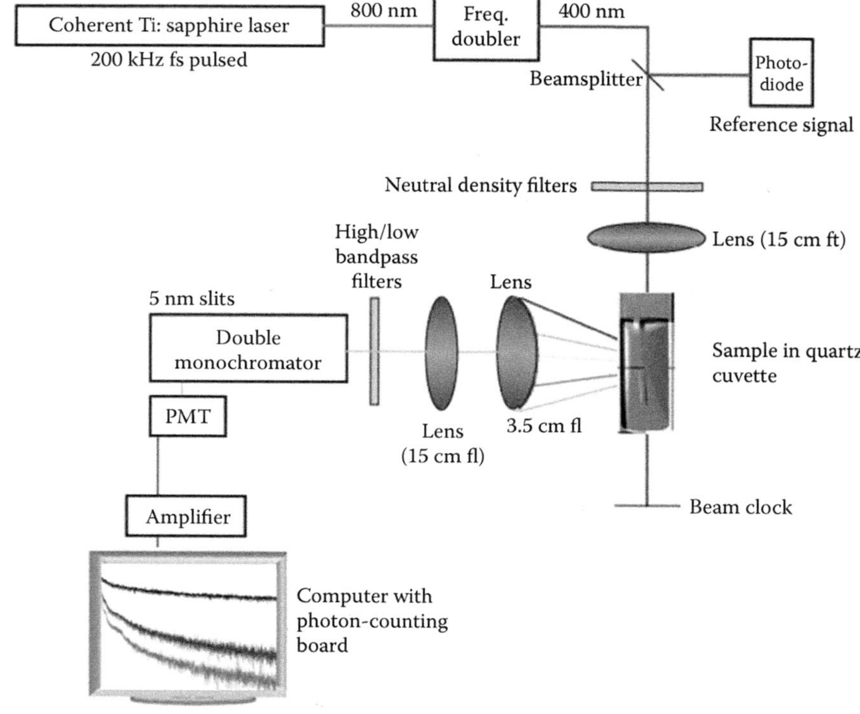


**Figure 16.18 IRF measured from the system in Figure 16.17.**

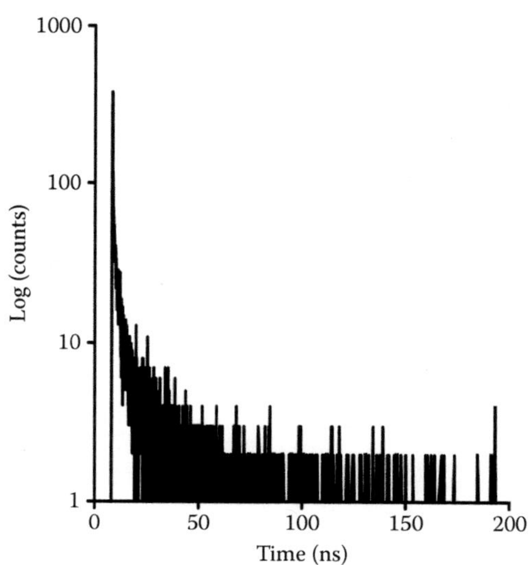


Deconvolution can be performed using TCSPC analysis packages, or may be coded at home using any of several mathematical algorithms. References to some of our favorites are at the end of the chapter.

The instrumentation for TCSPC is costly, finicky, and complex. Usually, time-resolved studies are done in laboratories dedicated to the subject, although "plug-and-play" TCSPC systems are becoming increasingly common and sometimes found in shared-use nanoscience or materials characterization facilities. Some of these systems come complete with a light source, usually a picosecond laser or pulsed LED; others contain all but the light source and are intended for use with picosecond or femtosecond lasers. The latter allow for optimum performance but require the researcher to purchase and maintain the laser. In some cases, monitoring of lifetimes at multiple wavelengths is possible; this can be useful for samples whose emission peak changes in the presence of conditions that change the lifetime. An example of this is in the study of fluorescent proteins, which often show substantial red shifts during maturation (from blue to green in GFP, from green to red in RFP). This technique is called time-resolved emission spectroscopy (TRES).

## 16.6 TIME-RESOLVED ABSORPTION*

Much as time-resolved fluorescence monitors the extinction of an excited state to a ground state over time by collecting the fluorescence of the sample, time-resolved absorption monitors the absorbance of photons by a system as a function of time. A time-resolved experiment can be thought of as two steady-state experiments: one where the system has been perturbed by a pump pulse ($OD_{on}$) and one lacking that perturbation ($OD_{off}$). The difference between these two signals ($\Delta OD$) is the quantity of interest in a time-resolved experiment, and expressed as

$$\Delta OD = OD_{on} - OD_{off}$$

$$OD_{on} = -\log\left(\frac{I_{on}}{I_{ref}}\right) \tag{16.23}$$

$$OD_{off} = -\log\left(\frac{I_{off}}{I_{ref}}\right)$$

where $I_{on}$, $I_{off}$, and $I_{ref}$ are the signal intensities for the pump, probe, and reference signal, respectively.

Time-resolved absorption experiments require pulsed-laser systems to create the pump and probe pulses. The pulse duration of a system determines the IRF as discussed for time-resolved emission, and so ultimately the timescale of the dynamics that can be investigated.

The pump pulse starts at $t = 0$, and the probe pulse is delayed by a defined time $\Delta t$. This is generally achieved by an optical delay stage, a crucial part of time-resolved experiments that sometimes must be specifically designed for the experiment in question. For example, it might be desirable to have a high repetition rate (> kHz) accompanied by a long delay range (> m). Designs for optical delay stages have been reported with these properties, but they are not widely available.

---

* Contributed by J. Saari, Kambhampati lab, Department of Chemistry, McGill University.

**Figure 16.19 Sample transient absorption data.**

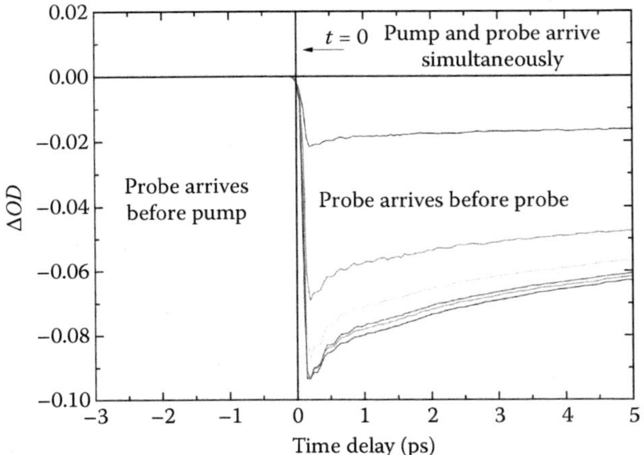


The pulse flight time must also be accounted for so that the pump and probe arrive with specific timing and can be distinguished. In ultrafast experiments, this is achieved by a fast-photodiode signal triggering an event upon pulse creation. This allows for the synchronization of all signals in the experiment. Once collected, data can be cast in the form of transient absorption spectra ($\Delta OD$ versus wavelength at a fixed $\Delta t$) or transient absorption measurements ($\Delta OD$ versus $\Delta t$ at a fixed wavelength). An example of transient absorption measurements is shown in **Figure 16.19**.

Time-resolved absorption systems are as diverse in cost and complexity as the timescales they can measure. Picosecond and nanosecond YAG, YLF, or excimer lasers combined with dye lasers or optical parametric oscillators/amplifiers are relatively inexpensive and provide for decent photochemistry and absorption instruments at these longer timescales. Titanium sapphire ultrafast systems are expensive on their own and also require the purchase of OPAs or OPOs for pump-bandwidth tuneability. Such systems also require vibration isolation, as optomechanics are susceptible to slight vibrations, changing their Poynting vectors. Detection at the femtosecond timescale also requires specialized equipment. Charge-coupled device cameras are often used for spectral acquisition, as are scanning monochromators; fast photodiodes can be used for transient acquisition. White-light continuum generation via the self-phase modulation of sapphire or salcite crystals allows for easy, stable, broadband visible probe sources in ultrafast systems. An example of a femtosecond absorption setup is given in **Figure 16.20**.

The number of possible applications of transient absorption spectroscopy is vast, and we refer you to textbooks on the subject for details. A few examples include the following:

- Studying the dynamics of electron-hole recombination, electron relaxation, and electron–phonon coupling in QDs.

- Studying the transient excited states of photosensitizing molecules, which are used in *photodynamic therapy* (PDT). In PDT, it is conventionally understood that excited photosensitizers transmit their energy to molecular oxygen via their triplet state; this leads to the generation of singlet

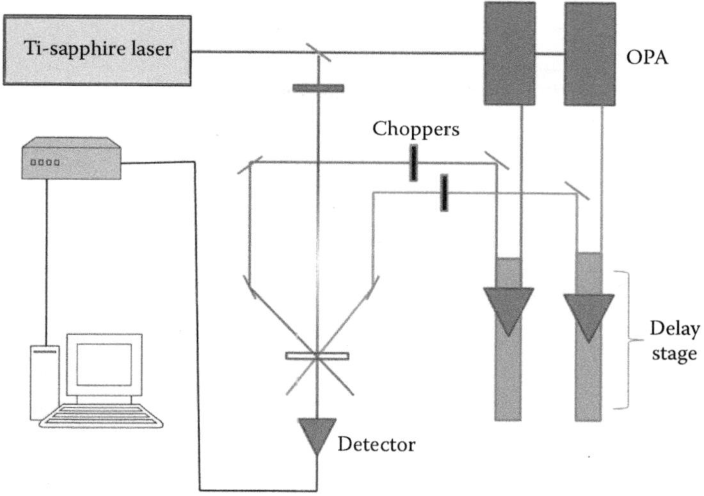

**Figure 16.20 Schematic of a setup for an ultrafast pump-probe transient absorption experiment.**



oxygen (a cytotoxic species). For this reason, ground state bleaching recovery measurements are useful in obtaining a triplet state yield, which help determine the efficacy of the photosensitizing drug.

- Studying the efficiency of photosynthetic reactions. As energy passes between molecules in the photosynthetic pathway, changes in their absorption spectra can be detected.

## 16.7  INFRARED SPECTROSCOPY

Mid-IR (2.5–15 μm wavelengths) excites molecular vibrations, so IR spectroscopy probes the vibrational states of a polyatomic molecule, yielding a distinct (though often messy) fingerprint. The selection rule arises from the transition dipole moment between the initial and final states, which for a vibrational change only is given by

$$\left\langle v_f \middle| \hat{\mu} \middle| v_i \right\rangle = \left\langle v_f \middle| x \middle| v_i \right\rangle \left( \frac{d\mu}{dx} \right) \tag{16.24}$$

where $dx$ is the displacement of one of the charges of the dipole relative to the other upon the vibrational motion. The take-home message from this general selection rule is that the dipole moment must change for a particular vibration to be *IR active*. Specific transition matrix elements may be calculated based upon the possible vibrations about a given bond; however, for complex molecules, the environment of the bond affects the energy, and so the signal will occur across a possible band corresponding to an organic functional group (Table 16.4). Interpretation of the spectra can be tricky, as the primary peaks must be assigned to a vibration (or a contaminant; Figure 16.21). This is a standard technique used in organic chemistry for identification of compounds; it also has many applications in forensics, geology, and other fields. IR spectrometers are being proposed for Mars exploration as they can identify minerals and traces of organic compounds that might be signatures of life. Extensive databases of IR spectra are available online.

## Table 16.4

IR Energy Bands of Stretching and Bending Vibrations of Selected Functional Groups

| Functional Group | Energy Range for Stretch | Peak Size | Assignment (Stretch) | Energy Range for Bend | Peak Size | Assignment (Bend) |
|---|---|---|---|---|---|---|
| Alkane ($CH_3$, $CH_2$, CH) | 2850–3000 | +++ | Alkane | 1350–1470<br>1370–1390<br>720–725 | ++<br>++<br>+ | $CH_{2,3}$ deformation<br>$CH_3$ deformation<br>$CH_2$ rocking |
| Alkene (double bond) | 3020–3100<br>1630–1680 | ++<br>++/+ | $=CH$ and $=CH_2$<br>C=C | 880–995<br>780–850<br>675–730 | +++<br>++<br>++ | $=CH$ and $=CH_2$<br>out of plane bend<br>RCH=CHR |
| Alkyne (triple bond) | 3300<br>2100–2250 | +++<br>+/++ | C–H<br>C≡ | 600–700 | +++ | C–H deformation |
| Aromatic | 3030<br>1600/1500 | ++/+++<br>+/++ | C–H<br>C=C in ring | 690–900 | ++/+++ | Ring bend/pucker |
| Alcohol | 3580–3650<br>3200–3550<br>970–1250 | ++/+++<br>+++<br>+++ | Free OH<br>H bonded OH<br>CO | 1330–1430<br>650–770 | ++<br>++/+ | In plane OH bend<br>Out of plane bend |
| Amine | 3400–3500<br>3300–3400<br>1000–1250 | +<br>+<br>++ | 1° amine<br>2° amine<br>C–N | 1550–1650<br>660–900 | ++/+++<br>++ | $NH_2$ scissoring<br>$NH_2$ and NH wag |
| Aldehyde/ketone | 2690–2840<br>1720–1740<br>1710–1720<br>1690<br>1675<br>1745<br>1780 | ++<br>+++<br>+++<br>+++<br>+++<br>+++<br>+++ | C–H aldehyde<br>C=O aldehyde<br>C=O ketone<br>Aryl ketone<br>Unsaturated<br>Cyclopentanone<br>Cyclobutanone | 1350–1360<br>1400–1450<br>1100 | +++<br>+++<br>++ | –$CH_3$ bend<br>–$CH_2$ bend<br>C–C–C bend |
| Carboxylic acids, etc. | 2500–3300<br>1705–1720<br>1210–1320<br>1785–1815<br>1750/1820<br>1040–1100<br>1735–1750<br>1000–1300<br>1630–1695 | ++<br>+++<br>++/+++<br>+++<br>+++<br>+++<br>+++<br>+++<br>+++ | OH acid<br>H-bonded C=O<br>O–C acid<br>C=O acyl halide<br>C=O anhydride<br>O–C anhydride<br>C=O ester<br>O–C ester<br>C=O amide | 1395–1440<br>1590–1650<br>1500–1560 | ++<br>++<br>++ | C–O–H bend<br>N–H bend 1° amide<br>N–H bend 2° amide |
| Nitrile | 2240–2260 | ++ | C≡N | | | |
| Sulfur compounds | 2250–2600<br>700–900<br>500–540<br>1050–1200<br>1030–1060<br>1345<br>1350–1450 | +<br>+++<br>+<br>+++<br>+++<br>+++<br>+++ | SH<br>S–OR<br>S–S<br>C=S<br>S=O sulfoxide<br>S=O sulfonate<br>S=O sulfate | | | |
| Phosphorous compounds | 2280–2440<br>2550–2700<br>900–1050<br>1100–1200<br>1230–1260<br>1100–1200<br>1200–1275 | ++<br>++ | P–H<br>(O=)PO–H<br>P–OR<br>P=O ester<br>P=O phosphonate<br>P=O phosphate<br>Phosphoramide | 950–1250 | + | P–H bend |
| Silicon compounds | 2100–2360<br>1000–1110<br>1250 ± 10 | +++<br>+++<br>+++ | S–H<br>Si–OR<br>Si–$CH_3$ | | | |

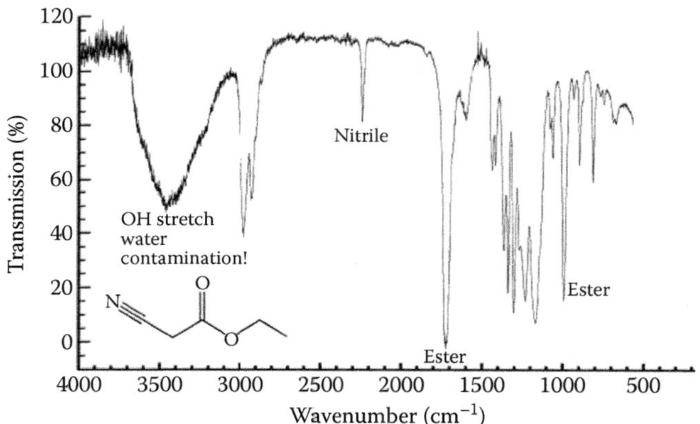



**Figure 16.21 Example IR spectrum for ethyl cyanoacetate.** Both the ester and nitrile can be clearly identified. The ester has multiple peaks, corresponding to the C=O and the C–O stretches. Notice that this sample has an unexpected O–H stretch, which is characteristic of $H_2O$ contamination. $CO_2$ is another common contaminant that should be recognized and ignored.

A variety of IR spectrometers are commercially available, and they can be found in individual investigators' labs as well as in shared facilities. A quartz–tungsten–halogen lamp can provide wavelengths up to 2500 nm. To go more deeply into the IR, most spectrometers use a relatively inexpensive igniter consisting of ceramic wires resistively heated to ~1500 K. These emit a classic black-body spectrum spanning ~1000–7000 cm$^{-1}$. Many sources can be air-cooled, whereas some must be water-cooled. Nearly all commercial IR spectrometers are *Fourier transform infrared (FTIR)* spectrometers, which have a Michaelson interferometer in the place of the diffraction grating seen in dispersive instruments (**Figure 16.22**). This allows all frequencies to be measured simultaneously, and is faster and more sensitive than scanning through all the wavelengths (compare the wavelength range with IR as opposed to visible spectroscopy and you will appreciate how slow scanning could be). The quality of the interferometer is the greatest determiner of the performance of the instrument. Data are reported as % transmittance ($T$) relative to a blank; absorbance may be calculated as $A = -\log(T)$.

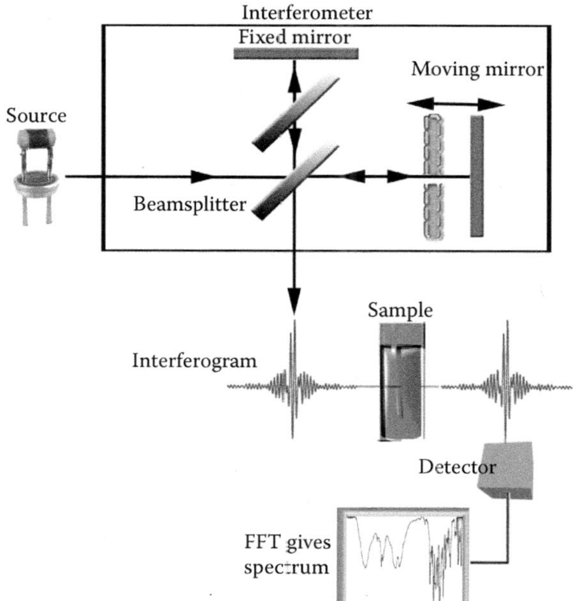



**Figure 16.22 Schematic of an FTIR spectrometer, showing a diagram of a Michaelson interferometer.** The mirror in the interferometer is the instrument's only moving part.

**Table 16.5**

IR Transmission Properties and Solvent Compatibilities of Salt Windows

| | NaCl | KBr | CsI | AgCl | Ge | ZnSe |
|---|---|---|---|---|---|---|
| Transmission Range (cm⁻¹) | 625–40,000 | 400–40,000 | 200–40,000 | 360–25,000 | 525–5500 | 454–20,000 |
| % Transmission | 91.5 (4 mm) | 90.5 (4 mm) | 92.0 (2 mm) | 84.0 (3 mm) | 50.0 (2 mm) | 65.0 (1 mm) |
| Refractive Index | 1.49 | 1.52 | 1.74 | 1.98 | 4.0 | 2.4 |
| Attacking Solvents | Water | Water | Water | Ammonium salts | $H_2SO_4$ | Strong acids, alkalis, chlorinated solvents |
| Cleaning Solvents | Anhydrous | Anhydrous | Anhydrous | Acetone Dichloromethane | Acetone Alcohols $H_2O$ | Acetone Alcohols $H_2O$ |

*Note:* The % transmission is given for a window of (thickness in cm).

Liquid samples for FTIR are deposited as thin films on or solutions between salt plates (NaCl, KBr, or other) as the salts do not contribute to the IR spectrum (it is important, however, that your sample not dissolve the plate). Solid samples or powders pressed into a pellet with salt (usually KBr) may also be used. **Table 16.5** gives the IR windows and solvent compatibility of commonly used salt plates.

Several variations on FTIR are readily available. Some instruments permit focusing on a microscopic area through an IR objective lens; still more sophisticated instruments allow simultaneous imaging with spectral measurements (these techniques are called IR microscopy). Add-ons are also possible for *attenuated total reflectance* (ATR) spectroscopy, which uses evanescent waves and is independent of sample thickness, and for diffuse reflectance infrared Fourier transform (DRIFT) spectroscopy, which is used primarily on surfaces.

In molecular biophysics, FTIR should be used to classify products of synthesis and/or conjugation. It is difficult to identify molecules bound to the surface of nanoparticles, as the effective concentration is low. However, processes such as ligand exchange can readily be seen (**Figure 16.23**), as can the major peaks of compounds used for functionalization (see sample experiment in **Section 16.11**).

IR spectroscopy may also be performed with time resolution (see **Advanced Topic 16.1**).

**Figure 16.23 Changes in FTIR spectrum of quantum dots under cap exchange from mercaptosuccinic acid (MSA) to pyridine.** (a) Schematic of exchange. (b) Before and after spectra (offset for clarity). The most obvious feature of the MSA spectrum is the very broad O–H stretch. This disappears upon pyridine cap exchange. The C–H stretch of pyridine is the most obvious feature in the "after" spectrum. Some of the other pyridine modes are also visible; these are much more pronounced in pure pyridine than the particle-bound form.

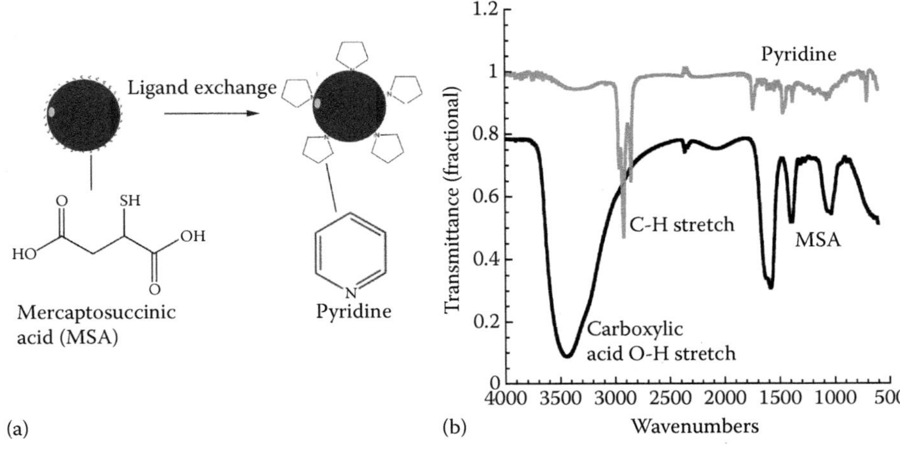

## ADVANCED TOPIC 16.1:    TIME-RESOLVED IR SPECTROSCOPY

Femtosecond time-resolved mid-IR spectroscopy is an emerging technique for observation of conformational changes in biomolecules in their native environments. In these large molecules, the vibrational modes involve the concerted motion of large numbers of atoms, such as protein backbones and side chains. Results are reported as a *difference* in spectrum between a defined State 1 and State 2, which corrects for nonspecific changes.

There are several challenges facing this technique in terms of instrumentation, experimental design, and data analysis. As of this writing, only one company offers an off-the-shelf time-resolved IR system, and it is very costly (>$1 million). The few systems that are operational are found in specialized laboratories and were designed by laser physicists. The detectors are much more costly than those for visible light, and the systems significantly more difficult to set up because the light beams cannot be seen.

Once the system is built, the key to good data is to be able to switch between states in a fashion fast enough to capture protein kinetics. Virtually all transient IR experiments have been done on photoactivatable proteins such as rhodopsin and heme proteins, because state switching can be obtained with a fast light pulse. Other light-activated processes, such as the formation of excited states in DNA in response to ultraviolet, have also been studied. For molecules that are not light-activated, some other means must be found to turn them on very rapidly at a time that can be determined.

Data interpretation is tricky for all but the smallest molecules. While static structures obtained from FTIR, crystallography, and other methods can help with band assignment, time-resolved spectra show unstable intermediate states that will need to be identified. Theoretical techniques have been used to help elucidate and confirm the reaction mechanisms observed: the energy of particular vibrational states is calculated, and it is confirmed that the proposed mechanisms agree with the energies observed in the spectra. This has been done successfully with proteins such as cytochrome P450BM3 and NADPH:protochlorophyllide (Pchlide) oxidoreductase.

For larger proteins, an even newer emerging technique is two-dimensional time-resolved infrared spectroscopy (2D IR). This is the ultrafast vibrational analog of 2D NMR spectroscopic methods, coherent spectroscopy (COSY), and nuclear Overhauser effect spectroscopy (NOESY; see the section on NMR). It has the advantages over NMR of being several orders of magnitude faster and of not being restricted to smaller proteins. In ultrafast 2D IR, a sequence of femtosecond duration mid-IR pulses is applied to the system under study. By controlling the frequency, phase, and time delay between these pulses, it is possible to generate signals (i.e., 2D vibrational spectra) that are sensitive to the coupling between various vibrational oscillators on the same (or different) molecules and changes in molecular environment that surrounds important chemical groups.

Time-resolved IR spectroscopy has recently been implemented at the nanoscale by using a commercial scanning near-field optical microscope (SNOM) in pump-probe mode. This allowed the measurement of spectral signatures of mid-IR surface plasmons in InAs with femtosecond time resolution.

### SUGGESTED READING

Bruijn, J.R., van der Loop, T.H., and Woutersen, S. (2016). Changing hydrogen-bond structure during an aqueous liquid-liquid transition investigated with time-resolved and two-dimensional vibrational spectroscopy. *Journal of Physical Chemistry Letters*, 7, 795–799.

Chen, H.H., Bobroff, V., Delugin, M., Pineau, R., Noreen, R., Seydou, Y., Banerjee, S., Chatterjee, J., Javerzat, S., and Petibois, C. (2016). The future of infrared spectroscopy in biosciences: In vitro, time-resolved, and 3D. *Acta Physica Polonica A*, 129, 255–259.

Courtney, T.L., Fox, Z.W., Slenkamp, K.M., and Khalil, M. (2015). Two-dimensional vibrational-electronic spectroscopy. *Journal of Chemical Physics*, 143, 154201.

Diem, M. Time-resolved methods in vibrational spectroscopy. In *Modern Vibrational Spectroscopy and Micro-Spectroscopy*, John Wiley & Sons, Ltd., Chichester, UK, 2015.

(*Continued*)

**ADVANCED TOPIC 16.1 (CONTINUED):    TIME-RESOLVED IR SPECTROSCOPY**

Doorley, G.W., Wojdyla, M., Watson, G.W., Towrie, M., Parker, A.W., Kelly, J.M., and Quinn, S.J. (2013). Tracking DNA excited states by picosecond-time-resolved infrared spectroscopy: Signature band for a charge-transfer excited state in stacked adenine–thymine systems. *Journal of Physical Chemistry Letters*, 4, 2739–2744.

Fayer, M.D. *Ultrafast Infrared Vibrational Spectroscopy*. Taylor & Francis, Boca Raton, 2013.

Groot, M.L., van Wilderen, L.J., and Di Donato, M. (2007). Time-resolved methods in biophysics. 5. Femtosecond time-resolved and dispersed infrared spectroscopy on proteins. *Photochemical and Photobiological Sciences*, 6, 501–507.

Lewis, N.H.C., and Fleming, G.R. (2016). Two-dimensional electronic-vibrational spectroscopy of chlorophyll a and b. *Journal of Physical Chemistry Letters*, 7, 831–837.

Wagner, M., McLeod, A.S., Maddox, S.J., Fei, Z., Liu, M., Averitt, R.D., Fogler, M.M., Bank, S.R., Keilmann, F., and Basov, D.N. (2014). Ultrafast dynamics of surface plasmons in InAs by time-resolved infrared nanospectroscopy. *Nano Letters*, 14, 4529–4534.

# 16.8  NUCLEAR MAGNETIC RESONANCE

## Introduction

NMR is a ubiquitous technique with a vast literature, even more so because it includes techniques labeled with the more "patient-friendly" name magnetic resonance imaging (MRI). Synthetic chemists make use of NMR on a daily basis to investigate the structure and purity of compounds, and NMR spectrometers are available at nearly every institution on a departmental or fee-for-use basis, many times with trained technician assistance. References are given to some very thorough texts and classic papers at the end of this chapter, and we encourage anyone delving into synthetic chemistry to consult them. Here we will give only a brief introduction to the technique, followed by a couple of examples of how it can be used on the particular types of materials discussed in this book—in particular, nanoparticles.

NMR is based upon the interaction of nuclear spins with a magnetic field. The ground state of all nuclei, except those with even numbers of protons *and* neutrons, has nonzero spin (either integer or half-integer). The Hamiltonian for a nucleus with magnetic moment $\mu$ in an external magnetic field $B$ is given by

$$H = -\gamma B \cdot \hat{I}, \tag{16.25}$$

with energy levels for a field $B = B_0 \hat{z}$

$$E_{m_I} = -\gamma \hbar B_0 m_I \tag{16.26}$$

The value $\gamma$ is the *magnetogyric ratio*, an empirical property of each nucleus that may also be expressed as the product of the nuclear magneton $\mu_N$ and the *nuclear g-factor* $g_I$:

$$\hbar \gamma = g_I \mu_N \tag{16.27}$$

If $\gamma > 0$, which is usually the case, then a spin $-1/2$ nucleus in a magnetic field will show an energy splitting of the two possible spin states proportional to $\hbar\gamma B_0$,

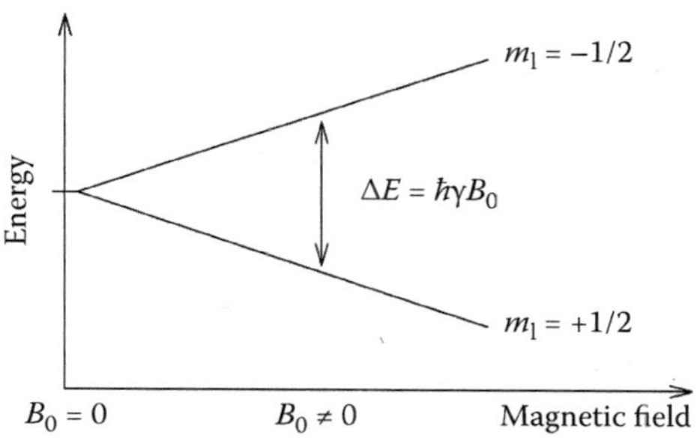

Figure 16.24 **Splitting of degenerate-energy nuclear spin states in a magnetic field.** Note that for some nuclei, $\gamma < 0$, so the sign of the shift will be inverted.

with the spin +1/2 state at lower energy (**Figure 16.24**). The resonance condition is satisfied when the applied radiation has energy $h\nu = \Delta E$, which corresponds to radiofrequency energies. NMR thus essentially consists of applying a strong magnetic field (usually on the order of 10 T) and observing the resonant frequency of molecules containing magnetic nuclei. The most common nuclei studied by NMR are protons (spin ½), as well as a few other spin -1/2 nuclei such as $^{13}$C and $^{19}$F. The techniques are then referred to as "$^1$H NMR," "$^{13}$C NMR," etc.

Although simple in principle, NMR can be tricky to interpret, as the positions of the resonances are affected by the structure of the molecule. This is, of course, what makes the technique so useful. The local magnetic field felt by a particular nucleus is influenced by the chemical structure in ways that can be classified and predicted. The difference between the measured resonance frequency $\nu$ and that of a reference standard $\nu_0$ is expressed as a dimensionless quantity known as the *chemical shift* $\delta$, usually expressed as parts per million:

$$\delta = \frac{\nu - \nu_0}{\nu_0} \times 10^6 \tag{16.28}$$

For protons, tetramethylsilane is usually the reference (it has a lot of protons). A characteristic chemical shift is then seen in a proton attached to any of the major functional groups: carboxylate, alcohol, methyl, aromatic, etc. (**Figure 16.25**). Similar principles apply to the other nuclei.

*Fine structure* results from the splitting of the peaks at a specific chemical shift caused by *scalar coupling*, which is a special case of spin–spin coupling. The

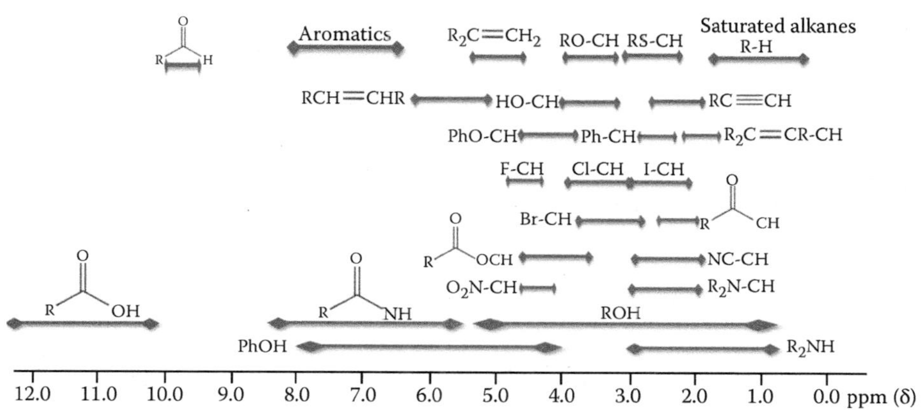

Figure 16.25 **Chemical shifts for protons in different functional groups relative to tetramethylsilane.** The boldface indicates the proton being measured.

**Figure 16.26 Examples of spectra with (a) splitting much smaller than chemical shift (1,3-dichloropropane) and (b) splitting comparable to chemical shift (para-nitrotoluene).**

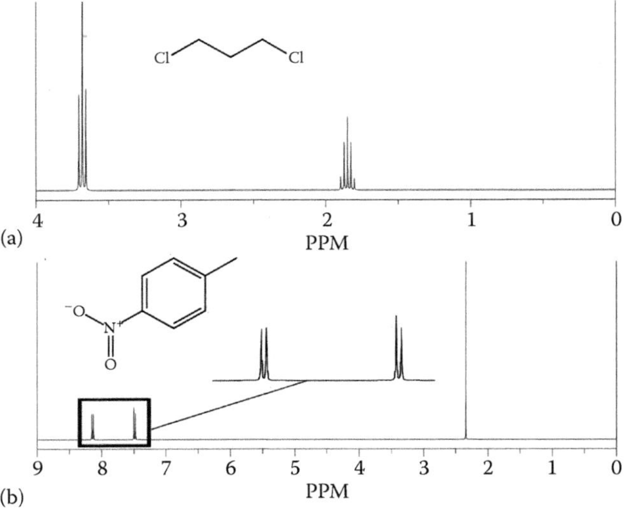

coupling constant is called $J$ and is in units of frequency and independent of the field strength. When the splitting is much smaller than the chemical shifts between groups ($J << \delta$), the spectra can be relatively easy to interpret. When the splitting is on the order of the chemical shift ($J \sim \delta$), it can be difficult to assign peaks to particular functional groups (**Figure 16.26**).

In general, calculating NMR peaks from first principles is difficult. Rather than attempt to model the spectra, they are usually compared to spectra that have been reported and deposited in the *NMR spectra database.* Many chemistry software packages also have spectral-predicting tools that work reasonably well for a wide range of molecules.

## Example: Examining QD surfaces with liquid-phase NMR

One of the most widely tested, and now widely accepted, models for the instability of thiol-capped QDs in aqueous suspension suggests that oxidation of the surface ligands leads to formation of dithiols, which detach from the QD surface eventually leading to precipitation (**Figure 16.27**).

The original article reporting this method used liquid-phase NMR to confirm this mechanism. Liquid-phase NMR of thiol-capped QDs is rarely used, as it generally gives broad peaks that are of little use because the concentration of the ligands is too low for good data, even with very high nanoparticle concentrations. What the authors of this study did to improve their signal was to add an *excess* of the thiol ligands (mercaptopropionic acid [MPA]) to the solution containing the QDs, and watch for oxidation over a period of many hours. In essence, they were measuring the ability of the QDs to act as a photocatalyst for the oxidation of the thiol compound. Free thiol without QDs was used as a comparison. The final solutions tested contained 2 mg of MPA and 50 μL of QDs at 1–3 μM in 2 mL of $D_2O$ (deuterated water is used to avoid an overwhelming signal from the protons of $H_2O$, and is routinely used in liquid NMR). The results showed the appearance of a dithiol signal in the QD-containing solutions essentially immediately. Oxidation occurred in MPA-only solutions as well, but much more slowly (**Figure 16.28**).

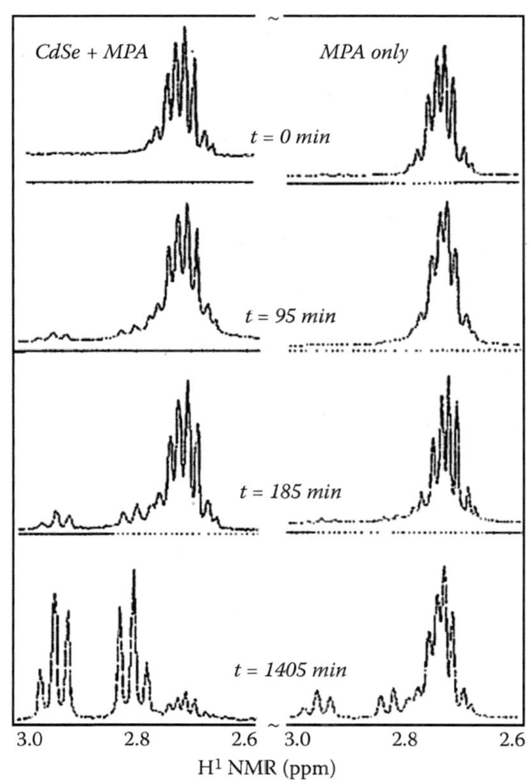



**Figure 16.27 Probable mechanism for photoinstability of alkanethiol-capped QDs in aqueous solution.** Exposure to light and oxygen causes formation of S–S bonds between ligands, which detach from the QD. If there are more free thiols in solution, they can then attach to the QD and be oxidized in turn.



**Figure 16.28 Photocatalytic oxidation of thiol ligands on the surface of CdSe nanocrystals.** At $t$ = 185 min (the third row of the NMR spectra), another control sample containing only MPA-coated CdSe nanocrystals with the same concentration as the one mentioned above started to be photooxidized and precipitated out of the solution. The last row corresponds to the precipitation point of the CdSe-MPA + free MPA solution. (Adapted with permission from Aldana, J., Wang, Y. A., and Peng, X, Photochemical instability of CdSe nanocrystals coated by hydrophilic thiols, *Journal of the American Chemical Society* 123, 8844–8850. Copyright 2001 American Chemical Society.)

## Solid-state NMR*

Solid-state NMR is rather more complicated than solution NMR, because a number of interactions of the magnetic nuclei that are reduced to zero in solution, due to very rapid molecular reorientation, come into play and cannot be neglected. In studies of CdSe QDs, two of these interactions are particularly significant: the internuclear dipole–dipole interaction and the chemical shift anisotropy. These can cause considerable line broadening as both interactions have orientation dependence. Consequently, powder samples, which consist of many tiny particles with different orientations, give rise to a distribution of frequencies with a characteristic lineshape. Techniques such as $^1$H high power decoupling (HPDEC) and magic angle spinning (MAS) are usually applied to reduce the effects of these interactions and produce a high-resolution spectrum. At the same time, they can be exploited in numerous ways: to enhance signal strength via cross-polarization (CP) say from $^1$H to $^{13}$C; by using the dipolar coupling to identify neighboring atoms; and also to provide new information not available from the liquid-phase spectrum, such as local symmetries in the lattice and dynamic processes.

$^{113}$Cd is a reasonably sensitive spin ½ NMR nucleus for which MAS spectra can be obtained both by CP and HPDEC (**Figure 16.29**). Direct excitation of the $^{113}$Cd requires long recycle times, typically 120 s or more, because the $^{113}$Cd relaxes very slowly, but this will provide quantitative spectra provided the recycle time is long enough to allow all types of Cd to fully relax. $^1$H HPDEC is applied to reduce any broadening arising from dipolar coupling to neighboring $^1$H.

The CP experiment is not quantitative and shows the Cd closest to $^1$H (surface Cd) at greater intensity than Cd further away (core Cd). Indeed, for larger QDs, some of the core Cd may not show at all in the CP spectrum. Core Cd have a chemical shift very similar to that of bulk CdSe, and the absence of spinning sidebands (ssb) is consistent with a similar symmetric tetrahedral coordination of four Se. The chemical shift for surface Cd is quite different, and the line has numerous ssb indicating a substantial chemical shift anisotropy, which means that this kind of Cd site is quite asymmetric. This is as expected, as the surface Cd must be attached to the capping ligand and fewer than four Se. The line corresponding to the isotropic frequency of the surface Cd can be determined by spinning at different speeds; only the ssb change positions. $^1$H-$^{113}$Cd CP has the advantage that the recycle time depends on the $^1$H relaxation time ($T_1$), which usually is much faster than for $^{113}$Cd, typically of the order of seconds. Integration of the HPDEC spectra for core Cd and surface Cd (ssb included) gives a ratio of the two types.

In mixed systems, for example, Cd/Zn/Se/S or Cd/Se/Te, the sensitivity of the chemical shift of $^{113}$Cd to its local environment can be used to study whether alloying has occurred or if a core/shell structure with different components has been produced in the course of preparation of the QD.

$^{77}$Se is also a spin ½ NMR nucleus, though less sensitive than $^{113}$Cd, and the same techniques can be applied. For both nuclei in CdSe QDs, long collection times are necessary, unless the very expensive option of isotopic enrichment is used,

---

* Section contributed by C. I. Ratcliffe, Steacie Institute for Molecular Sciences, National Research Council, Ottawa, Canada.

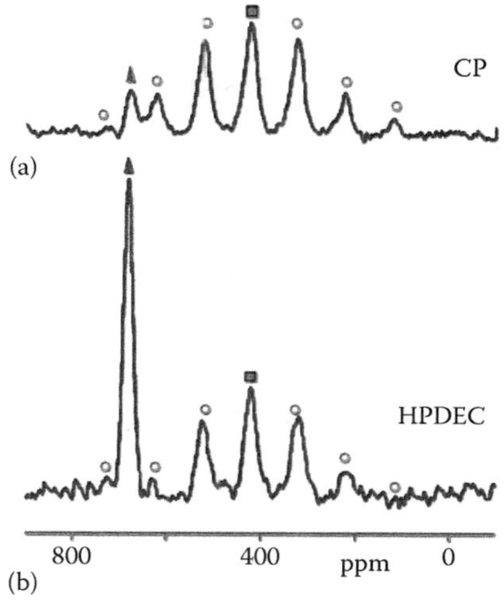

**Figure 16.29** $^{113}$**Cd MAS NMR spectra at 66.58 MHz of CdSe nanoparticles with a diameter of about 2.4 nm capped with myristate ions (CH$_3$(CH$_2$)$_{12}$COO$^-$).** Spin rate 6.6 kHz. (a) CP: 1 ms cross-polarization, 2 s recycle. This shows suppression of the core Cd signal. (b) HPDEC: 120 s recycle. This shows quantitative signals. Core Cd at 680 ppm (triangle). Surface Cd at 422 ppm (filled square for isotropic shift and circles for ssb).

in order to build up sufficient signal to noise. There are several reasons for this: (i) NMR is an insensitive technique so relatively large amounts of material are required, and unfortunately, it can require a lot of effort to produce enough QD sample to fill the MAS sample rotor. In practice, this often means that less than the optimal amount of material is available. (ii) Long recycle times are necessary to obtain quantitative spectra. (iii) The spectral lines, even under MAS, are relatively broad compared to macrocrystalline bulk CdSe, presumably because of slight variations in local environments in the QDs.

$^{13}$C CP/MAS spectra can be used to study the organic capping ligands, and by using dipolar dephasing, a technique that is also known as nonquaternary suppression, it is possible to identify carbons that either have no attached $^1$H or that are undergoing dynamics such that their coupling to $^1$H is very much reduced. In this way, it is possible to determine if the capping ligand is firmly attached, as in the case of carboxylate ions or alkylamines, or whether it is more labile—for example, TOP (trioctylphosphine or its oxide, TOPO) hops rapidly from site to site on the surface of the QD.

## Pulse techniques and MRI

Much greater sensitivity can be obtained over traditional NMR by using Fourier transform NMR, which measures the radiation emitted by nuclear spins as they return to equilibrium after a pulse, a process called *spin relaxation*. At the radiofrequencies used by NMR, relaxation is nonradiative (thermal). There are two types: *spin–lattice (longitudinal) relaxation*, also called T1; and *spin–spin (transverse) relaxation*, called T2.

Spin–lattice relaxation is so called because it occurs due to interactions of the nuclear spins with the surroundings (lattice), which act to return them to thermal equilibrium. These interactions may be of several different types, which can be distinguished by their strength, field, dependence, and other variables, but they all result from field fluctuations caused by molecular motion. Truly isolated spins would show negligible rates of T1 relaxation. A common way to measure T1 is to apply a *180° pulse* or π pulse, which flips the magnetization vector along the z-axis.

Relaxation is then proportional to the difference between the initial $z$ component of magnetization, $M_0$, and the measured value $M_z(t)$:

$$M_z(t) - M_0 \propto e^{-t/T_1} \tag{16.29}$$

T1 relaxation times range from 0.1 to 100 s, and the pulse repetition rate and/or the angle of the pulse should be adjusted to allow for accurate integration.

Spin–spin relaxation refers to random variations in the precession frequency of different spins leading to decoherence of the component of the magnetization vector perpendicular to the applied field (called $M_{xy}$ or $M_T$). T2 times are similar to T1 times when molecular motions are fast (high temperatures, low viscosity solutions); if motion is slowed, T2 times are slower. Typical times are on the order of seconds.

MRI uses the techniques of pulsed NMR to map the distribution of protons in a cell or entire organism. It is widely used in medicine because unlike x-rays, MRI can image soft tissue such as tumors and nervous tissue (brain and spinal cord). The radiofrequencies used are also nonionizing and therefore considered entirely safe, except that caution must be taken not to wear any metals near the powerful magnets used. People with some types of metal implants, especially those in the eyes or brain, cannot have MRI. This includes those who have been wounded by bullets or shrapnel. Other types of implants, including bone staples and spinal fixation hardware, are not contraindications to MRI but may blur the signal in the area of the implant.

The field in an MRI instrument varies linearly along the $z$-axis as $B = B_0 + Gz$. This allows for imaging of a thick specimen on a slice-by-slice basis, since the protons will be resonant at frequencies

$$\nu = \frac{\gamma}{2\pi}(B_0 + Gz) \tag{16.30}$$

giving a signal proportional to the number of protons at a specific spatial location $z$. Slice selection is followed by projection reconstruction of the entire image.

MR images are usually either *T1-weighted* or *T2-weighted*, meaning they distinguish areas that differ in T1 or T2 relaxation times. The types of image taken depend upon the time between radiofrequency pulses (repetition time or TR) and the time between application of the pulses and measurement of the MR signal (echo time or TE). T1-weighted images have both a short TR (<1 s) and a short TE (<30 ms). They are good for imaging anatomic structures and for distinguishing fat, protein-containing fluid, and hemorrhage, all of which appear bright. T2 images have a long TR (>1.5 s) and long TE (>45 ms). Water appears bright on T2-weighted images, so these types of images are best for distinguishing tissue hydration states, especially pathological collections of water that occur in tumors or abscesses. The way to tell whether an image is T1 or T2 is to look for a heavily hydrated area, such as CSF. It will appear dark on a T1-weighted image and bright on a T2-weighted image (**Figure 16.30**). The administration of *contrast agents* is also frequently used in MRI. Pathological structures such as tumors that are heavily hydrated also tend to take up contrast agent more than surrounding normal tissue, so T2 contrast agents are very useful in imaging tumors. The following section gives a recipe for a simple iron-based T2 contrast agent.

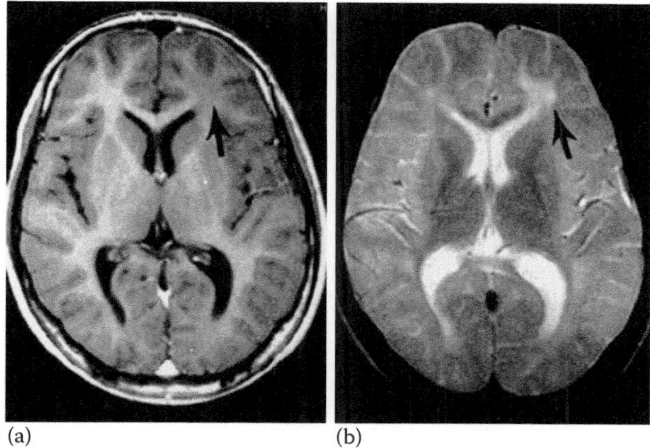

**Figure 16.30 MR images of the brain.** (a) T1-weighted. (b) T2-weighted. (Image from Bonthius, D.J., Stanek, N., and Grose, C. Subacute sclerosing panencephalitis, a measles complication, in an internationally adopted child. *Emerging Infectious Diseases* 6(4), 2000. Available from http://www.cdc.gov/ncidod/EID/vol6no4/contents.htm)

(a)                    (b)

## Paramagnetic nanoparticles as MR contrast agents

Tissues that are histologically distinct may be magnetically similar and thus hard to distinguish with MRI. Contrast agents act to enhance targeted tissues or cells by changing either T1 or T2. The efficacy of a contrast agent depends upon the degree of time constant change per unit concentration and is known as *relaxivity* $r_{1,2}$:

$$\frac{1}{T_{1,2}} = \frac{1}{T_0} + r_{1,2}[C]$$
(16.31)

where $T_0$ is the value of T1 or T2 without the contrast agent, and $[C]$ is the concentration of the agent. The unit of relaxivity is $M^{-1}\,s^{-1}$.

T1 contrast agents generally require direct contact with the aqueous environment of the tissue, whereas T2 agents do not. Iron oxide nanoparticles, coated with dextrans for biocompatibility, have a long and successful history as T2 MRI contrast agents. They have received approval for *in vivo* use in humans in the United States and Europe and have been used for identification of cancer metastases to lymph nodes.

In addition to the usual issues, the use of MRI contrast agents in biological research faces different barriers than does their use in humans. *In situ* examination of samples may be performed in a small-bore, high-field magnet (11–13 T as opposed to 1.5 T for medical imaging). This increases the potential resolution of imaging but puts increasing demands on the contrast agent. Rather than be restricted to the extracellular space, the agent should be intracellular without disrupting internal structures or organelles. Because developing tissues divide rapidly, the contrast agent must be passed from cell to cell during mitosis with even distribution of signal. To meet both of these requirements, the particles should be as small as possible, must not show aggregation under intracellular conditions, and should not be toxic even at concentrations many-fold higher than what is required for detection (in order to allow for dilution during cell division).

Even with the highest-field magnets, resolution of MRI is limited to microns. The ability to directly couple MR data to higher-resolution images would bridge the resolution gap—for this, multifunctional agents need to be designed that allow for not only MR contrast but also another imaging modality. In cell biology, where samples are often thin and transparent, fluorescence microscopy provides the best complement. The preparation of fluorescent paramagnetic particles has been reported in many studies using different approaches: doping of inherently fluorescent particles such as QDs to make them paramagnetic; attaching two particles with different properties together via a chemical bridge; and more. Table 16.6 gives a summary of multifunctional particles used for combined MR/ other imaging, and a simple method for preparation of fluorescent iron oxide nanoparticles is provided in **Practical Tips 16.2**.

The usefulness of an MR contrast agent can be estimated in an ordinary NMR spectrometer by measuring T2 contrast as a function of concentration and then calculating the relaxivity. This will give an estimate of the intravital concentration needed to observe the agent in cells or organisms. It should be on the order of nanomolars (**Figure 16.31**).

To test the suitability of an agent for generating MR contrast deep within intact organisms, they need to be injected and imaged using MRI (**Figure 16.32**).

NMR is an emerging technique in structural biology (see **Advanced Topic 16.2**).

**Table 16.6**

A Few Examples of Multifunctional Nanoparticles for Combined Imaging with References

| Type of Particle | Types of Imaging | Reference |
| --- | --- | --- |
| Iron oxide core/ gold shell | IR, NMR/MRI, visible, electron microscopy | Jin, Y., Jia, C., Huang, S. W., O'Donnell, M., and Gao, X. (2010). Multifunctional nanoparticles as coupled contrast agents. *Nature Communications* 1:DOI: 10.1038/ncomms1042. |
| Mn-doped CdS | Fluorescence, x-ray, NMR | Santra, S., Yang, H., Holloway, P.H., Stanley, J.T., and Mericle, R.A. (2005). Synthesis of water-dispersible fluorescent, radio-opaque, and paramagnetic CdS:Mn/ZnS quantum dots: A multifunctional probe for bioimaging. *Journal of the American Chemical Society* 127, 1656–1657. |
| Gd chelate coated Au | MRI and CT scanning | Alric, C., Taleb, J., Le Duc, G., Mandon, C., Billotey, C., Le Meur-Herland, A., Brochard, T., Vocanson, F., Janier, M., Perriat, P., Roux, S., and Tillement, O. (2008). Gadolinium chelate coated gold nanoparticles as contrast agents for both X-ray computed tomography and magnetic resonance imaging. *Journal of the American Chemical Society* 130, 5908–5915. |
| Si particles with Gd | MRI and fluorescence | Rieter, W.J., Kim, J.S., Taylor, K.M., An, H., Lin, W., and Tarrant, T. (2007). Hybrid silica nanoparticles for multimodal imaging. *Angewandte Chemie International Edition* 46, 3680–3682. |

## PRACTICAL TIPS 16.2:   FLUORESCENT PARAMAGNETIC IRON OXIDE NANOPARTICLES

There are many ways to prepare iron oxide nanoparticles. This is a particularly easy method to make them fluorescent and water-soluble.

Synthesis of hydrophobic particles:

- Add 0.2 mL of $Fe(CO)_5$ to 10 mL octyl ether and 478 mL oleic acid at 100°C.
- Reflux at 280°C for 1 h; solution should be black.
- Allow to cool, then add 0.34 g trimethyl-N-oxide to fully oxidize the nanoparticles.
- Heat to 125°C for 2 h under an inert atmosphere.
- Reflux at 280°C for 1 h.
- Allow to cool, then precipitate the iron with ethanol.
- Recover the nanoparticles by centrifugation and dispersal in hexane.

### SOLUBILIZATION WITH FLUORESCENTLY LABELED LIPID MICELLES

To avoid aggregation, perform micelle encapsulation in dilute solutions.

- Add 100 µL of 10-fold diluted nanoparticles to 900 µL of chloroform.
- Add 1 mg of mPEG-2000 PE, 0.25 mg of dipalmitoylphosphatidylcholine (DPPC), and 34 µL of a 5.4 mM solution of a lipid-soluble dye of the desired color, such as DiI, DiO, or DiD (addition of greater amounts of lipid leads to aggregation) (see the note below for full names of these dyes).
- Allow the organic solvents to evaporate.
- Resuspend in 1 mL $H_2O$.
- Purify by filtration using a 3000 molecular weight cutoff spin column according to manufacturer instructions. Adjust the final volume to 1 mL, which will give an effective iron concentration of ~700 µM.

The particles as synthesized are ~7 nm in diameter, with a hydrodynamic radius of about 50 nm after micelle coating (**Figure P16.2.1**).

Note:

DiD = 1,1′-dioctadecyl-3,3,3′,3′-tetramethylindodicarbocyanine perchlorate
DiI = 1,1′-dioctadecyl-3,3,3′,3′-tetramethylindocarbocyanine perchlorate
DiO = 3,3′-dioctadecyloxacarbocyanine perchlorate

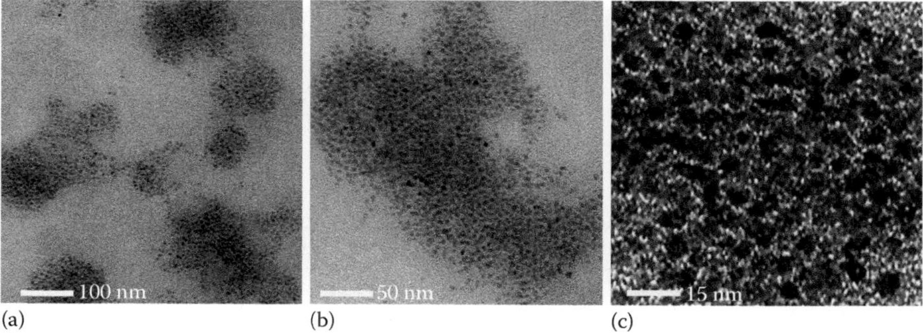


**Figure P16.2.1 Appearance of phosphotungstic acid (PTA)-stained iron nanocrystals with micelle coat under TEM at different magnifications.** (a) Low power image showing typical clustering of micelle-coated particles. (b) Higher resolution showing individual particles with lighter-colored spacing indicating the presence of the lipid micelle. (c) High-resolution image of negatively stained particles showing light rings of lipid around the dark iron crystal.

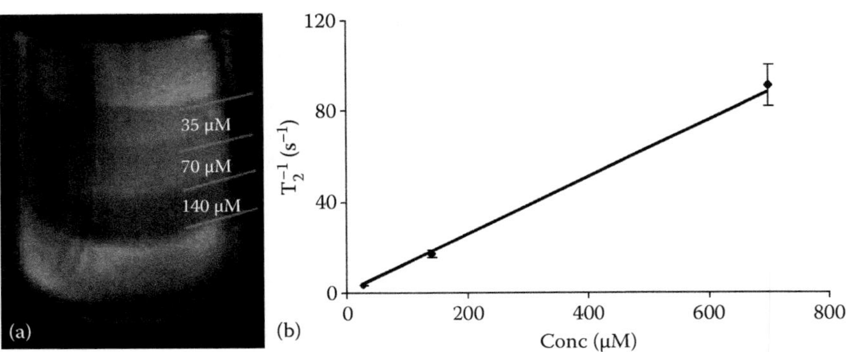

**Figure 16.31 NMR contrast of iron nanoparticles in agarose slab (particles made as in Practical Tips 16.2).** A 1% agarose solution was mixed with iron nanoparticles to the final iron concentrations listed in the image. The solutions were then poured into a vial and allowed to solidify to form three sections with different iron concentrations. T2 times were measured with a CPMG sequence of two echoes with interecho time of 10 ms using a 300 MHz NMR spectrometer (Varian). (a) T2-weighted magnetic resonance image of the vial, showing increasing T2 contrast as the concentration of iron nanoparticles increases. (b) T2 relaxivity curve measured at 300 MHz. The relaxivity is linearly dependent on concentration as expected. The value of $T_0$ for curve fitting is that for water (3 s).

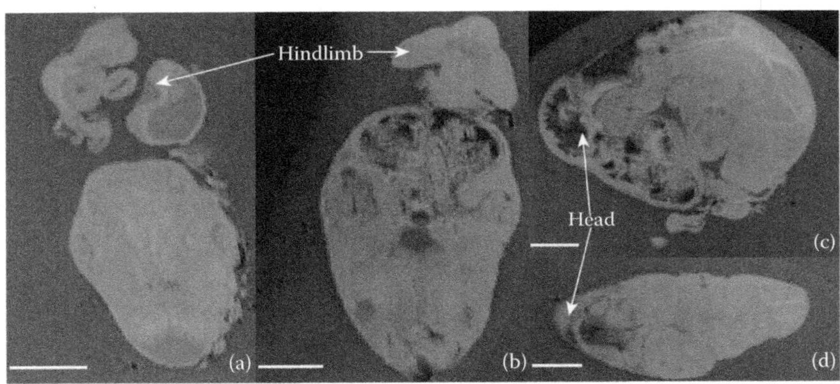

**Figure 16.32 MR images of quail embryos taken with an 11.7 T Bruker MRI.** Voxel resolution was $39 \times 78 \times 78$ µm; field of view = $0.5 \times 0.5 \times 0.5$ cm; array = $128 \times 64 \times 64$; repetition time (TR) = 400 ms; echo time (TE) = 8.78 ms, number of echoes = 2. Scale bar = 1 mm. (a) Axial section of uninjected quail embryo. No darkened areas can be seen. (b) Axial section of quail embryo injected with 1 µL of iron nanoparticles. The enhanced T2 contrast can be seen in the upper right of the image, in the cranial region of the quail. (c) Sagittal section of embryo in (b). (d) Coronal section of embryo in (b). The T2 contrast can easily be seen in the right portion of the embryo; this is where the injection was made.

## 16.9 ELECTRON PARAMAGNETIC RESONANCE SPECTROSCOPY

### Basic principles

Electron paramagnetic resonance (EPR) spectroscopy, also called electron spin resonance (ESR) spectroscopy, is a technique very similar in principle to NMR, but it acts only on paramagnetic species—that is, those with an unpaired electron: radicals and triplet states. When an unpaired electron is placed in a magnetic field

## ADVANCED TOPIC 16.2:    NMR IN STRUCTURAL BIOLOGY

NMR spectroscopy is an important emerging technique for structure determination of relatively small, unstructured proteins that are not suitable for x-ray crystallography (difficult to crystallize). As can be readily appreciated, the number of atoms in even a small protein leads to NMR spectra that are highly complex and where the peaks are impossible to resolve. To solve this problem, "multidimensional" NMR techniques have been developed in which two pulses are applied to the sample and the spectra plotted along two axes. *Correlation spectroscopy* (COSY) uses two consecutive 90° pulses to determine spin–spin coupling.

Probably the most useful technique for protein structure determination is nuclear Overhauser effect spectroscopy (*NOESY*). The intensity of the NOESY signal between two nuclei drops off as a dipole–dipole interaction ($r^6$), allowing for measurements of internuclear distances. This can yield a secondary structure of a protein, especially when the protein is labeled at key residues with $^{15}N$ or $^{13}C$. The tertiary structure of the protein is then modeled from the secondary structure using either *distance geometry* or *simulating annealing*.

As of this writing, NMR structure determination is routine for proteins up to 25 kD, and possible for structures to ~80 kD. The data analysis is complex and outside the scope of this work; interested readers are encouraged to explore the recent literature.

### SUGGESTED READING

Curry, S. (2015). Structural biology: A century-long journey into an unseen world. *Interdisciplinary Science Reviews*, 40, 308–328.

Marion, D. (2013). An introduction to biological NMR spectroscopy. *Molecular and Cellular Proteomics*, 12, 3006–3025.

### DATABASE

Biological Magnetic Resonance Data Bank (http://www.bmrb.wisc.edu/)

$B_0$, the two spin states become nondegenerate in energy due to the Zeeman effect, with the spin –½ state lower in energy than the spin +½ state by a value $\Delta E$:

$$E_{m_s} = g_e \mu_\beta B_0 m_s,$$
$$\Delta E = g_e \mu_\beta B_0 \tag{16.32}$$

where $\mu_\beta$ is the Bohr magneton, and $g_e$ is the (dimensionless) magnetic moment of the electron. The resonance condition is satisfied when the applied radiation has energy $h\nu = \Delta E$, which, for electronic spin conditions, corresponds to microwave energies (contrast with NMR; **Figure 16.33**). At resonance, an electron in the lower energy level has a high probability of absorbing this radiation and transitioning to the higher energy level, generating an absorbance feature in the microwave spectrum.

For a free electron, $g_e = 2.0023193$. However, the magnetic moment of an electron in a molecule is no longer that of a free electron, but depends upon the strength of the magnetic field in the molecule in response to currents generated through the molecular framework. The quantity $g_e$ is then replaced by simply $g$, known as the *g-factor*, Landé *g* factor, or *g*-value. The *g* factor is related to *spin-orbit coupling*

**Figure 16.33 Splitting of degenerate-energy electron spin states in a magnetic field.** Note that the −1/2 spin state is the lower energy.

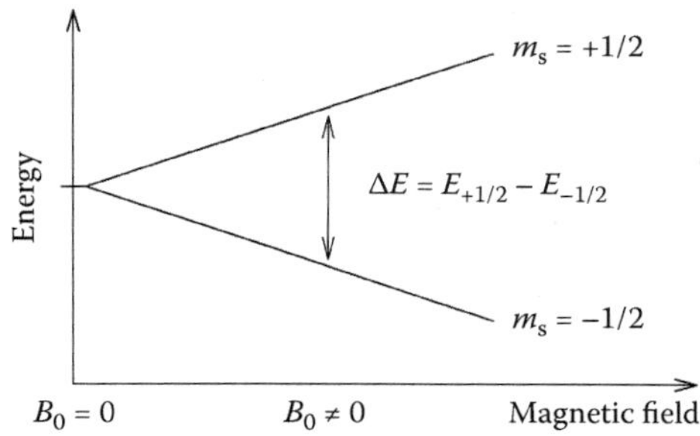



and is given as a function of orbital angular momentum $L$ and total angular momentum $J$ as

$$g = 1 + \frac{S(S+1) - L(L+1) + J(J-1)}{2J(J+1)} \tag{16.33}$$

For most radicals, $L$ is close to 0 and so $g$ is close to the free-electron value. The exception is with transition metals, which have high spin-orbit coupling constants and resulting $g$ values that vary a good deal from 2, and which are also anisotropic (that is, they depend upon the orientation of the molecule with respect to the $B$ field). We will not discuss this situation further here, but some references at the end of the chapter provide more in-depth calculations for the curious.

EPR spectroscopy consists of searching for the absorption peaks that occur when electrons are excited into the more-energetic spin state. Resonance can be achieved by either varying the magnetic field (constant frequency) or varying the frequency at constant $B_0$. Commercial EPR spectrometers usually do the former (in contrast to NMR). The result is a peak at a given value of $B_0$ indicating the presence of a radical. Most commonly, phase-sensitive detection is used, and the resulting absorption spectrum is actually the first derivative of the microwave absorption with respect to the field (**Figure 16.34**).

The key to molecular recognition using EPR spectroscopy is *hyperfine structure*. This occurs because the nuclei in the molecule also show energy level splitting

**Figure 16.34 Example of an EPR spectrum, showing the absorbance and the appearance of its first derivative.** Spectra are presented as the first derivative of the absorption vs. field.

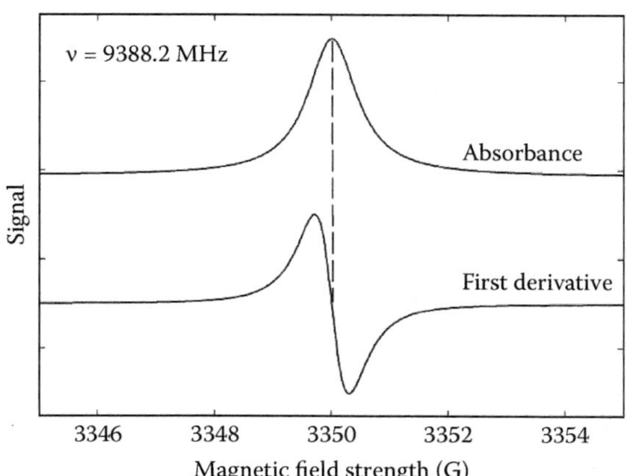

due to the Zeeman effect, in a similar way as the unpaired electron but to a lesser degree:

$$E_{m_I} = g_N \mu_N B_0 m_I \tag{16.34}$$

where $g_N$ is the nuclear $g$ factor, $\mu_N$ is the nuclear Bohr magneton, and $m_I$ is one of the projections of the nuclear spin (for a nucleus of spin $I$, there are $2I + 1$ possible projections). The electron–nuclear interaction depends on the projections of both electron and nuclear spins via the *hyperfine coupling constant A*, or its normalized value $a = A/g_e m_B$. This changes **Equation 16.31** to become

$$E = g_e \mu_B m_s (B_0 + Sa_i m_{I_i}) - g_N \mu_N B_0 m_I \tag{16.35}$$

Let us take the simplest example of a hydrogen nucleus, where the spins may be ±1/2. So the spectrum will be split into two lines, one representing spin up where the radicals resonate at

$$h\nu = g\mu_B \left( B_0 + \frac{1}{2} a \right) \tag{16.36a}$$

and the other spin down, where they resonate at

$$h\nu = g\mu_B \left( B_0 - \frac{1}{2} a \right) \tag{16.36b}$$

More complex nuclei can show even more spin states. For example, $^{14}$N has $I = 1$, so will split the spectrum into three lines (**Figure 16.35a**). More complex molecules can have more than one equivalent nucleus, be it a proton or a more complex nucleus. This means that some of the hyperfine lines will overlap each other, resulting in greater intensity of the same line. If a radical contains $N$ equivalent spin-$I$ nuclei, this results in a spectrum of $2NI + 1$ lines with intensities following a binomial distribution (**Figure 16.35b**).

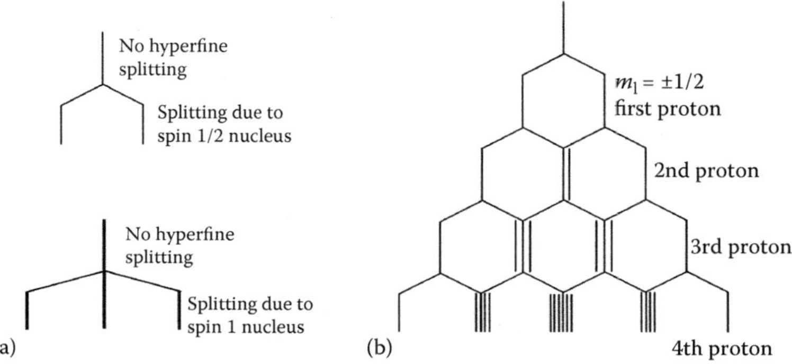


**Figure 16.35 Using stick diagrams to illustrate hyperfine splitting.** (a) Splitting depends upon the possible spin projections of the nucleus. (b) Example for a molecule with four equivalent hydrogen nuclei (protons). The fold degeneracy at each energy determines the peak height in the spectrum.

**Table 16.7**

Magnetic Properties of Some Common Nuclei

| Nucleus | Spin (I) | g value, $g_N$ | Magnetic moment ($\mu/\mu_N$) | Hyperfine Coupling $a$/mT (Isotropic/Anisotropic) |
|---|---|---|---|---|
| $^1$H | ½ | 5.5857 | 2.79285 | 50.8 |
| $^2$H | 1 | 0.85745 | 0.85745 | 7.8 |
| $^{13}$C | ½ | 1.4046 | 0.7024 | 113.0 (2s)/6.6 (2p) |
| $^{14}$N | 1 | 0.40356 | 0.40376 | 55.2 (2s)/4.8 (2p) |
| $^{17}$O | 5/2 | −0.7572 | −1.89379 | |
| $^{19}$F | ½ | 5.2567 | 2.62887 | 1720 (2s)/108.4 (2p) |
| $^{31}$P | ½ | 2.2634 | 1.1316 | 364 (3s)/20.6 (3p) |
| $^{33}$S | 3/2 | 0.4289 | 0.6438 | |
| $^{35}$Cl | 3/2 | 0.5479 | 0.8219 | 168 (3s)/10.0 (3p) |
| $^{37}$Cl | 3/2 | 0.4561 | 0.6841 | 140 (3s)/8.4 (3p) |

*Source:* Atkins, P., and De Paula, J. *Atkins' Physical Chemistry*. Oxford University Press, New York, 2006.

The case of aromatics is particular. A proton bound to a carbon of an aromatic ring has a hyperfine coupling that depends upon the *spin polarization* (or *spin density*) of the carbon atom $\rho_\pi$:

$$a = Q\rho_\pi \tag{16.37}$$

where $Q$ is a constant of order 2.5 mT, depending slightly upon the ring.

Some nuclear spins, g values, and hyperfine coupling constants are given in **Table 16.7**. The quantum chemistry programs cited at the end of **Chapter 1** and of this chapter can be used to calculate hyperfine couplings and predict EPR spectra; textbooks are available for guidance with these calculations. For most applications in molecular biophysics, simulations are not needed, as a small number of well-characterized radicals are used.

## Spin probes and spin traps

Most biomolecules are EPR silent, and radicals produced during biological reactions tend to be short-lived. The greatest use of EPR spectroscopy in biological applications thus makes use of *spin probes* (or spin labels) and *spin traps*.

A spin probe is a stable radical that can be conjugated to a specific part of a biomolecule in order to measure motility. The most common spin probes in biology are *nitroxide* radicals (**Figure 16.36a**). Spin-labeling was used to demonstrate for the first time that the interiors of lipid membranes are highly fluid, and is also used to measure association kinetics of biomolecules. Both of these types of experiments make use of the fact that electron exchange between radicals leads to spectral broadening (**Figure 16.36b**).

Spin-labeling has proven very useful for characterizing very flexible molecules or unstable heterodimers, structures that are difficult to crystallize (as discussed in **Chapter 6**). The process of spin-labeling is similar to that of other types of targeted biofunctionalization: mutation of target residues to cysteines, followed by the

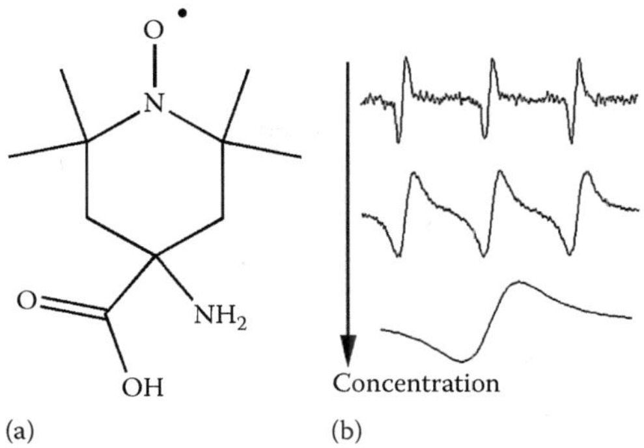

O•

N

O

NH$_2$

OH

(a)

Concentration

(b)

**Figure 16.36  Spin labeling.** (a) Example of a nitroxide spin label. This one is an artificial amino acid, 2,2,6,6-tetramethylpiperidine-1-oxyl-4-amino-4-carboxylic acid. (b) Effect of increasing concentration on spectral linewidth. This effect is undesirable in pure compounds, but can be used to measure distances between spin probes on labeled molecules.

attachment of a thiol-reactive spin probe (often methanethiolsulfonate [MTSSL]). This technique can probe length scales of about 0.8–8 nm on timescales between 10 ps and 1 µs, ideal ranges for protein dimerization and conformational changes. While the modification itself is straightforward, the choice of sites to modify, the selection of experimental parameters, and the interpretation of the data can all be tricky. The technique has been applied to many different soluble and membrane proteins, nucleic acids, peptides, and other macromolecules, and the current literature should be consulted before designing an experiment of this kind. For protein studies, it helps to have a crystal structure, even if it is incomplete or represents only a monomer. Along with electron exchange, line broadening can also occur due to the proximity of other paramagnetic species, including oxygen.

Another important application of EPR in biology is in the study of free radical generation. A spin trap either (a) converts an unstable radical into a stable, EPR-active species, or (b) causes an unstable radical to combine with an EPR-active species to yield an EPR-silent product. An example of (a) is the EPR-silent 2,2,6,6-tetramethylpiperidine (TMP). The reaction of TMP with singlet oxygen/superoxide yields the formation of a stable nitroxide radical via oxidation (**Figure 16.37a**). An example of (b) is 2,2,6,6-tetramethyl-1-piperidinyloxy (TEMPO), a stable free radical that can be oxidized by holes, OH radicals, or any other oxidative species that have a redox potential ≥ +0.75 V vs. NHE. Thus, the *disappearance* of TEMPO radical EPR spectra upon irradiation indicates photogenerated oxidative species (**Figure 16.37b**).

The kinetics of TMP and TEMPO are often used in biology to determine mechanisms of radical formation (**Figure 16.38**). **Practical Tips 16.3** describes

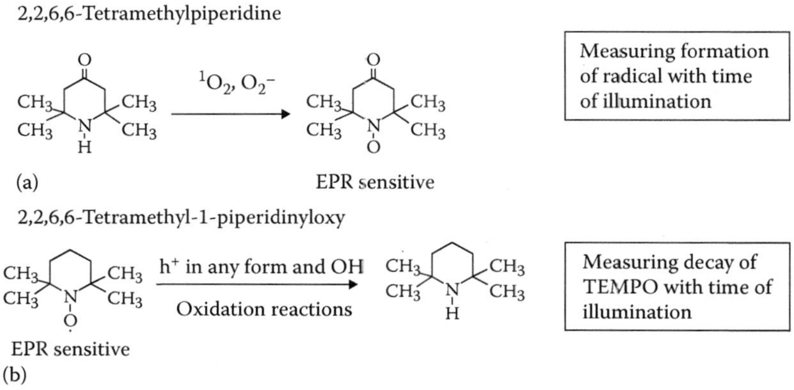

2,2,6,6-Tetramethylpiperidine

CH$_3$  CH$_3$
CH$_3$  N  CH$_3$
       H

$^1$O$_2$, O$_2^-$

CH$_3$  CH$_3$
CH$_3$  N  CH$_3$
       O

(a)                              EPR sensitive

2,2,6,6-Tetramethyl-1-piperidinyloxy

CH$_3$  CH$_3$
CH$_3$  N  CH$_3$
       O

h$^+$ in any form and OH

Oxidation reactions

CH$_3$  CH$_3$
CH$_3$  N  CH$_3$
       H

EPR sensitive

(b)

Measuring formation of radical with time of illumination

Measuring decay of TEMPO with time of illumination

**Figure 16.37  Example of spin traps.** (a) TMP. (b) TEMPO.

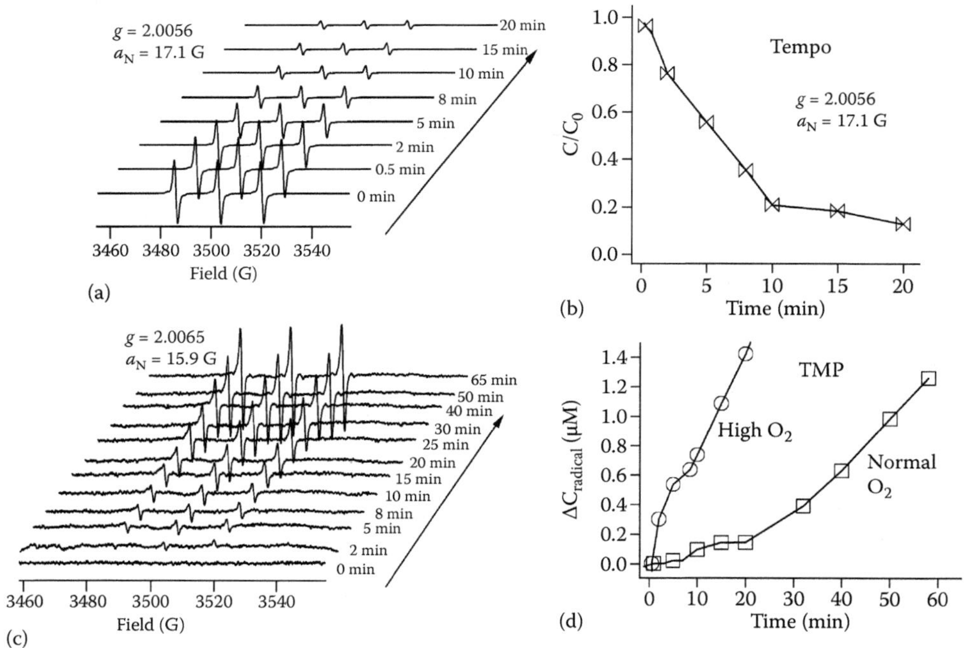

**Figure 16.38 Measuring ROS with TMP and TEMPO.** (a) TEMPO data showing the decay of the radical signal with time. (b) Plot of signal vs. time for the data in (a). (c) TMP data showing the growth of the radical signal with time. (d) Plot of signal vs. time for the data in (c), along with a similar curve taken under increased-oxygen conditions.

how to use spin-trap EPR to investigate the production of reactive oxygen species from water-soluble QDs.

Another commonly used spin trap is $N$-tert-butyl-$\alpha$-phenylnitrone (PBN) (**Figure 16.39a**). It has been used to demonstrate free radical production from amyloid-$\beta$, suggesting a link between reactive oxygen species and the pathophysiology of Alzheimer's disease (**Figure 16.39b**). PBN also shows neuroprotective effects against seizures and excitotoxicity. The role of free radicals in aging, cognitive decline, and many forms of neurodegenerative diseases remains largely unexplored and promises to be an exciting avenue of research for the future.

**Figure 16.39 (a) $N$-tert-butyl-$\alpha$-phenylnitrone (PBN). (b) Using PBN to detect ROS from amyloid-$\beta$.** The control is 50 mM PBN without amyloid. (Adapted with permission from Varadarajan, S. et al., *Journal of Structural Biology* 130, 184–208, 2000.)

## PRACTICAL TIPS 16.3:   MEASUREMENT OF REACTIVE OXYGEN SPECIES FROM QDS USING EPR

Many studies have reported formation (or nonformation) of different types of reactive oxygen species (ROS), particularly singlet oxygen, from quantum dots. There is a lot of disagreement in the literature because of several factors: heterogeneity of quantum dots; different types of ROS, of very different biological importance; the complexity of the QD exciton's interaction with water and oxygen; and the wavelength and power of the illumination used to generate the excitons that give rise to the ROS.

There are many different mechanisms for ROS production and several different forms of ROS. Free radicals may be generated from photoexcited nanoparticles by either the *reductive pathway* (involving the electron transferring to an acceptor, $A$) or the oxidative pathway (involving the hole transferring to a donor, $D$) (**Figure P16.3.1a**):

$$A + e^-_{CB} \rightarrow A^-$$
$$D + h^+_{VB} \rightarrow D^+. \tag{1}$$

If the radicals formed interact with water or oxygen, ROS can result. However, the radicals might also recombine rapidly, such as in the "electron shuttling" seen with quinones, for example by the process

$$A + e^-_{CB} \rightarrow A^-$$
$$A^- + h^+_{VB} \rightarrow A. \tag{2}$$

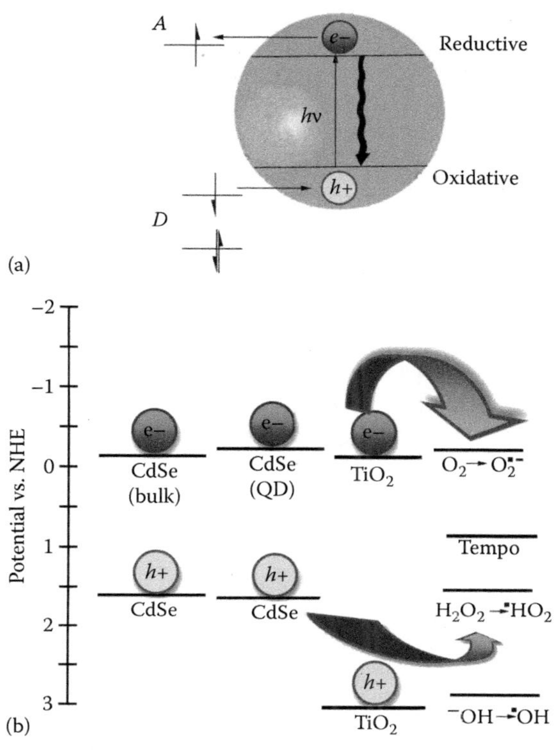


**Figure P16.3.1   Mechanisms and energy levels involved in QD redox processes.** (a) When a nanoparticle is excited by light more energetic than the bad gap, an electron–hole pair is formed. The electron may interact with an acceptor $A$ (the reductive process) and/or the hole with a donor $D$ (the oxidative process). (b) Approximate energy levels (vs. NHE) in aqueous solution for bulk CdSe (bandgap 1.7 eV) and a yellow CdSe QD (bandgap 2.1 eV as measured from absorbance peak). TiO₂ is shown for comparison, as are the energies of TEMPO, oxygen, peroxide, and hydroxylate ions.

(*Continued*)

## PRACTICAL TIPS 16.3 (CONTINUED):  MEASUREMENT OF REACTIVE OXYGEN SPECIES FROM QDS USING EPR

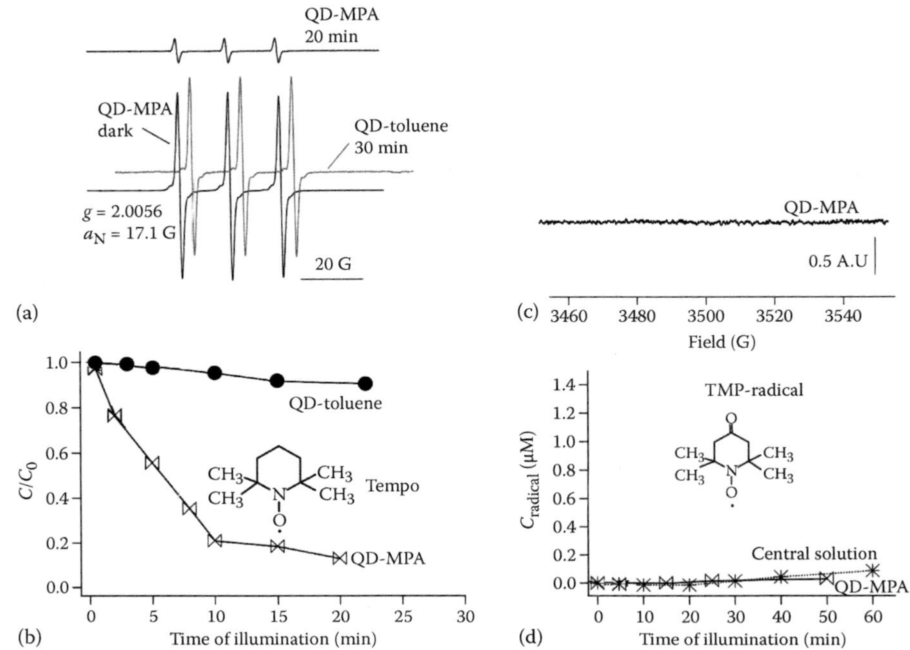


**Figure P16.3.2 EPR spectroscopy using TMP and TEMPO radicals as spin traps.** (a) Spectra of TEMPO radicals showing initial intensity of QD-MPA vs. substantial decay at 20 min. There is no decay for QDs in toluene (gray line, spectrum is shifted for better visibility). (b) Decay of TEMPO radical relative concentrations with time of illumination for QD-MPA and QDs in toluene. (c) QD-MPA shows no signal from the TMP radical after 30 min of illumination. (d) Signal vs. time of QD-MPA and control solution with TMP. There is no significant signal.

$TiO_2$ nanoparticles create very reactive holes. However, CdSe QD holes are much less reactive, and their ability to act via the oxidative pathway is questionable (**Figure P16.3.2b**). Spin-trap EPR using TMP and TEMPO can be used to distinguish the oxidative from the reductive pathway. A TMP signal should indicate the reductive pathway, whereas a disappearance of TEMPO indicates the oxidative pathway.

The data shown were collected on a Bruker Elexys E580 spectrometer at room temperature, with a power of 66.32 mW and a modulation amplitude of 1.0 Gauss. Illumination was with a 300 W Xe lamp using a cutoff filter of 400 nm long pass, intensity ~ 100 mW/cm². The changes in spin-trap concentration over time were determined by measuring EPR spectra at certain time intervals, while solutions were under continuous illumination. Typically, the accumulation of a single spectrum (sweep time) was 42 s in all experiments. The concentration of radicals was determined after double integration of spectra, and normalized to the 10 μM TEMPO radical. The g tensor values were calibrated for homogeneity and accuracy by comparing to a coal standard ($g = 2.00285 \pm 0.00005$). The concentration of TMP was 0.1 M for all solutions; the concentration of TEMPO was varied, usually ~33 μM. All solutions were in air.

**Figure P16.3.2** shows the results from 1 mM CdSe/ZnS QD solutions in $H_2O$ (QDs solubilized with mercaptopropionic acid) or in toluene. The QDs were orange emitting (~590 nm emission peak). It can be seen that there is a distinct TEMPO signal only when the solution is illuminated, and thus the oxidative pathway is possible with these QDs. However, there is no signal from TMP upon prolonged illumination, so the reductive pathway is not taking place.

## Instrumentation

EPR spectrometers consist of a microwave source (typically a *klystron*, may also be a *Gunn diode*) connected via a waveguide to a high-Q resonant cavity. The microwave power cannot be easily adjusted at the source, so is controlled with an attenuator. Samples are placed in the middle of the cavity, where the microwave power is maximum and perpendicular to the B field generated by an electromagnet surrounding the cavity. Most EPR spectrometers are reflection spectrometers, although some have a transmission geometry. Microwaves reflected back from the cavity are routed to a detector diode; the signal is a decrease in current at the detector corresponding to absorption of microwaves by the sample (**Figure 16.40**).

Microwave bands are defined by the Radio Society of Great Britain according to their frequencies (**Table 16.8**). Most commercial EPR spectrometers are at X-band frequencies, but sometimes at L, S, and Q. The choice depends mainly upon the availability of microwave components, and the fact that electromagnets cannot generate reliable fields greater than ~ 1 T. High-frequency high-field EPR (>70 GHz) can be performed at a few sites, including the National High Magnetic Field Laboratory (Tallahassee, FL).

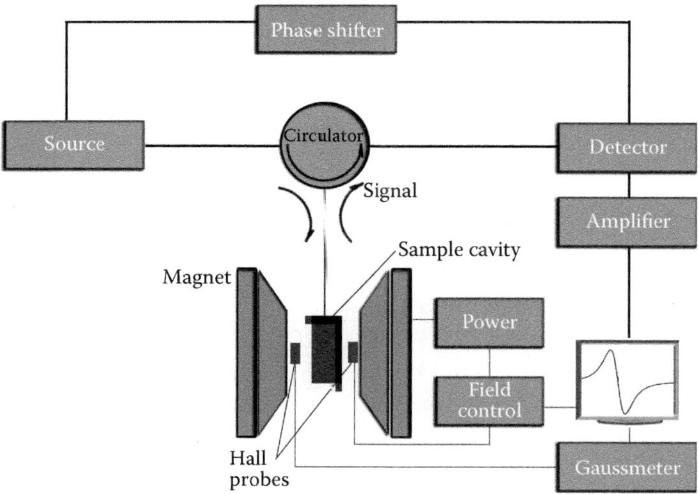


**Figure 16.40 Schematic of an EPR spectrometer.**

**Table 16.8**

Microwave Bands

| Letter | Frequency Range | Typical EPR Frequency (GHz) | Typical Wavelength (mm) | Typical B Field (T) |
|---|---|---|---|---|
| L | 1–2 | 1.5 | 200 | 0.054 |
| S | 2–4 | 3.0 | 100 | 0.11 |
| X | 8–12 | 9.5 | 30 | 0.34 |
| $K_u$ | 12–18 | 17 | 17 | 0.6 |
| Q | 30–50 | 36 | 8 | 1.28 |
| V | 50–75 | 70 | 4 | 2.5 |
| W | 75–110 | 95 | 3 | 3.39 |
| D | 110–170 | 149 | 2 | 5 |

**Table 16.9**

Some Examples of Glass-Forming Solvents

| Solvent | Melting Temperature (K) | Glass Transition (K) |
|---|---|---|
| Ethanol | 155.7 | 97.2 |
| Methanol | 175.2 | 102.6 |
| Isopropanol | 146.6 | 109 |
| Toluene | 178 | 117.2 |
| Ethylene glycol | 255.6 | 154.2 |
| Glycerol | 291.2 | 190.9 |
| Polystyrene | 513 | 373 |

Samples for EPR can be gaseous, crystalline, liquid, or solid. Typical concentrations in liquid samples are tens to hundreds of micromolars to avoid spectral broadening. Water is a great absorber of microwaves, so aqueous solutions are usually held in thin capillary tubes and must be centered carefully in the cavity. Oxygen can also distort spectra since it is paramagnetic; the solubility of oxygen in water is low, however, so this is usually only an issue in nonaqueous solvents. In crystalline samples, an EPR labeled molecule should be diluted 1:100 to $1:10^5$ with its unlabeled counterpart. If no diamagnetic counterpart exists, the alternative is to freeze the molecule in a glass-forming solvent. Some of these solvents with their glass transition temperatures are given in Table 16.9.

Choice of a spectrometer depends upon the sensitivity and tunability required. Higher-end instruments have cavities with higher $Q$ and multifrequency capabilities; the highest end offer X-band to W-band frequencies within a single instrument. Other options that we will not discuss in detail, but which can be very helpful in biological applications, include *ENDOR spectroscopy*, pulsed EPR, and Fourier transform EPR. All of these techniques are discussed in detail in the textbooks referenced at the end of the chapter. EPR spectrometers are costly and are not owned by most individual labs. Many large universities have at least one EPR spectrometer, usually owned by a department or in a shared facility. They may also be found at national laboratories in nanomaterials characterization facilities, and can be made available to external researchers on a fee-for-use or proposal basis.

## 16.10 X-RAY SPECTROSCOPY

Synchotron or home source x-ray spectroscopy for crystallography is discussed in Chapter 6. The only type of x-ray spectroscopy we will cover in this chapter is *energy-dispersive x-ray spectroscopy*, also called EDS or EDX. This is a very commonly available technique in electron microscopy facilities; an EDX system may be attached to a scanning or transmission electron microscope, or to a microprobe. EDX provides an elemental map of a sample by detecting x-rays that are emitted at different energies in response to bombardment by the microscope's electron beam. These energies are characteristic of particular elements because they represent the difference in energy between a shell from which the electron was ejected by the beam and the shell from which another electron enters to fill the hole. The peaks are referred to by the names of the shells from which the electron originated, e.g., "Na K" (Figure 16.41).

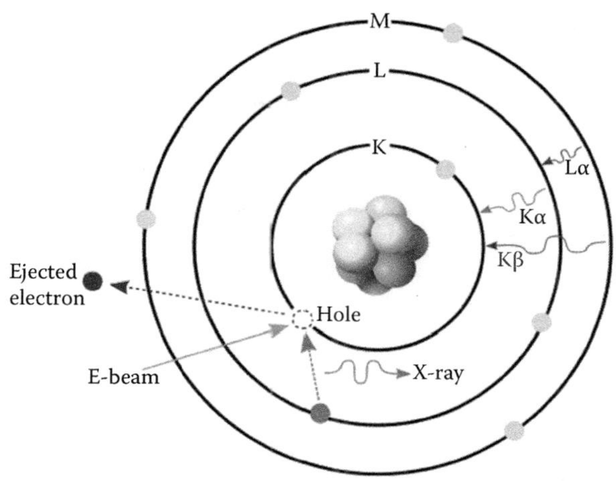


**Figure 16.41 Schematic of EDX spectroscopy.** The electron microscope e-beam ejects an electron, and an electron from a higher shell moves to fill the hole, releasing the excess energy in the form of an x-ray.

There are a few caveats to the use of EDX:

- In the case of inhomogeneous samples, it is important to be aware of the instrument spot size to make sure that the signal comes from only the desired region of the sample.

- EDX cannot see low-energy x-rays since they are absorbed by the detector window, so it is completely silent to elements smaller than atomic number 4; for practical purposes, it is usually used only for elements heavier than carbon.

- Some elements have characteristic x-rays with overlapping energies. Examples are Ti and V; Mn and Fe; and S with Mo and Pb. Sometimes it is possible just to guess which elements "make sense" in the sample; other times it is trickier. Table 16.10 gives a list of x-ray energies for all the elements, showing overlaps.

**Table 16.10**

X-ray Energies (keV) and Overlaps for the Elements through Atomic Number 100

| Element | K Line (Overlap) | L Line (Overlap) | M Line (Overlap) |
|---|---|---|---|
| Li | 0.052 | | |
| Be | 0.110 | | |
| B | 0.185 | | |
| C | 0.282 | | |
| N | 0.392 (Ti) | | |
| O | 0.523 (V, Cr) | | |
| F | 0.677 (Mn, Fe) | | |
| Ne | 0.851 (Ni, La) | | |
| Na | 1.041 (Zn) | | |
| Mg | 1.254 (As, Tb) | | |
| Al | 1.487 (Br) | | |
| Si | 1.740 (Ta, W) | | |

(Continued)

**Table 16.10 (Continued)**

X-ray Energies (keV) and Overlaps for the Elements through Atomic Number 100

| Element | K Line (Overlap) | L Line (Overlap) | M Line (Overlap) |
|---|---|---|---|
| P | 2.015 (Zr, Ir) | | |
| S | 2.308 (Mo, Pb) | | |
| Cl | 2.622 | | |
| Ar | 2.957 (Ag) | | |
| K | 3.313 (In) | | |
| Ca | 3.691 | 0.341 | |
| Sc | 4.090 | 0.395 (N) | |
| Ti | 4.510 (Ba) | 0.452 (N) | |
| V | 4.952 (Ti, Cr) | 0.510 (O) | |
| Cr | 5.414 (V, Pm) | 0.571 (O) | |
| Mn | 5.898 (Cr) | 0.636 (F) | |
| Fe | 6.403 (Mn) | 0.704 (F) | |
| Co | 6.930 (Er) | 0.775 | |
| Ni | 7.477 | 0.849 (Ne, La) | |
| Cu | 8.047 | 0.928 (Pr) | |
| Zn | 8.638 (Re) | 1.009 (Na, Nd) | |
| Ga | 9.251 | 1.096 (Sm) | |
| Ge | 9.885 | 1.186 (Gd) | |
| As | 10.543 (Pb) | 1.282 (As, Dy) | |
| Se | 11.221 | 1.379 (Ho) | |
| Br | 11.923 | 1.480 (Al) | |
| Kr | 12.648 | 1.587 (Lu) | |
| Rb | 13.394 | 1.694 (Si, Ta) | |
| Sr | 14.164 | 1.806 (W) | |
| Y | 14.957 (Cm) | 1.922 (Os) | |
| Zr | 15.774 | 2.042 (P, Pt) | |
| Nb | 16.614 | 2.166 (Hg) | |
| Mo | 17.478 | 2.293 (S, Pb) | |
| Tc | 18.410 | 2.424 (Bi) | |
| Ru | 19.278 | 2.558 | |
| Rh | 20.214 | 2.696 | |
| Pd | 21.175 | 2.838 | |
| Ag | 22.162 | 2.984 (Ar, Th) | |
| Cd | 23.172 | 3.133 | |
| In | 24.207 | 3.287 (K) | |
| Sn | 25.270 | 3.444 | |
| Sb | 26.357 | 3.605 | |
| Te | 27.471 | 3.769 | |
| I | 28.610 | 3.937 | |
| Xe | 29.802 | 4.111 (Sc) | |
| Cs | 30.970 | 4.286 | |
| Ba | 32.191 | 4.467 (Ti) | |

(Continued)

**Table 16.10 (Continued)**

X-ray Energies (keV) and Overlaps for the Elements through Atomic Number 100

| Element | K Line (Overlap) | L Line (Overlap) | M Line (Overlap) |
|---|---|---|---|
| La | 33.440 | 4.651 | 0.833 (Ne) |
| Ce | 34.717 | 4.840 | 0.883 |
| Pr | 36.023 | 5.034 | 0.929 (Cu) |
| Nd | 37.359 | 5.230 | 0.978 (Zn) |
| Pm | 38.649 | 5.431 (Cr) | |
| Sm | 40.124 | 5.636 | 1.081 (Ga) |
| Eu | 41.529 | 5.846 | 1.131 |
| Gd | 42.983 | 6.059 | 1.185 (Ge) |
| Tb | 44.470 | 6.275 | 1.240 (Mg) |
| Dy | 45.985 | 6.495 | 1.293 (As) |
| Ho | 47.528 | 6.720 | 1.348 (Se) |
| Er | 49.099 | 6.948 (Co) | 1.406 |
| Tm | 50.730 | 7.181 | 1.462 |
| Yb | 52.360 | 7.414 | 1.521 |
| Lu | 54.063 | 7.654 | 1.581 (Kr) |
| Hf | 55.757 | 7.898 | 1.645 |
| Ta | 57.524 | 8.145 | 1.710 (Si, Rb) |
| W | 59.310 | 8.396 | 1.775 (Si, Sr) |
| Re | 61.131 | 8.651 (Zn) | 1.843 |
| Os | 62.991 | 8.910 | 1.910 (Y) |
| Ir | 64.886 | 9.173 | 1.980 (P) |
| Pt | 66.820 | 9.441 | 2.051 (Zr) |
| Au | 68.794 | 9.711 | 2.123 |
| Hg | 70.821 | 9.987 | 2.195 (Nb) |
| Tl | 72.860 | 10.266 | 2.271 |
| Pb | 74.957 | 10.549 (As) | 2.346 (S, Mo) |
| Bi | 77.097 | 10.836 | 2.423 (Tc) |
| Po | 79.296 | 11.128 | |
| At | 81.525 | 11.424 | |
| Rn | 83.800 | 11.724 | |
| Fr | 86.119 | 12.029 | |
| Ra | 88.485 | 12.338 | |
| Ac | 90.894 | 12.650 (Kr) | |
| Th | 93.334 | 12.966 | 2.996 (Ag) |
| Pa | 95.851 | 13.291 | 3.082 |
| U | 98.428 | 13.613 | 3.171 |
| Np | 101.005 | 13.945 | |
| Pu | 103.653 | 14.279 | |
| Am | 106.351 | 14.618 | |
| Cm | 109.098 | 14.961 (Y) | |
| Bk | 111.896 | 15.309 | |
| Cf | 114.745 | 15.661 | |
| Es | 117.646 | 10.018 | |
| Fm | 120.598 | 16.379 | |

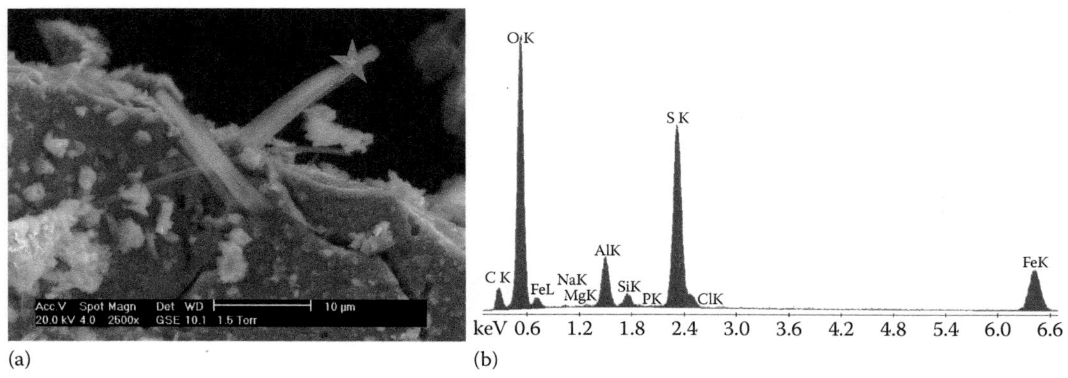

**Figure 16.42 EDX analysis of the mineral pyrite.** (a) SEM image. The star indicates the area where the spectrum was taken. (b) Spectrum with assignments. (Data courtesy R.E. Mielke, UCSB.)

EDX spectra are given as peaks vs. energy in keV and are usually sharp and readily distinguishable (**Figure 16.42**).

# 16.11　EXAMPLE EXPERIMENT: CHARACTERIZATION OF CDSE/ZNS NANOPARTICLE BIOCONJUGATE USING UV–VIS, FLUORESCENCE EMISSION, TIME-RESOLVED EMISSION, FTIR, AND EPR SPECTROSCOPY

## UV–Vis and fluorescence emission

You have made a batch of CdSe/ZnS QDs as discussed in **Chapter 11**. They are coated with oleic acid (OA) and dissolved in toluene and have a lovely orange color, like artificial cheese powder. You wish to solubilize them into water by coating them with MPA, and then conjugate them to dopamine via an amide bond (**Figure 16.43a**). The first thing to do is to take the UV–Vis spectra of the QDs as is. This allows you to observe the energetic position of the first exciton peak and to estimate the size and QD concentration based upon the formulas in **Chapter 11**. From **Figure 16.43b**, we find the absorbance at the exciton peak to be 0.6 at a position of 575 nm. Using **Equation 11.4** now gives a core size of ~3.7 nm and a concentration of about 3.3 µM (compare with the core/shell size of 4.1 nm measured by TEM).

Now dilute the QDs in toluene to get an emission spectrum. It does not matter if you know the exact concentration for this; it is just important to get them diluted enough to get a good spectrum. The solution should look clear or nearly clear, not yellow. The emission peak measured is at 590 nm—the author's lab calls this "Quantum Dot Orange" since it is such an easy color to make for both CdSe and CdTe (**Figure 16.43b**). In order to evaluate the quantum yield $Q$, a reference standard should be chosen, such as Rhodamine 6G. A plot of integrated emission vs. absorbance at a chosen value gives $Q$ according to **Equation 6.5** (**Figure 16.43c**).

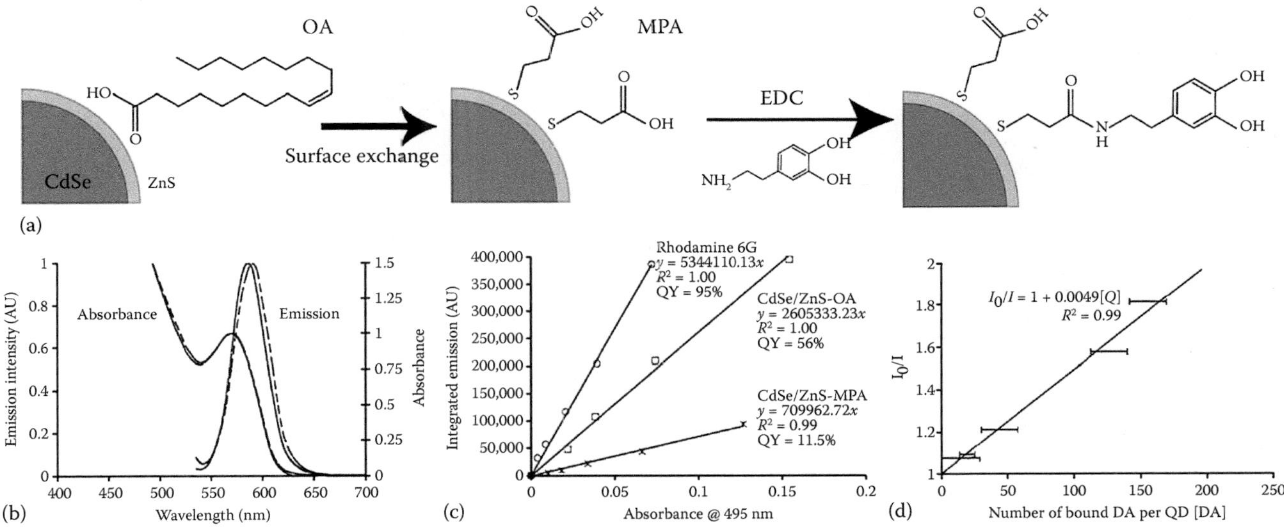



**Figure 16.43 UV–Vis analysis of CdSe/ZnS QD preparation and dopamine conjugate.** (a) Schematic of QDs as prepared, coated with OA in toluene; after cap exchange with MPA and solubilization into water; and after EDC coupling to dopamine at varying ratios, where some or all of the MPA groups are attached to dopamine via its amino terminus. (b) Absorbance and normalized emission for QDs before (solid) and after (dashed) solubilization. (c) Integrated emission vs. absorbance at 495 nm for determination of $Q$ relative to rhodamine 6G. Points of different concentrations are fit to a straight line. (d) Stern–Volmer plot of quenching of QD peak emission vs. number of dopamine molecules bound per particle. (Figures from S. Clarke, PhD thesis 2008.)

The absorbance, emission, and $Q$ experiments should be repeated after the solubilization procedure. It can be seen from **Figure 16.43b and c** that there is a slight red shift in the emission spectrum after cap exchange, and a significant reduction in $Q$: from 56% to 11.5%. This is a typical result with CdSe/ZnS QDs, for which thiol compounds act as quenchers.

Conjugation of dopamine to QDs leads to additional quenching. We discussed in **Chapter 11.4** how to control and estimate the number of molecules bound to a QD via an amide bond. Using this, we can create a sort of Stern–Volmer plot of quenching vs. QD concentration (**Figure 16.43d**). Note that this is a bit different from usual quenching plots, as we know that all of the dopamine molecules are bound to QDs rather than free in solution.

## FTIR

We do not have an FTIR, so we go to the lab next door to see if we can borrow theirs. It turns out they have a micro-FTIR, which focuses to a small spot on a window of choice. This is good, as it allows us to deposit a thick layer of QDs onto a small area, without having to use a huge amount of material. Next we have to buy (or borrow) a window. Looking at the table of compatibilities, and realizing that the samples are mostly aqueous, we decide on a ZnSe window.

First, let us see what we expect the solubilized particles to look like by predicting the FTIR spectrum of mercaptosuccinic acid. Using the software ChemDraw, we find the spectrum shown in **Figure 16.44a**. Comparing this with the spectrum of the solubilized QDs, we find that there are a few similarities. The SH feature has

**Figure 16.44 Predicted and measured FTIR spectra from QD conjugates.** (a) Predicted MSA spectrum vs. QD-MSA measured spectrum. The QD spectrum is dominated by two carboxylate peaks and an OH stretch. (b) Predicted and measured spectra of dopamine (DA) alone. (c) Measured DA alone vs. QD-DA. The QD-DA shows some characteristic features of the DA spectrum, such as the CH bend and aryl oxygen stretch.

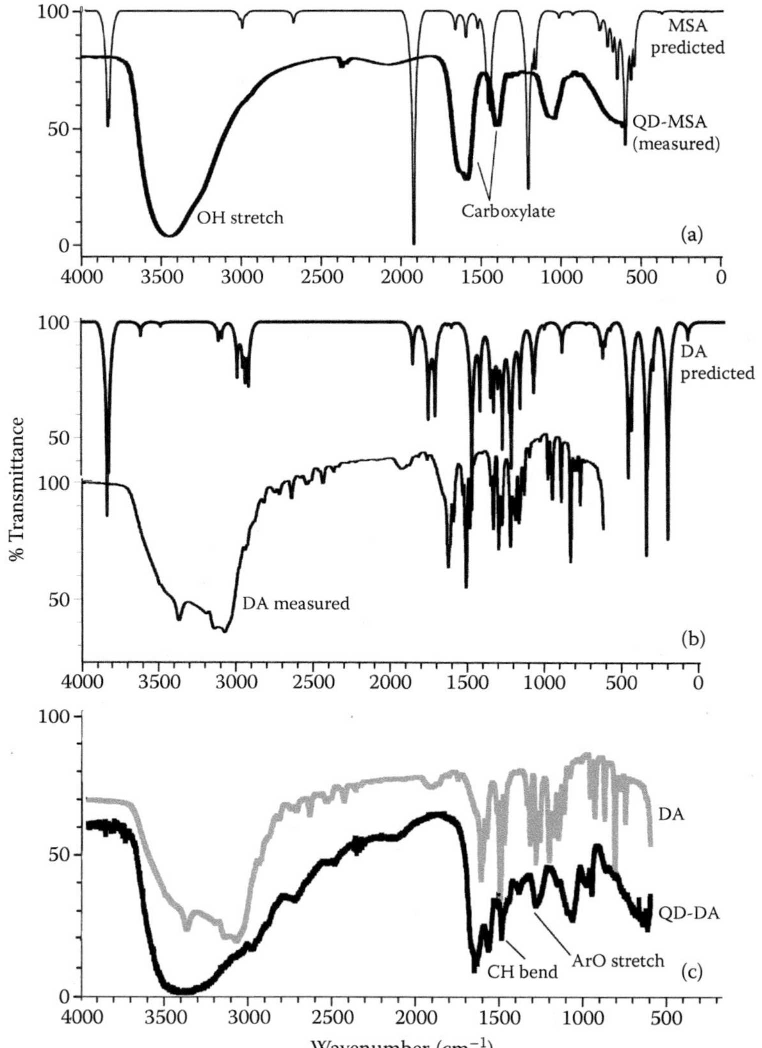



disappeared in the QDs, which is good, and there is a large OH stretch probably caused by adsorbed water. The only really recognizable MSA feature in the QDs are the two carboxylate vibrations, one in the 1540–1650 range, and the other in the 1360–1450 range. Note that both are asymmetric. This is a typical appearance of solubilized nanoparticles, where we are seeing only what is on the surface.

Next we look at the predicted and measured spectra of dopamine (DA) alone. Many of the features of the predicted spectrum are seen; because of the ZnSe window, we cannot see those below ~450 cm$^{-1}$. The features that are not in the predicted spectra are consistent with those in the Sigma-Aldrich reported spectrum, and probably involve water, including hydrogen bonding interactions (**Figure 16.44b**).

Finally, compare the spectrum of the dopamine-conjugated QDs to that of DA alone (**Figure 16.44c**). Note that the carboxyl peaks of the solubilized QDs have disappeared, but that the DA peaks in the 3300 range cannot be resolved because of the massive O–H stretch. Two of the dopamine peaks, the aryl oxygen stretch at 1245 and the CH bend at 1500 wavenumbers, are readily seen on the conjugated QDs.

## TCSPC

Does binding dopamine change QD emission lifetimes? This is an interesting question, since this is not really the typical experiment to tell the difference between static and dynamic quenching. We *know* that the quenchers are not free to diffuse, since we have complexed them to the QD. Therefore, we will be seeing the emission only from QD-dopamine, and if the lifetime is different from that of the QDs alone, it suggests energy transfer (such as FRET) or direct electron transfer from the photoexcited hole of the QD to the dopamine, which is energetically favored (**Figure 16.45a**).

We do TCSPC on an apparatus much like the one pictured in **Figure 16.17**. QD concentrations for our TCSPC experiments range from 20 to 100 nm (bear in mind that the dopamine conjugates are quenched, so more is required than in unconjugated QD solutions). Concentrations much higher than 100 nM cause scattering and are not recommended. The frequency-doubled laser light at 400 nm is used to excite as well as "photoage" the QDs.

Neutral density filters should be placed before the focusing lens to obtain an optimum peak pulse intensity; for us, this was $10^7$ W/cm². Emission should be collected at the emission peak of the QD conjugate at the magic angle with respect to the 400 nm vertical excitation laser polarization and focused into a monochromator. The PMT is mounted on the exit slit of the monochromator.

In order to hold excitation power constant across different samples, the monochromator slit width is changed to obtain discriminated count rates between ~500 and 10,000 per second (or below 0.04 of the laser repetition rate) to avoid pulse pileup. Controls containing dopamine alone, EDC alone, and dopamine and EDC should be run once to ensure that they show no fluorescence signal. The resulting spectra should be plotted against each other, normalized or not, revealing the dramatic lifetime change upon dopamine conjugation (**Figure 16.45b**).

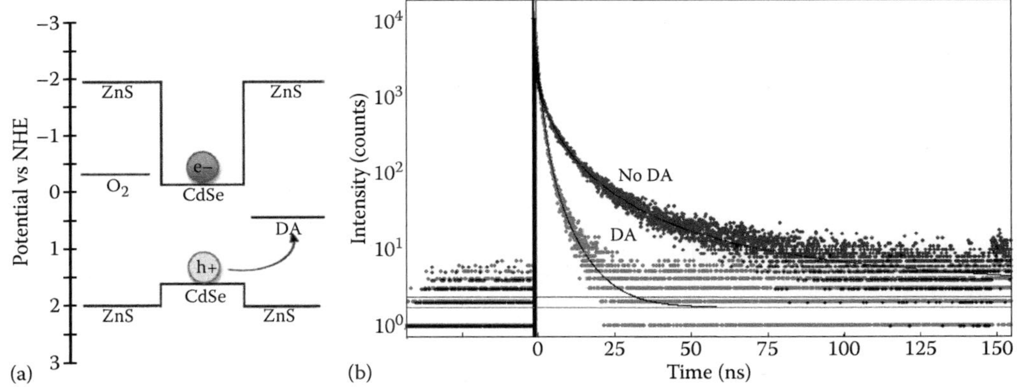


**Figure 16.45 TCSPC on quantum dots with attached dopamine.** (a) Electrochemical energies of the CdSe/ZnS QD electron and hole vs. the redox potential of dopamine (DA). A hole could be transferred to dopamine, oxidizing it (arrow). (b) TCSPC plots for QDs with and without DA, showing dramatic lifetime change. Fits are to stretched exponentials (**Figure 16.17**) with average lifetimes of 1.4 ns before conjugation, and 0.3 ns with dopamine.

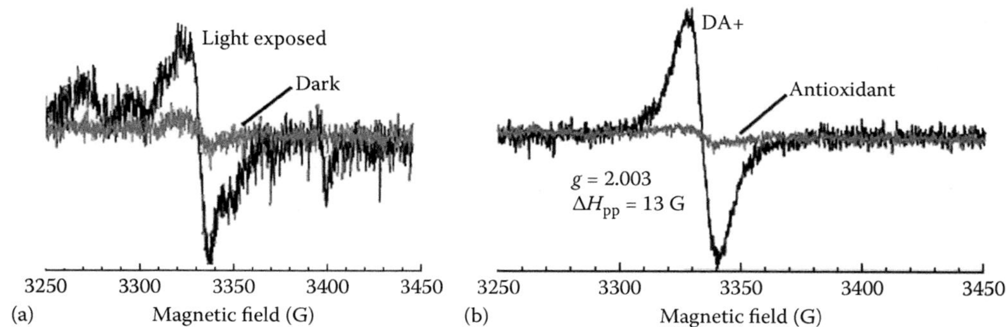


**Figure 16.46  (a) EPR signal from thiol-solubilized QDs in water, with and without UV light exposure. (b) EPR signal from QD-dopamine in water under UV irradiation, showing appearance of the dopamine radical.** When antioxidants are added, there is almost no signal.

### EPR

EPR can be used to investigate the mechanism of fluorescence quenching of QD-DA conjugates. EPR spectroscopy provides an unambiguous identification of the species involved in the charge separation processes by revealing changes in local symmetry and hyperfine couplings along the pathway of charge carriers. X-band EPR experiments were conducted on a Bruker ESP300E spectrometer equipped with a Varian cavity and a variable-temperature cryostat (Air Products) cooled to liquid helium temperature. The microwave frequency was determined after each measurement using a Hewlett-Packard 5352B frequency counter. The $g$ tensors were calibrated for homogeneity and accuracy by comparison to a coal standard, $g = 2.00285 \pm 0.00005$.

Samples were excited directly into the cavity using a 300-W Xe lamp (ILC) with a 355 nm cutoff filter. The filter was used to avoid possible excitation of dopamine.

Xe lamp illumination of solubilized QDs alone gives a spectrum corresponding to the formation of ˙CHCH radicals from the solubilizing thiol (**Figure 16.46a**). Cutoff illumination ($\lambda > 355$ nm) of dopamine-conjugated QDs leads to the formation of a carbon centered radical with $g$ tensor of 2.003 (**Figure 16.46b**). This signal corresponds to the radical dopamine⁺, with spin density on the pendant side chain, indicating charge transfer between excited QD and attached dopamine. When antioxidants are added or oxygen is removed, the intensity of the dopamine⁺ decreases drastically, confirming that the presence of the antioxidant suppresses the oxidation of dopamine. This confirms the mechanism suggested by the TCSPC experiments.

## 16.12  FINAL COMMENTS

Spectroscopy of different energies remains the most quantitative way to analyze molecules and materials at the angstrom scale. Some techniques are indispensable and must be mastered; others cannot be mastered in a lifetime. You will be limited by your financial resources and by the instrumentation available at your institution, but do not be limited by your imagination. New techniques and better instruments are becoming available all the time, and new applications are being discovered in the biological sciences at an increasingly rapid rate.

Although this chapter did not have room to explore any of the techniques in great depth, and some techniques were omitted entirely, I hope that it will help you to get started. One of the first things to do is take an inventory of the instruments available to your group, as the presence of a rare type of spectrometer may open experimental possibilities previously unimagined.

## End-of-Chapter Questions

1. In **Chapter 2**, we discussed quantifying plasmid DNA using an extinction coefficient of 6600 $M^{-1}$ $cm^{-1}$. Explain why this is less accurate for short sequences, such as primers, and give a formula for obtaining the extinction coefficient of a DNA oligomer.

2. Derive **Equation 16.2**.

3. Derive the corresponding equation for static quenching.

4. Derive the corresponding equation for when both dynamic and static quenching occur in the same sample.

5. If a quencher has a "sphere of action" of volume $V$, where any fluorophore will be quenched if it is found with a quencher within that volume, show that **Equation 16.2** becomes

$$\frac{I_0}{I} = (1 + K_D[Q])e^{QVN/1000}$$ where $N$ is the number of quenchers.

6. Derive **Equation 16.10**.

7. Calculate the theoretical values of anisotropy for fluorescein (MW 332, density 1.54 $g/cm^3$) in ethanol and in propylene glycol at 25°C. Use the following values: $r_0 = 0.36$, $\tau_f = 4.4$ ns, $\eta$ (ethanol) = 1.194 cP, $\eta$ (propylene glycol) = 32 cP.

8. Explain how you would use anisotropy measurements to determine the binding constants between a fluorophore and a (nonfluorescent) protein, in the cases where (a) binding does not change the fluorophore quantum yield and (b) binding reduces the quantum yield by a factor of 6. Discuss how the anisotropy measurements would change for a molecule that is a prolate ellipsoid instead of a sphere.

9. Derive **Equation 16.16b** when $f = 1$.

10. Correct **Equation 16.16b** for the case in which the donor shows some emission in $\lambda_{em}$.

11. Discuss how you might use FRET to design a genetically encoded sensor for $[Ca^{2+}]$.

12. So-called "homo-FRET" involves FRET between identical molecules. Discuss how this works and when it is advantageous.

13. Suggest a few fluorescent protein FRET pairs and give the parameters that make these pairs a good match.

14. One way to become very confused with FRET is when apparent donor "quenching" is not quenching at all, but rather absorbance of some of the excitation light by the acceptor. How can this be detected and controlled for?

15. Discuss how time-resolved measurements might be done in the frequency domain. What sort of light source would be required?

16. A population of fluorophores all have a single-exponential lifetime decay with lifetime $\tau_{slow}$. Now a fraction $n$ of the fluorophores are quenched and show a lifetime $\tau_{fast}$. What does the decay look like now? What is the average lifetime? Values $\tau_{fast} = 0.1$ ns, $\tau_{slow} = 2$ ns and plot $n = 0.2, 0.4, 0.6$.

17. For measuring quenching due to a conformational change, is it better to use amplitude- or intensity-weighted lifetime? Justify your answer.

18. What general class of molecules is infrared inactive? Why?

19. How many vibrational modes are possible for (a) $H_2O$; (b) $CO_2$; (c) $C_2H_2$; (d) $CH_2Cl_2$? Which of these modes are infrared active?

20. Predict the fine structure in the NMR spectrum of ethanol.

21. Calculate the resonance frequencies of (a) a proton and (b) a $^{19}F$ nucleus if $B_0 = 15$ T.

22. Predict the EPR spectrum of a radical that contains three equivalent $^{14}N$ nuclei.

**23.** Explain why the concentration dependence shown in **Figure 16.36b** is true.

**24.** Predict the EPR spectrum of the phenyl radical.

**25.** What magnetic fields are needed for Q-band EPR?

**26.** Imagine that the TMP and TEMPO curves shown in **Figure 16.38b and d** come from mercaptopropionic acid–capped CdSe quantum dots (for B) and dopamine-conjugated CdSe (for D) (they really do!). Describe the reactions of photogenerated electrons and holes with water and oxygen that might explain these data.

**27.** You want to design a high-resolution MRI for small animals. Discuss the relationship between signal-to-noise and field strength, coil size, and acquisition times. Why are neuroanatomy experiments performed with fixed brain tissue rather than live specimens?

## Background Reading and Resources

### Books

There are of course hundreds of books on spectroscopy, from every level from pre-undergraduate to advanced graduate. These are just a few of my favorites, with a slight bias toward the theoretical background behind the different types of spectroscopy. The books on MRI are deliberately aimed at research scientists rather than clinicians.

Atkins, P., and De Paula, J. *Atkins' Physical Chemistry*, Edn. 10. Oxford University Press, New York, 2014.

Becker, W. *Advanced Time-Correlated Single Photon Counting Techniques*. Springer Series in Chemical Physics, Springer, New York, 2010.

Becker, W. *Advanced Time-Correlated Single Photon Counting Applications*. Springer Series in Chemical Physics, Springer, New York, 2015.

Bishop, D.M. *Group Theory and Chemistry*. Dover, Mineola, 1993.

Brustolon, M., and Giamello, E. (eds.). *Electron Paramagnetic Resonance: A Practitioner's Toolkit*. John Wiley & Sons, Hoboken, NJ, 2009.

Cazes, J., and Ewing, G.W. *Ewing's Analytical Instrumentation Handbook*. Edn. 3. Wiley, New York, 2009.

Cook, D.B. *Handbook of Computational Quantum Chemistry*. Dover, New York, 2005.

de Graaf, R.A. *In Vivo NMR Spectroscopy: Principles and Techniques*. John Wiley & Sons, Hoboken, NJ, 2007.

Gregoriou, V.G., and Braiman, M.S. (eds.). *Vibrational Spectroscopy of Biological and Polymeric Materials*. CRC Press, Boca Raton, FL, 2006.

Guo, J. (ed.). *X-rays in Nanoscience: Spectroscopy, Spectromicroscopy, and Scattering Techniques*. Wiley-VCH, Weinheim, Germany, 2010.

Hollas, J.M. *Modern Spectroscopy*. John Wiley & Sons, Hoboken, NJ, 2010.

Kaupp, M., Bühl, M., and Malkin, V.G. (eds.). *Calculation of NMR and EPR Parameters: Theory and Applications*. Wiley-VCH Verlag GmbH & Co, Weinheim, Germany, 2004.

Lakowicz, J.R. *Principles of Fluorescence Spectroscopy*. Edn. 3. Springer Science + Business Media, LLC, New York, 2011.

McQuarrie, D.A., and Simon, J.D. *Physical Chemistry: A Molecular Approach*. John Wiley & Sons, Hoboken, NJ, 2004.

Rieger, P.H. *Electron Spin Resonance: Analysis and Interpretation*. Royal Society of Chemistry, Cambridge, UK, 2007.

Salzer, R., and Siesler, H.W. (eds.). *Infrared and Raman Spectroscopic Imaging*. Wiley-VCH, Weinheim, Germany, 2009.

Szabo, A., and Ostlund, N.S. *Modern Quantum Chemistry: Introduction to Advanced Electronic Structure Theory*. Dover, New York, 1996.

Turro, N.J., Scaiano, J.C., and Ramamurthy, V. *Principles of Molecular Photochemistry: An Introduction*. University Science Books, Sausalito, CA, 2009.

Weil, J., Bolton, H., and Wertz, J. *Electron Paramagnetic Resonance: Elemental Theory and Practical Applications*. John Wiley and Sons, New York, 1994.

Weishaupt, D., Koechli, V.D., and Marincek, B. *How Does MRI Work?: An Introduction to the Physics and Function of Magnetic Resonance Imaging*. Springer, New York, 2006.

Wilson, E.B., Decius, J.C., and Cross, P.C. *Vibrations: The Theory of Infrared and Raman Vibrational Spectra*. Dover, New York, 1980.

### Journal articles

*UV–Vis absorption and fluorescence, steady-state and time-resolved*

Baba, M., Suzuki, M., Ganeev, R.A., Kuroda, H., Ozaki, T., Hamakubo, T., Masuda, K., Hayashi, M., Sakihama, T., Kodama, T., and Kozasa, T. (2007). Decay time shortening of fluorescence from donor-acceptor pair proteins using ultrafast time-resolved fluorescence resonance energy transfer spectroscopy. *Journal of Luminescence* 127, 355–361.

Beljonne, D., Curutchet, C., Scholes, G.D., and Silbey, R.J. (2009). Beyond forster resonance energy transfer in biological and nanoscale systems. *Journal of Physical Chemistry B* 113, 6583–6599.

Bohmer, M., and Enderlein, J. (2003). Fluorescence spectroscopy of single molecules under ambient conditions: Methodology and technology. *Chemphyschem* 4, 793–808.

Bonsma, S., Purchase, R., Jezowski, S., Gallus, J., Konz, F., and Volker, S. (2005). Green and red fluorescent proteins: Photo- and thermally induced dynamics probed by site-selective spectroscopy and hole burning. *Chemphyschem* 6, 838–849.

Borst, J.W., Hink, M.A., van Hoek, A., and Visser, A. (2005). Effects of refractive index and viscosity on fluorescence and anisotropy decays of enhanced cyan and yellow fluorescent proteins. *Journal of Fluorescence* 15, 153–160.

Borst, W.L., and Liu, L.I. (1999). Time-autocorrelated two-photon counting technique for time-resolved fluorescence measurements. *Review of Scientific Instruments* 70, 41–49.

Carlson, H.J., and Campbell, R.E. (2009). Genetically encoded FRET-based biosensors for multiparameter fluorescence imaging. *Current Opinion in Biotechnology* 20, 19–27.

Cash, T.P., Pan, Y., and Simon, M.C. (2007). Reactive oxygen species and cellular oxygen sensing. *Free Radical Biology and Medicine* 43, 1219–1225.

Chudakov, D.M., Lukyanov, S., and Lukyanov, K.A. (2005). Fluorescent proteins as a toolkit for in vivo imaging. *Trends in Biotechnology* 23, 605–616.

Cox, G., Matz, M., and Salih, A. (2007). Fluorescence lifetime imaging of coral fluorescent proteins. *Microscopy Research and Technique* 70, 243–251.

Day, R.N., and Schaufele, F. (2008). Fluorescent protein tools for studying protein dynamics in living cells: A review. *Journal of Biomedical Optics* 13(3), 031202.

de Almeida, R.F.M., Loura, L.M.S., and Prieto, M. (2009). Membrane lipid domains and rafts: Current applications of fluorescence lifetime spectroscopy and imaging. *Chemistry and Physics of Lipids* 157, 61–77.

Evers, T.H., van Dongen, E., Faesen, A.C., Meijer, E.W., and Merkx, M. (2006). Quantitative understanding of the energy transfer between fluorescent proteins connected via flexible peptide linkers. *Biochemistry* 45, 13183–13192.

Felekyan, S., Kuhnemuth, R., Kudryavtsev, V., Sandhagen, C., Becker, W., and Seidel, C.A.M. (2005). Full correlation from picoseconds to seconds by time-resolved and time-correlated single photon detection. *Review of Scientific Instruments* 76, 083104.

Frommer, W.B., Davidson, M.W., and Campbell, R.E. (2009). Genetically encoded biosensors based on engineered fluorescent proteins. *Chemical Society Reviews* 38, 2833–2841.

Gambin, Y., and Deniz, A.A. (2010). Multicolor single-molecule FRET to explore protein folding and binding. *Molecular Biosystems* 6, 1540–1547.

Gouanve, F., Schuster, T., Allard, E., Meallet-Renault, R., and Larpent, C. (2007). Fluorescence quenching upon binding of copper ions in dye-doped and ligand-capped polymer nanoparticles: A simple way to probe the dye accessibility in nano-sized templates. *Advanced Functional Materials* 17, 2746–2756.

Gradinaru, C.C., Marushchak, D.O., Samim, M., and Krull, U.J. (2010). Fluorescence anisotropy: From single molecules to live cells. *Analyst* 135, 452–459.

Greulich, K.O. (2005). Fluoresence spectroscopy onn single biomolecules. *Chemphyschem* 6, 2458–2471.

Haas, E. (2005). The study of protein folding and dynamics by determination of intramolecular distance distributions and their fluctuations using ensemble and single-molecule FRET measurements. *Chemphyschem* 6, 858–870.

Heberle, F.A., Buboltz, J.T., Stringer, D., and Feigenson, G.W. (2005). Fluorescence methods to detect phase boundaries in lipid bilayer mixtures. *Biochimica et Biophysica Acta—Molecular Cell Research* 1746, 186–192.

Heyduk, T. (2002). Measuring protein conformational changes by FRET/LRET. *Current Opinion in Biotechnology* 13, 292–296.

Hoffmann, B., Zimmer, T., Klocker, N., Kelbauskas, L., Konig, K., Benndorf, K., and Biskup, C. (2008). Prolonged irradiation of enhanced cyan fluorescent protein or Cerulean can invalidate Forster resonance energy transfer measurements. *Journal of Biomedical Optics* 13(3), 031205.

Hosoi, H., Mizuno, H., Miyawaki, A., and Tahara, T. (2006). Competition between energy and proton transfer in ultrafast excited-state dynamics of an oligomeric fluorescent protein red kaede. *Journal of Physical Chemistry B* 110, 22853–22860.

Huebsch, N.D., and Mooney, D.J. (2007). Fluorescent resonance energy transfer: A tool for probing molecular cell-biomaterial interactions in three dimensions. *Biomaterials* 28, 2424–2437.

Jares-Erijman, E.A., and Jovin, T.M. (2003). FRET imaging. *Nature Biotechnology* 21, 1387–1395.

Jones, M., Kumar, S., Lo, S.S., and Scholes, G.D. (2008). Exciton trapping and recombination in type IICdSe/CdTe nanorod heterostructures. *Journal of Physical Chemistry C* 112, 5423–5431.

Jones, M., Nedeljkovic, J., Ellingson, R.J., Nozik, A.J., and Rumbles, G. (2003). Photoenhancement of luminescence in colloidal CdSe quantum dot solutions. *Journal of Physical Chemistry B* 107, 11346–11352.

Joo, C., Balci, H., Ishitsuka, Y., Buranachai, C., and Ha. (2008). Advances in single-molecule fluorescence methods for molecular biology. *Annual Review of Biochemistry* 77, 51–76.

Kerppola, T.K. (2008). Biomolecular fluorescence complementation (BiFC) analysis as a probe of protein interactions in living cells. *Annual Review of Biophysics* 37, 465–487.

Kikuchi, K. (2010). Design, synthesis and biological application of chemical probes for bio-imaging. *Chemical Society Reviews* 39, 2048–2053.

Klostermeier, D., and Millar, D.P. (2001). Time-resolved fluorescence resonance energy transfer: A versatile tool for the analysis of nucleic acids. *Biopolymers* 61, 159–179.

Kulzer, F., and Orrit, M. (2004). Single-molecule optics. *Annual Review of Physical Chemistry* 55, 585–611.

Lakowicz, J.R., and Berndt, K. (1990). Frequency-domain measurements of photon migration in tissues. *Chemical Physics Letters* 166, 246–252.

Laptenok, S.P., Borst, J.W., Mullen, K.M., van Stokkum, I.H.M., Visser, A., and van Amerongen, H. (2010). Global analysis of Forster resonance energy transfer in live cells measured by fluorescence lifetime imaging microscopy exploiting the rise time of acceptor fluorescence. *Physical Chemistry Chemical Physics* 12, 7593–7602.

Levitt, J.A., Matthews, D.R., Ameer-Beg, S.M., and Suhling, K. (2009). Fluorescence lifetime and polarization-resolved imaging in cell biology. *Current Opinion in Biotechnology* 20, 28–36.

Li, Y.S., Zhou, X.Y., and Ye, D.Y. (2008). Molecular beacons: An optimal multifunctional biological probe. *Biochemical and Biophysical Research Communications* 373, 457–461.

Lu, H., Schops, O., Woggon, U., and Niemeyer, C.M. (2008). Self-assembled donor comprising quantum dots and fluorescent proteins for long-range fluorescence resonance energy transfer. *Journal of the American Chemical Society* 130, 4815–4827.

Lukyanov, K.A., Chudakov, D.M., Lukyanov, S., and Verkhusha, V.V. (2005). Photoactivatable fluorescent proteins. *Nature Reviews Molecular Cell Biology* 6, 885–891.

Lymperopoulos, K., Kiel, A., Seefeld, A., Stohr, K., and Herten, D.P. (2010). Fluorescent probes and delivery methods for single-molecule experiments. *Chemphyschem* 11, 43–53.

Marras, S.A.E. (2008). Interactive fluorophore and quencher pairs for labeling fluorescent nucleic acid hybridization probes. *Molecular Biotechnology* 38, 247–255.

Marshall, R.A., Aitken, C.E., Dorywalska, M., and Puglisi, J.D. (2008). Translation at the single-molecule level. *Annual Review of Biochemistry* 77, 177–203.

McLoskey, D., Birch, D.J.S., Sanderson, A., Suhling, K., Welch, E., and Hicks, P.J. (1996). Multiplexed single-photon counting.1. A time-correlated fluorescence lifetime camera. *Review of Scientific Instruments* 67, 2228–2237.

Medintz, I.L., and Mattoussi, H. (2009). Quantum dot-based resonance energy transfer and its growing application in biology. *Physical Chemistry Chemical Physics* 11, 17–45.

Muller-Taubenberger, A., and Anderson, K.I. (2007). Recent advances using green and red fluorescent protein variants. *Applied Microbiology and Biotechnology* 77, 1–16.

Phillips, D. (1994). Luminescence lifetimes in biological-systems. *Analyst* 119, 543–550.

Phillips, D., Drake, R.C., Oconnor, D.V., and Christensen, R.L. (1985). Time correlated single-photon counting (TCSPC) using laser excitation. *Analytical Instrumentation* 14, 267–292.

Piston, D.W., and Kremers, G.J. (2007). Fluorescent protein FRET: The good, the bad and the ugly. *Trends in Biochemical Sciences* 32, 407–414.

Piszczek, G. (2006). Luminescent metal-ligand complexes as probes of macromolecular interactions and biopolymer dynamics. *Archives of Biochemistry and Biophysics* 453, 54–62.

Pons, T., and Mattoussi, H. (2009). Investigating biological processes at the single molecule level using luminescent quantum dots. *Annals of Biomedical Engineering* 37, 1934–1959.

Posokhov, Y.O., and Ladokhin, A.S. (2006). Lifetime fluorescence method for determining membrane topology of proteins. *Analytical Biochemistry* 348, 87–93.

Rajapakse, H.E., Gahlaut, N., Mohandessi, S., Yu, D., Turner, J.R., and Miller, L.W. (2010). Time-resolved luminescence resonance energy transfer imaging of protein-protein interactions in living cells. *Proceedings of the National Academy of Sciences of the United States of America* 107, 13582–13587.

Ramadass, R., Becker, D., Jendrach, M., and Bereiter-Hahn, J. (2007). Spectrally and spatially resolved fluorescence lifetime imaging in living cells: TRPV4-microfilament interactions. *Archives of Biochemistry and Biophysics* 463, 27–36.

Rasnik, I., McKinney, S.A., and Ha, T. (2005). Surfaces and orientations: Much to FRET about? *Accounts of Chemical Research* 38, 542–548.

Roy, R., Hohng, S., and Ha, T. (2008). A practical guide to single-molecule FRET. *Nature Methods* 5, 507–516.

Royer, C.A., and Scarlata, S.F. (2008). Fluorescence approaches to quantifying biomolecular interactions. *Fluorescence Spectroscopy* 450, 79–106.

Sanders, J.C., Ottaviani, M.F., Vanhoek, A., Visser, A., and Hemminga, M.A. (1992). A small protein in model membranes—A time-resolved fluorescence and ESR study on the interaction of M13 coat protein with lipid bilayers. *European Biophysics Journal with Biophysics Letters* 21, 305–311.

Saxena, A.M., Krishnamoorthy, G., Udgaonkar, J.B., and Periasamy, N. (2007). Identification of intermediate species in protein-folding by quantitative analysis of amplitudes in time-domain fluorescence spectroscopy. *Journal of Chemical Sciences* 119, 61–69.

Schlothauer, J., Hackbarth, S., and Roder, B. (2009). A new benchmark for time-resolved detection of singlet oxygen luminescence—Revealing the evolution of lifetime in living cells with low dose illumination. *Laser Physics Letters* 6, 216–221.

Schuler, B. (2005). Single-molecule fluorescence spectroscopy of protein folding. *Chemphyschem* 6, 1206–1220.

Schuttrigkeit, T.A., Zachariae, U., von Feilitzsch, T., Wiehler, J., von Hummel, J., Steipe, B., and Michel-Beyerle, M.E. (2001). Picosecond time-resolved FRET in the fluorescent protein from Discosoma Red (wt-DsRed). *Chemphyschem* 2, 325–328.

Seward, H.E., and Bagshaw, C.R. (2009). The photochemistry of fluorescent proteins: Implications for their biological applications. *Chemical Society Reviews* 38, 2842–2851.

Shanker, N., and Bane, S.L. (2008). Basic aspects of absorption and fluorescence spectroscopy and resonance energy transfer methods. *Methods in Cell Biology*, 84, 213–242.

Shimozono, S., Hosoi, H., Mizuno, H., Fukano, T., Tahara, T., and Miyawaki, A. (2006). Concatenation of cyan and yellow fluorescent proteins for efficient resonance energy transfer. *Biochemistry* 45, 6267–6271.

Spriet, C., Trinel, D., Waharte, F., Deslee, D., Vandenbunder, B., Barbillat, J., and Heliot, L. (2007). Correlated fluorescence lifetime and spectral measurements in living cells. *Microscopy Research and Technique* 70, 85–94.

Stortelder, A., Buijs, J.B., Bulthuis, J., Gooijer, C., and van der Zwan, G. (2004). Fast-gated intensified charge-coupled device camera to record time-resolved fluorescence spectra of tryptophan. *Applied Spectroscopy* 58, 705–710.

Subramaniam, V., Hanley, Q.S., Clayton, A.H.A., and Jovin, T.M. (2003). Photophysics of green and red fluorescent proteins: Implications for quantitative microscopy. *Biophotonics A* 360, 178–201.

Suhling, K., McLoskey, D., and Birch, D.J.S. (1996). Multiplexed single-photon counting. 2. The statistical theory of time-correlated measurements. *Review of Scientific Instruments* 67, 2238–2246.

Szollosi, J., and Alexander, D.R. (2003). The application of fluorescence resonance energy transfer to the investigation of phosphatases. *Protein Phosphatases* 366, 203–224.

Tian, Z.Y., Wu, W.W., and Li, A.D.Q. (2009). Photoswitchable fluorescent nanoparticles: Preparation, properties and applications. *Chemphyschem* 10, 2577–2591.

Umezawa, Y. (2005). Genetically encoded optical probes for imaging cellular signaling pathways. *Biosensors and Bioelectronics* 20, 2504–2511.

Wagner, M.K., Li, F., Li, J.J., Li, X.F., and Le, X.C. (2010). Use of quantum dots in the development of assays for cancer biomarkers. *Analytical and Bioanalytical Chemistry* 397, 3213–3224.

Wang, Y.X., and Wang, N. (2009). FRET and mechanobiology. *Integrative Biology* 1, 565–573.

Wessels, J.T., Yamauchi, K., Hoffman, R.M., and Wouters, F.S. (2010). Advances in cellular, subcellular, and nanoscale imaging in vitro and in vivo. *Cytometry Part A* 77A, 667–676.

Wrozowa, T., Ciesielska, B., Komar, D., Karolczak, J., Maciejewski, A., and Kubicki, J. (2004). Measurements of picosecond lifetimes by time correlated single photon counting method: The effect of the refraction index of the solvent on the instrument response function. *Review of Scientific Instruments* 75, 3107–3121.

Xia, Z.Y., and Rao, J.H. (2009). Biosensing and imaging based on bioluminescence resonance energy transfer. *Current Opinion in Biotechnology* 20, 37–44.

Yan, Y.L., and Marriott, G. (2003). Analysis of protein interactions using fluorescence technologies. *Current Opinion in Chemical Biology* 7, 635–640.

Zhou, R.B., Schlierf, M., and Ha, T. (2010). Force fluorescence spectroscopy at the single-molecule level. *Methods in Enzymology* 475, 405–426; *Single Molecule Tools B* 474.

## Infrared

Alexandre, M.T.A., Domratcheva, T., Bonetti, C., van Wilderen, L., van Grondelle, R., Groot, M.L., Hellingwerf, K.J., and Kennis, J.T.M. (2009). Primary reactions of the LOV2 domain of phototropin studied with ultrafast mid-infrared spectroscopy and quantum chemistry. *Biophysical Journal* 97, 227–237.

Arkin, I.T. (2006). Isotope-edited IR spectroscopy for the study of membrane proteins. *Current Opinion in Chemical Biology* 10, 394–401.

Arrondo, J.L.R., and Goni, F.M. (1999). Structure and dynamics of membrane proteins as studied by infrared spectroscopy. *Progress in Biophysics and Molecular Biology* 72, 367–405.

Ataka, K., Kottke, T., and Heberle, J. (2010). Thinner, smaller, faster: IR techniques to probe the functionality of biological and biomimetic systems. *Angewandte Chemie-International Edition* 49, 5416–5424.

Barth, A. (2000). The infrared absorption of amino acid side chains. *Progress in Biophysics and Molecular Biology* 74, 141–173.

Barth, A. (2007). Infrared spectroscopy of proteins. *Biochimica Et Biophysica Acta—Bioenergetics* 1767, 1073–1101.

Berera, R., van Grondelle, R., and Kennis, J.T.M. (2009). Ultrafast transient absorption spectroscopy: Principles and application to photosynthetic systems. *Photosynthesis Research* 101, 105–118.

Bista, R.K., Bruch, R.F., and Covington, A.M. (2010). Vibrational spectroscopic studies of newly developed synthetic biopolymers. *Biopolymers* 93, 403–417.

Bonetti, C., Alexandre, M.T.A., van Stokkum, I.H.M., Hiller, R.G., Groot, M.L., van Grondelle, R., and Kennis, J.T.M. (2010). Identification of excited-state energy transfer and relaxation pathways in the peridinin-chlorophyll complex: An ultrafast mid-infrared study. *Physical Chemistry Chemical Physics* 12, 9256–9266.

Borodko, Y., Lee, H.S., Joo, S.H., Zhang, Y.W., and Somorjai, G. (2010). Spectroscopic study of the thermal degradation of PVP-capped Rh and Pt nanoparticles in H-2 and O-2 environments. *Journal of Physical Chemistry C* 114, 1117–1126.

Bunaciu, A.A., Aboul-Enein, H.Y., and Fleschin, S. (2010). Application of Fourier transform infrared spectrophotometry in pharmaceutical drugs analysis. *Applied Spectroscopy Reviews* 45, 206–219.

Chu, H.A., Hillier, W., and Debus, R.J. (2004). Evidence that the C-terminus of the D1 polypeptide of photosystem II is ligated to the manganese ion that undergoes oxidation during the S-1 to S-2 transition: An isotope-edited FTIR study. *Biochemistry* 43, 3152–3166.

Deflores, L.P., Ganim, Z., Nicodemus, R.A., and Tokmakoff, A. (2009). Amide I'-II' 2D IR spectroscopy provides enhanced protein secondary structural sensitivity. *Journal of the American Chemical Society* 131, 3385–3391.

Della Ventura, G., Bellatreccia, F., Marcelli, A., Guidi, M.C., Piccinini, M., Cavallo, A., and Piochi, M. (2010). Application of micro-FTIR imaging in the Earth sciences. *Analytical and Bioanalytical Chemistry* 397, 2039–2049.

Downes, A., Mouras, R., and Elfick, A. (2010). Optical spectroscopy for noninvasive monitoring of stem cell differentiation. *Journal of Biomedicine and Biotechnology* 2010, 101864.

Ede, N.J., and Wu, Z.M. (2003). Beyond Rf tagging. *Current Opinion in Chemical Biology* 7, 374–379.

Ganim, Z., Chung, H.S., Smith, A.W., Deflores, L.P., Jones, K.C., and Tokmakoff, A. (2008). Amide I two-dimensional infrared spectroscopy of proteins. *Accounts of Chemical Research* 41, 432–441.

Goormaghtigh, E., Ruysschaert, J.M., and Raussens, V. (2006). Evaluation of the information content in infrared spectra for protein secondary structure determination. *Biophysical Journal* 90, 2946–2957.

Gordon, L.M., Mobley, P.W., Pilpa, R., Sherman, M.A., and Waring, A.J. (2002). Conformational mapping of the N-terminal peptide of HIV-1 gp41 in membrane environments using C-13-enhanced Fourier transform infrared spectroscopy. *Biochimica Et Biophysica Acta—Biomembranes* 1559, 96–120.

Haupts, U., Tittor, J., and Oesterhelt, D. (1999). Closing in on bacteriorhodopsin: Progress in understanding the molecule. *Annual Review of Biophysics and Biomolecular Structure* 28, 367–399.

Iwaki, M., Puustinen, A., Wikstrom, M., and Rich, P.R. (2004). ATR-FTIR spectroscopy and isotope labeling of the P-M intermediate of *Paracoccus denitrificans* cytochrome c oxidase. *Biochemistry* 43, 14370–14378.

Jackson, M., and Mantsch, H.H. (1995). The use and misuse of FTIR spectroscopy in the determination of protein-structure. *Critical Reviews in Biochemistry and Molecular Biology* 30, 95–120.

Jayaraman, V. (2004). Spectroscopic and kinetic methods for ligand–protein interactions of glutamate receptor. *Energetics of Biological Macromolecules E* 380, 170–187.

Jung, C. (2000). Insight into protein structure and protein-ligand recognition by Fourier transform infrared spectroscopy. *Journal of Molecular Recognition* 13, 325–351.

Karyakin, A., Motiejunas, D., Wade, R.C., and Jung, C. (2007). FTIR studies of the redox partner interaction in cytochrome P450: The Pdx-P450cam couple. *Biochimica et Biophysica Acta—General Subjects* 1770, 420–431.

Kazarian, S.G., and Chan, K.L.A. (2006). Applications of ATR-FTIR spectroscopic imaging to biomedical samples. *Biochimica et Biophysica Acta—Biomembranes* 1758, 858–867.

Kosumi, D., Abe, K., Karasawa, H., Fujiwara, M., Cogdell, R. J., Hashimoto, H., and Yoshizawa, M. (2010). Ultrafast relaxation kinetics of the dark S-1 state in all-trans-beta-carotene explored by one- and two-photon pump-probe spectroscopy. *Chemical Physics* 373, 33–37.

Kretlow, A., Wang, Q., Kneipp, J., Lasch, P., Beekes, M., Miller, L., and Naumann, D. (2006). FTIR-microspectroscopy of prion-infected nervous tissue. *Biochimica et Biophysica Acta—Biomembranes* 1758, 948–959.

Lawrie, G., Keen, I., Drew, B., Chandler-Temple, A., Rintoul, L., Fredericks, P., and Grondahl, L. (2007). Interactions between alginate and chitosan biopolymers characterized using FTIR and XPS. *Biomacromolecules* 8, 2533–2541.

Lim, M., and Anfinrud, P.A. Ultrafast time-resolved IR studies of protein–ligand interactions. In G. U. Nienhaus (ed.), *Methods in Molecular Biology*, vol. 305: Protein–Ligand Interactions: Methods and Applications. Humana Press, Totowa, NJ, 243–258, 2005.

Lorenz-Fonfria, V.A., Granell, M., Leon, X., Leblanc, G., and Padros, E. (2009). In-plane and out-of-plane infrared difference spectroscopy unravels tilting of helices and structural changes in a membrane protein upon substrate binding. *Journal of the American Chemical Society* 131, 15094.

Mackanos, M.A., and Contag, C.H. (2010). Fiber-optic probes enable cancer detection with FTIR spectroscopy. *Trends in Biotechnology* 28, 317–323.

Marshall, D., and Rich, P.R. (2009). Studies of complex I by Fourier transform infrared spectroscopy. *Methods in Enzymology* 456, 53–74.

Martin, I., Goormaghtigh, E., and Ruysschaert, J. M. (2003). Attenuated total reflection IR spectroscopy as a tool to investigate the orientation and tertiary structure changes in fusion proteins. *Biochimica et Biophysica Acta—Biomembranes* 1614, 97–103.

Materazzi, S., and Vecchio, S. (2010). Evolved gas analysis by infrared spectroscopy. *Applied Spectroscopy Reviews* 45, 241–273.

Nabedryk, E., and Breton, J. (2008). Coupling of electron transfer to proton uptake at the Q(B) site of the bacterial reaction center: A perspective from FTIR difference spectroscopy. *Biochimica et Biophysica Acta—Bioenergetics* 1777, 1229–1248.

Neault, J.F., Diamantoglou, S., Nafisi, S., and Tajmir-Riahi, H.A. (2008). Conformational analysis of Na,K-ATPase in drug-protein complexes. *Journal of Photochemistry and Photobiology B—Biology* 91, 167–174.

Niedzwiedzki, D.M., Collins, A.M., LaFountain, A.M., Enriquez, M.M., Frank, H.A., and Blankenship, R.E. (2010). Spectroscopic studies of carotenoid-to-bacteriochlorophyll energy transfer in LHRC photosynthetic complex from *Roseiflexus castenholzii*. *Journal of Physical Chemistry B* 114, 8723–8734.

Noguchi, T. (2010). Fourier transform infrared spectroscopy of special pair bacteriochlorophylls in homodimeric reaction centers of heliobacteria and green sulfur bacteria. *Photosynthesis Research* 104, 321–331.

Petersen, P.B., and Tokmakoff, A. (2010). Source for ultrafast continuum infrared and terahertz radiation. *Optics Letters* 35, 1962–1964.

Prati, S., Joseph, E., Sciutto, G., and Mazzeo, R. (2010). New advances in the application of FTIR microscopy and spectroscopy for the characterization of artistic materials. *Accounts of Chemical Research* 43, 792–801.

Rich, P.R., and Iwaki, M. (2007). Methods to probe protein transitions with ATR infrared spectroscopy. *Molecular Biosystems* 3, 398–407.

Ritter, E., Elgeti, M., and Bartl, F.J. (2008). Activity switches of rhodopsin. *Photochemistry and Photobiology* 84, 911–920.

Ryan, S.E., Demers, C.N., Chew, J.P., and Baenziger, J.E. (1996). Structural effects of neutral and anionic lipids on the nicotinic acetylcholine receptor. An infrared difference spectroscopy study. *Journal of Biological Chemistry* 271, 24590–24597.

Saulou, C., Jamme, F., Maranges, C., Fourquaux, I., Despax, B., Raynaud, P., Dumas, P., and Mercier-Bonin, M. (2010). Synchrotron FTIR microspectroscopy of the yeast *Saccharomyces cerevisiae* after exposure to plasma-deposited nanosilver-containing coating. *Analytical and Bioanalytical Chemistry* 396, 1441–1450.

Schmidt, B., Hillier, W., McCracken, J., and Ferguson-Miller, S. (2004). The use of stable isotopes and spectroscopy to investigate the energy transducing function of cytochrome c oxidase. *Biochimica et Biophysica Acta—Bioenergetics* 1655, 248–255.

Siebert, F. (1995). Infrared-spectroscopy applied to biochemical and biological problems. *Biochemical Spectroscopy* 246, 501–526.

Vogel, R., and Siebert, F. (2003). Fourier transform IR spectroscopy study for new insights into molecular properties and activation mechanisms of visual pigment rhodopsin. *Biopolymers* 72, 133–148.

Zhang, B., and Yan, B. (2010). Analytical strategies for characterizing the surface chemistry of nanoparticles. *Analytical and Bioanalytical Chemistry* 396, 973–982.

## NMR, MRI, EPR

Altieri, A.S., and Byrd, R.A. (2004). Automation of NMR structure determination of proteins. *Current Opinion in Structural Biology* 14, 547–553.

Benveniste, H., and Blackband, S. (2002). MR microscopy and high resolution'small animal MRI: Applications in neuroscience research. *Progress in Neurobiology* 67, 393–420.

Bren, K.L., Gray, H.B., Banci, L., Bertini, I., and Turano, P. (1995). Paramagnetic H-1-NMR spectroscopy of the cyanide derivative of MET80ALA-ISO-1-cytochrome-C. *Journal of the American Chemical Society* 117, 8067–8073.

Brezova, V., Dvoranova, D., and Stasko, A. (2007). Characterization of titanium dioxide photoactivity following the formation of radicals by EPR spectroscopy. *Research on Chemical Intermediates* 33, 251–268.

Cooper, D.R., Dimitrijevic, N.M., and Nadeau, J.L. (2010). Photosensitization of CdSe/ZnS QDs and reliability of assays for reactive oxygen species production. *Nanoscale* 2, 114–121.

Doty, F.D., Entzminger, G., Kulkarni, J., Pamarthy, K., and Staab, J.P. (2007). Radio frequency coil technology for small-animal MRI. *NMR in Biomedicine* 20, 304–325.

Felli, I.C., and Brutscher, B. (2009). Recent advances in solution NMR: Fast methods and heteronuclear direct detection. *Chemphyschem* 10, 1356–1368.

Fox, G.B., McGaraughty, S., and Lao, Y.P. (2006). Pharmacological and functional magnetic resonance imaging techniques in CNS drug discovery. *Expert Opinion on Drug Discovery* 1, 211–224.

Fritzinger, B., Capek, R.K., Lambert, K., Martins, J.C., and Hens, Z. (2010). Utilizing self-exchange to address the binding of carboxylic acid ligands to CdSe quantum dots. *Journal of the American Chemical Society* 132, 10195–10201.

Gawrisch, K., Eldho, N.V., and Polozov, I.V. (2002). Novel NMR tools to study structure and dynamics of biomembranes. *Chemistry and Physics of Lipids* 116, 135–151.

Gawrisch, K., and Koenig, B.W. (2002). Lipid–peptide interaction investigated by NMR. *Peptide–Lipid Interactions* 52, 163–190.

Grishaev, A., and Llinas, M. (2005). Protein structure elucidation from minimal NMR data: The CLOUDS approach. *Nuclear Magnetic Resonance of Biological Macromolecules C* 394, 261–295.

Huang, Y.P.J., Moseley, H.N.B., Baran, M.C., Arrowsmith, C., Powers, R., Tejero, R., Szyperski, T., and Montelione, G.T. (2005). An integrated platform for automated analysis of protein NMR structures. *Nuclear Magnetic Resonance of Biological Macromolecules, Part C* 394, 111.

Jennings, L.E., and Long, N.J. (2009). 'Two is better than one'-probes for dual-modality molecular imaging. *Chemical Communications* 3511–3524.

Kay, L.E. (1995). Pulsed field gradient multi-dimensional NMR methods for the study of protein structure and dynamics in solution. *Progress in Biophysics and Molecular Biology* 63, 277–299.

Keifer, P.A. (1999). NMR tools for biotechnology. *Current Opinion in Biotechnology* 10, 34–41.

Krishna, N.R., and Jayalakshmi, V. (2006). Complete relaxation and conformational exchange matrix analysis of STD-NMR spectra of ligand–receptor complexes. *Progress in Nuclear Magnetic Resonance Spectroscopy* 49, 1–25.

Lipovsky, A., Tzitrinovich, Z., Friedmann, H., Applerot, G., Gedanken, A., and Lubart, R. (2009). EPR study of visible light-induced ROS generation by nanoparticles of ZnO. *Journal of Physical Chemistry C* 113, 15997–16001.

Lokteva, I., Radychev, N., Witt, F., Borchert, H., Parisi, J., and Kolny-Olesiak, J. (2010). Surface treatment of CdSe nanoparticles for application in hybrid solar cells: The effect of multiple ligand exchange with pyridine. *Journal of Physical Chemistry C* 114, 12784–12791.

Lu, Z.R., Mohs, A.M., Zong, Y., and Feng, Y. (2006). Polydisulfide Gd(III) chelates as biodegradable macromolecular magnetic resonance imaging contrast agents. *International Journal of Nanomedicine* 1, 31–40.

Montelione, G.T., and Szyperski, T. (2010). Advances in protein NMR provided by the NIGMS Protein Structure Initiative: Impact on drug discovery. *Current Opinion in Drug Discovery and Development* 13, 335–349.

Moseley, H.N.B., and Montelione, G.T. (1999). Automated analysis of NMR assignments and structures for proteins. *Current Opinion in Structural Biology* 9, 635–642.

O'Connell, M.R., Gamsjaeger, R., and Mackay, J.P. (2009). The structural analysis of protein-protein interactions by NMR spectroscopy. *Proteomics* 9, 5224–5232.

Oschkinat, H., Muller, T., and Dieckmann, T. (1994). Protein-structure determination with 3-dimensional and 4-dimensional NMR-spectroscopy. *Angewandte Chemie International Edition* 33, 277–293.

Pattabiraman, N. (2002). Analysis of ligand-macromolecule contacts: Computational methods. *Current Medicinal Chemistry* 9, 609–621.

Rajendran, V., Lehnig, M., and Niemeyer, C.M. (2009). Photocatalytic activity of colloidal CdS nanoparticles with different capping ligands. *Journal of Materials Chemistry* 19, 6348–6353.

Reynolds, W.F., and Enriquez, R.G. (2002). Choosing the best pulse sequences, acquisition parameters, postacquisition processing strategies, and probes for natural product structure elucidation by NMR spectroscopy. *Journal of Natural Products* 65, 221–244.

Shimizu, H. (2003). Studies on molecular interactions of protein and their carbohydrate ligands by NMR. *Trends in Glycoscience and Glycotechnology* 15, 221–233.

Spasojevic, I. (2010). Electron paramagnetic resonance—A powerful tool of medical biochemistry in discovering mechanisms of disease and treatment prospects. *Journal of Medical Biochemistry* 29, 175–188.

Sugar, I.P., and Xu, Y. 1992. Computer-simulation of 2D-NMR (NOESY) spectra and polypeptide structure determination. *Progress in Biophysics and Molecular Biology* 58, 61–84.

Swartz, H.M., Khan, N., and Khramtsov, V.V. (2007). Use of electron paramagnetic resonance spectroscopy to evaluate the redox state in vivo. *Antioxidants and Redox Signaling* 9, 1757–1771.

Van der Linden, A., Van Camp, N., Ramos-Cabrer, P., and Hoehn, M. (2007). Current status of functional MRI on small animals: Application to physiology, pathophysiology, and cognition. *NMR in Biomedicine* 20, 522–545.

Wang, G.S. (2008). NMR of membrane-associated peptides and proteins. *Current Protein and Peptide Science* 9, 50–69.

Wise, R.G., and Tracey, I. (2006). The role of fMRI in drug discovery. *Journal of Magnetic Resonance Imaging* 23, 862–876.

Wishart, D. (2005). NMR spectroscopy and protein structure determination: Applications to drug discovery and development. *Current Pharmaceutical Biotechnology* 6, 105–120.

Yan, G.L., Chen, J., and Hua, Z.Z. (2009). Roles of $H_2O_2$ and OH center dot radical in bactericidal action of immobilized $TiO_2$ thin-film reactor: An ESR study. *Journal of Photochemistry and Photobiology A—Chemistry* 207, 153–159.

Yee, A., Gutmanas, A., and Arrowsmith, C.H. (2006). Solution NMR in structural genomics. *Current Opinion in Structural Biology* 16, 611–617.

*X-ray*

Asakura, K., Niimi, H., and Kato, M. (2010). Energy filtered X-ray photoemission electron microscopy. *Advances in Imaging and Electron Physics* 162, 1–43.

Bakshi, M.S., Jaswal, V.S., Kaur, G., Simpson, T.W., Banipal, P.K., Banipal, T.S., Possmayer, F., and Petersen, N.O. (2009). Biomineralization of BSA-chalcogenide bioconjugate nano- and microcrystals. *Journal of Physical Chemistry C* 113, 9121–9127.

Fang, Z., Li, Y., Zhang, H., Zhong, X.H., and Zhu, L.Y. (2009). Facile synthesis of highly luminescent UV-blue-emitting ZnSe/ZnS core/shell nanocrystals in aqueous media. *Journal of Physical Chemistry C* 113, 14145–14150.

Fay, F., Linossier, I., Langlois, V., Haras, D., and Vallee-Rehel, K. (2005). SEM and EDX analysis: Two powerful techniques for the study of antifouling paints. *Progress in Organic Coatings* 54, 216–223.

George, G.N., and Pickering, I.J. X-ray absorption spectroscopy in biology and chemistry. In V. Tsakanov and H. Wiedemann (eds.), *Brilliant Light in Life and Material Sciences*, p. 97-119, 2007.

Hahner, G. (2006). Near edge X-ray absorption fine structure spectroscopy as a tool to probe electronic and structural properties of thin organic films and liquids. *Chemical Society Reviews* 35, 1244–1255.

Hofer, F., and Pabst, M.A. Characterization of deposits in human lung tissue by a combination of different methods of analytical electron microscopy. *Micron* 29, 7–15, 1998.

Hornyak, G.L., Peschel, S., Sawitowski, T., and Schmid, G. (1998). TEM, STM and AFM as tools to study clusters and colloids. *Micron* 29, 183–190.

Kalko, S.G., Chagoyen, M., Jimenez-Lozano, N., Verdaguer, N., Fita, I., and Carazo, J.M. (2000). The need for a shared database infrastructure: Combining X-ray crystallography and electron microscopy. *European Biophysics Journal with Biophysics Letters* 29, 457–462.

Lee, M.R. (2010). Transmission electron microscopy (TEM) of Earth and planetary materials: A review. *Mineralogical Magazine* 74, 1–27.

Loukanov, A., Kamasawa, N., Danev, R., Shigemoto, R., and Nagayama, K. (2010). Immunolocalization of multiple membrane proteins on a carbon replica with STEM and EDX. *Ultramicroscopy* 110, 366–374.

Lund, P.K., Morningstar, D.A., and Mathies, J.C. (1964). X-ray spectroscopy in biology + medicine. 6. Use of ultra-soft x-ray source for nondestructive detection of microgram amounts of sodium + magnesium. *Biochemical and Biophysical Research Communications* 14, 177.

Menzel, A., Kewish, C.M., Kraft, P., Henrich, B., Jefimovs, K., Vila-Comamala, J., David, C., Dierolf, M., Thibault, P., Pfeiffer, F., and Bunk, O. (2010). Scanning transmission X-ray microscopy with a fast, framing pixel detector. *Ultramicroscopy* 110, 1143–1147.

Moore, K.T. (2010). X-ray and electron microscopy of actinide materials. *Micron* 41, 336–358.

Opella, S.J., DeSilva, T.M., and Veglia, G. (2002). Structural biology of metal-binding sequences. *Current Opinion in Chemical Biology* 6, 217–223.

Petibois, C., and Guidi, M.C. (2008). Bioimaging of cells and tissues using accelerator-based sources. *Analytical and Bioanalytical Chemistry* 391, 1599–1608.

Quiney, H.M. (2010). Coherent diffractive imaging using short wavelength light sources. *Journal of Modern Optics* 57, 1109–1149.

Ratner, B.D., and Castner, D.G. 1994. Advances in X-ray photoelectron-spectroscopy instrumentation and methodology—Instrument evaluation and new techniques with special reference to biomedical studies. *Colloids and Surfaces B—Biointerfaces* 2, 333–346.

Ruckert, B., and Kolb, U. (2005). Distribution of molecularly imprinted polymer layers on macroporous silica gel particles by STEM and EDX. *Micron* 36, 247–260.

Schreiner, M., Melcher, M., and Uhlir, K. (2007). Scanning electron microscopy and energy dispersive analysis: Applications in the field of cultural heritage. *Analytical and Bioanalytical Chemistry* 387, 737–747.

Vazina, A.A. Biological application of synchrotron radiation: From Artem Alikhanyan to nowadays. In V. Tsakanov and H. Wiedemann (eds.), *Brilliant Light in Life and Material Sciences*, Springer Science, Berlin, Germany 121–131, 2007.

## Online resources and software

### Databases

These databases make spectral techniques such as FTIR, Raman, and NMR useful, providing libraries against which new or unknown compounds may be compared. When citing a chemical compound retrieved from a database, always include the CAS Registry Number (RN), Beilstein/Gmelin Registry Number (BRN/GRN), or chemical name and the date the database was accessed.

### Free databases (key: N=NMR, R=Raman)

*Spectral Database System*. IR, 1H-NMR, 13C-NMR, mass, and EPR spectra of organic compounds.

*Sigma Aldrich*. FT-IR/Raman and NMR spectra of most compounds in the catalog.

*NIST Chemistry WebBook*. IR, mass, UV/VIS spectra, and ion energetics data.

*Environmental Protection Agency*. Several hundred spectra, primarily of pollutants.

*Public database of NMR spectra*. Growing set of NMR data, with links to others.

*Caltech Mineral Spectroscopy Server*. IR spectra of minerals.

*University of Northern Colorado Protein Infrared Database*. IR spectra of proteins.

*WebSpectra UCLA*. Introduction to IR and NMR with problems

### Commercial databases

(Many of these are available for site license and may be at your institution.)

*Bio-Rad Spectral Databases*. World's largest FTIR database (Sadtler); also NMR, UV–Vis, and more. Available for license.

*FDM Reference Spectra Databases*. FTIR, Raman, mass spec libraries for purchase. Free trial available.

*SciFinder (IR, NMR)*. Massive database including not only spectra, but chemical reactions, literature, and more.

*Beilstein CrossFire (MDL)*. Two of the most complete databases in a single search tool: Beilstein, containing information on over 8 million organic compounds, and Gmelin, containing information on over 1 million inorganic and organometallic compounds.

*ChemNetBase* (includes the CRC handbook, *Properties of Organic Compounds*). Spectral peaks of various types for >25,000 important organic substances.

## Free software

*Bio-Rad Know It All Academic Edition*. Free Windows software package for drawing chemical structures, importing spectra, and performing IR and Raman functional group analysis.

*Math NMR*. Mathematica package for NMR spin and spatial tensor manipulations. Arbitrary spin systems, commutators, projection operators, rotations, Redfield matrix elements, matrix decomposition into basis operators, change of basis, coherence filtering, and the manipulation of Hamiltonians. Offered from the lab of Alexej Jerschow, NYU.

*MA-Table*. From microanalyst.net. EDX software.

*Spekwin*. Freeware with import functions for many spectrometer file types.

*Essential FTIR*. Free software for spectral analysis with search tool.

*TARQUIN*. Analysis tool for automatically determining the quantities of molecules present in NMR spectroscopic data. The intended purpose is to aid the characterization of tumors.

*rNMR*. Open-source software for NMR data analysis. rNMR simplifies repetitive resonance assignments and quantification tasks common in metabolomics, titrations, and small molecule analyses. Versions of rNMR are available for Windows, Macintosh, Linux, and Unix (courtesy of the University of Wisconsin-Madison).

*matNMR*. Processing toolbox for NMR and EPR spectra under MATLAB (all platforms). It can read in data from most commercial spectrometers, can process any 1D or 2D spectrum, and uses all graphical capabilities that MATLAB has to offer (Jacco van Beek, ETH Zurich).

*RMN*. NMR data processing program for the Macintosh. (Dr. Philip J. Grandinetti, Ohio State).

*MARS*. Program for robust automatic backbone assignment of $^{13}C/^{15}N$ labeled proteins (Markus Zweckstetter, Max Planck Institute for Biophysical Chemistry, Germany).

*3DiCSI*. 3D Interactive Chemical Shift Imaging (3DiCSI) is a comprehensive software program for multidimensional CSI data visualization, spectral processing/analysis, spectral localization, and quantification. Free for research (Qi Zhao, Columbia University).

*AMBER*. General-purpose molecular dynamics and simulation package, with many features relevant to NMR structure determination.

## Commercial software (also see instrument manufacturers)

*Bio-Rad KnowItAll*. Spectral databases and software for NMR, FTIR, UV–Vis, and more. Integrates library searching with analysis.

*Advanced Chemistry Development*. Commercial software and databases for visible, IR, NMR, and other spectroscopies.

*ScienceSoft NMRanalyst*. Computerized analysis of phase sensitive 1D through 3D NMR FIDs and spectra. Generates accurate spin system descriptions including signal integrals and coupling constants. (Sold by Varian as "Full Reduction of Entire Datasets," FRED(tm)).

*ACD/Labs NMR Prediction Software*. ACD/HNMR and ACD/CNMR enable you to calculate the proton/$^{13}C$ NMR spectra for any organic structure to a high accuracy. Prediction is based on an internal data file with over a million experimental chemical shifts and coupling constants. ACD/Labs also offers NMR prediction modules for $^{15}N$, $^{19}F$, and $^{31}P$.

*Cambridgesoft Chem3D*. Predicts spectra via multiple computational chemistry packages. Those packages include MOPAC, GAMESS, Jaguar, and Gaussian. Chem3D can be used to predict NMR, IR, and UV–Vis spectra.

The computational chemistry packages that include spectra predictions are as follows:

- MOPAC:
  - NMR spectra
  - IR spectra
- GAMESS:
  - 1H-NMR spectra
  - 13C-NMR spectra
  - IR/Raman spectra

- Jaguar:
  - IR spectra
- Gaussian:
  - NMR spectra
  - IR/Raman spectra
  - UV/Vis spectra

## Instrument manufacturers

This is by no means an exhaustive list, especially of more common instruments such as UV–Vis. We encourage you to comparison shop. Please note that most manufacturers also provide software for analysis, usually as part of the instrument package. We did not list these packages separately under software, but they should be assumed to exist.

*Hamamatsu.* Optics and photonics: lasers, flash lamps, LEDs, PMTs, photodiodes, cameras, streak cameras, much more; also provides full systems for time-resolved emission (Vis and NIR) and absorption (UV–Vis).

*CVI Melles Griot.* Laser and custom optics specialist: custom and standard lenses, mirrors, prisms, filters, and other optical components; gas, diode-pumped solid state, and semiconductor diode lasers; optomechanics; shutters, controllers; isolation tables; more.

*Becker & Hickel.* Specialist in photon counting. Offers components such as photon counting boards, amplifiers, detectors, and lasers; also offers a large selection of complete lifetime-measurement systems, including spectrometers, fluorescence lifetime imaging microscopes preassembled with confocal or widefield fluorescence microscopes or for installation into a separate instrument.

*Horiba Jobin Yvon.* Very wide range of high-performance fluorescence spectrometers, including time- and frequency-domain lifetime spectrometers. TCSPC modules available as add-ons to steady-state spectrofluorometers, or as dedicated instruments. NanoLog instrument measures near-IR emission for work with semiconductor materials. Many other types of instruments available for targeted applications:

analytical Raman; x-ray fluorescence; surface plasmon resonance imaging; FTIR microscopy; and more.

*Bruker.* FTIR, near IR, Raman, NMR, MRI, EPR spectrometers; IR microscopes; also scanning probe microscopies, x-ray instruments, gas chromatography/mass spec. Also offers analytical services to the United States and Europe, including FTIR, NMR, and EPR.

*PicoQuant.* Pulsed light sources, photon counting instrumentation, full plug-and-play TCSPC systems.

*Newport.* All components for time-resolved experiments: lasers, vibration tables, monochromators, etc. Also offers plug-and-play transient absorption systems in the visible and near IR (requiring only a Ti:sapphire laser).

*Ultrafast Systems LLC.* Turnkey systems for transient absorption (femtosecond and subnanosecond) and for TCSPC/fluorescence upconversion. Also accessories and optical components.

*Agiltron.* Wide array of components; IR detectors; fiber optics; Raman spectrometers.

*ThermoFisher Scientific.* UV–Vis and fluorescence, including ultrasmall volume "Nanodrop" spectrometers and diode array spectrometers; x-ray; FTIR; wide choice of small and/or entry-level instruments for teaching and a specialist in high-throughput analysis.

*Varian.* UV–Vis–NIR, fluorescence, NMR, FTIR spectrometers; FTIR imaging; MRI; x-ray crystallography systems; chromatography.

*Edinburgh Instruments.* Wide choice of steady-state and time-resolved spectrometers, including lifetime plate reader. Lasers for time-resolved measurements. Many accessories. Also offer a TCSPC "teaching kit" for introducing the technique in an educational setting.

*PerkinElmer.* Wide range of fluorescence, UV–Vis, UV–Vis–NIR, FTIR, and Raman spectrometers.

*Labomed.* Specialist in UV–Vis and UV–Vis–NIR spectrometers and accessories (e.g., cuvettes).

*Jeol.* Usually known for electron microscopes, but also manufacturers of NMR and EPR spectrometers,

NMR magnets, NMR processing software, and mass spectrometers.

*Jasco*. UV–Vis, fluorescence, Raman, polarization spectrometers.

*Molecular Devices*. Plate readers for biological applications—UV–Vis, fluorescence, polarization, luminescence/phosphorescence, some with automated liquid handling.

*Ocean Optics*. Miniature, portable field spectrometers particularly for field applications.

*SpinCore Technologies*. NMR system, components for NMR, EPR.

*CRAIC Technologies*. Specialist in UV–Vis–NIR and Raman microanalysis (spectra and images).

*Princeton Instruments*. Wide range of imaging and spectroscopy cameras; monochromators; specialist in Raman and laser-induced breakdown spectroscopy (LIBS).

I hope you have enjoyed reading this book, found it practical, and will get stains on it, rip pages out, and write to me with your corrections, additions, comments, and questions. This is a work in progress that will be updated as new techniques emerge and are adopted by more and more researchers, and your feedback will help keep it up-to-date for years to come.

# CHAPTER 17

# Introduction to Nanofabrication

Orad Reshef

## 17.1  INTRODUCTION

This chapter aims to give a practical introduction to some of the techniques of nanofabrication, written by someone who recently completed a PhD using these techniques. It begins by discussing patterning and pattern transfer, and then techniques of deposition by spin coating, evaporation, and chemical vapor deposition (CVD). Techniques of characterization specific to nanoscale devices are then discussed in two sections: the first dealing with *metrology* and the second with spectroscopy, with some spectroscopic techniques that were not mentioned in **Chapters 14** and **16**. Finally, the chapter concludes with some practical tips for achieving and maintaining a clean, high-quality sample, noting where cutting corners can have disastrous results.

### The planar process

Most nanofabrication techniques in use today were developed to be compatible with the *planar process* (**Figure 17.1**). This process, which was initially developed in the 1950s by Fairchild Semiconductor (a predecessor of Intel), is a top-down approach to fabricating nanostructures. A sample begins with a substrate, a wafer or chip, which acts as the foundation for the sample. A layer of material is deposited onto the substrate, forming a *thin film*. A layer of *resist*, a chemically sensitive polymer, is subsequently deposited onto the thin film. The desired shape is then patterned into the resist using *lithography* techniques. Finally, the design is transferred from the resist to the deposited film using a *pattern transfer* technique. The resist is dissolved away, and a protective layer called a *cladding* is deposited and polished. This procedure may be repeated multiple times to form a vertical stack of two-dimensional structured layers. In this chapter, we will discuss each of these steps in detail to produce a nanostructure composed of a single two-dimensional layer.

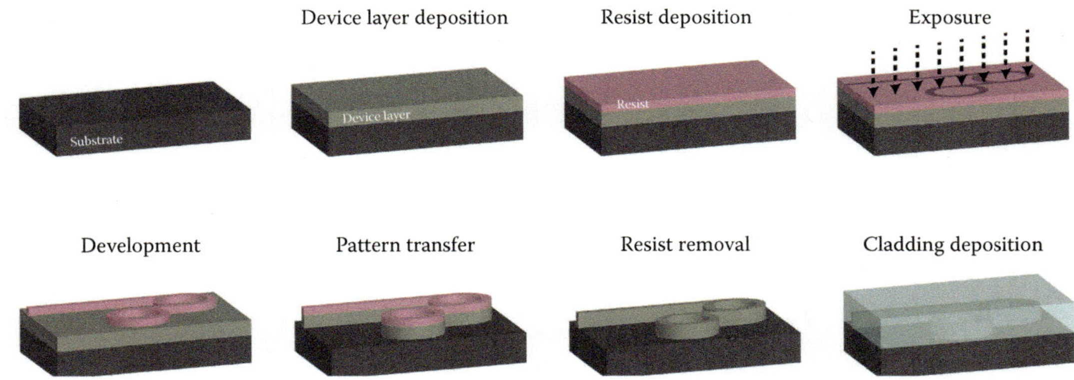

**Figure 17.1  The planar process can produce arbitrary shapes that are confined to the 2-D plane of the device layer.**

## 17.2  PATTERNING

### Resist

The first step to structuring a pattern into a surface is to deposit a layer of resist. A resist is a soft material that is sensitive to a lithography (i.e., pattern writing) step but is insensitive to subsequent processing steps. A pattern can be written into a resist and then transferred via a pattern transfer method into a surface of choice. Where the resist is exposed, the chemical bonds are changed from the surrounding area, so that it reacts differently to a developer solvent (**Figure 17.2**). Resist can be positive-tone, where regions that are exposed during the lithography

**Figure 17.2**  (a) In a positive-tone resist, exposing a circle yields a hole in the polymer, whereas in a negative-tone resist, exposing a circle yields a polymer cylinder. (b) Negative-tone resist (SU-8) structures after development. (SEM courtesy of Victor White.)

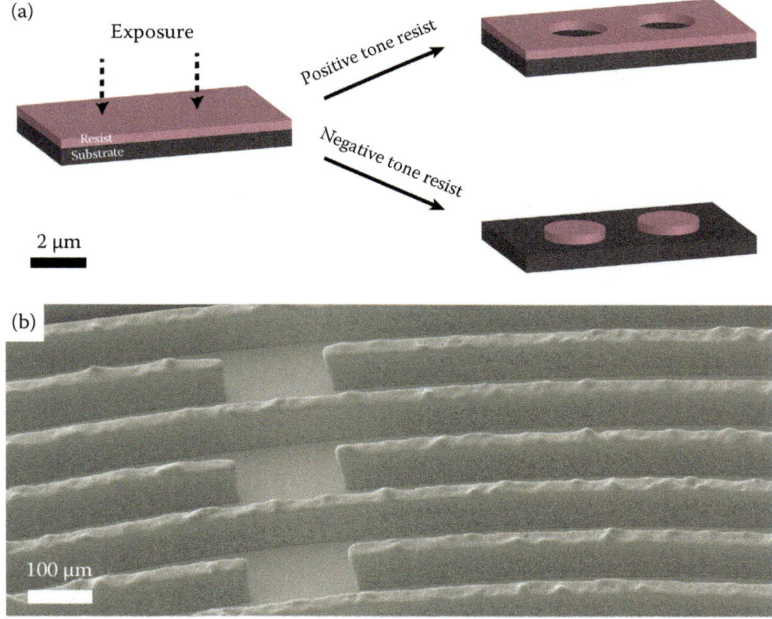

step are dissolved away, or negative-tone, where the affected regions are the only ones that remain.

In research contexts, the workhorse resist is polymethyl methacrylate (PMMA) due to its cost and versatility. It is a positive-tone resist for electron beam lithography (EBL). It can be produced at varying concentrations and molecular weights, which yield a broad range of viscosities and exposure sensitivities. Thus, it can be reliably deposited at different thicknesses, ranging from below 100 nm up to over 1000 nm. Its chemical resistivity is fair. We compare this polymer to a proprietary alternative positive-tone resist, ZEP520A (by Zeon Chemicals), which is much more chemically inert and can be exposed using much smaller doses, dramatically reducing exposure times. This resist is known to also yield smoother features with shorter exposure times at the expense of a much higher price.

## Lithography

There are many different methods with which a pattern can be defined in a resist. To a certain degree, the desired structure and application will dictate what lithography tool to use. *Photolithography*, where the exposure is provided by light, is the most used lithography method by industry due to its speed. A common variant of photolithography is called *contact lithography*; a wafer is coated with resist, a mask with a desired pattern is pressed against it, and the entire wafer is flooded with light at the same time. In other variants, the mask never physically touches the resist, preventing defects and damage. The image can even be focused on the surface of the resist using a lens following transmission through the mask, which provides an additional reduction in feature size (e.g., by a factor of 5). Advanced photolithography tools are even capable of maskless photolithography by projecting an image directly on the resist. The entire exposure process takes a fraction of a second, regardless of the number of features in the pattern. A whole *wafer* can be up to 300 mm in diameter and is subsequently diced into its individual chips, resulting in hundreds or thousands of chips being fabricated per minute. A *stepper* can be used to repeatedly expose a smaller pattern many times on the surface of a wafer.

Photolithography suffers from many limitations. First, the smallest feature sizes are limited by the exposure wavelength, which is usually fixed by the tool you have available. Thus, without the absolute state-of-the-art equipment, it is challenging to reliably achieve feature sizes smaller than a wavelength (e.g., 500 nm), which is a major obstacle to many applications. Second, before the exposure step, most photolithography tools require a mask to be fabricated, which dramatically slows down the prototyping and iteration cycle.

For these reasons, *EBL* is a popular tool in research and development. EBL can routinely achieve feature sizes down to 10 nm without need of a mask. It works by raster-scanning a collimated beam of electrons onto the resist. The major drawback in EBL is its operating speed—the writing process scales with the exposed area due to the rastering of a focused beam, so the total write time for an entire wafer can grow to be prohibitively long. For this reason, mass production is universally done in photolithography once a design is perfected and ready for the market.

These lithographic techniques produce 2-D structures that are compatible with the planar process. Though the pattern in the plane of the sample can have arbitrary shapes, the cross-sectional profiles out of the plane are regularly rectangular, with near-right-angled corners. Any three-dimensional structures have to be produced with some ingenuity that takes advantage of anisotropic material-dependent deposition or etching properties. A true arbitrary three-dimensional polymer structure can be produced using grayscale lithography, which makes use of resist that is only partially exposed, or direct laser writing techniques, where a laser beam is focused within the body of the resist. These methods are highly specialized and are beyond the scope of this chapter; however, they are rapidly becoming accessible and commercialized.

## Nanoimprint lithography

*Nanoimprint lithography* is a fast and cheap method to scale up devices that are created using EBL. A negative template is first created using other lithographic methods, and then it is stamped into an imprint resist. This method can ultimately yield the mass production of devices with a resolution on the order of tens of nanometers.

## Focused ion beam

The final lithographic technique we will discuss is *focused ion beam* (FIB), where a collimated beam of inert ions (e.g., Ga$^+$) mills away the material. Using an FIB, the desired structure is directly chiseled out of the material without the need for a resist, making it the most versatile of all lithography techniques. Because it is also a rastering technique, it suffers from the same issues with production speed as EBL. It also produces rough surfaces compared to the other methods and is known to be relatively challenging to master. Nevertheless, it can yield very small features and can structure materials that are often otherwise challenging to etch. It is most often used to prepare extremely thin samples (on the order of 100 nm wide) for characterization, for example, for transmission electron microscopy.

# 17.3 PATTERN TRANSFER

## Wet etching

Etching a material is the process of chemically removing it. In a wet etch, the entire sample is immersed in a reactive solvent. Ideally, this solvent will target a specific material, leaving all the other materials on the sample untouched. For example, to remove a layer of gold sitting on a glass substrate, the gold should be removed completely, while the glass should be kept in pristine condition. This chemical selectivity is thus important to consider during the planning stages of a fabrication process, before even beginning the deposition of any materials. There exist a lot of literature with tables of materials and solvents and their etch rates that have been established experimentally. They should be heavily consulted before proceeding with any etching. Some references are given at the end of the chapter to articles giving these data.

A *wet etch* is typically isotropic, which means the etchant reacts equally in all directions, ignoring any lattice planes in the material. Thus, wet etches are used almost solely to remove a selected material from a sample indiscriminately. However, the isotropy of the etching process could also be exploited to undercut a material and achieve freestanding structures (**Figure 17.3**).

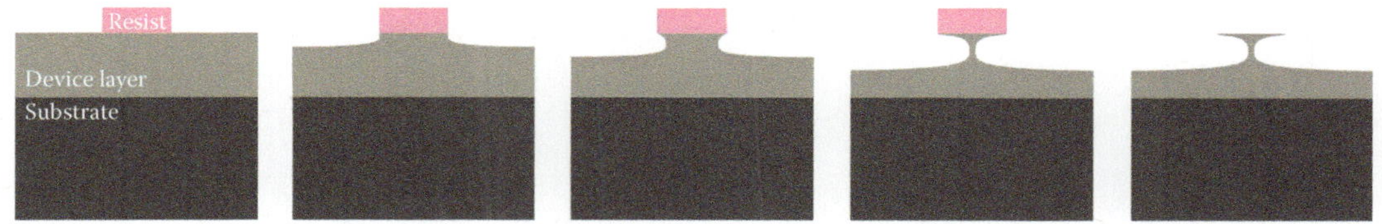

**Figure 17.3 A wet etch is used to etch under the resist and form a suspended structure.**

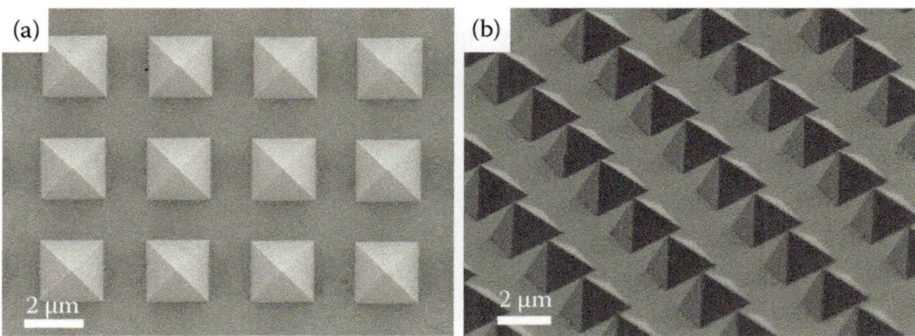

**Figure 17.4 Gold nanopyramids fabricated using silicon and an anisotropic wet etch.** (Figure courtesy of Nabiha Saklayen.)

Besides isotropic wet etches, there also exist anisotropic wet etches that are much more material specific. For example, potassium hydroxide (KOH) etches the (100) silicon surface along the (111)-oriented crystal lattice planes, which can be used to yield sloped walls. These properties have been used to create negative templates for nanopyramids (**Figure 17.4**). Anisotropic etches thus provide a degree of freedom for creating three-dimensional structures.

After completing a wet etch, the sample is rinsed with an inert solvent (e.g., distilled water). Best practices dictate rinsing the sample multiple times in separate flasks of water. This might seem excessive, but there is a sound reason for it—the concentration of the etchant could be high enough that only rinsing in a single cup will leave some residue that is sufficiently concentrated to continue etching the surface.

## Dry etching

As the name suggests, a *dry etch* is similar to a wet etch, but instead of using liquids, gases are used to selectively remove materials. It turns out that this small change complicates things dramatically. First, a dry etch procedure cannot be done in the open air or a fume hood; at the very least, a vacuum chamber is necessary. Second, since the gases must be flowed at a consistent rate, extra equipment is necessary to control their flow and the chamber pressure. All of these considerations drastically increase the cost of the etch.

However, this increased cost and complexity afford a finer control of etching properties and more sophistication in the etch. A voltage applied on the gases can excite them to form a plasma, which is accelerated toward the sample. This bombardment adds a unidirectional physical component to the etch, in addition to the chemical component provided by the choice of reacting gases. Together, standard dry etching tools provide enough flexibility to obtain perfectly vertical sidewalls or sidewalls with tailored angles (**Figure 17.5**).

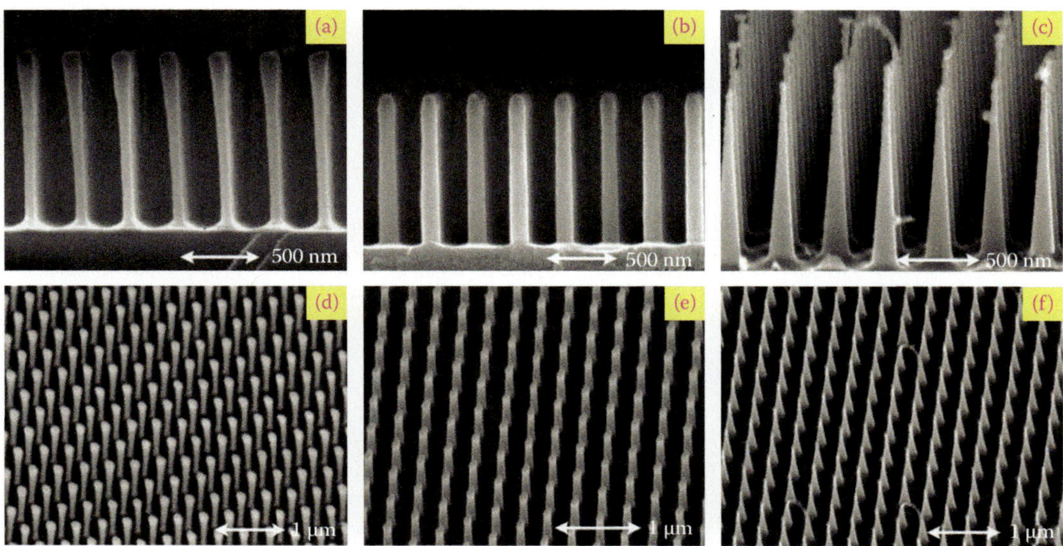

**Figure 17.5** Dry etching can produce sidewalls with (a, d) positive or (b, e) negative slope angles. (c, f) With some optimization, even vertical sidewalls can be achieved. (Figure courtesy of Yung-Jr Hung and Brian Thibeault.)

There are dozens of variables that can be tuned while etching, which include the pressure of the chamber, the flow rate of the gases, as well as various temperatures, powers, and voltages. References are given at the end of the chapter to some key articles summarizing these variables. Though finding an acceptable etch recipe sounds daunting, the manufacturer always delivers the tool with a set of standard recipes for the most commonly used materials, such as Si or GaAs. However, even with the same equipment and process parameters, results will be variable and will have to be tweaked to optimize the etch to exhibit the desired properties. Beyond this, a design of experiments (i.e., the Taguchi method) could be employed to find an ideal etch without exploring the entire parameter space of variables.

Two more practical points on etches are apropos. First, there needs to be excellent thermal contact between the sample and the sample holder to allow for the dissipation of heat accumulated by the sample due to ion bombardment. This is paramount to maintaining a consistent etch between uses. We recommend using a thermal grease instead of thermal paste; grease or oil is much cleaner because it dissolves completely in a solvent and does not redeposit on the sample. The appropriate amount of grease for a given sample size will need to be determined empirically. Too much grease will seep out from under the sample, and the ions in the etch will sputter grease onto the face of the sample. Too little grease will not be enough to spread out under every part of the sample, and only the small region of the sample on top of the grease will etch properly.

Second, the etch in the very center of the chip will be a little different from the etch on the edges of the sample. This is because the complete physics of a dry etch process includes contributions from secondary ions that recoil from reacting elsewhere on the chip in addition to the primary ions that come directly from above. Since the regions on the edges have access to half as many of these secondary ions, the etch rate on the etch can vary significantly from the center. If all the features of interest lie in the center of the sample, perhaps this is not an important consideration. However, if there are any structures that reach the very edge of the chip, they will be etched to a different depth and at a different quality. A quick fix to this problem is to make the starting substrate larger. Alternatively, a secondary set of dummy samples may be placed surrounding the sample. As these samples are etched away, they can produce the secondary ions to mimic the center of the chip on its edges.

## Lift-off

Many materials are very challenging or even impossible to etch reliably. Often, there are practical reasons not to etch, such as the lack of the necessary gases for a given reaction, or a ban on certain materials by the owner of the tool for fear of contamination. The layer *under* the one being etched might also need to be protected at all costs, and so etching is too risky. Whatever the reason, there exists an entirely different pattern transfer procedure called a *lift-off* process, illustrated in **Figure 17.6**. The negative image of a pattern is written into a resist, and a layer of the desired material is deposited directly on top of it. In the regions where the resist is dissolved away, the deposited material makes direct contact with the substrate. All remaining resist is then dissolved away, leaving behind the desired material. Using this method, virtually any material that can be evaporated can be structured.

There are many reasons a dry etch is typically the preferred method to structure a material. First, lift-off processes are relatively slow and unscalable when compared to dry etches. Next, using a negative-tone resist directly will always result in smaller features than lifting off material from a positive-tone resist. Features yielded in an etch can be optimized to be smoother than those lifted off, as well. Finally, when etching, the starting material can be atomically smooth and epitaxial, and can also be thoroughly characterized prior to being structured.

As in etch procedures, the lift-off procedure is highly material specific, and so a recipe needs to be developed for every individual scenario. There are certain rules of thumb:

- The deposition method needs to be nonconformal, that is, the material should not coat the sides of the resist in the exposed areas. Otherwise, the structured region remains in contact with the region designed to be lifted off. This causes the entirety of the deposited material to be removed with the resist, including the desired structures.

- The thickness of the resist needs to be at least two or three times thicker than the desired thickness of the final material.

- During the lift-off procedure, the sample needs to be agitated gently in the solvent. Anything more aggressive that uses stir bars or ultrasonication

**Figure 17.6 In a standard material lift-off procedure, material is deposited on top of the resist.** The resist is subsequently removed, leaving the desired structure on the substrate.

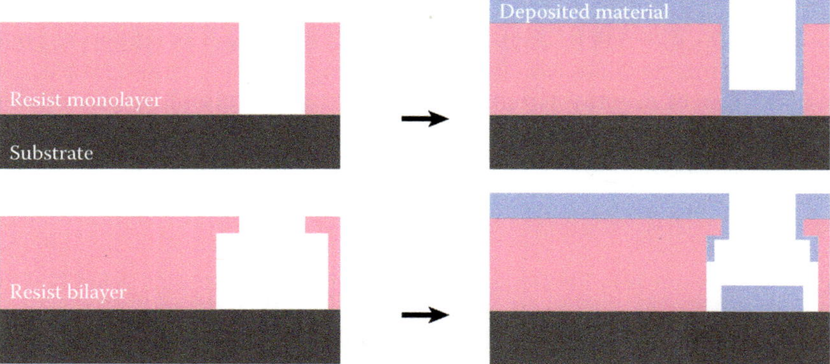

**Figure 17.7 Cross section of a sample, comparing a resist bilayer to a resist monolayer in the lift-off process.** When using a bilayer with an overhang, the desired structure is isolated from the material being lifted off.



should be avoided if possible. However, sometimes, the adhesion between the deposited material and the substrate is really good, and sonication can still be employed to accelerate the process.

- It is preferable to set the sample to soak in the solvent on a hot plate overnight if there is time. This method yields excellent reproducible results for most cases.

It sometimes remains challenging to successfully lift off a material. A contributing factor could be that the deposition method is inherently a little too conformal, making it impossible to avoid continuous material deposition along the resist sidewalls, despite following the rules of thumb outlined previously. To avoid this, a resist bilayer may be used (**Figure 17.7**). Instead of depositing only a single layer of resist, two layers are deposited; however, the bottom layer is chosen to be a resist that is slightly more sensitive to the exposure. This produces an overhang in the resist, which helps isolate the desired structure on the substrate. Using a resist bilayer is not necessary; however, it does act as a cure-all to most lift-off problems.

## Template stripping

Another way to structure a material for which there is not an etcher or an etch recipe available is called *template stripping*. It is a slow, manual process that is not particularly scalable, but it can achieve excellent results for research purposes. It begins with a structured surface called a template. The pattern in the template needs to be the negative pattern of what is desired in the final result. A layer of the desired material (e.g., gold) is deposited on top of the template. This is followed by an adhesion layer, which is subsequently pressed into a separate sample (e.g., a glass slide), which will act as the new substrate. Finally, the template is peeled off of the sample, yielding the desired structure (**Figure 17.8**). Template stripping is notable for producing smooth structures in any material that can be deposited

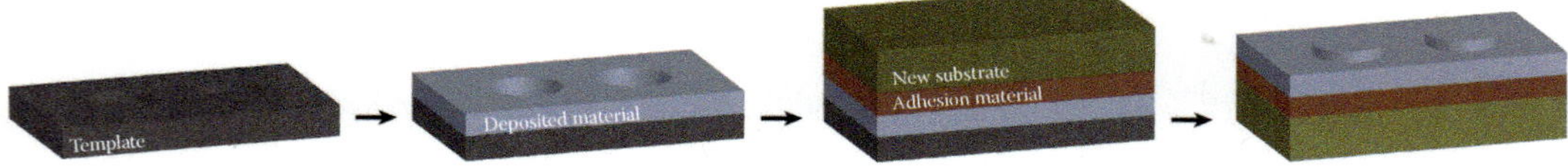



**Figure 17.8 The template-stripping process.**

since the roughness of the final surface is dictated only by the roughness of the template. If the procedure is performed carefully, the master template can even be reused to produce multiple samples.

# 17.4 MATERIAL DEPOSITION

The deposition method to use is in part dictated by the material in question and the desired film properties. Not every method is capable of depositing every material, and the properties of the material once deposited (e.g., the crystal structure, grain size, conformality, surface conductivity, etc.) vary dramatically depending on the tool that is used. Additionally, just because an available facility owns a particular tool does not mean that all of the necessary raw materials (e.g., the gases or pure material target) are available. Finally, no single technique is optimal for all materials—every material has a deposition method it "prefers" for a particular application—some materials can only be grown epitaxially (i.e., in a perfect crystal lattice) using only one particular deposition method. For these reasons, a lot of homework needs to be done when selecting the deposition technique for a process. In this section, we will present the most commonly used deposition techniques.

For most of these devices, the deposition rate can be monitored *in situ* using a quartz crystal microbalance (see **Chapter 10**) or an interferometer. In newer tools, a feedback loop can be set up so that the deposition rate is held constant, yielding a consistent film within a few nanometers (i.e., a few atomic layers) of the desired thickness. However, not all tools have these capabilities. In these cases, a test run is performed on a so-called witness sample (e.g., an unstructured substrate like a glass slide) to establish a deposition rate.

## Spin coating

The cheapest and simplest tool to deposit a thin film of material is the *spin-coater*. A small amount of viscous solvent is pipetted onto a sample or a wafer, which is then spun at several thousand revolutions per minute. The faster the spin, the thinner the layer produced. This is almost always followed by a baking step, e.g., 60 s on a hot plate set at 180°C. It is also good practice to bake a sample *before* spinning anything on it (e.g., 5 min above 100°C) to remove any moisture that has adsorbed onto the surface and to ensure a consistent spin.

This is the method by which resist is deposited. Commercial resists are shipped with a spin curve, a recipe that predicts the film thickness as a function of the spin speed. It is recommended to use the speed in the middle of the curve for the best consistency. For example, for the spin curve in **Figure 17.9**, this corresponds to 2500 rpm. A spun-on layer is usually only uniform in the center of the sample, with a raised ridge at the edges of the sample, known as an edge bead. The more symmetric the sample, the larger the uniform region in the middle of the sample can be. Circular samples such as wafers produce the most uniform layers with the smallest edge beads, the corners in a square sample are typically unusable, and rectangular samples will have even larger unusable areas.

As a final note, we recommend doing a spin test before using a sample in the spin-coater. A spinner has a vacuum safety lock that will prevent it from spinning if it does not sense a sample. The spin should also stop if the sample is launched

**Figure 17.9 Spin curve for SU-8 2000, a negative-tone photoresist.**

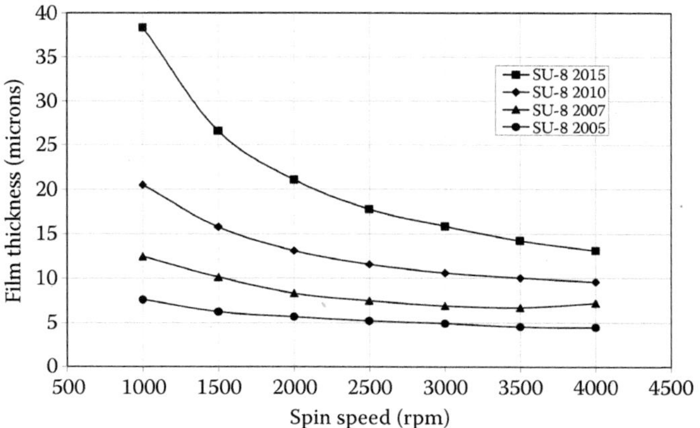


off of the spinner. Thus, the standard spinner test is as follows: with no sample mounted, start the spinner. It should not spin at this point. Place your gloved finger on the spinner to block the suction hole. This should prompt the spinner to start spinning. When you remove your finger, it should stop spinning. If any of these parts of the test fail, it is likely because there is some blockage in the spinner caused by some resist that has leaked into its pipes, and you have just saved your sample.

## Sputter deposition

To *sputter* a material, a sample is placed facing a pure target of a material in a vacuum chamber. An inert gas (e.g., argon) is flowed into the chamber, bombarding the target, which in turn emits charged ions of the target material. These ions are accelerated toward the sample with either a direct current (DC) or radiofrequency (RF) bias similar to what was discussed previously in etching systems, redepositing the ions on the sample and forming a thin film. The sample can be rotated on its axis to achieve a more uniform film. The mechanism is entirely physical and does not depend on any material-dependent properties, meaning most materials can be deposited using sputter deposition provided there is a high-quality target available.

The sample can be heated during deposition to encourage the formation of a particular crystal structure. There is often the option of flowing an additional gas to react with the target material. For example, oxygen gas can be flowed alongside argon in a chamber with a titanium target to produce titanium dioxide. This is called reactive sputtering, and it is a suitable method to obtain a compound material when a particular target for that material (in this case, a titanium dioxide target) is unavailable. A chamber can also come with multiple targets. For example, a titanium target and a gold target could be installed in the chamber, and both could be sputtered simultaneously. This makes it relatively straightforward to deposit a stack of materials (e.g., titanium/gold/titanium/gold, etc.).

Beyond this, the physics of the sputtering process is very complex and can lead to counterintuitive effects such as hysteresis. For example, a higher-powered plasma does not necessarily produce a faster deposition rate. More so than for any other of the other deposition methods described in this chapter, we highly recommend reading some literature if serious about developing a sputtering process.

## Thermal and electron beam evaporation

Evaporation techniques are relatively simple and affordable. A sample is placed facedown above a crucible (often called a "boat") filled with the desired material in a vacuum chamber. The boat is then heated up until the material evaporates, leaving a thin film on the sample. This is called *thermal evaporation* (TE). Alternatively, the material is placed in a crucible and is melted with a focused electron beam. This is called *electron beam evaporation* (EBE). Both techniques are similar and yield films of comparable quality. However, the region heated in the crucible is much smaller when using EBE, leading to a much more directional deposition. EBE can achieve much higher temperatures than TE as well, allowing for the deposition of materials with higher boiling points. A more important consideration is that the material quality depends on the tool that is selected, and it is not always straightforward to predict which tool will perform better for a given material. Finally, compared to sputtering, there are fewer parameters available to control when using evaporation tools. Thus, there are fewer degrees of freedom available to optimize a deposition.

## Chemical vapor deposition

In *CVD*, gases are flowed over a sample. These gases react on the surface of the sample to yield a film. The CVD tool itself is in fact almost identical to a dry etcher, save for the choice of gases and the recipes employed. Thus, it is just as complicated to develop a reliable deposition recipe because of the large number of tunable parameters during deposition as it is to obtain a reliable etch recipe. In this case, just as in the case for the dry etcher, the manufacturer ships the tool with a few recipes for commonly used materials. In general, it is a very complex tool with many sensitive components, making it more fragile and prone to downtime than the other deposition tools listed so far. It is also capable of depositing a shorter list of materials since a recipe needs to be developed for each one individually. However, once a good recipe is established, the CVD is a reliable turnkey machine with repeatable depositions, constant deposition rates, and little room for human error.

## Atomic layer deposition

*Atomic layer deposition* (ALD) is a relatively new deposition technique, which is derived from CVD. The principle behind ALD is the same as the one behind CVD, with one key difference—instead of flowing the gases continuously, the gases are injected in intermittent spurts. This allows for the deposition of individual atomic layers upon the surface, making ALD an extremely precise and controllable deposition technique, albeit an incredibly slow one. ALD also has the distinction of being the most conformal (i.e., least directional) deposition method. Using ALD, the sidewalls of a structure are uniformly coated as well as the top. Trenches with very high aspect ratios (50:1) can be filled with ease. ALD conforms so well that even the backside of the sample gets coated during a deposition!

Just as for CVD, there is but a short list of materials that can be deposited (mostly oxides). However, the extreme conformality and accuracy of this deposition technique has caused ALD to become an attractive tool for industrial applications, and expanding this list of materials has become an active area of research. For example, in 2016, researchers managed to deposit gold for the first time, a material without a naturally occurring oxide.

# 17.5 METROLOGY

Once a structure has been fabricated, it needs to be characterized. For feature sizes on the nanoscale, this is something surprisingly nontrivial to accomplish. Most features are too small to resolve using optical microscopy. Light microscopes are also incapable of measuring heights or thicknesses. To this end, the entire field of metrology has been developing advanced microscopy and profilometry techniques. In this section, we discuss the most useful and accessible metrology tools that can be expected to be available in every cleanroom.

## Scanning electron microscopy

*Scanning electron microscopy* (SEM) is the simplest and most reliable way to image a nanostructured surface. It produces high-quality, high-resolution images of features down to tens of nanometers, far smaller than what is physically possible using an optical microscope. (Every picture in this chapter was taken using an SEM.) Turnkey commercial SEMs are now so reliable that anyone can learn how to load samples and take stunning images of nanoscale structures with very little practice or preparation.

An SEM works by focusing a collimated beam of electrons (often the very same beam that is used in EBL) onto the surface of a sample in a vacuum chamber. These electrons raster-scan the surface, reflecting off of it as well as sometimes ejecting a secondary electron from the material on the surface of the sample. Detectors sense these electrons to reconstitute an image. Different types of detectors can be employed in an SEM. Some are more sensitive to relative heights on a sample, while others target the secondary electrons and so are more capable of distinguishing constituent materials. Most SEMs have access to multiple detectors that can be toggled while imaging. The beam rasters the surface so quickly that the image appears in a fraction of a second; the scan speed and averaging methods can also be altered to find a reasonable balance between a short wait time and an increased signal-to-noise ratio.

There is another important consideration when imaging a sample using an SEM. The electron beam is constantly pumping electrons onto the sample. In principle, the beam could damage the sample, although this is rarely an issue save for specific cases. The real problem is that these electrons need an escape route from the sample. Otherwise, the surface of the sample collects these charges, deflecting subsequent electrons and severely distorting the resulting image. This charging effect happens on the order of a fraction of a second, making it challenging to obtain a well-focused image.

One solution to this problem is just to set the acceleration voltage of the electron beam very low (~1 kV). This often works at the cost of a lower-resolution, grainier image. Some SEMs can flow gas onto the sample during imaging, which acts to compensate for the charging effect. An alternative solution is to deposit a conductive coating onto the surface of the sample. Usually, sample preparation rooms will have a small sputterer for precisely this purpose, with just the ability to deposit a thin (5–10 nm) layer of gold. This surface needs to have a direct connection between the surface of the sample and the "ground," i.e., the conductive sample holder. Often, this deposited surface will spoil the sample— for example, gold can diffuse into silicon and disrupt its conductive and optical properties. It also cannot be deposited and removed from a biological sample in a biocompatible manner. Thus, if a sample is being prepared to be imaged in the SEM, it can be considered a destructive measurement.

## Profilometry

At its heart, a *profilometer* is meant to supplement the SEM in characterizing microscopic devices. While an SEM excels at measuring lateral dimensions, it falls short at measuring heights and depths. There is a vast library of profilometry tools available. The simplest profilometer is a cheap tool consisting of a needle on a cantilever that operates using the same principle as a record player or gramophone. A profilometer can make coarse measurements for a range of step heights, down to tens of nanometers and up to tens of microns. It has very poor lateral resolution, so it cannot measure the width of any features on a sample with any reliability. Also, since the stylus makes contact with the surface of the sample, it will often blemish that region of a sample. If this is to be avoided, there exist noncontact profilometers that work on the principle optical interference, but these depend on certain optical properties of the materials that might be beyond control.

## Atomic force microscopy

An *atomic force microscope* (AFM) is a high-resolution stylus profilometer, one type of *scanning probe microscope* (SPM). It has the capability to raster-scan a surface and provides stunning true three-dimensional images with the highest available resolution, down to a fraction of a nanometer (**Figure 17.10**). AFMs also come with a noncontact mode, which leaves the sample surface untouched and clean. The main drawback of the AFM is how slow it is: because it scans by physically moving around a needle, it can take several minutes to complete an image. Also, unlike an SEM, an AFM has no way of distinguishing materials and is only capable of measuring relative heights. Either way, commercial AFMs have come a long way to becoming turnkey and user friendly, just like the SEM.

The AFM tip itself is always user replaceable because of its fragility. Tips come in a range of sizes and qualities (**Figure 17.11**). It is critical to know the properties of the tip used during a measurement since the results will be heavily influenced by tip choice. Tips also come with different conductive and mechanical properties. They range in price from a few dollars to several hundreds of dollars each, so knowing the required parameters is essential before purchasing a set.

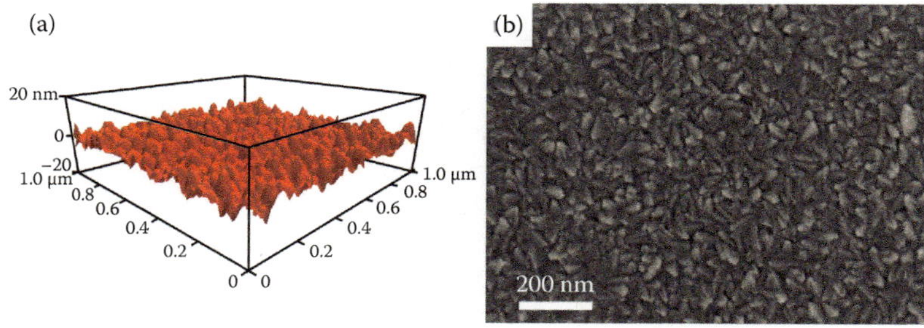


**Figure 17.10** (a) AFM is used to supplement (b) SEM to measure profile data and extract a surface roughness measurement. (Figure courtesy of Jonathan D. B. Bradley.)

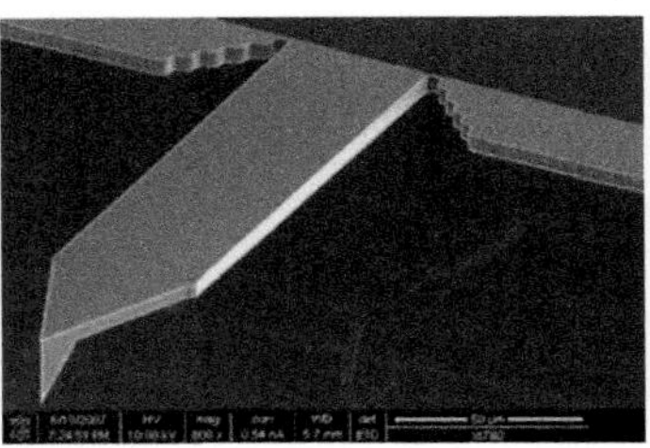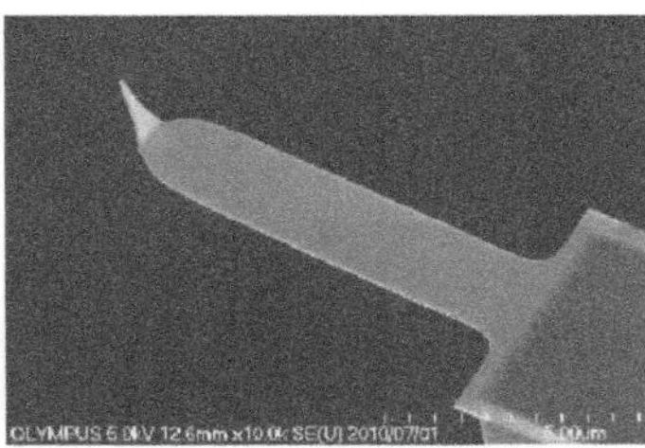

**Figure 17.11 SEM images of different kinds of commercially available AFM tips.**

## 17.6 THIN-FILM CHARACTERIZATION

Just as fabricated surfaces should be characterized once they are structured, thin films should also be evaluated to make sure that they possess the desired properties. In fact, it should not even be taken for granted that a tool will deposit precisely the material that it claims to! Here, we present several metrology techniques that are specifically tailored to investigating unstructured thin films.

### Ellipsometry

An *ellipsometer* is a metrology tool that is used to characterize thin films. It works by focusing a laser on the surface of a sample and then measuring the coherent reflection for different input and output polarizations of light. This information could be used to back-calculate the film thicknesses and optical constants (e.g., absorption and refractive index information). In a fabrication setting, they are most often used to determine the thickness of films for the optical properties have already been established. For example, most resists are delivered with optical constants to be used with an ellipsometer to characterize your resist thickness before exposure.

The simplest ellipsometer only performs a measurement at a single optical wavelength and is known as a single-point ellipsometer. This tool can only reliably measure thicknesses below one optical wavelength, ~1 μm. This is a known physical limitation of the tool that has to do with fundamental properties of the propagation of light. On the other hand, spectroscopic ellipsometers take multiple measurements at different wavelengths and incident angles. They can use this extra information in sophisticated modeling software to unambiguously determine the thickness of a thin film. With software that is advanced enough, an ellipsometer can provide a lot more information, such as the thickness of interim layers in a stack (e.g., silicon on silica on silicon) or the surface roughness, all without ever touching the surface of the sample.

### X-ray photoelectron spectroscopy

*X-ray photoelectron spectroscopy* (XPS) is one of many tools that can be used to characterize a thin film. A calibrated x-ray source illuminates the sample, dislodging electrons from the surface. These electrons are counted as a function of the x-ray photon energy, which corresponds to the ionization energy of the electron, yielding a spectrum. This spectrum can help you figure out the elemental composition of a deposited material. For example, suppose you reactively sputter titanium in an oxygen atmosphere. An XPS measurement will tell you what the titanium-to-oxygen ratio is, and can help you determine if you have deposited $TiO_2$ or $Ti_2O_3$. The measurement is inherently nondestructive, but the tool often includes inert atoms, which can be used to bombard and thin the material and probe spectra as a function of depth.

## Raman spectroscopy

*Raman spectroscopy* is one of a number of spectroscopic techniques that operate entirely on optical processes, where a laser illuminates a sample and the optical spectrum that is emitted by the sample is recorded. Several of these techniques were discussed in **Chapter 16**. Raman uses inelastic scattering from vibrations in a molecule to give information about vibrational states. Infrared, visible, and ultraviolet (UV) Raman are all used to probe different types of chemical bonds. As an example, both UV and visible Raman spectroscopy have been used to distinguish between different phases of $TiO_2$ such as rutile, anatase, or amorphous. Spectra must be cross-referenced with literature values to confirm identity of a sample; increasingly large databases of Raman spectra are becoming available. Raman spectrometers are usually available in materials characterization facilities and departments of geology.

## X-ray diffraction

In *x-ray diffraction* (XRD), a monochromatic beam of x-rays is used to illuminate a sample. It propagates through the material and diffracts into multiple beams. In this measurement, the sample behaves like a diffraction grating, and the distance between the atoms within the sample act as the spacing in the grating. The angles at which the output beams propagate inform us about the crystallinity of the sample. This is how crystal structures are determined, which won the Nobel Prize in Physics in both 1914 and 1915, among dozens of other Nobel Prizes in Physics, Chemistry, and Medicine that made use of XRD. See also **Chapter 6** for applications of XRD to protein structure determination.

For microfabrication purposes, XRD provides information about the crystallinity of deposited materials that is complementary to the other spectroscopy tools discussed in this chapter. One thing that the other tools *cannot* do is quantify the level of crystallinity of a sample. This is called a *rocking curve analysis*: a sample is tilted, and the width of the XRD order is measured as a function of angle. The highest-quality materials with the most ordered crystals will produce the sharpest peaks. Thus, XRD is an indispensable tool for developing a recipe to deposit material epitaxially.

## Four-point probe

A *four-point probe* is an ohmmeter that can be used to measure the resistance or conductivity of a thin film. It yields resistance values in units of $\Omega/sq$ (i.e., unitless area). When assessing multiple deposition recipes for a metal or conductor, the sample with the lowest resistance will likely have the purest, highest-quality material. It is a foolproof device that is found in every cleanroom and that provides yet another angle from which a deposition recipe can be assessed.

## 17.7 KEEPING A SAMPLE CLEAN

To conclude this chapter, we will provide some commonsense advice (often called *lore*) that has been acquired over time, through many failures and some personal experience. Following these guidelines should increase your yield and make you a more successful "fabber."

### Yield yield yield

You will *never* get to the point where every device you fabricate will be a success. Something unexpected always goes wrong, from accidentally dropping the sample facedown to an etcher malfunctioning and entirely ruining a sample. The solution to this problem is to prepare multiple samples in parallel. Prepare four samples in parallel, and only process one or two through the final step. If you need to develop a particular step in the recipe, say an etch recipe, then prepare 5 or 10 samples and try something different on each sample. *Never* just process only a single sample at a time.

### Minimum feature sizes

As a rule of thumb, smaller features are, in general, harder to make. If your application permits it, it is to your benefit to redesign your pattern with larger and less dense structures. This will greatly increase your yield and reduce your fabrication cost, as well. Minimum feature sizes above 5 μm are preferred, though even keeping above 1 μm makes a substantial difference.

### Mind your tolerances

It is paramount to understand what the tolerances are for every fabrication and metrology step. It is still impossible to obtain a structure that is accurate to within 1 nm in every dimension, as well as to perfectly characterize that structure. Try and design your device to be insensitive to these kinds of fabrication imperfections, with tolerances to dimension variations of up to at least 5%. Doing so will greatly increase your yield, as well.

### Dedicated labware

Once you have developed a complete and working fabrication process, from the deposition of the device layer, through patterning and structuring, to deposition of the final protective cladding layer, you will likely repeat this entire procedure multiple times for different devices. At this point, you should purchase new glassware and dedicate them to the various steps of the process. For example, you should have a set of "lift-off" glassware, a set of "substrate cleaning" glassware, etc. This ensures that earlier steps in the process are not contaminated by dirtier steps downstream. You should also have dedicated tweezers for certain processes. For instance, your "gold" tweezers should never be used to process silicon samples, as gold dissolves in silicon, as mentioned previously. The cost of these elements will add up, but in the long run, they will dwarf in comparison to having to restart the entire process multiple times.

## "Nothing goes in the bottle"

A bottle of resist can be relatively expensive to replace, so you should make every effort to keep it from getting contaminated. This is why you should follow the "nothing goes *in* the bottle" rule. Instead of pipetting directly from the resist bottle, decant some resist into a smaller bottle, and pipette for your daily process from there.

## AMI wash, RCA clean, piranha etch

There are a few standard cleaning processes that use commonly accessible solvents and acids. The simplest is called an AMI rinse, where the sample is ultrasonicated in acetone, methanol, and isopropanol for 5 min each. It is imperative that isopropanol be the last thing to touch the sample since it evaporates cleanly without leaving a residue. Following this, the sample is blow-dried using a dry nitrogen gun and baked on a hot plate. This is likely to be enough to clean most samples. More aggressive processes (specific to silicon substrates) are the RCA clean and the piranha clean, which aggressively target surface contaminants such as oxide layers and organics.

## Descumming

Cleanrooms often have a small dedicated dry etcher that flows oxygen plasma over a sample for the purpose of removing resist and organic material. This step is called descumming, ashing, stripping, or plasma cleaning. The multiple names stem from the fact that the operating principle can take many forms, but the general idea is the same. For example, in ozone ashing, oxygen is gently flowed over a sample under an intense deep UV source. The UV ionizes oxygen to create ozone as well as any carbon compounds on the surface. They bond to form $CO_2$ that drifts away with the airflow. A descummer should be used as part of a sample preparation, for example, after every AMI wash.

## Take your time

Do not rush through your process. Cleanroom fabrication work is expensive and requires exacting care and patience. If you have allotted a certain amount of time to do something, and you are running behind, it would be better to stop what you are doing and come back another day than to rush through and risk spoiling a sample that might have taken a few days to prepare. A staffer once said to me, "If you don't have the time, what are you even doing here?" In the same vein, do not skimp on cleaning thoroughly in the earlier steps of the process. Otherwise, something will go wrong down the line, and you will not know what caused it.

## Be nice to the cleanroom staff

The cleanroom staff's job is actually to maintain equipment, not to help you with your research. However, solving your obscure research problem is often the most exciting part of their day, so they might end up giving you a whole lot of unpaid help anyway. These technicians will be intimately familiar with the abilities and history of their equipment and thus often can come up with creative solutions. Be sure to ask for their advice and take advantage of their experience. Be incredibly gracious with them; at the end of the day, you will be dependent on their happiness to get your work done.

## Before experimenting with a new recipe, recreate something you KNOW will work

During your first few weeks in the cleanroom, you are probably not going to be developing a process from scratch. Instead, you will likely be asked to adapt an established process from a senior student or postdoc. Before experimenting with your own process, first recreate the old work. Otherwise, you will not know when you inevitably fail if the cause was your adjustment to the procedure or some misunderstanding of the original recipe. Also, while you are working toward perfecting a process, take images of your structure during the intermediate steps. Ideally, you can image your sample with an SEM as well as an optical microscope. If something goes wrong, you can track which step caused the problem.

Once you have finally recreated the old recipe, note that if you now tweak an earlier part of the process, you will often need to reoptimize the subsequent steps in the process, as well. This is why experienced fabbers might end up working with inefficient processes that already work over untested new processes.

## Keep your toolbox properly outfitted

Start collecting spare pieces of labware in a toolbox—you never know when you will need them in a pinch. Extra pens, sample holders, razor blades, Scotch tape, double-sided carbon tape, kapton tape (for use in vacuum chambers), diamond scribes, spare silicon wafers, crystal bond... these are all incredibly cheap items that can save a fabrication day.

## Spinning resist on small samples

If you have a sample that is too small to spin resist on reliably, mount it onto a larger silicon wafer with a small piece of tape, and spin the sample when on top of the wafer. Touch the tape with a gloved hand a few times to reduce its stickiness, or else you might find yourself in trouble when trying to remove your sample after the spin.

## Keep your surroundings clean

A droplet of clear liquid on the wet bench can be anything, including water (which is harmless) or hydrogen fluoride (which is not). Just because *you* know that it is water does not mean that everyone else can figure it out. Everyone has access to some pretty scary chemicals in a cleanroom! Do yourself and everyone else in the cleanroom a huge favor and keep every bench and tabletop clean and sparkling. Also, label every piece of glassware and bottle with the chemical they contain. All of these little steps will add a little bit of time to your recipe development but will save time in the long run while making the entire process cleaner and safer as a result.

## Double check

- The tone of your resist

- The safety procedures (Safety Data Sheet, called the SDS) for every new chemical you use

- Where to put your waste chemical when you are done with it *before* you pour any out

- If your tools are chemically compatible with the chemical you are about to use (e.g., HF dissolves glass, so use a plastic container and plastic tweezers)

# 17.8  FINAL COMMENTS

I hope you have enjoyed reading this book, found it practical, and will get stains on it, rip pages out, and write to me with your corrections, additions, comments, and questions. This is a work in progress that will be updated as new techniques emerge and are adopted by more and more researchers, and your feedback will help keep it up-to-date for years to come.

## Background Reading

### Books

There are few books currently available on the subject of nanofabrication.

Ayers, J.E., Kujofsa, T., Rago, P., and Raphael, J. *Heteroepitaxy of Semiconductors: Theory, Growth, and Characterization*. Edn 2. CRC Press, Boca Raton, FL, 2016.

Gerlach, G., Dotzel, W., and Muller, D. *Introduction to Microsystem Technology*. Limited edition. John Wiley & Sons, West Sussex, England, 2008.

Plummer, J.D., Deal, M.D., and Griffin, P.B. *Silicon VLSI Technology: Fundamentals, Practice, and Modeling*. Edn 1. Pearson, Prentice Hall Upper Saddle River, New Jersey, 2000.

### Journal articles

This is by no means an exhaustive list but should give some idea of the methods, applications, and, in some cases, history of the techniques in this chapter.

Berg, S., and Nyberg, T. (2005). Fundamental understanding and modeling of reactive sputtering processes. *Thin Solid Films* 476(2), 215–230.

Bradley, J.D.B., Evans, C.C., Choy, J.T., Reshef, O., Deotare, P.B., Parsy, F., Phillips, K.C., Lončar, M., and Mazur, E. (2012). Submicrometer-wide amorphous and polycrystalline anatase TiO2 waveguides for microphotonic devices. *Optics Express* 20(21), 23821–23831.

Burek, M.J., de Leon, N.P., Shields, B.J., Hausmann, B.J.M., Chu, Y., Quan, Q., Zibrov, A.S., Park, H., Lukin, M.D., and Lončar, M. (2012). Free-standing mechanical and photonic nanostructures in single-crystal diamond. *Nano Letters* 12(12), 6084–6089.

Griffiths, M.B.E., Pallister, P.J., Mandia, D.J., and Barry, S.T. (2016). Atomic layer deposition of gold metal. *Chemistry of Materials* 28(1), 44–46.

Hung, Y.J., Lee, S.L., Thibeault, B.J., and Coldren, L.A. (2011). Fabrication of highly ordered silicon nanowire arrays with controllable sidewall profiles for achieving low-surface reflection. *IEEE Journal of Selected Topics in Quantum Electronics* 17(4), 869–877.

Kontio, J.M., Husu, H., Simonen, J., Huttunen, M.J., Tommila, J., Pessa, M., and Kauranen M. (2009). Nanoimprint fabrication of gold nanocones with ~10 nm tips for enhanced optical interactions. *Optics Letters* 34, 1979–1981.

Kudryashov, V., Yuan, X.-C., Cheong, W.-C., Radhakrishnan, K. (2003). Grey scale structures formation in SU-8 with e-beam and UV. *Microelectronic Engineering* 67–68, 306–311.

Nagpal, P., Lindquist, N.C., Oh, S.-H., and Norris, D.J. (2009). Ultra-smooth patterned metals for plasmonics and metamaterials. *Science* 325(5940), 594–597.

Saklayen, N., Huber, M., Madrid, M., Nuzzo, V., Vulis, D.I., Shen, W., Nelson, J., McClelland, A.A., Heisterkamp, A., and Mazur, E. (2017). Intracellular Delivery Using Nanosecond-Laser Excitation of Large-Area Plasmonic Substrates. *ACS Nano*, 11(4), 3671–3680.

Van Laer, R., Kuyken, B., Van Thourhout, D., and Baets, R. (2015). Interaction between light and highly confined hypersound in a silicon photonic nanowire. *Nature Photonics* 16(February), 1–5.

Vogel, N., Zieleniecki, J., and Koper, I. (2012). As flat as it gets: Ultrasmooth surfaces from template-stripping procedures. *Nanoscale* 4(13), 3820–3832.

Vu, K.T., and Madden, S.J. (2011). Reactive ion etching of tellurite and chalcogenide waveguides using hydrogen, methane, and argon. *Journal of Vacuum Science and Technology A: Vacuum, Surfaces, and Films* 29(1), 11023.

Wilkinson, C.D.W., and Rahman, M. (2004). Dry etching and sputtering. *Philosophical Transactions of the Royal Society A: Mathematical, Physical and Engineering Sciences* 362(1814), 125–138.

Williams, K.R., and Muller, R.S. (1996). Etch rates for micromachining processing. Journal of Microelectromechanical Systems 5(4), 256–269.

Williams, K.R., Gupta, K., and Wasilik, W. (2003). Etch rates for micromachining processing—Part II. *Journal of Micro-electromechanical Systems* 12(6), 761–778.

## Equipment and supplies

### Chemicals

### Thermal grease

Santovac 5 (SantoLubes)

### Positive-tone resist

ZEP520A (Zeon Chemicals)

PMMA (MicroChem)

Shipley (Dow)

### Negative-rone resist

SU-8 (MicroChem)

HSQ (Dow)

Fox (Dow)

XR (Dow)

ma-N 2400 (Microresist Technology)

### Instruments

J. A. Woollam Co., Inc, Gaertner Scientific. Ellipsometers and measuring microscopes.

Raith nanofabrication, JEOL, Elionix. Electron beam lithography.

JEOL, Zeiss. SEMs.

Park Systems, Asylum Research, Veeco, Bruker. AFMs, surface profilometers.

AJA International, Kurt J. Lesker. Thin-film deposition.

SPP Process Technology Systems Ltd. Etching and chemical vapor deposition.

Ultratech/CNT. Atomic layer deposition.

# Glossary

**180° pulse (π pulse):** Nuclear magnetic resonance (NMR) pulse that flips the magnetization vector along the $z$-axis.

**3′ end (pronounced "three prime"):** The end of a DNA or RNA strand that terminates in the hydroxyl group of the third carbon in the sugar (deoxyribose or ribose). It is also called the tail end, and the direction toward the 3′ end is called *downstream*.

**5′ end (pronounced "five prime"):** The upstream end of DNA or RNA, which terminates in the phosphate group of the fifth carbon of the sugar.

**Aberration:** Failure of a mirror, refracting surface, lens, etc. to produce exact correspondence between an object and its image (causes blurring of the image).

**Absorbed dose:** A measurement of the energy per unit mass deposited in a target by ionizing radiation. This does not take into account relative health effects of different types of radiation. The unit is J/kg or gray (Gy). The previously used CGS unit is the rad.

**Acetylcholine (ACh):** An ester of acetic acid and choline. It functions as a neurotransmitter in both the peripheral and central nervous systems, being released by both parasympathetic and sympathetic neurons, somatic neurons, and some central nervous system neurons.

**Acid-fast stain:** Colorimetric bacterial stain used to stain mycobacteria and other structures that have a mycolic acid content.

**Acousto-optical tunable filter (AOTF):** An anisotropic crystal bonded to a piezoelectric transducer that diffracts light in a wavelength-dependent fashion in response to a radio-frequency (RF) signal.

**Actin:** Globular protein found in all eukaryotic cells; it is the monomeric subunit of two types of filamentous structures involved in the cytoskeleton and contractile apparatus in muscle cells.

**Action potential:** A brief, regenerative change in a cell's membrane potential that propagates along the axon of nerve cells as well as over the surface of some muscle and glandular cells.

**Acute slice:** A brain section used for experimentation immediately after preparation.

**Adaptive immune system:** The part of the immune system that recognizes specific pathogens. In vertebrates, antibody- and cell-mediated immunity, facilitated by B and T cells, both make up the adaptive immune system. In prokaryotes, CRISPR-Cas plays this role.

**Additive screen:** The addition of various substances in addition to altering common crystallization parameters to enhance the formation of macromolecular crystals.

**Adeno-associated virus:** Small parvovirus with single-stranded DNA (4.8 kb genome). Many properties, including the fact that it does not cause any known disease, make it ideal for gene therapy.

**Adenovirus:** A family of double-stranded DNA viruses that are nonenveloped and medium-sized (26–45 kb genome) and that can be used as vectors to deliver foreign genes to a wide variety of cells in culture and in experimental animals. Limited use in gene therapy due to strong immune response.

**Adult stem cell:** An undifferentiated cell found in a tissue or organ that can renew itself and differentiate to yield some or all of the specialized cells types of that tissue or organ.

**Affinity chromatography:** Type of column chromatography based upon a specific affinity of the resin in the column to a protein that is to be purified.

**Agar:** A mixture of agarose and agaropectin, which is used to make solid nutrient media. Not suitable for gel electrophoresis; not to be confused with purified agarose.

**Agarose:** Polysaccharide that is a linear polymer of the disaccharide agarobiose. It gels into a porous matrix with favorable properties for gel electrophoresis, supported bilayers, and other applications.

**Agonist:** A substance that binds to a receptor to induce a biochemical response.

**Airy disk:** The intensity distribution produced by Fraunhofer diffraction around a circular aperture.

**Alamethicin:** A peptide antibiotic, produced by the fungus *Trichoderma viride*, that forms voltage-dependent ion channels in membranes that show multiple subconductance states.

**Alcohol:** An organic compound containing a hydroxyl functional group (chemical formula ROH).

**Alkanethiol:** An organic compound consisting of carbon and hydrogen atoms linked by single bonds and a thiol functional group. Common chain lengths used in biotechnology range from 1 to 16 carbons.

**Alkylating agent:** A reactive agent that can transfer alkyl groups (hydrocarbon chains missing one hydrogen; formula $C_nH_{2n+1}$) from one molecule to another.

**Alpha helix:** Folding pattern in proteins in which an amino acid chain folds into a right-handed helix, which is stabilized by internal hydrogen bonding.

**Alternative splicing:** A process by which different mRNAs are produced from the same transcript through variations in the splicing pattern of the transcript.

**Amide bond:** A covalent bond joining a carboxylic acid and an amine.

**Amine:** Nitrogen-containing organic compound that is a derivative of ammonia.

**Amino acid:** A molecule containing both amine and carboxyl functional groups and a side chain that can be neutral, positively charged, or negatively charged. There are 22 naturally occurring amino acids in eukaryotes that link together to form peptides and proteins.

**Aminoglycosides:** A group of antibiotics effective against certain types of bacteria.

**Amorphous:** Lacking a distinct form or shape.

**Amplification:** The production of many DNA copies from one master region of DNA.

**Analog mode:** Mode of photomultiplier tube detection that measures the average photocurrent and incorporates the noise into the signal.

**Annexin V:** A cellular protein that binds to externalized phosphatidylserine; used as a marker for apoptosis.

**Anodic decomposition:** Breakdown of a compound under oxidizing (positive) potentials. Occurs with most semiconductors.

**Antibiotic:** Substance that destroys or injures bacteria.

**Antibody:** A type of protein produced by the body's immune system when it detects harmful substances (antigens); each type of antibody is unique and can bind to one specific type of antigen.

**Antibonding orbital:** A type of molecular orbital that occurs when overlap of the individual atomic orbitals is such that the electrons are repulsive and act to destabilize the molecule. Energy level is higher than the ground state of a single atom.

**Antigen:** A substance that prompts the production of antibodies. They are parts of bacteria, viruses, or other microorganisms that bind to specific antibodies produced by the immune system.

**Antiparallel:** Parallel strands oriented in opposite directions.

**Aptamer:** A small sequence of DNA or a peptide that binds a specific target molecule.

**Archaerhodopsin-3:** Yellow light–driven outward proton pump. Isolated from the organism *Halorubrum sodomense.*

**Aromatic group:** Planar ring that is made of either carbon only or carbon, oxygen, nitrogen, or sulfur.

**Array synthesizer:** Instrument that synthesizes specific DNA oligonucleotide sequences in individually addressable regions to create array chips containing hundreds of thousands to millions of different oligo sequences.

**Astrocyte:** Type of glial cell involved in numerous regulatory processes including regulation of blood flow, provision of energy metabolites to neurons, maintenance of extracellular balance of ions, fluids, neurotransmitters, etc.

**Atomic absorption spectroscopy:** A method of determining atomic composition of a material by atomizing it and measuring its light absorbance relative to a standard, and then determining concentration using the Beer–Lambert law. Mostly used for heavy elements.

**Atomic force microscope:** A type of scanning probe microscope that measures force between a probe (usually a sharp tip that may be chemically functionalized) and the sample by deflection of the probe. Has atomic-scale resolution; forces measured may be van der Waals forces, electrostatic forces, and many others.

**Atomic layer deposition:** A type of chemical vapor deposition in which precursors are present sequentially in the chamber, so that the interactions are self-limited, resulting in thin, conformal films.

**Attenuated total reflectance (ATR) spectroscopy:** Type of infrared (IR) spectroscopy that uses evanescent waves and is independent of sample thickness.

**Auger therapy:** A type of radiotherapy relying on low-energy (Auger) electrons.

**Autofluorescence:** Fluorescence of other substances other than the desired structure, creating difficulties in studying the fluorescently labeled structure of interest.

**Autoimmune:** Relating to a disease caused by the production of antibodies or lymphocytes against the host's own tissues.

**Avidin:** Protein produced in the oviducts of egg-laying animals, which protects the developing embryo from bacterial invasion. It binds biotin in the strongest noncovalent interaction known.

**Bacterial expression vector:** Plasmid designed to express a cloned gene in bacteria. Contains a resistance gene with its promoter, as well as a separate bacterial promoter and cloning site for insertion of foreign DNA.

**Bacteriophage:** A virus that infects bacteria, often called simply phage. Phages may be the most common organisms on Earth, though only a few dozen have been well characterized, and even fewer are used in routine molecular cloning.

**Bacteriorhodopsin:** A transmembrane protein that functions to pump protons across the cellular membrane of archaea.

**Bar code:** A short DNA sequence that serves as a marker for sequencing. Used to identify species as well as artificially introduced sequences.

**bCNG:** Bacterial cyclic nucleotide–gated ion channels.

**Beta-galactosidase:** See LacZ.

**Beta sheet:** Structural motif in proteins in which different sections of the polypeptide chain run alongside each other joined together by hydrogen bonding between atoms of the polypeptide backbone.

**Bidentate thiol:** Thiol that can donate two pairs of electrons to a metal atom.

**Bilayer chamber:** Vessel for lipid bilayer recordings; usually has two vertical, equally sized reservoirs, which are filled with electrolyte solutions.

**Biodistribution:** The way in which a drug or nanomedicine accumulates in organs or tumors of animals or humans. Biodistribution studies are required for any agent proposed for the market; for anticancer agents, studies should include healthy animals as well as tumor-bearing animals.

**Biofunctionalization:** Modification of a nonbiological surface to add biological functionality.

**Biomolecular quenching constant:** Rate constant describing the chemical process of biomolecular quenching.

**Biosafety containment:** The isolation of dangerous biological agents in an enclosed cabinet, room, and/or facility to protect against the infection of scientific workers or release into surrounding environment. Levels range from 1 (minimal risk) to 4 (maximum risk).

**Biosensor:** Any device that detects, records, or can transduce data pertaining to biological molecules or chemicals.

**Biotin:** Essential vitamin in the B family, also called vitamin B7 or vitamin H, that binds avidin with the strongest noncovalent interaction known. It is a small molecule (molecular weight, 244) and so may be added to other molecules without greatly affecting their size.

**Biotinylation:** Linking of biotin to any other molecule.

**Birefringence:** The double refraction of light in a transparent, molecularly ordered material.

**Blastocyst:** Early embryonic stage consisting of a mass of developing cells surrounding a central cavity.

**Bleed-through:** Entry of unwanted light into an image. Examples may be excitation light entering a fluorescence image, or the tail end green fluorescence entering into the image of red fluorescence.

**Blunt end:** The end of a molecule of DNA that has been cut by an enzyme so that the complementary strands are of equal length. Lowers the efficiency of ligation reactions.

**Bonding orbital:** A molecular orbital in which the greatest electron density is between the nuclei of the constituent atoms. Energy level is lower than that of the ground state of a single atom.

**Bone marrow transplantation:** Procedure in which normal bone marrow stem cells are infused into a person; transplanted cells can give rise to all blood cell types and can replace marrow damaged by disease or deliberately ablated with chemotherapy (e.g., for leukemia).

**Brachytherapy:** Radiotherapy in which the radioactive source is placed inside the body near or inside the tumor target. Also called curietherapy.

**Bradford assay:** A colorimetric assay for protein that makes use of the shift in spectrum of Coomassie blue when it binds to protein.

**Budding yeast:** Yeast that divide by budding off a smaller cell. *Saccharomyces cerevisiae* is an example of a budding yeast.

**Burst:** A group of closely spaced openings of an ion channel.

**Capsid:** The outer shell of a virus, made up of monomeric protein subunits.

**Carboxylate:** A salt or ester of a carboxylic acid (chemical formula $RCOO^-$).

**Carboxylic acid:** Organic acid containing at least one carboxyl group (chemical formula RCOOH).

**Cas9:** CRISPR-associated protein 9, an RNA-guided DNA endonuclease associated with CRISPR-based adaptive immunity in *Streptococcus pyogenes* and other prokaryotes. The most commonly used Cas in molecular biology.

**Catalyst, catalyze:** A substance that increases the rate of a reaction without being consumed by the reaction itself.

**Cation–pi interaction:** Monopole–quadrupole interaction between a positively charged ion (cation) and the electron-rich face of an aromatic pi orbital.

**cDNA (complementary DNA):** Synthetic DNA transcribed from a specific DNA using the enzyme reverse transcriptase. This eliminates the introns.

**Cell:** Basic unit of life.

**Cell-attached:** Type of patch recording in which the pipette contacts the cell membrane directly, forming a high resistance seal.

**Cell line:** A particular type of cell that has been immortalized for continuous growth and reproduction under cell culture conditions.

**Cell membrane:** A selectively permeable phospholipid bilayer that acts as a barrier between the cell interior and the extracellular fluid surrounding the cell.

**Cell wall:** Rigid layer surrounding the cell membrane in plants, bacteria, archaea, and fungi. Animal cells are distinguished by their lack of cell walls.

**Centromere:** The central constriction point of a eukaryotic chromosome that joins sister chromatids.

**Chalcogens:** Elements that form group 16 of the periodic table of elements (oxygen, sulfur, etc.).

**Channelrhodopsin:** Light-gated channel protein that nonselectively conducts cations when stimulated by visible light.

**Chaperone:** Protein that helps guide the proper folding of other proteins, or helps them avoid misfolding.

**Chemical shift:** In NMR, the difference between the measured resonance frequency and that of a reference standard expressed as a dimensionless quantity.

**Chemical vapor deposition:** A method of thin film production where volatile gases react or decompose on the wafer to produce the required film. Unwanted by-products are removed by flowing gas through the chamber.

**Chemically competent:** Bacterial cells that have been made **competent** via exposure to divalent cations (usually $Ca^{2+}$) followed by heat shock.

**Chemotherapy, chemotherapeutic:** Chemical agents (drugs) used to treat cancer.

**Chiral, chirality:** An object whose mirror image is not superimposable.

**Choline:** A B vitamin and essential nutrient; precursor to the neurotransmitter acetylcholine.

**Chloramphenicol acetyltransferase (CAT):** Enzyme that catalyzes the transfer of an acetyl group from acetyl-CoA to the antibiotic chloramphenicol. Enzymatic activity is used as a reporter gene.

**Chromatic aberration:** Distortion caused by lenses that have a slightly different index of refraction at different wavelengths, which causes rays of different colors to come to a focus at different points along the optical axis.

**Chromosome:** Highly coiled condensed form of chromatin formed in cell nucleus of eukaryotes. In prokaryotes, the packaging is much less, and there is no nuclear membrane surrounding the usually circular chromosome.

***cis* face:** Face of a Golgi stack where substances enter from the endoplasmic reticulum for processing.

**Cladding:** Outer protective layers on a device. In semiconductor devices, usually has a higher bandgap than the inner layer.

**Cloning:** See Molecular cloning.

**Clearance:** In pharmacology, measurement of the plasma volume from which a drug is completely removed per unit time (mL/min). Includes clearance by the liver, kidney, and lung.

**Cloning strain:** A strain (usually of *Escherichia coli*) optimized for efficient uptake and production of foreign DNA.

**Cloning vector:** Small DNA molecule used to carry the fragment of DNA to be cloned into the recipient cell, enabling the DNA fragment to be replicated.

**Clonogenic assay:** An assay in which cells that are able to divide and form colonies are quantified.

**Closed complex:** Structure formed between RNA polymerase and DNA.

**Closed tip:** Technique that eliminates the air interface at the tip of the microinjection needle by pulling a needle that is sealed and then breaking it.

**$CO_2$ Incubator:** Airtight chamber that enables the creation of a sterile, temperature-controlled environment with a specific mix of gases for cell growth.

**Cocrystallization:** Adding ligand to the protein solution before crystallizing.

**Codon:** A three-nucleotide sequence that codes for a specific amino acid or a termination (*stop*).

**Codon optimization:** Method of improving the expression of a gene in a different system. For instance, one can determine the optimal codons for bacterial expression of a mammalian gene and synthesize an appropriate DNA sequence that efficiently expresses that mammalian gene in bacteria.

**Collisional quenching:** Occurs when a quencher diffuses to an excited-state fluorophore, causing the latter to return to its ground state without fluorescence.

**Colloid, colloidal:** A system in which small particles (solid, liquid, or gas) are suspended within a continuous medium (solid, liquid, or gas; gas–gas colloids are not possible).

**Colony:** A cluster of genetically identical bacteria derived from a single bacterium growing on a solid medium.

**Column chromatography:** A method of purifying chemicals in a solution by passing the solution through a column partially filled with a *stationary phase* that separates components differently. The speed with which different fractions elute from the column is determined by their chemical composition and so can be used as a separation technique. Affinity chromatography is one type of column chromatography.

**Column refolding:** A method of protein purification where the protein is denatured, and then renatured on the purification column.

**Compartmental model:** A model that describes the electrical properties of a neuron as a series of isopotential compartments.

**Competent:** The ability of a cell to take up extracellular DNA from its environment.

**Complement, complementary:** A property of double-stranded nucleic acids meaning that the base pairs of each strand can be joined by two or three hydrogen bonds. A is complementary to T (or U), and G to C.

**Concatemer:** A long continuous DNA molecule containing multiple repeats of the same sequence.

**Confluency:** Measure of the coverage of available space by adherent cells.

**Conjugate:** Specific chemical group that is attached to a protein or useful sequence.

**Conjugate focal planes:** Consists of two sets of planes, referred to as the image-forming conjugate set and the illumination conjugate set, that are simultaneously in focus.

**Conjugation:** (1) Alternating single and double bonds leading to electron delocalization. (2) The direct transfer of genetic material between bacteria via a cell-to-cell bridge. (3) The formation of a chemical compound formed by the union of two others (*Alexa 633–streptavidin conjugate*).

**Contrast agent:** Substance used to stain or highlight a certain structure for imaging purposes.

**Coordinating:** Of a solvent, ability to complex with metal ions.

**Core–shell quantum dots:** Nanocrystallites that have been coated (and passivated) with higher-bandgap inorganic materials.

**Correlation spectroscopy (COSY):** 2-D NMR technique in which two consecutive 90° pulses are applied to the sample to determine spin–spin coupling.

**Cosmid:** A hybrid cloning vector that can replicate autonomously like a plasmid and be packaged into a phage.

**Cotransfection:** Simultaneous transfection with two unrelated nucleic acid molecules. Usually, one encodes a reporter gene, which is easily assayed and acts as a marker.

**Covalent bond:** A chemical bond formed by the sharing of one or more electrons between atoms.

**CRISPR-associated proteins (Cas):** Proteins that play a role in CRISPR-related gene editing. At least 45 families have been identified, though the endonuclease Cas9 is the one most commonly used in molecular biology.

**Cross-linker:** A reagent that catalyzes the linkage of one reactive group to another.

**Cryoprotection:** Process of protecting a cell, tissue or organism from cold-induced damage.

**Cryo-TEM:** Electron microscopic imaging of samples ultrafast-cooled to cryogenic temperatures.

**Crystal dye:** Small-molecule dyes that can penetrate into the solvent tracts of protein crystals.

**Crystal monitor:** A quartz crystal used to measure the thickness of deposited thin films.

**Current clamp:** Type of electrophysiological recording in which an amplifier is used to set the current through the electrode and the resulting voltage of the cell membrane is measured.

**Cystinuria:** A genetic disease causing high levels of the amino acid cystine in the urine.

**Cytoplasm:** Fluid within the cell membrane containing all organelles.

**Cytoskeleton:** System of protein filaments into the cytoplasm that gives the cell its shape and capacity for directed movement.

**Darkfield:** Illumination technique whereby specimens are illuminated by oblique light.

**DAVID (Database for Annotation, Visualization and Integrated Discovery):** A National Institutes of Health (NIH)–funded, web-based tool for genomics that provides a "comprehensive set of functional annotation tools for investigators to understand biological meaning behind large list of genes."

**Dehydration synthesis:** Chemical reaction in which two smaller molecules are joined to form a larger molecule and a single water molecule is lost in the process.

**Deletion:** A mutation in which one or more nucleotide pairs are deleted.

**Denaturation:** Change in conformation of a protein or nucleic acid caused by heating or exposure to a chemical, usually resulting in loss of biological function.

**Denature:** Process of disturbing the secondary and tertiary structure of proteins or nucleic acids.

**Deoxyribose:** Sugar backbone of DNA derived from the sugar ribose by loss of one oxygen atom.

**Depolarize:** Change in membrane potential toward a more positive value.

**Descanning:** Return of fluorescence emission through the galvanometer mirror system.

**Dialysis:** Separation of substances in solution by unequal diffusion through semipermeable membranes.

**Dichroic:** Filter that selectively passes light of a small range of colors while reflecting the rest.

**Dideoxynucleotides:** Block further polymerization when added to the end of a DNA strand. This property is utilized for DNA sequencing.

**Differentiation:** Process by which a cell undergoes a change to a specialized cell type.

**Differential interference contrast (DIC):** Method of deriving contrast in an unstained specimen from differences in index of refraction of specimen components.

**Diffracted wave (D-wave):** Light that passes through the specimen.

**Diffraction:** A class of optical phenomena caused by the bending of a light wave when it encounters an obstacle (slit, edge, etc.).

**Digital holographic microscopy:** A microscopic technique in which the wave front information of an object is recorded on a digital sensor.

**Digital polymerase chain reaction (PCR):** A quantitative PCR technique in which a sample is separated into multiple "partitions," only some of which contain the target molecule.

**Diode array:** A collection of photosensitive diodes placed in a side-by-side arrangement.

**Dipole:** The separation of positive and negative charge around a molecule due to nonuniform electron distributions.

**Dipole approximation:** Approximation where the electric field inside a nanoparticle is assumed to be constant and in phase with the external electric field, so the nanoparticle can be treated as a dipole emitter.

**Directed evolution:** Method of evolving proteins or nucleic acids by introducing random mutations and utilizing the concept of natural selection to optimize their activity.

**Direct immunofluorescence:** Technique of antibody labeling in which the primary antibody is directly labeled with a fluorophore.

**Dissociated:** In cell culture, refers to cells that have been physically or enzymatically separated so that they grow individually in a monolayer.

**Distance geometry:** Mathematical basis for a geometric theory of molecular conformation.

**Disulfide:** Covalent bond between two thiols.

**DNA (deoxyribonucleic acid):** A double-stranded nucleic acid that contains the genetic information for cell growth, division, and function.

**DNA helicase:** Enzyme involved in opening the DNA helix into its single strands for DNA replication.

**DNA library:** Collection of cloned DNA fragments.

**DNA ligase:** Enzyme that joins the ends of two strands of DNA together with a covalent bond to make a continuous DNA strand.

**DNA polymerase:** Enzyme that synthesizes DNA by joining nucleotides together using a DNA template as a guide.

**DNA primase:** Enzyme that synthesizes a short strand of RNA on a DNA template producing a primer for DNA synthesis.

**DNA replication:** Process by which a copy of a DNA molecule is made.

**DNA shuffling:** Method for generating highly recombined genes and evolved enzymes.

**DNAses:** Enzymes that hydrolyze DNA to nucleotides.

**Dry etch:** A plasma of reactive gases such as fluorocarbons, chlorine, oxygen, or boron trichloride that is used to remove material in a masked pattern for microfabrication. Reactive-ion etching is one type of dry etch. Dry etching is usually anisotropic.

**Dwell time:** Amount of time an ion channel remains in a certain conductance state.

**Dynamic light scattering:** Also known as photon correlation spectroscopy or quasi-elastic light scattering. Technique used to determine the size of particles in solution.

**Electrically competent:** Bacterial cells specially treated so that they become **competent** via exposure to an electric field.

**Electron beam evaporation:** A type of thin film deposition in which a target anode is vaporized by an electron beam, and the anode material condenses onto a substrate.

**Electron beam lithography:** A type of lithography in which an electron beam is used to draw a pattern in an electron-sensitive resist. Features down to 10 nm are possible with this method.

**Electron diffraction:** Interference effects owing to the wavelike nature of an electron beam when passing near matter. Can be used as a form of spectroscopy.

**Electronegativity (symbol $\chi$):** Measure of the affinity of an atom for electrons. It is reported as dimensionless Pauling units that range from 0.7 to 4.0.

**Electroporation cuvettes:** Specialized containers to hold cells during electroporation. Contain parallel metal plates. May explode if reused.

**Electroporator:** An appliance that exposes a cell solution to an electric field, causing a significant increase in the permeability of the cell plasma membrane enabling the cells to take up extracellular DNA.

**Ellipsometer:** A specialized tool for measuring the dielectric properties of thin films (complex refractive index, dielectric function) at one or more wavelengths. The machine measures changes in polarization upon reflection from or transmission through the sample, and then uses a model to calculate the dielectric properties; the models are preprogrammed into ellipsometer software.

**Emission filter:** A filter for fluorescence microscopy that selectively passes the emission wavelength and eliminates any trace of the wavelengths used for excitation.

**Enantiomer:** One possible handedness of a chiral molecule.

**Enantiomeric excess:** A measure of the preponderance of one handedness of a molecule over the other in a mixture, expressed as the absolute difference of the mole fractions of each enantiomer.

**Endonuclease:** Enzyme that cleaves nucleic acids within the polynucleotide chain. Restriction enzymes are endonucleases, as is Cas9.

**Endoplasmic reticulum (ER):** Membrane-bound compartment where lipids and proteins are synthesized. Rough ER contains ribosomes on its surface, and smooth ER does not.

**Endotoxin:** Toxic substance bound to bacterial cell wall and released when cell ruptures or deteriorates. Almost always synonymous with the lipopolysaccharide (LPS) of Gram-negative bacteria.

**Energy of hydration:** Amount of energy released when a mole of an ion is dissolved in an infinitely large amount of water.

**Energy of solvation:** The change in energy when an ion or molecule is transferred from a vacuum (or the gas phase) to a solvent.

**Energy-dispersive x-ray spectroscopy (EDS or EDX):** Describes the elemental makeup of a sample by detecting x-rays that are emitted at different energies in response to bombardment by the microscope's electron beam.

**Enhanced permeability and retention (EPR) effect:** The phenomenon by which nanoparticles accumulate in tumors (or infected, inflamed tissue) because rapidly growing, leaky capillaries found in such tissues permit nanomaterials easy entry but not exit (relative to healthy tissues).

**Enhancer:** Regulatory DNA sequence to which gene regulatory proteins bind, increasing the rate of transcription of a structural gene.

**Envelope plasmid:** Encodes an envelope gene of a different virus for pseudotyping.

**Epifluorescence:** Fluorescence microscopy performed using reflected light.

**Epigenome:** Chemical changes to DNA that do not involve changing the DNA sequence. Examples are DNA methylation, histone modification, and siRNA expression.

**Ester:** Covalent bond formed by joining a carboxylic acid and an alcohol.

**Exchange chromatography (ion exchange chromatography):** Enables the purification of charged molecules by utilizing the charge–charge interactions between the proteins in the sample and the charges immobilized on the resin.

**Excitable cell:** A cell that can be stimulated to fire action potentials.

**Excitation balancer:** Used to adjust intensity at given wavelengths.

**Excitation filter:** Used in fluorescence microscopy for selection of the excitation wavelength of light from a light source.

**Excitotoxicity:** Neuronal toxicity caused by overstimulation of excitatory (depolarizing) receptors, usually for the neurotransmitter glutamate.

**Exonuclease:** Enzyme that cleaves nucleotides one at a time from the ends of polynucleotides.

**Expression leak:** Of promoters—a promoter that expresses while in its "off" state.

**External beam radiotherapy:** A type of radiotherapy where the radiation is delivered from a source located outside the body; it must penetrate the tissue to the target (usually a cancerous tumor).

**External conversion:** Process by which a molecule returns to its ground state, with the energy of the excited state being lost as heat or transferred to other molecules by collisions.

**Extinction:** Total attenuation of incident radiation (sum of the scattered and absorbed light).

**Extremophiles:** Organisms that survive in conditions that are destructive to most life on Earth, including extremes of heat, cold, salt concentration, pH, and more.

**Fab fragment:** Region on an antibody that binds to antigens.

**Fastidious:** Of bacteria, having complex nutritional requirements.

**Fatty acid:** A carboxylic acid with long hydrocarbon chains. The two groups of fatty acids, unsaturated and saturated, are characterized by the presence of one or more double bonds between the carbons in the hydrocarbon chain. If no double bonds exist, the carbon atoms are "saturated" with hydrogen atoms.

**Feeder layer:** Cells used to support the growth of a variety of cultured cell types; the cells serve as a basal layer and supply metabolites without any growth or division of their own. (They are inactivated with gamma irradiation.)

**Field curvature aberration:** Distortion of microscopic image caused by the center of the field of view being in a different focal plane than the periphery.

**Filter cube:** Fluorescence filters packaged into a cube so that all components are held at correct angles and so that all filters can be easily replaced.

**Fine structure:** A small splitting of spectral lines attributed to an interaction between the electron spin and the orbital angular momentum.

**Fingerprinting:** A technique for identifying organisms on the individual, species, genus, or higher level based upon patterns of DNA or RNA. For example, may be used to identify how many organisms in a hot spring are eukaryotes, archaea, or bacteria.

**Fission yeast:** Yeast that divide to give two-equal sized cells. *Schizosaccharomyces pombe* is an example.

**Fluorescence lifetime imaging microscopy (FLIM):** A type of fluorescence microscopy coupled to **time-correlated single-photon counting** so that each pixel of the image displays a fluorescence lifetime decay curve.

**Fluorescence localization after photobleaching (FLAP):** A variation on FRAP in which two fluorophores are used; only one is selectively bleached, and the ratio of the two is used to determine diffusion.

**Fluorescence loss in photobleaching (FLIP):** A variation on FRAP in which a select region is bleached over and over, and loss of fluorescence throughout the entire cell is measured.

**Fluorescence quenching:** Process in which the emission from a fluorophore is reduced.

**Fluorescence recovery after photobleaching (FRAP):** A microscopic technique for measuring diffusion that involves photobleaching a selected area and measuring the fluorescence recovery.

**Fluorescence resonance energy transfer (FRET):** Distance-dependent interaction between two fluorescent molecules in which the donor, initially in its electronic excited state, transfers energy to the acceptor without the emission of a photon.

**Focused-ion beam (FIB):** A technique using a focused beam of ions for site-specific analysis, deposition, and ablation of materials.

**Folic acid (folate when deprotonated):** One of the B vitamins, essential for cell growth so a major target for anticancer drugs.

**Förster distance:** Represents the molecular separation at which FRET energy transfer is 50% efficient.

**Fourier transform infrared (FTIR) spectrometer:** Type of IR spectrometer that has a Michaelson interferometer in the place of the diffraction grating, allowing all frequencies to be measured simultaneously.

**Four-point probe:** A device for measuring bulk resistivity in thin films, containing four thin collinear tungsten wires that all contact the surface. Current flows between the outer probes, and voltage is measured between the inner probes.

**Frameshift mutation:** The insertion or deletion of a nucleotide pair(s), which causes a disruption of the translational reading frame and subsequent abnormalities in the translation of the gene downstream of the mutation.

**French press (French pressure cell):** Device for rupturing bacteria by exposing them to high pressure. Useful for large volumes.

**FRET efficiency:** The percentage of the excitation photons that contribute to FRET; strongly dependent on the distance between the donor and acceptor fluorophores.

**Functional groups:** Groups of atoms found within molecules that are involved in the chemical reactions characteristic of those molecules.

**Fundamental anisotropy:** The anisotropy (feature whereby a certain property is exhibited at different values when measured along different axes) that is measured in the absence of rotational diffusion.

**Fusion protein:** Product of joining two genes together genetically and expressing as a single protein. Frequently done with green fluorescent protein (GFP).

**Gating:** Property of ion channels in which they are open only in the presence of a specific stimulus.

**Gel electrophoresis:** Method of molecular separation in which molecules (DNA, RNA, or protein) are separated in a gel mold based on molecular size, done by applying an electric field in a specified direction.

**Gene:** Region of DNA that carries information for a discrete hereditary characteristic, usually corresponding to a single protein or single RNA.

**Gene expression profiling:** The analysis of the varying expression of genes in different organisms.

**Gene therapy:** Experimental technique that uses gene delivery to treat or prevent disease. Genes may be stably or transiently delivered to an individual's cells or tissues using liposomes, viral vectors, or any of the other techniques used for gene delivery in experimental models.

**Genome assembly:** An algorithm that assembles a genome from thousands or millions of short individual DNA sequences.

**Genome mapping:** Determination of the sequence of genes and their relative distances from each other on a given chromosome. Also called *physical mapping* when done by molecular biology techniques.

**Gigaseal:** Very high-resistance seal formed between perfectly clean glass and a cell membrane; the basis of the patch clamp technique. Can easily reach tens of gigaohms.

**Glial cell:** Supporting nonneural cell of the nervous system. Different types perform different essential roles, from physical support to production of myelin to immune support.

**Glycerol:** Three-carbon carbohydrate; forms backbone of triglyceride.

**Glycosylation:** Reaction in which one or more sugars is added to a protein or lipid molecule.

**Goldman–Hodgkin–Katz:** Equation used to predict the resting membrane potential for a cell with multiple ionic species.

**Golgi apparatus:** Organelle in which proteins and lipids transferred from the endoplasmic reticulum are modified and sorted.

**Gradient gel:** Polyacrylamide gel with a higher concentration of acrylamide at the bottom than the top. Helps to separates proteins that comigrate on a single-concentration gel.

**Gram negative:** Bacterial classification characterized by thin peptidoglycan layer in cell wall and an outer membrane containing lipopolysaccharide (LPS). Stain red under the Gram stain procedure.

**Gram positive:** Bacterial classification characterized by thick peptidoglycan layer in cell wall. Stain violet under Gram stain.

**Gram stain:** Colorimetric bacterial stain used to distinguish Gram-positive bacteria (violet) from Gram-negative bacteria (red).

**Gramicidin:** A family of linear pentadecapeptides from *Bacillus brevis* that form cation-selective channels in membranes (slightly selective for potassium over sodium).

**Green fluorescent protein:** A protein that exhibits bright green fluorescence when exposed to blue light, originally isolated from the jellyfish *Aequorea victoria*.

**Gunn diode:** Device that is capable of converting direct current power into radio-frequency power when it is coupled to the appropriate resonator. Also referred to as *transferred electron device* (TED).

**Halorhodopsin:** Light-driven inward chloride pump. Isolated from different species of *Halobacteria*.

**Hanging drop:** Type of vapor diffusion technique of protein crystallization in which the protein drop is placed hanging upside down over the reservoir solution.

**Headstage:** Part of an electrophysiology amplifier that interfaces directly with the recording and ground electrodes. It serves as a stable holder for the recording pipette and electrode and an interface to the preamp.

**Heat fix:** Process by which bacteria are fixed (killed and adhered to the slide) by application of heat from a flame or specialized apparatus.

**HeLa:** A cervical cancer cell line, the first cell line to be successfully immortalized. Its name comes from Henrietta Lacks, the owner of the tumor (which killed her).

**Helper:** When referring to a virus, refers to a virus that assists in the replication of another defective virus within the same host cell.

**Hematopoetic stem cells:** Give rise to all blood cell types.

**Hemocytometer:** Thick glass microscope slide with a gridded chamber designed to allow for counting the number of cells per volume in a cell suspension.

**HER2:** Human epidermal growth factor receptor 2, a growth-promoting receptor found on some breast cancers. Tumors are tested for HER2 in order to determine the best course of treatment, since specific drugs exist for HER2-positive cancers.

**Heterocyclic:** Organic compound containing carbon and at least one element other than carbon in a ring structure.

**Heterozygote:** Having different alleles of a given gene. One allele may be dominant over the other (for example, brown eyes over blue), or the two may be codominant (for example, type AB blood).

**Hologram:** A photographic recording of a light field that can be used to reconstruct a three-dimensional image.

**Homochirality:** A group of molecules that all have the same handedness. Also see **enantiomeric excess**.

**Homologous recombination:** A type of genetic recombination that occurs when two very similar (or identical) pieces of DNA "cross over" and exchange sequences. It is used in nature to repair errors and create genetic diversity during meiosis, and in molecular biology as a gene targeting technique.

**Homology directed repair:** A mechanism used by cells to repair double-strand breaks in DNA using an unbroken homologous strand present in the cell. Important for cancer prevention.

**Homozygous:** Having identical alleles for any given gene. In diploid organisms (those with two copies of each chromosome), a homozygote has two identical copies. Most mammals are diploid; many plants have three or more copies (polyploid).

**Host:** An organism that sustains another organism or foreign material inside of it.

**Hybridoma:** A type of cell line created by fusing an immortal line with a nonimmortalized cell type.

**Hydrogen bond:** An attractive force between a hydrogen atom bound to an electronegative atom and another electronegative atom of a different molecule.

**Hydrolysis:** Cleavage of a covalent bond with accompanying addition of water.

**Hydrophilic:** Property of a molecule or surface causing it to have an affinity for water and be able to bind to water through hydrogen bonding.

**Hydrophobic:** Property of a molecule causing it to be repelled by water. Nonpolar molecules are hydrophobic.

**Hyperfine coupling constant:** Mathematical interpretation of the projections of both electron and nuclear spins, which dictates the electron–nuclear interaction.

**Hyperfine structure:** A term for several small spectral shifts that result from the interactions of molecular nuclei with external electric and magnetic fields.

**Hyperpolarize:** Change in membrane potential toward a more negative value.

**Hyperspectral imaging:** Imaging in wavelength as well as space ($xy\lambda$, $xyz\lambda$) by spectral scanning, spatial scanning, spatiospectral scanning, or nonscanning (snapshot) methods.

**Hyperthermia therapy:** The use of local or generalized heating in order to kill cancer cells or make them more susceptible to radiation and/or chemotherapy.

**Immunocompetent:** An animal with a normal immune response.

***In vivo*:** Refers to an experiment done within an entire living organism (Latin for "in the living").

***In vivo* excision:** Process that converts a lambda clone to a phagemid clone within *E. coli*.

**Inclusion bodies:** Insoluble aggregates of misfolded proteins.

**Indirect immunofluorescence:** Technique of antibody staining involving two sets of antibodies: a primary antibody against the antigen of interest, and a secondary dye-coupled antibody that recognizes the primary.

**Infection:** Transfection via viral vectors.

**Infinity corrected (of a microscope lens):** The lens projects an image that is collimated to infinity, and the image is brought into focus at the point of the eyepiece or the photo tube through the use of a convergence lens.

**Inner cell mass:** Portion of the blastocyst that becomes the embryo.

**Insert:** The DNA fragment to be cloned that is included in a DNA vector. May be a gene, or a gene plus promoter, multiple genes, etc.

**Insertion:** A mutation in which one or more nucleotide pairs are added.

**Inside-out patch:** Type of patch clamp recording in which the pipette is separated from the cell with a piece of the inside of the membrane attached to the pipette and exposed to the solution in the surrounding bath.

**Instrument response function:** The curve obtained from a zero-lifetime sample used to estimate the time resolution of an instrument.

**Integrase:** Retroviral enzyme that inserts the viral genome (**provirus**) into the DNA of the host cell.

**Integration complex:** Complex of retroviral provirus and other viral and host enzymes and proteins. In the case of lentiviruses, can pass into the nucleus.

**Intercalating agent:** A mutagen that can insert itself between the stacked bases at the center of the DNA double helix. Intercalating agents such as ethidium bromide are commonly used as fluorescent tags in gel electrophoresis.

**Interface diffusion:** A method of protein crystallization in which protein and reagents gradually diffuse under the influence of a concentration gradient.

**Intermediate filament:** Fibrous protein filament that forms ropelike networks in animal cells.

**Intron:** A noncoding segment of a eukaryotic gene, removed from the mRNA posttranscriptionally during splicing. Introns may serve to regulate gene expression timing and extent by controlling packaging in the chromosome.

**Inverse FRAP (iFRAP):** A FRAP variant in which an entire cell *outside* of a region of interest is photobleached.

***In vitro*:** An experiment done outside a living organism—in a test tube or a cell line (Latin for "in glass").

***In vitro* transcription:** Generation of RNA from DNA using enzymes only, no cells.

**Inward rectifier channel:** Ion channel that conducts more current at negative potentials.

**Ion channel:** A protein that allows a flow of ions in or out of the cell as determined by the cell's electrochemical gradient.

**Ionic bond:** A chemical bond formed by the complete transfer of one or more electrons from one atom to another.

**Ionophore:** Pore-forming peptide.

**IPTG (Isopropyl β-D-1-thiogalactopyranoside):** A chemical that triggers expression from the *lac* promoter. Used to induce expression.

**IR active:** Vibration associated with a change in dipole moment.

**Isomerization:** The process by which the atoms of a molecule are rearranged, transforming the molecule to another, without changing its chemical formula.

**Jablonski diagram:** State diagram in which molecular electronic states are grouped according to multiplicity into horizontally displaced columns.

**Junction potential:** Potential difference at the boundary between dissimilar solutions.

**Ketone:** Organic compound in which a carbonyl group is bound to two carbon atoms (chemical formula RCOR).

**Klystron:** An evacuated electron tube used to generate or amplify microwave radiation by varying the velocity of an electron beam. Used as a microwave source for EPR and many other applications.

**Knockout:** Inactivation of a particular gene.

**Lac operon:** A set of genes and regulatory elements from *E. coli* that are needed for transport and metabolism of lactose. One element of this operon is the inducible *lac* promoter, used commonly in molecular biology vectors as it can be induced with IPTG.

**LacZ:** Common reporter gene encoding for beta-galactosidase. Used in bacteria and mammalian cells.

**Ladder:** A mixture of molecules (DNA, RNA, or protein) of various known sizes, run alongside a sample of molecules of unknown size in gel electrophoresis.

**Lagging strand:** One of the two newly synthesized strands of DNA found at a replication fork. The lagging strand is made in discontinuous lengths that are later joined covalently.

**Laser tweezers:** Also called an optical trap, a highly focused laser beam that "traps" dielectric particles by creating an electric field gradient. An Nd:YAG laser (1064 nm) is usually used for biological specimens.

**Lawn:** A continuous growth of bacteria on a solid medium where individual colonies cannot be distinguished.

**Leading strand:** One of the two newly synthesized strands of DNA found at a replication fork. The leading strand is made by continuous synthesis in the 5′-to-3′ direction.

**Lectin:** One of a class of proteins that bind specific surface carbohydrates.

**Lentivirus:** Retrovirus with the unique ability of being able to infect nondividing cells.

**Lift-off:** A type of patterning in which the inverse pattern is formed in the resist. Used when direct etching of structural material would damage the layer below.

**Ligand:** In biochemistry, a substance that is able to bind to a site on a target biomolecule to serve a biological purpose. Binding is nearly always reversible.

**Ligand-gated:** Type of ion channel that opens in response to the binding of a specific chemical or molecule.

**Ligation:** Joining of two molecules of DNA end to end through phosphodiester bonds catalyzed by the enzyme DNA ligase.

**Lipid bilayer recording:** An experimental setup to measure the electric properties of ion channels consisting of two chambers separated by an artificial cell membrane in which ion channels can be implanted to simulate a cell.

**Lithography:** A microfabrication technique in which a masked pattern is used to create a "chemical stencil" that serves as a template for further process steps.

**London dispersion interactions:** Weak intermolecular forces that arise from the interaction between temporary dipole moments in molecules.

**Long-term depression (LTD), long-term potentiation (LTP):** Reduction or enhancement (respectively) in signal transmission between neurons that are used as cellular models of learning and memory.

**Luciferase:** Enzyme that catalyzes a luminescent reaction. There are several forms, used in biosensors and as reporter genes. Require a substrate (luciferin) to produce luminescence.

**Lyophilized, lyophilization:** The process by which water from a given sample is extracted. By this technique, the sample remains stable and is easily stored at room temperature.

**Lysis:** Rupture of a cell's plasma membrane leading to the release of cytoplasm and cell death.

**Lysosome:** Membrane-bound organelle in eukaryotic cells containing digestive enzymes, which are typically most active at an acidic pH.

**Magnetogyric ratio:** Ratio of a nucleus's magnetic dipole moment to its angular momentum.

**Maxiprep:** Processes of extracting DNA from a large (150–250 mL) volume of bacterial culture.

**Mean luminance:** A measurement that is a function of source brightness, arc geometry, and the angular spread of the light.

**Mechanosensitive:** Type of ion channel that exhibits gating in response to mechanical stress. An example is MscL, the *mechanosensitive channel (large)*.

**Membrane potential:** Potential difference across the cell membrane, measured in mV.

**Mesenchymal cell (mesenchymal stem cell):** A cell that can differentiate into a variety of cell types (bone, cartilage, fat, and connective tissue).

**Messenger RNA (mRNA):** RNA molecule that specifies the amino acid sequence of a protein and relays that information to other parts of the cell.

**Metabolism:** Refers to the biochemical processes in which energy is transformed within cells.

**Metrology:** The science of measurement.

**Microbatch:** Crystallization technique in which aqueous protein solutions are dispensed under liquid oil and as the water evaporates through the oil, the concentrations of protein and precipitant increase until the nucleation point is reached.

**Microfilament:** Helical protein filament formed by polymerization of globular actin molecules. Forms part of the contractile apparatus of skeletal muscle.

**Micrometastasis (plural micrometastases):** Microscopic spread of a cancerous tumor.

**Microplate reader:** Laboratory instrument designed to read fluorescence or another biological signal from multiwell culture plates.

**MicroRNAs:** Small, noncoding RNA molecules that regulate gene expression by targeting mRNA and therefore are important modulators in many cellular pathways.

**Microtubule:** Hollow cylindrical structure composed of the protein tubulin.

**Midiprep:** Processes of extracting DNA from a medium (50–100 mL) volume of bacterial culture.

**Miniprep:** Processes of extracting DNA from a small (3-5–mL) volume of bacterial culture. Usually performed quickly, leading to DNA of quality suitable only for screening, not transfection, spectroscopy, or sequencing.

**Missense mutation:** Nucleotide pair substitution that leads to the replacement of one amino acid by another in the translated protein.

**Mitochondrion (plural mitochondria):** Membrane-bound organelle that performs oxidative phosphorylation and produces most of the ATP in eukaryotic cells.

**Mobility:** Ability of a charged particle to move in response to an electric field.

**Molar extinction coefficient $\varepsilon$:** Measure of the ability of a molecule to absorb light. Reported at a specific wavelength, e.g., $\varepsilon_{485}$.

**Molecular beacons:** Molecular beacons are oligonucleotides that, in the presence of their target sequence, bind and fluoresce and therefore report the presence of specific nucleic acids in a solution.

**Molecular cloning:** The process of creating and amplifying specific DNA segments by growing a clone of carrier cells (i.e., *E. coli*) into which the gene has been introduced by recombinant DNA techniques.

**Molecular orbital calculation:** Mathematical investigation concerning atomic interactions using molecular orbital theory.

**Monodispersity:** Collection of particles with the same size and shape or mass.

**Monogenetic:** Caused by a single gene. Also called monogenic. Monogenetic diseases include cystic fibrosis and Tay–Sachs disease.

**Monomer, monomeric:** (1) Single molecule that is able to form polymers with others. (2) A protein made of a single polypeptide chain.

**Monosaccharide:** The most basic unit of carbohydrates, with the chemical formula $C_x(H_2O)_y$.

**Monovalent:** (1) An element that can form only one chemical bond. (2) A binding protein with only one site of attachment (e.g., monovalent antibody, monovalent streptavidin). (3) Having activity against one specific antigen (e.g., monovalent vaccine).

**Mother liquor:** In the vapor diffusion method of protein crystallization, the drop that contains a mixture of the reservoir solution and protein sample.

**Motor protein:** Protein that uses energy derived from nucleoside triphosphate hydrolysis to propel itself along a linear track. Examples are myosin, kinesin, and dynein.

**Mounting medium:** Material used to mount microscopic specimens before examination. Chosen based on color, viscosity, refractive index, etc.

**MscS, MscL (mechanosensitive channel [small, large]):** Ion channels from bacteria that open in response to mechanical stress.

**Multielectrode array (MEA):** An array of several to several thousand electrodes designed to be interfaced with neuronal networks.

**Multiple cloning site (polylinker):** Short fragment of DNA containing multiple unique restriction enzyme sites.

**Multiphoton microscopy:** A type of fluorescence microscopy in which the energy of the excitation light is lower than the energy of the excited state, so that two or more photons must hit the sample simultaneously in order to excite the fluorophore.

**Multipotent:** A cell that has the potential to differentiate into cell types from a small number of lineages.

**Mutagenesis:** The use of methods to change a genetic sequence in either a targeted or random fashion.

**Mutagenic PCR:** Process of producing multiple copies of the same gene with random mutations by amplifying with an error-prone polymerase.

**Mutagenic primer:** A primer containing a single targeted error relative to the template strand. The error is near the center of the primer and does not prevent binding and amplification. Used to insert a point mutation.

**Mutation:** Change in the nucleotide sequence of a region of DNA.

***Mycoplasma*:** Genus of bacteria without a cell wall (thus resistant to ampicillin and similar antibiotics). Common contaminants of cell cultures, difficult to detect and treat.

**Nanoimprint lithography:** A method of lithography that can scale up production of a pattern made by low-throughput methods (such as electron beam lithography) by stamping the pattern into an imprint resist.

**Nanomedicine:** Application of nanotechnology to medicine; usually refers to therapeutic and diagnostic nanoparticles, though may apply to more futuristic concepts such as biomachines.

**Native (state, configuration):** The properly folded tertiary structure of a protein, as opposed to denatured or unfolded protein.

**Negative (bright) phase contrast:** A technique of phase contrast microscopy in which details of specimen appear bright on a darker background.

**Negative-sense RNA:** Viral RNA that is complementary to the sequence of mRNA. Must be replicated into its complement before it can be transcribed.

**Nernst equation:** Equation for the reduction potential of a half cell or electromotive force (EMF) of a full electrochemical cell. In physiology, gives the equilibrium potential of an ion across a membrane as a function of concentration gradient.

**Neural stem cell:** Self-renewing, multipotent cells that generate the cells of the nervous system (neurons, oligodendrocytes, astrocytes).

**Neurotransmitter:** A chemical substance that relays signals between a neuron and another cell across the synapse. Examples include glutamate, dopamine, and nitric oxide.

**Nickase:** An endonuclease that only cuts one strand of double-stranded DNA.

**Nicotinamide adenine dinucleotide (NADH):** Coenzyme with central role in cellular metabolism and energy production.

**Nitrogenous base:** In molecular biology, one of five nitrogen-containing bases that make up nucleotides (adenine, guanine, cytosine, uracil, thymine). More generally, any nitrogen-containing molecule that acts as a base.

**Noncoding RNA:** Functional RNA molecule not translated into a protein.

**Nonhomologous end joining (NHEJ):** An innate mechanism in all organisms that repairs DNA double-strand breaks without the need for a template.

**Nonpolar molecule:** A molecule in which the electrons are symmetrically distributed and neither end is more or less charged than the other.

**Nonradiative recombination:** In semiconductors, when an electron in the conduction band recombines with a hole in the valence band and the excess energy is emitted in the form of heat in the semiconductor crystal lattice instead of as fluorescence.

**Nonsense mutation:** Nucleotide pair substitution that changes a codon for an amino acid into a termination codon. (Translation is signaled to end prematurely.)

**Northern blot:** The transfer of electrophoretically separated RNA molecules from a gel onto an absorbent sheet, which is then immersed in a solution containing a labeled probe that can bind to the RNA of interest.

**Nuclear g-factor:** Ratio that characterizes the magnetic moment and spin angular momentum of a nucleus.

**Nuclear Overhauser effect spectroscopy (NOESY):** Form of two-dimensional NMR, useful for structural studies. Its strength is a function of the distance between two protons.

**Nuclease-deficient Cas9:** A Cas9 that binds DNA but does not cut it.

**Nucleic acid:** Polymer of nucleotides (DNA, RNA).

**Nucleoside:** Nitrogenous base linked to a sugar (ribose or deoxyribose).

**Nucleotide:** Composed of a nitrogenous base, a five-carbon sugar and one to three phosphate groups. Nucleotide monomers join together to make nucleic acids (DNA or RNA).

**Nucleus:** Large membrane-bound organelle containing the DNA of a eukaryotic cell.

**Numerical aperture:** Measure of a lens's ability to gather light and resolve fine specimen detail at a fixed object distance.

**Ocular:** Perceived or related to the eye. A name for the eyepiece of a microscope.

**Off-target activity:** In gene editing, refers to editing anywhere in the genome other than the desired location. Characteristic of all systems, but a greater problem with CRISPR than zinc finger nucleases.

**Oligonucleotide:** A short nucleic acid polymer.

**Oncogene:** Gene that when mutated or expressed at abnormally high levels contributes to converting a normal cell into a cancer cell.

**Open complex:** Structure formed after the closed complex when 12–17 base pairs of double-stranded DNA are split into single strands to permit transcription.

**Optical path difference (OPD):** Difference between the indices of refraction of the medium and the specimen.

**Optical section:** A chosen plane in a thick specimen, made visible by use of a confocal scanning microscope.

**Optogenetics:** The use of genetically encoded actuator and reporter proteins that allow the use of light to either control or report on the activity of molecular processes in specific cell populations.

**Orbital:** Quantum mechanical function describing the electron probability density around an atom (atomic orbital) or molecule (molecular orbital).

**Organelles:** Subcellular compartments that perform specialized functions. Membrane-bound organelles are characteristic of eukaryotic cells.

**Organic molecule:** A molecule normally found or produced in living systems, typically consisting of carbon, hydrogen, oxygen, or nitrogen atoms. Must contain at least one carbon and at least one C–H bond. (e.g., cyanide and carbon dioxide are inorganic.)

**Organometallic synthesis:** Thermolysis or reduction of organometallic precursors in the presence of ligand molecules.

**Organotypic slice:** Brain slice cultured for several weeks. Cell morphology, anatomical relations, and network connections are maintained.

**Origin of replication:** Point of a specific sequence where DNA replication is initiated.

**Osmometer:** An instrument that measures the freezing point of a solution to determine osmolarity.

**Ostwald ripening:** Process by which smaller particles dissolve and their material is consumed by larger particles. As the larger crystals grow, the area around them is depleted of smaller crystals.

**Outside-out patch:** A type of patch clamp recording in which the pipette is pulled from a whole-cell patch and the membrane is folded back over the pipette end.

**Overlap integral:** In FRET, describes the degree of overlap between the donor emission spectrum and the acceptor absorption spectrum.

**Packaging cell line:** Cells used to produce viral vectors. May be ordinary cells that are transiently transfected with viral plasmids, but often are special lines stably transfected to produce one or more viral proteins.

**PALM (photoactivated localization microscopy):** A type of superresolution fluorescence microscopy that uses photoactivatable fluorescent proteins.

**PANTHER (Protein Analysis through Evolutionary Relationships):** A web-based classification tool "designed to classify proteins (and their genes) in order to facilitate high-throughput analysis."

**Passaging (cell passaging):** Subculturing cells into fresh medium to prevent overgrowth. Applies to bacteria and suspension or adherent mammalian cells.

**Passivate:** To decrease the reactivity of a surface by chemical treatment.

**Patch clamping:** Direct measurement of electrical properties of cells via connection with an electrode filled with a particular ionic solution.

**Pattern transfer:** Means by which patterns that have been lithographically printed are transferred onto the substrate.

**Pellet:** Densely packed concentration of cells in a tube or rotor after centrifugation.

**Peptide:** A short polymer formed by the linking of amino acids. The association between one amino acid molecule and the next is called a peptide bond. Proteins are polypeptide molecules.

**Perforated patch:** Type of whole-cell recording in which the pipette contains a chemical that degrades the cell membrane upon contact and reduces the resistance from the pipette to the cell.

**Perfusion chamber:** A holder for live cells that permits a constant flow of electrolyte solution.

**Periplasm:** Space between the cell membrane and cell wall in Gram-negative bacteria.

**Perrin equation:** Formula that describes the relationship between the observed fluorescence polarization, the limiting polarization, the fluorescence lifetime of the fluorophore, and its rotational relaxation time.

**Phage display:** A technique for high-throughput screening of proteins by expressing them on the surface of bacteriophage.

**Phalloidin:** A mushroom toxin (from *Amanita phalloides*) that targets the cytoskeleton.

**Pharmacokinetic/pharmacodynamic (PK/PD) studies:** A pharmacokinetic study measures the drug concentration-time courses in selected body fluids resulting from administration of a certain dose. A pharmacodynamic study measures the observed effect resulting from a certain drug concentration. Both are essential in the drug approval process.

**Phase contrast:** A light microscopic technique that translates phase difference into amplitude difference to enhance the contrast of unstained, relatively transparent specimens.

**Phase problem:** Any problem in experimental physics that occurs when information is lost because only the intensity but not the phase of emitted light is recorded by a detector. More specifically used in x-ray crystallography since x-ray diffraction gives the Fourier transform of the unit cell's electron density. If phases are known, the electron density may be calculated.

**Phosphate:** A salt or ester of phosphoric acid.

**Phosphodiester:** Covalent bond between a phosphate and two other molecules.

**Phosphorescence:** A type of photoluminescence where emission occurs from a triplet state via "forbidden" transitions. Much slower than fluorescence; may occur over a time course of minutes to hours. Seen in some "glow-in-the-dark" materials that must be preexposed to light.

**Phosphorylation:** Reaction in which a phosphate group is covalently coupled to another molecule.

**Photoactivatable:** In fluorescence microscopy, a probe (dye or fluorescent protein) that may be switched on with a specific wavelength of light. The process is usually irreversible, such as with caged compounds.

**Photobleaching:** Occurs when a fluorophore permanently loses the ability to fluoresce due to chemical damage and covalent modification.

**Photoconvertible:** In fluorescence microscopy, a probe (dye or fluorescent protein) that may be switched from one fluorescent state to another with a specific wavelength of light. The process is usually irreversible.

**Photolithography:** Lithography in which light is used to transfer a pattern from a mask to a light-sensitive photoresist. Ultraviolet (UV) light is usually used, and so photolithography may be called *UV lithography* or *optical lithography*.

**Photoluminescence, photoluminescent:** Emission of light that is caused by the irradiation of a substance with other light; property of a substance that emits photoluminescence.

**Photoswitchable:** In fluorescence microscopy, a probe (dye or fluorescent protein) that may be reversibly switched from one fluorescent state to another and back again with a specific wavelength or wavelengths of light.

**Photon counting mode:** Sensitive spectrofluorometer mode in which photons are counted one by one at the photomultiplier tube.

**Photothermal therapy:** A type of hyperthermia therapy in which light is used to heat target tissue to a therapeutic range of temperature.

**Pinhole aperture:** Small opening through which light travels.

**Pipette puller:** Device used to pull patch pipettes and/or sharp electrodes from glass capillary tubes.

**Planar process:** A step-by-step and layer-upon-layer method of fabricating elements on a wafer substrate.

**Plaque:** Regions of cell destruction formed within a cell culture by viral infection. Each plaque consists of genetically identical viruses.

**Plaque purification:** Process of isolating virus from a plaque.

**Plasmid:** Circular DNA molecule that replicates independently from the genome. Modified plasmids are used extensively as vectors for DNA cloning.

**Plasmon:** Dipolar oscillation of conduction electrons.

**Plasmon resonance:** Wavelength (or frequency) of a beam of light that matches the plasmon on the surface of the material it encounters.

**Pluripotent:** A cell that can give rise to any fetal or adult cell type but cannot form embryonic tissue.

**Point mutation:** The change (or addition or deletion) of a single base pair in a DNA sequence.

**Point scanning:** A type of scanning in confocal microscopy that employs a dot-shaped aperture, which moves with constant velocity along the scan line.

**Polar:** Molecule with an uneven distribution of charge along the molecule, leading to an electric dipole moment.

**Polyacrylamide gel:** See SDS-PAGE.

**Polyadenylated, poly-A:** Long sequence of adenine nucleotides added to the 3′ end of a nascent mRNA molecule as transcription finishes. In eukaryotes, mRNA must be polyadenylated to be stable and be exported from the nucleus.

**Polydimethylsiloxane (PDMS):** A polymeric organosilicon (silicone) very frequently used in soft lithography for biological applications, since it is easy to mold, optically clear, and nontoxic to cells.

**Polymer, polymerize:** Any process in which small molecules, or monomers, combine chemically to produce a chainlike molecule, called a polymer.

**Polynucleotide:** Polymer of nucleotides.

**Polysaccharide:** Carbohydrate structure formed by repeating units of monosaccharides or disaccharides linked by glycosidic bonds.

**Positive (dark) phase contrast:** A technique of phase contrast microscopy in which details of specimen appear dark against a lighter background.

**Positive-sense RNA:** Viral RNA that is identical in sequence to mRNA. Positive-sense viral genomes are transcribed directly into protein by the host cell without need for viral polymerases.

**Posttranslational processing:** Changes to a protein's primary structure (polypeptide) made after it is synthesized and before, during, or after folding. These changes can include cleavage of the initial methionine and any signal sequences; addition of groups such as phosphates; formation of disulfide bonds; and others.

**Precipitant:** An agent that causes the formation of a precipitate.

**Preclinical:** In drug development, preclinical studies determine efficacy and toxicity of a drug with specific *in vivo* (animal) and *in vitro* tests.

**Primary amine:** A derivative of ammonia in which only one of the hydrogen atoms in the ammonia molecule has been replaced (chemical formula $RNH_2$). The $-NH_2$ group is highly reactive and often used for coupling.

**Primary cells:** Cells taken directly from a living organism for culture.

**Primary structure:** Linear sequence of monomer units in a polymer, such as the amino acid sequence of a protein.

**Primer dimer:** Primer molecules that have joined to each other due to complementary bases. Potentially inhibits proper PCR amplification of DNA.

**Primer:** An RNA or DNA oligonucleotide that can serve as a template for DNA synthesis when annealed to a longer DNA molecule. Used extensively in PCR experiments.

**Probe hybridization:** The use of a small, labeled molecule of RNA or DNA ("probe") to identify complementary nucleic acid sequences by hybridization with a target.

**Probe sonicator:** An instrument with a thin metal probe designed to be dipped into a flask or tube and disrupt a solution with sound waves. The frequency and intensity are usually controllable.

**Profilometer:** A device for measuring the 3-D surface profile of a sample. May be as simple as a stylus (contact profilometer) or any one of many noncontact methods. Spatial resolution varies greatly depending upon the exact technique.

**Promoter:** Nucleotide sequence in DNA to which RNA polymerase binds to begin transcription.

**Promoter clearance:** Step in transcription in which RNA polymerase detaches from the promoter and binds to nonspecific DNA downstream.

**Protease:** Enzyme that degrades protein by hydrolyzing some of the peptide bonds between amino acids.

**Protease cleavage site:** Protease recognition sequence placed into a cloning vector between two expression sequences (e.g., an affinity tag and a gene of interest). After purification of the protein, protease enzymes are used to cut the sequences apart.

**Protein:** Large polymer consisting of one or more sequences of amino acid subunits joined by peptide bonds.

**Proteome:** The ensemble of the expressed proteins in a given cell or organism at a given time.

**Protoplast:** Cell (plant, bacterial, fungal) in which the cell wall has been removed. In bacteria, protoplast is usually used to refer to Gram-positive bacteria, where only a single cell membrane remains.

**Protospacer adjacent motif:** A 2- to 6-base-pair sequence in a virus or plasmid that allows it to be targeted by the Cas proteins of the CRISPR system. "Protospacers" refers to invading sequences that then became part of the CRISPR locus ("spacers"). The PAM ensures that the Cas proteins do not cleave the CRISPR locus itself.

**Provirus:** Virus integrated into the genetic material of the host, passively replicated, and passed along to the host's offspring.

**Pseudotyped:** A viral particle whose outer shell originates from a virus different from that of the genome and replication apparatus.

**Pulsed field gel electrophoresis:** Laboratory technique that allows separation of much larger pieces of DNA than traditional gel electrophoresis. The direction of current is altered at a regular interval rather than proceeding in a constant direction.

**Purine:** An organic heterocycle consisting of a pyrimidine ring fused to an imidazole. The bases adenine and guanine are purines, as are other important compounds such as caffeine and uric acid.

**Pyrimidine:** An organic heterocycle with nitrogens at positions 1 and 3 of the six-membered ring. The bases thymine, uracil, and cytosine are pyrimidine derivatives.

**Quantum dot:** A nanoscale structure of semiconductor material that can confine particles (usually electrons) in all three dimensions. Photoluminescence is characteristic.

**Quantum yield:** Quantitative measure of fluorescence emission efficiency expressed as the ratio of the number of photons emitted to the number of photons absorbed.

**Quaternary structure:** Three-dimensional relationship of the different polypeptide chains in a multisubunit protein or protein complex.

**Quenching:** Process involving collisional energy transfer with radiationless return of the fluorophore to the ground state.

**Radiotherapy:** The use of ionizing radiation to kill unwanted cells, usually for cancer therapy.

**Racemic, racemization, racemize:** A racemic mixture contains an equal fraction of each handedness of a chiral compound. Racemization refers to the gradual conversion of a homochiral mixture into a racemic mixture, which may be used as a tool for dating organic matter.

**Radiative recombination:** Form of spontaneous emission from a semiconductor in which a photon is emitted with wavelength corresponding to the energy released.

**Radiosensitizer:** A drug or other agent that makes cells more sensitive to radiation therapy.

**Raman scattering:** Scattering in which light is scattered inelastically, meaning the photon energy is either increased or decreased. This results from an interaction of the incident electric field with that of a molecule, resulting in a change in the rotational, vibrational, or electronic energy of the molecule.

**Raman spectroscopy:** A type of spectroscopy that uses characteristic Raman scattering patterns to identify chemical species.

**Ratiometric dye:** Dyes that shift their excitation or emission spectra upon binding to a substrate, so that the ratio of the bound to the unbound spectrum can be used to probe substrate concentration.

**Ratiometric imaging:** In fluorescence microscopy, imaging the emission (or absorbance) at one specific wavelength relative to that of a second wavelength.

**Rayleigh criterion:** Two neighboring points are just distinguishable if the main intensity peak from one point is located on the first dark fringe of the second point's Airy pattern.

**Reading frame:** Phase in which nucleotides are read in sets of three to encode a protein.

**Real-time PCR:** Specialized PCR technique that allows for quantization of the amplification of DNA as the PCR experiment progresses.

**RecA:** Protein present in *E. coli* essential for the repair and maintenance of DNA.

**Receptor:** A protein molecule to which a mobile *signaling* molecule or ligand may bind, usually eliciting a conformational change that initiates a cellular response.

**Recombine, recombination:** Breaking of a DNA molecule and rejoining to a different molecule. Occurs spontaneously when two or more parts of the sequence are similar (homologous recombination) and can be useful or unwanted depending upon the experiment.

**Reconstruction:** In holography, using a recorded interferometric pattern to reconstruct a 3-D image. Requires some assumptions regarding the nature of the propagator.

**Reflected light:** In microscopy, when incident illumination is reflected from the specimen surface into the objective.

**Regulatory region:** DNA sequence to which a gene regulatory protein binds to control the rate of assembly of the transcriptional complex at the promoter. Promoters and repressors are examples.

**Relaxivity:** A measure of the efficacy of an MRI contrast agent. Dependent on the degree of time constant change per unit concentration.

**Replication fork:** Y-shaped region of a replicating DNA molecule at which the two strands of the DNA are being separated and the daughter strands are being formed.

**Reporter gene:** A gene whose phenotype expression is simple to monitor; used to study tissue-specific promoter and enhancer activities in transgenes.

**Repressor sequence:** Regulatory region of DNA where repressor proteins bind to prevent transcription of an adjacent gene.

**Resist:** As a general term, something added to part of a structure to create a pattern by protecting the part from being affected by a subsequent stage in the process. Often used in lithography as a shorter term for *photoresist*, a light-sensitive polymer.

**Resolution (optical resolution):** The ability to resolve detail in an image.

**Resonance:** Delocalization of electrons in certain molecules. In the structural model, two or more equivalent structures with different bond placements are possible; the "real" structure is a hybrid of all of these. Molecular orbital theory describes resonance automatically in terms of highly delocalized electrons.

**Restriction enzyme:** An endonuclease that recognizes a specific nucleotide sequence in DNA and breaks the DNA molecule at points containing that sequence.

**Restriction mapping:** The mapping of a DNA sequence to indicate the positions of target sequences of specific restriction enzymes.

**Restriction sites:** The specific nucleotide sequences recognized by restriction enzymes.

**Reticuloendothelial system:** A network of cells having both endothelial and reticular attributes; found in the blood, general connective tissue, spleen, liver, lungs, bone marrow, lymph nodes, and elsewhere. These cells are well known for taking up colloidal particles, including nanomedicines.

**Retrovirus:** RNA-containing virus that replicates in a cell by first making an RNA–DNA intermediate and then a double-stranded DNA molecule that becomes integrated into the host's DNA.

**Reversal potential:** Potential difference measured across the cell membrane when the current of a given ion reverses direction.

**Reverse transcriptase:** Enzyme that catalyzes the synthesis of a DNA strand from an RNA template. Evolved only in the **retroviruses**.

**Reverse transcriptase PCR (rt-PCR):** DNA amplification in which the DNA to be amplified is initially copied from a strand of RNA by the enzyme reverse transcriptase.

**Ribose:** Sugar backbone of RNA.

**Ribosome:** Cytoplasmic particle that mediates the linking of amino acids to form proteins.

**RNA polymerase:** Enzyme that synthesizes RNA by joining nucleotides together using a DNA template as a guide.

**RNA virus:** A virus that uses RNA exclusively as its genetic material.

**RNA-dependent RNA polymerase:** Polymerase that makes RNA from RNA.

**RNA-induced silencing complex (RISC):** A multiprotein complex that incorporates an siRNA and allows it to bind its target; one protein in RISC, called Argonaute, then cleaves the target mRNA.

**RNAses:** Enzymes that hydrolyze RNA to nucleotides.

**Salt bridges:** Glass or plastic V or U shapes containing electrolyte embedded in agar; connect two solutions by allowing ions to migrate from one end to the other.

**Salting out:** Forcing a protein out of solution by saturation with a neutral salt.

**Satellite colony:** Bacteria that does not contain the desired plasmid.

**Saturated:** Fatty acid with only single bonds between the carbon atoms. The molecule is therefore saturated with hydrogen atoms.

**Scalar coupling:** Indirect (electron-mediated) nuclear spin interaction between spins in the same molecule. Results in NMR fine structure.

**Scan head:** Part of a confocal microscope responsible for rasterizing the excitation scans and collecting the photon signals from the specimen.

**Scanning cysteine accessibility mutagenesis (SCAM):** A method of determining protein topology by mutating selected amino acids into reactive cysteines and then determining the accessibility of those reactive sites to ligands.

**Scanning electron microscopy:** A type of microscopy in which a focused beam of electrons is raster-scanned across a sample, producing secondary and reflected electrons and photons; the secondary electrons are the signal most usually used. Spatial resolution can be <1 nm.

**Scanning probe microscope:** A type of microscope that forms images of a surface by scanning with a physical probe. Atomic force microscopy and scanning tunneling microscopy are the most common, but there are many others.

**SDS-PAGE (sodium dodecyl sulfate–polyacrylamide gel electrophoresis):** A type of gel electrophoresis used to separate proteins (and sometimes very small DNA fragments) based upon pores in gels of polyacrylamide at different concentrations.

**Secondary amine:** A derivative of ammonia in which two of the hydrogen atoms in the ammonia molecule have been replaced (chemical formula $R_2NH$). The –NH group is much less reactive than a primary amine.

**Secondary structure:** Regular local folding pattern of a polymeric molecule.

**Secretory pathway:** Method by which proteins are exported from the cell.

**Seeding:** A technique to initiate crystallization from saturated or supersaturated solutions.

**Selectable marker:** Gene included in a DNA construct to impart a trait into target cells that allows for artificial selection. This makes it possible to select cells according to whether or not they contain the construct. Selectable markers are most often antibiotic-resistance genes. (In contrast, a *screenable marker* can be seen but not selected—for example, GFP.)

**Selective plate:** Dish containing nutrient agar as well as the antibiotic whose resistance gene is expressed in the culture. Foreign bacteria (which do not contain the resistance gene) are unable to proliferate.

**Selectivity:** Specificity toward the type of ion allowed through an ion channel.

**Self-inactivating lentiviral vector:** A lentiviral vector with a deletion in the 3′ long terminal repeat (LTR) that inactivates any promoter activity of the lentivirus itself after genome integration.

**Senescence:** Phenomenon observed in primary cell cultures in which cell proliferation slows down and finally halts irreversibly.

**Sequencer:** Laboratory apparatus that automatically determines the sequence (see **sequencing**) of a DNA sample.

**Sequencing gel:** Gel used to analyze a sample of DNA and determine the order of nucleotide bases in the sample.

**Sequencing primer:** Oligonucleotide used for DNA sequencing. Must anneal to the target DNA segment in a unique and predictable location.

**Sequencing:** DNA sequencing refers to the elucidation of the order of nucleotide bases in a segment of DNA.

**Series resistance:** In electrophysiology, the access resistance to the cell caused by the resistance of the recording pipette and the hole in the membrane.

**Serum:** Blood plasma with clotting factors removed. Contains electrolytes, antibodies, hormones, and many unknown factors.

**Sharp electrode:** A type of electrophysiological recording in which a sharp micropipette is pushed through the cell membrane so that the interior electrical properties can be recorded.

**Short hairpin RNA:** An artificial RNA molecule with a sharp hairpin turn. This type of RNA can be expressed on a plasmid and so is useful for RNA interference experiments.

**Shuttle vector:** A vector constructed in such a way that it can replicate in at least two different host species. DNA inserted in a shuttle vector can be manipulated and investigated in at least two different cell types.

**Side chain:** Branch of an amino acid that determines its identity.

**Silent mutation:** A mutation that has no effect on the amino acid composition of the transcript (i.e., it changes a codon for alanine into a different codon that also encodes alanine).

**Simulating annealing:** A probabilistic method for finding the global minimum of a cost function that may possess several local minima.

**Single guide RNA:** A single artificial RNA sequence that plays the roles of both the trans-activating CRISPR RNA (tracrRNA) and the CRISPR RNA (crRNA) in a CRISPR experiment.

**Single-strand binding protein:** Protein that binds to the single strands of opened-up DNA during transcription, to prevent it from rewinding or self-binding.

**Site-directed mutagenesis:** The alteration of a specific site on a cloned DNA segment.

**Sitting drop:** Type of vapor diffusion technique of protein crystallization in which each well contains a post where the drop is placed, which is elevated above the bottom of the reservoir.

**Small interfering RNA (siRNA):** A class of short (20–25 base pairs) double-stranded RNA molecules that bind specific mRNAs and degrade them, thus inhibiting transcription of the target gene.

**Solid cancers:** Cancers that form tumors—as opposed to blood cancers, which are circulating individual cells.

**Southern blot:** The transfer of electrophoretically separated fragments of DNA from a gel to an absorbent sheet, which is then immersed in a solution containing a labeled probe that can bind to a fragment of interest.

**Space clamp:** Describes the differences in spatial distribution of voltage control in a complex cell given that a voltage-clamp electrode is a point source and that there is finite resistance between areas of the cell.

**Sparse matrix screen:** Testing a large combination of conditions (varying: pH, buffer, additive, and precipitant) in an initial effort to crystallize a protein.

**Spectroscopy:** The measurement or analysis of the interaction between radiation (electromagnetic, particle) and matter.

**Spermidine:** A polymer formed from putrescine found in almost all tissues in association with nucleic acids. It is though to help stabilize some membranes and nucleic acid structures.

**Spherical aberration:** Effect that causes the outer rays exiting from the curved lens surface to converge to a point on a different plane from the rays passing through the lens center, limiting resolution.

**Spheroplast:** Cell (plant, bacterial, fungal) in which the cell wall has been almost completely removed. Often used to refer to Gram-negative bacteria, which retain two membranes after cell wall digestion.

**Spin-coater:** A machine for spin-coating resist onto wafers. Should feature sensitive, reproducible speeds and resistance to chemicals (usually a reservoir made of Teflon).

**Spin polarization:** Expectation value of the spin operator. It demonstrates the average spin direction of the ensemble.

**Spin probe:** A stable radical that can be conjugated to a specific part of a biomolecule in order to measure motility.

**Spin relaxation:** Return of nuclear spins to equilibrium after a pulse.

**Spin trap:** A compound that either converts an unstable radical into a stable, EPR-active species or causes an unstable radical to combine with an EPR-active species to yield an EPR-silent product. Often used in biology to determine mechanisms of radical formation.

**Spin-lattice (longitudinal) relaxation:** The return of the longitudinal magnetization to its equilibrium value along the $z$-axis.

**Spin–orbit coupling:** The interaction of the electron spin magnetic moment with the magnetic moment due to the orbital motion of the electron.

**Spin–spin (transverse) relaxation:** The return of the transverse magnetization to its equilibrium value.

**Sputter:** Sputtering is a process where a beam of ions is used to bombard a target and cause it to eject particles; the kinetic energy of the incoming particles must be much greater than thermal energies.

**Squamous cell carcinoma (SCC):** A cancer of squamous cells, which are a particular type of epithelial cell found in the skin, mouth, and other areas. SCC is one of the most common types of skin cancer.

**Stab:** The process by which a bacterial culture is made by piercing a solid medium with an inoculating needle.

**Stain:** Material used to highlight certain biological structures and improve microscopic contrast.

**Static quenching:** Quenching that occurs due to the formation of a nonfluorescent complex or conjugate between a fluorophore and a quencher.

**Stem cell:** Undifferentiated cell that can continuously divide indefinitely and differentiate into specialized cell types.

**Sterile technique:** Procedures whereby cell cultures are manipulated without contaminating the cultures or release into surrounding environment.

**Stern–Volmer constant:** The product of the biomolecular quenching constant and the fluorophore lifetime.

**Sticky end:** The end of a molecule of DNA that has been cut by an enzyme so that one strand of the double helix is slightly longer than its complementary strand. This aids in ligation reactions.

**Stokes shift:** The difference (usually in frequency units) between the spectral positions of the band maxima of the absorption and luminescence arising from the same electronic transition.

**STORM (stochastic optical reconstruction microscopy):** A type of superresolution fluorescence microscopy that uses photoactivatable dyes.

**Streptavidin:** Tetrameric protein that binds with high affinity to biotin. Similar to avidin.

**Strong confinement regime:** Occurs when a nanocrystal is smaller than the excitonic Bohr radius. The confinement energy dominates.

**Structural model:** Classical model of organic chemistry based upon the theory of valence.

**Sulfhydryl (thiol, mercaptan):** Organic compound containing a functional group consisting of a sulfur–hydrogen bond (chemical formula RSH).

**Supercoiling, supercoiled:** Occurs when a DNA molecule relieves the helical stress by twisting around itself. Expressed as the sum of the number of helical turns in the DNA (twist) and the number of times the double helix crosses over itself (writhe).

**Supersaturated:** Solution in which the dissolved material is present in higher amounts than could normally be dissolved in passive conditions. Slight disturbance causes crystallization.

**Surface-enhanced Raman (SER):** Spectroscopy making use of the fact that when a free molecule in solution is adjacent to a metal surface, either by physisorption or the formation of chemical bonds, the apparent Raman signal can be enhanced many orders of magnitude.

**Surfactant:** Substance that can lower the surface tension of a liquid. Aids in spreading and dispersion.

**Surround wave (S-wave):** Undiffracted light that has not interacted with the specimen.

**Tags (purification and immunohistochemistry):** Small sequences placed on a protein so that when the protein is expressed, it will be marked with the tag. Purification tags aid in specific binding to purification columns; immunohistochemistry tags are recognized by antibodies. Many tags are both.

**Telomeres:** Repetitive, noncoding regions of DNA on the ends of eukaryotic chromosomes that protect the rest of the DNA from degradation during replication.

**Template stripping:** A low-cost, manual microfabrication technique where a negative image of the desired pattern is formed on a template.

**Terminally differentiated:** Specialized cells that are unable to divide or differentiate further.

**Terminator:** Signal in DNA that halts transcription.

**Tertiary amine:** A derivative of ammonia in which all three hydrogen atoms in the ammonia molecule have been replaced (chemical formula $R_3N$). These groups are not reactive enough for typical coupling reactions.

**Tertiary structure:** The folded, three-dimensional structure of proteins, largely determined by the amino acids of which it is composed.

**Tetravalent:** (1) Atom having four valence electrons available to form bonds with other atoms. (2) Protein having four ligand-binding sites (e.g., avidin).

**Thermal cycler:** Laboratory apparatus used in PCR experiments to raise and lower the temperature of DNA in discrete, preprogrammed steps.

**Thermal evaporation:** A simple, low-cost method of thin film deposition in which the film material is heated up in a chamber and allowed to condense onto the substrate.

**Thin film:** A layer of material that may be a single monolayer (nanometers) in thickness up to a few micrometers.

**Thioester:** Covalent bond joining a carboxylic acid and a thiol.

**Time-correlated single-photon counting:** Method of measuring fluorescence lifetimes in the time domain. Pulsed excitation of the sample is monitored by a fast photodiode, resulting in the production of a single photon. Summing over tens of thousands of pulses yields a probability histogram of counts versus time.

**Time-gated imaging:** In fluorescence microscopy, the use of a gated detector synchronized to the excitation source to permit collection of fluorescence emission only after a specified delay time. Allows separation of short-lived from long-lived fluorophores.

**Tissue culture treated:** Modified glass or plastic surface that becomes hydrophilic and negatively charged when covered with medium to allow for more effective cell adhesion.

**Topoisomerase:** Enzyme that binds to DNA and reversibly breaks a phosphodiester bond in one or both strands. This relaxes supercoiling.

**Total RNA:** RNA isolated from cells or tissue (in contrast to purifying purely mRNA).

**Trafficking:** Delivery of proteins to their specific sites.

*trans* **face:** Face of a Golgi stack where substances exit in the form of smaller detached vesicles.

**Transcription, transcribe:** First stage of protein synthesis in which genetic information is copied from DNA to RNA.

**Transcription bubble:** A small area (12–17 base pairs) of melted DNA within which transcription occurs.

**Transcription factor:** A protein that binds to specific DNA sequences controlling the transcription of genetic information from DNA to mRNA.

**Transcriptome:** Collection of all the transcripts (RNA molecules) in a given cell.

**Transduction, transduce:** The movement of genes from a bacterial donor to a bacterial recipient with a phage as the vector.

**Transfection, transfect:** Introduction of foreign DNA or RNA into a complex eukaryotic cell.

**Transfer RNA (tRNA):** Set of small RNA molecules used in protein synthesis as an interface between mRNA and amino acids. Each type of tRNA molecule is covalently linked to a particular amino acid.

**Transferrin:** An iron-binding glycoprotein found in all vertebrates.

**Transformation:** The insertion of foreign DNA, usually as a plasmid, into a bacterial or yeast cell.

**Transformation efficiency:** A quantitative assessment of the competence of cells defined by the number of gene-expressing colonies obtained per microgram of DNA.

**Transgenic:** An organism whose genome has been modified by externally applied new DNA.

**Transient transfection:** The introduction of foreign DNA into cells in which the gene is only expressed in the cell to which it was originally inserted and only for a brief period of time.

**Transition:** (1) A type of nucleotide pair substitution in which a purine replaces another purine or in which a pyrimidine replaces another pyrimidine. (2) Movement of electrons to a higher or lower energy state requiring the absorbance or emission of a precise amount of energy.

**Translation, translate:** Assembly of amino acids to form proteins according to the genetic instructions contained in RNA.

**Transmembrane domain:** Membrane-spanning region of a protein.

**Transmitted light:** Light that passes through a microscopic specimen and its mounting medium.

**Transversion:** A type of nucleotide pair substitution in which a pyrimidine replaces a purine or vice versa.

**Trioctylphosphine oxide (TOPO):** Chemical compound commonly used as a solvent for extraction and separation processes.

**Tropism:** Of a virus, specificity for a particular host or tissue.

**Trypsin:** Serine protease produced in the pancreas, which hydrolyzes proteins. Trypsins are used to resuspend cells adherent to the cell culture dish wall during cell passaging or harvesting.

**Tube length:** In microscopy, the length from the specimen to the intermediate image plane.

**Tumor necrosis factor (TNF):** A group of cytokines that induce apoptosis. There is a whole family of TNFs; TNF-$\alpha$ is the most well known.

**Type strain:** One strain of a species or cell type is designated as the type strain, often one of the first strains studied and usually more fully characterized than other strains.

**Unilamellar vesicles (small, large, giant):** Vesicles with one outer wall rather than several. Small ones are usually hundreds of nanometers in size, large up to 1 micron, and giant several microns. Also called SUVs, LUVs, GUVs.

**Unnatural amino acid:** Synthetic variation of an amino acid (caged, fluorescently tagged, etc.).

**Unsaturated:** Fatty acid containing at least one double or triple bond between carbon atoms.

**Valence:** The number of electrons surrounding a given atom; defines the number of bonds the atom can form.

**Valinomycin:** Dodecadepsipeptide from *Streptomyces* species that forms a potassium-selective channel in membranes.

**Van der Waals interactions:** A general term for intermolecular interactions between uncharged molecules. Includes London dispersion forces and dipole–dipole interactions.

**Vapor diffusion:** Type of protein crystallization in which a drop composed of a sample and reagent is placed in vapor equilibration with a liquid reservoir of reagent. To achieve equilibrium, water vapor leaves the drop and moves into the reservoir, and the sample undergoes an increase in relative supersaturation.

**Vector:** A biological entity used to transfer genetic material into a cell.

**Viral vector:** A virus that has been modified in a laboratory environment for the purpose of introducing genetic material into a cell.

**Voltage clamp:** Type of electrophysiological recording in which an amplifier is used to set the cell membrane potential to a particular value and the resulting currents are measured.

**Voltage-gated:** Type of ion channel that opens in response to a particular transmembrane potential.

**Wafer:** A thin slice of a semiconductor, often crystalline silicon, used in microfabrication. The wafer serves as a substrate for lithographic patterning. Typically 6 inches in diameter, though diameters from 1 to 17.7 inches are available.

**Western blotting, Western blot:** A protein immunoblot for detecting specific proteins after SDS-PAGE.

**Wet etch:** A liquid chemical that selectively removes layers from a wafer for microfabrication. May be isotropic or anisotropic.

**Whole cell (patch):** Type of non-single-channel patch clamp recording in which the cell membrane is punctured to create a hole in the membrane that is then encircled by the pipette orifice.

**Wright's stain:** A colorimetric stain used to distinguish blood cell types and stain DNA.

**Xenograft:** A tissue graft from one species to another. For example, growing a human tumor in a mouse is a xenograft.

***Xenopus* oocytes:** Eggs of *Xenopus laevis*, the African clawed frog. They have many applications in biophysics, particularly for electrophysiology.

**X-gal:** 5-bromo-4-chloro-3-indolyl-$\beta$-galactopyranoside, the most commonly used substrate for $\beta$-galactosidase. Yields a blue product.

**X-ray diffraction:** A method of determining crystal structure by bombarding a crystal with monochromatic x-rays with wavelength similar to the spacing of the atoms in the crystal (~0.1 nm or 1 Å) and studying the diffraction pattern that results from elastic scattering. Used for any crystalline material, including protein crystals and semiconductors.

**X-ray photoelectron spectroscopy:** A form of spectroscopy that measures the number and energy of electrons emitted from a thin film under x-ray irradiation. Probes only to ~10 nm deep.

**Zeta potential:** The electrokinetic potential of a colloidal dispersion; a reasonably accurate measure of the charge on nanoparticles.

**Zinc finger nuclease:** Artificial restriction enzymes designed to cut DNA at a specific sequence. Commonly used prior to the advent of CRISPR.

# Appendix A: Common Solutions

**Ammonium acetate, 10 M**

Dissolve 385.4 g ammonium acetate in 150 mL $H_2O$

Adjust volume to 500 mL with $H_2O$

Filter-sterilize

**BSA (bovine serum albumin), 10% (100 mg/mL)**

Dissolve 10 g BSA in 100 mL $H_2O$

Filter-sterilize with low-protein-binding 0.22 μm filter and store at 4°C

**DTT (dithiothreitol), 1 M**

Dissolve 15.45 g DTT in 100 mL $H_2O$

Filter-sterilize, do not autoclave, and store at –20°C

**EDTA, 0.5 M (pH 8.0)**

Add 186.1 g ethylenediamine tetraacetic acid ($Na_2EDTA \cdot 2H_2O$) to 700 mL $H_2O$

Stir while adding 50 mL of 10 M NaOH

Adjust volume to 1 L with $H_2O$

*EDTA will not dissolve readily at pH < 7. Heating may also assist in dissolving.*

**HCl, 1 M**

Add 86.2 mL of concentrated HCl to 913.8 mL $H_2O$

Do not sterilize

**HBS (HEPES-buffered saline), 2×**

Dissolve the following in 90 mL $H_2O$:

  1.6 g NaCl (0.27 M)

  74.6 mg KCl (10 mM)

  21.3 mg sodium phosphate dibasic ($Na_2HPO_4$; 1.5 mM)

  0.18 g glucose (10 mM)

  1.07 g HEPES (45 mM)

Adjust pH to desired volume with 0.5 N NaOH

Adjust volume to 100 mL with $H_2O$

Filter-sterilize and store in aliquots at $-20°C$

**KCl, 1 M**

Dissolve 74.6 g KCl in 1 L $H_2O$

**β-ME (β-mercaptoethanol), 50 mM**

For 1 M stock: add 0.5 mL of 14.3 M β-ME to 6.6 mL $H_2O$

For 50 mM working solution: dilute 5 mL of 1 M β-ME in 95 mL $H_2O$

Do not autoclave; store in dark or in foil-wrapped container at 4°C

**$MgCl_2$, 1 M**

Dissolve 20.3 g $MgCl_2 \cdot 6H_2O$ in 100 mL $H_2O$

**$MgSO_4$, 1 M**

Dissolve 24.6 g $MgSO_4 \cdot 7H_2O$ in 100 mL $H_2O$

**NaCl, 5 M**

Dissolve 292 g NaCl in 1 L $H_2O$

**NaOH, 10 M**

Dissolve 400 g NaOH in 450 mL $H_2O$

Adjust volume to 1 L with $H_2O$

Do not sterilize

**PBS (phosphate-buffered saline), pH ~7.3**

Dissolve the following in 800 mL $H_2O$:

8.0 g NaCl (137 mM)

0.2 g KCl (2.7 mM)

2.16 g sodium phosphate dibasic heptahydrate ($Na_2HPO_4 \cdot 7H_2O$; 8 mM)

0.2 g potassium phosphate monobasic ($KH_2PO_4$; 1.5 mM)

Adjust volume to 1 L with $H_2O$

Filter-sterilize and store at 4°C

**Potassium acetate buffer, 0.1 M**

Solution A: 11.55 mL glacial acetic acid/L $H_2O$ (0.2 M)

Solution B: 19.6 g potassium acetate ($KC_2H_3O_2$)/L $H_2O$ (0.2 M)

Refer to **Table A1.1** for the desired pH: combine the indicated volumes of
solutions A and B, and then adjust volume to 100 mL with $H_2O$

Table A1.1

Recipes for Potassium and Sodium Acetate Buffers at Various pH Values

| Desired pH | Solution A (mL) | Solution B (mL) |
| --- | --- | --- |
| 3.6 | 46.3 | 3.7 |
| 3.8 | 44.0 | 6.0 |
| 4.0 | 41.0 | 9.0 |
| 4.2 | 36.8 | 13.2 |
| 4.4 | 30.5 | 19.5 |
| 4.6 | 25.5 | 24.5 |
| 4.8 | 20.0 | 30.0 |
| 5.0 | 14.8 | 35.2 |
| 5.2 | 10.5 | 39.5 |
| 5.4 | 8.8 | 41.2 |
| 5.6 | 4.8 | 45.2 |

Filter-sterilize if needed, and store up to 3 months at room temperature

*For intermediate pH values, prepare at the closest higher pH and then titrate with solution A*

## Potassium phosphate buffer, 0.1 M

Solution A: 27.2 g potassium phosphate monobasic ($KH_2PO_4$)/L $H_2O$ (0.2 M)

Solution B: 34.8 g potassium phosphate dibasic ($K_2HPO_4$)/L $H_2O$ (0.2 M)

Refer to Table A1.2 for the desired pH: combine the indicated volumes of solutions A and B, and then adjust volume to 200 mL with $H_2O$

Filter-sterilize if needed, and store up to 3 months at room temperature

*For intermediate pH values, prepare at the closest higher pH and then titrate with solution A*

Table A1.2

Recipes for Potassium and Sodium Phosphate Buffers at Various pH Values

| Desired pH | Solution A (mL) | Solution B (mL) |
|---|---|---|
| 5.7 | 93.5 | 6.5 |
| 5.8 | 92.0 | 8.0 |
| 5.9 | 90.0 | 10.0 |
| 6.0 | 87.7 | 12.3 |
| 6.1 | 85.0 | 15.0 |
| 6.2 | 81.5 | 18.5 |
| 6.3 | 77.5 | 22.5 |
| 6.4 | 73.5 | 26.5 |
| 6.5 | 68.5 | 31.5 |
| 6.6 | 62.5 | 37.5 |
| 6.7 | 56.5 | 43.5 |
| 6.8 | 51.0 | 49.0 |
| 6.9 | 45.0 | 55.0 |
| 7.0 | 39.0 | 61.0 |
| 7.1 | 33.0 | 67.0 |
| 7.2 | 28.0 | 72.0 |
| 7.3 | 23.0 | 77.0 |
| 7.4 | 19.0 | 81.0 |
| 7.5 | 16.0 | 84.0 |
| 7.6 | 13.0 | 87.0 |
| 7.7 | 10.5 | 90.5 |
| 7.8 | 8.5 | 91.5 |
| 7.9 | 7.0 | 93.0 |
| 8.0 | 5.3 | 94.7 |

**SDS (sodium dodecyl sulfate or sodium lauryl sulfate), 20%**

Dissolve 20 g SDS in 100 mL $H_2O$ with stirring (heat if necessary)

Filter-sterilize with 0.45 µm filter

**Sodium acetate, 3 M**

Dissolve 408 g sodium acetate trihydrate ($NaC_2H_3O_2 \cdot 3H_2O$) in 800 mL $H_2O$

Adjust pH to 5.2 with 3 M acetic acid

Adjust volume to 1 L with $H_2O$

Filter-sterilize

**Sodium acetate buffer, 0.1 M**

Solution A: 11.55 mL glacial acetic acid/L $H_2O$ (0.2 M)

Solution B: 27.2 g sodium acetate trihydrate ($NaC_2H_3O_2 \cdot 3H_2O$) /L $H_2O$ (0.2 M)

Refer to Table A1.1 for the desired pH: combine the indicated volumes of solutions A and B, and then adjust volume to 100 mL with $H_2O$

Filter-sterilize if needed, and store up to 3 months at room temperature

*For intermediate pH values, prepare at the closest higher pH then titrate with solution A*

**Sodium phosphate buffer, 0.1 M**

Solution A: 27.6 g sodium phosphate monobasic monohydrate ($NaH_2PO_4 \cdot H_2O$)/L $H_2O$ (0.2 M)

Solution B: 53.65 g sodium phosphate dibasic heptahydrate ($Na_2HPO_4 \cdot 7H_2O$)/L $H_2O$ (0.2 M)

Refer to Table A1.2 for the desired pH: combine the indicated volumes of solutions A and B, and then adjust volume to 200 mL with $H_2O$

Filter-sterilize if needed, and store up to 3 months at room temperature

*For intermediate pH values, prepare at the closest higher pH then titrate with solution A*

**TAE (Tris-acetate EDTA) buffer, 50×**

Dissolve the following in 900 mL $H_2O$:

242 g Tris base

>    57.1 mL glacial acetic acid
>
>    37.2 g $Na_2EDTA \cdot 2H_2O$ (2 mM)
>
>    Adjust volume to 1 L with $H_2O$

### TBS (Tris-buffered saline)

>    Dissolve the following in 800 mL $H_2O$:
>
>    8.0 g NaCl (137 mM)
>
>    0.2 g KCl (2.7 mM)
>
>    3.0 g Tris base (24.8 mM)
>
>    Adjust pH as desired (typically to pH 8) with 1 M HCl
>
>    Adjust volume to 1 L with $H_2O$

### TE (Tris-EDTA) buffer

>    Combine:
>
>    10 mL of 1 M Tris $\cdot$ Cl (10 mM)
>
>    2 mL 0.5 M EDTA, pH 8 (1 mM)
>
>    Adjust volume to 1 L with $H_2O$

### Tris $\cdot$ Cl, 1 M

>    Dissolve 121 g Tris base (tris(hydroxymethyl)aminomethane) in 800 mL $H_2O$
>
>    Adjust to desired pH with concentrated HCl
>
>    Adjust volume to 1 L with $H_2O$
>
>    Store up to 6 months at 4°C or room temperature

# Appendix B: Common Media



*NOTE: All ingredients should be bacteriological grade, and media should be autoclaved prior to use; all $H_2O$ should be distilled.*

**2× LB (Lysogeny Broth)**

Dissolve the following in 800 mL $H_2O$:

10 g tryptone

5 g yeast extract

10 g NaCl

Adjust to desired pH with NaOH

Adjust volume to 1 L with $H_2O$

*Low-salt variant: Use 1/2 (5 g) NaCl*

**M9 minimal media**

First, prepare 5× M9 salts solution:

Dissolve the following in 800 mL $H_2O$:

64 g sodium phosphate dibasic heptahydrate ($Na_2HPO_4 \cdot 7H_2O$)

15 g potassium phosphate monobasic ($KH_2PO_4$)

2.5 g NaCl

5.0 g ammonium chloride ($NH_4Cl$)

Adjust volume to 1 L with $H_2O$

Autoclave or filter-sterilize

Next, add/dissolve in 700 mL $H_2O$:

200 mL of 5× M9 salts solution

2 mL of sterile 1 M $MgSO_4$

20 mL of sterile filtered 20% (w/w) glucose

100 μL of sterile 1 M $CaCl_2$

Adjust volume to 1 L with $H_2O$

## M17

Dissolve the following in 800 mL $H_2O$:

5 g neopeptone

5 g soytone

5 g beef extract

2.5 g yeast extract

0.5 ascorbic acid

1 g 2-glycerolphosphate pentahydrate, disodium salt

Adjust volume to 960 mL with $H_2O$

Autoclave to sterilize, and let cool to room temperature

Add 40 mL of 25% glucose

Add 1 mL of 1 M $MgSO_4$

## Nutrient broth

Dissolve the following in 800 mL $H_2O$:

5 g peptone

3 g meat extract

Adjust pH to 7.0

Adjust volume to 1 L with $H_2O$

## SOC

Dissolve the following in 800 mL $H_2O$:

2% vegetable peptone (or tryptone)

0.5% yeast extract

10 mM NaCl

2.5 mM KCl

10 mM $MgCl_2$

10 mM MgSO

Adjust volume to 1 L with $H_2O$

*SOB variant: add 20 mM sterile filtered glucose (do not autoclave glucose)*

## Terrific broth

Dissolve the following in 800 mL $H_2O$:

12 g tryptone

24 g yeast extract

4 mL glycerol

Adjust volume to 1 L with $H_2O$

## $2\times$ YT

Dissolve the following in 800 mL $H_2O$:

16 g tryptone

10 g yeast extract

5 g NaCl

Adjust volume to 1 L with $H_2O$

# Appendix C: Restriction Endonucleases

Shown are the enzyme names with their recognition sequence. Cut points are indicated as "–".

*NOTE: Some commonly used enzymes are shown in bold.*

AatII

    5′...GACGT-C...3′

    3′...C-TGCAG...5′

Acc65I

    5′...G-GTACC...3′

    3′...CCATG-G...5′

AccI

    5′...GT-MKAC...3′

    3′...CAKM-TG...5′

AciI

    5′...C-CGC...3′

    3′...GGC-G...5′

AclI

    5′...AA-CGTT...3′

    3′...TTGC-AA...5′

AcuI

    5′...CTGAAG(N)$_{16}$-...3′

    3′...GACTTC(N)$_{14}$-...5′

AfeI

    5′...AGC-GCT...3′

    3′...TCG-CGA...5′

AflII

    5′...C-TTAAG...3′

    3′...GAATT-C...5′

AflIII

    5′...A-CRYGT...3′

    3′...TGYRC-A...5′

AgeI

    5′...A-CCGGT...3′

    3′...TGGCC-A...5′

AhdI

    5′...GACNNN-NNGTC...3′

    3′...CTGNN-NNNCAG...5′

AleI

    5′...CACNN-NNGTG...3′

    3′...GTGNN-NNCAC...5′

AluI

    5′...AG-CT...3′

    3′...TC-GA...5′

AlwI

$5'...GGATC(N)_4-...3'$

$3'...CCTAG(N)_5-...5'$

AlwNI

$5'...CAGNNN-CTG...3'$

$3'...GTC-NNNGAC...5'$

ApaI

$5'...GGGCC-C...3'$

$3'...C-CCGGG...5'$

ApaLI

$5'...G-TGCAC...3'$

$3'...CACGT-G...5'$

ApeKI

$5'...G-CWGC...3'$

$3'...CGWC-G...5'$

ApoI

$5'...R-AATTY...3'$

$3'...YTTAA-R...5'$

AscI

$5'...GG-CGCGCC...3'$

$3'...CCGCGC-GG...5'$

AseI

$5'...AT-TAAT...3'$

$3'...TAAT-TA...5'$

AsiSI

$5'...GCGAT-CGC...3'$

$3'...CGC-TAGCG...5'$

AvaI

    5′...C-YCGRG...3′

    3′...GRGCY-C...5′

AvaII

    5′...G-GWCC...3′

    3′...CCWG-G...5′

AvrII

    5′...C-CTAGG...3′

    3′...GGATC-C...5′

BaeGI

    5′...GKGCM-C...3′

    3′...C-MCGKG...5′

BaeI

    5′...-$_{10}$(N)AC(N)$_4$GTAYC(N)$_{12}$-...3′

    3′...-$_{15}$(N)TG(N)$_4$CATRG(N)$_7$-...5′

**BamHI**

    5′...G-GATCC...3′

    3′...CCTAG-G...5′

BanI

    5′...G-GYRCC...3′

    3′...CCRYG-G...5′

BanII

    5′...GRGCY-C...3′

    3′...C-YCGRG...5′

BbsI

    5′...GAAGAC(N)$_2$-...3′

    3′...CTTCTG(N)$_6$-...5′

BbvCI

5′...CC-TCAGC...3′

3′...GGAGT-CG...5′

BbvI

5′...GCAGC(N)$_8$-...3′

3′...CGTCG(N)$_{12}$-...5′

BccI

5′...CCATC(N)$_4$-...3′

3′...GGTAG(N)$_5$-...5′

BceAI

5′...ACGGC(N)$_{12}$-...3′

3′...TGCCG(N)$_{14}$-...5′

BcgI

5′...-$_{10}$(N)CGA(N)$_6$TGC(N)$_{12}$-...3′

3′...-$_{12}$(N)GCT(N)$_6$ACG(N)$_{10}$-...5′

BciVI

5′...GTATCC(N)$_6$-...3′

3′...CATAGG(N)$_5$-...5′

BclI

5′...T-GATCA...3′

3′...ACTAG-T...5′

BfaI

5′...C-TAG...3′

3′...GAT-C...5′

BfuAI

5′...ACCTGC(N)$_4$-...3′

3′...TGGACG(N)$_8$-...5′

BfuCI

5′...-GATC...3′

3′...CTAG-...5′

BglI

5′...GCCNNNN-NGGC...3′

3′...CGGN-NNNNCCG...5′

**BglII**

5′...A-GATCT...3′

3′...TCTAG-A...5′

BlpI

5′...GC-TNAGC...3′

3′...CGANT-CG...5′

BmgBI

5′...CAC-GTC...3′

3′...GTG-CAG...5′

BmrI

5′...ACTGGG(N)$_5$-...3′

3′...TGACCC(N)$_4$-...5′

BmtI

5′...GCTAG-C...3′

3′...C-GATCG...5′

BpmI

5′...CTGGAG(N)$_{16}$-...3′

3′...GACCTC(N)$_{14}$-...5′

Bpu10I

5′...CC-TNAGC...3′

3′...GGANT-CG...5′

**BpuEI**

    5′...CTTGAG(N)$_{16}$-...3′

    3′...GAACTC(N)$_{14}$-...5′

**BsaAI**

    5′...YAC-GTR...3′

    3′...RTG-CAY...5′

**BsaBI**

    5′...GATNN-NNATC...3′

    3′...CTANN-NNTAG...5′

**BsaHI**

    5′...GR-CGYC...3′

    3′...CYGC-RG...5′

**BsaI**

    5′...GGTCTC(N)$_{1}$-...3′

    3′...CCAGAG(N)$_{5}$-...5′

**BsaJI**

    5′...C-CNNGG...3′

    3′...GGNNC-C...5′

**BsaWI**

    5′...W-CCGGW...3′

    3′...WGGCC-W...5′

**BsaXI**

    5′...-$_{9}$(N)AC(N)$_{5}$CTCC(N)$_{10}$-...3′

    3′...-$_{12}$(N)TG(N)$_{5}$GAGG(N)$_{7}$-...5′

**BseRI**

    5′...GAGGAG(N)$_{10}$-...3′

    3′...CTCCTC(N)$_{8}$-...5′

BseYI

    5′...C-CCAGC...3′

    3′...GGGTC-G...5′

BsgI

    5′...GTGCAG(N)$_{16}$-...3′

    3′...CACGTC(N)$_{14}$-...5′

BsiEI

    5′...CGRY-CG...3′

    3′...GC-YRGC...5′

BsiHKAI

    5′...GWGCW-C...3′

    3′...C-WCGWG...5′

BsiWI

    5′...C-GTACG...3′

    3′...GCATG-C...5′

BslI

    5′...CCNNNNN-NNGG...3′

    3′...GGNN-NNNNNCC...5′

BsmAI

    5′...GTCTC(N)$_1$-...3′

    3′...CAGAG(N)$_5$-...5′

BsmBI

    5′...CGTCTC(N)$_1$-...3′

    3′...GCAGAG(N)$_5$-...5′

BsmFI

    5′...GGGAC(N)$_{10}$-...3′

    3′...CCCTG(N)$_{14}$-...5′

**BsmI**

5′...GAATGCN-...3′

3′...CTTAC-GN...5′

**BsoBI**

5′...C-YCGRG...3′

3′...GRGCY-C...5′

**Bsp1286I**

5′...GDGCH-C...3′

3′...C-HCGDG...5′

**BspCNI**

5′...CTCAG(N)$_{10}$-...3′

3′...GAGTC(N)$_8$-...5′

and

5′...CTCAG(N)$_9$-...3′

3′...GAGTC(N)$_7$-...5′

**BspDI**

5′...AT-CGAT...3′

3′...TAGC-TA...5′

**BspEI**

5′...T-CCGGA...3′

3′...AGGCC-T...5′

**BspHI**

5′...T-CATGA...3′

3′...AGTAC-T...5′

**BspMI**

5′...ACCTGC(N)$_4$-...3′

3′...TGGACG(N)$_8$-...5′

BspQI

    5′...GCTCTTC(N)$_1$-...3′

    3′...CGAGAAG(N)$_4$-...5′

BsrBI

    5′...CCG-CTC...3′

    3′...GGC-GAG...5′

BsrDI

    5′...GCAATGNN-...3′

    3′...CGTTAC-NN...5′

BsrFI

    5′...R-CCGGY...3′

    3′...YGGCC-R...5′

**BsrGI**

    5′...T-GTACA...3′

    3′...ACATG-T...5′

BsrI

    5′...ACTGGN-...3′

    3′...TGAC-CN...5′

BssHII

    5′...G-CGCGC...3′

    3′...CGCGC-G...5′

BssKI

    5′...-CCNGG...3′

    3′...GGNCC-...5′

BssSI

    5′...C-ACGAG...3′

    3′...GTGCT-C...5′

BstAPI

5′...GCANNNN-NTGC...3′

3′...CGTN-NNNNACG...5′

BstBI

5′...TT-CGAA...3′

3′...AAGC-TT...5′

BstEII

5′...G-GTNACC...3′

3′...CCANTG-G...5′

BstNI

5′...CC-WGG...3′

3′...GGW-CC...5′

BstUI

5′...CG-CG...3′

3′...GC-GC...5′

**BstXI**

5′...CCANNNNN-NTGG...3′

3′...GGTN-NNNNNACC...5′

BstYI

5′...R-GATCY...3′

3′...YCTAG-R...5′

BstZ17I

5′...GTA-TAC...3′

3′...CAT-ATG...5′

Bsu36I

5′...CC-TNAGG...3′

3′...GGANT-CC...5′

BtgI

5′...C-CRYGG...3′

3′...GGYRC-C...5′

BtgZI

5′...GCGATG(N)$_{10}$-...3′

3′...CGCTAC(N)$_{14}$-...5′

BtsCI

5′...GGATGNN-...3′

3′...CCTAC-NN...5′

BtsI

5′...GCAGTGNN-...3′

3′...CGTCAC-NN...5′

Cac8I

5′...GCN-NGC...3′

3′...CGN-NCG...5′

ClaI

5′...AT-CGAT...3′

3′...TAGC-TA...5′

CspCI

5′...-$_{10-11}$(N)CAA(N)$_5$GTGG(N)$_{12-13}$-...3′

3′...-$_{12-13}$(N)GTT(N)$_5$CACC(N)$_{10-11}$-...5′

CviAII

5′...C-ATG...3′

3′...GTA-C...5′

CviKI-1

5′...RG-CY...3′

3′...YC-GR...5′

CviQI

5′...G-TAC...3′

3′...CAT-G...5′

DdeI

5′...C-TNAG...3′

3′...GANT-C...5′

DpnI

5′...GA$_M$-TC...3′

3′...CT-A$_M$G...5′

*NOTE: DpnI will cleave fully adenomethylated dam sites and hemi-adenomethylated dam sites more slowly.*

DpnII

5′...-GATC...3′

3′...CTAG-...5′

DraI

5′...TTT-AAA...3′

3′...AAA-TTT...5′

DraIII

5′...CACNNN-GTG...3′

3′...GTG-NNNCAC...5′

DrdI

5′...GACNNNN-NNGTC...3′

3′...CTGNN-NNNNCAG...5′

EaeI

5′...Y-GGCCR...3′

3′...RCCGG-Y...5′

EagI

5′...C-GGCCG...3′

3′...GCCGG-C...5′

EarI

5′...CTCTTC(N)$_1$-...3′

3′...GAGAAG(N)$_4$-...5′

EciI

5′...GGCGGA(N)$_{11}$-...3′

3′...CCGCCT(N)$_9$-...5′

Eco53KI

5′...GAG-CTC...3′

3′...CTC-GAG...5′

EcoNI

5′...CCTNN-NNNAGG...3′

3′...GGANNN-NNTCC...5′

EcoO109I

5′...RG-GNCCY...3′

3′...YCCNG-GR...5′

EcoP15I

5′...CAGCAG(N)$_{25}$-...3′

3′...GTCGTC(N)$_{27}$-...5′

**EcoRI**

5′...G-AATTC...3′

3′...CTTAA-G...5′

**EcoRV**

5′...GAT-ATC...3′

3′...CTA-TAG...5′

FatI

5′...-CATG...3′

3′...GTAC-...5′

FauI

5′...CCCGC(N)$_4$-...3′

3′...GGGCG(N)$_6$-...5′

Fnu4HI

5′...GC-NGC...3′

3′...CGN-CG...5′

FokI

5′...GGATG(N)$_9$-...3′

3′...CCTAC(N)$_{13}$-...5′

FseI

5′...GGCCGG-CC...3′

3′...CC-GGCCGG...5′

FspI

5′...TGC-GCA...3′

3′...ACG-CGT...5′

HaeII

5′...RGCGC-Y...3′

3′...Y-CGCGR...5′

HaeIII

5′...GG-CC...3′

3′...CC-GG...5′

HgaI

5′...GACGC(N)$_5$-...3′

3′...CTGCG(N)$_{10}$-...5′

HhaI

5′...GCG-C...3′

3′...C-GCG...5′

HincII

5′...GTY-RAC...3′

3′...CAR-YTG...5′

**HindIII**

5′...A-AGCTT...3′

3′...TTCGA-A...5′

HinfI

5′...G-ANTC...3′

3′...CTNA-G...5′

HinP1I

5′...G-CGC...3′

3′...CGC-G...5′

HpaI

5′...GTT-AAC...3′

3′...CAA-TTG...5′

HpaII

5′...C-CGG...3′

3′...GGC-C...5′

HphI

5′...GGTGA(N)$_8$-...3′

3′...CCACT(N)$_7$-...5′

Hpy166II

5′...GTN-NAC...3′

3′...CAN-NTG...5′

Hpy188I

    5′...TCN-GA...3′

    3′...AG-NCT...5′

Hpy188III

    5′...TC-NNGA...3′

    3′...AGNN-CT...5′

Hpy99I

    5′...CGWCG-...3′

    3′...-GCWGC...5′

HpyAV

    5′...CCTTC(N)$_6$-...3′

    3′...GGAAG(N)$_5$-...5′

HpyCH4III

    5′...ACN-GT...3′

    3′...TG-NCA...5′

HpyCH4IV

    5′...A-CGT...3′

    3′...TGC-A...5′

HpyCH4V

    5′...TG-CA...3′

    3′...AC-GT...5′

KasI

    5′...G-GCGCC...3′

    3′...CCGCG-G...5′

KpnI

    5′...GGTAC-C...3′

    3′...C-CATGG...5′

MboI

5′...-GATC...3′

3′...CTAG-...5′

MboII

5′...GAAGA(N)$_8$-...3′

3′...CTTCT(N)$_7$-...5′

MfeI

5′...C-AATTG...3′

3′...GTTAA-C...5′

MluI

5′...A-CGCGT...3′

3′...TGCGC-A...5′

MlyI

5′...GAGTC(N)$_5$-...3′

3′...CTCAG(N)$_5$-...5′

MmeI

5′...TCCRAC(N)$_{20}$-...3′

3′...AGGYTG(N)$_{18}$-...5′

MnlI

5′...CCTC(N)$_7$-...3′

3′...GGAG(N)$_6$-...5′

MscI

5′...TGG-CCA...3′

3′...ACC-GGT...5′

MseI

5′...T-TAA...3′

3′...AAT-T...5′

MslI

    5′...CAYNN-NNRTG...3′

    3′...GTRNN-NNYAC...5′

MspA1I

    5′...CMG-CKG...3′

    3′...GKC-GMC...5′

MspI

    5′...C-CGG...3′

    3′...GGC-C...5′

MwoI

    5′...GCNNNNN-NNGC...3′

    3′...CGNN-NNNNNCG...5′

NaeI

    5′...GCC-GGC...3′

    3′...CGG-CCG...5′

NarI

    5′...GG-CGCC...3′

    3′...CCGC-GG...5′

NciI

    5′...CC-SGG...3′

    3′...GGS-CC...5′

**NcoI**

    5′...C-CATGG...3′

    3′...GGTAC-C...5′

NdeI

    5′...CA-TATG...3′

    3′...GTAT-AC...5′

NgoMIV

5′...G-CCGGC...3′

3′...CGGCC-G...5′

**NheI**

5′...G-CTAGC...3′

3′...CGATC-G...5′

NlaIII

5′...CATG-...3′

3′...-GTAC...5′

NlaIV

5′...GGN-NCC...3′

3′...CCN-NGG...5′

NmeAIII

5′...GCCGAG(N)$_{20\text{-}21}$-...3′

3′...CGGCTC(N)$_{18\text{-}19}$-...5′

**NotI**

5′...GC-GGCCGC...3′

3′...CGCCGG-CG...5′

NruI

5′...TCG-CGA...3′

3′...AGC-GCT...5′

NsiI

5′...ATGCA-T...3′

3′...T-ACGTA...5′

NspI

5′...RCATG-Y...3′

3′...Y-GTACR...5′

PacI

    5′...TTAAT-TAA...3′

    3′...AAT-TAATT...5′

PaeR7I

    5′...C-TCGAG...3′

    3′...GAGCT-C...5′

PciI

    5′...A-CATGT...3′

    3′...TGTAC-A...5′

PflFI

    5′...GACN-NNGTC...3′

    3′...CTGNN-NCAG...5′

PflMI

    5′...CCANNNN-NTGG...3′

    3′...GGTN-NNNNACC...5′

PhoI

    5′...GG-CC...3′

    3′...CC-GG...5′

PleI

    5′...GAGTC(N)$_4$-...3′

    3′...CTCAG(N)$_5$-...5′

PmeI

    5′...GTTT-AAAC...3′

    3′...CAAA-TTTG...5′

PmlI

    5′...CAC-GTG...3′

    3′...GTG-CAC...5′

PpuMI

5′...RG-GWCCY...3′

3′...YCCWG-GR...5′

PshAI

5′...GACNN-NNGTC...3′

3′...CTGNN-NNCAG...5′

PsiI

5′...TTA-TAA...3′

3′...AAT-ATT...5′

PspGI

5′...-CCWGG...3′

3′...GGWWCC-...5′

PspOMI

5′...G-GGCCC...3′

3′...CCCGG-G...5′

PspXI

5′...VC-TCGAGB...3′

3′...BGAGCT-CV...5′

**PstI**

5′...CTGCA-G...3′

3′...G-ACGTC...5′

PvuI

5′...CGAT-CG...3′

3′...GC-TAGC...5′

PvuII

5′...CAG-CTG...3′

3′...GTC-GAC...5′

RsaI

    5′...GT-AC...3′

    3′...CA-TG...5′

RsrII

    5′...CG-GWCCG...3′

    3′...GCCWG-GC...5′

SacI

    5′...GAGCT-C...3′

    3′...C-TCGAG...5′

SacII

    5′...CCGC-GG...3′

    3′...GG-CGCC...5′

SalI

    5′...G-TCGAC...3′

    3′...CAGCT-G...5′

SapI

    5′...GCTCTTC(N)$_1$-...3′

    3′...CGAGAAG(N)$_4$-...5′

Sau3AI

    5′...-GATC...3′

    3′...CTAG-...5′

Sau96I

    5′...G-GNCC...3′

    3′...CCNG-G...5′

SbfI

    5′...CCTGCA-GG...3′

    3′...GG-ACGTCC...5′

ScaI

    5′...AGT-ACT...3′

    3′...TCA-TGA...5′

ScrFI

    5′...CC-NGG...3′

    3′...GGN-CC...5′

SexAI

    5′...A-CCWGGT...3′

    3′...TGGWCC-A...5′

SfaNI

    5′...GCATC(N)$_5$-...3′

    3′...CGTAG(N)$_9$-...5′

SfcI

    5′...C-TRYAG...3′

    3′...GAYRT-C...5′

SfiI

    5′...GGCCNNNN-NGGCC...3′

    3′...CCGGN-NNNNCCGG...5′

SfoI

    5′...GGC-GCC...3′

    3′...CCG-CGG...5′

SgrAI

    5′...CR-CCGGYG...3′

    3′...GYGGCC-RC...5′

**SmaI**

    5′...CCC-GGG...3′

    3′...GGG-CCC...5′

SmlI

    5′...C-TYRAG...3′

    3′...GARYT-C...5′

SnaBI

    5′...TAC-GTA...3′

    3′...ATG-CAT...5′

SpeI

    5′...A-CTAGT...3′

    3′...TGATC-A...5′

SphI

    5′...GCATG-C...3′

    3′...C-GTACG...5′

SspI

    5′...AAT-ATT...3′

    3′...TTA-TAA...5′

StuI

    5′...AGG-CCT...3′

    3′...TCC-GGA...5′

StyD4I

    5′...-CCNGG...3′

    3′...GGNCC-...5′

StyI

    5′...C-CWWGG...3′

    3′...GGWWC-C...5′

SwaI

    5′...ATTT-AAAT...3′

    3′...TAAA-TTTA...5′

Taq$^{\alpha}$I

5′...T-CGA...3′

3′...AGC-T...5′

TfiI

5′...G-AWTC...3′

3′...CTWA-G...5′

TliI

5′...C-TCGAG...3′

3′...GAGCT-C...5′

TseI

5′...G-CWGC...3′

3′...CGWC-G...5′

Tsp45I

5′...-GTSAC...3′

3′...CASTG-...5′

Tsp509I

5′...-AATT...3′

3′...TTAA-...5′

TspMI

5′...C-CCGGG...3′

3′...GGGCC-C...5′

TspRI

5′...NNCASTGNN-...3′

3′...-NNGTSACNN...5′

Tth111I

5′...GACN-NNGTC...3′

3′...CTGNN-NCAG...5′

**XbaI**

5′...T-CTAGA...3′

3′...AGATC-T...5′

XcmI

5′...CCANNNNN-NNNNTGG...3′

3′...GGTNNNN-NNNNNACC...5′

**XhoI**

5′...C-TCGAG...3′

3′...GAGCT-C...5′

XmaI

5′...C-CCGGG...3′

3′...GGGCC-C...5′

XmnI

5′...GAANN-NNTTC...3′

3′...CTTNN-NNAAG...5′

ZraI

5′...GAC-GTC...3′

3′...CTG-CAG...5′

# Appendix D: Common Enzymes

**DNA Ligase (e.g., T4)**—for ligation reactions

**Phosphatases**

> **CIP (calf intestinal phosphatase), SAP (shrimp alkaline phosphatase), Antarctic**—for dephosphorylating nucleic acids (prior to ligation)
>
> **Lambda**—for dephosphorylating proteins

**Polymerase (e.g., *Taq*)**—for polymerization of new DNA/RNA (used in polymerase chain reaction [PCR])

# Appendix E: Fluorescent Dyes and Quenchers

*NOTE: Many dye properties, especially time constants and quantum yields, are unknown or not given by the manufacturer.*

| Dye | Abs. Peak (nm) | Emission Max (nm) | Molar Ext. Coefficient | Quantum Yield | Molecular Weight (g/mol) | Notes |
|---|---|---|---|---|---|---|
| 2′,7′-Dichloro-fluorescein | 495 (pH 4)<br>504 (pH 8) | 529 | 38,000 (pH 4)<br>107,000 (pH 8) | | 445.21 | AX, pH |
| 4′,5′-Dichloro-2′,7′-dimethoxy-fluorescein (JOE) | 522 | 550 | 75,000 | | 602.34 | AX, DNA sequencing |
| 4-Dimethylaminophenylazophenyl | 419 | – | 34,000 | | 320.35 | TX, Q |
| Acridine Orange | 500 (dsDNA)<br>460 (ssDNA/RNA) | 526 (dsDNA)<br>650 (ssDNA/RNA) | 53,000 | | 301.82 | DNA/RNA |
| Alexa Fluor 350 | 346 | 442 | 19,000 | | 410.35 | AX, TX |
| Alexa Fluor 405 | 402 | 421 | 35,000 | | 1028.26 | AX, TX |
| Alexa Fluor 430 | 434 | 539 | 15,000 | | 701.75 | AX |
| Alexa Fluor 488 | 495 | 519 | 73,000 | 0.92 | 643.41 | AX, TX |
| Alexa Fluor 514 | 518 | 540 | 80,000 | | 713.69 | AX |
| Alexa Fluor 532 | 531 | 554 | 81,000 | 0.61 | 723.77 | AX, TX |
| Alexa Fluor 555 | 555 | 565 | 155,000 | 0.10 | ~1250 | AX, TX |
| Alexa Fluor 546 | 556 | 575 | 112,000 | 0.79 | 1079.39 | AX, TX |
| Alexa Fluor 568 | 578 | 603 | 88,000 | 0.69 | 791.80 | AX, TX |
| Alexa Fluor 594 | 590 | 617 | 92,000 | 0.66 | 819.85 | AX, TX |
| Alexa Fluor 610 | 612 | 628 | 144,000 | | 1284.82 | AX |
| Alexa Fluor 633 | 632 | 647 | 159,000 | | ~1200 | AX, TX |
| Alexa Fluor 647 | 650 | 668 | 270,000 | 0.33 | ~1250 | AX, TX |
| Alexa Fluor 660 | 663 | 690 | 132,000 | 0.37 | ~1100 | AX, TX |

*(Continued)*

| Dye | Abs. Peak (nm) | Emission Max (nm) | Molar Ext. Coefficient | Quantum Yield | Molecular Weight (g/mol) | Notes |
|---|---|---|---|---|---|---|
| Alexa Fluor 680 | 679 | 702 | 183,000 | 0.36 | ~1150 | AX, TX |
| Alexa Fluor 700 | 702 | 723 | 205,000 | 0.25 | ~1400 | AX |
| Alexa Fluor 750 | 749 | 775 | 290,000 | 0.12 | ~1300 | AX, TX |
| Alexa Fluor 790 | 782 | 805 | 260,000 | | ~1750 | AX |
| AMCA | 349 | 448 | 19,000 | | 443.46 | AX, pH |
| Aminocoumarin | 350 | 445 | | | 330 | - |
| Anilinonaphthalene | 402 | 421 | 27,000 | | ~416-505 | TX |
| Atto 390 | 390 | 470 | 24,000 | | | AX, TX |
| Atto 425 | 436 | 484 | 45,000 | | | AX, TX |
| Atto 465 | 454 | 505 | 75,000 | | | AX, TX |
| Atto 488 | 504 | 521 | 105,000 | | | AX, TX |
| Atto 520 | 520 | 542 | 110,000 | | | AX, TX |
| Atto 532 | 533 | 560 | 115,000 | | | AX, TX |
| Atto 550 | 554 | 576 | 115,000 | | | AX, TX |
| Atto 565 | 561 | 585 | 120,000 | | | AX, TX |
| Atto 590 | 598 | 634 | 120,000 | | | AX, TX |
| Atto 594 | 601 | 627 | 12,000 | | | AX, TX |
| Atto 610 | 605 | 646 | 110,000 | | | AX, TX |
| Atto 620 | 620 | 641 | 120,000 | | | AX |
| Atto 635 | 635 | 659 | 120,000 | | | AX |
| Atto 647 | 645 | 673 | 120,000 | | | AX, TX |
| Atto 647N | 644 | 667 | 150,000 | | | AX, TX |
| Atto 655 | 665 | 690 | 110,000 | | | AX, TX |
| Atto 680 | 680 | 702 | 110,000 | | | AX, TX |
| Atto 700 | 504 | 714 | 120,000 | | | AX, TX |
| Atto 725 | 725 | 752 | | | | AX |
| Benzophenone | 282 | – | | | 182.2 | TX |
| Bimane | 380 | 458 | | | 282.31 | AX, TX |
| BODIPY FL | ~503 | ~513 | >80,000 | ~1 | 292.09 | AX<br>TX |
| BODIPY R6G | 528 | 550 | 70,000 | | 437.21 | AX |
| BODIPY TMR | 542 | 574 | 60,000 | | 608.45 | AX, TX |
| BODIPY TR | 589 | 617 | 68,000 | | 634.46 | AX, TX |
| BODIPY 493/503 | 500 | 509 | 79,000 | | 417.22 | AX |
| | 493 | 503 | 62,000 | | 341 | TX |
| BODIPY 499/508 | 499 | 508 | 88,000 | | 419.24 | TX |
| BODIPY 507/545 | 508 | 543 | 69,000 | | 431.03 | TX |
| BODIPY 530/550 | 534 | 554 | 77,000 | | 513.31 | AX |
| BODIPY 558/568 | 558 | 569 | 97,000 | | 443.23 | AX |
| BODIPY 564/570 | 565 | 571 | 142,000 | | 463.25 | AX |
| BODIPY 576/589 | 576 | 590 | | | | AX |
| BODIPY 577/618 | 577 | 618 | | | | TX |
| BODIPY 581/591 | 584 | 592 | 136,000 | | 489.28 | AX |

*(Continued)*

| Dye | Abs. Peak (nm) | Emission Max (nm) | Molar Ext. Coefficient | Quantum Yield | Molecular Weight (g/mol) | Notes |
|---|---|---|---|---|---|---|
| BODIPY 630/650 | 625 | 640 | 101,000 | | 660.50 | AX, TX |
| BODIPY 650/665 | 646 | 660 | 102,000 | | 643.45 | AX |
| Cascade Blue dye | 400 | 420 | | | 607.42 | AX |
| Cascade Yellow dye | 402 | 545 | | | 563.54 | AX |
| CF 350 | 347 | 448 | 18,000 | | 496 | AX, TX |
| CF 405S | 404 | 431 | 33,000 | | 1169 | AX, TX |
| CF 405M | 408 | 452 | 41,000 | | 503 | AX, TX |
| CF 485 | ~485 | 513 | | | | |
| CF 488A | 490 | 515 | 70,000 | | ~910 | AX, TX |
| CF 555 | ~555 | | | | | AX, TX |
| CF 568 | 562 | 583 | 100,000 | | 714 | AX, TX |
| CF 594 | 593 | 614 | 115,000 | | ~730 | AX, TX |
| CF 620R | 617 | 639 | 115,000 | | 738 | AX, TX |
| CF 633 | 630 | 650 | 100,000 | | ~820 | AX, TX |
| CF 640R | 642 | 662 | 105,000 | | 832 | AX |
| CF 647 | 650 | 665 | 240,000 | | ~2980 | AX |
| CF 660C | 667 | 685 | 200,000 | | 3112 | AX |
| CF 660R | 663 | 682 | 100,000 | | 888 | AX |
| CF 680 | 681 | 698 | 210,000 | | 3241 | AX, TX |
| CF 680R | 680 | 701 | 140,000 | | 912 | AX, TX |
| CF 750 | 755 | 777 | | | | AX |
| CF 770 | 770 | 797 | | | | AX |
| CF 790 | | | | | | AX |
| Cy2 | 489 | 506 | | | ~714 | |
| Cy3 | 550 | 570 615 | | 0.04 | ~766 | |
| Cy3B | 558 | 572 620 | | 0.67 | ~658 | |
| Cy3.5 | 581 | 594 640 | | 0.15 | ~1102 | |
| Cy5 | 625 649 | 670 | | 0.28 | ~792 | |
| Cy5.5 | 675 | 694 | | 0.23 | ~1128 | |
| Cy7 | 743 | 767 | | 0.28 | ~818 | |
| DAPI | 358 | 461 | | | 277.32 | DNA |
| Dabcyl | 453 | – | 32,000 | | | AX |
| Dansyl | 340 | 520 | | | | AX |
| Dapoxyl dye | 373 | 551 | | | | AX |
| Dialkylaminocoumarin | 375 435 | 470 475 | | | | AX |
| Dibromobimane | 394 | 490 | | | | TX |
| Diethylaminocoumarin | 334 | 470 | | | | TX |
| Dimethylaminocoumarin | 376 | 465 | | | | TX |
| Dimethylaminonaphthalene | 391 | 500 | | | | TX |

*(Continued)*

| Dye | Abs. Peak (nm) | Emission Max (nm) | Molar Ext. Coefficient | Quantum Yield | Molecular Weight (g/mol) | Notes |
|---|---|---|---|---|---|---|
| Dipicrylamine | ~410 | – | | | 439.21 | Q |
| Di-4-ANEPPS | 468 | 635 | | | ~481 | MP |
| Di-8-ANEPPS | 468 | 635 | | | ~593 | MP |
| DiA | 456 | 590 | | | | |
| DiD (DilC$_{18}$(5)) | 644 | 665 | | | | |
| DiI (DilC$_{18}$(3)) | 549 | 565 | | | | |
| DiO (DiOC$_{18}$(3)) | 484 | 501 | | | | |
| DiR (DilC$_{18}$(7)) | 750 | 779 | | | | |
| DyLight 350 | 353 | 432 | 15,000 | | 874.1<br>899.15 | AX<br>TX |
| DyLight 405 | 400 | 420 | 30,000 | | 793<br>818 | AX<br>TX |
| DyLight 488 | 493 | 518 | 70,000 | | 1011<br>800<br>1088.01 | AX<br>TX<br>ZX |
| DyLight 549 | 562 | 576 | 150,000 | | 982<br>1007<br>1273.32 | AX<br>TX<br>ZX |
| DyLight 594 | 593 | 618 | 80,000 | | 1078<br>1059 | AX<br>TX |
| DyLight 633 | 638 | 658 | 170,000 | | 1066<br>1091 | AX<br>TX |
| DyLight 649 | 654 | 673 | 250,000 | | 1008<br>1033<br>1299.36 | AX<br>TX<br>ZX |
| DyLight 680 | 692 | 712 | 140,000 | | 950<br>972 | AX<br>TX |
| DyLight 680B | 679 | 698 | 180,000 | | 1196.16<br>1221.21 | AX<br>TX |
| DyLight 750 | 752 | 778 | 210,000 | | 1034<br>1059 | AX<br>TX |
| DyLight 800 | 777 | 794 | 270,000 | | 1050<br>1075 | AX<br>TX |
| Eosin | 522 | 550 | | | | AX, TX |
| Ethidium bromide | 518 | 605 | | | | |
| Fluorescein | 494 | 518 | 93,000 | | 332.31 | AX, TX |
| Hydroxycoumarin | 385 | 445 | | | 331 | AX |
| Lissamine rhodamine B | 570 | 590 | | | | AX |
| Lucifer Yellow | 425 | 528 | | | | TX |
| Malachite green | 630 | – | 76,000 | | | AX |
| Marina Blue dye | 365 | 460 | | | 367.26 | AX |
| Methoxycoumarin | 360 | 410 | | | 317 | AX |
| Monobromobimane | 394 | 490 | | | | TX |
| Monochlorobimane | 394 | 490 | | | | TX |

*(Continued)*

| Dye | Abs. Peak (nm) | Emission Max (nm) | Molar Ext. Coefficient | Quantum Yield | Molecular Weight (g/mol) | Notes |
|---|---|---|---|---|---|---|
| Napthalene | 336 | 490 | | | | TX |
| Napthofluorescein | 605 | 675 | | | | AX |
| NBD | 465 | 535 | | | | AX, TX |
| Oregon Green 488 | 496 | 524 | | | | AX, TX |
| Oregon Green 514 | 511 | 530 | | | | AX |
| Pacific Blue dye | 410 | 455 | | | 339.21 | AX, TX |
| Pacific Orange dye | 400 | 551 | 23,000 | | ~750 | AX |
| | | | 25,000 | | ~800 | TX |
| Phenanthroline | 270 | – | | | | TX, Q |
| Propidium iodide | 536 | 617 | | | | |
| PyMPO | 415 | 570 | | | 564.39 | AX, TX |
| Pyrene | 345 | 378 | | | 385.42 | AX, TX |
| QSY 35 | 475 | – | 23,000 | | | AX, TX, Q |
| QSY 7 | 560 | – | 90,000 | | | AX, TX, Q |
| QSY 9 | 562 | – | 88,000 | | | AX, TX, Q |
| QSY 21 | 661 | – | 90,000 | | | AX, Q |
| Rhodamine 110 | 496 | 520 | | | | |
| Rhodamine 123 | 507 | 529 | | ~0.9 | 380.82 | |
| Rhodamine 6G | 525 | 555 | | ~0.95 | 479.02 | AX |
| Rhodamine B | 543 | 610 | 106,000 | ~0.5-0.7 | 479.02 | |
| Rhodamine Green dye | 502 | 527 | >75,000 | | ~508-621 | AX |
| Rhodamine Red dye | 570 | 590 | 119,000 | | 680.79 | AX, TX |
| RH 414 | 500 | 635 | | | | |
| Stilbene | 339 | 384 | | | | TX |
| Sulfonerhodamine | 555 | 580 | 88,000 | | 840.47 | TX |
| Tetramethyl-rhodamine (TMR) | 555 | 580 | | | | AX, TX |
| Texas Red dye | 595 | 615 | | | | AX, TX |
| Tracy 645 | 638 | 656 | | | | AX |
| Tracy 652 | 648 | 670 | | | | AX |
| Tracy 690 | 688 | 715 | | | | AX |
| TRITC | 547 | 572 | | | 444 | |
| X-rhodamine | 580 | 605 | | | | AX |

*Notes:*  AX, available in amine-reactive form (carboxylic acid, sulfosuccinimidyl ester, or 4-sulfotetrafluorophenyl ester derivatives); DNA/RNA, reacts with nucleic acids of these types; dsDNA, double-stranded DNA; MP, membrane potential probe; pH, pH-dependent properties; Q, quencher (absorbs but does not emit); ssDNA, single-stranded DNA; TX, available in thiol-reactive form (maleimide, alkyl halide, haloacetamine, bromomethyl, cystine or thiolsulfate derivatives); ZX, phosphine derivatives. The following dyes are patented:
CF dyes (Biotium), BODIPY and Alexa Fluor dyes (Invitrogen), DyLight Fluor (Thermo Scientific, Pierce), Atto and Tracy (Sigma Aldrich).

# Appendix F: Fluorescent Proteins

| Protein | Excitation Max (nm) | Emission Max (nm) | Molar Ext. Coefficient | Quantum Yield | *In Vivo* Structure | Relative Brightness (% of EGFP) |
|---|---|---|---|---|---|---|
| GFP (wt) | 395/475 | 509 | 21,000 | 0.77 | Monomer* | 48 |
| EGFP | 484 | 507 | 56,000 | 0.66 | Monomer* | 100 |
| Emerald | 487 | 509 | 57,500 | 0.68 | Monomer* | 116 |
| Superfolder GFP | 485 | 510 | 83,300 | 0.65 | Monomer* | 160 |
| Azami Green | 492 | 505 | 55,000 | 0.74 | Monomer | 121 |
| mWasabi | 493 | 509 | 70,000 | 0.80 | Monomer | 167 |
| TagGFP | 482 | 505 | 58,200 | 0.59 | Monomer* | 110 |
| TurboGFP | 482 | 502 | 70,000 | 0.53 | Dimer | 102 |
| AcGFP | 480 | 505 | 50,000 | 0.55 | Monomer* | 82 |
| ZsGreen | 493 | 505 | 43,000 | 0.91 | Tetramer | 117 |
| T-Sapphire | 399 | 511 | 44,000 | 0.60 | Monomer* | 79 |
| EBFP | 383 | 445 | 29,000 | 0.31 | Monomer* | 27 |
| EBFP2 | 383 | 448 | 32,000 | 0.56 | Monomer* | 53 |
| Azurite | 384 | 450 | 26,200 | 0.55 | Monomer* | 43 |
| mTagBFP | 399 | 456 | 52,000 | 0.63 | Monomer | 98 |
| ECFP | 439 | 476 | 32,500 | 0.40 | Monomer* | 39 |
| mECFP | 433 | 475 | 32,500 | 0.40 | Monomer | 39 |
| Cerulean | 433 | 475 | 43,000 | 0.62 | Monomer* | 79 |
| CyPet | 435 | 477 | 35,000 | 0.51 | Monomer* | 53 |
| AmCyan1 | 458 | 489 | 44,000 | 0.24 | Tetramer | 31 |
| Midori-ishi Cyan | 472 | 495 | 27,300 | 0.90 | Dimer | 73 |
| TagCFP | 458 | 480 | 37,000 | 0.57 | Monomer | 63 |
| mTFP1 (teal) | 462 | 492 | 64,000 | 0.85 | Monomer | 162 |
| EYFP | 514 | 527 | 83,400 | 0.61 | Monomer* | 151 |
| Topaz | 514 | 527 | 94,500 | 0.60 | Monomer* | 169 |
| Venus | 515 | 528 | 92,200 | 0.57 | Monomer* | 156 |
| mCitrine | 516 | 529 | 77,000 | 0.76 | Monomer | 174 |

(Continued)

| Protein | Excitation Max (nm) | Emission Max (nm) | Molar Ext. Coefficient | Quantum Yield | *In Vivo* Structure | Relative Brightness (% of EGFP) |
|---|---|---|---|---|---|---|
| YPet | 517 | 530 | 104,000 | 0.77 | Monomer* | 238 |
| TagYFP | 508 | 524 | 64,000 | 0.60 | Monomer | 118 |
| PhiYFP | 525 | 537 | 124,000 | 0.39 | Monomer* | 144 |
| ZsYellow1 | 529 | 539 | 20,200 | 0.42 | Tetramer | 25 |
| mBanana | 540 | 553 | 6,000 | 0.70 | Monomer | 13 |
| Kusabira Orange | 548 | 559 | 51,600 | 0.60 | Monomer | 92 |
| Kusabira Orange2 | 551 | 565 | 63,800 | 0.62 | Monomer | 118 |
| mOrange | 548 | 562 | 71,000 | 0.69 | Monomer | 146 |
| mOrange2 | 549 | 565 | 58,000 | 0.60 | Monomer | 104 |
| dTomato | 554 | 581 | 69,000 | 0.69 | Dimer | 142 |
| dTomato-Tandem | 554 | 581 | 138,000 | 0.69 | Monomer | 283 |
| TagRFP | 555 | 584 | 100,000 | 0.48 | Monomer | 142 |
| TagRFP-T | 555 | 584 | 81,000 | 0.41 | Monomer | 99 |
| DsRed | 558 | 583 | 75,000 | 0.79 | Tetramer | 176 |
| DsRed2 | 563 | 582 | 43,800 | 0.55 | Tetramer | 72 |
| DsRed Express (T1) | 555 | 584 | 38,000 | 0.51 | Tetramer | 58 |
| DsRed-monomer | 556 | 586 | 35,000 | 0.10 | Monomer | 10 |
| mTangerine | 568 | 585 | 38,000 | 0.30 | Monomer | 34 |
| mRuby | 558 | 605 | 112,000 | 0.35 | Monomer | 117 |
| mApple | 568 | 592 | 75,000 | 0.49 | Monomer | 109 |
| mStrawberry | 574 | 596 | 90,000 | 0.29 | Monomer | 78 |
| AsRed2 | 576 | 592 | 56,200 | 0.05 | Tetramer | 8 |
| mRFP1 | 584 | 607 | 50,000 | 0.25 | Monomer | 37 |
| JRed | 584 | 610 | 44,000 | 0.20 | Dimer | 26 |
| mCherry | 587 | 610 | 72,000 | 0.22 | Monomer | 47 |
| HcRed1 | 588 | 618 | 20,000 | 0.015 | Dimer | 1 |
| mRaspberry | 598 | 625 | 86,000 | 0.15 | Monomer | 38 |
| dKeima-Tandem | 440 | 620 | 28,800 | 0.24 | Monomer | 21 |
| HcRed-Tandem | 590 | 637 | 160,000 | 0.04 | Monomer | 19 |
| mPlum | 590 | 649 | 41,000 | 0.10 | Monomer | 12 |
| AQ143 | 595 | 655 | 90,000 | 0.04 | Tetramer | 11 |

*Note:* EGFP, enhanced green fluorescent protein.
*Forms weak dimer.

# Index

Page numbers followed by f and t indicate figures and tables, respectively.